AF905581

Applied Mathematical Sciences

Volume 164

The mathematization of all sciences, the fading of traditional scientific boundaries, the impact of computer technology, the growing importance of computer modeling and the necessity of scientific planning all create the need both in education and research for books that are introductory to and abreast of these developments. The purpose of this series is to provide such books, suitable for the user of mathematics, the mathematician interested in applications, and the student scientist. In particular, this series will provide an outlet for topics of immediate interest because of the novelty of its treatment of an application or of mathematics being applied or lying close to applications. These books should be accessible to readers versed in mathematics or science and engineering, and will feature a lively tutorial style, a focus on topics of current interest, and present clear exposition of broad appeal. A compliment to the Applied Mathematical Sciences series is the Texts in Applied Mathematics series, which publishes textbooks suitable for advanced undergraduate and beginning graduate courses.

More information about this series at http://www.springer.com/series/34

George C. Hsiao · Wolfgang L. Wendland

Boundary Integral Equations

Second Edition

Springer

George C. Hsiao
Department of Mathematical Sciences
University of Delaware
Newark, USA

Wolfgang L. Wendland
IANS
Universität Stuttgart
Stuttgart, Germany

ISSN 0066-5452 ISSN 2196-968X (electronic)
Applied Mathematical Sciences
ISBN 978-3-030-71129-0 ISBN 978-3-030-71127-6 (eBook)
https://doi.org/10.1007/978-3-030-71127-6

Mathematics Subject Classification: 31A10, 31B10, 35J05, 35J20, 35J35, 35Q61, 45B05, 47G30, 74B05, 76D07, 78A25, 78A45

Responsible Editor: Martin Peters
This Springer imprint is published by the registered company Springer Nature Switzerland AG.
The registered company address is: Gewerbestrasse 11, 6330 Cham, Switzerland

To our families for their love and understanding

Preface to the Second Edition

The first edition of this book was published in August, 2008. Since then, various theoretical and computational techniques have been developed for boundary integral equation methods. In particular, fundamental results for solutions of the boundary integral equations in electromagnetics have matured substantially. The widespread scattering of these results throughout the literature has motivated us to update our monograph and collect these important developments and fundamental results to facilitate accessibility.

In this second edition, in addition to making minor corrections, correcting typographical errors, updating references, as well as improving or simplifying some of the presentations, we have added two new chapters. One of the new chapters is on *Maxwell equations*. The other new chapter is on *Remarks on Pseudodifferential Operators Related to Time–harmonic Maxwell Equations.* The chapter on Maxwell equations includes an introduction to basic electromagnetic theory with emphasis on the reduction of time-harmonic solutions to corresponding boundary integral equations. The chapter on the pseudodifferential operator approach to Maxwell equations discusses, based on the symbol analysis of the electromagnetic boundary integral operators, the coerciveness, strong ellipticity and solutions for boundary integral equations in standard Sobolev spaces.

We also added a new section on *Function Spaces* $H(div, \Omega)$ *and* $H(curl, \Omega)$ which gives a brief summary of some basic vector function spaces and related topics needed for the two new chapters. Finally, for convenience, we added a new appendix to the second edition that contains *Vector Identities, Green's Formulae, Surface Differential Operators and Stokes Formulae.* The second edition of the monograph now contains a total of twelve chapters and two appendices. We are very pleased that this second edition includes theory and applications of boundary integral equations arising in potential flow, acoustics, elasticity, Stokes flow and electromagnetics. Because of the breadth of these applications, we hope that this second edition will also attract a broader readership.

We would like to take this opportunity to express our deepest gratitude to the "Mathematisches Forschungsinstitut Oberwolfach" (MFO), which supported us twice through the Research in Pairs Program (RiP 1634p, and RiP 1837q). Thanks go to Greg Silber for proof reading of the manuscript of the two added chapters. Heartfelt thanks are also due to Gisela Wendland for her highly skilled LaTex typesetting and for preparing the manuscript.

Newark, Delaware — *George C. Hsiao*
Stuttgart, Germany 2020 — *Wolfgang L. Wendland*

Preface to the First Edition

This book is devoted to the mathematical foundation of boundary integral equations. The combination of finite element analysis on the boundary with these equations has led to very efficient computational tools, the boundary element methods (see e. g., the authors [202] and Schanz and Steinbach (eds.) [373]). Although we do not deal with the boundary element discretizations in this book, the material presented here gives the mathematical foundation of these methods. In order to avoid over generalization we have confined ourselves to the treatment of elliptic boundary value problems.

The central idea of eliminating the field equations in the domain and reducing boundary value problems to equivalent equations only on the boundary requires the knowledge of corresponding fundamental solutions, and this idea has a long history dating back to the work of Green [155] and Gauss [141, 142]. Today the resulting boundary integral equations still serve as a major tool for the analysis and construction of solutions to boundary value problems.

As is well known, the reduction to equivalent boundary integral equations is by no means unique, and there are primarily two procedures for this reduction, the 'direct' and 'indirect' approaches. The direct procedure is based on Green's representation formula for solutions of the boundary value problem, whereas the indirect approach rests on an appropriate layer ansatz. In our presentation we concentrate on the direct approach although the corresponding analysis and basic properties of the boundary integral operators remain the same for the indirect approaches. Roughly speaking, one obtains two kinds of boundary integral equations with both procedures, those of the first kind and those of the second kind.

The basic mathematical properties that guarantee existence of solutions to the boundary integral equations and also stability and convergence analysis of corresponding numerical procedures hinge on Gårding inequalities for the boundary integral operators on appropriate function spaces. In addition, contraction properties allow the application of Carl Neumann's classical successive iteration procedure to a class of boundary integral equations of the second kind. It turns out that these basic features are intimately related to the variational forms of the underlying elliptic boundary value problems and the potential energies of their solution fields, allowing us to consider the boundary integral equations in the form of variational problems on the boundary manifold of the domain.

On the other hand, the Newton potentials as the inverses of the elliptic partial differential operators are particular pseudodifferential operators on the domain or in the Euclidean space. The boundary potentials (or Poisson operators) are just Newton potentials of distributions with support on the boundary manifold and the boundary integral operators are their traces there. Therefore, it is rather natural to consider the boundary integral operators as pseudodifferential operators on the boundary manifold. Indeed, most of the boundary integral operators in applications can be recast as such pseudodifferential operators provided that the boundary manifold is smooth enough.

With the application of boundary element methods in mind, where strong ellipticity is the basic concept for stability, convergence and error analysis of corresponding discretization methods for the boundary integral equations, we are most interested in establishing strong ellipticity in terms of Gårding's inequality for the variational formulation as well as strong ellipticity of the pseudodifferential operators generated by the boundary integral equations. The combination of both, namely the variational properties of the elliptic boundary value and transmission problems as well as the strongly elliptic pseudodifferential operators provides us with an efficient means to analyze a large class of elliptic boundary value problems.

This book contains 10 chapters and an appendix. For the reader's benefit, Figure 0.0.1 gives a sketch of the topics contained in this book. Chapters 1 through 4 present various examples and background information relevant to the premises of this book.

In Chapter 5, we discuss the variational formulation of boundary integral equations and their connection to the variational solution of associated boundary value or transmission problems. In particular, continuity and coerciveness properties of a rather large class of boundary integral equations are obtained, including those discussed in the first and second chapters. In Chapter 4, we collect basic properties of Sobolev spaces in the domain and their traces on the boundary, which are needed for the variational formulations in Chapter 5.

Chapter 7 presents an introduction to the basic theory of classical pseudodifferential operators. In particular, we present the construction of a parametrix for elliptic pseudodifferential operators in subdomains of $\mathbb{R}^n$. Moreover, we give an iterative procedure to find Levi functions of arbitrary order for general elliptic systems of partial differential equations. If the fundamental solution exists then Levi's method based on Levi functions allows its construction via an appropriate integral equation.

In Chapter 8, we show that every pseudodifferential operator is an Hadamard's finite part integral operator with integrable or nonintegrable kernel plus possibly a differential operator of the same order as that of the pseudodifferential operator in case of nonnegative integer order. In addition, we formulate the necessary and sufficient Tricomi conditions for the integral operator kernels to define pseudodifferential operators in the domain by using the asymptotic expansions of the symbols and those of pseudohomogeneous kernels. We close Chapter 8 with a presentation of the transformation formulae and invariance properties under the change of coordinates.

Chapter 9 is devoted to the relation between the classical pseudodifferential operators and boundary integral operators. For smooth boundaries, all of our examples in Chapters 1 and 2 of boundary integral operators belong to the class of classical pseudodifferential operators on compact manifolds having symbols of the rational type. If the corresponding class of pseudodifferential operators in the form of Newton potentials is applied to tensor product distributions with support on the boundary manifold, then they generate, in a natural way, boundary integral operators which again are classical pseudodifferential operators on the boundary manifold. Moreover, for these operators associated with regular elliptic boundary value problems, it turns out that the corresponding Hadamard's finite part integral operators are invariant under the change of coordinates, as considered in Chapter 3. This approach also provides the jump relations of the potentials. We obtain these properties by using only the Schwartz kernels of the boundary integral operators. However, these are covered by Boutet de Monvel's work in the 1960's on regular elliptic problems involving the transmission properties.

The last two chapters, 10 and 11, contain concrete examples of boundary integral equations in the framework of pseudodifferential operators on the boundary manifold. In Chapter 10, we provide explicit calculations of the symbols corresponding to typical boundary integral operators on closed surfaces in $\mathbb{R}^3$. If the fundamental solution is not available then the boundary value problem can still be reduced to a coupled system of domain and boundary integral equations. As an illustration we show that these coupled systems can be considered as some particular Green operators of the Boutet de Monvel algebra. In Chapter 11, the special features of Fourier series expansions of boundary integral operators on closed curves are exploited.

We conclude the book with a short Appendix on differential operators in local coordinates with minimal differentiability. Here, we avoid the explicit use of the normal vector field as employed in Hadamard's coordinates in Chapter 3. These local coordinates may also serve for a more detailed analysis for Lipschitz domains.

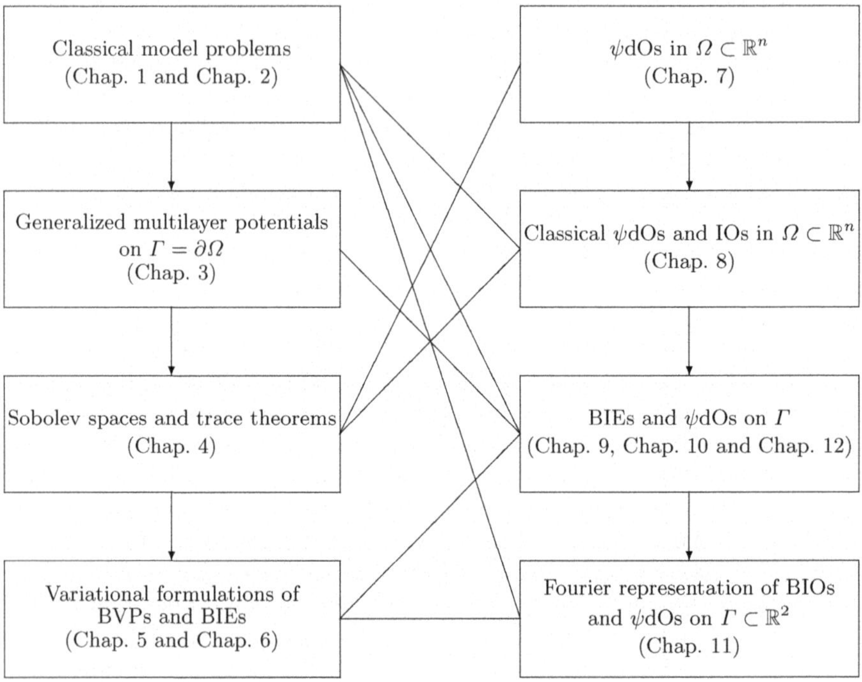

Abbreviations:

$\Omega \subset \mathbb{R}^n$	– *A given domain with compact boundary* Γ
BVPs	– *Boundary value problems*
BIEs	– *Boundary integral equations*
ψ*dOs*	– *Pseudodifferential operators*
IOs	– *Integral operators*
BIOs	– *Boundary integral operators*

Fig. 0.1. A schematic sketch of the topics and their relations

Our original plan was to finish this book project about 10 years ago. However, many new ideas and developments in boundary integral equation methods appeared during these years which we have attempted to incorporate.

Nevertheless, we regret to say that the present book is by no means complete. For instance, we only slightly touch on the boundary integral operator methods involving Lipschitz boundaries which have recently become more important in engineering applications. We do hope that we have made a small step forward to bridge the gap between the theory of boundary integral equation methods and their applications. We further hope that this book will lead to better understanding of the underlying mathematical structure of these methods and will serve as a mathematical foundation of the boundary element methods.

In closing, we would also like to mention some other relevant books related to boundary integral methods such as the classical books on potential theory by Kellogg [224] and Günter [162], the mathematical books on boundary integral equations by Hackbusch [167], Jaswon and Symm [213], Kupradze [251, 252, 253], Schatz, Thomée and Wendland [374], Mikhlin [301, 303, 302], Nedelec [328, 331], Colton and Kress [75, 77], Mikhlin, Morozov and Paukshto [304], Mikhlin and Prössdorf [305], Dautray and Lions [97], Chen and Zhou [65], Gatica and Hsiao [139], Kress [247], McLean [290], Yu, De–hao [442], Steinbach [399], Freeden and Michel [127], Kohr and Pop [233], Sauter and Schwab [371], as well as the Encyclopedia articles by Maz'ya [289], Prössdorf [354], Agranovich [6] and the authors [204]. For engineering books on boundary integral equations, we suggest the books by Brebbia [43], Crouch and Starfield [92], Brebbia, Telles and Wrobel [44], Manolis and Beskos [281], Balaš, Sladek and Sladek [23], Pozrikidis [353], Power and Wrobel [352], Bonnet [37], Gaul, Kögel and Wagner [140].

Acknowledgements

We are very grateful for the continuous support and encouragement by our students, colleagues, and friends. During the course of preparing this book we have benefitted from countless discussions with so many excellent individuals including Martin Costabel, Gabriel Gatica, Olaf Steinbach, Ernst Stephan and the late Siegfried Prössdorf; to name a few. Moreover, we are indebted to our reviewers of the first draft of the book manuscript for their critical reviews and helpful suggestions which helped us to improve our presentation. We would like to extend our thanks to Greg Silber, Clemens Förster and Gülnihal Meral for careful and critical proof reading of the manuscript.

We take this opportunity to acknowledge our gratitude to our universities, the University of Delaware at Newark, DE. U.S.A. and the University of Stuttgart in Germany; the Alexander von Humboldt Foundation, the Fulbright Foundation, and the German Research Foundation DFG for repeated support within the Priority Research Program on Boundary Element Methods and within the Collaborative Research Center on Multifield Problems, SFB 404 at the University of Stuttgart, the MURI program of AFOSR at the University of Delaware, and the Nečas Center in Prague. We express in particular our gratitude to the Oberwolfach Research Institute in Germany which supported us three times through the Research in Pairs Program, where we enjoyed the excellent research environment and atmosphere. It is also a pleasure to acknowledge the generous attitude, the unfailing courtesy, and the ready cooperation of the publisher.

Last, but by no means least, we are gratefully indebted to Gisela Wendland for her highly skilled hands in the LaTeX typing and preparation of this manuscript.

Newark, Delaware *George C. Hsiao*
Stuttgart, Germany 2008 *Wolfgang L. Wendland*

Table of Contents

1. Introduction

This chapter serves as a basic introduction to the reduction of elliptic boundary value problems to boundary integral equations. We begin with model problems for the Laplace equation. Our approach is the direct formulation based on Green's formula, in contrast to the indirect approach based on a layer ansatz. For ease of reading, we begin with the interior and exterior Dirichlet and Neumann problems of the Laplacian and their reduction to various forms of boundary integral equations, without detailed analysis. (For the classical results see e.g. Günter [162] and Kellogg [224].) The *Laplace equation*, and more generally, the *Poisson equation* ,

$$-\Delta v = f \quad \text{in } \Omega \text{ or } \Omega^c$$

already models many problems in engineering, physics and other disciplines (Dautray and Lions [95] and Tychonoff and Samarski [422]). This equation appears, for instance, in conformal mapping (Gaier [133, 134]), electrostatics (Gauss [141], Martensen [283] and Stratton [408]), stationary heat conduction (Günter [162]), in plane elasticity as the membrane state and the torsion problem (Szabo [411]), in Darcy flow through porous media (Bear [25] and Liggett and Liu [267]) and in potential flow (Glauert [150], Hess and Smith [178], Jameson [212] and Lamb [258]), to mention a few.

The approach here is based on the relation between the Cauchy data of solutions via the Calderón projector. As will be seen, the corresponding boundary integral equations may have eigensolutions in spite of the uniqueness of the solutions of the original boundary value problems. By appropriate modifications of the boundary integral equations in terms of these eigensolutions, the uniqness of the boundary integral equations can be achieved. Although these simple, classical model problems are well known, the concepts and procedures outlined here will be applied in the same manner for more general cases.

1.1 The Green Representation Formula

For the sake of simplicity, let us first consider, as a model problem, the Laplacian in two and three dimensions. As usual, we use $x = (x_1, \ldots, x_n) \in$

G. C. Hsiao und W. L. Wendland, *Boundary Integral Equations*,
Applied Mathematical Sciences 164,
https://doi.org/10.1007/978-3-030-71127-6_1

$\mathbb{R}^n$ ($n = 2$ or 3) to denote the Cartesian co–ordinates of the points in the Euclidean space $\mathbb{R}^n$. Furthermore, for $x, y \in \mathbb{R}^n$, we set

$$x \cdot y = \sum_{j=1}^{n} x_j y_j \quad \text{and} \quad |x| = (x \cdot x)^{\frac{1}{2}}$$

for the inner product and the Euclidean norm, respectively. We want to find the solution u satisfying the differential equation

$$-\Delta v := -\sum_{j=1}^{n} \frac{\partial^2 v}{\partial x_j^2} = f \text{ in } \Omega. \tag{1.1.1}$$

Here $\Omega \subset \mathbb{R}^n$ is a bounded, simply connected domain, and its boundary Γ is sufficiently smooth, say twice continuously differentiable, i.e. $\Gamma \in C^2$. (Later this assumption will be reduced.) As is known from classical analysis, a classical solution $v \in C^2(\Omega) \cap C^1(\overline{\Omega})$ can be represented by boundary potentials via the Green representation formula and the *fundamental solution* E of (1.1.1). For the Laplacian, $E(x, y)$ is given by

$$E(x,y) = \begin{cases} -\frac{1}{2\pi} \log |x-y| & \text{for } n = 2 \ , \\ \frac{1}{4\pi} \frac{1}{|x-y|} & \text{for } n = 3 \ . \end{cases} \tag{1.1.2}$$

The presentation of the solution reads

$$v(x) = \int\limits_{y \in \Gamma} E(x,y) \frac{\partial v}{\partial n}(y) ds_y - \int\limits_{y \in \Gamma} v(y) \frac{\partial E(x,y)}{\partial n_y} ds_y + \int\limits_{\Omega} E(x,y) f(y) dy \tag{1.1.3}$$

for $x \in \Omega$ (see Mikhlin [302, p. 220ff.]) where $\boldsymbol{n}_y$ denotes the exterior normal to Γ at $y \in \Gamma$, ds_y the surface element or the arclength element for $n = 3$ or 2, respectively, and

$$\frac{\partial v}{\partial n}(y) := \lim_{\tilde{y} \to y \in \Gamma, \tilde{y} \in \Omega} \operatorname{grad} v(\tilde{y}) \cdot \boldsymbol{n}_y \, . \tag{1.1.4}$$

The notation $\partial / \partial n_y$ will be used if there could be misunderstanding due to more variables.

In the case when $f \not\equiv 0$ in (1.1.1), one may also use the decomposition in the following form:

$$v(x) = v_p(x) + u(x) := \int\limits_{R^n} E(x,y) f(y) dy + u(x) \tag{1.1.5}$$

where u now solves the Laplace equation

$$-\Delta u = 0 \quad \text{in} \quad \Omega \, . \tag{1.1.6}$$

Here v_p denotes a particular solution of (1.1.1) in Ω or Ω^c and f has been extended from Ω (or Ω^c) to the entire R^n. Moreover, for the extended f we assume that the integral defined in (1.1.5) exists for all $x \in \overline{\Omega}$ (or $\overline{\Omega^c}$). Clearly, with this particular solution, the boundary conditions for u are to be modified accordingly.

Now, without loss of generality, we restrict our considerations to the solution u of the Laplacian (1.1.6) which now can be represented in the form:

$$u(x) = \int\limits_{y \in \Gamma} E(x,y)\frac{\partial u}{\partial n}(y) - \int\limits_{y \in \Gamma} u(y)\frac{\partial E(x,y)}{\partial n_y} ds_y \,. \tag{1.1.7}$$

For given boundary data $u|_\Gamma$ and $\frac{\partial u}{\partial n}|_\Gamma$, the representation formula (1.1.7) defines the solution of (1.1.6) everywhere in Ω. Therefore, the pair of boundary functions belonging to a solution u of (1.1.6) is called the *Cauchy data*, namely

$$\text{Cauchy data of}\, u := \begin{pmatrix} u|_\Gamma \\ \frac{\partial u}{\partial n}|_\Gamma \end{pmatrix} . \tag{1.1.8}$$

In solid mechanics, the representation formula (1.1.7) can also be derived by the principle of virtual work in terms of the so–called weighted residual formulation. The Laplacian (1.1.6) corresponds to the equation of the equilibrium state of the membrane without external body forces and vertical displacement u. Then, for fixed $x \in \Omega$, the terms

$$u(x) + \int\limits_{y \in \Gamma} u(y)\frac{\partial E(x,y)}{\partial n_y} ds_y$$

correspond to the virtual work of the point force at x and of the resulting boundary forces $\partial E(x,y)/\partial n_y$ against the displacement field u, which are equal to the virtual work of the resulting boundary forces $\frac{\partial u}{\partial n}|_\Gamma$ acting against the displacement $E(x,y)$, i.e.

$$\int\limits_{y \in \Gamma} E(x,y)\frac{\partial u}{\partial n}(y) ds_y \,.$$

This equality is known as *Betti's principle* (see e.g. Ciarlet [68], Fichera [119] and Hartmann [173, p.159]). Corresponding formulas can also be obtained for more general elliptic partial differential equations than (1.1.6), as will be discussed in Chapter 2.

1.2 Boundary Potentials and Calderón's Projector

The representation formula (1.1.7) contains two boundary potentials, the *simple layer potential*

$$V\sigma(x) \quad := \quad \int\limits_{y\in\Gamma} E(x,y)\sigma(y)\,ds_y, \qquad x \in \Omega \cup \Omega^c\,, \tag{1.2.1}$$

and the *double layer potential*

$$W\varphi(x) \quad := \quad \int\limits_{y\in\Gamma} (\frac{\partial}{\partial n_y} E(x,y))\varphi(y) ds_y,\ x \in \Omega \cup \Omega^c\,. \tag{1.2.2}$$

Here, σ and φ are referred to as the densities of the corresponding potentials. In (1.1.7), for the solution of (1.1.6), these are the Cauchy data which are not both given for boundary value problems. For their complete determination we consider the Cauchy data of the left– and the right–hand sides of (1.1.7) on Γ; this requires the limits of the potentials for x approaching Γ and their normal derivatives. This leads us to the following definitions of boundary integral operators, provided the corresponding limits exist. For the potential equation (1.1.6), this is well known from classical analysis (Mikhlin [302, p. 360] and Günter [162, Chap.II]):

$$V\sigma(x) \quad := \quad \lim_{z\to x\in\Gamma} V\sigma(z) \qquad \text{for } x \in \Gamma\,, \tag{1.2.3}$$

$$K\varphi(x) \quad := \quad \lim_{z\to x\in\Gamma, z\in\Omega} W\varphi(z) + \tfrac{1}{2}\varphi(x) \qquad \text{for } x \in \Gamma\,, \tag{1.2.4}$$

$$K'\sigma(x) \quad := \quad \lim_{z\to x\in\Gamma, z\in\Omega} \operatorname{grad}_z V\sigma(z)\cdot \boldsymbol{n}_x - \tfrac{1}{2}\sigma(x) \quad \text{for } x \in \Gamma\,, \tag{1.2.5}$$

$$D\varphi(x) \quad := \quad -\lim_{z\to x\in\Gamma, z\in\Omega} \operatorname{grad}_z W\varphi(z)\cdot \boldsymbol{n}_x \qquad \text{for } x \in \Gamma\,. \tag{1.2.6}$$

To be more explicit, we quote the following standard results without proof.

Lemma 1.2.1. *Let $\Gamma \in C^2$ and let σ and φ be continuous. Then the limits in* (1.2.3)–(1.2.5) *exist uniformly with respect to all $x \in \Gamma$ and all σ and φ with $\sup_{x\in\Gamma} |\sigma(x)| \le 1$, $\sup_{x\in\Gamma} |\varphi(x)| \le 1$. Furthermore, these limits can be expressed by*

$$V\sigma(x) \quad = \quad \int\limits_{y\in\Gamma\setminus\{x\}} E(x,y)\sigma(y) ds_y \qquad \text{for } x \in \Gamma, \tag{1.2.7}$$

$$K\varphi(x) \quad = \quad \int\limits_{y\in\Gamma\setminus\{x\}} \frac{\partial E}{\partial n_y}(x,y)\varphi(y) ds_y \ \text{ for } x \in \Gamma\,, \tag{1.2.8}$$

$$K'\sigma(x) \quad = \quad \int\limits_{y\in\Gamma\setminus\{x\}} \frac{\partial E}{\partial n_x}(x,y)\sigma(y) ds_y \ \text{ for } x \in \Gamma\,. \tag{1.2.9}$$

We remark that here all of the above boundary integrals are improper with weakly singular kernels in the following sense (see [302, p. 158]): The kernel $k(x,y)$ of an integral operator of the form

$$\int\limits_{\Gamma} k(x,y)\varphi(y) ds_y$$

is called *weakly singular* if there exist constants c and $\lambda < n-1$ such that

$$|k(x,y)| \le c|x-y|^{-\lambda} \text{ for all } x,y \in \Gamma\,. \tag{1.2.10}$$

For the Laplacian, for $\Gamma \in C^2$ and $E(x,y)$ given by (1.1.2), one even has

$$|E(x,y)| \le c_\lambda |x-y|^{-\lambda} \text{ for any } \lambda > 0 \text{ for } n=2 \text{ and } \lambda = 1 \text{ for } n=3\,, \tag{1.2.11}$$

$$\frac{\partial E}{\partial n_y}(x,y) = \frac{1}{2(n-1)\pi}\frac{(x-y)\cdot \boldsymbol{n}_y}{|x-y|^n}\,, \tag{1.2.12}$$

$$\frac{\partial E}{\partial n_x}(x,y) = \frac{1}{2(n-1)\pi}\frac{(y-x)\cdot \boldsymbol{n}_x}{|x-y|^n} \text{ for } x,y \in \Gamma\,. \tag{1.2.13}$$

In case $n=2$, both kernels in (1.2.12), (1.2.13) are continuously extendable to a C^0-function for $y \to x$ (see Mikhlin [302]), in case $n=3$ they are weakly singular with $\lambda = 1$ (see Günter [162, Sections II.3 and II.6]). For other differential equations, as e.g. for elasticity problems, the boundary integrals in (1.2.7)–(1.2.9) are strongly singular and need to be defined in terms of Cauchy principal value integrals or even as finite part integrals in the sense of Hadamard. In the classical approach, the corresponding function spaces are the *Hölder spaces* which are defined as follows:

$$C^{m+\alpha}(\Gamma) := \{\varphi \in C^m(\Gamma) \,\big|\, ||\varphi||_{C^{m+\alpha}(\Gamma)} < \infty\}$$

where the norm is defined by

$$||\varphi||_{C^{m+\alpha}(\Gamma)} := \sum_{|\beta|\le m} \sup_{x\in\Gamma} |\partial^\beta \varphi(x)| + \sum_{|\beta|=m} \sup_{\substack{x,y\in\Gamma\\ x\ne y}} \frac{|\partial^\beta\varphi(x) - \partial^\beta\varphi(y)|}{|x-y|^\alpha}$$

for $m \in \mathbb{N}_0$ and $0<\alpha<1$. Here, ∂^β denotes the covariant derivatives

$$\partial^\beta := \partial_1^{\beta_1}\cdots\partial_{n-1}^{\beta_{n-1}}$$

on the $(n-1)$–dimensional boundary surface Γ where $\beta \in \mathbb{N}_0^{n-1}$ is a multi–index and $|\beta| = \beta_1 + \ldots + \beta_{n-1}$ (see Millman and Parker [306]).

Lemma 1.2.2. *Let $\Gamma \in C^2$ and let φ be a Hölder continuously differentiable function. Then the limit in* (1.2.6) *exists uniformly with respect to all $x \in \Gamma$ and all φ with $\|\varphi\|_{C^{1+\alpha}} \le 1$. Moreover, the operator D can be expressed as a composition of tangential derivatives and the simple layer potential operator V:*

$$D\varphi(x) = -\frac{d}{ds_x} V\Big(\frac{d\varphi}{ds}\Big)(x) \quad \textit{for } n=2 \tag{1.2.14}$$

and

$$D\varphi(x) = -(\boldsymbol{n}_x \times \nabla_x)\cdot V(\boldsymbol{n}_y \times \nabla_y\varphi)(x) \quad \textit{for } n=3\,. \tag{1.2.15}$$

For the classical proof see Maue [284] and Günter [162, p.73ff].

Note that the differential operator $(\boldsymbol{n}_y \times \nabla_y)\varphi$ defines the tangential derivatives of $\varphi(y)$ within Γ which are Hölder–continuous functions on Γ. Often this operator is also called the *surface curl* (see Giroire and Nedelec [149, 329]). In the following, we give a brief derivation for these formulae based on classical results of potential theory with Hölder continuous densities $\frac{d\varphi}{ds}(y)$ and $(\boldsymbol{n}_y \times \nabla_y)\varphi(y)$, respectively. Note that $\frac{d}{ds_x}$ and $(\boldsymbol{n}_x \times \nabla_x)$ in (1.2.14) and (1.2.15), respectively, are *not* interchanged with integration over Γ. Later on we will discuss the connection of such an interchange with the concept of Hadamard's finite part integrals. For $n = 2$, note that, for $z \in \Omega,\ z \neq y \in \Gamma$,

$$
\begin{aligned}
-\boldsymbol{n}_x \cdot \nabla_z \int\limits_\Gamma (\boldsymbol{n}_y \cdot \nabla_y E(z,y))\ \varphi(y) ds_y \\
= \ - \int\limits_\Gamma \boldsymbol{n}_x \cdot \nabla_z \big(\boldsymbol{n}_y \cdot \nabla_y E(z,y)\big)\varphi(y) ds_y, \\
= \ \int\limits_\Gamma \boldsymbol{n}_x \cdot \big(\nabla_y \nabla_y^\top E(z,y)\big) \cdot \boldsymbol{n}_y \varphi(y) ds\,.
\end{aligned}
$$

Here,

$$
\boldsymbol{n}_x = \begin{pmatrix} \dfrac{dx_2}{ds} \\ -\dfrac{dx_1}{ds} \end{pmatrix},
$$

hence, with

$$
\boldsymbol{t}_x = \begin{pmatrix} \dfrac{dx_1}{ds} \\ \dfrac{dx_2}{ds} \end{pmatrix} = \boldsymbol{n}_x^\perp
$$

where $\boldsymbol{a}^\perp := \binom{-a_2}{a_1}$ is defined as counterclockwise rotation by $\frac{\pi}{2}$.

An elementary computation shows that

$$
\boldsymbol{n}_x^\top A \boldsymbol{n}_y = -\boldsymbol{t}_x^\top A^\top \boldsymbol{t}_y + (\text{trace } A)\boldsymbol{t}_x \cdot \boldsymbol{t}_y
$$

for any 2×2–matrix A. Hence,

$$
\begin{aligned}
-\boldsymbol{n}_x \cdot \nabla_z W\varphi(z) &= \int\limits_\Gamma \boldsymbol{n}_x \cdot \big(\nabla_y \nabla_y^\top E(z,y)\big) \cdot \boldsymbol{n}_y \varphi(y) ds_y \\
&= \int\limits_\Gamma \{\boldsymbol{t}_x \cdot \Delta_y E(z,y) \boldsymbol{t}_y - (\boldsymbol{t}_x \cdot \nabla_y)(\boldsymbol{t}_y \cdot \nabla_y E(z,y))\}\varphi(y) ds_y\,.
\end{aligned}
$$

Since $y \neq z$ and $\Delta_y E(z,y) = 0$, the second term on the right takes the form

$$
\begin{aligned}
-\boldsymbol{n}_x \cdot \nabla_z W\varphi(z) &= \int_\Gamma (\boldsymbol{t}_x \cdot \nabla_z)(\boldsymbol{t}_y \cdot \nabla_y E(z,y))\varphi(y)ds_y \\
&= (\boldsymbol{t}_x \cdot \nabla_z) \int_\Gamma (\frac{d}{ds_y} E(z,y))\varphi(y)ds_y
\end{aligned}
$$

and, after integration by parts,

$$
\begin{aligned}
&= -\boldsymbol{t}_x \cdot \nabla_z \int_\Gamma E(z,y)\frac{d\varphi}{ds_y}(y)ds_y \\
&= -\boldsymbol{t}_x \cdot \nabla_z \{V(\frac{d\varphi}{ds})\,(z)\}\,.
\end{aligned}
$$

First note that $\nabla_z V(\frac{d\varphi}{ds})(z)$ is a Hölder continuous function for $z \in \Omega$ which admits a Hölder continuous extension up to Γ (Günter [162, p.68]). The definition of derivatives at the boundary gives us

$$
\nabla_x V(\frac{d\varphi}{ds})(x) = \lim_{z\to x} \nabla_z V(\frac{d\varphi}{ds})(z)
$$

which yields

$$
\frac{d}{ds_x} V(\frac{d\varphi}{ds})(x) = \boldsymbol{t}_x \cdot \nabla_x V(\frac{d\varphi}{ds})(x) = \lim_{z\to x} \boldsymbol{t}_x \cdot \nabla_z V(\frac{d\varphi}{ds})(z)\,,
$$

i.e. (1.2.14).

Similarly, for $n = 3$, we see that

$$
-\boldsymbol{n}_x \cdot \nabla_z W\varphi(z) = \int_\Gamma \boldsymbol{n}_x \cdot (\nabla_y \nabla_y^T E(z,y)) \cdot \boldsymbol{n}_y \varphi(y)ds_y
$$

and with the formulae of vector analysis

$$
\begin{aligned}
&= \int_\Gamma \{(\boldsymbol{n}_y \cdot \nabla_y)(\boldsymbol{n}_x \cdot \nabla_y)E(z,y)\}\varphi(y)ds_y \\
&= -\int_\Gamma \{(\boldsymbol{n}_y \times \nabla_y)\cdot(\boldsymbol{n}_x \times \nabla_y)E(z,y)\}\varphi(y)ds_y \\
&\quad + \int_\Gamma \{(\boldsymbol{n}_x \cdot \boldsymbol{n}_y)\Delta_y E(z,y)\}\varphi(y)ds_y\,,
\end{aligned}
$$

where the last term vanishes since $z \notin \Gamma$. Now, with elementary vector analysis,

$$\begin{aligned}
-\boldsymbol{n}_x \cdot \nabla_z W\varphi(z) &= -\int_\Gamma \{\boldsymbol{n}_y \cdot (\nabla_y \times (\boldsymbol{n}_x \times \nabla_y))E(z,y)\}\varphi(y)ds_y \\
&= -\int_\Gamma \boldsymbol{n}_y \cdot \{\nabla_y \times ((\boldsymbol{n}_x \times \nabla_y)E(z,y)\varphi(y))\}ds_y \\
&\quad -\int_\Gamma \boldsymbol{n}_y \cdot \{(\boldsymbol{n}_x \times \nabla_y)E(z,y) \times \nabla_y\varphi(y)\}ds_y\,,
\end{aligned}$$

where $\varphi(y)$ denotes any $C^{1+\alpha}$–extension from Γ into $\mathbb{R}^3$. The first term on the right–hand side vanishes due to the Stokes theorem, whereas the second term gives

$$\begin{aligned}
-\boldsymbol{n}_x \cdot \nabla_z W\varphi(z) &= -\int_\Gamma \boldsymbol{n}_y \cdot \{\nabla_y\varphi(y) \times (\boldsymbol{n}_x \times \nabla_z)E(z,y)\}ds_y \\
&= -\int_\Gamma (\boldsymbol{n}_x \times \nabla_z E(z,y)) \cdot (\boldsymbol{n}_y \times \nabla_y\varphi(y))ds_y \\
&= -(\boldsymbol{n}_x \times \nabla_z) \cdot \int_\Gamma E(z,y)(\boldsymbol{n}_y \times \nabla_y)\varphi(y)ds_y\,.
\end{aligned}$$

Since $(\boldsymbol{n}_y \times \nabla_y)\varphi(y)$ defines tangential derivatives of φ,

$$\nabla_z \cdot V((\boldsymbol{n}_y \times \nabla_y)\ \varphi)(z)$$

defines a Hölder continuous function for $z \in \Omega$ which admits a Hölder continuous limit for $z \to x \in \Gamma$ due to Günter [162, p.68] implying (1.2.15).

From (1.2.15) we see that the hypersingular integral operator (1.2.6) can be expressed in terms of a composition of differentiation and a weakly singular operator. This, in fact, is a regularization of the hypersingular distribution, which will also be useful for the variational formulation and related computational procedures.

A more elementary, but different regularization can be obtained as follows (see Giroire and Nedelec [149]). From the definition (1.2.6), we see that

$$\begin{aligned}
D\varphi(x) \;=\; \lim_{\Omega \ni z \to x \in \Gamma} \Big\{ &-\boldsymbol{n}_x \cdot \nabla_z \int_\Gamma \frac{\partial E}{\partial n_y}(z,y)\,(\varphi(y) - \varphi(x))\,ds_y \\
&-\boldsymbol{n}_x \cdot \nabla_z \int_\Gamma \frac{\partial E}{\partial n_y}(z,y)\varphi(x)ds_y \Big\}.
\end{aligned}$$

If we apply the representation formula (1.1.7) to $u \equiv 1$, then we obtain Gauss' well known formula

$$\int_\Gamma \frac{\partial E}{\partial n_y}(z,y)ds_y = -1 \;\text{ for all } z \in \Omega\,.$$

This yields

$$\nabla_z \int_\Gamma \frac{\partial E}{\partial n_y}(z,y)\varphi(x)ds_y = 0 \text{ for all } z \in \Omega ,$$

hence, we find the simple regularization

$$D\varphi(x) = \lim_{\Omega \ni z \to x \in \Gamma} -\boldsymbol{n}_x \cdot \int_\Gamma \nabla_z \frac{\partial E}{\partial n_y}(z,y)\left(\varphi(y) - \varphi(x)\right) ds_y . \qquad (1.2.16)$$

In fact, the limit in (1.2.16) can be expressed in terms of a Cauchy principal value integral,

$$\begin{aligned} D\varphi(x) &= -\text{ p.v.} \int_\Gamma \Big(\frac{\partial}{\partial n_x}\frac{\partial}{\partial n_y}E(x,y)\Big)\left(\varphi(y)-\varphi(x)\right) ds_y \\ &:= -\lim_{\varepsilon \to 0+} \int_{y\in\Gamma \wedge |y-x|\geq\varepsilon} \Big(\frac{\partial}{\partial n_x}\frac{\partial}{\partial n_y}E(x,y)\Big)\left(\varphi(y)-\varphi(x)\right) ds_y \\ &= \lim_{\varepsilon \to 0+} \int_{y\in\Gamma \wedge |y-x|\geq\varepsilon} \frac{1}{2(n-1)\pi}\Big\{\frac{\boldsymbol{n}_x\cdot\boldsymbol{n}_y}{r^n} + n\frac{(y-x)\cdot\boldsymbol{n}_y(x-y)\cdot\boldsymbol{n}_x}{r^{2+n}}\Big\} \\ &\qquad \times \big(\varphi(y)-\varphi(x)\big)ds_y . \end{aligned} \qquad (1.2.17)$$

The derivation of (1.2.17) from (1.2.16), however, requires detailed analysis (see Günter [162, Section II,10]).

Since the boundary values for the various potentials are now characterized, we are in a position to discuss the relations between the Cauchy data on Γ by taking the limit $x \to \Gamma$ and the normal derivative of the left– and right–hand sides in the representation formula (1.1.7). For any solution of (1.1.6), this leads to the following relations between the Cauchy data:

$$u(x) = (\frac{1}{2}I - K)u(x) + V\frac{\partial u}{\partial n}(x) \qquad (1.2.18)$$

and

$$\frac{\partial u}{\partial n}(x) = Du(x) + (\frac{1}{2}I + K')\frac{\partial u}{\partial n}(x) . \qquad (1.2.19)$$

Consequently, for any solution of (1.1.6), the Cauchy data $(u, \frac{\partial u}{\partial n})^\top$ on Γ are reproduced by the operators on the right–hand side of (1.2.18), (1.2.19), namely by

$$\mathcal{C}_\Omega := \begin{pmatrix} \frac{1}{2}I - K & , & V \\ D & , & \frac{1}{2}I + K' \end{pmatrix} \qquad (1.2.20)$$

This operator is called the *Calderón projector* (with respect to Ω) (Calderón [59]). The operators in $\mathcal{C}_\Omega$ have mapping properties in the classical Hölder function spaces as follows:

Theorem 1.2.3. *Let $\Gamma \in C^2$ and $0 < \alpha < 1$, a fixed constant . Then the boundary potentials V, K, K', D define continuous mappings in the following spaces,*

$$V \; : \; C^\alpha(\Gamma) \;\to C^{1+\alpha}(\Gamma)\,, \tag{1.2.21}$$
$$K, K' \; : \; C^\alpha(\Gamma) \;\to C^\alpha(\Gamma),\; C^{1+\alpha}(\Gamma) \to C^{1+\alpha}(\Gamma)\,, \tag{1.2.22}$$
$$D \; : \; C^{1+\alpha}(\Gamma) \to C^\alpha(\Gamma)\,. \tag{1.2.23}$$

For the proofs see Mikhlin and Prössdorf [305, Sections IX, 4 and 7].

Remark 1.2.1: The double layer potential operator K and its adjoint K' for the Laplacian have even stronger continuity properties than those in (1.2.22), namely K, K' map continuously $C^\alpha(\Gamma) \to C^{1+\alpha}(\Gamma)$ and $C^{1+\alpha}(\Gamma) \to C^{1+\beta}(\Gamma)$ for any $\alpha \le \beta < 1$ (Mikhlin and Prössdorf [305, Sections IX, 4 and 7] and Colton and Kress [75, Chap.2]). Because of the compact imbeddings $C^{1+\alpha}(\Gamma) \hookrightarrow C^\alpha(\Gamma)$ and $C^{1+\beta}(\Gamma) \hookrightarrow C^{1+\alpha}(\Gamma), K$ and K' are compact. These smoothing properties of K and K' do not hold anymore, if K and K' correspond to more general elliptic partial differential equations than (1.1.6). This is e.g. the case in linear elasticity. However, the continuity properties (1.2.22) remain valid.

With Theorem 1.2.3, we now are in a position to show that $\mathcal{C}_\Omega$ indeed is a projection. More precisely, there holds:

Lemma 1.2.4. *Let $\Gamma \in C^2$. Then $\mathcal{C}_\Omega$ maps $C^{1+\alpha}(\Gamma) \times C^\alpha(\Gamma)$ into itself continuously. Moreover,*

$$\mathcal{C}_\Omega^2 \; = \; \mathcal{C}_\Omega\,. \tag{1.2.24}$$

Consequently, we have the following identities:

$$VD \; = \; \tfrac{1}{4}I - K^2\,, \tag{1.2.25}$$
$$DV \; = \; \tfrac{1}{4}I - {K'}^2\,, \tag{1.2.26}$$
$$KV \; = \; VK'\,, \tag{1.2.27}$$
$$DK \; = \; K'D\,. \tag{1.2.28}$$

These relations will show their usefulness in our variational formulation later on, and as will be seen in the next section, the Calderón projector leads in a direct manner to boundary integral equations for boundary value problems.

1.3 Boundary Integral Equations

As we have seen from (1.2.18) and (1.2.19), the Cauchy data of a *solution* of the differential equation in Ω are related to each other by these two equations. As is well known, for regular elliptic boundary value problems, only half of the Cauchy data on Γ is given. For the remaining part, the equations (1.2.18), (1.2.19) define an overdetermined system of boundary integral equations which may be used for determining the complete Cauchy data. In general, any combination of (1.2.18) and (1.2.19) can serve as a boundary integral equation for the missing part of the Cauchy data. Hence, the boundary integral equations associated with one particular boundary condition are by no means uniquely determined. The 'direct' approach for formulating boundary integral equations becomes particularly simple if one considers the Dirichlet problem or the Neumann problem. In what follows, we will always prefer the direct formulation.

1.3.1 The Dirichlet Problem

In the Dirichlet problem for (1.1.6), the boundary values

$$u|_\Gamma = \varphi \quad \text{on}\, \Gamma \tag{1.3.1}$$

are given. Hence,

$$\sigma = \frac{\partial u}{\partial n}|_\Gamma \tag{1.3.2}$$

is the missing Cauchy datum required to satisfy (1.2.18) and (1.2.19) for any solution u of (1.1.6). In the direct formulation, if we take the first equation (1.2.18) of the Calderón projection then σ is to be determined by the boundary integral equation

$$V\sigma(x) = \tfrac{1}{2}\varphi(x) + K\varphi(x)\,,\; x \in \Gamma\,. \tag{1.3.3}$$

Explicitly, we have

$$\int\limits_{y\in\Gamma} E(x,y)\sigma(y)ds_y = f(x)\,,\; x \in \Gamma \tag{1.3.4}$$

where f is given by the right–hand side of (1.3.3) and E is given by (1.1.2), a weakly singular kernel. Hence, (1.3.4) is a *Fredholm integral equation of the first kind*. In the case $n = 2$ and a boundary curve Γ with conformal radius equal to 1, the integral equation (1.3.4) has exactly one eigensolution, the so–called *natural charge* $e(y)$ (Plemelj [349]). However, the modified equation

$$\int\limits_{y\in\Gamma} E(x,y)\sigma(y)ds_y - \omega = f(x)\,,\; x \in \Gamma \tag{1.3.5}$$

together with the normalizing condition

$$\int\limits_{y\in\Gamma} \sigma ds = \Sigma \tag{1.3.6}$$

is always solvable for σ and the constant ω for given f and given constant Σ [198]. Later on we will come back to this modification.

For $\Sigma = 0$ it can be shown that $\omega = 0$. Hence, with $\Sigma = 0$, this modified formulation can also be used for solving the interior Dirichlet problem.

Alternatively to (1.3.4), if we take the second equation (1.2.19) of the Calderón projector, we arrive at

$$\tfrac{1}{2}\sigma(x) - K'\sigma(x) = D\varphi(x) \quad \text{for } x \in \Gamma\,. \tag{1.3.7}$$

In view of (1.2.9) and (1.2.11), the explicit form of (1.3.7) reads

$$\sigma(x) - 2 \int\limits_{y\in\Gamma\setminus\{x\}} \frac{\partial E}{\partial n_x}(x,y)\sigma(y)ds_y = g(x) \quad \text{for } x \in \Gamma\,, \tag{1.3.8}$$

where $\frac{\partial E}{\partial n_x}(x,y)$ is weakly singular due to (1.2.13), provided Γ is smooth. $g = 2D\varphi$ is defined by the right–hand side of (1.3.7). Therefore, in contrast to (1.3.4), this is a *Fredholm integral equation of the second kind.*

This simple example shows that for the same problem we may employ different boundary integral equations. In fact, (1.3.8) is one of the celebrated integral equations of classical potential theory — the adjoint to the Neumann–Fredholm integral equation of the second kind with the double layer potential — which can be obtained by using the double layer ansatz in the indirect approach. In the classical framework, the analysis of integral equation (1.3.8) has been studied intensively for centuries, including its numerical solution. For more details and references, see, e.g., Atkinson [16], Bruhn et al. [46], Dautray and Lions [95, 97], Jeggle [214, 215], Kellogg [224], Kral et al [239, 240, 241, 242, 243], Martensen [282, 283], Maz'ya [289], Neumann [335, 336, 337], Radon [360] and [432]. In recent years, increasing efforts have also been devoted to the integral equation of the first kind (1.3.4) which — contrary to conventional belief — became a very rewarding and fundamental formulation theoretically as well as computationally. It will be seen that this equation is particularly suitable for the variational analysis.

1.3.2 The Neumann Problem

In the Neumann problem for (1.1.6), the boundary condition reads as

$$\frac{\partial u}{\partial n}|_\Gamma = \psi \ \text{ on } \Gamma \tag{1.3.9}$$

with given ψ. For the interior problem (1.1.6) in Ω, the normal derivative ψ needs to satisfy the necessary compatibility condition

$$\int\limits_\Gamma \psi ds = 0 \tag{1.3.10}$$

for any solution of (1.1.6), (1.3.9) to exist. Here, $u_{|\Gamma}$ is the missing Cauchy datum required to satisfy (1.2.18) and (1.2.19) for any solution u of the Neumann problem (1.1.6), (1.3.9). If we take the first equation (1.2.18) of the Calderón projector, then $u_{|\Gamma}$ is determined by the solution of the boundary integral equation

$$\tfrac{1}{2}u(x) + Ku(x) = V\psi(x),\; x \in \Gamma\,. \tag{1.3.11}$$

This is a classical Fredholm integral equation of the second kind on Γ, namely

$$u(x) + 2\int\limits_{y\in\Gamma\setminus\{x\}} \frac{\partial E}{\partial n_y}(x,y)u(y)ds_y = 2\int\limits_{y\in\Gamma\setminus\{x\}} E(x,y)\psi(y)ds_y =: f(x)\,. \tag{1.3.12}$$

For the Laplacian we have (1.2.12), which shows that the kernel of the integral operator in (1.3.12) is continuous for $n = 2$ and weakly singular for $n = 3$. It is easily shown that $u_0 = 1$ defines an eigensolution of the homogeneous equation corresponding to (1.3.12); and that $f(x)$ in (1.3.12) satisfies the classical orthogonality condition if and only if ψ satisfies (1.3.10). Classical potential theory provides that (1.3.12) is always solvable if (1.3.10) holds and that the null–space of (1.3.12) is one–dimensional, see e.g. Mikhlin [303, Chap.17, 11].

Alternatively, if we take the second equation (1.2.19) of the Calderón projector, we arrive at the equation

$$Du(x) = \tfrac{1}{2}\psi(x) - K'\psi(x) \;\; \text{for}\, x \in \Gamma\,. \tag{1.3.13}$$

This is a *hypersingular boundary integral equation* of the first kind for $u_{|\Gamma}$ which also has the one–dimensional null–space spanned by $u_0|_\Gamma = 1$, as can easily be seen from (1.2.14) and (1.2.15) for $n = 2$ and $n = 3$, respectively. Although this integral equation (1.3.13) is not one of the standard types, we will see that, nevertheless, it has advantages for the variational formulation and corresponding numerical treatment.

1.4 Exterior Problems

In many applications such as electrostatics and potential flow, one often deals with exterior problems which we will now consider for our simple model equation.

1.4.1 The Exterior Dirichlet Problem

For boundary value problems exterior to Ω, i.e. in $\Omega^c = \mathbb{R}^n \setminus \overline{\Omega}$, infinity belongs to the boundary of Ω^c and, therefore, we need additional growth or radiation conditions for u at infinity. Moreover, in electrostatic problems, for instance, the total charge Σ on Γ will be given. This leads to the following *exterior Dirichlet problem*, defined by the differential equation

$$-\Delta u = 0 \text{ in } \Omega^c\,, \tag{1.4.1}$$

the boundary condition

$$u|_\Gamma = \varphi \text{ on } \Gamma \tag{1.4.2}$$

and the additional growth condition

$$u(x) = \frac{1}{2\pi}\Sigma \log|x| + \omega + o(1) \text{ as } |x| \to \infty \text{ for } n = 2 \tag{1.4.3}$$

or

$$u(x) = -\frac{1}{4\pi}\Sigma\frac{1}{|x|} + \omega + O(|x|^{-2}) \text{ as } |x| \to \infty \text{ for } n = 3\,. \tag{1.4.4}$$

The Green representation formula now reads

$$u(x) = Wu(x) - \left(V\frac{\partial u}{\partial n}\right)(x) + \omega \text{ for } x \in \Omega^c\,, \tag{1.4.5}$$

where the direction of $\boldsymbol{n}$ is defined as before and the normal derivative is now defined as in (1.1.7) but with $z \in \Omega^c$. In the case $\omega = 0$, we may consider the Cauchy data from Ω^c on Γ, which leads with the boundary data of (1.4.5) on Γ to the equations

$$\begin{pmatrix} u \\ \frac{\partial u}{\partial n} \end{pmatrix} = \begin{pmatrix} \frac{1}{2}I + K & , & -V \\ -D & , & \frac{1}{2}I - K' \end{pmatrix} \begin{pmatrix} u \\ \frac{\partial u}{\partial n} \end{pmatrix} \text{ on } \Gamma\,. \tag{1.4.6}$$

Here, the boundary integral operators V, K, K', D are related to the limits of the boundary potentials from Ω^c similar to (1.2.3)–(1.2.6), namely

$$V\sigma(x) = \lim_{z\to x} V\sigma(z)\; x \in \Gamma\,, \tag{1.4.7}$$

$$K\varphi(x) = \lim_{z\to x, z\in\Omega^c} W\varphi(z) - \tfrac{1}{2}\varphi(x)\,,\; x \in \Gamma\,, \tag{1.4.8}$$

$$K'\sigma(x) = \lim_{z\to x, z\in\Omega^c} \operatorname{grad}_z V(z)\cdot n_x + \tfrac{1}{2}\sigma(x)\,,\; x \in \Gamma\,, \tag{1.4.9}$$

$$D\varphi(x) = -\lim_{z\to x, z\in\Omega^c} \operatorname{grad}_z W\varphi(z)\cdot n_x\,,\; x \in \Gamma. \tag{1.4.10}$$

Note that in (1.4.8) and (1.4.9) the signs at $\frac{1}{2}\varphi(x)$ and $\frac{1}{2}\sigma(x)$ are different from those in (1.2.4) and (1.2.5), respectively.

For any solution u of (1.4.1) in Ω^c with $\omega = 0$, the Cauchy data on Γ are reproduced by the right–hand side of (1.4.6), which therefore defines the *Calderón projector* $\mathcal{C}_{\Omega^c}$ for the Laplacian with respect to the exterior domain Ω^c. Clearly,

$$\mathcal{C}_{\Omega^c} = \mathcal{I} - \mathcal{C}_\Omega\,, \tag{1.4.11}$$

where $\mathcal{I}$ denotes the identity matrix operator.

For the solution of (1.4.1), (1.4.2) and (1.4.3), we obtain from (1.4.5) a modified boundary integral equation,

$$V\sigma(x) - \omega = -\tfrac{1}{2}\varphi(x) + K\varphi(x) \;\text{ for } x \in \Gamma\,. \tag{1.4.12}$$

This, again, is a *first kind integral equation* for $\sigma = \frac{\partial u}{\partial n}|_\Gamma$, the unknown Cauchy datum. However, in addition, the constant ω is also unknown. Hence, we need an additional constraint, which here is given by

$$\int\limits_\Gamma \sigma ds = \Sigma\,. \tag{1.4.13}$$

This is the same modified system as (1.3.5), (1.3.6), which is always uniquely solvable for (σ, ω).

If we take the normal derivative at Γ on both sides of (1.4.5), we arrive at the following *Fredholm integral equation of the second kind* for σ, namely

$$\tfrac{1}{2}\sigma(x) + K'\sigma(x) = -D\varphi(x) \;\text{ for } x \in \Gamma\,. \tag{1.4.14}$$

This is the classical integral equation associated with the exterior Dirichlet problem which has a one–dimensional space of eigensolutions. Here, the special right–hand side of (1.4.14) always satisfies the orthogonality condition in the classical Fredholm alternative. Hence, (1.4.14) always has a solution, which becomes unique if the additional constraint of (1.4.13) is included.

1.4.2 The Exterior Neumann Problem

Here, in addition to (1.4.1), we require the Neumann condition

$$\frac{\partial u}{\partial n}|_\Gamma = \psi \;\text{ on } \Gamma \tag{1.4.15}$$

where ψ is given. Moreover, we again need a condition at infinity. We choose the growth condition (1.4.3) or (1.4.4), respectively, where the constant Σ is given by

$$\Sigma = \int\limits_\Gamma \psi ds$$

from (1.4.15), where ω is now an additional parameter, which can be prescribed arbitrarily according to the special situation. The representation formula (1.4.5) remains valid. Often $\omega = 0$ is chosen in (1.4.3), (1.4.4) and (1.4.5). The direct approach with $x \to \Gamma$ in (1.4.5) now leads to the boundary integral equation

$$-\tfrac{1}{2}u(x) + Ku(x) = V\psi(x) - \omega \;\text{ for } x \in \Gamma\,. \tag{1.4.16}$$

For any given ψ and ω, this is the *classical Fredholm integral equation of the second kind* which has been studied intensively (Günter [162]). (See also

Atkinson [16], Dieudonné [100], Kral [240, 241], Maz'ya [289], Mikhlin [301, 303]) (1.4.16) is uniquely solvable for $u|_\Gamma$.

If we apply the normal derivative to both sides of (1.4.5), we find the *hypersingular integral equation of the first kind,*

$$Du(x) = -\tfrac{1}{2}\psi(x) - K'\psi(x) \text{ for } x \in \Gamma . \tag{1.4.17}$$

This equation has the constants as an one–dimensional eigenspace. The special right–hand side in (1.4.17) satisfies an orthogonality condition in the classical Fredholm alternative, which is also valid for (1.4.17), e.g., in the space of Hölder continuous functions on Γ, as will be shown later. Therefore, (1.4.17) always has solutions $u|_\Gamma$. Any solution of (1.4.17) inserted into the right hand side of (1.4.5) together with any choice of ω will give the desired unique solution of the exterior Neumann problem.

For further illustration, we now consider the historical example of the two–dimensional potential flow of an inviscid incompressible fluid around an airfoil. Let $\boldsymbol{q}_\infty$ denote the given traveling velocity of the profile defining a uniform velocity at infinity and let $\boldsymbol{q}$ denote the velocity field. Then we have the following exterior boundary value problem for $\boldsymbol{q} = (q_1, q_2)^\top$:

$$(\nabla \times \boldsymbol{q})_3 = \frac{\partial q_2}{\partial x_1} - \frac{\partial q_1}{\partial x_2} = 0 \text{ in } \Omega^c , \tag{1.4.18}$$

$$\operatorname{div}\boldsymbol{q} = \frac{\partial q_1}{\partial x_1} + \frac{\partial q_2}{\partial x_2} = 0 \text{ in } \Omega^c , \tag{1.4.19}$$

$$\boldsymbol{q} \cdot \boldsymbol{n}_{|\Gamma} = 0 \text{ on } \Gamma , \tag{1.4.20}$$

$$\lim_{\Omega^c \ni x \to TE} |\boldsymbol{q}(x)| = |\boldsymbol{q}|_{|TE} \text{ exists at the trailing edge } TE , \tag{1.4.21}$$

$$\boldsymbol{q} - \boldsymbol{q}_\infty = o(1) \text{ as } |x| \to \infty . \tag{1.4.22}$$

Here, the airfoil's profile Γ is given by a simply closed curve with one corner point at the trailing edge TE. Moreover, Γ has a C^∞– parametrization $\boldsymbol{x}(s)$ for the arc length $0 \le s \le L$ with $\boldsymbol{x}(0) = \boldsymbol{x}(L) = TE$, whose periodic extension is only piecewise C^∞. With Bernoulli's law, the condition (1.4.21) is equivalent to the Kutta–Joukowski condition, which requires bounded and equal pressure at the trailing edge. (See also Ciavaldini et al [70]). The origin 0 of the co–ordinate system is chosen within the airfoil with TE on the x_1–axis and the line $\overline{0\,TE}$ within Ω, as shown in Figure 1.4.1.

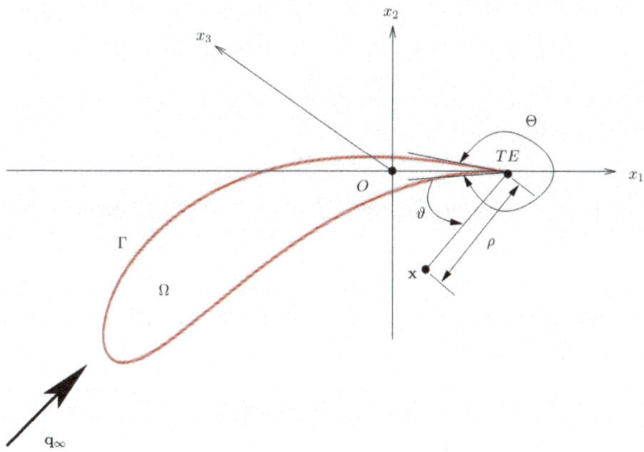

Figure 1.4.1: Airfoil in two dimensions

As before, the exterior domain is denoted by $\Omega^c := \mathbb{R}^2\backslash\overline{\Omega}$. Since the flow is irrotational and divergence–free, $\boldsymbol{q}$ has a potential which allows the reformulation of (1.4.18)–(1.4.22) as

$$\boldsymbol{q} = \boldsymbol{q}_\infty + \frac{\omega}{2\pi}\frac{1}{|x|^2}\begin{pmatrix}-x_2\\ x_1\end{pmatrix} + \nabla u \tag{1.4.23}$$

where u is the solution of the exterior Neumann boundary value problem

$$-\Delta u = 0 \quad \text{in } \Omega^c\,, \tag{1.4.24}$$

$$\frac{\partial u}{\partial n}_{|\Gamma} = -\boldsymbol{q}_\infty \cdot \boldsymbol{n}_{|\Gamma} - \frac{\omega_0}{2\pi}\frac{1}{|x|^2}\begin{pmatrix}-x_2\\ x_1\end{pmatrix}\cdot \boldsymbol{n} \quad \text{on } \Gamma\,, \tag{1.4.25}$$

$$u(x) = o(1) \ \text{ as } |x| \to \infty\,. \tag{1.4.26}$$

In this formulation, $u \in C^2(\Omega^c) \cap C^0(\overline{\Omega^c})$ is the unknown disturbance potential, ω_0 is the unknown circulation around Γ which will be determined by the additional Kutta–Joukowski condition, that is

$$\lim_{\Omega^c \ni x \to TE} |\nabla u(x)| = |\nabla u|_{|TE} \quad \text{exists}\,. \tag{1.4.27}$$

We remark that condition (1.4.27) is a direct consequence of condition (1.4.21). By using conformal mapping, the solution u was constructed by Kirchhoff [226], see also Goldstein [153].

We now reduce this problem to a boundary integral equation. As in (1.4.5) and in view of (1.4.26), the solution admits the representation

$$u(x) = Wu(x) + V\left(\boldsymbol{q}_\infty \cdot \boldsymbol{n} + \frac{\omega_0}{2\pi|y|^2}\begin{pmatrix}-y_2\\ y_1\end{pmatrix}\cdot \boldsymbol{n}\right)(x) \ \text{ for } x \in \Omega^c\,. \tag{1.4.28}$$

Since

$$\int_{\Gamma} \boldsymbol{q}_{\infty} \cdot \boldsymbol{n}(y) ds_y = \int_{\Omega} (\mathrm{div} \boldsymbol{q}_{\infty}) dx = 0\,,$$

and, by Green's theorem for

$$\int_{\Gamma} \frac{1}{|y|^2} \begin{pmatrix} -y_2 \\ y_1 \end{pmatrix} \cdot \boldsymbol{n}(y) ds_y = \int_{|y|=R} \frac{1}{|y|^2} \begin{pmatrix} -y_2 \\ y_1 \end{pmatrix} \cdot \boldsymbol{n}(y) ds_y = 0\,,$$

it follows that every solution u represented by (1.4.28) satisfies (1.4.26) for any choice of ω_0. We now set

$$u = u_0 + \omega_0 u_1\,, \tag{1.4.29}$$

where the potentials u_0 and u_1 are solutions of the exterior Neumann problems

$$-\Delta u_i = 0 \quad \text{in } \Omega^c \quad \text{with } i = 0, 1, \quad \text{and}$$

$$\frac{\partial u_0}{\partial n}|_{\Gamma} = -\boldsymbol{q}_{\infty} \cdot \boldsymbol{n}_{|\Gamma}\,, \quad \frac{\partial u_1}{\partial n}|_{\Gamma} = -\frac{1}{2\pi} |x|^{-2} \begin{pmatrix} -x_2 \\ x_1 \end{pmatrix} \cdot \boldsymbol{n}_{|\Gamma}\,,$$

respectively, and

$$u_i = o(1) \quad \text{as } |x| \to \infty, \ i = 0, 1.$$

Hence, these functions admit the representations

$$u_0(x) = W u_0(x) + V(\boldsymbol{q}_{\infty} \cdot \boldsymbol{n})(x)\,, \tag{1.4.30}$$

$$u_1(x) = W u_1(x) + V \left(\frac{1}{2\pi} \frac{1}{|y|^2} \begin{pmatrix} -y_2 \\ y_1 \end{pmatrix} \cdot \boldsymbol{n} \right)(x) \tag{1.4.31}$$

for $x \in \Omega^c$, whose boundary traces are the unique solutions of the boundary integral equations

$$\frac{1}{2} u_0(x) - K u_0(x) = V(\boldsymbol{q}_{\infty} \cdot \boldsymbol{n})(x)\,, x \in \Gamma, \tag{1.4.32}$$

$$\frac{1}{2} u_1(x) - K u_1(x) = V \left(\frac{1}{2\pi} \frac{1}{|y|^2} \begin{pmatrix} -y_2 \\ y_1 \end{pmatrix} \cdot \boldsymbol{n} \right)(x)\,, x \in \Gamma\,. \tag{1.4.33}$$

Note that K is defined by (1.5.2) which is valid at TE, too. The right–hand sides of (1.4.32) and (1.4.33) are both Hölder continuous functions on Γ. Due to the classical results by Carleman [62] and Radon [360], there exist unique solutions u_0 and u_1 in the class of continuous functions. A more detailed analysis shows that the derivatives of these solutions possess singularities at the trailing edge TE. More precisely, one finds (e.g. in the book by Grisvard [156, Theorem 5.1.1.4 p.255]) that the solutions admit local singular expansions of the form

$$u_i(x) = \alpha_i \varrho^{\frac{\pi}{\Theta}} \cos\left(\frac{\pi}{\Theta} \vartheta \right) + O\left(\varrho^{\frac{2\pi}{\Theta} - \varepsilon} \right), \quad i = 0, 1\,, \tag{1.4.34}$$

where Θ is the exterior angle of the two tangents at the trailing edge, ϱ denotes the distance from the trailing edge to x, ϑ is the angle from the lower trailing edge tangent to the vector $(\boldsymbol{x} - TE)$, where ε is any positive number. Consequently, the gradients are of the form

$$\nabla u_i(x) = \alpha_i \frac{\pi}{\Theta} \varrho^{\frac{\pi}{\Theta}-1} \boldsymbol{e}_\vartheta + O\left(\varrho^{\frac{2\pi}{\Theta}-1-\varepsilon}\right), \quad i = 0, 1, \tag{1.4.35}$$

where $\boldsymbol{e}_\vartheta$ is a unit vector with angle $(1 - \frac{\pi}{\Theta})\vartheta$ from the lower trailing edge tangent, for both cases, $i = 0, 1$.

Hence, from equations (1.4.27) and (1.4.29) we obtain, for $\varrho \to 0$, the condition for ω_0,

$$\alpha_0 + \omega_0 \alpha_1 = 0. \tag{1.4.36}$$

The solution u_1 corresponds to $\boldsymbol{q}_\infty = 0$, i.e. the pure circulation flow, which can easily be found by mapping Ω conformally onto the unit circle in the complex plane. The mapping has the local behavior as in (1.4.34) with $\alpha_1 \neq 0$ since TE is mapped onto a point on the unit circle (see Lehman [261]). Consequently, ω_0 is uniquely determined from (1.4.36). We remark that this choice of ω_0 shows that

$$\nabla u = O\left(\varrho^{\frac{2\pi}{\Theta}-1-\varepsilon}\right)$$

due to (1.4.35) and, hence, the singularity vanishes for $\Theta < 2\pi$ at the trailing edge TE; which indeed is then a stagnation point for the disturbance velocity ∇u. This approach can be generalized to two–dimensional transonic flow problems (Coclici et al [74]).

1.5 Remarks

For applications in engineering, the strong smoothness assumptions for the boundary Γ need to be relaxed allowing corners and edges. Moreover, for crack and screen problems as in elasticity and acoustics, respectively, Γ is not closed but only a part of a curve or a surface. To handle these types of problems, the approach in the previous sections needs to be modified accordingly.

To be more specific, we first consider *Lyapounov boundaries*. Following Mikhlin [303, Chap.18], a Lyapounov curve in $\mathbb{R}^2$ or Lyapunov surface Γ in $\mathbb{R}^3$ (Günter [162]) satisfies the following two conditions:

1. There exists a normal $\boldsymbol{n}_x$ at any point x on Γ.
2. There exist positive constants a and $\kappa \leq 1$ such that for any two points x and ξ on Γ with corresponding vectors $\boldsymbol{n}_x$ and $\boldsymbol{n}_\xi$ the angle ϑ between them satisfies

$$|\vartheta| \leq a r^\kappa \quad \text{where}\, r = |x - \xi|\,.$$

In fact, it can be shown that for $0 < \kappa < 1$, a Lyapounov boundary coincides with a $C^{1,\kappa}$ boundary curve or surface [303, Chap.18].

For a Lyapounov boundary, all results in Sections 1.1–1.4 remain valid if C^2 is replaced by $C^{1,\kappa}$ accordingly. These non–trivial generalizations can be found in the classical books on potential theory. See, e.g., Günter [162], Mikhlin [301, 303, 302] and Smirnov [392].

In applications, one often has to deal with boundary curves with corners, or with boundary surfaces with corners and edges. The simplest generalization of the previous approach can be obtained for piecewise Lyapounov curves in $\mathbb{R}^2$ with finitely many corners where $\Gamma = \cup_{j=1}^{N} \overline{\Gamma}_j$ and each Γ_j being an open arc of a particular closed Lyapounov curve. The intersections $\overline{\Gamma}_j \cap \overline{\Gamma}_{j+1}$ are the corner points where $\Gamma_{N+1} := \Gamma_1$. In this case it easily follows that there exists a constant C such that

$$\int\limits_{\Gamma\setminus\{x\}} |\frac{\partial E}{\partial n_y}(x,y)|ds \leq C \quad \text{for all } x \in \mathbb{R}^2 . \tag{1.5.1}$$

This property already ensures that for continuous φ on Γ, the operator K is well defined by (1.2.4) and that it is a continuous mapping in $C^0(\Gamma)$. However, (1.2.8) needs to be modified and becomes

$$\begin{aligned} K\varphi(x) \; := \; & \int\limits_{\Gamma\setminus\{x\}} \varphi(y)\frac{\partial E}{\partial n_y}(x,y)ds_y \\ & - \Big(\frac{1}{2} + \int\limits_{\Gamma\setminus\{x\}} \frac{\partial E}{\partial n_y}(x,y)ds_y\Big)\varphi(x) , \end{aligned} \tag{1.5.2}$$

where the last expression takes care of the corner points and vanishes if x is not a corner point. Here K is not compact anymore as in the case of a Lyapounov boundary, however, it can be shown that K can be decomposed into a sum of a compact operator and a contraction, provided the corner angles are not 0 or 2π. This decomposition is sufficient for the classical Fredholm alternative to hold for (1.3.11) with continuous u, as was shown by Radon [360]. For the most general two–dimensional case we refer to Kral [240].

For the Neumann problem, one needs a generalization of the normal derivative in terms of the so–called *boundary flow*, which originally was introduced by Plemelj [349] and has been generalized by Kral [241]. It should be mentioned that in this case the adjoint operator K' to K is no longer a bounded operator on the space of continuous functions (Netuka [334]). The simple layer potential $V\sigma$ is still Hölder continuous in $\mathbb{R}^2$ for continuous σ. However, its normal derivative needs to be interpreted in the sense of boundary flow.

This situation is even more complicated in the three–dimensional case because of the presence of edges and corners. Here, for continuous φ it is still not clear whether the Fredholm alternative for equation (1.3.11) remains valid even for general piecewise Lyapounov surfaces with finitely many corners and edges; see, e.g., Angell et al [13], Burago et al [55, 56], Kral et al [242, 243], Maz'ya [289] and [432].

On the other hand, as we will see, in the variational formulation of the boundary integral equations, many of these difficulties can be circumvented for even more general boundaries such as Lipschitz boundaries (see Section 5.6).

To conclude these remarks, we consider Γ to be an oriented, open part of a closed curve or surface $\widetilde{\Gamma}$ (see Figure 1.5.1). The *Dirichlet problem* here is to find the solution u of (1.4.1) in the domain $\Omega^c = \mathbb{R}^n \setminus \overline{\Gamma}$ subject to the boundary conditions

$$u_+ = \varphi_+ \text{ on } \Gamma_+ \quad \text{and} \quad u_- = \varphi_- \text{ on } \Gamma_- \tag{1.5.3}$$

where Γ_+ and Γ_- are the respective sides of Γ and u_+ and u_- the corresponding traces of u. The functions φ_+ and φ_- are given with the additional requirement that

$$\varphi_+ - \varphi_- = 0$$

at the endpoints of Γ for $n = 2$, or at the boundary edge of Γ for $n = 3$. In the latter case we require $\partial\Gamma$ to be a C^∞– smooth curve. Similar to the regular exterior problem, we again require the growth condition (1.4.3) for $n = 2$ and (1.4.4) for $n = 3$. For a sufficiently smooth solution, the Green representation formula has the form

$$u(x) = W_\Gamma[u](x) - V_\Gamma\left[\frac{\partial u}{\partial n}\right](x) + \omega \text{ for } x \in \Omega^c\,, \tag{1.5.4}$$

where W_Γ, V_Γ are the corresponding boundary potentials with integration over Γ only,

$$W_\Gamma\varphi(x) := \int\limits_{y\in\Gamma}\Big(\frac{\partial}{\partial n_y}E(x,y)\Big)\varphi(y)ds_y\,,\ x \notin \overline{\Gamma}\,; \tag{1.5.5}$$

$$V_\Gamma\sigma(x) := \int\limits_{y\in\Gamma} E(x,y)\sigma(y)ds_y\,,\ x \notin \overline{\Gamma}\,. \tag{1.5.6}$$

$$[u] = u_+ - u_-\,,\ \sigma := \left[\frac{\partial u}{\partial n}\right] = \frac{\partial u_+}{\partial n} - \frac{\partial u_-}{\partial n} \text{ on } \Gamma \tag{1.5.7}$$

with $\boldsymbol{n}$ the normal to Γ pointing in the direction of the side Γ_+. If we substitute the given boundary values φ_+, φ_- into (1.5.4), the missing Cauchy datum σ is now the jump of the normal derivative across Γ. Between this unknown datum and the behaviour of u at infinity viz.(1.4.3) we arrive at

$$\int\limits_\Gamma \sigma ds = \Sigma\,. \tag{1.5.8}$$

By taking x to Γ_+ (or Γ_-) we obtain (in both cases) the boundary integral equation of the first kind for σ on Γ,

$$V_\Gamma\sigma(x) - \omega = -\tfrac{1}{2}\big(\varphi_+(x) + \varphi_-(x)\big) + K_\Gamma(\varphi_+ - \varphi_-)(x) =: f(x) \quad (1.5.9)$$

where K_Γ is defined by

$$K_\Gamma\varphi(x) = \int\limits_{y\in\Gamma\setminus\{x\}} \Big(\frac{\partial}{\partial n_y}E(x,y)\Big)\varphi(y)ds_y \ \text{ for } x \in \Gamma\,. \quad (1.5.10)$$

As for the previous case of a closed curve or surface Γ, respectively, the system (1.5.8), (1.5.9) admits a unique solution pair $(\sigma,\ \omega)$ for any given φ_+, φ_- and Σ. Here, however, σ will have singularities at the endpoints of Γ or the boundary edge of Γ for $n = 2$ or 3, respectively, and our classical approach, presented here, requires a more careful justification in terms of appropriate function spaces and variational setting.

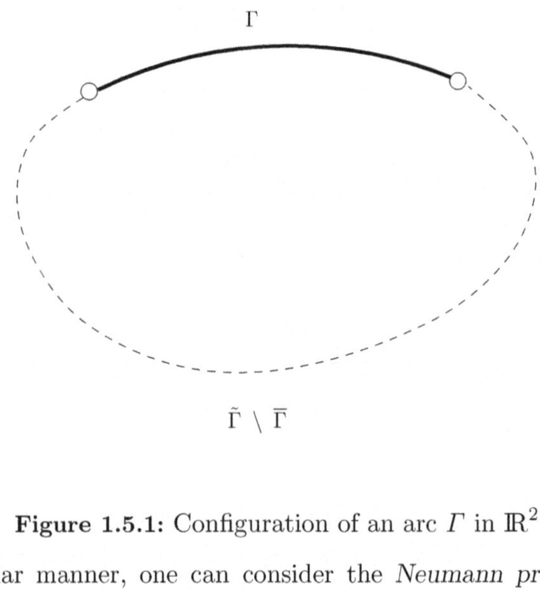

Figure 1.5.1: Configuration of an arc Γ in $\mathbb{R}^2$

In a similar manner, one can consider the *Neumann problem*: Find u satisfying (1.4.1) in Ω^c subject to the boundary conditions

$$\frac{\partial u_+}{\partial n} = \psi_+ \ \text{ on } \Gamma_+ \ \text{ and } \ \frac{\partial u_-}{\partial n} = \psi_- \ \text{ on } \Gamma_- \quad (1.5.11)$$

where ψ_+ and ψ_- are given smooth functions. By applying the normal derivatives $\partial/\partial n_x$ to the representation formula (1.5.4) from both sides of Γ, it is not difficult to see that the missing Cauchy datum $\varphi =: [u] = u_+ - u_-$ on Γ satisfies the hypersingular boundary integral equation of the first kind for φ,

$$D_\Gamma\varphi = -\tfrac{1}{2}(\psi_+ + \psi_-) - K'_\Gamma[\psi] \ \text{ on } \Gamma\,, \quad (1.5.12)$$

where the operators D_Γ and K'_Γ again are given by (1.2.6) and (1.2.5) with W and V replaced by W_Γ and V_Γ, respectively. As we will see later on in the framework of variational problems, this integral equation (1.5.12) is uniquely solvable for φ with $\varphi = 0$ at the endpoints of Γ for $n = 2$ or at the boundary edge $\partial\Gamma$ of Γ for $n = 3$, respectively. Here, Σ in (1.4.3) or (1.4.4) is already given by (1.5.8) and ω can be chosen arbitrarily. For further analysis of these problems see [210], Stephan et al [403, 406], Costabel et al [79, 85].

2. Boundary Integral Equations

In Chapter 1 we presented basic ideas for the reduction of boundary value problems of the Laplacian to various forms of boundary integral equations based on the direct approach. This reduction can be easily extended to more general partial differential equations. Here we will consider, in particular, the Helmholtz equation, the Lamé system, the Stokes equations and the biharmonic equation.

For the Helmholtz equation, we also investigate the solution's asymptotic behavior for small wave numbers and the relation to solutions of the Laplace equation by using the boundary integral equations.

For the Lamé system of elasticity, we first present the boundary integral equations of the first kind as well as of the second kind. Furthermore, we study the behavior of the solution and the boundary integral equations for incompressible materials. As will be seen, this has a close relation to the Stokes system and its boundary integral equations. In the two–dimensional case, both the Stokes and the Lamé problems can be reduced to solutions of biharmonic boundary value problems which, again, can be solved by using boundary integral equations based on the direct approach.

In this chapter we consider these problems for domains whose boundaries are smooth enough, mostly Lyapounov boundaries, and the boundary charges belonging to Hölder spaces. Later on we shall consider the boundary integral equations again on Sobolev trace spaces which is more appropriate for stability and convergence of corresponding discretization procedures.

2.1 The Helmholtz Equation

A slight generalization of the Laplace equation is the well–known *Helmholtz equation*

$$-(\Delta + k^2)u = 0 \quad \text{in} \quad \Omega \text{ (or } \Omega^c) \,. \tag{2.1.1}$$

This equation arises in connection with the propagation of waves, in particular in acoustics (Filippi [122], Kupradze [251] and Wilcox [438]) and electromagnetics (Ammari [8], Cessenat [63], Colton and Kress [75], Jones [220], Müller [317] and Neledec [331]). In acoustics, k with $\mathrm{Im}\, k \geq 0$ denotes the complex *wave number* and u corresponds to the *acoustic pressure* field.

G. C. Hsiao und W. L. Wendland, *Boundary Integral Equations*,
Applied Mathematical Sciences 164,
https://doi.org/10.1007/978-3-030-71127-6_2

The reduction of boundary value problems for (2.1.1) to boundary integral equations can be carried out in the same manner as for the Laplacian in Chapter 1. For (2.1.1), in the exterior domain, one requires at infinity the *Sommerfeld radiation conditions*,

$$u(x) = \mathrm{O}(|x|^{-(n-1)/2}) \text{ and } \frac{\partial u}{\partial |x|}(x) - iku(x) = o\left(|x|^{-(n-1)/2}\right) \tag{2.1.2}$$

where i is the imaginary unit. (See the book by Sommerfeld [395] and the further references therein; see also Neittaanmäki and Roach [333] and Wilcox [438]). These conditions select the outgoing waves; they are needed for uniqueness of the exterior Dirichlet problem as well as for the Neumann problem.

The pointwise condition (2.1.2) can be replaced by a more appropriate and weaker version of the radiation condition given by Rellich [363, 364],

$$\lim_{R\to\infty} \int\limits_{|x|=R} |\frac{\partial u}{\partial n}(x) - iku(x)|^2 ds = 0\,. \tag{2.1.3}$$

This form is to be used in the variational formulation of exterior boundary value problems.

The fundamental solution $E(x,y)$ to (2.1.1), subject to the radiation condition (2.1.2) for fixed $y \in R^n$ is given by

$$E_k(x,y) = \begin{cases} \frac{i}{4} H_0^{(1)}(kr) & \text{in } \mathbb{R}^2\,, \\ & \qquad \text{with } r = |x-y| \\ \frac{e^{ikr}}{4\pi r} & \text{in } \mathbb{R}^3\,, \end{cases} \tag{2.1.4}$$

where $H_0^{(1)}$ denotes the modified Bessel function of the first kind. We note, that for $n = 2, E(x,y)$ has a branch point for $\mathbb{C} \ni k \to 0$. Therefore, in the following we confine ourselves first to the case $k \neq 0$. In terms of these fundamental solutions, which obviously are symmetric, the representation of solutions to (2.1.1) in Ω or in Ω^c with (2.1.2) assumes the same forms as (1.1.7) and (1.4.5), namely

$$\begin{aligned} u(x) &= \pm\Big\{ \int\limits_{y\in\Gamma} E_k(x,y)\frac{\partial u}{\partial n}(y) ds_y - \int\limits_{y\in\Gamma} u(y)\frac{\partial E_k(x,y)}{\partial n_y} ds_y \Big\} \\ &= \pm\Big\{ V_k \frac{\partial u}{\partial n}(x) - W_k u(x) \Big\} \end{aligned} \tag{2.1.5}$$

for all $x \in \Omega$ or Ω^c respectively, where the $\pm$sign corresponds to the interior and the exterior domain. Here, V_k is defined by

$$\begin{aligned} V_k\sigma(x) &:= \int\limits_{y\in\Gamma} E_k(x,y)\sigma(y) ds_y \\ &= V\sigma(x) + S_k\sigma(x) + \{\delta_{n3}\tfrac{ik}{4\pi} - \delta_{n2}(\log k + \gamma_0)\} \int\limits_{\Gamma} \sigma ds\,, \end{aligned} \tag{2.1.6}$$

where V is given by (1.2.1) with (1.1.2) and S_k is a k–dependent remainder defined by

$$S_k\sigma(x) = \begin{cases} \int\limits_{y\in\Gamma} \Big\{\frac{i}{4}\, H_0^{(1)}(k|x-y|) + \frac{1}{2\pi}\log|x-y|\Big\}\sigma(y)ds_y, \\ -\frac{1}{4\pi}\int\limits_{y\in\Gamma}\Big\{\sum\limits_{m=2}^{\infty}\frac{(m-1)}{m!}(ik|x-y|)^m\Big\}\frac{1}{|x-y|}\sigma(y)ds_y \end{cases} \tag{2.1.7}$$

for $n=2$ or 3, respectively. The potential W_k is defined by

$$W_k\varphi(x) := \int\limits_{y\in\Gamma}\frac{\partial E_k(x,y)}{\partial n_y}\varphi(y)ds_y = W\varphi(x) + R_k\varphi(x) \tag{2.1.8}$$

where W is given by (1.2.2) with (1.1.2) and R_k is a k–dependent remainder defined by

$$R_k\varphi(x) = \begin{cases} \int\limits_{y\in\Gamma\setminus\{x\}} \frac{\partial}{\partial n_y}\{\frac{i}{4}H_0^{(1)}(k|x-y|) + \frac{1}{2\pi}\log|x-y|\}\varphi(y)ds_y\,, \\ -\frac{1}{4\pi}\int\limits_{y\in\Gamma\setminus\{x\}}\Big\{\sum\limits_{m=2}^{\infty}\frac{(m-1)}{m!}(ik|x-y|)^m\Big\}\Big(\frac{\partial}{\partial n_y}\frac{1}{|x-y|}\Big)\varphi(y)ds_y \end{cases} \tag{2.1.9}$$

for $n=2$ or 3, respectively. Note that S_k and R_k have bounded kernels for all $y\in\Gamma$ and $x\in\mathbb{R}^n$ and for $k\neq 0$ in the case $n=2$. Hence, $S_k\sigma(x)$ and $R_k\varphi(x)$ are well defined for all $x\in\mathbb{R}^n$. Moreover, the properties of the operators V and W, as given by Lemmata 1.2.1, 1.2.2, 1.2.4 and Theorem 1.2.3 for the Laplacian, remain valid for V_k and W_k. Since the kernels of V_k and W_k, S_k and V_k depend analytically on $k\in\mathbb{C}\setminus\{0\}$ for $n=2$ and $k\in\mathbb{C}$ for $n=3$, the solutions of the corresponding boundary integral equations will depend analytically on the wave number k, as well.

Here again we consider the interior and *exterior Dirichlet problem* for (2.1.1), where

$$u_{|\Gamma} = \varphi \quad \text{on } \Gamma \text{ is given .} \tag{2.1.10}$$

In acoustics, (2.1.10) models a "soft" boundary for the interior and a "soft scatterer" for the exterior problem. As in Section 1.3.1, here the missing Cauchy datum on Γ is $\frac{\partial u}{\partial n}_{|\Gamma} = \sigma$.

The boundary integral equation for the interior Dirichlet problem reads

$$V_k\sigma(x) = \frac{1}{2}\varphi(x) + K_k\varphi(x), \quad x\in\Gamma\,, \tag{2.1.11}$$

where

$$K_k\varphi(x) := \int\limits_{y\in\Gamma\setminus\{x\}} \frac{\partial E_k(x,y)}{\partial n_y}\varphi(y)ds_y = K\varphi(x) + R_k\varphi(x) \quad \text{for } x\in\Gamma \tag{2.1.12}$$

with K given by (1.2.8) with (1.2.12) or (1.2.13), respectively. As before, (2.1.11) is a Fredholm boundary integral equation of the first kind for σ on Γ. In the classical Hölder continuous function space $C^\alpha, 0<\alpha<1$, this integral equation has been studied by Colton and Kress in [75, Chap.3]. In particular, for $k\neq 0$ and $\varphi\in C^{1+\alpha}(\Gamma)$, (2.1.11) is uniquely solvable with $\sigma\in C^\alpha(\Gamma)$, except for certain values of $k\in\mathbb{C}$ which are the *exceptional* or *irregular frequencies* of the boundary integral operator V_k. For any irregular frequency k_0, the operator V_{k_0} has a nontrivial nullspace $ker\ V_{k_0} = span\ \{\sigma_{0j}\}$. The eigensolutions σ_{0j} are related to the eigensolutions $\widetilde{u}_{0j}$ of the *interior Dirichlet problem for the Laplacian*,

$$\begin{aligned} -\Delta\widetilde{u}_0 &= k_0^2\widetilde{u}_0 \text{ in } \Omega\,, \\ \widetilde{u}_{0|\Gamma} &= 0 \text{ on } \Gamma\,. \end{aligned} \tag{2.1.13}$$

That is,

$$\sigma_{0j} = \frac{\partial\widetilde{u}_{0j}}{\partial n|_\Gamma}\,. \tag{2.1.14}$$

Moreover, the solutions are real–valued and

$$\dim ker V_{k_0} = \text{ dimension of the eigenspace of (2.1.13)}\,.$$

As is known, see e.g. Hellwig [177, p.229], the eigenvalue problem (2.1.13) admits denumerably infinitely many eigenvalues $k_{0\ell}^2$. They are all real and have at most finite multiplicity. Moreover, they can be ordered according to size $0<k_{01}^2<k_{02}^2<\cdots$ and have $+\infty$ as their only limit point. For any of the corresponding eigensolutions $\widetilde{u}_{0j}$, (2.1.14) can be obtained from (2.1.5) applied to $\widetilde{u}_{0j}$. In this case, when k_0 is an eigenvalue, the interior Dirichlet problem (2.1.1), (2.1.10) admits solutions in $C^\alpha(\Gamma)$ if and only if the given boundary values $\varphi\in C^{1+\alpha}(\Gamma)$ satisfy the *orthogonality conditions*

$$\int\limits_\Gamma \varphi\sigma_0 ds = \int\limits_\Gamma \varphi\frac{\partial\widetilde{u}_0}{\partial n}ds = 0 \quad \text{for all } \sigma_0\in kerV_{k_0}\,. \tag{2.1.15}$$

Correspondingly, for $\varphi\in C^{1+\alpha}(\Gamma)$, the boundary integral equation (2.1.11) has solutions $\sigma\in C^\alpha(\Gamma)$ if and only if (2.1.15) is satisfied.

For the *exterior Dirichlet problem*, i.e. (2.1.1) in Ω^c with the Sommerfeld radiation conditions (2.1.2) and boundary condition (2.1.10), from (2.1.5) again we obtain a boundary integral equation of the first kind,

$$V_k\sigma(x) = -\tfrac{1}{2}\varphi(x) + K_k\varphi(x)\,,\ x\in\Gamma\,, \tag{2.1.16}$$

which differs from (2.1.11) only by a sign in the right–hand side. Hence, the exceptional values k_0 are the same as for the interior Dirichlet problem, namely the eigenvalues of (2.1.13). If $k \neq k_0$, (2.1.16) is always uniquely solvable for $\sigma \in C^\alpha(\Gamma)$ if $\varphi \in C^{1+\alpha}(\Gamma)$. For $k = k_0$, in contrast to the interior Dirichlet problem, the exterior Dirichlet problem remains uniquely solvable. However, (2.1.16) now has eigensolutions, and the right–hand side always satisfies the orthogonality conditions

$$\int\limits_{x\in\Gamma} \big(-\tfrac{1}{2}\varphi(x) + K_{k_0}\varphi(x)\big)\sigma_0(x)ds_x$$

$$= \int\limits_{x\in\Gamma} \varphi(x)\{-\tfrac{1}{2}\sigma_0(x) + K'_{k_0}\sigma_0(x)\}ds_x = 0 \ \text{ for all } \sigma_0 \in \ker V_{k_0}\,,$$

since σ_0 is real valued and the simple layer potential $V_{k_0}\sigma_0(x)$ vanishes identically for $x \in \Omega^c$. The latter implies

$$\frac{\partial}{\partial n_x} V_{k_0}\sigma_0(x) = -\frac{1}{2}\sigma_0(x) + K'_{k_0}\sigma_0(x) = 0 \ \text{ for } x \in \Gamma\,,$$

where

$$K'_k\sigma(x) := \int\limits_{y\in\Gamma\setminus\{x\}} \Big(\frac{\partial}{\partial n_x}E_k(x,y)\Big)\sigma(y)ds_y = K'\sigma(x) + R'_k\sigma(x)$$

with

$$R'_k\sigma(x) = \begin{cases} \displaystyle\int\limits_{y\in\Gamma\setminus\{x\}} \frac{\partial}{\partial n_x}\Big\{\frac{i}{4}H_0^{(1)}(k|x-y|) + \frac{1}{2\pi}\log|x-y|\Big\}\sigma(y)ds_y\,, \\ \displaystyle -\frac{1}{4\pi}\int\limits_{y\in\Gamma\setminus\{x\}} \Big\{\sum_{m=2}^{\infty}\frac{(m-1)}{m!}(ik|x-y|)^m\Big\}\Big(\frac{\partial}{\partial n_x}\frac{1}{|x-y|}\Big)\sigma(y)ds_y\,, \end{cases}$$

for $n = 2$ and $n = 3$, respectively. Accordingly, the representation formula (2.1.5) with $u_{|\Gamma} = \varphi$ and $\frac{\partial u}{\partial n}_{|\Gamma} = \sigma$ will generate a *unique* solution for any σ solving (2.1.16).

Alternatively, both, the interior and exterior Dirichlet problem can also be solved by the Fredholm integral equations of the second kind as (1.3.7) and (1.4.14) for the Laplacian. In order to avoid repetition we summarize the different direct formulations of the interior and exterior Dirichlet and Neumann problems which will be abbreviated by (IDP), (EDP), (INP) and (ENP), accordingly, in Table 2.1.1. The Neumann data in (INP) and (ENP) will be denoted by ψ.

In addition to the previously defined integral operators V_k, K_k, K'_k we also introduce the hypersingular integral operator, D_k for the Helmholtz equation, namely

$$D_k\varphi(x) := -\frac{\partial}{\partial n_x}\int\limits_{\Gamma\setminus\{x\}} \frac{\partial E_k(x,y)}{\partial n_y}\varphi(y)ds_y = D\varphi(x) + \frac{\partial}{\partial n_x}R_k\varphi(x) \ \text{ on } \Gamma\,, \tag{2.1.17}$$

Table 2.1.1. Summary of the boundary integral equations for the Helmholtz equation and the related eigenvalue problems

BVP	BIE	Eigensolutions $\tilde{u}_0$ or $\tilde{u}_1$ for BVP and Exceptional values k_0, k_1	Eigensolutions for BIE, σ_0, σ_1 or u_0, u_1	Solvability Conditions for given φ, ψ
IDP	$(1) V_k\sigma = (\frac{1}{2}I + K_k)\varphi$	(D_0):		
	$(2) (\frac{1}{2}I - K_k')\sigma = D_k\varphi$	$-(\Delta + k_0^2)\tilde{u}_0 = 0$ in Ω,	$\sigma_0 = \frac{\partial \tilde{u}_0}{\partial n}_{\mid\Gamma}$	$\int\limits_\Gamma \sigma_0 \varphi ds = 0$
EDP	$(1) V_k\sigma = (-\frac{1}{2}I + K_k)\varphi$	$\tilde{u}_{0\mid\Gamma} = 0$ on Γ		
	$(2) (\frac{1}{2}I + K_k')\sigma = -D_k\varphi$	(N_0) :	$V_{k_1}\sigma_1 = \tilde{u}_1$ on Γ	None
INP	$(1) D_k u = (\frac{1}{2}I - K_k')\psi$	$-(\Delta + k_1^2)\tilde{u}_1 = 0$ in Ω		
	$(2) (\frac{1}{2}I + K_k)u = V_k\psi$	$\frac{\partial \tilde{u}_1}{\partial n}_{\mid\Gamma} = 0$ on Γ	$u_1 = \tilde{u}_{1\mid\Gamma}$ on Γ	$\int\limits_\Gamma u_1 \psi ds = 0$
ENP	$(1) D_k u = -(\frac{1}{2}I + K_k')\psi$			
	$(2) (\frac{1}{2}I - K_k)u = -V_k\psi$	(D_0)	$D_{k_0}u_0 = \frac{\partial \tilde{u}_0}{\partial n}_{\mid\Gamma}$	None

where D is the hypersingular integral operator (1.4.10) of the Laplacian and R_k is the remainder in (2.1.9).

Note that the relations between the eigensolutions of the BIEs and the interior eigenvalue problems of the Laplacian are given explicitly in column three of Table 2.1.1. We also observe that for the exterior boundary value problems, the exceptional values k_0 and k_1 of the corresponding boundary integral operators depend on the type of boundary integral equations derived by the direct formulation. For instance, we see that for (EDP), k_0 are the exceptional values for V_k whereas k_1 are those for $(\frac{1}{2}I + K_k')$. Similar relations hold for (ENP).

In Table 2.1.1, the second column contains all of the boundary integral equations (BIE) obtained by the direct approach. As we mentioned earlier, for the *exterior* boundary value problems, the solvability conditions of the corresponding boundary integral equations at the exceptional values are always satisfied due to the special forms of the corresponding right-hand sides. For the indirect approach, this is not the case anymore; see Colton and Kress [75, Chap. 3]. There are various ways to modify the boundary integral equations so that some of the exceptional values will not belong to the spectrum of the boundary integral operator anymore. In this connection, we refer to the work by Brakhage and Werner [42] Colton and Kress [75], Jones [220], Kleinman and Kress [228] and Ursell [423], to name a few.

2.1.1 Low Frequency Behaviour

Of particular interest is the case $k \to 0$ which corresponds to the low–frequency behaviour. This case also determines the large–time behaviour of the solution to time–dependent problems if (2.1.1) is obtained from the wave equation by the Fourier–Laplace transformation (see e.g. MacCamy [274, 275] and Werner [436]). As will be seen, some of the boundary value problems will exhibit a singular behaviour for $k \to 0$. The main results are summarized in the Table 2.1.2 below.

To illustrate the singular behaviour we begin with the explicit asymptotic expansions of the boundary integral equations in Table 2.1.1. Our presentation here follows [203]. In particular, we begin with the fundamental solution for small kr having the series expansions:

<u>For $n = 2$:</u>

$$E_k(x,y) = \frac{i}{4}H_0^{(1)}(kr) = E(x,y) - \frac{1}{2\pi}(\log k + \gamma_0) + S_k(x,y)\,, \tag{2.1.18}$$

where

$$\gamma_0 = c_0 - \log 2 - i\frac{\pi}{2} \quad \text{with } c_0 \approx 0.5772\,, \quad \text{Euler's constant,}$$

and

$$\begin{aligned}
S_k(x,y) &= \frac{i}{4}H_0^{(1)}(kr) - \{E(x,y) + \frac{1}{2\pi}(\log(kr)+\gamma_0)\} \\
&= -\frac{1}{2\pi}\{\log(kr)\sum_{m=1}^{\infty} a_m(kr)^{2m} + \sum_{m=1}^{\infty} b_m(kr)^{2m}\}, \\
a_m &= \frac{(-1)^m}{2^{2m}(m!)^2}, \qquad b_m = (\gamma_0 - 1 - \frac{1}{2} - \cdots - \frac{1}{m})a_m .
\end{aligned}$$

For $n = 3$:

$$E_k(x,y) = E(x,y) + \frac{ik}{4\pi} + S_k(x,y)\,,$$

where

$$S_k(x,y) = \frac{1}{4\pi r}(e^{ikr} - 1 - ikr) = \frac{1}{4\pi r}\sum_{m=2}^{\infty}\frac{(ikr)^m}{m!}\,. \tag{2.1.19}$$

Correspondingly, for the double layer potential kernel we obtain

$$\frac{\partial}{\partial n_y}E_k(x,y) = \frac{\partial}{\partial n_y}E(x,y) + R_k(x,y)\,,$$

where

$$R_k(x,y) = \Big\{\sum_{m=1}^{\infty}\big(a_m(1+2m\log(kr)) + 2mb_m\big)(rk)^{2m}\Big\}\frac{\partial}{\partial n_y}E(x,y) \quad \text{for } n=2\,, \tag{2.1.20}$$

and

$$R_k(x,y) = -\Big\{\sum_{m=2}^{\infty}\frac{(m-1)}{m!}(ikr)^m\Big\}\frac{\partial}{\partial n_y}E(x,y) \quad \text{for } n=3\,. \tag{2.1.21}$$

The kernel of the adjoint operator R'_k can be obtained by interchanging the variables x and y. Hence, R'_k has the same asymptotic behaviour as R_k for $k \to 0$.

Similarly, for the hypersingular kernel we have, for $n = 2$,

$$\begin{aligned}
\frac{\partial R_k}{\partial n_x}(x,y) &= -4\pi\sum_{m=2}^{\infty}(m-1)c_m(k)(rk)^{2m}\frac{\partial E}{\partial n_x}\frac{\partial E}{\partial n_y}(x,y) \qquad (2.1.22)\\
&\quad + \frac{1}{2\pi}\sum_{m=1}^{\infty}c_m(k)(rk)^{2m}\frac{n_x\cdot n_y}{r^2}\,,\\
&\quad + \log r\sum_{m=1}^{\infty}2ma_m(rk)^{2m}\Big\{\frac{1}{2\pi}\frac{\boldsymbol{n}_x\cdot\boldsymbol{n}_y}{r^2} - 4\pi(m-1)\frac{\partial E}{\partial n_x}\frac{\partial E}{\partial n_y}\Big\}
\end{aligned}$$

where

$$c_m(k) = (1 + 2m \log k)a_m + 2mb_m\,,$$

and, for $n = 3$,

$$\begin{aligned}\frac{\partial R_k}{\partial n_x}(x,y) = &-\frac{1}{4\pi}\sum_{m=2}^{\infty}\frac{(m-1)}{m!}(ikr)^m\frac{\boldsymbol{n}_x\cdot\boldsymbol{n}_y}{r^3}\\ &-4\pi\sum_{m=2}^{\infty}(3+m)\frac{(m-1)}{m!}r(ikr)^m\frac{\partial E}{\partial n_x}\frac{\partial E}{\partial n_y}(x,y)\,.\end{aligned} \tag{2.1.23}$$

Note that the kernel $\dfrac{\partial R_k}{\partial n_x}(x,y)$ is symmetric.

As can be seen from the above expansions, the term $\log k$ appears in (2.1.18) explicitly which shows that V_k is a singular perturbation of V whereas the other operators are regular perturbations of the corresponding operators of the Laplacian as $k \to 0$.

IDP: Let us consider first the simplest case, i.e. Equation (2) for (IDP) in Table 2.1.1,

$$\left(\tfrac{1}{2}I - K' - R_k'\right)\sigma = D_k\varphi \quad \text{on } \Gamma\,. \tag{2.1.24}$$

Since, for given $\varphi \in C^{1+\alpha}(\Gamma)$, the equation

$$(\tfrac{1}{2}I - K')\widetilde{\sigma} = D\varphi$$

has a unique solution $\widetilde{\sigma} \in C^{\alpha}(\Gamma)$, we may rewrite (2.1.24) as

$$\begin{aligned}\sigma &= (\frac{1}{2}I - K')^{-1}R_k'\sigma + (\frac{1}{2}I - K')^{-1}D_k\varphi\,,\\ &= \widetilde{\sigma} + (\frac{1}{2}I - K')^{-1}R_k'\sigma + (\frac{1}{2}I - K')^{-1}\frac{\partial}{\partial n_x}R_k\varphi\,, \qquad (2.1.25)\\ &= \widetilde{\sigma} + \begin{cases} O(k^2 \log k) & \text{for } n = 2\,,\\ O(k^2) & \text{for } n = 3\,,\end{cases}\end{aligned}$$

where the last expressions can be obtained from the expansions (2.1.20) and (2.1.22) in case $n = 2$ and from (2.1.21) and (2.1.23) in case $n = 3$ (see MacCamy [274]).

The analysis for the integral equation of the first kind (1) for (IDP) in Table 2.1.1 is more involved, depending on $n = 2$ or 3. For $n = 2$, from (2.1.11) with the expansion (2.1.18) of (2.1.7) we have

$$V\sigma + \omega + S_k\sigma = \frac{1}{2}\varphi + K\varphi + R_k\varphi \tag{2.1.26}$$

where

$$\omega = -\frac{1}{2\pi}(\log k + \gamma_0)\int\limits_{\Gamma}\sigma ds\,. \tag{2.1.27}$$

Similar to (2.1.25), we seek the solution of (2.1.26) and (2.1.27) in the asymptotic form,

$$\begin{aligned} \sigma &= \widetilde{\sigma} + \alpha_1(k)\widetilde{\sigma}_1 + \sigma_R\,, \\ \omega &= \widetilde{\omega} + \alpha_1(k)\widetilde{\omega}_1 + \omega_R\,, \end{aligned} \tag{2.1.28}$$

where $\widetilde{\sigma}, \widetilde{\omega}$ correspond to the solution of the interior Dirichlet problem for the Laplacian (1.1.6), (1.3.1). Hence, $\widetilde{\sigma}, \widetilde{\omega}$ satisfy (1.3.3), namely

$$V\widetilde{\sigma} + \widetilde{\omega} = \frac{1}{2}\varphi + K\varphi \tag{2.1.29}$$

subject to the constraints

$$\int_\Gamma \widetilde{\sigma} ds = 0 \quad \text{and} \quad \widetilde{\omega} = 0\,.$$

The first perturbation terms $\widetilde{\sigma}_1, \widetilde{\omega}_1$ are independent of k with the coefficient $\alpha_1(k) = o(1)$ for $k \to 0$. The functions σ_R, ω_R are the remainders which are of order $o(\alpha_1(k))$. To construct $\widetilde{\sigma}_1$ and $\widetilde{\omega}_1$, we employ equation (2.1.26) inserting (2.1.28). For $k \to 0$ we arrive at

$$\begin{aligned} V\widetilde{\sigma}_1 + \widetilde{\omega}_1 &= 0\,, \\ \int_\Gamma \widetilde{\sigma}_1 ds &= 1\,, \end{aligned} \tag{2.1.30}$$

where we appended the last normalizing condition for $\widetilde{\sigma}_1$, since from the previous results in Section 1.3 we know that $\int_\Gamma \widetilde{\sigma}_1 ds = 0$ would yield the trivial solution $\widetilde{\sigma}_1 = 0, \widetilde{\omega}_1 = 0$. Inserting (2.1.28) into (2.1.27), it follows from $\int_\Gamma \widetilde{\sigma}_1 ds = 1$ with $\widetilde{\omega} = 0$ that $\alpha_1(k) = O(\sigma_R)$. Hence, without loss of generality, we may set $\alpha_1(k) = 0$ in (2.2.28). Now (2.1.26) and (2.1.27) with (2.1.28) imply that the remaining terms σ_R, ω_R satisfy the equations

$$\begin{aligned} V\sigma_R + \omega_R + S_k\sigma_R &= R_k\varphi - S_k\widetilde{\sigma}\,, \\ \int_\Gamma \sigma_R ds + \omega_R 2\pi(\log k + \gamma_0)^{-1} &= 0\,; \end{aligned} \tag{2.1.31}$$

which are *regular* perturbations of equations (1.3.5), (1.3.6) due to the expansions for S_k in (2.1.18). The right–hand side of (2.1.31) is of order $O(k^2 \log k)$ because of the expansions of S_k and of R_k in (2.1.20). Therefore, σ_R and ω_R are, indeed, of order $O(k^2 \log k)$ as in (2.1.25), which was already obtained with the integral equations (2.1.24) of the second kind.

For $n = 3$, the integral equation of the first kind takes the form

$$V\sigma + ik\frac{1}{4\pi}\int_\Gamma \sigma ds + S_k\sigma = \frac{1}{2}\varphi + K\varphi + R_k\varphi\,, \tag{2.1.32}$$

where the kernels of the integral operators S_k and R_k are given by (2.1.19) and (2.1.21), respectively. Both are of order $O(k^2)$. Hence, (2.1.32) is a regular perturbation of equation (1.3.3). If we insert (2.1.28) with $\alpha_1(k) = k$ then the function $\widetilde{\sigma}$ is given by the solution of

$$V\widetilde{\sigma} = \tfrac{1}{2}\varphi + K\varphi\,,$$

which corresponds to the interior Dirichlet problem of the Laplacian. Hence,

$$\int_\Gamma \widetilde{\sigma} ds = 0 \quad \text{and} \quad \widetilde{\sigma}_1 \equiv 0$$

since it is the solution of

$$V\widetilde{\sigma}_1 + \frac{i}{4\pi}\int\limits_\Gamma \widetilde{\sigma} ds = 0\,.$$

Therefore, the solution of (2.1.32) is of the form $\sigma = \widetilde{\sigma} + O(k^2)$.

EDP: By using the indirect formulation of boundary integral equations, this case was also analyzed by Hariharan and MacCamy in [171]. In case $n = 2$, for the exterior Dirichlet problem (EDP), Equation (1) in Table 2.1.1 has the form

$$V\sigma + \omega + S_k\sigma = -\tfrac{1}{2}\varphi + K\varphi + R_k\varphi \tag{2.1.33}$$

with ω again defined by (2.1.27). Again, the solution admits the asymptotic expansion (2.1.28) with $\widetilde{\sigma}, \widetilde{\omega}$ being the solution of

$$V\widetilde{\sigma} + \widetilde{\omega} = -\tfrac{1}{2}\varphi + K\varphi\,, \quad \int\limits_\Gamma \widetilde{\sigma} ds = 0\,. \tag{2.1.34}$$

Hence, $\widetilde{\sigma}, \widetilde{\omega}$ correspond to the exterior Dirichlet problem in Section 1.4.1. Therefore, in contrast to the (IDP), here $\widetilde{\omega} \neq 0$, in general. Again, $\widetilde{\sigma}_1, \widetilde{\omega}_1$ are solutions of equations (2.1.30). In contrast to the (IDP), the coefficient $\alpha_1(k)$ is explicitly given by

$$\alpha_1(k) = -\widetilde{\omega}\left\{\frac{1}{\pi}(\log k + \gamma_0) + \widetilde{\omega}_1\right\}^{-1},$$

which is not identically equal to zero, in general. The remainders σ_R, ω_R satisfy equations similar to (2.1.31) and are of order $O(k^2 \log k)$.

For the case $n = 3$, the integral equation(1) for (EDP) is of the form

$$V\sigma + ik\frac{1}{4\pi}\int\limits_\Gamma \sigma ds + S_k\sigma = -\tfrac{1}{2}\varphi + K\varphi + R_k\varphi$$

and σ is of the form

$$\sigma = \widetilde{\sigma} + k\widetilde{\sigma}_1 + \sigma_R\,.$$

This yields

$$V\widetilde{\sigma} = -\tfrac{1}{2}\varphi + K\varphi$$

for $\widetilde{\sigma}$ corresponding to the (EDP) for the Laplacian . For the next term $\widetilde{\sigma}_1$ we have

$$V\widetilde{\sigma}_1 = -\frac{i}{4\pi}\int\limits_{\Gamma}\widetilde{\sigma}ds\,.$$

Therefore, $\widetilde{\sigma}_1$ is proportional to the *natural charge* q which is the unique eigensolution of

$$\tfrac{1}{2}q + K'q = 0 \;\text{ on }\; \Gamma \;\text{ with }\; \int\limits_{\Gamma} qds = 1\,.$$

Note that the corresponding simple layer potential satisfies

$$Vq(x) = c_0 = \;\text{const. for all}\; x \in \overline{\Omega}\,. \tag{2.1.35}$$

For $n = 3$, a simple contradiction argument shows $c_0 \neq 0$. Hence,

$$\widetilde{\sigma}_1(x) = -\Big(\frac{i}{4\pi c_0}\int\limits_{\Gamma}\widetilde{\sigma}ds\Big)q(x)\,.$$

The remainder term σ_R is of order k^2 which follows easily from

$$V\sigma_R + S_k\sigma_R = -S_k(\widetilde{\sigma} + k\widetilde{\sigma}_1) - k^2\frac{i}{4\pi}\int\limits_{\Gamma}\widetilde{\sigma}_1 ds - R_k\varphi\,,$$

which is a regular perturbation of (1.4.12).

For the integral equation of the second kind (2) of the (EDP) and $k = 0$, the homogeneous reduced integral equation reads

$$\frac{1}{2}q + K'q = 0 \tag{2.1.36}$$

with the natural charge q as an eigensolution. We therefore modify boundary integral equation (2) by using a method by Wielandt [437] (see also Werner [435]). Using Equation (1), we obtain a normalization condition for σ,

$$\int\limits_{\Gamma}\sigma ds = l_k(\varphi) + \widetilde{l}_k(\sigma)\,,$$

where the linear functionals l_k and $\widetilde{l_k}$ are given by

$$l_k(\varphi) = \begin{cases} \left(\frac{1}{2\pi}(\log k + \gamma_0) - c_0\right)^{-1} \left\{ \int\limits_\Gamma \varphi q ds + \int\limits_\Gamma (R_k\varphi) q ds \right\} & \text{for } n = 2 \ , \\ (c_0 + k\frac{i}{4\pi})^{-1} \left\{ \int\limits_\Gamma \varphi q ds + \int\limits_\Gamma (R_k\varphi) q ds \right\} & \text{for } n = 3 \ ; \end{cases}$$

and

$$\widetilde{l}_k(\sigma) = \begin{cases} \left(\frac{1}{2\pi}(\log k + \gamma_0) - c_0\right)^{-1} \int\limits_\Gamma (S_k\sigma) q ds & \text{for } n = 2 \ , \\ \left(c_0 + k\frac{i}{4\pi}\right)^{-1} \int\limits_\Gamma (S_k\sigma) q ds & \text{for } n = 3 \ . \end{cases}$$

The modified boundary integral equation of the second kind reads

$$(\tfrac{1}{2}I + K')\sigma + q \int\limits_\Gamma \sigma ds + R'_k\sigma - q\widetilde{l}_k(\sigma) = -D_k\varphi + ql_k(\varphi) \,, \tag{2.1.37}$$

which is a regular perturbation of the equation

$$(\tfrac{1}{2}I + K')\widetilde{\sigma} + q \int\limits_\Gamma \widetilde{\sigma} ds = -D\varphi + ql_0(\varphi) \,.$$

The latter is uniquely solvable. The next term $\widetilde{\sigma}_1$ satisfies the equation

$$(\tfrac{1}{2}I + K')\widetilde{\sigma}_1 + q \int\limits_\Gamma \widetilde{\sigma}_1 ds = \int\limits_\Gamma \varphi q ds \,.$$

It is not difficult to see that both boundary integral equations (2.1.33) and (2.1.37) provide the same asymptotic solutions

$$\sigma = \begin{cases} \widetilde{\sigma} + \alpha_1(k)\widetilde{\sigma}_1 + O(k^2 \log k) & \text{for } n = 2 \ , \\ \widetilde{\sigma} + k\widetilde{\sigma}_1 + O(k^2) & \text{for } n = 3 \ . \end{cases}$$

<u>INP:</u> Since the boundary integral equation of the second kind (see (2) in Table 2.1.1) for the (INP) and $k = 0$ has the constant functions as eigensolutions, we need an appropriate modification of (2). This modification can be derived from the Helmholtz equation (2.1.1) in Ω together with the Neumann boundary condition

$$\frac{\partial u}{\partial n}|_\Gamma = \psi \ \text{ on } \Gamma \,, \tag{2.1.38}$$

namely from the solvability condition (Green's formula for the Laplacian)

$$\int\limits_\Omega u(x) dx = -\frac{1}{k^2} \int\limits_\Gamma \psi ds \,. \tag{2.1.39}$$

This condition can be rewritten in terms of the representation formula (2.1.5),

$$\int_{\Omega} W_k u(x)dx = \int_{\Omega} V_k\psi(x)dx + \frac{1}{k^2}\int_{\Gamma} \psi ds\,, \tag{2.1.40}$$

the left–hand side of which actually depends on the boundary values $u_{|\Gamma}$. It is not difficult to replace the domain integration on the left–hand side by appropriate boundary integrals. With the help of (2.1.40), the boundary integral equation of the second kind (see (2) in Table 2.1.1) for (INP) can be modified,

$$\begin{aligned}\frac{1}{2}u + Ku - \frac{1}{|\Omega|}\int_{\Omega} Wu(x)dx + R_k u - \frac{1}{|\Omega|}\int_{\Omega} R_k u(x)dx \\ = -\frac{1}{|\Omega|k^2}\int_{\Gamma}\psi ds + V\psi - \frac{1}{|\Omega|}\int_{\Omega} V\psi(x)dx \\ + S_k\psi - \frac{1}{|\Omega|}\int_{\Omega} S_k\psi(x)dx \quad \text{on } \Gamma\end{aligned} \tag{2.1.41}$$

where $|\Omega| = meas(\Omega)$. Note, that the constant term of order $\log k$ in V_k for $n = 2$ and of order k for $n = 3$, respectively, has been canceled in (2.1.41) due to (2.1.40).

The boundary integral operator on the left–hand side in (2.1.41) is a regular perturbation of the reduced modified operator

$$A := \tfrac{1}{2}I + K - \frac{1}{|\Omega|}\int_{\Omega} W \bullet dx \quad \text{on } \Gamma$$

due to (2.1.20) and (2.1.21). Moreover, we will see that the reduced homogeneous equation

$$Av_0 = 0 \text{ on } \Gamma$$

admits only the trivial solution $v_0 = 0$, since for the equivalent equation,

$$\tfrac{1}{2}v_0 + Kv_0 = \frac{1}{|\Omega|}\int_{\Omega} Wv_0 dx = \kappa = \text{ const}\,, \tag{2.1.42}$$

the orthogonality condition for the original reduced operator on the left–hand side reads

$$\int_{\Gamma} q\kappa ds = \kappa = 0\,.$$

This implies from (2.1.42) that $v_0 = const$ as the eigensolution discussed in Section 1.3.2. Hence, with (1.1.7), we have

$$0 = \kappa = \frac{1}{|\Omega|}\int_{\Omega} Wv_0 dx = -v_0\,.$$

Consequently, A^{-1} exists in the classical function space $C^\alpha(\Gamma)$ due to the Fredholm alternative.

As for the asymptotic behaviour of u, it is suggested from (2.1.40) or (2.1.41) that u admits the form

$$u = \frac{1}{k^2}\alpha + \widetilde{u} + u_R\,. \tag{2.1.43}$$

The first term α satisfies the equation

$$A\alpha = \tfrac{1}{2}\alpha + K\alpha - \frac{1}{|\Omega|}\int\limits_{\Omega} W\alpha dx = -\frac{1}{|\Omega|}\int\limits_{\Gamma}\psi ds$$

which has the unique solution

$$\alpha = -\frac{1}{|\Omega|}\int\limits_{\Gamma}\psi ds = \text{ const}\,. \tag{2.1.44}$$

The second term $\widetilde{u}$ is the unique solution of the equation

$$A\widetilde{u} = V\psi - \frac{1}{|\Omega|}\int\limits_{\Omega} V\psi dx - \alpha\lim_{k\to 0}\frac{1}{k^2}\Big\{R_k 1 - \frac{1}{|\Omega|}\int\limits_{\Omega} R_k 1 dx\Big\}\,. \tag{2.1.45}$$

Now we investigate the last term in (2.1.45).
For $n = 2$, we have

$$R_k 1 = \frac{|\Omega|}{2\pi}(\log k + \gamma_0)k^2 + \frac{k^2}{2\pi}\int\limits_{\Omega}\log|x-y|dy + O(k^4\log k).$$

This implies

$$\begin{aligned}&\frac{1}{k^2}\Big\{R_k 1 - \frac{1}{|\Omega|}\int\limits_{\Omega} R_k 1 dx\Big\},\\ &\quad = \frac{1}{2\pi}\Big\{\int\limits_{\Omega}\log|x-y|dy - \frac{1}{|\Omega|}\int\limits_{\Omega}\int\limits_{\Omega}\log|x-y|dydx\Big\} + O(k^2\log k)\,.\end{aligned}$$

Hence, the limit on the right–hand side in (2.1.45) exists.
For $n = 3$, the limit is a well defined function in $C^\alpha(\Gamma)$ due to (2.1.21).

The remainder u_R satisfies the equation

$$Au_R + R_k u_R - \frac{1}{|\Omega|}\int\limits_{\Omega} R_k u_R(x)dx = f_R \text{ on } \Gamma \tag{2.1.46}$$

where

$$f_R = \begin{cases} O(k^2 \log k) & \text{for } n = 2 \ , \\ O(k^2) & \text{for } n = 3 \ . \end{cases}$$

Hence, u_R is uniquely determined by (2.1.46) and is of the same order as f_R.

For the first kind hypersingular equation (1) in (INP) we use the same normalization condition (2.1.40) by subtracting it from the equation and obtain the modified hypersingular equation

$$\begin{aligned} Du - \int\limits_{\Omega} Wu(z)dz - \frac{\partial}{\partial n_x} R_k u - \int\limits_{\Omega} R_k u dz \\ = (\tfrac{1}{2} - K')\psi - R_k'\psi - \int\limits_{\Omega} V\psi dz - \int\limits_{\Gamma} S_k \psi dz \qquad (2.1.47) \\ + |\Omega| \Big(\delta_{n2} \frac{1}{2\pi} (\log k + \gamma_0) - \delta_{n3} \frac{ik}{4\pi} \Big) \int\limits_{\Gamma} \psi ds - \frac{1}{k^2} \int\limits_{\Gamma} \psi ds \, . \end{aligned}$$

Again, the operator on the left-hand side is a regular perturbation of the reduced operator

$$Bu := Du - \int\limits_{\Omega} Wu(z)dz,$$

which can be shown to be invertible, in the same manner as for A. From the previous analysis, we write the solution in the form (2.1.43), namely

$$u = \frac{\alpha}{k^2} + \widetilde{u} + u_R \, .$$

We note that

$$B1 = |\Omega|,$$

and therefore (2.1.47) for $k \to 0$ again yields (2.1.44) for α. We further note that for $n = 2$, it can be verified from (2.1.20) that

$$\int\limits_{\Omega} R_k \left(\frac{\alpha}{k^2} \right) dz = \frac{1}{2\pi} |\Omega|^2 \alpha \log k + O(1) \ \text{ for } \ k \to 0 \, ,$$

which shows that the choice of α from (2.1.44) cancels the term $\frac{1}{2\pi} |\Omega| \log k \int_{\Gamma} \psi ds$ on the right–hand side of (2.1.47).

Consequently in both cases $n = 2$ and $n = 3$, $\widetilde{u}$ can be obtained as the unique solution of the equation

$$B\widetilde{u} = \frac{1}{2}\psi - K'\psi - \int\limits_{\Omega} V\psi dz + \alpha \chi_n \, ,$$

where

$$\chi_n = \begin{cases} -\frac{\gamma_0}{2\pi}|\Omega|^2 + \lim\limits_{k\to 0}\left\{\frac{1}{k^2}\int\limits_\Omega R_k 1 dz - \frac{|\Omega|^2}{2\pi}\log k\right\}, & n=2\ , \\ \lim\limits_{k\to 0}\frac{1}{k^2}\{\frac{\partial}{\partial n_x}R_k 1 + \int\limits_\Omega R_k 1 dz\}, & n=3\ . \end{cases}$$

Similarly, again we can show that u_R is the unique solution of the equation

$$Bu_R - \frac{\partial}{\partial n_x}R_k u_R - \int\limits_\Omega R_k u_R(z)dz = f_R$$

with

$$\begin{aligned} f_R &= -R_k'\psi - \int\limits_\Omega S_k\psi dx + \frac{\partial}{\partial n_x}R_k\widetilde{u} + \int\limits_\Omega R_k\widetilde{u}dx \\ &\quad + \frac{\alpha}{k^2}\left\{\frac{\partial}{\partial n_x}R_k 1 + \int\limits_\Omega R_k 1 dx - \delta_{n2}\frac{|\Omega|^2}{2\pi}k^2\log k + \delta_{n3}\frac{|\Omega|^2}{4\pi}ik^3\right\} \\ &\quad - \alpha\lim_{k\to 0}\frac{1}{k^2}\left\{\frac{\partial}{\partial n_x}R_k 1 + \int\limits_\Omega R_k 1 dx - \delta_{n2}\frac{|\Omega|^2}{2\pi}k^2\log k\right\}. \end{aligned}$$

For $n = 2$, the above relations show that $f_R = O(k^2\log k)$, which implies that $u_R = O(k^2\log k)$, as well.

In case $n = 3$, by using the Gaussian theorem for the first two terms from (2.1.21) we obtain explicitly

$$R_k 1 = \int\limits_\Gamma R_k(x,y)ds_y = -\frac{1}{4\pi}k^2\int\limits_\Omega \frac{1}{|x-y|}dy - \frac{i}{4\pi}k^3|\Omega| + O(k^4)\,.$$

Hence,

$$\begin{aligned} \frac{\partial}{\partial n_x}R_k 1 &= -\frac{1}{4\pi}k^2\int\limits_\Omega \frac{\partial}{\partial n_x}\frac{1}{|x-y|}dy + O(k^4)\,, \\ \int\limits_\Omega R_k 1 dx &= -\frac{1}{4\pi}k^2\int\limits_\Omega\int\limits_\Omega \frac{1}{|x-y|}dydx - \frac{i}{4\pi}k^3|\Omega|^2 + O(k^4)\,. \end{aligned}$$

Together with (2.1.44) this yields

$$f_R = O(k^2)$$

which implies that u_R is of the same order as $k \to 0$.

ENP: Finally, let us consider the boundary integral equations of (ENP) and begin with the simplest case, i.e. the integral equation of the second kind (see (2) in Table 2.1.1),

$$\frac{1}{2}u - Ku - R_k u = -V\psi + \left\{\delta_{n2}\left(\frac{1}{2\pi}\log k + \gamma_0\right) - \delta_{n3}\frac{ik}{4\pi}\right\}\int\limits_\Gamma \psi ds - S_k\psi\,. \tag{2.1.48}$$

The operator on the left–hand side is a regular perturbation of the reduced invertible operator $\frac{1}{2}I - K$ (see Section 1.4.2).

For $n = 2$, $S_k\psi$ is of order $O(k^2 \log k)$ in view of (2.1.19). Hence, for $u_{|\Gamma}$ we use the expansion

$$u = \frac{1}{2\pi}(\log k + \gamma_0)\alpha + \widetilde{u} + u_R\,.$$

The highest order term

$$\alpha = \int\limits_\Gamma \psi ds$$

is the unique solution of the reduced equation

$$\tfrac{1}{2}\alpha - K\alpha = \int\limits_\Gamma \psi ds\,;$$

the second term $\widetilde{u}$ is the unique solution of the reduced equation

$$\tfrac{1}{2}\widetilde{u} - K\widetilde{u} = -V\psi\,. \tag{2.1.49}$$

For the remainder u_R we obtain the equation

$$\tfrac{1}{2}u_R - Ku_R - R_k u_R = \frac{\alpha}{2\pi}(\log k + \gamma_0)R_k 1 + R_k\widetilde{u} - S_k\psi\,.$$

The dominating term on the right–hand side is defined by

$$R_k 1 = \frac{1}{2\pi}|\Omega| k^2 \log k + O(k^2)\,,$$

therefore u_R is of order $O((k \log k)^2)$.

For $n = 3$, $S_k\psi$ in (2.1.48) is of order $O(k^2)$ in view of (2.1.19). Hence, now u is of the form

$$u = \widetilde{u} - \alpha\frac{ik}{4\pi} + u_R$$

where $\widetilde{u}$ is the unique solution of the reduced equation (2.1.49) and $\alpha = \int_\Gamma \psi ds$. The remainder u_R satisfies the equation

$$\tfrac{1}{2}u_R - Ku_R - R_k u_R = R_k\widetilde{u} - S_k\psi - \frac{ik}{4\pi}R_k\alpha\,,$$

and, therefore, is of the order $O(k^2)$.

The hypersingular boundary integral equation (1) of the first kind in Table 2.1.2 reads

$$Du - \frac{\partial}{\partial n_x} R_k u = -\Big(\frac{1}{2} I + K'\Big)\psi - R'_k \psi\,, \tag{2.1.50}$$

where D has the constant functions as eigensolutions.

Multiplying (2.1.48) by the natural charge q, integrating over Γ and using (2.1.36) and $V_q = c_0$, we obtain the relation

$$\begin{aligned} &\int\limits_\Gamma uqds - \int\limits_\Gamma q(R_k u)ds \\ &= -\Big\{c_0 - \delta_{n2}\frac{1}{2\pi}(\log k + \gamma_0) + \delta_{n3}\frac{ik}{4\pi}\Big\}\int\limits_\Gamma \psi ds - \int\limits_\Gamma q(S_k\psi)ds \end{aligned} \tag{2.1.51}$$

with c_0 given by (2.1.35). Now (2.1.51) can be used as a normalizing condition by combining (2.1.50) and (2.1.51); we arrive at the modified equation

$$\begin{aligned} Du + \int\limits_\Gamma uqds - \int\limits_\Gamma (R_k u)qds - \frac{\partial}{\partial n_x} R_k u & \\ = \Big(-c_0 + \delta_{n2}\frac{1}{2\pi}(\log k + \gamma_0) - \delta_{n3}\frac{ik}{\pi}\Big)\int\limits_\Gamma \psi ds & \\ - \big(\tfrac{1}{2}I + K'\big)\psi - R'_k\psi - \int\limits_\Gamma q(S_k\psi)ds\,. & \end{aligned} \tag{2.1.52}$$

Again, we assume u in the form

$$u = \delta_{n2}\frac{1}{2\pi}(\log k + \gamma_0 - 2\pi c_0)\alpha - \delta_{n3}\frac{ik}{4\pi}\alpha + \widetilde{u} + u_R\,,$$

where $k \to 0$ yields

$$\alpha = \int\limits_\Gamma \psi ds\,.$$

The term $\widetilde{u}$ is the unique solution of

$$D\widetilde{u} + \int\limits_\Gamma \widetilde{u}qds = -\delta_{n3}c_0\int\limits_\Gamma \psi ds - (\tfrac{1}{2}I + K')\psi\,.$$

The remainder u_R satisfies the equation

$$Du_R + \int\limits_\Gamma u_R qds - \int\limits_\Gamma q(R_k u_R)ds - \frac{\partial}{\partial n_x} R_k u_R = f_R\,,$$

where

$$f_R = \alpha\Big\{\delta_{n2}(\log k + \gamma_0 - 2\pi c_0) - \delta_{n3}\frac{ik}{4\pi}\Big\}\Big\{\frac{\partial}{\partial n_x}R_k 1 + \int\limits_\Gamma q(R_k 1)ds\Big\}$$
$$+ \frac{\partial}{\partial n_x}R_k\widetilde{u} + \int\limits_\Gamma q(R_k\widetilde{u})ds - R'_k\psi - \int\limits_\Gamma q(S_k\psi)ds$$

is of order $O\big((k\log k)^2\big)$ for $n = 2$ and of order $O(k^2)$ for $n = 3$, which is in agreement with the result from the previous analysis of the boundary integral equation of the second kind.

With the solutions of the BIEs available, we now summarize the asymptotic behaviour of the solutions of the BVPs for small k by substituting the boundary densities into the representation formula (2.1.5). In all the cases we arrive at the following asymptotic expression

$$u(x) = \pm[V\widetilde{\sigma}(x) - W\widetilde{u}(x)] + C(k;x) + R(k;x) \tag{2.1.53}$$

where the $\pm$ sign corresponds to the interior and exterior domain and $x \in \Omega$ or Ω^c as in (2.1.5). For the Dirichlet problems, $\widetilde{u}_{|\Gamma} = \varphi$, and for the Neumann problems, $\widetilde{\sigma}_{|\Gamma} = \psi$ on Γ, are the given boundary data, respectively, whereas the missing densities are the solutions of the corresponding BIEs presented above. In Formula (2.1.53), $C(k;x)$ denotes the highest order terms of the perturbations in Ω or Ω^c; whereas $R(k;x)$ denotes the remaining boundary potentials. The behaviour of C and R for $k \to 0$ is summarized in Table 2.1.2 below.

Table 2.1.2. Low frequency characteristics

BVP	$C(k;x)$	$R(k;x)$	n
IDP	0	$O((k\log k)^2)$	$n=2$
		$O(k^2)$	$n=3$
EDP	$-\widetilde{\omega}$	$O((k\log k)^2)$ [1]	$n=2$
	$-k\{V\widetilde{\sigma}_1(x) + \frac{i}{4\pi}\int\limits_\Gamma \widetilde{\sigma}ds\}$	$O(k^2)$	$n=3$
INP	$-\{\frac{1}{k^2} - \frac{1}{2\pi}\int\limits_\Omega \log\lvert x-y\rvert dy\}\frac{1}{\lvert\Omega\rvert}\int\limits_\Gamma \psi ds$	$O(k^2\log k)$	$n=2$
	$-\{\frac{1}{k^2} + \frac{1}{4\pi}\int\limits_\Omega \frac{1}{\lvert x-y\rvert}dy\}\frac{1}{\lvert\Omega\rvert}\int\limits_\Gamma \psi ds$	$O(k^2)$	$n=3$
ENP	$\frac{1}{2\pi}(\log k + \gamma_0)\int\limits_\Gamma \psi ds$	$O((k\log k)^2)$	$n=2$
	$-\frac{ik}{4\pi}\int\limits_\Gamma \psi ds$	$O(k^2)$	$n=3$

[1] For this case a sharper result with $O(k^2)$ is given by MacCamy [274]

The remainders $R(x;k)$ are of the orders as given in the table, uniformly in $x \in \Omega$ for the interior problems and in compact subsets of $\overline{\Omega^c}$ only, for the exterior problems.

2.2 The Lamé System

In linear elasticity for isotropic materials, the governing equations are

$$-\Delta^* \boldsymbol{v} := -\mu \Delta \boldsymbol{v} - (\mu + \lambda)\,\mathrm{grad}\mathrm{div}\, \boldsymbol{v} = \boldsymbol{f} \text{ in } \Omega \text{ (or } \Omega^c) , \tag{2.2.1}$$

where $\boldsymbol{v}$ is the desired displacement field and $\boldsymbol{f}$ is a given body force. The parameters μ and λ are the *Lamé constants* which characterize the elastic material. (See e.g. Ciarlet [68], Fichera [119], Gurtin [165], Kupradze et al [253] and Leis [263]).

For $n = 3$, one also has the relation $\lambda = 2\mu\nu/(1-2\nu)$ with $0 \le \nu < \frac{1}{2}$, the *Poisson ratio.* The latter relation is also valid for the *plane strain* problem in two dimensions, where (2.2.1) is considered with $n = 2$. In the special case of so–called *generalized plane stress* problems one still has (2.2.1) with $n = 2$ for the first two displacement components $(v_1, v_2)^\top$ but with a modified $\overline{\lambda} = 2\mu\overline{\nu}/(1 - 2\overline{\nu})$ and a modified Poisson ratio $\overline{\nu} = \nu/(1 + \nu)$. In this case and in what follows, we will keep the same notation for λ.

For the Lamé system, the *fundamental solution* is given by

$$E(x, y) = \frac{\lambda+3\mu}{4\pi(n-1)\mu(\lambda+2\mu)} \left\{ \gamma_n(x, y)\boldsymbol{I} + \frac{\lambda+\mu}{\lambda+3\mu} \frac{1}{r^n} (\boldsymbol{x} - \boldsymbol{y})(\boldsymbol{x} - \boldsymbol{y})^\top \right\} , \tag{2.2.2}$$

a *matrix*–valued function, where $\boldsymbol{I}$ is the identity matrix, $r = |x - y|$ and

$$\gamma_n(x, y) = \begin{cases} -\log r & \text{for } n = 2 \ , \\ \frac{1}{r} & \text{for } n = 3 \ . \end{cases} \tag{2.2.3}$$

The boundary integral equations for the so–called fundamental boundary value problems are based on the Green representation formula, which in elasticity also is termed the *Betti–Somigliana representation formula.* For interior problems, we have the representation

$$\boldsymbol{v}(x) = \int\limits_\Gamma E(x, y) T\boldsymbol{v}(y) ds_y - \int\limits_\Gamma \left(T_y E(x, y)\right)^\top \boldsymbol{v}(y) ds_y + \int\limits_\Omega E(x, y)\boldsymbol{f}(y) dy \tag{2.2.4}$$

for $x \in \Omega$. Here the traction on Γ is defined by

$$T\boldsymbol{v}_{|\Gamma} = \left(\lambda(\mathrm{div}\ \boldsymbol{v})\boldsymbol{n} + 2\mu \frac{\partial \boldsymbol{v}}{\partial n} + \mu \boldsymbol{n} \times \mathit{curl}\boldsymbol{v} \right)\Big|_\Gamma \tag{2.2.5}$$

for $n = 3$ which reduces to the case $n = 2$ by setting $u_3 = 0$ and the third component of the normal $n_3 = 0$. The subscript y in $T_y E(x, y)$ again denotes differentiations in (2.2.4) with respect to the variable y.

The last term in the representation (2.2.4) is the *volume potential* (or *Newton potential*) due to the body force $\boldsymbol{f}$ defining a particular solution $\boldsymbol{v}_p$ of (2.2.1). As in Section 2.1, we decompose the solution in the form

$$\boldsymbol{v} = \boldsymbol{v}_p + \boldsymbol{u}$$

where $\boldsymbol{u}$ now satisfies the homogeneous Equation (2.2.1) with $\boldsymbol{f} = \boldsymbol{0}$ and has a representation (2.2.4) with $\boldsymbol{f} = \boldsymbol{0}$, i.e.

$$\boldsymbol{u}(x) = V\boldsymbol{\sigma}(x) - W\boldsymbol{\varphi}(x)\,. \tag{2.2.6}$$

Here V and W are the simple and double layer potentials, now defined by

$$V\boldsymbol{\sigma}(x) := \int\limits_{\Gamma} E(x,y)\boldsymbol{\sigma}(y)ds_y\,, \tag{2.2.7}$$

$$W\boldsymbol{\varphi}(x) := \int\limits_{\Gamma} \left(T_y(x,y)E(x,y)\right)^{\top} \boldsymbol{\varphi}(y)ds_y\,; \tag{2.2.8}$$

and where in (2.2.6) the boundary charges are the Cauchy data $\boldsymbol{\varphi}(x) = \boldsymbol{u}(x)_{|\Gamma}$, $\boldsymbol{\sigma}(x) = T\boldsymbol{u}(x)_{|\Gamma}$ of the solution to

$$-\Delta^*\boldsymbol{u} = \boldsymbol{0} \;\text{ in }\; \Omega\,. \tag{2.2.9}$$

For linear problems, because of the above decomposition, in the following, we shall consider, without loss of generality, only the case of the homogeneous equation (2.2.9), i.e., $\boldsymbol{f} = \boldsymbol{0}$.

For the exterior problems, the representation formula for $\boldsymbol{v}$ needs to be modified by taking into account growth conditions at infinity. For $\boldsymbol{f} = \boldsymbol{0}$, the growth conditions are

$$\boldsymbol{u}(x) = -E(x,0)\boldsymbol{\Sigma} + \boldsymbol{\omega}(x) + O(|x|^{1-n}) \;\text{ as }\; |x| \to \infty\,, \tag{2.2.10}$$

where $\boldsymbol{\omega}(x)$ is a rigid motion defined by

$$\boldsymbol{\omega}(x) = \begin{cases} \boldsymbol{a} + b(-x_2, x_1)^{\top} & \text{for } n = 2\;, \\ \boldsymbol{a} + \boldsymbol{b} \times \boldsymbol{x} & \text{for } n = 3\;. \end{cases} \tag{2.2.11}$$

Here, $\boldsymbol{a}, \boldsymbol{b}$ and $\boldsymbol{\Sigma}$ are constant vectors; the former denote translation and rotation, respectively. The representation formula for solutions of

$$-\Delta^*\boldsymbol{u} = \boldsymbol{0} \;\text{ in }\; \Omega^c \tag{2.2.12}$$

with the growth condition (2.2.10) has the form

$$\boldsymbol{u}(x) = -V\boldsymbol{\sigma}(x) + W\boldsymbol{\varphi}(x) + \boldsymbol{\omega}(x) \tag{2.2.13}$$

with the Cauchy data $\boldsymbol{\varphi} = \boldsymbol{u}_{|\Gamma}$ and $\boldsymbol{\sigma} = T\boldsymbol{u}_{|\Gamma}$ and with

$$\boldsymbol{\Sigma} = \int\limits_{\Gamma} \boldsymbol{\sigma} ds\,. \tag{2.2.14}$$

2.2.1 The Interior Displacement Problem

The Dirichlet problem for the Lamé equations (2.2.9) in Ω is called the *displacement problem* since here the boundary displacement

$$\boldsymbol{u}_{|\Gamma} = \boldsymbol{\varphi} \text{ on } \Gamma \tag{2.2.15}$$

is prescribed. The missing Cauchy datum on Γ is the boundary traction $\boldsymbol{\sigma} = T\boldsymbol{u}_{|\Gamma}$. Applying the trace and the traction operator T (2.2.5) to both sides of the representation formula (2.2.4), we obtain the overdetermined system of boundary integral equations

$$\boldsymbol{\varphi}(x) = (\tfrac{1}{2}I - K)\boldsymbol{\varphi}(x) + V\boldsymbol{\sigma}(x)\,, \tag{2.2.16}$$

$$\boldsymbol{\sigma}(x) = D\boldsymbol{\varphi} + (\tfrac{1}{2}I + K')\boldsymbol{\sigma}(x) \text{ on } \Gamma\,. \tag{2.2.17}$$

Here, the boundary integral operators are defined as in (1.2.3) –(1.2.6), however, it is understood that the fundamental solution now is given by (2.2.2) and the differentiation $\frac{\partial}{\partial n_{|\Gamma}}$ (1.2.4)–(1.2.6) is to be replaced by the traction operator $T_{|\Gamma}$ (2.2.5). Explicitly, we also have:

Lemma 2.2.1. *Let $\Gamma \in C^2$ and let $\boldsymbol{\varphi} \in C^\alpha(\Gamma), \boldsymbol{\sigma} \in C^\alpha(\Gamma)$ with $0 < \alpha < 1$. Then, for the case of elasticity, the limits (1.2.3)–(1.2.5) exist uniformly with respect to all $x \in \Gamma$ and all $\boldsymbol{\varphi}$ and $\boldsymbol{\sigma}$ with $||\boldsymbol{\varphi}||_{C^\alpha} \le 1$, $||\boldsymbol{\sigma}||_{C^\alpha} \le 1$. Furthermore, these limits can be expressed by*

$$V\boldsymbol{\sigma}(x) = \int\limits_{y\in\Gamma\setminus\{x\}} E(x,y)\boldsymbol{\sigma}(y)ds_y, \qquad x \in \Gamma\,, \tag{2.2.18}$$

$$K\boldsymbol{\varphi}(x) = \text{p.v.} \int\limits_{y\in\Gamma\setminus\{x\}} \left(T_y E(x,y)\right)^\top \boldsymbol{\varphi}(y)ds_y\,, \quad x \in \Gamma\,, \tag{2.2.19}$$

$$K'\boldsymbol{\sigma}(x) = \text{p.v.} \int\limits_{y\in\Gamma\setminus\{x\}} \left(T_x E(x,y)\right)\boldsymbol{\sigma}(y)ds_y\,, \quad x \in \Gamma\,. \tag{2.2.20}$$

These results are originally due to Giraud [145], see also Kupradze [252, 253] and the references therein. We remark that the integral in (2.2.18) is a weakly singular improper integral, whereas the integrals in (2.2.19) and (2.2.20) are to be defined as *Cauchy principal value integrals*, i.e.

$$\text{p.v.} \int\limits_{y\in\Gamma\setminus\{x\}} \left(T_y E(x,y)\right)^\top \boldsymbol{\varphi}(y)ds_y = \lim_{\varepsilon\to 0} \int\limits_{|y-x|\ge\varepsilon>0\wedge y\in\Gamma} \left(T_y E(x,y)\right)^\top \boldsymbol{\varphi}(y)ds_y\,,$$

since the operators K and K' have Cauchy singular kernels:

$$\begin{aligned}
&(T_y E(x,y))^\top \\
&= \frac{\mu}{2(n-1)\pi(\lambda+2\mu)}\Big\{\Big(I+\frac{n(\lambda+\mu)}{\mu|x-y|^2}(x-y)(x-y)^\top\Big)\frac{\partial\gamma_n}{\partial n_y}(x,y) \\
&\quad +\frac{1}{|x-y|^n}\left((x-y)\boldsymbol{n}_y^\top-\boldsymbol{n}_y(x-y)^\top\right)\Big\}^\top , \qquad (2.2.21)
\end{aligned}$$

$$\begin{aligned}
&(T_x E(x,y)) \\
&= \frac{\mu}{2(n-1)\pi(\lambda+2\mu)}\Big\{\Big(I+\frac{n(\lambda+\mu)}{\mu|x-y|^2}(x-y)(x-y)^\top\Big)\frac{\partial\gamma_n}{\partial n_x}(x,y) \\
&\quad -\frac{1}{|x-y|^n}\left((x-y)\boldsymbol{n}_x^\top-\boldsymbol{n}_x(x-y)^\top\right)\Big\}. \qquad (2.2.22)
\end{aligned}$$

We note that in case $n=2$, the last term in (2.2.21) can also be written in the form

$$\frac{1}{|x-y|^2}\left((x-y)\boldsymbol{n}_y^\top-\boldsymbol{n}_y(x-y)^\top\right)=\begin{pmatrix}0 & , & 1\\ -1 & , & 0\end{pmatrix}\frac{d}{ds_y}\log|x-y|\,. \qquad (2.2.23)$$

The last terms in the kernels (2.2.21) and (2.2.22) are of the order $O(r^{-n})$ as $r=|x-y|\to 0$ and, in addition, satisfy the *Mikhlin condition* (see Mikhlin [302, Chap.5] or [305, Chap.IX]) which is necessary and sufficient for the existence of the above Cauchy principal value integrals for any $x\in\Gamma$. We shall return to this condition later on. The proof of Lemma 2.2.1 can be found in [302, Chap.45 p.210 ff].

It turns out that there exists a close relation between the single and double layer operators of the Laplacian and the Lamé system based on the following lemma and the *Günter derivatives*

$$m_{jk}(\partial_y,\boldsymbol{n}(y)) := n_k(y)\frac{\partial}{\partial y_j}-n_j(y)\frac{\partial}{\partial y_k}=-m_{kj}(\partial_y,\boldsymbol{n}(y)) \qquad (2.2.24)$$

(see Kupradze et al [253, (4.7), p. 99]) which are particular tangential derivatives; in matrix form

$$\mathcal{M}(\partial_y,\boldsymbol{n}(y)) := \big(\big(m_{jk}(\partial_y,\boldsymbol{n}(y))\big)\big)_{3\times 3}\,. \qquad (2.2.25)$$

Lemma 2.2.2. *For the Günter derivatives there hold the identities*

$$m_{j\ell}(\partial_y,\boldsymbol{n}(y))\gamma_n = \frac{n_j(y_\ell-x_\ell)-n_\ell(y_j-x_j)}{|x-y|^n}, \qquad (2.2.26)$$

$$\begin{aligned}
&\sum_{\ell=1}^n m_{j\ell}(\partial_y,\boldsymbol{n}(y))\frac{(y_\ell-x_\ell)(y_k-x_k)}{|x-y|^n} \\
&\qquad = -\Big(\delta_{jk}-\frac{n(y_j-x_j)(y_k-x_k)}{|x-y|^2}\Big)\frac{\partial}{\partial n_y}\gamma_n\,. \qquad (2.2.27)
\end{aligned}$$

The proof is straight forward by direct computation.

Inserting the identities (2.2.26) and (2.2.27) into (2.2.21) we arrive at

$$
\begin{aligned}
\big(T_y E(x,y)\big)^\top &= \frac{1}{2(n-1)\pi}\Big\{\frac{\partial}{\partial n_y}\gamma_n(x,y) + \mathcal{M}\big(\partial_y,(\boldsymbol{n}(y)\big)\gamma_n(x,y)\Big\} \\
&\quad + 2\mu\big(\mathcal{M}\big(\partial_y,\boldsymbol{n}(y)\big)E(x,y)\big)^\top . \qquad (2.2.28)
\end{aligned}
$$

In the same manner we find

$$
\begin{aligned}
\big(T_x E(x,y)\big) &= \frac{1}{2(n-1)\pi}\Big\{\frac{\partial}{\partial n_x}\gamma_n(x,y) - \mathcal{M}\big(\partial_x,(\boldsymbol{n}(x)\big)\gamma_n(x,y)\Big\} \\
&\quad + 2\mu\big(\mathcal{M}\big(\partial_x,\boldsymbol{n}(x)\big)E(x,y)\big) . \qquad (2.2.29)
\end{aligned}
$$

As a consequence, we may write the double layer potential operator in (2.2.19) after integration by parts in the form of the Stokes theorem as

$$
\begin{aligned}
K\varphi(x) &= \frac{1}{2(n-1)\pi}\Big\{\int\limits_\Gamma \Big(\frac{\partial}{\partial n_y}\gamma_n(x,y)\Big)\varphi(y)ds_y \\
&\quad - \int\limits_\Gamma \gamma_n(x,y)\mathcal{M}\big(\partial_y,\boldsymbol{n}(y)\big)\varphi(y)ds_y\Big\} \qquad (2.2.30) \\
&\quad + 2\mu\int\limits_\Gamma E(x,y)\mathcal{M}\big(\partial_y,\boldsymbol{n}(y)\big)\varphi(y)ds_y ,
\end{aligned}
$$

see Kupradze et al [253, Chap. V Theorem 6.1].

Finally, the hypersingular operator D in (2.2.17) is defined by

$$
\begin{aligned}
D\varphi(x) &= -T_x W\varphi(x) \\
&:= -\lim_{z\to x\in\Gamma, z\notin\Gamma}\Big(\lambda\big(\mathrm{div}_z W\varphi(z)\big) + 2\mu\big(\mathrm{grad}_z W\varphi(z)\big)\cdot\boldsymbol{n}_x \\
&\qquad + \mu\boldsymbol{n}_x \times \mathrm{curl}_z W\varphi(z)\Big) . \qquad (2.2.31)
\end{aligned}
$$

Lemma 2.2.3. *Let $\Gamma \in C^2$ and let φ be a Hölder continuously differentiable function. Then in the case of elasticity, the limits* (1.2.6) *exist uniformly with respect to all $x \in \Gamma$ and all φ with $\|\varphi\|_{C^{1+\alpha}} \le 1$. Moreover, the operator D can be expressed as a composition of tangential differential operators and the simple layer operators of the Laplacian and the Lamé system:*

For $n = 2$:

$$
D\varphi(x) = -\frac{d}{ds_x}\widetilde{V}\Big(\frac{d\varphi}{ds}\Big)(x) \qquad (2.2.32)
$$

where

$$
\widetilde{V}\boldsymbol{\chi}(x) := \frac{\mu(\lambda+\mu)}{\pi(\lambda+2\mu)}\int\limits_\Gamma\Big(-\log|x-y| + \frac{(x-y)(x-y)^\top}{|x-y|^2}\Big)\boldsymbol{\chi}(y)ds_y . \qquad (2.2.33)
$$

For $n = 3$:

$$D\varphi(x) = -\frac{\mu}{4\pi}(\boldsymbol{n}_x \times \nabla_x) \cdot \int\limits_{\Gamma} \frac{1}{|x-y|}(\boldsymbol{n}_y \times \nabla_y)\varphi(y)ds_y$$
$$-\mathcal{M}(\partial_x, \boldsymbol{n}(x)) \int\limits_{\Gamma} \Big\{4\mu^2 E(x,y) - \frac{\mu}{2\pi}\frac{1}{|x-y|}I\Big\}\mathcal{M}(\partial_y, \boldsymbol{n}(y))\varphi(y)ds_y$$
$$+\frac{\mu}{4\pi}\Big(\sum_{\ell,k=1}^{3} m_{\ell k}(\partial_x, \boldsymbol{n}(x)) \int\limits_{\Gamma} \frac{1}{|x-y|}(m_{kj}(\partial_y, \boldsymbol{n}(y))\varphi_\ell)(y)ds_y\Big)_{j=1,2,3}. \tag{2.2.34}$$

Proof: The proof for $n = 2$ was given by Bonnemay [36] and Nedelec [330]; here we follow the proof given by Houde Han [169, 170] for $n = 3$. The proof is based on a different representation of the traction operator T by employing Günter's derivatives $\mathcal{M}$; more precisely we have:

$$T(\partial_x, \boldsymbol{n}(x))\boldsymbol{u}(x) = (\lambda + \mu)(\text{div}\boldsymbol{u})\boldsymbol{n}(x) + \mu\Big(\frac{\partial \boldsymbol{u}}{\partial n} + \mathcal{M}\boldsymbol{u}\Big). \tag{2.2.35}$$

We note

$$\mathcal{M}\boldsymbol{u} = \frac{\partial \boldsymbol{u}}{\partial n} - (\text{div}\boldsymbol{u})\boldsymbol{n}(x) + \boldsymbol{n}(x) \times \text{curl}\,\boldsymbol{u}\,. \tag{2.2.36}$$

Now we apply T in the form (2.2.35) to the three individual terms in the double layer potential K in (2.2.30) successively, and begin with

$$T(\partial_z, \boldsymbol{n}(x))\frac{\partial}{\partial n_y}\gamma_n(z,y)\boldsymbol{I}$$
$$= \mu(\boldsymbol{n}(x)\cdot\nabla_z)\frac{\partial}{\partial n_y}\gamma_n(z,y)\boldsymbol{I} + \mu\mathcal{M}(\partial_z, \boldsymbol{n}(x))\frac{\partial}{\partial n_y}\gamma_n(z,y)\boldsymbol{I}$$
$$+ (\lambda+\mu)\boldsymbol{n}(x)\Big(\text{grad}_z\frac{\partial}{\partial n_y}\gamma_n(z,y)\Big)^{\top}.$$

Next we apply T to the remaining two terms by using (2.2.29) to obtain

$$T(\partial_z, \boldsymbol{n}(x))\Big(2\mu E(z,y) - \frac{1}{2(n-1)\pi}\gamma_n(z,y)\boldsymbol{I}\Big)$$
$$= \mathcal{M}(\partial_z, \boldsymbol{n}(x))\Big\{4\mu^2 E(x,y) - \frac{2\mu}{2(n-1)\pi}\gamma_n(x,y)\boldsymbol{I}\Big\}$$
$$+\frac{2\mu}{2(n-1)\pi}\boldsymbol{n}(x)\cdot\nabla_z\gamma_n(z,y)\boldsymbol{I} - T(\partial_z, \boldsymbol{n}(x))\frac{1}{4\pi}\gamma_n(z,y)\boldsymbol{I}\,.$$

Now we apply (2.2.35) to the last term in this expression and find

$$
\begin{aligned}
&T(\partial_z, \boldsymbol{n}(x)) \frac{1}{2(n-1)\pi} \gamma_n(z,y) \boldsymbol{I} \\
&= \frac{1}{2(n-1)\pi} (\lambda+\mu) \boldsymbol{n}(x) \big(\nabla_z \gamma_n(z,y)\big)^\top + \frac{\mu}{2(n-1)\pi} \boldsymbol{n}(x) \cdot \nabla_z \gamma_n(z,y) \boldsymbol{I} \\
&\qquad\qquad + \frac{\mu}{2(n-1)\pi} \mathcal{M}(\partial_z, \boldsymbol{n}(x)) \gamma_n(z,y) \boldsymbol{I} \,.
\end{aligned}
$$

Hence,

$$
\begin{aligned}
&T(\partial_z, \boldsymbol{n}(x)) \Big(2\mu E(z,y) - \frac{1}{2(n-1)\pi} \gamma_n(z,y) \boldsymbol{I} \Big) \\
&= \mathcal{M}(z, \boldsymbol{n}(x)) \Big\{ 4\mu^2 E(z,y) - \frac{3\mu}{2(n-1)\pi} \gamma_n \boldsymbol{I} \Big\} - \frac{\lambda+\mu}{2(n-1)\pi} \boldsymbol{n}(x) \big(\nabla_z \gamma_n(x,z)\big)^\top \\
&\quad + \frac{\mu}{2(n-1)\pi} \boldsymbol{n}(x) \cdot \nabla_z(z,y) \boldsymbol{I} \,.
\end{aligned}
$$

Collecting these results we obtain

$$
\begin{aligned}
&T(\partial_z, \boldsymbol{n}(x)) (K\boldsymbol{\varphi})(z) \\
&= \frac{\mu}{2(n-1)\pi} \boldsymbol{n}(x) \cdot \nabla_z \int\limits_\Gamma \frac{\partial}{\partial n_y} \gamma_n(z,y) \boldsymbol{\varphi}(y) ds_y \\
&\quad + \mathcal{M}(\partial_z, \boldsymbol{n}(x)) \int\limits_\Gamma \Big\{ 4\mu^2 E(z,y) - \frac{3\mu}{2(n-1)\pi} \gamma_n(z,y) \Big\} \mathcal{M}(\partial_y, \boldsymbol{n}(y)) \boldsymbol{\varphi}(y) ds_y \\
&\quad + \frac{\lambda+\mu}{2(n-1)\pi} \Big\{ \boldsymbol{n}(x) \int\limits_\Gamma \Big(\nabla_z \frac{\partial}{\partial n_y} \gamma_n(z,y) \Big) \cdot \boldsymbol{\varphi}(y) ds_y \\
&\qquad\qquad - \boldsymbol{n}(x) (\nabla_z \int\limits_\Gamma \gamma_n(z,y) \cdot \big(\mathcal{M}(\partial_y, \boldsymbol{n}(y)) \boldsymbol{\varphi}(y) \big) ds_y \Big\} \\
&\quad + \frac{\mu}{2(n-1)\pi} \Big\{ \mathcal{M}(\partial_z, \boldsymbol{n}(x)) \int\limits_\Gamma \frac{\partial}{\partial n_y} \gamma_n(z,y) \boldsymbol{\varphi}(y) ds_y \\
&\qquad\qquad + \boldsymbol{n}(x) \cdot \nabla_z \int\limits_\Gamma \gamma_n(z,y) \mathcal{M}(\partial_y, \boldsymbol{n}(y)) \boldsymbol{\varphi}(y) ds_y \Big\}
\end{aligned}
$$

$$
\begin{aligned}
&= \frac{\mu}{2(n-1)\pi} \boldsymbol{n}(x) \cdot \nabla_z \int\limits_\Gamma \Big(\frac{\partial}{\partial n_y} \gamma_n(z,y) \Big) \boldsymbol{\varphi}(y) ds_y \\
&\quad + \mathcal{M}(\partial_z, \boldsymbol{n}(x)) \int\limits_\Gamma \Big\{ 4\mu^2 E(z,y) - \frac{3\mu}{2(n-1)\pi} \gamma_n(z,y) \Big\} \mathcal{M}(\partial_y, \boldsymbol{n}(y)) \boldsymbol{\varphi}(y) ds_y \\
&\quad + \frac{(\lambda+\mu)}{2(n-1)\pi} \mathcal{J}_1 \boldsymbol{\varphi}(x) + \frac{\mu}{2(n-1)\pi} \mathcal{J}_2 \boldsymbol{\varphi}(x)
\end{aligned}
$$

where

$$\mathcal{J}_1\boldsymbol{\varphi}(x) := \int\limits_{\Gamma} \{\boldsymbol{n}(x)\Big(\nabla_z \frac{\partial}{\partial n_y}\gamma_n(z,y)\cdot\boldsymbol{\varphi}(y)\Big)$$
$$- \boldsymbol{n}(x)(\nabla_z\gamma_n(z,y))\cdot(\mathcal{M}(\partial_y,\boldsymbol{n}(y))\boldsymbol{\varphi}(y))\}ds_y$$

and

$$\mathcal{J}_2\boldsymbol{\varphi}(x) := \int\limits_{\Gamma} \Big\{\mathcal{M}(\partial_z,\boldsymbol{n}(x))\frac{\partial}{\partial n_y}\gamma_n(z,y)$$
$$+ \big(\boldsymbol{n}(x)\cdot\nabla_z\gamma_n(z,y)\big)\mathcal{M}(\partial_y,\boldsymbol{n}(y))\Big\}\boldsymbol{\varphi}(y)ds_y\,.$$

The product rule gives

$$\begin{aligned}
&(\mathcal{M}(\partial_y,\boldsymbol{n}(y))\nabla_z\gamma_n(z,y))\cdot\boldsymbol{\varphi}(y)\\
&\quad = \sum_{k,\ell=1}^{n}\Big(m_{k\ell}(\partial_y,\boldsymbol{n}(y))\frac{\partial}{\partial z_\ell}\gamma_n(z,y)\Big)\varphi_k(y)\\
&\quad = \sum_{k,\ell=1}^{n}\Big(m_{k\ell}(\partial_y,\boldsymbol{n}(y))\Big(\frac{\partial}{\partial z_\ell}\gamma_n(z,y)\varphi_k(y)\Big)\\
&\qquad\qquad -\big(m_{k\ell}(\partial_y,\boldsymbol{n}(y))\varphi_k(y)\big)\frac{\partial}{\partial z_\ell}\gamma_n(z,y)\Big)\\
&\quad = \sum_{k,\ell=1}^{n}\Big(-m_{\ell k}(\partial_y,\boldsymbol{n}(y))\Big(\frac{\partial}{\partial z_\ell}\gamma_n(z,y)\varphi_k(y)\Big)\\
&\qquad\qquad +\Big(\frac{\partial}{\partial z_\ell}\gamma_n(z,y)\Big)\big(m_{\ell k}(\partial_y,\boldsymbol{n}(y))\big)\varphi_k(y)\Big)\,.
\end{aligned}\tag{2.2.37}$$

The symmetry of $\gamma_n(z,y)$ and $\Delta_y\gamma_n(z,y)=0$ for $z\neq y$ implies

$$\begin{aligned}
\sum_{k=1}^{n} m_{jk}(y)\frac{\partial\gamma_n(z,y)}{\partial z_k} &= \sum_{k=1}^{n} n_k(y)\frac{\partial^2}{\partial y_j\partial z_k}\gamma_n(z,y) - n_j(y)\sum_{k=1}^{n}\frac{\partial^2}{\partial y_k\partial z_k}\gamma_n(z,y)\\
&= \sum_{k=1}^{n} n_k(y)\frac{\partial^2}{\partial y_k\partial z_j}\gamma_n(z,y) + n_j(y)\Delta_y\gamma_n(z,y)\\
&= \frac{\partial}{\partial z_j}\,\frac{\partial\gamma_n(z,y)}{\partial n_y}\,,
\end{aligned}$$

i.e.,

$$\nabla_z\frac{\partial\gamma_n(z,y)}{\partial n_y} = \mathcal{M}(\partial_y(y))\nabla_z\gamma_n(z,y)\,. \tag{2.2.38}$$

Using (2.2.37) and (2.2.38) for the first term of the integrand of $\mathcal{J}_1$, we find

$$\begin{aligned}\mathcal{J}_1\varphi(x) &= \boldsymbol{n}(x)\int\limits_{\Gamma}\{(\mathcal{M}(\partial_y,\boldsymbol{n}(y))\nabla_z\gamma_n(z,y))\cdot\varphi(y)\\ &\quad -\nabla_z\gamma_n(z,y)\cdot(\mathcal{M}(\partial_y(y))\varphi(y))\}ds_y\\ &= -\boldsymbol{n}(x)\int\limits_{\Gamma}\sum_{\ell,k=1}^{n} m_{\ell k}(\partial_y,\boldsymbol{n}(y))\Big(\frac{\partial}{\partial z_k}\gamma_n(z,y)\varphi_k(y)\Big)ds_y = 0\end{aligned}$$

for $x\notin\Gamma$ due to the Stokes theorem.

For the integral $\mathcal{J}_2$ we use

$$\begin{aligned}&\mathcal{M}(\partial_z,\boldsymbol{n}(x))\frac{\partial}{\partial n_y}\gamma_n(z,y)-\mathcal{M}(\partial_y,\boldsymbol{n}(y))(\nabla_z\cdot\boldsymbol{n}(x))\gamma_n(z,y)\boldsymbol{I}\\ &=\{\mathcal{M}(\partial_y,\boldsymbol{n}(y))\mathcal{M}(\partial_z,\boldsymbol{n}(x))-\mathcal{M}(\partial_z,\boldsymbol{n}(x))\mathcal{M}(\partial_y,\boldsymbol{n}(y))\}\gamma_n(z,y)\boldsymbol{I}\end{aligned}$$

to obtain

$$\begin{aligned}\mathcal{J}_2\varphi(x) &= \int\limits_{\Gamma}\Big(\mathcal{M}(\partial_z,\boldsymbol{n}(x))\gamma_n(z,y)\boldsymbol{I}\mathcal{M}(\partial_y,\boldsymbol{n}(y))\Big)^{\top}\varphi(y)ds_y\\ &\quad +\mathcal{M}(\partial_z,\boldsymbol{n}(x))\int\limits_{\Gamma}\gamma_n(z,y)\mathcal{M}(\partial_y,\boldsymbol{n}(y))\varphi(y)ds_y\,.\end{aligned}$$

The final result is then with (2.2.36):

$$\begin{aligned}&-T(\partial_z,\boldsymbol{n}(x))\int\limits_{\Gamma}K(z,y)\varphi(y)ds_y\\ &=-\frac{\mu}{2(n-1)\pi}(\boldsymbol{n}(x)\times\nabla_z)\cdot\int\limits_{\Gamma}\gamma_n(z,y)(\boldsymbol{n}(y)\times\nabla_y)\varphi(y)ds_y\\ &\quad -\mathcal{M}(\partial_z,\boldsymbol{n}(x))\int\limits_{\Gamma}\Big\{4\mu^2E(z,y)-\frac{2\mu}{2(n-1)\pi}\gamma_n(z,y)\boldsymbol{I}\Big\}\mathcal{M}(\partial_y,\boldsymbol{n}(y))\varphi ds_y\\ &\quad +\frac{\mu}{2(n-1)\pi}\int\limits_{\Gamma}\Big(\mathcal{M}(\partial_z,\boldsymbol{n}(x))\gamma_n(z,y)\mathcal{M}(\partial_y,\boldsymbol{n}(y))\Big)^{\top}\varphi(y)ds_y\,,\end{aligned}$$

where $z\notin\Gamma$.

As can be seen, this is a combination of applications of tangential derivatives to weakly singular operators operating on tangential derivatives of φ. Therefore, the limits $z\to x\in\Gamma$ with $z\in\Omega$ or $z\in\Omega^c$ exist, which leads to the desired result involving Cauchy singular integrals (see Hellwig [177, p.197]). Note that for $n=2$ we have $\big(n(x)\times\nabla_x\big)|_\Gamma=\frac{d}{ds_x}$. ■

The singular behaviour of the hypersingular operator now can be regularized as above. This facilitates the computational algorithm for the Galerkin method (Of et al [340]).

As for the hypersingular integral operator associated with the Laplacian in Section 1.2, here we can apply a more elementary, but different regularization. Based on (2.2.31), we get

$$\begin{aligned} D\varphi(x) &= -\lim_{\Omega\ni z\to x\in\Gamma}\{T_z\int_\Gamma (T_yE(x,y)^\top(\varphi(y)-\varphi(x))ds_y \\ &\qquad + T_z\int_\Gamma (T_yE(x,y))^\top\varphi(x)ds_y\}. \end{aligned}$$

Here, in the neighborhood of Γ, the operator T_z is defined by (2.2.22) where we identify $\boldsymbol{n}_z = \boldsymbol{n}_x$ for $z \notin \Gamma$. If we apply the representation formula (2.2.4) to any constant vector field $\boldsymbol{a}$, representing a rigid displacement, then we obtain

$$\boldsymbol{a} = -\int_\Gamma (T_yE(z,y))^\top \boldsymbol{a} ds_y \text{ for } z\in\Omega,$$

which yields for $z \in \Omega$ in the neighborhood of Γ

$$T_z\int_\Gamma (T_yE(z,y))^\top ds_y\varphi(x) = \boldsymbol{0}.$$

Hence,

$$D\varphi(x) = -\lim_{\Omega\ni z\to x\in\Gamma} T_z\int_\Gamma (T_yE(z,y))^\top(\varphi(y)-\varphi(x))ds_y,$$

from which it can finally be shown that

$$D\varphi(x) = -p.v.\int_\Gamma T_x\left(T_yE(x,y)\right)^\top\left(\varphi(y)-\varphi(x)\right)ds_y, \tag{2.2.39}$$

(see Kupradze et al [253, p.294], Schwab et al [380]).

If the boundary potential operators in (1.2.18) and (1.2.19) are replaced by the corresponding elastic potential operators, then the Calderón projector $\mathcal{C}_\Omega$ for solutions $\boldsymbol{u}$ of (2.2.9) in Ω again is given in the form of (1.2.20) with the corresponding elastic potential operators.

Also, Theorem 1.2.3. and Lemma 1.2.4. remain valid for the corresponding elastic potentials V, K, K' and D.

For the solution of the interior displacement problem we now may solve the Fredholm boundary integral equation of the first kind

$$V\boldsymbol{\sigma} = \tfrac{1}{2}\varphi + K\varphi \text{ on } \Gamma, \tag{2.2.40}$$

or the Cauchy singular integral equation of the second kind

$$\tfrac{1}{2}\boldsymbol{\sigma} - K'\boldsymbol{\varphi} = D\boldsymbol{\varphi} \text{ on } \Gamma\,, \tag{2.2.41}$$

which both are equations for $\boldsymbol{\sigma}$.

The first kind integral equation (2.2.40) may have eigensolutions for special Γ similar to (1.3.4) for the Laplacian (see Steinbach [400]). Again we can modify (2.2.40) by including rigid motions (2.2.11). More precisely, we consider the system

$$\begin{aligned} V\boldsymbol{\sigma} - \boldsymbol{\omega}_0 = \tfrac{1}{2}\boldsymbol{\varphi} + K\boldsymbol{\varphi} \text{ on } \Gamma\,, \\ \int\limits_\Gamma \boldsymbol{\sigma} ds = \mathbf{0} \end{aligned} \tag{2.2.42}$$

together with

$$\begin{aligned} \int\limits_\Gamma (-\sigma_1 x_2 + \sigma_2 x_1) ds &= 0 \quad \text{for } n=2 \text{ or} \\ \int\limits_\Gamma (\boldsymbol{\sigma} \times \boldsymbol{x}) ds &= \mathbf{0} \quad \text{for } n=3\,, \end{aligned} \tag{2.2.43}$$

where $\boldsymbol{\sigma}$ and the unknown constant vector $\boldsymbol{\omega}_0$ are to be determined. As was shown in [205], the rotation $\boldsymbol{b}$ in (2.2.11) can be prescribed as $\boldsymbol{b} = \mathbf{0}$; in this case the side conditions (2.2.43) will not be needed. For $n = 3$, many more choices in $\boldsymbol{\omega}_0$ can be made (see [205] and Mikhlin et al [304]). The modified system (2.2.42) and (2.2.43) is always uniquely solvable in the Hölder space, $\boldsymbol{\sigma} \in C^\alpha(\Gamma)$ for given $\boldsymbol{\varphi} \in C^{1+\alpha}(\Gamma)$.

For the special Cauchy singular integral equation of the second kind (2.2.41), Mikhlin showed in [300] that the Fredholm alternative, originally designed for compact operators K, remains valid here. Therefore, (2.2.41) admits a unique classical solution $\boldsymbol{\sigma} \in C^\alpha(\Gamma)$ provided $\boldsymbol{\varphi} \in C^{1+\alpha}(\Gamma), 0 < \alpha < 1$. (See Kupradze et al [253, Chapt. VI], Mikhlin et al [305, p.382 ff]).

2.2.2 The Interior Traction Problem

The Neumann problem for the Lamé system (2.2.9) in Ω is called the *traction problem*, since here the boundary traction

$$T\boldsymbol{u}_{|\Gamma} = \boldsymbol{\psi} \text{ on } \Gamma \tag{2.2.44}$$

is given, whereas the missing Cauchy datum $\boldsymbol{u}_{|\Gamma}$ needs to be determined. Corresponding to (2.2.16) and (2.2.17), we have the overdetermined system

$$\left(\tfrac{1}{2}I + K\right)\boldsymbol{u} = V\boldsymbol{\psi}\,, \tag{2.2.45}$$

$$D\boldsymbol{u} = \left(\tfrac{1}{2}I - K'\right)\boldsymbol{\psi} \text{ on } \Gamma \tag{2.2.46}$$

for the unknown boundary displacemant $\boldsymbol{u}_{|\Gamma}$.

As for the Neumann Problem for the Laplacian, here $\boldsymbol{\psi}$ needs to satisfy certain equilibrium conditions for a solution to exist. These can be obtained from the Betti formula, which for $\boldsymbol{u}$ and any rigid motion $\boldsymbol{\omega}(x)$ reads

$$\begin{aligned} 0 &= \int\limits_{\Omega} (\boldsymbol{\omega} \cdot \Delta^* \boldsymbol{u}) - (\boldsymbol{u} \cdot \Delta^* \boldsymbol{\omega}) dx \\ &= \int\limits_{\Gamma} \boldsymbol{\omega} \cdot T\boldsymbol{u} ds - \int\limits_{\Gamma} \boldsymbol{u} \cdot T\boldsymbol{\omega} ds \,. \end{aligned}$$

This implies, with $T\boldsymbol{\omega} = \boldsymbol{0}$, the necessary compatibility conditions for the given traction $\boldsymbol{\psi}$, namely

$$\int\limits_{\Gamma} \boldsymbol{\omega} \cdot \boldsymbol{\psi} ds = 0 \quad \text{for all rigid motions } \boldsymbol{\omega} \tag{2.2.47}$$

given by (2.2.11). This condition turns out to be also sufficient for the existence of $\boldsymbol{u}$ in the classical Hölder function spaces. If $\boldsymbol{\psi} \in C^{\alpha}(\Gamma)$ with $0 < \alpha < 1$ is given satisfying (2.2.47), then the right–hand side, $V\boldsymbol{\psi}$ in (2.2.45), automatically satisfies the orthogonality conditions from Fredholm's alternative; and the Cauchy singular integral equation (2.2.45) admits a solution$\boldsymbol{u} \in C^{1+\alpha}(\Gamma)$. The solution, however, is unique only up to all rigid motions $\boldsymbol{\omega}$, which are eigensolutions. For further details see [205].

The hypersingular integral equation of the first kind (2.2.46) also has eigensolutions which again are given by all rigid motions (2.2.11). As will be seen, the classical Fredholm alternative even holds for (2.2.46), and the right–hand side $\frac{1}{2}\boldsymbol{\psi} - K'\boldsymbol{\psi}$ satisfies the corresponding orthogonality conditions, provided, $\boldsymbol{\psi} \in C^{\alpha}(\Gamma)$ satisfies the equilibrium conditions (2.2.47). In both cases, the integral equations , together with appropriate side conditions, can be modified so that the resulting equations are uniquely solvable (see [204]).

2.2.3 Some Exterior Fundamental Problems

In this section we shall summarize the approach from [205]. For the *exterior displacement problem* for $\boldsymbol{u}$ satisfying the Lamé system (2.2.12) and the Dirichlet condition (2.2.15),

$$\boldsymbol{u}_{|\Gamma} = \boldsymbol{\varphi} \text{ on } \Gamma \,,$$

we require at infinity appropriate conditions according to (2.2.10); namely that there exist $\boldsymbol{\Sigma}$ and some rigid motion $\boldsymbol{\omega}$ such that

$$\boldsymbol{u}(x) + E(x,0)\boldsymbol{\Sigma} - \boldsymbol{\omega} = O(|x|^{1-n}) \quad \text{as} \quad |x| \to \infty \,.$$

In general, the constants in $\boldsymbol{\Sigma}$ and $\boldsymbol{\omega}$ are related to each other. However, some of them can still be specified.

For $n = 2$, we consider the following two cases. In the first case,

$$b \text{ in } \boldsymbol{\omega} = \boldsymbol{a} + b(-x_2, x_1)^T \text{ is given}.$$

In addition, the total forces $\boldsymbol{\Sigma}$ in (2.2.14) are also given, often as $\boldsymbol{\Sigma} = \mathbf{0}$ due to equilibrium. Then $\boldsymbol{a}$ in $\boldsymbol{\omega}$ is an additional unknown vector. The representation formula (2.2.13) for the Dirichlet problem (2.2.12) with (2.2.14) yields the modified boundary integral equation of the first kind

$$V\boldsymbol{\sigma} - \boldsymbol{a} = -\tfrac{1}{2}\varphi(x) + K\varphi(x) + b(-x_2, +x_1)^\top \text{ on } \Gamma, \quad (2.2.48)$$

$$\int_\Gamma \boldsymbol{\sigma} ds = \boldsymbol{\Sigma}. \quad (2.2.49)$$

Here, $\varphi, \boldsymbol{\Sigma}$ and b are given and $\boldsymbol{\sigma}, \boldsymbol{a}$ are the unknowns. As we will see, these equations are always uniquely solvable. In particular, for any given $\varphi \in C^{1+\alpha}(\Gamma)$, $\boldsymbol{\Sigma}$ and b, one finds in the classical Hölder–spaces $\boldsymbol{\sigma} \in C^\alpha(\Gamma)$.

In the second case, in addition to the total force $\boldsymbol{\Sigma}$, the total momentum

$$\int_\Gamma (-x_2\sigma_1 + x_1\sigma_2) ds_x = \Sigma_3$$

is also given, whereas b is now an additional unknown. Now the modified boundary integral equation of the first kind reads

$$V\boldsymbol{\sigma}(x) - \boldsymbol{a} - b(-x_2, +x_1)^\top = \frac{1}{2}\varphi(x) + K\varphi(x) \text{ on } \Gamma, \quad (2.2.50)$$

$$\int_\Gamma \boldsymbol{\sigma} ds = \boldsymbol{\Sigma}, \ \int_\Gamma (-x_2\sigma_1 + x_1\sigma_2) ds_x = \Sigma_3, \quad (2.2.51)$$

where $\varphi \in C^{1+\alpha}(\Gamma), \boldsymbol{\Sigma}, \Sigma_3$ are given and $\boldsymbol{\sigma}, \boldsymbol{a}$ and b are to be determined. The system (2.2.50), (2.2.51) always has a unique solution $\boldsymbol{\sigma} \in C^\alpha(\Gamma), \boldsymbol{a}, b$.

Both these problems can also be reduced to Cauchy singular integral equations by applying the traction operator to (2.2.13). This yields the singular integral equation

$$\tfrac{1}{2}\boldsymbol{\sigma}(x) + K'\boldsymbol{\sigma}(x) = -D\varphi(x) \text{ for } x \in \Gamma, \quad (2.2.52)$$

with the additional equation (2.2.49) in the first case or the additional equations (2.2.51) in the second case, respectively. The operator $\frac{1}{2}I + K'$ is adjoint to $\frac{1}{2}I + K$ in (2.2.17). Due to Mikhlin [300], for these special operators, Fredholm's classical alternative is still valid in the space $C^\alpha(\Gamma)$. Since

$$\tfrac{1}{2}\boldsymbol{\omega} + K\boldsymbol{\omega} = 0 \text{ on } \Gamma$$

for all rigid motions $\boldsymbol{\omega}$, the adjoint equation (2.2.52) has an $3(n-1)$–dimensional eigenspace, as well. Moreover, $D\boldsymbol{\omega} = \mathbf{0}$ for all rigid motions;

hence, the right–hand side of (2.2.52) always satisfies the orthogonality conditions for any given $\varphi \in C^{1+\alpha}(\Gamma)$. This implies that equation (2.2.52) always admits a solution $\boldsymbol{\sigma} \in C^{\alpha}(\Gamma)$ which is not unique. If, for $n = 2$, the total force and total momentum in addition are prescribed by (2.2.51), i.e. in the second case, then these equations determine $\boldsymbol{\sigma}(x)$ uniquely. For finding $\boldsymbol{a}$ and b we first determine three vector–valued functions $\boldsymbol{\lambda}_j(x), j = 1, 2, 3$, satisfying on Γ the equations

$$\begin{aligned}
\int\limits_{\Gamma} \boldsymbol{a} \cdot \boldsymbol{\lambda}_j ds &= a_j \quad \text{and} \quad \int\limits_{\Gamma} (-x_2, x_1) \cdot \boldsymbol{\lambda}_j(x) ds = 0 \ \text{ for } j = 1, 2\,, \\
\int\limits_{\Gamma} \boldsymbol{a} \cdot \boldsymbol{\lambda}_3 ds &= 0 \quad \text{and} \quad \int\limits_{\Gamma} (-x_2, x_1) \cdot \boldsymbol{\lambda}_3 ds = 1\,.
\end{aligned} \tag{2.2.53}$$

Since $\boldsymbol{\sigma}(x)$ on Γ is already known from solving (2.2.52), equation (2.2.50) can now be used to find $\boldsymbol{a}$ and b; namely

$$\left.\begin{array}{ll} \text{for } j = 1, 2 & : a_j \\ \text{for } j = 3 & : b \end{array}\right\} = \int\limits_{\Gamma} V\boldsymbol{\sigma}(x) \cdot \boldsymbol{\lambda}_j(x) d_x - \frac{1}{2} \int\limits_{\Gamma} \varphi \cdot \boldsymbol{\lambda}_j ds - \int\limits_{\Gamma} (K\varphi) \cdot \boldsymbol{\lambda}_j ds\,. \tag{2.2.54}$$

If, as in the first case, b and $\boldsymbol{\Sigma}$ are given, then the additional equations

$$\begin{aligned}
\int\limits_{\Gamma} \boldsymbol{\sigma} ds &= \boldsymbol{\Sigma} \quad \text{and} \\
\int\limits_{\Gamma} \boldsymbol{\lambda}_3(x) \cdot V\boldsymbol{\sigma}(x) ds_x &= b - \frac{1}{2} \int\limits_{\Gamma} \boldsymbol{\lambda}_3 \cdot \varphi ds + \int\limits_{\Gamma} \boldsymbol{\lambda}_3 \cdot K\varphi ds
\end{aligned}$$

determine $\boldsymbol{\sigma}$ uniquely; and a_1, a_2 can be found from (2.2.54), afterwards.

In the case $n = 3$, there are many more possible choices of additional conditions. To this end, we write the rigid motions (2.2.11) in the form

$$\boldsymbol{\omega}(x) = \sum_{j=1}^{3} a_j \boldsymbol{m}_j(x) + \sum_{j=4}^{6} b_{j-3} \boldsymbol{m}_j(x) =: \sum_{j=1}^{6} \omega_j \boldsymbol{m}_j(x)$$

where $\boldsymbol{m}_j(x)$ is the j–th column vector of the matrix

$$\begin{pmatrix} 1, & 0, & 0, & 0, & x_3\,, & -x_2 \\ 0, & 1, & 0, & -x_3\,, & 0\,, & x_1 \\ 0, & 0, & 1, & x_2\,, & -x_1\,, & 0\,. \end{pmatrix}. \tag{2.2.55}$$

Let $\mathcal{J} \subset \mathcal{F} := \{1, 2, 3, 4, 5, 6\}$ denote any fixed set of indices in $\mathcal{F}$. Then we may prescribe a_{j-3}, b_j for $j \in \mathcal{J}$, i.e., some of the parameters in $\boldsymbol{\omega}$ subject to the behavior of (2.2.10) at infinity. If $\mathcal{J}$ is a proper subset of $\mathcal{F}$ then we must include additional normalization conditions,

$$\int_\Gamma \boldsymbol{m}_k(y) \cdot \boldsymbol{\sigma}(y) ds_y = \Sigma_k \quad \text{for } k \in \mathcal{F} \setminus \mathcal{J}\,. \tag{2.2.56}$$

By taking the representation formula (2.2.13) on Γ, we obtain from the direct formulation the boundary integral equation of the first kind on Γ,

$$V\boldsymbol{\sigma}(x) - \sum_{k \in \mathcal{F} \setminus \mathcal{J}} \omega_k \boldsymbol{m}_k(x) = -\frac{1}{2}\boldsymbol{\varphi}(x) + K\boldsymbol{\varphi}(x) + \sum_{j \in \mathcal{J}} \omega_j \boldsymbol{m}_j(x)\,, \tag{2.2.57}$$

together with the additional equations,

$$\int_\Gamma \boldsymbol{m}_k(y) \cdot \boldsymbol{\sigma}(y) ds_y = \Sigma_k \quad \text{for } k \in \mathcal{F} \setminus \mathcal{J}\,. \tag{2.2.58}$$

In the right–hand side of (2.2.57), the ω_j are given, whereas the ω_k in the left–hand side are unknown. For given $\boldsymbol{\varphi} \in C^\alpha(\Gamma), 0 < \alpha < 1$, and given constants $\omega_j, j \in \mathcal{J}$, the unknowns are $\boldsymbol{\sigma} \in C^\alpha(\Gamma)$ together with ω_k for $k \in \mathcal{F} \setminus \mathcal{J}$.

Again, we may take the traction of (2.2.13) on Γ to obtain a Cauchy singular boundary integral equation instead of (2.2.57) and (2.2.58), namely (2.2.52) together with (2.2.58). Since (2.2.52) is always solvable for any given $\boldsymbol{\varphi} \in C^{1+\alpha}(\Gamma)$ due to the special form of the right–hand side, and since the eigenspace of $\frac{1}{2}I + K'$ is the linear space of all rigid motions, the linear equations (2.2.58) need to be completed by including (card $\mathcal{J}$) additional equations which resembles the required behavior of $\boldsymbol{u}(x)$ at infinity for those ω_j given already with $j \in \mathcal{J}$. For these constraints, we again choose the vector–valued functions, $\boldsymbol{\lambda}_\ell$ on Γ, $\ell \in \mathcal{F}$, (e.g. linear combinations of $\boldsymbol{m}_\ell|_\Gamma$) which are orthonormalized to $\boldsymbol{m}_j|_\Gamma$, i.e.,

$$\int_\Gamma \boldsymbol{m}_j(x) \cdot \boldsymbol{\lambda}_\ell(x) ds_x = \delta_{j,\ell}, \quad j, \ell \in \mathcal{F}\,. \tag{2.2.59}$$

Now, the complete system of equations for the modified Dirichlet problem can be formulated as

$$\begin{aligned} \frac{1}{2}\boldsymbol{\sigma}(x) + K'\boldsymbol{\sigma}(x) - \sum_{\ell \in \mathcal{F}} \alpha_\ell \boldsymbol{m}_\ell(x) &= -D\boldsymbol{\varphi}(x) \quad \text{for } x \in \Gamma \\ \int_\Gamma \boldsymbol{m}_k \cdot \boldsymbol{\sigma}_k(y) ds_y &= \Sigma_k, \; k \in \mathcal{F} \setminus \mathcal{J}\,, \end{aligned} \tag{2.2.60}$$

together with

$$\int_\Gamma \boldsymbol{\lambda}_j(x) \cdot V\boldsymbol{\sigma} ds = \int_\Gamma \boldsymbol{\lambda}_j \cdot \left\{\frac{1}{2}\boldsymbol{\varphi}(x) + K'\boldsymbol{\varphi}(x)\right\} ds + \omega_j \quad \text{for } j \in \mathcal{J}\,.$$

The desired displacement $\boldsymbol{u}(x)_{|\Gamma}$ then is obtained from (2.2.13) with $\boldsymbol{\sigma}(x)$ determined from (2.2.60) and ω_k given by

$$\omega_k = \int_\Gamma \boldsymbol{\lambda}_k \cdot V\boldsymbol{\sigma} ds + \int_\Gamma \boldsymbol{\lambda}_k \cdot \left\{ \frac{1}{2}\boldsymbol{\varphi} - K\boldsymbol{\varphi} \right\} ds \text{ for } k \in \mathcal{F} \setminus \mathcal{J}\,.$$

Note that we have included the extra unknown term $\sum \alpha_\ell \boldsymbol{m}_\ell(x)$ in (2.2.60) so that the number of unknowns and equations coincide. One can show that in fact $\alpha_\ell = 0$ for $\ell \in \mathcal{F}$. The last set of equations has been obtained from (2.2.50) with (2.2.59).

For the *exterior traction problem*, the Neumann datum is given by

$$T\boldsymbol{u}|_\Gamma = \boldsymbol{\psi} \text{ on } \Gamma\,,$$

and the total forces and momenta by

$$\int_\Gamma \boldsymbol{\psi} \cdot \boldsymbol{m}_k ds = \Sigma_k, \quad k \in \mathcal{F}.$$

Here, the standard direct approach with $x \to \Gamma$ in (2.2.13) yields the Cauchy singular boundary integral equation for $\boldsymbol{u}_{|\Gamma}$,

$$\tfrac{1}{2}\boldsymbol{u}(x) - K\boldsymbol{u}(x) = V\boldsymbol{\psi}(x) + \boldsymbol{\omega}(x) \text{ for } x \in \Gamma\,. \tag{2.2.61}$$

As is well known, (2.2.61) is always uniquely solvable for any given $\boldsymbol{\psi} \in C^\alpha(\Gamma)$ and given $\boldsymbol{\omega}$ with $\boldsymbol{u} \in C^{1+\alpha}(\Gamma)$; we refer for the details to Kupradze [252, p.118].

If we apply T to (2.2.13) then we obtain the hypersingular boundary integral equation

$$D\boldsymbol{u}(x) = -\tfrac{1}{2}\boldsymbol{\psi}(x) - K'\boldsymbol{\psi}(x) \text{ for } x \in \Gamma\,. \tag{2.2.62}$$

It is easily seen that the rigid motions $\boldsymbol{\omega}(y)$ on Γ are eigensolutions of (2.2.62). Therefore, in order to guarantee unique solvability of the boundary integral equation , we modify (2.2.62) by including restrictions and adding unknowns, e.g.,

$$\begin{aligned} D\boldsymbol{u}_0(x) &= -\frac{1}{2}\boldsymbol{\psi}(x) - K'\boldsymbol{\psi}(x) + \sum_{\ell=1}^{3(n-1)} \alpha_\ell \boldsymbol{m}_\ell(x) \text{ for } x \in \Gamma \text{ and} \\ \int_\Gamma \boldsymbol{m}_k(y) &\cdot \boldsymbol{u}_0(y) ds_y = 0\,, \quad k = 1, \cdots, 3(n-1)\,. \end{aligned} \tag{2.2.63}$$

Also here, the unknowns α_ℓ are introduced for obtaining a quadratic system. For the Neumann problem, they all vanish because of the special form of the right–hand side. As we will see, the system (2.2.63) is always uniquely solvable; for any given $\psi \in C^{\alpha}(\Gamma)$ we find exactly one $\boldsymbol{u}_0 \in C^{1+\alpha}(\Gamma)$. In (2.2.63), the additional compatibility conditions can also be incorporated into the first equation of (2.2.63) which yields the stabilized uniquely solvable version

$$\begin{aligned}\widetilde{D}\boldsymbol{u}_0(x) &:= D\boldsymbol{u}_0(x) + \sum_{k=1}^{3(n-1)} \int_{\Gamma} \boldsymbol{m}_k(y)\boldsymbol{u}_0(y)ds_y \boldsymbol{m}_k(x) \\ &= -\tfrac{1}{2}\psi(x) - K'\psi(x) \quad \text{for } x \in \Gamma\,.\end{aligned} \tag{2.2.64}$$

Once $\boldsymbol{u}_0$ is known, the actual displacement field $\boldsymbol{u}(x)$ is then given by

$$\boldsymbol{u}(x) = -V\psi(x) + W\boldsymbol{u}_0(x) \quad \text{for } x \in \Omega^c\,. \tag{2.2.65}$$

Note that the actual boundary values of $\boldsymbol{u}|_\Gamma$ may differ from $\boldsymbol{u}_0$ by a rigid motion. $\boldsymbol{u}|_\Gamma$ can be expressed via (2.2.65) in the form

$$\boldsymbol{u}(x)|_\Gamma = \frac{1}{2}\boldsymbol{u}_0(x) + K\boldsymbol{u}_0(x) - V\psi(x)\,. \tag{2.2.66}$$

In concluding this section we remark that the boundary integral equations on Hölder continuous charges are considered in most of the more classical works on this topic with applications in elasticity (e.g., Ahner et al [7], Balaš et al [23], Bonnemay [37], Kupradze [251, 252], Muskhelishvili [319] and Natroshvili [322]). However, as mentioned before, we shall come back to these equations in Chapters 5–11.

Now we extend our approach to incompressible materials.

2.2.4 The Incompressible Material

If the elastic material becomes *incompressible*, the Poisson ratio $\nu := \lambda/2(\lambda+\mu)$ tends to $1/2$ or $\lambda = 2\mu\nu/(1-2\nu) \to \infty$ for $n = 3$ and for the plane strain case where $n = 2$. However, for the plane stress case we have $\overline{\nu} \to 1/3$ and $\overline{\lambda} \to 2\mu$ if the material is incompressible; and our previous analysis remains valid without any restriction.

In order to analyze the incompressible case, we now rewrite the Lamé equation (2.2.1) in the form of a system

$$\begin{aligned}-\Delta\boldsymbol{u} + \nabla p &= 0\,, \\ \operatorname{div}\,\boldsymbol{u} &= -cp \quad \text{where } p = -\frac{\lambda+\mu}{\mu}\operatorname{div}\boldsymbol{u}\end{aligned} \tag{2.2.67}$$

and $c = 1-2\nu = \frac{\mu}{\lambda+\mu} \to 0+$ (see Duffin and Noll [108]). This system corresponds to the *Stokes system*. In terms of the parameter c, the fundamental solution (2.2.2) now takes the form

$$E(x,y) = \frac{1+2c}{4\pi(n-1)\mu(1+c)} \left\{ \gamma_n(x,y)\boldsymbol{I} + \tfrac{1}{(1+2c)r^n}(x-y)(x-y)^\top \right\} \tag{2.2.68}$$

which is well defined for $c = 0$, as well. Consequently, the Betti–Somigliana representation formula (2.2.4) remains valid, where the double layer potential kernel now reads as

$$\begin{aligned} (T_y E(x,y))^\top &= \frac{1}{2\pi(n-1)(1+c)} \Big\{ \big(c\boldsymbol{I} + \frac{n}{r^2}(x-y)(x-y)^\top\big)\frac{\partial\gamma}{\partial n_y} \\ &\quad + \frac{c}{r^n}\big((x-y)\boldsymbol{n}_y^\top - \boldsymbol{n}_y(x-y)^\top\big) \Big\}^\top , \end{aligned} \tag{2.2.69}$$

which again is well defined for the limiting case $c = 0$. However, the limiting case of the differential equations (2.2.63) leads to the more complicated Stokes system involving a mixed variational formulation (see Brezzi and Fortin [45]) whereas the associated boundary integral equations remain valid. In fact, from (2.2.65) and (2.2.66) it seems that the case $c = 0$ does not play any exceptional role. However, for small $c > 0$, a more detailed analysis is required. We shall return to this point after discussing the Stokes problem in Section 2.3.

2.3 The Stokes Equations

The linearized and stationary equations of the *incompressible viscous fluid* are modeled by the Stokes system consisting of the equations in the form

$$\begin{aligned} -\mu\Delta\boldsymbol{u} + \nabla p &= \boldsymbol{f}\,, \\ \operatorname{div}\,\boldsymbol{u} &= 0 \quad \text{in } \Omega \text{ (or } \Omega^c). \end{aligned} \tag{2.3.1}$$

Here $\boldsymbol{u}$ and p are the velocity and pressure of the fluid flow, respectively, which are the unknowns; $\boldsymbol{f}$ corresponds to a given forcing term, while μ is the given *dynamic viscosity* of the fluid. We have already seen this system previously in (2.2.67) for the elastic material when it becomes incompressible, although for viscous flow with given *fluid density* ρ, one introduces

$$\nu := \frac{\mu}{\rho} \gg 1$$

which is usually referred to as the *kinematic viscosity* of the fluid, not the Poisson ratio as in the case of elasticity. The fundamental solution of the Stokes system (2.3.1) is defined by the pair of distributions $\boldsymbol{v}^k$, and q^k satisfying

$$\begin{aligned} -\{\mu\Delta_x \boldsymbol{v}^k(x,y) - \nabla_x q^k(x,y)\} &= \delta(x,y)\boldsymbol{e}^k\,, \\ \operatorname{div}_x \boldsymbol{v}^k &= 0\,, \end{aligned} \tag{2.3.2}$$

where $\boldsymbol{e}^k$ denotes the unit vector along the x_k-axis, $k = 1, ., n$ with $n = 2$ or 3 (see Ladyženskaya [256]). By using the Fourier transform, we may obtain

the fundamental solution explicitly:
For $n = 2$,

$$\begin{aligned} \boldsymbol{v}^k(x,y) &= \frac{1}{4\pi\mu}\Big\{\log\frac{1}{|x-y|}\boldsymbol{e}^k + \sum_{j=1}^{2}\frac{(x_k-y_k)(x_j-y_j)\boldsymbol{e}^j}{|x-y|^2}\Big\}, \\ q^k(x,y) &= \frac{\partial}{\partial x_k}\{-\frac{1}{2\pi}\log\frac{1}{|x-y|}\}; \end{aligned} \tag{2.3.3}$$

and for $n = 3$,

$$\begin{aligned} \boldsymbol{v}^k(x,y) &= \frac{1}{8\pi\mu}\Big\{\frac{1}{|x-y|}\boldsymbol{e}^k + \sum_{j=1}^{3}\frac{(x_k-y_k)(x_j-y_j)\boldsymbol{e}^j}{|x-y|^3}\Big\}, \\ q^k(x,y) &= \frac{\partial}{\partial x_k}\{-\frac{1}{4\pi}\frac{1}{|x-y|}\}. \end{aligned} \tag{2.3.4}$$

We note that from their explicit forms, $\boldsymbol{v}^k(x,y)$ and $q^k(x,y)$ also satisfy the adjoint system in the y–variables, namely

$$\begin{aligned} -\{\mu\Delta_y\boldsymbol{v}(x,y) + \nabla_y q^k(x,y)\} &= \delta(x,y)\boldsymbol{e}^k\,, \\ -\mathrm{div}_y\,\boldsymbol{v}^k &= 0\,. \end{aligned} \tag{2.3.5}$$

This means that we may use the same fundamental solution for the Stokes system and for its adjoint depending on which variables are differentiated. As will be seen, we do not need to switch the variables x and y in the representation of the solution of (2.3.1) from Green's formula (see [256]). For the flow $(\boldsymbol{u}, p)$, we define the *stress operators* as in elasticity,

$$\begin{aligned} T(\boldsymbol{u}) &:= -p\boldsymbol{n} + \mu(\nabla\boldsymbol{u} + \nabla\boldsymbol{u}^\top)\boldsymbol{n} \\ &= \boldsymbol{\sigma}(\boldsymbol{u},p)\boldsymbol{n}\,, \\ T'(\boldsymbol{u}) &:= p\boldsymbol{n} + \mu(\nabla\boldsymbol{u} + \nabla\boldsymbol{u}^\top)\boldsymbol{n} \\ &= \boldsymbol{\sigma}(\boldsymbol{u},-p)\boldsymbol{n}\,, \end{aligned} \tag{2.3.6}$$

where

$$\boldsymbol{\sigma} := -pI + \mu(\nabla\boldsymbol{u} + \nabla\boldsymbol{u}^\top)$$

denotes the *stress tensor* in the viscous flow. We remark that it is understood that the stress operator T is always defined for the pair $(\boldsymbol{u}, p)$. For smooth $(\boldsymbol{u}, p)$ and $(\boldsymbol{v}, q)$, we have the second Green formula

$$\begin{aligned} &\int_\Omega \boldsymbol{u}\cdot\{-\mu\Delta\boldsymbol{v} - \nabla q)\}\,dy - \int_\Omega\{-\mu\Delta\boldsymbol{u} + \nabla p\}\cdot\boldsymbol{v}dy \\ &\qquad = \int_\Gamma [T(\boldsymbol{u})\cdot\boldsymbol{v} - \boldsymbol{u}\cdot T'(\boldsymbol{v})]\,ds \end{aligned} \tag{2.3.7}$$

provided
$$\operatorname{div} \boldsymbol{u} = \operatorname{div} \boldsymbol{v} = 0\,.$$

Now replacing $(\boldsymbol{v}, q)$ by $(\boldsymbol{v}^k, q^k)$, and by following standard arguments, the velocity component of the nonhomogeneous Stokes system (2.3.1) has the representation

$$u_k(x) = \int\limits_\Gamma \left[T_y(\boldsymbol{u}) \cdot \boldsymbol{v}^k(x,y) - \boldsymbol{u} \cdot T_y'(\boldsymbol{v}^k)(x,y)\right] ds_y + \int\limits_\Omega \boldsymbol{f} \cdot \boldsymbol{v}^k dy \quad \text{for } x \in \Omega\,. \tag{2.3.8}$$

To obtain the representation of the pressure, we may simply substitute the relation
$$\frac{\partial p}{\partial x_k} = \mu \Delta u_k + f_k$$

into (2.3.8). To simplify the representation, we now introduce the *fundamental velocity tensor* $E(x,y)$ and its *associated pressure vector* $Q(x,y)$, respectively, as
$$E(x,y) = [\boldsymbol{v}^1, \cdot, \boldsymbol{v}^n], \quad \text{and} \quad Q(x,y) = [q^1, \cdot, q^n]\,,$$

which satisfy
$$\begin{aligned} -\mu \Delta_x E + \nabla_x Q &= \delta(x,y) \boldsymbol{I}\,, \\ \operatorname{div}_x E &= \boldsymbol{0}^\top\,. \end{aligned} \tag{2.3.9}$$

As a result of (2.3.4) and (2.3.5), we have explicitly

$$\begin{aligned} E(x,y) &= \frac{1}{4(n-1)\pi\mu}\Big(\gamma_n \boldsymbol{I} + \frac{(x-y)(x-y)^\top}{|x-y|^n}\Big)\,, \\ Q(x,y) &= \frac{1}{2(n-1)\pi}(-\nabla_x \gamma_n)^\top = \frac{1}{2(n-1)\pi}(\nabla_y \gamma_n)^\top \end{aligned} \tag{2.3.10}$$

with
$$\gamma_n(x,y) = \begin{cases} -\log|x-y| & \text{for } n = 2 \\ \frac{1}{|x-y|} & \text{for } n = 3 \end{cases},$$

as in (2.2.2). In terms of $E(x,y)$ and $Q(x,y)$, we finally have the representation for solutions in the form:

$$\begin{aligned} \boldsymbol{u}(x) &= \int\limits_\Gamma E(x,y) T(\boldsymbol{u})(y) ds_y - \int\limits_\Gamma \left(T_y' E(x,y)\right)^\top \boldsymbol{u}(y) ds_y \\ &\quad + \int\limits_\Omega E(x,y) \boldsymbol{f}(y) dy \quad \text{for } x \in \Omega\,, \end{aligned} \tag{2.3.11}$$

$$\begin{aligned} p(x) &= \int\limits_\Gamma Q(x,y) \cdot T(\boldsymbol{u})(y) ds_y - 2\mu \int\limits_\Gamma \Big(\frac{\partial Q}{\partial n_y}(x,y)\Big) \cdot \boldsymbol{u}(y) ds_y \\ &\quad + \int\limits_\Omega Q(x,y) \cdot \boldsymbol{f}(y) dy \quad \text{for } x \in \Omega\,. \end{aligned} \tag{2.3.12}$$

It is understood that the representation of p is unique only up to an additive constant. Also, as was explained before,

$$T'_y E(x,y) := \boldsymbol{\sigma}\big(E(x,y), -Q(x,y)\big)\boldsymbol{n}(y)\,.$$

2.3.1 Hydrodynamic Potentials

The last terms in the representation (2.3.11) and (2.3.12) corresponding to the body force $\boldsymbol{f}$ define a particular solution $(\boldsymbol{U}, P)$ of the nonhomogeneous Stokes system (2.3.1). As in elasticity, if we decompose the solution in the form

$$\boldsymbol{u} = \boldsymbol{u}_c + \boldsymbol{U}\,, \quad p = p_c + P,$$

then the pair $(\boldsymbol{u}_c, p_c)$ will satisfy the corresponding homogeneous system of (2.3.1). Hence, in the following, without loss of generality, we shall confine ourselves only to the homogeneous Stokes system. The solution of the homogeneous system now has the representation from (2.3.11) and (2.3.12) with $\boldsymbol{f} = \boldsymbol{0}$, i.e.,

$$\boldsymbol{u}(x) = V\boldsymbol{\tau}(x) - W\boldsymbol{\varphi}(x)\,, \tag{2.3.13}$$

$$p(x) = \Phi\boldsymbol{\tau}(x) - \Pi\boldsymbol{\varphi}(x). \tag{2.3.14}$$

(The subscript c has been suppressed.) Here the pair (V, Φ) and (W, Π) are the respective simple– and double layer hydrodynamic potentials defined by

$$\left.\begin{aligned} V\boldsymbol{\tau}(x) &:= \int\limits_\Gamma E(x,y)\boldsymbol{\tau}(y)ds_y\,, \\ \Phi\boldsymbol{\tau}(x) &:= \int\limits_\Gamma Q(x,y)\cdot\boldsymbol{\tau}(y)ds_y\,; \end{aligned}\right. \tag{2.3.15}$$

$$\left.\begin{aligned} W\boldsymbol{\varphi}(x) &:= \int\limits_\Gamma \big(T'_y(E(x,y))\big)^\top \boldsymbol{\varphi}(y)ds_y\,, \\ \Pi\boldsymbol{\varphi}(x) &:= 2\mu\int\limits_\Gamma \Big(\frac{\partial}{\partial n_y}Q(x,y)\Big)\cdot\boldsymbol{\varphi}(y)ds_y \quad \text{for } x \notin \Gamma\,. \end{aligned}\right. \tag{2.3.16}$$

In (2.3.13) and (2.3.14) the boundary charges are the Cauchy data $\boldsymbol{\varphi} = \boldsymbol{u}(x)|_\Gamma$ and $\boldsymbol{\tau}(x) = T\boldsymbol{u}(x)|_\Gamma$ of the solution to the Stokes equations

$$\left.\begin{aligned} -\mu\Delta\boldsymbol{u} + \nabla p &= \boldsymbol{0}\,, \\ \text{div}\ \boldsymbol{u} &= 0 \quad \text{in } \Omega\,. \end{aligned}\right. \tag{2.3.17}$$

For the exterior problems, the representation formula for $\boldsymbol{u}$ needs to be modified by taking into account the growth conditions at infinity. Here proper growth conditions are

$$\boldsymbol{u}(x) = \begin{cases} \boldsymbol{\Sigma} \ \log|x| + O(1) & \text{for } n = 2 \ , \\ O(|x|^{-1}) & \text{for } n = 3 \ ; \end{cases} \tag{2.3.18}$$

$$p(x) = O(|x|^{1-n}) \quad \text{as} \quad |x| \to \infty. \tag{2.3.19}$$

In the two–dimensional case $\boldsymbol{\Sigma}$ is a given constant vector. The representation formula for solutions of the Stokes equations (2.3.17) in Ω^c with the growth conditions (2.3.18) and (2.3.19) has the form

$$\boldsymbol{u}(x) = -V\boldsymbol{\tau}(x) + W\boldsymbol{\varphi}(x) + \boldsymbol{\omega}\,, \tag{2.3.20}$$

$$p(x) = -\boldsymbol{\Phi}\boldsymbol{\tau}(x) + \boldsymbol{\Pi}\boldsymbol{\varphi}(x) \tag{2.3.21}$$

with the Cauchy data $\boldsymbol{\varphi} = \boldsymbol{u}_{|\Gamma}$ and $\boldsymbol{\tau} = T(\boldsymbol{u})_{|\Gamma}$ satisfying

$$\boldsymbol{\Sigma} = \int\limits_{\Gamma} \boldsymbol{\tau} ds\,; \tag{2.3.22}$$

and $\boldsymbol{\omega}$ is an unknown constant vector which vanishes when $n = 3$.

2.3.2 The Stokes Boundary Value Problems

We consider two boundary value problems for the Stokes system (2.3.17) in Ω as well as in Ω^c. In the first problem (the Dirichlet problem), the boundary trace of the velocity

$$\boldsymbol{u}|_{\Gamma} = \boldsymbol{\varphi} \qquad \text{on } \Gamma \tag{2.3.23}$$

is specified, and in the second problem (the Neumann problem), the hydrodynamic boundary traction

$$T(\boldsymbol{u})|_{\Gamma} = \boldsymbol{\tau} \qquad \text{on } \Gamma \tag{2.3.24}$$

is given. As consequences of the incompressible flow equations and the Green formula for the interior problem, the prescribed Cauchy data need to satisfy, respectively, the *compatibility conditions*

$$\begin{aligned} &\int\limits_{\Gamma} \boldsymbol{\varphi} \cdot \boldsymbol{n}\, ds = 0\,, \\ &\int\limits_{\Gamma} \boldsymbol{\tau} \cdot (\boldsymbol{a} + \boldsymbol{b} \times \boldsymbol{x})\, ds = \boldsymbol{0} \quad \text{for all } \boldsymbol{a} \in \mathbb{R}^n \quad \text{and} \quad \boldsymbol{b} \in \mathbb{R}^{1+2(n-2)}\,, \end{aligned} \tag{2.3.25}$$

with $\boldsymbol{b} \times \boldsymbol{x} := b(x_2, -x_1)^\top$ for $n = 2$.

For the exterior problem we require the decay conditions (2.3.18) and (2.3.19). We again solve these problems by using the direct method of boundary integral equations.

Since the pressure p will be completely determined once the Cauchy data for the velocity are known, in the following, it suffices to consider only the boundary integral equations for the velocity $\boldsymbol{u}$. We need, of course, the representation formula for p implicitly when we deal with the stress operator. In analogy to elasticity, we begin with the representation formula (2.3.13) for the velocity $\boldsymbol{u}$ in Ω and (2.3.20) and (2.3.21) in Ω^c. Applying the trace operator and the stress operator T to both sides of the representation formula, we obtain the overdetermined system of boundary integral equations (the Calderón projection) for the interior problem

$$\boldsymbol{\varphi}(x) = (\tfrac{1}{2}I - K)\boldsymbol{\varphi}(x) + V\boldsymbol{\tau}(x)\,, \tag{2.3.26}$$

$$\boldsymbol{\tau}(x) = D\boldsymbol{\varphi} + (\tfrac{1}{2}I + K')\boldsymbol{\tau}(x) \text{ on } \Gamma\,. \tag{2.3.27}$$

Hence, the Calderón projector for Ω can also be written in operator matrix form as

$$\mathcal{C}_\Omega = \begin{pmatrix} \frac{1}{2}I - K & V \\ D & \frac{1}{2}I + K' \end{pmatrix}.$$

Here V, K, K' and D are the four corresponding basic boundary integral operators of the Stokes flow. Hence, the Calderón projector $\mathcal{C}_\Omega$ for the interior domain has the same form as in (1.2.20) with the corresponding hydrodynamic potential operators.

For the exterior problem, the Calderón projector on solutions having the decay properties (2.3.18) and (2.3.19) with $\boldsymbol{\Sigma}$ given by (2.3.22) is also given by (1.4.11), i.e.,

$$\mathcal{C}_{\Omega^c} = \mathcal{I} - \mathcal{C}_\Omega = \begin{pmatrix} \frac{1}{2}I + K & -V \\ -D & \frac{1}{2}I - K' \end{pmatrix}. \tag{2.3.28}$$

As always, the solutions of both Dirichlet problems as well as both Neumann problems in Ω and Ω^c can be solved by using the boundary integral equations of the first as well as of the second kind by employing the relations between the Cauchy data given by the Calderón projectors.

The four basic operators appearing in the Calderón projectors for the Stokes problem are defined in the same manner as in elasticity (see Lemmata 2.3.1 and 2.2.3) but with appropriate modifications involving the pressure terms. More specifically, the double layer operator is defined as

$$\begin{aligned} K\boldsymbol{\varphi}(x) &:= \frac{1}{2}\boldsymbol{\varphi}(x) + \lim_{\Omega\ni z\to x\in\Gamma} \int\limits_\Gamma \big(T'_y E(z,y)\big)^\top \boldsymbol{\varphi}(y)ds_y \\ &= \int\limits_{\Gamma\setminus\{x\}} \big(T'_y E(x,y)\big)^\top \boldsymbol{\varphi}(y)ds_y \\ &= \int\limits_{\Gamma\setminus\{x\}} \sum_{i,j,k,=1}^n \Big\{Q^k(x,y)\delta_{ij} + \mu\Big(\frac{\partial E_{ik}(x,y)}{\partial y_j} + \frac{\partial E_{jk}(x,y)}{\partial y_i}\Big)\Big\} n_j(y)\varphi_i(y)ds_y \\ &= \frac{n}{2(n-1)\pi} \int\limits_{\Gamma\setminus\{x\}} \frac{\big((x-y)\cdot\boldsymbol{n}(y)\big)\big((x-y)\cdot\boldsymbol{\varphi}(y)\big)(x-y)}{|x-y|^{n+2}} ds_y \end{aligned} \tag{2.3.29}$$

having a weakly singular kernel for $\Gamma \subset C^2$, and, hence, defines a continuous mapping $K : C^\alpha(\Gamma) \to C^{1+\alpha}(\Gamma)$ (see Ladyženskaya [256, p.35] where the fundamental solution and the potentials carry the opposite sign). The hypersingular operator D is now defined by

$$\begin{aligned} D\varphi &= -T_x W\varphi(x) \\ &:= -\lim_{\Omega\ni z\to x\in\Gamma} T_z(\partial_z, x)\big(W\varphi(z)\big) \qquad (2.3.30) \\ &= \lim_{\Omega\ni z\to x\in\Gamma} \big\{\big(\Pi\varphi(z)\big)\boldsymbol{n}(x) - \mu\big(\nabla_z W\varphi(z) + \big(\nabla_z W\varphi(z)\big)^\top\big)\boldsymbol{n}(x)\big\}\,. \end{aligned}$$

With the standard regularization this reads

$$D\varphi(x) = -\text{p.v.}\int_\Gamma \big\{T_x\big(T'_y E(x,y)\big)^\top\big\}\big(\varphi(y)-\varphi(x)\big)ds_y \qquad (2.3.31)$$

$$\begin{aligned} &= \frac{-\mu}{2(n-1)\pi}\text{p.v.}\int_\Gamma \Big\{2\frac{1}{|x-y|^n}\boldsymbol{n}(y)\cdot\big(\varphi(y)-\varphi(x)\big) \\ &\qquad + \frac{n}{|x-y|^{n+2}}\boldsymbol{n}(y)\cdot\boldsymbol{n}(x)\big[(x-y)\cdot\big(\varphi(y)-\varphi(x)\big)\big](\boldsymbol{x}-\boldsymbol{y})\Big\}ds_y \\ &+ \frac{\mu}{2(n-1)\pi}\int_\Gamma \Big\{\frac{2n(n+2)}{|x-y|^{n+4}}\big[(x-y)\cdot\boldsymbol{n}(x)\big] \\ &\qquad \times\big[(x-y)\cdot\boldsymbol{n}(y)\big]\big[(x-y)\cdot\big(\varphi(y)-\varphi(x)\big)\big](\boldsymbol{x}-\boldsymbol{y}) \\ &\qquad - \frac{n}{|x-y|^{n+2}}\Big(\big[(x-y)\cdot\boldsymbol{n}(x)\big]\big[(x-y)\cdot\big(\varphi(y)-\varphi(x)\big)\big]\boldsymbol{n}(y) \\ &\qquad + \boldsymbol{n}(x)\cdot\big(\varphi(y)-\varphi(x)\big)\big[(y-x)\cdot\boldsymbol{n}(y)\big](\boldsymbol{x}-\boldsymbol{y}) \\ &\qquad + \big[(x-y)\cdot\boldsymbol{n}(x)\big]\big[(x-y)\cdot\boldsymbol{n}(y)\big]\big(\varphi(y)-\varphi(x)\big)\Big)\Big\}ds_y\,. \end{aligned}$$

Again, as in Lemma 2.2.3, the hypersingular operator can be reformulated.

Lemma 2.3.1. *Kohr et al* [234] *Let $\Gamma \in C^2$ and let φ be a Hölder continuously differentiable function. Then the operator D in* (2.3.30) *can be expressed as a composition of tangential differential operators and simple layer potentials as in* (2.2.32)–(2.2.34) *where in the case $n = 2$ set $\frac{\lambda+\mu}{\lambda+2\mu} = 1$ in* (2.2.33) *and in the case $n = 3$ take $E(x, y)$ from* (2.3.10) *in* (2.2.34).

Now let us assume that the boundary $\Gamma = \bigcup_{\ell=1}^{L} \Gamma_\ell$ consists of L separate, mutually non intersecting compact boundary components $\Gamma_1, \ldots, \Gamma_L$.

Before we exemplify the details of solvability of the boundary integral equations, we first summarize some basic properties of their eigenspaces.

Theorem 2.3.2. *(See also Kohr and Pop* [233].*) Let $n = 3$. Then we have the following relations.*

i) *The normal vector fields $\boldsymbol{n}_\ell \in C^\alpha(\Gamma)$ where $\boldsymbol{n}_\ell|_{\Gamma_j} = \boldsymbol{0}$ for $\ell \neq j$ generate exterior to Ω on $\Gamma = \bigcup_{\ell=1}^{L} \Gamma_\ell$ the L–dimensional eigenspace or kernel of the simple layer operator V as well as of $(\frac{1}{2}I - K')$. Then the operator $(\frac{1}{2}I - K)$ also has an L–dimensional eigenspace generated by $\boldsymbol{\varphi}_{0\ell} \in C^{1+\alpha}(\Gamma)$ with $\boldsymbol{\varphi}_{0\ell}|_{\Gamma_j} = \boldsymbol{0}$ for $\ell \neq j$ satisfying the equations*

$$\boldsymbol{n}_\ell = D\boldsymbol{\varphi}_{0\ell} \quad \textit{for} \ \ \ell = \overline{1, L}\,. \tag{2.3.32}$$

Any eigenfunction $\sum_{j=1}^{L} \gamma_j \boldsymbol{n}_j$ generates a solution

$$\boldsymbol{0} \equiv \sum_{j=1}^{L} \gamma_j V \boldsymbol{n}_j \quad \textit{and} \ \ p_0 = \gamma_1 \quad \textit{in} \ \ \Omega$$

(see Kohr and Pop [233], Reidinger and Steinbach [361]).
ii) *On each component Γ_ℓ of the boundary, the boundary integral operators $D|_{\Gamma_\ell}$ as well as $(\frac{1}{2}I + K)|_{\Gamma_\ell}$ have the 6–dimensional eigenspace $\boldsymbol{v}_\ell = (\boldsymbol{a}_\ell + \boldsymbol{b}_\ell \times \boldsymbol{x})|_\Gamma$ for all $\boldsymbol{a}_\ell \in \mathbb{R}^3$ with $\boldsymbol{b}_\ell \in \mathbb{R}^3$.*

If $\boldsymbol{v}_{j,\ell}$ with $j = 1, \dots, 6$ and $\ell = 1, \dots, L$ denotes a basis of this eigenspace then there exist $6L$ linearly independent eigenvectors $\boldsymbol{\tau}_{j,\ell} \in C^\alpha(\Gamma)$ of the adjoint operator $(\frac{1}{2}I + K')|_{\Gamma_\ell}$; and there holds the relation

$$\boldsymbol{v}_{j,\ell} = V|_{\Gamma_\ell} \boldsymbol{\tau}_{j,\ell} \tag{2.3.33}$$

between these two eigenspaces.

Any of the eigenfunctions $v_{j,\ell}$ on Γ generates a solution

$$\boldsymbol{u}_{0j,\ell}(x) = -\int\limits_{\Gamma_\ell} K(x,y)\boldsymbol{v}_{j,\ell}(x)ds_x = \begin{cases} \boldsymbol{0} & \text{for } x \in \overline{\Omega} \text{ if } \ell = \overline{2, L}, \\ v_{j,1}(x) & \text{for } x \in \Omega \text{ if } \ell = 1 \end{cases}$$

$$p_{0j,\ell}(x) = \begin{cases} 0 & \text{for } x \in \overline{\Omega} \text{ if } \ell = \overline{2, L}, \\ \mathrm{div}_x \frac{\mu}{\pi(n-1)} \int\limits_{\Gamma_1} \left(\frac{\partial}{\partial n_y}\gamma_n(x,y)\right) \boldsymbol{v}_{j,1}(y)ds_y & \text{for } x \in \Omega \text{ if } \ell = 1 \end{cases}.$$

In the case $n = 2$, the operator V needs to be replaced by

$$\widetilde{V}\boldsymbol{\tau} := V\boldsymbol{\tau} + \alpha\Big(\int_\Gamma \boldsymbol{\tau} ds\Big) \quad \textit{with} \ \ \alpha > 0$$

an appropriately large chosen scaling constant α and $\boldsymbol{a} + \boldsymbol{b} \times \boldsymbol{x}$ replaced by $\boldsymbol{a} + b(x_2, -x_1)^\top$ and 6 by 3 in **ii)**.

Proof: Let $n = 3$ and, for brevity, $L = 1$.
i) It is shown by Ladyženskaya in [256, p.61] that $\boldsymbol{n}$ is the only eigensolution of $(\frac{1}{2}I - K')$. Therefore, due to the classical Fredholm alternative, the adjoint

operator $(\frac{1}{2}I - K)$ has only one eigensolution φ_0, as well. It remains to show that $\boldsymbol{n}$ is also the only linear independent eigensolution to V and satisfies (2.3.32).

As we will show later on, for V, the Fredholm theorems are also valid and $V : C^\alpha(\Gamma) \to C^{\alpha+1}(\Gamma)$ has the Fredholm index zero. Let $\boldsymbol{\tau}_0$ be any solution of $V\boldsymbol{\tau}_0 = \boldsymbol{0}$. Then the single layer potential

$$\boldsymbol{u}_0 = V\boldsymbol{\tau}_0 \quad \text{with} \quad p_0 = \Phi\boldsymbol{\tau}_0$$

is a solution of the Stokes system in Ω as well as in Ω^c with $\boldsymbol{u}_0|_\Gamma = \boldsymbol{0}$. Then $\boldsymbol{u}_0 \equiv \boldsymbol{0}$ in $\mathbb{R}^3$ and the associated pressure is zero in Ω^c and $p_0 = \beta =$ constant in Ω. As a consequence we have from the jump relations

$$T_x(\boldsymbol{u}_{0-}, p_0) - T_x(\boldsymbol{u}_{0+}, 0) = \big[\sigma(\boldsymbol{u}_0, p_0)\boldsymbol{n}\big]|_\Gamma = -\beta\boldsymbol{n}\,,$$

therefore $\boldsymbol{\tau}_0 = -\beta\boldsymbol{n}$.

On the other hand, $V\boldsymbol{n}|_\Gamma = \boldsymbol{0}$ follows from the fact that $\boldsymbol{u} := V\boldsymbol{n}$ and $p := \Pi\boldsymbol{n}$ is the solution of the exterior homogeneous Neumann problem of the Stokes system since

$$T(V\boldsymbol{u})|_\Gamma = (-\tfrac{1}{2}I + K')\boldsymbol{n} = \boldsymbol{0}\,.$$

and therefore vanishes identically (see [256, Theorem 1 p.60]).

In order to show (2.3.32), we consider the solution of the exterior Dirichlet Stokes problem with $\boldsymbol{u}^+|_\Gamma = \varphi_0 \neq 0$ which admits the representation

$$\boldsymbol{u}(x) = W\varphi_0 - V\boldsymbol{\tau}\,.$$

Then it follows that the corresponding simple layer term has vanishing boundary values,

$$-V\boldsymbol{\tau}|_\Gamma = \varphi_0 - (\tfrac{1}{2}I + K)\varphi_0 = (\tfrac{1}{2}I - K)\varphi_0 = \boldsymbol{0}\,.$$

Hence, $\boldsymbol{\tau} = \beta\boldsymbol{n}$ with some constant $\beta \in \mathbb{R}$. Application of $T_x|_\Gamma$ gives

$$\boldsymbol{\tau} = \beta\boldsymbol{n} = -D\varphi_0 + (\tfrac{1}{2}I - K')\beta\boldsymbol{n} = -D\varphi_0\,.$$

The case $\beta = 0$ would imply $\boldsymbol{\tau} = \boldsymbol{0}$ and then $\boldsymbol{u}(x)$ solved the homogeneous Neumann problem which has only the trivial solution [256, p.60] implying $\varphi_0 = \boldsymbol{0}$ which is excluded. So, $\beta \neq 0$ and scaling of φ_0 implies (2.3.32).

ii) For the operator $(\frac{1}{2}I + K)$ having the eigenspace $(\boldsymbol{a} + \boldsymbol{b} \times \boldsymbol{x})|_\Gamma$ of dimension 6 we refer to [256, p.62]. Hence, the adjoint operator $(\frac{1}{2}I + K')$ also has a 6–dimensional eigenspace due to the classical Fredholm theory since K is a compact operator. For the operator D let us consider the potential $\boldsymbol{u}(x) = W\boldsymbol{v}(x)$ in Ω^c with $\boldsymbol{v} = \boldsymbol{a} + \boldsymbol{b} \times \boldsymbol{x}|_\Gamma$. Then $\boldsymbol{u}$ is a solution of the Stokes problem and on the boundary we find

$$\boldsymbol{u}^+|_\Gamma = (\tfrac{1}{2}I + K)\boldsymbol{v} = \boldsymbol{0}\,.$$

Therefore $\boldsymbol{u}(x) = \mathbf{0}$ for all $x \in \overline{\Omega^c}$ and, hence,

$$TW\boldsymbol{v} = -D\boldsymbol{v} = \mathbf{0} \quad \text{and} \quad \boldsymbol{v} \in \ker D\,.$$

Conversely, if $D\boldsymbol{v} = \mathbf{0}$ then let $\boldsymbol{u}$ be the solution of the interior Dirichlet problem with $\boldsymbol{u}^-|_\Gamma = \boldsymbol{v}$ which has the representation

$$\boldsymbol{u}(x) = V\boldsymbol{\tau} - W\boldsymbol{\tau} \quad \text{for } x \in \Omega$$

with an appropriate $\boldsymbol{\tau}$. Then applying T we find

$$\boldsymbol{\tau} = (\tfrac{1}{2}I + K')\boldsymbol{\tau} + D\boldsymbol{v} = (\tfrac{1}{2}I + K')\boldsymbol{\tau}\,.$$

Therefore $\boldsymbol{\tau}$ satisfies $(\frac{1}{2}I - K')\boldsymbol{\tau} = \mathbf{0}$ which implies $\boldsymbol{\tau} = \beta\boldsymbol{n}$ with some $\beta \in \mathrm{I\!R}$. Hence,

$$\boldsymbol{u}(x) = \beta V\boldsymbol{n}(x) - W\boldsymbol{v}(x) = -W\boldsymbol{v}(x)$$

and its trace yields

$$(\tfrac{1}{2}I + K)\boldsymbol{v} = \mathbf{0} \quad \text{on } \Gamma\,.$$

Therefore $\boldsymbol{v} = \boldsymbol{a} + \boldsymbol{b} \times \boldsymbol{x}$ with some $\boldsymbol{a}, \boldsymbol{b} \in \mathrm{I\!R}^3$.

Now let $\boldsymbol{\tau}_0 \in \ker(\frac{1}{2}I + K')$, $\boldsymbol{\tau}_0 \neq \mathbf{0}$. Then $\boldsymbol{u}(x) := V\boldsymbol{\tau}_0(x)$ in Ω is a solution of the homogeneous Neumann problem in Ω since $T\boldsymbol{u}|_\Gamma = (\frac{1}{2}I + K')\boldsymbol{\tau}_0 = \mathbf{0}$. Therefore $\boldsymbol{u} = \boldsymbol{a} + \boldsymbol{b} \times \boldsymbol{x}$ with some $\boldsymbol{a}, \boldsymbol{b} \in \mathrm{I\!R}^3$ and

$$V\boldsymbol{\tau}_0 \in \ker(\tfrac{1}{2}I + K)\,.$$

The mapping $V : \ker(\frac{1}{2}I + K') \to \ker(\frac{1}{2}I + K)$ is also injective since for $\boldsymbol{\tau}_0 \neq \mathbf{0}$,

$$V\boldsymbol{\tau}_0 = 0 \quad \text{would imply} \quad \boldsymbol{\tau}_0 = \beta\boldsymbol{n}$$

and, hence,

$$(\tfrac{1}{2}I + K')\boldsymbol{\tau}_0 = \mathbf{0} = \beta(\tfrac{1}{2}I + K')\boldsymbol{n} = \beta\boldsymbol{n} - (\tfrac{1}{2}I - K')\beta\boldsymbol{n} = \beta\boldsymbol{n}\,.$$

So, $\beta = 0$, which is a contradiction. The case $L > 1$ we leave to the reader (see [207]).

For $n = 2$ the proof follows in the same manner and we omit the details. ∎

In the Table 2.3.3 below we summarize the boundary integral equations of the first and second kind for solving the four fundamental boundary value problems together with the corresponding eigenspaces as well as the compatibility conditions. We emphasize that, as a consequence of Theorem 2.3.2, the orthogonality conditions for the right–hand side given data in the boundary integral equations will be automatically satisfied provided the given Cauchy data satisfy the compatibility conditions if required because of the direct approach.

In the case of $n = 2$ in Table 2.3.3, replace V by $\widetilde{V}$ and $\boldsymbol{b} \times \boldsymbol{x}$ by $(bx_2, -x_1)^\top$, $b \in \mathrm{I\!R}$, $\boldsymbol{a} \in \mathrm{I\!R}^2$.

Table 2.3.3. Boundary Integral Equations for the 3–D Stokes Problem

BVP	BIE	Eigenspaces of BIO $\ell, k = \overline{1, L};\ j = \overline{1, 3(n-1)}$	Compability conditions for given $\boldsymbol{\varphi}, \boldsymbol{\psi}$
IDP	$(1) V\boldsymbol{\tau} = (\frac{1}{2}I + K)\boldsymbol{\varphi}$	$\ker V = \ker(\frac{1}{2} - K') = \text{span}\ \{\boldsymbol{n}_\ell\}$	$\int\limits_\Gamma \boldsymbol{\varphi} \cdot \boldsymbol{n} ds = 0$
	$(2) (\frac{1}{2}I - K')\boldsymbol{\tau} = D\boldsymbol{\varphi}$		
INP	$(1) D\boldsymbol{u} = (\frac{1}{2}I - K')\boldsymbol{\psi}$	$\ker D = \ker(\frac{1}{2}I + K) = \text{span}\ \{\boldsymbol{v}_{j,\ell}\}$	$\int\limits_\Gamma \boldsymbol{\psi} \cdot (\boldsymbol{a} + \boldsymbol{b} \times \boldsymbol{x}) ds = 0$
	$(2) (\frac{1}{2}I + K)\boldsymbol{u} = V\boldsymbol{\psi}$	$\boldsymbol{v}_{j,\ell}$ basis of $\{\boldsymbol{v}\|_{\Gamma_\ell} = \boldsymbol{a} + \boldsymbol{b} \times \boldsymbol{x}\|_{\Gamma_\ell}, \boldsymbol{v}\|_{\Gamma_k} = \boldsymbol{0}, \ell \neq k\}$	for all $\boldsymbol{a}, \boldsymbol{b} \in \mathbb{R}^3$.
EDP	$(1) V\boldsymbol{\tau} = (-\frac{1}{2}I + K)\boldsymbol{\varphi}$	$\ker V = \text{span}\ \{\boldsymbol{n}_\ell\}$	None
	$(2) (\frac{1}{2}I + K')\boldsymbol{\tau} = -D\boldsymbol{\varphi}$	$\ker(\frac{1}{2}I + K') = \text{span}\ \{\boldsymbol{\tau}_{j,\ell}\}$, $V\boldsymbol{\tau}_{j,\ell} = \boldsymbol{v}_{j,\ell} \in \ker(\frac{1}{2}I + K)$	
ENP	$(1) D\boldsymbol{u} = -(\frac{1}{2} + K')\boldsymbol{\psi}$	$\ker D = \text{span}\ \{\boldsymbol{v}_{j,\ell}\}$	None
	$(2) (-\frac{1}{2}I + K)\boldsymbol{u} = V\boldsymbol{\psi}$	$\ker(\frac{1}{2}I - K) = \text{span}\ \{\boldsymbol{u}_{0\ell}\}$, $D\boldsymbol{u}_{0\ell} = \boldsymbol{n}_\ell \in \ker V$	

	Modified Equations I, $\ell = \overline{1,L}\,,\;\; j = \overline{1,3(n-1)}$	Modified Equations II
IDP	$(1) V\boldsymbol{\tau} + \sum\limits_{\ell=1}^{L} \omega_\ell \boldsymbol{n}_\ell = (\frac{1}{2}I + K)\boldsymbol{\varphi}$ and $\int\limits_{\Gamma_\ell} \boldsymbol{\tau} \cdot \boldsymbol{n}_\ell ds = 0$	$V\boldsymbol{\tau} + \sum\limits_{\ell=1}^{L} \int\limits_{\Gamma} \boldsymbol{\tau} \cdot \boldsymbol{n}_\ell ds \boldsymbol{n}_\ell = (\frac{1}{2}I + K)\boldsymbol{\varphi}$
	$(2) (\frac{1}{2}I - K')\boldsymbol{\tau} + \sum\limits_{\ell=1}^{\ell} \omega_\ell \boldsymbol{n}_\ell = D\boldsymbol{\varphi}$ and $\int\limits_{\Gamma_\ell} \boldsymbol{\tau} \cdot \boldsymbol{n}_\ell ds = 0$	$(\frac{1}{2}I - K')\boldsymbol{\tau} + \sum\limits_{\ell=1}^{L} \int\limits_{\Gamma_\ell} \boldsymbol{\tau} \cdot \boldsymbol{n}_\ell ds \boldsymbol{n}_\ell = D\boldsymbol{\varphi}$
INP	$(1) D\boldsymbol{u} + \sum\limits_{\ell=1}^{L} \sum\limits_{j=1}^{3(n-1)} \omega_{j,\ell} \boldsymbol{v}_{j,\ell} = (\frac{1}{2}I - K')\boldsymbol{\psi}$ and $\int\limits_{\Gamma_\ell} \boldsymbol{u} \cdot \boldsymbol{v}_{j,\ell} ds = 0$	$D\boldsymbol{u} + \sum\limits_{\ell=1}^{L} \sum\limits_{j=1}^{3(n-1)} \int\limits_{\Gamma_\ell} \boldsymbol{u} \cdot \boldsymbol{v}_{j,\ell} ds \boldsymbol{v}_{j,\ell} = (\frac{1}{2}I - K')\boldsymbol{\tau}$
	$(2) (\frac{1}{2}I + K)\boldsymbol{u} + \sum\limits_{\ell=1}^{L} \sum\limits_{j=1}^{3(n-1)} \omega_{j,\ell} \boldsymbol{v}_{j,\ell} = V\boldsymbol{\psi}$ and $\int\limits_{\Gamma_\ell} \boldsymbol{u} \cdot V\boldsymbol{\tau}_{j,\ell} ds = \int\limits_{\Gamma_\ell} \boldsymbol{u} \cdot \boldsymbol{v}_{j,\ell} ds = 0$	$(\frac{1}{2}I + K)\boldsymbol{u} + \sum\limits_{\ell=1}^{L} \sum\limits_{j=1}^{3(n-1)} \int\limits_{\Gamma_\ell} \boldsymbol{u} \cdot \boldsymbol{v}_{j,\ell} ds \boldsymbol{v}_j = V\boldsymbol{\psi}$
EDP	$(1) V\boldsymbol{\tau} + \sum\limits_{\ell=1}^{L} \omega_\ell \boldsymbol{n}_\ell = (-\frac{1}{2}I + K)\boldsymbol{\varphi}$ and $\int\limits_{\Gamma_\ell} \boldsymbol{\tau} \cdot \boldsymbol{n}_\ell ds = 0$	$V\boldsymbol{\tau} + \sum\limits_{\ell=1}^{L} \int\limits_{\Gamma_\ell} \boldsymbol{\tau} \cdot \boldsymbol{n}_\ell ds \boldsymbol{n}_\ell = (-\frac{1}{2}I + K)\boldsymbol{\varphi}$
	$(2) (\frac{1}{2}I - K')\boldsymbol{\tau} + \sum\limits_{\ell=1}^{L} \sum\limits_{j=1}^{3(n-1)} \omega_{j,\ell} \boldsymbol{\tau}_{j,\ell} = -D\boldsymbol{\varphi}$ and $\int\limits_{\Gamma_\ell} \boldsymbol{\tau} \cdot \boldsymbol{v}_{j,\ell} ds = 0$	$(\frac{1}{2}I - K')\boldsymbol{\tau} + \sum\limits_{\ell=1}^{L} \sum\limits_{j=1}^{3(n-1)} \int\limits_{\Gamma_\ell} \boldsymbol{\tau} \cdot \boldsymbol{v}_{j,\ell} ds \boldsymbol{\tau}_{j,\ell} = -D\boldsymbol{\varphi}$
ENP	$(1) D\boldsymbol{u} + \sum\limits_{\ell=1}^{L} \sum\limits_{j=1}^{3(n-1)} \omega_{j,\ell} \boldsymbol{v}_{j,\ell} = -(\frac{1}{2}I + K')\boldsymbol{\psi}$ and $\int\limits_{\Gamma_\ell} \boldsymbol{u} \cdot \boldsymbol{v}_{j,\ell} ds = 0,$	$D\boldsymbol{u} + \sum\limits_{\ell=1}^{L} \sum\limits_{j=1}^{3(n-1)} \int\limits_{\Gamma_\ell} \boldsymbol{u} \cdot \boldsymbol{v}_{j,\ell} ds \boldsymbol{v}_{j,\ell} = -(\frac{1}{2}I + K')\boldsymbol{\psi}$
	$(2) (\frac{1}{2}I - K)\boldsymbol{u} + \sum\limits_{\ell=1}^{L} \omega_\ell \boldsymbol{n}_\ell = -V\boldsymbol{\psi}$ and $\int\limits_{\Gamma_\ell} \boldsymbol{u} \cdot D\boldsymbol{u}_{0\ell} ds = \int\limits_{\Gamma_\ell} \boldsymbol{u} \cdot \boldsymbol{n}_\ell ds = 0$	$(\frac{1}{2}I - K)\boldsymbol{u} + \sum\limits_{\ell=1}^{L} \int\limits_{\Gamma_\ell} \boldsymbol{u} \cdot \boldsymbol{n}_\ell ds \boldsymbol{n}_\ell = -V\boldsymbol{\psi}$

Table 2.3.4. Modified Integral Equations for the 3–D Stokes Problems

Since each of the integral equations in Table 2.3.3 has a nonempty kernel, we now modify these equations in the same manner as in elasticity by incorporating eigenspaces to obtain uniquely solvable boundary integral equations. Again, in order not to be repetitious, we summarize the modified equations in Table 2.3.4.

A few comments are in order.

In the two–dimensional case, it should be understood that V should be replaced by $\widetilde{V}$ and that $\ker D = \text{span}\,\{\boldsymbol{v}_{j,\ell}\}$ with $\boldsymbol{v}_{j,\ell}$ a basis of $\{\boldsymbol{a}+b\binom{x_2}{-x_1}\}|_{\Gamma_\ell}$ with $\boldsymbol{a} \in \mathbb{R}^2$, $b \in \mathbb{R}$. Moreover, as in elasticity in Section 2.2, one has to incorporate $\int\limits_\Gamma \boldsymbol{\sigma} ds$ appropriately, in order to take into account the decay conditions (2.3.18).

For exterior problems, special attention has to be paid to the behavior at infinity. In particular, $\boldsymbol{u}$ has the representation (2.3.20), i.e.,

$$\boldsymbol{u} = W\boldsymbol{\varphi} - V\boldsymbol{\tau} + \boldsymbol{\omega} \quad \text{in } \Omega^c .$$

Then the Dirichlet condition leads on Γ to the system

$$V\boldsymbol{\tau} - \boldsymbol{\omega} = -(\tfrac{1}{2}I - K)\boldsymbol{\varphi} \quad \text{and} \quad \int\limits_\Gamma \boldsymbol{\tau} ds = \boldsymbol{\Sigma}\,, \tag{2.3.34}$$

where in the last equation $\boldsymbol{\Sigma}$ is a given constant vector determining the logarithmic behavior of $\boldsymbol{u}$ at infinity (see (2.3.18)). For uniqueness, this system is modified by adding the additional conditions

$$\int\limits_\Gamma \boldsymbol{\tau}\cdot\boldsymbol{n}_\ell ds = 0 \;\; \ell = \overline{1,L}\,.$$

Then the system (2.3.34) is equivalent to the uniquely solvable system

$$\begin{aligned} V\boldsymbol{\tau} - \boldsymbol{\omega} + \sum_{\ell=1}^{L} \omega_{3\ell}\boldsymbol{n}_\ell &= -(\tfrac{1}{2}I - K)\boldsymbol{\varphi}\,, \\ \int\limits_\Gamma \boldsymbol{\tau}\cdot\boldsymbol{n}_\ell ds = 0\,,\; \int\limits_\Gamma \boldsymbol{\sigma} ds &= \boldsymbol{\Sigma}\,,\; \overline{\ell = 1, L}\,, \quad \text{or} \end{aligned} \tag{2.3.35}$$

$$V\boldsymbol{\tau} - \boldsymbol{\omega} + \sum_{\ell=1}^{L} \int\limits_\Gamma \boldsymbol{\tau}\cdot\boldsymbol{n}_\ell ds\boldsymbol{n}_\ell = -(\tfrac{1}{2}I - K)\boldsymbol{\varphi}\,, \quad \int\limits_\Gamma \boldsymbol{\sigma} ds = \boldsymbol{\Sigma}\,. \tag{2.3.36}$$

These last two versions (2.3.35) and (2.3.36) correspond to mixed formulations and have been analyzed in detail in Fischer et al [124] and [196, 197, 199, 202].

In the same manner, appropriate modifications are to be considered for other boundary co nditions and the time harmonic unsteady problems and corresponding boundary integral equations as well [199, 202], Kohr et al [233, 232] and Varnhorn [424]. There, as in this section, the boundary integral equations are considered for charges in Hölder spaces. We shall come back to these problems in a more general setting in Chapter 5.

Note that for the interior Neumann problem, the modified integral equations will provide specific uniquely determined solutions of the integral equations, whereas the solution of the original Stokes Neumann problem still has the nullspace $\boldsymbol{a} + \boldsymbol{b} \times \boldsymbol{x}$ for $n = 3$ and $\{\boldsymbol{a} + b(x_2, -x_1)^\top\}$ for $n = 2$.

Finally, the second versions of the modified integral equations (II) in Table 2.3.4 are often referred to as *stabilized versions* in scientific computing. Clearly, the two versions are always equivalent Fischer et al [123].

2.3.3 The Incompressible Material — Revisited

With the analysis of the Stokes problems available, we now return to the interior elasticity problems in Section 2.2.4 for almost incompressible materials, i.e., for small $c \geq 0$, but restrict ourselves to the case that Γ is one connected compact manifold (see also [207], and Steinbach [398]). The case of $\Gamma = \bigcup_{\ell=1}^{L} \Gamma_\ell$ as in Theorem 2.3.2 is considered in [207].

For the interior displacement problem, the unknown boundary traction $\boldsymbol{\sigma}$ satisfies the boundary integral equation (2.2.40) of the first kind,

$$V_{el}\boldsymbol{\sigma} = (\tfrac{1}{2}I + K_{el})\boldsymbol{\varphi} \quad \text{on } \Gamma\,. \tag{2.3.37}$$

where the index el indicates that these are the operators in elasticity where the kernel $E_{el}(x, y)$ can be expressed via (2.2.68). Then with the simple layer potential operator V_{st} of the Stokes equation and its kernel given in (2.3.10) we have the relation

$$V_{el} = \frac{1}{1+c} V_{st} + \frac{2c}{1+c}\,\frac{1}{\mu} V_\Delta I \tag{2.3.38}$$

where V_Δ denotes the simple layer potential operator (1.2.1) of the Laplacian. Inserting (2.3.38) into (2.3.37) yields the equation

$$V_{st}\boldsymbol{\sigma} = (1+c)(\tfrac{1}{2}I + K_{el})\boldsymbol{\varphi} - c\frac{2}{\mu} V_\Delta \boldsymbol{\sigma}\,, \tag{2.3.39}$$

which corresponds to the equation (1) of the interior Stokes problem in Table 2.3.3.

As was shown in Theorem 2.3.2, the solution of (2.3.39) can be decomposed in the form

$$\boldsymbol{\sigma} = \boldsymbol{\sigma}_0 + \alpha \boldsymbol{n} \quad \text{with } \int_\Gamma \boldsymbol{\sigma}_0 \cdot \boldsymbol{n} ds = 0 \quad \text{and } \alpha \in \mathbb{R}\,. \tag{2.3.40}$$

Hence,

$$\begin{aligned} V_{st}\boldsymbol{\sigma}_0 &= (1+c)(\tfrac{1}{2}I + K_{el})\boldsymbol{\varphi} - c\frac{2}{\mu} V_\Delta(\boldsymbol{\sigma}_0 + \alpha\boldsymbol{n})\,, \\ &\int_\Gamma \boldsymbol{\sigma}_0 \cdot \boldsymbol{n} ds = 0\,. \end{aligned} \tag{2.3.41}$$

A necessary and sufficient condition for the solvability of this system is the orthogonality condition

$$\int\limits_{\Gamma} \Big\{(1+c)(\tfrac{1}{2}I + K_{e\ell})\varphi - c\frac{2}{\mu}V_{\Delta}\boldsymbol{\sigma}_0 - \alpha c\frac{2}{\mu}V_{\Delta}\boldsymbol{n}\Big\} \cdot \boldsymbol{n} ds = 0\,.$$

Now we combine (2.2.68) with (2.3.29) and obtain the relation

$$(1+c)K_{e\ell}\varphi = K_{st}\varphi + c(K_{\Delta}\varphi + L_1\varphi) \tag{2.3.42}$$

between the double layer potential operators of the Lamé and the Stokes systems where K_{Δ} is the double layer potential operator (1.2.8) of the Laplacian and L is the linear Cauchy singular integral operator defined by

$$L_1\varphi = \frac{1}{2\pi(n-1)}\ \text{p.v.}\!\!\int\limits_{\Gamma\setminus\{x\}} \Big(\frac{\boldsymbol{n}(y)\cdot\varphi(y)(x-y) - (x-y)\cdot\varphi(y)\boldsymbol{n}(y)}{|x-y|^n}\Big) ds_y\,. \tag{2.3.43}$$

Therefore the orthogonality condition becomes

$$\int\limits_{\Gamma}\{(\tfrac{1}{2}I + K_{st})\varphi\}\cdot\boldsymbol{n}ds + c\Big[\int\limits_{\Gamma}\{(\tfrac{1}{2}I + K_{\Delta} + L_1)\varphi\}\cdot\boldsymbol{n}ds - \frac{2}{\mu}\int\limits_{\Gamma}(V_{\Delta}\boldsymbol{\sigma}_0)\cdot\boldsymbol{n}ds - \alpha\frac{2}{\mu}\beta_{\Delta}\Big] = 0$$

where $\beta_{\Delta} := \int\limits_{\Gamma}(V_{\Delta}\boldsymbol{n})\cdot\boldsymbol{n}ds$. Since $\int\limits_{\Gamma}\boldsymbol{n}ds = 0$, it can be shown that $\beta_{\Delta} > 0$ (see [201], [204, Theorem 3.7]). In the first integral, however, we interchange orders of integration and obtain

$$\int\limits_{\Gamma}\big((\tfrac{1}{2}I + K_{st})\varphi\big)\cdot\boldsymbol{n}ds = \int\limits_{\Gamma}\varphi\cdot\big((\tfrac{1}{2}I + K'_{st})\boldsymbol{n}\big)ds = \int\limits_{\Gamma}\varphi\cdot\boldsymbol{n}ds$$

from Theorem 2.3.2. Hence, the orthogonality condition implies that α must be chosen as

$$\alpha = \frac{1}{c}\frac{\mu}{2\beta_{\Delta}}\int\limits_{\Gamma}\varphi\cdot\boldsymbol{n}ds + \frac{\mu}{2\beta_{\Delta}}\int\limits_{\Gamma}\{(\tfrac{1}{2}I + K_{\Delta} + L_1)\varphi\}\cdot\boldsymbol{n}ds - \frac{1}{\beta_{\Delta}}\int\limits_{\Gamma}(V_{\Delta}\boldsymbol{\sigma}_0)\cdot\boldsymbol{n}ds\,. \tag{2.3.44}$$

Replacing α from (2.3.44) in (2.3.41) we finally obtain the corresponding stabilized equation,

$$V_{st}\boldsymbol{\sigma}_0 + \int\limits_{\Gamma}\boldsymbol{\sigma}_0\cdot\boldsymbol{n}ds\boldsymbol{n} + cB\boldsymbol{\sigma}_0 = \boldsymbol{f} \tag{2.3.45}$$

where the linear operator B is defined by

$$B\boldsymbol{\sigma}_0 := \frac{2}{\mu}\Big(V_{\Delta}\boldsymbol{\sigma}_0 - \frac{1}{\beta_{\Delta}}\int\limits_{\Gamma}(V_{\Delta}\boldsymbol{\sigma}_0)\cdot\boldsymbol{n}dsV_{\Delta}\boldsymbol{n}\Big)$$

and the right–hand side $\boldsymbol{f}$ is given by

$$\boldsymbol{f} = (\tfrac{1}{2}I + K_{st})\boldsymbol{\varphi} - \frac{1}{\beta_\Delta}\int_\Gamma \varphi \cdot \boldsymbol{n} ds V_\Delta \boldsymbol{n}$$
$$+c\Big[(\tfrac{1}{2}I + K_\Delta + L_1)\boldsymbol{\varphi} - \frac{1}{\beta_\Delta}\int_\Gamma \{(\tfrac{1}{2}I + K_\Delta + L_1)\boldsymbol{\varphi}\} \cdot \boldsymbol{n} ds V_\Delta \boldsymbol{n}\Big] .$$

Since for $c = 0$ the equation (2.3.45) is uniquely solvable, the regularly perturbed equation (2.3.45) for small c but $c \neq 0$ is still uniquely solvable.

With $\boldsymbol{\sigma}_0$ available, α can be found from (2.3.44) and, finally, the boundary traction $\boldsymbol{\sigma}$ is given by (2.3.40). Then the representation formula (2.2.6) provides us with the elastic displacement field $\boldsymbol{u}$ and the solution's behavior for the elastic, but almost incompressible materials, which one may expand with respect to small $c \geq 0$, as well. In particular we see that, for the almost incompressible material

$$\boldsymbol{u}_{e\ell} = \boldsymbol{u}_{st} + \tfrac{1}{\beta_\Delta}\int_\Gamma \boldsymbol{\varphi} \cdot \boldsymbol{n} ds V_\Delta \boldsymbol{n} + O(c) \quad \text{as} \quad c \to 0$$

where $\boldsymbol{u}_{st}$ is the unique solution of the Stokes problem with

$$\boldsymbol{u}_{st}|_\Gamma = \boldsymbol{\varphi} - \tfrac{1}{\beta_\Delta}\int_\Gamma \boldsymbol{\varphi} \cdot \boldsymbol{n} ds V_\Delta \boldsymbol{n} + O(c) .$$

We also have the relation

$$\int_\Gamma \boldsymbol{\varphi} \cdot \boldsymbol{n} ds = \int_\Omega \mathrm{div} \boldsymbol{u} dx = -c \int_\Omega p dx .$$

This shows that only if the given datum $\int_\Gamma \boldsymbol{\varphi} \cdot \boldsymbol{n} ds = O(c)$ then we have

$$\boldsymbol{u}_{e\ell} = \boldsymbol{u}_{st} + O(c) .$$

Next, we consider the interior traction problem for the almost incompressible material. For simplicity, we now employ Equation (2.2.46),

$$D_{e\ell}\boldsymbol{u} = (\tfrac{1}{2}I - K'_{e\ell})\boldsymbol{\psi} \quad \text{on} \quad \Gamma \tag{2.3.46}$$

where now $\boldsymbol{\psi}$, the boundary stress, is given on Γ satisfying the compatibility conditions (2.2.47), and the boundary displacement $\boldsymbol{u}$ is the unknown.

·With (2.3.38), i.e.,

$$E_{e\ell}(x,y) = \frac{1}{1+c}E_{st}(x,y) + \frac{c}{1+c}\,\frac{1}{2(n-1)\pi\mu}\gamma_n(x,y)I \tag{2.3.47}$$

and with Lemma 2.3.1 we obtain for the hypersingular operators

$$D_{e\ell}\boldsymbol{\varphi} = D_{st}\boldsymbol{\varphi} + cL_2\boldsymbol{\varphi} \tag{2.3.48}$$

where for $n = 3$

$$L_2\varphi(x) = \mathcal{M}_x \int\limits_{\Gamma\setminus\{x\}} 4\mu^2 \frac{1}{1+c}\Big(E_{st}(x,y) - \frac{1}{2(n-1)\mu\pi}\gamma_n(x,y)I\Big)\mathcal{M}_y\varphi(y)ds_y\,, \tag{2.3.49}$$

and for $n = 2$ the differential operators can be replaced as $\mathcal{M}_x = \frac{d}{ds_x}$ and $\mathcal{M}_y = \frac{d}{ds_y}$. Hence, (2.3.46) can be written as

$$D_{st}\boldsymbol{u} = (\tfrac{1}{2}I - K'_{e\ell})\boldsymbol{\psi} - cL_2\boldsymbol{u}\,. \tag{2.3.50}$$

In view of Theorem 2.3.2, one may decompose the solution $\boldsymbol{u}$ in the form

$$\boldsymbol{u}(x) = \boldsymbol{u}_0(x) + \sum_{j=1}^{M} \alpha_j \boldsymbol{m}_j(x) \tag{2.3.51}$$

where

$$\int\limits_\Gamma \boldsymbol{u}_0 \cdot \boldsymbol{m}_j ds = 0 \quad \text{for } j = 1, \dots, M \quad \text{with } M := \tfrac{1}{2}n(n+1)\,,$$

and $\boldsymbol{m}_j(x)$ are the traces of the rigid motions given in (2.2.55). These vector valued functions form a basis of the kernel to $D_{e\ell}$ as well as to D_{st} which implies also that

$$L_2\boldsymbol{m}_j = 0 \quad \text{for } j = 1, \dots, M \quad \text{and } c \in \mathbb{R}\,. \tag{2.3.52}$$

Substituting (2.3.51) into (2.3.50) yields the uniquely solvable system of equations

$$\begin{aligned} D_{st}\boldsymbol{u}_0 + cL_2\boldsymbol{u}_0 &= (\tfrac{1}{2}I - K'_{e\ell})\boldsymbol{\psi}\,, \\ \int\limits_\Gamma \boldsymbol{u}_0 \cdot \boldsymbol{m}_j ds &= 0 \quad \text{for } j = 1, \dots, M\,; \end{aligned} \tag{2.3.53}$$

or, in stabilized form

$$D_{st}\boldsymbol{u}_0 + \sum_{j=1}^{M} \int\limits_\Gamma \boldsymbol{u}_0 \cdot \boldsymbol{m}_j ds\, \boldsymbol{m}_j + cL_2\boldsymbol{u}_0 = (\tfrac{1}{2}I - K'_{e\ell})\boldsymbol{\psi}\,. \tag{2.3.54}$$

The right–hand side in (2.3.53) satisfies the orthogonality conditions

$$\int\limits_\Gamma \big((\tfrac{1}{2}I - K'_{e\ell})\boldsymbol{\psi}\big) \cdot \boldsymbol{m}_j ds = 0 \quad \text{for } j = 1, \dots, M$$

since the given $\boldsymbol{\psi}$ satisfies the compatibility conditions

$$\int\limits_\Gamma \boldsymbol{\psi} \cdot \boldsymbol{m}_j ds = 0 \quad \text{for } j = 1, \dots, M$$

and the vector valued function $\boldsymbol{m}_j$ satisfies

$$(\tfrac{1}{2}I + K_{e\ell})\boldsymbol{m}_j = \boldsymbol{0} \quad \text{on } \Gamma .$$

The equations (2.3.53) or (2.3.54) are uniquely solvable for every $c \in [0, \infty)$ and so, the general elastic solution $\boldsymbol{u}_{e\ell}$ for almost incompressible material has the form

$$\begin{aligned}\boldsymbol{u}_{e\ell}(x) &= V_{e\ell}\boldsymbol{\psi}(x) - W_{e\ell}\boldsymbol{u}_0(x) + \sum_{j=1}^{M} \alpha_j \boldsymbol{m}_j(x)\\ &= \frac{1}{1+c}\Big\{\boldsymbol{u}_{st} + \frac{2c}{\mu}V_{\Delta}\boldsymbol{\psi} - c(W_{\Delta}\boldsymbol{u}_0 + L_1\boldsymbol{u}_0\Big\}(x) + \sum_{j=1}^{M} \alpha_j \boldsymbol{m}_j(x)\end{aligned}$$

for $x \in \Omega$ with arbitrary $\alpha_j \in \mathbb{R}$ and where $\boldsymbol{u}_{st}$ is the solution of the Stokes problem with given boundary tractions $\boldsymbol{\psi}$, and L_1 is defined in (2.3.43). For $c \to 0$ we see that for the elastic Neumann problem

$$\boldsymbol{u}_{e\ell} = \boldsymbol{u}_{st} + O(c)$$

up to rigid motions, i.e., a regular perturbation with respect to the Stokes solution.

2.4 The Biharmonic Equation

In both problems, plane elasticity and plane Stokes flow, the systems of partial differential equations can be reduced to a single scalar 4th–order equation,

$$\Delta^2 u = 0 \quad \text{in} \quad \Omega \text{ (or } \Omega^c) \subset \mathbb{R}^2, \tag{2.4.1}$$

kwown as the *biharmonic equation.* In the elasticity case, u is the *Airy stress function,* whereas in the Stokes flow u is the *stream function* of the flow. The Airy function $W(x)$ is defined in terms of the stress tensor $\sigma_{ij}(\boldsymbol{u})$ for the displacement field $\boldsymbol{u}$ as

$$\sigma_{11}(\boldsymbol{u}) = \frac{\partial^2 W}{\partial x_2^2}, \quad \sigma_{12}(\boldsymbol{u}) = -\frac{\partial^2 W}{\partial x_1 \partial x_2}, \quad \sigma_{22}(\boldsymbol{u}) = \frac{\partial^2 W}{\partial x_1^2},$$

which satisfies the equilibrium equation $\mathrm{div}\boldsymbol{\sigma}(\boldsymbol{u}) = 0$ automatically for any smooth function W. Then from the stress-strain relation in the form of Hooke's law, it follows that

$$\Delta W = \sigma_{11}(\boldsymbol{u}) + \sigma_{22}(\boldsymbol{u}) = (\lambda + 2\mu)\mathrm{div}\boldsymbol{u}\,;$$

and thus, W satisfies (2.4.1) since $\Delta(\mathrm{div}\ \boldsymbol{u}) = 0$ from the Lamé system. On the other hand, the stream function u is defined in terms of the velocity $\boldsymbol{u}$ in the form

$$\boldsymbol{u} = (\nabla u)^{\perp}.$$

Here $\perp$ indicates the operation of rotating a vector counter–clockwise by a right angle. From the definition, the continuity equation for the velocity is satisfied for any choice of a smooth stream function u. One can verify directly that u satisfies (2.4.1) by taking the curl of the balance of momentum equation in the Stokes system. We note that in terms of the stream function, the vorticity is equal to $\omega \boldsymbol{k} = \nabla \times \boldsymbol{u} = \Delta u \boldsymbol{k}$, where $\boldsymbol{k}$ is the unit vector perpendicular to the $(x_1, x_2)-$ plane of the flow. For the homogeneous Stokes system, the vorticity is a harmonic function, and as a consequence, u satisfies the biharmonic equation (2.4.1).

To discuss boundary value problems for the biharmonic equation (2.4.1), it is best to begin with Green's formula for the equation in Ω. As is well known, for fourth-order differential equations, the Green formula generally varies and depends on the choice of boundary operators, i.e., how to apply the integration by parts formulae. In order to include boundary conditions arising for the thin plate, we rewrite $\Delta^2 u$ in terms of the Poisson ratio ν in the form

$$\Delta^2 u = \frac{\partial^2}{\partial x_1^2}\Big(\frac{\partial^2 u}{\partial x_1^2} + \nu\frac{\partial^2 u}{\partial x_2^2}\Big) + 2(1-\nu)\frac{\partial^2}{\partial x_1 \partial x_2}\Big(\frac{\partial^2 u}{\partial x_1 \partial x_2}\Big) + \frac{\partial^2}{\partial x_2^2}\Big(\frac{\partial^2 u}{\partial x_2^2} + \nu\frac{\partial^2 u}{\partial x_1^2}\Big).$$

Now integration by parts leads to the first Green formula in the form

$$\int_{\Omega} (\Delta^2 u)\, v dx = a(u,v) - \int_{\Gamma} \Big\{\frac{\partial v}{\partial n} Mu + vNu\Big\} ds\,, \tag{2.4.2}$$

where the bilinear form $a(u, v)$ is defined by

$$a(u,v) := \int_{\Omega} \Big\{\nu \Delta u\; \Delta v + (1-\nu)\sum_{i,j=1}^{2}\Big(\frac{\partial^2 u}{\partial x_i \partial x_j}\,\frac{\partial^2 v}{\partial x_i \partial x_j}\Big)\Big\} dx\,; \tag{2.4.3}$$

and M and N are differential operators defined by

$$Mu := \nu \Delta u + (1-\nu)\big((\boldsymbol{n}(z)\cdot\nabla_x)^2 u)|_{z=x}\big)\,, \tag{2.4.4}$$

$$Nu := -\Big\{\frac{\partial}{\partial n}\Delta u + (1-\nu)\frac{d}{ds}\big((\boldsymbol{n}(z)\cdot\nabla_x)(\boldsymbol{t}(z)\cdot\nabla_x)u(x)\big)\Big\}_{|z=x} \tag{2.4.5}$$

where $\boldsymbol{t} = \boldsymbol{n}^{\perp}$ is the unit tangent vector, i.e. $t_1 = -n_2\,,\ t_2 = n_1$. Then

$$Mu = \nu\Delta u + (1-\nu)\Big[\frac{\partial^2 u}{\partial x_1^2}n_1^2 + 2\frac{\partial^2 u}{\partial x_1 \partial x_2}n_1 n_2 + \frac{\partial^2 u}{\partial x_2^2}n_2^2\Big]\,,$$

$$Nu = -\frac{\partial}{\partial n}\Delta u + (1-\nu)\frac{d}{ds}\Big\{\Big(\frac{\partial^2 u}{\partial x_1^2} - \frac{\partial^2 u}{\partial x_2^2}\Big)n_1 n_2 - \frac{\partial^2 u}{\partial x_1 \partial x_2}(n_1^2 - n_2^2)\Big\}.$$

For the interior boundary value problems for (2.4.1), the starting point is the representation formula

$$\begin{aligned} u(x) &= \int_\Gamma \{E(x,y)Nu(y) + \Big(\frac{\partial E}{\partial n_y}(x,y)\Big)Mu(y)\}ds_y \\ &\quad - \int_\Gamma \{(M_yE(x,y))\frac{\partial u}{\partial n_y} + (N_yE(x,y))u(y)\}ds_y \quad \text{for } x \in \Omega\,, \end{aligned} \tag{2.4.6}$$

where $E(x,y)$ is the fundamental solution for the biharmonic equation given by

$$E(x,y) = \frac{1}{8\pi}|x-y|^2 \log|x-y| \tag{2.4.7}$$

which satisfies

$$\Delta_x^2 E(x,y) = \delta(x-y) \quad \text{in} \quad \mathbb{R}^2\,.$$

As in case of the Laplacian, we may rewrite u in the form

$$u(x) = V(Mu, Nu) - W\Big(u, \frac{\partial u}{\partial n}\Big) \tag{2.4.8}$$

where

$$V(Mu, Nu) := \int_\Gamma \{E(x,y)Nu(y) + \Big(\frac{\partial E}{\partial n_y}(x,y)\Big)Mu(y)\}ds_y\,, \tag{2.4.9}$$

$$W\Big(u, \frac{\partial u}{\partial n}\Big) := \int_\Gamma \{(M_yE(x,y))\frac{\partial u}{\partial n}(y) + (N_yE(x,y))u(y)\}ds_y \tag{2.4.10}$$

are the simple and double layer potentials, respectively, and $u|_\Gamma, \frac{\partial u}{\partial n}|_\Gamma, Mu|_\Gamma$ and $Nu|_\Gamma$ are the (modified) Cauchy data. This representation formula (2.4.6) suggests two basic types of boundary conditions:
The **Dirichlet boundary condition**, where $u|_\Gamma$ and $\frac{\partial u}{\partial n}|_\Gamma$ are prescribed on Γ, and the **Neumann boundary condition**, where $Mu|_\Gamma$ and $Nu|_\Gamma$ are prescribed on Γ. In thin plate theory, where u stands for the deflection of the middle surface of the plate, the Dirichlet condition specifies the displacement and the angle of rotation of the plate at the boundary, whereas the Neumann condition provides the bending moment and shear force at the boundary. Clearly, various linear combinations will lead to other mixed boundary conditions, which will not be discussed here.

From the bilinear form (2.4.3), we see that

$$a(u,v) = 0 \quad \text{for} \quad v \in \mathcal{R} := \{v = c_1x_1 + c_2x_2 + c_3 \mid \text{ for all } c_1, c_2, c_3 \in \mathbb{R}\}\,. \tag{2.4.11}$$

This implies that the Neumann data need to satisfy the compatibility condition

$$\int_\Gamma \Big\{\frac{\partial v}{\partial n}Mu + vNu\Big\}ds = 0 \quad \text{for all} \quad v \in \mathcal{R}\,. \tag{2.4.12}$$

We remark that looking at (2.4.2), one might think of choosing Δu and $-\frac{\partial}{\partial n}\Delta u$ as the Neumann boundary conditions which correspond to the Poisson ratio $\nu = 1$. This means that the compatibility condition (2.4.12) requires that it should hold for all harmonic functions v. However, the space of harmonic functions in Ω has *infinite* dimension, and this does not lead to a regular boundary value problem in the sense of Agmon [3, p. 151].

As for the exterior boundary value problems, in order to ensure the uniqueness of the solution of (2.4.1) in Ω^c, we need to augment (2.4.1) with an appropriate radiation condition (see (2.3.18)). We require that

$$u(x) = \Big(A_0 r + \frac{\boldsymbol{A}_1 \cdot \boldsymbol{x}}{|x|}\Big)\, r \,\log r + O(r) \quad \text{as} \quad r = |x| \to \infty \tag{2.4.13}$$

for given constant A_0 and constant vector $\boldsymbol{A}_1$. Under the condition (2.4.9), we then have the representation formula for the solution of (2.4.1) in Ω^c,

$$u(x) = -V(Mu, Nu) + W\Big(u, \frac{\partial u}{\partial n}\Big) + p(x), \tag{2.4.14}$$

where $p \in \mathcal{R}$ is a polynomial of degree less than or equal to one.

Before we formulate the boundary integral equations we first summarize some classical basic results.

Theorem 2.4.1. (Gakhov [135], Mikhlin [298, 299, 301] and Muskhelishvili [319]). *Let* $\Gamma \in C^{2,\alpha}$, $0 < \alpha < 1$.
i) *Let*

$$\boldsymbol{\varphi} = (\varphi_1, \varphi_2)^\top = \Big(u|_\Gamma\, ,\, \frac{\partial u}{\partial n}|_\Gamma\Big) \in C^{3+\alpha}(\Gamma) \times \Gamma^{2+\alpha}(\Gamma)$$

be given. Then there exists a unique solution $u \in C^{3+\alpha}(\overline{\Omega}) \cap C^4(\Omega)$ *of the interior Dirichlet problem satisfying the Dirichlet conditions*

$$u|_\Gamma = \varphi_1 \quad \textit{and} \quad \frac{\partial u}{\partial n}|_\Gamma = \varphi_2\,. \tag{2.4.15}$$

For given $A_0 \in \mathbb{R}$ *and* $\boldsymbol{A}_1 \in \mathbb{R}^2$ *there also exists a unique solution* $u \in C^{3+\alpha}(\Omega^c \cup \Gamma) \cap C^4(\Omega^c)$ *of the exterior Dirichlet problem which behaves at infinity as in* (2.4.13).
ii) *For given*

$$\boldsymbol{\psi} = (\psi_1, \psi_2)^\top \in C^{1+\alpha}(\Gamma) \times C^\alpha(\Gamma)$$

satisfying the compatibility conditions (2.4.12), *i.e.,*

$$\int\limits_\Gamma \Big\{\psi_1 \frac{\partial v}{\partial n} + \psi_2 v\Big\} ds = 0 \quad \textit{for all} \quad v \in \mathcal{R}\,, \tag{2.4.16}$$

the interior Neumann problem consisting of (2.4.1) *and the Neumann conditions*

$$Mu|_\Gamma = \psi_1 \quad and \quad Nu|_\Gamma = \psi_2 \tag{2.4.17}$$

has a solution $u \in C^{3+\alpha}(\overline{\Omega}) \cap C^4(\Omega)$ *which is unique up to a linear function* $p \in \mathcal{R}$.

If, for the exterior Neumann problem in Ω^c, *in addition to* ψ *the linear function* $p \in \mathcal{R}$ *is given, then it has a unique solution* $u \in C^{3+\alpha}(\Omega^c \cup \Gamma) \cap C^4(\Omega^c)$ *with the behaviour* (2.4.13), (2.4.14) *where*

$$A_0 = -\int\limits_\Gamma \psi_2 ds \quad and \quad \boldsymbol{A}_1 = \int\limits_\Gamma (\psi_1 \boldsymbol{n} + \psi_2 \boldsymbol{x}) ds\,. \tag{2.4.18}$$

As a consequence of Theorem 2.4.1 one has the useful identity of Gaussian type,

$$-W\Big(p, \frac{\partial p}{\partial n}\Big) = \begin{cases} p & \text{for} \quad x \in \Omega\,, \\ \frac{1}{2}p & \text{for} \quad x \in \Gamma\,, \\ 0 & \text{for} \quad x \in \Omega^c\,, \end{cases} \qquad \text{for any } \; p \in \mathcal{R}\,. \tag{2.4.19}$$

2.4.1 Calderón's Projector

(See also [208].) In order to obtain the boundary integral operators as x approaches Γ, from the simple– and double–layer potentials in the representation formulae (2.4.8) and (2.4.14), we need explicit information concerning the kernels of the potentials. A straightforward calculation gives

$$\begin{aligned} V(Mu, Nu)(x) &= \int\limits_\Gamma E(x,y) Nu(y) ds_y + \int\limits_\Gamma \Big(\frac{\partial E}{\partial n_y}(x,y)\Big) Mu(y) ds_y \\ &= \frac{1}{8\pi}\int\limits_\Gamma |x-y|^2 \log|x-y| Nu(y) ds_y \qquad\qquad (2.4.20) \\ &\quad + \frac{1}{8\pi}\int\limits_\Gamma \boldsymbol{n}(y)\cdot(y-x)(2\log|x-y|+1) Mu(y) ds_y \end{aligned}$$

$$\begin{aligned} W\Big(u, \frac{\partial u}{\partial n}\Big)(x) &= \int\limits_\Gamma (M_y E(x,y)) \frac{\partial u(y)}{\partial n} ds_y + \int\limits_\Gamma (N_y E(x,y)) u(y) ds_y \\ &= \frac{1}{8\pi}\int\limits_\Gamma \Big\{(2\log|x-y|+1) + \nu(2\log|x-y|+3) \\ &\qquad + 2(1-\nu)\frac{((y-x)\cdot \boldsymbol{n}(y))^2}{|x-y|^2}\Big\}\frac{\partial u(y)}{\partial n} ds_y \qquad (2.4.21) \\ &\quad + \frac{1}{2\pi}\int\limits_\Gamma \Big\{\frac{\partial}{\partial n_y}\log\Big(\frac{1}{|x-y|}\Big) \\ &\qquad - \frac{1}{2}(1-\nu)\frac{d}{ds_y}\Big(\frac{(x-y)\cdot \boldsymbol{t}(y)(x-y)\cdot \boldsymbol{n}(y)}{|x-y|^2}\Big)\Big\} u(y) ds_y\,. \end{aligned}$$

This leads to the following 16 boundary integral operators.

$$
\mathcal{K}\begin{pmatrix}\boldsymbol{\varphi}\\ \boldsymbol{\sigma}\end{pmatrix} = \lim_{\Omega \ni z \to x \in \Gamma} \begin{pmatrix} -W(\varphi_1,0)(z) - \frac{1}{2}\varphi_1(z) & -W(0,\varphi_2)(z) & V(\sigma_1,0)(z) & V(0,\sigma_2)(z) \\ -\boldsymbol{n}_x \cdot \nabla_z W(\varphi_1,0)(z) & -\boldsymbol{n}_x \cdot \nabla_z W(0,\varphi_2)(z) - \frac{1}{2}\varphi_2(x) & \boldsymbol{n}_x \cdot \nabla_z V(\sigma_1,0)(z) & \boldsymbol{n}_x \cdot \nabla_z V(0,\sigma_2)(z) \\ -M_z W(\varphi_1,0)(z) & -M_z W(0,\varphi_2)(z) & M_z V(\sigma_1,0)(z) - \frac{1}{2}\sigma_1(x) & M_z V(0,\sigma_2)(z) \\ -N_z W(\varphi_1,0)(z) & -N_z W(0,\varphi_2)(z) & N_z V(\sigma_1,0)(z) & N_z V(0,\sigma_2)(z) - \frac{1}{2}\sigma_2(x) \end{pmatrix}
$$

$$
= \begin{pmatrix} -K_{11} & V_{12} & V_{13} & V_{14} \\ D_{21} & K_{22} & V_{23} & V_{24} \\ D_{31} & D_{32} & -K_{33} & V_{34} \\ D_{41} & D_{42} & D_{43} & K_{44} \end{pmatrix} \tag{2.4.22}
$$

where we write

$$
\begin{pmatrix}\boldsymbol{\varphi}\\ \boldsymbol{\sigma}\end{pmatrix} = (\varphi_1, \varphi_2, \sigma_1, \sigma_2)^\top = \left(u, \frac{\partial u}{\partial n}, Mu, Nu\right)^\top |_\Gamma \,.
$$

Then the Calderon projector associated with the bi–Laplacian is defined by

$$
\mathcal{C}_\Omega := \frac{1}{2}\mathcal{I} + \mathcal{K} = \begin{pmatrix} \frac{1}{2}I - K_{11} & V_{12} & V_{13} & V_{14} \\ D_{21} & \frac{1}{2}I + K_{22} & V_{23} & V_{24} \\ D_{31} & D_{32} & \frac{1}{2}I - K_{33} & V_{34} \\ D_{41} & D_{42} & D_{43} & \frac{1}{2}I + K_{44} \end{pmatrix} \tag{2.4.23}
$$

Some more explanations are needed here. In order to maintain consistency with our notations for the Laplacian, we have adopted the notations V_{ij}, K_{ij} and D_{ij} for the weakly and hypersingular boundary integral operators according to our terminology. These boundary integral operators are obtained by taking limits of the operations $\nabla_z(\bullet) \cdot n_x, M_z, N_z$, respectively on the corresponding potentials V and W as $\Omega \ni z \to x \in \Gamma$. As in the case of the Laplacian, for any solution of (2.4.1), the Cauchy data $(u, \frac{\partial u}{\partial n}, Mu, Nu)_\Gamma$ on Γ are reproduced by the matrix operators in (2.4.23), and $\mathcal{C}_\Omega$ is the Calderón projector corresponding to the bi–Laplacian. In the classical Hölder function spaces, we have the following lemma.

Lemma 2.4.2. *Let* $\Gamma \in C^{2,\alpha}$, $0 < \alpha < 1$. *Then* $\mathcal{C}_\Omega$ *maps* $\prod_{k=0}^{3} C^{3+\alpha-k}(\Gamma)$ *into itself continuously. Moreover,*

$$\mathcal{C}_\Omega^2 = \mathcal{C}_\Omega \,. \tag{2.4.24}$$

As a consequence of this lemma, one finds the following specific identities:

$$\begin{aligned}
V_{12}D_{21} + V_{13}D_{31} + V_{14}D_{41} &= (\tfrac{1}{4}I - K_{11}^2)\,,\\
D_{21}V_{12} + V_{23}D_{32} + V_{24}D_{42} &= (\tfrac{1}{4}I - K_{22}^2)\,,\\
D_{31}V_{13} + D_{32}V_{23} + V_{34}D_{43} &= (\tfrac{1}{4}I - K_{33}^2)\,,\\
D_{41}V_{14} + D_{42}V_{24} + D_{43}V_{34} &= (\tfrac{1}{4}I - K_{44}^2)\,.
\end{aligned}$$

Clearly, from (2.4.24) one finds 12 more identities between these operators.

In the same manner as in the case for the Laplacian, for any solution u of (2.4.1) in Ω^c with $p = 0$, we may introduce the Calderón projection C_{Ω^c} for the exterior domain for the biharmonic equation. Then clearly, we have

$$\mathcal{C}_{\Omega^c} = \mathcal{I} - \mathcal{C}_\Omega \,,$$

where $\mathcal{I}$ denotes the identity matrix operator. This relation then provides the corresponding boundary integral equations for exterior boundary value problems. As will be seen, the boundary integral operators in $\mathcal{C}_\Omega$ are pseudodifferential operators on Γ and their orders are summarized systematically in the following:

$$Ord(\mathcal{C}_\Omega) := \begin{pmatrix} 0 & -1 & -3 & -3 \\ +1 & 0 & -1 & -3 \\ +1 & +1 & 0 & -1 \\ +3 & +1 & +1 & 0 \end{pmatrix} \tag{2.4.25}$$

The orders of these operators can be calculated from their symbols and provide the mapping properties in the Sobolev spaces to be discussed in Chapter 11.

2.4.2 Boundary Value Problems and Boundary Integral Equations

We begin with the boundary integral equations for the Dirichlet problems. For the integral equations of the first kind we employ the second and the first row of $\mathcal{C}_\Omega$ which leads to the following system for the *interior Dirichlet problem*,

$$\mathbf{V}\boldsymbol{\sigma} := \begin{pmatrix} V_{23} & V_{24} \\ V_{13} & V_{14} \end{pmatrix} \begin{pmatrix} \sigma_1 \\ \sigma_2 \end{pmatrix} = \begin{pmatrix} -D_{21} & \frac{1}{2}I - K_{22} \\ \frac{1}{2}I + K_{11} & -V_{12} \end{pmatrix} \begin{pmatrix} \varphi_1 \\ \varphi_2 \end{pmatrix} =: \boldsymbol{f}_i \,. \tag{2.4.26}$$

The solution of the interior Dirichlet problem has associated Cauchy data $\boldsymbol{\sigma}$ which satisfy the three compatibility conditions:

$$\int_\Gamma (\sigma_1 \boldsymbol{n} + \sigma_2 \boldsymbol{x}) ds_x = \mathbf{0}\,, \; -\int_\Gamma \sigma_2 ds = 0\,. \tag{2.4.27}$$

As we shall see in Chapter 11 , $\mathbf{V}$ is known as a strongly elliptic operator for which the classical Fredholm alternative holds. Hence uniqueness will imply the existence of exactly one solution $\boldsymbol{\sigma} \in C^{1+\alpha}(\Gamma) \times C^\alpha(\Gamma)$.

For the exterior Dirichlet problem, by using $\mathcal{C}_{\Omega^c}$ and the representation (2.4.13) we obtain the system with integral equations of the first kind,

$$\mathbf{V}\boldsymbol{\sigma} + R\boldsymbol{\omega} = -\begin{pmatrix} D_{21} & \frac{1}{2}I + K_{22} \\ \frac{1}{2}I - K_{11} & V_{12} \end{pmatrix} \begin{pmatrix} \varphi_1 \\ \varphi_2 \end{pmatrix} =: \boldsymbol{f}_e \,,$$

$$\int_\Gamma (\sigma_1 \boldsymbol{n} + \sigma_2 \boldsymbol{x}) = \boldsymbol{A}_1\,, \; -\int_\Gamma \sigma_2 ds = A_0 \tag{2.4.28}$$

where

$$R(x) = -\begin{pmatrix} 0 & n_1 & n_2 \\ 1 & x_1 & x_2 \end{pmatrix} \quad \text{and} \quad \boldsymbol{\omega} = (\omega_0, \omega_1, \omega_2)^\top \in \mathbb{R}^3\,.$$

Lemma 2.4.3. *The homogeneous system corresponding to* (2.4.28) *has only the trivial solution in* $C^{1+\alpha}(\Gamma) \times \mathbb{C}^\alpha(\Gamma) \times \mathbb{R}^3$.

Proof: Let $\boldsymbol{\sigma}_0, \boldsymbol{\omega}_0$ be any solution of

$$\mathbf{V}\boldsymbol{\sigma}_0 + R\boldsymbol{\omega}_0 = 0 \quad \text{on } \Gamma\,, \; \int_\Gamma \Big(\frac{\partial v}{\partial n}, v\Big) \boldsymbol{\sigma}_0 ds = 0 \quad \text{for all } v \in \Re \tag{2.4.29}$$

and consider the solution of (2.4.1),

$$u_0(x) := V\boldsymbol{\sigma}_0(x) + p_0(x) \quad \text{with } p_0(x) = \mathring{\omega}_0 + \mathring{\omega}_1 x_1 + \mathring{\omega}_2 x_2 \quad \text{for } x \in \Omega^c\,.$$

Then $A_0 = 0\,,\; \boldsymbol{A}_1 = \mathbf{0}$ because of (2.4.29) and (2.4.18), hence $u_0 = O(|x|)$at infinity due to (2.4.13) which implies $u_0(x) = p_0(x)$ for all $x \in \Omega^c \cup \Gamma$. On the other hand, $u_0(x)$ is also a solution of (2.4.1) in Ω and is continuous across Γ where $u_0|_\Gamma = 0$. Hence, due to Theorem 2.4.1, $u_0(x) = 0$ for all x in Ω. Consequently, $Mu_0^\pm|_\Gamma = 0$ and $Nu_0^\pm|_\Gamma = 0$. Then the jump relations corresponding to $\mathcal{C}_{\Omega^c} - \mathcal{C}_\Omega$ imply $\boldsymbol{\sigma}_0 = ([Mu]|_\Gamma\,,\ [Nu]|_\Gamma)^\top = \mathbf{0}$ on Γ and $0 = u_0^-|_\Gamma = u_0^+|_\Gamma = p_0$ implies $p_0(x) = 0$ for all x, i.e., $\overset{\circ}{\boldsymbol{\omega}} = 0$. ■

As a consequence, both, interior and exterior Dirichlet problems lead to the same uniquely solvable system (2.4.28) where only the right–hand sides are different and, for the interior Dirichlet problem, $\boldsymbol{\omega} = 0$.

Clearly, the solution of the Dirichlet problems can also be treated by using the boundary integral equations of the second kind. To illustrate the idea we consider again the interior Dirichlet problem where $u|_\Gamma = \varphi_1$ and $\frac{\partial u}{\partial n}|_\Gamma = \varphi_2$. From the representation formula (2.4.6) we obtain the following system for the unknown $\boldsymbol{\sigma} = (Mu, Nu)^\top$ on Γ:

$$\begin{pmatrix} \frac{1}{2}I + K_{33} & -V_{34} \\ -D_{43} & \frac{1}{2}I - K_{44} \end{pmatrix} \begin{pmatrix} \sigma_1 \\ \sigma_2 \end{pmatrix} = \begin{pmatrix} D_{31} & D_{32} \\ D_{41} & D_{42} \end{pmatrix} \begin{pmatrix} \varphi_1 \\ \varphi_2 \end{pmatrix} =: \boldsymbol{D}\boldsymbol{\varphi}\,. \tag{2.4.30}$$

This system (2.4.30) of integral equations has a unique solution. As we shall see in Chapter 11, for $0 \le \nu < 1$ the Fredholm alternative is still valid for these integral equations and $\boldsymbol{\sigma} \in C^{1+\alpha}(\Gamma) \times C^\alpha(\Gamma)$. So, uniqueness implies existence.

Lemma 2.4.4. *Let $\overset{\circ}{\boldsymbol{\sigma}} \in C^\alpha(\Gamma) \times C^{1+\alpha}(\Gamma)$ be the solution of the homogeneous system*

$$\begin{aligned} (\tfrac{1}{2}I + K_{33})\overset{\circ}{\sigma}_1 - V_{34}\overset{\circ}{\sigma}_2 &= 0 \\ -D_{34}\overset{\circ}{\sigma}_1 + (\tfrac{1}{2}I - K_{44})\overset{\circ}{\sigma}_2 &= 0 \quad \text{on } \Gamma\,. \end{aligned} \tag{2.4.31}$$

Then $\overset{\circ}{\boldsymbol{\sigma}} = \mathbf{0}$.

Proof: For the proof we consider the simple layer potential

$$u_0(x) = V\overset{\circ}{\boldsymbol{\sigma}}$$

which is a solution of (2.4.1) for $x \notin \Gamma$. Then for $x \in \Omega$ we obtain with (2.4.31):

$$\begin{aligned} Mu_0^-|_\Gamma &= (\tfrac{1}{2}I - K_{33})\overset{\circ}{\sigma}_1 + V_{34}\overset{\circ}{\sigma}_2 &= \overset{\circ}{\sigma}_1\,, \\ Nu_0^-|_\Gamma &= (\tfrac{1}{2}I + K_{44})\overset{\circ}{\sigma}_2 + D_{43}\overset{\circ}{\sigma}_1 &= \overset{\circ}{\sigma}_2\,. \end{aligned}$$

Then the Green formula (2.4.2) implies

$$\int_\Gamma \Big(\overset{\circ}{\sigma}_1 \frac{\partial v}{\partial u} + \overset{\circ}{\sigma}_2 v\Big) ds = 0 \quad \text{for all } v \in \Re\,. \tag{2.4.32}$$

For $x \in \Omega^c$, we find $Mu_0^+|_\Gamma = 0$ and $Nu_0^+|_\Gamma = 0$ due to (2.4.31). Then Theorem 2.4.1 implies with (2.4.32) that

$$u_0(x) = p(x) \quad \text{for } x \in \Omega^c \cup \Gamma \quad \text{with some } p \in \Re\,.$$

But $u_0(x)$ is continuously differentiable across Γ and satisfies (2.4.1) in Ω with boundary conditions $u_0^-|_\Gamma = p|_\Gamma$ and $\frac{\partial u_0^-}{\partial n}|_\Gamma = \frac{\partial p}{\partial n}|_\Gamma$. Hence, with Theorem 2.4.1 we find

$$V\overset{\circ}{\boldsymbol{\sigma}}(x) = u_0(x) = p(x) \quad \text{for } x \in \mathbb{R}^2\,.$$

Then

$$\overset{\circ}{\sigma}_1 = [Mu_0]|_\Gamma = 0 \quad \text{and} \quad \overset{\circ}{\sigma}_2 = [Nu_0]|_\Gamma = 0\,.$$

■

We now conclude this section by summarizing the boundary integral equations associated with the two boundary value problems of the biharmonic equation considered here in the following Tables 2.4.5 and 2.4.6. However, the missing details will not be pursued here. We shall return to these equations in later chapters.

We remark that in Table 2.4.5 we did not include orthogonality conditions for the right–hand sides in the equations INP (1) and (2), EDP (2) and ENP (1) since due to the direct approach it is known that the right–hand sides always lie in the range of the operators. Hence, we know that the solutions exist due to the basic results in Theorem 2.4.1, and, moreover, the classical Fredholm alternative holds for the systems in Table 2.4.5. From this table we now consider the modified systems so that the latter will always be uniquely solvable. The main idea here is to incorporate additional side conditions as well as eigensolutions. These modifications are collected in Table 2.4.6. In particular, we have augmented the systems by including additional unknowns $\boldsymbol{\omega} \in \mathbb{R}^3$ in the same manner as in Section 2.2 for the Lamé system. Note that in Table 2.4.6 the matrix valued function S is defined by

$$S(x) := \big(\overset{\circ}{\boldsymbol{\sigma}}{}^1(x), \overset{\circ}{\boldsymbol{\sigma}}{}^2(x), \overset{\circ}{\boldsymbol{\sigma}}{}^3(x)\big)$$

where the columns of S are the three linearly independent eigensolutions of the operator on the left–hand side of EDP (2) in Table 2.4.5. If we solve the exterior Neumann problem with the system ENP (1) in Table 2.4.6, then we obtain a particular solution with $p(x) = 0$ in Ω^c, and for given $p(x) \neq 0$, the latter is to be added to the representation formula (2.4.14). For the interior Neumann problem, the modified boundary integral equation INP (1) and (2) provide a particular solution which presents the general solution only up to linear polynomials.

Note that here we needed $\Gamma \in C^{2,\alpha}$ and even jumps of the curvature are excluded. For piecewise $\Gamma \in C^{2,\alpha}$–boundary, Green's formula, the representation formula as well as the boundary integral equations need to be modified appropriately by including certain functionals at the discontinuity points (Knöpke [230]).

	Systems of BIEs	Side conditions	Eigenspaces
IDP	(1) $\begin{pmatrix} V_{23} & V_{24} \\ V_{13} & V_{14} \end{pmatrix} \begin{pmatrix} \sigma_1 \\ \sigma_2 \end{pmatrix} = \begin{pmatrix} -D_{21} & \frac{1}{2}I - K_{22} \\ \frac{1}{2}I + K_{11} & -V_{12} \end{pmatrix} \begin{pmatrix} \varphi_1 \\ \varphi_2 \end{pmatrix}$	$\int_\Gamma \left(\sigma_1 \frac{\partial v}{\partial n} + \sigma_2 v\right) ds = 0 \, *$ for all $v \in \Re$	none for $\overset{\circ}{\boldsymbol{\sigma}}$ satisfying *
	(2) $\begin{pmatrix} \frac{1}{2}I + K_{33} & -V_{34} \\ -D_{43} & \frac{1}{2}I - K_{44} \end{pmatrix} = \begin{pmatrix} D_{31} & D_{32} \\ D_{41} & D_{42} \end{pmatrix} \begin{pmatrix} \varphi_1 \\ \varphi_2 \end{pmatrix}$	none	none
INP	(1) $\begin{pmatrix} D_{41} & D_{42} \\ D_{31} & D_{32} \end{pmatrix} \begin{pmatrix} u \\ \frac{\partial u}{\partial n} \end{pmatrix} = \begin{pmatrix} -D_{43} & \frac{1}{2}I - K_{44} \\ \frac{1}{2}I + K_{33} & -V_{34} \end{pmatrix} \begin{pmatrix} \psi_1 \\ \psi_2 \end{pmatrix}$	$\int_\Gamma \left(\psi_1 \frac{\partial v}{\partial n} + \psi_2 v\right) ds = 0$ for all $v \in \Re$	$\begin{pmatrix} p \\ \frac{\partial p}{\partial n} \end{pmatrix}, p \in \Re$
	(2) $\begin{pmatrix} \frac{1}{2}I + K_{11} & -V_{12} \\ -D_{21} & \frac{1}{2}I - K_{22} \end{pmatrix} \begin{pmatrix} u \\ \frac{\partial u}{\partial n} \end{pmatrix} = \begin{pmatrix} V_{13} & V_{14} \\ V_{23} & V_{24} \end{pmatrix} \begin{pmatrix} \psi_1 \\ \psi_2 \end{pmatrix}$		
EDP	(1) $\begin{pmatrix} V_{23} & V_{24} \\ V_{13} & V_{14} \end{pmatrix} \begin{pmatrix} \sigma_1 \\ \sigma_2 \end{pmatrix} - \begin{pmatrix} \frac{\partial}{\partial n} p \\ p \end{pmatrix} = -\begin{pmatrix} D_{21} & \frac{1}{2}I + K_{22} \\ \frac{1}{2}I - K_{11} & V_{12} \end{pmatrix} \begin{pmatrix} \varphi_1 \\ \varphi_2 \end{pmatrix}$	$-\int_\Gamma \sigma_2 ds = A_0$ $\int_\Gamma (\sigma_1 \boldsymbol{n} + \sigma_2 \boldsymbol{x}) ds = \boldsymbol{A}_1 .$	none for $\overset{\circ}{\boldsymbol{\sigma}}$ satisfying *
	(2) $\begin{pmatrix} \frac{1}{2}I - K_{33} & V_{34} \\ D_{43} & \frac{1}{2}I + K_{44} \end{pmatrix} \begin{pmatrix} \sigma_1 \\ \sigma_2 \end{pmatrix} = -\begin{pmatrix} D_{31} & D_{32} \\ D_{41} & D_{42} \end{pmatrix} \begin{pmatrix} \varphi_1 \\ \varphi_2 \end{pmatrix}$	$A_0, \boldsymbol{A}_1$ given	$\overset{\circ}{\boldsymbol{\sigma}}{}^1, \overset{\circ}{\boldsymbol{\sigma}}{}^2, \overset{\circ}{\boldsymbol{\sigma}}{}^3$
ENP	(1) $\begin{pmatrix} D_{41} & D_{42} \\ D_{31} & D_{32} \end{pmatrix} \begin{pmatrix} u \\ \frac{\partial u}{\partial n} \end{pmatrix} = -\begin{pmatrix} D_{43} & \frac{1}{2}I + K_{44} \\ \frac{1}{2}I - K_{33} & V_{34} \end{pmatrix} \begin{pmatrix} \psi_1 \\ \psi_2 \end{pmatrix}$	none, $p(x)$ given	$\begin{pmatrix} p \\ \frac{\partial p}{\partial n} \end{pmatrix}, p \in \Re$
	(2) $\begin{pmatrix} \frac{1}{2}I - K_{11} & V_{12} \\ D_{21} & \frac{1}{2}I + K_{22} \end{pmatrix} \begin{pmatrix} u \\ \frac{\partial u}{\partial n} \end{pmatrix} = -\begin{pmatrix} V_{13} & V_{14} \\ V_{23} & V_{24} \end{pmatrix} \begin{pmatrix} \psi_1 \\ \psi_2 \end{pmatrix} + \begin{pmatrix} p(x) \\ \frac{\partial}{\partial n} p(x) \end{pmatrix}$		none

Table 2.4.5. Boundary Integral Equations for the Biharmonic Equation

BVP	
IDP	(1) $\begin{pmatrix} V_{23} & V_{24} \\ V_{13} & V_{14} \end{pmatrix} \begin{pmatrix} \sigma_1 \\ \sigma_2 \end{pmatrix} + R\boldsymbol{\omega} = \begin{pmatrix} -D_{21} & \frac{1}{2}I - K_{22} \\ \frac{1}{2}I + K_{11} & -V_{12} \end{pmatrix} \begin{pmatrix} \varphi_1 \\ \varphi_2 \end{pmatrix}$ $\int_\Gamma \sigma_2 ds = 0\,,\ \int_\Gamma (\sigma_1 \boldsymbol{n} + \sigma_2 \boldsymbol{x}) ds_x = 0$
	(2) $\begin{pmatrix} \frac{1}{2}I + K_{33} & -V_{34} \\ -D_{43} & \frac{1}{2}I - K_{44} \end{pmatrix} \begin{pmatrix} \sigma_1 \\ \sigma_2 \end{pmatrix} = \begin{pmatrix} D_{31} & D_{32} \\ D_{41} & D_{42} \end{pmatrix} \begin{pmatrix} \varphi_1 \\ \varphi_2 \end{pmatrix}$
INP	(1) $\begin{pmatrix} D_{41} & D_{42} \\ D_{31} & D_{32} \end{pmatrix} \begin{pmatrix} u \\ \frac{\partial u}{\partial n} \end{pmatrix} + R\boldsymbol{\omega} = \begin{pmatrix} -D_{43} & \frac{1}{2}I - K_{44} \\ \frac{1}{2}I + K_{33} & -V_{34} \end{pmatrix} \begin{pmatrix} \psi_1 \\ \psi_2 \end{pmatrix}$ $\int_\Gamma u ds = 0\,,\ \int_\Gamma \left(u x_1 + n_1 \frac{\partial u}{\partial n}\right) ds = 0\,,\ \int_\Gamma \left(u x_2 + n_2 \frac{\partial u}{\partial n}\right) ds = 0$
	(2) $\begin{pmatrix} \frac{1}{2}I + K_{11} & -V_{12} \\ -D_{21} & \frac{1}{2}I - K_{22} \end{pmatrix} \begin{pmatrix} u \\ \frac{\partial u}{\partial n} \end{pmatrix} + S\boldsymbol{\omega} = \begin{pmatrix} V_{13} & V_{14} \\ V_{23} & V_{24} \end{pmatrix} \begin{pmatrix} \psi_1 \\ \psi_2 \end{pmatrix}$ $\int_\Gamma u ds = 0\,,\ \int_\Gamma \left(u x_1 + n_1 \frac{\partial u}{\partial n}\right) ds = 0\,,\ \int_\Gamma \left(u x_2 + n_2 \frac{\partial u}{\partial n}\right) ds = 0$
EDP	(1) $\begin{pmatrix} V_{23} & V_{24} \\ V_{13} & V_{14} \end{pmatrix} \begin{pmatrix} \sigma_1 \\ \sigma_2 \end{pmatrix} + R\boldsymbol{\omega} = -\begin{pmatrix} D_{21} & \frac{1}{2}I + K_{22} \\ \frac{1}{2}I - K_{11} & V_{12} \end{pmatrix} \begin{pmatrix} \varphi_1 \\ \varphi_2 \end{pmatrix}$ $-\int_\Gamma \sigma_2 ds = A_0\,,\ \int_\Gamma (\sigma_1 \boldsymbol{n} + \sigma_2 \boldsymbol{x}) ds = \boldsymbol{A}_1$
	(2) $\begin{pmatrix} \frac{1}{2}I - K_{33} & V_{34} \\ D_{43} & \frac{1}{2}I + K_{44} \end{pmatrix} \begin{pmatrix} \sigma_1 \\ \sigma_2 \end{pmatrix} + \begin{pmatrix} 1 & x_1 & x_2 \\ 0 & n_1 & n_2 \end{pmatrix} \boldsymbol{\omega} = -\begin{pmatrix} D_{31} & D_{32} \\ D_{41} & D42 \end{pmatrix} \begin{pmatrix} \varphi_1 \\ \varphi_2 \end{pmatrix}$ $-\int_\Gamma \sigma_2 ds = A_0\,,\ \int_\Gamma (\sigma_1 \boldsymbol{n} + \sigma_2 \boldsymbol{x}) ds = \boldsymbol{A}_1$
ENP	(1) $\begin{pmatrix} D_{41} & D_{42} \\ D_{31} & D_{32} \end{pmatrix} \begin{pmatrix} u_p \\ \frac{\partial u_p}{\partial n} \end{pmatrix} + R\boldsymbol{\omega} = -\begin{pmatrix} D_{43} & \frac{1}{2}I + K_{44} \\ \frac{1}{2}I - K_{33} & V_{34} \end{pmatrix} \begin{pmatrix} \psi_1 \\ \psi_2 \end{pmatrix}$ $\int_\Gamma u_p ds = 0\,,\ \int_\Gamma \left(u_p n_1 + \frac{\partial u_p}{\partial n} x_1\right) ds = 0\,,\ \int_\Gamma \left(u_p n_2 + \frac{\partial u_p}{\partial n} x_2\right) ds = 0$
	(2) $\begin{pmatrix} \frac{1}{2}I - K_{11} & V_{12} \\ D_{21} & \frac{1}{2}I + K_{22} \end{pmatrix} \begin{pmatrix} u \\ \frac{\partial u}{\partial n} \end{pmatrix} = -\begin{pmatrix} V_{13} & V_{14} \\ V_{23} & V_{24} \end{pmatrix} \begin{pmatrix} \psi_1 \\ \psi_2 \end{pmatrix} + \begin{pmatrix} p \\ \frac{\partial}{\partial n} p \end{pmatrix}$

Table 2.4.6. Modified Systems for the Biharmonic Equation

2.5 Remarks

Very often in applications, on different parts of the boundary, different boundary conditions are required or, as in classical crack mechanics (Cruse [93]), the boundaries are given as transmission conditions on some bounded manifold, the crack surface, in the interior of the domain. A similar situation can be found for screen problems (see also Costabel and Dauge [85] and Stephan [402]).

As an example of *mixed boundary conditions* let us consider the Lamé system with given Dirichlet data on $\Gamma_D \subset \Gamma$ and given Neumann data on $\Gamma_N \subset \Gamma$ where $\Gamma = \Gamma_D \cup \Gamma_N \cup \gamma$ with the set of collision points γ of the two boundary conditions (which might also be empty if Γ_D and Γ_N are separated components of Γ) (see e.g. Fichera [120], [209], Kohr et al [234], Maz'ya [289], Stephan [404]) where meas $(\Gamma_D) > 0$.

For $n = 2$, where Γ is a closed curve, we assume that either $\gamma = \emptyset$ or consists of finitely many points; for $n = 3$, the set γ is either empty or a closed curve and as smooth as Γ. For the Lamé system (2.2.1) the classical mixed boundary value problem reads:

Find $u \in C^2(\Omega) \cap C^\alpha(\overline{\Omega})$, $0 < \alpha < 1$, *satisfying*

$$\begin{aligned} -\Delta^* \boldsymbol{u} &= \boldsymbol{f} \quad \textit{in } \Omega \quad \textit{with} \\ \gamma_0 \boldsymbol{u} &= \varphi_D \quad \textit{on } \Gamma_D \quad \textit{and } Tu = \boldsymbol{\psi}_N \quad \textit{on } \Gamma_N \,. \end{aligned} \tag{2.5.1}$$

For reformulating this problem with boundary integral equations we first extend φ_D from Γ_D and $\boldsymbol{\psi}_N$ from Γ_N onto the complete boundary Γ such that

$$\boldsymbol{\varphi}_D = \boldsymbol{\varphi}|_{\Gamma_D} \quad \text{and} \quad \boldsymbol{\psi}_N = \boldsymbol{\psi}|_{\Gamma_N} \tag{2.5.2}$$

with $\boldsymbol{\varphi} \in C^\alpha(\Gamma)$, $0 < \alpha < 1$ and appropriate $\boldsymbol{\psi}$. Then

$$\gamma_0 \boldsymbol{u} = \boldsymbol{\varphi} + \widetilde{\boldsymbol{\varphi}} \,, \; T\boldsymbol{u} = \boldsymbol{\psi} + \widetilde{\boldsymbol{\psi}} \tag{2.5.3}$$

where now

$$\widetilde{\boldsymbol{\varphi}} \in C_0^\alpha(\Gamma_N) = \{\boldsymbol{\varphi} \in C^\alpha(\Gamma) \mid \operatorname{supp} \boldsymbol{\varphi} \subset \Gamma_N\} \tag{2.5.4}$$

and $\widetilde{\boldsymbol{\psi}}$ with supp $\widetilde{\psi} \subset \overline{\Gamma}_D$ are the yet unknown Cauchy data to be determined. With (2.5.3), the representation formula (2.2.4) reads

$$\begin{aligned} \boldsymbol{v}(x) = & \int_\Gamma E(x,y)\boldsymbol{\psi}(y)ds_y - \int_\Gamma \big(T_y E(x,y)\big)^\top \boldsymbol{\varphi}(y)ds_y \\ & + \int_\Gamma E(x,y)\widetilde{\boldsymbol{\psi}}(y)ds_y - \int_\Gamma \big(T_y E(x,y)\big)^\top \widetilde{\boldsymbol{\varphi}}(y)ds_y \\ & + \int_\Omega E(x,y)\boldsymbol{f}(y)dy \quad \text{for } x \in \Omega \,. \end{aligned} \tag{2.5.5}$$

As is well known, even if $\psi \in C^\alpha(\Gamma)$ then $\widetilde{\psi}$ will have singularities at γ which need to be taken into account either by $\{\text{dist}\,(x,\gamma)\}^{-\frac{1}{2}}\widetilde{\psi}_1$ with $\widetilde{\psi}_1 \in C^\alpha(\Gamma_D)$ or by adding singular functions at γ.

Taking the trace and the traction of (2.5.5) on Γ leads with (2.5.3) to the system of boundary integral equations

$$\begin{aligned} V\widetilde{\psi}(x) - K\widetilde{\varphi}(x) &= \tfrac{1}{2}\varphi(x) + K\varphi(x) - V\psi(x) - N\boldsymbol{f}(x) &\quad \text{for } x \in \Gamma_D\,,\\ K'\widetilde{\psi}(x) + D\widetilde{\varphi}(x) &= \tfrac{1}{2}\psi(x) - K'\psi(x) - D\varphi(x) - T_x N\boldsymbol{f}(x) &\quad \text{for } x \in \Gamma_N\,. \end{aligned} \tag{2.5.6}$$

As will be seen in Chapter 5, the system (2.5.6) is uniquely solvable for $\widetilde{\varphi} \in C_0^\alpha(\Gamma_N)$ and $\widetilde{\psi}$ either in the space with the weight $\{\text{dist}\,(x,\gamma)\}^{-\frac{1}{2}}$ or in an augmented space according to the asymptotic behaviour of the solution and involving the stress intensity factors (Stephan et al [406] provided meas$(\Gamma_D) > 0$.

In a similar manner one might also use the system of integral equations of the second kind

$$\begin{aligned} \tfrac{1}{2}\widetilde{\varphi}(x) + K\widetilde{\varphi}(x) - V\widetilde{\psi}(x) &= V\psi(x) - \tfrac{1}{2}\varphi(x) - K\varphi(x) + N\boldsymbol{f}(x)\\ &\qquad \text{for } x \in \Gamma_N\,,\\ \tfrac{1}{2}\widetilde{\psi}(x) - K'\widetilde{\psi}(x) - D\widetilde{\varphi}(x) &= -\tfrac{1}{2}\psi(x) + K'\psi(x) + D\varphi(x) + T_x N\boldsymbol{f}(x)\\ &\qquad \text{for } x \in \Gamma_D\,. \end{aligned} \tag{2.5.7}$$

For the Laplacian and the Helmholtz equation and mixed boundary value problems as well as for the Stokes system one may proceed in the same manner. As will be seen in Chapter 5, the variational formulation for the mixed boundary conditions provides us with the right analytical tools for showing the well–posedness of the formulation (2.5.6) (see e.g., Kohr et al [234], Sauter and Schwab [371] and Steinbach [399]). In the engineering literature, usually the system (2.5.7) is used for discretization and then the equations corresponding to (2.5.7) are obtained by assembling the discrete given and unknown Cauchy data appropriately (see e.g., Bonnet [37], Brebbia et al [43, 44] and Gaul et al [140]).

For crack and insertion problems let us again consider just the example of classical linear theory without volume forces. Let us consider a bounded open domain $\Omega \subset \mathbb{R}^n$ with $n = 2$ or 3 enclosing a given bounded crack or insertion surface as an oriented piece of a curve $\Gamma_c \in C^\alpha$, if $n = 2$ or, if $n = 3$, as an open piece of an oriented surface $\Gamma_c \in C^\alpha$, with a simple, closed boundary curve $\partial\Gamma_c = \gamma \in C^\alpha$. Further the crack should not reach the boundary $\partial\Omega = \Gamma$ of Ω, i.e., $\overline{\Gamma}_c \subset \Omega$. The annulus $\Omega_c := \Omega \setminus F_c$ is not a Lipschitz domain anymore but if we distinguish the two sides of Γ_c assigning with + the points near Γ_c on the side of the normal vector $\boldsymbol{n}_c$ given due to the orientation of Γ_c and the points of the opposite side with −, the traces from either side are still defined. For the crack or insertion problem, an elastic field $\boldsymbol{u} \in C^2(\Omega_c)$ is sought which satisfies the homogeneous Lamé system

$$-\Delta^* \boldsymbol{u} = \boldsymbol{0} \quad \text{in} \quad \Omega_c\,, \tag{2.5.8}$$

in $C^\alpha(\Omega_c \cup \Gamma)$ and up to Γ_c from either side with possibly different traces at Γ_c,

$$\gamma_0^\pm \boldsymbol{u}|_{\Gamma_c} = \boldsymbol{\varphi}^\pm \quad \text{and} \quad T_c^\pm \boldsymbol{u} = \boldsymbol{\psi}^\pm \tag{2.5.9}$$

where we have the transmission properties

$$[\gamma_0 \boldsymbol{u}]|_{\Gamma_c} := (\gamma_0^+ \boldsymbol{u} - \gamma_0^- \boldsymbol{u})|_{\Gamma_c} = [\boldsymbol{\varphi}]|_{\Gamma_c} = (\boldsymbol{\varphi}^+ - \boldsymbol{\varphi}^-)|_{\Gamma_c} \in C_0^\alpha(\Gamma_c) \tag{2.5.10}$$

and

$$[T_c \boldsymbol{u}]|_{\Gamma_c} := (T_c^+ \boldsymbol{u} - T_c^- \boldsymbol{u})|_{\Gamma_c} = [\boldsymbol{\psi}]|_{\Gamma_c} := (\boldsymbol{\psi}^+ - \boldsymbol{\psi}^-)|_{\Gamma_c} \in C_1^\alpha(\Gamma_c) \tag{2.5.11}$$

with

$$C_0^\alpha(\Gamma_c) := \{\boldsymbol{v} \in C^\alpha(\overline{\Gamma}_c) \,|\, (\gamma_0^+ \boldsymbol{v} - \gamma_0^- \boldsymbol{v})|_\gamma = \boldsymbol{0}\}\,, \tag{2.5.12}$$

and

$$C_1^\alpha(\Gamma_c) := \{\boldsymbol{\psi} = \{\text{dist}\,(x-\gamma)\}^{-\frac{1}{2}} \boldsymbol{\psi}_1(x) \,|\, \boldsymbol{\psi}_1 \in C^\alpha(\overline{\Gamma}_c)\}\,. \tag{2.5.13}$$

For the classical *insertion problem* with Dirichlet conditions $\gamma_0 \boldsymbol{u} = \boldsymbol{\varphi} \in C^\alpha(\Gamma)$ on Γ the functions $\boldsymbol{\varphi}^\pm \in C_0^\alpha(\Gamma_c)$ are given. The unknown field $\boldsymbol{u}$ then has to satisfy the boundary conditions

$$\begin{aligned} \gamma_0 \boldsymbol{u}|_\Gamma = \boldsymbol{\varphi} \quad \text{on} \quad \Gamma \quad \text{and with} \quad (\boldsymbol{\varphi}^+ - \boldsymbol{\varphi}^-)|_\gamma = \boldsymbol{0}\,, \\ \gamma_0^+ \boldsymbol{u}|_{\Gamma_c} = \boldsymbol{\varphi}^+ \quad \text{and} \quad \gamma_0^- \boldsymbol{u}|_{\Gamma_c} = \boldsymbol{\varphi}^- \quad \text{on} \quad \Gamma_c \end{aligned} \tag{2.5.14}$$

as well as the transmission conditions (2.5.10) and (2.5.11).

By extending Γ_c up to the boundary Γ ficticiously and applying the Green formula to the two ficticiously separated subdomains of Ω one finds the representation formula

$$\begin{aligned} \boldsymbol{u}(x) = &\int_\Gamma E(x,y)\boldsymbol{\psi}(y)ds_y - \int_\Gamma \big(T_y E(x,y)\big)^\top \boldsymbol{\varphi}(y)ds_y \\ &- \int_{y\in\Gamma_c} E(x,y)[\boldsymbol{\psi}]|_{\Gamma_c}(y)ds_y + \int_{\Gamma_c} \big(T_y^c E(x,y)\big)^\top [\boldsymbol{\varphi}]|_{\Gamma_c}(y)ds_y \end{aligned} \tag{2.5.15}$$

for $x \in \Omega_c$.

By taking the traces of (2.5.15) at Γ and at Γ_c one obtains the following system of equations on Γ and on Γ_c:

$$\begin{aligned}
&\int\limits_{y\in\Gamma} E(x,y)\psi(y)ds_y - \int\limits_{y\in\Gamma_c} E(x,y)[\psi](y)ds_y \\
&\quad = \tfrac{1}{2}\varphi(x) + K\varphi(x) - \int\limits_{y\in\Gamma_c} \big(T_y^c E(x,y)\big)^\top [\varphi] ds_y \quad \text{for } x\in\Gamma\,, \\
&\quad - \int\limits_{y\in\Gamma} E(x,y)\psi(y)ds_y + \int\limits_{\Gamma_c} E(x,y)[\psi](y)ds_y \\
&\quad = -\tfrac{1}{2}\big(\varphi^+(x)+\varphi^-(x)\big) + \int\limits_{\Gamma_c} \big(T_y^c E(x,y)\big)^\top [\psi] ds_y \quad \text{for } x\in\Gamma_c\,.
\end{aligned} \tag{2.5.16}$$

This is a coupled system for $\psi \in C^\alpha(\Gamma)$ on Γ and $[\psi] \in C_1^\alpha(\Gamma_c)$ on Γ_c, which, in fact, is uniquely solvable for any given triple $(\varphi, \varphi^+, \varphi^-)$ with the required properties.

For the classical *crack problem* with Dirichlet conditions on Γ, e.g. as $\psi^+ \in C^\alpha(\overline{\Gamma}_c)$ and $\psi^- \in C^\alpha(\overline{\Gamma}_c)$ are given with $(\psi^+ - \psi^-)|_\gamma = \mathbf{0}$; the desired fields $\boldsymbol{u}$ has to satisfy (2.5.8) and the boundary conditions

$$\gamma_0 \boldsymbol{u}|_\Gamma = \varphi \quad \text{on } \Gamma \quad \text{and } T_c^+ \boldsymbol{u}|_{\Gamma_c} = \psi^+\,,\ T_c^- \boldsymbol{u}|_{\Gamma_c} = \psi^- \quad \text{on } \Gamma_c \tag{2.5.17}$$

as well as the transmission conditions (2.5.10), (2.5.11).

Again from the representation formula (2.5.15) we now obtain the coupled system

$$\begin{aligned}
&\int\limits_{y\in\Gamma} E(x,y)\psi(y)ds_y + \int\limits_{y\in\Gamma_c} \big(T_y^c E(x,y)\big)^\top [\varphi](y)ds_y \\
&\quad = \tfrac{1}{2}\varphi(x) + K\varphi(x) + \int\limits_{y\in\Gamma_c} E(x,y)[\psi]|_{\Gamma_c}(y)ds_y \quad \text{for } x\in\Gamma\,, \\
&D_c[\varphi](x) - \int\limits_{y\in\Gamma} T_x^c E(x,y)\psi(y)ds_y \\
&\quad = \tfrac{1}{2}\big(\psi^+(x)+\psi^-(x)\big) - K^c\,|\,([\psi]|_{\Gamma_c})(x) \\
&\qquad - \int\limits_{y\in\Gamma} T_x^c\big(T_y E(x.y)\big)^\top \varphi(y)ds_y \quad \text{for } x\in\Gamma_c
\end{aligned} \tag{2.5.18}$$

for the unknowns $\psi \in C^\alpha(\Gamma)$ and $[\varphi] \in C_0^\alpha(\Gamma_c)$. As it turns out, this system always has a unique solution for any given triple $(\varphi, \psi^+, \psi^-)$ with the required properties.

The desired displacement field in Ω_c is in both cases given by (2.5.15).

3. Representation Formulae, Local Coordinates and Direct Boundary Integral Equations

In order to generalize the direct approach for the reduction of more general boundary value problems to boundary integral equations than those presented in Chapters 1 and 2 we consider here the $2m$–th order positive elliptic systems with real C^∞–coefficients. We begin by collecting all the necessary machinery. This includes the basic definitions and properties of classical function spaces and distributions, the Fourier transform and the definition of Hadamard's finite part integrals which, in fact, represent the natural regularization of homogeneous distributions and of the hypersingular boundary integral operators. For the definition of boundary integral operators one needs the appropriate representation of the boundary manifold Γ involving local coordinates. Moreover, the calculus of vector fields on Γ requires some basic knowledge in classical differential geometry. For this purpose, a short excursion into differential geometry is included. Once the fundamental solution is available, the representation of solutions to the boundary value problems is based on general Green's formulae which are formulated in terms of distributions and multilayer potentials on Γ. The latter leads us to the direct boundary integral equations of the first and second kind for interior and exterior boundary value problems as well as for transmission problems. As expected, the hypersingular integral operators are given by direct values in terms of Hadamard's finite part integrals.

The results obtained in this chapter will serve as examples of the class of pseudodifferential operators to be considered in Chapters 7–11.

3.1 Classical Function Spaces and Distributions

For rigorous definitions of classical as well as generalized function spaces we first collect some standard results and notations.

Multi–Index Notation

Let $\mathbb{N}_0$ be the set of all non–negative integers and let $\mathbb{N}_0^n$ be the set of all ordered n–tuples $\alpha = (\alpha_1, \cdots, \alpha_n)$ of non–negative integers $\alpha_i \in \mathbb{N}_0$. Such an n–tuple α is called a *multi–index*. For all $\alpha \in \mathbb{N}_0^n$, we denote by $|\alpha| = \alpha_1 + \alpha_2 + \cdots + \alpha_n$ the order of the multi-index α. If $\alpha, \beta \in \mathbb{N}_0^n$, we define $\alpha + \beta = (\alpha_1 + \beta_1, \cdots, \alpha_n + \beta_n)$. The notation $\alpha \leq \beta$ means that $\alpha_i \leq \beta_i$ for $1 \leq i \leq n$. We set

G. C. Hsiao und W. L. Wendland, *Boundary Integral Equations*,
Applied Mathematical Sciences 164,
https://doi.org/10.1007/978-3-030-71127-6_3

$$\alpha! = \alpha_1! \cdots \alpha_n!$$

and

$$\binom{\beta}{\alpha} = \frac{\beta!}{(\beta-\alpha)!\alpha!} = \binom{\beta_1}{\alpha_1} \cdots \binom{\beta_n}{\alpha_n}.$$

We use the usual compact notation for the partial derivatives: If $\alpha = (\alpha_1, \cdots, \alpha_n) \in \mathbb{N}_0^n$, we denote by $D^\alpha u$ the partial derivatives

$$D^\alpha u = \frac{\partial^{|\alpha|} u}{\partial x_1^{\alpha_1}, \partial x_2^{\alpha_2} \ldots \partial x_n^{\alpha_n}}$$

of order $|\alpha|$. In particular, if $|\alpha| = 0$, then $D^\alpha u = u$.

Functions with Compact Support

Let u be a function defined on an open subset $\Omega \subset \mathbb{R}^n$. The *support of* u, denoted by $\operatorname{supp} u$, is the closure in $\mathbb{R}^n$ of the set

$$\{x \in \Omega : u(x) \neq 0\}.$$

In other words, the support of u is the smallest closed subset of Ω outside of which u vanishes. We say that u has a *compact support* in Ω if its support is a compact (i.e. closed and bounded) subset of Ω. By the notation $K \Subset \Omega$, we mean not only that $K \subset \Omega$ but also that K is a compact subset of Ω. Thus, if u has a compact support in Ω, we may write $\operatorname{supp} u \Subset \Omega$.

The Spaces $C^m(\Omega)$ and $C_0^\infty(\Omega)$

We denote by $C^m(\Omega), m \in \mathbb{N}_0$, the space of all real– (or complex–)valued functions defined in an open subset $\Omega \subset \mathbb{R}^n$ having continuous derivatives of order $\leq m$. Thus, for $m = 0, C^0(\Omega)$ is the space of all continuous functions in Ω which will be simply denoted by $C(\Omega)$. We set

$$C^\infty(\Omega) = \bigcap_{m \in \mathbb{N}_0} C^m(\Omega),$$

the space of functions defined in Ω having derivatives of all orders, i.e., the space of functions which are infinitely differentiable.

We define $C_0^\infty(\Omega)$ to be the space of all infinitely differentiable functions which, together with all of their derivatives, have compact support in Ω. We denote by $C_0^m(\Omega)$ the space of functions $u \in C^m(\Omega)$ with $\operatorname{supp} u \Subset \Omega$. The spaces $C_0^m(\Omega)$ as well as $C_0^\infty(\Omega)$ are linear function spaces.

On the linear function space $C_0^\infty(\Omega)$ one can introduce the notion of convergence $\varphi_j \to \varphi$ in $C_0^\infty(\Omega)$ if to the sequence of functions $\{\varphi_j\}_{j=1}^\infty$ there exists a common compact subset $K \Subset \Omega$ with $\operatorname{supp} \varphi_j \subset K$ for all $j \in \mathbb{N}$ and $D^\alpha \varphi_j \to D^\alpha \varphi$ uniformly for every multi–index α. This notation of convergence defines a locally convex topology on $C_0^\infty(\Omega)$ (see Treves [419]). With this concept of convergence, $C_0^\infty(\Omega)$ is a locally convex topological vector space called $\mathcal{D}(\Omega)$.

Similarly, one can define $C^m(\overline{\Omega})$ to be the space of functions in $C^m(\Omega)$ which, together with their derivatives of order $\leq m$, have continuous extensions to $\overline{\Omega} = \Omega \cup \partial\Omega$. If Ω is bounded and $m < \infty$, then $C^m(\overline{\Omega})$ is a Banach space with the norm

$$\|u\|_{C^m(\overline{\Omega})} = \sum_{|\alpha| \leq m} \sup_{x \in \overline{\Omega}} |D^\alpha u(x)|.$$

The Spaces $C^{m,\alpha}(\Omega)$

In Section 1.2, we have already introduced Hölder spaces, however, for completeness, we restate the definition here again. Let Ω be a subset of $\mathbb{R}^n$ and $0 < \alpha \leq 1$. A function u defined on Ω is said to be *Hölder continuous* with exponent α in Ω if $0 < \alpha < 1$ and if there exists a constant $c \geq 0$ such that

$$|u(x) - u(y)| \leq c|x - y|^\alpha$$

for all $x, y, \in \Omega$. The quantity

$$[u]_{\alpha;\Omega} := \sup_{\substack{x,y \in \Omega \\ x \neq y}} \frac{|u(x) - u(y)|}{|x - y|^\alpha}$$

is called the *Hölder modulus* of u. For $\alpha = 1, u$ is called *Lipschitz continuous* and $[u]_{1;\Omega}$ is called the *Lipschitz modulus*. We say that u is *locally Hölder* or *Lipschitz continuous* with exponent α on Ω if it is Hölder or Lipschitz continuous with exponent α on every compact subset of Ω, respectively. By $C^{m,\alpha}(\Omega), m \in \mathbb{N}_0, 0 < \alpha \leq 1$, we denote the space of functions in $C^m(\Omega)$ whose m–th order derivatives are locally Hölder or Lipschitz continuous with exponent α on the open subset $\Omega \subset \mathbb{R}^n$. We remark that Hölder continuity may be viewed as a fractional differentiability.

Further, by $C^{m,\alpha}(\overline{\Omega})$ we denote the subspace of $C^m(\overline{\Omega})$ consisting of functions which have m–th order Hölder or Lipschitz continuous derivatives of exponent α in $\overline{\Omega}$. If Ω is bounded, we define the *Hölder* or *Lipschitz norm* by

$$\|u\|_{C^{m,\alpha}(\overline{\Omega})} := \|u\|_{C^m(\overline{\Omega})} + \sum_{|\beta|=m} \sup_{\substack{y \in \overline{\Omega} \\ x \neq y}} \frac{|D^\beta u(x) - D^\beta u(y)|}{|x - y|^\alpha} \tag{3.1.1}$$

The space $C^{m,\alpha}(\overline{\Omega})$ equipped with the norm $\|\cdot\|_{C^{m,\alpha}(\overline{\Omega})}$ is a Banach space.

Distributions

Let v be a linear functional on $\mathcal{D}(\Omega)$. Then we denote by $\langle v, \varphi \rangle$ the image of $\varphi \in \mathcal{D}(\Omega)$.

Definition 3.1.1. *A linear functional v on $\mathcal{D}(\Omega)$ is called a* distribution *if, for every compact set $K \Subset \Omega$, there exist constants $c_K \geq 0$ and $N \in \mathbb{N}_0$ such that*

$$|\langle v, \varphi \rangle| \leq c_K \sum_{|\alpha| \leq N} \|D^\alpha \varphi\|_{C^0(\Omega)} := c_K \sum_{|\alpha| \leq N} \sup_{\alpha \in \Omega} |D^\alpha \varphi(x)| \tag{3.1.2}$$

for all $\varphi \in \mathcal{D}(\Omega)$ with $\operatorname{supp} \varphi \subset K$. *The vector space of distributions is called $\mathcal{D}'(\Omega)$. For $\Omega = \mathbb{R}^n$ we set $\mathcal{D} := \mathcal{D}(\mathbb{R}^n)$ and $\mathcal{D}' := \mathcal{D}'(\mathbb{R}^n)$.*

For a distribution $v \in \mathcal{D}'(\Omega)$, (3.1.2) implies that $\langle v, \varphi_j \rangle \to \langle v, \varphi \rangle$ for $v \in \mathcal{D}'(\Omega)$ if $\varphi_j \to \varphi$ in $\mathcal{D}(\Omega)$. Correspondingly, we equip $\mathcal{D}'(\Omega)$ by the weak topology; i.e., $v_j \rightharpoonup v$ in $\mathcal{D}'(\Omega)$ if and only if $\langle v_j, \varphi \rangle \to \langle v, \varphi \rangle$ as $j \to \infty$ for every $\varphi \in \mathcal{D}(\Omega)$.

For $v, w \in \mathcal{D}'(\Omega)$, we say that $v = w$ on an open set $\Theta \subset \Omega$ if and only if $\langle v, \varphi \rangle = \langle w, \varphi \rangle$ for all $\varphi \in C_0^\infty(\Theta)$. The *support of a distribution* $v \in \mathcal{D}'(\Omega)$ is the complement of the largest open set on which $v = 0$.

The function space $C^\infty(\Omega)$ equipped with the family of seminorms $\| \bullet \|_{C^m(K)}$ for all compact subsets $K \Subset \Omega$ is denoted by $\mathcal{E}(\Omega)$. The space of *distributions with compact supports* in Ω is denoted by $\mathcal{E}'(\Omega)$ and for $\Omega = \mathbb{R}^n$ we set $\mathcal{E} := \mathcal{E}(\mathbb{R}^n)$ and $\mathcal{E}' := \mathcal{E}'(\mathbb{R}^n)$. As a consequence of this definition, one obtains the following characterization of $\mathcal{E}'(\Omega)$.

Proposition 3.1.1. *A linear functional w on $\mathcal{E}$ is an element of $\mathcal{E}'(\Omega)$ if and only if there exist a compact set $K \Subset \Omega$ and constants c and $N \in \mathbb{N}_0$ such that*

$$|\langle w, \varphi \rangle| \leq c \|\varphi\|_{C^N(K)} \quad \textit{for all} \quad \varphi \in \mathcal{E}(\Omega) . \tag{3.1.3}$$

We remark that the standard operations, such as the multiplication by a function and differentiation can be extended to distributions in the following way. Let T be any linear continuous operator on $\mathcal{D}(\Omega)$, i.e., $T\varphi_j \to T\varphi$ in $\mathcal{D}(\Omega)$ if $\varphi_j \to \varphi$ in $\mathcal{D}(\Omega)$. Then the operator T on $\mathcal{D}'(\Omega)$ is defined by

$$\langle Tv, \varphi \rangle = \langle v, T'\varphi \rangle \quad \text{for } v \in \mathcal{D}'(\Omega) \quad \text{and} \quad \varphi \in \mathcal{D}(\Omega) \tag{3.1.4}$$

where the *dual operator* T' is given by

$$\int_{\mathbb{R}^n} (T\varphi)\psi dx = \int_{\mathbb{R}^n} \varphi(T'\psi) dx \quad \text{for all } \varphi, \psi \in \mathcal{D}(\Omega) . \tag{3.1.5}$$

Therefore, the linear functional (Tv) in (3.1.4) is continuous on $\mathcal{D}(\Omega)$ provided T' is continuous on $\mathcal{D}(\Omega)$.

If $f \in \mathcal{E}(\Omega)$, then for $v \in \mathcal{D}'(\Omega)$ we can define the *product* $(fv) \in \mathcal{D}'(\Omega)$ via the continuous linear functional given by

$$\langle (fv), \varphi \rangle := \langle v, f\varphi \rangle . \tag{3.1.6}$$

For the *differential operator* $D^\alpha = T$, which is continuous on $\mathcal{D}(\Omega)$, $(-1)^{|\alpha|} D^\alpha = T'$ is also continuous on $\mathcal{D}(\Omega)$, hence, for $v \in \mathcal{D}'(\Omega)$ we define $D^\alpha v$ by

$$\langle D^\alpha v, \varphi \rangle := \langle v, (-1)^{|\alpha|} D^\alpha \varphi \rangle \quad \text{for all} \quad \varphi \in \mathcal{D}(\Omega)\,. \tag{3.1.7}$$

Next, we collect the basic definitions of the rapidly decreasing functions and corresponding tempered distributions $\mathbb{R}^n$.

Definition 3.1.2. *A function $\varphi \in C^\infty(\mathbb{R}^n)$ is called* r a p i d l y d e c r e a s i n g *if*

$$\|\varphi\|_{\alpha,\beta} := \sup_{x \in \mathbb{R}^n} |x^\alpha D^\beta \varphi| < \infty \tag{3.1.8}$$

for all pairs of multi–indices α, β.

The vector space of the rapidly decreasing functions equipped with this family of seminorms is called S c h w a r t z s p a c e *$\mathcal{S}$.*

A sequence of functions $\{\varphi_k\}_{k=1}^\infty \subset \mathcal{S}$ converges to $\varphi \in \mathcal{S}$ if and only if

$$\|\varphi - \varphi_j\|_{\alpha,\beta} \to 0 \quad as \ \ j \to 0$$

for all pairs of multi–indices α, β.

Definition 3.1.3. *By $\mathcal{S}' \subset \mathcal{D}'$ we denote the space of all continuous linear functionals on $\mathcal{S}$, i.e., to $v \in \mathcal{S}'$ there exist constants $c_v \geq 0$ and $N \in \mathbb{N}_0$ such that*

$$|\langle v, \varphi \rangle| \leq c_v \sum_{|\alpha|,|\beta| \leq N} \sup |x^\alpha D^\beta \varphi| \quad \textit{for all} \quad \varphi \in \mathcal{S}\,. \tag{3.1.9}$$

The distribution space $\mathcal{S}'$ is usually referred to as the space of *tempered distributions* on $\mathbb{R}^n$. A sequence $\{v_j\}_{j=1}^\infty \subset \mathcal{S}'$ is said to *converge* to $v \in \mathcal{S}'$ in $\mathcal{S}'$ if and only if

$$\langle v_j, \varphi \rangle \to \langle v, \varphi \rangle \quad \text{for all} \quad \varphi \in \mathcal{S} \quad \text{as} \quad j \to \infty\,. \tag{3.1.10}$$

Remark 3.1.1: For the above spaces we have the following inclusions:

$$\mathcal{D} \subset \mathcal{E}'\,,\ \mathcal{S} \subset \mathcal{S}'\,,\ \mathcal{E} \subset \mathcal{D}'\,,\ \mathcal{D} \subset \mathcal{S} \subset \mathcal{E} \quad \text{and} \quad \mathcal{E}' \subset \mathcal{S}' \subset \mathcal{D}'\,. \tag{3.1.11}$$

As will be seen, the Fourier transform will play a crucial role in the analysis of integral– and pseudodifferential operators. It is defined as follows.

Fourier Transform

On the function space $\mathcal{D}$ of functions in $C_0^\infty(\mathbb{R}^n)$, the *Fourier transform* is defined by

$$\mathcal{F}_{x \mapsto \xi} u(x) := \widehat{u}(\xi) := (2\pi)^{-n/2} \int\limits_{\mathbb{R}^n} e^{-ix \cdot \xi} u(x) dx \quad \text{for} \quad u \in \mathcal{D}\,. \tag{3.1.12}$$

Clearly, $\mathcal{F}:\mathcal{D}\to\mathcal{E}$ is a continuous linear mapping. By using Definition (3.1.4), we extend the Fourier transform to distributions $v\in\mathcal{E}'$ by

$$\langle \mathcal{F}v,\varphi\rangle := \langle v,\mathcal{F}\varphi\rangle \quad \text{for all } \varphi\in\mathcal{D}. \tag{3.1.13}$$

Then the linear map $\mathcal{F}:\mathcal{E}'\to\mathcal{D}'$ is also continuous. If $u\in\mathcal{S}$ then $\mathcal{F}$ is also well defined by (3.1.12) and the following theorem is valid.

Theorem 3.1.2. *The Fourier transform is a topological isomorphism from $\mathcal{S}$ onto itself and the following identity holds:*

$$u(x) = (2\pi)^{-n/2}\int\limits_{\mathbb{R}^n} e^{ix\cdot\xi}\mathcal{F}u(\xi)d\xi \quad \textit{for all } \varphi\in\mathcal{S}. \tag{3.1.14}$$

This is the inverse Fourier transform $\mathcal{F}^*v$ for $v=\mathcal{F}u$.
In addition, we have for $u,v\in\mathcal{S}$:

$$\mathcal{F}_{x\mapsto\xi}\big(D_x^\alpha u(x)\big) = (i\xi)^\alpha\widehat{u}(\xi) \quad \textit{and} \quad \big(D_\xi^\beta\widehat{u}(\xi)\big) = \mathcal{F}_{x\mapsto\xi}\big((-ix)^\beta u(x)\big) \tag{3.1.15}$$

$$\langle u,\mathcal{F}v\rangle = \langle \mathcal{F}u,v\rangle \quad \textit{and} \quad \langle \mathcal{F}u,\mathcal{F}v\rangle = \langle u,v\rangle\,. \tag{3.1.16}$$

(Formula (3.1.16) is called *Parseval's formula.*)
Similar to the extension of $\mathcal{F}$ from $\mathcal{D}$ to $\mathcal{E}'$, we now extend $\mathcal{F}$ from $\mathcal{S}$ to $\mathcal{S}'$ by the formula

$$\langle \mathcal{F}u,\varphi\rangle := \langle u,\mathcal{F}\varphi\rangle \quad \text{for all } \varphi\in\mathcal{S} \tag{3.1.17}$$

which defines a continuous linear functional on $\mathcal{S}$ since $\mathcal{F}:\mathcal{S}\to\mathcal{S}$ is an isomorphism. The inverse Fourier transform $\mathcal{F}^*v$ for $v\in\mathcal{S}'$ is then defined by

$$\langle \mathcal{F}^*v,\varphi\rangle = \langle v,\mathcal{F}^*\varphi\rangle \quad \text{for all } \varphi\in\mathcal{S}\,. \tag{3.1.18}$$

With these definitions,
Theorem 3.1.2 **remains valid for $\mathcal{S}'$.**
A useful characterization of $\mathcal{E}'$ in terms of the Fourier transformed distributions is given by the fundamental Paley–Wiener–Schwartz theorem.

Theorem 3.1.3.

(i) *If $u\in\mathcal{E}'(\mathbb{R}^n)$ then its Fourier transform*

$$\widehat{u}(\xi) = \mathcal{F}_{x\mapsto\xi}u := (2\pi)^{-n/2}\int\limits_{\mathbb{R}^n} u(x)e^{-ix\cdot\xi}dx$$

is analytic in $\xi\in\mathbb{C}^n$. Here, the integration is understood in the distributional sense.

(ii) *For $u\in\mathcal{E}'(\mathbb{R}^n)$ and $\operatorname{supp} u\subset\{x\in\mathbb{R}^n \mid |x|\le a\}$ with $a>0$, there exist constants $c,N\ge 0$, such that*

$$|\widehat{u}(\xi)| \le c(1+|\xi|)^N e^{a|Im\xi|} \quad \textit{for } \xi\in\mathbb{C}^n\,.$$

(iii) *If* $u \in C_0^\infty(\mathbb{R}^n)$ *and* $\operatorname{supp} u \subset \{x \in \mathbb{R}^n \mid |x| \le a\}$ *then for every* $m \in \mathbb{N}_0$ *exists a constant* $c_m \ge 0$ *such that*

$$|\widehat{u}(\xi)| \le c_m(1+|\xi|)^{-m} e^{a|Im\xi|} \quad \textit{for all } \xi \in \mathbb{C}^n .$$

(see e.g. Schwartz [382] or Friedlander [128, Theorem 10.2.1]*).*

3.2 Hadamard's Finite Part Integrals

We now consider a subclass of $\mathcal{S}'$ which contains distributions defined by functions having isolated singularities of finite order. In order to give the precise definition of such distributions, we introduce the concept of Hadamard's finite part integrals. Since our boundary integral operators have kernels defined by functions with singularities of this type, we confine ourselves to the following classes of functions.

Definition 3.2.1.
A function $h_q(z)$ *is a* $C^\infty(\mathbb{R}^n \setminus \{0\})$ *pseudohomogeneous function of degree* $q \in \mathbb{R}$ *if*

$$\begin{array}{rcll} h_q(tz) & = & t^q h_q(z) \quad \textit{for every } t>0 \quad \textit{and } z \ne 0 & \textit{if } q \notin \mathbb{N}_0\,; \\ h_q(z) & = & f_q(z) + \log|z| p_q(z) & \textit{if } q \in \mathbb{N}_0\,, \end{array} \qquad (3.2.1)$$

where $p_q(z)$ *is a homogeneous polynomial in* z *of degree* q *and where the function* $f_q(z)$ *satisfies*

$$f_q(tz) = t^q f_q(z) \quad \textit{for every } t>0 \quad \textit{and } z \ne 0 .$$

In short, we denote the class of pseudohomogeneous functions of degree $q \in \mathbb{R}$ *by* Ψhf_q.

We note that for $q > -n$, the integral

$$\int_\Omega h_q(z)u(z)dz \qquad (3.2.2)$$

for $u \in C_0^\infty(\Omega)$ is well defined as an improper integral. For $q \le -n$, however, $h_q(z)$ is non integrable except that $u(y)$ and its derivatives up to the order $\kappa := [-n-q]$ vanish at the origin[1]. Hence, (3.2.2) defines a homogeneous distribution on $C_0^\infty(\Omega \setminus \{0\})$. In order to extend this distribution to all of $C_0^\infty(\Omega)$, we use the Hadamard finite part concept which is the most natural extension. It is based on the idea of integration by parts. Consider first the family of regular integrals:

[1] As usual, $[p] := \max\{m \in \mathbb{Z} \mid m \le p\}$ with $\mathbb{Z}$ the integers, denotes the Gaussian bracket for $p \in \mathbb{R}$.

$$I_\varepsilon^q(u) := \int\limits_{\Omega\setminus\Omega_0} h_q(z)u(z)dz + \int\limits_{\Omega_0\setminus\{|z|<\varepsilon\}} h_q(z)u(z)dz \tag{3.2.3}$$

depending on $\varepsilon > 0$ where $u \in C_0^\infty(\Omega)$ and Ω_0 is any fixed star–shaped domain with respect to the origin, satisfying $\overline{\Omega_0} \subset \Omega$. The first integrals on the right–hand side has C^∞–integrands and exists. With the decomposition of the second integral we have

$$\begin{aligned} I_\varepsilon^q(u) = & \int\limits_{\Omega_0\setminus\{|z|<\varepsilon\}} h_q(z)\Big\{u(z) - \sum_{|\alpha|\le\kappa} \frac{1}{\alpha!} z^\alpha \frac{\partial^\alpha u}{\partial x^\alpha}(0)\Big\}dz \\ & + \sum_{|\alpha|\le\kappa} \frac{1}{\alpha!}\frac{\partial^\alpha u}{\partial x^\alpha}(0) \int\limits_{\Omega_0\setminus\{|z|<\varepsilon\}} h_q(z)z^\alpha dz\,. \end{aligned} \tag{3.2.4}$$

We note that the first integral on the right–hand side has a singularity of order $|z|^{-\varrho}$ where

$$0 \le \varrho := -(q+n) - \kappa < 1\,.$$

We shall see that the last term on the right–hand side can be written in the form

$$\sum_{0\le|\alpha|\le\kappa} d_\alpha \frac{\partial^\alpha u}{\partial x^\alpha}(0) + \sum_{j=0}^{\kappa} C_j(u)\varepsilon^{-j-\varrho} \tag{3.2.5}$$

if $\varrho > 0$. For $\varrho = 0$ one gets the same expansion as in (3.2.5) except that $C_0(u)\varepsilon^0$ is replaced by $C_0(u)\log\varepsilon$. The constants $C_j(u)$ and d_α will be given explicitly later on. We note that for $\varepsilon \to 0$, the terms $C_0 \log\varepsilon + \sum_{j=1}^{\kappa} C_j\varepsilon^{-j-\varrho}$ diverge. These are the terms causing the non–existence of the limit value of the integral on the left–hand side in (3.2.4). More precisely, the coefficients d_α and the functionals $C_j(u)$ in (3.2.5) are computed from (3.2.4). To illustrate the idea, we now compute a typical term where $q + |\alpha| + n = -\varrho - j < 0$ and $|\alpha| = \kappa - j$:

$$\sum_{|\alpha|=\kappa-j} \frac{1}{\alpha!}\frac{\partial^\alpha u}{\partial x^\alpha}(0) \int\limits_{\Omega_0\setminus\{|z|<\varepsilon\}} h_q(z) z^\alpha dz$$

$$= \sum_{|\alpha|=\kappa-j} \frac{1}{\alpha!}\frac{\partial^\alpha u}{\partial x^\alpha}(0) \int\limits_{|\widehat{\Theta}|=1} \int\limits_{r=\varepsilon}^{R(\widehat{\Theta})} r^q h_q(\widehat{\Theta}) r^{|\alpha|} \widehat{\Theta}^\alpha r^{n-1} dr d\omega$$

$$= \sum_{|\alpha|=\kappa-j} \frac{1}{\alpha!}\frac{\partial^\alpha u}{\partial x^\alpha}(0) \Big\{ \int\limits_{|\widehat{\Theta}|=1} \frac{(R(\widehat{\Theta})^{q+|\alpha|+n} - \varepsilon^{q+|\alpha|+n})}{(q+|\alpha|+n)} \widehat{\Theta}^\alpha h_q(\widehat{\Theta}) d\omega \Big\}$$

$$= -\sum_{|\alpha|=\kappa-j} \frac{1}{\alpha!} \int\limits_{|\widehat{\Theta}|=1} \frac{R(\widehat{\Theta})^{-j-\varrho}}{(j+\varrho)} \widehat{\Theta}^\alpha h_q(\widehat{\Theta}) d\omega \frac{\partial^\alpha u}{\partial x^\alpha}(0)$$

$$+ \varepsilon^{-j-\varrho} \sum_{|\alpha|=\kappa-j} \frac{1}{\alpha!} \int\limits_{|\widehat{\Theta}|=1} \frac{\widehat{\Theta}^\alpha}{(j+\varrho)} h_q(\widehat{\Theta}) d\omega \frac{\partial^\alpha u}{\partial x^\alpha}(0)$$

$$=: \sum_{|\alpha|=\kappa-j} \Big\{ d_\alpha - \int\limits_{\Omega\setminus\Omega_0} h_q(z) z^\alpha dz \Big\} \frac{\partial^\alpha u}{\partial x^\alpha}(0) + C_j(u)\varepsilon^{-j-\varrho}$$

provided $j + \varrho > 0$. The coefficients d_α and the functional C_j are defined in an obvious manner. Here, $\widehat{\Theta} = z/|z|$, and $d\omega$ is the surface element on the unit sphere. $R(\widehat{\Theta})$ describes the boundary of Ω_0 in polar coordinates. For $j + \varrho = 0$ one may proceed similarly; we omit the details.

The Hadamard finite part integral as an extension of the homogeneous distribution is now defined by neglecting the divergent terms in (3.2.4) given in (3.2.5) (see also Sellier [386]).

Definition 3.2.2. *Hadamard's finite part integral is defined by*

$$\text{p.f.} \int\limits_{\Omega} h_q(z) u(z) dz := \int\limits_{\Omega\setminus\Omega_0} h_q(z) u(z) dz$$
$$+ \int\limits_{\Omega_0} h_q(z) \Big\{ u(z) - \sum_{|\alpha|\le\kappa} \frac{1}{\alpha!} z^\alpha \frac{\partial^\alpha u}{\partial x^\alpha}(0) \Big\} dz + \sum_{|\alpha|\le\kappa} d_\alpha \frac{\partial^\alpha u}{\partial x^\alpha} \tag{3.2.6}$$

where, for $0 < \varrho < 1$ *and for all multi-indices* α *with* $|\alpha| = \kappa - j$ *and* $j = 0, \ldots, \kappa$,

$$d_\alpha = -\frac{1}{(j+\varrho)} \frac{1}{\alpha!} \int\limits_{|\widehat{\Theta}|=1} R^{-j-\varrho}(\widehat{\Theta}) \widehat{\Theta}^\alpha h_q(\widehat{\Theta}) d\omega \tag{3.2.7}$$

if $n \ge 2$ *and where for the case* $n = 1$ *we set* $\Omega_0 := [-R_0, R_0]$ *and identify*

$$\int\limits_{|\widehat{\Theta}|=1} R^{-j-\varrho}(\widehat{\Theta}) \widehat{\Theta}^\alpha h_q(\widehat{\Theta}) d\omega := R_0^{-j-\varrho} \big(h_q(1) + (-1)^\alpha h_q(-1)\big) . \tag{3.2.8}$$

If $\varrho = 0$ and $|\alpha| < \kappa$, i.e., $j = 1, \dots, \kappa$, then d_α is still given by (3.2.7), (3.2.8), respectively.

For $\varrho = 0$ and $|\alpha| = \kappa$, i.e. $j = 0$, then

$$d_\alpha := \frac{1}{\alpha!} \int\limits_{|\widehat{\Theta}|=1} \log R(\widehat{\Theta}) \widehat{\Theta}^\alpha h_q(\widehat{\Theta}) d\omega \quad \text{if } n \geq 2 \tag{3.2.9}$$

and for $n = 1$ we identify

$$\int\limits_{|\widehat{\Theta}|=1} \log R(\widehat{\Theta}) \widehat{\Theta}^\alpha h_q(\widehat{\Theta}) d\omega := (log R_0)\big(h_q(1) + (-1)^\alpha h_q(-1)\big) . \tag{3.2.10}$$

The Hadamard finite part integral is also often denoted by

$$\oint\limits_{\Omega} h_q(z) u(z) dz = \text{ p.f.} \int\limits_{\Omega} h_q(z) u(z) dz .$$

Alternatively, by subtracting the singular terms in (3.2.5), the value of Hadamard's finite part integral is also given by the limiting procedure

$$\begin{aligned} &\text{p.f.} \int\limits_{\Omega} h_q(z) u(z) dz = \\ &\lim_{\varepsilon \to 0} \Big\{ \int\limits_{\Omega \setminus \{|z| < \varepsilon\}} h_q(z) u(z) dz - \sum_{j=0}^{\kappa} C_j(u) \varepsilon^{-j-\varrho} \Big\} \end{aligned} \tag{3.2.11}$$

where $C_0(u)\varepsilon^0$ is to be replaced by $C_0(u) \log \varepsilon$ for $\varrho = 0$ and where the functionals $C_j(u)$ are given as follows:
For $0 < \varrho < 1$ we have for $j = 0, \dots, \kappa$:

$$C_j(u) = \frac{1}{(j + \varrho)} \sum_{|\alpha| = \kappa - j} \frac{1}{\alpha!} \frac{\partial^\alpha u}{\partial x^\alpha}(0) \int\limits_{|\widehat{\Theta}|=1} \widehat{\Theta}^\alpha h_q(\widehat{\Theta}) d\omega . \tag{3.2.12}$$

If $\varrho = 0$, then for $j = 0$ we have

$$C_0(u) = - \sum_{|\alpha| = \kappa} \frac{1}{\alpha!} \frac{\partial^\alpha u}{\partial x^\alpha} \int\limits_{|\widehat{\Theta}|=1} \widehat{\Theta}^\alpha h_q(\widehat{\Theta}) d\omega , \tag{3.2.13}$$

whereas for $j = 1, \dots, \kappa$, the functionals are still given by (3.2.12) with $\varrho = 0$.

In the sequel it is understood, that for $n = 1$ the surface integral in (3.2.13) should always be replaced by

$$\int\limits_{|\widehat{\Theta}|=1} \widehat{\Theta}^\alpha h_q(\widehat{\Theta}) d\omega := \big(h_q(1) + (-1)^\alpha h_q(-1)\big) . \tag{3.2.14}$$

Remarks 3.2.1: We observe that the functionals C_j in (3.2.12) and (3.2.13) are independent of the choice of Ω_0. Thus, the equation (3.2.11) of Hadamard's finite part integral is independent of the choice of Ω_0.

On the other hand, since Definition 3.2.2 and (3.2.11) are equivalent, non of the coefficients d_α in (3.2.7) dependent on the choice of Ω_0.

The advantage of Definition 3.2.2 is that it provides a constructive procedure for the evaluation of the Hadamard finite part integrals in terms of the weakly singular and regular integrals. Furthermore, one has the freedom to choose Ω_0 as one wishes. This freedom is very handy from a practical computational point of view. Moreover, we may decompose the integral

$$I(\varepsilon) := \int\limits_{\Omega\setminus\{|z|<\varepsilon\}} h_q(z)u(z)dz$$

in the form

$$I(\varepsilon) = \widetilde{I} + \sum_{j=0}^{\kappa} C_j \varepsilon^{-j-\varrho} + o(1)\,, \tag{3.2.15}$$

(where for $\varrho = 0$ we replace ε^0 by $\log\varepsilon$) and where $\widetilde{I}$ is the finite part of the integral due to (3.2.11). Conversely, if such an expansion is available then the finite part is simply given by $\widetilde{I}$.

Definition 3.2.3. *A Hadamard finite part integral is called the Cauchy principal value integral if in* (3.2.15)

$$C_j = 0 \quad \textit{for} \ j = 0, \ldots, \kappa\,. \tag{3.2.16}$$

In particular, we obtain a Cauchy principal value integral if

$$\int\limits_{|\widehat{\Theta}|=1} \widehat{\Theta}^\alpha h_q(\widehat{\Theta}) d\omega = 0 \tag{3.2.17}$$

for all multiindices α *with* $|\alpha| = \kappa - j\,,\ j = 0, \ldots, \kappa$. *We then denote by*

$$\text{p.v.} \int\limits_{\Omega} h_q(z)u(z)dz = \lim_{\varepsilon\to 0} \int\limits_{\Omega\setminus\{|z|<\varepsilon\}} h_q(z)u(z)dz \tag{3.2.18}$$

the Cauchy principal value of the integral, whenever it exists.

With the definition of Hadamard's finite part integral available, every pseudohomogeneous function $h_q(z)$ defines a distribution.

Lemma 3.2.1. *Let* $h_q(z) \in \psi h f_q$ *then*

$$\text{p.f.} \int\limits_{\mathbb{R}^n} h_q(z)u(z)dz =: \langle h_q, u\rangle \quad \textit{for all} \ \ u \in \mathcal{D} \tag{3.2.19}$$

defines a distribution $h_q \in \mathcal{D}'$. This distribution can be extended to $u \in \mathcal{S}$, hence, $h_q \in \mathcal{S}'$.

Proof: The first claim follows from Definition 3.2.2 since the right–hand side of (3.2.6) can be bounded by $c_K \|u\|_{C^\kappa(K)}$ for $u \in C_0^\infty(K)$ for any compact set $K \Subset \mathbb{R}^n$.

If $u \in \mathcal{S}$, then we write

$$\begin{aligned} \text{p.f.} \int_{\mathbb{R}^n} h_q(z)u(z)dz &= \int_{|z|\leq 1} h_q(z)\Big\{u(z) - \sum_{|\alpha|\leq\kappa} \frac{1}{\alpha!} z^\alpha \frac{\partial^\alpha u}{\partial z^\alpha}(0)\Big\}dz \\ &\quad + \sum_{|\alpha|\leq\kappa} d_\alpha \frac{\partial^\alpha u}{\partial z^\alpha}(0) + \int_{|z|\geq 1} h_q(z)u(z)dz\,, \end{aligned}$$

where the d_α are defined by (3.2.7) with $\Omega = \Omega_0 = \{z \mid |z| \leq 1\}$. Let $q \notin \mathbb{N}_0$. Then the integrand of the first integral is of order $|z|^{\kappa+1+q}$, hence, this integral is dominated by $c\|u\|_{C^{\kappa+1}(\mathbb{R}^n)}$ for $\kappa > -q-n-1$. The last integral is dominated by $c \sup |z^{q+n+1}u(z)|$. Hence,

$$|\langle h_q, u\rangle| \leq c \sup_{|\alpha|,|\beta|\leq N} |z^\alpha D^\beta(z)|$$

for $N \geq |q+n|+1$; and Definition 3.1.3 ensures $h_q \in \mathcal{S}'$.

If $q \in \mathbb{N}_0$ then the first integral exists already for $\kappa = 0$, whereas the additional logarithmic growth does not effect the estimate of the integral for $|z| \geq 1$. ∎

Remark 3.2.2: Let $\chi \in L^\infty(\mathbb{R}^n)$ with compact $\operatorname{supp}\chi =: K$. Then

$$\int_{\mathbb{R}^n} \chi(z)h_q(z)u(z)dz =: \langle \chi h_q, u\rangle \quad \text{for} \quad u \in \mathcal{E} \tag{3.2.20}$$

defines a distribution $\chi h_q \in \mathcal{E}'$ since χh_q also has support K.

As an example of the Hadamard's finite part integral we consider the hypersingular operator D of the Laplacian defined in (1.2.16) for $n = 2$. We shall show that for $\Gamma \in C^{2+\alpha}$ and $\varphi \in C^{1+\alpha}$, $0 < \alpha < 1$, we are able to express the hypersingular operator in terms of a finite part integral, more precisely,

$$\begin{aligned} D\varphi(x) &= -\lim_{z\to x\in\Gamma, z\in\Omega} n_x \bullet \nabla_z \int_\Gamma \Big(\frac{\partial}{\partial n_y} E(z,y)\Big)\varphi(y)ds_y \\ &= -\,\text{p.f.} \int_\Gamma \frac{\partial^2 E}{\partial n_x \partial n_y}(x,y)\varphi(y)ds_y\,,\ x \in \Gamma\,, \end{aligned}$$

where the finite part integral is defined by the limit

$$\text{p.f.} \int\limits_{\Gamma} \frac{\partial^2 E}{\partial n_x \partial n_y}(x,y)\varphi(y)ds_y = \lim_{\varepsilon \to 0} \Big\{ \int\limits_{\Gamma_\varepsilon} \frac{\partial^2 E}{\partial n_x \partial n_y}(x,y)\varphi(y)ds_y - H(x;\varepsilon;\sigma) \Big\}$$

by using Formula (3.2.11) with $\Gamma_\varepsilon = \Gamma \cap \{y \,|\, |x-y| \geq \varepsilon\}$ and,

$$H(x;\varepsilon;\sigma) = \int\limits_{\Gamma_\varepsilon} \frac{d}{ds_x}\frac{d}{ds_y}\big(E(x,y)\varphi(y)\big)ds_y = -\frac{\varphi(x)}{\pi\varepsilon} + O(\varepsilon^\alpha) \,. \tag{3.2.21}$$

To show (3.2.21), we first use formula (1.2.14), i.e.,

$$D\varphi(x) = -\frac{d}{ds_x} V\Big(\frac{d\varphi}{ds_y}\Big)(x)$$

and, with the well known result for singular integrals of Cauchy type, (see e.g. Muskhelishvili [320, p. 31]) we can interchange $\frac{d}{ds_x}$ with integration, to obtain the principal value integral

$$D\sigma(x) = -\text{p.v.} \int\limits_{\Gamma} \frac{d}{ds_x} E(x,y) \frac{d\varphi}{ds_y}(y)ds_y = -\lim_{\varepsilon \to 0} \int\limits_{\Gamma_\varepsilon} \frac{d}{ds_x} E(x,y)\frac{d\varphi}{ds_y}(y)ds_y \,.$$

Here we see that the above limit depends on the special choice of Γ_ε. If Γ_ε were not chosen symmetrically with respect to x and the arclength distance from x, the interchange of $\frac{d}{ds_x}$ and the integration needs to be modified. Now, for fixed $\varepsilon > 0$ we have

$$\int\limits_{\Gamma_\varepsilon} \frac{d}{ds_x} E(x,y) \frac{d\varphi}{ds_y} ds_y = \int\limits_{\Gamma_\varepsilon} \frac{d}{ds_x}\frac{d}{ds_y}\Big(E(x,y)\varphi(y)\Big)ds_y - \int\limits_{\Gamma_\varepsilon} \Big(\frac{d}{ds_x}\frac{d}{ds_y}E(x,y)\Big)\varphi(y)ds_y \,.$$

Note that with our assumptions, the limit on the left–hand side exists for $\varepsilon \to 0$, so does the difference on the right–hand side whereas each of the two terms blows up individually. In particular, the first term can be simplified with integration by parts, i.e.

$$\int\limits_{\Gamma_\varepsilon} \frac{d}{ds_x}\frac{d}{ds_y}\big(E(x,y)\varphi(y)\big)ds_y = \frac{d}{ds_x} E(x,y)\varphi(y)\Big|_{p_1}^{p_2}$$

where $\boldsymbol{p}_1 = y(s_x + \varepsilon_1)$ and $\boldsymbol{p}_2 = y(s_x - \varepsilon_2)$ with $\varepsilon_1 > 0$ and $\varepsilon_2 > 1$ which are defined by $|\boldsymbol{x} - \boldsymbol{p}_1| = \varepsilon = |\boldsymbol{x} - \boldsymbol{p}_2|$. If we use the Taylor expansion of the curve Γ about x, we have

$$\boldsymbol{p}_1 = \boldsymbol{x} + \varepsilon \boldsymbol{t}_x - \kappa_x \boldsymbol{n}_x \frac{\varepsilon^2}{2} + O(\varepsilon^{2+\alpha}) \quad \text{and}$$

$$\boldsymbol{p}_2 = \boldsymbol{x} - \varepsilon \boldsymbol{t}_x - \kappa_x \boldsymbol{n}_x \frac{\varepsilon^2}{2} + O(\varepsilon^{2+\alpha})$$

where κ_x denotes the curvature of Γ at x. Hence, from (1.1.2) we find

$$\begin{aligned} \frac{d}{ds_x} E(x;y)\varphi(x)(y)\Big|_{p_1}^{p_2} &= -\frac{1}{2\pi} \frac{\boldsymbol{t}_x \cdot (x-y)\varphi(y)}{|x-y|^2}\Big|_{p_1}^{p_2} \\ &= -\frac{1}{2\pi\varepsilon^2}\Big\{\big(2\varepsilon + O(\varepsilon^{2+\alpha})\big)\big(\varphi(x) + O(\varepsilon^{1+\alpha})\big)\Big\} \\ &= -\frac{\varphi(x)}{\varepsilon\pi} + O(\varepsilon^{\alpha}), \end{aligned}$$

which is (3.2.21).

Alternatively, the same result can be achieved according to Definition 3.2.2 by chosing $\Omega_0 = \Gamma \cap \{y \,|\, |y-x| < \varepsilon\}$. Then $n = 1$, $q = -2$ and $\kappa = 1$; $d_0 = -\frac{1}{2\pi}$ is given by (3.2.7) with (3.2.8) and $d_1 = 0$ is given by (3.2.9); and the regularized part in (3.2.6) satisfies

$$\int\limits_{|x-y|\le\varepsilon} h_{-2}(x-y)\{\ldots\} ds_y = O(\varepsilon^{\alpha}).$$

3.3 Local Coordinates

In order to introduce the notion of a boundary surface Γ of a bounded domain $\Omega \subset \mathbb{R}^n$, we shall first define its regularity in the classes $C^{k,\kappa}, k \in \mathbb{N}_0, \kappa \in [0,1]$ by following Nedelec [326] and Grisvard [156], see also Adams [1, p.67] and Wloka [439]. Given a point $y = (y_1, \cdots, y_n) \in \mathbb{R}^n$, we shall write

$$y = (y', y_n), \tag{3.3.1}$$

where

$$y' = (y_1, \cdots, y_{n-1}) \in \mathbb{R}^{n-1}. \tag{3.3.2}$$

Definition 3.3.1. *A bounded domain Ω in $\mathbb{R}^n$ is said to be of class $C^{k,\kappa}$ (in short $\Omega \in C^{k,\kappa}$) if the following properties are satisfied:*

i. *There exists a finite number p of orthogonal linear transformations $T_{(r)}$ (i.e. $n \times n$ orthogonal matrices) and the same number of points $x_{(r)} \in \Gamma$ and functions $a_{(r)}(y'), r = 1, \ldots, p$, defined on the closures of the $(n-1)$-dimensional ball,*

$$Q = \{y' \in \mathbb{R}^{n-1} \,|\, |y'| < \delta\}$$

where $\delta > 0$ is a fixed constant. For each $x \in \Gamma$ there is at least one $r \in \{1, \ldots, p\}$ such that

$$x = x_{(r)} + T_{(r)}(y', a_{(r)}(y')). \tag{3.3.3}$$

ii. *The functions* $a_{(r)} \in C^{k,\kappa}(Q)$.
iii. *There exists a positive number* ε *such that for each* $r \in \{1,\dots,p\}$ *the open set*

$$B_{(r)} := \{x = x_{(r)} + T_{(r)}y \,|\, y = (y', y_n), y' \in Q \text{ and } |y_n| < \varepsilon\}$$

is the union of the sets

$$\begin{aligned}\mathcal{U}^-_{(r)} &= B_{(r)} \cap \Omega = \{x = x_{(r)} + T_{(r)}y \,|\, y = (y', y_n), y' \in Q \\ &\quad \text{and } a_{(r)}(y') - \varepsilon < y_n < a_{(r)}(y')\}\,, \\ \mathcal{U}^+_{(r)} &= B_{(r)} \cap (\mathbb{R}^n \setminus \overline{\Omega}) = \{x = x_{(r)} + T_{(r)}y \,|\, y = (y', y_n), y' \in Q \\ &\quad \text{and } a_{(r)}(y') < y_n < a_{(r)}(y') + \varepsilon\}\,, \end{aligned} \qquad (3.3.4)$$

and

$$\Gamma_{(r)} = B_{(r)} \cap \partial\Omega = \{x = x_{(r)} + T_{(r)}(y', a_{(r)}(y')) \,|\, y' \in Q\}\,. \qquad (3.3.5)$$

The boundary surface $\Gamma = \partial\Omega$ is said to be in the class $C^{k,\kappa}$ if $\Omega \in C^{k,\kappa}$; and in short we also write $\Gamma \in C^{k,\kappa}$. The collection of the above local parametric representations is called a *finite atlas* and each particular mapping (3.3.3) a *chart of* Γ. For the geometric interpretation see Figure 3.3.1.

Remark 3.3.1: Note that we have a special open covering of Γ by the open sets $B_{(r)}$,

$$\Gamma \subset \bigcup_{r=1}^{p} B_{(r)}.$$

Moreover, the mappings

$$y = (y', y_n) = \Phi_{(r)}(x) \text{ where } x \in B_{(r)} \text{ and}$$

$$y' := \Big(T^{-1}_{(r)}(x - x_{(r)})\Big)', y_n = \Big(T^{-1}_{(r)}(x - x_{(r)})\Big)_n - a_{(r)}\Big((T^{-1}_{(r)}(x - x_{(r)}))'\Big)$$

are one–to–one mappings from $B_{(r)}$ onto $Q \times (-\varepsilon, \varepsilon) \subset \mathbb{R}^n$ having the inverse mappings

$$x = \Psi_{(r)}(y) := x_{(r)} + T_{(r)}(y', a_{(r)}(y') + y_n)$$

from $Q \times (-\varepsilon, \varepsilon)$ onto $B_{(r)}$ as considered by Grisvard in [156] (see also Wloka [439]). Correspondingly, we write $\Gamma \in C^\infty$ if $\Gamma \in \cap_{k=1}^{\infty} C^k$.

In the special case, when $\Gamma \in C^{0,1}$, the boundary is called a *Lipschitz boundary* with a *strong Lipschitz property* and Ω is called a *strong Lipschitz domain* (see Adams [1, p.67]). That means in particular that the local parametric representation (3.3.3) is given by a Lipschitz continuous function $a_{(r)}$. Such a boundary can contain conical points or edges (see e.g. in Kufner et al

[249]). Note that the domains shown in Figure 3.3.2 are not strong Lipschitz domains.

It was shown by Gagliardo [132] and Chenais [66] that every strong Lipschitz domain has the uniform cone property and conversely, every domain having the uniform cone property, is a strong Lipschitz domain (Grisvard [156, Theorem 1.2.2.2]).

A domain $\Omega \in C^{1,\kappa}$ with $0 < \kappa < 1$ is a *Lyapounov surface* as introduced in Section 1.5. In fact, we have $C^{2,0} \subset C^{1,1} \subset C^{1,\alpha} \subset C^{0,1}$ where $0 < \alpha \leq 1$.

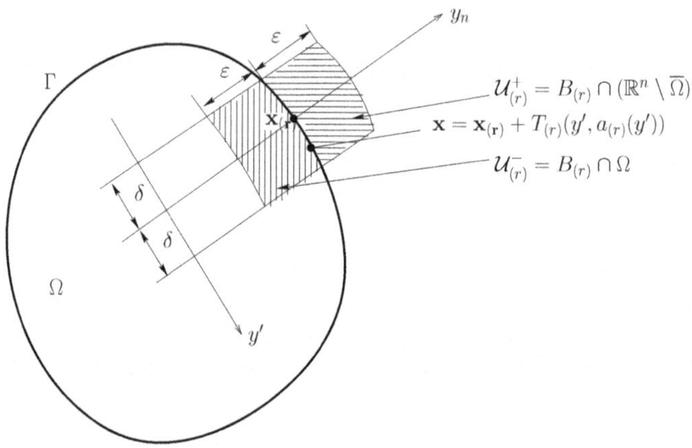

Figure 3.3.1: The local coordinates of a $C^{k,\kappa}$ domain

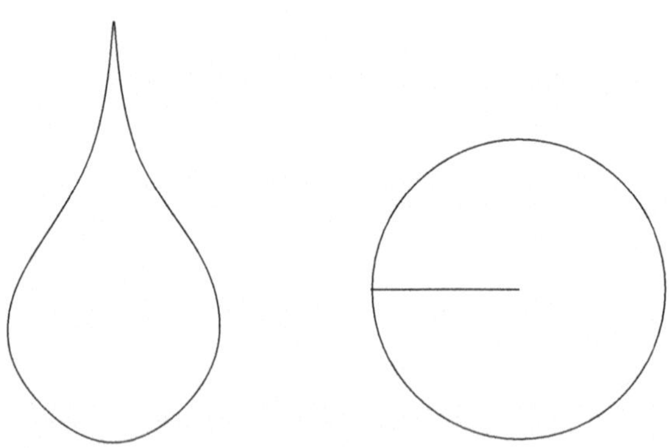

Figure 3.3.2: Two non–Lipschitzian domains

If $f \in C^\infty(\mathbb{R}^n)$ then, with our definitions, the restriction of f to the boundary $\Gamma \in C^\infty$ will be denoted by $f_{|_\Gamma}$, and it defines by $f(x(y'))$ a C^∞–

function of y' in Q. The mapping $\gamma_0 : f \mapsto f_{|_\Gamma}$ defines the linear continuous trace operator from $C^\infty(\mathbb{R}^n)$ into $C^\infty(\Gamma)$.

These local coordinates are necessary for obtaining explicit representations of the corresponding boundary integral operators on Γ. For this purpose we are going to rewrite the representation formulae.

According to (3.3.3), let Γ be given locally by the regular parametric representation

$$\Gamma : x_{|\Gamma} = y(\sigma') \tag{3.3.6}$$

in the vicinity of a fixed point $x_{(r)} \in \Gamma$, where $\sigma' = (\sigma_1, \ldots, \sigma_{n-1})$ are the parameters and $x_{(r)} = y(0)$. In the following, we require only that the tangent vectors

$$\frac{\partial y}{\partial \sigma_\lambda}(\sigma') \text{ for } \lambda = 1, \ldots, n-1 \text{ are linearly independent}$$

and sufficiently smooth but not necessarily orthonormal. We note that in (3.3.3) a particular representation was given by

$$y(\sigma') = x_{(r)} + T_{(r)}(\sigma', a_{(r)}(\sigma'))\,.$$

For $x \notin \Gamma$ but near to Γ, we introduce a spatial transformation

$$x(\sigma', \sigma_n) := y(\sigma') + \sigma_n n(\sigma')\,, \tag{3.3.7}$$

where $n(\sigma') = \boldsymbol{n}$ is the exterior normal to Γ at $y(\sigma')$ and σ_n is an additional parameter. (The coordinates (σ_1', σ_n) and (3.3.7) are often called "Hadamard coordinates".) For Γ sufficiently smooth, it can easily be shown that (3.3.7) defines a diffeomorphism between $x \in \mathbb{R}^n$ and $\sigma = (\sigma', \sigma_n) \in \mathbb{R}^n$ in appropriate neighbourhoods of $x = x_{(r)}$ and $\sigma = 0$. However, note that for $x(\sigma) \in C^{\ell,\kappa}$, the surface Γ must be given in $C^{\ell+1,\kappa}$. We shall come back to this point more specifically in Section 3.7.

3.4 Short Excursion to Elementary Differential Geometry

For the local analysis of differential as well as boundary integral operators on the boundary manifold we need some corresponding calculus in the form of elementary differential geometry. For straightforward introductions to this area we refer to the presentations in Berger and Gostiaux [30], Bishop and Goldberg [32], Klingenberg [229], Millman and Parker [306] and Sokolnikoff [394]. We begin with the differential operator $\frac{\partial}{\partial x_q}$ which can be transformed by

$$\frac{\partial u}{\partial x_q} = \sum_{\ell=1}^{n} \frac{\partial \sigma_\ell}{\partial x_q} \frac{\partial u}{\partial \sigma_\ell} \tag{3.4.1}$$

where the Jacobian $\frac{\partial \sigma_\ell}{\partial x_q}$ is to be expressed in terms of the transformation (3.3.7). To this end, let us define the coefficients of the associated Riemannian metric

$$g_{ik}(\sigma) = g_{ki} := \frac{\partial \boldsymbol{x}}{\partial \sigma_i} \bullet \frac{\partial \boldsymbol{x}}{\partial \sigma_k} \text{ for } i, k, = 1, \ldots, n\,. \tag{3.4.2}$$

The diffeomorphism defined by (3.3.7) is said to be regular if the determinant does not vanish,

$$g := \det\,(g_{ik}) \neq 0\,. \tag{3.4.3}$$

For a regular diffeomorphism, the inverse $g^{m\ell}$ of g_{ik} is well defined by

$$\sum_{m=1}^{n} g_{jm} g^{m\ell} = \sum_{k=1}^{n}\sum_{m=1}^{n} \frac{\partial x_k}{\partial \sigma_m} g^{m\ell} \frac{\partial x_k}{\partial \sigma_j} = \delta_{j\ell}\,; \tag{3.4.4}$$

and from

$$\sum_{k=1}^{n} \frac{\partial \sigma_\ell}{\partial x_k}\frac{\partial x_k}{\partial \sigma_j} = \delta_{j\ell}\,,$$

one gets

$$\frac{\partial \sigma_\ell}{\partial x_k} = \sum_{m=1}^{n} \frac{\partial x_k}{\partial \sigma_m} g^{m\ell}\,. \tag{3.4.5}$$

Hence, (3.4.1) becomes

$$\frac{\partial u}{\partial x_q} = \sum_{\ell,m=1}^{n} \frac{\partial x_q}{\partial \sigma_m} g^{m\ell} \frac{\partial u}{\partial \sigma_\ell}\,. \tag{3.4.6}$$

This equation can be rewritten as follows.

Lemma 3.4.1. *Let* (3.3.7) *be a regular, sufficiently smooth transformation. Then*

$$\frac{\partial u}{\partial x_q} = \frac{1}{\sqrt{g}} \sum_{m=1}^{n} \frac{\partial}{\partial \sigma_m}\Big(\sqrt{g} \sum_{\ell=1}^{n} \frac{\partial x_q}{\partial \sigma_\ell} g^{\ell m} u\Big)\,. \tag{3.4.7}$$

Proof: In order to show (3.4.7) we need to introduce the Christoffel symbols G^r_{pq} of the diffeomorphism (3.3.7) defined by[2]

$$\frac{\partial^2 x_i}{\partial \sigma_q \partial \sigma_p} = \sum_{r=1}^{n} G^r_{pq} \frac{\partial x_i}{\partial \sigma_r} \text{ for } i, p, q = 1, \ldots, n\,. \tag{3.4.8}$$

For the Christoffel symbols, the following relations are easily established (Sokolnikoff [394, p.80 ff.]):

[2] In Millman and Parker [306, (32.6)] one finds the additional term $\left\{ {i \atop st} \right\}_{\!y} \frac{\partial x_s}{\partial \sigma_p} \frac{\partial x_t}{\partial \sigma_q}$ which can be shown to be zero after some manipulations.

$$\begin{aligned}
\frac{\partial g_{\ell k}}{\partial \sigma_q} &= \sum_{m=1}^{n} G_{\ell q}^{m} g_{mk} + \sum_{m=1}^{n} G_{kq}^{m} g_{m\ell}\,, & (3.4.9)\\
\sum_{q=1}^{n} \frac{\partial g^{pq}}{\partial \sigma_q} &= -\sum_{q,\ell=1}^{n} G_{q\ell}^{p} g^{\ell q} - \sum_{q,\ell=1}^{n} G_{q\ell}^{q} g^{\ell p}\,, & (3.4.10)\\
\sum_{p,q=1}^{n} \frac{\partial}{\partial \sigma_q}\left(\frac{\partial x_i}{\partial \sigma_p} g^{pq}\right) &= -\sum_{p,q,r=1}^{n} \frac{\partial x_i}{\partial \sigma_p} G_{rq}^{r} g^{pq}\,, & (3.4.11)\\
\frac{\partial g}{\partial \sigma_q} &= 2g \sum_{r=1}^{n} G_{rq}^{r}\,. & (3.4.12)
\end{aligned}$$

The right–hand side of (3.4.6) can be written as

$$\begin{aligned}
\frac{\partial u}{\partial x_q} &= \frac{1}{\sqrt{g}} \sum_{\ell,m=1}^{n} \frac{\partial}{\partial \sigma_\ell}\left(\sqrt{g}\frac{\partial x_q}{\partial \sigma_m} g^{m\ell} u\right)\\
&\quad - \frac{1}{\sqrt{g}} \sum_{\ell,m=1}^{n} \left(\frac{\partial}{\partial \sigma_\ell}\big(\sqrt{g}\frac{\partial x_q}{\partial \sigma_m} g^{m\ell}\big)\right) u\,.
\end{aligned}$$

Then the second term on the right–hand side vanishes, since

$$\begin{aligned}
&\frac{1}{\sqrt{g}} \sum_{\ell,m=1}^{n} \frac{\partial}{\partial \sigma_\ell}\left(\sqrt{g}\frac{\partial x_q}{\partial \sigma_m} g^{m\ell}\right)\\
&= \frac{1}{\sqrt{g}} \sum_{\ell,m=1}^{n} \left\{\frac{1}{2\sqrt{g}} \frac{\partial g}{\partial \sigma_\ell} \frac{\partial x_q}{\partial \sigma_m} g^{m\ell} + \sqrt{g}\frac{\partial^2 x_q}{\partial \sigma_\ell \partial \sigma_m} g^{m\ell} + \sqrt{g}\frac{\partial x_q}{\partial \sigma_m} \frac{\partial}{\partial \sigma_\ell} g^{m\ell}\right\}\\
&= \frac{1}{\sqrt{g}} \sum_{\ell,m=1}^{n} \frac{1}{2\sqrt{g}} 2g \sum_{r=1}^{n} G_{r\ell}^{r} \frac{\partial x_q}{\partial \sigma_m} g^{m\ell} + \sum_{r,m,\ell=1}^{n} G_{\ell m}^{r} \frac{\partial x_q}{\partial \sigma_r} g^{m\ell}\\
&\quad - \sum_{m,\ell,t=1}^{n} \frac{\partial x_q}{\partial \sigma_m} G_{\ell t}^{m} g^{t\ell} - \sum_{m,\ell,t=1}^{n} \frac{\partial x_q}{\partial \sigma_m} G_{\ell t}^{\ell} g^{tm}\\
&= 0\,, \qquad (3.4.13)
\end{aligned}$$

which gives (3.4.7). ■

Our coordinates (3.3.7) are closely related to the surface Γ which corresponds to $\sigma_n = 0$, i.e. $x = x(\sigma', 0) = y(\sigma')$.

In the following, we refer to these coordinates (3.3.7) as *Hadamard's tubular coordinates*. The metric coefficients of the surface Γ are given by

$$\gamma_{\nu\mu}(\sigma') = \gamma_{\mu\nu}(\sigma') := \frac{\partial y}{\partial \sigma_\nu} \cdot \frac{\partial y}{\partial \sigma_\mu} \quad \text{for } \nu, \mu = 1, \ldots, n-1\,. \qquad (3.4.14)$$

As usual, we denote by

$$\gamma = \det\,(\gamma_{\nu\mu}) \tag{3.4.15}$$

the corresponding determinant and by $\gamma^{\lambda\kappa}$ the coefficients of the inverse to $\gamma_{\nu\mu}$. Then

$$ds = \gamma^{\frac{1}{2}} d\sigma' \tag{3.4.16}$$

is the surface element of Γ in terms of the local coordinates.

In order to express g_{ik} in terms of the surface geometry we will need the Weingarten equations (Klingenberg [229], Millman and Parker [306, p. 126]),

$$\frac{\partial n}{\partial \sigma_\lambda} = -\sum_{\nu=1}^{n-1} L_\lambda^\nu \frac{\partial y}{\partial \sigma_\nu} \quad \text{for } \lambda = 1, \dots, n-1 \tag{3.4.17}$$

where the coefficients L_λ^ν of the *second fundamental form* of Γ describe the curvature of Γ and are given by

$$L_\lambda^\mu = \sum_{\nu=1}^{n-1} L_{\lambda\nu}\gamma^{\nu\mu}\,,\, L_{\lambda\nu} = \frac{\partial^2 y}{\partial\sigma_\lambda \partial\sigma_\nu} \bullet n(\sigma')\,. \tag{3.4.18}$$

In addition, we have

$$2H = \sum_{\lambda=1}^{n-1} L_\lambda^\lambda = -(\nabla \bullet n) \tag{3.4.19}$$

with the *mean curvature* H of Γ and

$$K = \det\,(L_\lambda^\mu)\,, \quad \sum_{\lambda=1}^{n-1} L_{\nu\lambda}L_\mu^\lambda = 2HL_{\nu\mu} - K\gamma_{\nu\mu} \tag{3.4.20}$$

where K is the *Gaussian curvature* of Γ. With (3.4.2) we obtain from (3.3.7)

$$g_{\nu\mu} = \left(\frac{\partial y}{\partial\sigma_\nu} + \sigma_n \frac{\partial n}{\partial \sigma_\nu}\right) \bullet \left(\frac{\partial y}{\partial\sigma_\mu} + \sigma_n \frac{\partial n}{\partial \sigma_\mu}\right)$$

and with (3.4.17) and (3.4.20),

$$\begin{aligned} g_{\nu\mu} &= \gamma_{\nu\mu} - 2\sigma_n L_{\nu\mu} + \sigma_n^2 \sum_{\lambda=1}^{n-1} L_{\nu\lambda}L_\mu^\lambda \\ &= (1 - \sigma_n^2 K)\gamma_{\nu\mu} - 2\sigma_n(1 - \sigma_n H)L_{\nu\mu} \end{aligned} \tag{3.4.21}$$

for $\nu, \mu = 1, \dots, n-1$. Furthermore, we see immediately

$$\begin{aligned} g_{\nu n} &= \frac{\partial x}{\partial \sigma_\nu} \bullet n(\sigma') = \left(\frac{\partial y}{\partial\sigma_\nu} - \sigma_n \sum_{\kappa=1}^{n-1} L_\nu^\kappa \frac{\partial y}{\partial \sigma_\kappa}\right) \bullet n(\sigma') \\ &= 0 \quad \text{for } \nu = 1, \dots, n-1\,; \text{ and} \\ g_{nn} &= 1\,. \end{aligned} \tag{3.4.22}$$

From (3.4.21) one obtains

$$\det\left(\sum_{\mu=1}^{n-1} g_{\nu\mu}\gamma^{\mu\lambda}\right) = (1 - 2\sigma_n H + \sigma_n^2 K)^2$$

and, hence,

$$g = \gamma(1 - 2\sigma_n H + \sigma_n^2 K)^2 . \tag{3.4.23}$$

Equations (3.4.21), (3.4.22) and (3.4.23) imply in particular that for $\sigma_n = 0$ one has

$$g_{\nu\mu}(\sigma', 0) = \gamma_{\nu\mu}(\sigma') \quad \text{and} \quad g^{\nu\mu}(\sigma', 0) = \gamma^{\nu\mu}(\sigma') . \tag{3.4.24}$$

Similar to (3.4.22), by computing the left–hand side in (3.4.8), we find that some of the Christoffel symbols in (3.4.8) take special values, namely:

$$\begin{aligned} G^r_{nn} &= 0 \quad \text{for } r = 1, \ldots, n; \\ G^n_{\lambda n} &= 0 \quad \text{for } \lambda = 1, \ldots, n-1 . \end{aligned} \tag{3.4.25}$$

We also find for our special coordinates (3.3.7) with (3.4.25) and (3.4.17),

$$\sum_{\nu=1}^{n-1} G^\nu_{n\lambda}\frac{\partial x}{\partial \sigma_\nu} = \frac{\partial^2 x}{\partial \sigma_n \partial \sigma_\lambda} = \frac{\partial n}{\partial \sigma_\lambda} = -\sum_{\nu=1}^{n-1} L^\nu_\lambda \frac{\partial y}{\partial \sigma_\nu} .$$

As a consequence from (3.3.7) with (3.4.17) we have

$$\sum_{\nu=1}^{n-1} G^\nu_{n\lambda}\left(\frac{\partial y}{\partial \sigma_\nu} - \sigma_n \sum_{\kappa=1}^{n-1} L^\kappa_\nu \frac{\partial y}{\partial \sigma_\kappa}\right) = -\sum_{\nu=1}^{n-1} L^\nu_\lambda \frac{\partial y}{\partial \sigma_\nu}$$

from which we obtain the relations

$$-L^\nu_\lambda = G^\nu_{n\lambda} - \sigma_n \sum_{\varrho=1}^{n-1} G^\varrho_{n\lambda} L^\nu_\varrho \tag{3.4.26}$$

for $\nu, \lambda = 1, \ldots, n-1$. Note that on Γ, for $\sigma_n = 0$ we have

$$G^\nu_{n\lambda}(\sigma', 0) = -L^\nu_\lambda(\sigma') .$$

For the Jacobian $\frac{\partial \sigma_\ell}{\partial x_k}$ of our special diffeomorphism (3.3.7), we find from (3.4.5) and (3.4.22) the equations

$$\frac{\partial \sigma_\lambda}{\partial x_k} = \sum_{\mu=1}^{n-1} \frac{\partial x_k}{\partial \sigma_\mu} g^{\mu\lambda} + n_k g^{n\lambda} \quad \text{for } k = 1, \ldots, n\,, \lambda = 1, \ldots, n-1 . \tag{3.4.27}$$

Multiplying by n_k we obtain

$$0 = \sum_{k=1}^{n} \frac{\partial \sigma_\lambda}{\partial x_k} \frac{\partial x_k}{\partial \sigma_n} = \sum_{k=1}^{n} \frac{\partial \sigma_\lambda}{\partial x_k} n_k = 1 \cdot g^{n\lambda} .$$

Hence,

$$g^{n\lambda} = 0 \quad \text{and} \quad \frac{\partial \sigma_\lambda}{\partial x_k} = \sum_{\mu=1}^{n-1} \frac{\partial x_k}{\partial \sigma_\mu} g^{\mu\lambda} \tag{3.4.28}$$

for $k = 1, \ldots, n$; $\lambda = 1, \ldots, n-1$. Moreover, with $\frac{\partial x_k}{\partial \sigma_n} = n_k$ from (3.3.7), the relation

$$\frac{\partial \sigma_n}{\partial x_k} = \sum_{\mu=1}^{n-1} \frac{\partial x_k}{\partial \sigma_\mu} g^{\mu n} + n_k g^{nn} = n_k g^{nn}$$

yields

$$g^{nn} = 1 \quad \text{and} \quad \frac{\partial \sigma_n}{\partial x_k} = n_k \quad \text{for} \quad k = 1, \ldots, n . \tag{3.4.29}$$

We also note that for any function u,

$$\frac{\partial u}{\partial \sigma_n} = \sum_{k=1}^{n} \frac{\partial u}{\partial x_k} \frac{\partial x_k}{\partial \sigma_n} = \sum_{k=1}^{n} \frac{\partial u}{\partial x_k} n_k = \frac{\partial u}{\partial n} . \tag{3.4.30}$$

With (3.4.28), (3.3.7) and (3.4.30) one obtains from (3.4.7), in addition to (3.4.19), the equation

$$\nabla \cdot n = \frac{1}{\sqrt{g}} \frac{\partial}{\partial \sigma_n} \sqrt{g} = \frac{1}{\sqrt{g}} \frac{\partial}{\partial n} \sqrt{g} . \tag{3.4.31}$$

Finally, by collecting (3.4.27), (3.4.30) we obtain from (3.4.7)

$$\frac{\partial u}{\partial x_q} = (\mathcal{D}_q + n_q \frac{\partial}{\partial n}) u , \tag{3.4.32}$$

where

$$\begin{aligned} \mathcal{D}_q := & \sum_{\lambda,\mu=1}^{n-1} \frac{\partial x_q}{\partial \sigma_\mu} g^{\mu\lambda} \frac{\partial}{\partial \sigma_\lambda} \\ = & \sum_{\lambda,\mu=1}^{n-1} \left(\frac{\partial y_q}{\partial \sigma_\mu} - \sigma_n \sum_{\kappa=1}^{n-1} L_\mu^\kappa \frac{\partial y_q}{\partial \sigma_\kappa} \right) g^{\mu\lambda} \frac{\partial}{\partial \sigma_\lambda} \end{aligned} \tag{3.4.33}$$

is a tangential operator on the surface $\sigma_n = \text{const}$, parallel to Γ. We note that the tangential operator $\mathcal{D}_q$ can be written in terms of the Günter derivatives (2.2.24) in the tubular coordinates, i.e.,

$$m_{jk}\big(\partial_x, n(\sigma')\big) := n_k(\sigma') \frac{\partial}{\partial x_j} - n_j(\sigma') \frac{\partial}{\partial x_k}$$

as

$$\mathcal{D}_q = -\sum_{q=1}^{n} n_q(\sigma')m_{qk}(\partial_x, n(\sigma')) \tag{3.4.34}$$

Now, from (3.4.32), (3.4.33) we have on Γ, i.e. for $\sigma_n = 0$,

$$\frac{\partial u}{\partial x_q}|_\Gamma = \sum_{\mu,\lambda=1}^{n-1} \frac{\partial y_q}{\partial \sigma_\mu}\gamma^{\mu\lambda}\frac{\partial u}{\partial \sigma_\lambda} + n_q\frac{\partial u}{\partial n}|_\Gamma\,, \tag{3.4.35}$$

or from (3.4.7),

$$\frac{\partial u}{\partial x_q} = \frac{1}{\sqrt{g}}\Big\{\sum_{\mu,\lambda=1}^{n-1}\frac{\partial}{\partial\sigma_\lambda}\Big(\sqrt{g}\frac{\partial x_q}{\partial\sigma_\mu}g^{\mu\lambda}u\Big) + \frac{\partial}{\partial n}(\sqrt{g}n_q u)\Big\}\,,$$

and with (3.4.23),

$$\begin{aligned}\frac{\partial u}{\partial x_q} &= \frac{1}{\sqrt{g}}\sum_{\mu,\lambda=1}^{n-1}\frac{\partial}{\partial\sigma_\lambda}\Big(\sqrt{g}\frac{\partial x_q}{\partial\sigma_\mu}g^{\mu\lambda}u\Big)\\ &\quad + 2\frac{\sqrt{\gamma}}{\sqrt{g}}(\sigma_n K - H)n_q u + n_q\frac{\partial u}{\partial n}\,,\end{aligned}$$

which becomes on Γ, i.e. for $\sigma_n = 0$,

$$\frac{\partial u}{\partial x_q}|_\Gamma = \frac{1}{\sqrt{\gamma}}\sum_{\mu,\lambda=1}^{n-1}\frac{\partial}{\partial\sigma_\lambda}\Big(\sqrt{\gamma}\frac{\partial y_q}{\partial\sigma_\mu}\gamma^{\mu\lambda}u\Big) - 2Hn_q u + n_q\frac{\partial u}{\partial n}\,. \tag{3.4.36}$$

Correspondingly, the so–called *surface gradient* $\mathrm{Grad}_\Gamma u$ is defined by

$$\mathrm{Grad}_\Gamma u = \sum_{\lambda,\mu=1}^{n-1}\frac{\partial y}{\partial\sigma_\mu}\gamma^{\mu\lambda}\frac{\partial u}{\partial\sigma_\lambda}\,, \tag{3.4.37}$$

which yields from (3.4.32) and (3.4.33) with $\sigma_n = 0$ the relation

$$\nabla u_{|\Gamma} = \mathrm{Grad}_\Gamma u + n\frac{\partial u}{\partial n} \quad \text{on } \Gamma\,. \tag{3.4.38}$$

Again, with the Günter derivatives (2.2.25) we have the relation

$$\mathrm{Grad}_\Gamma u = -(n^\top \mathcal{M}|_\Gamma)^\top u \quad \text{on } \Gamma\,. \tag{3.4.39}$$

In order to rewrite a differential operator Pu along Γ we further need higher order derivatives:

$$D^\alpha u = \Big(\mathcal{D} + n(\sigma')\frac{\partial}{\partial n}\Big)^\alpha u := \prod_{q=1}^{n}\Big(\mathcal{D}_q(\sigma',\sigma_n) + n_q(\sigma')\frac{\partial}{\partial n}\Big)^{\alpha_q} u \tag{3.4.40}$$

where $\frac{\partial}{\partial n} = \frac{\partial}{\partial \sigma_n}$ due to (3.4.30). Note that only the operators $n_p(\sigma')\frac{\partial}{\partial n}$ and $n_q(\sigma')\frac{\partial}{\partial n}$ commute, whereas the tangential differential operators $\mathcal{D}_p$ and $\mathcal{D}_q$ and the operators $n_p(\sigma')\frac{\partial}{\partial n}$ mutually are noncommutative. However, due to the Meinardi–Codazzi compatibility conditions, the operators

$$\frac{\partial}{\partial x_q} = \Big(\mathcal{D}_q + n_q \frac{\partial}{\partial n}\Big) \quad \text{and} \quad \frac{\partial}{\partial x_p} = \Big(\mathcal{D}_p + n_p \frac{\partial}{\partial n}\Big)$$

do commute. Similar to the arrangement in (3.4.40) we now write

$$\sum_{k=0}^{|\alpha|} \mathcal{C}_{\alpha,k} \frac{\partial^k u}{\partial n^k} := \Big(\mathcal{D} + n\frac{\partial}{\partial n}\Big)^{\alpha} u = D_x^{\alpha} u \tag{3.4.41}$$

with tangential differential operators

$$\mathcal{C}_{\alpha,k} u = \sum_{|\gamma| \le |\alpha| - k} c_{\alpha,\gamma,k}(\sigma', \sigma_n) \Big(\frac{\partial}{\partial \sigma'}\Big)^{\gamma} u\,, \tag{3.4.42}$$

for $\gamma = (\gamma_1, \dots, \gamma_{n-1})\,, k = 0, \dots, |\alpha|$ and

$$\mathcal{C}_{\alpha,k} u = 0 \quad \text{for} \quad k \le -1 \quad \text{and for} \quad k > |\alpha|\,. \tag{3.4.43}$$

For convenience in calculations we set

$$c_{\alpha,\gamma,k}(\sigma', \sigma_n) = 0 \text{ for } \quad \text{any} \quad \gamma_j < 0 \quad \text{or for } |\gamma| > |\alpha| - k\,. \tag{3.4.44}$$

Then, in particular, $c_{\alpha,\gamma,k} = 0$ for $k > |\alpha|$ in accordance with (3.4.42). The tangential operators $\mathcal{C}_{\alpha,k}$ can be introduced from (3.4.40) and (3.4.41) recursively as follows: Set

$$\mathcal{C}_{0,0} = 1\,, \text{ hence } \mathcal{C}_{0,0} u = u\,. \tag{3.4.45}$$

Then we have for

$$\beta = (\alpha_1, \dots, \alpha_\ell + 1, \dots, \alpha_n) \text{ with } |\beta| = |\alpha| + 1$$

the recursion formulae

$$\mathcal{C}_{\beta,k} = \mathcal{D}_\ell \mathcal{C}_{\alpha,k} + n_\ell \Big(\frac{\partial}{\partial \sigma_n} \mathcal{C}_{\alpha,k}\Big) + n_\ell \mathcal{C}_{\alpha,k-1} \tag{3.4.46}$$

and

$$\begin{aligned} c_{\beta,\varrho,k}(\sigma', \sigma_n) = & \sum_{\lambda,\mu=1}^{n-1} \frac{\partial x_\ell}{\partial \sigma_\mu} g^{\mu\lambda} \Big(\frac{\partial c_{\alpha,\varrho,k}}{\partial \sigma_\lambda} + c_{\alpha,(\varrho_j - \delta_{j\lambda}),k}\Big) \\ & + n_\ell \Big(\frac{\partial c_{\alpha,\varrho,k}}{\partial \sigma_n} + c_{\alpha,\varrho,k-1}\Big)\,. \end{aligned} \tag{3.4.47}$$

In particular, we have

$$\mathcal{C}_{\beta,|\beta|} = c_{\beta,0,|\beta|} = n^{\beta}(\sigma') = \prod_{\ell=1}^{n} n_{\ell}(\sigma')^{\beta_{\ell}} . \tag{3.4.48}$$

Finally, inserting (3.4.40)–(3.4.42) we can write along Γ

$$\begin{aligned} Pu(x) &= \sum_{|\alpha|\le 2m} a_{\alpha}(x) D^{\alpha} u(x) = \sum_{|\alpha|\le 2m} a_{\alpha}(x)\left(\mathcal{D} + n\frac{\partial}{\partial n}\right)^{\alpha} u \\ &= \sum_{|\alpha|\le 2m} a_{\alpha}(x) \sum_{k=0}^{|\alpha|} \mathcal{C}_{\alpha,k} \frac{\partial^k u}{\partial n^k} \\ &=: \sum_{k=0}^{2m} \mathcal{P}_k \frac{\partial^k u}{\partial n^k} , \end{aligned} \tag{3.4.49}$$

where the tangential operators $\mathcal{P}_k$ of orders $2m - k$ are given by

$$\begin{aligned} \mathcal{P}_k &= \sum_{|\alpha|\le 2m} a_{\alpha}(x) \mathcal{C}_{\alpha,k} \\ &= \sum_{|\alpha|\le 2m} \sum_{|\gamma|\le|\alpha|-k} a_{\alpha}(x) c_{\alpha,\gamma,k}(\sigma',\sigma_n)\left(\frac{\partial}{\partial\sigma'}\right)^{\gamma} \\ &= \sum_{|\gamma|\le 2m-k} \sum_{|\alpha|\le 2m} a_{\alpha}(x) c_{\alpha,\gamma,k}(\sigma',\sigma_n)\left(\frac{\partial}{\partial\sigma'}\right)^{\gamma} \end{aligned} \tag{3.4.50}$$

in view of (3.4.44). With (3.4.43) and (3.4.48) we find

$$\mathcal{P}_{2m} = \sum_{|\alpha|\le 2m} a_{\alpha}(x)\mathcal{C}_{\alpha,2m} = \sum_{|\alpha|=2m} a_{\alpha}(x(\sigma',\sigma_n)) n^{\alpha}(\sigma') . \tag{3.4.51}$$

The operator

$$\mathcal{P}_0 = \sum_{|\gamma|\le 2m} \sum_{|\alpha|\le 2m} a_{\alpha}(y) c_{\alpha,\gamma,0}(\sigma',\sigma_n)\left(\frac{\partial}{\partial\sigma'}\right)^{\gamma} \tag{3.4.52}$$

is known as the *Beltrami operator* along Γ associated with P and Γ.

3.4.1 Second Order Differential Operators in Divergence Form

As a special example let us consider a second order partial differential operator in divergence form,

$$Pu(x) = \sum_{i,k=1}^{n} \frac{\partial}{\partial x_i}\left(a_{ik}(x)\frac{\partial u}{\partial x_k}\right) + \sum_{k=1}^{n} b_k(x)\frac{\partial u}{\partial x_k} + c(x)u(x) . \tag{3.4.53}$$

Then (3.4.6) yields

$$\sum_{i,k=1}^{n} \frac{\partial}{\partial x_i}\left(a_{ik}\frac{\partial u}{\partial x_k}\right) = \sum_{i,k,\ell,m=1}^{n} \frac{\partial}{\partial x_i}\left\{a_{ik}\frac{\partial x_k}{\partial \sigma_m} g^{m\ell}\frac{\partial u}{\partial \sigma_\ell}\right\}.$$

Now applying formula (3.4.7) to the term $\{\cdots\}$, we obtain

$$\begin{aligned}
&\sum_{i,k=1}^{n} \frac{\partial}{\partial x_i}\left(a_{ik}(x)\frac{\partial u}{\partial x_k}\right) \\
&= \frac{1}{\sqrt{g}}\sum_{p,\ell=1}^{n} \frac{\partial}{\partial \sigma_p}\Big(\Big\{\sqrt{g}\sum_{i,k,q,m=1}^{n} a_{ik}(x)\frac{\partial x_i}{\partial \sigma_q}\frac{\partial x_k}{\partial \sigma_m} g^{m\ell}g^{qp}\Big\}\frac{\partial u}{\partial \sigma_\ell}\Big).
\end{aligned}$$

By using the properties $g^{n\lambda} = 0$ and $g^{nn} = 1$ in (3.4.28) and (3.4.29), formula (3.4.54) takes the form

$$\begin{aligned}
&\sum_{i,k=1}^{n} \frac{\partial}{\partial x_i}\left(a_{ik}(x)\frac{\partial u}{\partial x_k}\right) \\
&= \sum_{\kappa,\lambda=1}^{n-1} \frac{1}{\sqrt{g}}\frac{\partial}{\partial \sigma_\kappa}\Big(\Big\{\sqrt{g}\sum_{i,k=1}^{n}\sum_{\mu,\nu=1}^{n-1} a_{ik}\frac{\partial x_i}{\partial \sigma_\nu}\frac{\partial x_k}{\partial \sigma_\mu} g^{\mu\lambda}g^{\nu\kappa}\Big\}\frac{\partial u}{\partial \sigma_\lambda}\Big) \qquad (3.4.54)\\
&\quad + \sum_{\kappa=1}^{n-1} \frac{1}{\sqrt{g}}\frac{\partial}{\partial \sigma_\kappa}\Big(\Big\{\sqrt{g}\sum_{i,k=1}^{n}\sum_{\nu=1}^{n-1} a_{ik}\frac{\partial x_i}{\partial \sigma_\nu} n_k g^{\nu\kappa}\Big\}\frac{\partial u}{\partial n}\Big) \\
&\quad + \frac{1}{\sqrt{g}}\frac{\partial}{\partial n}\Big(\sum_{\lambda=1}^{n-1}\Big\{\sqrt{g}\sum_{i,k=1}^{n}\sum_{\mu=1}^{n-1} a_{ik}n_i\frac{\partial x_k}{\partial \sigma_\mu} g^{\mu\lambda}\Big\}\frac{\partial u}{\partial \sigma_\lambda}\Big) \\
&\quad + \frac{1}{\sqrt{g}}\frac{\partial}{\partial n}\Big(\Big\{\sqrt{g}\sum_{i,k=0}^{n} a_{ik}n_i n_k\Big\}\frac{\partial u}{\partial n}\Big).
\end{aligned}$$

On the other hand, (3.4.6) gives

$$\sum_{k=1}^{n} b_k\frac{\partial u}{\partial x_k} = \sum_{\lambda=1}^{n-1}\left\{\sum_{k=1}^{n}\sum_{\mu=1}^{n-1} b_k\frac{\partial x_k}{\partial \sigma_\mu}g^{\mu\lambda}\right\}\frac{\partial u}{\partial \sigma_\lambda} + \left(\sum_{k=1}^{n} b_k n_k\right)\frac{\partial u}{\partial n}. \qquad (3.4.55)$$

Substituting (3.4.54) and (3.4.55) into (3.4.53) yields

$$Pu = \mathcal{P}_0 u + \mathcal{P}_1\frac{\partial u}{\partial n} + \mathcal{P}_2\frac{\partial^2 u}{\partial n^2}. \qquad (3.4.56)$$

For ease of reading, at the end of this section we shall show how these tangential operators $\mathcal{P}_k$ for $k = 0, 1, 2$, can be derived explicitly:

$$\mathcal{P}_2 = \sum_{i,k=1}^{n} a_{ik}(x(\sigma', \sigma_n))n_i(\sigma')n_k(\sigma') . \tag{3.4.57}$$

$$\begin{aligned}
\mathcal{P}_1 = {} & \sum_{\lambda=1}^{n-1}\Big(\sum_{i,k=1}^{n}\sum_{\nu=1}^{n-1} a_{ik}\frac{\partial x_i}{\partial \sigma_\nu}n_k g^{\nu\lambda}\Big)\frac{\partial}{\partial \sigma_\lambda} \\
& + \Big(2\frac{\sqrt{\gamma}}{\sqrt{g}}(\sigma_n K - H) - \sum_{\kappa=1}^{n-1} G^{\kappa}_{\kappa n}\Big)\sum_{i,k=1}^{n} a_{ik}n_i n_k \\
& - \sum_{i,k=1}^{n}\sum_{\varrho,\lambda,\nu=1}^{n-1} a_{ik}\frac{\partial x_i}{\partial \sigma_\varrho}L^{\nu}_{\lambda}\frac{\partial y_k}{\partial \sigma_\nu}g^{\varrho\lambda} \\
& + \sum_{i,k=1}^{n}\frac{\partial a_{ik}}{\partial n}n_i n_k + \sum_{i,k,\ell=1}^{n}\sum_{\varrho,\nu=1}^{n-1}\frac{\partial a_{ik}}{\partial x_\ell}\frac{\partial x_\ell}{\partial \sigma_\varrho}\frac{\partial x_i}{\partial \sigma_\nu}n_k g^{\varrho\nu} + \sum_{k=1}^{n} b_k n_k ,
\end{aligned} \tag{3.4.58}$$

$$\begin{aligned}
\mathcal{P}_0 = {} & \sum_{\kappa,\lambda=1}^{n-1}\frac{1}{\sqrt{g}}\frac{\partial}{\partial \sigma_\kappa}\Big(\Big\{\sqrt{g}\sum_{i,k=1}^{n}\sum_{\mu,\nu=1}^{n-1} a_{ik}\frac{\partial x_i}{\partial \sigma_\nu}\frac{\partial x_k}{\partial \sigma_\mu}g^{\nu\kappa}g^{\mu\lambda}\Big\}\frac{\partial}{\partial \sigma_\lambda}\Big) \\
& + \sum_{\lambda=1}^{n-1}\Big\{\sum_{i,k=1}^{n} a_{ik}n_i\Big(\sum_{\mu=1}^{n-1}\Big\{\frac{2\sqrt{\gamma}}{\sqrt{g}}(\sigma_n K - H)\frac{\partial x_k}{\partial \sigma_\mu} \\
& - \sum_{\nu=1}^{n-1}\Big(\frac{\partial y_k}{\partial \sigma_\nu}L^{\nu}_{\mu} + \frac{\partial x_k}{\partial \sigma_\nu}G^{\nu}_{\mu n}\Big)\Big\}g^{\mu\lambda} - \sum_{\mu,\lambda=1}^{n-1}\frac{\partial x_k}{\partial \sigma_\nu}g^{\mu\nu}G^{\lambda}_{\mu n}\Big) \\
& + \sum_{\mu=1}^{n-1}\Big\{\sum_{k=1}^{n}\Big(\sum_{i=1}^{n}\frac{\partial a_{ik}}{\partial n}n_i + b_k\frac{\partial x_k}{\partial \sigma_\mu}\Big)\Big\}g^{\mu\lambda}\Big\}\frac{\partial}{\partial \sigma_\lambda} + c(x) \\
& + c(x) .
\end{aligned} \tag{3.4.59}$$

On Γ, where $\sigma_n = 0$, the operators further simplify due to (3.4.24) and (3.4.26) and the definition of H. Moreover, we have the relation

$$\frac{\partial}{\partial \sigma_n}g^{\mu\lambda}\big|_{\sigma_n=0} = \sum_{\nu=1}^{n-1}\gamma^{\mu\nu}L^{\lambda}_{\nu} + \gamma^{\lambda\nu}L^{\mu}_{\nu} = 2L^{\lambda\mu} . \tag{3.4.60}$$

The operators $\mathcal{P}_k$ on Γ now read explicitly as

$$\mathcal{P}_2 = \sum_{i,k=1}^{n} a_{ik}(y) n_i(y) n_k(y)\,, \tag{3.4.61}$$

$$\begin{aligned}
\mathcal{P}_1 = & \sum_{\lambda=1}^{n-1}\Big(\sum_{i,k=1}^{n}\sum_{\nu=1}^{n-1} a_{ik}(y)\frac{\partial y_i}{\partial \sigma_\nu} n_k(y)\gamma^{\nu\lambda}\Big)\frac{\partial}{\partial \sigma_\lambda} \\
& - \sum_{i,k=1}^{n}\sum_{\varrho,\nu=1}^{n-1} a_{ik}\frac{\partial y_i}{\partial \sigma_\varrho} L^{\varrho\nu}\frac{\partial y_k}{\partial \sigma_\nu} \\
& + \sum_{i,k=1}^{n}\Big\{\frac{\partial a_{ik}}{\partial n} n_i n_k + \sum_{\varrho,\nu=1}^{n-1}\sum_{\ell=1}^{n}\frac{\partial a_{ik}}{\partial x_\ell}\frac{\partial y_\ell}{\partial \sigma_\varrho}\frac{\partial y_i}{\partial \sigma_\nu} n_k \gamma^{\nu\varrho}\Big\} \\
& + \sum_{k=1}^{n} b_k n_k\,,
\end{aligned} \tag{3.4.62}$$

$$\begin{aligned}
\mathcal{P}_0 = & \sum_{\kappa,\lambda=1}^{n-1}\frac{1}{\sqrt{\gamma}}\frac{\partial}{\partial \sigma_\kappa}\Big(\Big\{\sqrt{\gamma}\sum_{i,k=1}^{n}\sum_{\mu,\nu=1}^{n-1} a_{ik}\frac{\partial y_i}{\partial \sigma_\nu}\frac{\partial y_k}{\partial \sigma_\mu}\gamma^{\nu\kappa}\gamma^{\mu\lambda}\Big\}\frac{\partial}{\partial \sigma_\lambda}\Big) \\
& + \sum_{\lambda=1}^{n-1}\Big\{\sum_{i,k=1}^{n} a_{ik} n_i \sum_{\mu=1}^{n-1}\frac{\partial y_k}{\partial \sigma_\mu}(-2H\gamma^{\mu\lambda} + L^{\mu\lambda}) \\
& + \sum_{\mu=1}^{n-1}\sum_{k=1}^{n}\Big(\sum_{i=1}^{n}\frac{\partial a_{ik}}{\partial n} n_i + b_k\frac{\partial y_k}{\partial \sigma_\mu}\Big)\gamma^{\mu\lambda}\Big\}\frac{\partial}{\partial \sigma_\lambda} \\
& + c(y)\,.
\end{aligned} \tag{3.4.63}$$

Remark 3.4.1: In case of the Laplacian we have $a_{ik} = \delta_{ik}, b_k = 0, c = 0$. Then the expressions (3.4.60) and (3.4.62), (3.4.63) become

$$\begin{aligned}
\mathcal{P}_2 &= 1\,, \\
\mathcal{P}_1 &= -\sum_{\varrho,\nu=1}^{n-1}\gamma_{\varrho\nu}L^{\varrho\nu} = -\sum_{\varrho=1}^{n-1} L^{\varrho}_{\varrho} = -2H\,, \\
\mathcal{P}_0 &= \sum_{\lambda,\kappa=0}^{n-1}\frac{1}{\sqrt{\gamma}}\frac{\partial}{\partial \sigma_\kappa}\Big(\Big\{\sqrt{\gamma}\sum_{\nu=1}^{n-1}\gamma_{\mu\nu}\gamma^{\nu\kappa}\gamma^{\mu\lambda}\Big\}\frac{\partial}{\partial \sigma_\lambda}\Big) \\
&= \frac{1}{\sqrt{\gamma}}\sum_{\lambda,\kappa=1}^{n-1}\frac{\partial}{\partial \sigma_\kappa}\Big(\sqrt{\gamma}\gamma^{\kappa\lambda}\frac{\partial}{\partial \sigma_\lambda}\Big) = \Delta_\Gamma\,,
\end{aligned} \tag{3.4.64}$$

which is the *Laplace–Beltrami* operator on Γ. (See e.g. in Leis [262, p.38]).

Derivation of $\mathcal{P}_1$ and $\mathcal{P}_0$:

We begin with the derivation of the $\mathcal{P}_1$ by collecting the corresponding terms from (3.4.54) and (3.4.55) arriving at

$$\begin{aligned}
\mathcal{P}_1 &= \sum_{\lambda=1}^{n-1}\Big(\sum_{i,k=1}^{n}\sum_{\nu=1}^{n-1} a_{ik}(x(\sigma',\sigma_n))\frac{\partial x_i}{\partial\sigma_\nu}n_k(\sigma')g^{\nu\lambda}\Big)\frac{\partial}{\partial\sigma_\lambda}\\
&\quad+\frac{1}{\sqrt{g}}\Big\{\frac{\partial}{\partial n}\Big(\sqrt{g}\sum_{i,k=1}^{n}a_{ik}n_in_k\Big)\\
&\qquad+\sum_{\kappa=1}^{n-1}\frac{\partial}{\partial\sigma_\kappa}\Big(\sqrt{g}\sum_{i,k=1}^{n}\sum_{\nu=1}^{n-1}a_{ik}\frac{\partial x_i}{\partial\sigma_\nu}n_kg^{\nu\kappa}\Big)\Big\}\\
&\quad+\sum_{k=1}^{n}b_kn_\kappa\,.
\end{aligned}$$

The following formulae are needed when we apply the product rule to the second term in the above expression.

$$\begin{aligned}
\frac{\partial}{\partial n}\sqrt{g} &= 2\sqrt{\gamma}(\sigma_nK-H) && \text{from (3.4.23)},\\
\frac{\partial}{\partial\sigma_\kappa}\sqrt{g} &= \sqrt{g}\sum_{r=1}^{n}G^r_{r\kappa} && \text{from (3.4.11)},\\
\frac{\partial^2x_i}{\partial\sigma_p\partial\sigma_q} &= \sum_{r=1}^{n}G^r_{pq}\frac{\partial x_i}{\partial\sigma_r} && \text{from (3.4.8)},\\
\frac{\partial n_k}{\partial\sigma_\kappa} &= -\sum_{\nu=1}^{n-1}L^\nu_\kappa\frac{\partial y_k}{\partial\sigma_\nu} && \text{from (3.4.17)},\\
\sum_{\kappa=1}^{n-1}\frac{\partial g^{\nu\kappa}}{\partial\sigma_\kappa}+\frac{\partial g^{\nu n}}{\partial\sigma_n} &= -\sum_{q,\ell=1}^{n}G^\nu_{q\ell}g^{\ell q}-\sum_{q,\ell=1}^{n}G^q_{q\ell}g^{\ell\nu} && \text{from (3.4.10)}\ .
\end{aligned}$$

As a consequence, we find

$$
\begin{aligned}
\mathcal{P}_1 &= \sum_{\lambda=1}^{n-1}\left(\sum_{i,k=1}^{n}\sum_{\mu=1}^{n-1} a_{ik}\frac{\partial x_i}{\partial \sigma_\mu} n_k g^{\mu\lambda}\right)\frac{\partial}{\partial \sigma_\lambda} \\
&\quad + 2\frac{\sqrt{\gamma}}{\sqrt{g}}(\sigma_n K - H)\sum_{i,k=1}^{n} a_{ik} n_i n_k + \sum_{i,k=1}^{n}\frac{\partial a_{ik}}{\partial n} n_i n_k \\
&\quad + \sum_{\nu,\kappa=1}^{n-1}\sum_{i,k}^{n}\left(a_{ik}\sum_{r=1}^{n} G^r_{r\kappa} + \sum_{\ell=1}^{n}\frac{\partial a_{ik}}{\partial n_\ell}\frac{\partial x_\ell}{\partial \sigma_\kappa}\right)\frac{\partial x_i}{\partial \sigma_\nu} n_k g^{\nu\kappa} \\
&\quad + \sum_{i,k,r=1}^{n}\sum_{\kappa,\nu=1}^{n-1} a_{ik}\frac{\partial x_i}{\partial \sigma_r} n_k G^r_{\kappa\nu} g^{\nu\kappa} \\
&\quad - \sum_{i,k=1}^{n}\sum_{\mu,\kappa,\nu=1}^{n-1} a_{ik}\frac{\partial x_i}{\partial \sigma_\nu} L^\mu_\kappa \frac{\partial y_k}{\partial \sigma_\mu} g^{\kappa\nu} \\
&\quad - \sum_{i,k,\ell,q=1}^{n}\sum_{\nu=1}^{n-1} a_{ik}\frac{\partial x_i}{\partial \sigma_\nu} n_k (G^\nu_{q\ell} g^{\ell q} + G^q_{q\ell} g^{\ell\nu}) \\
&\quad + \sum_{k=1}^{n} b_k n_k .
\end{aligned}
$$

In addition, we see that with (3.4.28) and (3.4.29), i.e. $g^{n\kappa} = 0$ and $g^{nn} = 1$ there holds

$$
\frac{\partial g^{nn}}{\partial \sigma_n} + \sum_{\kappa=1}^{n-1}\frac{\partial g^{n\kappa}}{\partial \sigma_\kappa} = 0 = \sum_{p=0}^{n}\frac{\partial g^{np}}{\partial \sigma_p}
$$

and (3.4.10) gives

$$
\begin{aligned}
0 &= -\sum_{\kappa,\nu=0}^{n-1} G^n_{\kappa\nu} g^{\nu\kappa} - \sum_{\kappa=1}^{n-1} G^n_{\kappa n} g^{n\kappa} - G^n_{nn} g^{nn} \\
&\quad - \sum_{\ell=1}^{n} G^n_{n\ell} g^{\ell n} - \sum_{\kappa,\lambda=1}^{n-1} G^\kappa_{\kappa\lambda} g^{\lambda n} - \sum_{\kappa=1}^{n-1} G^n_{\kappa n} g^{nn} \\
&= -\sum_{\kappa,\nu=1}^{n-1} G^n_{\kappa\nu} g^{\nu\kappa} - G^n_{nn} - \sum_{\ell=1}^{n} G^n_{n\ell} g^{\ell n} - \sum_{\kappa=1}^{n-1} G^\kappa_{\kappa n} .
\end{aligned}
$$

Using (3.4.25), we obtain

$$
\sum_{\kappa,\nu=1}^{n-1} G^n_{\kappa\nu} g^{\nu\kappa} = -\sum_{\kappa=1}^{n-1} G^\kappa_{\kappa n} . \tag{3.4.65}
$$

A simple computation with the help of (3.4.65) and (3.4.25) yields

$$\begin{aligned}
\mathcal{P}_1 &= \sum_{\lambda=1}^{n-1}\Big(\sum_{i,k=1}^{n}\sum_{\mu=1}^{n-1} a_{ik}\frac{\partial x_i}{\partial\sigma_\mu}n_k g^{\mu\lambda}\Big)\frac{\partial}{\partial\sigma_\lambda}\\
&\quad + 2\frac{\sqrt{\gamma}}{\sqrt{g}}(\sigma_n K - H)\sum_{i,k=1}^{n} a_{ik}n_i n_k + \sum_{i,k=1}^{n}\frac{\partial a_{ik}}{\partial\sigma_n}n_i n_k\\
&\quad + \sum_{i,k=1}^{n} a_{ik}\Bigg\{\sum_{\varrho,\nu,\kappa=1}^{n-1}\frac{\partial x_i}{\partial\sigma_\varrho}n_k G^{\varrho}_{\kappa\nu}g^{\nu\kappa} - \sum_{\kappa=1}^{n-1} n_i n_k G^{\kappa}_{\kappa n}\\
&\qquad\qquad - \sum_{\varrho,\nu,\kappa=1}^{n-1}\frac{\partial x_i}{\partial\sigma_\varrho}n_k G_{\kappa\varrho}g^{\nu\kappa} - \sum_{\varrho=1}^{n-1}\frac{\partial x_i}{\partial\sigma_\varrho}n_k G^{\varrho}_{nn}\\
&\qquad\qquad - \sum_{\varrho,\kappa,\nu=1}^{n-1}\frac{\partial x_i}{\partial\sigma_\varrho}L^{\nu}_{\kappa}\frac{\partial y_k}{\partial\sigma_\nu}g^{\varrho\kappa}\Bigg\}\\
&\quad + \sum_{i,k,\ell=1}^{n}\sum_{\nu,\varrho=1}^{n-1}\frac{\partial a_{ik}}{\partial x_\ell}\frac{\partial x_\ell}{\partial\sigma_\varrho}\frac{\partial x_i}{\partial\sigma_\nu}n_k g^{\varrho\nu} + \sum_{k=1}^{n} b_k n_k\,.
\end{aligned}$$

With (3.4.25) and rearranging terms, the result (3.4.58) follows.

For $\mathcal{P}_0$ we collect from (3.4.54) the terms

$$\begin{aligned}
\mathcal{P}_0 &= \sum_{\kappa,\lambda=1}^{n-1}\frac{1}{\sqrt{g}}\frac{\partial}{\partial\sigma_\kappa}\Big(\Big\{\sqrt{g}\sum_{i,k=1}^{n}\sum_{\mu,\nu=1}^{n-1} a_{ik}\frac{\partial x_i}{\partial\sigma_\nu}\frac{\partial x_k}{\partial\sigma_\mu}g^{\mu\lambda}g^{\nu\kappa}\Big\}\frac{\partial}{\partial\sigma_\lambda}\Big)\\
&\quad + \frac{1}{\sqrt{g}}\sum_{\lambda=1}^{n-1}\Big(\frac{\partial}{\partial\sigma_n}\Big\{\sqrt{g}\sum_{i,k=1}^{n}\sum_{\mu=1}^{n-1} a_{ik}n_i\frac{\partial x_k}{\partial\sigma_\mu}g^{\mu\lambda}\Big\}\Big)\frac{\partial}{\partial\sigma_\lambda}\\
&\quad + \sum_{\lambda=1}^{n-1}\Big\{\sum_{k=1}^{n}\sum_{\mu=1}^{n-1} b_k\frac{\partial x_k}{\partial\sigma_\mu}g^{\mu\lambda}\Big\}\frac{\partial}{\partial\sigma_\lambda} + c(x)\,.
\end{aligned}$$

With the product rule and (3.4.23) we obtain

$$
\begin{aligned}
\mathcal{P}_0 &= \sum_{\kappa,\lambda=1}^{n-1} \frac{1}{\sqrt{g}} \frac{\partial}{\partial\sigma_\kappa} \Big(\Big\{ \sqrt{g} \sum_{i,k=1}^{n} \sum_{\mu,\nu=1}^{n-1} a_{ik} \frac{\partial x_i}{\partial\sigma_\nu} \frac{\partial x_k}{\partial\sigma_\mu} g^{\mu\lambda} g^{\nu\kappa} \Big\} \frac{\partial}{\partial\sigma_\lambda} \Big) \\
&\quad + \sum_{\lambda=1}^{n-1} \Big\{ \frac{2\sqrt{\gamma}}{\sqrt{g}} (\sigma_n K - H) \sum_{i,k=1}^{n} \sum_{\mu=1}^{n-1} a_{ik} n_i \frac{\partial x_k}{\partial\sigma_\mu} g^{\mu\lambda} \\
&\quad + \sum_{i,k=1}^{n} \sum_{\mu=1}^{n-1} a_{ik} n_i \Big(\frac{\partial n_k}{\partial\sigma_\mu} g^{\mu\lambda} + \frac{\partial x_k}{\partial\sigma_\mu} \Big(\frac{\partial}{\partial\sigma_n} g^{\mu\lambda} \Big) \Big) \\
&\quad + \sum_{i,k=1}^{n} \sum_{\mu=1}^{n-1} \frac{\partial a_{ik}}{\partial n} n_i \frac{\partial x_k}{\partial\sigma_\mu} g^{\mu\lambda} \\
&\quad + \sum_{k=1}^{n} \sum_{\mu=1}^{n-1} b_k \frac{\partial x_k}{\partial\sigma_\mu} g^{\lambda\mu} \Big\} \frac{\partial}{\partial\sigma_\lambda} + c(x) .
\end{aligned}
$$

In the third term we insert $\frac{\partial n_k}{\partial\sigma_\mu}$ from (3.4.17), whereas for $\frac{\partial}{\partial\sigma_n} g^{\mu\lambda}$ we use (3.4.10), which yields

$$
\begin{aligned}
\frac{\partial}{\partial\sigma_n} g^{\mu\lambda} &= -\sum_{\ell,p=1}^{n} g^{\mu\ell} \left(\frac{\partial}{\partial\sigma_n} g_{\ell p} \right) g^{p\lambda} \\
&= -\sum_{\ell=1}^{n} (g^{\mu\ell} G^{\lambda}_{\ell n} + g^{\lambda\ell} G^{\mu}_{\ell n}) \\
&= -\sum_{\nu=1}^{n-1} (g^{\mu\nu} G^{\lambda}_{\nu n} + g^{\lambda\nu} G^{\mu}_{\nu n})
\end{aligned}
$$

following from (3.4.25). On the surface Γ where $\sigma_n = 0$ we obtain again (3.4.60).

3.5 Distributional Derivatives and Abstract Green's Second Formula

Let Γ be sufficiently smooth, say, $\Gamma \in C^\infty$ and let $\chi_\Omega(x)$ be the characteristic function for Ω, namely

$$
\chi_\Omega(x) := \begin{cases} 1 & \text{for } x \in \Omega \text{ and} \\ 0 & \text{otherwise.} \end{cases}
$$

We begin with the distributional derivatives of χ_Ω, and claim

$$
\frac{\partial}{\partial x_i} \chi_\Omega(x) = -n_i(x) \delta_\Gamma \tag{3.5.1}
$$

where $n_i(x)$ is the i–th component of the exterior normal $\boldsymbol{n}(x)$ to Ω at $x \in \Gamma = \partial\Omega$ extended to a $C_0^\infty(\mathbb{R}^n)$–function, δ_Γ is the Dirac distribution of Γ defined by

$$\langle \delta_\Gamma, \varphi \rangle_{\mathbb{R}^n} = \int\limits_{\mathbb{R}^n} \delta_\Gamma(x)\varphi(x)dx := \int\limits_\Gamma \varphi(x)ds_x \tag{3.5.2}$$

for every $\varphi \in C_0^\infty(\mathbb{R}^n)$. Equation (3.5.1) follows immediately from the definition of the derivative of distributions and the Gaussian divergence theorem as follows:

$$\begin{aligned}\Big\langle \Big(\frac{\partial}{\partial x_i}\chi_\Omega\Big), \varphi \Big\rangle_{\mathbb{R}^n} &= -\int\limits_{\mathbb{R}^n} \chi_\Omega \frac{\partial \varphi}{\partial x_i} dx = -\int\limits_\Omega \frac{\partial \varphi}{\partial x_i} dx \\ = -\int\limits_\Gamma \varphi(x) n_i(x) ds_x &= -\langle \delta_\Gamma, n_i \varphi \rangle_{\mathbb{R}^n} = -\langle n_i \delta_\Gamma, \varphi \rangle_{\mathbb{R}^n}\end{aligned}$$

from (3.1.6).

Now we define the distribution $\delta_\Gamma^{(1)}(x) := \frac{\partial}{\partial n}\delta_\Gamma$ given by

$$\langle \delta_\Gamma^{(1)}, \varphi \rangle_{\mathbb{R}^n} = \langle (\boldsymbol{n} \cdot \nabla)\delta_\Gamma, \varphi \rangle_{\mathbb{R}^n} = \Big\langle \frac{\partial}{\partial n}\delta_\Gamma, \varphi \Big\rangle_{\mathbb{R}^n}$$

which, by the definition of derivatives of distributions becomes

$$\Big\langle \frac{\partial}{\partial n}\delta_\Gamma, \varphi \Big\rangle_{\mathbb{R}^n} = \langle \delta_\Gamma^{(1)}, \varphi \rangle_{\mathbb{R}^n} = \langle -\delta_\Gamma, \nabla \cdot (\boldsymbol{n}\varphi) \rangle_{\mathbb{R}^n} \quad \text{for all } \varphi \in C_0^\infty(\mathbb{R}^n)\,. \tag{3.5.3}$$

Consequently, we have

$$\langle \delta_\Gamma^{(1)}, \varphi \rangle_{\mathbb{R}^n} = -\int\limits_\Gamma \Big(\frac{\partial}{\partial n} + (\nabla \cdot \boldsymbol{n})\Big)\varphi ds_x = -\Big\langle \delta_\Gamma, \frac{\partial}{\partial n}'\varphi \Big\rangle_{\mathbb{R}^n} \tag{3.5.4}$$

which also defines the transpose of the operator $\frac{\partial}{\partial n}$,

$$\frac{\partial}{\partial n}'\varphi = -\Big(\frac{\partial}{\partial n} + \nabla \cdot \boldsymbol{n}\Big)\varphi = -\Big(\frac{\partial}{\partial n} - 2H\Big)\varphi \text{ on } \Gamma\,, \tag{3.5.5}$$

where $H = -\frac{1}{2}\nabla \cdot \boldsymbol{n}$ is the mean curvature of Γ (see (3.4.19)). With these operators, the following relations of corresponding distributions can be derived:

$$\nabla \chi_\Omega(x) = -\boldsymbol{n}(x)\delta_\Gamma(x)\,, \tag{3.5.6}$$

$$\Delta \chi_\Omega = \frac{\partial}{\partial n}'\delta_\Gamma\,, \tag{3.5.7}$$

$$\Delta(\varphi\chi_\Omega) = (\Delta\varphi)\chi_\Omega - 2\frac{\partial \varphi}{\partial n}\delta_\Gamma + \varphi\Big(\frac{\partial}{\partial n}'\delta_\Gamma\Big) \tag{3.5.8}$$

for any $\varphi \in C_0^\infty$. Here, the right–hand side in (3.5.7) is the composition of the differential operator $\frac{\partial}{\partial n}'$ given in (3.5.5) with the distribution δ_Γ, and it defines again a distribution. Correspondingly, (3.5.8) yields the *Green identity* in the form:

$$\begin{aligned}\langle \varphi\chi_\Omega, \Delta\psi\rangle_{\mathbb{R}^n} - \langle(\Delta\varphi)\chi_\Omega, \psi\rangle_{\mathbb{R}^n} &= \Big\langle \varphi\Big(\frac{\partial}{\partial n}'\delta_\Gamma\Big), \psi\Big\rangle_{\mathbb{R}^n} - 2\Big\langle\frac{\partial\varphi}{\partial n}\delta_\Gamma, \psi\Big\rangle_{\mathbb{R}^n}\\ &= \Big\langle \delta_\Gamma\frac{\partial\psi}{\partial n}, \varphi\Big\rangle_{\mathbb{R}^n} - \Big\langle\frac{\partial\varphi}{\partial n}\delta_\Gamma, \psi\Big\rangle_{\mathbb{R}^n},\end{aligned}$$

or, in the traditional form,

$$\int_\Omega (\varphi\Delta\psi - \psi\Delta\varphi)dx = \int_\Gamma \Big(\varphi\frac{\partial\psi}{\partial n} - \psi\frac{\partial\varphi}{\partial n}\Big)ds_x\,. \tag{3.5.9}$$

In order to formulate the Green identities for higher order partial differential operators, we also need a relation between the distribution's derivatives $\frac{\partial}{\partial x_i}\delta_\Gamma$ and $\delta_\Gamma^{(1)}$ which is given by

$$\frac{\partial}{\partial x_i}\delta_\Gamma = n_i(x)\delta_\Gamma^{(1)}\,. \tag{3.5.10}$$

The derivation of this relation is not straight forward and needs the concept of local coordinates near Γ, corresponding formulae from differential geometry which are given by (3.4.29) and (3.4.30), and also the Stokes theorem on Γ.

For u, a sufficiently smooth p–vector valued function, e.g. $u \in C^\infty(\mathbb{R}^n)$ and for a general $2m$–th order differential operator P, we have

$$\begin{aligned} P(u\chi_\Omega) &= \sum_{|\alpha|\le 2m} a_\alpha(x) D^\alpha(u\chi_\Omega)\\ &= \sum_{|\alpha|\le 2m} a_\alpha(x) \sum_{0\le\beta\le\alpha}\frac{\alpha!}{\beta!(\alpha-\beta)!}D^{\alpha-\beta}uD^\beta\chi_\Omega \\ &= Pu\chi_\Omega + \sum_{1\le|\alpha|\le 2m} a_\alpha(x)\sum_{0<\beta\le\alpha}\frac{\alpha!}{\beta!(\alpha-\beta)!}D^{\alpha-\beta}uD^\beta\chi_\Omega\,,\end{aligned} \tag{3.5.11}$$

where

$$D^\beta\chi_\Omega = -\sum_{\nu=0}^{|\beta|-1} c_{\beta\nu}\delta_\Gamma^{(\nu)} \text{ for } |\beta|\ge 1 \text{ and } D^0\chi_\Omega = \chi_\Omega\,.$$

The distribution $\delta_\Gamma^{(\nu+1)}$ is defined recursively by

$$\delta_\Gamma^{(\nu+1)} = (\boldsymbol{n}\cdot\nabla)\delta_\Gamma^{(\nu)}$$

according to (3.5.3). Similarly to (3.5.10) is can be shown that

$$\frac{\partial}{\partial x_i}\delta_\Gamma^{(\nu)} = n_i(x)\delta_\Gamma^{(\nu+1)} . \tag{3.5.12}$$

The coefficients $c_{\beta\nu}$ are given by the recursion relations

$$D^\beta \chi_\Omega = \frac{\partial}{\partial x_\ell} D^\alpha \chi_\Omega \text{ for } \beta = (\alpha_1, \dots, \alpha_\ell + 1, \dots, \alpha_n) ,$$

and by

$$\begin{aligned} c_{\beta,0} &= \frac{\partial}{\partial x_\ell} c_{\alpha,0} , \\ c_{\beta,\nu} &= c_{\alpha,\nu-1} n_\ell + \frac{\partial}{\partial x_\ell} c_{\alpha,\nu} \text{ for } \nu = 1, \dots, |\alpha| - 1 , \\ c_{\beta,|\beta|-1} &= c_{\beta,|\alpha|} = c_{\alpha,|\alpha|-1} n_\ell . \end{aligned} \tag{3.5.13}$$

For $|\alpha| = 1$ with $\alpha_\ell = 1$ and $\alpha_j = 0$ for $j \neq \ell$ we find from (3.5.10)

$$c_{\alpha,0} = n_\ell . \tag{3.5.14}$$

Hence, collecting and rearranging terms in (3.5.11), we obtain

$$P(u\chi_\Omega) = (Pu)\chi_\Omega - \sum_{1 \le |\alpha| \le 2m} a_\alpha(x) \sum_{0 < \beta \le \alpha} \frac{\alpha!}{\beta!(\alpha - \beta)!} D^{\alpha-\beta} u \sum_{\nu=0}^{|\beta|-1} c_{\beta,\nu} \delta_\Gamma^{(\nu)} . \tag{3.5.15}$$

By setting

$$c_{\beta,\nu} := 0 \text{ for } \nu \ge |\beta| , \tag{3.5.16}$$

we can rewrite equation (3.5.15) in the form

$$P(u\chi_\Omega) = (Pu)\chi_\Omega - \sum_{\nu=0}^{2m-1} Q_\nu u \delta_\Gamma^{(\nu)} \tag{3.5.17}$$

where

$$\begin{aligned} Q_\nu u &= \sum_{1 \le |\alpha| \le 2m} a_\alpha(x) \sum_{0 \le \gamma < \alpha} \frac{\alpha!}{(\alpha - \gamma)!\gamma!} c_{\alpha-\gamma,\nu}(x) D^\gamma u \\ &=: \sum_{|\mu| \le 2m-\nu-1} q_{\mu,\nu}(x) D^\mu u \end{aligned} \tag{3.5.18}$$

and where the coefficients $q_{\mu,\nu}$ are defined in an obvious manner by (3.5.18). In fact, for $m = 1$ and

$$a_\alpha(x) = \begin{cases} 1 & \text{for } \alpha = \alpha^{(j)} \text{ with } \alpha_i^{(j)} = 2\delta_{ij} \text{ and } i, j = 1, \dots, n, \\ 0 & \text{otherwise.} \end{cases}$$

we recover formula (3.5.8) from (3.5.17). The latter is the *second Green's formula* in distributional form, namely by multiplying (3.5.17) with $\phi \in (C_0^\infty(\mathbb{R}^n))^p$ and integrating, we get

$$\langle P(u\chi_\Omega), \phi\rangle_{\mathbb{R}^n} - \langle (Pu)\chi_\Omega, \phi\rangle_{\mathbb{R}^n} = -\sum_{\nu=0}^{2m-1} \langle (Q_\nu u)\delta_\Gamma^{(\nu)}, \phi\rangle_{\mathbb{R}^n}$$

and

$$\begin{aligned}
&\int_\Omega uP^\top \phi\, dy - \int_\Omega \phi Pu\, dy && (3.5.19)\\
&= -\sum_{\nu=0}^{2m-1} \int_\Gamma \Big(\frac{\partial}{\partial n}'\Big)^\nu (\phi\, Q_\nu u) ds_y\\
&= -\sum_{\nu=0}^{2m-1}\sum_{k=0}^{\nu} \binom{\nu}{k} \int_\Gamma \Big\{\Big(\frac{\partial}{\partial n}'\Big)^k \phi\Big\}\Big\{\Big(\frac{\partial}{\partial n}'\Big)^{\nu-k} Q_\nu u\Big\} ds_y\\
&= \sum_{\nu=0}^{2m-1}\sum_{k=0}^{\nu} \binom{\nu}{k} \int_\Gamma \Big\{\Big(\frac{\partial}{\partial n}'\Big)^k Q_\nu^{(\top)}\phi\Big\}\Big\{\Big(\frac{\partial}{\partial n}'\Big)^{\nu-k} u\Big\} ds_y\,,
\end{aligned}$$

where $Q_\nu^{(\top)}$ corresponds to the transposed differential operator $P^\top$ of P via (3.5.18). Obviously, formula (3.5.19) is also valid for any $\phi \in \big(C^\infty(\overline{\Omega})\big)$.

3.6 The Green Representation Formula

In this section we consider a scalar differential equation or a $2m$th–order uniformly strongly elliptic system, in the sense of Petrovski, in the domain Ω or $\Omega^c = \mathbb{R}^n \setminus \overline{\Omega}$, respectively,

$$Pu := \sum_{|\alpha|\le 2m} a_\alpha(x) D^\alpha u = f \tag{3.6.1}$$

with real $p \times p$ coefficient matrices a_α.

Definition 3.6.1. (see Mikhlin [300], Miranda [307, p.244] and Stephan et al [405]). *The linear operator P is called* s t r o n g l y e l l i p t i c *in Ω if there exists a smooth matrix–valued function $\Theta(x)$ and $\gamma_0(x) > 0$ such that*

$$Re\{\zeta^\top \Theta(x) \sum_{|\alpha|=2m} a_\alpha(x)\xi^\alpha \overline{\zeta}\} \ge \gamma_0 |\xi|^{2m}|\zeta|^2 \tag{3.6.2}$$

for all $\xi \in \mathbb{R}^n$ and all $\zeta \in \mathbb{C}^p$ is satisfied where p denotes the dimension of the vector valued functions u . If, in addition, $\gamma_0 > 0$ is a constant independent of x and (3.6.2) holds for all $x \in \Omega$, then P is called u n i f o r m l y s t r o n g l y e l l i p t i c.

In many applications one has $\Theta(x) \equiv 1$ with the unity matrix I. Note that strongly elliptic real differential operators are of even order and are properly elliptic (see Wloka [439, Def. 9.24]).

The formula (3.5.19) can be employed to obtain a general Green representation formula for the solution of (3.6.1) by using a fundamental solution.

Definition 3.6.2. *Let $x \in \mathbb{R}^n$ be any chosen point. Then the distribution $E(x,y)$ is called a fundamental solution of the differential operator $P^\top$ (in $\mathbb{R}^n$) if it satisfies the equation*

$$P_y^\top E(x,y) = \delta_x(y) := \delta(y-x) \tag{3.6.3}$$

in the distributional sense where $\delta_x(y)$ is the Dirac distribution given by

$$\langle \delta_x, \varphi \rangle_{\mathbb{R}^n} = \varphi(x) \text{ for all } \varphi \in C_0^\infty(\mathbb{R}^n)\,.$$

(As usual, the notation $P_y^\top$ stands for differentiation with respect to y.)

For a general differential operator $P^\top$, the existence of a fundamental solution is by no means trivial, see Section 7.3.

For strongly elliptic operators it can be shown with the Green formula (3.5.19) that (3.6.3) implies

$$P_x E(x,y)^\top = \delta_y(x) \text{ for any fixed } y \in \mathbb{R}^n\,. \tag{3.6.4}$$

Hence, if $E_x(y) := E(x,y)$ is the fundamental solution of $P_y^\top$ then $E_y(x) = E(x,y)^\top$ is the fundamental solution of P_x.

Lemma 3.6.1. *Let P be a uniformly strongly elliptic differential operator of order $2m$ with real C^∞ coefficients $a_\alpha(x)$. Then for every compact region $\overline{\Omega} \Subset \mathbb{R}^n$ with $\partial\Omega = \Gamma \in C^\infty$ there exists a local fundamental solution $E(x,y)$ which is a C^∞ function of all variables for $x \neq y$ and $x, y \in \overline{\Omega}$.*

We shall come back to this topic in Section 7.3 and also refer for the interested reader to the proofs given by Hörmander [188, III Theorem 17.1.1].

Hence, in what follows we shall always assume that a fundamental solution exists, either for $\overline{\Omega}$ or the the whole space $\mathbb{R}^n$. With the fundamental solution E of $P^\top$ as test function ϕ in (3.5.19) we obtain a desired *representation formula* for any fixed $x \in \Omega$:

$$\begin{aligned} u(x) &= \int\limits_\Omega E(x,y)^\top f(y)dy \\ &+ \sum_{\nu=0}^{2m-1}\sum_{k=0}^{\nu}\binom{\nu}{k}\int\limits_\Gamma \Big\{\Big(\frac{\partial}{\partial n}'\Big)^k Q_\nu^{(\top)} E(x,y)\Big\}^\top \Big\{\Big(\frac{\partial}{\partial n}'\Big)^{\nu-k} u\Big\} ds_y\,. \end{aligned} \tag{3.6.5}$$

This formula, however, still needs to be justified since $E(x,y)$ is not in $C_0^\infty(\mathbb{R}^n)$. Therefore, let us introduce a cut–off function $\chi \in C_0^\infty(\Omega)$ with the properties

$$\chi(y) = \begin{cases} 1 & \text{for } |y-x| \le \delta_0\,, \\ 0 & \text{for } |y-x| \ge 2\delta_0 \end{cases}$$

for fixed $x \in \Omega$ where $\delta_0 > 0$ is an appropriately chosen constant depending on x and Ω. Then

$$E(x,y) = \chi E + (1-\chi)E$$

and χE is a well defined distribution whereas $\phi(y) := (1-\chi)E \in C^\infty(\mathbb{R}^n)$ can be used in formula (3.5.19). This yields

$$\int_\Omega (1-\chi)E^\top (Pu)dy - \int_\Omega uP^\top((1-\chi)E)dy \tag{3.6.6}$$
$$+ \sum_{\nu=0}^{2m-1}\sum_{k=0}^{\nu}\binom{\nu}{k}\int_\Gamma \Big\{\Big(\frac{\partial}{\partial n_y}'\Big)^k Q_\nu^{(\top)} E(x,y)\Big\}^\top \Big\{\Big(\frac{\partial}{\partial n}'\Big)^{\nu-k} u\Big\} ds_y = 0$$

since $\chi(y) \equiv 0$ along Γ. On the other hand, by definition of distributional derivatives and by $\chi \in C_0^\infty(\Omega)$, we have

$$\langle \chi E, Pu\rangle = \langle P^\top(\chi E), u\rangle\,,$$

i.e.

$$\int_\Omega (\chi E)^\top Pudy - \int_\Omega uP^\top(\chi E)dy = 0\,. \tag{3.6.7}$$

Since $\chi \equiv 1$ for $|x-y| \le \delta_0$ we find with $P_y^\top E(x,y) = 0$ for $y \ne x$,

$$\begin{aligned} \int_\Omega uP_y^\top(\chi E)dy &= \int_{|x-y|\le\delta_0} uP_y^\top E(x,y)dy + \int_{\delta_0\le|x-y|\wedge y\in\Omega} uP^\top(\chi E)dy \\ &= \langle P^\top E, u\rangle_{\mathbb{R}^n} + \int_{\delta_0\le|x-y|\wedge y\in\Omega} uP^\top(\chi E)dy\,, \end{aligned}$$

formula (3.6.7) becomes

$$\int_\Omega (\chi E)^\top Pudy - \int_{\delta_0\le|x-y|\wedge y\in\Omega} uP^\top(\chi E)dy = u(x)\,. \tag{3.6.8}$$

Now, formula (3.6.5) follows immediately by adding (3.6.6) and (3.6.8) since $(1-\chi) \equiv 0$ for $|x-y| \le \delta_0$ and $P_y^\top E = 0$ for $x \ne y$.

In order to use the Cauchy data $(\frac{\partial}{\partial n})^\nu u_{|\Gamma}$ in the representation formula (3.6.5), we need to replace $(\frac{\partial}{\partial n}')^\lambda$ by using (3.5.5) which yields

$$u(x) = \int_\Omega E(x,y)^\top f(y)dy + \sum_{\lambda=0}^{2m-1}\int_\Gamma \{T_{\lambda y}E(x,y)\}^\top \Big\{\Big(\frac{\partial}{\partial n}\Big)^\lambda u(y)\Big\} ds_y \tag{3.6.9}$$

where the boundary operators $T_{\lambda y}$ are defined by the equations

$$\sum_{\lambda=0}^{2m-1} (T_\lambda \phi)^\top \Big(\frac{\partial}{\partial n}\Big)^\lambda u_{|\Gamma} = \tag{3.6.10}$$
$$\sum_{\nu=0}^{2m-1}\sum_{k=0}^{\nu}\sum_{\ell=0}^{\nu-k} \binom{\nu}{k}(-1)^\ell \binom{\nu-k}{\ell}\Big(\Big(\frac{\partial}{\partial n}\Big)^{\nu-k-\ell} 2H\Big)\Big(\Big(\frac{\partial}{\partial n}\Big)'^k Q_\nu^{(\top)}\phi\Big)^\top \Big(\frac{\partial}{\partial n}\Big)^\ell u_{|\Gamma}\,.$$

This formula will be used for defining general boundary potentials and Calderón projectors as introduced in Chapters 1 and 2 for special differential operators.

Now, let $x \in \Omega^c$. Then (3.6.3) yields

$$P_y^\top E(x,y) = 0 \text{ for } y \in \Omega$$

and from (3.5.19), in exactly the same manner as before, one obtains the equation

$$\begin{aligned} 0 &= \int_\Omega E(x,y)^\top Pu(y)dy \\ &\quad + \sum_{\nu=0}^{2m-1}\sum_{k=0}^{\nu}\binom{\nu}{k}\int_\Gamma \Big\{\Big(\frac{\partial}{\partial n_y}\Big)'^k Q_{\nu y}^{(\top)} E(x,y)\Big\}^\top \Big\{\Big(\frac{\partial}{\partial n}\Big)'^{\nu-k} u\Big\} ds_y \\ &= \int_\Omega E(x,y)^\top Pu(y)dy \\ &\quad + \sum_{\lambda=0}^{2m-1}\int_\Gamma (T_{\lambda y}E(x,y))^\top \Big(\Big(\frac{\partial}{\partial n}\Big)^\lambda u\Big) ds_y \text{ for every } x\in\Omega^c \end{aligned} \tag{3.6.11}$$

and for any $u \in C^\infty(\overline{\Omega})$.

For exterior problems one can proceed correspondingly. For $u \in C^\infty(\mathbb{R}^n)$, one has with (3.5.17) the equation

$$\begin{aligned} P(u\chi_{\Omega^c}) &= P(u(1-\chi_\Omega)) \\ &= Pu - P(u\chi_\Omega) \\ &= Pu\chi_{\overline{\Omega}^c} + \sum_{\nu=0}^{2m-1}(Q_\nu u)\delta_\Gamma^{(\nu)}\,. \end{aligned} \tag{3.6.12}$$

Hence, if $R > 0$ is chosen large enough such that

$$B_R := \{y \in \mathbb{R}^n \,\big|\, |y| < R\} \supset \overline{\Omega}$$

then (3.5.19) for $\overline{\Omega}^c \cap B_R$ gives

$$\int\limits_{\overline{\Omega}^c \cap B_R} u P^\top \phi \, dy - \int\limits_{\overline{\Omega}^c \cap B_R} \phi P u dy$$

$$= \sum_{\nu=0}^{2m-1} \int\limits_{\Gamma} \Big(\frac{\partial}{\partial n}'\Big)^{\nu} (\phi \, Q_\nu u) ds_y - \sum_{\nu=0}^{2m-1} \int\limits_{\partial B_R} \Big(\frac{\partial}{\partial n}'\Big)^{\nu} (\phi \, Q_\nu) ds_y$$

for every test function $\phi \in C^\infty(\mathbb{R}^n)$ and every R large enough. If $\phi \in C_0^\infty(\mathbb{R}^n)$ has compact support then the last integral over ∂B_R vanishes for $\operatorname{supp} \phi \subset B_R$. In order to obtain the Green representation formula in Ω^c, we again use $E(x, y)$ instead of $\phi(y)$ and in exactly the same manner as before we now find

$$\begin{aligned} u(x) &= \int\limits_{\overline{\Omega}^c \cap B_R} E(x,y)^\top P u(y) dy \\ &\quad - \sum_{\lambda=0}^{2m-1} \int\limits_{\Gamma} \{T_{\lambda y} E(x,y)\}^\top \Big\{\Big(\frac{\partial}{\partial n}\Big)^{\lambda} u(y)\Big\} ds_y \qquad (3.6.13) \\ &\quad + \sum_{\lambda=0}^{2m-1} \int\limits_{\Gamma_R} \{T_{\lambda y} E(x,y)\}^\top \Big\{\Big(\frac{\partial}{\partial n}\Big)^{\lambda} u(y)\Big\} ds_y \text{ for } x \in \Omega^c \cap B_R \end{aligned}$$

and for every $u \in C^\infty(\overline{\Omega}^c)$ and every R large enough, $\Gamma_R = \partial B_R$.

If the right–hand side f of the differential equation

$$Pu = f \text{ in } \Omega^c \qquad (3.6.14)$$

has compact support then the limit of the volume integral in (3.6.13) exists for $R \to \infty$ and, in this case, for any solution of (3.6.14) the existence of the limit

$$M(x; u) := \lim_{R\to\infty} \sum_{\lambda=0}^{2m-1} \int\limits_{\Gamma_R} \{T_{\lambda y} E(x,y)\}^\top \Big\{\Big(\frac{\partial}{\partial n}\Big)^{\lambda} u(y)\Big\} ds_y \qquad (3.6.15)$$

follows from (3.6.13). Since all the operators in (3.6.13) are linear, for any fixed $x \in \Omega^c$, the limit in (3.6.15) defines a linear functional of u. Its value depends on the behaviour of u at ∞, i.e., on the radiation conditions imposed. For instance, from the exterior Dirichlet and Neumann problems for the Helmholtz equation subject to the Sommerfeld condition we have

$$M(x; u) = \omega(u)$$

as was explained in Chapter 2.2, whereas for exterior problems of elasticity we have

$$M(x; u) = \sum_{j=1}^{3(n-1)} \omega_j(u) \boldsymbol{m}_j(x)$$

for $n = 2$ and 3 dimensions due to (2.2.11).

An immediate consequence of (3.6.4) is

$$P_x M(x;u) = 0 \text{ for all } x \in \mathbb{R}^n \tag{3.6.16}$$

and for every solution of (3.6.14) with f having compact support. Collecting the above arguments, we find in this case the representation formula for *exterior problems*,

$$\begin{aligned} u(x) &= \int\limits_{\Omega^c} E(x,y)^\top f(y)dy + M(x;u) \\ &\quad - \sum_{\lambda=0}^{2m-1} \int\limits_{\Gamma} \{T_{\lambda y} E(x,y)\}^\top \Big\{\Big(\frac{\partial}{\partial n}\Big)^\lambda u(y)\Big\} ds_y \text{ for } x \in \Omega^c . \end{aligned} \tag{3.6.17}$$

Finally, we find from (3.6.4) in the same manner the identity

$$\begin{aligned} 0 = &\int\limits_{\Omega^c} E(x,y)^\top f(y)dy + M(x;u) \\ &- \sum_{\lambda=0}^{2m-1} \int\limits_{\Gamma} \{T_{\lambda y} E(x,y)\}^\top \Big\{\Big(\frac{\partial}{\partial n}\Big)^\lambda u(y)\Big\} ds_y \text{ for } x \in \Omega . \end{aligned} \tag{3.6.18}$$

3.7 Green's Representation Formulae in Local Coordinates

The differential operators T_λ in the representation formulae (3.6.9), (3.6.13) and (3.6.17) and in the identities (3.6.11) and (3.6.18) can be rewritten in terms of local coordinates in the neighbourhood of Γ by using the parametric representation of Γ. This will allow us to obtain appropriate properties of the corresponding Calderón projections as in the special cases considered in Chapters 1 and 2.

In order to simplify the Green representation formulae introduced in Section 3.6 we introduce tensor distributions subordinate to the local coordinates $\sigma = (\sigma', \sigma_n)$ or $\tau = (\tau', \tau_n)$ of the previous sections.

For simplicity, we present here the case of Hadamard's tubular local coordinates (σ', σ_n). Let $\Omega_1 \subset \mathbb{R}^{n-1}$ and $\Omega_2 \subset \mathbb{R}$ be open subsets. If $\varphi \in C_0^\infty(\Omega_1)$ and $\psi \in C_0^\infty(\Omega_2)$, the *tensor product* $\varphi \otimes \psi$ of φ and ψ is defined by

$$(\varphi \otimes \psi)(\sigma', \sigma_n) := \varphi(\sigma')\psi(\sigma_n) \tag{3.7.1}$$

which is a $C_0^\infty(\Omega_1 \times \Omega_2)$–function. The corresponding space $C_0^\infty(\Omega_1) \otimes C_0^\infty(\Omega_2)$ is a sequentially dense linear subspace of $C_0^\infty(\Omega_1 \times \Omega_2)$, that is, for every $\Phi \in C_0^\infty(\Omega_1 \times \Omega_2)$, there exists a sequence $\{\Phi_j\}$ in $C_0^\infty(\Omega_1) \otimes C_0^\infty(\Omega_2)$

such that $\Phi_j = \varphi_j(\sigma') \otimes \psi_j(\sigma_n) \to \Phi$ in $C_0^\infty(\Omega_1 \times \Omega_2)$ (Friedlander [128, Section 4.3] and Schwartz [382]).

The tensor product can be extended to distributions; and the resulting distributional tensor product then satisfies

$$\langle u \otimes v\,,\, \varphi \otimes \psi \rangle_{\mathbb{R}^n} = \langle u, \varphi \rangle_{\mathbb{R}^{n-1}} \langle v, \psi \rangle_{\mathbb{R}} \tag{3.7.2}$$

where u and v are distributions on Ω_1 and Ω_2, respectively. By density arguments, the equation (3.7.1) extends to

$$\begin{aligned} \langle (u \otimes v), \Phi \rangle_{\mathbb{R}^n} &= \big\langle u, \langle v, \Phi(\sigma', \bullet) \rangle_{\mathbb{R}} \big\rangle_{\mathbb{R}^{n-1}} \\ &= \big\langle v, \langle u, \Phi(\bullet, \sigma_n) \rangle_{\mathbb{R}^{n-1}} \big\rangle_{\mathbb{R}} \end{aligned} \tag{3.7.3}$$

for all $\Phi \in C_0^\infty(\Omega_1 \times \Omega_2)$. The basic properties of the tensor product are

$$\operatorname{supp}(u \otimes v) = \operatorname{supp}(u) \times \operatorname{supp}(v) \tag{3.7.4}$$

and

$$D_{\sigma'}^\alpha D_{\sigma_n}^\beta (u \otimes v) = (D^\alpha u) \otimes (D^\beta v)\,. \tag{3.7.5}$$

In order to make use of (3.7.5) we first use the separation of the differential operator P in the neighbourhood of Γ in the form (3.4.49).

We begin with the following jump formula:

Lemma 3.7.1. *For $k \in \mathbb{N}$ and for u and Γ sufficiently smooth one has the jump formula*

$$\frac{\partial^k}{\partial n^k}(u \chi_\Omega) = \left(\frac{\partial^k u}{\partial n^k} \right) \chi_\Omega - \sum_{\ell=0}^{k-1} \frac{\partial^\ell u}{\partial n^\ell}\Big|_\Gamma \otimes \delta_\Gamma^{(k-\ell-1)}\,. \tag{3.7.6}$$

Proof: The proof is done by induction with respect to k. For $k = 1$ we have from (3.5.6)

$$\begin{aligned} \frac{\partial}{\partial n}(u \chi_\Omega) &= \frac{\partial u}{\partial n} \chi_\Omega + u \boldsymbol{n}(x) \cdot \nabla \chi_\Omega \\ &= \frac{\partial u}{\partial n} \chi_\Omega - u \delta_\Gamma(x) \\ &= \frac{\partial u}{\partial n} \chi_\Omega - (u|_\Gamma \otimes \delta_\Gamma)\,. \end{aligned}$$

Now, if (3.7.6) holds for k, we find

$$\begin{aligned} \frac{\partial}{\partial n}\Big(\frac{\partial^k}{\partial n^k}(u \chi_\Omega) \Big) &= \frac{\partial}{\partial n}\Big\{ \frac{\partial^k u}{\partial n^k} \chi_\Omega - \sum_{\ell=0}^{k-1} \Big(\frac{\partial^\ell u}{\partial n^\ell}\Big|_\Gamma \otimes \delta_\Gamma^{k-\ell-1} \Big) \Big\} \\ &= \Big(\frac{\partial^{k+1} u}{\partial n^{k+1}} \Big) \chi_\Omega - \Big(\frac{\partial^k u}{\partial n^k}\Big|_\Gamma \otimes \delta_\Gamma \Big) \\ &\quad - \frac{\partial}{\partial \sigma_n} \sum_{\ell=0}^{k-1} \Big(\frac{\partial^\ell u}{\partial \sigma_n^\ell}\Big|_\Gamma \otimes \delta_\Gamma^{(k-\ell-1)} \Big) \end{aligned}$$

and, with (3.7.5), the desired formula (3.7.6) for $k+1$:

$$\frac{\partial^{k+1}}{\partial n^{k+1}}(u\chi_\Omega) = \Big(\frac{\partial^{k+1} u}{\partial n^{k+1}}\Big)\chi_\Omega - \sum_{\ell=0}^{k}\Big(\frac{\partial^\ell u}{\partial n^\ell}\Big|_\Gamma \otimes \delta_\Gamma^{k-\ell}\Big)\,.$$

■

In terms of tensor products, Formula (3.5.8) can now be rewritten in the form of the jump relation

$$\Delta(u\chi_\Omega) = (\Delta u)\chi_\Omega + \Big(u|_\Gamma \otimes \Big(\frac{\partial}{\partial n}'\delta_\Gamma\Big)\Big) - 2\Big(\frac{\partial u}{\partial n}\Big|_\Gamma \otimes \delta_\Gamma\Big)\,.$$

Now we consider the jump relation for any differential operator P as in (3.4.49), namely

$$Pu(x(\sigma)) = \sum_{k=0}^{2m}\mathcal{P}_k\frac{\partial^k u}{\partial n^k} = \sum_{k=0}^{2m}\mathcal{P}_k\Big(\sigma;\frac{\partial}{\partial\sigma'}\Big)\frac{\partial^k u}{\partial\sigma_n^k}$$

in the neighbourhood of Γ where $\mathcal{P}_k$ is the differential operator (3.4.50) of order $2m-k$ acting along the parallel surfaces σ_n =const. By virtue of Lemma 3.7.1 we have

$$\begin{aligned}
P(u\chi_\Omega) &= \sum_{k=0}^{2m}\mathcal{P}_k\Big(\sigma;\frac{\partial}{\partial\sigma'}\Big)\frac{\partial^k}{\partial\sigma_n^k}(u\chi_\Omega)\\
&= \mathcal{P}_0(u\chi_\Omega)\\
&\quad+\sum_{k=1}^{2m}\mathcal{P}_k\Big(\sigma;\frac{\partial}{\partial\sigma'}\Big)\Big\{\Big(\frac{\partial^k u}{\partial n^k}\Big)\chi_\Omega - \sum_{\ell=0}^{k-1}\Big(\frac{\partial^\ell u}{\partial n^\ell}\Big|_\Gamma\otimes\delta_\Gamma^{(k-\ell-1)}\Big)\Big\}\\
&= \Big(\sum_{k=0}^{2m}\mathcal{P}_k\frac{\partial^k u}{\partial n^k}\Big)\chi_\Omega - \sum_{k=1}^{2m}\sum_{\ell=0}^{k-1}\Big(\mathcal{P}_k\Big(\sigma;\frac{\partial}{\partial\sigma'}\Big)\frac{\partial^\ell u}{\partial n^\ell}\Big|_\Gamma\Big)\otimes\delta_\Gamma^{(k-\ell-1)}\,.
\end{aligned}$$

By interchanging the order of summation between k and ℓ and afterwards setting $p = k-\ell-1$ we arrive at

$$P(u\chi_\Omega) = (Pu)\chi_\Omega - \sum_{\ell=0}^{2m-1}\sum_{p=0}^{2m-\ell-1}\Big(\Big(\mathcal{P}_{p+\ell+1}\frac{\partial^\ell u}{\partial n^\ell}\Big)\Big|_\Gamma\otimes\delta_\Gamma^{(p)}\Big) \tag{3.7.7}$$

similar to (3.5.17), the equation (3.7.7) is again the second Green formula in distributional form, which yields

$$\begin{aligned}
&\int_\Omega uP^\top\varphi dx - \int_\Omega \varphi Pu dx\\
&\quad= -\sum_{\ell=0}^{2m-1}\sum_{p=0}^{2m-\ell-1}\int_\Gamma\Big(\mathcal{P}_{p+\ell+1}\frac{\partial^\ell u}{\partial n^\ell}\Big)\Big(\frac{\partial}{\partial n}'\Big)^p\varphi ds
\end{aligned} \tag{3.7.8}$$

for any $\varphi \in C^\infty(\bar{\Omega})$. This formula evidently is much simpler than (3.5.19). With (3.5.5) and (3.4.31) one also obtains

$$\Big(\frac{\partial}{\partial n}'\Big)^p \varphi = \Big(-\frac{1}{\sqrt{\gamma}}\frac{\partial}{\partial n}\sqrt{g}\Big)^p \varphi\,. \tag{3.7.9}$$

In particular, the tangential operators $\mathcal{P}_k$ appear in the same way as in the differential equation (3.4.49) operating only on the Cauchy data of u.

As in Section 3.6, one may replace φ in (3.7.8) by the fundamental solution $E(x,y)$ of the differential operator $P^\top$, satisfying equations (3.6.3) and (3.6.4). This yields the representation formulae

$$u(x) = \int\limits_\Omega E(x,y)^\top f(y)dy \tag{3.7.10}$$
$$- \sum_{\ell=0}^{2m-1}\sum_{p=0}^{2m-\ell-1} \int\limits_\Gamma \Big\{\Big(\frac{\partial}{\partial n_y}'\Big)^p E(x,y)\Big\}^\top \Big\{\mathcal{P}_{p+\ell+1}\frac{\partial^\ell u}{\partial n^\ell}(y)\Big\}ds_y \text{ for } x\in\Omega$$

and

$$0 = \int\limits_\Omega E(x,y)^\top f(y)dy \tag{3.7.11}$$
$$- \sum_{\ell=0}^{2m-1}\sum_{p=0}^{2m-\ell-1} \int\limits_\Gamma \Big\{\Big(\frac{\partial}{\partial n_y}'\Big)^p E(x,y)\Big\}^\top \Big\{\mathcal{P}_{p+\ell+1}\frac{\partial^\ell u}{\partial n^\ell}(y)\Big\}ds_y \text{ for } x\in\Omega^c\,.$$

For the representation formula of the solution of $Pu = f$ in Ω^c, we again require the existence of the limit

$$M(x;u) := \tag{3.7.12}$$
$$-\lim_{R\to\infty}\sum_{\ell=0}^{2m-1}\sum_{p=0}^{2m-\ell-1} \int\limits_\Gamma \Big\{\Big(\frac{\partial}{\partial n_y}'\Big)^p E(x,y)\Big\}^\top \Big\{\mathcal{P}_{p+\ell+1}\frac{\partial^\ell u}{\partial n^\ell}(y)\Big\}ds_y\,.$$

If $f \in C_0^\infty(\mathbb{R}^n)$, then the existence of $M(x;u)$ follows from the representation formula (3.7.10) with Ω replaced by $\Omega^c \cap B_R$ as in Section 4.2; and (3.6.16) remains the same for the present case.

The exterior representation formulae (3.6.17) and (3.6.18) now read as

$$u(x) = \int\limits_{\Omega^c} E(x,y)^\top f(y)dy + M(x;u) \tag{3.7.13}$$
$$+ \sum_{\ell=0}^{2m-1}\sum_{p=0}^{2m-\ell-1} \int\limits_\Gamma \Big\{\Big(\frac{\partial}{\partial n_y}'\Big)^p E(x,y)\Big\}^\top \Big\{\mathcal{P}_{p+\ell+1}\frac{\partial^\ell u}{\partial n^\ell}(y)\Big\}ds_y$$

for $x \in \Omega^c$ and

$$0 = \int_{\Omega^c} E(x,y)^\top f(y)dy + M(x;u) \qquad (3.7.14)$$
$$+ \sum_{\ell=0}^{2m-1} \sum_{p=0}^{2m-\ell-1} \int_\Gamma \Big\{\Big(\frac{\partial}{\partial n_y}{}'\Big)^p E(x,y)\Big\}^\top \Big\{\mathcal{P}_{p+\ell+1}\frac{\partial^\ell u}{\partial n^\ell}(y)\Big\}ds_y$$

for $x \in \Omega$.

We note that the representation formulae hold equally well for systems with p equations and p unknowns, such as in elasticity where $p = n$.

We remark that similar formulae can be obtained when using the local coordinates $\tau = (\tau', \tau_n)$ as in the Appendix.

3.8 Multilayer Potentials

In this section we follow closely Costabel et al [90].

From the representation formulae such as (3.6.9), (3.7.10) and (3.7.13), the boundary integrals all define boundary potentials. For simplicity, we assume $\Gamma \in C^\infty$ and consider first (3.7.10) by defining the corresponding multiple layer potentials:

$$(\mathcal{K}_p\phi)(x) := \int_\Gamma \Big\{\Big(\frac{\partial}{\partial n_y}{}'\Big)^p E(x,y)\Big\}^\top \phi(y)ds_y \text{ for } x \in \Omega \text{ or } x \in \Omega^c \quad (3.8.1)$$

for the density $\phi \in C^\infty(\Gamma)$. Using the tensor product of distributions, we can write (3.8.1) as

$$\begin{aligned}\mathcal{K}_p\phi(x) &:= \int_{\mathbb{R}^n} E(x,y)^\top \Big(\phi \otimes \Big(\frac{\partial}{\partial n}\Big)^p \delta_\Gamma\Big)dy \\ &= \langle E(x,\bullet)\,,\, \phi\Big|_\Gamma \otimes \delta_\Gamma^{(p)}\rangle_{\mathbb{R}^n}\end{aligned} \qquad (3.8.2)$$

for $x \notin \Gamma$ since then $E(x,y)$ is a smooth function of $y \in \Gamma$ and x is in its vicinity. As we will see in Chapter 9, this defines a pseudodifferential operator whose mapping properties in Sobolev spaces then follow from corresponding general results. We shall come back to this point later on.

Clearly, the boundary potentials (3.8.1) are in $C^\infty(\mathbb{R}^n \setminus \Gamma)$ for any given $\phi \in C^\infty(\Gamma)$. Moreover, for $x \notin \Gamma$ we have

$$P\mathcal{K}_p\phi(x) = \int_\Gamma P_x\Big\{\Big(\frac{\partial'}{\partial n_y}\Big)^p E^\top(x,y)\Big\}\phi(y)ds_y = 0 \qquad (3.8.3)$$

since here $x \neq y$ for $y \in \Gamma$ and; therefore, integration and differentiation can be interchanged. Moreover,

$$P \int_\Omega E(x,y)^\top f(y)dy = f(x) \quad \text{for all } x \in \Omega\,. \tag{3.8.4}$$

In terms of the above potentials, the representation formula (3.7.10) now reads for $u \in C^\infty(\overline{\Omega})$ and $x \in \Omega$:

$$u(x) = \int_\Omega E(x,y)^\top f(y)dy - \sum_{\ell=0}^{2m-1} \sum_{p=0}^{2m-\ell-1} \mathcal{K}_p\Big(\mathcal{P}_{p+\ell+1}\frac{\partial^\ell u}{\partial n^\ell}\Big)(x)\,. \tag{3.8.5}$$

In addition, as we already have seen in Chapters 1 and 2, these potentials admit boundary traces as x approaches the boundary Γ from Ω or Ω^c, respectively, which are not equal in general. More specifically, the following lemma is needed and will be proved in Chapter 9, Theorem 9.5.8.

Lemma 3.8.1. *Let E be the fundamental solution of P with C^∞–coefficients defined in Lemma* 3.6.1. *Let $\Gamma \in C^\infty$ and $\varphi \in C^\infty(\Gamma)$. Then the multilayer potential $\mathcal{K}_p\varphi(x)$ can be extended from $x \in \Omega$ up to the boundary $x \in \overline{\Omega}$ in $C^\infty(\overline{\Omega})$ defining a continuous mapping*

$$\mathcal{K}_p : C^\infty(\Gamma) \to C^\infty(\overline{\Omega})\,.$$

Similarly, $\mathcal{K}_p\varphi(x)$ can be extended from $x \in \Omega^c$ to $\overline{\Omega^c}$ defining a continuous mapping

$$\mathcal{K}_p : C^\infty(\Gamma) \to C^\infty(\overline{\Omega^c})\,.$$

By using the traces from Ω of both sides in equation (3.8.5) we obtain the relation

$$\begin{aligned}\gamma u(x) =& \gamma \int_\Omega E(x,y) f(y)dy \\ & - \sum_{\ell=0}^{2m-1} \sum_{p=0}^{2m-\ell-1} \gamma\mathcal{K}_p\Big(\mathcal{P}_{p+\ell+1}\frac{\partial^\ell u}{\partial n^\ell}\Big)(x) \text{ for } x \in \Gamma\,,\end{aligned} \tag{3.8.6}$$

where the trace operator γu is defined by

$$\gamma u(x) = (\gamma_0, \ldots, \gamma_{2m-1})u(x) := \lim_{\Omega \ni z \to x \in \Gamma} \Big(u(z), \frac{\partial u(z)}{\partial n}, \ldots, \frac{\partial^{2m-1}u(z)}{\partial n^{2m-1}}\Big)^\top \tag{3.8.7}$$

for $x \in \Gamma$ and $u \in C^{2m}(\overline{\Omega})$. Of course, $\gamma\mathcal{K}_p$ needs to be analyzed carefully since it contains the jump relations (see e.g. (1.2.3) – (1.2.6)).

The second term on the right–hand side of (3.8.6) defines a boundary integral operator

$$\mathcal{Q}_\Omega \mathcal{P}\psi(x) := \sum_{\ell=0}^{2m-1} \sum_{p=0}^{2m-1-\ell} \gamma\mathcal{K}_p(\mathcal{P}_{p+\ell+1}\psi_\ell)(x) \tag{3.8.8}$$

where both, $\mathcal{P}$ and $\mathcal{Q}_\Omega$ themselves may be considered as matrix–valued operators with

$$\mathcal{P} = \begin{pmatrix} \mathcal{P}_1 & \mathcal{P}_2 & \mathcal{P}_3 & \dots & \mathcal{P}_{2m} \\ \mathcal{P}_2 & \mathcal{P}_3 & & & 0 \\ \mathcal{P}_3 & & & & 0 \\ \vdots & & & & \vdots \\ \mathcal{P}_{2m} & 0 & 0 & \dots & \end{pmatrix} \tag{3.8.9}$$

and

$$\mathcal{Q}_\Omega = \begin{pmatrix} \gamma_0\mathcal{K}_0 & \dots & \gamma_0\mathcal{K}_{2m-1} \\ \vdots & \vdots & \\ \gamma_{2m-1}\mathcal{K}_0 & \dots & \gamma_{2m-1}\mathcal{K}_{2m-1} \end{pmatrix} = \begin{pmatrix} \gamma_0 \\ \vdots \\ \gamma_{2m-1} \end{pmatrix} (\mathcal{K}_0, \cdots \mathcal{K}_{2m-1}) \,. \tag{3.8.10}$$

Again, the boundary integral operator

$$\mathcal{C}_\Omega \psi := -\mathcal{Q}_\Omega \mathcal{P} \psi \tag{3.8.11}$$

defines the *Calderón projector* associated with the partition of the differential operator P in (3.4.49) for the domain Ω (see Calderón [59] and Seeley [384]). For the Laplacian, it was given explicitly in (1.2.20).

If ψ is specifically chosen to be $\psi = \gamma u$ as the trace of a solution to the homogeneous partial differential equation

$$Pu = \sum_{|\alpha| \le 2m} a_\alpha(x) D^\alpha u = 0 \tag{3.8.12}$$

in Ω, then (3.8.6) yields with (3.8.11) the identity

$$\gamma u = \mathcal{C}_\Omega \gamma u \,. \tag{3.8.13}$$

Consequently, because of (3.8.6),

$$u = -\sum_{\ell=0}^{2m-1} \sum_{p=0}^{2m-\ell-1} \mathcal{K}_p(\mathcal{P}_{p+\ell+1}\psi_\ell)$$

is a special solution of (3.8.12), hence we find

$$\mathcal{C}_\Omega^2 \psi = \mathcal{C}_\Omega \gamma u = \gamma u = \mathcal{C}_\Omega \psi \tag{3.8.14}$$

for every $\psi \in (C^\infty(\overline{\Omega}))^{2m}$. Hence, $\mathcal{C}_\Omega$ is a projection of general boundary functions $\psi|_\Gamma$ onto the Cauchy data γu of solutions to the homogeneous differential equation (3.8.12).

On the other hand, the relation (3.8.13) gives a *necessary compatibility condition* for the Cauchy data of any solution u of the differential equation (3.8.12). For solutions of the inhomogeneous partial differential equation

$$Pu = f \tag{3.8.15}$$

in Ω we have the corresponding necessary compatibility condition

$$\gamma u = \gamma \int_{\Omega} E(x,y)^\top f(x) dy + \mathcal{C}_\Omega \gamma u \,. \tag{3.8.16}$$

In the same manner, for the exterior domain we may consider the corresponding Calderón projector

$$\mathcal{C}_{\Omega^c}\psi := \mathcal{Q}_{\Omega^c}\mathcal{P}\psi := \sum_{\ell=0}^{2m-1}\sum_{p=0}^{2m-1-\ell} \gamma_c \mathcal{K}_p(\mathcal{P}_{p+\ell+1}\psi_\ell) \tag{3.8.17}$$

where γ_c now denotes the trace operator with respect to Ω^c, i.e.,

$$(\gamma_{c0},\ldots,\gamma_{c,2m-1})u(x) = \gamma_c u(x) := \lim_{\Omega^c \ni z \to x \in \Gamma} \Big(u(z), \frac{\partial u(z)}{\partial n}, \ldots, \frac{\partial^{2m-1} u}{\partial n^{2m-1}}(z)\Big)^\top$$

on Γ. Here, $\mathcal{Q}_{\Omega^c}$ is defined by (3.8.10) with γ replaced by γ_c. For solutions satisfying the radiation condition (3.7.12) we find the necessary compatibility condition for its trace:

$$\gamma_c u = \gamma_c \int_{\Omega^c} E(x,y)^\top f(y) dy + \mathcal{C}_{\Omega^c}\gamma_c u + \mathcal{C}_{\Omega^c}\gamma_c M \,. \tag{3.8.18}$$

The following theorem insures that $\mathcal{C}_{\Omega^c}$ is also a projection.

Theorem 3.8.2. (Costabel et al [90, Lemma1.1])
For the multiple layer potential operators $\mathcal{K}_j$ and $\mathcal{K}_{jc}$ as well as for $\mathcal{Q}_\Omega$ and $\mathcal{Q}_{\Omega^c}$ (cf. (3.8.10)) there holds

$$\mathcal{Q}_{\Omega^c} - \mathcal{Q}_\Omega = \mathcal{P}^{-1} \,. \tag{3.8.19}$$

Equivalently, the Calderón projectors satisfy

$$\mathcal{C}_\Omega + \mathcal{C}_{\Omega^c} = \mathcal{I} \,. \tag{3.8.20}$$

Proof: As was shown in (3.4.51), $\mathcal{P}_{2m} = \sum_{|\alpha|=2m} a_\alpha \boldsymbol{n}^\alpha$ is a matrix–valued function and the assumption of uniform strong ellipticity (3.6.2) for the operator $P^\top$ assures

$$|\Theta(y)\mathcal{P}_{2m}(y)| \ge \gamma_0 |\boldsymbol{n}|^{2m} = \gamma_0 > 0 \,,$$

hence, the existence of the inverse matrix–valued function $\mathcal{P}_{2m}^{-1}$ exists along Γ. Therefore, the existence of the inverse of $\mathcal{P}$ in (3.8.9) follows immediately from that of $\mathcal{P}_{2m}^{-1}$ since $\mathcal{P}$ has triangular form. Moreover, $\mathcal{P}^{-1}$ is a lower triangular matrix of similar type as $\mathcal{P}$ consisting of tangential differential operators $\mathcal{P}^{(j)}$ of orders $2m-j$;

$$\mathcal{P}^{-1} = \begin{pmatrix} 0 & \dots & 0 & 0 & \mathcal{P}_{2m}^{-1} \\ \dots & & & & \vdots \\ 0 & & & & \mathcal{P}^{(3)} \\ 0 & & & \mathcal{P}^{(3)} & \mathcal{P}^{(2)} \\ \mathcal{P}_{2m}^{-1} & \dots & \mathcal{P}^{(3)} & \mathcal{P}^{(2)} & \mathcal{P}^{(1)} . \end{pmatrix} \tag{3.8.21}$$

Therefore, (3.8.19) and (3.8.20) are equivalent in view of (3.8.11) and (3.8.17).

Next we shall show (3.8.19). For this purpose we consider the potentials

$$v(x) := \mathcal{K}_k\varphi(x) = \langle \varphi|_\Gamma \otimes \delta_\Gamma^{(k)}, E(x, \bullet)\rangle \tag{3.8.22}$$

for $x \in \Omega$ and Ω^c, respectively. For $\varphi \in C^\infty(\mathbb{R}^n)$, the multilayer potential v is in $C^\infty(\Omega)$ up to the boundary and also in $C^\infty(\Omega^c)$ up to the boundary, as well, due to Lemma 3.8.1.

Then, from the formula (3.7.7) we find, after changing the orders of summation,

$$P(v\chi_\Omega) = (Pv)\chi_\Omega - \sum_{p=0}^{2m-1}\sum_{\ell=0}^{2m-p-1} (\mathcal{P}_{p+\ell+1}\gamma_\ell v)_{|\Gamma} \otimes \delta_\Gamma^{(p)}$$

and, similarly,

$$P(v\chi_{\Omega^c}) = (Pv)\chi_{\Omega^c} + \sum_{p=0}^{2m-1}\sum_{\ell=0}^{2m-p-1} (\mathcal{P}_{p+\ell+1}\gamma_{c\ell} v)_{|\Gamma} \otimes \delta_\Gamma^{(p)} ,$$

since

$$\frac{\partial}{\partial x_i}\chi_{\Omega^c} = n_i\delta_\Gamma .$$

Adding these two equations yields, with (3.8.3), i.e., $Pv = 0$ in $\Omega \cup \Omega^c$, and with any test function
$\psi \in C_0^\infty(\mathbb{R}^n)$, the equation

$$\langle P(v\chi_\Omega) + P(v\chi_{\Omega^c}), \psi\rangle = \sum_{p=0}^{2m-1}\sum_{\ell=0}^{2m-p-1} \langle \mathcal{P}_{p+\ell+1}(\gamma_{c\ell} v - \gamma_\ell v) \otimes \delta_\Gamma^{(p)}, \psi\rangle .$$

By the definition of derivatives of distributions, the left–hand side can be rewritten as

$$\int\limits_\Omega vP^\top\psi dx + \int\limits_{\Omega^c} vP^\top\psi dx = \int\limits_{\mathbb{R}^n} vP^\top\psi dx ,$$

where the last equality holds since v was in $C^\infty(\Omega)$ up to the boundary and in $C^\infty(\Omega^c) \cap C_0^\infty(\mathbb{R}^n)$ up to the boundary, too. With (3.8.22), this implies

$$\langle P(v\chi_\Omega) + P(v\chi_{\Omega^c}), \psi\rangle = \langle Pv, \psi\rangle = \langle \varphi_{|\Gamma} \otimes \delta_\Gamma^{(k)}, \psi\rangle \tag{3.8.23}$$

since E is the fundamental solution of P. The last step will be justified at the end of the proof. Therefore, we obtain with (3.8.10),

$$\begin{aligned}\varphi_{|\Gamma} \otimes \delta_\Gamma^{(k)} &= \sum_{p=0}^{2m-1} \sum_{\ell=0}^{2m-p-1} \mathcal{P}_{p+\ell+1}(\gamma_{c\ell}\mathcal{K}_k\varphi - \gamma_\ell\mathcal{K}_k\varphi) \otimes \delta_\Gamma^{(p)} \\ &= \sum_{p=0}^{2m-1} \sum_{\ell=0}^{2m-p-1} \mathcal{P}_{p+\ell+1}(\mathcal{Q}_{\Omega^c\ell+1,k+1}\varphi - \mathcal{Q}_{\Omega\ell+1,k+1}\varphi) \otimes \delta_\Gamma^{(p)} .\end{aligned}$$

Hence, from the definitions (3.8.9) and (3.8.10) we find the proposed equation

$$\mathcal{I} = \mathcal{P}(\mathcal{Q}_{\Omega^c} - \mathcal{Q}_\Omega) .$$

Since we have shown in (3.8.21) that $\mathcal{P}$ is invertible, this yields (3.8.19).

To complete the proof, we return to the justification of (3.8.23). Let Φ and ψ belong to $C_0^\infty(\mathbb{R}^n)$. Then, since $E(x,y)$ is a fundamental solution of P satisfying (3.6.3) and (3.6.4) , we obtain the relation

$$\begin{aligned}\int_{\mathbb{R}^n} \langle \phi, E(x,\bullet)\rangle P_x^\top \psi(x)dx &= \langle P \int_{\mathbb{R}^n} E(\bullet,y)^\top \phi(y)dy, \psi\rangle = \langle \phi, \psi\rangle \\ &= \langle \psi, \phi\rangle = \int_{\mathbb{R}^n} \langle \psi, E(x,\bullet)\rangle P_x^\top \phi(x)dx ,\end{aligned} \tag{3.8.24}$$

where the last identity follows by symmetry. Since (3.8.24) defines for any $\phi \in C^\infty$ a bounded linear functional on $\psi \in C_0^\infty(\mathbb{R}^n)$, we can extend (3.8.24) to any distribution $\phi \in \mathcal{D}'$. In particular, we may choose

$$\phi := \phi_{|\Gamma} \otimes \delta_\Gamma^{(k)} .$$

Then (3.8.24) implies

$$\langle v, P^\top \psi\rangle = \int_{\mathbb{R}^n} \langle \phi_{|\Gamma} \otimes \delta_\Gamma^{(k)}, E(x,\bullet)\rangle P_x^\top \psi(x)dx = \langle \phi_\Gamma \otimes \delta_\Gamma^{(k)}, \psi\rangle ,$$

i.e. (3.8.23). This completes the proof. ∎

As a consequence of Theorem 3.8.2 we have

Corollary 3.8.3. *For the "rigid motions" $M(x;u)$ in (3.7.12) and with the exterior Calderón projector, we find*

$$\mathcal{C}_{\Omega^c}\gamma_c M(\bullet;u) = 0 . \tag{3.8.25}$$

Proof: Since $M(x;u)$ is a C^∞–solution of

$$PM = 0 \quad \text{in } \mathbb{R}^n\,,$$

which follows from (3.7.12), we find

$$\mathcal{C}_\Omega \gamma M = \gamma M = \gamma_c M\,.$$

Applying $\mathcal{C}_{\Omega^c}$ to this equation yields (3.8.25) since $\mathcal{C}_{\Omega^c}\mathcal{C}_\Omega = 0$ because of (3.8.20). ■

Remark 3.8.1: From the definition of $\mathcal{Q}_\Omega$ we now define the *finite part multiple layer potentials* on Γ,

$$\mathcal{D}_{jk}\phi_k \;:=\; \text{p.f.}\int\limits_\Gamma \Big\{\Big(\frac{\partial}{\partial n_x}\Big)^{j-1}\Big(\frac{\partial}{\partial n_y}{}'\Big)^{k-1} E(x,y)^\top\Big\}\phi_k(y)ds_y\,, \qquad j,k = 1,\dots,2m \tag{3.8.26}$$

and

$$\begin{aligned}\mathcal{D}\phi \;&:=\; \Big(\sum_{k=1}^{2m}\mathcal{D}_{1,k}\phi_k,\dots,\sum_{k=1}^{2m}\mathcal{D}_{2m,k}\phi_k\Big)^\top \\ &= \mathcal{Q}_\Omega\phi + \frac{1}{2}\mathcal{P}^{-1}\phi = \mathcal{Q}_{\Omega^c}\phi - \frac{1}{2}\mathcal{P}^{-1}\phi \quad \text{on } \Gamma\end{aligned} \tag{3.8.27}$$

for any given $\phi = (\phi_1,\dots,\phi_{2m})^\top \in C^\infty(\Gamma)$. Correspondingly, we have

$$\mathcal{D}(\mathcal{P}\psi) = -\mathcal{C}_\Omega\psi + \frac{1}{2}\psi = \mathcal{C}_{\Omega^c}\psi - \frac{1}{2}\psi \quad \text{on } \Gamma\,. \tag{3.8.28}$$

As we will see, (3.8.27) or (3.8.28) are the *jump relations* for multilayer potentials.

3.9 Direct Boundary Integral Equations

As we have seen in the examples in elasticity and the biharmonic equation in Chapter 2, the boundary conditions are more general than parts of the Cauchy γu data appearing in the Calderón projector directly. Correspondingly, for more general boundary conditions, the boundary potentials in applications will also be more complicated.

3.9.1 Boundary Value Problems

Let us consider a more general boundary value problem for strongly elliptic equations associated with (3.6.1) in Ω (and later on also in Ω^c). As before,

Ω is assumed to be a bounded domain with C^∞–boundary $\Gamma = \partial\Omega$. The linear boundary value problem here consists of the uniformly strongly elliptic system of partial differential equations

$$Pu = \sum_{|\alpha|\le 2m} a_\alpha(x) D^\alpha u = f \quad \text{in } \Omega \tag{3.9.1}$$

with real C^∞ coefficient $p \times p$ matrices $a_\alpha(x)$ (which automatically is properly elliptic as mentioned before, see Wloka [439, Sect. 9]) together with the boundary conditions

$$R\gamma u = \varphi \quad \text{on } \Gamma \tag{3.9.2}$$

where $R = (R_{jk})$ with $j = 1, \dots, m$ and $k = 1, \dots, 2m$ is a rectangular matrix of tangential differential operators along Γ having orders

$$\text{order } (R_{jk}) = \mu_j - k + 1 \quad \text{where } 0 \le \mu_j \le 2m - 1$$
$$\text{and } \mu_j < \mu_{j+1} \quad \text{for } 1 \le j \le m - 1\,.$$

If $\mu_j - k + 1 < 0$ then $R_{jk} := 0$. Hence, component–wise, the boundary conditions (3.9.2) are of the form

$$\sum_{k=1}^{2m} R_{jk}\gamma_{k-1}u = \sum_{k=1}^{\mu_j+1} R_{jk}\gamma_{k-1}u = \varphi_j \text{ for } j = 1, \dots, m\,. \tag{3.9.3}$$

For p–vector–valued functions u each R_{jk} is a $p \times p$ matrix of operators. In the simplest case μ_j is constant for all components of $u = (u_1, \dots, u_p)^\top$, and let us consider here first this simple case. We shall return to the more general case later on.

In order to have *normal boundary conditions* (see e.g. Lions and Magenes [270]), we require the following crucial assumption.

Fundamental assumption: *There exist complementary tangential boundary operators $S = (S_{jk})$ with $j = 1, \dots, m$ and $k = 1, \dots, 2m$, having the orders*

$$\textit{order } (S_{jk}) = \nu_j - k + 1\,, \quad \textit{where } \nu_j \ne \mu_\ell$$
$$\textit{for all } j, \ell = 1, \dots, m\,,\ 0 \le \nu_j \le 2m - 1\,;\ \nu_j > \nu_{j+1} \quad \textit{for } 1 \le j \le m - 1$$

such that the square matrix of tangential differential operators

$$\mathcal{M} \;:=\; \begin{pmatrix} R \\ S \end{pmatrix} \tag{3.9.4}$$

admits an inverse

$$\mathcal{N} \;:=\; \mathcal{M}^{-1}\,, \tag{3.9.5}$$

which is a matrix of tangential differential operators.

Since $\mathcal{M}$ is a rearranged triangular matrix this means that the operators on the rearranged diagonal, i.e. R_{j,μ_j+1} and S_{j,ν_j+1} are multiplications with non vanishing coefficients.

Definition 3.9.1. *The boundary value problem is called a* regular elliptic boundary value problem *if P in* (3.9.1) *is elliptic and the boundary conditions in* (3.9.2) *are normal and satisfy the Lopatinski–Shapiro conditions.*

In order to have a *regular elliptic boundary value problem* (see Hörmander [187], Wloka [439]) we require in addition to the fundamental assumption, i.e. normal boundary conditions, that the *Lopatinski–Shapiro condition* is fulfilled (see [439, Chap.11]). This condition reads as follows:

Definition 3.9.2. *The boundary conditions* (3.9.3) *for the differential operator in* (3.9.1) *are said to satisfy the Lopatinski–Shapiro conditions if the initial value problem defined by the constant coefficient ordinary differential equations*

$$\sum_{k=0}^{2m} \mathcal{P}_k^{(2m)}(x, i\xi')\Big(\frac{d}{d\sigma_n}\Big)^k \chi(\sigma_n) = 0 \;\; \textit{for} \;\; 0 < \sigma_n \tag{3.9.6}$$

together with the initial conditions

$$\sum_{|\lambda|=\mu_j-k-1} r_{jk\lambda}(x)(i\xi')^\lambda \Big(\frac{d}{d\sigma_n}\Big)^{k-1} \chi_{|\sigma_n=0} = 0 \;\; \textit{for} \; j = 1, \dots, m \tag{3.9.7}$$

and the radiation condition $\chi(\sigma_n) = o(1)$ as $\sigma_n \to \infty$ admits only the trivial solution $\chi(\sigma_n) = 0$ for every fixed $x \in \Gamma$ and every $\xi' \in \mathbb{R}^{n-1}$ with $|\xi'| = 1$.

Here, the coefficients in (3.9.6) are given by the principal part of P in local coordinates via (3.4.50) as

$$\mathcal{P}_k^{(2m)}(x, \xi') := \sum_{|\gamma| \le 2m-k} \; \sum_{|\alpha|=2m} a_\alpha(x) c_{\alpha,\gamma,k}(\sigma', 0)(i\xi')^\gamma \, ;$$

and the coefficients in (3.9.7), from the tangential boundary operators in (3.9.2) in local coordinates are given by

$$R_{jk} = \sum_{|\lambda| \le \mu_j - k - 1} r_{jk\lambda}(x)\Big(\frac{\partial}{\partial\sigma'}\Big)^\lambda .$$

The case of more general boundary conditions in which the orders $\mu_{j\ell}$ for the different components u_ℓ of u are different, with corresponding given components φ_{jt} of φ_j, was treated by Grubb in [158]. Here the boundary conditions have the form

$$\sum_{\ell=1}^{p} \sum_{k=1}^{\mu_{j\ell}+1} \sum_{|\lambda|=\mu_{j\ell}-k-1} r_{jk\lambda}^{t\ell}(x)(i\xi')^\lambda \gamma_{k-1} u_\ell = \varphi_{jt} \, , \tag{3.9.8}$$

$$t = 1, \dots, q_j \le p \, , \; j = 1, \dots, m \quad \text{and} \quad \sum_{j=1}^{m} q_j = m \cdot p \, .$$

Here, for each fixed ℓ we have $\mu_{j\ell} \in \{0,\dots,2m-1\}$ which are ordered as $\mu_{j\ell} < \mu_{j+1,\ell}$, then we also need complementing orders $\nu_{j\ell} \in \{0,\dots,2m-1\}\setminus \bigcup_{j=1}^{2m-1}\{\mu_{j\ell}\}$ with $\nu_{j\ell} > \nu_{j+1,\ell}$ and complementing boundary conditions $S^{t\ell}_{jk}$. For a system of *normal boundary conditions* S the boundary conditions in R together with the complementing boundary conditions should be such that the *rearranged* triangular matrix $\widetilde{\mathcal{M}} = ((M_{jk}))_{2m\times 2m}$ with $M_{jk} = 0$ for $j < k$ and tangential operators

$$M_{jk} = ((M^{t\ell}_{jk}))_{q_j\times p}$$

now define a complete system of boundary conditions

$$\sum_{\ell=1}^{p}\sum_{k\le j} M^{t\ell}_{jk}\gamma_{k-1}u_\ell = \varphi_{jt}\,,\ t = \overline{1,q_j}\,,\ j = \overline{1,2m}$$

where the diagonal matrices $M_{jj} = ((M_{jj}))_{q_j\times p}$, whose entries are functions, define surjective mappings onto the q_j–vector valued functions, and where $q_j \le p$. Then there exists the right inverse $\mathcal{N}$ to the triangular matrix $\widetilde{\mathcal{M}}$ and, hence, to $\mathcal{M}$.

If we require the stronger fundamental assumption that $\mathcal{M}^{-1}$ exists, the boundary conditions still will be a normal system. In this case, in the Shapiro–Lopatinski condition, the initial conditions (3.9.7) are now to be replaced by

$$\sum_{\ell=1}^{p}\sum_{k=1}^{\mu_{j\ell}+1}\sum_{|\lambda|=\mu_{j\ell}-k-1} r^{t\ell}_{jk\lambda}(x)(i\xi')^{\lambda}\Big(\frac{d}{d\sigma_n}\Big)^{k-1}\chi_\ell|_{\sigma_n=0} = 0 \qquad (3.9.9)$$
$$\text{for } t = 1,\dots,q_j \text{ with } q_j \le p \text{ and } j = 1,\dots,m\,.$$

In regard to the Shapiro–Lopatinski condition for general elliptic systems in the sense of Agmon–Douglis–Nirenberg, we refer to the book by Wloka et al [440, Sections 9.3,10.1], where one can find an excellent presentation of these topics.

Now, for the special cases for $m = 1$ we consider the following **second order systems**:

In this case, only the following two combinations of boundary conditions are possible:

i)

$$\begin{aligned} R &= (R_{00}, 0) && \text{with order } (R_{00}) = 0 \\ S &= (S_{00}, S_{01}) && \text{with order } (S_{00}) = 1 \text{ and order } (S_{01}) = 0\,. \end{aligned} \qquad (3.9.10)$$

For (3.9.5) to be satisfied, we require that the inverse matrices R^{-1}_{00} and S^{-1}_{01} exist along Γ. Then $\mathcal{M}$ and $\mathcal{N}$ are given by

$$\mathcal{M} = \begin{pmatrix} R_{00} & , & 0 \\ S_{00} & , & S_{01} \end{pmatrix} ; \mathcal{N} = \begin{pmatrix} R_{00}^{-1} & , & 0 \\ -S_{01}^{-1} S_{00} R_{00}^{-1} & , & S_{01}^{-1} \end{pmatrix} \tag{3.9.11}$$

ii)

$$\begin{aligned} R &= (R_{00}, R_{01}) \quad \text{with order } (R_{00}) &&= 1 \quad \text{and order } (R_{01}) = 0\,, \\ S &= (S_{00}, 0) \quad \text{with order } (S_{00}) &&= 0\,. \end{aligned} \tag{3.9.12}$$

Similarly, here for (3.9.5) one needs the existence of R_{01}^{-1} and S_{00}^{-1}. Then $\mathcal{M}$ and $\mathcal{N}$ are given by

$$\mathcal{M} = \begin{pmatrix} R_{00} & , & R_{01} \\ S_{00} & , & 0 \end{pmatrix} ; \mathcal{N} = \begin{pmatrix} 0 & , & S_{\infty}^{-1} \\ R_{01}^{-1} & , & -R_{01}^{-1} R_{00} S_{00}^{-1} \end{pmatrix} \tag{3.9.13}$$

In general, the verification of the fundamental assumption is nontrivial (see e.g. Grubb [158, 159]), however, in most applications, the fundamental assumption is fulfilled.

Because of (3.9.4) one may express the Cauchy data $\gamma u = (\gamma_0 u, \ldots, \gamma_{2m-1} u)$ by the given boundary functions φ_j in (3.9.3) and a set of complementary boundary functions λ_j via

$$\sum_{k=1}^{\nu_j+1} S_{jk} \gamma_{k-1} u = \lambda_j \quad \text{with } j = 1, \ldots, m \text{ on } \Gamma\,. \tag{3.9.14}$$

If we require (3.9.5) or normal boundary conditions, the Cauchy data γu can be recovered from φ and λ by the use of $\mathcal{N}$,

$$\gamma_{\ell-1} u = \sum_{j=1}^{m} \mathcal{N}_{\ell j} \varphi_j + \sum_{p=m+1}^{2m} \mathcal{N}_{\ell p} \lambda_{p-m} \text{ for } \ell = 1, \ldots, 2m\,; \tag{3.9.15}$$

in short,

$$\gamma u = \mathcal{N}_1 \varphi + \mathcal{N}_2 \lambda \tag{3.9.16}$$

with rectangular tangential differential operators $\mathcal{N}_1$ and $\mathcal{N}_2$ from (3.9.15). Then

$$R\mathcal{N}_1 = I\,,\ R\mathcal{N}_2 = 0\,,\ S\mathcal{N}_1 = 0\,,\ S\mathcal{N}_2 = I \tag{3.9.17}$$

since $\mathcal{N}$ is the right inverse. This enables us to insert the boundary data φ and λ with given φ into the representation formula (3.8.5), i.e., for $x \in \Omega$ we obtain:

$$\begin{aligned} u(x) &= \int_{\Omega} E(x, y)^\top f(y) dy \qquad (3.9.18) \\ &\quad - \sum_{\ell=0}^{2m-1} \sum_{p=0}^{2m-\ell-1} \mathcal{K}_p \mathcal{P}_{p+\ell+1} \Big\{ \sum_{j=1}^{m} \mathcal{N}_{\ell+1,j} \varphi_j + \sum_{j=1}^{m} \mathcal{N}_{\ell+1,m+j} \lambda_j \Big\}(x)\,. \end{aligned}$$

By taking the traces on both sides of (3.9.18) we arrive with (3.8.6) or (3.8.10) at

$$\gamma u = \mathcal{N}\binom{\varphi}{\lambda} = \gamma F - \mathcal{Q}_\Omega \mathcal{P}\mathcal{N}\binom{\varphi}{\lambda} = \gamma F + \mathcal{C}_\Omega \mathcal{N}\binom{\varphi}{\lambda}$$

where

$$F(x) := \int\limits_\Omega E(x,y)^\top f(y) dy\,.$$

Now, applying the boundary operator $\mathcal{M}$, we find

$$\mathcal{M}\gamma u = \binom{\varphi}{\lambda} = \mathcal{M}\gamma F + \mathcal{M}\mathcal{C}_\Omega \mathcal{M}^{-1}\binom{\varphi}{\lambda}\,.$$

If $f = 0$, the right–hand side again is a projector for $\binom{\varphi}{\lambda} = \mathcal{M}\gamma u$, which are generalized Cauchy data and boundary data to any solution u of the homogeneous equation $Pu = 0$ in Ω. We call $\widetilde{\mathcal{C}}_\Omega$, defined by

$$\widetilde{\mathcal{C}}_\Omega \binom{\varphi}{\lambda} := \mathcal{M}\mathcal{C}_\Omega \mathcal{M}^{-1}\binom{\varphi}{\lambda} = \binom{R}{S}\mathcal{C}_\Omega(\mathcal{N}_1\varphi + \mathcal{N}_2\lambda)\,, \tag{3.9.19}$$

the *modified Calderón projector* associated with P, Ω and $\mathcal{M}$, which satisfies

$$\widetilde{\mathcal{C}}^2_\Omega = \widetilde{\mathcal{C}}_\Omega\,. \tag{3.9.20}$$

In terms of the finite–part integrals defined in (3.8.26), we have a relation corresponding to (3.8.28) for the modified Calderón projector as well:

$$\mathcal{M}\mathcal{D}\mathcal{P}\mathcal{M}^{-1}\chi = -\,\widetilde{\mathcal{C}}_\Omega \chi + \frac{1}{2}\chi \quad \text{on } \Gamma\,. \tag{3.9.21}$$

For the boundary value problem (3.9.1) and (3.9.3), φ is given. Consequently, for the representation of u in (3.9.18) the missing boundary datum λ is to be determined by the boundary integral equations

$$\binom{\varphi}{\lambda} = \widetilde{\mathcal{C}}_\Omega \binom{\varphi}{\lambda} + \mathcal{M}\gamma F \quad \text{on } \Gamma\,. \tag{3.9.22}$$

As for the Laplacian, (3.9.22) is an overdetermined system for λ in the sense that only one of the two sets of equations is sufficient to determine λ.

Applying $R\mathcal{M}^{-1} = \binom{I}{0}$ and $S\mathcal{M}^{-1} = \binom{0}{I}$ to (3.9.22) and employing (3.9.21) leads to the following two choices of boundary integral equations in terms of finite part integral operators:

Integral equations of the "first kind":

$$\begin{aligned} \mathcal{A}_\Omega\lambda := -R\mathcal{Q}_\Omega\mathcal{P}\mathcal{N}_2\lambda &= \varphi + R\mathcal{Q}_\Omega\mathcal{P}\mathcal{N}_1\varphi - R\gamma F \\ = -R\mathcal{D}\mathcal{P}\mathcal{N}_2\lambda \quad &= \tfrac{1}{2}\varphi + R\mathcal{D}\mathcal{P}\mathcal{N}_1\varphi - R\gamma F \text{ on } \Gamma\,. \end{aligned} \tag{3.9.23}$$

Integral equations of the "second kind":

$$\begin{aligned} \mathcal{B}_\Omega \lambda := \lambda + S\mathcal{Q}_\Omega \mathcal{P}\mathcal{N}_2 \lambda &= -S\mathcal{Q}_\Omega \mathcal{P}\mathcal{N}_1 \varphi + S\gamma F \\ = \tfrac{1}{2}\lambda + S\mathcal{D}\mathcal{P}\mathcal{N}_2 \lambda &= -S\mathcal{D}\mathcal{N}_1 \varphi + S\gamma F \text{ on } \Gamma . \end{aligned} \tag{3.9.24}$$

Note that in the above equations the tangential differential operators R and S are applied to finite part integral operators. These operators can be shown to commute with the finite part integration (see Corollary 9.3.3).

We now have shown how the boundary value problem could be reduced to each of these boundary integral equations. Consequently, the solution λ of one of the boundary integral equations generates the solution of the boundary value problem with the help of the representation formula (3.9.18). More precisely, we now establish the following equivalence theorem.

This theorem assures that every solution of the boundary integral equation of the first kind generates solutions of the interior as well as of the exterior boundary value problem and that every solution of the boundary value problems can be obtained by solving the first kind integral equations. This equivalence, however, does not guarantee that the numbers of solutions of boundary value problems and corresponding boundary integral equations coincide.

Theorem 3.9.1. *Assume the fundamental assumption holds. Let $f \in C^\infty(\overline{\Omega})$ and $\varphi \in C^\infty(\Gamma)$. Then every solution $u \in C^\infty(\overline{\Omega})$ of the boundary value problem* (3.9.1), (3.9.2) *defines by* (3.9.14) *a solution λ of both integral equations* (3.9.23) *and* (3.9.24). *Conversely, let $\lambda \in C^\infty(\Gamma)$ be a solution of the integral equation of the first kind* (3.9.23). *Then λ together with φ defines via* (3.9.18) *a solution $u \in C^\infty(\overline{\Omega})$ of the boundary value problem* (3.9.1), (3.9.2) *and, in addition, λ solves also the the second kind equation* (3.9.24).

If $\lambda \in C^\infty(\Gamma)$ is a solution of the integral equation of the second kind (3.9.24) *and if, in addition, the operator $S\mathcal{D}\mathcal{P}\mathcal{N}_1$ is injective, then the potential u defined by λ and φ via* (3.9.18) *is a C^∞–solution of the boundary value problem* (3.9.1), (3.9.2) *and, in addition, λ solves also the first kind equation* (3.9.23).

Proof: i.) If $u \in C^\infty(\overline{\Omega})$ is a solution to (3.9.1) and (3.9.2) then the previous derivation shows that $\lambda \in C^\infty(\Gamma)$ and that λ satisfies both boundary integral equations.

ii.) If $\lambda \in C^\infty(\Gamma)$ is a solution of the integral equation of the first kind (3.9.23) then the potentials in (3.9.18) define a C^∞–solution u of the partial differential equation (3.9.1) due to Lemma 3.8.1 and (3.8.3), (3.8.4). Applying to both sides of (3.9.18) the trace operator yields

$$\gamma u = \gamma F + \mathcal{C}_\Omega \mathcal{M}^{-1} \begin{pmatrix} \varphi \\ \lambda \end{pmatrix} \quad \text{on } \Gamma . \tag{3.9.25}$$

Applying R on both sides gives with (3.9.5)

$$R\gamma u = R\gamma F + RC_\Omega(\mathcal{N}_1\varphi + \mathcal{N}_2\lambda)\,;$$

and (3.9.23) for λ yields with (3.8.11)

$$R\gamma u = \varphi\,.$$

Hence, $\gamma u = \mathcal{M}^{-1}\binom{\varphi}{\lambda}$ and the application of $S\mathcal{M}$ to (3.9.25) yields (3.9.24) for λ, too.

iii.) If $\lambda \in C^\infty(\Gamma)$ solves (3.9.24), i.e.

$$\lambda + S\mathcal{Q}_\Omega\mathcal{P}(\mathcal{N}_1\varphi + \mathcal{N}_2\lambda) = S\gamma F\,, \tag{3.9.26}$$

then the potentials in (3.9.18) again define a C^∞–solution u of the partial differential equation (3.9.1) in Ω, and for its trace holds (3.9.25) on Γ. Applying S yields with (3.9.26) and (3.8.11) the equation

$$S\gamma u = \lambda \quad \text{on } \Gamma\,.$$

On the other hand, the solution u defined by (3.9.18) satisfies on Γ

$$\mathcal{M}\gamma u = \mathcal{M}\gamma F + \widetilde{\mathcal{C}}_\Omega\mathcal{M}\gamma u$$

whose second set of components reads

$$\lambda = S\gamma u = S\gamma F - S\mathcal{Q}_\Omega\mathcal{P}(\mathcal{N}_1 R\gamma u + \mathcal{N}_2\lambda)$$

in view of (3.9.23) and (3.8.11). If we subtract (3.9.26) we obtain

$$SC_\Omega\mathcal{N}_1(R\gamma u - \varphi) = 0\,.$$

On the other hand, with (3.8.28),

$$SC_\Omega\mathcal{N}_1 = S\{\tfrac{1}{2}\mathcal{I} - \mathcal{D}\mathcal{P}\}\mathcal{N}_1 = -S\mathcal{D}\mathcal{P}\mathcal{N}_1$$

due to (3.9.17). Therefore

$$S\mathcal{D}\mathcal{P}\mathcal{N}_1(\varphi - R\gamma u) = 0$$

which implies $\varphi = R\gamma u$ due to the assumed injectivity. Now, the first equation (3.9.23) then follows as in ii). ∎

Now let us consider the *exterior boundary value problem* for the equation (3.6.14), i.e.,

$$Pu = f \quad \text{in } \Omega^c \tag{3.9.27}$$

together with the boundary condition

$$R\gamma_c u = \varphi \quad \text{on } \Gamma \tag{3.9.28}$$

and the radiation condition — which means that $M(x;u)$ defined by (3.6.15) is also *given*. Here, the solution can be represented by (3.7.13),

$$u(x) = \int_{\Omega^c} E(x,y)^\top f(y)dy + M(x;u)$$
$$+ \sum_{\ell=0}^{2m-1} \sum_{p=0}^{2m-\ell-1} \int_{\Gamma} \Big\{ \Big(\frac{\partial}{\partial n_y}{}'\Big)^p E(x,y)\Big\}^\top \mathcal{P}_{p+\ell+1}\gamma_{c\ell}u(y)ds_y .$$

As for the interior problem, we require the *fundamental assumption* and obtain with the exterior Cauchy data $R\gamma_c u = \varphi$ and $S\gamma_c u = \lambda$, the modified *representation formula* for $x \in \Omega^c$, similar to (3.9.18):

$$u(x) = F_c(x) + M(x;u) \tag{3.9.29}$$
$$+ \sum_{\ell=0}^{2m-1} \sum_{p=0}^{2m-\ell-1} \mathcal{K}_p \mathcal{P}_{p+\ell+1} \Big\{ \sum_{j=1}^{m} \mathcal{N}_{\ell+1,j}\varphi_j + \sum \mathcal{N}_{\ell+1,m+j}\lambda_j \Big\}(x)$$

where

$$F_c(x) = \int_{\Omega^c} E(x,y)^\top f(y)dy .$$

In the same manner as for the interior problem, we obtain the set of boundary integral equations

$$\begin{pmatrix}\varphi\\ \lambda\end{pmatrix} = \widetilde{\mathcal{C}}_{\Omega^c}\begin{pmatrix}\varphi\\ \lambda\end{pmatrix} + \mathcal{M}\gamma_c F_c + \mathcal{M}\gamma_c M(\bullet;u) \tag{3.9.30}$$

where

$$\widetilde{\mathcal{C}}_{\Omega^c} = \mathcal{I} - \widetilde{\mathcal{C}}_{\Omega} = \mathcal{M}\mathcal{C}_{\Omega^c}\mathcal{M}^{-1} \tag{3.9.31}$$

and $\mathcal{C}_{\Omega^c}$ is given in (3.8.17). These are the ordinary and modified exterior Calderón projectors, respectively, associated with P, Ω^c and $\mathcal{M}$.

As before, we obtain two sets of boundary integral equations for λ.

Integral equation of the "first kind":

$$\begin{aligned} \mathcal{A}_{\Omega^c}\lambda &:= -R\mathcal{Q}_{\Omega^c}\mathcal{P}\mathcal{N}_2\lambda = -\varphi + R\mathcal{Q}_{\Omega^c}\mathcal{P}\mathcal{N}_1\varphi + R\gamma_c F_c + R\gamma_c M , \\ &= -R\mathcal{D}\mathcal{P}\mathcal{N}_2\lambda = -\tfrac{1}{2}\varphi + R\mathcal{D}\mathcal{P}\mathcal{N}_1\varphi + R\gamma_c F_c + R\gamma_c M . \end{aligned} \tag{3.9.32}$$

Integral equation of the "second kind":

$$\begin{aligned} \mathcal{B}_{\Omega^c}\lambda &:= \lambda - S\mathcal{Q}_{\Omega^c}\mathcal{P}\mathcal{N}_2\lambda = S\mathcal{Q}_{\Omega^c}\mathcal{P}\mathcal{N}_1\varphi + S\gamma_c F_c + S\gamma_c M , \\ &= \tfrac{1}{2}\lambda - S\mathcal{D}\mathcal{P}\mathcal{N}_2\lambda = S\mathcal{D}\mathcal{P}\mathcal{N}_1\varphi + S\gamma_c F_c + S\gamma_c M . \end{aligned} \tag{3.9.33}$$

In comparison to equations (3.9.23) and (3.9.32), we see that $\mathcal{A}_{\Omega}$ and $\mathcal{A}_{\Omega^c}$ are the same for interior and exterior problems. Hence, we simply denote this operator in the sequel by

$$\mathcal{A} := \mathcal{A}_\Omega = \mathcal{A}_{\Omega^c} = -R\mathcal{D}\mathcal{P}\mathcal{N}_2\,.$$

It is not difficult to see that under the same assumptions as for the interior problem, **the equivalence Theorem 3.9.1 remains valid** for the exterior problem (3.9.27), (3.9.28) with the radiation condition provided $f \in C^\infty(\overline{\Omega^c})$ and f has compact support in $\mathbb{R}^n$.

The following theorem shows the relation between the solutions of the homogeneous boundary value problems and the homogeneous boundary integral equations of the first kind.

Theorem 3.9.2. *Under the previous assumptions $\Gamma \in C^\infty$ and $a_\alpha \in C^\infty$, we have for the eigenspaces of the interior and exterior boundary value problems and for the eigenspace of the boundary integral operator $\mathcal{A}$ on Γ, respectively, the relation:*

$$\ker\mathcal{A} = \mathcal{X} \tag{3.9.34}$$

where

$$\ker\mathcal{A} = \{\lambda \in C^\infty(\Gamma) \,\big|\, \mathcal{A}\lambda = 0 \text{ on } \Gamma\}$$

and

$$\begin{aligned}\mathcal{X} = \text{span}\ &\{S\gamma u \,\big|\, u \in C^\infty(\overline{\Omega}) \wedge Pu = 0 \ \textit{in } \Omega \textit{ and } R\gamma u = 0 \text{ on } \Gamma\}\\ &\cup \{S\gamma_c u_c \,\big|\, u_c \in C^\infty(\overline{\Omega^c}) \wedge Pu_c = 0 \ \textit{in } \Omega^c\,,\ R\gamma_c u_c = 0 \text{ on } \Gamma\\ &\qquad \textit{and the radiation condition } M(x; u_c) = 0\}\end{aligned}$$

Proof:

i.) Suppose $\lambda \in \ker\mathcal{A}$, i.e. $\mathcal{A}\lambda = 0$ on Γ. Then define the potential

$$u(x) := \sum_{\ell=0}^{2m-1}\sum_{p=0}^{2m-\ell-1} \mathcal{K}_p\mathcal{P}_{p+\ell+1}(\mathcal{N}_2\lambda)_\ell$$

where $(\psi)_\ell$ denotes the ℓ–th component of the vector ψ. Now, u is a solution of

$$Pu = 0 \quad \text{in } \Omega \text{ and in } \Omega^c\,.$$

Then we find in Ω

$$\begin{aligned} R\gamma u &= R\mathcal{Q}_\Omega\mathcal{P}\mathcal{N}_2\lambda = -\mathcal{A}\lambda = 0 \quad \text{and}\\ S\gamma u &= S\mathcal{Q}_\Omega\mathcal{P}\mathcal{N}_2\lambda\,. \end{aligned} \tag{3.9.35}$$

Similarly, we find in Ω^c,

$$\begin{aligned} R\gamma_c u &= R\mathcal{Q}_{\Omega^c}\mathcal{P}\mathcal{N}_2\lambda = -\mathcal{A}\lambda = 0 \quad \text{and}\\ S\gamma_c u &= S\mathcal{Q}_{\Omega^c}\mathcal{P}\mathcal{N}_2\lambda\,. \end{aligned} \tag{3.9.36}$$

Subtracting (3.9.36) from (3.9.35) gives

$$\begin{aligned} R\gamma u &= R\gamma_c u = 0 \quad \text{and} \\ S\gamma u - S\gamma_c u &= S(\mathcal{Q}_\Omega - \mathcal{Q}_{\Omega^c})\mathcal{P}\mathcal{N}_2\lambda = S\mathcal{P}^{-1}\mathcal{P}\mathcal{N}_2\lambda = \lambda\,. \end{aligned}$$

This shows that $\lambda \in \mathcal{X}$.
ii.) Let u be the eigensolution of $Pu = 0$ in Ω with $R\gamma u = 0$ on Γ. Then u admits the representation (3.9.18), namely

$$u(x) = -\sum_{\ell=0}^{2m-1}\sum_{p=0}^{2m-\ell-1} \mathcal{K}_p\mathcal{P}_{p+\ell+1}\{\mathcal{N}_2 S\gamma u\}_\ell \text{ in } \Omega\,.$$

By taking traces and applying the boundary operator R we obtain

$$0 = R\gamma u = \mathcal{A}(S\gamma u)\,.$$

This implies that $S\gamma u \in \ker\mathcal{A}$.

In the same manner we proceed for an eigensolution of the exterior problem and conclude that $S\gamma_c u \in \ker\mathcal{A}$ as well.

This completes the proof. ∎

Corollary 3.9.3. *The injectivity of the operator*

$$S\mathcal{D}\mathcal{P}\mathcal{N}_1$$

in the equivalence Theorem 3.9.1 *is guaranteed if and only if the interior and exterior homogeneous boundary value problems*

$$\begin{aligned} Pu &= 0 \text{ in } \Omega \quad \text{and} & Pu_c &= 0 \text{ in } \Omega^c\,, \\ S\gamma u &= 0 \quad \text{and} & S\gamma_c u_c &= 0 \text{ on } \Gamma \text{ with } M(x;u_c) = 0 \end{aligned}$$

admit only the trivial solutions.

Proof: For the proof we exchange the rôles of R and S, λ and φ, respectively. Then the operator $\mathcal{A}$ will just be $S\mathcal{D}\mathcal{P}\mathcal{N}_1$ and Theorem 3.9.2 implies the Corollary. ∎

3.9.2 Transmission Problems

In this section we consider transmission problems of the following rather general type:

Find u satisfying the differential equation

$$Pu = f \quad \text{in } \Omega \cup \Omega^c$$

and the transmission conditions

$$R\gamma u = \varphi^- \quad \text{and} \quad [R\gamma u]_\Gamma = (R\gamma_c u - R\gamma u) = [\varphi] \quad \text{on } \Gamma \tag{3.9.37}$$

where $f \in C_0^\infty(\mathbb{R}^n)$, φ^-, $[\varphi] \in C^\infty(\Gamma)$ are given functions and where u satisfies the radiation condition (3.6.15).

We begin with some properties of the modified Calderón projectors. Obviously, they enjoy the same properties as the ordinary Calderón projectors, namely

$$\widetilde{\mathcal{C}}_\Omega^2 = \widetilde{\mathcal{C}}_\Omega, \quad \widetilde{\mathcal{C}}_\Omega + \widetilde{\mathcal{C}}_{\Omega^c} = \mathcal{I}, \quad \widetilde{\mathcal{C}}_{\Omega^c}^2 = \widetilde{\mathcal{C}}_{\Omega^c}\,. \tag{3.9.38}$$

In terms of the boundary potentials and boundary operators R and S, the explicit forms of $\widetilde{\mathcal{C}}_\Omega$ and $\widetilde{\mathcal{C}}_{\Omega^c}$ are given by

$$\begin{aligned} \widetilde{\mathcal{C}}_\Omega \begin{pmatrix}\varphi\\ \lambda\end{pmatrix} &= -\begin{pmatrix} R\mathcal{Q}_\Omega\mathcal{P}\mathcal{N}_1, R\mathcal{Q}_\Omega\mathcal{P}\mathcal{N}_2\\ S\mathcal{Q}_\Omega\mathcal{P}\mathcal{N}_1, S\mathcal{Q}_\Omega\mathcal{P}\mathcal{N}_2\end{pmatrix}\begin{pmatrix}\varphi\\ \lambda\end{pmatrix},\\ \widetilde{\mathcal{C}}_{\Omega^c} \begin{pmatrix}\varphi\\ \lambda\end{pmatrix} &= \begin{pmatrix} R\mathcal{Q}_{\Omega^c}\mathcal{P}\mathcal{N}_1, R\mathcal{Q}_{\Omega^c}\mathcal{P}\mathcal{N}_2\\ S\mathcal{Q}_{\Omega^c}\mathcal{P}\mathcal{N}_1, S\mathcal{Q}_{\Omega^c}\mathcal{P}\mathcal{N}_2\end{pmatrix}\begin{pmatrix}\varphi\\ \lambda\end{pmatrix}. \end{aligned} \tag{3.9.39}$$

If we denote the jump of the boundary potentials u across Γ by

$$[\mathcal{Q}\mathcal{P}\mathcal{N}_j] = \mathcal{Q}_{\Omega^c}\mathcal{P}\mathcal{N}_j - \mathcal{Q}_\Omega\mathcal{P}\mathcal{N}_j$$

then, as a consequence of (3.8.27), we have the following jump relations for any $\varphi, \lambda \in C^\infty(\Gamma)$ across Γ:

$$\begin{aligned} [R\mathcal{Q}\mathcal{P}\mathcal{N}_1\varphi] &= \varphi\,, \ [R\mathcal{Q}\mathcal{P}\mathcal{N}_2\lambda] &= 0\,,\\ [S\mathcal{Q}\mathcal{P}\mathcal{N}_1\varphi] &= 0\,, \ [S\mathcal{Q}\mathcal{P}\mathcal{N}_2\lambda] &= \lambda\,. \end{aligned} \tag{3.9.40}$$

From the representation formulae (3.7.10), (3.7.11) and (3.7.13), (3.7.14) we arrive at the representation formula for the transmission problem,

$$\begin{aligned} u(x) &= \int\limits_{\mathbb{R}^n} E(x,y)^\top f(y)dy + M(x;u)\\ &+ \sum_{\ell=0}^{2m-1}\sum_{p=0}^{2m-\ell-1} \int\limits_\Gamma \Big\{\Big(\frac{\partial}{\partial n_y}\Big)^{\prime p} E(x,y)\Big\}^\top \mathcal{P}_{p+\ell+1}\big(\gamma_{c\ell}u(y) - \gamma_\ell u(y)\big)ds_y\\ &= \int\limits_{\mathbb{R}^n} E(x,y)^\top f(y)dy + M(x;u)\\ &+ \sum_{\ell=0}^{2m-1}\sum_{p=0}^{2m-\ell-1} \mathcal{K}_p\mathcal{P}_{p+\ell+1}\{\mathcal{N}_1[\varphi] + \mathcal{N}_2[\lambda]\}_\ell \quad \text{for } x \in \Omega\cup\Omega^c\,,\ x \notin \Gamma\,. \end{aligned} \tag{3.9.41}$$

Now, we apply γ and γ_c to both sides of (3.9.41) and obtain after applying $\mathcal{M}$:

$$\begin{aligned}
\mathcal{M}\gamma u &= \mathcal{M}\gamma F + \mathcal{M}\gamma M + \mathcal{M}\gamma \sum_{\ell=0}^{2m-1}\sum_{p=0}^{2m-\ell-1} \mathcal{K}_p \mathcal{P}_{p+\ell+1}\{\mathcal{N}_1[\varphi] + \mathcal{N}_2[\lambda]\}_\ell \\
&= \mathcal{M}\gamma F + \mathcal{M}\gamma M + \mathcal{M}\mathcal{Q}_\Omega \mathcal{P}\{\mathcal{N}_1[\varphi] + \mathcal{N}_2[\lambda]\}, \qquad (3.9.42)\\
\mathcal{M}\gamma_c u &= \mathcal{M}\gamma_c F + \mathcal{M}\gamma_c M + \mathcal{M} \sum_{\ell=0}^{2m-1}\sum_{p=0}^{2m-\ell-1} \mathcal{K}_p \mathcal{P}_{p+\ell+1}\{\mathcal{N}_1[\varphi] + \mathcal{N}_2[\lambda]\}_\ell \\
&= \mathcal{M}\gamma_c F + \mathcal{M}\gamma_c M + \mathcal{M}\underset{\sim}{\mathcal{Q}}_{\Omega^c} \mathcal{P}\{\mathcal{N}_1[\varphi] + \mathcal{N}_2[\lambda]\} \qquad (3.9.43)
\end{aligned}$$

where

$$[\varphi] = R(\gamma_c u - \gamma u) \quad \text{and} \quad [\lambda] = S(\gamma_c u - \gamma u). \qquad (3.9.44)$$

Inserting (3.8.27) into (3.9.42) and (3.9.43), we arrive at the boundary integral equations

$$\varphi^{\mp} = R\gamma F + R\gamma M + R\mathcal{D}\mathcal{P}\{\mathcal{N}_1[\varphi] + \mathcal{N}_2[\lambda]\} \mp \frac{1}{2}[x] \qquad (3.9.45)$$

and

$$\lambda^{\mp} = S\gamma F + S\gamma M + S\mathcal{D}\mathcal{P}\{\mathcal{N}_1[\varphi] + \mathcal{N}_2[\lambda]\} \mp \tfrac{1}{2}[x], \qquad (3.9.46)$$

where we use the properties $\gamma_c F = \gamma F$ and $\gamma_c M = \gamma M$ due to the C^∞–continuity of F and M.

Since φ^- and $[\varphi]$ in (3.9.45) are given we may use the boundary integral equation of the first kind for finding the unknown $[\lambda]$:

$$\begin{aligned}
\mathcal{A}[\lambda] &:= -R\mathcal{D}\mathcal{P}\mathcal{N}_2[\lambda] \qquad (3.9.47)\\
&= R\gamma F + R\gamma M(\bullet; u) - \varphi^- - \tfrac{1}{2}[\varphi] + R\mathcal{D}\mathcal{P}\mathcal{N}_1[\varphi].
\end{aligned}$$

Once $[\lambda]$ is known, λ^- may be obtained from the equation (3.9.46).

The solution u of the transmission problem is then given by the representation formula (3.9.41).

We notice that in all the exterior problems as well as in the transmission problem, the corresponding boundary integral equations as (3.9.32), (3.9.33) and (3.9.47) contain the term $M(x; u)$ which describes the behavior of the solution at infinity and may or may not be known a priori. In elasticity, $M(x; u)$ corresponds to the rigid motions. In order to guarantee the unique solvability of these boundary integral equations, additional compatibility conditions associated with $M(x; u)$ need to be appended.

3.10 Remarks

In all of the derivations of Sections 3.4–3.9 we did not care for less regularity of Γ by assuming $\Gamma \in C^\infty$. It should be understood that with appropriate bookkeeping of the respective orders of differentiation, all of the calculations presented in this chapter can be carried out under weaker assumptions for $C^{k,\kappa}$ boundaries according to the particular formulae and derivations. In order not to be overwhelmingly descriptive we leave these details to the reader.

As far as generalizations to mixed and screen boundary value problems are concerned, as indicated in Section 2.5, one has to modify the representation formulae as e.g. (3.9.18) by splitting the generalized Cauchy data $\varphi_{jD} = \varphi_j + \widetilde{\varphi}_j$ and $\lambda_{jN} = \lambda_j + \widetilde{\lambda}_j$ in order to obtain systems of boundary integral equations instead of (3.9.23) and (3.9.26). Also, singularities of the charges at the collision points need to be incorporated which is crucial for the corresponding analysis.

In the next chapter, we also include one of the very important applications of integral equations, namely those for the general Maxwell equations in electromagnetic theory, which have drawn much attention in recent years.

4. Sobolev Spaces

In order to study the variational formulations of boundary integral equations and their numerical approximations, one needs proper function spaces. The Sobolev spaces provide a very natural setting for variational problems. This chapter contains a brief summary of the basic definitions and results of the L^2–theory of Sobolev spaces which will suffice for our purposes. A more general discussion on these topics may be found in the standard books such as Adams [1], Grisvard [156], Lions and Magenes [270], Maz'ya [288], Tartar [414], and also in McLean [290].

4.1 The Spaces $H^s(\Omega)$

The Spaces $L^p(\Omega)(1 \le p \le \infty)$

We denote by $L^p(\Omega)$ for $1 \le p < \infty$, the space of equivalence classes of Lebesgue measurable functions u on the open subset $\Omega \subset \mathbb{R}^n$ such that $|u|^p$ is integrable on Ω. We recall that two Lebesgue measurable functions u, v on Ω are said to be *equivalent* if they are equal almost everywhere in Ω, i.e. $u(x) = v(x)$ for all x outside a set of Lebesgue measure zero (see Kufner and Fučik [249]). The space $L^p(\Omega)$ is a Banach space with the norm

$$\|u\|_{L^p(\Omega)} := \Big(\int\limits_\Omega |u(x)|^p dx \Big)^{1/p} .$$

In particular, for $p = 2$, we have the space of all square integrable functions $L^2(\Omega)$ which is also a Hilbert space with the inner product

$$(u, v)_{L^2(\Omega)} := \int\limits_\Omega u(x)\overline{v(x)}dx \quad \text{for all } u, v \in L^2(\Omega) .$$

A Lebesgue measurable function u on Ω is said to be *essentially bounded* if there exists a constant $c \ge 0$ such that $|u(x)| \le c$ almost everywhere, i.e. (a.e.) in Ω. We define

$$\operatorname*{ess\,sup}_{x\in\Omega} |u(x)| = \inf\{c \in \mathbb{R} \,|\, |u(x)| \le c \text{ a.e.in } \Omega\}.$$

G. C. Hsiao und W. L. Wendland, *Boundary Integral Equations*,
Applied Mathematical Sciences 164,
https://doi.org/10.1007/978-3-030-71127-6_4

By $L^\infty(\Omega)$, we denote the space of equivalence classes of essentially bounded, Lebesgue measurable functions on Ω. The space $L^\infty(\Omega)$ is a Banach space equipped with the norm

$$\| u \|_{L^\infty(\Omega)} := \operatorname*{ess\,sup}_{x\in\Omega} |u(x)|.$$

We now introduce the Sobolev spaces $H^s(\Omega)$. Here and in the rest of this chapter $\Omega \subset \mathbb{R}^n$ is a domain. For simplicity we begin with $s = m \in \mathbb{N}_0$ and define these spaces by the completion of $C^m(\Omega)$–functions. Alternatively, Sobolev spaces may be defined in terms of distributions and their generalized derivatives (or weak derivatives), (see, e.g. Adams [1], Hörmander [187], McLean [290]). However, it is our belief that from a computational point of view. The following approach is more constructive.

Let us first introduce the function space

$$C^m_*(\Omega) := \{u \in C^m(\Omega) \mid \|u\|_{W^m(\Omega)} < \infty\}$$

where

$$\|u\|_{W^m(\Omega)} := \Big\{ \sum_{|\alpha|\le m} \int_\Omega |D^\alpha u|^2 dx \Big\}^{1/2}. \tag{4.1.1}$$

Then we define the Sobolev space of order m to be the completion of $C^m_*(\Omega)$ with respect to the norm $\| \cdot \|_{W^m(\Omega)}$. By this we mean that for every $u \in W^m(\Omega)$ there exists a sequence $\{u_k\}_{k\in\mathbb{N}} \subset C^m_*(\Omega)$ such that

$$\lim_{k\to\infty} \|u - u_k\|_{W^m(\Omega)} = 0. \tag{4.1.2}$$

We recall that two Cauchy sequences $\{u_k\}$ and $\{v_k\}$ in $C^m_*(\Omega)$ are said to be equivalent if and only if $\lim_{k\to\infty} \|u_k - v_k\|_{W^m(\Omega)} = 0$. This implies that $W^m(\Omega)$, in fact, consists of all equivalence classes of Cauchy sequences and that the limit u in (4.1.2) is just a representative for the class of equivalent Cauchy sequences $\{u_k\}$. The space $W^m(\Omega)$ is a Hilbert space with the inner product defined by

$$(u, v)_m := \sum_{|\alpha|\le m} \int_\Omega D^\alpha u \overline{D^\alpha v} dx. \tag{4.1.3}$$

Clearly, for $m = 0$ we have $W^0(\Omega) = L^2(\Omega)$.

The same approach can be used for defining the Sobolev space $W^s(\Omega)$ for non–integer real positive s. Let

$$s = m + \sigma \quad \text{with } m \in \mathbb{N}_0 \quad \text{and } 0 < \sigma < 1; \tag{4.1.4}$$

and let us introduce the function space

$$C^s_*(\Omega) := \{u \in C^m(\Omega) \mid \|u\|_{W^s(\Omega)} < \infty\}$$

where

$$\|u\|_{W^s(\Omega)} := \left\{ \|u\|^2_{W^m(\Omega)} + \sum_{|\alpha|=m} \int_\Omega \int_\Omega \frac{|D^\alpha u(x) - D^\alpha u(y)|^2}{|x-y|^{n+2\sigma}} dxdy \right\}^{1/2}, \tag{4.1.5}$$

the *Slobodetskii norm*. Note that the second part in the definition (4.1.5) of the norm in $W^s(\Omega)$ gives the L^2–version of fractional differentiability, which is compatible to the pointwise version in $C^{m,\alpha}(\Omega)$. In the same manner as for the case of integer order, the Sobolev space $W^s(\Omega)$ of order s is the completion of the space $C^s_*(\Omega)$ with respect to the norm $\|\cdot\|_{W^s(\Omega)}$. Again, $W^s(\Omega)$ is a Hilbert space with respect to the inner product

$$(u,v)_s := \tag{4.1.6}$$

$$(u,v)_m + \sum_{|\alpha|=m} \int_\Omega \int_\Omega \frac{\big(D^\alpha u(x) - D^\alpha u(y)\big)\overline{\big(D^\alpha v(x) - D^\alpha v(y)\big)}}{|x-y|^{n+2\sigma}} dx\, dy\,.$$

Clearly, for $m = 0$ we have $W^0(\Omega) = L^2(\Omega)$.

Note, that all of the definitions above are also valid for $\Omega = \mathbb{R}^n$. In this case, Meyers and Serrin [294] have shown that the space $C_0^\infty(\mathbb{R}^n)$ is dense in $W^s(\mathbb{R}^n)$ which implies that for $\Omega = \mathbb{R}^n$. The Sobolev spaces defined via distributions are the same as $W^s(\mathbb{R}^n)$. We therefore denote

$$H^s(\mathbb{R}^n) := W^s(\mathbb{R}^n) \quad \text{for} \quad s \geq 0\,. \tag{4.1.7}$$

Instead of the functions in $C^m(\Omega)$ let us now consider the function space of restrictions,

$$C^\infty(\overline{\Omega}) := \{u = \widetilde{u}|_\Omega \quad \text{with} \quad \widetilde{u} \in C_0^\infty(\mathbb{R}^n)\}$$

and introduce for $s \geq 0$ the norm

$$\|u\|_{H^s(\Omega)} := \inf\{\|\widetilde{u}\|_{H^s(\mathbb{R}^n)} \,|\, u = \widetilde{u}|_\Omega\}\,. \tag{4.1.8}$$

Now we define $H^s(\Omega)$ to be the completion of $C^\infty(\overline{\Omega})$ with respect to the norm $\|\cdot\|_{H^s(\Omega)}$, which means that to every $u \in H^s(\Omega)$ there exists a sequence $\{u_k\}_{k\in\mathbb{N}} \subset C^\infty(\overline{\Omega})$ such that

$$\lim_{k\to\infty} \|u - u_k\|_{H^s(\Omega)} = 0\,.$$

Since $C_0^\infty(\Omega) \subset C^\infty(\overline{\Omega})$ where for any $u \in C_0^\infty(\Omega)$ the trivial extension $\widetilde{u}$ by zero outside of Ω is in $C_0^\infty(\mathbb{R}^n)$, we define the space $\widetilde{H}^s(\Omega)$ for $s \geq 0$ to be the completion of $C_0^\infty(\Omega)$ with respect to the norm

$$\|u\|_{\widetilde{H}^s(\Omega)} := \|\widetilde{u}\|_{H^s(\mathbb{R}^n)}\,. \tag{4.1.9}$$

This definition implies that

$$\widetilde{H}^s(\Omega) \subset H^s(\Omega) \text{ for } s \geq 0.$$

For negative s, we define $H^s(\Omega)$ by duality with respect to the inner product $(\cdot,\cdot)_{L^2(\Omega)}$. More precisely, for $s < 0$ we define the norm by

$$\|u\|_s = \sup_{0 \neq \varphi \in \widetilde{H}^{-s}(\Omega)} |(\varphi, u)_{L^2(\Omega)}| \,/\, \|\varphi\|_{\widetilde{H}^{-s}(\Omega)} \,. \tag{4.1.10}$$

As usual, we denote the completion of $L^2(\Omega)$ with respect to (4.1.10) by

$$H^s(\Omega) = (\widetilde{H}^{-s}(\Omega))' \quad \text{for} \quad s < 0\,, \tag{4.1.11}$$

the dual space of $\widetilde{H}^{-s}(\Omega)$. These are the Sobolev spaces of negative order.

Extending (4.1.11), we also define the spaces $\widetilde{H}^s(\Omega)$ for $s < 0$ by the completion of $L^2(\Omega)$ with respect to the norm

$$\|u\|_{\widetilde{H}^s(\Omega)} := \sup_{0 \neq \psi \in H^{-s}(\Omega)} |(\psi, u)_{L^2(\Omega)}| \,/\, \|\psi\|_{H^{-s}(\Omega)} \,. \tag{4.1.12}$$

This completion is denoted by

$$\widetilde{H}^s(\Omega) = \big(H^{-s}(\Omega)\big)'\,, \tag{4.1.13}$$

which is the dual of $H^{-s}(\Omega)$. In fact, one can show that

$$\widetilde{H}^s(\Omega) = \{u \in H^s(\mathbb{R}^n) \mid \operatorname{supp} u \subset \overline{\Omega}\}\,, \tag{4.1.14}$$

see Hörmander [187]. These are also subspaces of the Sobolev spaces $H^s(\Omega)$.

For $s > 0$, we have the inclusions:

$$\widetilde{H}^s(\Omega) \subset H^s(\Omega) \subset L^2(\Omega) \subset \widetilde{H}^{-s}(\Omega) \subset H^{-s}(\Omega) \tag{4.1.15}$$

with continuous injections (i.e. injective linear mappings). We remark that the elements of the Sobolev spaces of negative order, $H^{-s}(\Omega)$ and $\widetilde{H}^{-s}(\Omega)$, define bounded, i.e. continuous, linear functionals on the Sobolev spaces $\widetilde{H}^s(\Omega)$ and $H^s(\Omega)$, respectively. As will be seen, these functionals can be represented in two different ways.

In what follows, let V denote the Hilbert space $\widetilde{H}^s(\Omega)$ or $H^s(\Omega)$ with $0 \leq s$, and let V' denote its dual. That is, we have $V \subset L^2 \subset V'$. We set the value of the continuous linear functional ℓ at v by

$$\langle \ell, v\rangle := \ell(v) \text{ for } \ \ell \in V' \text{ and } \ v \in V \quad \text{with} \ \ \langle \ell, v\rangle = (v, \overline{\ell})_{L^2} \tag{4.1.16}$$

provided $\ell \in L^2$. Here the bilinear form $\langle \ell, v\rangle$ is called the *duality pairing* on $V' \times V$. If $\|u\|_V$ is the norm on V, the space V' is supplied with the *dual norm*

$$\|\ell\|_{V'} := \sup_{0 \neq v \in V} \frac{|\langle \ell, v\rangle|}{\|v\|_V}.$$

Since V is a Hilbert space, by the Riesz representation theorem, for each $\ell \in V'$ there exists a unique $\lambda \in V$ depending on ℓ such that

$$\langle \ell, v\rangle = \ell(v) = (v, \lambda)_V \text{ and } \|\ell\|_{V'} = \|\lambda\|_V, \tag{4.1.17}$$

where $(v, \lambda)_V$ denotes the inner product in V. This is one way to represent the element of V', although it is not very practical, since this representation involves here the complicated inner product $(\bullet, \bullet)_V$ of the Hilbert space V.

The other way to represent $\ell \in V'$ is to identify ℓ with an element $\overline{u}_\ell$ in $L^2(\Omega)$ and to consider the value $\ell(v)$ at $v \in V$ as the L^2-inner product $(v, u_\ell)_{L^2(\Omega)}$ provided $\ell \in V'$ is also bounded on L^2. In other words, we express the bilinear form $\langle \ell, v\rangle$ on $V' \times V$ as an inner product $(v, u_\ell)_{L^2(\Omega)}$. In fact, the definitions (4.1.10) and (4.1.12) for the dual spaces $V' = H^s(\Omega)$ and $\widetilde{H}^s(\Omega)$ for $s < 0$ are based on this representation. In case $\ell \in L^2(\Omega) \subset V'$, we simply let $\overline{u}_\ell = \ell \in L^2(\Omega)$ and obtain

$$\langle \ell, v\rangle = (v, u_\ell)_{L^2(\Omega)} = \int\limits_\Omega v(x)\overline{u_\ell(x)}\, dx = \int\limits_\Omega v(x)\ell(x) dx$$

for all $v \in V$. On the other hand, if $\ell \in V'$ but $\ell \notin L^2(\Omega)$, then we define

$$\langle \ell, v\rangle = \lim_{k\to\infty} (v, u_{\ell_k})_{L^2(\Omega)} = \lim_{k\to\infty} \int\limits_\Omega v(x)\overline{u_{\ell_k}(x)} dx \tag{4.1.18}$$

for all $v \in V$, where $\{u_{\ell_k}\} \subset L^2(\Omega)$ is a sequence such that

$$\lim_{k\to\infty} \|\overline{u}_{\ell_k} - \ell\|_{V'} = 0.$$

We know that $\{\overline{u}_{\ell_k}\}$ exists and (4.1.18) makes sense, since V' is the completion of $L^2(\Omega)$ with respect to the norm $\|\cdot\|_{V'}$, defined by (4.1.10) and (4.1.11), respectively.

The second way is more practical; and as will be seen, our variational formulations of boundary integral equations in Chapter 5 are based on this representation. In what follows, we write

$$\langle u, v\rangle_{L^2(\Omega)} = (u, \overline{v})_{L^2(\Omega)} \tag{4.1.19}$$

for the duality pairing of $(u, v) \in V' \times V$; and the L^2–inner product on the right–hand side is understood in the sense of (4.1.18) for $u \notin L^2(\Omega)$.

We remark that both representations are equivalent and are defined by the same linear functional. Hence, the mapping $\ell \mapsto \lambda$ from $\widetilde{H}^{-s}(\Omega)$ onto $H^s(\Omega)$ is a well defined *isomorphism*. By an isomorphism we mean a continuous one–to–one linear mapping from $\widetilde{H}^{-s}(\Omega)$ onto $H^s(\Omega)$ whose inverse is also continuous.

Now we consider the special case $\Omega = \mathbb{R}^n$. In this case, for any $s \in \mathbb{R}$, in addition we have

$$\widetilde{H}^s(\mathbb{R}^n) = H^s(\mathbb{R}^n)\,. \tag{4.1.20}$$

Moreover, these Sobolev spaces can be characterized via the Fourier transformation. For any $u \in C_0^\infty(\mathbb{R}^n)$, the Fourier transformed $\hat{u}$ is defined by (3.1.12), and there holds

$$\mathcal{F}^*_{\xi\mapsto x}\widehat{u} := u(x) = (2\pi)^{-n/2}\int\limits_{\mathbb{R}^n} e^{ix\cdot\xi}\hat{u}(\xi)d\xi\,, \tag{4.1.21}$$

with the *inverse Fourier transform* of $\hat{u}$. The following properties are well known (see e.g. Petersen [346, p.79]; i.e. *the Parseval–Plancherel formula*

$$(u,v)_{L^2(\mathbb{R}^n)} = \int\limits_{\mathbb{R}^n} u(x)\overline{v(x)}dx = \int\limits_{\mathbb{R}^n} \hat{u}(\xi)\overline{\hat{v}(\xi)}d\xi = (\hat{u},\hat{v})_{L^2(\mathbb{R}^n)}\,, \tag{4.1.22}$$

$$\int\limits_{\mathbb{R}^n} u(x)\overline{\hat{v}(x)}dx = \int\limits_{\mathbb{R}^n} \hat{u}(\xi)\overline{v(\xi)}d\xi\,. \tag{4.1.23}$$

Formula (4.1.22) implies

$$\|\hat{u}\|_{L^2(\mathbb{R}^n)} = \|u\|_{L^2(\mathbb{R}^n)} \tag{4.1.24}$$

for any $u \in C_0^\infty(\mathbb{R}^n)$. A simple application of (4.1.24) to the identity

$$\begin{aligned}(2\pi)^{-n/2}\int\limits_{\mathbb{R}^n} e^{-ix\cdot\xi}(D^\alpha u(x))dx &= (-1)^{|\alpha|}(2\pi)^{-n/2}\int\limits_{\mathbb{R}^n}(D_x^\alpha e^{-ix\cdot\xi})u(x)dx\\ &= (i\xi)^\alpha\hat{u}(\xi)\end{aligned}$$

for any $\alpha \in \mathbb{N}_0^n$ shows that

$$(\xi^\alpha\hat{u}(\xi)) \in L^2(\mathbb{R}^n) \quad \text{for every } \alpha \in \mathbb{N}_0^n \text{ and } u \in C_0^\infty(\mathbb{R}^n).$$

This implies that we have

$$|||u|||_s^2 := (2\pi)^{-n}\int\limits_{\mathbb{R}^n}(1+|\xi|^2)^s|\hat{u}(\xi)|^2d\xi < \infty \tag{4.1.25}$$

for every $s \in \mathbb{R}$ and $u \in C_0^\infty(\mathbb{R}^n)$. Obviously, $|||\cdot|||_s$ is a norm. Now let $\mathcal{H}^s(\mathbb{R}^n)$ be the completion of $C_0^\infty(\mathbb{R}^n)$ with respect to the norm $|||\cdot|||_s$. Then the space $\mathcal{H}^s(\mathbb{R}^n)$ again is a Hilbert space with the inner product defined by

$$((u,v))_s := (2\pi)^{-n}\int\limits_{\mathbb{R}^n}(1+|\xi|^2)^s\hat{u}(\xi)\overline{\hat{v}(\xi)}d\xi\,. \tag{4.1.26}$$

From Parseval's formula, (4.1.24) it follows that

$$H^0(\mathbb{R}^n) = \mathcal{H}^0(\mathbb{R}^n) = L^2(\mathbb{R}^n).$$

In fact, one can show that $H^s(\mathbb{R}^n)$ is equivalent to $\mathcal{H}^s(\mathbb{R}^n)$ for every $s \in \mathbb{R}$, see e.g. Petersen [346, p.235]. Therefore, in what follows, we shall not distinguish between the spaces $H^s(\mathbb{R}^n)$ and $\mathcal{H}^s(\mathbb{R}^n)$ anymore. The corresponding norms $\|\cdot\|_s$ and $|||\cdot|||_s$ will be employed interchangeably.

As a further consequence of Parseval's equality (4.1.24), we see that the Fourier transform $\mathcal{F}u(\xi) := \hat{u}(\xi)$ as a bounded linear mapping from $C_0^\infty(\mathbb{R}^n)$ into $L^2(\mathbb{R}^n)$ extends continuously to $L^2(\mathbb{R}^n)$ and the extension defines an isomorphism $L^2(\mathbb{R}^n) \to L^2(\mathbb{R}^n)$.

Here again, by an isomorphism we mean a continuous one–to–one mapping from $L^2(\mathbb{R}^n)$ onto itself whose inverse is also continuous. In fact, the inverse $\mathcal{F}^{-1}$ is given by (4.1.21). Further properties of $\mathcal{F}$ on the Sobolev spaces and on distributions will be discussed in Chapter 7.

Now we summarize some of the relevant results and properties of $H^s(\Omega)$ without proofs. Proofs can be found e.g. in Adams [1], Kufner and Fučik [249], McLean [290] and Petersen [346].

Lemma 4.1.1. Generalized Cauchy-Schwarz inequality (Adams [1, p.50]).

The $H^s(\Omega)$–scalar product extends to a continuous bilinear form on $H^{s+t}(\Omega) \times \widetilde{H}^{s-t}(\Omega)$. Moreover, we have

$$|(u,v)_{H^s(\Omega)}| \le \|u\|_{H^{s+t}(\Omega)} \|v\|_{\widetilde{H}^{s-t}(\Omega)} \tag{4.1.27}$$

for all $(u,v) \in H^{s+t}(\Omega) \times \widetilde{H}^{s-t}(\Omega)$ and all $s,t \in \mathbb{R}$. This inequality also holds for $\Omega = \mathbb{R}^n$ where $\widetilde{H}^{s-t}(\mathbb{R}^n) = H^{s-t}(\mathbb{R}^n)$.

Lemma 4.1.2. (Petersen [346, Chap.4, Lemma 2.2])

For $s < t$, the inclusion $H^t(\mathbb{R}^n) \subset H^s(\mathbb{R}^n)$ is continuous and has dense image.

Theorem 4.1.3. Rellich's Lemma (Petersen [346, Chap.4, Theorem 1.12])

Let $s < t$ and $\Omega \subset \mathbb{R}^n$ be a bounded domain. Then the imbedding

$$\widetilde{H}^t(\Omega) \overset{c}{\hookrightarrow} H^s(\mathbb{R}^n)$$

is bounded and compact.

In the following, the symbol $\overset{c}{\hookrightarrow}$ will be used for compact imbedding.

Theorem 4.1.4. Sobolev imbedding theorem (Petersen [346, Chap.4, Theorem 2.13])

Let $m \in \mathbb{N}_0$ and $s > m + \frac{n}{2}$. Then for every $u \in H^s(\mathbb{R}^n)$ and every compact $K \Subset \mathbb{R}^n$ we have $u|_K \in C^{m+\alpha}(K)$ and

$$\|u\|_{C^{m,\alpha}(K)} \le C(K,m,\alpha)\|u\|_{H^s(\mathbb{R}^n)}$$

if $0 < \alpha \le s - m - \frac{n}{2}$ and $\alpha < 1$.

Before we can state further properties we need some mild restrictions on the domain Ω under consideration.

The domain Ω is said to have the *uniform cone property* if there exists a locally finite open covering $\{\mathcal{U}_j\}$ of $\partial\Omega = \Gamma$ and a corresponding sequence $\{\mathcal{C}_j\}$ of finite cones, each congruent to some fixed finite cone $\mathcal{C}$ such that:

(i) *For some finite number M, every $\mathcal{U}_j$ has diameter less than M.*
(ii) *For some constant $\delta > 0$ we have*
$\bigcup_{j=1}^{\infty} \mathcal{U}_j \supset \Omega_\delta := \{x \in \Omega \,|\, \inf_{y\in\Gamma} |x - y| < \delta\}$.
(iii) *For every index j we have $\bigcup_{x\in\Omega\cap\mathcal{U}_j}(x + \mathcal{C}_j) =: Q_j \subset \Omega$.*
(iv) *For some finite number R, every collection of $R + 1$ of the sets Q_j has empty intersection.*

We remark that the uniform cone property is closely related to the smoothness of Γ. In particular, any C^2–boundary Γ will possess the uniform cone property. It was shown by Gagliardo [132] and Chenais [66] that every Lipschitz domain has the uniform cone property. Conversely, Chenais [66] showed that every bounded domain having the uniform cone property, is a Lipschitz domain.

Theorem 4.1.5. Strong extension property
(Calderon and Zygmund [60], McLean [290], Wloka [439])
Let Ω be bounded and satisfy the uniform cone property.

i. Let $m \in \mathbb{N}_0$. Then there exists a continuous, linear so–called extension operator

$$F_\Omega : H^s(\Omega) \to H^s(\mathbb{R}^n) \quad \text{for } m \le s < m + 1$$

satisfying

$$\|F_\Omega u\|_{H^s(\mathbb{R}^n)} \le c_\Omega(s)\|u\|_{H^s(\Omega)} \quad \text{for all } u \in H^s(\Omega)\,. \tag{4.1.28}$$

ii. Let Γ be a $C^{k,\kappa}$–boundary with $k+\kappa \ge 1$, $k \in \mathbb{N}$, and let $0 \le s < k+\kappa$ if $k+\kappa \notin \mathbb{N}$ or $0 \le s \le k+\kappa$ if $k+\kappa \in \mathbb{N}$. Then there exists a continuous, linear extension operator F_Ω which is independent of s satisfying (4.1.28).

As a consequence, whenever the extension operator F_Ω exists, we have $W^s(\Omega) = H^s(\Omega)$ and the norms

$$\|u\|_{W^s(\Omega)} \quad \text{and} \quad \|u\|_{H^s(\Omega)} \tag{4.1.29}$$

are equivalent (Wloka [439, Theorem 5.3]).

Theorem 4.1.6. Rellich's Lemma, Sobolev imbedding Theorem
Adams [1]: *Let Ω be bounded and satisfy the uniform cone property. Then the following imbeddings are continuous and compact:*

i.

$$\widetilde{H}^s(\Omega) \overset{c}{\hookrightarrow} \widetilde{H}^t(\Omega) \quad \text{for } -\infty < t < s < \infty\,, \tag{4.1.30}$$

ii.

$$H^s(\Omega) \overset{c}{\hookrightarrow} H^t(\Omega) \quad \textit{for } -\infty < t < s < \infty\,. \tag{4.1.31}$$

iii.

$$H^s(\Omega) \overset{c}{\hookrightarrow} C^{m,\alpha}(\overline{\Omega}) \quad \text{for } m \in \mathbb{N}_0, 0 \le \alpha < 1, m+\alpha < s - \tfrac{n}{2}\,. \tag{4.1.32}$$

For $s = m + \alpha + \frac{n}{2}$ and $0 < \alpha < 1$, the imbeddding

$$H^s(\Omega) \hookrightarrow C^{m,\alpha}(\overline{\Omega}) \tag{4.1.33}$$

is continuous.

Classical Sobolev spaces

For $m \in \mathbb{N}_0$, let $H_0^m(\Omega)$ be the completion of $C_0^m(\Omega)$ with respect to the norm $\|\cdot\|_{H^m(\Omega)}$. Then we have

$$\begin{aligned} H_0^m(\Omega) &= \widetilde{H}^m(\Omega) \quad \text{and} \\ H^{-m}(\Omega) &= \Big(\widetilde{H}^m(\Omega)\Big)' = (H_0^m(\Omega))'\,. \end{aligned} \tag{4.1.34}$$

In the case $0 < s \notin \mathbb{N}_0$, for $\widetilde{H}^s(\Omega)$ there are two cases to be considered. If $s = m + \sigma$ with $|\sigma| < \frac{1}{2}$ and $m \in \mathbb{N}_0$ then there holds

$$\widetilde{H}^s(\Omega) = \overline{C_0^m(\Omega)}^{\|\cdot\|_{H^s(\mathbb{R}^n)}} = H_0^s(\Omega) := \overline{C_0^m(\Omega)}^{\|\cdot\|_{H^s(\Omega)}}\,, \tag{4.1.35}$$

where the completion of the space $C_0^m(\Omega)$ with respect to the norms $\|\cdot\|_{H^s(\Omega)}$ and $\|\cdot\|_{H^s(\mathbb{R}^n)}$, respectively, is taken.

The spaces $H_{00}^s(\Omega)$

If $s = m + \frac{1}{2}$, then the space $\widetilde{H}^s(\Omega)$ is *strictly* contained in $H_0^s(\Omega)$:

$$\widetilde{H}^s(\Omega) = \overline{C_0^m(\Omega)}^{\|\cdot\|_{H^s(\mathbb{R}^n)}} \overset{\bullet}{\subset} H_0^s(\Omega)\,. \tag{4.1.36}$$

In this case, Lions and Magenes [270] characterize $\widetilde{H}^s(\Omega)$ by using the norm

$$\|u\|_{H_{00}^s(\Omega)} = \Big\{\|u\|^2_{H^s(\Omega)} + \sum_{|\alpha|=m} \|\rho^{-\frac{1}{2}} D^\alpha u\|^2_{L^2(\Omega)}\Big\}^{\frac{1}{2}} \tag{4.1.37}$$

where $\rho = \operatorname{dist}(x, \partial\Omega)$ for $x \in \Omega$. It is shown in Lions and Magenes [270, p.66] that the norms

$$\|u\|_{H^s_{00}(\Omega)} \text{ and } \|\widetilde{u}\|_{H^s(\mathbb{R}^n)} \text{ with } \widetilde{u} = \begin{cases} u & \text{for } x \in \Omega \ , \\ 0 & \text{otherwise} \ , \end{cases}$$

are equivalent.

In variational problems it is sometimes convenient to express equivalent norms in different forms. In the following we present a corresponding general theorem from which various well known equivalent norms can be deduced.

Theorem 4.1.7. Various Equivalent Norms
Triebel [421, Theorem 6.28.2]

Let Ω be a bounded domain having the uniform cone property, $m \in \mathbb{N}$, and let $q(v)$ be a nonnegative continuous quadratic functional on $H^m(\Omega)$; i.e.,

$$\begin{aligned} q : H^m(\Omega) &\to \mathbb{R} \quad \textit{with } q(\lambda v) = |\lambda|^2 q(v) \quad \textit{for all } \lambda \in \mathbb{C}, \ v \in H^m(\Omega)\,, \\ 0 \le q(v) &\le c\|v\|^2_{H^m(\Omega)} \quad \textit{and} \\ q(\wp) &> 0 \quad \textit{for all polynomials } \wp \textit{ of degree less than } m\,. \end{aligned} \tag{4.1.38}$$

Then the norms $\|v\|_{H^m(\Omega)}$ and $\{\sum\limits_{|\alpha|=m} \|D^\alpha v\|^2_{L^2(\Omega)} + q(v)\}^{\frac{1}{2}}$ are equivalent on $H^m(\Omega)$.

In particular, let us consider the following special choices of q:

i. $q(v) := \sum\limits_{|\beta|<m} \left| \int_\Omega D^\beta v dx \right|^2$. Then we have the

Poincaré inequality:

$$\|v\|_{H^\ell(\Omega)} \le \|v\|_{H^m(\Omega)} \le c\Big\{ \sum_{|\alpha|=m} \|D^\alpha v\|^2_{L^2(\Omega)} + \sum_{|\beta|<m} \left| \int_\Omega D^\beta v dx \right|^2 \Big\}^{\frac{1}{2}} \tag{4.1.39}$$

for $0 \le \ell \le m$.

For equivalent inequalities see Amrouche, Ciarlet and Mardare [11].

ii. $q(v) := \int\limits_\Gamma |v|^2 ds$. Here we have the

Friedrichs inequality:

$$\|v\|_{H^m(\Omega)} \le c\Big\{ \sum_{|\alpha|=m} \|D^\alpha v\|^2_{L^2(\Omega)} + \int_\Gamma |v|^2 ds \Big\}^{\frac{1}{2}}. \tag{4.1.40}$$

When $v \in H^m_0(\Omega)$ then we have in particular

$$\|v\|_{H^\ell(\Omega)} \le \|v\|_{H^m(\Omega)} \le c\Big\{ \sum_{|\alpha|=m} \|D^\alpha v\|^2_{L^2(\Omega)} \Big\}^{\frac{1}{2}} \quad \text{for } 0 \le \ell \le m\,. \tag{4.1.41}$$

We often shall also need the local spaces defined as

$$H^s_{\mathrm{loc}}(\Omega) := \{u \in \mathcal{D}'(\Omega) \,|\, \forall \varphi \in C_0^\infty(\Omega) \,:\, \varphi u \in H^s(\mathbb{R}^n)\}\,. \tag{4.1.42}$$

Theorem 4.1.8. *The multiplication by φ is continuous; i.e.*

$$\|\varphi u\|_{H^s(\mathbb{R}^n)} \le 2^{|s|/2} \|u\|_{H^s(\mathbb{R}^n)} \int\limits_{\mathbb{R}^n} (1 + |\xi|^2)^{|s|/2} |\widehat{\varphi}(\xi)| d\xi \tag{4.1.43}$$

where $\widehat{\varphi}(\xi) = \mathcal{F}_{x\mapsto\xi}\varphi(x)$.

The space $H^s_{\mathrm{loc}}(\Omega)$ is a Frechet space since with an exhaustive sequence of compact subsets K_j of Ω one may choose a sequence $\{\varphi_j\} \subset C_0^\infty(\Omega)$ with $\varphi_j(x) = 1$ on K_j. Then the family of seminorms $\|\varphi_j u\|_{H^s(\mathbb{R}^n)}$ could be used to define the corresponding Frechet space.

We also introduce the space

$$H^s_{\mathrm{comp}}(\Omega) = \{u \in \mathcal{E}'(\Omega) \,|\, \text{ to } u \text{ there exists a compact } K \Subset \Omega \text{ and } u \in \widetilde{H}^s(K)\}\,. \tag{4.1.44}$$

The spaces $H^s_{\mathrm{loc}}(\Omega)$ and $H^{-s}_{\mathrm{comp}}(\Omega)$ are dual to each other with respect to the bilinear pairing defined by the bilinear form

$$\langle u, v\rangle_{L^2(\Omega)} = \int\limits_{\Omega} u(x)v(x)dx \quad \text{for } v \in H^s_{\mathrm{loc}}(\Omega)\,,\ u \in H^{-s}_{\mathrm{comp}}(\Omega)\,. \tag{4.1.45}$$

4.2 The Trace Spaces $H^s(\Gamma)$

In the study of boundary value problems we need to speak about the values which certain elements of $H^1(\Omega)$ take on the boundary Γ of Ω. If $u \in H^s(\Omega)$ is continuous up to the boundary Γ, then one can say that the value which u takes on Γ is the restriction to Γ (of the extension by continuity to $\overline{\Omega}$) of the function u, which will be denoted by $u|_\Gamma$. In general, however, the elements of $H^s(\Omega)$ are defined except for a set of n–dimensional zero measure and it is meaningless therefore to speak of their restrictions to Γ (which has an n–dimensional zero measure). We therefore need a new concept, the concept of the trace of a function on Γ, which can substitute and generalize the restriction $u|_\Gamma$ whenever the latter is inapplicable in the classical sense. In this connection, we need the boundary spaces. In the following, we define the boundary spaces in three different ways. However, it turns out that for Sobolev indices s with $|s| < k + \kappa - \frac{1}{2}$, these spaces are equivalent due to the trace theorem.

Let us begin with the definition of the integration on Γ. We introduce partitions of unity subordinate to the covering of Γ by the sets $B_{(r)} \subset \mathbb{R}^n$. We say, a family of functions $\alpha_{(r)} \in C_0^\infty(\mathbb{R}^n), r = 1, \dots, p$, is a *partition of unity*, if $\alpha_{(r)} : \mathbb{R}^n \to [0,1]$ has compact support $\operatorname{supp} \alpha_{(r)} \Subset B_{(r)}$ satisfying

$$\sum_{r=1}^{p} \alpha_{(r)}(x) = 1 \text{ for all } \; x \in \mathcal{U}_\Gamma ,$$

where $\mathcal{U}_\Gamma \supset \Gamma$ is an appropriate n–dimensional open neighbourhood of Γ. For $\Gamma \in C^{k,\kappa}$ and $k + \kappa \geq 1$, the tangent vectors are well defined almost everywhere for $x \in \Gamma \cap B_{(r)}$, so as in the surface element ds_x. Then, for f given on Γ, we define the *surface integral* by

$$\int_\Gamma f ds := \sum_{r=1}^{p} \int_{\Gamma \cap B_{(r)}} f(x) \alpha_{(r)}(x) ds_x .$$

By using the local representation (3.3.3) of Γ, the integrals on the right-hand side are reduced to integrals of the function $(f\alpha_{(r)}) \circ \big(x_{(r)} + T_{(r)}(y', a_{(r)}(y')\big)$ over $Q \subset \mathbb{R}^{n-1}$. This definition is intrinsic in the sense that it does neither depend on the special local coordinate representation of Γ nor on the particular partition of unity considered.

Now let $L^2(\Gamma)$ be the completion of $C^0(\Gamma)$, the space of all continuous functions on Γ, with respect to the norm

$$\|u\|_{L^2(\Gamma)} := \Big\{ \int_\Gamma |u(x)|^2 ds_x \Big\}^{\frac{1}{2}} . \tag{4.2.1}$$

As is well known, $L^2(\Gamma)$ is a Hilbert space with the scalar product

$$(u, v)_{L^2(\Gamma)} := \int_\Gamma u(x) \overline{v(x)} ds_x . \tag{4.2.2}$$

For a Lipschitz domain Ω one can show that there exists a unique linear mapping $\gamma_0 : H^s(\Omega) \to L^2(\Gamma)$ such that if $u \in C^0(\overline{\Omega})$ then $\gamma_0 u = u|_\Gamma$. If $u \in H^1(\Omega)$, we will call $\gamma_0 u$ the *trace* of u on Γ and the mapping γ_0 the *trace operator* (of order 0). However, in order to characterize all those elements in $L^2(\Gamma)$ which can be the trace of elements of $H^1(\Omega)$, we need to introduce the so-called *trace spaces* $H^s(\Gamma)$, and for $s = 0$, we simply set $H^0(\Gamma) = L^2(\Gamma)$.

Natural trace space $\mathbf{H}^s(\Gamma)$

The simplest way to define the trace spaces on Γ is to use extensions of functions defined on Γ to Sobolev spaces defined in Ω. For $s > 0$ let us introduce the linear space

$$
\begin{aligned}
C_{(s)}(\Gamma) \;:=\; & \{\varphi \in C^0(\Gamma) |\ \textit{to}\ \varphi\ \textit{there exists}\ \widetilde{\varphi} \in H^{s+\frac{1}{2}}(\Omega) \\
& \textit{such that}\ \gamma_0\widetilde{\varphi} := \widetilde{\varphi}_{|\Gamma} = \varphi\ \textit{on}\ \Gamma\}\,.
\end{aligned}
$$

Then the *natural trace space* $\mathbf{H}^s(\Gamma)$ is defined to be the completion of $C_{(s)}(\Gamma)$ with respect to the norm

$$\|u\|_{\mathbf{H}^s(\Gamma)} := \inf_{\gamma_0\widetilde{u}=u} \|\widetilde{u}\|_{H^{s+\frac{1}{2}}(\Omega)}\,. \tag{4.2.3}$$

However, with this definition we note that it is difficult to define a scalar product associated with (4.2.3). On the other hand, the trace theorem holds *by definition*, namely

$$\|\gamma_0\widetilde{u}\|_{\mathbf{H}^s(\Gamma)} \leq \|\widetilde{u}\|_{H^{s+\frac{1}{2}}(\Omega)} \quad \text{for every } \widetilde{u} \in H^{s+\frac{1}{2}}(\Omega) \tag{4.2.4}$$

and for any $s > 0$. As we will show later on, $\mathbf{H}^s(\Gamma)$ is actually a Hilbert space, although the inner product can not be deduced from the above definition (4.2.3).

For $s < 0$, we can define the space $\mathbf{H}^s(\Gamma)$ as the dual of $\mathbf{H}^{-s}(\Gamma)$ with respect to the $L^2(\Gamma)$ scalar product; i.e. the completion of $L^2(\Gamma)$ with respect to the norm

$$\|u\|_{\mathbf{H}^s(\Gamma)} := \sup_{\|\varphi\|_{\mathbf{H}^{-s}(\Gamma)}=1} |\langle \varphi, u\rangle_{L^2(\Gamma)}|. \tag{4.2.5}$$

These are the boundary spaces of negative orders.

The Fichera trace spaces $\mathcal{H}^s(\Gamma)$

Alternatively, one may define the Sobolev spaces on the boundary in terms of boundary norms due to Fichera [121], instead of the domain norms as in (4.2.3). We begin with the simplest case; for $0 < s < 1$ we define $\mathcal{H}^s(\Gamma)$ to be the completion of

$$C_s{}^0(\Gamma) := \{\varphi \in C^0(\Gamma) |\ \|\varphi\|_{\mathcal{H}^s(\Gamma)} < \infty\}$$

with respect to the norm

$$\|u\|_{\mathcal{H}^s(\Gamma)} := \left\{ \|u\|^2_{L^2(\Gamma)} + \int_\Gamma\int_\Gamma \frac{|u(x)-u(y)|^2}{|x-y|^{n-1+2s}} ds_x ds_y \right\}^{\frac{1}{2}}. \tag{4.2.6}$$

Again, $\mathcal{H}^s(\Gamma)$ is a Hilbert space equipped with the inner product

$$(u,v)_{\mathcal{H}^s(\Gamma)} := (u,v)_{L^2(\Gamma)} + \int_\Gamma\int_\Gamma \frac{(u(x)-u(y))\,\overline{(v(x)-v(y))}}{|x-y|^{n-1+2s}} ds_x ds_y\,. \tag{4.2.7}$$

Defining $\mathcal{H}^s(\Gamma)$ for $s \geq 1$, it is more involved. In the following, let $m \in \mathbb{N}$ be fixed and let

$$\begin{aligned}\mathcal{M} &:= \{u \in C^m(\overline{\Omega}) \,|\, (\boldsymbol{t}(z)\cdot\nabla_x)^\alpha u(x)|_{z=x\in\Gamma} = 0 \text{ for all} \\ &\qquad 0 \le |\alpha| \le m \text{ and for all tangential vectors } \boldsymbol{t}_{|\Gamma}\}\,.\end{aligned}$$

This defines an equivalence relation on $C^m(\overline{\Omega})$. With $\mathcal{M}$ we can define the cosets of any $u \in C^m(\overline{\Omega})$ by

$$[u] = \{v \in C^m(\overline{\Omega}) | \, v - u \in \mathcal{M}\}\,.$$

Now let us consider the quotient space

$$C^m(\overline{\Omega})/\mathcal{M} := \{[u] | u \in C^m(\overline{\Omega})\}\,.$$

As usual, the quotient space $C^m(\overline{\Omega})/\mathcal{M}$ becomes a linear vector space by the proper extension of linear operations from $C^m(\overline{\Omega})$ to $C^m(\overline{\Omega})/\mathcal{M}$, i.e.

$$\alpha[u] + \beta[v] = [\alpha u + \beta v]$$

for all $\alpha, \beta \in \mathbb{C}$ and $u, v \in C^m(\overline{\Omega})$. On $C^m(\overline{\Omega})$, we introduce the Hermitian bilinear form

$$((u,v))_m := \sum_{|\alpha|\le m} \int_\Gamma D^\alpha u \overline{D^\alpha v}\, ds \tag{4.2.8}$$

and the associated semi–norm

$$|||u|||_m := ((u,u))_m^{\frac{1}{2}}\,. \tag{4.2.9}$$

If $|||\cdot|||_m$ were a norm on $C^m(\overline{\Omega})$, then the completion of the quotient space $C^m(\overline{\Omega})/\mathcal{M}$, would provide us with the desired Hilbert space on Γ. However, since $\{u \in C^m(\overline{\Omega}) \,|\, |||u|||_m = 0\} \ne \{0\}$, we need to introduce an additional quotient space in which $|||\cdot|||_m$ will become a norm. To be more precise, let

$$\begin{aligned}\mathcal{N} &:= \{u \in C^m(\overline{\Omega}) \,|\, D^\alpha u_{|\Gamma} = 0 \textit{ for all } \alpha \textit{ with } 0 \le |\alpha| \le m\} \\ &= \{u \in C^m(\overline{\Omega}) \,|\, |||u|||_m = 0\}\,.\end{aligned}$$

Then $\mathcal{N}$ defines an additional equivalence relation on $C^m(\overline{\Omega})$. Moreover, on the quotient space $C^m(\overline{\Omega})/\mathcal{N}$, given by the collection of cosets,

$$[u]_\mathcal{N} := \{v \in C^m(\overline{\Omega}) \,|\, |||v - u|||_m = 0\}\,,$$

$((\bullet,\bullet))_m$ and $|||\bullet|||_m$ define, by construction, a scalar product and an associated norm. The corresponding completion

$$\mathcal{B}^m(\Gamma) := \overline{C^m(\overline{\Omega})/\mathcal{N}}^{|||\cdot|||_m}$$

then defines a Hilbert space with the scalar product (4.2.8) and norm (4.2.9); for any $\dot{u}, \dot{v} \in \mathcal{B}^m(\Gamma)$ we find sequences $u_n, v_n \in C^m(\overline{\Omega})$ such that $\dot{u} = \lim_{n\to\infty}[u_n]_\mathcal{N}$ and $\dot{v} = \lim_{n\to\infty}[v_n]_\mathcal{N}$ in $\mathcal{B}^m(\Gamma)$. Thus, we may define

$$
\begin{aligned}
(\dot{u},\dot{v})_m := (\dot{u},\dot{v})_{\mathcal{B}^m(\Gamma)} &= \lim_{n\to\infty} ([u_n]_{\mathcal{N}},[v_n]_{\mathcal{N}})_{\mathcal{B}^m(\Gamma)} \\
&= \lim_{n\to\infty} ((u_n,v_n))_m \,, \qquad (4.2.10)\\
\|\dot{u}\|_m := \|\dot{u}\|_{\mathcal{B}^m(\Gamma)} &= \lim_{n\to\infty} \|[u_n]_{\mathcal{N}}\|_{\mathcal{B}^m(\Gamma)} = \lim_{n\to\infty} |||u_n|||_m \,. \qquad (4.2.11)
\end{aligned}
$$

In order to identify $C^m(\overline{\Omega})/\mathcal{M}$ with a subspace of $\mathcal{B}^m(\Gamma)$, we map $\mathcal{M}$ into $C^m(\overline{\Omega})/\mathcal{N}$ by

$$\mathcal{M} \ni u \longmapsto [u]_{\mathcal{N}} \in \mathcal{M}/\mathcal{N}\,,$$

where

$$\mathcal{M}/\mathcal{N} := \{[u]_{\mathcal{N}} \mid u \in \mathcal{M}\} \subset C^m(\overline{\Omega})/\mathcal{N} \subset \mathcal{B}^m\,.$$

One can easily show that the quotient spaces

$$C^m(\overline{\Omega})/\mathcal{M} \simeq \left(C^m(\overline{\Omega})/\mathcal{N}\right)/(\mathcal{M}/\mathcal{N})$$

are algebraically isomorphic, i.e., there exists a linear one–to–one correspondence between these two spaces. Now we denote by $\overline{\mathcal{M}}$ the completion of $\mathcal{M}/\mathcal{N}$ in $\mathcal{B}^m(\Gamma)$, and define the trace spaces by

$$\mathcal{H}^m(\Gamma) := \mathcal{B}^m(\Gamma)/\overline{\mathcal{M}}\,. \qquad (4.2.12)$$

For any $u \in \mathcal{H}^m(\Gamma)$, by the definition of the quotient space (4.2.12) this means that u has the form $u = [\dot{u}]$ for some $\dot{u} \in \mathcal{B}^m(\Gamma)$, and the norm of u is defined by

$$\|u\|_{\mathcal{H}^m(\Gamma)} := \inf_{\dot{v}\in\overline{\mathcal{M}}} \|\dot{u}+\dot{v}\|_m\,. \qquad (4.2.13)$$

Since $\overline{\mathcal{M}}$ is complete, we can find some $\dot{u}^* \in \overline{\mathcal{M}}$ such that

$$\|u\|_{\mathcal{H}^m(\Gamma)} = \|\dot{u}+\dot{u}^*\|_m = \inf_{\dot{v}\in\overline{\mathcal{M}}} \|\dot{u}+\dot{v}\|_m\,. \qquad (4.2.14)$$

Moreover, for

$$\overset{\circ}{u} := \dot{u}+\dot{u}^* \in \overline{\mathcal{M}}^{\perp} \subset \mathcal{B}^m(\Gamma) \quad \text{there holds } \; (\overset{\circ}{u},\dot{v})_m = 0 \text{ for all } \dot{v} \in \overline{\mathcal{M}}\,.$$

Therefore, $\mathcal{H}^m(\Gamma)$ and $\overline{\mathcal{M}}^{\perp}$ are isomorphic.

By the definition of completion, there exists a sequence $\overset{\circ}{u}_n \in C^m(\overline{\Omega})$ such that

$$\lim_{n\to\infty} \|[\overset{\circ}{u}_n]_{\mathcal{N}} - \overset{\circ}{u}\|_m = 0$$

and

$$\lim_{n,k\to\infty} \|[\overset{\circ}{u}_n]_{\mathcal{N}} - [\overset{\circ}{u}_k]_{\mathcal{N}}\|_m = \lim_{n,k\to\infty} |||\,\overset{\circ}{u}_n - \overset{\circ}{u}_k|||_m = 0\,. \qquad (4.2.15)$$

In particular, we see that

$$\|u\|_{\mathcal{H}^m(\Gamma)} = \lim_{n\to\infty} |||\overset{\circ}{u}_n|||_m\,. \tag{4.2.16}$$

Similarly, for $v \in \mathcal{H}^m(\Gamma)$ we can find $\overset{\circ}{v} \in \overline{\mathcal{M}}^{\perp}$ and a sequence $\overset{\circ}{v}_n \in C^m(\overline{\Omega})$ with the corresponding properties. Then we can define the inner product by

$$(u,v)_{\mathcal{H}^m(\Gamma)} := (\overset{\circ}{u},\overset{\circ}{v})_m = \lim_{n\to\infty} ((\overset{\circ}{u}_n,\overset{\circ}{v}_n))_m \tag{4.2.17}$$

so that $\mathcal{H}^m(\Gamma)$ becomes a Hilbert space.

Remark 4.2.1: As we mentioned earlier, the space $\mathcal{H}^m(\Gamma)$ may be considered as the completion of the quotient space $C^m(\overline{\Omega})/\mathcal{M}$ with respect to the norm

$$\|[u]\|_m = \inf_{\varphi\in\mathcal{M}} |||u+\varphi|||_m \tag{4.2.18}$$

corresponding to (4.2.14). However, in order to define the inner product $(\cdot,\cdot)_m$ in (4.2.17), it is necessary to introduce the quotient space $C^m(\overline{\Omega})/\mathcal{N}$. In addition, note that, in general, the infimum in (4.2.18) will not be attained in $\mathcal{M}$ since $\mathcal{M}$ is not complete.

To define $\mathcal{H}^s(\Gamma)$ for $s \notin \mathbb{N}, s>0$, we write $s = m+\sigma$ where $-\frac{1}{2} \le \sigma < \frac{1}{2}, m \in \mathbb{N}$, and proceed in a similar manner as before. More precisely, we first replace the scalar product (4.2.8) by

$$\begin{aligned}((u,v))_s &:= ((u,v))_{[s]}\\ &+ \sum_{|\alpha|=[s]} \int\limits_{\Gamma}\int\limits_{\Gamma} \frac{(D^\alpha u(x)-D^\alpha u(y))\overline{(D^\alpha v(x)-D^\alpha v(y))}}{|x-y|^{n-1+2(s-[s])}} ds_x ds_y\end{aligned} \tag{4.2.19}$$

where $[s] := \max\{\ell \in \mathbb{Z}\,|\,\ell \le s\}$ denotes the Gaussian bracket. Correspondingly, we define the associated semi–norm

$$|||u|||_s := ((u,u))_s^{\frac{1}{2}} \tag{4.2.20}$$

on $\widetilde{C}^s_*(\overline{\Omega})$ which is a subspace of $C^m(\overline{\Omega})$ defined by

$$\widetilde{C}^s_*(\overline{\Omega}) = \{u \in C^m(\overline{\Omega})\,|\,|||u|||_s < \infty\}\,.$$

The subspace $\mathcal{M}$ is now replaced by

$$\begin{aligned}\mathcal{M}^s \;=\; &\{u \in \widetilde{C}^s_*(\overline{\Omega})\,|\,(\boldsymbol{t}(z)\cdot\nabla_x)^\alpha u(x)|_{z=x\in\Gamma} = 0\\ &\text{for all } \;|\alpha| \le m \text{ and all tangent vectors } \boldsymbol{t}_{|\Gamma}\}\,;\end{aligned}$$

and the quotient space is replaced by $\widetilde{C}^s_*(\overline{\Omega})/\mathcal{M}^s$. Furthermore, let

$$\mathcal{N}^s := \{u \in \widetilde{C}^s_*(\overline{\Omega})\,|\,|||u|||_s = 0\}\,;$$

and the completion of the quotient space $\widetilde{C}^s_*(\overline{\Omega})/\mathcal{N}^s$, given by the cosets

$$[u]_{\mathcal{N}^s} := \{v \in \widetilde{C}^s_*(\overline{\Omega}) \mid |||u - v|||_s = 0\},$$

is denoted by $\mathcal{B}^s(\Gamma)$. We denote by $\overline{\mathcal{M}}^s$ the completion of

$$\mathcal{M}^s/\mathcal{N}^s := \{[u]_{\mathcal{N}^s} \mid u \in \mathcal{M}^s\} \subset \mathcal{B}^s(\Gamma)$$

in the Hilbert space $\mathcal{B}^s(\Gamma)$. Then, finally we define

$$\mathcal{H}^s(\Gamma) := \mathcal{B}^s(\Gamma)/\overline{\mathcal{M}^s} \tag{4.2.21}$$

as the new boundary Hilbert space. Note that any element $u \in \mathcal{H}^s(\Gamma)$ can again be approximated in $\mathcal{H}^s(\Gamma)$ by sequences $\overset{\circ}{u}_n \in \widetilde{C}^s_*(\overline{\Omega})$; and the details are exactly the same as in the case of integer $s = m$.

For negative s, we define $\mathcal{H}^s(\Gamma)$ by duality with respect to the $L^2(\Gamma)$–inner product, $(\cdot,\cdot)_{L^2(\Gamma)}$. Similar to (4.1.10), the norm is given by

$$\|u\|_{\mathcal{H}^s(\Gamma)} := \sup_{0 \neq \varphi \in \mathcal{H}^{-s}(\Gamma)} |(\varphi, u)_{L^2(\Gamma)}| \,/\, \|\varphi\|_{\mathcal{H}^{-s}(\Gamma)}\,; \tag{4.2.22}$$

and the completion of $L^2(\Gamma)$ with respect to (4.2.22) is denoted by

$$\begin{aligned} \mathcal{H}^s(\Gamma) &= \left(\mathcal{H}^{-s}(\Gamma)\right)', \\ &= \{\ell : \mathcal{H}^{-s}(\Gamma) \to \mathbb{C} \mid \ell \textit{ is linear and continuous }\}, \end{aligned} \tag{4.2.23}$$

i.e., the dual space of $\mathcal{H}^{-s}(\Gamma)$. We write again

$$\langle \ell, v\rangle = (\overline{v}, \ell)_{L^2(\Gamma)}$$

for $\ell \in \mathcal{H}^s(\Gamma)$ and $v \in \mathcal{H}^{-s}(\Gamma)$, and the $L^2(\Gamma)$-inner product $(\cdot,\cdot)_{L^2(\Gamma)}$ is understood to be similar in a sense to (4.1.18) for $\ell \notin L^2(\Gamma)$.

Remark 4.2.2: In principle, to replace the cosets, one can define $\mathcal{H}^s(\Gamma)$ in terms of functions on Γ with covariant derivatives of orders α with $|\alpha| \leq m$ from differential geometry which we tried to avoid here. Nevertheless, properties concerning the geometry of Γ will be discussed later.

Standard trace spaces $\mathbf{H}^s(\Gamma)$

In this approach we identify the boundary Γ with $\mathbb{R}^{n-1}$ by means of the local parametric representations of the boundary. Roughly speaking, we define the trace space to be isomorphic to the Sobolev space $H^s(\mathbb{R}^{n-1})$. This approach is the one which can be found in standard text books in connection with the trace theorem see Aubin [17, p.198ff].

Let us recall the parametric representation (3.3.3) of Γ, namely

$$x = x_{(r)} + T_{(r)}\left(y', a_{(r)}(y')\right) \text{ for } y' \in Q, r = 1, \dots, p.$$

Then, for s with $0 \le s < k + \kappa$ or for noninteger $k + \kappa$ or $0 \le s \le k + \kappa$ for integer $k + \kappa$ we define here the Sobolev space $\mathsf{H}^s(\Gamma)$ by the boundary space

$$\{u \in L^2(\Gamma) \,|\, u\left(x_{(r)} + T_{(r)}(y', a_{(r)}(y'))\right) \in H^s(Q), r = 1, \ldots, p\}$$

equipped with the norm

$$\|u\|_{\mathsf{H}^s(\Gamma)} := \Big\{ \sum_{r=1}^{p} \|u\left(x_{(r)} + T_{(r)}(y', a_{(r)}(y'))\right)\|^2_{H^s(Q)} \Big\}^{\frac{1}{2}} . \tag{4.2.24}$$

This space is a Hilbert space with the inner product

$$(u, v)_{\mathsf{H}^s(\Gamma)} := \sum_{r=1}^{p} \big(u(x_{(r)} + T_{(r)}(y', a_{(r)}(y'))), v(x_{(r)} + T_{(r)}(y', a_{(r)}(y')))\big)_{H^s(Q)} . \tag{4.2.25}$$

Note that the above restrictions of s are necessary since otherwise the differentiations with respect to y' required in (4.2.24) and (4.2.25) may not be well defined. Also note that $\mathsf{H}^s(Q)$ is the Sobolev space in the $(n-1)$–dimensional domain Q as in Section 4.1. In an additional step, these definitions, (4.2.24) and (4.2.25), can be rewritten in terms of the Sobolev space $H^s(\mathbb{R}^{n-1})$ by using the partition of unity on Γ subordinate to the open covering $B_{(r)}$. More precisely, we take the functions $\alpha_{(r)} \in C_0^\infty(\mathbb{R}^n), \operatorname{supp} \alpha_{(r)} \Subset B_{(r)}$ introduced previously, with

$$\sum_{r=1}^{p} \alpha_{(r)}(x) = 1 \tag{4.2.26}$$

in some neighbourhood of Γ. For u given on Γ, we define the extended function on $\mathbb{R}^{n-1}$ by

$$(\widetilde{\alpha_{(r)} u})(y') := \begin{cases} (\alpha_{(r)} u)\left(x_{(r)} + T_{(r)}(y', a_{(r)}(y'))\right) & \text{for } y' \in Q \;, \\ 0 & \text{otherwise}\,. \end{cases}$$

Then $\mathsf{H}^s(\Gamma)$ is given by all u on Γ for which $\widetilde{\alpha_{(r)} u} \in H^s(\mathbb{R}^{n-1}), r = 1, \ldots, p$. The corresponding norm now reads

$$\|u\|_{\mathsf{H}^s(\Gamma)} := \{\sum_{r=1}^{p} \|(\widetilde{\alpha_{(r)} u})\|^2_{H^s(\mathbb{R}^{n-1})}\}^{\frac{1}{2}} \tag{4.2.27}$$

and is associated with the scalar product

$$(u, v)_{\mathsf{H}^s(\Gamma)} := \sum_{r=1}^{p} (\widetilde{\alpha_{(r)} u}, \widetilde{\alpha_{(r)} v})_{H^s(\mathbb{R}^{n-1})} . \tag{4.2.28}$$

Clearly, for $1 \le s \le k+\kappa$, the integrals in (4.2.24)–(4.2.28) contain derivatives of u on Γ with respect to the local coordinates, i.e. derivatives of

$$u\left(x_{(r)} + T_{(r)}(y', a_{(r)}(y'))\right)$$

with respect to $y_j, j = 1, \cdots, n-1$. These derivatives are, in fact, the covariant derivatives of u with respect to Γ (see Bishop and Goldberg [32] and Millman and Parker [306]).

The particular charts, the pushed forward functions $\widetilde{\alpha_{(r)} u}$ are defined on $\mathbb{R}^{n-1}$ having compact supports in Q and in (4.2.27), (4.2.28) and using $H^s(\mathbb{R}^{n-1})$, we can introduce via L^2–duality the whole scale of Sobolev spaces for $\mathsf{H}^s(\Gamma)$ all s with $-k - \kappa \underset{(-)}{\leq} s \underset{(-)}{\leq} k + \kappa$.

Up to now, we have introduced three different boundary spaces. The connection between these boundary spaces and the Sobolev spaces on the domain Ω are given through the *trace theorem* in Ω. To this end, let us introduce the *trace operators* $\gamma_j, j = 0, 1, \cdots, m-1, 1 \leq m \in \mathbb{N}$,

$$\gamma_j : C^{m-1}(\overline{\Omega}) \to C^{m-1}(\Gamma)$$

defined by

$$\gamma_j u = \frac{\partial^j u}{\partial n^j}|_\Gamma := (n \cdot \nabla)^j u_{|\Gamma}, \quad j = 0, 1, \cdots, m-1\,.$$

In particular, we have, as before, for $u \in C^o(\overline{\Omega})$,

$$\gamma_0 u = u|_\Gamma,$$

and γ_0 is a linear operator which acts on a continuous function $u \in C^o(\overline{\Omega})$ to produce its restriction to the boundary Γ as a continuous function $\gamma_0 u$ on Γ. What we are really interested in, is the problem of how to define $\gamma_0 u$ when u belongs to $L^2(\Omega)$ or, more generally, to one of the Sobolev spaces $H^m(\Omega)$. As we mentioned earlier, functions belonging to $H^m(\Omega)$ are only defined uniquely on Ω and not on $\overline{\Omega}$ since Γ is a set of n–dimensional measure zero and functions in $H^m(\Omega)$ differing on a set of measure zero are regarded as being identical. However, the trace theorem enables us to define unambiguously $\gamma_0 u$ (or more generally $\gamma_j u$), provided that u is smooth enough, e.g. in $H^1(\Omega)$ (or $u \in H^m(\overline{\Omega})$) and Γ is sufficiently smooth. We now state the *Trace Theorem*.

Theorem 4.2.1. Trace Theorem (Costabel [80], Gagliardo [131], Grisvard [156], Lions [268], Miyazaki [312], Nečas [326, Chap.2,5], Ding [102], Wloka [439, p.130], Dindos and Mitrea [101], Mikhailov [297], Mitrea and Wright [310], Wendland [433] and on manifolds Große and Schneider [157])

i. Let $\Omega \in C^{k,\kappa}$ where $k+\kappa \geq 1$ and $s \in \mathbb{R}$ with $\frac{1}{2} < s < \max\{k+\kappa, \frac{3}{2}\}$ for non integer $k + \kappa$ or $\frac{1}{2} < s \leq k + \kappa$ for integer $k + \kappa$. Then there exists a linear continuous trace operator γ_0 with

$$\gamma_0 \;:\; H^s(\Omega) \to H^{s-\frac{1}{2}}(\Gamma) \tag{4.2.29}$$

which is an extension of

$$\gamma_0 u = u|_\Gamma \quad for \;\; u \in C^0(\overline{\Omega})\,.$$

The adjoint operator

$$\gamma_0^* \,:\, H^{-s+\frac{1}{2}}(\Gamma) \to H^{-s}(\mathbb{R}^n) \tag{4.2.30}$$

given by

$$(\gamma_0^*\sigma, \Phi)_{L^2(\mathbb{R}^n)} = \langle \sigma, \gamma_0\Phi\rangle_{L^2(\Gamma)}$$

satisfies

$$c_0\|\sigma\|_{H^{-s+\frac{1}{2}}(\Gamma)} \le \|\gamma_0^*\sigma\|_{H^{-s}(\mathbb{R}^n)} \le c_1\|\sigma\|_{H^{-s+\frac{1}{2}}(\Gamma)} \tag{4.2.31}$$

with some positive constants c_0 and c_1 (depending on Γ and s).

ii. Let $\Omega \in C^{k,1}$ and $s \in \mathbb{R}$ and $j,k \in \mathbb{N}_0$ with $\frac{1}{2}+j < s \le k+1$ for non integer $k+\kappa$ or $\frac{1}{2}+j < s \le k+\kappa$ for integer s. Then there exists a linear continuous trace operator γ_j with

$$\gamma_j : H^s(\Omega) \to H^{s-j-\frac{1}{2}}(\Gamma) \tag{4.2.32}$$

which is an extension of

$$\gamma_j u = \frac{\partial^j u}{\partial n^j}|_\Gamma \quad for \quad u \in C^\ell(\overline{\Omega}) \quad with \quad s+j \le \ell \in \mathbb{N}\,.$$

Moreover, for these s the three boundary spaces are equivalent:

$$\mathbf{H}^{s-\frac{1}{2}}(\Gamma) \simeq \mathcal{H}^{s-\frac{1}{2}}(\Gamma) \simeq \mathsf{H}^{s-\frac{1}{2}}(\Gamma) \tag{4.2.33}$$

In view of the trace theorem, for $\Omega \in C^{k,1}$ we identify once for all the boundary spaces

$$H^s(\Gamma) := \mathbf{H}^s(\Gamma) = \mathcal{H}^s(\Gamma) = \mathsf{H}^s(\Gamma) \tag{4.2.34}$$

for $0 < |s| \le k + \frac{1}{2}$, take $H^0(\Gamma) = L^2(\Gamma)$ and call $H^s(\Gamma)$ the *trace spaces.* We note that for $0 < s \le k+\frac{1}{2}$, the equivalence is due to the trace theorem. For $-k-\frac{1}{2} \le s < 0$ the equivalence follows from the common definition via $L^2(\Gamma)$–duality.

Since in these cases the boundary spaces are equivalent to the spaces $H^s(\mathbb{R}^{n-1})$, we have with the imbedding theorem in $\mathbb{R}^{n-1}$ also on Γ:

Theorem 4.2.2. *Let Γ be a $C^{k,1}$ boundary, $k \in \mathbb{N}_0$ and let $|t|\,,\;|s| \le k+\frac{1}{2}$. Then the imbeddings*

$$H^s(\Gamma) \overset{c}{\hookrightarrow} H^t(\Gamma) \qquad for \;\; t < s \tag{4.2.35}$$

$$H^s(\Gamma) \overset{c}{\hookrightarrow} C^{m,\alpha}(\Gamma) \quad for \;\; m+\alpha < s - \tfrac{n}{2} + \tfrac{1}{2}\,, m \in \mathbb{N}_0, 0 \le \alpha < 1 \tag{4.2.36}$$

are compact.

Remarks 4.2.3:

i. In case of a Lipschitz domain, i.e. $k = 0, \kappa = 1$, we see that for $\frac{1}{2} < s \leq 1$, the trace operator γ_0 exists with

$$\gamma_0 : H^s(\Omega) \to H^{s-\frac{1}{2}}(\Gamma)$$

and

$$\gamma_0 u = u|_\Gamma \text{ for } u \in C^0(\overline{\Omega}).$$

However, as shown by Costabel [80], this even extends to $\frac{1}{2} < s < \frac{3}{2}$ and in (Wendland [433]) for $\frac{1}{2} < s < k + \frac{3}{2}, k \in \mathbb{N}$, which indicates that the equivalence (4.2.33) can also be extended for $0 \leq |s| < k + \kappa$. However, as shown in (Wloka [439, p.130]), for $0 < \kappa < 1$ it seems that one needs the stronger restrictions $\frac{1}{2} + j < s < k + \kappa - 1$ for (4.2.32) to hold.

ii. For $\Omega \in C^{k,1}$ and fixed s with $\frac{1}{2} < s \leq k+1$ let m be the largest integer in $\mathbb{N}$ with $m < s + \frac{1}{2}$. Here, we may collect the trace operators $\gamma_j u$ for $j = 0, \cdots, m-1$ and define

$$\gamma u := (\gamma_0 u, \ldots, \gamma_{m-1} u)^\top . \tag{4.2.37}$$

Then

$$\gamma : H^s(\Omega) \to \prod_{j=0}^{m-1} H^{s-j-\frac{1}{2}}(\Gamma) \tag{4.2.38}$$

is continuous and linear and is an extension of

$$\gamma u = \Big(u, \frac{\partial u}{\partial n}, \cdots \frac{\partial^{m-1} u}{\partial n^{m-1}}\Big)^\top |_\Gamma \text{ for } u \in C^{m-1}(\overline{\Omega}) .$$

The continuity of γ can be expressed through the inequality

$$\sum_{j=0}^{m-1} \|\gamma_j u\|_{H^{s-j-\frac{1}{2}}(\Gamma)} \leq c\|u\|_{H^s(\Omega)}$$

with a constant c independent of u.

iii. The first approach shows that the spaces $\mathbf{H}^{s-\frac{1}{2}}(\Gamma)$ are just the quotient spaces

$$\mathbf{H}^{s-\frac{1}{2}}(\Gamma) = H^s(\Omega)/H^s_0(\Omega) . \tag{4.2.39}$$

Here, $H^s_0(\Omega)$ denotes the completion of $C^\infty_0(\Omega)$ in $H^s(\Omega)$. Hence, due to (4.2.34) we also have with $c_2, c_1 > 0$,

$$c_1\|u\|_{H^{s-\frac{1}{2}}(\Gamma)} \leq \inf_{\gamma_0 v = u} \|v\|_{H^s(\Omega)} \leq c_2\|u\|_{H^{s-\frac{1}{2}}(\Gamma)} . \tag{4.2.40}$$

iv. The trace theorem for $\gamma_0 : L^p_s(\Omega) \to B^{p,p}_{1-\frac{1}{p}}$ with $1 < p < \infty$ and $\frac{1}{p} < s < 1 + \frac{1}{p}$ and L^p_s- and Besov spaces can be found in (Jerison et al., [216], Mikhailov [297], Mitrea et al. [310]).

From the trace theorem, it is clear that the concept of the trace on Γ is a natural generalization of the concept of the restriction on Γ. We shall adopt the convention that the trace of the function u will be denoted by the symbol $u|_\Gamma$ or simply by the same symbol u. If $\varphi = \varphi(x)$ is a function defined on Γ, then the statement: $u = \varphi$ on Γ in the sense of traces will express the fact that $\gamma_0 u = \varphi$ on Γ. In addition to the trace theorem, we also have the *inverse trace theorem.*

Theorem 4.2.3. Inverse Trace Theorem (McLean [290], Neças [326, p.104], Wloka [439]). *For $\Omega \in C^{k,\kappa}$ let s be fixed with $\frac{1}{2} < s \leq k + \kappa$. Let m be the largest integer with $m < s + \frac{1}{2}$. Then there exists a linear bounded (i.e. continuous) right inverse $\mathcal{Z}$ to γ with*

$$\mathcal{Z} : \prod_{j=0}^{m-1} H^{s-j-\frac{1}{2}}(\Gamma) \to H^s(\Omega) \text{ and } \gamma(\mathcal{Z}\boldsymbol{\varphi}) = \boldsymbol{\varphi} \tag{4.2.41}$$

$$\text{for all } \boldsymbol{\varphi} \in \prod_{j=0}^{m-1} H^{s-j-\frac{1}{2}}(\Gamma)\,.$$

The continuity of $\mathcal{Z}$ can be expressed through the inequality

$$\|\mathcal{Z}\boldsymbol{\varphi}\|_{H^s(\Omega)} \leq c \sum_{j=0}^{m-1} \|\varphi_j\|_{H^{s-j-\frac{1}{2}}(\Gamma)} \tag{4.2.42}$$

with c independent of φ_j, the components of $\boldsymbol{\varphi}$.

Remark 4.2.4: The trace theorem for $\Omega \in C^{k,1}$ together with the inverse trace theorem implies that the trace operator γ in (4.2.38) is *surjective*. The kernel of γ is $H^s_0(\Omega)$, dense in $H^0(\Omega) = L^2(\Omega)$. The proof of the trace theorem usually is reduced to the case when $\Omega = \mathbb{R}^n_+ = \{x \in \mathbb{R}^n \,|\, x_n > 0\}$ and $\Gamma = \mathbb{R}^{n-1} = \{x \in \mathbb{R}^n \,|\, x_n = 0\}$ is its boundary (Aubin [17, p.198 ff.]), which corresponds to our spaces $\mathsf{H}^s(\Gamma)$. Note that in this case one also has

$$\gamma H^s(\Omega) = \gamma H^s_{\mathrm{loc}}(\Omega^c)\,.$$

Remark 4.2.5: For $s > k+1$ and $\Omega \in C^{k,1}$, the equivalence of our three boundary spaces in (4.2.34) is no longer valid since the trace theorem does not hold anymore. However, the definition of the boundary spaces $\mathbf{H}^{s-\frac{1}{2}}(\Gamma)$ provides the *trace theorem by definition* and we therefore define

$$H^{s-\frac{1}{2}}(\Gamma) := \mathbf{H}^{s-\frac{1}{2}}(\Gamma) \text{ for } s > k+1$$

to be the trace spaces.

For these s, we can see that the trace space $H^{s-\frac{1}{2}}(\Gamma)$ is in fact the same as the subspace of $L^2(\Gamma)$ defined by

$$\{u \in L^2(\Gamma) \mid \text{There exists (at least) one extension } \widetilde{u} \in H^s(\Omega) \text{ with } \gamma_0\widetilde{u} = u\} \tag{4.2.43}$$

equipped with the norm

$$\|u\|_{H^{s-\frac{1}{2}}(\Gamma)} := \inf_{\gamma_0\widetilde{u}=u} \|\widetilde{u}\|_{H^s(\Omega)}\,; \tag{4.2.44}$$

since $C^0(\Gamma)$ is dense in $L^2(\Gamma)$ and $s > 1$, we have $H^s(\Omega) \subset H^1(\Omega) \cap C^0(\overline{\Omega})$. Moreover, for any $\widetilde{u} \in H^s(\Omega)$ we have $\gamma_0\widetilde{u} \in H^{\frac{1}{2}}(\Gamma) \subset L^2(\Gamma)$. Consequently, *every* $\widetilde{u} \in H^s(\Omega)$ is admissible in (4.2.43) as well as in (4.2.3).

4.2.1 Trace Spaces for Periodic Functions on a Smooth Curve in $\mathbb{R}^2$

In the special case $n = 2$ and $\Gamma \in C^\infty$, the trace spaces can be identified with spaces of periodic functions defined by Fourier series. For simplicity, let Γ be a simple, closed curve. Then Γ admits a global parametric representation

$$\Gamma:\ x = x(t) \text{ for } t \in [0,1] \text{ with } x(0) = x(1) \tag{4.2.45}$$

satisfying

$$\Big|\frac{dx}{dt}\Big| \ge \gamma_0 > 0 \text{ for all } t \in \mathbb{R}\,.$$

Clearly, any function on Γ can be identified with a function on $[0,1]$ and its 1–periodic continuation to the real axis. In particular, $x(t)$ in (4.2.45) can be extended periodically. Then any function f on Γ can be identified with $f \circ x$ which is 1–periodic. Since $\Gamma \in C^\infty$, we may use either of the definitions of the trace spaces defined previously. In particular, for $s = m \in \mathbb{N}_0$, the norm of $u \in \mathsf{H}^m(\Gamma)$ in the standard trace spaces is given by (4.2.27) with $n = 2$; namely

$$\|u\|_{\mathsf{H}^m(\Gamma)} = \Big\{\sum_{r=1}^{p} \sum_{0\le\beta\le m} \int_{\mathbb{R}} \big|(\alpha_{(r)}u)^{(\beta)}(t)\big|^2 dt\Big\}^{\frac{1}{2}}.$$

For 1–periodic functions $u(t)$, this norm can be shown to be equivalent to a weighted norm in terms of the Fourier coefficients of the functions. As is well known, any 1–periodic function can be represented in the form

$$u\,(x(t)) = \sum_{j=-\infty}^{+\infty} \widehat{u}_j e^{2\pi ijt},\ t \in \mathbb{R} \tag{4.2.46}$$

where $\widehat{u}_j$ are the Fourier coefficients defined by

$$\widehat{u}_j = \int_0^1 e^{-2\pi ijt} u(x(t))\,dt,\; j \in \mathbb{Z}. \tag{4.2.47}$$

By using the Parseval equality, the following Lemma can be established.

Lemma 4.2.4. *Let $u(t)$ be 1–periodic: Then, for $m \in \mathbb{N}_0$, the norms*

$$\|u\|_{H^m(\Gamma)} \text{ and } \|u\|_{\mathcal{S}^m} := \Big\{\sum_{k\in\mathbb{Z}} |\widehat{u}(k)|^2 (k+\delta_{k0})^{2m}\Big\}^{\frac{1}{2}}$$

are equivalent.

Proof:

i. The definition (4.2.27) implies that

$$\|u\|^2_{H^m(\Gamma)} \le c \sum_{0\le\beta\le m} \sum_{r=1}^{p} \int_{\mathrm{supp}(\alpha_{(r)})} |u(x(t))^{(\beta)}|^2 dt,$$

from which by making use of the 1–periodicity of $u\circ x$, we arrive at

$$\|u\|^2_{H^m(\Gamma)} \le c \sum_{0\le\beta\le m} \int_0^1 |u(x(t))^{(\beta)}|^2 dt.$$

An application of the Parseval equality then yields

$$\begin{aligned}\sum_{0\le\beta\le m} \int_0^1 |u(x(t))^{(\beta)}|^2 dt &= \sum_{0\le\beta\le m} \sum_{j\in\mathbb{Z}} |2\pi j|^{2\beta} |\widehat{u}(j)|^2 \\ &\le (2\pi)^{2m}(m+1) \sum_{j\in\mathbb{Z}} |j+\delta_{0j}|^{2m} |\widehat{u}(j)|^2\end{aligned}$$

which gives

$$\|u\|_{H^m(\Gamma)} \le c\|u\|_{\mathcal{S}^m}.$$

ii. Conversely, again by the Parseval equality, we have

$$\begin{aligned}\|u\|^2_{\mathcal{S}^m} &= |\widehat{u}(0)|^2 + \frac{1}{(2\pi)^m} \sum_{j\in\mathbb{Z}} |j2\pi|^{2m} |\widehat{u}(j)|^2 \\ &= \Big(\int_0^1 (u\circ x)dt\Big)^2 + \int_0^1 |(u\circ x)^{(m)}|^2 dt \\ &= \Big(\int_0^1 \sum_{r=1}^{p} \alpha_{(r)} u dt\Big)^2 + \int_0^1 \sum_{r=1}^{p} \alpha_{(r)} |(u\circ x)^{(m)}|^2 dt \\ &\le c \sum_{r=1}^{p} \sum_{0\le\beta\le m} \int_{\mathbb{R}} |(\alpha_{(r)} u)^{(\beta)}(t)|^2 dt.\end{aligned}$$

This shows that

$$\|u\|_{\mathcal{S}^m} \le c\|u\|_{H^m(\Gamma)} \,.$$

■

In general, for $0 < s \notin \mathbb{N}$, we write $s = m + \sigma$ with $0 < \sigma < 1$. We now define for periodic functions the *Fourier series norm*,

$$\|u\|_{\mathcal{S}^s} := \{\sum_{k\in\mathbb{Z}} |\widehat{u}(k)|^2 |k + \delta_{k0}|^{2s}\}^{\frac{1}{2}} \,. \tag{4.2.48}$$

Following Kufner and Kadlec [250, p.250ff], we can again show the equivalence between the norms $\|\cdot\|_{H^s(\Gamma)}$ and $\|\cdot\|_{\mathcal{S}^s}$.

Lemma 4.2.5. *Let $u(t)$ be 1–periodic. Then for $0 < s \notin \mathbb{N}$, the norms $\|u\|_{H^s(\Gamma)}$ and $\|u\|_{\mathcal{S}^s}$ are equivalent.*

Proof: In view of (4.2.27), (4.1.5) with $\Omega = \mathbb{R}$ and Lemma 4.2.4, it suffices to prove that the double integral

$$I(u) := \int_0^1\int_0^1 \frac{|u^{(m)}(t) - u^{(m)}(\tau)|^2}{|\sin\pi(t-\tau)|^{1+2\sigma}} dt d\tau$$

converges if and only if $\|u^{(m)}\|_{\mathcal{S}^\sigma}$ is bounded. Of course, we may restrict ourselves to $m = 0$, since the case $m \in \mathbb{N}$ reduces to $m = 0$, for $v := u^{(m)}$.

Using periodicity and interchanging the order of integration we obtain

$$\begin{aligned} I(v) &= \int_0^1\int_0^1 \frac{|v(t+\tau) - v(t)|^2}{|\sin\pi\tau|^{1+2\sigma}} dt d\tau \\ &= \int_0^1 |w(\tau)|^2 |\sin\pi\tau|^{-1-2\sigma} d\tau \end{aligned}$$

where

$$w(\tau) = \|v(\bullet + \tau) - v(\bullet)\|_{L^2(0,1)} \,.$$

With the identity

$$(v(\bullet+\tau) - v(\bullet))\widehat{\ } = -\widehat{v}(j)(1 - e^{2\pi i j\tau})$$

and the Parseval equality one finds

$$w^2(\tau) = 4\sum_{j\in\mathbb{Z}} |\widehat{v}(j)|^2 (\sin\pi j\tau)^2 \,.$$

Therefore,

$$
\begin{aligned}
I(v) &= 4\sum_{j\in\mathbb{Z}} |\widehat{v}(j)|^2 \int_0^1 \frac{(\sin\pi j\tau)^2}{|\sin\pi\tau|^{1+2\sigma}}d\tau \\
&= 8\sum_{j\in\mathbb{Z}} |\widehat{v}(j)|^2 \int_0^{\frac{1}{2}} \frac{(\sin\pi j\tau)^2}{(\sin\pi\tau)^{1+2\sigma}}d\tau\,.
\end{aligned}
$$

Since for $\tau \in (0, \frac{1}{2})$ we have

$$
\frac{\tau}{2} \le \sin\pi\tau \le \pi\tau\,,
$$

it follows that

$$
\begin{aligned}
8\pi^{-1-2\sigma}\int_0^{\frac{1}{2}} \frac{(\sin\pi j\tau)^2}{\tau^{1+2\sigma}}d\tau &\le 8\int_0^{\frac{1}{2}} \frac{(\sin\pi j\tau)^2}{(\sin\pi\tau)^{1+2\sigma}}d\tau \\
&\le 8\cdot 2^{1+2\sigma}\int_0^{\frac{1}{2}} \frac{(\sin\pi j\tau)^2}{\tau^{1+2\sigma}}d\tau\,.
\end{aligned}
$$

We now rewrite the last integral on the right-hand side in the form:

$$
\begin{aligned}
\int_0^{\frac{1}{2}} \frac{(\sin\pi j\tau)^2}{\tau^{1+2\sigma}}d\tau &= (\pi j)^{2\sigma}\int_0^{j\frac{\pi}{2}} \frac{\sin^2 s}{s^{1+2\sigma}}ds \\
&=: (\pi j)^{2\sigma} c_{\sigma j}\,,
\end{aligned}
$$

and we see that

$$
c_{\sigma 1} \le c_{\sigma j} \le c_{\sigma\infty} = \int_0^{\infty} \frac{\sin^2 s}{s^{1+2\sigma}}ds < \infty\,.
$$

As a consequence, we find

$$
\begin{aligned}
&\pi^{-1}\cdot 8\cdot c_{\sigma 1}\sum |\widehat{v}(j)|^2|j|^{2\sigma} \\
&\le I(v) \le 8\cdot 2^{1+2\sigma}\cdot \pi^{2\sigma}c_{\sigma\infty}\sum |\widehat{v}(j)|^2|j|^{2\sigma}
\end{aligned}
$$

which shows the desired equivalence of $\|v\|_{H^\sigma(\Gamma)}$ and $\|u\|_{\mathcal{S}^\sigma}$. With $u^{(m)} = v$ and Lemma 4.2.4, the desired equivalence of $\|u\|_{H^s(\Gamma)}$ and $\|u\|_{\mathcal{S}^s}$ for $s = m+\sigma$ and $0 < \sigma < 1$ follows easily. ∎

For $s \ge 0$, the trace space $H^s(\Gamma)$ can also be considered to be the completion of $C^\infty(\Gamma)$ with respect to the Fourier series norm $\|\bullet\|_{\mathcal{S}^s}$ which, in fact, is generated by the scalar product

$$(u,v)_{\mathcal{S}^s} := \sum_{j\in\mathbb{Z}} |j+\delta_{j0}|^{2s}\widehat{u}(j)\overline{\widehat{v}(j)}\,. \tag{4.2.49}$$

By applying the Schwarz inequality to the right–hand side of (4.2.49), we immediately obtain the inequality

$$|(u,v)| \le \Big\{\sum_{j\in\mathbb{Z}} |j+\delta_{j0}|^{2(s+\varrho)}|\widehat{u}(j)|^2\Big\}^{\frac12}\Big\{\sum_{j\in\mathbb{Z}} |j+\delta_{j0}|^{2(s-\varrho)}|\widehat{v}(j)|^2\Big\}^{\frac12} \tag{4.2.50}$$

for *any* real ϱ. In terms of the norms (4.2.48), the last inequality can be written as

$$|(u,v)_{\mathcal{S}^s}| \le \|u\|_{\mathcal{S}^{s+\varrho}}\|v\|_{\mathcal{S}^{s-\varrho}} \tag{4.2.51}$$

which shows that the spaces $H^{s+\varrho}(\Gamma)$ and $H^{s-\varrho}(\Gamma)$ form a duality pair with respect to the inner product (4.2.49), if these spaces are supplied with the norms $\|\cdot\|_{\mathcal{S}^{s+\varrho}}$ and $\|\cdot\|_{\mathcal{S}^{s-\varrho}}$, respectively, for $0\le\varrho\le s$.

Moreover, the inequality (4.2.50) for *any* $\varrho\in\mathbb{R}$ indicates that the norm of the dual space gives for *negative* s the norm of the dual space of $H^{-s}(\Gamma)$ with respect to the $(\bullet,\bullet)_{\mathcal{S}^0}$–inner product. Therefore, for negative s, we can also use the norm (4.2.48) for the dual spaces of $H^{-s}(\Gamma)$.

Corollary 4.2.6. *Let u be a 1–periodic distribution. Then, for $s\in\mathbb{R}$, the norms $\|u\|_{H^s(\Gamma)}$ and $\|u\|_{\mathcal{S}^s}$ are equivalent.*

Remarks 4.2.6: Since for 1–periodic functions on $\mathbb{R}$, the Fourier series representation coincides with (4.1.22), the Fourier transformation of u on $\mathbb{R}$, the $\mathcal{S}^s$–norm (4.2.48) is identical to the one in terms of Fourier transform (4.1.25).

From the explicit expression (4.2.48) of the norm one can obtain elementarily the Sobolev imbedding theorems (4.1.31), (4.1.32) for the trace spaces where $n=1$ and Ω is replaced by Γ.

4.2.2 Trace Spaces on Curved Polygons in $\mathbb{R}^2$

Let $\Gamma\in C^{0,1}$ be a *curved polygon* which is composed of N simple C^∞–arcs $\Gamma_j, j=1,\cdots,N$ such that their closures $\overline{\Gamma}_j$ are C^∞. The curve $\overline{\Gamma}_{j+1}$ follows $\overline{\Gamma}_j$ according to the positive orientation. We denote by Z_j the vertex being the end point of Γ_j and the starting point of Γ_{j+1}.

For $s\in\mathbb{R}$ let $\underset{\sim}{H}^s(\Gamma_j)$ be the standard Sobolev spaces on the pieces $\Gamma_j, j=1,\cdots,N$ which are defined as follows. Without loss of generality, we assume that for each of the Γ_j we have a parametric representation

$$x = x_{(j)}(t) \text{ for } t\in\overline{\Omega}_j=[a_j,b_j]\subset\mathbb{R}$$

with $x_{(j)}(a_j)=Z_{j-1}, x_{(j)}(b_j)=Z_j, \Omega_j=(a_j,b_j), j=1,\cdots,N$ where $x_{(j)}\in C^\infty(\overline{\Omega})$. Then, we define the space

$$\underset{\sim}{H}^s(\Gamma_j) = \{\varphi \,|\, \varphi\,(x_{(j)}(\bullet)) \in H^s(\Omega_j)\}$$

to be equipped with the norm

$$\|\varphi\|_{\underset{\sim}{H}^s(\Gamma_j)} := \|\varphi \circ x_{(j)}\|_{H^s(\Omega_j)},$$

where $\|\cdot\|_{H^s(\Omega_j)}$ is defined as in Section 4.1. Then $\underset{\sim}{H}^s(\Gamma_j)$ is a Hilbert space with inner product

$$(\varphi, \psi)_{\underset{\sim}{H}^s(\Gamma_j)} = (\varphi \circ x_{(j)}, \psi \circ x_{(j)})_{H^s(\Omega_j)}. \tag{4.2.52}$$

From the formulation of the trace theorem, for $s = m + \sigma$ with $m \in \mathbb{N}$ and $|\sigma| < \frac{1}{2}$, we now introduce the boundary spaces

$$\mathbf{T}^{m+\sigma-\frac{1}{2}}(\Gamma) := \prod_{\ell=0}^{N} \prod_{j=1}^{m-1} \underset{\sim}{H}^{m-j+\sigma-\frac{1}{2}}(\Gamma_\ell).$$

We note that the restriction of a smooth function in $\mathbb{R}^2$ to the curve Γ automatically satisfies compatibility conditions of the derivatives at each vertex z_j, and this is not the case generally for functions in $\mathbf{T}^{m+\sigma-\frac{1}{2}}(\Gamma)$. Here we see that

$$\mathbf{T}^{m+\sigma-\frac{1}{2}}(\Gamma) \neq \prod_{j=0}^{m-1} \mathbf{H}^{m-j+\sigma-\frac{1}{2}}(\Gamma),$$

since the latter, i.e. the natural trace space, does implicitly include the appropriate compatibility conditions by construction.

The compatibility conditions can be found in Grisvard's book [156]. These are the following conditions: Let

$$P_\alpha = \sum_{|\alpha| \le d} a_\alpha D^\alpha = \sum_{j=0}^{\alpha} P_{\ell,j} \frac{\partial^j}{\partial n^j}|_{\Gamma_\ell}$$

be any partial differential operator in $\mathbb{R}^2$ of order d having constant coefficients a_α where the sums on the right–hand side correspond to the local decomposition of P_d in terms of the normal derivatives and the tangential differential operators $P_{\ell,j}$ of orders $d - j$ along Γ_ℓ for $\ell = 1, \cdots, N$.

Further let $\mathcal{P}_d$ denote the class of all these operators of order d. Then the *compatibility conditions* read for all $P_d \in \mathcal{P}_d$,

$$(C_1) \qquad \sum_{j=0}^{d} P_{\ell,j} f_{\ell,j}(z_\ell) = \sum_{j=0}^{d} P_{\ell+1,j} f_{\ell+1,j}(z_\ell)$$

for $d \le m - 2$ if $\sigma \le 0$ or $d \le m - 1$ if $\sigma > 0$,

$$(C_2) \qquad \int_0^\delta |\sum_{j=0}^d P_{\ell,j} f_{\ell,j}(x_{(\ell)}(t)) - P_{\ell+1,j} f_{\ell+1,j}(x_{\ell+1}(t))|^2 \frac{dt}{t} < \infty$$

for $\sigma = 0$, $d = m-1$ and some $\delta > 0$. Here, the functions

$$f_{\ell,j} \in \underset{\sim}{H}^{m-j+\sigma-\frac{1}{2}}(\Gamma_\ell) \quad \text{for} \quad j = 0, \ldots, m-1\,.$$

By requiring the functions in $\mathbf{T}^{m+\sigma-\frac{1}{2}}(\Gamma)$ to satisfy the compatibility conditions we now have the following version of the trace theorem.

Theorem 4.2.7. Trace theorem Grisvard [156, Theorem 1.5-2-8]: *Let*

$$\boldsymbol{P}^{m-\frac{1}{2}+\sigma}(\Gamma) := \{\boldsymbol{f} \in \mathbf{T}^{m-\sigma-\frac{1}{2}}(\Gamma) | \boldsymbol{f} = (f_0, \cdots, f_{m-1})^\top \wedge f_{j|\Gamma_\ell} =: f_{\ell,j}\}$$

satisfy (C_1) and for $\sigma = 0$ also (C_2). Then

$$\prod_{j=0}^{m-1} \mathbf{H}^{m-j-\frac{1}{2}+\sigma}(\Gamma) = \boldsymbol{P}^{m-\frac{1}{2}+\sigma}(\Gamma).$$

Alternatively, the mapping

$$u \longmapsto \gamma u_{|\Gamma_\ell} = (u, \frac{\partial u}{\partial n}, \cdots, \frac{\partial u^{m-1}}{\partial n^{m-1}})_{|\Gamma_\ell}$$

is a linear continuous and surjective mapping from $H^{m+\sigma}(\Omega)$ onto $\boldsymbol{P}^{m-\frac{1}{2}+\sigma}(\Gamma)$.

In order to illustrate the idea of the compatibility conditions (C_1) and (C_2), we now consider the special example for the polygonal region consisting of only two pieces Γ_1 and Γ_2 at the vertex z as shown in Figure 4.2.2. We consider two cases $m = 1$ and $m = 2$:

1. $m = 1, \sigma > 0$. Then $d = m - 1 = 0, \mathcal{P}_d = \mathcal{P}_0 = \{f_0 = a_0 D^0 \,|\, a_0 \in \mathbb{R}\}$ and $P_{\ell,0} = a_0, \ell = 1, 2$. The condition (C_1) takes the form

$$P_{1,0} f_{1,0}(z_1) = P_{2,0} f_{2,0}(z_1)$$

   which reduces simply to the condition

$$f_{1,0}(z_1) = f_{2,0}(z_1).$$

2. $m = 2, \sigma > 0$. Then $d \geq m - 1 = 1$ and

$$\mathcal{P}_d = \mathcal{P}_1 = \{P_1 = a_0 D^0 + a_{10}\partial_1 + a_{01}\partial_2 \,|\, a_0, a_{10}, a_{01} \in \mathbb{R}\}.$$

The differential operator P_1 restricted to Γ_1 can be written in terms of tangential and normal derivatives,

$$\begin{aligned}
P_1|_{\Gamma_1} &= a_0 + a_{10}\Big(\cos\alpha\frac{\partial}{\partial s} - \sin\alpha\frac{\partial}{\partial n}\Big) + a_{01}\Big(\sin\alpha\frac{\partial}{\partial s} + \cos\alpha\frac{\partial}{\partial n}\Big)\\
&= a_0 + (a_{10}\cos\alpha + a_{01}\sin\alpha)\frac{\partial}{\partial s} + (-a_{10}\sin\alpha + a_{01}\cos\alpha)\frac{\partial}{\partial n}\\
&= P_{1,0} + P_{1,1}\frac{\partial}{\partial n}|_{\Gamma_1},
\end{aligned}$$

where $P_{1,j}, j = 0,1$ are the differential operators tangential to Γ_1, explicitly given by

$$\begin{aligned}
P_{1,0} &= a_0 + (a_{10}\cos\alpha + a_{01}\sin\alpha)\frac{\partial}{\partial s}|_{\Gamma_1},\\
P_{1,1} &= (-a_{10}\sin\alpha + a_{01}\cos\alpha).
\end{aligned}$$

The differential operator P_1 restricted to Γ_2 is

$$P_1|_{\Gamma_2} = a_0 + a_{10}\frac{\partial}{\partial s}|_{\Gamma_2} - a_{01}\frac{\partial}{\partial n}|_{\Gamma_2} = P_{2,0} + P_{2,1}\frac{\partial}{\partial n}|_{\Gamma_2}$$

with

$$\begin{aligned}
P_{2,0} &= a_0 + a_{10}\frac{\partial}{\partial s}|_{\Gamma_2},\\
P_{2,1} &= -a_{01}.
\end{aligned}$$

The compatibility condition (C_1) now reads

$$P_{1,0}f_{1,0}(z_1) + P_{1,1}f_{1,1}(z_1) = P_{2,0}f_{2,0}(z_1) + P_{2,1}f_{2,1}(z_1)$$

which reduces explicitly to the relation:

$$\begin{aligned}
&a_0\,(f_{1,0}(z_1) - f_{2,0}(z_1))\\
&\quad + a_{10}\Big(\cos\alpha\frac{\partial}{\partial s}f_{1,0}(z_1) - \sin\alpha f_{1,1}(z_1) - \frac{\partial}{\partial s}f_{2,0}(z_1)\Big)\\
&\quad + a_{01}\Big(\sin\alpha\frac{\partial f_{1,0}}{\partial s}(z_1) + \cos\alpha f_{1,1}(z_1) + f_{2,1}(z_1)\Big) = 0.
\end{aligned}$$

Since a_0, a_{10}, a_{01} are arbitrary, this yields three linearly independent conditions:

$$\begin{aligned}
f_{1,0}(z_1) - f_{2,0}(z_1) &= 0,\\
\cos\alpha\frac{\partial f_{1,0}}{\partial s}(Z_1) - \sin\alpha f_{1,1}(z_1) - \frac{\partial f_{2,0}}{\partial s}(z_1) &= 0,\\
\sin\alpha\frac{\partial f_{1,0}}{\partial s}(z_1) + \cos\alpha f_{1,1}(z_1) + f_{2,1}(z_1) &= 0.
\end{aligned}$$

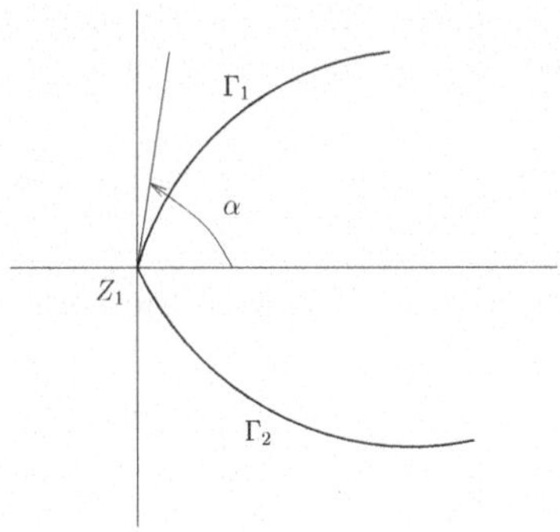

Fig. 4.1. Polygonal vertex

4.3 The Trace Spaces on an Open Surface

In some applications we need trace spaces on an open connected part $\Gamma_0 \overset{\bullet}{\subset} \Gamma$ of a closed surface Γ. To simplify the presentation let us assume $\Gamma \in C^{k,1}$ with $k \in \mathbb{N}_0$. In the two–dimensional case $\Gamma_0 \overset{\bullet}{\subset} \Gamma = \partial\Omega$ with $\Omega \subset \mathbb{R}^2$, the boundary of Γ_0 denoted by $\gamma = \partial\Gamma_0$ consists just of two endpoints $\gamma = \{z_1, z_2\}$.

In the three–dimensional case let us assume that the boundary $\partial\Gamma_0$ of Γ_0 is a closed curve γ. In what follows we assume s satisfying $|s| \leq k+1$. In this case, all the trace spaces $H^s(\Gamma)$ coincide. Similar to Section 4.1 let us introduce the spaces of trivial extensions from $\overline{\Gamma}_0$ to Γ of functions u defined on $\overline{\Gamma}_0$ by zero outside of $\overline{\Gamma}_0$ which will be denoted by $\widetilde{u}$. Then we define

$$\widetilde{H}^s(\Gamma_0) := \{u \in H^s(\Gamma) \big| \, u_{|\Gamma \setminus \overline{\Gamma}_0} = 0\} = \{u \in H^s(\Gamma) \big| \, \operatorname{supp} u \subset \overline{\Gamma}_0\} \quad (4.3.1)$$

as a subspace of $H^s(\Gamma)$ equipped with the corresponding norm

$$\|u\|_{\widetilde{H}^s(\Gamma_0)} = \|\widetilde{u}\|_{H^s(\Gamma)} \, . \quad (4.3.2)$$

By definition, $\widetilde{H}^s(\Gamma_0) \subset H^s(\Gamma)$.

For $s \geq 0$ we also introduce the spaces

$$H^s(\Gamma_0) = \{u = v_{|\Gamma_0} \big| \, v \in H^s(\Gamma)\} \quad (4.3.3)$$

equipped with the norm

$$\|u\|_{H^s(\Gamma_0)} = \inf_{v \in H^s(\Gamma) \wedge v_{|\Gamma_0} = u} \|v\|_{H^s(\Gamma)} \, . \quad (4.3.4)$$

Clearly, $\widetilde{H}^s(\Gamma_0) \subset H^s(\Gamma_0)$.

For $s < 0$, the dual spaces $H^s(\Gamma_0)$ with respect to the $L^2(\Gamma_0)$–inner product are well defined by the completion of $L^2(\Gamma_0)$ with respect to the norm

$$\|u\|_{H^s(\Gamma_0)} := \sup_{0 \neq \varphi \in \widetilde{H}^{-s}(\Gamma_0)} |\int_{\Gamma_0} \varphi \bar{u} ds| / \|\varphi\|_{\widetilde{H}^s(\Gamma_0)} . \tag{4.3.5}$$

Correspondingly, we also have $\widetilde{H}^s(\Gamma_0)$ with the norm

$$\|u\|_{\widetilde{H}^s(\Gamma_0)} := \sup_{0 \neq \varphi \in H^{-s}(\Gamma_0)} | \int_{\Gamma_0} \varphi \bar{u} ds \,| / \|\varphi\|_{H^{-s}(\Gamma)} . \tag{4.3.6}$$

The following lemma shows the relations between these spaces in full analogy to (4.1.11) and (4.1.13).

Lemma 4.3.1. *Let $\Gamma \in C^{k,1}$, $k \in \mathbb{N}_0$. For $|s| \le k+1$ we have*

$$(H^s(\Gamma_0))' = \widetilde{H}^{-s}(\Gamma_0) \tag{4.3.7}$$

and

$$\left(\widetilde{H}^s(\Gamma_0)\right)' = H^{-s}(\Gamma_0) . \tag{4.3.8}$$

We also have the inclusions for $s > 0$:

$$\widetilde{H}^s(\Gamma_0) \subset H^s(\Gamma_0) \subset L^2(\Gamma_0) \subset \widetilde{H}^{-s}(\Gamma_0) \subset H^{-s}(\Gamma_0) . \tag{4.3.9}$$

Similar to the spaces $H^s_{00}(\Omega)$ in Section 4.1, the inclusions in (4.3.9) are strict for $s \ge \frac{1}{2}$ and

$$\widetilde{H}^s(\Gamma_0) = H^s(\Gamma_0) \text{ for } |s| < \frac{1}{2}.$$

For $s > \frac{1}{2}$, we note that $u \in \widetilde{H}^s(\Gamma_0)$ satisfies $u_{|_\gamma} = 0$. Hence, we can introduce the space $H^s_0(\Gamma_0)$ as the completion of $\widetilde{H}^s(\Gamma_0)$ with respect to the norm $\|\cdot\|_{H^s(\Gamma_0)}$. Then the spaces $\widetilde{H}^s(\Gamma_0)$ can also be defined by

$$\widetilde{H}^s(\Gamma_0) = \begin{cases} H^s_0(\Gamma_0) & \text{for } s \neq m + \frac{1}{2}, m \in \mathbb{N}_0 \;, \\ H^s_{00}(\Gamma_0) & \text{for } s = m + \frac{1}{2}, m \in \mathbb{N}_0 \,. \end{cases} \tag{4.3.10}$$

Here, as in Section 4.1, the space $H^{m+\frac{1}{2}}_{00}(\Gamma_0)$ is defined by

$$H^{m+\frac{1}{2}}_{00}(\Gamma_0) = \{u \,|\, u \in H^{m+\frac{1}{2}}_0(\Gamma_0), \varrho^{-\frac{1}{2}} D^\alpha u \in L^2(\Gamma_0) \text{ for } |\alpha| = m\} ,$$

where $\varrho =$ dist (x, γ) for $x \in \Gamma_0$, and D^α stands for the covariant derivatives in Γ (see e.g. Berger and Gostiaux [30]). It is worth pointing out that again

$$\widetilde{H}^{m+\frac{1}{2}}(\Gamma_0) = H^{m+\frac{1}{2}}_{00}(\Gamma_0) \neq H^{m+\frac{1}{2}}_0(\Gamma_0) . \tag{4.3.11}$$

4.4 The Weighted Sobolev Spaces $H^m(\Omega^c;\lambda)$ and $H^m(\mathbb{R}^n;\lambda)$

In applications one often deals with exterior boundary-value problems. In order to ensure the uniqueness of the solutions to the problems, appropriate growth conditions at infinity must be incorporated into the solution spaces. For this purpose, a class of weighted Sobolev spaces in the exterior domain $\Omega^c := \mathbb{R}^n \setminus \overline{\Omega}$ or in the whole $\mathbb{R}^n$ is often used.

For simplicity we shall confine ourselves only to the case of the weighted Sobolev spaces of order $m \in \mathbb{N}$. We begin with the subspace of $C^m(\Omega^c)$,

$$C^m(\Omega^c;\lambda) := \{u \in C^m(\Omega^c) \big| \|u\|_{H^m(\Omega^c;\lambda)} < \infty\},$$

where $\lambda \in \mathbb{N}_0$ is given. Here, $\|\cdot\|_{H^m(\Omega^c;\lambda)}$ is the weighted norm defined by

$$\begin{aligned}\|u\|_{H^m(\Omega^c;\lambda)} &:= \Big\{ \sum_{0\le|\alpha|\le\kappa} \|\varrho^{-(m-|\alpha|-\lambda)}\varrho_0^{-1} D^\alpha u\|^2_{L^2(\Omega^c)} \\ &\quad + \sum_{\kappa+1\le|\alpha|\le m} \|\varrho^{-(m-|\alpha|-\lambda)} D^\alpha u\|^2_{L^2(\Omega^c)} \Big\}^{\frac{1}{2}},\end{aligned} \tag{4.4.1}$$

where $\varrho = \varrho(|x|)$ and $\varrho_0 = \varrho_0(|x|)$ are the weight functions defined by

$$\varrho(|x|) = (1+|x|^2)^{\frac{1}{2}}, \; \varrho_0(|x|) = \log(2+|x|^2), \; x \in \mathbb{R}^n,$$

and the index κ in (4.4.1) is chosen depending on n and λ such that

$$\kappa = \begin{cases} m - (\frac{n}{2}+\lambda) & \text{if } \frac{n}{2}+\lambda \in \{1,\dots,m\} \;, \\ -1 & \text{otherwise} \;. \end{cases}$$

The weighted Sobolev space $H^m(\Omega^c;\lambda)$ is the completion of $C^m(\Omega^c;\lambda)$ with respect to the norm $\|\cdot\|_{H^m(\Omega^c;\lambda)}$. Again, $H^m(\Omega^c;\lambda)$ is a Hilbert space with the inner product

$$\begin{aligned}(u,v)_{H^m(\Omega^c;\lambda)} &:= \sum_{0\le|\alpha|\le\kappa} (\varrho^{-(m-|\alpha|-\lambda)}\varrho_0^{-1}D^\alpha u\,,\,\varrho^{-(m-|\alpha|-\lambda)}\varrho_0^{-1}D^\alpha v)_{L^2(\Omega^c)} \\ &\quad + \sum_{\kappa+1\le|\alpha|\le m} (\varrho^{-(m-|\alpha|-\lambda)}D^\alpha u, \varrho^{-(m-|\alpha|-\lambda)}D^\alpha v)_{L^2(\Omega^c)} \,.\end{aligned} \tag{4.4.2}$$

In a complete analogy to $H_0^m(\Omega)$, we may define

$$H_0^m(\Omega^c;\lambda) = \overline{C_0^m(\Omega^c;\lambda)}^{\|\cdot\|_{H^m(\Omega^c;\lambda)}} \tag{4.4.3}$$

where $C_0^m(\Omega^c;\lambda)$ is the subspace

$$C_0^m(\Omega^c;\lambda) = \{u \in C_0^m(\Omega^c) \;\big|\; \|u\|_{H^m(\Omega^c;\lambda)} < \infty\}\,.$$

Then the following result can be shown (see Nedelec [327, Lemma 2.2], [328, p.16ff.])

Lemma 4.4.1. (a) *The semi–norm* $|u|_{H^m(\Omega^c;\lambda)}$ *defined by*

$$|u|_{H^m(\Omega^c;\lambda)} := \Big\{ \sum_{|\alpha|=m} \|\varrho^\lambda D^\alpha u\|^2_{L^2(\Omega^c)} \Big\}^{\frac{1}{2}} \tag{4.4.4}$$

is a norm on $H_0^m(\Omega^c;\lambda)$*, and is equivalent to the norm* $\|u\|_{H^m(\Omega^c;\lambda)}$ *on* $H_0^m(\Omega^c;\lambda)$.
(b) *In general,* $|u|_{H^m(\Omega^c;\lambda)}$ *is a norm on the quotient space* $H^m(\Omega^c;\lambda)/\mathcal{P}_\eta$*, and is equivalent to the quotient norm*

$$\|\dot{u}\|_{H^m(\Omega^c;\lambda)} := \inf_{p\in\mathcal{P}_\eta} \|u+p\|_{H^m(\Omega^c;\lambda)} \tag{4.4.5}$$

for any representative $\dot{u}$ *in the coset. Here* $\mathcal{P}_\eta$ *is the set of all polynomials of degree less than or equal to* $\eta = \min\{\eta', m-1\}$ *where* η' *is the non–negative integer defined by*

$$\eta' = \begin{cases} m - \frac{n}{2} - \lambda & \textit{for } \frac{n}{2} + \lambda \in \{0,1,\cdots,m\}\ , \\ [m - \frac{n}{2} - \lambda] & \textit{otherwise}\ . \end{cases}$$

We remark that the definition of $H^m(\Omega^c;\lambda)$ carries over to cases when Ω^c is the entire $\mathbb{R}^n$, in which case we have the weighted Sobolev space $H^m(\mathbb{R}^n;\lambda)$. It is also worth pointing out that the space $H^m(\Omega^c;\lambda)$ is contained in the space

$$H^m_{\text{loc}}(\Omega^c) := \{u \in H^m(K)\,|\quad \text{for every}\quad K \Subset \Omega^c\}\,, \tag{4.4.6}$$

the latter can also be characterized by the statement

$$u \in H^m_{\text{loc}}(\Omega^c) \text{ iff } \varphi u \in H^m(\Omega^c) \text{ for every } \varphi \in C_0^\infty(\mathbb{R}^n).$$

Consequently, the trace theorem can also be applied to φu and, hence, to all these spaces.

To conclude this section, we now consider two special cases for $H^m(\Omega^c;\lambda)$. For $n=2$, we let $m=1$ and $\lambda=0$. Then we see that the weighted Sobolev space $H^1(\Omega^c;0)$ is equipped with the norm

$$\|u\|_{H^1(\Omega^c;0)} := \{\|\rho^{-1}\varrho_0^{-1}u\|_{L^2(\Omega^c)} + \sum_{|\alpha|=1} \|D^\alpha u\|^2_{L^2(\Omega^c)}\}^{\frac{1}{2}}\,, \tag{4.4.7}$$

and the corresponding inner product is

$$(u,v)_{H^1(\Omega^c;0)} = (\varrho^{-1}\varrho_0^{-1}u, \varrho^{-1}\varrho_0^{-1}v)_{L^2(\Omega^c)} + \sum_{|\alpha|=1} (D^\alpha u, D^\alpha v)_{L^2(\Omega^c)}\,. \tag{4.4.8}$$

Similarly, for $n=3, \lambda=0, m=1$, the norm$\|\cdot\|_{H^1(\Omega^c:0)}$ is given by

$$\|u\|_{H^1(\Omega^c;0)} := \{\|\varrho^{-1}u\|^2_{L^2(\Omega^2)} + \sum_{|\alpha|=1} \|D^\alpha u\|^2_{L^2(\Omega^c)}\}^{\frac{1}{2}} \tag{4.4.9}$$

and the associated inner product is

$$(u,v)_{H^1(\Omega^c;0)} = (\varrho^{-1}u, \varrho^{-1}v)_{L^2(\Omega^c)} + \sum_{|\alpha|=1} (D^\alpha u, D^\alpha v)_{L^2(\Omega^c)} \,. \tag{4.4.10}$$

As will be seen, these are most frequently used weighted Sobolev spaces which are connected with the energy spaces of potential functions for the two– and three–dimensional Laplacian, respectively.

An Exterior Equivalent Norm

In analogy to the interior Friedrichs inequality (4.1.40) we now give the corresponding version for $m = 1$ and the weighted Sobolev space $H^1(\Omega^c\,;\,0)$.

Lemma 4.4.2. Nedelec [328] *Let Ω^c be an unbounded domain with Lipschitz boundary. Then the norms*

$$\|u\|_{H^1(\Omega^c\,;\,0)} \quad \textit{and} \quad \{\|\nabla u\|^2_{L^2(\Omega^c)} + \int\limits_\Gamma |v|^2 ds\}^{\frac{1}{2}} \quad \textit{are equivalent} \;. \tag{4.4.11}$$

Moreover, the exterior Friedrichs–Poincaré inequality

$$\|u\|^2_{H^1(\Omega^c\,;\,0)} \le c\Big\{\|\nabla u\|^2_{L^2(\Omega^c)} + \|u\|^2_{H^1(\Omega^c_{R_0})}\Big\} \tag{4.4.12}$$

is valid where $\Omega^c_{R_0} := \{x \in \Omega^c \,|\, |x| < R_0\}$ and $R_0 > 0$ is chosen large enough to guarantee $\overline{\Omega} \subset \{x \in \mathbb{R}^n \,|\, |x| < R_0\}$.

Proof: Let $\varphi \in C^\infty(\mathbb{R}^n)$ be a cut–off function with

$$0 \le \varphi(x) \le 1 \quad \text{and} \quad \varphi(x) = 0 \quad \text{for} \quad |x| \le R_0\,, \; \varphi(x) = 1 \quad \text{for} \quad |x| \ge 2R_0\,.$$

Let $B^c_{R_0} := \{x \in \mathbb{R}^n \,|\, |x| > R_0\}$. Then

$$\|u\|_{H^1(\Omega^c\,;\,0)} \le \|\varphi u\|_{H^1(B^c_{R_0}\,;\,0)} + c\|u\|_{H^1(\Omega^c_{2R_0})} \,.$$

Since $\varphi u \in H^1_0(\Omega^c\,;\,0)$, we can apply Lemma 4.4.1 part (a) and obtain with (4.4.1),

$$\|u\|_{H^1(\Omega^c\,;\,0)} \le c\|\nabla(\varphi u)\|_{L^2(B^c_{R_0})} + c\|u\|^2_{H^1(\Omega^c_{2R_0})} \,.$$

Because of the properties of φ the product rule implies

$$\|u\|^2_{H^1(\Omega^c\,;\,0)} \le c_1\|\nabla u\|^2_{L^2(\Omega^c)} + c_2\|u\|^2_{H^1(\Omega^c_{2R_0})} \,.$$

In $\Omega^c_{2R_0}$, the norms $\{\|\nabla u\|^2_{L^2(\Omega^c)} + \|u\|^2_{H^1(\Omega^c_{2R_0})}\}^{\frac{1}{2}}$ and $\{\|\nabla u\|_{L^2(\Omega^c)} + \int\limits_\Gamma |u|^2 ds\}^{\frac{1}{2}}$ are equivalent due to (4.1.40) which yields

$$\|u\|^2_{H^1(\Omega^c\,;\,0)} \le c\Big(\|\nabla u\|^2_{L^2(\Omega^c)} + \int\limits_\Gamma |u|^2 ds\Big)\,; \tag{4.4.13}$$

and with the trace theorem, (4.2.32) for ψu with $\psi(x) = 1 - \varphi$ we find

$$\int\limits_\Gamma |u|^2 ds \le \|\psi u\|_{H^1(\Omega^c\,;\,0)} \le c\|u\|_{H^1(\Omega^c\,;\,0)}\,,$$

i. e.

$$\|\nabla u\|^2_{L^2(\Omega^c)} + \int\limits_\Gamma |u|^2 ds \le c\|u\|^2_{H^1(\Omega^c\,;\,0)}\,.$$

Hence, $\|u\|_{H^1(\Omega^c\,;\,0)}$ and (4.4.11) are equivalent.

The inequality (4.4.12) follows from (4.4.13) with the trace theorem, (4.2.32) for $\Omega^c_{R_0}$. ∎

4.5 Function Spaces H(div , Ω) and H(curl , Ω)

In this section, we will give a brief summary of some vector function spaces and related topics which are needed for treating Maxwell's equations in Chapters 6 and 12. For a detailed description of the subject, we refer to the fundamental papers Buffa and Ciarlet [49] and [50] for Lipschitz polyhedral domains and the seminal paper Buffa, Costabel and Sheen [52] in the case of general Lipschitz boundaries (see also Claeys and Hiptmair [71]). For smooth domains, most of relevant results are available in the monographs Cessenat [63] and Dautray and Lions [96].

Unless otherwise specified, through out the section, we assume that $\Omega \subset \mathbb{R}^3$ is a bounded domain with a Lipschitz–continuous boundary Γ and a unit outward normal to Ω defined almost everywhere on $\Gamma = \partial\Omega$. We begin with two basic function spaces for Maxwell's equations:

$$\mathbf{H}(\mathrm{div}\,,\Omega) := \{\boldsymbol{u} \in \boldsymbol{L}^2(\Omega) := L^2(\Omega)^3 |\, \mathrm{div}\,\boldsymbol{u} \in L^2(\Omega)\} \tag{4.5.1}$$

endowed with the norm

$$\|\boldsymbol{u}\|_{\mathbf{H}(\mathrm{div}\,,\Omega)} := \big\{\|\boldsymbol{u}\|^2_{\boldsymbol{L}^2(\Omega)} + \|\mathrm{div}\,\boldsymbol{u}\|^2_{L^2(\Omega)}\big\}^{1/2}.$$

and

$$\mathbf{H}(\mathbf{curl}\,,\Omega) := \{\boldsymbol{u} \in \boldsymbol{L}^2(\Omega) |\, \mathbf{curl}\, u \in \boldsymbol{L}^2(\Omega)\} \tag{4.5.2}$$

with the norm

$$\|\boldsymbol{u}\|_{\mathbf{H}(\mathbf{curl}\,,\Omega)} := \big\{\|\boldsymbol{u}\|^2_{\boldsymbol{L}^2(\Omega)} + \|\mathbf{curl}\, u\|^2_{\boldsymbol{L}^2(\Omega)}\big\}^{1/2},$$

Clearly, both $\mathbf{H}(\mathrm{div}\,,\Omega)$ and $\mathbf{H}(\mathbf{curl}\,,\Omega)$ are Hilbert spaces. But note that their imbeddings into $L^2(\Omega)$ are not compact as was shown by Murat [318]. Consequently, the intersection space

$$\mathbf{X}(\Omega) := \mathbf{H}(\operatorname{div},\Omega) \cap \mathbf{H}(\mathbf{curl},\Omega) \hookrightarrow L^2(\Omega)^3 \tag{4.5.3}$$

with the norm

$$\|\boldsymbol{u}\|_{\mathbf{X}(\Omega)} := \{\|\boldsymbol{u}\|^2_{\mathbf{H}(\operatorname{div},\Omega)} + \|\boldsymbol{u}\|^2_{\mathbf{H}(\mathbf{curl},\Omega)}\}^{\frac{1}{2}}$$

is in $L^2(\Omega)^3$ **not** compactly imbedded. However, the subspace

$$\mathbf{H}^1_0(\Omega) = \{\boldsymbol{u} \in H^1(\Omega)^3 \quad \text{with } \gamma_0\boldsymbol{u} = \mathbf{0} \quad \text{on } \Gamma\}$$

is compactly imbedded into $\boldsymbol{L}_2(\Omega)$. And in Amrouche, Ciarlet and Mardare [11] it is shown that the Poincaré inequality (4.1.39) is equivalent to the right–invertibility of the div–operator on $\mathbf{H}^1_0(\Omega)$.

Lemma 4.5.1. *The mapping*

$$\operatorname{div}\mathbf{H}^1_0(\Omega) \to L^2_0(\Omega) = \{f \in L_2(\Omega) \quad \textit{with} \int\limits_\Omega f d\boldsymbol{x} = 0\} \tag{4.5.4}$$

is onto. Consequently, for each $f \in L_2\Omega)$ there exists a unique element $\boldsymbol{u}_f$ satisfying

$$(\boldsymbol{u}_f, \boldsymbol{v}_0)_{\boldsymbol{L}_2(\Omega)} = 0 \quad \textit{for all } \boldsymbol{v}_0 \in \operatorname{Ker}(\operatorname{div}).$$

If Ω is starlike with respect to an open ball $B \subset \Omega$ then with a cut–off function $\psi \in C^\infty_0(\mathbb{R}^3)$ with supp $\psi \subset B$, $\psi(\boldsymbol{x}) \geq 0$ and $\int \psi(\boldsymbol{x})d\boldsymbol{x} = 1$, the Bogovskii operator [35]

$$\boldsymbol{B}f(\boldsymbol{x}) := \int\limits_\Omega f(\boldsymbol{y}) \frac{\boldsymbol{x}-\boldsymbol{y}}{|\boldsymbol{x}-\boldsymbol{y}|^3} \Big\{ \int\limits_{|\boldsymbol{x}-\boldsymbol{y}|}^\infty \psi\Big(\boldsymbol{y} + r\frac{\boldsymbol{x}-\boldsymbol{y}}{|\boldsymbol{x}-\boldsymbol{y}|}\Big) r^2 dr \Big\} d\boldsymbol{y} \tag{4.5.5}$$

defines a right–inverse to the div–operator on L^2_0 (see Costabel and McIntosh [86]). Since a Lipschitz domain Ω has the strong cone property the bounded Lipschitz domain $\Omega \subset \mathbb{R}^3$ can be decomposed into a finite number of subdomains where each of the subdomains is starlike to some ball and on each of the subdomains the corresponding Bogovskii operators provide a local right–inverse to the div–operator.

In particular, we are interested in the traces of functions in the spaces $\mathbf{H}(\operatorname{div},\Omega)$ and $\mathbf{H}(\mathbf{curl},\Omega)$ because of boundary integral formulations of solutions of Maxwell equations.

We begin with $\boldsymbol{u} \in \mathbf{H}(\operatorname{div},\Omega)$ and $\varphi \in H^1(\Omega)$. Then from generalized Green's formula, we see that

$$\begin{aligned}(\boldsymbol{u}, \mathbf{grad}\,\varphi) + (\operatorname{div}\boldsymbol{u}, \varphi) &= \int\limits_\Gamma (\boldsymbol{u}|_\Gamma \cdot \boldsymbol{n})\varphi|_\Gamma d\Gamma \\ &=: \langle(\boldsymbol{u}\cdot\boldsymbol{n})|_\Gamma\,,\,\varphi|_\Gamma\rangle\end{aligned} \tag{4.5.6}$$

where $\langle \cdot, \cdot \rangle$ denotes the duality pairing between $H^{-1/2}(\Gamma)$ and $H^{1/2}(\Gamma)$ (see McLean [290, Chapt. 3]). This means we may define a normal trace operator

$$\gamma_{\boldsymbol{n}} : \mathbf{H}(\mathrm{div}\,, \Omega) \longrightarrow H^{-1/2}(\Gamma)\,,\ \boldsymbol{u} \mapsto \gamma_{\boldsymbol{n}}(\boldsymbol{u}) := \boldsymbol{u}|_{\Gamma} \cdot \boldsymbol{n}. \tag{4.5.7}$$

Indeed, based on the density argument, the following theorem is available in Dautray and Lions [96] and Girault and Raviart [147].

Theorem 4.5.2. (Trace Theorem for $\mathbf{H}(\mathrm{div}\,, \Omega)$). *Let $\Omega \subset \mathbb{R}^3$ be a bounded domain with Lipschitz boundary Γ. Then*

(a) *The space $C^\infty(\overline{\Omega})^3$ is dense in $\mathbf{H}(\mathrm{div}\,, \Omega)$.*
(b) *The normal trace map $\gamma_{\boldsymbol{n}} : \boldsymbol{u} \mapsto \boldsymbol{u}|_{\Gamma} \cdot \boldsymbol{n}$ defined on $C^\infty(\overline{\Omega})^3$ extends by continuity to a continuous linear map — still denoted by $\gamma_{\boldsymbol{n}}$ — from $\mathbf{H}(\mathrm{div}\,, \Omega)$ onto $H^{-1/2}(\Gamma)$.*
(c) *The Green's formula* (4.5.6) *holds for functions $\boldsymbol{u} \in \mathbf{H}(\mathrm{div}\,, \Omega)$ and $\varphi \in H^{1/2}(\Gamma)$.*
(d) *The kernel* Ker $(\gamma_{\boldsymbol{n}})$ *of this mapping $\gamma_{\boldsymbol{n}}$ is the space*

$$\mathbf{H}_0(\mathrm{div}\,, \Omega) := \big\{\boldsymbol{u} \in \mathbf{H}(\mathrm{div}\,, \Omega) : \boldsymbol{u} \cdot \boldsymbol{n}|_{\Gamma} = \mathbf{0}\big\}. \tag{4.5.8}$$

On the other hand, for the space $\mathbf{H}(\mathbf{curl}, \Omega)$ two tangential traces of functions are involved. We collect some of basic results, give here an overview of their developments, and point out the corresponding relevant mathematical ingredients employed. In the following, let

$$\boldsymbol{L}_t^2(\Gamma) := \big\{\boldsymbol{\mu} \in \boldsymbol{L}^2(\Gamma) : \boldsymbol{\mu} \cdot \boldsymbol{n} = 0\big\}.$$

We define the tangential trace operator:

$$\gamma_t : C^\infty(\overline{\Omega})^3 \longrightarrow \boldsymbol{L}_t^2(\Gamma)\,,\ \boldsymbol{u} \mapsto \gamma_t \boldsymbol{u} := \boldsymbol{n} \times \boldsymbol{u}|_{\Gamma} \tag{4.5.9}$$

and the tangential projection operator:

$$\begin{aligned} \pi_t : C^\infty(\overline{\Omega})^3 \longrightarrow \boldsymbol{L}_t^2(\Gamma)\ :\ \boldsymbol{u} \mapsto \pi_t(\boldsymbol{u}) &:= -\boldsymbol{n} \times (\boldsymbol{n} \times \boldsymbol{u}|_{\Gamma}) \\ &= \boldsymbol{u}|_{\Gamma} - (\boldsymbol{u}|_{\Gamma} \cdot \boldsymbol{n})\boldsymbol{n}. \end{aligned} \tag{4.5.10}$$

Note that various notations have been adopted for the tangential trace operator γ_t and for the tangential projection operator π_t; these notions are not uniform. Geometrically, $\gamma_t \boldsymbol{u}$ is tangent to Γ, while $\pi_t \boldsymbol{u}$ is the projection onto the tangent plane to Γ of the restriction $\boldsymbol{u}|_{\Gamma}$ of $\boldsymbol{u}$ to Γ (i.e. onto the tangent bundle). These traces are rotated ninety degrees about the normal vector field with respect to each other.

Remark 4.5.1: Let γ_0 be the standard trace operator such that $\gamma_0 : \mathbf{H}^1(\Omega) := H^1(\Omega)^3 \longrightarrow \mathbf{H}^{1/2}(\Gamma) := H^{1/2}(\Gamma)^3\,,\ \gamma_0 : \boldsymbol{u} \mapsto \gamma_0(\boldsymbol{u}) = \boldsymbol{u}|_{\Gamma}$ and with the adjoint

$$\gamma_0^* : (\mathbf{H}^{-\frac{1}{2}}(\Gamma)) \to \widetilde{\mathbf{H}}^{-1}(\Omega) : \boldsymbol{\sigma} \mapsto \gamma_0^*(\boldsymbol{\sigma}) = \boldsymbol{\sigma} \otimes \delta_{\Gamma}.$$

Since γ_0 is surjective, let $\boldsymbol{\mathcal{Z}}$ be one of its right inverses. Then the composite operator $\pi_t \circ \boldsymbol{\mathcal{Z}}$ (resp. $\gamma_t \circ \boldsymbol{\mathcal{Z}}$) may also be used to act only on the trace functions (see, e.g Buffa, Costabel and Sheen [52]). By density of $C^\infty(\overline{\Omega})^3|_\Gamma$ into $\boldsymbol{L}^2(\Gamma)$, the operators π_t and γ_t can be extended to linear continuous operators in $\boldsymbol{L}^2(\Gamma)$.

Tangential trace spaces for $\mathbf{H}^1(\Omega)$
A simple integration by parts shows that

$$(\boldsymbol{u}, \mathbf{curl}\, \boldsymbol{v})_{\boldsymbol{L}^2(\Omega)} - (\mathbf{curl}\, \boldsymbol{u}, \boldsymbol{v})_{\boldsymbol{L}^2(\Omega)} = \int_\Gamma (\boldsymbol{n} \times \boldsymbol{u}) \cdot \boldsymbol{v}\, d\Gamma \tag{4.5.11}$$
$$= \langle \gamma_t \boldsymbol{u}, \pi_t\, \boldsymbol{v} \rangle_t, \quad \text{for all } \boldsymbol{u}, \boldsymbol{v} \in \mathbf{H}^1(\Omega),$$

where the bracket in the right hand side of (4.5.11) is the inner product of $\boldsymbol{L}_t^2(\Gamma)$. This motives us to define the spaces

$$\mathbf{H}_\perp^{1/2}(\Gamma) := \gamma_t(\mathbf{H}^1(\Omega)) = \{\boldsymbol{\varphi} \in \boldsymbol{L}_t^2(\Gamma) : \boldsymbol{\varphi} = \gamma_t \boldsymbol{u},\ \boldsymbol{u} \in \mathbf{H}^1(\Omega)\}, \tag{4.5.12}$$
$$\mathbf{H}_{||}^{1/2}(\Gamma) := \pi_t(\mathbf{H}^1(\Omega)) = \{\boldsymbol{\varphi} \in \boldsymbol{L}_t^2(\Gamma) : \boldsymbol{\varphi} = \pi_t \boldsymbol{u},\ \boldsymbol{u} \in \mathbf{H}^1(\Omega)\}. \tag{4.5.13}$$

These spaces are dense in $\boldsymbol{L}_t^2(\Gamma)$ and they are different for non–smooth boundaries. The spaces are equipped with the norms:

$$\|\boldsymbol{\varphi}\|_{\mathbf{H}_\perp^{1/2}(\Gamma)} := \inf\{\|\boldsymbol{u}\|_{\mathbf{H}^1(\Omega))} : \gamma_t \boldsymbol{u} = \boldsymbol{\varphi}\} \quad \text{and}$$
$$\|\boldsymbol{\varphi}\|_{\mathbf{H}_{||}^{1/2}(\Gamma)} := \inf\{\|\boldsymbol{u}\|_{\mathbf{H}^1(\Omega))} : \pi_t \boldsymbol{u} = \boldsymbol{\varphi}\},$$

which ensure the mappings $\gamma_t : \mathbf{H}^1(\Omega)) \to \mathbf{H}_\perp^{1/2}(\Gamma)$ and $\pi_t : \mathbf{H}^1(\Omega)) \to \mathbf{H}_{||}^{1/2}(\Gamma)$ are bounded and surjective. We may also define their dual spaces :

$$\mathbf{H}_\perp^{-1/2}(\Gamma) := \mathbf{H}_\perp^{1/2}(\Gamma)', \quad \text{and} \quad \mathbf{H}_{||}^{-1/2}(\Gamma) := \mathbf{H}_{||}^{1/2}(\Gamma)', \tag{4.5.14}$$

equipped, respectively with the norms:

$$\|\boldsymbol{\lambda}\|_{\mathbf{H}_\perp^{-1/2}(\Gamma)} := \sup_{\mathbf{0} \neq \boldsymbol{\varphi} \in \mathbf{H}_\perp^{1/2}(\Gamma)} \frac{|\langle \boldsymbol{\lambda}, \boldsymbol{\varphi} \rangle_t|}{\|\boldsymbol{\varphi}\|_{\mathbf{H}_\perp^{1/2}(\Gamma)}},$$
$$\|\boldsymbol{\lambda}\|_{\mathbf{H}_{||}^{-1/2}(\Gamma)} := \sup_{\mathbf{0} \neq \boldsymbol{\varphi} \in \mathbf{H}_{||}^{1/2}(\Gamma)} \frac{|\langle \boldsymbol{\lambda}, \boldsymbol{\varphi} \rangle_t|}{\|\boldsymbol{\varphi}\|_{\mathbf{H}_{||}^{1/2}(\Gamma)}}.$$

Here the bracket $\langle\ ,\ \rangle_t$ denotes the two possible extensions of the $\boldsymbol{L}_t^2(\Gamma)$ inner product to the duality pairings $\mathbf{H}_\perp^{-1/2}(\Gamma) \times \mathbf{H}_\perp^{1/2}(\Gamma)$ and $\mathbf{H}_{||}^{-1/2}(\Gamma) \times \mathbf{H}_{||}^{1/2}(\Gamma)$. In order to see their extensions to the tangential traces for the space $\mathbf{H}(\mathbf{curl}\, , \Omega)$, we need some informations concerning various differential operators defined on the surface Γ.

Tangential differential operators (see Buffa and Ciarlet [49, 50], Buffa, Costabel and Sheen [52])
In the following we need differential operators $\mathbf{grad}_\Gamma$ and $\mathbf{curl}_\Gamma$ as well as their corresponding adjoint operators defined on the surface Γ, that is, a closed Lipschitz surface without boundary. For definiteness, we define the operators :

$$\mathbf{grad}_\Gamma := (\pi_t \mathbf{grad}) \circ \mathcal{Z}^{-1} : H^1(\Gamma) \longrightarrow \boldsymbol{L}_t^2(\Gamma) \quad \text{and} \tag{4.5.15}$$

$$\mathbf{curl}_\Gamma := -(\gamma_t \mathbf{grad} \ \circ \mathcal{Z}^{-1} : H^1(\Gamma) :\longrightarrow \boldsymbol{L}_t^2(\Gamma). \tag{4.5.16}$$

The adjoint operators of $-\mathbf{grad}_\Gamma$ and $\mathbf{curl}_\Gamma$ are respectively,

$$\mathrm{div}_\Gamma : \boldsymbol{L}_t^2(\Gamma) \to H_*^{-1}(\Gamma), \quad \text{and } \mathrm{curl}_\Gamma : \boldsymbol{L}_t^2(\Gamma) \to H_*^{-1}(\Gamma), \tag{4.5.17}$$

which are defined by the following duality pairings :

$$\langle \mathrm{div}_\Gamma \boldsymbol{\lambda}, \varphi \rangle = -\int_\Gamma \boldsymbol{\lambda} \cdot \mathbf{grad}_\Gamma \varphi \, d\Gamma \quad \text{and}$$

$$\langle \mathrm{curl}_\Gamma \boldsymbol{\lambda}, \varphi \rangle = \int_\Gamma \boldsymbol{\lambda} \cdot \mathbf{curl}_\Gamma \varphi \, d\Gamma$$

for all $\varphi \in H^1(\Gamma)$ and for all $\boldsymbol{\lambda} \in \boldsymbol{L}_t^2(\Gamma)$, i.e.
$\mathrm{div}_\Gamma^* = -\mathbf{grad}_\Gamma$ and $\mathrm{curl}_\Gamma^* = \mathbf{curl}_\Gamma$.
In the equation (4.5.17) we denote

$$H_*^{-1}(\Gamma) = \{\varphi \in H^{-1}(\Gamma) | \int_\Gamma \varphi d\Gamma = 0\}.$$

The operators $\mathbf{grad}_\Gamma$ and $\mathbf{curl}_\Gamma$ can be restricted to more regular spaces. If $H^{3/2}(\Gamma) = \gamma_0(H^2(\Omega))$, and $H^{-3/2}(\Gamma) = H^{3/2}(\Gamma)'$, then we see that

$$\mathbf{grad}_\Gamma : H^{3/2}(\Gamma) \longrightarrow \mathbf{H}_{||}^{1/2}(\Gamma), \quad \text{and } \mathbf{curl}_\Gamma : H^{3/2}(\Gamma) \longrightarrow \mathbf{H}_{\perp}^{1/2}(\Gamma). \tag{4.5.18}$$

Thus, by $\boldsymbol{L}_t^2(\Gamma)$ duality

$$\mathrm{div}_\Gamma : \mathbf{H}_{||}^{-1/2}(\Gamma) \longrightarrow H^{-3/2}(\Gamma), \quad \text{and } \mathrm{curl}_\Gamma : \mathbf{H}_{\perp}^{-1/2}(\Gamma) \longrightarrow H^{-3/2}(\Gamma). \tag{4.5.19}$$

Now in order to define suitable extensions of the operators π_t and γ_t, we make use of the **Green formula**: For all $\boldsymbol{u} \in \mathbf{H}(\mathbf{curl}, \Omega)$ and for all $\boldsymbol{v} \in \mathbf{H}^1(\Omega)$, we see that

$$\begin{aligned} \int_\Omega \Big\{ (\mathbf{curl}\, \boldsymbol{u}) \cdot \boldsymbol{v} - \boldsymbol{u} \cdot \mathbf{curl}\, \boldsymbol{v} \Big\} d\Omega &= \langle \gamma_t(\boldsymbol{u}), \pi_t(\boldsymbol{v}) \rangle_t \\ &= -\langle \pi_t(\boldsymbol{u}), \gamma_t(\boldsymbol{v}) \rangle_t. \end{aligned} \tag{4.5.20}$$

Since the operators $\pi_t : \mathbf{H}^1(\Omega) \longrightarrow \mathbf{H}_{||}^{1/2}(\Gamma)$ and $\gamma_t : \mathbf{H}^1(\Omega) \longrightarrow \mathbf{H}_{\perp}^{1/2}(\Gamma)$ are surjective, based on (4.5.20) and the closed graph theorem (see Bishop [32] and Ammari [8], Ammari and Nedelec [9]), we obtain that the operators

$$\gamma_t : \mathbf{H}(\mathbf{curl}\,, \Omega) \longrightarrow \mathbf{H}_{||}^{-1/2}(\Gamma) \quad \text{and} \quad \pi_t : \mathbf{H}(\mathbf{curl}\,, \Omega) \longrightarrow \mathbf{H}_{\perp}^{-1/2}(\Gamma) \tag{4.5.21}$$

are linear and continuous. We are now in a position to describe the ranges of these operators. By making use of (4.5.20), this shows that for all $\boldsymbol{u} \in \mathbf{H}(\mathbf{curl}\,, \Omega)$ there holds $\gamma_t(\boldsymbol{u}) \in \mathbf{H}_{||}^{-1/2}(\Gamma)$ and

$$\begin{aligned} \operatorname{div}_\Gamma(\gamma_t(\boldsymbol{u})) &= -(\mathbf{curl}\ \boldsymbol{u}) \cdot \boldsymbol{n} \in H^{-1/2}(\Gamma), \\ \|\operatorname{div}_\Gamma\Big(\gamma_t(\boldsymbol{u})\Big)\|_{H^{-1/2}(\Gamma)} &\le c\|\boldsymbol{u}\|_{\mathbf{H}(\mathbf{curl}\,,\Omega)}, \end{aligned} \tag{4.5.22}$$

$(\mathbf{curl}\,\boldsymbol{u}) \cdot \boldsymbol{n} \in H^{-1/2}(\Gamma)$ since $\mathbf{curl}\ \boldsymbol{u} \in \mathbf{H}(\operatorname{div}, \Omega)$. From (4.5.20) again, we obtain similarly for all $\boldsymbol{u} \in \mathbf{H}(\mathbf{curl}\,, \Omega)$ that $\pi_t(\boldsymbol{u}) \in \mathbf{H}_{\perp}^{-1/2}(\Gamma)$ and

$$\begin{aligned} \operatorname{curl}_\Gamma\left(\pi_t(\boldsymbol{u})\right) &= -\operatorname{div}_\Gamma(\boldsymbol{n} \times \pi_t(\boldsymbol{u})) = -\operatorname{div}_\Gamma(\boldsymbol{n} \times \boldsymbol{u}), \\ \operatorname{curl}_\Gamma\Big(\pi_t(\boldsymbol{u})\Big) &= \mathbf{curl}\ \boldsymbol{u} \cdot \boldsymbol{n} \in H^{-1/2}(\Gamma), \\ \|\operatorname{curl}_\Gamma\Big(\pi_t(\boldsymbol{u})\Big)\|_{H^{-1/2}(\Gamma)} &\le c\|\boldsymbol{u}\|_{\mathbf{H}(\mathbf{curl},\Omega)}. \end{aligned} \tag{4.5.23}$$

Thus, we arrive at two trace spaces, namely $\mathbf{H}_{||}^{-1/2}(\operatorname{div}_\Gamma, \Gamma)$ and $\mathbf{H}_{\perp}^{-1/2}(\operatorname{curl}_\Gamma, \Gamma)$ for the space $\mathbf{H}(\mathbf{curl}\,, \Omega)$ in the following main theorem due to Buffa, Costabel and Sheen in [52]. We state the theorem and refer details of the proof to [52].

Theorem 4.5.3. (Trace Theorem for $\mathbf{H}(\mathbf{curl}\,, \Omega)$)
Let

$$\mathbf{H}_{||}^{-1/2}(\operatorname{div}_\Gamma, \Gamma) := \{\boldsymbol{\lambda} \in \mathbf{H}_{||}^{-1/2}(\Gamma) \,|\, \operatorname{div}_\Gamma(\boldsymbol{\lambda}) \in H^{-1/2}(\Gamma)\}, \tag{4.5.24}$$

and

$$\mathbf{H}_{\perp}^{-1/2}(\operatorname{curl}_\Gamma, \Gamma) := \{\boldsymbol{\lambda} \in \mathbf{H}_{\perp}^{-1/2}(\Gamma) |\, \operatorname{curl}_\Gamma(\boldsymbol{\lambda}) \in \mathrm{H}^{-1/2}(\Gamma)\}, \tag{4.5.25}$$

be the trace spaces for $\mathbf{H}(\mathbf{curl}\,, \Omega)$ equipped with graph norms:

$$\|\boldsymbol{\lambda}\|_{\mathbf{H}_{||}^{-1/2}(\operatorname{div}_\Gamma,\Gamma)} := \|\boldsymbol{\lambda}\|_{\mathbf{H}_{||}^{-1/2}(\Gamma)} + \|\operatorname{div}_\Gamma\boldsymbol{\lambda}\|_{H^{-1/2}(\Gamma)}, \quad \textit{and} \tag{4.5.26}$$

$$\|\boldsymbol{\lambda}\|_{\mathbf{H}_{\perp}^{-1/2}(\operatorname{curl}_\Gamma,\Gamma)} := \|\boldsymbol{\lambda}\|_{\mathbf{H}_{\perp}^{-1/2}(\Gamma)} + \|\operatorname{curl}_\Gamma\boldsymbol{\lambda}\|_{\mathrm{H}^{-1/2}(\Gamma)}, \tag{4.5.27}$$

respectively, Then the operators

$$\begin{aligned} \gamma_t &: \mathbf{H}(\mathbf{curl}\,, \Omega) \to \mathbf{H}_{||}^{-1/2}(\operatorname{div}_\Gamma, \Gamma) \quad \textit{and} \\ \pi_t &: \mathbf{H}(\mathbf{curl}\,, \Omega) \to \mathbf{H}_{\perp}^{-1/2}(\operatorname{curl}_\Gamma, \Gamma) \end{aligned} \tag{4.5.28}$$

are linear, continuous, surjective and possess continuous right inverses. Moreover, the $\boldsymbol{L}_t^2(\Gamma)$ *inner product can be extended to define a duality pairing* $\langle\ ,\ \rangle$ *between the spaces* $\mathbf{H}_{||}^{-1/2}(\mathrm{div}_\Gamma, \Gamma)$ *and* $\mathbf{H}_{\perp}^{-1/2}(\mathrm{curl}_\Gamma, \Gamma)$.

Remark 4.5.2: In Theorem 4.5.3, we have replaced the original notations V_π' and V_γ' adopted in Buffa, Costabel and Sheen [52] by $\mathbf{H}_{||}^{-1/2}(\Gamma)$ and $\mathbf{H}_{\perp}^{-1/2}(\Gamma)$.

As a consequence of the theorem, Green's formula (4.5.20) can be extended to functions $\boldsymbol{u}, \boldsymbol{v} \in \mathbf{H}(\,\mathbf{curl}\,, \Omega)$, if the boundary integral of the right–hand side is interpreted as $\langle \gamma_t(\boldsymbol{u}), \pi_t(\boldsymbol{v})\rangle_t$, that is, the duality pairing between $\mathbf{H}_{\perp}^{-1/2}(\,\mathbf{curl}\,_\Gamma, \Gamma)$ and $\mathbf{H}_{||}^{-1/2}(\mathrm{div}_\Gamma, \Gamma)$ with respect to the pivot space index duality pairing $\boldsymbol{L}_t^2(\Gamma)$. The extension of $\boldsymbol{L}_t^2(\Gamma)$ dualities to the pair $\big(\mathbf{H}_{||}^{-1/2}(\mathrm{div}_\Gamma, \Gamma)\,,\ \mathbf{H}_{\perp}^{-1/2}(\mathrm{curl}_\Gamma, \Gamma)\big)$ is a surprising and deep result since we were used to dualities for pairs of Sobolev spaces like $\big(\mathbf{H}^s(\Gamma), \mathbf{H}^{-s}(\Gamma)\big)$ with $0 \le s \le 1$. We follow Buffa and Ciarlet [49, 50] and Buffa, Costabel and Sheen [52] to proceed; first based on a Hodge decomposition of functions $\mathbf{H}(\,\mathbf{curl}\,, \Omega)$, and then based on Hodge decompositions of trace spaces $\mathbf{H}_{||}^{-1/2}(\mathrm{div}_\Gamma, \Gamma)$ and $\mathbf{H}_{\perp}^{-1/2}(\,\mathrm{curl}_\Gamma, \Gamma)$.

Hodge decomposition of vector fields

We recall that there exists the decomposition: $\mathbf{H}(\,\mathbf{curl}\,, \Omega) = \mathbf{H}^1(\Omega) + \mathbf{grad}\, H^1(\Omega)$ (see, e.g. Amrouche, Bernardi, Dauge and Girault [10], Buffa and Ciarlet [49], [50]). Then for $\boldsymbol{u}, \boldsymbol{v} \in \mathbf{H}(\,\mathbf{curl}\,, \Omega)$, we assume that

$$\boldsymbol{u} = \boldsymbol{\phi} + \mathbf{grad}\, p, \quad \text{and} \quad \boldsymbol{v} = \boldsymbol{\psi} + \mathbf{grad}\, q \tag{4.5.29}$$

where $\boldsymbol{\phi}, \boldsymbol{\psi} \in \mathbf{H}^1(\Omega)$ and $p, q \in H^1(\Omega)$. We obtain

$$\begin{aligned}\int_\Omega \Big\{(\,\mathbf{curl}\ \boldsymbol{u}) \cdot \boldsymbol{v} - \boldsymbol{u} \cdot \mathbf{curl}\ \boldsymbol{v}\Big\}\, d\Omega &= \int_\Omega \Big\{(\,\mathbf{curl}\ \boldsymbol{\phi}) \cdot \boldsymbol{\psi} - \boldsymbol{\phi} \cdot \mathbf{curl}\ \boldsymbol{\psi}\Big\} d\Omega \\ &+ \int_\Omega \mathbf{curl}\ \boldsymbol{\phi} \cdot \mathbf{grad}\ q\, d\Omega - \int_\Omega \mathbf{grad}\ p \cdot \mathbf{curl}\ \boldsymbol{\psi}\, d\Omega.\end{aligned}$$

Applying formula (4.5.20) three times yields

$$\int_\Omega \Big\{(\,\mathbf{curl}\ \boldsymbol{u}) \cdot \boldsymbol{v} - \boldsymbol{u} \cdot \mathbf{curl}\ \boldsymbol{v}\Big\}\, d\Omega = \langle \gamma_t(\boldsymbol{u}), \pi_t(\boldsymbol{v})\rangle_t, \tag{4.5.30}$$

where the boundary term can be defined as

$$\begin{aligned}\big\langle \gamma_t(\boldsymbol{u}), \pi_t(\boldsymbol{v})\big\rangle_t &= \int\limits_\Gamma \Big\{\gamma_t(\boldsymbol{\phi})\cdot\pi_t(\boldsymbol{\psi})\Big\} d\,\Gamma + \big\langle\,\mathbf{grad}_\Gamma\, q, \gamma_t(\boldsymbol{\phi})\big\rangle \\ &\quad - \big\langle\,\mathbf{curl}_\Gamma\, p, \pi_t(\boldsymbol{\psi})\big\rangle \\ &= \int\limits_\Gamma \Big\{\gamma_t(\boldsymbol{\phi})\cdot\pi_t(\boldsymbol{\psi})\Big\} d\,\Gamma - \big\langle\,\mathrm{div}_\Gamma(\gamma_t(\boldsymbol{\phi})), \gamma_0\, q\big\rangle \\ &\quad - \big\langle\,\mathrm{curl}_\Gamma\, \pi_t(\boldsymbol{\psi}), \gamma p\big\rangle \end{aligned} \tag{4.5.31}$$

Hence, the integration by parts formula (4.5.30) holds for any fields $\boldsymbol{u},\boldsymbol{v}\in \mathbf{H}(\,\mathbf{curl}\,,\Omega)$ and from the surjectivity of the trace operators in Theorem 4.5.3, it is then possible to define the following duality for any $\boldsymbol{u},\boldsymbol{v}\in\mathbf{H}(\,\mathbf{curl}\,,\Omega)$:

$$\begin{aligned}\big\langle \gamma_t(\boldsymbol{u}), \pi_t(\boldsymbol{v})\big\rangle_t &= \big\langle \gamma_t(\boldsymbol{\phi}), \pi_t(\boldsymbol{\psi})\big\rangle_{\boldsymbol{L}^2_t(\Gamma)} \\ &\quad - \big\langle\,\mathrm{div}_\Gamma(\gamma_t(\boldsymbol{\phi})), \gamma\, q\big\rangle - \big\langle\,\mathrm{curl}_\Gamma(\pi_t(\boldsymbol{\psi})), \gamma\, p\big\rangle\end{aligned} \tag{4.5.32}$$

These three dualities on the right–hand side are well defined and continuous with respect to the $\mathbf{H}(\,\mathbf{curl}\,,\Omega)$ norms. However, this duality pairing based on a decomposition of vector fields in $\mathbf{H}(\,\mathbf{curl}\,,\Omega)$ is by no means an intrinsic characterization of the spaces $\mathbf{H}_{||}^{-1/2}(\mathrm{div}_\Gamma,\Gamma)$ and $\mathbf{H}_{\perp}^{-1/2}(\,\mathrm{curl}_\Gamma\,,\Gamma)$.

We now consider the duality pairing between $\mathbf{H}_{||}^{-1/2}(\mathrm{div}_\Gamma,\Gamma)$ and $\mathbf{H}_{\perp}^{-1/2}(\,\mathrm{curl}_\Gamma,\Gamma)$ based on Hodge decompositions of tangential vector fields in the trace spaces directly.

Hodge decompositions of tangential vector fields

The following orthogonal Hodge decompositions of the tangential spaces have been introduced in Buffa and Ciarlet [50]:

$$\mathbf{H}_{||}^{-1/2}(\mathrm{div}_\Gamma,\Gamma) = \mathbf{grad}_\Gamma\,\mathcal{H}(\Gamma)\oplus\mathbf{curl}_\Gamma H^{1/2}(\Gamma)\quad\text{and} \tag{4.5.33}$$

$$\mathbf{H}_{\perp}^{-1/2}(\,\mathrm{curl}_\Gamma,\Gamma) = \mathbf{curl}_\Gamma\,\mathcal{H}(\Gamma)\oplus\mathbf{grad}_\Gamma H^{1/2}(\Gamma), \tag{4.5.34}$$

where

$$\mathcal{H}(\Gamma) := \Big\{\alpha\in H^1\Gamma)\quad\text{with}\quad \Delta_\Gamma\alpha\in H^{-1/2}(\Gamma)\quad\text{and}\quad\langle\alpha,1\rangle = 0\Big\}.$$

A duality can now be defined between $\mathbf{H}_{||}^{-1/2}(\mathrm{div}_\Gamma,\Gamma)$ and $\mathbf{H}_{\perp}^{-1/2}(\,\mathrm{curl}_\Gamma,\Gamma)$ since the pair of spaces $\mathbf{H}_{||}^{-1/2}(\mathrm{div}_\Gamma,\Gamma)$ and $\mathbf{H}_{\perp}^{-1/2}(\,\mathrm{curl}_\Gamma,\Gamma)$ form a duality pairing. That is, there exists a bounded sesquilinear form

$$\langle\cdot,\cdot\rangle_t : \mathbf{H}_{||}^{-1/2}(\mathrm{div}_\Gamma,\Gamma)\times\mathbf{H}_{\perp}^{-1/2}(\,\mathrm{curl}_\Gamma,\Gamma)\to\mathbb{C}$$

that extends the $\boldsymbol{L}^2_t(\Gamma)$ inner product to the duality pairing of the two spaces such that the following inequalities hold:

$$c_1\|\boldsymbol{\lambda}\|_{\mathbf{H}_{||}^{-1/2}(\operatorname{div}_\Gamma,\Gamma)} \leq \sup_{0\neq\boldsymbol{\eta}\in\mathbf{H}_{\perp}^{-1/2}(\operatorname{curl}_\Gamma,\Gamma)} \frac{|\langle\boldsymbol{\lambda},\boldsymbol{\eta}\rangle_t|}{\|\boldsymbol{\eta}\|_{\mathbf{H}_{\perp}^{-1/2}(\operatorname{curl}_\Gamma,\Gamma)}} \leq c_2\|\boldsymbol{\lambda}\|_{\mathbf{H}_{||}^{-1/2}(\operatorname{div}_\Gamma,\Gamma)}$$

and

$$c_3\|\boldsymbol{\eta}\|_{\mathbf{H}_{\perp}^{-1/2}(\operatorname{curl}_\Gamma,\Gamma)} \leq \sup_{0\neq\boldsymbol{\lambda}\in\mathbf{H}_{||}^{-1/2}(\operatorname{div}_\Gamma,\Gamma)} \frac{|\langle\boldsymbol{\lambda},\boldsymbol{\eta}\rangle_t|}{\|\boldsymbol{\lambda}\|_{\mathbf{H}_{||}^{-1/2}(\operatorname{div}_\Gamma,\Gamma)}} \leq c_4\|\boldsymbol{\eta}\|_{\mathbf{H}_{\perp}^{-1/2}(\operatorname{curl}_\Gamma,\Gamma)},$$

where c_i are positive constants, independent of $\boldsymbol{\lambda}$ and $\boldsymbol{\eta}$. We now state without proof the following theorem:

Theorem 4.5.4. *Let $\boldsymbol{u} \in \mathbf{H}_{||}^{-1/2}(\operatorname{div}_\Gamma,\Gamma)$ and $\boldsymbol{v} \in \mathbf{H}_{\perp}^{-1/2}(\operatorname{curl}_\Gamma,\Gamma)$. Then there exist $\alpha_u,\alpha_v \in \mathcal{H}(\Gamma)$ and $\beta_u,\beta_v \in H^{1/2}(\Gamma)$ such that the following decompositions hold:*

$$\boldsymbol{u} = \mathbf{grad}_\Gamma\, \alpha_u + \mathbf{curl}_\Gamma\, \beta_u, \boldsymbol{v} = \mathbf{curl}_\Gamma \alpha_v + \mathbf{grad}_\Gamma\, \beta_v. \tag{4.5.35}$$

Let us define

$$\langle\boldsymbol{u},\boldsymbol{v}\rangle_t := -\langle\Delta_\Gamma\alpha_u,\beta_v\rangle_t - \langle\Delta_\Gamma\alpha_v,\beta_u\rangle_t. \tag{4.5.36}$$

Then there holds the integration by parts formula:

$$\int_\Omega \{\mathbf{curl}\ \boldsymbol{u}\cdot\boldsymbol{v} - \boldsymbol{u}\cdot\ \mathbf{curl}\ \boldsymbol{v}\}d\Omega = \big\langle\gamma_t(\boldsymbol{u}),\pi_t(\boldsymbol{v})\big\rangle_t \tag{4.5.37}$$

for all $\boldsymbol{u},\boldsymbol{v} \in \mathbf{H}(\mathbf{curl}\ ,\Omega)$, and the formula (4.5.37) *is consistent with the definition of $\langle\boldsymbol{u},\boldsymbol{v}\rangle_t$ in* (4.5.36).

The proof of this theorem can be found in Buffa and Ciarlet [50] in the case of Lipschitz polyhedra and in Buffa, Costabel and Sheen [52] in the case of general Lipschitz domains. We remark that the integration by parts formula (4.5.30) is consistent with the definition of $\langle\boldsymbol{u},\boldsymbol{v}\rangle_t$ in (4.5.36). The duality pairing

$$\mathbf{H}_{||}^{-1/2}(\operatorname{div}_\Gamma,\Gamma) = (\mathbf{H}_{\perp}^{-1/2}(\operatorname{curl}_\Gamma,\Gamma))' \tag{4.5.38}$$

with $\boldsymbol{L}_t^2(\Gamma)$ as a pivot space has already been stated in the case of a smooth domain with $C^{1,1}$ boundary Γ in Cessenat [63, p.40]. Note that from (4.5.37), we have

$$|\big\langle\gamma_t(\boldsymbol{u}),\pi_t(\boldsymbol{v})\big\rangle_t| \leq \|\boldsymbol{u}\|_{\mathbf{H}(\mathbf{curl}\ ,\Omega)}\|\boldsymbol{v}\|_{\mathbf{H}(\mathbf{curl}\ ,\Omega)}$$

and since the right–hand side of (4.5.37) depends only on the traces, this shows

$$|\big\langle\gamma_t(\boldsymbol{u}),\pi_t(\boldsymbol{v})\big\rangle_t| \leq \|\gamma_t(\boldsymbol{u})\|_{\mathbf{H}_{||}^{-1/2}(\operatorname{div}_\Gamma,\Gamma)}\ \|\pi_t(\boldsymbol{v})\|_{\mathbf{H}_{\perp}^{-1/2}(\operatorname{curl}_\Gamma,\Gamma)},$$

which implies that the duality is continuous. Furthermore, we have the duality property

$$\big\langle \gamma_t(\boldsymbol{u}), \pi_t(\boldsymbol{v})\big\rangle_t = -\big\langle \gamma_t(\boldsymbol{v}), \pi_t(\boldsymbol{u})\big\rangle_t, \quad \forall\, \boldsymbol{u}, \boldsymbol{v} \in \mathbf{H}(\mathbf{curl}\,, \Omega), \tag{4.5.39}$$

as a consequence of the Green formula (4.5.37). $\mathbf{grad}_{\,\Gamma}\alpha_u$ is already in $\gamma_t \boldsymbol{u}$ see (4.5.35),

$$\begin{aligned}
&\text{and } \langle \mathbf{grad}_{\,\Gamma}\alpha_u, \mathbf{curl}_{\,\Gamma}\alpha_v\rangle_t = \langle \mathrm{curl}_\Gamma \mathbf{grad}_{\,\Gamma}\alpha_u, \alpha_v\rangle_t = 0,\\
&\text{also } \langle \mathbf{curl}_{\,\Gamma}\beta_u, \mathbf{grad}_{\,\Gamma}\beta_v\rangle_t = 0, \text{(see V(A) or V(3) in Appendix B).}
\end{aligned}$$

The basic function spaces for the solutions of Maxwell's equations are $\mathbf{H}(\mathbf{curl}\,, \Omega)$ and $\mathbf{H}(\operatorname{div}, \Omega)$. However, as is well–known, one can not establish weak solutions of Maxwell equations by applying the Fredholm alternative directly to their variational formulations from either PDEs or BIEs, because of non–compactness of the imbeddings $\mathbf{H}(\operatorname{div}, \Omega) \hookrightarrow \boldsymbol{L}^2(\Omega)$ and $\mathbf{H}(\mathbf{curl}\,, \Omega) \hookrightarrow \boldsymbol{L}^2(\Omega)$ as well as $\mathbf{H}(\operatorname{div}, \Omega) \cap \mathbf{H}(\mathbf{curl}, \Omega) \hookrightarrow \boldsymbol{L}^2(\Omega)$ (see, e.g. Amrouche, Bernardi, Dauge and Girault [10]). Fortunately, these difficulties can be circumvented by introducing the idea of the before mentioned Hodge decompositions (or Helmholtz decompositions) of solution spaces of the PDEs or the trace spaces for the BIEs to reveal hidden compactness properties of the formulations as will be demonstrated in Chapter 12. It is worth mentioning that for the same space there may be different decompositions. To understand the developments for the Hodge decompositions of spaces, we need the concept of known as *de Rham diagram* in differential geometry to characterize the range and the kernel of the operators **grad**, **curl** and div in the space of square integrable fields in an open bounded domain $\Omega \subset \mathbb{R}^3$ with boundary Γ. In the exact sequence of the de Rham diagram, the co–domain of each previous operator fills out the kernel of the next operator in the sequence. This topic is out of the scope of our book. For interested readers, we refer to the references Bossavit [38, pp. 132–134], Cessenat [63, Appendix] and Dautray and Lions [96, pp.202, 313].

We comment that as mentioned in Dautray and Lions [96, p.209], it is important to make precise the connection between the spaces $\mathbf{H}(\operatorname{div}, \Omega)$, $\mathbf{H}(\mathbf{curl}\,, \Omega)$ and the Sobolev space $\mathbf{H}^1(\Omega)$. We now state the following theorem:

Theorem 4.5.5. *Let Ω be a bounded domain in $\mathbb{R}^3$ with a $C^{1,1}$ boundary Γ. Then the Sobolev spaces:*

$$\begin{cases} \mathbf{H}^1_t(\Omega) := \{\boldsymbol{u} \in \mathbf{H}^1(\Omega) | \gamma_n(\boldsymbol{u}) = 0\};\\[1ex] \mathbf{H}^1_n(\Omega) := \{\boldsymbol{u} \in \mathbf{H}^1(\Omega) | \gamma_t(\boldsymbol{u}) = \mathbf{0}\} \end{cases} \tag{4.5.40}$$

are such that

$$\begin{cases} \mathbf{H}^1_t(\Omega) = \mathbf{H}(\mathbf{curl}\,, \Omega) \cap \mathbf{H}_0(\operatorname{div}, \Omega);\\[1ex] \mathbf{H}^1_n(\Omega) = \mathbf{H}_0(\mathbf{curl}, \Omega) \cap \mathbf{H}(\operatorname{div}, \Omega) \end{cases} \tag{4.5.41}$$

where

$$\mathbf{H}_0(\mathbf{curl}\,,\Omega) := \{\boldsymbol{u} \in \mathbf{H}(\mathbf{curl}, \Omega) \quad \textit{and} \quad \gamma_t \boldsymbol{u} = \mathbf{0}\}, \quad \textit{and}$$
$$\mathbf{H}_0(\mathrm{div}, \Omega) := \{\boldsymbol{u} \in \mathbf{H}(\mathrm{div}, \Omega) \quad \textit{and} \quad \gamma_{\boldsymbol{n}} \boldsymbol{u} = 0\}.$$

On these spaces, the norm $\|\boldsymbol{u}\|_{\mathbf{H}^1(\Omega)}$ *is equivalent to the norm:*

$$\|\boldsymbol{u}\| := \Big\{ \int_\Omega (|\,\mathbf{curl}\ \boldsymbol{u}|^2 + |\mathrm{div}\,\boldsymbol{u}|^2 + |\boldsymbol{u}|^2) d\Omega \Big\}^{1/2}. \tag{4.5.42}$$

The proof of Theorem 4.5.5 is given in Dautray and Lions [96, Theorem 3]. In both cases, based on a Green formula after a tedious manipulation, it shows that there holds the following inequality:

$$\int_\Omega |\,\mathbf{grad}\ \boldsymbol{u}|^2 d\Omega \le \int_\Omega \left(|\,\mathbf{curl}\ \boldsymbol{u}|^2 + |\mathrm{div}\,\boldsymbol{u}|^2\right) d\Omega + c \int_\Gamma |\boldsymbol{u}|^2 ds_\Gamma$$

where c is a positive constant. Then to complete the proof, Peetre's lemma Peetre [345] is applied to the Sobolev spaces $\mathbf{H}^1_t(\Omega)$ as well as to $\mathbf{H}^1_n(\Omega)$. As a consequence of Theorem 4.5.5, one has

Corollary 4.5.6. *Under the hypotheses of Theorem* 4.5.5 on Ω , the spaces

$$\left.\begin{array}{l} \left\{\boldsymbol{u} \in \mathbf{H}(\,\mathbf{curl}\,,\Omega) \cap \mathbf{H}(\mathrm{div},\Omega) | \gamma_n(\boldsymbol{u}) \in H^{1/2}(\Gamma)\right\} \\[2ex] \left\{\boldsymbol{u} \in \mathbf{H}(\,\mathbf{curl}\,,\Omega) \cap \mathbf{H}(\mathrm{div},\Omega) | \gamma_t(\boldsymbol{u}) \in \mathbf{H}^{1/2}(\Gamma)\right\} \end{array}\right\} \tag{4.5.43}$$

are identified with the Sobolev space $\mathbf{H}^1(\Omega)$. Moreover, there exist positive constants c_1 and c_2 such that

$$\begin{cases} \|\boldsymbol{u}\|_{\mathbf{H}^1(\Omega)} \le c_1\Big(\|\boldsymbol{u}\|_{\boldsymbol{L}^2(\Omega)} + \|\,\mathbf{curl}\ \boldsymbol{u}\|_{\boldsymbol{L}^2(\Omega)} \\ \qquad\qquad + \|\mathrm{div}\,\boldsymbol{u}\|_{\boldsymbol{L}^2(\Omega)} + \|\gamma_n(\boldsymbol{u})\|_{H^{1/2}(\Gamma)}\Big) \\ \|\boldsymbol{u}\|_{\mathbf{H}^1(\Omega)} \le c_2\Big(\|\boldsymbol{u}\|_{\boldsymbol{L}^2(\Omega)} + \|\,\mathbf{curl}\ u\|_{\boldsymbol{L}^2(\Omega)} \\ \qquad\qquad + \|\mathrm{div}\,\boldsymbol{u}\|_{\boldsymbol{L}^2(\Omega)} + \|\gamma_t(\boldsymbol{u})\|_{\mathbf{H}_\perp^{1/2}(\Gamma)}\Big) \end{cases} \tag{4.5.44}$$

for all $\in \mathbf{H}^1(\Omega)$.

In closing this section, we include the following crucial results:

Theorem 4.5.7. *The embeddings of* $\mathbf{H}^1_t(\Omega)$ *and* $\mathbf{H}^1_n(\Omega)$ *into* $L^2(\Omega)^3$ *are compact* (see Theorem 2.8 in Amrouche, Bernardi, Dauge and Girault [10] and Picard, Weck and Witsch [348]).

This Theorem is proven by Ch. Weber and P. Werner [430]. As a consequence, there holds

Theorem 4.5.8. *The intersection space* $\mathbf{H}_t^1(\Omega) \cap \mathbf{H}_n^1(\Omega)$ *coincides with* $\left(H_0^1(\Omega)\right)^3$ (see Theorem 2.5 in Amrouche, Bernardi, Dauge and Girault [10]).

Theorem 4.5.8 is proven by V. Girault and P.A. Raviart [147].

5. Variational Formulations

In this chapter we will discuss the variational formulation for boundary integral equations and its connection to the variational solution of partial differential equations. We collect some basic theorems in functional analysis which are needed for this purpose. In particular, Green's theorems and the Lax–Milgram theorem are fundamental tools for the solvability of boundary integral equations as well as for elliptic partial differential equations. We will present here a subclass of boundary value problems for which the coerciveness property for some associated boundary integral operators follows directly from that of the variational form of the boundary and transmission problems. In this class, the solution of the boundary integral equations will be established with the help of the existence and regularity results of elliptic partial differential equations in variational form. This part of our presentation goes back to J.C. Nedelec and J. Planchard [332] and is an extension of the approach used by J.C. Nedelec [328] and the "French School" (see Dautray and Lions [95, Vol. 4]). It also follows closely the work by M. Costabel [79, 80, 84] and our works [90, 201, 401].

5.1 Partial Differential Equations of Second Order

Variational methods proved to be very successful for a class of elliptic boundary value problems which admit an equivalent variational formulation in terms of a coercive bilinear form. It is this class for which we collect some of the basic results. We begin with the second order differential equations which are the simplest examples in this class.

In this section let $\Omega \subset \mathbb{R}^n$ be a strong Lipschitz domain with strong Lipschitz boundary Γ, and let $\Omega^c = \mathbb{R}^n \setminus \overline{\Omega}$ denote the exterior domain. We consider the second order elliptic $p \times p$ system

$$Pu := -\sum_{j,k=1}^{n} \frac{\partial}{\partial x_j}\Big(a_{jk}(x)\frac{\partial u}{\partial x_k}\Big) + \sum_{j=1}^{n} b_j(x)\frac{\partial u}{\partial x_j} + c(x)u = f(x) \qquad (5.1.1)$$

in Ω and/or in Ω^c. The coefficients a_{jk}, b_j, c are $p \times p$ matrix–valued functions which, for convenience, are supposed to be C^∞–functions. For a_{jk} we

G. C. Hsiao und W. L. Wendland, *Boundary Integral Equations*,
Applied Mathematical Sciences 164,
https://doi.org/10.1007/978-3-030-71127-6_5

require the uniform strong ellipticity condition (3.6.2). We begin with the *first Green's formula* for (5.1.1) in Sobolev spaces.

Throughout this chapter we assume that a fundamental solution $E(x, y)$ to the system (5.1.1) is available (see Lemma 3.6.1 and Section 7.3).

First Green's Formula (Nečas [326, Chap.3])

Multiplying (5.1.1) by a test function v and applying the Gauss divergence theorem one obtains

$$\int_\Omega (Pu)^\top \overline{v} dx = a_\Omega(u, v) - \int_\Gamma (\partial_\nu u)^\top \overline{v} ds \tag{5.1.2}$$

with the *energy–sesquilinear form*

$$a_\Omega(u, v) := \int_\Omega \Big\{ \sum_{j,k=1}^n \Big(a_{jk}(x)\frac{\partial u}{\partial x_k}\Big)^\top \frac{\partial \overline{v}}{\partial x_j} + \sum_{j=1}^n \Big(b_j(x)\frac{\partial u}{\partial x_j}\Big)^\top \overline{v} + (cu)^\top \overline{v} \Big\} dx \tag{5.1.3}$$

and the conormal derivative

$$\partial_\nu u := \sum_{j,k=1}^n n_j a_{jk} \gamma_0 \frac{\partial u}{\partial x_k} \tag{5.1.4}$$

provided $u \in H^2(\Omega)$ and $v \in H^1(\Omega)$ (see Nečas [326]); or in terms of local coordinates, inserting (3.4.35) into (5.1.4), by

$$\partial_\nu u = \Big(\sum_{j,k=1}^n n_j a_{jk} n_k \Big) \frac{\partial u}{\partial n} + \sum_{j,k=1}^n \sum_{\mu,\varrho=1}^{n-1} n_j a_{jk} \frac{\partial y_k}{\partial \sigma_\mu} \gamma^{\mu\varrho} \frac{\partial u}{\partial \sigma_\varrho} . \tag{5.1.5}$$

For further generalization we introduce the function space

$$H^1(\Omega, P) := \{u \in H^1(\Omega) \,|\, Pu \in \widetilde{H}_0^{-1}(\Omega)\}$$

equipped with the graph norm

$$\|u\|_{H^1(\Omega,P)} := \|u\|_{H^1(\Omega)} + \|Pu\|_{\widetilde{H}^{-1}(\Omega)} .$$

Here $\widetilde{H}_0^{-1}(\Omega)$ is a subspace embedded into $\widetilde{H}^{-1} = \{f \in H^{-1}(\mathbb{R}^\kappa)$ with supp $f_1 \subseteq \overline{\Omega}\}$ which consists of all functions in the form:

$$f = f_1 + f_2 \quad \text{where } f_1 \in \widetilde{H}^{-1} \text{ with supp } f_1 \Subset \Omega \\ \text{and } f_2 \in \widetilde{H}^t(\Omega), \quad \text{for } t \geq -\frac{1}{2}, \tag{5.1.6}$$

equipped with the norm $\inf\{\|f_1\|_{\widetilde{H}^{-1}(\Omega)} + \|f_2\|_{\widetilde{H}^t(\Omega)}\}$.

Clearly, the following inclusions hold:

$$H_0^1(\Omega) \subset H^1(\Omega) \subset L^2(\Omega) \subset \widetilde{H}_0^{-1}(\Omega) \subset H^{-1}(\Omega) .$$

It is understood, that $u = (u_1, \dots, u_p)^\top \in (H^1(\Omega, P))^p =: H^1(\Omega, P)$. Here and in the sequel, as in the scalar case, we omit the superindex p.

Then first Green's formula (5.1.2) remains valid in the following sense.

Lemma 5.1.1 (Generalized First Green's Formula).

For fixed $u \in H^1(\Omega, P)$, the mapping

$$v \mapsto \langle \tau u, v \rangle_\Gamma := a_\Omega(u, \overline{\mathcal{Z}v}) - \int\limits_\Omega (Pu)^\top \mathcal{Z}v dx \tag{5.1.7}$$

is a continuous linear functional τu on $v \in H^{\frac{1}{2}}(\Gamma)$ that coincides for $u \in H^2(\Omega)$ with $\partial_\nu u$, i.e. $\tau u = \partial_\nu u$. The mapping $\tau : H^1(\Omega, P) \to H^{-\frac{1}{2}}(\Gamma)$ with $u \mapsto \tau u$ is continuous. Here, $\mathcal{Z}$ is a right inverse to the trace operator γ_0 (4.2.41).

In addition, there holds also the generalized first Green's formula

$$\int\limits_\Omega (Pu)^\top \overline{v} dx = a_\Omega(u, v) - \langle \tau u, \overline{\gamma_0 v} \rangle_\Gamma \tag{5.1.8}$$

for $u \in H^1(\Omega, P)$ and $v \in H^1(\Omega)$ where

$$\langle \tau u, \overline{\gamma_0 v} \rangle_\Gamma = (\tau u, \gamma_0 v)_{L^2(\Gamma)} = \int\limits_\Gamma (\tau u)^\top \overline{\gamma_0 v} ds \tag{5.1.9}$$

with $\tau u \in H^{-\frac{1}{2}}(\Gamma)$.

Note that the resulting operator $\tau u = \partial_\nu u$ in (5.1.7) does not depend on the special choice of the right inverse $\mathcal{Z}$ in (4.2.41); and that $\langle \cdot, \cdot \rangle_\Gamma$ denotes the duality pairing (5.1.9) with respect to $L^2(\Gamma)$.

If f is given in (5.1.6) then the definition of the generalized conormal derivative τ in (5.1.7) corresponds to T^+ in Mikhailov [297, Definition 3] and Costabel [80, Lemma 2.3].

For $f \in \widetilde{H}^{-1}$ set in (5.1.8) $\int_\Gamma (Pu)\mathcal{Z}v dx = \langle f, \mathcal{Z}u \rangle_\Omega$ and approximate u in $H^1(\Omega)$ and f in $\widetilde{H}^{-1}(\Omega)$ by distributions in $\widetilde{H}_0^{-1}(\Omega)$, characterized in (5.1.6). Then $r \in H^{-1/2}(\Omega)$ is the well defined limit but now depends not only on u in Ω but also on f, since the extension of $Pu \in H^{-1}$ to $f \in \widetilde{H}^{-1}(\Omega)$ is not unique (see McLean [290, Lemma 4.3] and Mikhailov [297, Lemma 6]).

Proof: First, let $u \in C^\infty(\overline{\Omega})$ be any given function. Then for any test function $v \in H^{\frac{1}{2}}(\Gamma)$ and its extension $\mathcal{Z}v \in H^1(\Omega)$ (see (4.2.41)) with $\mathcal{Z}$ a right inverse of γ_0, the first Green's formula (5.1.2) yields

$$\langle \partial_\nu u, v \rangle_\Gamma = a_\Omega(u, \overline{\mathcal{Z}v}) - \int\limits_\Omega (Pu)^\top \mathcal{Z}v dx \,, \tag{5.1.10}$$

where $\partial_\nu u$ is independent of $\mathcal{Z}$. Moreover, by applying the Cauchy–Schwarz inequality and the duality argument on the right–hand side of (5.1.10), we obtain the estimate

$$\begin{aligned}
&|\langle \partial_\nu u, v\rangle_\Gamma| \\
&\leq \; c\|u\|_{H^1(\Omega)}\|\mathcal{Z}v\|_{H^1(\Omega)} + \|Pu\|_{\widetilde{H}^{-1}(\Omega)}\|\mathcal{Z}v\|_{H^1(\Omega)} \\
&\leq \; c_{\mathcal{Z}}\{\|u\|_{H^1(\Omega)} + \|Pu\|_{\widetilde{H}^{-1}(\Omega)}\}\|v\|_{H^{\frac{1}{2}}(\Gamma)} = c_{\mathcal{Z}}\|u\|_{H^1(\Omega,P)}\|v\|_{H^{\frac{1}{2}}(\Gamma)}
\end{aligned}$$

where $c_{\mathcal{Z}}$ is independent of u and v. This shows that the right–hand side of (5.1.10) defines a bounded linear functional on $v \in H^{\frac{1}{2}}(\Gamma)$ which for every $u \in C^\infty(\overline{\Omega})$ is given by $\partial_\nu u$ satisfying

$$\|\partial_\nu u\|_{H^{-\frac{1}{2}}(\Gamma)} = \sup_{\|v\|_{H^{\frac{1}{2}}(\Gamma)}=1} |\langle \partial_\nu u, v\rangle_\Gamma| \leq c_{\mathcal{Z}}\|u\|_{H^1(\Omega,P)}\,, \tag{5.1.11}$$

where $c_{\mathcal{Z}}$ is independent of u. This defines a linear mapping from $u \in C^\infty(\overline{\Omega}) \subset H^1(\Omega, P)$ onto the linear functional on $H^{\frac{1}{2}}(\Gamma)$ given by $\tau u = \partial_\nu u \in C^\infty(\Gamma) \subset H^{-\frac{1}{2}}(\Gamma)$. This mapping $u \mapsto \tau u$ can now be extended continuously to $u \in H^1(\Omega, P)$ due to the estimate (5.1.11) by the well known completion procedure since $C^\infty(\overline{\Omega})$ is dense in $H^1(\Omega, P)$. Namely, if $u \in H^1(\Omega, P)$ is given then choose a sequence $u_k \in C^\infty(\overline{\Omega})$ such that $\|u - u_k\|_{H^1(\Omega,P)} \to 0$ for $k \to \infty$. Then the sequence $\tau u_k = \partial_\nu u_k$ defines a Cauchy sequence in $H^{-\frac{1}{2}}(\Gamma)$ whose limit defines τu. This procedure also proves the validity of the generalized Green's formula (5.1.7). ■

Note that the generalized first Green's formula (5.1.8) in particular holds for the H^1–solution u of $Pu = f$ with $f \in \widetilde{H}_0^{-1}(\Omega)$.

We also present the generalized first Green's formula for the exterior domain Ω^c. Here we use the local space

$$H^1_{\text{loc}}(\Omega^c, P) := \{u \in H^1_{\text{loc}}(\Omega^c) \,|\, Pu \in H^{-1}_{\text{comp}}(\Omega^c)\}$$

as defined in (4.1.44). The following version of Lemma 5.1.1 is valid in Ω^c.

Lemma 5.1.2 (Exterior Generalized First Green's Formula).
Let $u \in H^1_{\text{loc}}(\Omega^c, P)$. Then the linear functional $\tau_c u$ given by the mapping

$$\langle \tau_c u, \gamma_{c0} v\rangle_\Gamma := -a_{\Omega^c}(u, \overline{\mathcal{Z}_c v}) + \int_{\Omega^c} (Pu)^\top \mathcal{Z}_c v dx \;\; \textit{for all} \;\; v \in H^{\frac{1}{2}}(\Gamma)$$

belongs to $H^{-\frac{1}{2}}(\Gamma)$ and coincides with

$$\tau_c u = \partial_{c\nu} u = \sum_{j,k=0}^{n} n_j a_{jk} \gamma_{c0} \frac{\partial u}{\partial x_k} \tag{5.1.12}$$

provided $u \in C^\infty(\overline{\Omega^c})$. *Moreover,* $\tau_c : u \mapsto \tau_c u$ *is a linear continuous mapping from* $H^1_{\mathrm{loc}}(\Omega^c, P)$ *into* $H^{-\frac{1}{2}}(\Gamma)$. *In addition, there holds the exterior generalized first Green's formula*

$$\int_{\Omega^c} (Pu)^\top \overline{v} dx = a_{\Omega^c}(u,v) + \langle \tau_c u, \gamma_{c0}\overline{v}\rangle_\Gamma \tag{5.1.13}$$

for $u \in H^1_{\mathrm{loc}}(\Omega^c, P)$ *and for every*

$$v \in H^1_{\mathrm{comp}}(\overline{\Omega^c}) := \{v \in H^1_{\mathrm{loc}}(\Omega^c) \,\big|\, v \text{ has compact support in } \mathbb{R}^n\}.$$

Here $\mathcal{Z}_c$ denotes some right inverse to γ_{c0} (cf. (4.2.41)) which maps $v \in H^{\frac{1}{2}}(\Gamma)$ into $\mathcal{Z}_c v \in H^1_{\mathrm{comp}}(\overline{\Omega^c})$ where $\mathrm{supp}(\mathcal{Z}_c v)$ is contained in some fixed compact set $\Omega^c_R \subset \mathbb{R}^n$ containing Γ. Note that the distributions belonging to $H^{-1}_{\mathrm{comp}}(\Omega^c)$ cannot have a singular support on Γ due to the definition in (4.1.44). The sesquilinear form $a_{\Omega^c}(u,v)$ is defined as in (5.1.3) with Ω replaced by Ω^c.

Proof: The proof resembles word by word the proof of Lemma 5.1.1 when replacing Ω by Ω^c_R in the corresponding norms there. ∎

Before we formulate the variational formulation for the boundary value problems for P we return to the sesquilinear form $a_\Omega(u,v)$ given by (5.1.3). In the sequel, the properties of a_Ω and a_{Ω^c} will be used extensively. Let us begin with the general definition of sesquilinear forms.

The sesquilinear form $a(u,v)$ (Lions [268, Chap.II], see also Stummel [410] and Zeidler [444].)
We recall that $a(u,v)$ is called a *sesquilinear form* on the product space $\mathcal{H}\times\mathcal{H}$, where $\mathcal{H}$ is a Hilbert space with scalar product $(\cdot,\cdot)_\mathcal{H}$ and $\|u\|^2_\mathcal{H} = (u,u)_\mathcal{H}$, iff:

a: $\mathcal{H}\times\mathcal{H} \to \mathbb{C}$;
$a\ (\bullet, v)$ is linear for each fixed $v \in \mathcal{H}$; i.e. for all $c_1, c_2 \in \mathbb{C}$, $u_1, u_2 \in \mathcal{H}$:
$a\ (c_1u_1 + c_2u_2, v) = c_1 a(u_1, v) + c_2 a(u_2, v)$;
$a\ (u, \bullet)$ is antilinear for each fixed $u \in \mathcal{H}$; i.e. for all $c_1 c_2 \in \mathbb{C}, v_1, v_2 \in \mathcal{H}$:
$a\ (u, c_1v_1 + c_2v_2) = \overline{c_1} a(u, v_1) + \overline{c_2} a(u, v_2)$.

The sesquilinear form $a(\cdot,\cdot)$ is said to be *continuous*, if there exists a constant M such that for all $u, v \in \mathcal{H}$:

$$|a(u,v)| \le M\|u\|_\mathcal{H}\, \|v\|_\mathcal{H}\,. \tag{5.1.14}$$

5.1.1 Interior Problems

With the sesquilinear form $a_\Omega(u,v)$ available, we now are in a position to present the variational formulation of boundary value problems. For simplicity, we begin with the standard Dirichlet problem.

The interior Dirichlet problem
Given $f \in \widetilde{H}_0^{-1}(\Omega)$ and $\varphi \in H^{\frac{1}{2}}(\Gamma)$, find $u \in H^1(\Omega)$ with $\gamma_0 u = \varphi$ on Γ such that

$$a_\Omega(u,v) = \ell_f(\overline{v}) := \langle f, \overline{v}\rangle_\Omega \text{ for all } v \in H_0^1(\Omega)\,, \tag{5.1.15}$$

where $a_\Omega(u,v)$ is the sesquilinear form (5.1.3) *and the right–hand side is a given bounded linear functional on $H_0^1(\Omega)$.*

We notice that the standard boundary condition $\gamma_0 u = \varphi$ on Γ is imposed in the set of admissible functions for the solution. For this reason, the Dirichlet boundary condition is referred to as an *essential boundary condition*. It can be replaced by the slightly more general form

$$B\gamma u = B_{00}\gamma_0 u = \varphi \in H^{\frac{1}{2}}(\Gamma) \tag{5.1.16}$$

with order $(B_{00}) = 0$ corresponding to (3.9.10) with $R\gamma u = B\gamma u$ where B_{00} is an invertible smooth matrix–valued function on Γ. For this boundary condition, in order to formulate the variational formulation similar to (5.1.15), one only needs to modify the last term in the first Green's formula (5.1.8) according to (5.1.16):

$$\langle \tau u, \overline{\gamma_0 v}\rangle_\Gamma = \langle N\gamma u, \overline{B_{00}\gamma_0 v}\rangle_\Gamma \tag{5.1.17}$$

if we choose the complementary operator $S = N$ of (3.9.10) as

$$N\gamma u := (B_{00}^{-1})^* \tau u = N_{00}\gamma_0 u + N_{01}\frac{\partial u}{\partial n}|_\Gamma \tag{5.1.18}$$

with

$$\begin{aligned} N_{00} &:= (B_{00}^{-1})^* \sum_{j,k=1}^{n}\sum_{\mu,\rho=1}^{n-1}\Big(n_j a_{jk}\frac{\partial y_k}{\partial \sigma_\mu}\gamma^{\mu\rho}\Big)\frac{\partial}{\partial \sigma_\rho}\,,\\ N_{01} &:= (B_{00}^{-1})^*\Big(\sum_{j,k=1}^{n} n_j a_{jk} n_k\Big) \end{aligned}$$

satisfying the conditions required in (3.9.5) because of the strong ellipticity of P. This leads to the variational formulation:

The general interior Dirichlet problem

For given $f \in L^2(\Omega)$ and $\varphi \in H^{\frac{1}{2}}(\Gamma)$, find $u \in H^1(\Omega)$ with $B_{00}\gamma_0 u = \varphi$ on Γ such that

$$a_\Omega(u,v) = \ell_f(\overline{v}) := \langle f, \overline{v}\rangle_\Omega \quad \textit{for all} \;\; v \in \mathcal{H} := \{v \in H^1(\Omega) \,|\, B\gamma v = 0 \quad \textit{on} \;\; \Gamma\}\,. \tag{5.1.19}$$

Clearly, $\mathcal{H} = H_0^1(\Omega)$ since $B\gamma v = B_{00}\gamma_0 v = 0$ for $v \in H^1(\Omega)$ iff $\gamma_0 v = 0$ because the smooth matrix–valued function B_{00} is assumed to be invertible.

The interior Neumann problem

Given $f \in \widetilde{H}_0^{-1}(\Omega)$ and $\psi \in H^{-\frac{1}{2}}(\Gamma)$, find $u \in H^1(\Omega)$ such that

$$a_\Omega(u,v) = \ell_{f,\psi}(\overline{v}) := \langle f, \overline{v}\rangle_\Omega + \langle \psi, \overline{v}\rangle_\Gamma \;\; \textit{for all} \;\; v \in H^1(\Omega)\,. \tag{5.1.20}$$

where $a_\Omega(u,v)$ again is the sesquilinear form (5.1.3) *and the right–hand side is a given bounded linear functional on $H^1(\Omega)$ defined by f and ψ.*

In contrast to the Dirichlet problem, the Neumann boundary condition $\tau u = \psi$ is not required to be imposed on the solution space. Consequently, the Neumann boundary condition is referred to as the *natural boundary condition*. Both formulations (5.1.15) and (5.1.20) are derived from the first Green's formula (5.1.2).

The more general Neumann–Robin condition with $R\gamma u = N\gamma u$ in (3.9.13) is given by

$$N\gamma u := N_{00}\gamma_0 u + N_{01}\frac{\partial u}{\partial n} = \psi \in H^{-\frac{1}{2}}(\Gamma)\,. \tag{5.1.21}$$

Here N_{01} is now a smooth, invertible matrix–valued function and N_{00} is a tangential first order differential operator of the form

$$N_{00}\gamma_0 u = \sum_{\rho=1}^{n-1} c_{0\rho}\frac{\partial}{\partial\sigma_\rho}\gamma_0 u + c_{00}\gamma_0 u \quad \text{on } \Gamma \tag{5.1.22}$$

with smooth matrix–valued coefficients $c_{0\rho}$ and c_{00}. In contrast to the standard Neumann problem, we now require additionally that Γ is more regular than Lipschitz, namely $\Gamma \in C^{1,1}$. The latter guarantees the continuity of the mapping $N\gamma : H^1(\Omega, P) \to H^{-\frac{1}{2}}(\Gamma)$.

In (5.1.8) we now insert (5.1.5) with (5.1.21) and (5.1.22) to obtain

$$\tau u = \Big(\sum_{j,k=1}^{n} n_j a_{jk} n_k\Big) N_{01}^{-1}(\psi - N_{00}\gamma_0 u) + \sum_{j,k=1}^{n}\sum_{\mu,\rho=1}^{n-1} n_j a_{jk}\frac{\partial y_k}{\partial\sigma_\mu}\gamma^{\mu\rho}\frac{\partial}{\partial\sigma_\rho}\gamma_0 u$$

and

$$\langle \tau u, \overline{\gamma_0 v}\rangle_\Gamma = \langle N\gamma u, \overline{S\gamma v}\rangle_\Gamma + \langle \widetilde{N}\gamma_0 u, \overline{\gamma_0 v}\rangle_\Gamma\,, \tag{5.1.23}$$

where the complementary boundary operator is now chosen as

$$S\gamma v := B\gamma u := (N_{01}^{-1})^*\Big(\sum_{j,k=1}^{n} n_j a_{jk} n_k\Big)^*\gamma_0 v \tag{5.1.24}$$

and $\widetilde{N}$ is therefore defined by

$$\widetilde{N}\gamma_0 u := \Big\{ -\Big(\sum_{j,k=1}^{n} n_j a_{jk} n_k\Big) N_{01}^{-1} N_{00} + \sum_{j,k=1}^{n} \sum_{\mu,\rho=1}^{n-1} n_j a_{jk} \frac{\partial y_k}{\partial \sigma_\mu} \gamma^{\mu\rho} \frac{\partial}{\partial \sigma_\rho} \Big\} \gamma_0 u \,.$$

This operator can be written with (5.1.22) as

$$\widetilde{N}\gamma_0 u = \sum_{\rho=1}^{n-1} \widetilde{c}_{0\rho} \frac{\partial}{\partial \sigma_\rho} \gamma_0 u + \widetilde{c}_{00} \gamma_0 u \quad \text{on } \Gamma \,, \tag{5.1.25}$$

where

$$\widetilde{c}_{0\varrho} = \sum_{j,k=1}^{n} \sum_{\mu=1}^{n-1} n_j a_{jk} \Big(\frac{\partial y_k}{\partial \sigma_\mu} \gamma^{\mu\varrho} - n_k N_{01}^{-1} c_{0\rho} \Big) \,.$$

Following an idea of Weinberger (see the remarks in Agmon [3, p.147ff.], Fichera [121]), the corresponding term in (5.1.23) will now be incorporated into $a_\Omega(u,v)$ generating an additional skew–symmetric bilinear form. For fixed ρ, (3.3.7) gives, with the chain rule,

$$\frac{\partial u}{\partial \sigma_\rho} = \sum_{k=1}^{n} \Big(\frac{\partial y_k}{\partial \sigma_\rho} + \sigma_n \frac{\partial n_k}{\partial \sigma_\rho} \Big) \frac{\partial u}{\partial x_k} \quad \text{near } \Gamma \,.$$

By extending y_k and n_j to all of Ω smoothly, e.g. by zero sufficiently far away from Γ, we define the skew–symmetric real–valued coefficients

$$\beta_{kj\rho} := \Big(n_k \frac{\partial y_j}{\partial \sigma_\rho} - n_j \frac{\partial y_k}{\partial \sigma_\rho} \Big) = -\beta_{jk\rho} \,. \tag{5.1.26}$$

With

$$\sum_{k=1}^{n} n_k^2 = 1 \quad \text{and} \quad \sum_{k=1}^{n} n_k \frac{\partial y_k}{\partial \sigma_\varrho} |_\Gamma = 0$$

we have

$$\begin{aligned}
&\sum_{j=1}^{n} \sum_{\rho=1}^{n-1} \int_\Gamma \Big(\widetilde{c}_{0\rho} \frac{\partial y_j}{\partial \sigma_\rho} \frac{\partial u}{\partial x_j} \Big)^\top \overline{\gamma_0 v} ds \\
&= \sum_{k,j=1}^{n} \sum_{\varrho=1}^{n-1} \int_\Gamma \Big(\widetilde{c}_{0\rho} \Big(n_k \frac{\partial y_j}{\partial \sigma_\rho} - n_j \frac{\partial y_k}{\partial \sigma_\rho} \Big) n_k \frac{\partial u}{\partial x_j} \Big)^\top \overline{\gamma_0 v} ds \\
&= \int_\Gamma \sum_{j,k=1}^{n} \sum_{\rho=1}^{n-1} \Big(\widetilde{c}_{0\rho} \beta_{jk\rho} \frac{\partial u}{\partial x_j} \Big)^\top n_k \overline{\gamma_0 v} ds
\end{aligned}$$

and integration by parts shows that

$$\sum_{j=1}^{n}\sum_{\rho=1}^{n-1}\int\limits_{\Gamma}\Big(\widetilde{c}_{0\rho}\frac{\partial y_j}{\partial\sigma_\rho}\frac{\partial u}{\partial x_j}\Big)^\top \gamma_0\overline{v}ds$$

$$= \int\limits_{\Omega}\sum_{j,k=1}^{n}\Big(\Big(\frac{\partial}{\partial x_k}\sum_{\rho=1}^{n-1}\widetilde{c}_{0\rho}\beta_{kj\rho}\Big)\frac{\partial u}{\partial x_j}\Big)^\top\overline{v}dx$$

$$+\int\limits_{\Omega}\sum_{j,k=1}^{n}\Big(\sum_{\rho=1}^{n-1}\widetilde{c}_{0\rho}\beta_{kj\rho}\frac{\partial u}{\partial x_j}\Big)^\top\frac{\partial\overline{v}}{\partial x_k}dx\,.$$

Since the last term vanishes because of (5.1.26), this yields

$$\langle\widetilde{N}\gamma_0 u,\overline{\gamma_0 v}\rangle_\Gamma = \sum_{j,k=1}^{n}\sum_{\rho=1}^{n-1}\int\limits_{\Omega}\Big(\widetilde{c}_{0\rho}\beta_{kj\rho}\frac{\partial u}{\partial x_j}\Big)^\top\frac{\partial\overline{v}}{\partial x_k}dx$$
$$+\sum_{j=1}^{n}\int\limits_{\Omega}\Big\{\frac{\partial}{\partial x_j}\big((n_j\widetilde{c}_{00}u)^\top\overline{v}\big)+\sum_{k=1}^{n}\sum_{\rho=1}^{n-1}\Big(\Big(\frac{\partial}{\partial x_k}\widetilde{c}_{0\rho}\beta_{kj\rho}\Big)\frac{\partial u}{\partial x_j}\Big)^\top\overline{v}\Big\}dx \tag{5.1.27}$$

which motivates us to introduce the modified bilinear form

$$\widetilde{a}_\Omega(u,v) := a_\Omega(u,v) - \sum_{j,k=1}^{n}\int\limits_{\Omega}\Big(\Big(\sum_{\rho=1}^{n-1}\widetilde{c}_{0\rho}\beta_{jk\rho}\Big)\frac{\partial u}{\partial x_j}\Big)^\top\frac{\partial\overline{v}}{\partial x_k}dx$$
$$-\int\limits_{\Omega}\sum_{j=1}^{n}\Big\{\frac{\partial}{\partial x_j}\big((n_j\widetilde{c}_{00}u)^\top\overline{v}\big)+\sum_{\rho=1}^{n-1}\sum_{k=1}^{n}\Big(\Big(\frac{\partial}{\partial x_k}\widetilde{c}_{0\rho}\beta_{kj\rho}\Big)\frac{\partial u}{\partial x_j}\Big)^\top\overline{v}\Big\}dx\,. \tag{5.1.28}$$

Equations (5.1.23) and (5.1.28) in connection with the first Green's formula (5.1.8) lead to the variational formulation of the general Neumann–Robin problem.

The general interior Neumann–Robin problem

Let $\Gamma\in C^{1,1}$. Given $f\in\widetilde{H}_0^{-1}(\Omega)$ and $\psi\in H^{-\frac{1}{2}}(\Gamma)$ in (5.1.21), find $u\in H^1(\Omega)$ such that

$$\widetilde{a}_\Omega(u,v) = \langle f,\overline{v}\rangle_\Omega + \langle\psi,\overline{B\gamma v}\rangle_\Gamma \quad \text{for all } v\in H^1(\Omega) \tag{5.1.29}$$

where $\widetilde{a}_\Omega(u,v)$ is defined by (5.1.28) and the right–hand side is a bounded linear functional on $H^1(\Omega)$ defined by f and ψ.

The combined Dirichlet–Neumann problem

A class of more general boundary value problems combining both Dirichlet and Neumann conditions can be formulated as follows. First let us introduce a symmetric, real $(p\times p)$ projection matrix π on Γ which has the properties:

$$\operatorname{rank} \pi(x) = r \le p \quad \text{for all } x \in \Gamma\,, \; \pi = \pi^\top \quad \text{and} \quad \pi^2 = \pi\,, \tag{5.1.30}$$

where r is constant on Γ and $\pi(x) \in C^1(\Gamma)$. As a simple example consider $\pi = T(x) \begin{pmatrix} 1 & 0 & 0 \\ 0 & 0 & 0 \\ 0 & 0 & 0 \end{pmatrix} T^\top(x)$ for $p = 3$ where $T(x)$ is any smooth orthogonal matrix on Γ. The combined interior Dirichlet–Neumann problem reads:
For given $f \in \widetilde{H}_0^{-1}(\Omega)$, $\varphi_1 \in \pi H^{\frac{1}{2}}(\Gamma)$ and $\psi_2 \in (I - \pi)H^{-\frac{1}{2}}(\Gamma)$ find $u \in H^1(\Omega, P)$ satisfying

$$Pu = f \quad \text{in } \Omega\,, \; \pi B\gamma u = \varphi_1 \quad \text{and} \quad (I - \pi)N\gamma u = \psi_2 \quad \text{on } \Gamma\,. \tag{5.1.31}$$

Here I denotes the identity matrix on Γ.

The corresponding variational formulation then takes the form:
Find $u \in H^1(\Omega)$ with $\pi B\gamma u = \varphi_1$ on Γ such that

$$\widetilde{a}_\Omega(u, v) = \int_\Omega f^\top \overline{v} dy + \langle \psi_2, (I - \pi)\overline{B\gamma u}\rangle_\Gamma \tag{5.1.32}$$

$$\text{for all } v \in \mathcal{H}_{\pi B} := \{v \in H^1(\Omega) \,|\, \pi B\gamma v_{|_\Gamma} = 0\}\,.$$

Note that in the special case $\pi = I$, (5.1.32) reduces to the interior Dirichlet problem (5.1.16), while in the case $\pi = 0$, (5.1.32) becomes the interior Neumann–Robin problem (5.1.29).

Note that in the special case $B_{00} = I$ and $N_{00} = 0$, for the treatment of (5.1.31) we only need $\Gamma \in C^{0,1}$, i.e., to be Lipschitz.

5.1.2 Exterior Problems

For extending our variational approach also to exterior problems, we need to incorporate the radiation conditions. For motivation we begin with the "classical" representation formula (3.6.17) in terms of our boundary data $\gamma_{c0}u$ and $\tau_c u = \partial_{c\nu} u$ (see (5.1.12)), respectively,

$$u(x) = \int_{y \in \Omega^c} E^\top(x, y) f(y) dy - \int_\Gamma E^\top(x, y) \tau_c u ds_y \tag{5.1.33}$$
$$+ \int_\Gamma \Big(\partial_{c\nu y} E(x, y) + \sum_{j=1}^n n_j(y) b_j(y) E(x, y)\Big)^\top \gamma_{c0} u ds_y + M(x; u)$$

provided $f \in H^{-1}_{\text{comp}}(\Omega^c)$ has compact support in $\mathbb{R}^n$ (see (4.1.44)). Here, M is defined by

$$M(x;u) := \lim_{R\to\infty}\Big\{ \int\limits_{|y|=R} E^\top(x,y)\tau_c u ds \tag{5.1.34}$$
$$- \int\limits_{|y|=R} \Big(\partial_{\nu y}E(x,y) + \sum_{j=1}^n n_j(y)b_j(y)E(x,y)\Big)^\top \gamma_{c0}uds_y\Big\}.$$

From this definition, we see that M must satisfy the homogeneous differential equation

$$PM = 0 \tag{5.1.35}$$

in all of $\mathbb{R}^n$. Its behaviour at infinity is rather delicate, heavily depending on the nature of P there. In order to characterize M more systematically we require that M is a tempered distribution, and we further simplify P by assuming the following:

Conditions for the coefficients:

$$c \equiv 0 \textit{ for all } x\,,\ a_{jk}(x) = a_{jk}(\infty) = \text{ const. } \textit{and } b_j(x) = 0 \textit{ for } |x| \geq R_0\,. \tag{5.1.36}$$

In addition, we require that for P there holds the following

Unique continuation property: (See Hörmander [188, Chapter 17.2].)

$$\textit{If } Pu = 0 \quad \textit{in any domain } \omega \subset \mathbb{R}^n \quad \textit{and } u \equiv 0 \quad \textit{in some ball}$$
$$B_\rho(x_0) := \{x \textit{ with } |x - x_0| < \rho\} \subset \omega\,,\ \rho > 0 \quad \textit{then } u \equiv 0 \quad \textit{in } \omega\,. \tag{5.1.37}$$

Under these assumptions together with the restriction that uniqueness holds for both Dirichlet problems with P in Ω as well as in Ω^c, we now show that every tempered distribution $M \in \mathcal{S}'(\mathbb{R}^n)$ satisfying (5.1.35) admits the form

$$M_L(x;u) = \sum_{|\beta|\leq L} \alpha_\beta(u)p_\beta(x)\,, \tag{5.1.38}$$

where α_β are linear functionals on u and $p_\beta(x)$ are generalized polynomials behaving at infinity like homogeneous polynomials of degree $|\beta|$. (For the constant coefficient case see, e.g., Dautray and Lions [95, pp. 360ff.], [97, p. 119] and Miranda [307, p. 225].)

Let M be a tempered distribution satisfying (5.1.35). Because of assumption (5.1.36) we have $M(x;u) = p^\infty(x)$ for $|x| \geq R_0$ where p^∞ is a tempered distribution satisfying

$$P_\infty p^\infty(x) := \sum_{j,k=1}^n a_{jk}(\infty)\frac{\partial^2}{\partial x_j \partial x_k}p^\infty(x) = 0\,. \tag{5.1.39}$$

By taking the Fourier transform of (5.1.39) we obtain

$$\sum_{j,k=1}^n \xi_j a_{jk}(\infty)\xi_k \widehat{p^\infty}(\xi) = 0 \quad \text{for all } \ \xi \in \mathbb{R}^n \setminus \{0\}\,. \tag{5.1.40}$$

Since for each $\xi \in \mathbb{R}^n \setminus \{0\}$ the strong ellipticity (3.6.2) implies that $\big(\sum\limits_{j,k=1}^{n} \xi_j a_{jk}(\infty)\xi_k\big)^{-1}$ exists, it follows that

$$\widehat{p^\infty}(\xi) = 0 \quad \text{for all } \ \xi \ \in \mathbb{R}^n \setminus \{0\}\,. \tag{5.1.41}$$

Hence, every component of $\widehat{p^\infty}(\xi)$ is a distribution with only support $\{0\}$. The Schwartz theorem (Dieudonné [100, Theorem 17.7.3]) implies that $\widehat{p^\infty}(\xi)$ then is a finite linear combination of Dirac distributions at the origin. This yields:

Lemma 5.1.3. *For a strongly elliptic operator P with* (5.1.36) *in $\mathbb{R}^n$, every tempered distributional solution of* (5.1.39) *in $\mathbb{R}^n$ is a polynomial.*

To see that M_L is of the form (5.1.38), we first decompose every generalized polynomial of degree $|\beta|$ as

$$p_\beta(x) = p_\beta^\infty(x) + \widetilde{p}_\beta(x) \tag{5.1.42}$$

and require

$$Pp_\beta(x) = 0 \quad \text{in } \mathbb{R}^n \quad \text{and } \ \widetilde{p}_\beta(x) = 0 \quad \text{for all } \ |x| \geq R_0\,. \tag{5.1.43}$$

In order to find $\widetilde{p}_\beta(x)$ we need the following additional general
Assumption: *The Dirichlet problem*

$$Pw = 0 \quad in \ \ B_{R_0} := \{x \in \mathbb{R}^n \,|\, |x| < R_0\} \tag{5.1.44}$$

with $w \in H_0^1(B_{R_0})$ admits only the trivial solution $w \equiv 0$.

(For a more detailed discussion involving more general unbounded domains see Nazarov [325]).

Remark 5.1.1: In case that we have the unique continuation property (5.1.37) one can show that this assumption can always be fulfilled for some $R_0 > 0$ (see Miranda [307, Section 19]).

Now we solve the following variational problem:

Find $\widetilde{p}_\beta \in H_0^1(B_{R_0})$ such that

$$a_{\mathbb{R}^n}(\widetilde{p}_\beta, v) = -\int\limits_{B_{R_0}} (Pp_\beta^\infty)^\top \bar{v} dx \quad \textit{for all } \ v \in H_0^1(B_{R_0})\,. \tag{5.1.45}$$

This variational problem has a unique solution $\widetilde{p}_\beta \in H_0^1(B_{R_0})$ which can be extended by zero to all of $\mathbb{R}^n$ and $\widetilde{p}_\beta \in H^1_{\text{loc}}(\mathbb{R}^n)$. With the constructed $\widetilde{p}_\beta$, every polynomial β_β^∞ generates a generalized polynomial $p_\beta(x)$ via (5.1.42). This justifies the representation of M_L in (5.1.38).

On the other hand, since $E(x,y)$ represents in physics the potential of a point charge at y observed at x, the representation formula (5.1.33) suggests that one may assume for the behaviour of $u(x)$ at infinity a behaviour similar to that of $E(x,0)$. This motivates us, based on (5.1.33), to require, with given L and $M_L(x;u)$,

$$D^\alpha(u(x)-M_L(x;u)) = D^\alpha E^\top(x,0)q + o(1) \text{ for } |x|\to\infty \text{ and with } |\alpha|\leq 1\,, \tag{5.1.46}$$

where q is a suitable constant vector. Hence, we need to have some information concerning the growth of the fundamental solution at infinity. (We shall return to the question of fundamental solutions in Section 7.3.) Since for $|x|\geq R_0$, the equation (3.6.4) reduces to the homogeneous equation with constant coefficients, here the corresponding fundamental solution has the explicit form (see John [219, (3.87)]):

$$E_{jk}(x,y) = \frac{1}{8\pi^2}\Delta_y \int\limits_{|\xi|=1} P^{jk}(\xi)|(x-y)\cdot\xi|^2\log|(x-y)\cdot\xi|d\omega_\xi \tag{5.1.47}$$

for $n=2$ and $|x|\geq R_0$, where the integration is taken over the unit circle $|\xi|=1$ with arc length element $d\omega_\xi$ and where $P^{jk}(\xi)$ denotes the inverse of the matrix

$$((\widetilde{a}_{jk}(\xi)))_{n\times n} \text{ with } \widetilde{a}_{jk}(\xi) = a_{jk}(\infty)\xi_j\xi_k\,,$$

which is regular for all $\xi\in\mathbb{R}^2\setminus\{0\}$ due to the uniform ellipticity.

For $n=3$ and $|x|\geq R_0$, one has (see John [219, (3.86)])

$$E_{jk}(x,y) = -\frac{1}{16\pi^2}\Delta_y \int\limits_{|\xi|=1} P^{jk}(\xi)|(y-x)\cdot\xi|d\omega_\xi\,, \tag{5.1.48}$$

where $d\omega_\xi$ now denotes the surface element of the unit sphere.

For the radiation condition let us consider the case $n=2$ first. Here, (5.1.47) implies (see [219, (3.49)])

$$E_{jk} = O(\log|x|) \text{ and } D^\alpha E_{jk} = O\left(\frac{1}{|x|}\right) \text{ for } |\alpha|=1 \text{ as } |x|\to\infty\,.$$

Since $M_L(x;u)$ is of the form (5.1.38), no logarithmic terms belong to M_L. Consequently, it follows from (5.1.33) by collecting the coefficients of $E^\top$ that the constant vector q in (5.1.46) is related to u by

$$q = \Big\{\int\limits_{\Omega^c} f(y)dy - \int\limits_{\Gamma}(\tau_c u - (\boldsymbol{b}\cdot\boldsymbol{n})^\top\gamma_{c0}u)ds\Big\}\,. \tag{5.1.49}$$

The proper radiation conditions for u now will be

$$D^\alpha u = D^\alpha E^\top(x,0)q + D^\alpha M_L(x;u) + O\left(\frac{1}{|x|^{1+|\alpha|}}\right) \text{ for } |\alpha|\leq 1 \text{ as } |x|\to\infty\,. \tag{5.1.50}$$

The exterior Dirichlet problem

Now we can formulate the exterior Dirichlet problem in two versions. For given $f \in H^{-1}_{\text{comp}}(\Omega^c)$ with compact support in $\mathbb{R}^n$ and $\varphi \in H^{\frac{1}{2}}(\Gamma)$ we require that u belongs to $H^1_{\text{loc}}(\Omega^c) \cap \mathcal{S}'(\mathbb{R}^n)$ and satisfies

$$\begin{aligned} Pu &= f && \text{in } \Omega^c\,, \\ \gamma_{c0}u &= \varphi && \text{on } \Gamma\,, \end{aligned} \tag{5.1.51}$$

together with the radiation conditions (5.1.50). The space $H^{-1}_{\text{comp}}(\Omega^c)$ is defined as in (4.1.44) where Ω is replaced by $\Omega^c = \mathbb{R} \setminus \overline{\Omega}$. (5.1.51) allows the following two interpretations of the radiation conditions and leads to two different classes of problems.

First version: *In addition to f and φ, the constant vector q and the integer L in (5.1.50) are given a–priori. Find u as well as $M_L(x;u)$ to satisfy (5.1.51) and (5.1.50).*

Note that modifying the solution by adding the term $E^\top(x,0)q$ with given q, this problem can always be reduced to the case where $q = 0$. In the latter case, the corresponding variational formulation reads:

Find a tempered distribution u in $H^1_{\text{loc}}(\Omega^c) \cap \mathcal{S}'(\mathbb{R}^n)$ with $\gamma_{c0}u = \varphi$ on Γ which behaves as (5.1.50) with $q = 0$ and given M_L and satisfies

$$a_{\Omega^c}(u,v) = \langle f, \overline{v}\rangle_{\Omega^c} \text{ for all } v \in H^1_{\text{comp}}(\Omega^c) \text{ with } \gamma_{c0}v = 0 \text{ on } \Gamma\,, \tag{5.1.52}$$

where $a_{\Omega^c}(u,v)$ is defined by (5.1.3) with Ω replaced by Ω^c.

We remark that, in order to obtain a unique solution one requires side conditions.

Second version: *In addition to f and φ, the function $M_L(x;u)$ in the form (5.1.38) is given a–priori. Find u as well as q to satisfy (5.1.51) and (5.1.50).*

We may modify the solution by subtracting the given M_L and reduce the problem to the case $M_L = 0$. Under this assumption, the variational formulation of the second version now reads:

Find a tempered distribution $u \in H^1_{\text{loc}}(\Omega^c) \cap \mathcal{S}'(\mathbb{R}^n)$ with $\gamma_{c0}u = \varphi$ on Γ which behaves at infinity as (5.1.50) and satisfies

$$a_{\Omega^c}(u,v) = \langle f, \overline{v}\rangle_{\Omega^c} \text{ for all } v \in H^1_{\text{comp}}(\overline{\Omega^c}) \text{ with } \gamma_{c0}v = 0 \text{ on } \Gamma \tag{5.1.53}$$

subject to the constraint $M_L(x;u) = 0$ for $x \in \mathbb{R}^n$.

In view of (5.1.34) and our assumption (5.1.36), the radiation condition is

$$0 = \lim_{R\to\infty} \Big\{ \int\limits_{|y|=R} E^\top(x,y)\tau_R u ds_y - \int\limits_{|y|=R} (\partial_{\nu y}E(x,y))^\top \gamma_0 u(y) ds_y \Big\} \tag{5.1.54}$$

for every $x \in \Omega^c$. Alternatively, for $M_L(x;u) = 0$, in view of (5.1.50) and (5.1.49) we obtain from (5.1.33) together with the asymptotic behaviour of $E(x,y)$ for large $|x|$,

$$\lim_{|x|\to\infty} \Big[u(x) - E^\top(x,0)\Big\{ \int\limits_{\Omega^c} f(y)dy - \int\limits_{\Gamma} (\tau_c u - (\boldsymbol{b}\cdot\boldsymbol{n})\gamma_{c0}u)\, ds\Big\}\Big] = 0\,. \tag{5.1.55}$$

Since the representation formula (5.1.33) holds for *every* exterior domain, we may replace Ω^c by $\{y \in \mathbb{R}^2 \big| |y| > R \geq R_0\}$ with $\mathrm{supp}(f) \subset \{y \big| |y| \leq R\}$ and find instead of (5.1.55) the radiation condition

$$\lim_{|x|\to\infty} \Big\{u(x) + E^\top(x,0) \int\limits_{|y|=R} \tau_R u ds\Big\} = 0 \tag{5.1.56}$$

for every sufficiently large $R \geq R_0$.

In contrast to the interior Dirichlet problem, the above formulations in terms of local function spaces are not suitable for the standard Hilbert space treatment unless one can find an equivalent formulation in some appropriate Hilbert space. In the following, we will present such an approach for the first version only which is based on the finite energy of the potentials. The same approach, however, cannot directly be applied to the second version since the corresponding potentials do not, in general, have finite energy.

However, the following approach will be used for analyzing the corresponding integral equations for problems formulated in the second version. We will pursue this in Section 5.6.

The Hilbert space formulation for the first version of the general exterior Dirichlet problem

In view of the growth condition (5.1.50), in the first version with $q = 0$ we have to choose Hilbert spaces combining growth and $M(x;u)$ appropriately. Let us consider first the case of self–adjoint equations (5.1.1) with a symmetric energy–sesquilinear form $a_{\Omega^c}(u,v)$ where, in addition to the assumptions (5.1.36), $b_j(x) = 0$ for all $x \in \Omega^c$. Then

$$a_{\Omega^c}(u,v) = \overline{a_{\Omega^c}(v,u)} \quad \text{and} \quad a_{\Omega^c}(v,v) \geq 0 \quad \text{for all} \quad u,v \in C_0^\infty(\mathbb{R}^n)$$

due to the strong ellipticity condition (3.6.2).

For a fixed chosen L denote by $\mathcal{P}_L$ all the generalized polynomials $M_L(x)$ of the form

$$p(x) = \sum_{|\beta|\leq L} \alpha_\beta p_\beta(x)$$

satisfying (5.1.43).

By $\mathcal{E}_L$ let us denote the subspace of $\mathcal{P}_L$ of generalized polynomials with "finite energy", i. e.

$$\mathcal{E}_L := \{p \in \mathcal{P}_L \,|\, |a_{\Omega^c}(p,p)| < \infty\}\,. \tag{5.1.57}$$

Because of (5.1.35) one finds

$$a_\Omega(p,v) = 0 \quad \text{for all} \quad v \in H_0^1(\Omega)$$

and further

$$a_{\mathbb{R}^n}(p,v) = 0 \quad \text{for all} \quad v \in C_0^\infty(\mathbb{R}^n)\,. \tag{5.1.58}$$

Now define the pre–Hilbert space

$$\begin{aligned}\mathcal{H}_{E*}(\Omega^c) \quad := \quad & \{v = v_0 + p \,|\, p \in \mathcal{E}_L \textit{ and } v_0 \in H^1_{\text{comp}}(\overline{\Omega^c}) \\ & \textit{satisfying } \; a_{\Omega^c}(p, v_0) = 0 \quad \textit{for all } \; p \in \mathcal{E}_L\}\end{aligned} \tag{5.1.59}$$

equipped with the scalar product

$$(u,v)_{\mathcal{H}_E} := a_{\Omega^c}(u,v) + \int_\Gamma \gamma_{c0} u^\top \overline{\gamma_{c0} v} ds\,. \tag{5.1.60}$$

(See also [205], Leis [263, p.26].) Note that $C_0^\infty(\Omega^c) \subset \mathcal{H}_{E*}(\Omega^c)$ because of (5.1.58). For convenience, we tacitly assume $\Theta = 1$ in the uniform strong ellipticity condition (3.6.2). Then $(u,v)_{\mathcal{H}_E}$ has all the properties of an inner product; in particular, $(u,u)_{\mathcal{H}_E} \geq 0$ for every $u \in \mathcal{H}_{E*}(\Omega^c)$. Now, suppose

$$(v,v)_{\mathcal{H}_E} = a_{\Omega^c}(v_0 + p, v_0 + p) + \int_\Gamma |v_0 + p|^2 ds = 0\,,$$

which implies $(v_0 + p)_{|\Gamma} = 0$ since $a_{\Omega^c}(v_0 + p, v_0 + p) \geq 0$. The former allows us to extend v_0 by $-p$ into Ω and to define

$$\widetilde{v}_0 := \begin{Bmatrix} v_0 & \text{in } \Omega^c \\ -p & \text{in } \Omega \end{Bmatrix} \in H^1_{\text{comp}}(\mathbb{R}^n)\,, \quad \text{where } \; (\widetilde{v}_0 + p)_{|\Omega} = 0\,.$$

Then

$$0 = a_{\Omega^c}(v_0 + p, v_0 + p) = a_{\mathbb{R}^n}(\widetilde{v}_0 + p, \widetilde{v}_0 + p)$$

and (5.1.58) implies

$$a_{\mathbb{R}^n}(\widetilde{v}_0 + p, \widetilde{v}_0 + p) = a_{\mathbb{R}^n}(p,p) = 0\,.$$

Here, our assumption (5.1.44) together with the strong ellipticity of P implies the $H_0^1(B_{R_1})$–ellipticity of $a_{\mathbb{R}^n}(\cdot,\cdot)$ for any $R_1 > 0$, in particular for $\overset{\circ}{B}_{R_1} \supset \operatorname{supp}(\widetilde{v}_0)$. Hence, $\widetilde{v}_0 = 0$ in $\mathbb{R}^n$. Consequently, $p = 0$ in Ω and $Pp = 0$ in $\mathbb{R}^n$. The unique continuation property then implies $p = 0$ in $\mathbb{R}^n$. Hence, $(v,v)_{\mathcal{H}_E} = 0$ implies $v = 0$ in $\mathcal{H}_E$. Consequently,

$$\|v\|_{\mathcal{H}_E} := (v,v)^{\frac{1}{2}}_{\mathcal{H}_E} \tag{5.1.61}$$

defines a norm. Now, $\mathcal{H}_E(\Omega^c)$ is defined by the completion of $\mathcal{H}_{E*}(\Omega^c)$ with respect to the norm $\|\cdot\|_{\mathcal{H}_E}$.

In the more general case of a non–symmetric sesquilinear form $a_{\Omega^c}(u,v)$ (and for general exterior Neumann problems later-on) we take a slightly more general approach by using the *symmetric part* of a_{Ω^c}, i. e.

$$a^S_{\Omega^c}(u,v) := \tfrac{1}{2}\Big\{a_{\Omega^c}(u,v) + \overline{a_{\Omega^c}(v,u)}\Big\} \tag{5.1.62}$$

together with the requirement

$$a^S_{\Omega^c}(v,v) \ge 0 \quad \text{for all} \quad v \in C_0^\infty(\mathbb{R}^n)\,.$$

Note that again

$$(u,v)_{\mathcal{H}^S_E} := a^S_{\Omega^c}(u,v)$$

has all the properties of an inner product as before in view of condition (5.1.36). The corresponding space is now defined by

$$\begin{aligned}\mathcal{H}^S_{E*}(\Omega^c) := \{v_0 + p^S \,|\, p^S \in \mathcal{E}^S_L \quad &\textit{and} \quad v_0 \in H^1_{\text{comp}}(\overline{\Omega^c})\\ &\textit{satisfying} \;\; a^S_{\Omega^c}(p^S, v_0) = 0 \quad \textit{for all} \;\; p^S \in \mathcal{E}^S_L\}\end{aligned}$$

where $\mathcal{E}^S_L$ now denote all generalized polynomials of degree $\le L$ with finite energy $a^S_{\mathbb{R}^n}(p^S,p^S) < \infty$ satisfying

$$a^S_{\mathbb{R}^n}(p^S,v) = 0 \quad \text{for all} \quad v \in C_0^\infty(\mathbb{R}^n)\,.$$

Then by following the same arguments as in the symmetric case, one can again show that $(u,v)_{\mathcal{H}^S_{E*}}$ is an inner product on the space $\mathcal{H}^S_{E*}(\Omega^c)$.

For the formulation of the exterior Dirichlet problem we further introduce the closed subspace

$$\mathcal{H}_{E,0}(\Omega^c) := \{v \in \mathcal{H}_E(\Omega^c) \big| \gamma_{c0}v = 0 \,\text{on}\; \Gamma\}\,. \tag{5.1.63}$$

If $u \in \mathcal{H}_E(\Omega^c)$ and $v \in \mathcal{H}_{E,0}(\Omega^c)$ then $u = u_0 + p$ and $v = v_0 + r$ where $p, r \in \mathcal{E}_L$. Moreover, by applying Green's formula to the annular region $\Omega \cap B_R$ with $B_R := \{x \in \mathbb{R}^n \,|\, |x| < R\}$ for $R \ge R_0$ sufficiently large, we obtain the identity

$$\begin{aligned} a_{\Omega^c \cap B_R}(u,v) &= \int\limits_{\Omega^c \cap B_R} (Pu)^\top \bar{v}dx + \int\limits_{\partial B_R} (\tau_R p)^\top \overline{\gamma_{0_R} r} ds_R \\ &= \int\limits_{\Omega^c \cap B_R} (Pu)^\top \bar{v}dx + a_{B_R}(p,r)\,. \end{aligned}$$

For $R \to \infty$ we obtain

$$a_{\Omega^c}(u,v) = \int\limits_{\Omega^c} (Pu)^\top \bar{v}dx + a_{\mathbb{R}^n}(p,r)\,. \tag{5.1.64}$$

With a basis $\{q_s\}_{s=1}^{\widetilde{L}}$ of $\mathcal{E}_L$ where $\widetilde{L} = \dim\ \mathcal{E}_L$, the last term can be written more explicitly with $v = v_0 + r = v_0 + \sum_{\rho=1}^{\widetilde{L}} \kappa_\rho q_\rho$ as

$$a_{\mathbb{R}^n}(p, r) = \sum_{\rho=1}^{\widetilde{L}} \overline{\kappa_\rho} a_{\mathbb{R}^n}(p, q_\rho)\,.$$

Now, for the first version we formulate:

The general exterior Dirichlet problem. *For given* $q = 0, L$ *and* $\widetilde{L} = \dim \mathcal{E}_L$, $\varphi \in H^{\frac{1}{2}}(\Gamma)$, $f \in L^2_{\text{comp}}(\Omega^c)$, $d_\rho \in \mathbb{C}$ *for* $\rho = 1, \ldots, \widetilde{L}$*, find* $u = u_0 + p \in \mathcal{H}_E$ *with* $B\gamma_{c0}u = \varphi \in H^{\frac{1}{2}}(\Gamma)$ *on* Γ *satisfying*

$$a_{\Omega^c}(u, v) = \langle f, \overline{v}\rangle_{\Omega^c} + \sum_{\rho=1}^{\widetilde{L}} d_\rho \overline{\kappa_\rho} \quad \textit{for all } v = v_0 + \sum_{\rho=1}^{\widetilde{L}} \kappa_\rho q_\rho \in \mathcal{H}_{E,0} \quad (5.1.65)$$

subject to the side conditions

$$a_{\mathbb{R}^n}(p, q_\rho) = d_\rho\,,\ \rho = 1, \ldots, \widetilde{L}\,.$$

In this formulation, we need **in addition to the assumptions** (5.1.36) that

$$B_{00} = I \quad \text{for } |x| \geq R_0 \quad \text{where } I \quad \text{is the identity matrix} \quad . \qquad (5.1.66)$$

We note that in this formulation we have replaced the function space $\{v \in \mathcal{H}_E \mid B\gamma_c v = 0 \text{ on } \Gamma\}$ by $\mathcal{H}_{E,0}$. Since det $(B_{00})_{|\Gamma} \neq 0$, they are equivalent.

Remark 5.1.2: In our Hilbert space formulation, for the Laplacian in $\mathbb{R}^2$ and the special choice of $L = 0$, and in $\mathbb{R}^3$ and $\mathcal{E}_L = \{0\} = M_L$, the energy space $\mathcal{H}_E(\Omega)$ is equivalent to the weighted Sobolev space $H^1(\Omega^c; 0)$ corresponding to $W_0^1(\Omega')$ in Nedelec [328]. Hence, in these cases, one may also formulate the variational problems in the weighted Sobolev spaces as in Dautray and Lions [97, Chap.XI], Giroire [148] and Nedelec [328].

The exterior Neumann problem

The exterior Neumann problem is defined by

$$\begin{aligned} Pu &= f \quad \text{in } \Omega^c\,, \\ \tau_c u &= \psi \quad \text{on } \Gamma\,. \end{aligned} \qquad (5.1.67)$$

Again we require the solution to behave like (5.1.50) for $|x| \to \infty$. Since $\tau_c u = \psi$ is given, the equation (5.1.49) yields

$$q = \{\int_{\Omega^c} f dy - \int_\Gamma \psi ds + \int_\Gamma (\boldsymbol{b} \cdot \boldsymbol{n})^\top \gamma_{c0} u ds\}\,.$$

For the special case $\boldsymbol{b}\cdot\boldsymbol{n} = 0$ on Γ, the constant q is determined explicitly by the given data from (5.1.67). In order to find a solution in $\mathcal{H}_E(\Omega^c)$ for $n = 2$, the constant q in the growth condition (5.1.50) must vanish. This yields the *necessary compatibility condition*

$$q = 0 = \int\limits_{\Omega^c} f(y)dy - \int\limits_{\Gamma} \psi ds \quad \text{for } n = 2 \tag{5.1.68}$$

and for the given data in (5.1.67). Then the solution will take the form $u = u_0$ with $u_0(x) = O\left(\frac{1}{|x|}\right)$ for $|x| \to \infty$. In the three–dimensional case $n = 3$, the compatibility condition (5.1.68) is not needed to guarantee a solution of finite energy.

The behaviour at infinity motivates us to consider the following variational problem in the Hilbert space $\mathcal{H}_E(\Omega^c)$. For the solution we assume the form

$$u = u_0 + p + E^\top(x,0)q \quad \text{in } \Omega^c \tag{5.1.69}$$

where $u_0 + p \in \mathcal{H}_E$ and $q \in \mathbb{C}^p$ to be determined.

The exterior Neumann variational problem

Find $u_0 + p \in \mathcal{H}_E$ and $q \in \mathbb{C}^p$ satisfying

$$\begin{aligned} a_{\Omega^c}(u_0, v) &= \langle f, \overline{v}\rangle_{\Omega^c} - \langle \psi - \tau_c E^\top(x,0)q, \overline{v}\rangle_\Gamma + \sum_{\rho=1}^{\widetilde{L}} d_\rho \overline{\kappa_\rho} \\ &\qquad \text{for all } v = v_0 + \textstyle\sum_{\rho=1}^{\widetilde{L}} \kappa_\rho q_\rho \in \mathcal{H}_E \end{aligned} \tag{5.1.70}$$

and

$$q = \Big\{\int\limits_{\Omega^c} f(y)dy - \int\limits_{\Gamma} \psi ds\Big\} + \int\limits_{\Gamma} (\boldsymbol{b}\cdot\boldsymbol{n})^\top \gamma_{c0}(u_0 + p + E^\top(x,0)q)ds$$

subject to the side conditions

$$a_{\mathbb{R}^n}(p, q_\rho) = d_\rho \quad \text{for } \rho = 1, \dots, \widetilde{L}\,.$$

This formulation shows that the given data f, ψ do not have to satisfy the standard compatibility condition (5.1.68) and, in general, the total solution u in (5.1.69) does not have finite energy. If the latter is required, i.e., $q = 0$ in (5.1.70) then the compatibility condition between f and ψ can only be satisfied in an implicit manner resolving the bilinear equation with respect to $u_0 + p$ and with $q = 0$ and inserting p in the second equation of (5.1.70) with $q = 0$.

The general exterior Neumann–Robin problem

Here we extend the boundary conditions to $N_{01} = B_{00} = I$ and $N_{00} = 0$ on the artificial boundary Γ_R for $R \geq R_0$. Then we can proceed in the same manner as for the general interior Neumann–Robin problem except for the radiation conditions which need to be incorporated. In this case, the constant vector q in the radiation condition (5.1.50) is related to the solution u by

$$\begin{aligned} q &= \int\limits_{\Omega^c} f(y)dy - \int\limits_{\Gamma} (\sum_{j,k=1}^{n} n_j a_{jk} n_k) N_{01}^{-1} \psi ds \\ &+ \int\limits_{\Gamma} \Big(\boldsymbol{b} \cdot \boldsymbol{n} + \sum_{\rho=1}^{n-1} (\frac{\partial}{\partial \sigma_\rho} \widetilde{c}_{0\rho}) - \widetilde{c}_{00} \Big) \gamma_{c0} (u_0 + p + E^\top(\bullet, 0) q) ds \end{aligned}$$

where $\widetilde{c}_{0\rho}$ and $\widetilde{c}_{00}$ are given by (5.1.25), which modifies (5.1.49) of the standard Neumann problem.

For the special case

$$\boldsymbol{b} \cdot \boldsymbol{n} + \sum_{\rho=1}^{n-1} \frac{\partial}{\partial \sigma_\rho} \widetilde{c}_{0\rho} - \widetilde{c}_{00} = 0 \quad \text{on } \Gamma,$$

we recover, similar to (5.1.68), the necessary compatibility condition

$$q = 0 = \int\limits_{\Omega^c} f(y)dy - \int\limits_{\Gamma} (\Sigma n_j a_{jk} n_k) N_{01}^{-1} \psi ds$$

for the given data f and ψ and for a solution with finite energy.

The exterior combined Dirichlet–Neumann problem

We begin with the formulation of the boundary value problem in distributional form: *Given* $f \in H^{-1}_{\text{comp}}(\Omega^c)$, $\varphi_1 \in \pi H^{\frac{1}{2}}(\Gamma)$ *and* $\psi_2 \in (I-\pi) H^{-\frac{1}{2}}(\Gamma)$ *on* Γ *and* $d_\rho \in \mathbb{C}$, $\rho = 1, \ldots, \widetilde{L}$, *find* $u = u_0 + p + E^\top(x;0) q$ *with* $u_0 + p \in \mathcal{H}_E(\Omega^c)$ *and* $q \in \mathbb{C}^p$ *as the distributional solution of*

$$\begin{aligned} Pu &= f \ \text{ in } \Omega^c \\ \pi B \gamma_c u &= \varphi_1 \ \text{ and } \ (1-\pi) N \gamma_c u = \psi_2 \ \text{ on } \Gamma \end{aligned} \tag{5.1.71}$$

subject to the side conditions

$$a_{\mathbb{R}^n}(p, q_\rho) = d_\rho \quad \text{for } \rho = 1, \ldots, \widetilde{L} \quad \text{where } \mathcal{E}_L = \text{span}\{q_\varrho\}_{\varrho=1}^{\widetilde{L}},$$

and the constraint

$$q = \{ \int\limits_{\Omega^c} f(y)dy - \int\limits_{\Gamma} (\tau_c u - (\boldsymbol{b} \cdot \boldsymbol{n})^\top \gamma_{c0} u) ds \}.$$

The application of Lemma 5.1.2 then yields:

The variational formulation for the exterior problem
Find $u = u_0 + p + E^\top(\bullet;0)q$ *with* $u_0 + p \in \mathcal{H}(\Omega^c)$ *and* $\pi B\gamma_c u = \varphi_1$ *on* Γ *satisfying*

$$\widetilde{a}_{\Omega^c}(u_0+p,v) = \int\limits_{\Omega^c} f^\top \bar{v}dx - \langle \psi_2, (I-\pi)\overline{B\gamma_c v}\rangle_\Gamma \qquad (5.1.72)$$

$$+ \langle (I-\pi)N\gamma_c E^\top(\bullet,0)q\,,\ (I-\pi)\overline{B\gamma_c v}\rangle_\Gamma + \sum_{\rho=1}^{\widetilde{L}} d_s\overline{\kappa_\rho}$$

for all $v = v_0 + \sum_{\rho=1}^{\widetilde{L}} \kappa_\rho q_\rho \in \mathcal{H}_{E\pi B} := \{v \in \mathcal{H}_E(\Omega^c)\,|\,\pi B\gamma_c v = 0 \text{ on } \Gamma\}$, *subject to the side conditions*

$$a_{\mathbb{R}^n}(p,q_\rho) = d_\rho \quad \text{for } \rho = 1,\dots,\widetilde{L} \quad \text{where } \mathcal{E}_L = \operatorname{span}\{q_\varrho\}_{\varrho=1}^{\widetilde{L}}\,, \qquad (5.1.73)$$

and the constraint

$$q = \{\int\limits_{\Omega^c} f(y)dy - \int\limits_\Gamma (\tau_c - (\boldsymbol{b}\cdot\boldsymbol{n})^\top\gamma_{c0})(u_0+p+E^\top(\bullet;0)q)ds\}\,. \qquad (5.1.74)$$

Since $\tau_c u = N_{01}^{-1}(N\gamma_c u - N_{00}\gamma_{c0}u)$ due to (5.1.21) and $\gamma_{c0}u = B_{00}^{-1}B\gamma_c u$ due to (5.1.16), in view of the given boundary conditions in (5.1.71), the last constraint (5.1.74) can be reformulated as the linear equation of the form

$$\begin{aligned} q &= \Big\{\int\limits_{\Omega^c} f(y)dy - \int\limits_\Gamma \Big(N_{01}^{-1}(1-\pi)\psi_2 + (N_{01}^{-1}N_{00} - (\boldsymbol{b}\cdot\boldsymbol{n}))B_{00}^{-1}\pi\varphi_1\Big)ds\Big\} \\ &\quad - \int\limits_\Gamma \Big(N_{01}^{-1}\pi N\gamma_c + (N_{01}^{-1}N_{00} - (\boldsymbol{b}\cdot\boldsymbol{n}))B_{00}^{-1}(1-\pi)B\gamma_c\Big) \\ &\quad (u_0+p+E^\top(\cdot;0)q)ds\,. \end{aligned} \qquad (5.1.75)$$

Remark 5.1.3: In the special case $\pi = I$, one verifies that the combined problem reduces to the general Dirichlet problem (5.1.65). For $\pi = 0$, we recover the general exterior Neumann–Robin problem.

5.1.3 Transmission Problems

The combined Dirichlet–Neumann conditions
As we can see in the previous formulations of the boundary value problems, the boundary operator R is always given while the complementing boundary operator S could be chosen according to the restrictions (3.9.4) and (3.9.5). As generalization of these boundary value problems, we consider transmission

problems where both boundary operators are given to form a pair of mutually complementary boundary conditions.

We will classify the transmission problems in two main classes. In both classes we are given $f \in H^{-1}_{\text{comp}}(\mathbb{R}^n \setminus \Gamma)$ and $d_\rho \in \mathbb{C}$, $\rho = 1, \dots, \widetilde{L}$ and we are looking for a *distributional solution* u *with* $u_{|\Omega} \in H^1(\Omega, P)$ *and* $u_{|\Omega^c} = u_0 + p + E^\top(\bullet; 0)q$ *with* $u = u_0 + p \in \mathcal{H}_E(\Omega^c)$ *and* $q \in \mathbb{C}^p$ *satisfying*

$$Pu = f \quad \text{in } \mathbb{R}^n \setminus \Gamma \quad \text{and} \quad a_{\mathbb{R}^n}(p, q_\rho) = d_\rho \quad \text{for } \rho = 1, \dots, \widetilde{L}; \tag{5.1.76}$$

$$\text{and} \quad q = \int\limits_{\Omega^c} f dy - \int\limits_{\Gamma} (\tau_c u - (\boldsymbol{b} \cdot \boldsymbol{n}) \gamma_{c0} u) ds \tag{5.1.77}$$

subject to the following additional transmission conditions on Γ; $\mathcal{E}_L = \operatorname{span}\{a_\varrho\}_{\varrho=1}^{\widetilde{L}}$.

In the first class these transmission conditions are

$$1. \quad [\pi B \gamma u] = \varphi_1 \in \pi H^{\frac{1}{2}}(\Gamma) \quad \text{and} \quad [\pi N \gamma u] = \psi_1 \in \pi H^{-\frac{1}{2}}(\Gamma) \tag{5.1.78}$$

where φ_1 and ψ_1 are given on Γ together with one of the following conditions:

$$1.1 \quad \begin{array}{l} (I - \pi) B \gamma u = \varphi_{20} \in (I - \pi) H^{\frac{1}{2}}(\Gamma) \quad \text{and} \\ (I - \pi) N \gamma_c u = \psi_{21} \in (I - \pi) H^{-\frac{1}{2}}(\Gamma) \end{array} \tag{5.1.79}$$

$$1.2 \quad \begin{array}{l} (I - \pi) B \gamma_c u = \varphi_{21} \in (I - \pi) H^{\frac{1}{2}}(\Gamma) \quad \text{and} \\ (I - \pi) N \gamma u = \psi_{20} \in (I - \pi) H^{-\frac{1}{2}}(\Gamma) . \end{array} \tag{5.1.80}$$

For the corresponding variational formulations, for simplicity we confine ourselves to the cases with $d_\rho = 0$ and $p(x) \equiv 0$ in (5.1.69), (5.1.70). The corresponding exterior energy space will be denoted by $\mathcal{H}^0_E(\Omega^c)$. Then the variational equations for this class of transmission problems can be formulated as:

$$\begin{aligned} &\widetilde{a}_{\mathbb{R}^n \setminus \Gamma}(u, v) := \widetilde{a}_\Omega(u, v) + \widetilde{a}_{\Omega^c}(u, v) \\ &= \int\limits_{\mathbb{R}^n \setminus \Gamma} f^\top \bar{v} dx + \langle N \gamma_c E^\top(\cdot; 0) q, \overline{B \gamma_c v} \rangle_\Gamma + \ell(\bar{v}) \end{aligned} \tag{5.1.81}$$

for all test functions v as characterized in Table 5.1.1, and where the linear functional

$$\ell(\bar{v}) = \langle N \gamma u, \overline{B \gamma v} \rangle_\Gamma - \langle N \gamma_c u, \overline{B \gamma_c v} \rangle_\Gamma$$

will be specified depending on the particular transmission conditions (5.1.78)–(5.1.80), whereas $q \in \mathbb{C}^p$ must satisfy the constraint (5.1.77), i.e.,

$$q = \int\limits_{\Omega} f dy - \int\limits_{\Omega} (\tau_c u - (\boldsymbol{b} \cdot \boldsymbol{n})^\top \gamma_{c0} u) ds . \tag{5.1.82}$$

In this class with (5.1.78), the corresponding variational formulations now read:

Find $u \in H^1(\Omega) \times \mathcal{H}^0_E(\Omega^c)$ with the enforced constraints given in Table 5.1.1 *satisfying the variational equation* (5.1.81) *with* (5.1.82) *for all test functions v in the subspace of $H^1(\Omega) \times \mathcal{H}^0_E(\Omega^c)$ characterized in Table* 5.1.1 *with the functional $\ell(\bar{v})$ also specified in Table* 5.1.1.

Table 5.1.1. Relevant data for the variational equations in the first class

Boundary conditions	Constraints for u	$\ell(\bar{v})$	Subspace conditions
(5.1.78) (5.1.79)	$[\pi B\gamma u] = \varphi_1$, $(I-\pi)B\gamma u = \varphi_{20}$	$-\langle\psi_1, \pi\overline{B\gamma v}\rangle_\Gamma$ $-\langle\psi_{21}, (I-\pi)\overline{[B\gamma v]}\rangle_\Gamma$	$[\pi B\gamma v] = 0$, $(I-\pi)B\gamma v = 0$
(5.1.78) (5.1.80)	$[\pi B\gamma u] = \varphi_1$, $(I-\pi)B\gamma_c u = \varphi_{21}$	$-\langle\psi_1, \pi\overline{B\gamma v}\rangle_\Gamma$ $+\langle\psi_{20}, (I-\pi)\overline{B\gamma v}\rangle_\Gamma$	$[\pi B\gamma v] = 0$, $(I-\pi)B\gamma_c v = 0$

We remark that in all classes π, r and $(I-\pi), (p-r)$ can be interchanged throughout due to the properties (5.1.30).

In the second class, the transmission conditions are

$$2. \quad [\pi B\gamma u] = \varphi_1 \in \pi H^{\frac{1}{2}}(\Gamma) \quad \text{and} \quad [(I-\pi)N\gamma u] = \psi_2 \in (I-\pi)H^{-\frac{1}{2}}(\Gamma) \tag{5.1.83}$$

together with one of the following conditions:

$$2.1 \quad \begin{array}{l} \pi B\gamma u = \varphi_{10} \in \pi H^{\frac{1}{2}}(\Gamma) \quad \text{and} \\ (I-\pi)N\gamma u = \psi_{20} \in (I-\pi)H^{-\frac{1}{2}}(\Gamma)\,; \end{array} \tag{5.1.84}$$

$$2.2 \quad \begin{array}{l} \pi B\gamma_c u = \varphi_{11} \in \pi H^{\frac{1}{2}}(\Gamma) \quad \text{and} \\ (I-\pi)N\gamma_c u = \psi_{21} \in (I-\pi)H^{-\frac{1}{2}}(\Gamma)\,; \end{array} \tag{5.1.85}$$

$$2.3 \quad \begin{array}{l} \pi B\gamma u = \varphi_{10} \in \pi H^{\frac{1}{2}}(\Gamma) \quad \text{and} \\ (I-\pi)N\gamma_c u = \psi_{21} \in (I-\pi)H^{-\frac{1}{2}}(\Gamma)\,; \end{array} \tag{5.1.86}$$

$$2.4 \quad \begin{array}{l} \pi B\gamma_c u = \varphi_{11} \in \pi H^{\frac{1}{2}}(\Gamma) \quad \text{and} \\ (I-\pi)N\gamma u = \psi_{20} \in (I-\pi)H^{-\frac{1}{2}}(\Gamma)\,; \end{array} \tag{5.1.87}$$

$$2.5 \quad \begin{array}{l} (I-\pi)B\gamma u = \varphi_{20} \in (I-\pi)H^{\frac{1}{2}}(\Gamma) \quad \text{and} \\ \pi N\gamma u = \psi_{10} \in \pi H^{-\frac{1}{2}}(\Gamma)\,; \end{array} \tag{5.1.88}$$

$$2.6 \quad \begin{array}{l} (I-\pi)B\gamma_c u = \varphi_{21} \in (I-\pi)H^{\frac{1}{2}}(\Gamma) \quad \text{and} \\ \pi N\gamma_c u = \psi_{11} \in \pi H^{-\frac{1}{2}}(\Gamma)\,. \end{array} \tag{5.1.89}$$

We remark that in all the above cases, one may rewrite the equations (5.1.77) for q in terms of the given transmission and boundary conditions, accordingly.

Similar to the variational formulations for the first class of transmission problems, we may summarize these formulations for the cases 2.1–2.4 in Table 5.1.2.

Table 5.1.2. Relevant data for the variational equations in the second class

Boundary conditions	Constraints for u	$\ell(\bar{v})$	Subspace conditions
(5.1.83)	$\pi B\gamma u = \varphi_{10}$,	$-\langle\psi_2, (1-\pi)\overline{B\gamma_c v}\rangle_\Gamma$	$\pi B\gamma v = 0$,
(5.1.84)	$\pi B\gamma_c u = \varphi_1 + \varphi_{10}$	$-\langle\psi_{20}, (1-\pi)\overline{[B\gamma v]}\rangle_\Gamma$	$\pi B\gamma_c v = 0$
(5.1.83)	$\pi B\gamma_c u = \varphi_{11}$,	$-\langle\psi_2, (1-\pi)\overline{B\gamma v}\rangle_\Gamma$	$\pi B\gamma v = 0$,
(5.1.85)	$\pi B\gamma u = \varphi_{11} - \varphi_1$	$-\langle\psi_{21}, (1-\pi)\overline{[B\gamma v]}\rangle_\Gamma$	$\pi B\gamma_c v = 0$
(5.1.83)	$\pi B\gamma u = \varphi_{10}$,	$-\langle\psi_2, (1-\pi)\overline{B\gamma v}\rangle_\Gamma$	$\pi B\gamma v = 0$,
(5.1.86)	$\pi B\gamma_c u = \varphi_1 + \varphi_{10}$	$-\langle\psi_{20}, (1-\pi)\overline{[B\gamma v]}\rangle_\Gamma$	$\pi B\gamma_c v = 0$
(5.1.83)	$\pi B\gamma_c u = \varphi_{11}$,	$-\langle\psi_2, (1-\pi)\overline{B\gamma_c v}\rangle_\Gamma$	$\pi B\gamma v = 0$,
(5.1.87)	$\pi B\gamma u = -\varphi_1 + \varphi_{11}$	$-\langle\psi_{20}, (1-\pi)\overline{[B\gamma v]}\rangle_\Gamma$	$\pi B\gamma_c v = 0$

We note that for all the cases in Table 5.1.2 one could solve the transmission problems equally well by solving the interior and the exterior problems independently. However, in the remaining two cases 2.5, 2.6 we can solve the transmission problems only by solving first one of the interior or exterior problems since one needs the corresponding resulting Cauchy data for solving the remaining problem by making use of the transmission conditions in (5.1.83). To illustrate the idea we now consider case 2.5. Here, first find $u \in H^1(\Omega)$ with $(1-\pi)B\gamma u = \varphi_{20}$ satisfying the variational equation

$$\tilde{a}_\Omega(u, v) = \int_\Omega f^\top \bar{v} dx + \langle\psi_{10}, \pi\overline{B\gamma v}\rangle_\Gamma$$

$$\text{for all } \; v \in H^1(\Omega) \quad \text{with} \quad (1-\pi)B\gamma v = 0\,.$$

Then, with the help of (5.1.83), we have the data

$$(1-\pi)N\gamma_c u = \psi_3 := \psi_2 + (1-\pi)N\gamma u \quad \text{and}$$
$$\pi B\gamma_c u = \varphi_3 := \varphi_1 + \pi B\gamma u$$

for the exterior variational problem (5.1.72).

In the case 2.6, the exterior problem is to be solved first for providing the necessary missing Cauchy data for the interior problem.

Remark 5.1.4: We note that for the problems above we have restricted ourselves to the case $M_L = 0$ if exterior Neumann problems are involved. For the case $M_L \neq 0$, the formulation is more involved and will not be presented here.

5.2 Abstract Existence Theorems for Variational Problems

The variational formulation of boundary value problems for partial differential equations (as well as that of boundary integral equations) leads to sesquilinear variational equations in a Hilbert space $\mathcal{H}$ of the following form:

Find an element $u \in \mathcal{H}$ such that

$$a(u,v) = \ell(\overline{v}) \quad \text{for all} \quad v \in \mathcal{H}\,. \tag{5.2.1}$$

Here $a(v,u)$ is a continuous sesquilinear form on $\mathcal{H}$ and $\ell(v)$ is a given continuous linear functional. In order to obtain existence results, we often require further that a satisfies a *Gårding inequality* in the form

$$Re\{a(v,v) + (Cv,v)_{\mathcal{H}}\} \geq \alpha_0 \|v\|^2_{\mathcal{H}} \tag{5.2.2}$$

with a positive constant α_0 and a compact linear operator C from $\mathcal{H}$ into $\mathcal{H}$.

We recall the definition of a *linear compact operator* C between two Banach spaces $\mathcal{X}$ and $\mathcal{Y}$.

Definition 5.2.1. *A linear operator $C : \mathcal{X} \to \mathcal{Y}$ is compact iff the image $C\mathcal{B}$ of any bounded set $\mathcal{B} \subset \mathcal{X}$ is a relatively compact subset of $\mathcal{Y}$; i.e. every sequence of $C\mathcal{B}$ contains a convergent sub–sequence in $\mathcal{Y}$.*

Gårding's inequality plays a fundamental rôle in the variational methods for the solution of partial differential equations and boundary integral equations as well as for the stability and convergence analysis in finite and boundary elements [204] and Dautray and Lions [97], Nedelec [328], Sauter and Schwab [371] and Steinbach [399].

In this section we present the well known Lax–Milgram theorem which provides an existence proof for the $\mathcal{H}$–elliptic sesquilinear forms. Since these theorems are crucial in the underlying analysis of our subjects we present this fundamental part of functional analysis here.

Definition 5.2.2. *A continuous sesquilinear form $a(u,v)$ is called $\mathcal{H}$– elliptic if it satisfies the inequality*

$$|a(v,v)| \geq \alpha_0 \|v\|^2_{\mathcal{H}} \quad \textit{for all} \quad v \in \mathcal{H} \tag{5.2.3}$$

with $\alpha_0 > 0$.

Remark 5.2.1: In literature the term of $\mathcal{H}$–ellipticity is some times defined by the slightly stronger condition of *strong $\mathcal{H}$–ellipticity* (see Stummel [410]),

$$Re\, a(v,v) \geq \alpha_0 \|v\|^2_{\mathcal{H}} \quad \text{for all} \quad v \in \mathcal{H} \tag{5.2.4}$$

with $\alpha_0 > 0$. (See also Lions and Magenes [270, p.201].)

Clearly, (5.2.4) implies (5.2.3); in words, strong $\mathcal{H}$–ellipticity implies $\mathcal{H}$–ellipticity since

$$|a(v,v)| \geq Re\, a(v,v) \geq \alpha_0 \|v\|^2_{\mathcal{H}}\,.$$

Hence, when $C = 0$ in Gårding's inequality (5.2.2) then $a(u,v)$ is strongly $\mathcal{H}$–elliptic. In most applications, however, $C \neq 0$ but compact.

5.2.1 The Lax–Milgram Theorem

In this section we confine ourselves to the case of $\mathcal{H}$–ellipticity of $a(u,v)$. We begin with two elementary but fundamental theorems.

Theorem 5.2.1 (Projection theorem). *Let $\mathcal{M} \subset \mathcal{H}$ be a closed subspace of the Hilbert space $\mathcal{H}$ and let $\mathcal{M}^\perp := \{v \in \mathcal{H} \,|\, (v,u) = 0$ for all $u \in \mathcal{M}\}$. Then every $u \in \mathcal{H}$ can uniquely be decomposed into the direct sum*

$$u = u_0 + u_0^\perp \quad \textit{with} \quad u_0 \in \mathcal{M} \quad \textit{and} \quad u_0^\perp \in \mathcal{M}^\perp\,. \tag{5.2.5}$$

The operator defined by $\pi_{\mathcal{M}} : u \mapsto u_0 =: \pi_{\mathcal{M}} u$ is a linear continuous projection. In this case one also writes $\mathcal{H} = \mathcal{M} \oplus \mathcal{M}^\perp$.

Proof: The proof is purely based on geometric arguments in connection with the completeness of a Hilbert space. As shown in Figure 5.2.1, u_0 will be the closest point to u in the subspace $\mathcal{M}$.

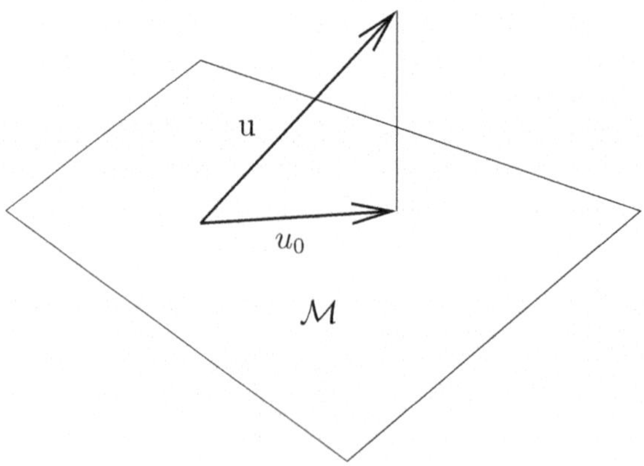

Figure 5.2.1: $\mathcal{H} = \mathcal{M} \oplus \mathcal{M}^\perp$

Let

$$d := \inf_{v \in \mathcal{M}} \|u - v\|_{\mathcal{H}}\,.$$

Then there exists a sequence $\{v_k\} \subset \mathcal{M}$ with

$$d = \lim_{k\to\infty} \|u - v_k\|_{\mathcal{H}} . \tag{5.2.6}$$

With the parallelogram equality

$$\|(u-v_k)+(u-v_\ell)\|_{\mathcal{H}}^2 + \|(u-v_k)+(u-v_\ell)\|_{\mathcal{H}}^2 = 2\|(u-v_k)\|_{\mathcal{H}}^2 + 2\|(u-v_\ell)\|_{\mathcal{H}}^2$$

we obtain together with $\frac{v_k + v_\ell}{2} \in \mathcal{M}$ the estimate

$$\begin{aligned} \|(v_k - v_\ell)\|_{\mathcal{H}}^2 &= 2\|(u-v_k)\|_{\mathcal{H}}^2 + 2\|(u-v_\ell)\|_{\mathcal{H}}^2 - 4\|u - \frac{v_k+v_\ell}{2}\|^2 \\ &\le 2\|(u-v_k)\|_{\mathcal{H}}^2 + 2\|(u-v_\ell)\|_{\mathcal{H}}^2 - 4d^2 . \end{aligned}$$

Hence, from (5.2.6) we obtain that

$$\|v_k - v_\ell\|_{\mathcal{H}}^2 \to 0 \quad \text{for} \quad k, \ell \to \infty .$$

Therefore, the sequence $\{v_k\}$ is a Cauchy sequence in $\mathcal{H}$. Thus, it converges to $u_0 \in \mathcal{H}$ due to completeness. Since $\mathcal{M}$ is closed, we find $u_0 = \lim_{k\to\infty} v_k \in \mathcal{M}$.

If $v_k \to u_0$ and $v_k' \to u_0'$ are two sequences in $\mathcal{M}$ with $\|u - v_k\|_{\mathcal{H}} \to d$ and $\|u - v_k'\|_{\mathcal{H}} \to d$ then with the same arguments as above we find

$$\|v_k' - v_k\|_{\mathcal{H}} \le 2\|u - v_k'\|_{\mathcal{H}} + 2\|u - v_k\|_{\mathcal{H}} - 4d^2 \to 0$$

for $k \to 0$. Hence, $u_0 = u_0'$; i.e. uniqueness of u_0 and, hence, of $u_0^\perp$, too. Thus, the mapping π is well defined.

Since for $u_0 \in \mathcal{M}$, $u_0 = \pi u_0 + 0$, one finds $\pi^2 u = \pi u_0 = u_0 = \pi u$ for all u. Hence, $\pi^2 = \pi$.

The linearity of π can be seen from

$$(\alpha\pi u_1 + \beta\pi u_2) \in \mathcal{M} \quad \text{and} \quad \alpha u_1^\perp + \beta u_2^\perp \in \mathcal{M}^\perp$$

for every $u_1, u_2 \in \mathcal{H}$ and $\alpha, \beta \in \mathbb{C}$, as follows. Since the decomposition

$$\alpha u_1 + \beta u_2 = \alpha(\pi u_1 + u_1^\perp) + \beta(\pi u_2 + u_2^\perp) = (\alpha\pi u_1 + \beta\pi u_2) + (\alpha u_1^\perp + \beta u_2^\perp)$$

is unique, there holds

$$\pi(\alpha u_1 + \beta u_2) = \alpha\pi u_1 + \beta\pi u_2 .$$

From

$$\|u\|_{\mathcal{H}}^2 = (u, u)_{\mathcal{H}} = (u_0 + u^\perp, u_0 + u^\perp)_{\mathcal{H}} = \|u_0\|_{\mathcal{H}}^2 + \|u^\perp\|_{\mathcal{H}}^2 ,$$

one immediately finds

$$\|\pi u\|_{\mathcal{H}} = \|u_0\|_{\mathcal{H}} \le \|u\|_{\mathcal{H}} \quad \text{and} \quad \|\pi\|_{\mathcal{H},\mathcal{H}} = 1$$

because of $\pi u_0 = u_0$ for $u_0 \in \mathcal{M}$. ∎

Theorem 5.2.2 (The Riesz representation theorem). *To every bounded linear functional $F(v)$ on the Hilbert space $\mathcal{H}$ there exists exactly one element $f \in \mathcal{H}$ such that*

$$F(v) = (v, f)_{\mathcal{H}} \quad \text{for all} \quad v \in \mathcal{H}\,. \tag{5.2.7}$$

Moreover,

$$\|f\|_{\mathcal{H}} = \sup_{\|v\|_{\mathcal{H}} \leq 1} |F(v)|\,.$$

Remark 5.2.2: For a Hilbert space $\mathcal{H}$ with the dual space $\mathcal{H}^*$ defined by all continuous linear functionals equipped with the norm

$$\|F\|_{\mathcal{H}^*} = \sup_{\|v\|_{\mathcal{H}} \leq 1} |F(v)|\,,$$

the *Riesz mapping* $j : \mathcal{H}^* \to \mathcal{H}$ defined by

$$jF := f$$

is an isometric isomorphism from $\mathcal{H}^*$ onto $\mathcal{H}$.

Proof: i.) Existence: We first show the existence of f for any given $F \in \mathcal{H}^*$. The linear space $\ker(F) := \{v \in \mathcal{H} \,|\, F(v) = 0\}$ is a closed subspace of $\mathcal{H}$ since F is continuous. Then either $\ker(F) = \mathcal{H}$ and $f = 0$, or $\mathcal{H} \setminus \ker(F) \neq \emptyset$. In the latter case there is a unique decomposition of $\mathcal{H}$ due to the Projection Theorem 5.2.1:

$$\begin{aligned} \mathcal{H} &= \ker(F) \oplus \{\ker(F)\}^{\perp} \quad \text{and} \\ u &= u_0 + u^{\perp} \quad \text{with} \quad u_0 \in \ker(F) \quad \text{and} \quad (u^{\perp}, v)_{\mathcal{H}} = 0 \quad \text{for all} \quad v \in \ker(F)\,. \end{aligned}$$

Since $\mathcal{H} \setminus \ker(F) \neq \emptyset$, there is an element $g \in \{\ker(F)\}^{\perp}$ with $F(g) \neq 0$ defining

$$z := \frac{1}{F(g)} g\,.$$

Then $F(z) = 1$ and for every $v \in \mathcal{H}$ we find

$$F(v - F(v)z) = F(v) - F(v)F(z) = 0\,, \text{ i.e. } v - F(v)z \in \ker(F)\,.$$

Moreover,

$$(v, z)_{\mathcal{H}} = (v - F(v)z + F(v)z, z)_{\mathcal{H}} = (F(v)z, z)_{\mathcal{H}} = F(v)\|z\|^2_{\mathcal{H}}\,.$$

Hence, $f := \frac{1}{\|z\|^2} z$ will be the desired element, since

$$(v, f)_{\mathcal{H}} = \left(v, \frac{z}{\|z\|^2}\right)_{\mathcal{H}} = \frac{1}{\|z\|^2}(v, z)_{\mathcal{H}} = F(v) \quad \text{for all} \quad v \in \mathcal{H}\,.$$

This implies also

$$\|F\|_{\mathcal{H}^*} = \sup_{\|v\|_{\mathcal{H}} \le 1} |F(v)| = \sup_{\|v\|_{\mathcal{H}} \le 1} |(v,f)_{\mathcal{H}}| = \|f\|_{\mathcal{H}} .$$

ii.) Uniqueness: Suppose $f_1, f_2 \in \mathcal{H}$ are two representing elements for $F \in \mathcal{H}^*$. Then

$$F(v) = (v, f_1)_{\mathcal{H}} = (v, f_2)_{\mathcal{H}} \text{ for all } v \in \mathcal{H}$$

implying

$$(v, f_1 - f_2)_{\mathcal{H}} = 0 \text{ for all } v \in \mathcal{H} .$$

The special choice of $v = f_1 - f_2$ yields

$$0 = \|f_1 - f_2\|^2_{\mathcal{H}} = (f_1 - f_2, f_1 - f_2)_{\mathcal{H}} , \text{ i.e. } f_1 = f_2 .$$

■

Theorem 5.2.3 (Lax–Milgram Lemma). *Let a be a continuous $\mathcal{H}$–elliptic sesquilinear form. Then to every bounded linear functional $\ell(v)$ on $\mathcal{H}$ there exists a unique solution $u \in \mathcal{H}$ of equation* (5.2.1).

Proof: For any fixed $u \in \mathcal{H}$, the relation

$$(Au)(v) := a(v, u)$$

defines a bounded linear functional $Au \in \mathcal{H}^*$ operating on $v \in \mathcal{H}$. By the Riesz representation theorem 5.2.2, there exists a unique element $jAu \in \mathcal{H}$ with

$$(v, jAu)_{\mathcal{H}} = a(v, u) \text{ for all } v \in \mathcal{H} .$$

The mappings $A : u \mapsto Au \in \mathcal{H}^*$ and $Bu := jA : \mathcal{H} \to \mathcal{H}$ are linear and bounded, since $a(v, u)$ is a continuous sesquilinear form (see(5.1.14)). Moreover,

$$M = \|B\|_{\mathcal{H},\mathcal{H}} = \|jA\|_{\mathcal{H},\mathcal{H}} = \|A\|_{\mathcal{H},\mathcal{H}^*} := \sup_{\|u\|_{\mathcal{H}} \le 1} \|Au\|_{\mathcal{H}^*} = \sup_{\|u\|=\|v\|=1} |a(v,u)| .$$

Obviously, the $\mathcal{H}$–ellipticity implies the invertibility of the operator B on its range. So, $B : \mathcal{H} \to$ range (B) is one–to–one and onto and there B^{-1} is bounded with

$$\|B^{-1}\|_{\text{range}(B),\mathcal{H}} \le \frac{1}{\alpha_0} .$$

This implies that range (B) is a closed subset of $\mathcal{H}$. Next we show that range $(B) = \mathcal{H}$. If not then there existed $z \in \mathcal{H} \setminus \text{range}(B)$ with $z \neq 0$ and there was $z_0 \in \big(\text{range}(B)\big)^{\perp}$ with $z \neq 0$ due to the Projection Theorem 5.2.1, but

$$a(z_0, z_0) = (Bz_0, z_0)_{\mathcal{H}} = 0 .$$

Then the $\mathcal{H}$–ellipticity implied $z_0 = 0$, a contradiction. Consequently, range $(B) = \mathcal{H}$ and the equation

$$a(u,v) = (Bu,v)_{\mathcal{H}} = \ell(v) = (j\ell,v)_{\mathcal{H}} \quad \text{for all} \quad v \in \mathcal{H}$$

has a unique solution

$$u = B^{-1}j\ell\,.$$

■

Note that the solution u then is bounded satisfying

$$\|u\|_{\mathcal{H}} \le \tfrac{1}{\alpha_0}\|j\ell\|_{\mathcal{H}} = \tfrac{1}{\alpha_0}\|\ell\|_{\mathcal{H}^*}\,. \tag{5.2.8}$$

As an immediate consequence we have

Corollary 5.2.4. *Let $a(u,v)$ be a continuous strongly $\mathcal{H}$–elliptic sesquilinear form on $\mathcal{H}\times\mathcal{H}$, i.e., satisfying* (5.2.4) *has a unique solution $u \in \mathcal{H}$.*

Clearly, since a is also $\mathcal{H}$–elliptic, we obtain u via the Lax–Milgram lemma, Theorem 5.2.3. However, for the strongly $\mathcal{H}$–elliptic sesquilinear form, we can reduce the problem to Banach's fixed point principle and present an alternative proof.

Proof: The equation (5.2.1) has a solution u if and only if u is a fixed point,

$$u = Q(u)\,,$$

of the mapping Q defined by

$$Q(w) := w - \rho(jAw - f) \ \text{ for } \ w \in \mathcal{H}$$

with the parameter $\rho \in (0, \frac{2\alpha_0}{M^2})$, where α_0 is the ellipticity constant in (5.2.4). With this choice of ρ (preferably $\rho = \frac{\alpha_0}{M^2}$), Q is a contractive mapping, since

$$\begin{aligned}
\|Q(w) - Q(v)\|^2_{\mathcal{H}} &= (w - v - \rho jA(w-v),\, w - v - \rho jA(w-v))_{\mathcal{H}}\\
&= \|w-v\|^2_{\mathcal{H}} - 2\rho Re\, a(w-v, w-v) + \rho^2\|jA(w-v)\|^2_{\mathcal{H}}\\
&\le \{1 - 2\rho\alpha_0 + \rho^2 M^2\}\|w-v\|^2_{\mathcal{H}}\\
&= q^2\|w-v\|^2_{\mathcal{H}}\,,
\end{aligned}$$

for $q^2 := (1 - 2\rho\alpha_0 + \rho^2 M^2) < 1$. Consequently, the successive approximation defined by the sequence $\{w_k\}$,

$$w_0 = 0 \ \text{ and } \ w_{k+1} := Q(w_k)\,,\ k = 0,1,\dots,$$

converges in $\mathcal{H}$ to the solution $u = \lim\limits_{k\to 0} w_k$. Moreover, we see that the solution u is unique. Since if u_1 and u_2 are two solutions, then

$$u_1 - u_2 = Q(u_1) - Q(u_2)$$

implies that

$$\|u_1 - u_2\|_{\mathcal{H}} \leq q\|u_1 - u_2\|_{\mathcal{H}} \,.$$

That is with $0 < q < 1$,

$$0 \leq (1-q)\|u_1 - u_2\|_{\mathcal{H}} \leq 0 \,,$$

which implies that $u_1 = u_2$. ∎

In applications one often obtains $\mathcal{H}$–ellipticity of a sesquilinear form $a(\cdot,\cdot)$ through Gårding's inequality (5.2.2) provided that *Re a* is semidefinite. More precisely, we have the following lemma.

Lemma 5.2.5. *Let the sesquilinear form $a(\cdot,\cdot)$ satisfy Gårding's inequality* (5.2.2) *and, in addition, have the property*

$$Re\, a(v,v) > 0 \text{ for all } v \in \mathcal{H} \quad \text{with } v \neq 0 \,.$$

Then $a(\cdot,\cdot)$ satisfies (5.2.4) *and, consequently, is strongly $\mathcal{H}$–elliptic.*

Proof: The proof rests on the well known weak compactness of the unit sphere in reflexive Banach spaces and in Hilbert spaces (Schechter [376, VIII Theorem 4.2]), i. e. every bounded sequence $\{v_j\}_{j\in\mathbb{N}} \subset \mathcal{H}$ with $\|v_j\|_{\mathcal{H}} \leq M$ contains a subsequence $v_{j'}$ with a weak limit $v_0 \in \mathcal{H}$ such that

$$\lim_{j'\to\infty} (g, v_{j'})_{\mathcal{H}} = (g, v_0)_{\mathcal{H}} \text{ for every } g \in \mathcal{H} \,.$$

We now prove the lemma by contradiction. If a were not strongly $\mathcal{H}$–elliptic then there existed a sequence $\{v_j\}_{j\in\mathbb{N}} \subset \mathcal{H}$ with $\|v_j\|_{\mathcal{H}} = 1$ and

$$\lim_{j\to\infty} Re\, a(v_j, v_j) = 0 \,.$$

Then $\{v_j\}$ contained a subsequence $\{v_{j'}\}$ converging weakly to $v_0 \in \mathcal{H}$. Gårding's inequality then implied

$$\begin{aligned}
\alpha_0 \|v_{j'} - v_0\|^2_{\mathcal{H}} &\leq Re\{a(v_{j'} - v_0, v_{j'} - v_0) + \Big(C(v_{j'} - v_0), v_{j'} - v_0\Big)_{\mathcal{H}}\} \\
&\leq Re\{a(v_{j'}, v_{j'}) - a(v_0, v_{j'}) - a(v_{j'}, v_0) \\
&\qquad + a(v_0, v_0) + (Cv_{j'}, v_{j'} - v_0)_{\mathcal{H}} - (Cv_0, v_{j'} - v_0)_{\mathcal{H}}\} \,.
\end{aligned}$$

Since C is compact, there existed a subsequence $\{v_{j''}\} \subset \{v_{j'}\} \subset \mathcal{H}$ such that $Cv_{j''} \to w \in \mathcal{H}$ for $j'' \to \infty$. Hence, due to the weak convergence $v_{j''} \rightharpoonup v_0$ we would have

$$\begin{aligned}
(Cv_{j''}, v_{j''})_{\mathcal{H}} - (Cv_{j''}, v_0)_{\mathcal{H}} &\to (w, v_0)_{\mathcal{H}} - (w, v_0)_{\mathcal{H}} = 0 \,, \\
a(v_0, v_{j''}) = \Big((jA)^* v_0, v_{j''}\Big)_{\mathcal{H}} &\to \Big((jA)^* v_0, v_0\Big)_{\mathcal{H}} = a(v_0, v_0)
\end{aligned}$$

and corresponding convergence of the remaining terms on the right–hand side. This yielded

$$0 \leq \overline{\lim}_{j'' \to \infty} \alpha_0 \|v_{j''} - v_0\|^2_{\mathcal{H}} \leq -Re\, a(v_0, v_0)$$

and, consequently, with $Re\, a(v_0, v_0) \geq 0$, we could find

$$\lim_{j'' \to \infty} \|v_{j''} - v_0\|_{\mathcal{H}} = 0 \quad \text{together with} \quad Re\, a(v_0, v_0) = 0\,.$$

The latter implied

$$v_0 = 0\,; \quad \text{however,} \quad \|v_0\|_{\mathcal{H}} = \lim_{j'' \to \infty} \|v_{j''}\|_{\mathcal{H}} = 1\,,$$

which is a contradiction. Consequently,

$$\inf_{\|v\|_{\mathcal{H}}=1} Re\, a(v, v) =: \alpha_0' > 0\,,$$

i. e., a is strongly $\mathcal{H}$–elliptic. ∎

In order to generalize the Lax–Milgram lemma to the case where the compact operator C is not zero, we need the following version of the classical Fredholm theorems in Hilbert spaces.

5.3 The Fredholm–Nikolski Theorems

5.3.1 Fredholm's Alternative

We begin with the Fredholm theorem for the standard equations of the second kind

$$Tu := (I - C)u = f \ \text{ in } \ \mathcal{H}\,, \tag{5.3.1}$$

where as before C is a compact linear operator on the Hilbert space $\mathcal{H}$ into itself. We adapt here the *Hilbert space adjoint* $C^* : \mathcal{H} \to \mathcal{H}$ of C which is defined by

$$(Cu, v)_{\mathcal{H}} = (u, C^*v)_{\mathcal{H}} \ \text{ for all } \ u, v \in \mathcal{H}\,.$$

In addition to (5.3.1), we also introduce the adjoint equation

$$T^*v := (I - C^*)v = g \ \text{ in } \ \mathcal{H}\,. \tag{5.3.2}$$

The range of T will be denoted by $\Re(T)$ and the nullspace of T by $\mathcal{N}(T) = \ker(T)$.

Theorem 5.3.1 (Fredholm's alternative). *For equation* (5.3.1) *and equation* (5.3.2), *respectively, the following alternative holds:*
Either

i.) $\mathcal{N}(T) = \{0\}$ *and* $\Re(T) = \mathcal{H}$

or

ii.) $0 < \dim \mathcal{N}(T) = \dim \mathcal{N}(T^*) < \infty$.
In this case, the inhomogeneous equations (5.3.1) *and* (5.3.2) *have solutions*

iff the corresponding right–hand sides satisfy the finitely many orthogonality conditions

$$(f, v_0)_{\mathcal{H}} = 0 \text{ for all } v_0 \in \mathcal{N}(T^*) \tag{5.3.3}$$

and

$$(u_0, g)_{\mathcal{H}} = 0 \text{ for all } u_0 \in \mathcal{N}(T), \tag{5.3.4}$$

respectively. If (5.3.3) *is satisfied then the general solution of* (5.3.1) *is of the form*

$$u = u^* + \sum_{\ell=1}^{k} c_\ell u_{0(\ell)}$$

where u^* is a particular solution of (5.3.1) *depending continuously on f:*

$$\|u^*\|_{\mathcal{H}} \le c\|f\|_{\mathcal{H}} \text{ with a suitable constant } c.$$

The $u_{0(\ell)}$ for $\ell = 1, \ldots, k$ are the linearly independent eigensolutions with $\mathcal{N}(T) = \text{span } \{u_{0(\ell)}\}_{\ell=1}^k$. *Similarly, if* (5.3.4) *is fulfilled then the solution of* (5.3.2) *has the representation*

$$v = v^* + \sum_{\ell=1}^{k} d_\ell v_{0(\ell)}$$

where v^* is a particular solution of (5.3.2) *depending continuously on g; and $v_{0(\ell)}$ for $\ell = 1, \ldots, k$ are the linearly independent eigensolutions with* $\mathcal{N}(T^*) = \text{span } \{v_{0(\ell)}\}_{\ell=1}^k$.

Our proof follows closely the presentation in Kantorowitsch and Akilow [222]. For ease of reading we collect the basic results in the following four lemmata and will give the proof at the end of this section.

Lemma 5.3.2. *The range $\Re(T)$ is a* **closed** *subspace of $\mathcal{H}$.*

Proof: Let $\{u_k\} \subset \Re(T)$ be a convergent sequence and suppose that $u_0 = \lim_{k\to\infty} u_k$. Our aim is to show that $u_0 \in \Re(T)$. First, note that by Projection Theorem 5.2.1, $\mathcal{N}(T)$ is closed and we have the unique decomposition

$$\mathcal{H} = \mathcal{N}(T) \oplus \mathcal{N}(T)^\perp .$$

Moreover, $\mathcal{R}(T) = T(\mathcal{N}(T)^\perp)$. Since $u_k \in \mathcal{R}(T)$, there exists $v_k \in \mathcal{N}(T)^\perp$ such that $u_k = Tv_k$.

Now we consider two cases.

(1) We assume that the sequence $\|v_k\|_{\mathcal{H}}$ is **bounded**. By the compactness of the operator C, we take a subsequence $\{v_{k'}\}$ for which $Cv_{k'} \to u'$ when $k' \to \infty$. We denote $u_{k'} = Tv_{k'} = v_{k'} - Cv_{k'}$. This implies that

$$v_{k'} = u_{k'} + Cv_{k'} \to v_0 = u_0 + u' .$$

Consequently,

$$Cv_{k'} \to Cv_0 \text{ and } u_0 = v_0 - Cv_0 = Tv_0$$

which implies $u_0 \in \mathcal{R}(T)$, that is, $\mathcal{R}(T)$ is closed.

(2) Suppose that $\|v_k\|_{\mathcal{H}}$ is **unbounded**. Then there exists a subsequence $\{v_{k'}\}$ with $\|v_{k'}\|_{\mathcal{H}} \to \infty$. Setting $\widetilde{v}_{k'} := \dfrac{v_{k'}}{\|v_{k'}\|_{\mathcal{H}}}$, we have $\widetilde{v}_{k'} \in \mathcal{N}(T)^\perp$ and $\|\widetilde{v}_{k'}\|_{\mathcal{H}} = 1$. With $u_{k'} = Tv_{k'}$ we obtain

$$\frac{u_{k'}}{\|v_{k'}\|_{\mathcal{H}}} = T\widetilde{v}_{k'} = \widetilde{v}_{k'} - C\widetilde{v}_{k'} \to 0 \text{ since } u_{k'} \to u_0 .$$

Now let $\{\widetilde{v}_{k''}\}$ be an other subsequence of $\widetilde{v}_{k'}$ such that $C\widetilde{v}_{k''} \to u''$. This implies the convergence $\widetilde{v}_{k''} \to u''$. Hence,

$$u'' = Cu'' \text{ with } \|u''\|_{\mathcal{H}} = 1 .$$

However, $\widetilde{v}_{k''} \in \mathcal{N}(T)^\perp$ and so is the limit $u'' \in \mathcal{N}(T)^\perp$. Hence, $u'' \in \mathcal{N}(T)^\perp \cap \mathcal{N}(T) = \{0\}$. This contradicts the fact that $\|u''\|_{\mathcal{H}} = 1$. Therefore, only case (1) is possible. ■

Lemma 5.3.3. *The sequence of sets $\mathcal{N}(T^j)$ for $j = 0, 1, 2, \dots$ with $T^0 = I =$ identity is an increasing sequence, i.e. $\mathcal{N}(T^j) \subseteq \mathcal{N}(T^{j+1})$. There exists a smallest index $m \in \mathbb{N}_0$ such that*

$$\mathcal{N}(T^j) = \mathcal{N}(T^m) \text{ for all } j \geq m \quad \text{and } \mathcal{N}(T^j) \dot{\subset} \mathcal{N}(T^{j+1}) \quad \text{for all } j < m .$$

Proof: The assertion $\mathcal{N}(T^j) \subseteq \mathcal{N}(T^{j+1})$ is clear. Observe that if there is an $m \in \mathbb{N}$ with

$$\mathcal{N}(T^{m+1}) = \mathcal{N}(T^m) ,$$

then, by induction, one has $\mathcal{N}(T^j) = \mathcal{N}(T^m)$ for *all* $j \geq m$. We now prove by contradiction that there exists such an m. Suppose that it would not exist. Then for every $j \in \mathbb{N}$, the set $\mathcal{N}(T^j)$ was a proper subspace of $\mathcal{N}(T^{j+1})$. Hence, there existed an element v_{j+1}

$$0 \neq v_{j+1} \in \mathcal{N}(T^{j+1}) \cap \mathcal{N}(T^j)^\perp .$$

Without loss of generality, we assume that

$$\|v_{j+1}\|_{\mathcal{H}} = 1 .$$

This would define a sequence $\{v_j\}_{j \in \mathbb{N}}$ with $v_j \in \mathcal{N}(T^j)$, and there existed a subsequence $\{v_{j'}\}$ for which

$$Cv_{j'} \to u_0 .$$

Now, for $m' > n'$, consider the difference

$$Cv_{m'} - Cv_{n'} = (v_{m'} - Tv_{m'}) - (v_{n'} - Tv_{n'}) = v_{m'} - z_{m'n'}$$

where

$$z_{m'n'} := v_{n'} + Tv_{m'} - Tv_{n'} \in \mathcal{N}(T^{m'-1})$$

since

$$T^{m'-1} z_{m'n'} = T^{m'-n'-1}(T^{n'} v_{n'}) + T^{m'} v_{m'} - T^{m'-n'}(T^{n'} v_{n'}) = 0\,.$$

Because of $T\mathcal{N}(T^{m'}) \subset \mathcal{N}(T^{m'-1})$ and $n' < m'$, this would imply that

$$(z_{m'n'}, v_{m'})_{\mathcal{H}} = 0\,.$$

Consequently,

$$\|v_{m'} - z_{m'n'}\|_{\mathcal{H}}^2 = \|v_{m'}\|_{\mathcal{H}}^2 + \|z_{m'n'}\|_{\mathcal{H}}^2 \geq 1\,.$$

On the other hand, we would have

$$\|v_{m'} - z_{m'n'}\|_{\mathcal{H}} = \|Cv_{m'} - Cv_{n'}\|_{\mathcal{H}} \leq \|Cv_{m'} - u_0\|_{\mathcal{H}} + \|Cv_{n'} - u_0\|_{\mathcal{H}}\,,$$

which implied that for $m' > n'$,

$$\|v_{m'} - z_{m'n'}\|_{\mathcal{H}} = \|Cv_{m'} - Cv_{n'}\|_{\mathcal{H}} \to 0 \ \text{ as } \ m', n' \to \infty\,.$$

Since $\|v_{m'} - z_{m'n'}\|_{\mathcal{H}} \geq 1$, we have a contradiction. ∎

Lemma 5.3.4. *The ranges* $\mathfrak{R}_j := \mathfrak{R}(T^j) = T^j\mathcal{H}$ *for* $j = 0, 1, 2, \ldots$ *with* $\mathfrak{R}_0 = \mathcal{H}$*, form a decreasing sequence of closed subspaces of* $\mathcal{H}$*, i.e.* $\mathfrak{R}_{j+1} \subseteq \mathfrak{R}_j$*. There exists a smallest index* $r \in \mathbb{N}_0$ *such that*

$$\mathfrak{R}_r = \mathfrak{R}_j \ \textit{ for all } j \geq r \quad \textit{and } \ \mathfrak{R}_{j+1} \dot{\subset} \mathfrak{R}_j \quad \textit{for all } \ j < r\,.$$

Proof: Lemma 5.3.2 guarantees that all $\mathfrak{R}_j$'s are closed subspaces. The inclusion $\mathfrak{R}_{j+1} \subseteq \mathfrak{R}_j$ is clear from the definition. The existence of r will be proved by contradiction.

Assume that for every j the space $\mathfrak{R}_{j+1}$ is a proper subspace of $\mathfrak{R}_j$. Then for every j there would exist an element $v_j \in \mathfrak{R}_j \cap \mathfrak{R}_{j+1}^{\perp}$ with $\|v_j\|_{\mathcal{H}} = 1$. Moreover, there existed a subsequence $\{v_{j'}\}$ such that

$$Cv_{j'} \to u_0 \in \mathcal{H} \ \text{ for } \ j' \to \infty\,,$$

since C is compact. Choose $m' > n'$ and consider the difference

$$(Cv_{n'} - u_0) - (Cv_{m'} - u_0) = v_{n'} - z_{n'm'}$$

where

$$z_{n'm'} := v_{m'} + Tv_{n'} - Tv_{m'} \in \Re_{n'+1}\,.$$

Since with $v_{m'} = T^{m'}u_{m'}$ and $v_{n'} = T^{n'}u_{n'}$ and with $T\Re_{n'} \subset \Re_{n'+1} \subseteq \Re_{m'}$ for $m' > n'$, there would hold

$$z_{n'm'} = T^{n'+1}(T^{m'-n'-1}u_{m'} - T^{m'-n'}u_{m'} + u_{n'})$$

whereas $v_{n'} \in \Re^{\perp}_{n'+1}$. Then, in the same manner as in the previous lemma, we would have

$$\|v_{n'} - z_{n'm'}\|^2_{\mathcal{H}} = \|v_{n'}\|^2_{\mathcal{H}} + \|z_{n'm'}\|^2_{\mathcal{H}} \geq 1$$

but

$$\|v_{n'} - z_{n'm'}\|_{\mathcal{H}} \leq \|Cv_{n'} - u_0\|_{\mathcal{H}} + \|Cv_{m'} - u_0\|_{\mathcal{H}} \to 0 \text{ for } m' > n' \to \infty\,;$$

a contradiction. ■

Lemma 5.3.5. *Let m and r be the indices in the previous lemmata. Then we have:*

i. $m = r$.

ii. $\mathcal{H} = \mathcal{N}(T^r) \oplus \Re_r$, *which means that every $u \in \mathcal{H}$ admits a unique decomposition*

$$u = u_0 + u_1 \quad \textit{where } u_0 \in \mathcal{N}(T^r) \textit{ and } u_1 \in \Re_r\,.$$

Proof: i.) We prove $m = r$ in two steps:

(a) Assume that $m < r$. Choose any $w \in \Re_{r-1}$, hence $w = T^{r-1}v$ for some $v \in \mathcal{H}$. This yields that

$$Tw = T^r v \in \Re_r = \Re_{r+1}$$

by definition of r. Then

$$Tw = T^{r+1}\widetilde{v} \text{ for some } \widetilde{v} \in \mathcal{H}\,,$$

and, hence,

$$T^r(v - T\widetilde{v}) = 0\,.$$

Now, since $m < r$ by assumption, we have

$$\mathcal{N}(T^r) = \mathcal{N}(T^{r-1}) = \cdots = \mathcal{N}(T^m)$$

and, thus,

$$T^{r-1}(v - T\widetilde{v}) = 0$$

implying

$$\Re_{r-1} \ni w = T^{r-1}v = T^r\widetilde{v} \in \Re_r\,.$$

Hence, we have the inclusions

$$\Re_{r-1} \subseteq \Re_r \subseteq \Re_{r-1} \text{ and } \Re_{r-1} = \Re_r$$

which contradicts the definition of r as to be the smallest index.

(b) Assume that $r < m$. Set $v \in \mathcal{N}(T^m)$. Then,

$$T^{m-1}v \in \mathfrak{R}_{m-1} = \cdots = \mathfrak{R}_r \text{ and } \mathfrak{R}_m = \mathfrak{R}_{m-1} = \mathfrak{R}_r\,,$$

since $m > r$. Hence,

$$T^{m-1}v = T^m\widetilde{v} \text{ for some } \widetilde{v} \in \mathcal{H}\,.$$

Moreover,

$$T^{m+1}\widetilde{v} = T^m v = 0\,, \text{ since } v \in \mathcal{N}(T^m)\,.$$

The latter implies that

$$\widetilde{v} \in \mathcal{N}(T^{m+1}) = \mathcal{N}(T^m) \text{ by the definition of } m\,.$$

Hence,

$$T^{m-1}v = T^m\widetilde{v} = 0$$

which implies $v \in \mathcal{N}(T^{m-1})$. Consequently, we have the inclusions

$$\mathcal{N}(T^m) \subseteq \mathcal{N}(T^{m-1}) \subseteq \mathcal{N}(T^m) \text{ and } \mathcal{N}(T^{m-1}) = \mathcal{N}(T^m)$$

which contradicts the definition of m.

The result $m = r$, i.e., **i.)** then follows from (a) and (b).

ii.) For showing $\mathcal{H} = \mathcal{N}(T^r) \oplus \mathfrak{R}_r$, let $u \in \mathcal{H}$ be given. Then $T^r u \in \mathfrak{R}_r = \mathfrak{R}_{2r}$ and $T^r u = T^{2r}\widetilde{u}$ form some $\widetilde{u} \in \mathcal{H}$. Let u_1 be defined by $u_1 := T^r\widetilde{u} \in \mathfrak{R}_r$, and let $u_0 := u - u_1$. For u_0, we find

$$T^r u_0 = T^r u - T^r u_1 = T^{2r}\widetilde{u} - T^{2r}\widetilde{u} = 0$$

and, hence, $u_0 \in \mathcal{N}(T^r)$. This establishes the decomposition: $u = u_0 + u_1$.

For the uniqueness of the decomposition, let $v \in \mathcal{N}(T^r) \cap \mathfrak{R}_r$. Then $v = T^r w$ for some $w \in \mathcal{H}$, and $T^r v = T^{2r} w = 0$, since $v \in \mathcal{N}(T^r)$. Hence, $w \in \mathcal{N}(T^{2r}) = \mathcal{N}(T^r)$ which implies that

$$v = T^r w = 0 \text{ and } \mathcal{N}(T^r) \cap \mathfrak{R}_r = \{0\}\,.$$

This completes the proof. ■

Theorem 5.3.6. *a) The operator T maps the subspace $\mathcal{H}' := T^r(\mathcal{H}) = \mathfrak{R}_r$ bijectively onto itself and $T : \mathcal{H}' \to \mathcal{H}'$ is an isomorphism.*
b) The space $\mathcal{H}'' := \mathcal{N}(T^r)$ is finite-dimensional and T maps $\mathcal{H}''$ into itself.
c) For the decomposition

$$u = u_0 + u_1 =: P_0 u + P_1 u$$

in Lemma 5.3.5*, the projection operators $P_0 : \mathcal{H} \to \mathcal{H}''$ and $P_1 : \mathcal{H} \to \mathcal{H}'$ are both bounded.*

d) The compact operator C admits a decomposition of the form

$$C = C_0 + C_1$$

where

$$C_0 : \mathcal{H} \to \mathcal{H}'' \text{ and } C_1 : \mathcal{H} \to \mathcal{H}' \text{ with } C_0C_1 = C_1C_0 = 0$$

and both operators are compact. Moreover, $T_1 := I - C_1$ is an isomorphism on $\mathcal{H}$.

Proof:

a) The definition of $\mathfrak{R}_r$ gives

$$T(\mathfrak{R}_r) = \mathfrak{R}_{r+1} = \mathfrak{R}_r\,,$$

i. e. surjectivity. The injectivity of T on $\mathfrak{R}_r$ follows from $r = m$ since

$$Tu_0 = 0$$

for $u_0 \in \mathfrak{R}_r$ implies $u_0 \in \mathcal{N}(T) \subset \mathcal{N}(T^r)$ and Lemma 5.3.5 **ii.)** yields $u_0 = 0$, i. e. injectivity.
To show that $T_{|\mathcal{H}'}$ is an isomorphism on $\mathcal{H}'$, it remains to prove that $(T_{|\mathcal{H}'})^{-1}$ on $\mathcal{H}'$ is bounded. In fact, this follows immediately by using Banach's closed graph theorem (Schechter [376, p. 65]) since $T_{|\mathcal{H}'}$ is continuous and bijective and $\mathcal{H}'$ is a closed subspace of the Hilbert space $\mathcal{H}$ due to Lemma 5.3.5. However, we present an elementary contradiction argument by taking advantage of the special form $T = I - C$ together with the compactness of C.
Suppose $(T_{|\mathcal{H}'})^{-1}$ is not bounded. Then there exists a sequence of elements $v_j \in \mathcal{H}'$ with $\|v_j\|_{\mathcal{H}} = 1$ such that $\|w_j\|_{\mathcal{H}} \to \infty$ where $w_j = (T_{|\mathcal{H}'})^{-1}v_j$.
For the corresponding sequence

$$u_j := \left(w_j \big/ \|w_j\|_{\mathcal{H}}\right) \in \mathcal{H}' \text{ with } \|u_j\|_{\mathcal{H}} = 1$$

one has

$$Tu_j = u_j - Cu_j = \frac{1}{\|w_j\|_{\mathcal{H}}}Tw_j = \left(v_j \big/ \|w_j\|_{\mathcal{H}}\right) \to 0 \text{ for } j \to \infty\,. \quad (5.3.5)$$

Because of the compactness of the operator C, there exists a subsequence $u_{j'} \in \mathcal{H}'$ such that

$$Cu_{j'} \to u_0 \text{ in } \mathcal{H}\,.$$

This implies from (5.3.5) that

$$u_{j'} = Cu_{j'} + v_{j'} \big/ \|w_{j'}\|_{\mathcal{H}} \to u_0\,,$$

hence, $u_0 \in \mathcal{H}'$ and, moreover,

$$Tu_0 = u_0 - Cu_0 = 0 \text{ with } \|u_0\|_{\mathcal{H}} = 1\,.$$

This contradicts the inectivity of $T_{|\mathcal{H}'}$.

b) Since

$$T^r = (I - C)^r =: I - \widetilde{C}$$

with a compact operator $\widetilde{C}$, the unit sphere $\{v \in \mathcal{H}'' \,|\, \|v\| \leq 1\}$ is compact and, hence, $\mathcal{H}''$ is finite–dimensional due to a well–known property of normal spaces, see e. g. Schechter [376, p.86].

c) For any $u \in \mathcal{H}$ define

$$\begin{aligned} u_1 &:= T_1^{-r} T^r u =: P_1 u \in \Re_r = \mathcal{H}' \text{ and} \\ u_0 &:= u - u_1 = u - P_1 u =: P_0 u \,. \end{aligned}$$

Then

$$T^r u_0 = T^r u - T^r P_1 u = T^r u - T_1^r T_1^{-r} T^r u = 0$$

and $P_0 u \in \mathcal{H}''$. Clearly, P_1 and P_0 are bounded linear operators by definition. Moreover, the decomposition

$$u = u_0 + u_1 = P_0 u + P_1 u \tag{5.3.6}$$

is unique from Lemma 5.3.5 and we have

$$P_0 P_1 = P_1 P_0 = 0 \,.$$

d) With (5.3.6) we define the compact operators

$$C_1 := C P_1 \text{ and } C_0 := C P_0 \,.$$

Since

$$C = I - T$$

we have with a) that

$$C_{|\Re_r} = C_{|\mathcal{H}'} = I_{|\mathcal{H}'} - T_{|\mathcal{H}'} : \mathcal{H}' \to \mathcal{H}' \,.$$

Hence, $C_1 : \mathcal{H} \to \mathcal{H}'$.

Similarly, for every $v_0 \in \mathcal{H}''$ we find that $w_0 := C v_0$ satisfies

$$T^r w_0 = T^r (C v_0) = T^r (I - T) v_0 = T^r v_0 - T T^r v_0 = 0 \,,$$

i. e.

$$C_{|\mathcal{H}''} : \mathcal{H}'' \to \mathcal{H}'' \text{ and } C_0 : \mathcal{H} \to \mathcal{H}'' \,.$$

The equation

$$C_1 C_0 = C_0 C_1 = 0$$

follows now immediately.

The invertibility of $T_1 = I - C_1$ follows from

$$T_1 u = T_1 u_1 + T_1 u_0 = T_1 u_1 + (I - C_1) u_0 = T_1 u_1 + u_0 = P_0 f + P_1 f$$

with

$$u = (T_{1_{|\mathcal{H}'}})^{-1} P_1 f + P_0 f =: (I - C_1)^{-1} f$$

where all the operators on the right–hand side are continuous. Therefore, $I - C_1$ is an isomorphism on $\mathcal{H}$. This completes the proof. ■

Proof of Theorem 5.3.1:
i.)

a) If the solution of (5.3.1) is unique then the homogeneous equation

$$Tu_0 = 0$$

admits only the trivial solution $u_0 = 0$. Hence, $r = m = 0, \mathcal{H}' = \mathcal{H}$ and $T(\mathcal{H}) = \mathcal{H}$ in Lemma 5.3.5. The latter implies $f \in T(\mathcal{H})$ which yields the existence of $u \in \mathcal{H}$ with (5.3.1), i. e. solvability and that $T : \mathcal{H} \to \mathcal{H}$ is an isomorphism.

b) Suppose that (5.3.1) admits a solution $u \in \mathcal{H}$ for every given $f \in \mathcal{H}$. Then $T(\mathcal{H}) = \mathcal{H}$ and Lemma 5.3.5 implies $m = r = 0$ and $\mathcal{N}(T) = \{0\}$, i. e. uniqueness. and $T : \mathcal{H} \to \mathcal{H}$ is an isomorphism.

ii.) First let us show $\dim \mathcal{N}(T) = \dim \mathcal{N}(T^*) < \infty$. Since $\mathcal{N}(T) \subset \mathcal{N}(T^r) = \mathcal{H}''$, it follows from Theorem 5.3.6 that $\dim \mathcal{N}(T) =: k < \infty$. Since with C also C^* being a compact operator on $\mathcal{H}$, the same arguments imply $\dim \mathcal{N}(T^*) =: k^* < \infty$.

Since a basis $u_{0(1)}, \dots, u_{0(k)}$ is linearly independent, the well known Hahn–Banach theorem (Schechter [375]) implies that there exist $\varphi_j \in \mathcal{H}$ with

$$(u_{0(\ell)}, \varphi_j) = \delta_{\ell j} \text{ for } \ell, j = 1, \dots, k\,.$$

Similarly, there exist $\psi_j \in \mathcal{H}$ with

$$(\psi_j, v_{0(\ell)}) = \delta_{j\ell} \text{ for } j, \ell = 1, \dots, k^*\,.$$

Now let us suppose $k < k^*$. Consider the new operator

$$\widetilde{T}w := Tw - \sum_{\ell=1}^{k} (w, \varphi_\ell)\psi_\ell\,.$$

If $w_0 \in \mathcal{N}(\widetilde{T})$ then

$$0 = (v_{0(j)}, \widetilde{T}w_0) = (T^* v_{0(j)}, w_0) - \sum_{\ell=1}^{k} (w_0, \varphi_\ell)(v_{0(j)}, \psi_\ell) = -(w_0, \varphi_j)$$
$$\text{for } j = 1, \dots, k\,.$$

Hence,

$$\widetilde{T}w_0 = Tw_0 - 0 = 0 \text{ and } w_0 = \sum_{\ell=1}^{k} \alpha_\ell u_{0(\ell)} \in \mathcal{N}(T)$$

where α_ℓ are some constants for $\ell = 1, \dots, k$. However,

$$0 = -(w_0, \varphi_j) = -\sum_{\ell=1}^{k} \alpha_\ell (u_{0(\ell)}, \varphi_j) = -\alpha_j \quad \text{for } j = 1, \ldots, k$$

and we find

$$w_0 = 0\,.$$

Hence, $\mathcal{N}(\widetilde{T}) = \{0\}$ and $\widetilde{T} : \mathcal{H} \to \mathcal{H}$ is an isomorphism due to **i.**). Then the equation

$$\widetilde{T}w^* = Tw^* - \sum_{\ell=1}^{k} (w^*, \varphi_\ell)\psi_\ell = \psi_{k+1}$$

has exactly one solution $w^* \in \mathcal{H}$ for which we find

$$\begin{aligned} 0 = (v_{0(k+1)}, \widetilde{T}w^*) &= (v_{0(k+1)}, Tw^*) - \sum_{\ell=1}^{k} (w^*, \varphi_\ell)(v_{0(k+1)}, \psi_\ell) \\ &= (v_{0(k+1)}, \psi_{k+1}) = 1\,, \end{aligned}$$

an obvious contradiction. Hence, $k^* \le k$.

In the same manner, by using $T^{**} = T$, we find $k \le k^*$ which shows

$$\dim \mathcal{N}(T) = \dim \mathcal{N}(T^*) = k\,.$$

Finally, if Equation (5.3.1) has a solution u then

$$(v_{0(\ell)}, f) = (v_{0(\ell)}, Tu) = (T^* v_{0(\ell)}, u) = 0 \quad \text{for } \ell = 1, \ldots, k\,.$$

Moreover, if u^* is any particular solution of (5.3.1) then $u_0 := u - u^* \in \mathcal{N}(T)$. Hence,

$$u = u^* + u_0 = u^* + \sum_{\ell=1}^{k} \alpha_\ell u_{0(\ell)}\,.$$

Conversely, if $(v_{0(\ell)}, f) = 0$ is satisfied for $\ell = 1, \ldots, k$ then solve the equation

$$\widetilde{T}u^* = Tu^* - \sum_{\ell=1}^{k} (u^*, \varphi_\ell)\psi_\ell = f$$

which admits exactly one solution $u^* \in \mathcal{H}$ due to **i.** and, furthermore,

$$\|u^*\|_{\mathcal{H}} \le \|\widetilde{T}^{-1}\|_{\mathcal{H},\mathcal{H}} \|f\|_{\mathcal{H}}$$

since $\widetilde{T}$ is an isomorphism. For this solution we see that

$$\begin{aligned} (v_{0(j)}, f) = 0 &= (v_{0(j)}, Tu^r - \sum_{\ell=1}^{k} (u^*, \varphi_\ell)\psi_\ell) = -(u^*, \varphi_j) \\ &\qquad \text{for every } j = 1, \ldots, k\,. \end{aligned}$$

Hence, u^* is a particular solution of (5.3.1). Clearly, any

$$u = u^* + u_0 \quad \text{with } u_0 \in \mathcal{N}(T)$$

solves (5.3.1). This completes the proof of Theorem 5.3.1. ∎

5.3.2 The Riesz–Schauder and the Nikolski Theorems

Fredholm's alternative can be generalized to operator equations between different Hilbert and Banach spaces which dates back to Riesz, Schauder and, more recently, to Nikolski. Here, for simplicity, we confine ourselves to the Hilbert space setting. Let

$$T : \mathcal{H}_1 \to \mathcal{H}_2$$

be a linear bounded operator between the two Hilbert spaces $\mathcal{H}_1$ and $\mathcal{H}_2$. If

$$T = T_I - T_C \tag{5.3.7}$$

where T_I is an isomorphism while T_C is compact from $\mathcal{H}_1$ into $\mathcal{H}_2$, then for T and its adjoint T^* defined by the relation

$$(Tu, v)_{\mathcal{H}_2} = (u, T^* v)_{\mathcal{H}_1} \quad \text{for all } u \in \mathcal{H}_1, v \in \mathcal{H}_2\,, \tag{5.3.8}$$

Fredholm's alternative remains valid in the following form.

Theorem 5.3.7 (Riesz–Schauder Theorem). *For the operator T of the form in* (5.3.7) *and its adjoint operator T^* defined by* (5.3.8) *the following alternative holds:*
Either

i. $\mathcal{N}(T) = \{0\}$ *and* $\Re(T) = \mathcal{H}_2$

or

ii. $0 < \dim \mathcal{N}(T) = \dim \mathcal{N}(T^*) < \infty$.

In this case, the equations

$$Tu = f \quad \textit{and} \quad T^* v = g$$

have solutions $u \in \mathcal{H}_1$ or $v \in \mathcal{H}_2$, respectively, iff the corresponding right–hand side satisfies the finitely many orthogonality conditions

$$\begin{array}{rll} & (f, v_0)_{\mathcal{H}_2} = 0 & \textit{for all } v_0 \in \mathcal{N}(T^*) \\ \textit{or} & (g, u_0)_{\mathcal{H}_1} = 0 & \textit{for all } u_0 \in \mathcal{N}(T)\,, \end{array}$$

respectively. If these conditions are fulfilled then the general solution is of the form

$$u = u^* + \widetilde{u}_0 \quad \textit{or} \quad v = v^* + \widetilde{v}_0 \tag{5.3.9}$$

with a particular solution u^* or v^* depending continuously on f or g; and any $\widetilde{u}_0 \in \mathcal{N}(T)$ or $\widetilde{v}_0 \in \mathcal{N}(T^*)$, respectively.

Proof: In order to apply Theorem 5.3.1 we define the compact operator

$$C := T_I^{-1} T_C \ \text{ in } \mathcal{H}_1\,.$$

Since T_I is an isomorphism, i. e. $T_I^{-1}\mathcal{H}_1 = \mathcal{H}_2$, one can show easily

$$C^* = T_C^*(T_I^*)^{-1}$$

from the definition of the adjoint operator,

$$(Cu, w)_{\mathcal{H}_1} = (u, C^*w)_{\mathcal{H}_1} \text{ for all } u, w \in \mathcal{H}_1 .$$

Then Fredholm's alternative, Theorem 5.3.1 is valid for the transformed equations

$$T_I^{-1}Tu = (I - C)u = T_I^{-1}f \text{ in } \mathcal{H}_1 \tag{5.3.10}$$

and

$$(I - C^*)w = g \text{ in } \mathcal{H}_1$$

where $w = T_I^* v$. Moreover, one verifies that the following relations hold:

$$\Re(T) = T_I \Re(I - C) \quad \text{and} \quad \mathcal{N}(T) = \mathcal{N}(I - C)\,; \tag{5.3.11}$$

$$\Re(T^*) = \Re(I - C^*) \quad \text{and} \quad \mathcal{N}(T^*) = (T_I^*)^{-1}\mathcal{N}(I - C^*)\,. \tag{5.3.12}$$

i. Equations (5.3.11) imply

$$\mathcal{N}(T) = \{0\} = \mathcal{N}(I - C)$$

and

$$\mathcal{H}_1 = \Re(I - C) \text{ is equivalent to } \mathcal{H}_2 = \Re(T)\,.$$

ii. Because of (5.3.3), Equation (5.3.10) admits a solution iff

$$(T_I^{-1}f, w_0)_{\mathcal{H}_1} = 0 \text{ for all } w_0 \in \mathcal{N}(I - C^*)\,.$$

The latter is equivalent to

$$(f, (T_I^*)^{-1}w_0)_{\mathcal{H}_2} = (f, v_0)_{\mathcal{H}_2} = 0$$

for all $v_0 = (T_I^*)^{-1}w_0$ with $w_0 \in \mathcal{N}(I - C^*)$. Then it follows from (5.3.12) that v_0 traces the whole eigenspace $\mathcal{N}(T^*) = (T_I^*)^{-1}\mathcal{N}(I - C^*)$. This proves the claims for the equation $Tu = f$. For the adjoint equation

$$T^*v = g$$

we can proceed in the same manner by making use of both equations in (5.3.12) and the second relation in (5.3.11). The representation formulae (5.3.9) for the solutions can be derived by using the relations (5.3.11) and (5.3.12) again together with the corresponding representations in Theorem 5.3.1. This completes the proof. ■

For the Riesz–Schauder Theorem 5.3.7, the assumptions on the operator T in (5.3.7) are not only sufficient but also necessary as will be shown in the following theorem.

Theorem 5.3.8 (Nikolski's Theorem). *Let $T : \mathcal{H}_1 \to \mathcal{H}_2$ be a bounded linear operator for which the Riesz–Schauder theorem* 5.3.7 *holds. Then T is of the form*

$$T = T_I - T_F \tag{5.3.13}$$

with some isomorphic operator $T_I; \mathcal{H}_1 \to \mathcal{H}_2$ and a finite–dimensional operator T_F.

The latter means that there exist two finite–dimensional orthogonal sets $\{\varphi_j\}_{j=1}^k \in \mathcal{H}_1$ and $\{\psi_j\}_{j=1}^k \in \mathcal{H}_2$ such that

$$T_F u = \sum_{j=1}^k (u, \varphi_j)_{\mathcal{H}_1} \psi_j \,. \tag{5.3.14}$$

Remark 5.3.1: Note that T_F is a compact operator and, hence, the representation (5.3.13) is a special case of the operators in (5.3.7) considered in the Riesz–Schauder Theorem 5.3.7.

Proof: Let $u_{0(\ell)} \in \mathcal{H}_1$ and $v_{0(\ell)} \in \mathcal{H}_2$ be orthogonal bases of the k–dimensional nullspaces $\mathcal{N}(T)$ and $\mathcal{N}(T^*)$ in the Riesz–Schauder theorem 5.3.7, respectively. Then define corresponding orthogonal systems $\varphi_j \in \mathcal{H}_1, \psi_j \in \mathcal{H}_2$ satisfying

$$(u_{0(\ell)}, \varphi_j)_{\mathcal{H}_1} = \delta_{\ell j} \text{ and } (\psi_j, v_{0(\ell)})_{\mathcal{H}_2} = \delta_{j\ell} \text{ for } j, \ell = 1, \dots, k \,.$$

With these elements we introduce the finite–dimensional operator T_F as in (5.3.14) and define the bounded linear operator

$$T_I := T + T_F \,.$$

It remains now to show that $T_I : \mathcal{H}_1 \to \mathcal{H}_2$.

i. Injectivity: Suppose $u \in \mathcal{H}_1$ with

$$T_I u = 0 \,.$$

Then u solves

$$Tu = -\sum_{j=1}^k (u, \varphi_j)_{\mathcal{H}_1} \psi_j \,.$$

Hence, the validity of the orthogonality conditions in Theorem 5.3.7 implies

$$-\sum_{j=1}^k (u, \varphi_j)_{\mathcal{H}_1} (\psi_j, v_{0(\ell)})_{\mathcal{H}_2} = -(u, \varphi_\ell)_{\mathcal{H}_1} = 0 \text{ for } \ell = 1, \dots, k \,. \tag{5.3.15}$$

Hence,

$$Tu = 0$$

and, therefore,

$$u = \sum_{\ell=1}^{k} \alpha_\ell u_{0(\ell)}$$

due to the assumed validity of the Riesz–Schauder theorem 5.3.7. However, equations (5.3.15) imply $\alpha_\ell = 0$ for $\ell = 1, \dots, k$, hence, $u \equiv 0$.

ii. Surjectivity: Given any $f \in \mathcal{H}_2$. We show that there exists a solution $w \in \mathcal{H}_1$ of

$$T_I w = f .$$

To this end consider

$$f' := f - \sum_{j=1}^{k} (f, v_{0(j)})_{\mathcal{H}_2} \psi_j$$

which satisfies

$$(f', v_{0(\ell)})_{\mathcal{H}_2} = 0 \text{ for } \ell = 1, \dots, k .$$

Hence, by the Riesz–Schauder theorem 5.3.7, there exists a family of solutions $w' \in \mathcal{H}_1$ of the form

$$w' = w^* + \sum_{j=1} \alpha_j u_{0(j)} \text{ with } \alpha_j \in \mathbb{C}$$

to

$$Tw' = f'$$

and

$$\|w^*\|_{\mathcal{H}_1} \le c \|f'\|_{\mathcal{H}_2} \le c' \|f\|_{\mathcal{H}_2} ,$$

the last inequality following from the construction of f'. In particular, for the special choice

$$\alpha_j := (f, v_{0(j)})_{\mathcal{H}_2} - (w^*, \varphi_j)_{\mathcal{H}_1}$$

this yields

$$\begin{aligned} T_I w' &= Tw' + T_F w' \\ &= f' + \sum_{j=1}^{k} (w^* + \sum_{\ell=1}^{k} \alpha_\ell u_{0(\ell)}, \varphi_j)_{\mathcal{H}_1} \psi_j \\ &= f - \sum_{j=1}^{k} (f, v_{0(j)})_{\mathcal{H}_2} \psi_j + \sum_{j=1}^{k} \sum_{\ell=1}^{k} (f, v_{0(\ell)})_{\mathcal{H}_2} (u_{0(\ell)}, \varphi_j)_{\mathcal{H}_1} \psi_j \\ &= f , \end{aligned}$$

which completes the proof of the surjectivity.

iii. Continuity of $(T_I)^{-1} : \mathcal{H}_2 \to \mathcal{H}_1$:

The definition of the α_j implies with the continuous dependence of w^* on f that

$$|\alpha_j| \le c'' \|f\|_{\mathcal{H}_2} .$$

Hence, we also have

$$\|w'\|_{\mathcal{H}_1} = \|(T_I)^{-1} f\|_{\mathcal{H}_1} \le c''' \|f\|_{\mathcal{H}_2}$$

which is the desired continuity. ■

As an easy consequence of Theorems 5.3.7 and 5.3.8 and Remark 5.3.1 we find the following result.

Corollary 5.3.9. *Let $T : \mathcal{H}_1 \to \mathcal{H}_2$ be a bounded linear operator. Then the following statements are equivalent:*

- $T = T_I - T_C$ *in* (5.3.7).
- *Fredholm's alternative in the form of the Riesz–Schauder theorem* 5.3.7 *is valid for* T.
- $T = \widetilde{T}_I - T_F$ *in* (5.3.13) *where* $\widetilde{T}_I$ *is an isomorphism (which can be different from* T_I*) and* T_F *is a finite–dimensional operator.*

5.3.3 Fredholm's Alternative for Sesquilinear Forms

In the context of variational problems one is lead to variational equations of the form (5.2.1) where the sesquilinear form $a(\cdot,\cdot)$ satisfies a Gårding inequality (5.2.2) and where, in general, the compact operator C does *not* vanish. Then Theorem 5.3.7 can be used to establish Fredholm's alternative for the variational equation (5.2.1) which generalizes the Lax–Milgram theorem 5.2.3. For this purpose, let us also introduce the *homogeneous variational problems* for the sesquilinear form a and its adjoint:

Find $u_0 \in \mathcal{H}$ *such that*

$$a(v, u_0) = 0 \text{ for all } v \in \mathcal{H} \tag{5.3.16}$$

and $v_0 \in \mathcal{H}$ *such that*

$$\overline{a^*(u, v_0)} = a(v_0, u) = 0 \text{ for all } u \in \mathcal{H} . \tag{5.3.17}$$

The latter is the *adjoint problem* to (5.3.16).

Theorem 5.3.10. *For the variational equation* (5.2.1) *in the Hilbert space* $\mathcal{H}$ *where the continuous sesquilinear form* $a(\cdot,\cdot)$ *satisfies Gårding's inequality* (5.2.2)*, there holds the alternative:*

Either

i. (5.2.1) *has exactly one solution* $u \in \mathcal{H}$ *for every given* $\ell \in \mathcal{H}^*$

or

ii. *The homogeneous problems* (5.3.16) *and* (5.3.17) *have finite–dimensional nullspaces of the same dimension* $k > 0$. *The nonhomogeneous problem* (5.2.1) *and its adjoint: Find* $v \in \mathcal{H}$ *such that*

$$\overline{a(v,w)} = \ell^*(w) \ \ \textit{for all} \ \ w \in \mathcal{H}$$

have solutions iff the orthogonality conditions

$$\ell(v_{0(j)}) = 0\,, \ \ \textit{respectively,} \ \ \ell^*(u_{0(j)}) = 0 \ \ \textit{for} \ \ j = 1, \ldots, k,$$

are satisfied where $\{u_{0(j)}\}_{j=1}^k$ *spans the eigenspace of* (5.3.16) *and* $\{v_{0(j)}\}_{j=1}^k$ *spans the eigenspace of* (5.3.17), *respectively.*

Proof: First let us rewrite the variational equation (5.2.1) in the form

$$a(v,u) = (v, jAu)_{\mathcal{H}} = (v,f)_{\mathcal{H}} = \ell(v)$$

where the mappings $A : \mathcal{H} \to \mathcal{H}^*$ and $jA : \mathcal{H} \to \mathcal{H}$ are linear and bounded and where $f \in \mathcal{H}$ represents $\ell \in \mathcal{H}^*$ due to the Riesz representation theorem 5.2.2. Then we can identify the variational equation (5.2.1) with the operator equation

$$(T_I - T_C)u = jAu = f \tag{5.3.18}$$

where

$$T_I := jA + C\,,\ T_C = C$$

with the compact operator C in Gårding's inequality (5.2.2). Because of the Lax–Milgram theorem 5.2.3, the $\mathcal{H}$–elliptic operator T_I defines an isomorphism from $\mathcal{H}$ onto $\mathcal{H}$.

Similarly, the **adjoint problem**:
Find $v \in \mathcal{H}$ *such that*

$$(w, (jA)^*)_{\mathcal{H}} = \overline{a(v,w)} = \ell^*(w) = (w,g)_{\mathcal{H}} \ \ \textit{for all} \ \ w \in \mathcal{H}\,,$$

can be rewritten as to find the solution $v \in \mathcal{H}$ *of the operator equation*

$$(T_I^* - T_C^*)v = (jA)^* v = g \tag{5.3.19}$$

with $T_I^* = (jA)^* + C^*$ and $T_C^* = C^*$.

The operators T_I and T_C satisfy all of the assumptions in the Riesz–Schauder theorem 5.3.7 with $\mathcal{H}_1 = \mathcal{H}_2 = \mathcal{H}$. An application of that theorem to equations (5.3.18) and (5.3.19) provides all claims in Theorem 5.3.10. ■

Remark 5.3.2: In the case that the Gårding inequality (5.2.2) is replaced by the weaker condition

$$|a(v,v) + (Cv,v)_{\mathcal{H}}| \geq \alpha_0 \|v\|_{\mathcal{H}}^2 \tag{5.3.20}$$

then Theorem 5.3.10 remains valid with the same proof.

5.3.4 Fredholm Operators

We conclude this section by including the following theorems on Fredholm operators on Hilbert spaces. (For general Banach spaces see e.g. Mikhlin and Prössdorf [305] where one may find detailed results and historical remarks concerning Fredholm operators.)

Definition 5.3.1. *Let $\mathcal{H}_1$ and $\mathcal{H}_2$ be two Hilbert spaces. The linear bounded operator $A : \mathcal{H}_1 \to \mathcal{H}_2$ is called a Fredholm operator if the following conditions are satisfied:*

i *The nullspace $\mathcal{N}(A)$ of A has finite dimension,*
ii *the range $\Re(A)$ of A is a closed subspace of $\mathcal{H}_2$,*
iii *the range $\Re(A)$ has finite codimension.*

The number

$$\text{index}(A) = \dim \mathcal{N}(A) - \text{codim } \Re(A) \tag{5.3.21}$$

is called the Fredholm index of A.

Theorem 5.3.11. *The linear bounded operator $A : \mathcal{H}_1 \to \mathcal{H}_2$ is a Fredholm operator if and only if there exist continuous linear operators $Q_1, Q_2 : \mathcal{H}_1 \to \mathcal{H}_2$ such that*

$$Q_1 A = I - C_1 \quad \textit{and} \quad AQ_2 = I - C_2 \tag{5.3.22}$$

with compact linear operators C_1 in $\mathcal{H}_1$ and C_2 in $\mathcal{H}_2$.

If both $B : \mathcal{H}_1 \to \mathcal{H}_2$ and $A : \mathcal{H}_2 \to \mathcal{H}_3$ are Fredholm operators then $A \circ B$ also is a Fredholm operator from $\mathcal{H}_1$ to $\mathcal{H}_3$ and

$$\text{index}(A \circ B) = \text{index}(A) + \text{index}(B)\,. \tag{5.3.23}$$

For a Fredholm operator A and a compact operator C, the sum $A + C$ is a Fredholm operator and

$$\text{index}(A + C) = \text{index}(A)\,. \tag{5.3.24}$$

The set of Fredholm operators is an open subset in the space of bounded linear operators and the index is a continuous function. The adjoint operator A^* of a Fredholm operator A is also a Fredholm operator and

$$\text{index}(A^*) = -\text{index}(A)\,. \tag{5.3.25}$$

The equation

$$Au = f \tag{5.3.26}$$

with $f \in \mathcal{H}_2$ is solvable if and only if

$$(f, v_0)_{\mathcal{H}_2} = 0 \quad \textit{for all} \quad v_0 \in \mathcal{N}(A^*) \subset \mathcal{H}_2\,. \tag{5.3.27}$$

Moreover,

$$\dim \mathcal{N}(A^*) = \text{codim } \Re(A) \quad \textit{and} \quad \dim \mathcal{N}(A) = \text{codim } \Re(A^*)\,.$$

5.4 Gårding's Inequality for Boundary Value Problems

In order to apply the abstract existence results for sesquilinear forms in Section 5.2 to the solution of the boundary value problems defined in Section 5.1, we need to verify the Gårding inequality for the sesquilinear form corresponding to the variational formulation of the boundary value problem. This is by no means a simple task. It depends on the particular underlying Hilbert space which depends on the boundary conditions together with the partial differential operators involved. In the most desirable situation, if the partial differential operators satisfy some strong, restrictive ellipticity conditions, the Gårding inequality can be established on the whole space and every regular boundary value problem with such a differential operator allows a variational treatment based on the validity of Gårding's inequality.

We shall present first the simple case of second order scalar equations and systems and then briefly present the cases of one $2m$–th order equation.

5.4.1 Gårding's Inequality for Second Order Strongly Elliptic Equations in Ω

In this subsection we shall consider systems of the form (5.1.1) and begin with the scalar case $p = 1$.

The case of one scalar second order equation

For the Gårding inequality, let us confine to the case of one real equation, i.e. the $a_{jk}(x)$ are real–valued functions and $\Theta = 1$ in (3.6.2). Then the uniform strong ellipticity condition takes the form

$$\sum_{j,k=1}^{n} a_{jk}(x)\xi_j\xi_k \geq \alpha_0|\xi|^2 \text{ for all } \xi \in \mathbb{R}^n \tag{5.4.1}$$

with $\alpha_0 > 0$. Without loss of generality, we may also assume $a_{jk} = a_{kj}$. In this case, we can easily show that Gårding's inequality is satisfied on the whole space $H^1(\Omega)$.

Lemma 5.4.1. *Let P in* (5.1.1) *be a scalar second order differential operator with real coefficients satisfying* (5.4.1). *Then $a_\Omega(u,v)$ given by* (5.1.3) *is a continuous sesquilinear form on $\mathcal{H} = H^1(\Omega)$ and satisfies the Gårding inequality*

$$Re\{a_\Omega(v,v) + (Cv,v)_{H^1(\Omega)}\} \geq \alpha_0\|v\|^2_{H^1(\Omega)} \text{ for all } v \in H^1(\Omega)\,, \tag{5.4.2}$$

where the compact operator C is defined by

$$(Cu,v)_{H^1(\Omega)} := -\int\limits_\Omega \Big\{\sum_{j=1}^{n} b_j(x)\frac{\partial u}{\partial x_j}(x) + (c(x)-\alpha_0)u(x)\Big\}^\top \overline{v}(x)dx \tag{5.4.3}$$

via the Riesz representation theorem in $H^1(\Omega)$. *Moreover, for the scalar operator* P, (5.4.2) *remains valid if* a_Ω *is replaced by* $\widetilde{a}_\Omega$, *cf.* (5.1.28) *and a compact operator* $\widetilde{C} = C +$ *additional lower order terms due to the obvious modifications in* (5.1.28).

Proof: The continuity of $a_\Omega(u,v)$ given by (5.1.3), in $H^1(\Omega) \times H^1(\Omega)$ follows directly from the Cauchy–Schwarz inequality and the boundedness of the coefficients of P. By definition, let

$$b_\Omega(u,v) := a_\Omega(u,v) + (Cu,v)_{H^1(\Omega)}$$
$$= \int\limits_\Omega \sum_{j,k=1}^n \Big(a_{jk}\frac{\partial u}{\partial x_k}\Big)^\top \frac{\partial \overline{v}}{\partial x_j} dx + \alpha_0 \int\limits_\Omega u(x)^\top \overline{v}(x) dx$$

with C given by (5.4.3). As can be easily verified, $b_\Omega(u,v) = \overline{b_\Omega(v,u)}$ and, hence, $b_\Omega(v,v)$ is real. Then, applying (5.4.1) to the first terms on the right for $u = v$ yields the inequality

$$Re\, b_\Omega(v,v) = b_\Omega(v,v) \ge \alpha_0 \Big\{ \int\limits_\Omega \sum_{j=1}^n \Big|\frac{\partial v}{\partial x_j}\Big|^2 dx + \int\limits_\Omega |v|^2 dx \Big\} = \alpha_0 \|v\|^2_{H^1(\Omega)} \,.$$

What remains to be shown, is the compactness of the operator C. Since

$$(u, C^* v)_{H^1(\Omega)} = (Cu, v)_{H^1(\Omega)}$$

we find from (5.4.3)

$$|(u, C^* v)_{H^1(\Omega)}| \le c' \|u\|_{H^1(\Omega)} \, \|v\|_{L^2(\Omega)}$$

where $c' = \max_{x \in \overline{\Omega} \wedge j=1,\dots,n} \{|b_j(x)|, |C(x) - \alpha_0|\}$ which implies that C^* satisfies

$$\|C^* v\|_{H^1(\Omega)} \le c' \|v\|_{L^2(\Omega)} \,.$$

Hence, C^* maps $L^2(\Omega)$ into $H^1(\Omega)$ continuously. Since strong Lipschitz domains enjoy the uniform cone property, Rellich's lemma (4.1.31) implies compactness of $C^* : H^1(\Omega) \to H^1(\Omega)$. Therefore, C becomes compact, too.

The Gårding inequality with $\widetilde{a}_\Omega$ instead of a_Ω follows from (5.4.2) since the two principal parts differ only by the skew–symmetric term (5.1.28) due to (5.1.26), which cancel each other for $u = v$, and lower order terms. ∎

The case of second order systems

In the scalar case, Lemma 5.4.1 shows that Gårding's inequality is valid on the whole space. However, for the second order uniformly strongly elliptic system (5.1.1) with $p > 1$, Gårding's inequality remains valid only on the subspace $H^1_0(\Omega)$. The latter is the well–known Gårding's theorem which we state without proof.

Theorem 5.4.2. (Gårding [137]) (see also Fichera [121, p.365], Nečas [326, Théorème 7.3. p.185], Wloka [439, Theorem 19.2])

For a bounded domain $\Omega \subset \mathbb{R}^n$ and P having continuous coefficients in $\overline{\Omega}$, the uniform strong ellipticity (3.6.2) *with $\Theta = I$ is necessary and sufficient for the validity of Gårding's inequality on $H_0^1(\Omega)$:*

$$Re\{a_\Omega(v,v) + (Cv,v)_{H_0^1(\Omega)}\} \geq \alpha_0 \|v\|^2_{H^1(\Omega)} \quad \text{for all} \quad v \in H_0^1(\Omega) \,. \tag{5.4.4}$$

On the whole space $H^1(\Omega)$, however, the validity of Gårding's inequality is by no means trivial. We now collect some corresponding results.

Definition 5.4.1. *The system* (5.1.1) *of second order is called* very strongly elliptic *at the point $x \in \mathbb{R}^n$, if there exists $\alpha_0(x) > 0$ such that*

$$Re \sum_{j,k=1}^{n} \zeta_j^\top a_{jk}(x) \overline{\zeta_k} \geq \alpha_0(x) \sum_{\ell=1}^{n} |\zeta_\ell|^2 \quad \text{for all} \quad \zeta_\ell \in \mathbb{C}^p \tag{5.4.5}$$

(*see* Nečas [326, p.186] *and for real equations* Ladyženskaya [257, p.296]).

The system (5.1.1) is called *uniformly very strongly elliptic* in Ω if $\alpha_0(x) \geq \alpha_0 > 0$ for all $x \in \Omega$ with a constant α_0.

Theorem 5.4.3. *If the system* (5.1.1) *is uniformly very strongly elliptic in Ω then Gårding's inequalities*

$$Re\{a_\Omega(v,v) + (Cv,v)_{H^1(\Omega)}\} \geq \alpha_0 \|v\|^2_{H^1(\Omega)} \tag{5.4.6}$$

holds for all $v \in H^1(\Omega)$, i.e. on the whole space $H^1(\Omega)$.

$$Re\{\widetilde{a}_\Omega(v,v) + (\widetilde{C}v,v)_{H^1(\Omega)}\} \geq \alpha_0 \|v\|^2_{H^1(\Omega)} \tag{5.4.7}$$

holds for all $v \in H^1(\Omega)$ provided $a_{jk} = a_{kj} = a^*_{jk}$.

The proof of (5.4.6) can be found in Nečas [326, p.186].

The very strong ellipticity is only a sufficient condition for the validity of (5.4.6). For instance, the important Lamé system is strongly elliptic but not very strongly elliptic. Nevertheless, as will be shown, here we still have Gårding's inequality on the whole space. This will lead to the *formally positive elliptic systems* which include the Lamé system in an isotropic medium as a particular case. To illustrate the idea, we begin with the general linear *system of elasticity in the form of the equations of equilibrium*,

$$(Pu)_j = -\sum_{k=1}^{n} \frac{\partial}{\partial x_k} \sigma_{jk}(x) = f_j(x) \quad \text{for} \quad j = 1, \cdots, n \,. \tag{5.4.8}$$

Here, $f_j(x)$ denotes the components of the given body force field and $\sigma_{jk}(x)$ is the stress tensor, related to the strain tensor

$$\varepsilon_{jk}(x) = \frac{1}{2}\Big(\frac{\partial u_j}{\partial x_k} + \frac{\partial u_k}{\partial x_j}\Big) \tag{5.4.9}$$

via Hooke's law

$$\sigma_{jk}(x) = \sum_{\ell,m=1}^{n} c_{jk\ell m}(x)\varepsilon_{\ell m}(x)\,. \tag{5.4.10}$$

The *elasticity tensor* $c_{jk\ell m}$ has the symmetry properties

$$c_{jk\ell m} = c_{kj\ell m} = c_{jkm\ell} = c_{m\ell jk}\,. \tag{5.4.11}$$

where $c_{jk\ell m}$ are called elastic constants. If the elastic constants do not depend on the position in the medium, the medium is called homogeneous. Otherwise the medium is called inhomogeneous

The stress and strain components and the elastic constants, related through Hook's law, depend on the orientation of the coordinate axes. The mediumis said to be isotropic, if its elastic constants $c_{jk\ell m}$ do not depend on the orientation of the coordinate axes or, in other words, if the elastic properties of the medium are the same in all directions. If the medium is not isotropic, it is called anisotropic. Therefore, in the case $n = 3$ or $n = 2$ there are only 21 or 6 different entries, respectively, the *elasticities* which define the symmetric positive definite $3(n-1) \times 3(n-1)$ *Sommerfeld matrix*

$$\widetilde{C}(x) = \begin{pmatrix} c_{1111} & c_{1122} & c_{1133} & c_{1123} & c_{1131} & c_{1112} \\ \cdot & c_{2222} & c_{2233} & c_{2223} & c_{2231} & c_{2212} \\ \cdot & \cdot & c_{3333} & c_{3323} & c_{3331} & c_{3312} \\ \cdot & \cdot & \cdot & c_{2323} & c_{2331} & c_{2312} \\ \cdot & \cdot & \cdot & \cdot & c_{3131} & c_{3112} \\ \cdot & \cdot & \cdot & \cdot & \cdot & c_{1212} \end{pmatrix} \quad \text{for } n = 3\,,$$

and

$$\widetilde{C}(x) = \begin{pmatrix} c_{1111} & c_{1122} & c_{1112} \\ \cdot & c_{2222} & c_{2212} \\ \cdot & \cdot & c_{1212} \end{pmatrix} \quad \text{for } n = 2\,.$$

For the *isotropic material*, the elasticities are explicitly given by

$$\begin{array}{lllllll} c_{1111} & = & c_{2222} & = & c_{3333} & = & 2\mu + \lambda\,, \\ c_{2323} & = & c_{3131} & = & c_{1212} & = & \mu\,, \\ c_{1122} & = & c_{1133} & = & c_{2233} & = & \lambda \text{ and} \\ c_{jkm\ell} & = & 0 & & \text{for all remaining indices} & . & \end{array}$$

Following Leis [263], we introduce the matrix $\mathcal{G}$ of differential operators generalizing the Nabla operator:

$$\mathcal{G} := \begin{pmatrix} \frac{\partial}{\partial x_1} & 0 & 0 & 0 & \frac{\partial}{\partial x_3} & \frac{\partial}{\partial x_2} \\ 0 & \frac{\partial}{\partial x_2} & 0 & \frac{\partial}{\partial x_3} & 0 & \frac{\partial}{\partial x_1} \\ 0 & 0 & \frac{\partial}{\partial x_3} & \frac{\partial}{\partial x_2} & \frac{\partial}{\partial x_1} & 0 \end{pmatrix}^{\top} \quad \text{for } n = 3\,,$$

whereas for $n = 2$, delete the columns containing $\frac{\partial}{\partial x_3}$ and the last row.

For convenience, we rewrite $\mathcal{G}$ in the form

$$\mathcal{G} = \sum_{k=1}^{n} \mathcal{G}_k \frac{\partial}{\partial x_k}$$

where, for each fixed $k = 1, \ldots, n$, $\mathcal{G}_k = ((\mathcal{G}_{\ell mk}))_{\substack{\ell=1,\ldots,3(n-1)\\ m=1,\cdots,n}}$ is a constant $3(n-1) \times n$ matrix of the same form as $\mathcal{G}$. In terms of the Sommerfeld matrix $\widetilde{C}$ and of $\mathcal{G}$, the equations of equilibrium (5.4.8) can be rewritten in the form

$$Pu = -\sum_{j,k=1}^{n} \frac{\partial}{\partial x_j}\left(a_{jk}\frac{\partial u}{\partial x_k}\right) := -\mathcal{G}^\top \widetilde{C}\mathcal{G}u = -\sum_{j,k=1}^{n} \frac{\partial}{\partial x_j}\mathcal{G}_j^\top \widetilde{C}\mathcal{G}_k \frac{\partial}{\partial x_k} u = f\,. \tag{5.4.12}$$

Then (5.1.2) reads

$$\int\limits_{\Omega} (Pu)^\top \bar{v} dx = a_\Omega(u,v) - \int\limits_{\Gamma} (\partial_\nu u)^\top \bar{v} ds$$

where the sesquilinear form a_Ω is given by

$$a_\Omega(u,v) = \int\limits_{\Omega} \Big(\sum_{j=1}^{n} \mathcal{G}_j \frac{\partial u}{\partial x_j}\Big)^\top \widetilde{C}\Big(\sum_{k=1}^{n} \mathcal{G}_k \frac{\partial \bar{v}}{\partial x_k}\Big) dx \tag{5.4.13}$$

and the boundary traction is the corresponding conormal derivative,

$$\partial_\nu u = \sum_{j,k=1}^{n} n_j \mathcal{G}_j^\top \widetilde{C}\mathcal{G}_k \frac{\partial u}{\partial x_k} = \Big(\sum_{k=1}^{n} n_k \sigma_{jk}\Big)_{j=1,\cdots,n}. \tag{5.4.14}$$

In this sesquilinear form, the elasticities $\widetilde{C}(x)$ form a symmetric $3(n-1) \times 3(n-1)$ matrix, which is assumed to be uniformly positive definite.

Consequently, we have

$$a_\Omega(u,u) \geq \int\limits_{\Omega} \alpha_0(x) \Big|\sum_{k=1}^{n} \mathcal{G}_k \frac{\partial u}{\partial x_k}\Big|^2 dx \geq \alpha_0 \int\limits_{\Omega} |\mathcal{G}u|^2_{\mathbb{R}^{3(n-1)}} dx$$

with a positive constant α_0, and where $|\cdot|_{\mathbb{R}^{3(n-1)}}$ denotes the $3(n-1)$–Euclidean norm. A simple computation shows

$$|\mathcal{G}u|^2_{\mathbb{R}^{3(n-1)}} = \sum_{j=1}^{n} \left|\frac{\partial u_j}{\partial x_j}\right|^2 + \sum_{\substack{j,k=1\\ j\neq k}}^{n} \left|\frac{\partial u_j}{\partial x_k} + \frac{\partial u_k}{\partial x_j}\right|^2 \geq \sum_{j,k=1}^{n} \varepsilon_{jk}^2\,.$$

Hence,

$$a_\Omega(u,u) \geq \alpha_0 \int_\Omega \sum_{j,k=1}^{n} \varepsilon_{jk}(x)^2 dx\,, \tag{5.4.15}$$

and Gårding's inequality in the form (5.4.6) on the whole space $H^1(\Omega)$ follows from the celebrated Korn's inequality given in the Lemma below, in combination with the compactness of the imbedding $H^1(\Omega) \hookrightarrow L^2(\Omega)$.

Lemma 5.4.4. (Korn's inequality) *For a bounded strong Lipschitz domain Ω there exists a constant $c > 0$ such that the inequality*

$$\int_\Omega \sum_{j,k=1}^{n} \varepsilon_{jk}(x)^2 dx \geq c\|u\|^2_{H^1(\Omega)} - \|u\|^2_{L^2(\Omega)} \tag{5.4.16}$$

holds for all $u \in H^1(\Omega)$.

The proof of Korn's inequality (5.4.16) is by no means trivial. A direct proof of (5.4.16) is given in Fichera [121], see also Ciarlet and Ciarlet [69], Kondratiev and Oleinik [236] and Nitsche [338]. However, as we mentioned previously, the elasticity system belongs to the class of systems called formally positive elliptic systems. For this class, under additional assumptions, one can establish Gårding's inequality on the whole space $H^1(\Omega)$ by a different approach which then also implies Korn's inequality.

Definition 5.4.2. (Nečas [326, Section 3.7.4], Schechter [375], Vishik [426]) *The system* (5.1.1) *is called formally positive elliptic if the coefficients of the principal part can be written as*

$$a_{jk}(x) = \sum_{r=1}^{t} \bar{\ell}_{rj}(x)\ell^\top_{rk}(x) \quad \textit{for } j,k = 1,\cdots,n \tag{5.4.17}$$

with $t \geq p$ complex vector–valued functions $\ell_{rk} = \big(\ell_{rk1}(x),\cdots,\ell_{rkp}(x)\big)^\top$ which satisfy the condition

$$\sum_{r=1}^{t}\Big|\sum_{j=1}^{n} \ell_{rj}(x)\xi_j\Big| \neq 0 \quad \textit{for all } 0 \neq \xi = (\xi_1,\cdots,\xi_n) \in \mathbb{R}^n \tag{5.4.18}$$

which is referred to as the collective ellipticity Schechter [375]. *Note that*

$$a_{jk} = a^*_{kj} \tag{5.4.19}$$

We now state the following theorem.

Theorem 5.4.5. (Nečas [326, Théorème 3.7.7]) *Let Ω be a bounded, strong Lipschitz domain and let the system* (5.1.1) *be formally positive elliptic with coefficients $\ell_{rj}(x)$ in* (5.4.17) *satisfying the additional rank condition:*

$$\operatorname{rank}\Big(\sum_{j=1}^{n} \ell_{rj}\xi_j\Big)_{r=1,\cdots,t} = p \quad \textit{for every } \xi = (\xi_1,\cdots,\xi_n) \in \mathbb{C}^n \setminus \{0\}\,. \tag{5.4.20}$$

Then the sesquilinear form a_Ω *in* (5.1.3) *as well as* $\widetilde{a}_\Omega$ *in* (5.1.28) *satisfy Gårding's inequalities* (5.4.6) *and* (5.4.7), *respectively, on the whole space* $H^1(\Omega)$.

We note that the rank condition (5.4.20) always implies the collective ellipticity condition (5.4.18); the converse, however, is not true in general. The Gårding inequality for $\widetilde{a}_\Omega$ follows from that for a_Ω with the same arguments as before in Theorem 5.4.3. The proof of Theorem 5.4.5 relies on the following crucial a–priori estimate:

$$\|u\|^2_{H^1(\Omega)} \;\le\; c\Big\{\|u\|^2_{L^2(\Omega)} + \sum_{r=1}^{t}\|\mathcal{L}_r u\|^2_{L^2(\Omega)}\Big\} \tag{5.4.21}$$

where

$$\mathcal{L}_r u \;=\; \sum_{k=1}^{n} \ell^{\top}_{rk}\frac{\partial u}{\partial x_k}\,,$$

which holds under the assumptions of Theorem 5.4.5 for the collectively elliptic system $\mathcal{L}_r$. For its derivation see Nečas [326, Théorème 3.7.6].

For the **elasticity system** (5.4.12) we now show that it satisfies all the above assumptions of Theorem 5.4.5. We note that the Sommerfeld matrix $\widetilde{C}(x)$ admits a unique square root which is again a $3(n-1)\times 3(n-1)$ positive definite symmetric matrix which will be denoted by

$$\widetilde{C}^{\frac12}(x) = ((\widetilde{c}^{(\frac12)}_{rq}))_{3(n-1)\times 3(n-1)}\,.$$

In particular, for isotropic materials we have explicitly

$$\widetilde{C}^{\frac12} \;=\; \frac{1}{3}\begin{pmatrix} \zeta+\eta & \zeta+\eta & \zeta+\eta & 0 & 0 & 0 & 0\\ \cdot & \zeta-\eta & \zeta-\eta & 0 & 0 & 0 & 0\\ \cdot & \cdot & \zeta-\eta & 0 & 0 & 0 & 0\\ \cdot & \cdot & \cdot & 3\sqrt{\mu} & 0 & 0 & 0\\ \cdot & \cdot & \cdot & \cdot & \cdot & 3\sqrt{\mu} & 0\\ \cdot & \cdot & \cdot & \cdot & \cdot & 0 & 3\sqrt{\mu} \end{pmatrix} \tag{5.4.22}$$

$$\text{for } n=3 \quad \text{with } \zeta=\sqrt{2\mu+n\lambda} \quad \text{and } \eta=\sqrt{2\mu}\,,$$

and where, for $n=2$ the rows and columns with $r,q = 3, 5$ and 6 are to be deleted. In terms of $\widetilde{C}^{\frac12}$, the system (5.4.12) takes the form

$$-\sum_{j,k=1}^{n}\frac{\partial}{\partial x_j}\Big(a_{jk}(x)\frac{\partial u}{\partial x_k}\Big) = -\sum_{j,k=1}^{n}\frac{\partial}{\partial x_j}\Big((\widetilde{C}^{\frac12}\mathcal{G}_j)^{\top}(\widetilde{C}^{\frac12}\mathcal{G}_k)\frac{\partial u}{\partial x_k}\Big) = f\,.$$

Then the product of the two matrices can also be written as

$$(\widetilde{C}^{\frac{1}{2}}(x)\mathcal{G}_j)^\top(\widetilde{C}^{\frac{1}{2}}(x)\mathcal{G}_k) = \sum_{r=1}^{3(n-1)} \ell_{rj}(x)\ell_{rk}^\top(x) = a_{jk}(x) \tag{5.4.23}$$

with the real valued column vector fields $\ell_{rj}(x)$ given by

$$\mathcal{G}_j^\top \widetilde{C}^{\frac{1}{2}}(x) = (\ell_{1j}(x), \cdots, \ell_{3(n-1),j}(x)) \quad \text{for } j = 1, \cdots, n\,.$$

In order to apply Theorem 5.4.5 to the elasticity system it remains to verify the rank condition (5.4.20) which will be shown by contradiction. Then (5.4.20) implies condition (5.4.18).

Suppose that the rank is less than $p = n$. Then there existed some complex vector $\xi = (\xi_1, \cdots, \xi_n) \neq 0$ and $\sum_{j=1}^n \ell_{rj}\xi_j = 0$ for all $r = 1, \cdots, 3(n-1)$. Then, clearly,

$$\sum_{r=1}^{3(n-1)} \sum_{j,k=1}^{n} (\ell_{rj}(x)\ell_{rk}^\top(x))\xi_j\bar{\xi}_k = 0$$

which would imply with (5.4.23) that the $n \times n$ matrix equation

$$\sum_{j,k=1}^{n} (\mathcal{G}_j\xi_j)^\top \widetilde{C}(x)(\mathcal{G}_k\bar{\xi}_k) = 0$$

holds and, hence, the positive definiteness of $\widetilde{C}(x)$ would imply

$$0 = \sum_{k=1}^{n} \mathcal{G}_k\bar{\xi}_k = \begin{cases} \begin{pmatrix} \bar{\xi}_1 & 0 & 0 & 0 & \bar{\xi}_3 & \bar{\xi}_2 \\ 0 & \bar{\xi}_2 & 0 & \bar{\xi}_3 & 0 & \bar{\xi}_1 \\ 0 & 0 & \bar{\xi}_3 & \bar{\xi}_2 & \bar{\xi}_1 & 0 \end{pmatrix}^\top & \text{for } n = 3\,, \\ \begin{pmatrix} \bar{\xi}_1 & 0 & \bar{\xi}_2 \\ 0 & \bar{\xi}_2 & \bar{\xi}_1 \end{pmatrix}^\top & \text{for } n = 2\,. \end{cases}$$

The latter yields $\xi = (\xi_1, \cdots, \xi_n) = 0$ which contradicts our supposition $\xi \neq 0$. Hence, the rank equals $p = n$. This completes the verification of all the assumptions required for the elasticity system to provide Gårding's inequality on $H^1(\Omega)$ for the associated sesquilinear form (5.4.13).

5.4.2 The Stokes System

The Stokes system (2.3.1) does not belong to the class of second order systems, elliptic in the sense of Petrovskii but nevertheless, in Section 2.3, Tables 2.3.3 and 2.3.4, we formulated boundary integral equations of various kinds for the hydrodynamic Dirichlet and traction problems. Here we formulate corresponding bilinear variational problems in the domain Ω and appropriate Gårding inequalities based on the first Green's identity for the Stokes system (see Dautray and Lions [95], Galdi [136], Kohr and Pop [233], Ladyženskaya [256], Temam [417], Varnhorn [424]).

Find $(\boldsymbol{u},p) \in H^1(\Omega) \times L^2(\Omega)$ *such that*

$$a_\Omega(\boldsymbol{u},\boldsymbol{v}) - b(p,\boldsymbol{v}) + b^\top(\boldsymbol{u},q) = \int\limits_\Omega \boldsymbol{f} \cdot \boldsymbol{v} dx + \int\limits_\Gamma T(\boldsymbol{u}) \cdot \boldsymbol{v} ds \tag{5.4.24}$$

is satisfied for $(\boldsymbol{v},q) \in H^1(\Omega) \times L^2(\Omega)$.

Here the bilinear forms in (5.4.24) are defined as

$$\begin{aligned} a_\Omega(\boldsymbol{u},\boldsymbol{v}) &:= \mu \sum_{j=1}^n \int\limits_\Omega \nabla u_j \cdot \nabla v_j dx + \mu \sum_{j,k=1}^n \int\limits_\Omega \frac{\partial u_j}{\partial x_k} \frac{\partial v_k}{\partial x_j} dx\,, \\ b(p,\boldsymbol{v}) &:= \int\limits_\Omega p \nabla \cdot \boldsymbol{v} dx\,,\; b^\top(\boldsymbol{u},q) := \int\limits_\Omega (\nabla \cdot \boldsymbol{u}) q dx\,, \end{aligned} \tag{5.4.25}$$

and μ is the fluid's dynamic viscosity.
Clearly, a_Ω is continuous on $H^1(\Omega) \times H^1(\Omega)$ and b on $L^2(\Omega) \times H^1(\Omega)$. In addition to (5.4.24) one has to append boundary conditions on Γ such as, e.g., the Dirichlet or traction conditions, and to specify the trial and test fields accordingly.

Since the bilinear form a_Ω can also be written as

$$a_\Omega(\boldsymbol{u},\boldsymbol{v}) = 2\mu \sum_{j,k=1}^n \int\limits_\Omega \varepsilon_{jk}(\boldsymbol{u}) \varepsilon_{jk}(\boldsymbol{v}) dx \tag{5.4.26}$$

with the strain tensor ε_{jk} given by (5.4.9), the Korn inequality (5.4.16) yields the Gårding inequality:

$$a_\Omega(\boldsymbol{v},\boldsymbol{v}) \geq c\|\boldsymbol{v}\|^2_{H^1(\Omega)} - (C\boldsymbol{v},\boldsymbol{v})_{H^1(\Omega)} \quad \text{for all } \boldsymbol{v} \in H^1(\Omega) \tag{5.4.27}$$

where $(C\boldsymbol{v},\boldsymbol{u})_{H^1(\Omega)} = \mu(\boldsymbol{v},\boldsymbol{u})_{L^2(\Omega)}$ and C is compact in $H^1(\Omega)$ due to the Rellich Lemma (4.1.31). Hence, (5.4.24) is to be treated as a typical saddle point problem (Brezzi and Fortin [45, II Theorem11]).

In order to formulate the boundary condition with the weak formulation (5.4.24) properly we again need an extension of the mapping (2.3.6), $(\boldsymbol{u},p) \mapsto T(\boldsymbol{u})$ to an appropriate subspace of $H^1(\Omega) \times L^2(\Omega)$ as in Lemma 5.1.1, and therefore define

$$H^1(\Omega, P_{st}) := \{(\boldsymbol{u},p) \in H^1(\Omega) \times L^2(\Omega) \mid \nabla p - \mu\Delta p \in \widetilde{H}_0^{-1}(\Omega)\} \tag{5.4.28}$$

equipped with the graph norm

$$\|(\boldsymbol{u},p)\|_{H^1(\Omega,P_{st})} := \|\boldsymbol{u}\|_{H^1(\Omega)} + \|p\|_{L^2(\Omega)} + \|\nabla p - \mu\Delta\boldsymbol{u}\|_{\widetilde{H}^{-1}(\Omega)}\,. \tag{5.4.29}$$

Lemma 5.4.6. *For given fixed $(\boldsymbol{u},p) \in H^1(\Omega, P_{st})$, the linear mapping*

$$\boldsymbol{v} \mapsto \int_\Gamma \boldsymbol{v}\cdot T(\boldsymbol{u})ds := a_\Omega(\mathcal{Z}\boldsymbol{v},\boldsymbol{u}) - b^\top(\mathcal{Z}\boldsymbol{v},p) - \int_\Omega \mathcal{Z}\boldsymbol{v}\cdot(\nabla p - \Delta\boldsymbol{u})dx \quad (5.4.30)$$

for $\boldsymbol{v} \in H^{\frac{1}{2}}(\Gamma)$ defines a continuous linear functional $T(\boldsymbol{u}) \in H^{-\frac{1}{2}}(\Gamma)$ where $\mathcal{Z}$ is a right inverse to the trace operator γ_0 in (4.2.41). The linear mapping $(\boldsymbol{u},p) \mapsto T(\boldsymbol{u})$ from $H^1(\Omega, P_{st})$ into $H^{-\frac{1}{2}}(\Gamma)$ is continuous, and there holds

$$\int_\Omega (\nabla p - \Delta\boldsymbol{u})\cdot\boldsymbol{v}dx + \int_\Gamma T(\boldsymbol{u})\cdot\boldsymbol{v}ds = a_\Omega(\boldsymbol{u},\boldsymbol{v}) - b(p,\boldsymbol{v}) \quad (5.4.31)$$
$$\textit{for } (\boldsymbol{u},p) \in H^1(\Omega, P_{st}) \textit{ and } \boldsymbol{v} \in H^1(\Omega).$$

Hence, $T(\boldsymbol{u})$ is an extension of (2.3.6) to $H^1(\Omega, P_{st})$.

The Interior Dirichlet Problem for the Stokes System

Here $\boldsymbol{f} \in \widetilde{H}_0^{-1}(\Omega)$ and $\boldsymbol{\varphi} \in H^{\frac{1}{2}}(\Gamma)$ are given where $\boldsymbol{\varphi}$ satisfies the necessary compatibility conditions

$$\int_{\Gamma_\ell} \boldsymbol{\varphi}\cdot\boldsymbol{n}_\ell ds = 0\,,\ \ell = 1,\ldots,\widetilde{L}\,. \quad (5.4.32)$$

Then we solve for the desired solenoidal field

$$\boldsymbol{u} \in H^1_{\text{div}}(\Omega) := \{\boldsymbol{w} \in H^1(\Omega)\,|\,\text{div}\boldsymbol{w} = 0 \text{ in } \Omega\} \quad (5.4.33)$$

the variational problem: *Find $\boldsymbol{u} \in H^1_{\text{div}}(\Omega)$ with $\boldsymbol{u}|_\Gamma = \boldsymbol{\varphi}$ as the solution of*

$$a_\Omega(\boldsymbol{u},\boldsymbol{v}) = \int_\Omega \boldsymbol{f}\cdot\boldsymbol{v}dx \quad \textit{for all } \boldsymbol{v} \in H^1_{0,\text{div}}(\Omega) := \{\boldsymbol{v} \in H^1_{\text{div}}(\Omega)\,|\,\boldsymbol{v}|_\Gamma = \boldsymbol{0}\}\,. \quad (5.4.34)$$

Since $H^1_{0,\text{div}}(\Omega)\dot{\subset}H^1_0(\Omega)\dot{\subset}H^1(\Omega)$ is a closed subspace of $H^1(\Omega)$ and a_Ω satisfies the Gårding inequality (5.4.27), for (5.4.34) Theorem 5.3.10 can be applied. Moreover, (5.4.34) is equivalent to the following variational problem with the Dirichlet bilinear form.

Lemma 5.4.7. *The variational problem* (5.4.34) *is equivalent to*
Find $\boldsymbol{u} \in H^1_{\text{div}}(\Omega)$ with $\boldsymbol{u}|_\Gamma = \boldsymbol{\varphi}$ of

$$\mu\sum_{j=1}^n \int_\Omega \nabla u_j\cdot\nabla v_j dx = \int_\Omega \boldsymbol{f}\cdot\boldsymbol{v}dx \quad \textit{for all } \boldsymbol{v} \in H^1_{0,\text{div}}(\Omega)\,. \quad (5.4.35)$$

The latter with $\varphi = \mathbf{0}$ only has the trivial solution. Hence, (5.4.35) as well as (5.4.34) have a unique solution, respectively.

Proof of Lemma 5.4.7: If $\boldsymbol{u} \in C^2(\Omega) \cap C^1(\overline{\Omega})$ with $\nabla \cdot \boldsymbol{u} = 0$ and $\boldsymbol{v} \in H^1_{0,\text{div}}(\Omega)$ then integration by parts yields

$$\begin{aligned} a_\Omega(\boldsymbol{u}, \boldsymbol{v}) &= \mu \sum_{j=1}^n \int_\Omega \nabla u_j \cdot \nabla v_j dx \\ &\quad + \mu \int_\Omega \Big\{ \sum_{j,k=1}^n \frac{\partial}{\partial x_j}\Big(\frac{\partial u_j}{\partial x_k} v_k\Big) - \sum_{k=1}^n \Big(\frac{\partial}{\partial x_k} \nabla \cdot \boldsymbol{u}\Big) v_k \Big\} dx \\ &= \mu \sum_{j=1}^n \int_\Omega \nabla u_j \cdot \nabla v_j dx + \mu \int_\Gamma \frac{\partial u_j}{\partial x_k} v_k n_k ds \\ &= \mu \sum_{j=1}^n \int_\Omega \nabla u_j \cdot \nabla v_j dx \end{aligned}$$

since $\boldsymbol{v}|_\Gamma = \mathbf{0}$. Since the bilinear forms on the left and the right hand side are continuous on $H^1_{\text{div}}(\Omega)$, by completion we obtain the proposed equality also for $\boldsymbol{u} \in H^1_{\text{div}}(\Omega)$. ■

With $\boldsymbol{u} \in H^1_{\text{div}}(\Omega)$ available we obtain $T(\boldsymbol{u})$ from (5.4.31), i.e.,

$$\int_\Gamma T(\boldsymbol{u}) \cdot \boldsymbol{v} ds = \int_\Omega \boldsymbol{f} \cdot \boldsymbol{v} ds + a_\Omega(\boldsymbol{u}, \boldsymbol{v}) \quad \text{for all } \boldsymbol{v} \in H^1_{\text{div}}(\Omega) \tag{5.4.36}$$

defines $T(\boldsymbol{u}) \in H^{-\frac{1}{2}}(\Gamma)$. Then the pressure p can be obtained from (2.3.12). The corresponding mapping properties of the potentials in (2.3.12) will be shown later on.

Clearly, for the solution we obtain the *a priori estimate*

$$\|\boldsymbol{u}\|_{H^1(\Omega)} \leq c\{\|\varphi\|_{H^{\frac{1}{2}}(\Gamma)} + \|\boldsymbol{f}\|_{\widetilde{H}^{-1}(\Omega)}\} . \tag{5.4.37}$$

The Interior Neumann Problem for the Stokes System

For the Neumann problem $\boldsymbol{f} \in \widetilde{H}_0^{-1}(\Omega)$ and $\boldsymbol{\psi} = T(\boldsymbol{u}) \in H^{-\frac{1}{2}}(\Gamma)$ are given satisfying the necessary compatibility conditions

$$\int_\Omega \boldsymbol{f} \cdot \boldsymbol{m}_k dx + \int_\Gamma \boldsymbol{\psi} \cdot \boldsymbol{m}_k ds = \mathbf{0} , \quad k = \overline{1, 3(n-1)} \tag{5.4.38}$$

where the $\boldsymbol{m}_k$ are the rigid motion basis in Ω. The desired solenoidal field $\boldsymbol{u} \in H^1_{\text{div}}(\Omega)$ will be obtained as the solution of the variational problem:

Find $\boldsymbol{u} \in H^1_{\text{div}}(\Omega)$ *satisfying*

$$a_\Omega(\boldsymbol{u},\boldsymbol{v}) = \int\limits_\Omega \boldsymbol{f}\cdot\boldsymbol{v}dx + \int\limits_\Gamma \boldsymbol{\psi}\cdot\boldsymbol{v}ds \quad \textit{for all } \boldsymbol{v}\in H^1_{\text{div}}(\Omega)\,. \tag{5.4.39}$$

Since $\Re = \text{span}\,\{\boldsymbol{m}_k\}$ is the null–space of $a_\Omega(\boldsymbol{u},\boldsymbol{v})$ we modify (5.4.38) to get the stabilized uniquely solvable bilinear equation:

$$\begin{aligned} &a_\Omega(\boldsymbol{u}_p,\boldsymbol{v}) + \sum_{k=1}^{3(n-1)} (\boldsymbol{u}_p,\boldsymbol{m}_k)_{L^2(\Omega)}\boldsymbol{m}_k \\ &= \int\limits_\Omega \boldsymbol{f}\cdot\boldsymbol{v}dx + \int\limits_\Gamma \boldsymbol{\psi}\cdot\boldsymbol{v}ds \quad \text{for all } \boldsymbol{v}\in H^1_{\text{div}}(\Omega)\,. \end{aligned} \tag{5.4.40}$$

The particular solution $\boldsymbol{u}_p$ of (5.4.40) is uniquely determined whereas the general solution then has the form

$$\boldsymbol{u} = \boldsymbol{u}_p + \sum_{k=1}^{3(n-1)} \boldsymbol{\alpha}_k\boldsymbol{m}_k \quad \text{for } n=2 \text{ or } 3\,,$$

with arbitrary $\boldsymbol{\alpha}_k \in R$. For $\boldsymbol{u}_p$ we have the *a priori estimate*

$$\|\boldsymbol{u}_p\|_{H^1(\Omega)} \le c\{\|\boldsymbol{\varphi}\|_{H^{-\frac{1}{2}}(\Gamma)} + \|\boldsymbol{f}\|_{\widetilde{H}^{-1}(\Omega)}\}\,. \tag{5.4.41}$$

With $\boldsymbol{u}$ and $\boldsymbol{\psi} = T(\boldsymbol{u})$ available, the pressure p also here can be calculated from (2.3.12).

5.4.3 Gårding's Inequality for Exterior Second Order Problems

Analogously to the interior problems, one may establish all properties for the sesquilinear form $a_{\Omega^c}(u,v)$ on $\mathcal{H}_E\times\mathcal{H}_E$ for later application of existence theory.

First we need the following inequalities between the exterior Dirichlet integrals and the $\mathcal{H}_E(\Omega^c)$–norm for the operator P. Here we require the conditions

$$b_j(x) = 0 \quad \text{for all } x\in\Omega^c \tag{5.4.42}$$

in addition to the assumptions (5.1.36).

Lemma 5.4.8. *Let P in* (5.1.1) *be uniformly strongly elliptic in Ω^c with coefficients satisfying the conditions* (5.1.36), (5.4.42). *Then the inequality*

$$\|\nabla v\|^2_{L^2(\Omega^c)} \le c\|v\|^2_{\mathcal{H}_{E,0}} \tag{5.4.43}$$

holds for all $v\in\mathcal{H}_{E,0}(\Omega^c)$.

In the scalar case, i.e., $p=1$ then also

$$\|\nabla v\|^2_{L^2(\Omega^c)} + \|\gamma_{c0}v\|^2_{L^2(\Gamma)} \le c\|v\|^2_{\mathcal{H}_{E,0}} \tag{5.4.44}$$

holds for all $v\in\mathcal{H}_E(\Omega^c)$. For a second order system, i.e., for $p>1$, (5.4.44) *holds for all $v\in\mathcal{H}_E(\Omega^c)$ provided P is very strongly elliptic or formally positive elliptic.*

Proof: For the proof of (5.4.43) note that the functions in $\mathcal{H}_{E,0}$ vanish on $\Gamma = \partial\Omega^c$ and can be extended by zero into Ω. The inequality then follows by Fourier transformation in $\mathbb{R}^n$ with the help of Parseval's equality and (3.6.2).

To establish inequality (5.4.44), we proceed as in Costabel et al [90] and introduce an auxiliary boundary $\Gamma_R : \{x \in \mathbb{R}^n \text{ with } |x| = R\}$ with R sufficiently large so that $|y| < R$ for all $y \in \Omega$. Let $\eta \in C^\infty(\mathbb{R}^n)$ be a cut–off function such that $0 \le \eta \le 1$, $\eta(x) = 1$ for $|x| \le R$ and $\eta(x) = 0$ for $|x| \ge 2R$. Define $\widetilde{v} := \eta v$ for $v \in \mathcal{H}_E$. Then, according to the corresponding assumptions on P, the application of Lemma 5.4.1 or Theorem 5.4.3 or Theorem 5.4.5 to the annular domains $\Omega^c_{jR} := \{x \in \Omega^c \,|\, |x| \le jR\}, j = 1, 2$, imply with the help of Lemma 5.2.5 the following inequalities:

There exist constants $\alpha_{01}, \alpha_{02} > 0$ such that

$$\alpha_{0j}\{\|\nabla v\|^2_{L^2(\Omega^c_{jR})} + \|\gamma_{c0} v\|^2_{L^2(\Gamma)}\} \le a_{\Omega^c_{jR}}(v, v) + \|\gamma_{c0} v\|^2_{L^2(\Gamma)} \tag{5.4.45}$$

for all $v \in H^1(\Omega^c_{jR})$ and for $j = 1$ and 2. Clearly, (5.4.45) yields for $\widetilde{v}$

$$\alpha_{02}\{\|\nabla \widetilde{v}\|^2_{L^2(\Omega^c)} + \|\gamma_{c0}\widetilde{v}\|^2_{L^2(\Gamma)}\} \le \|\widetilde{v}\|^2_{\mathcal{H}_E} \,. \tag{5.4.46}$$

Moreover, for $\widetilde{v}^* = (1-\eta)v = (v - \widetilde{v}) \in H^1_0(\Omega^c)$ we have $\nabla\widetilde{v}^* \in L_2(\Omega^c)$, and the Gårding inequality in the exterior is valid on the subspace $\mathcal{H}_{E,0}$,i.e.,

$$\alpha_{03}\|\nabla \widetilde{v}^*\|^2_{L^2(\Omega^c)} \le a_{\Omega^c}(\widetilde{v}^*, \widetilde{v}^*) \,. \tag{5.4.47}$$

From (5.4.46) and (5.4.47) we obtain

$$\begin{aligned}
&\alpha_{04}\{\|\nabla v\|^2_{L^2(\Omega^c)} + \|\gamma_{c0} v\|^2_{L^2(\Gamma)}\} \\
&\le \|\gamma_{c0} v\|^2_{L^2(\Gamma)} + \sum_{j,k=1}^n \int_{\Omega^c} \Big\{ \Big(a_{jk}\frac{\partial}{\partial x_k}(\eta v)\Big)^\top \Big(\frac{\partial}{\partial x_j}(\eta \bar{v})\Big) \\
&\qquad + \Big(a_{jk}\frac{\partial}{\partial x_k}(1-\eta)v\Big)^\top \Big(\frac{\partial}{\partial x_j}(1-\eta)\bar{v}\Big)\Big\} dx \\
&\le \|v\|^2_{\mathcal{H}_E} + c\|v\|^2_{H^1(\Omega^c_{2R}\setminus\Omega^c_R)} \,,
\end{aligned} \tag{5.4.48}$$

where $\alpha_{04} = \min\{\alpha_{02}, \alpha_{03}\} > 0$. To complete the proof of the inequality (5.4.44), we suppose the contrary. This implies the existence of a sequence $v_j \in \mathcal{H}_E(\Omega^c)$ with

$$\|\nabla v_j\|^2_{L^2(\Omega^c)} + \|\gamma_{c0} v_j\|^2_{L^2(\Gamma)} = 1 \quad \text{and} \quad \|v_j\|_{\mathcal{H}_E} \to 0 \tag{5.4.49}$$

which converges weakly in $H^1_{\text{loc}}(\Omega^c)$, i.e. there exists $v_0 \in H^1_{\text{loc}}(\Omega^c)$ and $v_j \rightharpoonup v_0$ in $H^1_{\text{loc}}(\Omega^c)$. With the compact imbedding $H^{\frac{1}{2}}(\Gamma) \hookrightarrow L^2(\Gamma)$, this yields

$$\|\gamma_{c0} v_j\|_{L^2(\Gamma)} \to 0 = \|\gamma_{c0} v_0\|_{L^2(\Gamma)}$$

due to (5.4.49). Then (5.4.45) for $j = 2$ yields with (5.4.49)

$$\alpha_{02}\{\|\nabla v_j\|^2_{L^2(\Omega^c_{2R})} + \|\gamma_{c0} v_j\|^2_{L^2(\Gamma)}\} \le \|v_j\|^2_{\mathcal{H}_E} \to 0$$

which implies that $\|v_j\|_{H^1(\Omega^c_{2R}\setminus\Omega^c_R)} \to 0$ together with $\|v_j\|_{\mathcal{H}_E} \to 0$. Hence, from (5.4.48) we find

$$\|\nabla v_j\|^2_{L^2(\Omega^c)} + \|\gamma_0 v_j\|^2_{L^2(\Gamma)} \to 0$$

which contradicts (5.4.49). Thus, (5.4.44) must hold. ■

Remark 5.4.1: From the proof we see that the result remains valid for the symmetrized sesquilinear form (5.1.62) since P is to be replaced by the self adjoint operator $\frac{1}{2}(P+P^*)$. For the latter, the additional conditions (5.4.42) are not required.

Lemma 5.4.9. *Under the assumptions of Lemma 5.4.8, the sesquilinear form $a_{\Omega^c}(u,v)$ is continuous on $\mathcal{H}_E$ and satisfies Gårding's inequality*

$$Re\{a_{\Omega^c}(v,v) + (Cv,v)_{\mathcal{H}_E}\} \ge \|v\|^2_{\mathcal{H}_E} \text{ for all } v \in \mathcal{H}_E(\Omega^c) \tag{5.4.50}$$

where C is defined by

$$(Cu,v)_{\mathcal{H}_E} := \int\limits_\Gamma u^\top \overline{v} ds - \sum_{j=1}^n \int\limits_{\Omega^c} \Big(b_j(x)\frac{\partial u}{\partial x_j}\Big)^\top \overline{v}(x)dx \tag{5.4.51}$$

via the Riesz representation theorem in $\mathcal{H}_E(\Omega^c)$ and is therefore a compact operator.

Proof:
i) Continuity: By definition (5.1.3) where Ω is replaced by Ω^c, assumptions (5.1.36) and (5.4.42) and the Schwarz inequality we have the estimate

$$|a_{\Omega^c}(u,v)| \le c\|\nabla u\|_{L^2(\Omega^c)}\{\|\nabla v\|_{L^2(\Omega^c)} + \|v\|_{L^2(\Omega^c_{R_0})}\}$$

where $\Omega^c_{R_0} = \{x \in \Omega^c \,\big|\, |x| < R_0\}$. The application of the following version of the Poincaré–Friedrichs inequality (4.1.40)

$$\|v\|_{L^2(\Omega^c_{R_0})} \le c(\Omega^c_{R_0})\{\|v\|_{L^2(\Gamma)} + \|\nabla v\|_{L^2(\Omega^c)}\},$$

together with (5.4.44) yields the desired estimate

$$|a_{\Omega^c}(u,v)| \le c\|u\|_{\mathcal{H}_E}\|v\|_{\mathcal{H}_E}.$$

ii) Gårding's inequality: The inequality (5.4.50) follows from the definition of C in (5.4.51) and the definition (5.1.60) of $(v,v)_{\mathcal{H}_E}$. It only remains to show that the mapping $C : \mathcal{H}_E \to \mathcal{H}_E$ is compact. To this end we consider the adjoint mapping C^* in the form

$$(Cu, v)_{\mathcal{H}_E} = (u, C^* v)_{\mathcal{H}_E} = (u, C_1^* v)_{\mathcal{H}_E} + (u, C_2^* v)_{\mathcal{H}_E}$$

where

$$(u, C_1^* v)_{\mathcal{H}_E} := \int\limits_{\Gamma} (\gamma_{c0} u)^\top \gamma_{c0} \overline{v} ds \,,$$

$$(u, C_2^* v)_{\mathcal{H}_E} := -\sum_{j=1}^{n} \int\limits_{\Omega^c_{R_0}} \Big(b_j(x) \frac{\partial u}{\partial x_j} \Big)^\top \overline{v}(x) dx \,.$$

The mapping C_1^* can be written as the composition of the mappings

$$\mathcal{H}_E(\Omega^c) \xrightarrow{\gamma_{c0}} H^{\frac{1}{2}}(\Gamma) \overset{i_c}{\hookrightarrow} L^2(\Gamma) \xrightarrow{\widetilde{C}_1^*} \mathcal{H}_E(\Omega^c)$$

where $\overset{i_c}{\hookrightarrow}$ denotes the compact imbedding i_c of the trace spaces and $\widetilde{C}_1^*$ is defined by the Riesz representation theorem from

$$(u, \widetilde{C}_1^* w)_{\mathcal{H}_E} := \int\limits_{\Gamma} (\gamma_{c0} u)^\top \overline{w} ds$$

with

$$|(u, \widetilde{C}_1^* w)|_{\mathcal{H}_E} \le \|\gamma_{c0} u\|_{L^2(\Gamma)} \|w\|_{L^2(\Gamma)} \le \|u\|_{\mathcal{H}_E} \|w\|_{L^2(\Gamma)} \,.$$

This implies the continuity of the mapping $\widetilde{C}_1^* : L^2(\Gamma) \to \mathcal{H}_E(\Omega^c)$. As a composition of continuous linear mappings with the compact mapping i_c, the mapping C_1^* is compact, and so is C_1.

The mapping C_2^* can be written as the composition of the mappings

$$\mathcal{H}_E(\Omega^c) \overset{v|_{\Omega^c_{R_0}}}{\longrightarrow} H^1(\Omega^c_{R_0}) \overset{i_c}{\hookrightarrow} L^2(\Omega^c_{R_0}) \xrightarrow{\widetilde{C}_2^*} \mathcal{H}_E(\Omega^c)$$

where the restriction to $\Omega^c_{R_0} \subset \Omega^c$ is continuous due to the equivalence of the spaces $\mathcal{H}_E(\Omega^c_{R_0})$ and $H^1(\Omega^c_{R_0})$ because of the boundedness of the domain $\Omega^c_{R_0}$ (see Triebel [421, Satz 28.5]). For the same reason, the imbedding $\overset{i_c}{\hookrightarrow}$ is compact. The mapping $\widetilde{C}_2^* : L^2(\Omega^c_{R_0}) \to \mathcal{H}_E(\Omega^c)$ is given by

$$(u, \widetilde{C}_2^* w)_{\mathcal{H}_E} := -\sum_{j=1}^{n} \int\limits_{\Omega^c_{R_0}} \Big(b_j(x) \frac{\partial u}{\partial x_j} \Big)^\top \overline{w} dx$$

and satisfies

$$|(u, \widetilde{C}_2^* w)_{\mathcal{H}_E}| \le c \|u\|_{\mathcal{H}_E} \|w\|_{L^2(\Omega^c_{R_0})} \,;$$

which implies its continuity. Thus, C_2^* is compact due to the compactness of i_c, and so is C_2. ∎

Alternatively, one also may use the weighted Sobolev spaces $H^1(\Omega^c;0)$ and $H^1_0(\Omega^c;0)$ (see (4.4.1) – (4.4.3)) instead of $\mathcal{H}_E$ and $\mathcal{H}_{E,0}$, respectively, by following Giroire [148] and Nedelec [328]; see also Dautray and Lions [97]. However, the following shows that the formulation (5.1.65) and Lemmata 5.4.8 and 5.4.9 both remain valid if the function spaces $\mathcal{H}_E$ and $\mathcal{H}_{E,0}$ are replaced by $H^1(\Omega^c;0)$ and $H^1_0(\Omega^c;0)$, respectively.

Lemma 5.4.10. *The weighted Sobolev space* $H^1(\Omega^c;0)$ *(and* $H^1_0(\Omega^c;0)$*) with the scalar product* (4.4.2) *is equivalent to the energy space* $\mathcal{H}_E$ *(and* $\mathcal{H}_{E,0}$*) associated with* $a_{jk} = \delta_{jk}$ *and also to the energy space* $\mathcal{H}_E$ *(and* $\mathcal{H}_{E,0}$*) given with the norm* (5.1.60) *(and by* (5.1.63)*) for general* a_{jk} *belonging to a formally positive elliptic system of second order.*

Proof: For the equivalence we have to show that there exist two positive constants c_1 and c_2 such that

$$c_1\|v\|_{H^1(\Omega^c;0)} \le \|v\|_{\mathcal{H}_E} \le c_2\|v\|_{H^1(\Omega^c;0)} \text{ for all } v \in H^1(\Omega^c;0)\,. \qquad (5.4.52)$$

Since the coefficients a_{jk} in (5.1.60) are supposed to be uniformly bounded and also to satisfy the uniform strong ellipticity condition (3.6.2) with $\Theta = 1$, it suffices to prove the lemma for $a_{jk} = \delta_{jk}$.

i) We show the right inequality in (5.4.52).

$$\begin{aligned} \|v\|^2_{\mathcal{H}_E} &:= \|v\|^2_{L^2(\Gamma)} + \|\nabla v\|^2_{L^2(\Omega^c)} \\ &= \|v\|^2_{L^2(\Gamma)} + \|\nabla v\|_{L^2(\Omega^c_{R_0})} + \|\nabla v\|^2_{L^2(|x|\ge R_0)} \end{aligned}$$

Using (4.1.40) which states that $\{\|v\|_{L^2(\Gamma)} + \|\nabla v\|_{L^2(\Omega^c_{R_0})}\}^{\frac12}$ and $\|v\|_{H^1(\Omega^c_{R_0})}$ are equivalent in the *bounded* domain $\Omega^c_{R_0}$, we find with some positive constant c_{R_0}

$$\|v\|^2_{\mathcal{H}_E} \le c_{R_0}\|v\|^2_{H^1(\Omega^c_{R_0})} + \|\nabla v\|^2_{L^2(|x|\ge R_0)}\,.$$

Since for the weight functions in (4.4.1) we have

$$\varrho(|x|)\varrho_0(|x|) \le \varrho(R_0)\varrho_0(R_0) \text{ for all } |x| \le R_0\,,$$

we find

$$\|v\|^2_{\mathcal{H}_E} \le c_{R_0}\varrho^2(R_0)\varrho_0^2(R_0)\|v\|^2_{H^1(\Omega^c_{R_0};0)} + \|v\|^2_{H^1(\Omega^c;0)}\,,$$

which implies the right inequality in (5.4.52)).

ii) For the second inequality we use the Friedrichs–Poincaré inequality (4.4.12) for weighted Sobolev spaces in exterior domains,

$$\|v\|^2_{H^1(\Omega^c;0)} \le c_1\|\nabla v\|^2_{L^2(\Omega^c)} + c_{2R_0}\|v\|^2_{H^1(\Omega^c_{R_0})}\,.$$

Using (4.1.40) and (4.1.41), again we find

$$\|v\|^2_{H^1(\Omega^c;0)} \le c_1\|\nabla v\|^2_{L^2(\Omega^c)} + c_2\{\|v\|^2_{L^2(\Gamma)} + \|\nabla u\|^2_{L^2(\Omega^c_{R_0})}\}$$

which gives the left inequality in (5.4.52).

Since $H^1(\Omega^c;0)$ and $\mathcal{H}_E$ are complete as Hilbert spaces and both contain the same dense subset $C_0^\infty(\mathbb{R}^n)\oplus\mathbb{C}^p$, it follows from inequality (5.4.52) that both spaces are equivalent. ∎

5.4.4 Gårding's Inequality for Second Order Transmission Problems

For the transmission problems 1.1, 1.2 and 2.1–2.4 (i.e., (5.1.79), (5.1.80) and (5.1.84)–(5.1.87)) the corresponding sesquilinear form is given by

$$\widetilde{a}(u,v) =: \widetilde{a}_\Omega(u,v) + \widetilde{a}_{\Omega^c}(u,v) \tag{5.4.53}$$

on $u,v \in H^1(\Omega)\times\mathcal{H}_E(\Omega^c)$. By making use of the Gårding inequalities (5.4.6), (5.4.7) in the bounded domain Ω as well as (5.4.50) for the exterior Ω^c, we obtain Gårding's inequality for these transmission problems.

Corollary 5.4.11. *Let P given in* (5.1.1) *be either scalar strongly elliptic or very strongly elliptic or positive elliptic satisfying in Ω^c the additional assumptions in Lemma* 5.4.8 *(i.e.,* (5.1.36) *and* (5.4.42)*). Then the sesquilinear form $\widetilde{a}(u,v)$ of the transmission problems* 1.1, 1.2, 2.1–2.4 *is continuous on $H^1(\Omega)\times\mathcal{H}_E(\Omega^c)$ and satisfies Gårding's inequality*

$$\begin{aligned}\mathrm{Re}\Big\{\widetilde{a}(v,v) + (C_1 v,v)_{H^1(\Omega)} + ((C_2 v,v)_{\mathcal{H}_E(\Omega^c)} + \int_\Gamma v\bar{v}^\top ds\Big\}& \\ \ge \alpha_0\{\|v\|^2_{H^1(\Omega)} + \|v\|^2_{\mathcal{H}_E(\Omega^c)}\}&\end{aligned} \tag{5.4.54}$$

for all $v\in H^1(\Omega)\times\mathcal{H}_E(\Omega^c)$.

As for the remaining transmission problems 2.5, 2.6 the Gårding inequalities for the interior and exterior problems both are already available from the previous sections. For 2.5 and 2.6 they are needed for solving the interior and exterior problems consecutively.

If we replace Ω^c by a bounded domain $\Omega^c_R = \Omega^c\cap B_R$ with $B_R := \{x\in\mathbb{R}^n \,|\, |x|<R\}$ and R sufficiently large such that $\overline{\Omega}\subset B_R$ then we may consider the previous transmission problems in Ω and Ω^c_R by replacing the radiation conditions by appropriate boundary conditions on $\Gamma_R = \partial B_R$, e.g., homogeneous Dirichlet conditions $B\gamma_0 u = 0$ on Γ_R. Then Gårding's inequality (5.4.54) remains valid if $\mathcal{H}_E(\Omega^c)$ is replaced by $H^1(\Omega^c_R)$.

5.5 Existence of Solutions to Strongly Elliptic Boundary Value Problems

We are now in the position to summarize some existence results for the boundary value and transmission problems in Section 5.1. These results are based on the corresponding variational formulations together with Gårding inequalities which have been established in Section 5.4. The latter implies the validity of Fredholm's alternative, Theorem 5.3.10. Hence, uniqueness implies existence of solutions in appropriate energy spaces. In many cases, uniqueness needs to be assumed.

In particular, the transmission problems play a fundamental role for establishing the mapping properties of the boundary integral operators in Section 5.6. These mapping properties will be needed to derive Gårding inequalities for the boundary integral operators which are a consequence of the Gårding inequalities for the sesquilinear forms associated with the transmission problems.

5.5.1 Interior Boundary Value Problems

We begin with the existence theorem for the general Dirichlet problem (5.1.19) in variational form.

Theorem 5.5.1. *Given $f \in \widetilde{H}_0^{-1}(\Omega)$ and $\varphi \in H^{\frac{1}{2}}(\Gamma)$, there exists a unique solution $u \in H^1(\Omega)$ with $B_{00}\gamma_0 u_{|\Gamma} = \varphi$ such that*

$$a_\Omega(u, v) = \langle f, \overline{v}\rangle_\Omega \quad \textit{for all} \;\; v \in \mathcal{H} = H_0^1(\Omega)\,, \tag{5.5.1}$$

provided the homogeneous adjoint equation

$$a_\Omega(u_0, v) = 0 \quad \textit{for all} \;\; v \in \mathcal{H} = H_0^1(\Omega) \tag{5.5.2}$$

has only the trivial solution $u_0 \equiv 0$.

Proof: The proof is an obvious consequence of Fredholm's alternative, Theorem 5.3.10 together with Gårding's inequality (5.4.6) from Theorem 5.4.3, where $u = \Phi + w$ with some $\Phi \in H^1(\Omega)$ satisfying $B_{00}\gamma_0\Phi = \varphi$ on Γ and with $w \in \mathcal{H}$; thus

$$a(v, w) := \overline{a_\Omega(w, v)} \quad \text{and} \quad \ell(v) := \langle f, v\rangle_\Omega - a_\Omega(\Phi, \overline{v})\,.$$

■

In the case of those Dirichlet problems where uniqueness does not hold, as a consequence of Theorem 5.3.10 (the second part of Fredholm's alternative) we have the following existence theorem.

Theorem 5.5.2. *Under the same assumptions as for Theorem 5.5.1, there exists a solution $u \in H^1(\Omega)$ of (5.1.19) in the form*

$$u = u_p + \sum_{j=1}^{k} \alpha_j u_{0(j)} \tag{5.5.3}$$

where u_p is a particular solution of (5.5.1) *and* $u_{0(j)} \in H_0^1(\Omega)$ *are linearly independent eigensolutions of*

$$a_\Omega(u_{0(j)}, v) = 0 \quad \textit{for all} \;\; v \in H_0^1(\Omega)$$

provided the orthogonality conditions

$$\ell(v_{0(j)}) = \langle f, v_{0(j)}\rangle_\Omega - a_\Omega(\Phi, \overline{v_{0(j)}}) = 0 \tag{5.5.4}$$

for all eigensolutions $v_{0(j)} \in H_0^1(\Omega)$ *of*

$$a_\Omega(w, v_{0(j)}) = 0 \quad \textit{for all} \;\; w \in H_0^1(\Omega)\,, \tag{5.5.5}$$

are satisfied; $j = 1, \ldots, k < \infty$.

In order to characterize these eigensolutions, we need the adjoint differential operator and a generalized second Green's formula corresponding to (5.1.1). The standard second Green's formula in Sobolev spaces reads as

$$\int\limits_\Omega \{u^\top(P^*\overline{v}) - (Pu)^\top \overline{v}\} dx = \int\limits_\Gamma \{(\partial_\nu u)^\top \overline{v} - u^\top \overline{(\widetilde{\partial}_\nu v)}\} ds - \int\limits_\Gamma \sum_{j=1}^n (n_j b_j u)^\top \overline{v} ds\,, \tag{5.5.6}$$

provided $u, v \in H^2(\Omega)$. The formal adjoint P^* is defined by

$$P^* v = -\sum_{j,k=1}^n \frac{\partial}{\partial x_j}\Big(a^*_{jk}\frac{\partial v}{\partial x_k}\Big) - \sum_{j=1}^n \frac{\partial}{\partial x_j}(b^*_j v) + c^* v$$

whose conormal derivative is

$$\widetilde{\partial}_\nu v = \sum_{k,j=1}^n n_j a^*_{jk} \frac{\partial v}{\partial x_k}\,. \tag{5.5.7}$$

The generalized second Green's formula can be derived in terms of the mapping τ and its adjoint. More precisely, we have the following lemma.

Lemma 5.5.3 (Generalized second Green's formula). *For every pair* $u, v \in H^1(\Omega, P)$ *there holds*

$$\begin{aligned} &\int\limits_\Omega \{u^\top(\overline{P^* v}) - (Pu)^\top \overline{v}\} dx \\ &= \langle \tau u, \overline{\gamma_0 v}\rangle_\Gamma - \langle \gamma_0 u, \overline{\widetilde{\tau} v}\rangle_\Gamma - \langle (\boldsymbol{n}\cdot\boldsymbol{b})\gamma_0 u, \overline{\gamma_0 v}\rangle_\Gamma\,, \end{aligned} \tag{5.5.8}$$

where $\widetilde{\tau} v = \widetilde{\partial}_\nu v$ *for* $v \in C^\infty$*; its extension* $\widetilde{\tau}$ *to* $H^{-\frac12}(\Gamma)$ *is defined by completion in the same manner as for* τ.

Proof: The proof follows from Lemma 5.1.1 with the operator P and P^*, subtraction of the corresponding equations (5.1.8), and from additional application of the divergence theorem in $H^1(\Omega)$. ■

Clearly, the functions $u_{0(j)}$ in (5.5.3) are nontrivial solutions of the homogeneous Dirichlet problem. It is also easy to see that the eigensolutions $v_{0(j)}$ of (5.5.5) are distributional solutions of the adjoint homogeneous boundary value problem

$$P^* v_{0(j)} = 0 \quad \text{in } \Omega \quad \text{satisfying } B_{00}\gamma_0 v_{0(j)} = 0 \quad \text{on } \Gamma . \tag{5.5.9}$$

Moreover, the orthogonality conditions (5.5.4) can be rewritten in the form

$$\int\limits_\Omega f^\top \overline{v}_{0(j)} dx - \int\limits_\Gamma \varphi^\top \overline{\Big((B_{00}^{-1})^* \widetilde{\tau} v_{0(j)} \Big)} ds = 0 . \tag{5.5.10}$$

For the general interior Neumann problem (5.1.29) we also have the following existence theorem:

Theorem 5.5.4. *Let $\Gamma \in C^{1,1}$ and let the differential operator P belong to one of the following cases: P is a scalar strongly elliptic operator viz. (5.4.1) or a very strongly elliptic system of operators viz. (5.4.5) or a formally positive elliptic system of operators viz. (5.4.17), (5.4.18). Then, given $f \in \widetilde{H}_0^{-1}(\Omega)$ and $\psi \in H^{-\frac{1}{2}}(\Gamma)$, there exists a unique solution $u \in H^1(\Omega)$ satisfying (5.1.29), i. e.*

$$\widetilde{a}_\Omega(u, v) = \langle f, \overline{v}\rangle_\Omega + \langle \psi, \overline{S\gamma v}\rangle_\Gamma \quad \textit{for all } v \in H^1(\Omega) , \tag{5.5.11}$$

with the bilinear form $\widetilde{a}_\Omega$ given by (5.1.28) associated with the operator P and $S = B$, provided the homogeneous equation

$$\widetilde{a}_\Omega(u_0, v) = 0 \quad \textit{for all } v \in H^1(\Omega)$$

admits only the trivial solution $u_0 \equiv 0$.

This theorem is again a direct consequence of Fredholm's alternative, Theorem 5.3.10, and Gårding's inequality (5.4.2), (5.4.7) on $\mathcal{H} = H^1(\Omega)$ (Lemma 5.4.1 and Theorems 5.4.3 and 5.4.5).

It is worth mentioning that the variational solution u is also in $H^1(\Omega, P)$. Hence the mappings $\tau : H^1(\Omega, P) \ni u \mapsto \tau u \in H^{-\frac{1}{2}}(\Gamma)$ as well as $\widetilde{N}\gamma_0 u$: $H^1(\Omega, P) \ni u \mapsto \widetilde{N}\gamma_0 u \in H^{-\frac{1}{2}}(\Gamma)$ via (5.1.27) are continuous. Therefore (5.1.23) implies that $R\gamma u = N\gamma u \in H^{-\frac{1}{2}}(\Gamma)$ is well defined and (5.5.11) yields

$$\langle N\gamma u - \psi, \overline{B\gamma v}\rangle_\Gamma = 0$$

for all $v \in H^1(\Omega)$. Hence, the variational solution of (5.5.11) satisfies the general Neumann condition in the form

$$N\gamma u = \psi \quad \text{in} \quad H^{-\frac{1}{2}}(\Gamma)\,. \tag{5.5.12}$$

In the case of non–uniqueness, we have a theorem, analogous to Theorem 5.5.2.

Theorem 5.5.5. *Under the same assumptions as for Theorem* 5.5.4, *there exists a solution* $u \in H^1(\Omega)$ *of* (5.5.11) *in the form*

$$u = u_p + \sum_{j=1}^{k} \alpha_j u_{0(j)} \tag{5.5.13}$$

where u_p *is a particular solution of* (5.5.11) *and* $u_{0(j)} \in H^1(\Omega)$ *are linearly independent eigensolutions of*

$$\widetilde{a}_\Omega(u_{0(j)}, v) = 0 \quad \textit{for all} \;\; v \in H^1(\Omega) \tag{5.5.14}$$

provided the necessary orthogonality conditions for f *and* ψ

$$\ell(v_{0(j)}) = \langle v_{0(j)}, f\rangle_\Omega + \langle \psi, \overline{B\gamma v_{0(j)}}\rangle_\Gamma = 0 \tag{5.5.15}$$

with all eigensolutions $v_{0(j)} \in H^1(\Omega)$ *of*

$$\widetilde{a}_\Omega(w, v_{0(j)}) = 0 \quad \textit{for all} \;\; w \in H^1(\Omega) \tag{5.5.16}$$

are satisfied; $j = 1, \ldots, k < \infty$.

In order to gain more insight into the eigensolutions $u_{0(j)}$, we recall the variational equation (5.1.29), i. e.

$$\widetilde{a}_\Omega(u, v) = \langle Pu, \bar{v}\rangle_\Omega + \langle N\gamma u, \overline{B\gamma v}\rangle_\Gamma \quad \text{for all} \;\; v \in H^1(\Omega)\,. \tag{5.5.17}$$

Hence, $u_{0(j)}$ must be a nontrivial distributional solution of the homogeneous Neumann problem,

$$Pu_{0(j)} = 0 \quad \text{in} \;\; \Omega\,, \;\; N\gamma u_{0(j)} = 0 \quad \text{on} \;\; \Gamma\,. \tag{5.5.18}$$

To characterize the eigensolutions $v_{0(j)}$ of the adjoint equation (5.5.16), we apply the generalized second Green's formula to (5.5.17) and obtain

$$\begin{aligned} \widetilde{a}_\Omega(w, v_{0(j)}) &= \langle Pw, \overline{v_{0(j)}}\rangle_\Omega + \langle N\gamma w, \overline{B\gamma v_{0(j)}}\rangle_\Gamma \\ &= \langle w, \overline{P^* v_{0(j)}}\rangle_\Omega + \langle N\gamma w, \overline{B\gamma v_{0(j)}}\rangle_\Gamma \\ - \langle \tau w, \overline{\gamma_0 v_{0(j)}}\rangle_\Gamma &+ \langle \gamma_0 w, \overline{\widetilde{\tau} v_{0(j)}}\rangle_\Gamma + \langle (\boldsymbol{b}\cdot\boldsymbol{n})\gamma_0 w, \overline{\gamma_0 v_{0(j)}}\rangle_\Gamma\,. \end{aligned} \tag{5.5.19}$$

We note that the relation (5.1.23),

$$\langle \tau w, \overline{\gamma_0 v_{0(j)}}\rangle_\Gamma = \langle N\gamma w, \overline{B\gamma v_{0(j)}}\rangle_\Gamma + \langle \widetilde{N}\gamma_0 w, \overline{\gamma_0 v_{0(j)}}\rangle_\Gamma\,,$$

implies that (5.5.19) takes the form

$$\begin{aligned}\widetilde{a}_\Omega(w, v_{0(j)}) &= \langle w, \overline{P^* v_{0(j)}}\rangle_\Omega \\ &\quad + \langle \widetilde{N}\gamma_0 w, \overline{\gamma_0 v_0}\rangle_\Gamma + \langle \gamma_0 w, \overline{\widetilde{\tau} v_{0(j)} + (\boldsymbol{b}\cdot\boldsymbol{n})\gamma_0 v_{0(j)}}\rangle_\Gamma \\ &= \langle w, \overline{P^* v_{0(j)}}\rangle_\Omega + \langle \gamma_0 w, \overline{(\widetilde{N}^*\gamma_0 + \widetilde{\tau} + (\boldsymbol{b}\cdot\boldsymbol{n})\gamma_0) v_{0(j)}}\rangle_\Gamma \,.\end{aligned}$$

Hence, it follows from (5.5.16) that the eigensolutions $v_{0(j)}$ are the nontrivial distributional solutions of the adjoint general Neumann problem

$$\begin{aligned} P^* v_{0(j)} &= 0 \quad \text{in } \Omega \quad \text{and} \\ \widetilde{\tau} v_{0(j)} + \widetilde{N}^*\gamma_0 v_{0(j)} + \overline{(\boldsymbol{b}\cdot\boldsymbol{n})}\gamma_0 v_{0(j)} &= 0 \quad \text{on } \Gamma \,. \end{aligned} \tag{5.5.20}$$

The operator $\widetilde{N}^*$ is derived from $\widetilde{R}$ in (5.1.25) via integration by parts and is of the form

$$\widetilde{N}^*\gamma_0 v_{0(j)} = -\sum_{j,k=1}^{n}\Big(\sum_{\mu,\varrho=1}^{n-1}\frac{\partial}{\partial\sigma_\varrho}(\gamma^{\mu\varrho}\frac{\partial y_k}{\partial\sigma_\mu}n_j a^*_{jk} v_{0(j)}) + N^*_{00}N^{-1*}_{01} n_j a^*_{jk} n_k v_{0(j)}\Big)_{|\Gamma} \,.$$

5.5.2 Exterior Boundary Value Problems

Theorem 5.5.6. *Given $f \in H^{-1}_{\mathrm{comp}}(\Omega^c)$ and $\varphi_1 \in \pi H^{\frac{1}{2}}(\Gamma)$ and $\psi_2 \in (I-\pi)H^{-\frac{1}{2}}(\Gamma)$ together with $d_\varrho \in \mathbb{C}$, $\varrho = 1, \ldots, \widetilde{L}$, there exists a unique solution $u = u_0 + p + E^\top(\bullet; 0)q$ with $u_0 + p \in \mathcal{H}_E(\Omega^c)$, $q \in \mathbb{C}^p$ such that Equation* (5.1.72) *holds for all $v = v_0 + \sum_{\varrho=1}^{\widetilde{L}}\kappa_\varrho q_\varrho \in H_{E\pi B}$ subject to the side conditions* (5.1.73) *and the constraint* (5.1.74)*, provided the homogeneous equation*

$$\widetilde{a}_{\Omega^c}(u_0 + p, v) = 0 \quad \text{for all } \; v \in \mathcal{H}_{E\pi B}$$

has only the trivial solution $u \equiv 0$.

If the homogeneous equation has nontrivial solutions, then the second part of Fredholm's alternative holds accordingly.

The proof follows with the same arguments as in the proof of Theorem 5.5.4 based on Gårding's inequality (5.4.50) and Fredholm's alternative. Details are omitted.

5.5.3 Transmission Problems

From the weak formulation (5.1.81) of the transmission problems and the corresponding Gårding's inequality (5.4.53) on all of $H^1(\Omega) \times \mathcal{H}^0_E(\Omega^c)$ it follows that we have the following existence result:

Theorem 5.5.7. *Given $f \in \widetilde{H}_0^{-1}(\Omega) \times H^{-1}_{\mathrm{comp}}(\Omega^c)$ and any set of boundary conditions in Table* 5.1.1 *with* (5.1.78) *or in Table* 5.1.2 *with* (5.1.83) *then* *there exists a unique solution* $u \in H^1(\Omega)\times\mathcal{H}^0_E(\Omega^c)$ *provided the corresponding homogeneous problem has only the trivial solution.*

If the corresponding homogeneous problem has nontrivial solutions then the second part of Fredholm's alternative holds accordingly.

We may also consider the above transmission problems in a bounded annular domain where Ω^c is replaced by $\Omega^c_R := \Omega^c \cap B_R$ with $R > 0$ large enough so that the ball B_R contains $\overline{\Omega}$ in its interior. In this case, no radiation conditions are needed and on the outer boundary $\partial B_R = \{x \in \mathbb{R}^n \,|\, |x| = R\}$, we may describe any of the regular boundary conditions and Theorem 5.5.6 remains valid with $f \in \widetilde{H}_0^{-1}(\Omega) \times \widetilde{H}_0^{-1}(\Omega^c_R)$ and $u \in H^1(\Omega) \times H^1_0(\Omega^c_R)$.

5.6 Solutions of Certain Boundary Integral Equations and Associated Boundary Value Problems

In this section we consider the weak formulation of the representation formulae for the solutions of boundary value problems, in particular the problems of Section 5.1. In order to establish existence and uniqueness results for the boundary integral equations we need to establish coerciveness properties of the integral operators on the boundary, which follow from the coerciveness of the corresponding interior, exterior and transmission problems in variational form as formulated in the previous section.

5.6.1 The Generalized Representation Formula for Second Order Systems

The representation formula can be derived from the generalized second Green's formula (5.1.8) in Sobolev spaces as in the classical approach by applying the generalized second formula (5.1.9) to the variational solution in

$$\Omega_\varepsilon = \Omega \setminus \{y \,\big|\, |y-x| \le \varepsilon\} \ \text{ and } \ \Omega^c_\varepsilon = \Omega^c \setminus \{y \,\big|\, |y-x| \le \varepsilon\}$$

with $\overline{v}(y)$ replaced by $E^\top(x,y)$ and then taking the limit $\varepsilon \to 0$ in connection with the explicit growth conditions of $E(x,y)$ and its derivatives when $y \to x$. Since we so far have not yet discussed the growth conditions of E at $y = x$, we therefore do not present this standard procedure here. Instead, we begin with the representation formula (3.6.9) and formula (3.6.18) for a classical solution with $\Gamma \in C^{1,1}$, $f \in C_0^\infty(\mathbb{R}^n)$ and for $M(x;u) = 0$,

$$\begin{aligned} u(x) = &\int\limits_\Omega E^\top(x,y) f(y)dy + \int\limits_\Gamma E^\top(x,y)\left\{\partial_\nu u - (\boldsymbol{b}\cdot\boldsymbol{n})\gamma_0 u\right\}(y)ds_y \\ &- \int\limits_\Gamma (\widetilde{\partial}_{\nu y} E(x,y))^\top \gamma_0 u(y) ds_y \quad \text{for } x \in \Omega \end{aligned} \tag{5.6.1}$$

and

$$0 = \int\limits_{\Omega^c} E^\top(x,y) f(y)dy - \int\limits_{\Gamma} E^\top(x,y) \{\partial_{c\nu} u - (\boldsymbol{b}\cdot\boldsymbol{n})\gamma_{c0} u\} ds_y \\ + \int\limits_{\Gamma} (\widetilde{\partial}_{c\nu y} E(x,y))^\top \gamma_{c0} u(y) ds_y \quad \text{for} \quad x \in \Omega \tag{5.6.2}$$

whereas we have

$$u(x) = \int\limits_{\Omega^c} E^\top(x,y) f(y)dy - \int\limits_{\Gamma} E^\top(x,y) \{\partial_{c\nu} u - (\boldsymbol{b}\cdot\boldsymbol{n})\gamma_{c0} u\} ds_y \\ + \int\limits_{\Gamma} (\widetilde{\partial}_{c\nu y} E(x,y))^\top \gamma_{c0} u(y) ds_y \quad \text{for} \quad x \in \Omega^c \tag{5.6.3}$$

together with

$$0 = \int\limits_{\Omega} E^\top(x,y) f(y)dy + \int\limits_{\Gamma} E^\top(x,y) \{\partial_{v} u - (\boldsymbol{b}\cdot\boldsymbol{n})\gamma_{0} u\} \\ - \int\limits_{\Gamma} \left(\widetilde{\partial}_{\nu y} E(x,y)\right)^\top \gamma_{0} u(y) ds_y \quad \text{for} \quad x \in \Omega^c .$$

By adding (5.6.1) and (5.6.2), we obtain

$$u(x) \;=\; \int\limits_{\mathbb{R}^n} E^\top(x,y) f(y)dy - \int\limits_{\Gamma} E^\top(x,y) [\partial_{\nu} u - (\boldsymbol{b}\cdot\boldsymbol{n})\gamma_{0} u] ds_y \\ + \int\limits_{\Gamma} (\widetilde{\partial}_{\nu y} E(x,y))^\top [\gamma_{0} u] ds_y \quad \text{for} \quad x \in \Omega\,, \tag{5.6.4}$$

where $[v] := v_c - v$ denotes the jump of v across Γ.

The same formula also holds for $x \in \Omega^c$; hence, (5.6.4) *is valid for all* $x \in \mathbb{R}^n \setminus \Gamma$. By density and completion arguments, one obtains the corresponding formulation in Sobolev spaces and for Lipschitz domains.

Theorem 5.6.1 (Generalized representation formula). *Let Ω be a bounded Lipschitz domain. Let $f \in \widetilde{H}_0^{-1}(\Omega) \otimes H^{-1}_{\text{comp}}(\Omega^c)$ with* $\operatorname{supp}(f) \Subset \mathbb{R}^n$ *and let u be a variational solution of $Pu = f$ in $\mathbb{R}^n \setminus \Gamma$, i.e., of* (5.1.81) *with $u|_\Omega \in H^1(\Omega, P)$ and $u|_{\Omega^c} \in H^1_{\text{loc}}(\Omega^c, P)$ satisfying the radiation condition $M(x;u) = 0$ (cf.* (3.6.15)*). Then $u(x)$ admits the representation*

$$u(x) = \langle E(x,\cdot), f\rangle_{\mathbb{R}^n} - \langle [\tau u - (\boldsymbol{b}\cdot\boldsymbol{n})\gamma_0 u], E(x,\cdot)\rangle_\Gamma \\ + \langle (\widetilde{\partial}_\nu E(x,\cdot)), [\gamma_0 u]\rangle_\Gamma \text{ for almost every } x \in \mathbb{R}^n \setminus \Gamma\,. \tag{5.6.5}$$

Remark 5.6.1: In case $f = 0$, formula (5.6.5) holds for *every* $x \in \mathbb{R}^n \setminus \Gamma$.

Note that for $u|_{\Omega^c} \in H^1_{\text{comp}}(\Omega^c, P)$ there holds $M(x; u) = 0$.

Proof of Theorem 5.6.1: For the proof we shall use the following estimates:

$$\|\langle E(x,\cdot), Pu\rangle_{\mathbb{R}^n\setminus\Gamma}\|_{H^1(\Omega^c_R)} \le c_{x,R}\left\{\|u\|_{H^1(\Omega,P)} + \|u\|_{H^1(\Omega^c_R,P)}\right\} \quad (5.6.6)$$

$$|\langle[\tau u - (\boldsymbol{b}\cdot\boldsymbol{n})\gamma_0 u], E(x,\cdot)\rangle_\Gamma| + |\langle(\widetilde{\partial}_\nu E(x,\cdot)), [\gamma_0 u]\rangle_\Gamma|$$
$$\le c_{x,R}\left\{\|u\|_{H^1(\Omega,P)} + \|u\|_{H^1(\Omega^c_R,P)}\right\} \quad \text{for } x \notin \Gamma\,. \quad (5.6.7)$$

In these estimates $\Omega^c_R = \Omega^c \cap \{y \in \mathbb{R}^n \mid |y| < R\}$ for R sufficiently large such that $\operatorname{supp}(f) \subset \Omega^c_R$. The estimate (5.6.6) is a standard a–priori estimate for Newton potentials (see e.g. (10.0.19) in Chapter 10). Since $E(x,\cdot)$ for fixed $x \notin \Gamma$ is a smooth function on the boundary Γ, the estimates (5.6.7) follow from Lemma 5.1.1 and Theorem 5.5.4.

With these two estimates, the representation (5.6.5) can be established by the usual completing procedure. More precisely, for smooth Γ and if u is a variational solution with the required properties we can approximate u by a sequence u_k with

$$u_k|_\Omega \in C^\infty(\overline{\Omega}) \text{ and } u_k|_{\Omega^c} \in C^\infty(\overline{\Omega^c})$$

satisfying (5.6.4) so that

$$\|u - u_k\|_{H^1(\Omega,P)} + \|u - u_k\|_{H^1_{\text{loc}}(\Omega^c,P)} \to 0 \text{ as } k \to \infty\,.$$

Then, because of (5.6.7), for any $x \notin \Gamma$ the two boundary potentials in (5.6.4) generated by u_k will converge to the corresponding boundary potentials $(u - F)(x)$ where $F(x) = \int\limits_{\mathbb{R}^n} E^\top(x,y)Pu(y)dy$ in (5.6.5). From (5.6.6), however, we only have

$$\langle E(x,y), Pu_k\rangle_{\mathbb{R}^n\setminus\Gamma} \to \langle E(x,\cdot), Pu\rangle_{\mathbb{R}^n\setminus\Gamma} = F(x)$$

for almost every $x \in \mathbb{R}^n \setminus \Gamma$.

If Ω is a strong Lipschitz domain with a strong Lipschitz boundary, then for fixed $x \notin \Gamma$, the generalized Green formulae can still be applied to the annular domains Ω_ε for any sufficiently small $\varepsilon > 0$. For the domain $B_\varepsilon(x) = \{y \mid |y - x| < \varepsilon\}$, the previous arguments can be used since $B_\varepsilon(x)$ has a smooth boundary. Note that the sum of the boundary integrals over ∂B_ε will be canceled. This completes the proof. ∎

From the representation formula (5.6.5) we may represent the solution of (5.1.19) or of (5.1.20) for $x \in \Omega$ by setting $u_{|\Omega^c} \equiv 0$ which coincides with our previous representation formulae in Ω. Similarly, if one is interested in the exterior problem we may set $u_{|\Omega} \equiv 0$ and obtain a corresponding representation of solutions in Ω^c. In fact, the representation formula (5.6.5) can be applied to the more general transmission problems which will be discussed below.

Based on the representation formula (5.6.5) we now consider associated boundary value problems.

5.6.2 Continuity of Some Boundary Integral Operators

Similar to the classical approach, the generalized Green's formulae lead us to the boundary integral equations in Sobolev spaces. For this purpose we need the mapping properties of boundary potentials including jump relations which can be derived from the theory of pseudo–differential operators as we shall discuss later on. For a restricted class of boundary value problems, however, these properties can be obtained by using solvability and regularity of associated transmission problems as in Section 5.1.3.

To illustrate the idea, let us consider the generalized representation formula (5.6.5) for the variational solution of the transmission problem (5.1.81) for second order equations with $f = 0$ and the radiation condition $M = 0$; namely

$$u(x) = -V[\tau u - (\boldsymbol{b}\cdot\boldsymbol{n})\gamma_0 u](x) + W[\gamma_0 u](x) \ \text{ for } x \in \mathbb{R}^n \setminus \Gamma\,, \qquad (5.6.8)$$

where $V\sigma(x)$ is the simple layer potential and $W\varphi(x)$ is the double layer potential corresponding to (5.6.5), namely

$$V\sigma(x) = \langle E(x,\cdot),\sigma\rangle_\Gamma\,,\ W\varphi(x) = \langle \widetilde{\partial}_\nu E(x,\cdot),\varphi\rangle_\Gamma\,. \qquad (5.6.9)$$

From this formula (5.6.8) we now establish the following mapping properties provided that Theorem 5.6.1 is valid. Otherwise, the proof presented below must be modified although the results remain valid (see Chapters 8 and 9).

Theorem 5.6.2. *The following operators are continuous:*

$$\begin{aligned}
V : H^{-\frac12}(\Gamma) &\to H^1(\Omega,P)\times H^1_{\text{loc}}(\Omega^c,P)\,,\\
\gamma_0 V \text{ and } \gamma_{c0}V : H^{-\frac12}(\Gamma) &\to H^{\frac12}(\Gamma)\,,\\
\tau V \text{ and } \tau_c V : H^{-\frac12}(\Gamma) &\to H^{-\frac12}(\Gamma)\,,\\[1ex]
W : H^{\frac12}(\Gamma) &\to H^1(\Omega,P)\times H^1_{\text{loc}}(\Omega^c,P)\,,\\
\gamma_0 W \text{ and } \gamma_{c0}W : H^{\frac12}(\Gamma) &\to H^{\frac12}(\Gamma)\,,\\
\tau W \text{ and } \tau_c W : H^{\frac12}(\Gamma) &\to H^{-\frac12}(\Gamma)\,.
\end{aligned}$$

Proof: For V we consider the transmission problem (5.1.76)–(5.1.78) with $\pi = I$ and $\varphi_1 = 0$, $\psi_1 = \sigma$:

Find $u \in H^1(\Omega,P)$ *and* $u \in \mathcal{H}^0_E(\Omega^c,P)$ *satisfying the differential equation*

$$Pu = 0 \ \text{ in } \Omega \text{ and in } \Omega^c\,,$$

together with the transmission conditions

$$[\gamma_0 u]_\Gamma = 0 \text{ and } [\partial_\nu u - (\boldsymbol{b}\cdot\boldsymbol{n})\gamma_0 u]_\Gamma = \sigma$$

with given $\sigma \in H^{-\frac{1}{2}}(\Gamma)$ *and some radiation condition.*

This is a special case of Theorem 5.6.1 which yields the existence of u. Moreover, u depends on σ continuously. The latter implies that the mappings $\sigma \mapsto u$ from $H^{-\frac{1}{2}}(\Gamma)$ to $H^1(\Omega, P)$ and to $\mathcal{H}^0_E(\Omega^c, P)$, respectively, are continuous. By the representation formula (5.6.8), it follows that the mapping $\sigma \mapsto u = V\sigma$ is continuous in the corresponding spaces. This together with the continuity of the trace operators

$$\gamma_0 : H^1(\Omega, P) \to H^{\frac{1}{2}}(\Gamma) \quad \text{and } \gamma_{c0} : H^1_{\text{loc}}(\Omega^c, P) \to H^{\frac{1}{2}}(\Gamma)$$

as well as of

$$\tau : H^1(\Omega, P) \to H^{-\frac{1}{2}}(\Gamma) \quad \text{and } \tau_c : H^1_{\text{loc}}(\Omega^c, P) \to H^{-\frac{1}{2}}(\Gamma)$$

implies the continuity properties of the operators associated with V.

Similarly, the solution of the transmission problem (5.1.76)–(5.1.78) with $\pi = I$ and $\psi_1 = 0$, $\varphi_1 = \varphi$:

Find $u \in H^1(\Omega, P)$ *and* $u \in \mathcal{H}^0_E(\Omega^c, P)$ *satisfying*

$$Pu = 0 \text{ in } \Omega \text{ and in } \Omega^c$$

with $\quad [\gamma_0 u]_{|\Gamma} = \varphi \in H^{\frac{1}{2}}(\Gamma)$ *and* $[\partial_\nu u - (\boldsymbol{b}\cdot\boldsymbol{n})\gamma_0 u]_\Gamma = 0;$

together with the representation formula (5.6.8),

$$u(x) = W\varphi(x) \text{ for } x \notin \Gamma$$

provides the desired continuity properties for the mappings associated with W. ∎

As a consequence of the mapping properties, we have the following jump relations.

Lemma 5.6.3. *Given* $(\sigma, \varphi) \in H^{-\frac{1}{2}}(\Gamma) \times H^{\frac{1}{2}}(\Gamma)$, *then the following jump relations hold:*

$$\begin{array}{rclrcl} [\gamma_0 V\sigma]_{|\Gamma} &=& 0, & [\tau V\sigma]_{|\Gamma} &=& -\sigma; \\ [\gamma_0 W\varphi]_{|\Gamma} &=& \varphi, & [\tau W\varphi]_{|\Gamma} &=& 0. \end{array}$$

Proof: We see from the representation formula (5.6.8) that $u(x) = -V\sigma(x)$ for $x \in \mathbb{R}^n \setminus \Gamma$ is the solution of the transmission problem (5.1.76)–(5.1.78) with $\pi = I$ and $\varphi_1 = 0$, $\psi_1 = \sigma$ with the transmission conditions

$$[\gamma_0 u]_\Gamma = 0 \text{ and } -\sigma = [\tau(u)]_\Gamma .$$

Inserting $u = -V\sigma$ gives the jump relations involving V.

Likewise, $u(x) = W\varphi(x)$ is the solution of the transmission problem (5.1.76)–(5.1.78) with $\pi = I$ and $\varphi_1 = \varphi$, $\psi_1 = 0$ satisfying

$$[\gamma_0 u]_\Gamma = \varphi \text{ and } [\tau(u)]_\Gamma = 0$$

which gives the desired jump relations involving W. ■

In accordance with the classical formulation (1.2.3)–(1.2.6), we now introduce the boundary integral operators on Γ for given $(\sigma, \varphi) \in H^{-\frac{1}{2}}(\Gamma) \times H^{\frac{1}{2}}(\Gamma)$ defined by:

$$\begin{array}{rclrcl} V\sigma & := & \gamma_0 V\sigma\,, & K\varphi & := & \gamma_0 W\varphi + \dfrac{1}{2}\varphi \\ K'\sigma & := & \tau V\sigma - \dfrac{1}{2}\sigma\,, & D\varphi & := & -\tau W\varphi\,. \end{array} \tag{5.6.10}$$

Clearly, Theorem 5.6.2 provides us the continuity of these boundary integral operators.

The corresponding Calderón projectors can now be defined in a weak sense by

$$\mathcal{C}_\Omega := \begin{pmatrix} \frac{1}{2}I - K & , & V \\ D & , & \frac{1}{2}I + K' \end{pmatrix} \text{ and } \mathcal{C}_{\Omega^c} := \mathcal{I} - \mathcal{C}_\Omega\,, \tag{5.6.11}$$

which are continuous mappings on $\big(H^{\frac{1}{2}}(\Gamma) \times H^{-\frac{1}{2}}(\Gamma)\big)$.

5.6.3 Continuity Based on Finite Regions

In deriving the continuity properties of the boundary integral operators we rely on the properties of the solutions to transmission problems in $\mathbb{R}^n \setminus \Gamma$. As we have seen, this complicates the analysis because of the solution's behaviour at infinity. Consequently, we needed restrictions on the coefficients of the differential operators. Alternatively, as we shall see, if one is only interested in the local properties of the solution, a similar analysis can be achieved by replacing the transmission problem in $\mathbb{R}^n \setminus \Gamma$ by one in a bounded domain $B_R \setminus \Gamma$. To be more precise, for illustration we consider the supplementary transmission problem:

Find $u \in H^1(\Omega, P) \times H^1(\Omega^c_R, P)$ satisfying

$$Pu = f_1 \quad \textit{in } \Omega \quad \textit{and } Pu = f_2 \quad \textit{in } \Omega^c_R \tag{5.6.12}$$

together with

$$[\gamma_0 u]|_\Gamma = \varphi \text{ and } [\partial_\nu u - (b\cdot n)^\top \gamma_0 u]|_\Gamma = \sigma \text{ and } \gamma_0 u|_{|x|=R} = 0 \tag{5.6.13}$$

with given $f_1 \in \widetilde{H}_0^{-1}(\Omega)$, $f_2 \in \widetilde{H}_0^{-1}(\Omega^c) \cap H^{-1}(B_R)$, $\varphi \in H^{\frac{1}{2}}(\Gamma)$, $\sigma \in H^{-\frac{1}{2}}(\Gamma)$.

The corresponding weak formulation then reads:

Find $u \in H^1(\Omega, P) \times H^1(\Omega_R^c, P)$ *with* $[\gamma_0 u]_\Gamma = \varphi$ *and* $\gamma_0 u|_{|x|=R} = 0$ *such that*

$$\widetilde{a}(u,v) = \widetilde{a}_\Omega(u,v) + \widetilde{a}_{\Omega_R^c}(u,v) = -\int\limits_\Gamma \sigma \bar{v} ds + \int\limits_\Omega f_1 \bar{v} dx + \int\limits_{\Omega_R^c} f_2 \bar{v} dx \tag{5.6.14}$$

for all
$v \in \mathcal{H} := \{v \in H^1(\Omega, P) \times H^1(\Omega_R^c, P)$ *with* $[\gamma_0 v]_\Gamma = 0$ *and* $\gamma_0 v|_{|x|=R} = 0\}$.

Since $\widetilde{a}(u,v)$ satisfies the Gårding inequality on $\mathcal{H}$, the Fredholm alternative is valid for the problem (5.6.14). For convenience let us assume that the homogeneous problem for (5.6.14) has only the trivial solution. Then the unique solution u of (5.6.14) satisfies an estimate

$$\|u\|_{H^1(\Omega)} + \|u\|_{H^1(\Omega_R^c)} \le c\{\|f_1\|_{\widetilde{H}^{-1}(\Omega)} + \|f_2\|_{\widetilde{H}^{-1}(\Omega^c)\cap H^{-1}(B_R)} + \|\sigma\|_{H^{-\frac{1}{2}}(\Gamma)} + \|\varphi\|_{H^{\frac{1}{2}}(\Gamma)}\} \tag{5.6.15}$$

and can be represented in the form

$$u(x) = \int\limits_\Omega E^\top(x,y) f_1(y) dy + \int\limits_{\Omega_R^c} E^\top(x,y) f_2(y) dy - \int\limits_\Gamma E^\top(x,y)\sigma(y) ds_y + \int\limits_\Gamma \big(\widetilde{\partial}_{\nu y} E(x,y)\big)^\top \varphi(y) ds_y \quad \text{for } x \in B_R \setminus \Gamma. \tag{5.6.16}$$

Now let us reexamine the continuity properties of the simple layer potential operator V again. Let $\sigma \in C^\infty(\Gamma)$ be given. Then consider

$$u(x) := V\sigma(x) = \int\limits_\Gamma E(x,y)^\top \sigma(y) ds_y \quad \text{for } x \notin \Gamma.$$

Also, let $\eta \in C_0^\infty(\mathbb{R}^n)$ be a cut–off function with

$$\eta(x) = 1 \quad \text{for } |x| \le R_1 < R \quad \text{and} \quad \operatorname{supp} \eta \subset B_R \tag{5.6.17}$$

where $R_1 > 0$ is chosen sufficiently large such that $\overline{\Omega} \subset B_{R_1}$. Now we define

$$\widetilde{u} := \eta(x) u(x).$$

Then $\widetilde{u}(x)$ is a solution of the transmission problem (5.6.12), (5.6.13) where $\varphi = 0$, $f_1 = 0$ and

$$f_2 := (P\eta - c\eta)V\sigma - \sum_{j,k=1}^n a_{jk}\Big\{\frac{\partial \eta}{\partial x_k}\frac{\partial V\sigma}{\partial x_j} + \frac{\partial \eta}{\partial x_j}\frac{\partial V\sigma}{\partial x_k}\Big\}. \tag{5.6.18}$$

Hence, $\operatorname{supp} f_2 \Subset B_R \setminus B_{R_1}$ and, since the coefficients of P are assumed to be $C^\infty(\mathbb{R}^n)$, we have $f_2 \in C_0^\infty(B_R \setminus B_{R_1})$ since $\operatorname{dist}(\Gamma, B_R \setminus B_{R_1}) \geq d_0 > 0$. (Note that this smoothness may be reduced depending on the smoothness of the coefficients of P.)

So, the estimate (5.6.15) implies for $\widetilde{u}$

$$\|V\sigma\|_{H^1(\Omega)} + \|\eta V\sigma\|_{H^1(\Omega_R^c)} \leq c\{\|f_2\|_{H^{-1}(B_R\setminus B_{R_1})} + \|\sigma\|_{H^{-\frac{1}{2}}(\Gamma)}\}.$$

Since $f_2 \in C_0^\infty(B_R \setminus B_{R_1})$ and because of $|x-y| \geq d_0 > 0$ for $x \in (B_R \setminus B_{R_1})$ and $y \in \Gamma$ from (5.6.18) it is clear that

$$\|f_2\|_{H^{-1}(B_R\setminus B_{R_1})} \leq c\|f_2\|_{L_2(B_R\setminus B_{R_1})} \leq c'\|\sigma\|_{H^{-\frac{1}{2}}(\Gamma)}.$$

Consequently, with Theorem 4.2.1, the trace theorem, we find

$$\|V\sigma\|_{H^{\frac{1}{2}}(\Gamma)} \leq c_1\{\|V\sigma\|_{H^1(\Omega)} + \|V\sigma\|_{H^1(\Omega_{R_1}^c)}\} \leq c_2\|\sigma\|_{H^{-\frac{1}{2}}(\Gamma)},$$

the first two continuity properties for V in Lemma 5.6.3, by the standard completion argument approximating $\sigma \in H^{-\frac{1}{2}}(\Gamma)$ by C^∞–functions.

Applying Lemma 5.1.1 to $\widetilde{u} \in H^1(\Omega, P)$ and Lemma 5.1.2 to $\widetilde{u} \in H^1(\Omega_R^c, P)$ we obtain the mapping properties of τV and $\tau_c V$ in Theorem 5.6.2.

For the properties of $W\varphi$ in Theorem 5.6.2 we may proceed in the same manner as for $V\varphi$ where $\psi_1 = 0$ and $\varphi_1 = \varphi \neq 0$ in the transmission problem (5.1.78).

5.6.4 Continuity of Hydrodynamic Potentials

Although the Stokes system (2.3.1) is not a second order Petrovskii elliptic system, its close relation with the Lamé system can be exploited to show continuity properties of the hydrodynamic potentials V, W and D in (2.3.15), (2.3.16), (2.3.31) (see Kohr et al [234]).

Lemma 5.6.4. *The mappings given by the hydrodynamic potentials define the following linear operators:*

$$\begin{aligned}
V \quad &: H^{-\frac{1}{2}}(\Gamma) \to H^1_{\text{div}}(\Omega, \Delta) \times H^1_{\text{div,loc}}(\Omega^c, \Delta),\\
\gamma_0 V \ \textit{and}\ \gamma_{c0} V &: H^{-\frac{1}{2}}(\Gamma) \to H^{\frac{1}{2}}(\Gamma),\\
W \quad &: H^{\frac{1}{2}}(\Gamma) \to H^1_{\text{div}}(\Omega, \Delta) \times H^1_{\text{div,loc}}(\Omega^c, \Delta),\\
\gamma_0 W \ \textit{and}\ \gamma_{c0} W &: H^{\frac{1}{2}}(\Gamma) \to H^{\frac{1}{2}}(\Gamma),\\
D \quad &: H^{\frac{1}{2}}(\Gamma) \to H^{-\frac{1}{2}}(\Gamma).
\end{aligned}$$

Here we denote

$$\begin{aligned} H^1_{\mathrm{div}}(\Omega,\Delta) &:= \{\boldsymbol{v}\in H^1(\Omega,\Delta) &&\mid \operatorname{div}\boldsymbol{v} &&= 0 \text{ in } \Omega\}, \\ H^1_{\mathrm{div,loc}}(\Omega^c,\Delta) &:= \{\boldsymbol{v}\in H^1_{\mathrm{loc}}(\Omega,\Delta) &&\mid \operatorname{div}\boldsymbol{v} &&= 0 \text{ in } \Omega^c\}. \end{aligned}$$

Proof: From (2.3.38) we see that for $V = V_{st}$ we have

$$V_{st} = (1+c)\Big\{V_{e\ell} - \frac{2c}{\mu}V_\Delta I\Big\}$$

which yields with Theorem 5.6.2 for $V_{e\ell}$with any $c > 0$ and for V_Δ the desired properties of V_{st} since the Stokes simple layer potential is solenoidal in Ω and in Ω^c.

Similarly, from (2.3.42) and (2.3.43) we obtain

$$W_{st} = (1+c)W_{e\ell} - c(W_\Delta + L_1)$$

and

$$L_1 = \frac{1}{c_1 - c_2}\{(1+c_1)W_{e\ell,c_1} - (1+c_2)W_{e\ell,c_2}\} - W_\Delta$$

with any $0 < c_2 < c_1$. Hence, the desired mapping properties again follow with Theorem 5.6.2 from those of $W_{e\ell,c_1}$, $W_{e\ell,c_2}$ and W_Δ.

For D_{st} we use (2.3.48) and (2.3.49),

$$D_{st} = D_{e\ell} - \frac{c}{1+c}L_{2,0}$$

where $L_{2,0}$ does not depend on c. Hence,

$$L_{2,0} = \frac{(1+c_1)(1+c_2)}{c_1 - c_2}\{D_{e\ell,c_1} - D_{e\ell,c_2}\}.$$

Theorem 5.6.2 for $D_{e\ell,c_1}$ and for $D_{e\ell,c_2}$ where $0 < c_2 < c_1$ yields that $L_{2,0} : H^{\frac{1}{2}}(\Gamma) \to H^{-\frac{1}{2}}(\Gamma)$ is continuous which then implies the desired continuity for D_{st} as well. ■

For the same reasons, the jump relations for the hydrodynamic potentials follow also from those of the Lamé system and the Laplacian.

Lemma 5.6.5. *Given $(\boldsymbol{\tau},\boldsymbol{\varphi}) \in H^{-\frac{1}{2}}(\Gamma) \times H^{\frac{1}{2}}(\Gamma)$, then the hydrodynamic potentials satisfy the following jump relations:*

$$\begin{aligned} [\gamma_0 V\boldsymbol{\tau}]|_\Gamma &= \mathbf{0}, & (T^c(V\boldsymbol{\tau}) - T(V\boldsymbol{\tau}))|_\Gamma &= -\boldsymbol{\tau} \\ [\gamma_0 W\boldsymbol{\varphi}]|_\Gamma &= \boldsymbol{\varphi}, & (T^c(W\boldsymbol{\varphi}) - T(W\boldsymbol{\varphi}))|_\Gamma &= \mathbf{0}. \end{aligned}$$

Proof: The relation $[\gamma_0 V\boldsymbol{\tau}]|_\Gamma = \mathbf{0}$ follows from (2.3.38) and Lemma 5.6.3 applied to $V_{e\ell}$ and V_Δ.

For smooth $\boldsymbol{\varphi}$ and $\boldsymbol{\tau}$ with (2.3.26) and (2.3.27) we obtain

$$2\varphi = \varphi + [\gamma_0 W\varphi]|_\Gamma - [\gamma_0 V\boldsymbol{\tau}]|_\Gamma \quad \text{on } \Gamma\,,$$

hence,

$$\varphi = [\gamma_0 W\varphi]|_\Gamma - \mathbf{0}\,.$$

With the continuity properties of $\gamma_0 W$ and $\gamma_{c0} W$, the proposition follows by completion arguments.

From Lemma 2.3.1 we obtain with (2.3.30):

$$\big(T^c(W\varphi) - T(W\varphi)\big)\big|_\Gamma = -\lim_{c\to 0}[D_{e\ell}\varphi]|_\Gamma = -\lim_{\lambda\to+\infty}[D_{e\ell}\varphi]|_\Gamma = \mathbf{0}\,.$$

For $T(V\boldsymbol{\tau})$ we invoke the second row of the Calderón projections (2.3.27) and (2.3.28) to obtain

$$2\boldsymbol{\tau} = -[D\varphi]|_\Gamma + \boldsymbol{\tau} + \big(T(V\boldsymbol{\tau}) - T^c(V\boldsymbol{\tau})\big)|_\Gamma$$

which is, because of $[D\varphi]|_\Gamma = \mathbf{0}$, the desired relation; first for smooth φ and $\boldsymbol{\tau}$ but then with Lemma 5.6.4 by completion in $H^{\frac{1}{2}}(\Gamma)$. ∎

For the final computation of the pressure p via (2.3.12) we need the mapping properties of the pressure operators Φ (2.3.15), Π (2.3.16) and of $\int\limits_\Omega Q(x,y)f(y)dy$.

Lemma 5.6.6. *The pressure operators define the continuous mappings*

$$\begin{aligned}
\Phi &: H^{-\frac{1}{2}}(\Gamma) &&\to L^2(\Omega)\,,\\
\Pi &: H^{\frac{1}{2}}(\Gamma) &&\to L^2(\Omega)\,,\\
\int\limits_\Omega Q(x,y)f(y)dy &: \widetilde{H}_0^{-1}(\Omega) &&\to L^2(\Omega)\,.
\end{aligned} \tag{5.6.19}$$

Proof: With $Q(x,y)$ given in (2.3.10) and γ_n in (2.2.3) as well as the simple layer potential V_Δ of the Laplacian in (1.2.1) we have

$$\Phi\boldsymbol{\tau} = -\text{div}V_\Delta\boldsymbol{\tau}\,.$$

Since Theorem 5.6.2 ensures the continuity $V_\Delta : H^{-\frac{1}{2}}(\Gamma) \to H^1(\Omega)$ differentiation yields the first of the assertions. Similarly,

$$\Pi\varphi = -2\mu\text{div}W_\Delta\varphi$$

with W_Δ for the Laplacian given in (1.2.2). Theorem 5.6.2 shows continuity of $W_\Delta : h^{\frac{1}{2}}(\Gamma) \to H^1(\Omega)$ and we get (5.6.19) for Π.

For the volume potential we have

$$\int\limits_\Omega Q(x,y)\boldsymbol{f}(y(dy = -\text{div}\int\limits_\Omega E_\Delta(x,y)\boldsymbol{f}(y)dy$$

where the Newton potential of the Laplacian is a pseudodifferential operator of order -2 in $\mathbb{R}^n$ of the type in (7.2.21). Since $\boldsymbol{f} \in \widetilde{H}_0^{-1}(\Omega)$ can also be seen as $\boldsymbol{f} \in H^{-1}(\mathbb{R}^n)$ if extended by zero we get

$$\int_\Omega E_\Delta(\cdot, y)\boldsymbol{f}(y)dy = \int_{\mathbb{R}^n} E_\Delta(\cdot, y)\boldsymbol{f}(y)dy \in H^1(\Omega)\,.$$

So, differentiation implies the last continuous mapping in (5.6.19). ■

5.6.5 The Equivalence Between Boundary Value Problems and Integral Equations

For the boundary value problems given in Section 5.5, from the corresponding representation formulae, together with the continuity properties of the boundary potentials given in Theorem 5.6.2 and the jump relations in Lemma 5.6.3, we arrive at boundary integral equations of the form

$$\mathcal{A}\lambda = \mathcal{B}\mu \quad \text{on } \Gamma \tag{5.6.20}$$

where μ is given and λ is unknown. The boundary integral operators $\mathcal{A}$ and $\mathcal{B}$ depend on the boundary value problem to be solved. In contrast to boundary integral equations in classical function spaces on Γ, here (5.6.20) should be understood as an *operator equation* in the corresponding Sobolev spaces on Γ; here $H^{-\frac{1}{2}}(\Gamma)$ or $H^{\frac{1}{2}}(\Gamma)$. For the boundary value problems (5.1.19), (5.1.52), (5.1.29) and (5.1.70), the specific forms of $\mathcal{A}, \mathcal{B}$ and the meaning of λ and μ together with the Sobolev spaces on Γ are listed in Table 5.6.3.

Theorem 5.6.7. *For given μ in the appropriate function space on Γ, the boundary value problems and the associated boundary integral equations* (5.6.20) *on Γ as listed in Table* 5.6.3 *are equivalent in the following sense:*

i. Every variational solution $u \in H^1(\Omega, P)$ (or $u \in H^1_{\mathrm{loc}}(\Omega^c, P)$, respectively) of the boundary value problem defines the boundary data μ and λ where λ is a solution of the associated boundary integral equations (5.6.20) *with $\mathcal{A}$ and $\mathcal{B}$ given in Table* 5.6.3.

ii. Conversely, every pair μ, λ where λ is a solution of the boundary integral equations (5.6.20) *defines, via the representation formula, a variational solution of the corresponding boundary value problem. Every variational solution of the boundary value problem can be obtained in this manner.*

Remark 5.6.2: Theorem 5.6.7 assures that every solution of the boundary integral equations (5.6.20) indeed generates solutions of the corresponding boundary value problem and that every solution of the boundary value problem can be obtained in this way. However, the theorem does *not imply* *that the number of solutions of both problems coincides.* It is possible that nontrivial solutions of the homogeneous boundary integral equations via the representation formula are mapped onto $u \equiv 0$.

Proof: Since the arguments of the proof are the same in each of the listed cases, we here consider only the first case to illustrate the main ideas:

i.) Let $u \in H^1(\Omega, P)$ be the variational solution of the interior Dirichlet boundary value problem (5.1.19) with given $\mu = \varphi = \gamma_0 u|_\Gamma = R\gamma u \in H^{\frac{1}{2}}(\Gamma)$ and $f = 0$. Then $\tau u|_\Gamma \in H^{-\frac{1}{2}}(\Gamma)$ is well defined. Hence, $\lambda := S\gamma u = \tau u|_\Gamma - \boldsymbol{b} \cdot \boldsymbol{n}\varphi$ is also well defined and $\lambda \in H^{-\frac{1}{2}}(\Gamma)$. As was shown in Section 5.6.1, the solution of (5.1.19) can be represented by

$$u(x) = V\lambda(x) - W\mu(x) \text{ for } x \in \Omega\,.$$

Taking the trace $\gamma_0|_\Gamma$ on both sides we obtain with Lemma 5.6.3 and with the operators $\mathcal{A} = V$ and $\mathcal{B} = \frac{1}{2}I + K$, as given in (5.6.10), the desired equation (5.6.20) between μ and λ on Γ.

ii.) Now suppose that $\mu \in H^{\frac{1}{2}}(\Gamma)$ and $\lambda \in H^{-\frac{1}{2}}(\Gamma)$ satisfy the boundary integral equation (5.6.20) in the form

$$V\lambda = \left(\tfrac{1}{2}I + K\right)\mu \text{ on } \Gamma\,.$$

Then the function

$$u(x) := V\lambda(x) - W\mu(x) \text{ for } x \in \Omega$$

satisfies the differential equation

$$Pu = P(V\lambda) - P(W\mu) = 0 \text{ in } \Omega$$

since $V\lambda$ and $W\mu$ are boundary potentials defined by the fundamental solution $E(x, y)$ of P. Moreover, both potentials are in $H^1(\Omega, P)$ because of Theorem 5.6.2. Therefore, u admits the trace $\gamma_0 u|_\Gamma$. Taking the trace on both sides of the representation formula we have with (5.6.10)

$$\gamma_0 u = \gamma_0 V\lambda - \gamma_0 W\mu = V\lambda - K\mu + \frac{1}{2}\mu \text{ on } \Gamma\,.$$

Now replacing $V\lambda$ by using the boundary integral equation (5.6.20), one obtains

$$\gamma_0 u = K\mu + \frac{1}{2}\mu - K\mu + \frac{1}{2}\mu = \mu = \varphi \text{ on } \Gamma\,.$$

Clearly, the remaining cases can be treated in the same manner. We omit the details. ∎

5.6.6 Variational Formulation of Direct Boundary Integral Equations

Because of the equivalence Theorem 5.6.7, solutions of the boundary value problems can now be constructed from the solutions of the boundary integral equations (5.6.20). It will be seen that the existence of their solutions can be established by applying Fredholm's alternative Theorem 5.3.10 to (5.6.20) in corresponding Sobolev spaces. Similar to the elliptic partial differential equations, we consider the *weak, bilinear formulation* of (5.6.20). For simplicity, let us consider the case that $\mathcal{A}$ is a boundary integral operator of the first kind with the mapping property

$$\mathcal{A} : H^{\alpha}(\Gamma) \to H^{-\alpha}(\Gamma) \text{ continuously }. \tag{5.6.21}$$

In the examples of Table 5.6.3, we have $\mathcal{A} = V$ where the Sobolev order is $\alpha = -\frac{1}{2}$ or $\mathcal{A} = D + \{\pm\frac{1}{2}I - K'\} \circ (\boldsymbol{b} \cdot \boldsymbol{n})^{\top}$ where $\alpha = \frac{1}{2}$. As for partial differential operators, the number 2α is called the *order* of $\mathcal{A}$.
The **variational formulation for the boundary integral equations of the first kind** (5.6.20) now reads:

Find $\lambda \in H^{\alpha}(\Gamma)$ such that

$$a_{\Gamma}(\chi, \lambda) := \langle \chi, \overline{\mathcal{A}\lambda} \rangle_{\Gamma} = \langle \chi, \overline{f} \rangle_{\Gamma} \quad \text{for all } \chi \in H^{\alpha}(\Gamma), \tag{5.6.22}$$

where $f = \mathcal{B}\mu \in H^{-\alpha}(\Gamma)$ is given.

Now we shall show that the sesquilinear form $a_{\Gamma}(\chi, \lambda)$ in (5.6.22) is continuous on $H^{\alpha}(\Gamma)$, the *energy space* for $\mathcal{A}$, and also satisfies Gårding's inequality (5.4.1) under suitable assumptions on the corresponding boundary value problems.

Theorem 5.6.8. *To the integral operators $\mathcal{A}$ of the first kind in Table* 5.6.3 *there exist compact operators $\mathcal{C}_{\mathcal{A}} : H^{\alpha}(\Gamma) \to H^{-\alpha}(\Gamma)$ and positive constants $\gamma_{\mathcal{A}}$ such that*

$$Re\big\{a_{\Gamma}(\lambda, \lambda) + \langle \lambda, \overline{\mathcal{C}_{\mathcal{A}}\lambda} \rangle_{\Gamma}\big\} \geq \gamma_{\mathcal{A}} \|\lambda\|^2_{H^{\alpha}(\Gamma)} \quad \text{for all } \lambda \in H^{\alpha}(\Gamma). \tag{5.6.23}$$

Boundary value problem	Given data	Unknown data	$\mathcal{A}$	$\mathcal{B}$	Type of BIE	Represen-tation
IDP (5.1.19)	$\mu \in H^{\frac{1}{2}}(\Gamma)$,	$\lambda \in H^{-\frac{1}{2}}(\Gamma)$,	V	$\frac{1}{2}I + K$	1st	(5.6.1)
	$\mu = \gamma_0 u\|_\Gamma = \varphi$	$\lambda = \tau u\|_\Gamma - (\boldsymbol{b}\cdot\boldsymbol{n})\|_\Gamma \varphi$	$\frac{1}{2}I - K'$	$D - (\boldsymbol{b}\cdot\boldsymbol{n})I$	2nd	
EDP (5.1.52)	$\mu \in H^{\frac{1}{2}}(\Gamma)$,	$\lambda \in H^{-\frac{1}{2}}(\Gamma)$,	V	$-\frac{1}{2}I + K$	1st	(5.6.3)
	$\mu = \gamma_{c0} u\|_\Gamma = \varphi$	$\lambda = \tau_c u\|_\Gamma - (\boldsymbol{b}\cdot\boldsymbol{n})\|_\Gamma \varphi$	$\frac{1}{2}I + K'$	$-D - (\boldsymbol{b}\cdot\boldsymbol{n})I$	2nd	
INP (5.1.29)	$\mu \in H^{-\frac{1}{2}}(\Gamma)$,	$\lambda \in H^{\frac{1}{2}}(\Gamma)$,	$D - (\frac{1}{2}I + K')\circ(\boldsymbol{b}\cdot\boldsymbol{n})$	$\frac{1}{2}I - K'$	1st	(5.6.1)
	$\mu = \tau u\|_\Gamma = \psi$	$\lambda = \gamma_0 u\|_\Gamma$	$\frac{1}{2}I + K + V\circ(\boldsymbol{b}\cdot\boldsymbol{n})$	V	2nd	
ENP (5.1.70)	$\mu \in H^{-\frac{1}{2}}(\Gamma)$,	$\lambda \in H^{\frac{1}{2}}(\Gamma)$,	$D + (\frac{1}{2}I - K')\circ(\boldsymbol{b}\cdot\boldsymbol{n})$	$-(\frac{1}{2}I + K')$	1st	(5.6.3)
	$\mu = \tau_c u\|_\Gamma = \psi$	$\lambda = \gamma_{c0} u\|_\Gamma$	$\frac{1}{2}I - K - V\circ(\boldsymbol{b}\cdot\boldsymbol{n})$	$-V$	2nd	

Table 5.6.3. Boundary value problems and their associated boundary integral equations

Proof:

i) Let $\mathcal{A} = V$, choose $\lambda \in H^{\frac{1}{2}}(\Gamma)$ and define

$$u(x) := -V\lambda(x) \text{ for } x \in \mathbb{R}^n .$$

Then $u \in H^1(\Omega, P)$ and $u \in H^1_{\text{loc}}(\Omega^c, P)$, respectively. Moreover, Lemma 5.6.3 yields

$$[\gamma_0 u]_{|\Gamma} = 0 \text{ and } [\tau u]_{|\Gamma} = \lambda .$$

By adding the generalized first Green formulae (5.1.7) and (5.1.13) we obtain

$$Re\Big\{ \int_\Gamma \lambda \overline{V\lambda} ds + \int_{\Omega^c} (P\eta V\lambda)^\top \overline{(\eta V\lambda)} dx \Big\}$$
$$= Re\Big\{ - \int_\Gamma [\tau u] \overline{\gamma_0 u} ds + \int_{\Omega^c} (P\eta u)^\top \overline{\eta u} dx \Big\} = Re\,\{a_\Omega(u,u) + a_{\Omega^c}(\eta u, \eta u)\}$$

where $\eta \in C_0^\infty(\mathbb{R}^n)$ with $\eta_{|\overline{\Omega}} \equiv 1$ and $\text{dist}(\{x|\eta(x) \neq 1\}, \overline{\Omega}) =: d_0 > 0$ is any fixed chosen cut–off function since the first Green's formula for Ω^c is valid for $v = \eta u$ having compact support.

On the other hand, from Theorem 5.6.2, Lemma 5.6.3 and Lemma 5.1.1, the mappings $\tau : H^1(\Omega, P) \to H^{-\frac{1}{2}}(\Gamma)$ and $\tau_c : H^1_{\text{loc}}(\Omega^c, P) \to H^{-\frac{1}{2}}(\Gamma)$ are continuous providing the inequality

$$\begin{aligned} \|\lambda\|^2_{H^{-\frac{1}{2}}(\Gamma)} = \|[\tau u]\|^2_{H^{-\frac{1}{2}}(\Gamma)} &\leq 2\|\tau u\|^2_{H^{-\frac{1}{2}}(\Gamma)} + 2\|\tau_c u\|^2_{H^{-\frac{1}{2}}(\Gamma)} \\ &\leq c\Big\{\|u\|^2_{H^1(\Omega)} + \|\eta u\|^2_{H^1(\Omega')} + \|P\eta u\|^2_{L^2(\Omega')}\Big\} \end{aligned}$$

where $\Omega' := \Omega^c \cap \text{supp}\,\eta$. Then Gårding's inequality (5.4.2) for a_Ω in the domain Ω and for $a_{\Omega'}$ in Ω' implies

$$\alpha_0 \Big\{\|u\|^2_{H^1(\Omega)} + \|\eta u\|^2_{H^1(\Omega')}\Big\}$$
$$\leq Re\,\big\{a_\Omega(u,u) + a_{\Omega'}(\eta u, \eta u) + (Cu,u)_{H^1(\Omega)} + (C'\eta u, \eta u)_{H^1(\Omega')}\big\}$$

with a positive constant α_0 and compact operators C and C' depending also on Ω and Ω'. Collecting the inequalities, we get

$$\alpha'\|\lambda\|^2_{H^{-\frac{1}{2}}(\Gamma)} \leq Re\{a_\Gamma(\lambda,\lambda)$$
$$+c_1\|P\eta u\|^2_{L^2(\Omega')} + c_2(Cu,u)_{H^1(\Omega)} + c_3(C'\eta u, \eta u)_{H^1(\Omega')} + (P\eta u, \eta u)_{L^2(\Omega')}\} .$$

Accordingly, we define the operator $\mathcal{C}_\mathcal{A} : H^{-\frac{1}{2}}(\Gamma) \to H^{\frac{1}{2}}(\Gamma)$ by the sesquilinear form

$$\langle \chi, \overline{\mathcal{C}_\mathcal{A}\lambda}\rangle_\Gamma = c(\chi,\lambda) := c_2(V\chi, CV\lambda)_{H^1(\Omega)} + c_3(\eta V\chi, C'\eta V\lambda)_{H^1(\Omega')}$$
$$+c_1(P\eta V\chi, P\eta V\lambda)_{L^2(\Omega')} + (P\eta V\chi, V\lambda)_{L^2(\Omega')} .$$

Boundary value problem	Given data	Unknown data	$\mathcal{A}$	$\mathcal{B}$	Type of BIE	Represen-tation
IDP (5.1.19)	$\mu \in H^{\frac{1}{2}}(\Gamma)$,	$\lambda \in H^{-\frac{1}{2}}(\Gamma)$,	V	$\frac{1}{2}I + K$	1st	(5.6.1)
	$\mu = \gamma_0 u\|_\Gamma = \varphi$	$\lambda = \tau u\|_\Gamma - (\boldsymbol{b} \cdot \boldsymbol{n})\|_\Gamma \varphi$	$\frac{1}{2}I - K'$	$D - (\boldsymbol{b} \cdot \boldsymbol{n})I$	2nd	
EDP (5.1.52)	$\mu \in H^{\frac{1}{2}}(\Gamma)$,	$\lambda \in H^{-\frac{1}{2}}(\Gamma)$,	V	$-\frac{1}{2}I + K$	1st	(5.6.3)
	$\mu = \gamma_{c0} u\|_\Gamma = \varphi$	$\lambda = \tau_c u\|_\Gamma - (\boldsymbol{b} \cdot \boldsymbol{n})\|_\Gamma \varphi$	$\frac{1}{2}I + K'$	$-D - (\boldsymbol{b} \cdot \boldsymbol{n})I$	2nd	
INP (5.1.29)	$\mu \in H^{-\frac{1}{2}}(\Gamma)$,	$\lambda \in H^{\frac{1}{2}}(\Gamma)$,	$D - (\frac{1}{2}I + K') \circ (\boldsymbol{b} \cdot \boldsymbol{n})$	$\frac{1}{2}I - K'$	1st	(5.6.1)
	$\mu = \tau u\|_\Gamma = \psi$	$\lambda = \gamma_0 u\|_\Gamma$	$\frac{1}{2}I + K + V \circ (\boldsymbol{b} \cdot \boldsymbol{n})$	V	2nd	
ENP (5.1.70)	$\mu \in H^{-\frac{1}{2}}(\Gamma)$,	$\lambda \in H^{\frac{1}{2}}(\Gamma)$,	$D + (\frac{1}{2}I - K') \circ (\boldsymbol{b} \cdot \boldsymbol{n})$	$-(\frac{1}{2}I + K')$	1st	(5.6.3)
	$\mu = \tau_c u\|_\Gamma = \psi$	$\lambda = \gamma_{c0} u\|_\Gamma$	$\frac{1}{2}I - K - V \circ (\boldsymbol{b} \cdot \boldsymbol{n})$	$-V$	2nd	

Table 5.6.4. Boundary value problems and their associated boundary integral equations

We note that the operator $\mathcal{C}_{\mathcal{A}}$ is well defined since each term on the right–hand side is a bounded sesquilinear form on $\chi, \lambda \in H^{-\frac{1}{2}}(\Gamma)$. Hence, by the Riesz representation theorem 5.2.2 there exists a linear mapping $j\mathcal{C}_{\mathcal{A}} : H^{-\frac{1}{2}} \to H^{-\frac{1}{2}}(\Gamma)$ such that

$$c(\chi, \lambda) = (\chi, j\mathcal{C}_{\mathcal{A}}\lambda)_{H^{-\frac{1}{2}}(\Gamma)} .$$

Since $j^{-1} : H^{-\frac{1}{2}}(\Gamma) \to \left(H^{-\frac{1}{2}}(\Gamma)\right)^* = H^{\frac{1}{2}}(\Gamma)$, this representation can be written in the desired form

$$\langle \chi, \overline{\mathcal{C}_{\mathcal{A}}\lambda} \rangle_\Gamma = c(\chi, \lambda) \quad \text{for all} \quad \chi, \lambda \in H^{-\frac{1}{2}}(\Gamma)$$

where $\mathcal{C}_{\mathcal{A}} = j^{-1} j\mathcal{C}_{\mathcal{A}}$. It remains to show that $\mathcal{C}_{\mathcal{A}}$ is compact.

Let us begin with the first sesquilinear form on the right–hand side,

$$\langle \chi, \overline{\mathcal{C}_1\lambda} \rangle_\Gamma := c_2(V\chi, CV\lambda)_{H^1(\Omega)} \quad \text{where } C : H^1(\Omega) \to H^1(\Omega)$$

is compact. Since $V : H^{-\frac{1}{2}}(\Gamma) \to H^1(\Omega)$ is continuous, $CV : H^{-\frac{1}{2}}(\Gamma) \to H^1(\Omega)$ is compact. Moreover, we have the inequality

$$\|\mathcal{C}_1\lambda\|_{H^{\frac{1}{2}}(\Gamma)} \leq \sup_{\|\chi\|_{H^{-\frac{1}{2}}(\Gamma)} \leq 1} |\langle \chi, \overline{\mathcal{C}_1\lambda} \rangle_\Gamma| \leq c\|CV\lambda\|_{H^1(\Gamma)} \cdot \|V\|_{H^{-\frac{1}{2}}(\Gamma), H^1(\Omega)} .$$

Hence, any bounded sequence $\{\lambda_j\}$ in $H^{-\frac{1}{2}}(\Gamma)$ provides a convergent subsequence $\{CV\lambda_{j'}\}$ in $H^1(\Omega)$, and the corresponding subsequence $\{\mathcal{C}_1\lambda_{j'}\}$ converges in $H^{\frac{1}{2}}(\Gamma)$ due to the above inequality. This shows that $\mathcal{C}_1 : H^{-\frac{1}{2}}(\Gamma) \to H^{\frac{1}{2}}(\Gamma)$ is a compact linear operator. By the same arguments with Ω replaced by Ω', the operator $\mathcal{C}_2$ corresponding to

$$\langle \chi, \overline{\mathcal{C}_2\lambda} \rangle_\Gamma = c_3(\eta V\chi, C'\eta V\lambda)_{H^1(\Omega')}$$

is also compact.

In the last two terms we observe that

$$\overline{\Omega''} := \text{supp}(P\eta u) \cap \overline{\Omega'} \subset \overline{\Omega'} \quad \text{and dist } (\overline{\Omega''}, \Gamma) =: d_0 > 0 .$$

Now consider the mapping

$$P\eta V\lambda(x) = \langle P_x\eta(x)E(x, \cdot), \lambda \rangle_\Gamma$$

and observe that the support of this function is in $\overline{\Omega''}$. Hence, this function is in $C^\infty(\Omega'')$ for $\lambda \in H^{-\frac{1}{2}}(\Gamma)$. This implies that

$$P\eta V : H^{-\frac{1}{2}}(\Gamma) \to H^1(\Omega'')$$

is continuous and because of the compact imbedding $H^1(\Omega'') \hookrightarrow L^2(\Omega'')$,

$$P\eta V : H^{-\frac{1}{2}}(\Gamma) \to L^2(\Omega'') \text{ is compact .}$$

As a consequence, the operator $\mathcal{C}_3$ defined by

$$\langle \chi, \overline{\mathcal{C}_3 \lambda} \rangle_\Gamma = c_1 (P\eta V\chi, P\eta V\lambda)_{L^2(\Omega'')} = c_1 (P\eta V\chi, P\eta V\lambda)_{L^2(\Omega')}$$

is compact since

$$\|\mathcal{C}_3 \lambda\|_{H^{\frac{1}{2}}(\Gamma)} \le c_1 \|P\eta V\lambda\|_{L^2(\Omega'')} \cdot \|P\eta V\|_{H^{-\frac{1}{2}}(\Gamma), L^2(\Omega'')} .$$

The last term can be written in the form

$$\langle \chi, \overline{\mathcal{C}_4 \lambda} \rangle_\Gamma = \langle \mathcal{C}_4^* \chi, \overline{\lambda} \rangle_\Gamma = (P\eta X\chi, V\lambda)_{L^2(\Omega'')}$$

and $\mathcal{C}_4^*$ is compact which follows from the estimate

$$\|\mathcal{C}_4^* \chi\|_{H^{\frac{1}{2}}(\Gamma)} \le \|P\eta V\chi\|_{L^2(\Omega'')} \|V\|_{H^{-\frac{1}{2}}(\Gamma), L^2(\Omega'')}$$

and the compactness of $P\eta V$ as previously shown. This completes the proof for $\mathcal{A} = V$.

ii) Now we consider

$$\mathcal{A} = D + \{\pm\tfrac{1}{2}I - K'\} \circ (\boldsymbol{b} \cdot \boldsymbol{n}) =: D + D_c$$

where $\alpha = \frac{1}{2}$. For any $\lambda \in H^{\frac{1}{2}}(\Gamma)$ define

$$u(x) := V \circ (\boldsymbol{b} \cdot \boldsymbol{n})\lambda(x) + W\lambda(x) \ \text{ for } x \in \mathbb{R}^n \setminus \Gamma \tag{5.6.24}$$

according to (5.6.5) and $\widetilde{u}(x) := \eta(x) u(x)$ for $x \in \Omega^c$ with η given as in case **i.**). Then Theorem 5.6.2 implies that $u \in H^1(\Omega, P)$ for $x \in \Omega$ and $\widetilde{u} \in H^1(\Omega', P)$ for $x \in \Omega'$. Moreover, Lemma 5.6.3 yields the jump relations

$$[\gamma_0 u]_\Gamma = \lambda , \ [\tau u]_\Gamma = -(\boldsymbol{b} \cdot \boldsymbol{n})\lambda$$

and the one sided jump relations from Ω,

$$\begin{aligned} \gamma_0 u &= V \circ (\boldsymbol{b} \cdot \boldsymbol{n})\lambda + (-\frac{1}{2}I + K)\lambda \ \text{ and} \\ \tau u &= (\frac{1}{2}I + K') \circ (\boldsymbol{b} \cdot \boldsymbol{n})\lambda - D\lambda \ \text{ on } \Gamma . \end{aligned}$$

As before, adding the generalized first Green's formulae (5.1.8) and (5.1.13), we obtain

$$\begin{aligned} \langle \lambda, \overline{D\lambda} \rangle_\Gamma &= a_\Omega(u, u) + a_{\Omega^c}(\eta u, \eta u) \\ &\quad - \int\limits_{\Omega''} (\eta u)^\top \overline{P\eta u} dx - \langle (\boldsymbol{b} \cdot \boldsymbol{n})\lambda, \overline{V \circ (\boldsymbol{b} \cdot \boldsymbol{n})\lambda} \rangle_\Gamma . \end{aligned}$$

Hence,

$$\begin{aligned} Re a_\Gamma(\lambda,\lambda) \;\geq\; & \alpha_0\Big\{\|u\|^2_{H^1(\Omega)} + \|\eta u\|^2_{H^1(\Omega')}\Big\} \\ & -Re\{(u,Cu)_{H^1(\Omega)} + (C\eta u, C'\eta u)_{H^1(\Omega')} + \int\limits_{\Omega''}(\eta u)^\top \overline{P\eta u}dx \\ & +\langle(\boldsymbol{b}\cdot\boldsymbol{n})\lambda, \overline{V\circ(\boldsymbol{b}\cdot\boldsymbol{n})\lambda}\rangle_\Gamma - \langle\lambda, \overline{D_c\lambda}\rangle_\Gamma\}\,. \end{aligned}$$

From the definition of λ we see that

$$\|\lambda\|^2_{H^{\frac12}(\Gamma)} = \|[\gamma_0 u]\|^2_{H^{\frac12}(\Gamma)} \leq c\Big\{\|u\|^2_{H^1(\Omega)} + \|\eta u\|^2_{H^1(\Omega')}\Big\}$$

following from the trace theorem in the form (4.2.40). This implies (5.6.23) with $\alpha = \frac12$ and

$$\begin{aligned} \langle\chi, \overline{\mathcal{C}_A\lambda}\rangle_\Gamma = & (v,Cu)_{H^1(\Omega)} + (\eta v, C'\eta u)_{H^1(\Omega')} \\ & + \int\limits_{\Omega''}(\eta v)^\top\overline{P\eta u}dx + \big(\chi, (\boldsymbol{b}\cdot\boldsymbol{n})V\circ(\boldsymbol{b}\cdot\boldsymbol{n})\lambda\big)_{L^2(\Gamma)} \\ & - \langle\chi, \overline{\{\pm\tfrac12 I - K'\}\circ(\boldsymbol{b}\cdot\boldsymbol{n})\lambda}\rangle_\Gamma \end{aligned}$$

where

$$\begin{aligned} v(x) &= V\circ(\boldsymbol{b}\cdot\boldsymbol{n})\chi(x) + W\chi(x) \text{ and} \\ u(x) &= V\circ(\boldsymbol{b}\cdot\boldsymbol{n})\lambda(x) + W\lambda(x) \text{ for } x\in\mathbb{R}^n\setminus\Gamma\,. \end{aligned}$$

The first three sesquilinear forms are defined by compact operators which can be analyzed as in case **i.)**. For the last two terms we have the compact imbeddings $i_{c1} : H^{\frac12}(\Gamma) \hookrightarrow L^2(\Gamma)$ and $i_{c2} : L^2(\Gamma) \hookrightarrow H^{-\frac12}(\Gamma)$. Then, with the continuity properties of $V : H^{-\frac12}(\Gamma) \to H^{\frac12}(\Gamma)$ and $(\boldsymbol{b}\cdot\boldsymbol{n}) : L^2(\Gamma) \to L^2(\Gamma)$ (note that $\boldsymbol{b}\cdot\boldsymbol{n} \in L^\infty(\Gamma)$), the following composition of mappings

$$\begin{aligned} H^{\frac12}(\Gamma) & \overset{i_{c1}}{\hookrightarrow} L^2(\Gamma) \overset{(\boldsymbol{b}\cdot\boldsymbol{n})}{\to} L^2(\Gamma) \overset{i_{c2}}{\hookrightarrow} H^{-\frac12}(\Gamma) \overset{V}{\to} H^{\frac12}(\Gamma) \\ & \overset{i_{c1}}{\hookrightarrow} L^2(\Gamma) \overset{(\boldsymbol{b}\cdot\boldsymbol{n})}{\to} L^2(\Gamma) \overset{i_{c2}}{\hookrightarrow} H^{-\frac12}(\Gamma) \end{aligned}$$

gives the compact mapping

$$i_{c2}\circ(\boldsymbol{b}\cdot\boldsymbol{n})\circ i_{c1}\circ V\circ i_{c2}\circ(\boldsymbol{b}\cdot\boldsymbol{n})\circ i_{c1} : H^{\frac12}(\Gamma)\to H^{-\frac12}(\Gamma)\,.$$

Similarly, we see that the mapping

$$\{\pm\tfrac12 I - K'\}\circ i_{c2}\circ(\boldsymbol{b}\cdot\boldsymbol{n})\circ i_{c1} : H^{\frac12}(\Gamma)\to H^{-\frac12}(\Gamma)$$

is compact. Collecting these results completes the proof. ∎

Integral equations of the second kind

Next, let us consider the cases that $\mathcal{A}$ is one of the boundary integral operators of the second kind in Table 5.6.3 with the mapping properties

$$\mathcal{A} : H^\alpha(\Gamma) \to H^\alpha(\Gamma) \ \text{ continuously} \tag{5.6.25}$$

where $\alpha = \frac{1}{2}$ or $-\frac{1}{2}$, respectively. The order of $\mathcal{A}$ now is equal to zero. Therefore, the *variational formulation* of (5.6.20) here requires the use of the $H^\alpha(\Gamma)$–scalar product instead of the L^2–duality:

Find $\lambda \in H^\alpha(\Gamma)$ *such that*

$$a_\Gamma(\chi, \lambda) := (\chi, \mathcal{A}\lambda)_{H^\alpha(\Gamma)} = (\chi, f)_{H^\alpha(\Gamma)} \ \text{ for all } \chi \in H^\alpha(\Gamma) \tag{5.6.26}$$

where $f = \mathcal{B}\mu \in H^\alpha(\Gamma)$ *is given.*

From Theorem 5.6.8 we have Gårding's inequalities available for the operators V and D on $H^{-\frac{1}{2}}(\Gamma)$ and $H^{\frac{1}{2}}(\Gamma)$, respectively. Moreover, we have the mapping properties $V : H^{-\frac{1}{2}}(\Gamma) \to H^{\frac{1}{2}}(\Gamma)$ and $D : H^{\frac{1}{2}}(\Gamma) \to H^{-\frac{1}{2}}(\Gamma)$. This allows us to modify the variational formulation (5.6.26) by replacing the $H^\alpha(\Gamma)$–scalar products in the following way:

Variational formulation of the boundary integral equations of the second kind

i.) *For* $\alpha = -\frac{1}{2}$ *and* $\mathcal{A} = (\frac{1}{2}I \pm K')$ *in the* IDP *and* EDP*:*
Find $\lambda \in H^{-\frac{1}{2}}(\Gamma)$ *such that*

$$a_\Gamma(\chi, \lambda) := \langle V\chi, \overline{\{\tfrac{1}{2}I \pm K'\}\lambda}\rangle_\Gamma = \langle V\chi, \overline{f}\rangle_\Gamma \ \text{ for all } \chi \in H^{-\frac{1}{2}}(\Gamma)\,. \tag{5.6.27}$$

ii.) *For* $\alpha = \frac{1}{2}$ *and* $\mathcal{A} = \{\frac{1}{2}I \pm K \pm V \circ (\boldsymbol{b} \cdot \boldsymbol{n})\}$ *in the* INP *and* ENP*:*
Find $\lambda \in H^{\frac{1}{2}}(\Gamma)$ *such that*

$$a_\Gamma(\chi, \lambda) := \langle D\chi, \overline{\{\tfrac{1}{2}I \pm K \pm V \circ (\boldsymbol{b} \cdot \boldsymbol{n})\}\lambda}\rangle_\Gamma = \langle D\chi, \overline{f}\rangle_\Gamma \tag{5.6.28}$$

for all $\chi \in H^{\frac{1}{2}}(\Gamma)$.

In the following, let us first consider Gårding's inequalities for the boundary sesquilinear forms (5.6.26) and (5.6.27), (5.6.28). As will be seen, the boundary integral equations of the second kind are also intimately related to the domain variational formulations. Whereas for the first kind boundary integral equations, the transmission problems played the decisive role, here, the boundary integral operators of the second kind associated with interior boundary value problems are related to the variational formulation of the differential equation in the exterior domain and, likewise, the boundary integral operator of the second kind associated with the exterior boundary value problems are related to the interior variational problems.

Theorem 5.6.9. *To the boundary sesquilinear forms* (5.6.27) and (5.6.28) *there exist compact operators* $\mathcal{C}_\mathcal{A} : H^\alpha(\Gamma) \to H^{-\alpha}(\Gamma)$ *and positive constants* $\gamma_\mathcal{A}$ *such that*

$$Re\{a_\Gamma(\lambda,\lambda) + c_\Gamma(\lambda,\lambda)\} \geq \gamma_\mathcal{A}\|\lambda\|^2_{H^\alpha(\Gamma)} \text{ for all } \lambda \in H^\alpha(\Gamma)\,, \tag{5.6.29}$$

where $c_\Gamma(\cdot,\cdot)$ *is a compact sesquilinear form with* $c_\Gamma(\chi,\lambda) = \langle \chi, \overline{\mathcal{C}_\mathcal{A}\lambda}\rangle_\Gamma$.

Proof:

i.) We begin with the IDP (5.1.19), and the equation in Table 5.6.3 where (5.6.20) reads

$$\mathcal{A}\lambda = (\tfrac{1}{2}I - K')\lambda = f := \{D - (\boldsymbol{b}\cdot\boldsymbol{n})\circ I\}\mu = \mathcal{B}\mu$$

for $\lambda \in H^{-\frac{1}{2}}(\Gamma)$ and given $\mu \in H^{\frac{1}{2}}(\Gamma)$ and $f \in H^{\frac{1}{2}}(\Gamma)$. Define

$$u(x) := V\lambda \text{ in } \mathbb{R}^n \text{ and take } \widetilde{u} = \phi u \text{ in } \Omega^c$$

as in the proof of Theorem 5.6.8. Then we obtain

$$\gamma_{c0}\widetilde{u} = V\lambda \text{ and } \tau_c\widetilde{u} = -(\tfrac{1}{2}I - K')\lambda \text{ on } \Gamma\,.$$

Hence, with the generalized Green's formula (5.1.13) we find

$$Re\langle V\lambda, \overline{\mathcal{A}\lambda}\rangle_\Gamma = -Re\langle \gamma_{c0}u, \overline{\tau_c u}\rangle_\Gamma = Re\,a_{\Omega^c}(\widetilde{u},\widetilde{u}) - Re\int\limits_{\Omega'} (P\widetilde{u})^\top \overline{\widetilde{u}}dx\,.$$

Since a_{Ω^c} satisfies Gårding's inequality (5.4.50) for $\widetilde{u}$ having compact support in $\overline{\Omega'}$, we obtain

$$Re\langle V\lambda, \overline{\mathcal{A}\lambda}\rangle_\Gamma \geq c_0\|\widetilde{u}\|^2_{H^1(\Omega')} - Re(C'\widetilde{u},\widetilde{u})_{H^1(\Omega')} - Re\int\limits_{\Omega'} (P\eta u)^\top \overline{\eta u}dx$$

and with the trace theorem, Theorem 4.2.1, (4.2.29),

$$Re\langle V\lambda, \overline{\mathcal{A}\lambda}\rangle_\Gamma \geq c_0\|V\lambda\|^2_{H^{\frac{1}{2}}(\Gamma)} - Re\{(C'\eta u, \eta u)_{H^1(\Omega')} + \int\limits_{\Omega''} (P\eta u)^\top \overline{\eta u}dx\}\,. \tag{5.6.30}$$

For V we already have established Gårding's inequality in Theorem 5.6.8, hence

$$Re\langle V\lambda, \overline{\lambda}\rangle_\Gamma + Re\langle \mathcal{C}_V\lambda, \overline{\lambda}\rangle_\Gamma \geq c_0\|\lambda\|^2_{H^{-\frac{1}{2}}(\Gamma)}$$

with $c_0 > 0$ [1] and $\mathcal{C}_V : H^{-\frac{1}{2}}(\Gamma) \to H^{\frac{1}{2}}(\Gamma)$ compactly. This inequality together with (4.2.22) yields the estimate

[1] Here and in the sequel, $c_0 > 0$ denotes a generic constant whose value may change from step to step.

$$c_0\|\lambda\|_{H^{-\frac{1}{2}}(\Gamma)} \le \|V\lambda\|_{H^{\frac{1}{2}}(\Gamma)} + \|\mathcal{C}_V\lambda\|_{H^{\frac{1}{2}}(\Gamma)}$$

and, consequently,

$$c_0^2\|\lambda\|^2_{H^{-\frac{1}{2}}(\Gamma)} \le (V\lambda, V\lambda)_{H^{\frac{1}{2}}(\Gamma)} + (\mathcal{C}_V\lambda, \mathcal{C}_V\lambda)_{H^{\frac{1}{2}}(\Gamma)}\,.$$

Inserting this inequality into (5.6.30) we obtain

$$Re\Big\{\langle V\lambda, \overline{\mathcal{A}\lambda}\rangle_\Gamma + c_\Gamma(\lambda,\lambda)\Big\} \ge c_0^2\|\lambda\|^2_{H^{-\frac{1}{2}}(\Gamma)}$$

where

$$c_\Gamma(\chi,\lambda) := (\eta V\chi, C'\eta V\lambda)_{H^1(\Omega')} + \int\limits_{\Omega''} (\eta V\chi)^\top \overline{(P\eta V\lambda)}dx + (\mathcal{C}_V\chi, \mathcal{C}_V\lambda)_{H^{\frac{1}{2}}(\Gamma)}\,.$$

As in the proof of Theorem 5.6.8, all three bilinear forms defining $c_\Gamma(\cdot,\cdot)$ are compact; and c_Γ can be represented in the proposed form.

ii.) For the EDP with the corresponding equation in Table 5.6.3 we have

$$\mathcal{A}\lambda = (\tfrac{1}{2}I + K')\lambda = f := -\{D + (\boldsymbol{b}\cdot\boldsymbol{n})\circ I\}\mu = \mathcal{B}\mu$$

and define now

$$u := V\lambda \text{ in } \mathbb{R}^n \text{ with } \gamma_0 u = V\lambda \text{ and } \tau u = (\tfrac{1}{2}I + K')\lambda \text{ on } \Gamma\,.$$

With the generalized first Green's formula (5.1.8) we now have

$$Re\langle V\lambda, \overline{\mathcal{A}\lambda}\rangle_\Gamma = Re\langle \gamma_0 u, \overline{\tau u}\rangle_\Gamma = Re\, a_\Omega(u,u)\,.$$

The remaining arguments of the proof are the same as in **i.)** but with the interior domain Ω instead of the exterior domain, $\Omega' \subset \Omega^c$.

iii) For the INP with the corresponding equations in Table 5.6.3 the operators in (5.6.20) now read

$$\mathcal{A}\lambda = \{\tfrac{1}{2}I + K + V\circ(\boldsymbol{b}\cdot\boldsymbol{n})\}\lambda = f := V\mu = \mathcal{B}\mu \text{ for } \lambda \in H^{\frac{1}{2}}(\Gamma)$$

with given $\mu \in H^{-\frac{1}{2}}(\Gamma)$ and $f \in H^{\frac{1}{2}}(\Gamma)$. Now we define

$$u := W\lambda \text{ in } \mathbb{R}^n \setminus \Gamma \text{ and } \widetilde{u} := \eta u \text{ in } \Omega^c\,.$$

Similarly, we obtain

$$Re\langle D\lambda, \overline{\mathcal{A}\lambda}\rangle_\Gamma = -Re\langle \tau_c\widetilde{u}, \overline{\gamma_{c0}\widetilde{u}}\rangle_\Gamma + Re\langle D\lambda, \overline{V\circ(\boldsymbol{b}\cdot\boldsymbol{n})\lambda}\rangle_\Gamma$$

$$= Re\; a_{\Omega^c}(\widetilde{u},\widetilde{u}) - Re\int\limits_{\Omega'} (P\eta u)^\top \overline{\eta u}dx + Re\langle D\lambda, \overline{V\circ(\boldsymbol{b}\cdot\boldsymbol{n})\lambda}\rangle_\Gamma\,.$$

Gårding's inequality (5.4.50) implies that

$$Re\langle D\lambda, \overline{\mathcal{A}\lambda}\rangle_\Gamma \geq c_0\|\widetilde{u}\|^2_{H^1(\Omega')} - Re(C'\widetilde{u}, \widetilde{u})_{H^1(\Omega')} - Re\int\limits_{\Omega''}(P\eta u)^\top \overline{\eta u}dx\,.$$

Now we need the continuity of the generalized conormal derivative $\tau_c\widetilde{u}$ from Lemma 5.1.2 which yields

$$\|\tau_c u\|^2_{H^{-\frac{1}{2}}[\Gamma)} = \|D\lambda\|^2_{H^{-\frac{1}{2}}(\Gamma)} \leq c\|\widetilde{u}\|^2_{H^1(\Omega')} + c(P\widetilde{u}, P\widetilde{u})_{L^2(\Omega')}$$

providing

$$Re\langle D\lambda, \overline{\mathcal{A}\lambda}\rangle_\Gamma \geq c_0\|D\lambda\|^2_{H^{-\frac{1}{2}}(\Gamma)} - Re\{(C'\eta u, \eta u)_{H^1(\Omega')} + \int\limits_{\Omega''}(P\eta u)^\top \overline{\eta u}dx$$
$$+c\int\limits_{\Omega''}(P\eta u)^\top \overline{P\eta u}dx - \langle D\lambda, \overline{V\circ(\boldsymbol{b}\cdot\boldsymbol{n})\lambda}\rangle_\Gamma\}\,.$$

Gårding's inequality for D as shown in Theorem 5.6.8 gives the estimate

$$\|D\lambda\|^2_{H^{-\frac{1}{2}}(\Gamma)} \geq c_0\|\lambda\|^2_{H^{\frac{1}{2}}(\Gamma)} - c(\mathcal{C}_D\lambda, \mathcal{C}_D\lambda)_{H^{-\frac{1}{2}}(\Gamma)}$$

where $\mathcal{C}_D : H^{\frac{1}{2}}(\Gamma) \to H^{-\frac{1}{2}}(\Gamma)$ is compact. Collecting the above estimates finally yields

$$Re\Big\{\langle D\lambda, \overline{\mathcal{A}\lambda}\rangle_\Gamma + c_\Gamma(\lambda,\lambda)\Big\} \geq c_0\|\lambda\|^2_{H^{\frac{1}{2}}(\Gamma)}$$

where

$$c_\Gamma(\chi,\lambda) = c(\mathcal{C}_D\chi, \mathcal{C}_D\lambda)_{H^{-\frac{1}{2}}(\Gamma)} + c(\eta W\chi, C'\eta W\lambda)_{H^1(\Omega')}$$
$$+c\int\limits_{\Omega''}(\eta W\chi)^\top \overline{P\eta W\lambda}dx + c\int\limits_{\Omega''}(P\eta W\chi)^\top \overline{(P\eta W\lambda)}dx - \langle D\chi, \overline{V\circ(\boldsymbol{b}\cdot\boldsymbol{n})\lambda}\rangle_\Gamma\,.$$

The compactness of all the sesquilinear forms on the right–hand side except the last one has already been shown in the proof of Theorem 5.6.8. The compactness of the last one follows again from the compact composition of the mappings

$$D\circ V\circ i_{c2}\circ(\boldsymbol{b}\cdot\boldsymbol{n})\circ i_{c1} : H^{\frac{1}{2}}(\Gamma) \to H^{-\frac{1}{2}}(\Gamma)\,.$$

iv) For the ENP (5.1.70), the operator $\mathcal{A}$ in Table 5.6.3 is of the form

$$\mathcal{A}\lambda = \{\tfrac{1}{2}I - K - V\circ(\boldsymbol{b}\cdot\boldsymbol{n})\}\lambda = f := -V\mu = \mathcal{B}\mu \quad \text{for } \lambda\in H^{\frac{1}{2}}(\Gamma)$$

with given $\mu \in H^{-\frac{1}{2}}(\Gamma)$ and $f\in H^{\frac{1}{2}}(\Gamma)$. As in case **iii)** take

$$u = W\lambda \text{ in } \Omega \text{ with } \gamma_0 u = -(\tfrac{1}{2}I - K)\lambda \text{ and } \tau u = -D\lambda \text{ on } \Gamma .$$

Then

$$\begin{aligned} Re\langle D\lambda, \overline{\mathcal{A}\lambda}\rangle_\Gamma &= Re\langle \tau u, \overline{\gamma_0 u}\rangle_\Gamma - Re\langle D\lambda, \overline{V \circ (\boldsymbol{b}\cdot\boldsymbol{n})\lambda}\rangle_\Gamma \\ &= a_\Omega(u,u) - Re\langle D\lambda, \overline{V \circ (\boldsymbol{b}\cdot\boldsymbol{n})\lambda}\rangle_\Gamma . \end{aligned}$$

The rest of the proof follows in the same manner as in case **iii)** with Ω^c replaced by the interior domain Ω. This completes the proof of Theorem 5.6.9. ■

Remark 5.6.3: For the bilinear form (5.6.26), Gårding's inequality

$$Re\{(\lambda, \mathcal{A}\lambda)_{H^\alpha(\Gamma)} + c_\Gamma(\lambda,\lambda)\} \geq \gamma_{\mathcal{A}} \|\lambda\|^2_{H^\alpha(\Gamma)} \tag{5.6.31}$$

with the operators $\mathcal{A}$ of the second kind in Table 5.6.3 follows immediately provided $K : H^{\frac{1}{2}}(\Gamma) \to H^{\frac{1}{2}}(\Gamma)$ and, consequently, the dual operator $K' : H^{-\frac{1}{2}}(\Gamma) \to H^{-\frac{1}{2}}(\Gamma)$, are compact. Here, the corresponding compact bilinear forms are defined by $c_\Gamma(\chi,\lambda) = \pm(\chi, \{K + V \circ (\boldsymbol{b}\cdot\boldsymbol{n})\}\lambda)_{H^{\frac{1}{2}}(\Gamma)}$ and $\pm(\chi, K'\lambda)_{H^{-\frac{1}{2}}(\Gamma)}$, respectively. However, the compactness of K or K' can only be established for a rather limited class of boundary value problems excluding the important problems in elasticity where K and K' are Cauchy singular integral operators, but including the classical double layer potential operators of the Laplacian, of the Helmholtz equation, of the Stokes problem and the like, provided Γ is at least $C^{1,\alpha}$ with $\alpha > 0$ excluding $C^{0,1}$–Lipschitz boundaries which are considered here. (See also Remark 1.2.1 and Chapter 8.)

Theorem 5.6.9 is valid without these limitations since it is based only on continuity, the trace theorems on Lipschitz domains and Gårding's inequality for the domain sesquilinear forms.

5.6.7 Positivity and Contraction of Boundary Integral Operators

Here we follow the presentation in Steinbach et al [401]. In [84], Costabel showed that these relations are rather general and have a long history.

As in Carl Neumann's classical method for the solution of the Dirichlet or Neumann problem in potential theory, we now show that for a certain class of boundary integral equations of the second kind Carl Neumann's method for solving these equations with the Neumann series can still be carried out in the corresponding Sobolev spaces.

For this purpose we consider the special class of boundary value problems leading to the boundary integral operators of the first kind with V and D which are positive in the sense that they satisfy Gårding inequalities and in addition

$$\langle V\lambda, \lambda\rangle \geq c_1^V \|\lambda\|^2_{H^{-\frac{1}{2}}(\Gamma)} \qquad \text{for all } \lambda \in H^{-\frac{1}{2}}(\Gamma) \tag{5.6.32}$$

$$\langle D\mu, \mu\rangle \geq c_1^D \|\mu\|^2_{H^{\frac{1}{2}}(\Gamma)} \qquad \text{for all } \mu \in H^{\frac{1}{2}}_{\Re}(\Gamma) \tag{5.6.33}$$

where

$$H^{\frac{1}{2}}_{\Re}(\Gamma) := \{\mu \in H^{\frac{1}{2}}(\Gamma) \mid \langle \mu, m\rangle = 0 \ \text{ for all } m \in \Re\} \tag{5.6.34}$$

and

$$\Re := \{\mu_0 \in H^{\frac{1}{2}}(\Gamma) \mid D\mu_0 = 0\}\,, \tag{5.6.35}$$

with constants $0 < c_1^V \leq \frac{1}{2}$, $0 < c_1^D \leq \frac{1}{2}$.

Since D satisfies a Gårding inequality the dimension of $\Re$ is finite. Moreover, these inequalities are satisfied if in the previous sections the corresponding energy forms of the underlying partial differential operators are finite (Costabel [84]).

With the Calderón projectors (5.6.11) we have also the relations (1.2.24)–(1.2.28) in the Sobolev spaces $H^{\frac{1}{2}}(\Gamma)$ and $H^{-\frac{1}{2}}(\Gamma)$, respectively, as in Theorem 5.6.2 for the four basic boundary integral operators V, K, K', D in (5.6.10).

In order to show the contraction properties of the operators $\frac{1}{2}I \pm K$ and $\frac{1}{2}I \pm K'$, we now introduce the norms

$$\|\lambda\|_V := \langle V\lambda, \lambda\rangle^{\frac{1}{2}} \quad \text{and} \quad \|\mu\|_{V^{-1}} := \langle V^{-1}\mu, \mu\rangle^{\frac{1}{2}} \tag{5.6.36}$$

which are equivalent to $\|\lambda\|_{H^{-\frac{1}{2}}(\Gamma)}$ and $\|\mu\|_{H^{\frac{1}{2}}(\Gamma)}$, respectively. More precisely, the following estimates are valid.

Lemma 5.6.10. *With the constants c_1^V and c_1^D in* (5.6.32) *and* (5.6.33) *we have*

$$c_1^V \|\mu\|^2_{V^{-1}} \leq \|\mu\|^2_{H^{\frac{1}{2}}(\Gamma)} \qquad \text{for all } \mu \in H^{\frac{1}{2}}(\Gamma)\,, \tag{5.6.37}$$

$$c_1^D \|\mu\|^2_{H^{\frac{1}{2}}(\Gamma)} \leq \langle S\mu, \mu\rangle \qquad \text{for all } \mu \in H^{\frac{1}{2}}_{\Re}(\Gamma) \tag{5.6.38}$$

and

$$\|V^{-\frac{1}{2}}\mu\|_{L^2(\Gamma)} = \|\mu\|_{V^{-1}} = \langle V^{-1}\mu, \mu\rangle^{\frac{1}{2}}\,. \tag{5.6.39}$$

Here

$$S = D + (\tfrac{1}{2}I + K')V^{-1}(\tfrac{1}{2}I + K) = V^{-1}(\tfrac{1}{2}I + K) \tag{5.6.40}$$

denotes the Steklov–Poincaré operator associated with Ω.

As we can see, for the solution u of the homogeneous equation (5.1.1) with $f = 0$ and given Dirichlet data $\mu \in H^{\frac{1}{2}}(\Gamma)$ on Γ, the Steklov–Poincaré operator in (5.6.40) maps μ into the Neumann data $\partial_\nu u = \lambda \in H^{-\frac{1}{2}}(\Gamma)$ on Γ. Therefore S is often also called the *Dirichlet to Neumann map*.

Proof: By definition,

$$\|V\lambda\|_{H^{\frac{1}{2}}(\Gamma)} = \sup_{0\neq\tau\in H^{-\frac{1}{2}}(\Gamma)} \frac{\langle V\lambda,\tau\rangle}{\|\tau\|_{H^{-\frac{1}{2}}(\Gamma)}} \geq \frac{|\langle V\lambda,\lambda\rangle|}{\|\lambda\|_{H^{-\frac{1}{2}}(\Gamma)}} \geq c_1^V \|\lambda\|_{H^{-\frac{1}{2}}(\Gamma)}$$

due to (5.6.32). On the other hand, with (4.1.27) and $\mu = V\lambda$ we have

$$\begin{aligned}\|V^{-1}\mu\|_{H^{-\frac{1}{2}}(\Gamma)} &= \sup_{0\neq\chi\in H^{\frac{1}{2}}(\Gamma)} \frac{|\langle V^{-1}\mu,\chi\rangle|}{\|\chi\|_{H^{\frac{1}{2}}(\Gamma)}} \\ &= \sup_{0\neq V\lambda\in H^{\frac{1}{2}}(\Gamma)} \frac{|\langle \mu,\lambda\rangle|}{\|V\lambda\|_{H^{\frac{1}{2}}(\Gamma)}} \leq \frac{1}{c_1^V}\|\mu\|_{H^{\frac{1}{2}}(\Gamma)}\end{aligned}$$

in view of the previous estimate. So, again with (4.1.27),

$$\langle V^{-1}\mu,\mu\rangle = \|\mu\|^2_{V^{-1}} \leq \|V^{-1}\mu\|_{H^{-\frac{1}{2}}(\Gamma)}\|\mu\|_{H^{\frac{1}{2}}(\Gamma)} \leq \frac{1}{c_1^V}\|\mu\|^2_{H^{\frac{1}{2}}(\Gamma)}$$

which establishes (5.6.37). The second inequality follows from the symmetric form of the Steklov–Poincaré operator, i.e.,

$$\begin{aligned}\langle S\mu,\mu\rangle &= \langle\big(D + (\tfrac{1}{2}I + K')V^{-1}(\tfrac{1}{2}I + K)\big)\mu,\mu\rangle \\ &= \langle D\mu,\mu\rangle + \|(\tfrac{1}{2}I + K)\mu\|^2_{V^{-1}} \geq c_1^D\|\mu\|^2_{H^{\frac{1}{2}}(\Gamma)} \quad \text{for } \mu\in H^{\frac{1}{2}}_{\Re}(\Gamma)\end{aligned}$$

which is the desired inequality (5.6.38).

The relation(5.6.39) is a simple consequence of the positivity and selfadjointness of V which defines an isomorphic mapping $H^{-\frac{1}{2}}(\Gamma) \to H^{\frac{1}{2}}(\Gamma)$, hence $V^{-1} : H^{-\frac{1}{2}}(\Gamma) \to H^{\frac{1}{2}}(\Gamma)$ is also positive and the square root $V^{-\frac{1}{2}} : H^{-\frac{1}{2}}(\Gamma) \to L^2(\Gamma)$ is well defined (Rudin [367, Theorem 12.33]). ∎

We are now in the position to formulate the contraction properties of the operators $(\frac{1}{2}I \pm K)$ and $(\frac{1}{2}I \pm K')$.

Theorem 5.6.11. *The operators* $\frac{1}{2}I + K$ *and* $\frac{1}{2}I + K'$ *are contractions on the corresponding energy spaces and*

$$\|(\tfrac{1}{2}I \pm K)\mu\|_{V^{-1}} \leq c_K\|\mu\|_{V^{-1}} \quad \textit{for all } \mu\in\begin{cases} H^{\frac{1}{2}}(\Gamma) \\ H^{-\frac{1}{2}}_{\Re}(\Gamma)\end{cases}, \tag{5.6.41}$$

$$\|(\tfrac{1}{2}I \pm K')\lambda\|_{V} \leq c_K\|\lambda\|_{V} \quad \textit{for all } \lambda\in\begin{cases} H^{-\frac{1}{2}}(\Gamma) \\ H^{-\frac{1}{2}}_{\Re}(\Gamma)\end{cases} \tag{5.6.42}$$

where

$$c_K = \tfrac{1}{2} + \sqrt{\tfrac{1}{4} - c_1^V c_1^D} < 1$$

and where the respective upper cases correspond to the + signs and the lower cases to the − signs. Here

$$H_{\Re}^{-\frac{1}{2}}(\Gamma) := \{\lambda \in H^{-\frac{1}{2}}(\Gamma) \,|\, \langle m, \lambda\rangle = 0 \quad \textit{for all} \;\; m \in \Re\}\,.$$

Moreover, we have

$$(1 - c_K)\|\mu\|_{V^{-1}} \le \|(\tfrac{1}{2}I \pm K)\mu\|_{V^{-1}} \quad \textit{for all} \;\; \mu \in H_{\Re}^{\frac{1}{2}}(\Gamma) \tag{5.6.43}$$

and

$$\|(\tfrac{1}{2}I - K)\mu_0\|_{V^{-1}} = \|\mu_0\|_{V^{-1}} \quad \textit{for} \;\; \mu_0 \in \Re\,. \tag{5.6.44}$$

Proof: We begin with the operator $(\frac{1}{2}I + K)$ and (5.6.41). Here,

$$\begin{aligned}\|(\tfrac{1}{2}I + K)\mu\|_{V^{-1}}^2 &= \langle V^{-1}(\tfrac{1}{2}I + K)\mu\,,\, (\tfrac{1}{2}I + K)\mu\rangle \\ &= \langle(\tfrac{1}{2}I + K')V^{-1}(\tfrac{1}{2}I + K)\mu, \mu\rangle = \langle S\mu, \mu\rangle - \langle D\mu, \mu\rangle\,.\end{aligned}$$

Hence, with (5.6.39) we have

$$\begin{aligned}\langle S\mu, \mu\rangle &= \langle V^{-\frac{1}{2}}VS\mu, V^{-\frac{1}{2}}\mu\rangle && \\ &\le \|V^{-\frac{1}{2}}VS\mu\|_{L^2}\|V^{-\frac{1}{2}}\mu\|_{L^2} &=& \|VS\mu\|_{V^{-1}}\|\mu\|_{V^{-1}} \\ && =& \|(\tfrac{1}{2}I + K)\mu\|_{V^{-1}}\|\mu\|_{V^{-1}}\end{aligned}$$

and with (5.6.33) and (5.6.37),

$$\langle D\mu, \mu\rangle \ge c_1^D\|\mu\|_{H^{\frac{1}{2}}}^2 \ge c_1^Dc_1^V\|\mu\|_{V^{-1}}^2\,.$$

This implies

$$\|(\tfrac{1}{2}I + K)\mu\|_{V^{-1}}^2 \le \|(\tfrac{1}{2}I + K)\mu\|_{V^{-1}}\|\mu\|_{V^{-1}} - c_1^Dc_1^V\|\mu\|_{V^{-1}}^2\,.$$

or

$$a^2 \le ab - c_1^Dc_1^Vb^2$$

with $0 \le a = \|(\frac{1}{2}I + K)\mu\|_{V^{-1}}$ and $0 \le b = \|\mu\|_{V^{-1}}$.

An elementary manipulation shows that

$$\tfrac{1}{2} - \sqrt{\tfrac{1}{4} - c_1^Vc_1^D} \le \tfrac{a}{b} \le \tfrac{1}{2} + \sqrt{\tfrac{1}{4} - c_1^Vc_1^D}$$

which leads to $a \le bc_K$, i.e., the desired estimate (5.6.41) with the + sign and to the estimate (5.6.43)

$$(1 - c_K)\|\mu\|_{V^{-1}} \le \|(\tfrac{1}{2}I + K)\mu\|_{V^{-1}} \quad \text{for all} \;\; \mu \in H_{\Re}^{\frac{1}{2}}(\Gamma)\,.$$

We also obtain

$$\begin{aligned}\|\mu\|_{V^{-1}} &= \|(\tfrac{1}{2}I - K)\mu + (\tfrac{1}{2}I + K)\mu\|_{V^{-1}} \\ &\le \|(\tfrac{1}{2}I - K)\mu\|_{V^{-1}} + c_K\|\mu\|_{V^{-1}}\end{aligned}$$

and

$$(1 - c_K)\|\mu\|_{V^{-1}} \le \|(\tfrac{1}{2}I - K)\mu\|_{V^{-1}} \quad \text{for all} \quad \mu \in H^{\frac{1}{2}}_{\Re}(\Gamma)\,,$$

i.e., (5.6.43) with the $-$ sign.

In order to show (5.6.41) for the case of the $-$ sign we proceed as follows.

$$\begin{aligned}\|(\tfrac{1}{2}I - K)\mu\|^2_{V^{-1}} &= \|(I - (\tfrac{1}{2}I + K))\mu\|^2_{V^{-1}} \\ &= \|\mu\|^2_{V^{-1}} + \|(I + K)\mu\|^2_{V^{-1}} - 2\langle V^{-1}(\tfrac{1}{2}I + K)\mu, \mu\rangle \\ &= \|\mu\|^2_{V^{-1}} + \|(\tfrac{1}{2}I + K)\mu\|^2_{V^{-1}} - 2\langle S\mu, \mu\rangle \\ &= \|\mu\|^2_{V^{-1}} - \|(\tfrac{1}{2}I + K)\mu\|^2_{V^{-1}} - 2\langle D\mu, \mu\rangle \\ &\le \|\mu\|^2_{V^{-1}}\{1 - (1 - c_K)^2 - 2c_1^D c_1^V\} = c_K^2\|\mu\|^2_{V^{-1}}\end{aligned}$$

for $\mu \in H^{\frac{1}{2}}_{\Re}$ which is (5.6.41).

Finally, the relation (5.6.44) follows from the fact that $\Re$ is also the kernel of $(\frac{1}{2}I + K)$, hence

$$(\tfrac{1}{2}I - K)\mu_0 = \mu_0 \quad \text{for all} \quad \mu_0 \in \Re$$

which implies (5.6.44).

The estimates (5.6.42) are a direct consequence of the norm definitions in (5.6.36):

$$\begin{aligned}\|(\tfrac{1}{2}I \pm K')\lambda\|^2_V &= \langle(\tfrac{1}{2} \pm K')\lambda, V\lambda\rangle \\ &= \langle V^{-1}\mu, (\tfrac{1}{2}I \pm K)\mu\rangle \\ &= \|(\tfrac{1}{2}I \pm K)\mu\|^2_{V^{-1}} \\ &\le c_K^2\|\mu\|^2_{V^{-1}} = c_K^2\langle V^{-1}V\lambda, V\lambda\rangle = c_K^2\|\lambda\|_{V^2}\end{aligned}$$

where $\mu = V\lambda$. ■

The Theorem 5.6.11 implies that the boundary integral equations of the second kind can be solved with appropriate Neumann's series in Carl Neumann's classical iterative scheme. In particular, the equations

$$\tfrac{1}{2}\mu \pm K\mu \;=\; f \;\text{ on } \Gamma \;\text{ with } f \in \left\{\begin{array}{l} H^{\frac{1}{2}}_{\Re}(\Gamma) \\ H^{\frac{1}{2}}(\Gamma)\end{array}\right. , \tag{5.6.45}$$

and

$$\tfrac{1}{2}\lambda \pm K'\lambda \;=\; g \;\text{ on } \Gamma \;\text{ with } g \in \left\{\begin{array}{l} H^{-\frac{1}{2}}_{\Re}(\Gamma) \\ H^{-\frac{1}{2}}(\Gamma)\end{array}\right. \tag{5.6.46}$$

can be solved by the convergent iterations

$$\mu^{(\ell+1)} := (\tfrac{1}{2}I \mp K)\mu^{(\ell)} + f \ \text{ in } \begin{cases} H^{\frac{1}{2}}_{\Re}(\Gamma) \\ H^{\frac{1}{2}}(\Gamma) \end{cases}, \tag{5.6.47}$$

and

$$\lambda^{(\ell+1)} := (\tfrac{1}{2}I \mp K')\lambda^{(\ell)} + g \ \text{ in } \begin{cases} H^{-\frac{1}{2}}_{\Re}(\Gamma) \\ H^{-\frac{1}{2}}(\Gamma) \end{cases} \tag{5.6.48}$$

for $\ell = 0, 1, \ldots$, in the respective Sobolev spaces.

5.6.8 The Solvability of Direct Boundary Integral Equations

With Gårding's inequality for the sesquilinear forms of direct boundary integral equations available, their solvability follows from the basic existence theorems for variational problems in Section 5.2. For the direct boundary integral equations these results of the previous chapters are summarized in the following theorem whose proof is clear from the previous presentation.

Theorem 5.6.12. *The variational boundary integral equations* (5.6.22) *of the first kind and* (5.6.26), (5.6.27) *and* (5.6.28) *of the second kind satisfy all the assumptions of Theorem* 5.3.10. *In particular, if one of these equations has the property that the corresponding homogeneous problem*

$$a_\Gamma(\chi, \lambda_0) = 0 \ \textit{ for all } \chi \in H^\alpha(\Gamma) \tag{5.6.49}$$

has only the trivial solution $\lambda_0 = 0$, then this equation is uniquely solvable and the solution λ together with the given data μ generates via the corresponding representation formula the unique variational solution of the corresponding boundary value problem.

On the other hand, if the homogeneous boundary variational form (5.6.49) *has nontrivial solutions then there are two cases:*

i. If the corresponding boundary value problem with the given data μ has a solution then the right–hand side of the boundary variational form defined by $f = \mathcal{B}\mu$ will always satisfy the required orthogonality conditions.
ii. The given data $f = \mathcal{B}\mu$ satisfy the orthogonality conditions.

In both cases, the orthogonality conditions read as

$$\begin{aligned} \langle \chi_0^*, \overline{\mathcal{B}\mu}\rangle_\Gamma &= 0 \quad \textit{for } (5.6.22) \textit{ or} \\ \langle V\chi_0^*, \overline{\mathcal{B}\mu}\rangle_\Gamma &= 0 \quad \textit{for } (5.6.27) \textit{ or} \\ \langle D\chi_0^*, \overline{\mathcal{B}\mu}\rangle_\Gamma &= 0 \quad \textit{for } (5.6.28) \end{aligned}$$

where χ_0^* are all the eigensolutions of the adjoint homogeneous boundary sesquilinear form

$$a_\Gamma(\chi_0^*, \nu) = 0 \ \ \textit{for all } \nu \in H^\alpha(\Gamma)\,.$$

The latter form a finite–dimensional subspace and the inhomogeneous boundary variational form has a solution but is not unique. Moreover, every pair (λ, μ) *again generates via the corresponding representation formula a variational solution of the corresponding boundary value problem.*

5.6.9 Positivity of the Boundary Integral Operators of the Stokes System

Based on the continuity of the hydraulic potential operators on the boundary Sobolev spaces and Gårding inequalities (5.4.27) for $a_\Omega(\boldsymbol{u},\boldsymbol{v})$ and $a_{\Omega^c}(\boldsymbol{u},\boldsymbol{v})$, respectively, the variational solutions in $H^1_{\text{div}}(\Omega)$ and $H^1_{0,\text{div}}(\Omega)$ of the traction or the Dirichlet problem, the jump relations in Lemma 5.6.5 and the generalized Green theorem, Lemma 5.4.6, one finally obtains the following Gårding inequalities for the hydrodynamic potentials given in Section 2.3.2, Table 2.3.4.

Theorem 5.6.13. *Given* $(\boldsymbol{\tau},\boldsymbol{\varphi}) \in H^{-\frac12}(\Gamma) \times H^{\frac12}(\Gamma)$. *Then the hydrodynamic boundary integral operators satisfy the following Gårding inequalities:*

$$\langle V\boldsymbol{\tau},\boldsymbol{\tau}\rangle_\Gamma + \sum_{\ell=1}^{L}\langle\boldsymbol{\tau},\boldsymbol{n}_\ell\rangle^2_{\Gamma_\ell} \geq \gamma_0\|\boldsymbol{\tau}\|^2_{H^{-\frac12}(\Gamma)}, \tag{5.6.50}$$

$$\langle D\boldsymbol{\varphi},\boldsymbol{\varphi}\rangle_\Gamma + \sum_{\ell=1}^{L}\sum_{k=1}^{3(n-1)}\langle\boldsymbol{\varphi},\boldsymbol{m}_{k,\ell}\rangle^2_{\Gamma_\ell} \geq \gamma_0\|\boldsymbol{\varphi}\|^2_{H^{\frac12}(\Gamma)}, \tag{5.6.51}$$

$$\langle V\boldsymbol{\tau}\,,\,(\tfrac12 I + K')\boldsymbol{\tau}\rangle_\Gamma + \sum_{\ell=1}^{L}\langle\boldsymbol{\tau},\boldsymbol{n}_\ell\rangle^2_{\Gamma_\ell} \geq \gamma_0\|\boldsymbol{\tau}\|^2_{H^{-\frac12}(\Gamma)}, \tag{5.6.52}$$

$$\langle D\boldsymbol{\varphi}\,,\,(\tfrac12 I - K)\boldsymbol{\varphi}\rangle_\Gamma + \sum_{\ell=1}^{L}\sum_{k=1}^{3(n-1)}\langle\boldsymbol{\varphi},\boldsymbol{m}_{k,\ell}\rangle^2_{\Gamma_\ell} \geq \gamma_0\|\boldsymbol{\varphi}\|^2_{H^{\frac12}(\Gamma)}. \tag{5.6.53}$$

Proof: Since the proof follows the same scheme as for the second order systems we sketch the proof here only for the operator V and (5.6.52). Here, the special field

$$(\boldsymbol{u},\mu) = (V\boldsymbol{\tau}_0, \Phi\boldsymbol{\tau}_0) \quad \text{where } \langle\boldsymbol{\tau}_0,\boldsymbol{n}_\ell\rangle = 0 \quad \text{for } \ell = \overline{1,L}$$

satisfies the homogeneous Stokes system where $\boldsymbol{f} = \boldsymbol{0} \in \widetilde{H}_0^{-1}$ in Ω as well as in Ω^c and, in addition, $\boldsymbol{u}$ is solenoidal. Moreover, for $V\boldsymbol{\tau}_0 = \boldsymbol{u}$ we have the jump relations in Lemma 5.6.5. Hence, $(\boldsymbol{u},p) \in H^1(\Omega, P_{st})$ and $(\boldsymbol{u},p) \in H^1(\Omega^c, P_{st})$ and the field decays at infinity. So, $(\boldsymbol{u},p)$ is a solution of the transmission problem and satisfies

$$\langle V\boldsymbol{\tau}_0, \boldsymbol{\tau}_0\rangle = a_\Omega(\boldsymbol{u},\boldsymbol{u}) + a_{\Omega^c}(\boldsymbol{u},\boldsymbol{u})$$

with the bilinear forms a_Ω and a_{Ω^c} given in (5.4.26). On solenoidal fields they satisfy Gårding inequalities (5.4.27) which here yield the form

$$a_\Omega(\boldsymbol{u},\boldsymbol{u}) + a_{\Omega^c}(\boldsymbol{u},\boldsymbol{u}) \geq c(\|\boldsymbol{u}\|^2_{H^1(\Omega)} + \|\boldsymbol{u}\|^2_{H^1(\Omega^c)})\,.$$

Then Lemma 5.4.6 for Ω and Ω^c implies finally

$$\begin{aligned}\langle V\boldsymbol{\tau}_0, \boldsymbol{\tau}_0\rangle &\geq c(\|\boldsymbol{u}\|^2_{H^1(\Omega)} + \|\boldsymbol{u}\|^2_{H^1(\Omega^c)})\\ &\leq c'\big(\|T(\boldsymbol{u})\|^2_{H^{-\frac{1}{2}}(\Gamma)} + \|T^c(\boldsymbol{u})\|^2_{H^{-\frac{1}{2}}(\Gamma)}\big)\\ &\geq c''\|\big(T(\boldsymbol{u}) - T^c(\boldsymbol{u})\big)\|^2_{H^{-\frac{1}{2}}(\Gamma)} = \gamma_0\|\boldsymbol{\tau}_0\|^2_{H^{-\frac{1}{2}}(\Gamma)}\,.\end{aligned}$$

For general $\boldsymbol{\tau} \in H^{-\frac{1}{2}}(\Gamma)$ write

$$\boldsymbol{\tau} = \boldsymbol{\tau}_0 + \sum_{\ell=1}^{L}\langle\boldsymbol{\tau}, \boldsymbol{n}_\ell\rangle\boldsymbol{n}_\ell$$

and with $V\boldsymbol{n}_\ell = \boldsymbol{0}$ for $\ell = \overline{1, L}$, the inequality (5.6.50) follows. ■

With Theorem 5.6.13 available, Theorem 5.6.6 can be carried through for the operators $\frac{1}{2}I \pm K$ and $\frac{1}{2}I \pm K'$ of the Stokes system.

We refer to further works on the Stokes system in Dautray and Lions [97], Fabes et al [114, 115], Fischer et al [123], Galdi [136], Kohr et al [233, 234], Ladyženskaya [256], Temam [417] and [434].

5.7 Partial Differential Equations of Higher Order

In the previous sections we presented a general procedure for treating second order equations and systems based on the generalized Green's formula by employing the direct approach for corresponding boundary integral equations. There the approach hinged on the intimate relations between Gårding inequalities for the boundary value problems of the underlying partial differential operators and corresponding integral operators.

In this section we demonstrate that the same approach can also be applied to higher order elliptic partial differential equations. To illustrate the idea we now consider boundary value problems for the *fourth order biharmonic equation* (2.4.1) as a simple model case. Here, the first Green's formula in $H^4(\Omega)$, in the interior domain is given by (2.4.2). For the boundary, we need at least $\Gamma \in C^{1,1}$. Moreover, we shall need the subspace of the dual space to $H^2(\Omega)$, which does not contain distributions in $\mathbb{R}^n$ having singular support on Γ. To this end, let $\widetilde{H}^{-2}(\Omega) := \big(H^2(\Omega)\big)'$ and

$$\widetilde{H}^{-2}_\Gamma(\Omega) := \{f \in \widetilde{H}^{-2}(\Omega)\,|\,(f,\varphi)_{L^2(\Omega)} = 0 \quad \text{for all } \ \varphi \in C_0^\infty(\Omega)\}\,.$$

Now let $\widetilde{H}^{-2}(\Omega)$ be decomposed as

$$\widetilde{H}^{-2}(\Omega) = \widetilde{H}_{\Gamma}^{-2}(\Omega) \oplus \widetilde{H}_0^{-2}(\Omega)$$

where $\widetilde{H}_0^{-2}(\Omega)$ and $\widetilde{H}_{\Gamma}^{-2}(\Omega)$ are orthogonal in the Hilbert space $\widetilde{H}^{-2}(\Omega)$.
For the generalized first Green's formula below, we introduce the space

$$H^2(\Omega, \Delta^2) := \{w \in H^2(\Omega) \,|\, \Delta^2 w \in \widetilde{H}_0^{-2}(\Omega)\}\,,$$

and throughout the section we need the boundary to be at least $\Gamma \in C^{1,1}$.

Lemma 5.7.1. *For fixed $u \in H^2(\Omega, \Delta^2)$, the mapping*

$$\gamma v = \Big(v, \frac{\partial v}{\partial n}\Big)^\top \mapsto \langle \boldsymbol{\tau} u, \gamma v\rangle_\Gamma := a_\Omega(u, \mathcal{Z}\gamma v) - \int\limits_\Omega \Delta^2 u \mathcal{Z}\gamma v dx \qquad (5.7.1)$$

defines a continuous linear functional $\boldsymbol{\tau} u$ on $\gamma v \in H^{\frac{3}{2}}(\Gamma) \times H^{\frac{1}{2}}(\Gamma)$ that for $u \in H^4(\Omega)$ is given by $\boldsymbol{\tau} u = (Nu, Mu)^\top$. The mapping

$$\boldsymbol{\tau} : H^2(\Omega, \Delta^2) \to H^{-\frac{3}{2}}(\Gamma) \times H^{-\frac{1}{2}}(\Gamma) \quad \textit{with } u \mapsto \boldsymbol{\tau} u$$

is continuous. Here $\mathcal{Z}$ is a right inverse to the trace operator γ.

Correspondingly, the generalized first Green's formula (5.7.1) remains valid in the space $H^2(\Omega_R^c, \Delta^2)$.
Similar to the transmission problem (5.1.76), we formulate in the domain B_R the **transmission problem:**

Find $u \in H^2(\Omega, \Delta^2) \times H^2(\Omega_R^c, \Delta^2)$ satisfying

$$\Delta^2 u = f_1 \quad \textit{in } \Omega \quad \textit{and} \quad \Delta^2 u = f_2 \quad \textit{in } \Omega_R^c \qquad (5.7.2)$$

together with the transmission conditions

$$[\gamma u]|_\Gamma = \boldsymbol{\varphi}\,,\ [\boldsymbol{\tau} u]|_\Gamma = \boldsymbol{\psi} \quad \text{on } \Gamma \qquad (5.7.3)$$

and the boundary condition

$$\gamma u|_{|x|=R} = \mathbf{0} \quad \text{on } \partial B_R = \{x \,|\, |x| = R\}$$

with given $f_1 \in \widetilde{H}_0^{-2}$, $f_2 \in H^{-2}_{\text{comp}} \cap H^{-2}(B_R)$, $\boldsymbol{\varphi} \in H^{\frac{3}{2}}(\Gamma) \times H^{\frac{1}{2}}(\Gamma)$, $\boldsymbol{\psi} \in H^{-\frac{3}{2}}(\Gamma) \times H^{-\frac{1}{2}}(\Gamma)$.

The corresponding weak formulation of the transmission problem then reads:
Find

$$u \in H^2(\Omega, \Delta^2) \times H^2(\Omega_R^c, \Delta^2) \quad \textit{with } [\gamma u]|_\Gamma = \boldsymbol{\varphi} \quad \textit{and } \gamma u|_{|x|=R} = \mathbf{0} \tag{5.7.4}$$

such that

$$\begin{aligned} a(u,v) &:= a_\Omega(u,v) + a_{\Omega_R^c}(u,v) \\ &= \int\limits_\Omega f_1 v dx + \int\limits_{\Omega_R^c} f_2 v dx - \int\limits_\Gamma \Big(\boldsymbol{v}, \frac{\partial v}{\partial n}\Big) \boldsymbol{\psi} ds \end{aligned} \tag{5.7.5}$$

for all $v \in \mathcal{H} := \{v \in H^2(\Omega, \Delta^2) \times H^2(\Omega_R^c, \Delta^2)$ *with* $[\gamma v]|_\Gamma = \mathbf{0}$ *and* $\gamma v|_{|x|=R} = \mathbf{0}\}$.

The bilinear forms a_Ω and $a_{\Omega_R^c}$ are given by (2.4.3). As S. Agmon has shown in [3], the sesquilinear form $a_\Omega(u,v)$ in (2.4.3) satisfies a Gårding inequality.

Lemma 5.7.2. (*See* [3]). *The bilinear form $a_\Omega(\cdot,\cdot)$ is coercive over $H^2(\Omega)$ iff $-3 < \nu < 1$. Here the coerciveness means that $a_\Omega(\cdot,\cdot)$ satisfies a Gårding inequality in the form:*

$$a_\Omega(v,v) + \lambda_0 \int\limits_\Omega v^2 dx \geq \alpha_0 \|v\|^2_{H^2(\Omega)} \quad \textit{for all } v \in H^2(\Omega)\,, \tag{5.7.6}$$

where $\alpha_0 > 0$ and $\lambda_0 > 0$ are constants.

Since $H^2(\Omega) \hookrightarrow L^2(\Omega)$ is compactly imbedded (see (4.1.31)),

$$\int\limits_\Omega v^2 dx =: (Cv, v)_{H^2(\Omega)}$$

where C is a compact linear operator in $H^2(\Omega)$. Clearly, this lemma is also valid for $a_{\Omega_R^c}(u,v)$ and Ω_R^c with fixed $R > 0$ sufficiently large.

Since a in (5.7.5) satisfies a Gårding inequality as (5.7.6) on $\mathcal{H}$, the Fredholm alternative for sesquilinear forms, Theorem 5.3.10 is valid. The homogeneous problem for (5.7.5) where $f_1 = 0\,,\ f_2 = 0\,,\ \boldsymbol{\varphi} = \mathbf{0}\,,\ \boldsymbol{\psi} = \mathbf{0}$ has only a trivial solution. Therefore (5.7.5) always has a unique solution u. Moreover, it satisfies the a priori estimate

$$\begin{aligned} \|u\|_{H^2(\Omega)} + \|u\|_{H^1(\Omega_R^c)} \leq c\{&\|f_1\|_{\widetilde{H}^{-2}(\Omega)} + \|f_2\|_{\widetilde{H}^{-2}(\Omega^c)\cap H^{-2}(B_R)} \\ &+ \|\boldsymbol{\varphi}\|_{H^{\frac{3}{2}}(\Gamma)\times H^{\frac{1}{2}}(\Gamma)} + \|\boldsymbol{\psi}\|_{H^{-\frac{3}{2}}(\Gamma)\times H^{-\frac{1}{2}}(\Gamma)}\}\,, \end{aligned} \tag{5.7.7}$$

and can be represented in the form

$$u(x) = W\boldsymbol{\varphi}(x) - V\boldsymbol{\psi}(x) + \int\limits_\Omega E(x,y) f_1(y) dy + \int\limits_{\Omega_R^c} E(x,y) f_2(y) dy\,. \tag{5.7.8}$$

For proving the continuity properties of the boundary integral operators in the Calderón projector (2.4.23) we again exploit the fact that $\mathbf{V}$ in (2.4.26) as well as $\boldsymbol{D}$ in (2.4.30) define potentials which are solutions of the transmission problem (5.7.4) with $f_1 = 0$ and $f_2 = 0$ together with the transmission conditions $\boldsymbol{\varphi} = \mathbf{0}$ for V and $\boldsymbol{\psi} = \mathbf{0}$ for W, respectively.

Theorem 5.7.3. *The following operators are continuous:*

$$
\begin{aligned}
& V : H^{-\frac{1}{2}}(\Gamma) \times H^{-\frac{3}{2}}(\Gamma) \to H^2(\Omega, \Delta^2) \times H^2_{\mathrm{loc}}(\Omega^c, \Delta^2)\,,\\
& \begin{pmatrix} \frac{\partial}{\partial n_c} \\ \gamma_{c0} \end{pmatrix} V = \begin{pmatrix} \frac{\partial}{\partial n} \\ \gamma_0 \end{pmatrix} V =: \mathbf{V} : H^{-\frac{1}{2}}(\Gamma) \times H^{-\frac{3}{2}}(\Gamma) \to H^{\frac{1}{2}}(\Gamma) \times H^{\frac{3}{2}}(\Gamma)\,,\\
& \begin{pmatrix} M_c \\ N_c \end{pmatrix} V \text{ and } \begin{pmatrix} M \\ N \end{pmatrix} V : H^{-\frac{1}{2}}(\Gamma) \times H^{-\frac{3}{2}}(\Gamma) \to H^{-\frac{1}{2}}(\Gamma) \times H^{-\frac{3}{2}}(\Gamma)\,,\\
& W : H^{\frac{1}{2}}(\Gamma) \times H^{\frac{3}{2}}(\Gamma) \to H^2(\Omega, \Delta^2) \times H^2_{\mathrm{loc}}(\Omega^c, \Delta^2)\,,\\
& \begin{pmatrix} \frac{\partial}{\partial n_c} \\ \gamma_{c0} \end{pmatrix} W \text{ and } \begin{pmatrix} \frac{\partial}{\partial n} \\ \gamma_0 \end{pmatrix} W : H^{\frac{1}{2}}(\Gamma) \times H^{\frac{3}{2}}(\Gamma) \to H^{\frac{1}{2}}(\Gamma) \times H^{\frac{3}{2}}(\Gamma)\,,\\
& \begin{pmatrix} M_c \\ N_c \end{pmatrix} W = \begin{pmatrix} M \\ N \end{pmatrix} W =: \boldsymbol{D} : H^{\frac{1}{2}}(\Gamma) \times H^{\frac{3}{2}}(\Gamma) \to H^{-\frac{1}{2}}(\Gamma) \times H^{-\frac{3}{2}}(\Gamma)\,.
\end{aligned}
$$

In the theorem, the corresponding operators are given by:

$$
V : \begin{pmatrix} Mu \\ Nu \end{pmatrix} \to u\,,\; W : \begin{pmatrix} \frac{\partial u}{\partial n} \\ u \end{pmatrix} \to u
$$

and

$$
\mathbf{V} = \begin{pmatrix} V_{23} & V_{24} \\ V_{13} & V_{14} \end{pmatrix}\,,\; \boldsymbol{D} = \begin{pmatrix} D_{32} & D_{31} \\ D_{42} & D_{41} \end{pmatrix} \tag{5.7.9}
$$

defined in (2.4.26), (2.4.30), respectively.

Proof: In fact, the proof can be carried out in the same manner as for Theorem 5.6.2 for second order systems. Therefore, here we only present the main ideas.

First we consider the continuity properties of V. To this end, we consider the related variational transmission problem (5.7.2), (5.7.3) with $\boldsymbol{\varphi} = \mathbf{0}$ which has a unique solution u in $H^2(\Omega, \Delta^2) \times H^2(\Omega^c_R, \Delta^2)$ satisfying (5.7.7) which can be represented in the form (5.7.8) with $\boldsymbol{\varphi} = \mathbf{0}$. In particular, we may consider $f_1 = 0\,,\ f_2 = 0$ which leads to the first three continuity properties in Theorem 5.7.3 by making use of the trace theorem 4.2.1.

Next, for the mapping properties of W, we again consider the transmission problem (5.7.2), (5.7.3) but now with $\boldsymbol{\psi} = \mathbf{0}$ and $\boldsymbol{\varphi} \in H^{\frac{1}{2}}(\Gamma) \times H^{\frac{3}{2}}(\Gamma)\,,\ f_1 = 0\,,\ f_2 = 0$. With representation formula (5.7.8) of the corresponding solution u in $H^2(\Omega, \Delta^2) \times H^2(\Omega^c_R, \Delta^2)$, i.e., with $u = W\boldsymbol{\varphi}$ and the trace theorem together with the definition of the operators M and N, i.e., of $\boldsymbol{\tau}$ based on the generalized Green's formula in Lemma 5.7.1, we obtain the last three continuity properties in Theorem 5.7.3. ∎

As a consequence of the mapping properties in Theorem 5.7.3 we find the jump relations as in Lemma 5.6.3.

Lemma 5.7.4. *Given $\boldsymbol{\varphi} \in H^{\frac{1}{2}}(\Gamma) \times H^{\frac{3}{2}}(\Gamma)$ and $\boldsymbol{\psi} \in H^{-\frac{1}{2}}(\Gamma) \times H^{-\frac{3}{2}}(\Gamma)$, the following jump relations hold:*

$$[\mathbf{V}\boldsymbol{\psi}]|_\Gamma = \mathbf{0}\,,\ [(M,N)^\top V\boldsymbol{\psi}]|_\Gamma = -\boldsymbol{\psi}\,,\ [(\tfrac{\partial}{\partial n},\gamma_0)^\top W\boldsymbol{\varphi}]|_\Gamma = \boldsymbol{\varphi}\,,\ [\boldsymbol{D}\boldsymbol{\varphi}]|_\Gamma = \mathbf{0} \tag{5.7.10}$$

We omit the proof since the arguments are the same as for Lemma 5.6.3.

In order to consider the boundary integral equations collected in Table 2.4.5 and Table 2.4.6 in the above Sobolev spaces and in variational form we now present the Gårding inequalities for the matrix operators $\mathbf{V}$ and $\boldsymbol{D}$ in (5.7.9).

Theorem 5.7.5. *With the real Sobolev spaces, the boundary integral matrix operators $\mathbf{V}$ and $\boldsymbol{D}$ satisfy Gårding's inequalities as*

$$\langle\boldsymbol{\psi},\mathbf{V}\boldsymbol{\psi}\rangle + \langle\boldsymbol{\psi}\boldsymbol{C}_V\boldsymbol{\psi}\rangle \geq c^V\|\boldsymbol{\psi}\|^2_{H^{-\frac{1}{2}}(\Gamma)\times H^{-\frac{3}{2}}(\Gamma)}\,, \tag{5.7.11}$$

$$\langle\boldsymbol{D}\boldsymbol{\varphi},\boldsymbol{\varphi}\rangle + \langle\boldsymbol{C}_D\boldsymbol{\varphi},\boldsymbol{\varphi}\rangle \geq c^D\|\boldsymbol{\varphi}\|^2_{H^{\frac{1}{2}}\times H^{\frac{3}{2}}(\Gamma)} \tag{5.7.12}$$

where c^V and c^D are positive constants and the linear operators $\boldsymbol{C}_V$ and $\boldsymbol{C}_D$ are compact operators on $H^{-\frac{1}{2}}(\Gamma) \times H^{-\frac{3}{2}}(\Gamma)$ and $H^{\frac{1}{2}}(\Gamma) \times H^{\frac{3}{2}}(\Gamma)$, respectively.

Proof: Since, again, the proof can be carried out in the same manner as for Theorem 5.6.8, we just outline the main ideas in this case without the details.

For (5.7.11) consider the function

$$u(x) := -\eta(x)V\boldsymbol{\psi}(x) \quad \text{for } \ x \in \mathbb{R}^2 \setminus \Gamma$$

with any $\boldsymbol{\psi} \in H^{-\frac{1}{2}}(\Gamma) \times H^{-\frac{3}{2}}(\Gamma)$ given , where $\eta \in C_0^\infty(\mathbb{R}^2)$ is a cut–off function as in (5.6.17). Then u is a solution of the transmission problem (5.7.2), (5.7.3) with $\boldsymbol{\varphi} = \mathbf{0}$ and $f_2 = \Delta^2(\eta V\boldsymbol{\psi}) \in C_0^\infty(\Omega_R^c)$. As in Section 5.6.6, the generalized first Green's formula yields with the Gårding inequality (5.7.6):

$$\begin{aligned}
\langle\boldsymbol{\psi},\mathbf{V}\boldsymbol{\psi}\rangle &= \int\limits_\Gamma \big(\tfrac{\partial u}{\partial n}\psi_1 + u\psi_2\big)ds \\
&= a_\Omega(u,u) + a_{\Omega_R^c}(u,u) - \int\limits_{R_1\leq|y|\leq R} u\Delta^2(\eta V\boldsymbol{\psi})dy \\
&\geq \gamma_0\{\|u\|^2_{H^2(\Omega)} + \|u\|^2_{H^2(\Omega_R^c)} - c_1\|\boldsymbol{\psi}\|^2_{H^{-1}(\Gamma)\times H^{-2}(\Gamma)}\}
\end{aligned}$$

Next, use Lemma 5.7.1 which yields

$$\|\boldsymbol{\psi}\|^2_{H^{-\frac{1}{2}}(\Gamma)\times H^{-\frac{3}{2}}(\Gamma)} \leq c_2\{\|u\|^2_{H^2(\Omega)} + \|u\|^2_{H^2(\Omega^c_R)} + \|\Delta^2 u\|^2_{\widetilde{H}^{-2}(\Omega)}\}$$

So,

$$\langle \boldsymbol{\psi}, \mathbf{V}\boldsymbol{\psi}\rangle \geq \tfrac{\gamma_0}{c_2}\|\boldsymbol{\psi}\|^2_{H^{-\frac{1}{2}}(\Gamma)\times H^{-\frac{3}{2}}(\Gamma)} - c_3\|\boldsymbol{\psi}\|^2_{H^{-1}(\Gamma)\times H^{-2}(\Gamma)}$$

since $\|\Delta^2 u\|^2_{\widetilde{H}^{-2}(\Omega)} \leq c\|\boldsymbol{\psi}\|_{H^{-1}(\Gamma)\times H^{-2}(\Gamma)}$. Then with the compact imbedding $H^{-\frac{1}{2}}(\Gamma)\times H^{-\frac{3}{2}}(\Gamma) \hookrightarrow H^{-1}(\Gamma)\times H^{-2}(\Gamma)$ the Gårding inequality (5.7.11) follows.

For (5.7.12) we use the function

$$u(x) := \eta(x)W\boldsymbol{\varphi}(x) \quad \text{for } x \in \mathbb{R}^2 \setminus \Gamma\,.$$

Then

$$\begin{aligned}\langle \boldsymbol{D}\boldsymbol{\varphi}, \boldsymbol{\varphi}\rangle &= \int\limits_\Gamma (\varphi_1 Mu + \varphi_2 Nu)ds \\ &= a_\Omega(u,u) + a_{\Omega^c_R}(u,u) - \int\limits_{R_1\leq|y|\leq R_2} u\Delta^2(\eta W\boldsymbol{\varphi})dy\,.\end{aligned}$$

Gårding inequalities (5.7.6) and the trace theorem lead to the desired estimate (5.7.12). ■

Remark 5.7.1: For the integral operator equations of the second kind as in Tables 2.4.5 and 2.4.6, corresponding boundary bilinear forms are related to the domain bilinear forms (2.4.3) and are given by

$$\left.\begin{array}{l} a_{\Omega^c}(W\boldsymbol{\varphi}, W\widetilde{\boldsymbol{\varphi}}) \\ a_\Omega(W\boldsymbol{\varphi}, W\widetilde{\boldsymbol{\varphi}}) \end{array}\right\} = \begin{array}{l}\langle(\tfrac{1}{2}I \pm K_{11})\varphi_1 \mp V_{12}\varphi_2\,,\ D_{41}\widetilde{\varphi}_1 + D_{42}\widetilde{\varphi}_2\rangle \\ +\langle \mp D_{21}\varphi_1 + (\tfrac{1}{2}I \mp K_{22})\varphi_2\,,\ D_{31}\widetilde{\varphi}_1 + D_{32}\widetilde{\varphi}_2\rangle\end{array} \tag{5.7.13}$$

and

$$\left.\begin{array}{l} a_{\Omega^c}(V\boldsymbol{\psi}, V\widetilde{\boldsymbol{\psi}}) \\ a_\Omega(V\boldsymbol{\psi}, V\widetilde{\boldsymbol{\psi}}) \end{array}\right\} = \begin{array}{l}\langle(\tfrac{1}{2}I \pm K_{33})\psi_1 \mp V_{34}\psi_2\,,\ V_{13}\widetilde{\psi}_1 + V_{14}\widetilde{\psi}_2\rangle \\ +\langle \mp D_{34}\psi_1 + (\tfrac{1}{2}I \mp K_{44})\varphi_2\,,\ V_{23}\widetilde{\psi}_1 + V_{24}\widetilde{\psi}_2\rangle\end{array} \tag{5.7.14}$$

Therefore, these bilinear forms also satisfy Gårding inequalities in the same manner as in Theorem 5.6.9 for second order partial differential equations. As we can see, the test functions are images under the operators of the first kind which serve as preconditioners.

With Gårding inequalities available, Fredholm's alternative, Theorem 5.3.10 can be applied to all the integral equations in Tables 2.4.5 and 2.4.6 in the corresponding Sobolev spaces. In fact, as formulations in Table 2.4.6 imply uniqueness, for these versions the existence of solutions in the Sobolev spaces is ensured.

Remark 5.7.2: As was pointed out by Costabel in [84] the positivity of the boundary integral operators $\mathbf{V}$ and $\boldsymbol{D}$ on appropriate Sobolev spaces as in Section 5.6.7 for the second order equations, the convergence of corresponding successive approximation for integral equations of the second kind, can be established. The details will not be pursued here.

Some further works on the biharmonic equation and boundary integral equations can be found in Costabel et al [87, 91], Giroire [148], Hartmann and Zotemantel [174], [208] and Knöpke [230].

5.8 Remarks

5.8.1 Assumptions on Γ

In this chapter we have for the most part, not been very specific concerning the assumptions on Γ which should at least be a strong Liptschitz boundary. For strongly elliptic second order elliptic equations and systems including the equations of linearized elasticity and also for the Stokes system, a strong Lipschitz boundary is sufficient as long as Dirichlet or Neumann conditions or boundary and transmission conditions of the form (5.1.21) with $B_{00} = I$ and $N_{00} = 0$ are considered (see Costabel [80], Fabes et al [114, 115], Nečas [326]). For the general Neumann–Robin conditions (5.1.21), however, one needs at least a $C^{1,1}$ boundary.

For higher order equations of order $2m'$, one needs as least $\Gamma \in C^{\ell,\kappa}$ with $\ell + \kappa = m'$ as $C^{1,1}$ for the biharmonic equation.

D. Mitrea, M. Mitrea and M. Taylor develop in Mitrea et al [309] the boundary integral equations for Lipschitz boundaries and for scalar strongly elliptic second order equations, in particular the Hodge Laplacian on Riemannian manifolds.

5.8.2 Higher Regularity of Solutions

Let us collect some facts regarding higher regularity of solutions without proofs. Higher regularity is often needed in applications and for the numerical approximation and corresponding error analysis (see e.g. [204], Sauter and Schwab [371], Steinbach [398]).

Without higher requirement for Γ, Theorem 5.6.2, Lemma 5.6.4 and Lemma 5.1.1 can be extended as follows due to the results by Fabes, Kenig, Verchota in [114, 115] and Costabel [80].

Theorem 5.8.1. *For $|\sigma| \le \frac{1}{2}$, the operators in Theorem* 5.6.2 *and in Lemma* 5.6.4 *can be extended to the following continuous mappings:*

$$\begin{aligned}
V \quad &: H^{-\frac{1}{2}+\sigma}(\Gamma) \to H^{1+\sigma}(\Omega) \times H^{1+\sigma}_{\text{loc}}(\Omega^c)\,,\\
\gamma_0 V \quad \textit{and}\ \gamma_{c0} V \quad &: H^{-\frac{1}{2}+\sigma}(\Gamma) \to H^{\frac{1}{2}+\sigma}(\Gamma)\,,\\
\tau W \quad \textit{and}\ \tau_c V \quad &: H^{-\frac{1}{2}+\sigma}(\Gamma) \to H^{-\frac{1}{2}+\sigma}(\Gamma)\,,\\
W \quad &: H^{\frac{1}{2}+\sigma}(\Gamma) \to H^{1+\sigma}(\Omega) \times H^{1+\sigma}_{\text{loc}}(\Omega^c)\,,\\
\gamma_0 W \quad \textit{and}\ \gamma_{c0} W \quad &: H^{\frac{1}{2}+\sigma}(\Gamma) \to H^{\frac{1}{2}+\sigma}(\Gamma)\,,\\
\tau W \quad \textit{and}\ \tau_c W \quad &: H^{\frac{1}{2}+\sigma}(\Gamma) \to H^{-\frac{1}{2}+\sigma}(\Gamma)\,.
\end{aligned}$$

For $0 \le \sigma < \frac{1}{2}$, the generalized conormal derivative satisfies $\tau u \in H^{-\frac{1}{2}+\sigma}(\Gamma)$ for $u \in H^{1+\sigma}(\Omega)$.

(For the Stokes system see Kohr et al [234].)

Moreover, the variational solutions in $H^1(\Omega)$ gain slightly more regularity and satisfy corresponding a priori estimates of the form

$$\|u\|_{H^{1+\sigma}(\Omega)} \le \{\|\varphi\|_{H^{\frac{1}{2}+\sigma}(\Gamma)} + \|f\|_{\widetilde{H}^{-1+\sigma}(\Omega)}\} \tag{5.8.1}$$

for the Dirichlet problem if $(\varphi, f) \in H^{\frac{1}{2}+\sigma}(\Gamma) \times \widetilde{H}_0^{-1+\sigma}(\Omega)$ or

$$\|u\|_{H^{1+\sigma}(\Omega)} \le c\{\|\psi\|_{H^{-\frac{1}{2}+\sigma}} + \|f\|_{\widetilde{H}^{-1+\sigma}(\Omega)}\} \tag{5.8.2}$$

for the Neumann problem if $(\psi, f) \in H^{-\frac{1}{2}+\sigma}(\Gamma) \times \widetilde{H}_0^{-1+\sigma}(\Omega)$ and $0 < \sigma \le \frac{1}{2}$.

Correspondingly, if for IDP and EDP in Table 5.1.2, $\mu \in H^{\frac{1}{2}+\sigma}(\Gamma)$ is given, then $\lambda \in H^{-\frac{1}{2}+\sigma}(\Gamma)$. If for INP and ENP $\mu \in H^{-\frac{1}{2}+\sigma}(\Gamma)$ is given then $\lambda \in H^{\frac{1}{2}+\sigma}(\Gamma)$.

If Γ is much smoother then one may obtain even higher regularity as will be seen in Chapters 8–11.

5.8.3 Mixed Boundary Conditions and Crack Problem

We return to the example of the *mixed boundary value problem* (2.5.1) for the Lamé system; but now with given $\boldsymbol{f} \in \widetilde{H}_0^{-1}(\Omega)$, $\boldsymbol{\varphi} \in H^{\frac{1}{2}}(\Gamma_D)$ and $\boldsymbol{\psi} \in H^{-\frac{1}{2}}(\Gamma_N)$. Again, we write with $\boldsymbol{\varphi} \in H^{\frac{1}{2}}(\Gamma)$ and $\boldsymbol{\psi} \in H^{-\frac{1}{2}}(\Gamma)$ according to (2.5.2),

$$\gamma_0 \boldsymbol{u} = \boldsymbol{\varphi} + \widetilde{\boldsymbol{\varphi}}\,,\ T\boldsymbol{u} = \boldsymbol{\psi} + \widetilde{\boldsymbol{\psi}} \quad \text{where } \widetilde{\boldsymbol{\varphi}} \in \widetilde{H}^{\frac{1}{2}}(\Gamma_N) \quad \text{and } \widetilde{\boldsymbol{\psi}} \in \widetilde{H}^{-\frac{1}{2}}(\Gamma_D) \tag{5.8.3}$$

are the yet unknown Cauchy data. The boundary integral equations (2.5.6) can now be written in variational form:

Find $(\widetilde{\boldsymbol{\psi}}, \widetilde{\boldsymbol{\varphi}}) \in \widetilde{H}^{-\frac{1}{2}}(\Gamma)$ satisfying

$$a_\Gamma\big((\widetilde{\psi},\widetilde{\varphi}),\,(\widetilde{\tau},\widetilde{\chi})\big) := \langle V\widetilde{\psi},\widetilde{\tau}\rangle - \langle K\widetilde{\varphi},\widetilde{\tau}\rangle_\Gamma + \langle K'\widetilde{\psi},\widetilde{\chi}\rangle + \langle D\widetilde{\varphi},\widetilde{\chi}\rangle_\Gamma$$
$$= \ell(\widetilde{\tau},\widetilde{\chi}) \quad \text{for all} \quad (\widetilde{\tau},\widetilde{\chi}) \in \widetilde{H}^{-\frac{1}{2}}(\Gamma)\times\widetilde{H}^{\frac{1}{2}}(\Gamma) \tag{5.8.4}$$

where the linear functional ℓ is given as

$$\begin{aligned}\ell(\widetilde{\tau},\widetilde{\chi}) &= \langle(\tfrac{1}{2}I+K)\varphi,\widetilde{\tau}\rangle - \langle V\psi,\widetilde{\tau}\rangle - \langle N\boldsymbol{f},\widetilde{\tau}\rangle \\ &\quad + \langle(\tfrac{1}{2}I-K')\psi,\widetilde{\chi}\rangle - \langle D\varphi,\chi\rangle - \langle T_x N\boldsymbol{f},\widetilde{\chi}\rangle\,.\end{aligned} \tag{5.8.5}$$

Since a_Γ is continuous and satisfies the Gårding inequality

$$\begin{aligned}a_\Gamma\big((\widetilde{\psi},\widetilde{\varphi}),\,(\widetilde{\psi},\widetilde{\varphi})\big) &= \langle V\widetilde{\psi},\widetilde{\psi}\rangle + \langle D\widetilde{\varphi},\widetilde{\varphi}\rangle \\ &\geq \gamma_0\{\|\widetilde{\psi}\|^2_{H^{-\frac{1}{2}}(\Gamma)} + \|\widetilde{\varphi}\|_{H^{\frac{1}{2}}(\Gamma)}\}\end{aligned} \tag{5.8.6}$$

for all $(\widetilde{\psi},\widetilde{\varphi}) \in \widetilde{H}^{-\frac{1}{2}}(\Gamma_D)\times\widetilde{H}^{\frac{1}{2}}(\Gamma_N) \subset H^{-\frac{1}{2}}(\Gamma)\times H^{\frac{1}{2}}(\Gamma)$ as in (5.6.32) and (5.6.33), a_Γ is $\widetilde{H}^{-\frac{1}{2}}(\Gamma_D)\times\widetilde{H}^{\frac{1}{2}}(\Gamma_N)$–elliptic and (5.8.4) is uniquely solvable. In fact, this formulation underlies the efficient and fast boundary element methods in Of et al [339, 340], Steinbach [398]. For the Stokes system, mixed boundary value problems are treated in [234] and for the biharmonic equation in Cakoni et al [58].

For the *classical insertion problem* (2.5.14), now $\varphi \in H^{\frac{1}{2}}(\Gamma)$ and $\varphi^\pm \in H^{\frac{1}{2}}(\Gamma_c)$ with $[\varphi]|_{\Gamma_c} \in \widetilde{H}^{\frac{1}{2}}(\Gamma_c)$ are given, and the variational form of (2.5.16) for the desired Cauchy data $\psi \in H^{-\frac{1}{2}}(\Gamma)$ and $[\psi]|_{\Gamma_c} \in \widetilde{H}^{-\frac{1}{2}}(\Gamma_c)$ reads:
Find $(\psi,[\psi]|_{\Gamma_c}) \in H^{-\frac{1}{2}}(\Gamma)\times\widetilde{H}^{-\frac{1}{2}}(\Gamma_c)$ *satisfying*

$$\begin{aligned}&a_\Gamma\big((\psi,[\varphi]|_{\Gamma_c}),\,(\tau,\widetilde{\tau})\big) := \\ &\langle V_\Gamma\psi,\tau\rangle_\Gamma - \langle V_{\Gamma_c}[\varphi]|_{\Gamma_c},\tau\rangle_\Gamma - \langle V_\Gamma\psi,\widetilde{\tau}\rangle_{\Gamma_c} + \langle V_{\Gamma_c}[\varphi]|_{\Gamma_c},\widetilde{\tau}\rangle_{\Gamma_c} \\ &\qquad = \ell(\tau,\widetilde{\tau}) \quad \text{for all} \quad (\tau,\widetilde{\tau}) \in H^{-\frac{1}{2}}(\Gamma)\times\widetilde{H}^{-\frac{1}{2}}(\Gamma_c)\end{aligned} \tag{5.8.7}$$

where the linear functional is given via (2.5.16). The bilinear form a_Γ is continuous and satisfies a Gårding inequality

$$\begin{aligned}a_\Gamma\big((\psi,\widetilde{\tau}),\,(\psi,\widetilde{\tau})\big) &\geq \gamma_0\{\|\psi\|^2_{H^{-\frac{1}{2}}(\Gamma)} + \|\widetilde{\tau}\|^2_{H^{-\frac{1}{2}}(\Gamma_c)} \\ &\quad c(\|\psi\|^2_{H^{-1}(\Gamma)} + \|\widetilde{\tau}\|^2_{H^{-1}(\Gamma_c)}\}\end{aligned} \tag{5.8.8}$$

Since, in addition, the homogeneous problem with $\varphi = \mathbf{0}$, $\psi = \mathbf{0}$ admits only the trivial solution, (5.8.7) is uniquely solvable.

For the *classical crack problem*, $\psi^\pm \in H^{-\frac{1}{2}}(\Gamma_c)$ with $[\psi]|_{\Gamma_c} \in \widetilde{H}^{-\frac{1}{2}}(\Gamma_c)$ and, in our example, $\varphi \in H^{\frac{1}{2}}(\Gamma)$ in (2.5.17) are given. Then the variational form of (2.5.18) reads:
Find $(\psi,[\varphi]|_{\Gamma_c}) \in H^{-\frac{1}{2}}(\Gamma)\times\widetilde{H}^{\frac{1}{2}}(\Gamma_c)$ *satisfying*

$$\begin{aligned}a_\Gamma\big((\psi,[\varphi]|_{\Gamma_c}),\,(\tau,\widetilde{\chi})\big) &:= \langle V_\Gamma\psi,\tau\rangle_\Gamma + \langle K_{\Gamma_c}[\varphi]|_{\Gamma_c},\tau\rangle_\Gamma \\ &\quad + \langle D_{\Gamma_c}[\varphi]|_{\Gamma_c},\widetilde{\chi}\rangle_{\Gamma_c} - \langle K'_\Gamma\psi,\widetilde{\chi}\rangle_{\Gamma_c} \\ &= \ell(\tau,\widetilde{\tau}) \quad \text{for all} \quad (\tau,\widetilde{\tau}) \in H^{-\frac{1}{2}}(\Gamma)\times\widetilde{H}^{\frac{1}{2}}(\Gamma_c)\,,\end{aligned} \tag{5.8.9}$$

where the linear functional is given via (2.5.18). Here a_Γ satisfies the Gårding inequality

$$a_\Gamma\big((\psi,\widetilde{\chi}),(\psi,\widetilde{\chi})\big) \geq \gamma_0\{\|\psi\|^2_{H^{-\frac{1}{2}}(\Gamma)} + \|\widetilde{\chi}\|^2_{\widetilde{H}^{\frac{1}{2}}(\Gamma)} - c(\|\psi\|^2_{H^{-1}(\Gamma)} + \|\widetilde{\chi}\|^2_{L^2}(\Gamma)\}\,. \tag{5.8.10}$$

Also here the homogeneous problem (5.8.9) with $\psi^{\pm}|_{\Gamma_c} = \mathbf{0}$ and $\varphi|_\Gamma = \mathbf{0}$, i.e., $\ell = 0$, admits only the trivial solution, hence (5.8.9) has a unique solution.

For more detailed analysis of these problems we refer to Costabel and Dauge [85], Duduchava et al [107], [209, 210], Natroshvili et al [324, 106], Stephan [403].

6. Electromagnetic Fields

6.1 Introduction

In this chapter we are concerned with the application of boundary integral methods to boundary value problems for time harmonic Maxwell equations in electromagnetic theory. In what follows in this chapter we assume that the results stated are formulated in classical function spaces if not specified otherwise. In particular, $\Gamma = \partial\Omega$ is assumed to be at least $C^{1+\alpha}$, i.e. Lyaponov. However, most of the analysis can be generalized to Lipschitz continuous boundaries.

The determination of the state of an electromagnetic system requires the knowledge of several physical quantities. We list these below:

$\mathcal{E}$, the electric field,
$\mathcal{H}$, the magnetic field,
$\mathcal{D}$, electric displacement field or flux,
$\mathcal{B}$, the magnetic induction or flux,
$\mathcal{J}$, the conduction current density,
ϱ, the charge density.

These are five vector fields and one scalar field depending on space and time. There are four basic laws which connect these quantities:

– *Gauss' law for electric fields* .
This states that for any subregion $\Omega \subset \mathbb{R}^3$ with boundary $\partial\Omega$, the electrical flux across Ω is equal to the total charge in Ω:

$$\int\limits_{\partial\Omega} \mathcal{D} \cdot ds = \int\limits_{\Omega} \varrho d. \tag{6.1.1}$$

– *Gauss' law for magnetic fields.*
The total magnetic flux across any closed surface $\partial\Omega$ is zero:

$$\int\limits_{\partial\Omega} \mathcal{B} \cdot ds = 0. \tag{6.1.2}$$

G. C. Hsiao und W. L. Wendland, *Boundary Integral Equations*,
Applied Mathematical Sciences 164,
https://doi.org/10.1007/978-3-030-71127-6_6

– *Faraday's law.*
The line integral of the electric field about any closed regular curve ∂S in the field is equal to the negative time rate of change of the magnetic flux through any surface S with the boundary ∂S; i.e.

$$\oint_{\partial S} \mathcal{E} \cdot d\ell = -\int_{S} \frac{\partial \mathcal{B}}{\partial t} \cdot ds. \tag{6.1.3}$$

– *Ampere–Maxwell law.*
The rate of change with time of the electrical flux and the electrical current through a surface S with boundary ∂S produces a circulating magnetic field around the closed path ∂S; i.e.,

$$\frac{\partial}{\partial t}\int_{S} \mathcal{D} \cdot ds + \int_{S} \mathcal{J} \cdot ds = \int_{\partial S} \mathcal{H} \cdot d\ell. \tag{6.1.4}$$

Equations (6.1.1) - (6.1.4) are the **Maxwell equations** in integral form [286] which are ingenious.

6.2 Maxwell Equations

Since the Maxwell equations in integral form are valid for arbitrary domains Ω and arbitrary surfaces S, by using the divergence theorem and the Stokes theorem, we obtain the **Maxwell differential equations** in $\Omega \times [0, T]$:

$$\mathbf{curl}\,\mathcal{E} = -\frac{\partial \mathcal{B}}{\partial t}, \tag{6.2.1}$$

$$\mathbf{curl}\,\mathcal{H} = \frac{\partial \mathcal{D}}{\partial t} + \mathcal{J}, \tag{6.2.2}$$

$$\operatorname{div} \mathcal{D} = \varrho, \tag{6.2.3}$$

$$\operatorname{div} \mathcal{B} = 0. \tag{6.2.4}$$

These equations are not linearly independent since Equations (6.2.1) imply (6.2.4) provided

$$(\operatorname{div}\mathcal{B})(\boldsymbol{x}, 0) = 0. \tag{6.2.5}$$

Also one may notice that from Equations (6.2.2) and (6.2.3) the charge density ϱ and the current density $\mathcal{J}$ satisfy the equation

$$\frac{\partial \varrho}{\partial t} + \operatorname{div} \mathcal{J} = 0. \tag{6.2.6}$$

This equation expresses the fundamental law that the electrical charge is conserved — any flow of charge must come from supply. Conversely, (6.2.6), (6.2.1) and (6.2.2) imply (6.2.3) provided

$$(\mathcal{D} - \varrho)(\boldsymbol{x}, 0) = 0. \tag{6.2.7}$$

This means, (6.2.1)–(6.2.3) are linearly independent, and under condition $(\operatorname{div} \mathcal{B})(\boldsymbol{x}, 0) = 0$, Equation (6.2.4) is a consequence. Alternatively, (6.2.1), (6.2.2) and (6.2.6) are linearly independent and under conditions (6.2.5) and (6.2.7) the equations (6.2.3) and (6.2.4) are the consequences.

6.3 Constitutive Equations

We have 16 unknowns but only 7 independent equations. Therefore, to characterize the electromagnetic fields completely, one needs 9 additional equations. These are constitutive relations taking into account the properties of media which must play a role in the complete system of equations.

Experiments show the relations

$$\mathcal{D}(\boldsymbol{x}, t) = \mathbf{F}\big(\mathcal{E}(\boldsymbol{x}, t)\big)\,,\ \mathcal{B}(\boldsymbol{x}, t) = \mathbf{G}\big(\mathcal{H}(\boldsymbol{x}, t)\big)\,,\ \mathcal{J}(\boldsymbol{x}, t) = \mathbf{H}\big(\mathcal{J}(\boldsymbol{x}, t)\big)$$

with appropriate functions $\mathbf{F}, \mathbf{G}$ and $\mathbf{H}$. We are going to consider linear isotropic homogeneous media. In this case, we adopt the following constitutive equations:

$$\mathcal{D} = \varepsilon\mathcal{E}\,,\ \mathcal{B} = \mu\mathcal{H}\,,\ \mathcal{J} = \sigma\mathcal{E} + \mathcal{J}_0\,, \tag{6.3.1}$$

where ε, μ and σ are constant; $\varepsilon \geq 0$ is the electric permittivity, $\mu > 0$ the magnetic permeability, $\sigma \geq 0$ the conductivity and $\mathcal{J}_0$ is a prescribed current density.

An energy estimate

For the Cauchy problem in $\Omega \times (0, T]$ with $\Gamma = \partial\Omega$ consisting of

$$\frac{\partial \mathcal{B}}{\partial t} + \frac{1}{\varepsilon}\operatorname{curl}\mathcal{D} = \mathbf{0}\,,$$

$$\frac{\partial \mathcal{D}}{\partial t} - \frac{1}{\mu}\mathbf{curl}\,\mathcal{B} + \frac{\sigma}{\varepsilon}\mathcal{D} = \mathbf{0}\,,$$

$$\operatorname{div}\mathcal{D} = 0\,,$$

$$\operatorname{div}\mathcal{B} = 0\,,$$

in terms of the constitutive equations for $\varrho = 0$ and $\mathcal{J}_0 = \mathbf{0}$, let us consider the following energy estimate:

$$\frac{\varepsilon}{2}\int_\Omega \frac{\partial}{\partial t}|\mathcal{B}|^2 d\boldsymbol{x} = -\int_\Omega (\mathbf{curl}\,\mathcal{D})\cdot\mathcal{B}d\boldsymbol{x} = \int_\Gamma \boldsymbol{n}\cdot(\mathcal{D}\times\mathcal{B})ds - \int_\Omega \mathcal{D}\cdot(\mathbf{curl}\,\mathcal{B})d\boldsymbol{x}, \tag{6.3.2}$$

$$\frac{\mu}{2}\int_\Omega \frac{\partial}{\partial t}|\mathcal{D}|^2 d\boldsymbol{x} = \int_\Omega (\mathbf{curl}\,\mathcal{B})\cdot\mathcal{D}d\boldsymbol{x} - \frac{\sigma}{\varepsilon}\int_\Omega |\mathcal{D}|^2 d\boldsymbol{x}.$$

Adding the equations in (6.3.2) yields the differential equation

$$\frac{d}{dt}\mathbb{E} = -\frac{\sigma}{\varepsilon}\int_{\Omega} |\mathcal{D}|^2 d\boldsymbol{x} + \int_{\Gamma} \boldsymbol{n}\cdot(\mathcal{B}\times\mathcal{D})ds$$

for the electromagnetic energy

$$\mathbb{E}(t) = \frac{1}{2}\int_{\Omega} \left\{\varepsilon|\mathcal{B}|^2 + \mu|\mathcal{D}|^2\right\} d\boldsymbol{x}.$$

Lemma 6.3.1. *Under the condition* $\boldsymbol{n}\cdot(\mathcal{B}\times\mathcal{D})_{|\Gamma} = 0$, *the homogeneous Cauchy problem only has the trivial solution* $\mathcal{B} = \mathcal{D} = \mathbf{0}$.

As a consequence of Lemma 6.3.1, it is suggested that the Cauchy problem with boundary conditions

$$(\boldsymbol{n}\times\mathcal{D})_{|\Gamma} = \Phi \quad \text{or} \quad (\boldsymbol{n}\times\mathcal{B})_{|\Gamma} = \Psi. \tag{6.3.3}$$

has at most one solution

Proof of of Lemma 6.3.1: Since $\mathbb{E}(0) = 0$

$$0 \le \frac{1}{2}\int_{\Omega} \mu|\mathcal{D}|^2 d\boldsymbol{x} \le \mathbb{E}(t) - \mathbb{E}(0) = -\frac{\sigma}{\varepsilon}\int_0\int_{\Omega} |\mathcal{D}|^2 d\boldsymbol{x}dt \le 0$$

it holds

$$\frac{1}{2}\int_{\Omega} \mu|\mathcal{D}|^2 d\boldsymbol{x} = 0,$$

hence $\mathcal{D} \equiv 0$. This implies

$$\frac{\partial\mathcal{B}}{\partial t} = 0 \quad \text{and} \quad \mathcal{B}(\boldsymbol{x},t) = \mathcal{B}(\boldsymbol{x},0) = 0$$

as proposed. ■

Lemma 6.3.2. *Let* $\varrho_0 \in L_2(\mathbb{R}^3)$ *have compact support and* $J_0 = \mathbf{0}$. *Then for* $\sigma > 0$, *and* $\varepsilon > 0$

$$\mathcal{D}(\boldsymbol{x},t) = e^{-\frac{\sigma t}{\varepsilon}}\mathbf{B}\varrho_0 \quad \textit{for} \quad t \ge \mathbf{0},$$

where $\mathbf{B}$ *denotes the Bogovskii operator in* $\mathbb{R}^3$.

Proof: Apply div to (6.2.2) and use (6.2.3) and (6.3.1) together with (6.2.3) to obtain

$$\frac{\partial\varrho}{\partial t} + \frac{\sigma}{\varepsilon}\varrho = 0.$$

Hence, with (6.2.3) ,

$$\mathrm{div}\mathcal{D} = \varrho = e^{-\frac{\sigma t}{\varepsilon}}\varrho_0,$$

and we obtain the proposed relation in terms of **B**, the Bogovskii operator as a right inverse to div which is a ψdO_{-1}. ■

Remark 6.3.1:

For positive constant ε and σ we have perfect electric and magnetic media Dautray and Lions [95]. Some special media are characterized in terms of the media parameters. So, substances for which σ is not negligibly small are called *conductors.*

- The perfectly conducting medium is given for $\sigma \to \infty$. Under the assumption that $\mathbf{curl}\,\mathcal{H}$ remains bounded, Equation (6.3.2) yields $\mathcal{E}(\boldsymbol{x}) = \mathbf{0}$ and then (6.3.2) implies $\mathcal{H}(\boldsymbol{x}) = \mathbf{0}$. That means, no electromagnetic fields exist in a perfectly conducting medium.
- If $\sigma = 0$ or negligibly small the medium is called *dielectric* or *insulator.* For $\sigma = 0$ we have *perfect dielectrics.* Then $\varrho = 0$ and $\mathcal{J} = \mathbf{0}$.

For a general conducting medium, $\sigma \neq 0$, if the conductivity increases with temperature over a wide range, the medium is called *semiconductor.* In free space, $\varepsilon = \varepsilon_0 = \frac{10^{-9}}{36\pi}$ Faraday/meter and $\mu = \mu_0 = 4\pi 10^{-7}$ Henry/meter. Moreover, $\varepsilon_0\mu_0 c^2 = 1$ with c the speed of light.

Remark 6.3.2:

The relation (6.2.6) of the conservation of charge with the last of the equations (6.3.1) yields

$$\frac{\partial \varrho}{\partial t} + \frac{\sigma}{\varepsilon}\varrho = 0 \quad \text{and} \quad \varrho(\boldsymbol{x}, 0) = \varrho_0(\boldsymbol{x}) \tag{6.3.4}$$

in view of (6.2.2) and (6.2.3). Hence, $\varrho = e^{-\frac{\sigma}{\varepsilon}t}\varrho_0$ where $\varrho_0(\boldsymbol{x})$ is given.

Remark 6.3.3: The generalized current concept:

It was Maxwell who first noted that Ampere's law for the statics:

$$\mathbf{curl}\,\mathbf{H} = \mathbf{J}$$

was incomplete for time-varying fields. He amended the law to include an electric displacement current $\partial D/\partial t$ in addition to the conduction current. In dielectrics, part of the term $\partial D/\partial t$ is a motion of bound particles and is thus a current in the true sense of the word. However, it is convenient to consider the entire $\partial D/\partial t$ term as a current. In view of the symmetry of Maxwell's equations, it also is convenient to consider the term $\partial \mathcal{B}/\partial t$ as a magnetic displacement current. To represent sources, we amend the field equations to include impressed currents, electric and magnetic. These are the currents we view as the cause of the field. The impressed currents represent energy sources. The symbols **J** and **M** will be used to denote electric and magnetic currents in general, with superscripts indicating the type of current. We define total currents:

$$\mathbf{J}^t = \frac{\partial D}{\partial t} + \mathbf{J}^c + \mathbf{J}^i$$
$$\mathbf{M}^t = \frac{\partial \mathcal{B}}{\partial t} + \mathbf{M}^i$$

where the superscripts t, c, and i denote total, conduction, and impressed currents. The symbols i and k will be used to denote net electric and magnetic currents, and the same superscripts will indicate the type. Thus the circuit form corresponding to the above equations is

$$i^t = \frac{\partial \psi^e}{\partial t} + i^c + i^i$$
$$k^t = \frac{\partial \psi^m}{\partial t} + k^i$$

The i and k are related to the $\mathbf{J}$ and $\mathbf{M}$ by

$$i = \int_{\Gamma} \mathbf{J} \cdot \boldsymbol{n} ds, \quad k = \int_{\Gamma} \mathbf{M} \cdot \boldsymbol{n} ds.$$

where these apply to any of the previous types of current.

Remark 6.3.4: The terminology of the electromagnetic quantities is by no means universal and one may find the following alternative terminologies for the corresponding electromagnetic quantities:

$\mathcal{E}$, the electric field is also called the electric field density;
$\mathcal{B}$, the magnetic induction is also called the magnetic field, magnetic field density or magnetic flux density;
$\mathcal{D}$, the electric displacement field is also called electric induction or electric flux density;
$\mathcal{H}$, the magnetic field is also called auxiliary magnetic field, magnetic field density, magnetizing field;
$\mathcal{J}$, the current density is also called the total current density consisting of free and bound current;
ϱ, the charge density is also called total charge density consisting of free and bound charge;
ε, the permittivity is also called the dielectric constant;
μ, the permeability is also called the magnetic constant.

6.4 Time Harmonic Fields

We consider only the monochromatic theory of electromagnetic waves, that is, not every thing is independent of time but rather that things vary periodically in time with a fixed frequency ω. Then with $i^2 = -1$

$$\begin{aligned} \mathcal{E}(\boldsymbol{x},t) &= \text{Re}\left\{\mathbf{E}(\boldsymbol{x})e^{-i\omega t}\right\}, \ \mathcal{D}(\boldsymbol{x},t) = \text{Re}\left\{\mathbf{D}(\boldsymbol{x})e^{-i\omega t}\right\}, \\ \mathcal{H}(\boldsymbol{x},t) &= \text{Re}\left\{\mathbf{H}(\boldsymbol{x})e^{-i\omega t}\right\}, \ \mathcal{B}(\boldsymbol{x},t) = \text{Re}\left\{\mathbf{B}(\boldsymbol{x})e^{-i\omega t}\right\}, \\ \mathcal{J}(\boldsymbol{x},t) &= \text{Re}\left\{\mathbf{J}(\boldsymbol{x})e^{-i\omega t}\right\}, \ \ \varrho(\boldsymbol{x},t) = \text{Re}\left\{\varrho(\boldsymbol{x})e^{-i\omega t}\right\}. \end{aligned}$$

By using these formulae the derivation with respect to time t transforms into multiplication by $-i\omega$.

The current density $\mathbf{J}$ commonly consists of two terms: one, $\mathbf{J}_0$ associated with external sources of electromagnetic disturbances and the other, $\mathbf{J}_c$ associated with conduction currents produced as a result of the electric field governed by Ohm's law.

The Maxwell equations (6.2.1)–(6.2.4) together with the constitutive equations (6.3.1) in a bounded domain Ω read now

$$\mathbf{curl}\,\mathbf{E}(\boldsymbol{x}) = i\omega\mu\mathbf{H}(\boldsymbol{x}), \tag{6.4.1}$$

$$\mathbf{curl}\,\mathbf{H}(\boldsymbol{x}) = (-i\omega\varepsilon + \sigma)\mathbf{E}(\boldsymbol{x}) + \mathbf{J}_0, \tag{6.4.2}$$

$$\text{div}\,(\varepsilon\,\mathbf{E})(\boldsymbol{x}) = \varrho, \tag{6.4.3}$$

$$\text{div}\,(\mu\mathbf{H})(\boldsymbol{x}) = 0. \tag{6.4.4}$$

Equation (6.2.6) then reduces to

$$\varrho(\boldsymbol{x}) = \text{div}\,\mathbf{J}_0(\boldsymbol{x})/(i\omega - \sigma/\varepsilon). \tag{6.4.5}$$

We remark that for time–harmonic fields ϱ and $\mathbf{J}_0$ are given source distributions satisfying the equation of conntinuity (6.4.5).

In the homogeneous isotropic medium, from the Maxwell equations we can eliminate either $\mathbf{E}$ or $\mathbf{H}$ and obtain the inhomogeneous Helmholtz equations

$$\Delta\mathbf{E} + k^2\mathbf{E} = \mathbf{grad}\ \varrho/\varepsilon - i\omega\mu\mathbf{J}_0 \quad \text{and} \quad \text{div}\,\mathbf{E} = \varrho/\varepsilon \tag{6.4.6}$$

or

$$\Delta\mathbf{H} + k^2\mathbf{H} = -\mathbf{curl}\,\mathbf{J}_0 \qquad \text{and} \quad \text{div}\,\mathbf{H} = 0, \tag{6.4.7}$$

with

$$k^2 = \omega^2\mu\varepsilon + i\omega\mu\sigma, \tag{6.4.8}$$

where we have used the vector identity

$$\Delta\,\boldsymbol{u} = \mathbf{grad}\ \text{div}\,\boldsymbol{u} - \mathbf{curl}\,\mathbf{curl}\,\boldsymbol{u} \tag{6.4.9}$$

for a vector field $\boldsymbol{u}$.

In what follows, we introduce the complex wave number $k \in \mathbb{C}$ on the branch of the square root of k^2 in (6.4.8) with $\operatorname{Re} k \geq 0$ and $\operatorname{Im} k \geq 0$.

Lossless and lossy media: A medium is generally referred to as a lossless medium, if the conductivity $\sigma = 0$ in the medium, and it is referred to as a lossy medium if $\sigma \neq 0$ in the medium. In the lossy medium, if the permittivity possesses an imaginary part, then the medium exhibits loss, and waves attenuate as they propagate through lossy media. Loss is incorporated into the permittivity via the conductivity σ which relates $\mathbf{J}_c$ to $\mathbf{E}$ by Ohm's law: $\mathbf{J}_c = \sigma \mathbf{E}$. The Maxwell–Ampere's law in a lossy medium now reads:

$$\mathbf{curl}\, \mathbf{H} = (\sigma - i\omega\epsilon)\mathbf{E} = -i\omega\ \epsilon_{eff}\mathbf{E}$$

where $\epsilon_{eff} := \epsilon\ (1 + i\sigma/\omega\epsilon)$ is called the effective permittivity and the quantity $\sigma/\omega\epsilon$ is called the loss tangent of the medium and is a measure of how successfully the medium conducts (see, e.g. Staelin, Morgenthaler and Kong [397, 237]). Mathematically, the conductor is considered good if $\sigma \gg \omega\epsilon$ and poor if the reverse is true.

6.4.1 Plane waves

In homogeneous, lossless and source free-media, plane waves

$$\mathbf{E}(x) = \boldsymbol{a}\ exp\{i\boldsymbol{k}\cdot\boldsymbol{x}\}, \quad \mathbf{H}(x) = \boldsymbol{b}\ exp\{i\boldsymbol{k}\cdot\boldsymbol{x}\} \tag{6.4.10}$$

are solutions of source–free Maxwell's equations, where $\boldsymbol{a}$ and $\boldsymbol{b}$ are wave amplitudes, $\boldsymbol{k}$ is called the propagation vector. They are related constant vectors to be specified. For the plane wave solution, Maxwell's equations become:

$$\begin{aligned} \boldsymbol{k}\times\mathbf{E} &= \omega\ \mu\mathbf{H}, \\ \boldsymbol{k}\times\mathbf{H} &= -\omega\ \epsilon\mathbf{E}, \\ \boldsymbol{k}\cdot\mu\mathbf{H} &= 0, \\ \boldsymbol{k}\cdot\ \epsilon\mathbf{E} &= 0. \end{aligned}$$

The propagation vector can be written as $\boldsymbol{k} = k\widehat{\boldsymbol{k}}$, where $k = \omega\sqrt{\mu\epsilon}$ is the magnitude of $\boldsymbol{k}$, the wave number and is real in a lossless medium; $\hat{\boldsymbol{k}}$ is the unit vector, which we may assume also real. The vector $\boldsymbol{a}$ is perpendicular to $\hat{\boldsymbol{k}}$, and the vector $\boldsymbol{b}$ is related to $\boldsymbol{a}$ as

$$\boldsymbol{b} = \frac{k}{\omega\mu}(\widehat{\boldsymbol{k}}\times\boldsymbol{a}) = Z^{-1}(\widehat{\boldsymbol{k}}\times\boldsymbol{a}), \quad Z = \sqrt{\frac{\mu}{\epsilon}} \quad \text{is the impedance} \ .$$

Geometrically, the unit vectors $\hat{\boldsymbol{a}}, \hat{\boldsymbol{b}}$ and $\hat{\boldsymbol{k}}$ form a right–hand orthonormal system (see Kong [237]).

When the medium is lossy, $k^2 = (i\omega\mu)(\sigma - i\omega\epsilon) = \omega^2\mu\epsilon_c$ is complex with $\epsilon_c = (\epsilon + i\sigma/\omega)$. The wave number k is also complex. In this case the wave propagating in a certain direction also decays exponentially in that direction. The imaginary part of the wave number $k = \omega\sqrt{\mu\epsilon_c}$ determines how fast the wave attenuates inside the medium. We define a penetration depth to be $1/Im(k)$. When the medium is very conductive so that $\epsilon_c \approx i\sigma/\omega$, we may approximate the wave number $k \approx (1+i)/\delta$. Here

$$\delta = \frac{1}{Im(k)} = \sqrt{\frac{2}{\omega\mu\sigma}}$$

which is called the skin depth of the conductor. More precisely, the distance the wave must travel in a lossy medium to reduce its value to $e^{-1} = 0.368$ of its original value is known as the penetration or skin depth δ (see, e.g. Balanis [21], Barkeshl [24]).

6.5 Electromagnetic potentials

It is a common practice in the analysis of electromagnetic boundary value problems to use auxiliary vector potentials as aids in obtaining solutions for the electric and magnetics fields. Two most commonly used electromagnetic vector potentials are the vector potential $\boldsymbol{A}$ and vector potential $\mathbf{Q}$. The vector potential $\boldsymbol{A}$ is used in the absence of the magnetic current density, to express the magnetic field $\mathbf{H}$, which is also called magnetic Hertz potential, while the vector potential $\mathbf{Q}$ is used to express the electric field $\mathbf{E}$ in the absence of the electric current density. The latter is called the electric Hertz potential.

To illustrate this idea, the magnetic flux density $\mathbf{B} = \mu\mathbf{H}$ is divergence free and, hence, can be expressed as

$$\mathbf{B} = \mathbf{curl}\,\boldsymbol{A} = \mu\mathbf{H}$$

where $\boldsymbol{A}$ is called the magnetic vector potential. In the homogeneous isotropic medium from (6.4.1), Faraday's law reads

$$\mathbf{curl}\,(\mathbf{E} - i\omega\boldsymbol{A}) = \mathbf{0}$$

which implies

$$\mathbf{E} = i\omega\boldsymbol{A} - \mathbf{grad}\,\varphi$$

where the scalar function φ is called the electric scalar potential. Then from the Ampere–Maxwell equation (6.4.2) we see that

$$\mathbf{curl}\,\mathbf{H} = \frac{1}{\mu}\mathbf{curl}\,\mathbf{curl}\,\boldsymbol{A}$$

which implies

$$\begin{aligned}\mathbf{curl}\,\mathbf{curl}\,\boldsymbol{A} &= (\sigma - i\omega\varepsilon)\mu\mathbf{E} + \mu\mathbf{J}_0,\\ &= (\sigma - i\omega\varepsilon)[i\omega\mu\boldsymbol{A} - \mu\,\mathbf{grad}\,\varphi] + \mu\mathbf{J}_0,\\ &= k^2\boldsymbol{A} - (\sigma - i\omega\varepsilon)\mu\,\mathbf{grad}\,\varphi + \mu\mathbf{J}_0.\end{aligned}$$

Hence,

$$\begin{aligned}-(\Delta + k^2)\boldsymbol{A} &= -\mathbf{grad}\ \mathrm{div}\ \boldsymbol{A} - (\sigma - i\omega\varepsilon)\mu\,\mathbf{grad}\,\varphi + \mu\mathbf{J}_0,\\ &= -\mathbf{grad}\left(\mathrm{div}\,\boldsymbol{A} + (\sigma - i\omega\varepsilon)\mu\varphi\right) + \mu\mathbf{J}_0.\end{aligned}$$

In order to determine $\boldsymbol{A}$ uniquely, we choose

$$\mathrm{div}\ \boldsymbol{A} = -(\sigma - i\omega\varepsilon)\mu\varphi$$

(see the **Remark 6.5.1** below concerning gauge conditions). This implies that

$$-(\Delta + k^2)\boldsymbol{A} = \mu\mathbf{J}_0.$$

Hence,

$$\boldsymbol{A} = \mu\int_{\Omega} G_k(\boldsymbol{x} - \boldsymbol{y})\mathbf{J}_0(\boldsymbol{y})d\boldsymbol{y},$$

where $G_k(\boldsymbol{x} - \boldsymbol{y}) = \frac{1}{4\pi}\frac{e^{ik|\boldsymbol{x}-\boldsymbol{y}|}}{|\boldsymbol{x}-\boldsymbol{y}|}$ is the fundamental solution of the Helmholtz equation. Consequently we see that

$$\mathbf{H}(\boldsymbol{x}) = \mathbf{curl}\int_{\Omega} G_k(\boldsymbol{x} - \boldsymbol{y})\mathbf{J}_0(\boldsymbol{y})d\boldsymbol{y}.$$

Now, for the electric field, we obtain

$$\begin{aligned}\mathbf{E} &= i\omega\boldsymbol{A} - \mathbf{grad}\ \varphi\\ &= i\omega\boldsymbol{A} + \frac{1}{\mu}\,\frac{1}{(\sigma - i\omega\varepsilon)}\mathbf{grad}\ \mathrm{div}\,\boldsymbol{A}\\ &= \frac{1}{\sigma - i\omega\varepsilon}(k^2\boldsymbol{I} + \mathbf{grad}\ \mathrm{div})\int_{\Omega} G_k(\boldsymbol{x} - \boldsymbol{y})\mathbf{J}_0(\boldsymbol{y})d\boldsymbol{y}.\end{aligned}$$

Moreover, from (6.4.3) we get

$$\mathrm{div}\,\mathbf{E} = \frac{\varrho}{\varepsilon}$$

which implies that

$$-\Delta\varphi + i\omega\,\mathrm{div}\boldsymbol{A} = \frac{\varrho}{\varepsilon}.$$

Thus

$$\begin{aligned}-\Delta\varphi - (i\omega\mu)(\sigma - i\omega\varepsilon)\varphi &= \frac{\varrho}{\varepsilon},\\ \text{or} \qquad -(\Delta + k^2)\varphi &= \frac{\varrho}{\varepsilon}.\end{aligned}$$

We have

$$\varphi(\boldsymbol{x}) = \int_{\Omega} G_k(\boldsymbol{x}-\boldsymbol{y})\frac{\varrho}{\varepsilon}(\boldsymbol{y})d\boldsymbol{y}\,,\ \boldsymbol{x}\in\Omega.$$

Note:

$$\frac{\varrho(\boldsymbol{x})}{\varepsilon} = \frac{-1}{(\sigma - i\omega\varepsilon)}\mathrm{div}_{\Gamma}\mathbf{J}_0(\boldsymbol{x})$$

from (6.4.5).

Similarly, for the electrical vector potential $\mathbf{Q}$ and the magnetic scalar potential ψ, the following Maxwell system in terms of a magnetic current density $\mathbf{M}_0$ and a magnetic charge density ζ has the form

$$\mathbf{curl}\,\mathbf{E} = i\omega\mu\mathbf{H} + \mathbf{M}_0 \tag{6.5.1}$$

$$\mathbf{curl}\,\mathbf{H} = (\sigma - i\omega\varepsilon)\mathbf{E}, \tag{6.5.2}$$

$$\mathrm{div}\,(\varepsilon\mathbf{E}) = 0, \tag{6.5.3}$$

$$\mathrm{div}\,(\mu\mathbf{H}) = \zeta. \tag{6.5.4}$$

We obtain

$$\mathbf{E} = \frac{1}{\varepsilon}\mathbf{curl}\,\mathbf{Q}\,,\ \mathbf{H} = \frac{1}{\varepsilon}(\sigma - i\omega\varepsilon)\mathbf{Q} - \mathbf{grad}\,\psi$$

where $\mathbf{Q}$ satisfies the inhomogeneous Helmholtz equation

$$-(\Delta + k^2)\mathbf{Q} = \varepsilon\mathbf{M}_0$$

and ψ the equation

$$-(\Delta\psi + k^2)\psi = \frac{\zeta}{\mu}.$$

By using the fundamental solution G_k we obtain

$$\mathbf{E}(\boldsymbol{x}) = \mathbf{curl}\int_{\Omega} G_k(\boldsymbol{x}-\boldsymbol{y})\mathbf{M}_0(\boldsymbol{y})d\boldsymbol{y}, \tag{6.5.5}$$

$$\mathbf{H}(\boldsymbol{x}) = \frac{1}{i\omega\mu}\{k^2\boldsymbol{I} + \mathbf{grad}\ \mathrm{div}\}\int_{\Omega} G_k(\boldsymbol{x}-\boldsymbol{y})\mathbf{M}_0(\boldsymbol{y})d\boldsymbol{y}. \tag{6.5.6}$$

Remark 6.5.1: Gauge condition. A fundamental theorem due to Helmholtz states that a vector field can be uniquely determined only by specifying both its curl and its divergence. Hence from equation

$$\mathbf{curl}\,\mathbf{A} = \mathbf{B}$$

to specify the **curl** of $\mathbf{A}$ exactly by the magnetic flux density $\mathbf{B}$ is not enough to determine $\mathbf{A}$ uniquely without specifying the divergence of $\mathbf{A}$.

Different choices for the divergence of $\mathbf{A}$ are referred to as choice of gauge. A physical problem may thus be described by a vector potential with different choice of gauge, which are said to be equivalent to within a gauge transformation, and each particular choice of $\nabla \cdot \mathbf{A}$ is called a gauge condition. In the previous formulation, we have chosen

$$\operatorname{div}\mathbf{A} = -(\sigma - i\omega\varepsilon)\mu\varphi$$

which is called **Lorentz's gauge.**

A best–known gauge condition used in electromagnetics is extremely simple, e. g. $\nabla \cdot \boldsymbol{A} = 0$, which is called the **Coulomb gauge** and with this choice, the boundary condition applicable to $\boldsymbol{A}$ can be further sharpened:

$$\int_{\partial\Omega} \boldsymbol{A} \cdot \boldsymbol{n} ds = 0.$$

In closing this section we now include for completeness the fundamental tensor of $\mathbf{curl}\,\mathbf{curl}\,\boldsymbol{A} - k^2\boldsymbol{A} = \mathbf{0}$, namely:

$$(\boldsymbol{I} + k^{-2}\mathbf{grad}_{\boldsymbol{x}}\operatorname{div}_{\boldsymbol{x}})\mathbf{G}_k(\boldsymbol{x}, \boldsymbol{y}). \tag{6.5.7}$$

where $\mathbf{G}_k(\boldsymbol{x}, \boldsymbol{y}) = G_k(\boldsymbol{x}, \boldsymbol{y})\boldsymbol{I}$.

6.6 Transmission and Boundary Conditions

We specify the distinct constitutive parameters and field quantities in two different media Ω_1 and Ω_2 in $\mathbb{R}^3$ with an interface Σ;

$$\begin{aligned} \mathbf{curl}\,\mathbf{E}_1 &= i\omega\mu_1\mathbf{H}_1\,,\ \ \mathbf{curl}\,\mathbf{H}_1 = (-i\omega\varepsilon_1 + \sigma_1)\mathbf{E}_1 + \mathbf{J}_{01} \quad \text{in } \Omega_1, \\ \mathbf{curl}\,\mathbf{E}_2 &= i\omega\mu_2\mathbf{H}_2\,,\ \ \mathbf{curl}\,\mathbf{H}_2 = (-i\omega\varepsilon_2 + \sigma_2)\mathbf{E}_2 + \mathbf{J}_{02} \quad \text{in } \Omega_2. \end{aligned} \tag{6.6.1}$$

Let $\boldsymbol{n}$ be the surface unit normal of Σ pointing from Ω_1 to Ω_2. The transmission conditions can be derived by using the integral form of the Maxwell equations (see Dautray and Lions [95, Section 4]) which are as follows:

$$\begin{aligned} (\varepsilon_2\mathbf{E}_2 - \varepsilon_1\mathbf{E}_1) \cdot \boldsymbol{n} &= \varrho_\Sigma, \\ (\mu_2\mathbf{H}_2 - \mu_1\mathbf{H}_1) \times \boldsymbol{n} &= -\mathbf{J}_\Sigma, \\ (\mu_2\mathbf{H}_2 - \mu_1\mathbf{H}_1) \cdot \boldsymbol{n} &= 0 \\ (\varepsilon_2\mathbf{E}_2 - \varepsilon_1\mathbf{E}_1) \times \boldsymbol{n} &= \mathbf{0} \quad \text{on } \Sigma. \end{aligned} \tag{6.6.2}$$

Here, ϱ_Σ and $\mathbf{J}_\Sigma$ denote the charge and current densities concentrated on Σ.

At the interface between two lossless dielectric media, it is usually assumed that both $\mathbf{J}_\Sigma$ and ϱ_Σ are identically zero. On the other hand, when Σ is the interface between two lossy dielectrics, the surface current $\mathbf{J}_\Sigma$ is identically zero because $\mathbf{E}$ is finite in both media. The surface charge density ϱ_Σ may or may not be zero.

In the special case where Ω_1 is a perfect conductor then $\mathbf{E}_1 = \mathbf{0}$ and $\mathbf{H}_1 = \mathbf{0}$ and (6.6.2) is reduced to

$$\begin{aligned} \varepsilon_2 \mathbf{E}_2 \cdot \boldsymbol{n} &= \varrho_\Sigma, \quad \mu_2 \mathbf{H}_2 \times \boldsymbol{n} = -\mathbf{J}_\Sigma; \\ \mu_2 \mathbf{H}_2 \cdot \boldsymbol{n} &= 0, \qquad \varepsilon_2 \mathbf{E}_2 \times \boldsymbol{n} = \mathbf{0}, \end{aligned} \tag{6.6.3}$$

where the surface charge ϱ_Σ and surface current density $\mathbf{J}_\Sigma$ in general are unknowns. The surface current and surface charge are not normally encountered except when one of the materials is a good conductor. At high frequencies there is a well known effect which confines current largely to surface regions. This is known as the skin effect MacCamy and Stephan [278, 279] in common conductors and is often sufficiently small to be neglected.

The case $\Omega_2 = \Omega^c = \mathbb{R}^3 \setminus \overline{\Omega}$ where Ω is a perfect conductor with boundary $\Gamma = \partial\Omega = \Sigma$ leads to the boundary conditions

$$\boldsymbol{n} \cdot \mathbf{H}|_\Gamma = 0 \quad \text{and} \quad \boldsymbol{n} \times \mathbf{E}|_\Gamma = \mathbf{0}. \tag{6.6.4}$$

6.7 Boundary Value Problems

We classify the boundary value problems into two classes according to applications. We begin with scattering problems followed by a discussion of eddy current problems.

6.7.1 Scattering problems

Let $\Omega \subset \mathbb{R}^3$ be a bounded domain with a regular $C^{1+\alpha}$ boundary $\Gamma = \partial\Omega$.

- *Interior problems*
  Find the electromagnetic complex–valued vector fields $\mathbf{E}$ and $\mathbf{H}$ in Ω having a perfect conducting boundary $\Gamma = \partial\Omega$ such that $(\mathbf{E}, \mathbf{H})^\top \in C^1(\Omega)^6 \cap C^0(\overline{\Omega})^6$ are satisfying the Maxwell equations (6.4.1)–(6.4.4) with $\mathbf{J}_0 = \mathbf{0}$ and $\varrho = 0$ together with the boundary conditions

$$(\boldsymbol{n} \times \mathbf{E})(\boldsymbol{x}) = \boldsymbol{f}(\boldsymbol{x}) \quad \text{for } \boldsymbol{x} \in \Gamma \tag{6.7.1}$$

  where $\boldsymbol{f}(\boldsymbol{x}) = -\boldsymbol{n} \times \mathbf{E}^{in}(\boldsymbol{x})$ for a given incident field $\mathbf{E}^{in}(\boldsymbol{x})$.

- *Exterior problems*
  Find the electromagnetic fields $\mathbf{E}$ and $\mathbf{H}$ in $\Omega^c := \mathbb{R}^3 \setminus \overline{\Omega}$ where $\partial\Omega = \Gamma$ is a perfectly conducting surface such that $(\mathbf{E}, \mathbf{H})^\top \in C^1(\Omega^c)^6 \cap C^0(\Omega^c \cup \Gamma)^6$ satisfy the Maxwell equations (6.4.1)–(6.4.4) in Ω^c with $\mathbf{J} = 0$ and $\varrho = 0$, the boundary conditions (6.7.1) and the **Silver–Müller radiation conditions**

$$\begin{aligned}\lim_{|\boldsymbol{x}|\to\infty}\left(\boldsymbol{x}\times(\mathbf{curl}\,\mathbf{E}(\boldsymbol{x})\right)+ik|\boldsymbol{x}|\mathbf{E}(\boldsymbol{x})&=\mathbf{0},\\ \lim_{|\boldsymbol{x}|\to\infty}\left(\boldsymbol{x}\times(\mathbf{curl}\,\mathbf{H}(\boldsymbol{x})\right)+ik|\boldsymbol{x}|\mathbf{H}(\boldsymbol{x})&=\mathbf{0}\end{aligned}\tag{6.7.2}$$

uniformly as $|\boldsymbol{x}|\to\infty$. They can also be written as

$$\begin{aligned}\frac{\boldsymbol{x}}{|\boldsymbol{x}|}\times\mathbf{H}+\frac{k}{\omega\mu}\mathbf{E}&=\mathfrak{o}(|\boldsymbol{x}|^{-1}),\\ \frac{\boldsymbol{x}}{|\boldsymbol{x}|}\times\mathbf{E}-\frac{\omega\mu}{k}\mathbf{H}&=\mathfrak{o}(|\boldsymbol{x}|^{-1}),\end{aligned}$$

uniformly as $|\boldsymbol{x}|\to\infty$.

- *Transmission problems*
  Let $\Omega^c := \mathbb{R}^3\setminus\overline{\Omega}$ be an exterior domain. Then the transmission problem reads: Given the functions $\boldsymbol{f}$ and $\boldsymbol{g}$ on the interface Γ. Find the two pairs of vector fields $(\mathbf{E},\mathbf{H})$ and $(\mathbf{E}^c,\mathbf{H}^c)$ satisfying the Maxwell equations Ω and Ω^c, respectively, and the transmission conditions

$$\boldsymbol{n}\times(\mathbf{E}-\mathbf{E}^c)=\boldsymbol{f}\,,\ \boldsymbol{n}\times(\mathbf{H}-\mathbf{H}^c)=\boldsymbol{g}\quad\text{on }\Gamma.\tag{6.7.3}$$

  In addition, the exterior field $(\mathbf{E}^c,\mathbf{H}^c)$ should satisfy the Silver–Müller radiation conditions (6.7.2).

6.7.2 Eddy current problems

Many applications in high electromagnetic energy problems are governed by the Maxwell equations in the low frequency region. In such a case the electrical displacement can be neglected in the Maxwell–Ampere's law (see Ammari, Buffa and Nedelec [12]). This leads to the eddy current model. As stated by Faraday's law, a time variation of the magnetic field generates an electric field. Therefore, a current density $\mathbf{J}=\sigma\mathbf{E}$ is induced in each of the conductors and is called the eddy current. The system of Maxwell equations obtained by disregarding the displacement current term then $i\omega\varepsilon_c\mathbf{E}$ is the eddy current model of the time–harmonic Maxwell equations, which can be justified by a dimensional analysis as in Rodriguez and Valli [366, p.7] if we let $L=\operatorname{diam}(\Omega^c)$ be the reference length, we see that

$$|\omega\varepsilon_c\mathbf{E}|=|\omega^2\varepsilon_c\mu_cL^2|\,|\mathbf{curl}\,\mathbf{H}|\quad\text{and}\quad|\omega\varepsilon_c\mathbf{E}|=\left|\frac{\omega\varepsilon}{\sigma}\right||\sigma\mathbf{E}|.$$

Hence, under the assumptions $|\omega^2\varepsilon_c\mu_cL^2|\ll 1$ and $|\sigma^{-1}\varepsilon_c\omega|\ll 1$, the electric field $\omega\varepsilon_c\mathbf{E}$ in the time–harmonic Ampere's equations can be neglected in comparison with **curl H** and $\sigma\mathbf{E}$, respectively (see, e.g. Rodriguez and Valli [366, p.7]).

(a) The eddy current problem in the bounded domain $\Omega\subset\mathbb{R}^3$
Let Ω^c with $\overline{\Omega^c}\subset\Omega^c$ be a conductor, and $\Omega_I := \Omega\setminus\overline{\Omega^c}$ be a dielectric

medium. Then the pair $(\mathbf{E},\mathbf{H})^t$ satisfies the **eddy current Maxwell equations:**

$$\left.\begin{aligned}\mathbf{curl}\,\mathbf{H}_c-\sigma_c\mathbf{E}_c=\mathbf{0},\\ \mathbf{curl}\,\mathbf{E}_c-i\omega\mu_c\mathbf{H}_c=\mathbf{0}\end{aligned}\right\}\quad \text{in}\ \ \Omega^c \tag{6.7.4}$$

and

$$\left.\begin{aligned}\mathbf{curl}\,\mathbf{H}=\mathbf{J}_0,\\ \mathbf{curl}\,\mathbf{E}-i\omega\mu\mathbf{H}=\mathbf{0}\quad\text{and}\\ \operatorname{div}(\varepsilon\mathbf{E})=0\end{aligned}\right\}\quad \text{in}\ \ \Omega_I \tag{6.7.5}$$

where the density $\mathbf{J}_0$ is given such that

$$\operatorname{div}\mathbf{J}_0=0\quad\text{in}\ \ \Omega_I. \tag{6.7.6}$$

In addition we need the **transmission conditions**

$$\begin{aligned}\boldsymbol{n}\times\mu_c\mathbf{H}_c-\boldsymbol{n}\times\mu\mathbf{H}=\mathbf{0}\quad&\text{and}\\ \boldsymbol{n}\times\mathbf{E}_c-\boldsymbol{n}\times\mathbf{E}=\mathbf{0}\quad&\text{on the interface}\ \ \Sigma:=\partial\Omega^c\cap\partial\Omega,\end{aligned} \tag{6.7.7}$$

and also the boundary conditions

$$\boldsymbol{n}\times\mathbf{E}=\mathbf{0}\quad\text{on}\ \ \partial\Omega. \tag{6.7.8}$$

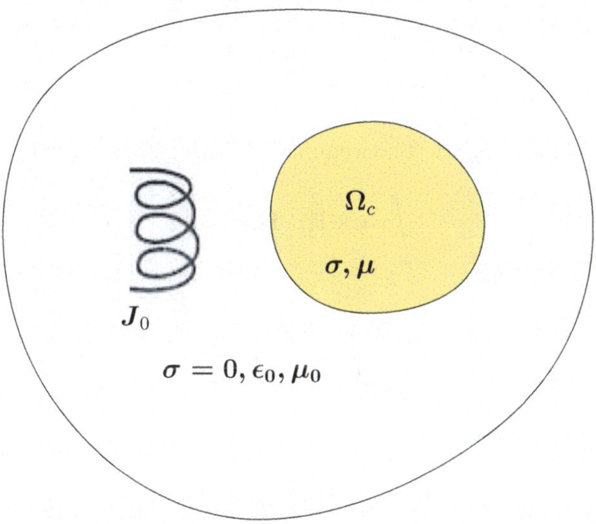

Fig. 6.1. The geometry of the bounded eddy current problem

(b) The eddy current problem in $\mathbb{R}^3$
Here $\Omega = \mathbb{R}^3$ and in addition to Equations (6.2.4)–(6.7.8) the electromagnetic field must satisfy the Silver–Müller radiation conditions (6.7.2) instead of boundary conditions on $\partial\Omega$.

6.8 Uniqueness

In the following, we include some of the uniqueness results of the boundary value problems. For presenting the proofs of uniqueness we follow E. Meister in [293, Section 5.4].

6.8.1 The cavity problem

In this problem let $\Omega \subset \mathbb{R}^3$ be a bounded Lipschitz domain and $(\mathbf{E},\mathbf{H})^\top \in C^1(\Omega)^6 \cap C^0(\overline{\Omega})^6$ be a solution of the Maxwell equations (6.4.1)–(6.4.4) satisfying the boundary conditions (6.7.1). Since the system as well as the boundary conditions are linear, for uniqueness, it suffices to consider the homogeneous problem with $\mathbf{J}_0 = \mathbf{0}$, $\varrho = 0$ in Ω, and $\boldsymbol{f} = \mathbf{0}$ on Γ.

Theorem 6.8.1. *The homogeneous cavity problem* (6.4.1)–(6.4.4), (6.7.1) *with* $\mathbf{J}_0 = \mathbf{0}$ *and* $\varrho = 0$ *in* Ω *and* $\boldsymbol{f} = \mathbf{0}$ *on* Γ *has*
a) for $\sigma > 0$ *in* Ω *only the trivial solution* $\mathbf{E} = \mathbf{0}$ *and* $\mathbf{H} = \mathbf{0}$.
b) If in addition $\sigma \equiv 0$ *then the homogeneous cavity problem in* Ω *has only the trivial solution if and only if* k^2 *is not an eigenvalue of the problem*

$$\Delta\mathbf{E} + k^2\mathbf{E} = \mathbf{0} \quad \text{and} \quad \operatorname{div}(\varepsilon\mathbf{E}) = 0 \quad \text{in} \;\; \Omega,$$
$$\text{and} \;\; \boldsymbol{n} \times \gamma_0\mathbf{E}(\boldsymbol{x}) = \mathbf{0} \quad \text{for} \;\; \boldsymbol{x} \in \Gamma,$$

Proof: (See Meister [293, Theorem 5.17]). From (6.7.1) with $\boldsymbol{f} = \mathbf{0}$ we get

$$\int_\Gamma (\boldsymbol{n} \times \overline{\mathbf{E}}) \cdot \mathbf{H} ds = 0$$

implying

$$0 = \int_\Gamma \boldsymbol{n} \cdot (\overline{\mathbf{E}} \times \mathbf{H}) ds$$

and with Gauss's theorem

$$0 = \int_\Omega \operatorname{div}(\overline{\mathbf{E}} \times \mathbf{H}) d\boldsymbol{x} = \int_\Omega (\mathbf{H} \cdot \mathbf{curl}\,\overline{\mathbf{E}}) d\boldsymbol{x} - \int_\Omega (\overline{\mathbf{E}} \cdot \mathbf{curl}\,\mathbf{H}) d\boldsymbol{x}.$$

Inserting (6.4.1) and (6.4.2) yields

$$0 = \int_\Omega \mathbf{H} \cdot (-i\omega\mu\overline{\mathbf{H}}) d\boldsymbol{x} - \int_\Omega \overline{\mathbf{E}} \cdot (-i\omega\varepsilon + \sigma)\mathbf{E} d\boldsymbol{x}$$

$$= i\omega \int_\Omega (\varepsilon|\mathbf{E}|^2 - \mu|\mathbf{H}|^2) d\boldsymbol{x} - \int_\Omega \sigma|\mathbf{E}|^2 d\boldsymbol{x}.$$

case a): Taking the real part of this relation gives

$$\int_\Omega \sigma|\mathbf{E}|^2 d\boldsymbol{x} = 0$$

which implies $\mathbf{E} = \mathbf{0}$ in $\overline{\Omega}$ since $\sigma > 0$. Then

$$-i\omega \int_\Omega \mu|\mathbf{H}|^2 d\boldsymbol{x} = 0$$

with $\mu > 0$ also $\mathbf{H} = \mathbf{0}$ in $\overline{\Omega}$ which completes the proof for $\sigma > 0$. If $\boldsymbol{n} \times \mathbf{H}|_\Gamma = \mathbf{0}$, vanishing of $\mathbf{E}$ and $\mathbf{H}$ in $\overline{\Omega}$ follow with the same arguments.

case b): Since $\sigma = 0$ there holds with constant $\varepsilon > 0$, $\operatorname{div}\mathbf{E} = 0$ due to (6.4.3) because $\mathbf{J}_0 = 0$ and $\varrho = 0$.

Therefore

$$\Delta\mathbf{E} + k^2\mathbf{E} = -\mathbf{curl\ curl\ E} + k^2\mathbf{E} = \mathbf{0}$$

and the first Green's identity yields

$$\int_\Omega k^2|\mathbf{E}|^2 dx = \int_\Omega \overline{\mathbf{E}}\mathbf{curl\ curl\ E}\, d\boldsymbol{x} = \int_\Omega \mathbf{curl}\ \overline{\mathbf{E}} \cdot \mathbf{curl\ E}\, d\boldsymbol{x} - \int_{\partial\Omega} \mathbf{curl}\ \overline{\mathbf{E}} \cdot (\boldsymbol{n} \times \mathbf{E}) ds.$$

Since $(\boldsymbol{n} \times \mathbf{E})|_\Gamma = \mathbf{0}$,

$$\int_\Omega \mathbf{curl}\ \overline{\mathbf{E}} \cdot \mathbf{curl\ E}\, d\boldsymbol{x} = \int_\Omega k^2|\mathbf{E}|^2 d\boldsymbol{x}. \tag{6.8.1}$$

As will be seen, the inhomogeneous problem for $\mathbf{E}$ with boundary condition (6.7.1) and $\boldsymbol{f} \in C^\alpha(\Gamma)$ defines a Fredholm mapping $\boldsymbol{f} \mapsto \mathbf{E} \in C^{1+\alpha}(\Omega) \cap C^\alpha(\overline{\Omega})$ of index zero. Then the homogeneous problem above only admits the trivial solution $\mathbf{E} = \mathbf{0}$ except for a countable set of real eigenvalues $0 < k_j^2 = \lambda_j^{-1} \leq \lambda_{j+1}^{-1}$, $\lambda_j^{-1} \to \infty$. The eigenvalue problem reads

$$\begin{aligned} -\Delta\mathbf{E} &= k^2\mathbf{E} \quad \text{and } \operatorname{div}\mathbf{E} = 0 && \text{in } \Omega \in C^{1,\alpha}, \\ \boldsymbol{n} \times \gamma_0\mathbf{E} &= \mathbf{0} && \text{on } \Gamma = \partial\Omega, \\ \mathbf{J}_0, \varrho &= 0 \qquad \text{and } \sigma \equiv 0 && \text{in } \Omega, \end{aligned} \tag{6.8.2}$$

where $k^2 = \varepsilon\mu\omega^2 > 0$. ∎

Theorem 6.8.2. (see also Meister [293, Theorem 5.20] and for Ω a Lipschitz domain see Clayes and Hiptmair [71],[72]).

Let Ω be simply connected with a $C^{1+\alpha}$ boundary Γ. Then (6.8.2) *has nontrivial eigensolutions for a countable number of eigenvalues* $0 < k_j^2 = \lambda_j^{-1} \le \lambda_{j+1}^{-1}$, $j \in \mathbb{N}$. *The linear operator associated with* (6.8.2) *is a Fredholm operator of index zero.*

Proof: From (6.8.1) we conclude that $k^2$is real and positiv. Moreover, any solution of (6.8.2) admits the representation

$$\mathbf{E}(\boldsymbol{y}) = k^2 \sum \int\limits_{\Omega} G_0(\boldsymbol{x}-\boldsymbol{y})\mathbf{E}(\boldsymbol{y})d\boldsymbol{y} \\ -\{\int\limits_{\Gamma} G_0(\boldsymbol{x}-\boldsymbol{y})(\gamma_t\,\mathbf{curl}\,\mathbf{E})ds_{\boldsymbol{y}} + \int\limits_{\Gamma}(\boldsymbol{n}\cdot\mathbf{E})(\boldsymbol{y})\mathbf{grad}_{\boldsymbol{y}}G_0(\boldsymbol{x}-\boldsymbol{y})ds_{\boldsymbol{y}} \\ + \int\limits_{\Gamma}(\gamma_t\mathbf{E})(\boldsymbol{y}) \times \mathbf{grad}_{\boldsymbol{y}}G_0(\boldsymbol{x}-\boldsymbol{y})ds_{\boldsymbol{y}}. \tag{6.8.3}$$

Since $\gamma_t\mathbf{E}|_\Gamma = \boldsymbol{n}\times\mathbf{E}|_\Gamma = \boldsymbol{n}\times\gamma_0\mathbf{E} = \mathbf{0}$ on Γ, the last Integral vanishes. Let $\varphi := \boldsymbol{n}\cdot\mathbf{E}$ and $\boldsymbol{v} := \gamma_t\,\mathbf{curl}\,\mathbf{E}$ on Γ. Then on the boundary Γ we obtain from (6.8.3) for $\boldsymbol{x}\in\Gamma$:

$$\varphi(\boldsymbol{x}) = k^2\boldsymbol{n}(\boldsymbol{x})\cdot\int\limits_{\Omega} G_0(\boldsymbol{x}-\boldsymbol{y})\mathbf{E}(\boldsymbol{y})d\boldsymbol{y} - \boldsymbol{n}(\boldsymbol{x})\cdot\int\limits_{\Gamma} G_0(\boldsymbol{x}-\boldsymbol{y})\boldsymbol{v}(\boldsymbol{y})ds_{\boldsymbol{y}} \\ -\boldsymbol{n}(\boldsymbol{x})\cdot\int\limits_{\Gamma}\varphi(\boldsymbol{y})\,\mathbf{grad}_{\boldsymbol{y}}G_0(\boldsymbol{x}-\boldsymbol{y})ds_{\boldsymbol{y}}, \tag{6.8.4}$$

$$\boldsymbol{v}(\boldsymbol{x}) = \gamma_t\,\mathbf{curl}\,\mathbf{E}(\boldsymbol{x}) = \boldsymbol{n}(\boldsymbol{x})\times\mathbf{curl}\,\mathbf{E}(\boldsymbol{x})|_\Gamma \\ = k^2\boldsymbol{n}(\boldsymbol{x})\times(\mathbf{curl}_{\boldsymbol{x}}\int\limits_{\Omega} G_0(\boldsymbol{x}-\boldsymbol{y})\mathbf{E}(\boldsymbol{y})d\boldsymbol{y})|_\Gamma \\ -\boldsymbol{n}(\boldsymbol{x})\times\mathbf{curl}_x\int\limits_{\Gamma} G_0(\boldsymbol{x}-\boldsymbol{y})\boldsymbol{v}(\boldsymbol{y})ds_{\boldsymbol{y}} \\ -\boldsymbol{n}(\boldsymbol{x})\times(\mathbf{curl}_{\boldsymbol{x}}\int\limits_{\Gamma}\varphi(\boldsymbol{y})\,\mathbf{grad}_{\boldsymbol{y}}G_0(\boldsymbol{x}-\boldsymbol{y})ds_{\boldsymbol{y}})|_\Gamma, \tag{6.8.5}$$

$$\mathbf{E}(\boldsymbol{x}) = k^2\int\limits_{\Omega} G_0(\boldsymbol{x}-\boldsymbol{y})\mathbf{E}(\boldsymbol{y})d\boldsymbol{y} - \int\limits_{\Gamma} G_0(\boldsymbol{x}-\boldsymbol{y})\boldsymbol{v}(\boldsymbol{y})ds_{\boldsymbol{y}} \\ -\int\limits_{\Gamma}\varphi(\boldsymbol{y})\,\mathbf{grad}_{\boldsymbol{y}}G_0(\boldsymbol{x}-\boldsymbol{y})ds_{\boldsymbol{y}} \quad \text{for } x\in\Omega. \tag{6.8.6}$$

This is a system of three integral equations for the triple $(\varphi,\boldsymbol{v},\mathbf{E}) \in \big(C^\alpha(\Gamma)\times (C^\alpha(\Gamma))^2\times C^{1+\alpha}(\overline{\Omega})\big)$.

In the first equation of the system, (6.8.4), use $\mathbf{grad}_{\boldsymbol{y}} G_0(\boldsymbol{x}-\boldsymbol{y}) = -\mathbf{grad}_{\boldsymbol{x}}(\boldsymbol{x}-\boldsymbol{y})$, then

$$-\boldsymbol{n}(\boldsymbol{x}) \cdot \int_\Gamma (\varphi(\boldsymbol{y}) \mathbf{grad}_{\boldsymbol{y}} G_0(\boldsymbol{x}-\boldsymbol{y}) ds_{\boldsymbol{y}} = \boldsymbol{n}(\boldsymbol{x}) \cdot \mathbf{grad}_{\boldsymbol{x}} \int_\Gamma G_0(\boldsymbol{x}-\boldsymbol{y})\varphi(\boldsymbol{y}) ds_{\boldsymbol{y}}$$

$$= \frac{\partial}{\partial n_{\boldsymbol{y}}} \int_\Gamma G_0(\boldsymbol{x}-\boldsymbol{y})\varphi(\boldsymbol{y}) ds_{\boldsymbol{y}}|_\Gamma = (\frac{1}{2}I + K')\varphi \quad \text{on } \Gamma$$

due to (1.2.5). Then (6.8.4) becomes

$$\begin{aligned}(\frac{1}{2}I - K')\varphi = k^2 \boldsymbol{n}(\boldsymbol{x}) \cdot & \int_\Omega G_0(\boldsymbol{x}-\boldsymbol{y})\mathbf{E}(\boldsymbol{y}) d\boldsymbol{y} \\ - \boldsymbol{n}(\boldsymbol{x}) \cdot & \int_\Gamma G_0(\boldsymbol{x}-\boldsymbol{y}) \boldsymbol{v}(\boldsymbol{y}) ds_{\boldsymbol{y}}.\end{aligned}$$

The operator $(\frac{1}{2}I - K')$ is invertible on $H^{-1/2}(\Gamma)$ (see (5.6.46) and (5.6.48)) and on $C^\alpha(\Gamma)$ as well, and (6.8.4) is equivalent to

$$\begin{aligned}\varphi = (\frac{1}{2}I - K')^{-1}\{k^2 \boldsymbol{n}(\boldsymbol{x}) \cdot & \int_\Omega G_0(\boldsymbol{x}-\boldsymbol{y})\mathbf{E}(\boldsymbol{y}) d\boldsymbol{y} \\ - \boldsymbol{n}(\boldsymbol{x}) \cdot & \int_\Gamma G_0(\boldsymbol{x}-\boldsymbol{y}) \boldsymbol{v}(\boldsymbol{y}) ds_{\boldsymbol{y}}\}.\end{aligned} \tag{6.8.7}$$

Similarly, for the last term in (6.8.5), we find

$$\boldsymbol{n}(\boldsymbol{x}) \times \mathbf{curl}_{\boldsymbol{x}} \int_\Gamma \varphi(\boldsymbol{y}) \, \mathbf{grad}_{\boldsymbol{y}} G_0(\boldsymbol{x}-\boldsymbol{y}) ds_{\boldsymbol{y}}$$

$$= -\boldsymbol{n}(\boldsymbol{x}) \times \int_\Gamma \mathbf{curl}_{\boldsymbol{x}} \, \mathbf{grad}_{\boldsymbol{x}} G_0(\boldsymbol{x}-\boldsymbol{y})\varphi(\boldsymbol{y}) ds_{\boldsymbol{y}} = 0$$

since tangential derivatives commute with surface integration. Then (6.8.5) takes the form

$$\begin{aligned}\boldsymbol{v}(\boldsymbol{x}) = k^2 \boldsymbol{n}(\boldsymbol{x}) \times \mathbf{curl}_{\boldsymbol{x}} & \int_\Omega G_0(\boldsymbol{x}-\boldsymbol{y})\mathbf{E}(\boldsymbol{y}) d\boldsymbol{y}|_\Gamma \\ - \boldsymbol{n}(\boldsymbol{x}) \times \mathbf{curl}_{\boldsymbol{x}} & \int_\Gamma G_0(\boldsymbol{x}-\boldsymbol{y}) \boldsymbol{v}(\boldsymbol{y}) ds_{\boldsymbol{y}} \quad \text{for } \boldsymbol{x} \in \Gamma.\end{aligned} \tag{6.8.8}$$

Using the local orthonormal vectors $(\boldsymbol{e}_1, \boldsymbol{e}_2, \boldsymbol{n})$ at $\boldsymbol{x} \in \Gamma$, the differential operator in (6.8.8) becomes with

$$\mathbf{curl}\,(w_1\boldsymbol{e}_1 + w_2\boldsymbol{e}_2) = -\partial_n w_2\boldsymbol{e}_1 + \partial_n w_1\boldsymbol{e}_2 + (\partial_1 w_2 - \partial_2 w_1)\boldsymbol{n},$$
$$\boldsymbol{n}\times\mathbf{curl}\,\boldsymbol{w} = -\partial_n w_1\boldsymbol{e}_1 - \partial_n w_2\boldsymbol{e}_2,$$

$$\begin{aligned}
\boldsymbol{v}(\boldsymbol{x}) &= k^2\boldsymbol{n}(\boldsymbol{x})\times\mathbf{curl}_{\boldsymbol{x}}\int\limits_{\Omega} G_0(\boldsymbol{x}-\boldsymbol{y})\mathbf{E}(\boldsymbol{y})d\boldsymbol{y}|_{\Gamma} + \frac{\partial}{\partial n_{\boldsymbol{x}}}\int\limits_{\Gamma} G_0(\boldsymbol{x}-\boldsymbol{y})\boldsymbol{v}(\boldsymbol{y})ds_{\boldsymbol{y}}\big|_{\Gamma}\\
&= k^2\boldsymbol{n}(\boldsymbol{x})\times\mathbf{curl}_{\boldsymbol{x}}\int\limits_{\Omega} G_0(\boldsymbol{x}-\boldsymbol{y})\mathbf{E}(\boldsymbol{y})d\boldsymbol{y}|_{\Gamma} + \big(\frac{1}{2}\boldsymbol{v}(\boldsymbol{x}) + K'\boldsymbol{v}(\boldsymbol{x})\big),\\
&= k^2\big(\frac{1}{2}I - K'\big)^{-1}\{\boldsymbol{n}\times\mathbf{curl}_{\boldsymbol{x}}\int\limits_{\Omega} G_0(\boldsymbol{x}-\boldsymbol{y})\mathbf{E}(\boldsymbol{y})d\boldsymbol{y}|_{\Gamma}\}\quad\text{for } x\in\Gamma
\end{aligned}\tag{6.8.9}$$

and $\boldsymbol{v}\in(C^{\alpha})^2\subset\big(H^{-1/2}(\Gamma)\big)^2$. Therefore the eigenvalue problem (6.8.2) is equivalent to the eigenvalue problem:

$$\boldsymbol{v}(\boldsymbol{x}) = k^2\big(\frac{1}{2}I - K'\big)^{-1}\{\boldsymbol{n}(\boldsymbol{x})\times\mathbf{curl}_{\boldsymbol{x}}\int\limits_{\Omega} G_0(\boldsymbol{x}-\boldsymbol{y})\mathbf{E}(\boldsymbol{y})d\boldsymbol{y}|_{\Gamma}\},\tag{6.8.10}$$

$$\begin{aligned}
&\varphi + \big(\frac{1}{2}I - K'\big)^{-1}\boldsymbol{n}(\boldsymbol{x})\cdot\int\limits_{\Gamma} G_0(\boldsymbol{x}-\boldsymbol{y})\boldsymbol{v}(\boldsymbol{y})ds_{\boldsymbol{y}}\\
&\quad = k^2\big(\frac{1}{2}I - K'\big)^{-1}\{\boldsymbol{n}(\boldsymbol{x})\cdot\int\limits_{\Omega} G_0(\boldsymbol{x}-\boldsymbol{y})\mathbf{E}(\boldsymbol{y})d\boldsymbol{y}\}
\end{aligned}$$

$$\begin{aligned}
&\mathbf{E}(\boldsymbol{y}) + \int\limits_{\Gamma} G_0(\boldsymbol{x}-\boldsymbol{y})\boldsymbol{v}(\boldsymbol{y})ds_{\boldsymbol{y}} + \int\limits_{\Gamma}\varphi(\boldsymbol{y})\mathbf{grad}_{\boldsymbol{y}}G_0(\boldsymbol{x}-\boldsymbol{y})ds_{\boldsymbol{y}}\\
&\quad = k^2\int\limits_{\Omega} G_0(\boldsymbol{x}-\boldsymbol{y})\mathbf{E}(\boldsymbol{y})d\boldsymbol{y}\quad\text{in }\Omega.
\end{aligned}$$

This is a triangular system of integral operators of the Riesz–Schauder type in the space $\mathfrak{X} = \Big(\big(C^{\alpha}(\Gamma)\big)^2\times C^{\alpha}(\Gamma)\times C^{1+\alpha}(\Omega)\big)^2\Big)$ where all the operators on the right hand side multiplied by k^2 are defining a compact mapping in $\mathfrak{X}$. Then there exists a countable number of eigenvalues of finite multiplicity $\lambda_j^{-1} = k_j\to\infty$ for $j\to\infty$ (see Zeidler [443]) and due to (6.8.1) they are real and strictly positive. ∎

Lemma 6.8.3. *Under the conditions of Theorem 6.8.2, for* $0 < k^2 < \lambda_1^{-1}$ *the problem* (6.8.2) *has only the trivial solution* $\mathbf{E} = \mathbf{0}$ *in* Ω.

As a consequence, for any $k^2\neq k_j^2$, (6.8.2) *has only the trivial solution* $\mathbf{E} = \mathbf{0}$ *in* Ω.

Proof: Since (6.8.2) with $k = 0$ is equivalent to the triangular system (6.8.10) with G_0 which is invertible, the analyticity with respect to k^2 assures invertibility up to $k^2 < \lambda_1^{-1}$ and Fredholm's theorem implies invertibility for $0 < \frac{1}{k^2}$ not an eigenvalue. ■

Corollary 6.8.4. *Theorem 6.8.2 and Lemma 6.8.3 are also true for Ω a bounded Lipschitz domain and $\mathfrak{X} := \big(H^{1/2}(\Gamma)\big)^2 \times H^{1/2}(\Gamma) \times \big(H^1(\Omega)\big)^3$ since the triangular operator matrix on the left of* (6.8.10) *in $\mathfrak{X}$ is continuous and continuously invertible, and because of $\int\limits_\Omega G_0(\boldsymbol{x}-\boldsymbol{y})\mathbf{E}(\boldsymbol{y})d\boldsymbol{y} \in \big(H^2(\Omega)\big)^3$, the operator vector on the right in* (6.8.10) *is mapping $\mathfrak{X}$ continuously into $\mathfrak{Y} := \big(H^{1-\varepsilon}(\Omega)\big)^2 \times H^{1-\varepsilon}(\Gamma) \times \big(H^{1-\varepsilon}(\Omega)\big)^3$ with $0 < \varepsilon < \frac{1}{2}$ and the imbedding $\mathfrak{Y} \hookrightarrow \mathfrak{X}$ is dense and compact, the propositions of Theorem* 6.8.2 *and Lemma* 6.8.3 *for the Riesz–Schauder system* (6.8.10) *again are valid.*

6.8.2 Exterior problems

Here the solution $(\mathbf{E},\mathbf{H})^\top \in C^1(\Omega^c)^6$ of the Maxwell equations (6.4.1)–(6.4.4) are considered in the exterior domain Ω^c satisfying one of the boundary conditions (6.7.1) together with the Silver–Müller radiation conditions (6.7.2) as $|\boldsymbol{x}| \to \infty$. Again, for the uniqueness it is sufficient to consider the homogeneous problem.

We consider two cases with $\sigma = 0$ or $\sigma > 0$. The case for $\sigma = 0$ is particularly relevant in scattering problems for perfect dielectrics.
For the case $\sigma = 0$ we shall need the following Theorem by Rellich ([293, Theorem 4.12]) for the uniqueness proof.

Theorem 6.8.5. *Let Φ in Ω^c be a regular solution of the homogeneous Helmholtz equation $\Delta\Phi + k^2\Phi = 0$ in Ω^c with $k > 0$ and satisfying the decay condition*

$$\int\limits_{|\boldsymbol{y}|=1} |\Phi(R\boldsymbol{y})|^2 do_1 = \mathfrak{o}(R^{-2}) \quad \textit{for } R \to \infty \tag{6.8.11}$$

then $\Phi = 0$ in Ω^c.

Theorem 6.8.6. *The homogeneous exterior problem* (6.4.1)–(6.4.4), (6.7.1) *with $\mathbf{J}_0 = \mathbf{0}$ and $\varrho = 0$ in Ω^c and $\boldsymbol{f} = \mathbf{0}$ on Γ satisfying the Silver Müller radiation conditions* (6.7.2) *for $\sigma > 0$ on Ω^c and also for $\sigma = 0$, only has the trivial solution $\mathbf{E} = \mathbf{0}$ and $\mathbf{H} = \mathbf{0}$.*

Proof: (See Meister [293, Theorem 5.18].) Let $R > 0$ be a parameter chosen so large that $\overline{\Omega} \subset B_R := \{\boldsymbol{x} \in \mathbb{R}^3 \text{ with } |\boldsymbol{x}| < R\}$ and let $\Omega^c_R := \Omega^c \cap B_R$. Then

$$\begin{aligned}\int\limits_{\Omega_R^c} \operatorname{div}(\mathbf{E}\times\overline{\mathbf{H}})d\boldsymbol{x} &= \int\limits_{|\boldsymbol{x}|=R} \frac{1}{R}\boldsymbol{x}\cdot(\mathbf{E}\times\overline{\mathbf{H}})do_R - \int\limits_{\Gamma} \boldsymbol{n}\cdot(\mathbf{E}\times\overline{\mathbf{H}})do_\Gamma \\ &= \int\limits_{|\boldsymbol{x}|=R} \overline{\mathbf{H}}\cdot(\frac{1}{R}\boldsymbol{x}\times\mathbf{E})do_R - \int\limits_{\Gamma} \overline{\mathbf{H}}\cdot(\boldsymbol{n}\times\mathbf{E})do_\Gamma \\ &= \int\limits_{|\boldsymbol{x}|=R} \overline{\mathbf{H}}\cdot(\frac{1}{R}\boldsymbol{x}\times\mathbf{E})do_R\end{aligned}$$

since $\boldsymbol{n}\times\mathbf{E}|_\Gamma = \mathbf{0}$.

Now we consider also the domain exterior to B_R where

$$\begin{aligned}\int\limits_{|\boldsymbol{x}|\geq R} \operatorname{div}(\mathbf{E}\times\overline{\mathbf{H}})d\boldsymbol{x} &= \int\limits_{|\boldsymbol{x}|\geq R} \overline{\mathbf{H}}\cdot\mathbf{curl}\,\mathbf{E}d\boldsymbol{x} - \int\limits_{|\boldsymbol{x}|\geq R} \mathbf{E}\cdot\mathbf{curl}\,\overline{\mathbf{H}}d\boldsymbol{x} \\ &= i\omega \int\limits_{|\boldsymbol{x}|\geq R} (\mu|\mathbf{H}|^2 - \varepsilon|\mathbf{E}|^2)d\boldsymbol{x} - \int\limits_{|\boldsymbol{x}|\geq R} \sigma|\mathbf{E}|^2 d\boldsymbol{x}\end{aligned}$$

due to (6.4.1), (6.4.2) with $\mathbf{J}_0 = \mathbf{0}$. Employing the Gauss theorem gives with the Silver–Müller radiation conditions (6.7.2),

$$\begin{aligned}\int\limits_{|\boldsymbol{x}|\geq R} \operatorname{div}(\mathbf{E}\times\overline{\mathbf{H}})d\boldsymbol{x} &= \int\limits_{|\boldsymbol{x}|=R} \overline{\mathbf{H}}\cdot(\frac{1}{R}\boldsymbol{x}\times\mathbf{E})do_R \\ &= k(\omega\varepsilon + i\sigma)^{-1} \int\limits_{|\boldsymbol{x}|=R} |\mathbf{H}|^2 do_R + \mathfrak{o}(1) \\ = \big\{(k_1\omega\varepsilon + k_2\sigma) + i(k_2\omega\varepsilon - k_1\sigma)\big\}&|\omega\varepsilon + i\sigma|^{-2} \int\limits_{|\boldsymbol{x}|=R} |\mathbf{H}|^2 do_R + \mathfrak{o}(1),\end{aligned}$$

$k = k_1 + ik_2$, for $R\to\infty$. Now insert the surface integral on $|\boldsymbol{x}| = R$ into the two previous equations to obtain for the real parts

$$0 \leq (k_1\omega\varepsilon + k_2\sigma)|\omega\varepsilon + i\sigma|^{-2} \int\limits_{|\boldsymbol{x}|=R} |\mathbf{H}|^2 do_R + \mathfrak{o}(1) = -\int\limits_{\Omega_R^c} \sigma|\mathbf{E}|^2 d\boldsymbol{x} \leq 0, \tag{6.8.12}$$

and for the imaginary parts

$$(k_2\omega\varepsilon - k_1\sigma)|\omega\varepsilon + i\sigma|^{-2} \int\limits_{|\boldsymbol{x}|=R} |\mathbf{H}|^2 do_R + \mathfrak{o}(1) = \int\limits_{\Omega_R^c} \{\mu|\mathbf{H}|^2 - \varepsilon|\mathbf{E}|^2\}d\boldsymbol{x}.$$

Hence, the relation (6.8.11) implies

$$\lim_{R\to\infty} \int\limits_{\Omega_R^c} \sigma|\mathbf{E}|^2 d\boldsymbol{x} = 0$$

which yields $\mathbf{E} = \mathbf{0}$ in Ω^c for $\sigma > 0$. Then (6.4.1) gives $\mathbf{H} = \mathbf{0}$ in Ω^c as proposed.

If $\sigma = 0$ then (6.4.8) gives $k^2 > 0$ and $\operatorname{Im} k = k_2 = 0$. Hence, (6.8.12) now implies with $\boldsymbol{y} := \frac{1}{R}\boldsymbol{x}$

$$\lim_{R\to\infty} \int\limits_{|\boldsymbol{x}|=R} |\mathbf{H}|^2 do_R = \lim_{R\to\infty} \int\limits_{|\boldsymbol{y}|=1} |\mathbf{H}(R\boldsymbol{y})|^2 do_1 R^2 = 0. \tag{6.8.13}$$

Since $\mathbf{J}_0 = \mathbf{0}$, $\mathbf{H}$ satisfies the homogeneous Helmholtz equation (6.4.7) in Ω^c and therefore is analytic up to $\boldsymbol{x} = \infty$ and must hence satisfy (6.8.12) which implies (6.8.1). The Rellich Theorem 6.8.5 then implies $\mathbf{H} = \mathbf{0}$ in Ω^c and $\mathbf{E} = \mathbf{0}$ follows from (6.4.2). ∎

6.8.3 The transmission problem

Theorem 6.8.7. (See Meister [293, Theorem 5.19]) *and* Koshlyakov, Smirnov and Gliner [238, p.91] *Let Ω be a bounded simply connected Lipschitz domain and $\Omega^c := \mathbb{R}^3 \setminus \overline{\Omega}$. $(\mathbf{E},\mathbf{H})^\top$ and $(\mathbf{E}^c,\mathbf{H}^c)^\top$ are solutions of the homogeneous Maxwell equations* (6.4.1)–(6.4.4) *with $\mathbf{J}_0 = \mathbf{0}$ in Ω and Ω^c, respectively, satisfying the transmission conditions* (6.7.3) *with $\boldsymbol{f} = \mathbf{0}$ and $\boldsymbol{g} = \mathbf{0}$ on the transmission surface $\Gamma = \partial\Omega$. In the exterior domain Ω^c the field $(\mathbf{E}^c,\mathbf{H}^c)$ satisfies the Silver–Müller radiation conditions* (6.7.2) *with $\boldsymbol{n}$ the exterior normal field on $\partial\Omega$.*

a) If $\sigma \geq 0$ in Ω and $\sigma^c > 0$ in Ω^c or
b) if $\sigma \geq 0$ in Ω and $\sigma^c = 0$ in Ω^c

then this homogeneous transmission problem only has the trivial solution $\mathbf{E} = \mathbf{0}\,,\ \mathbf{H} = \mathbf{0}\,,\ \mathbf{E}^c = \mathbf{0}\,,\ \mathbf{H}^c = \mathbf{0}$.

Proof: The Gauss Theorem in Ω and in Ω^c_R, respectively, gives

$$\int\limits_{\Omega} \operatorname{div}(\mathbf{E}\times\overline{\mathbf{H}})ds\boldsymbol{x} = \int\limits_{\Gamma} \boldsymbol{n}\cdot(\mathbf{E}\times\overline{\mathbf{H}})ds_\Gamma = \int\limits_{\Gamma} \overline{\mathbf{H}}\cdot(\boldsymbol{n}\times\mathbf{E})do_\Gamma,$$

$$\int\limits_{\Omega^c_R} \operatorname{div}(\mathbf{E}^c\times\overline{\mathbf{H}}^c)d\boldsymbol{x} = -\int\limits_{\Gamma} \overline{\mathbf{H}}^c\cdot(\boldsymbol{n}\times\mathbf{E}^c)ds_\Gamma + \int\limits_{|\boldsymbol{x}|=R} \frac{1}{R}\boldsymbol{x}\cdot(\mathbf{E}^c\times\overline{\mathbf{H}}^c)ds_R,$$

and with (6.4.1) and (6.4.2),

$$\operatorname{div}(\mathbf{E}\times\overline{\mathbf{H}}) = i\omega\varepsilon\mu|\mathbf{H}|^2 - (i\omega\varepsilon+\sigma)|\mathbf{E}|^2$$

we obtain

$$\int\limits_{\Omega} i\omega\mu|\mathbf{H}|^2 d\boldsymbol{x} - \int\limits_{\Omega} (i\omega\varepsilon + \sigma)|\mathbf{E}|^2 d\boldsymbol{x} = \int\limits_{\Gamma} \overline{\mathbf{H}} \cdot (\boldsymbol{n} \times \mathbf{E}) ds_\Gamma,$$

$$\int\limits_{\Omega_R^c} i\omega\mu^c|\mathbf{H}^c|^2 d\boldsymbol{x} - \int\limits_{\Omega_R^c} |\mathbf{E}^c|^2 (i\omega\varepsilon^c + \sigma^c) d\boldsymbol{x} = -\int\limits_{\Gamma} \overline{\mathbf{H}^c} \cdot (\boldsymbol{n} \times \mathbf{E}^c) ds_\Gamma + \int\limits_{|\boldsymbol{x}|=R} \frac{\boldsymbol{x}}{R} \cdot (\mathbf{E}^c \times \overline{\mathbf{H}^c}) ds_R.$$

Multiplying the first equation with $\frac{\mu\varepsilon}{\mu^c\varepsilon^c}$, adding to the second equation (6.7.3), and using the homogeneous transmission conditions (6.7.3) yields

$$\int\limits_{|\boldsymbol{x}|=R} \frac{1}{R}\boldsymbol{x} \cdot (\mathbf{E}^c \times \overline{\mathbf{H}}^c) ds_R = i\omega\Big\{ \int\limits_{\Omega_R^c} \mu^c|\mathbf{H}^c|^2 d\boldsymbol{x} - \int\limits_{\Omega_R^c} \varepsilon^c|\mathbf{E}^c|^2 d\boldsymbol{x} \Big\} - \int\limits_{\Omega_R^c} \sigma^c|\mathbf{E}^c|^2 d\boldsymbol{x}$$
$$+ i\omega\Big\{ \int\limits_{\Omega} (\mu|\mathbf{H}|^2 d\boldsymbol{x} - \varepsilon|\mathbf{E}|^2) \Big\} d\boldsymbol{x} - \frac{\mu\varepsilon}{\mu^c\varepsilon^c} \int\limits_{\Omega} \sigma|\mathbf{E}|^2 d\boldsymbol{x} \tag{6.8.14}$$

For the surface integral over the sphere $|\boldsymbol{x}| = R$, we employ the Silver-Müller radiation conditions (6.7.2) to obtain

$$\int\limits_{|\boldsymbol{x}|=R} \frac{1}{R}\boldsymbol{x} \cdot (\mathbf{E}^c \times \overline{\mathbf{H}}^c) do_R = \int\limits_{|\boldsymbol{x}|=R} k^c(\omega\varepsilon^c + i\sigma^c)^{-1}|\mathbf{H}^c|^2 do + \mathfrak{o}(1)$$
$$= \int\limits_{|\boldsymbol{x}|=R} \{(k_1^c\omega\varepsilon^c + k_2^c\sigma^c) + i(k_2^c\omega\varepsilon^c - k_1^c\sigma^c)\}|\omega\varepsilon^c + i\sigma^c|^{-2}|\mathbf{H}^c|^2 do + \mathfrak{o}(1) \tag{6.8.15}$$

for $R \to \infty$. Here $k^c = k_1^c + ik_2^c$.

Taking the real parts of (6.8.14) and (6.8.15) results in

$$0 \geq -\frac{\mu\varepsilon}{\mu^c\varepsilon^c} \int\limits_{\Omega} \sigma|\mathbf{E}|^2 d\boldsymbol{x} - \int\limits_{\Omega_R^c} \sigma^c|\mathbf{E}^c|^2 d\boldsymbol{x}$$
$$= \int\limits_{|\boldsymbol{x}|=R} (k_1^c\omega\varepsilon^c + k_2^c\sigma^c)|\omega\varepsilon^c + i\sigma^c|^{-2}|\mathbf{H}^c|^2 do + \mathfrak{o}(1) \tag{6.8.16}$$

and $R \to \infty$ yields

$$\frac{\mu\varepsilon}{\mu^c\varepsilon^c} \int\limits_{\Omega} \sigma|\mathbf{E}|^2 d\boldsymbol{x} + \int\limits_{\Omega^c} \sigma^c|\mathbf{E}^c|^2 d\boldsymbol{x} = 0.$$

Case a): In case of $\sigma^c > 0$ this implies

$$\mathbf{E}^c = \mathbf{0} \quad \text{in } \Omega^c$$

and Equation (6.4.1) gives $\mathbf{H}^c = \mathbf{0}$ in Ω^c. Then $\boldsymbol{n} \times \mathbf{E}|_\Gamma = \mathbf{0} = \boldsymbol{n} \times \mathbf{E}^c|_\Gamma$ and $\boldsymbol{n} \times \mathbf{H}|_\Gamma = \mathbf{0} = \boldsymbol{n} \times \mathbf{H}^c|_\Gamma$ and Theorem 6.8.1 in Subsection 6.8.1 implies also $\mathbf{E} = \mathbf{0}$ and $\mathbf{H} = \mathbf{0}$ in Ω.

Case b): In case $\sigma^c = 0$ we obtain from (6.8.16) now

$$\lim_{R\to\infty} \int\limits_{|\boldsymbol{x}|=R} |\mathbf{H}^c|^2 do = 0. \tag{6.8.17}$$

$\mathbf{H}^c$ satisfies the homogeneous Helmholtz equation (6.4.7) in Ω^c and is analytic around ∞. Then (6.8.17) implies $\mathbf{H}^c(\boldsymbol{x}) = \mathfrak{o}(|\boldsymbol{x}|^{-2})$ for $|\boldsymbol{x}| \to \infty$. So, the Rellich Theorem 6.8.5 implies $\mathbf{H}^c = \mathbf{0}$ in Ω^c. Now, Equation (6.4.2) yields $\mathbf{E}^c = \mathbf{0}$ in Ω^c implying $\boldsymbol{n} \times \mathbf{E}^c|_\Gamma = 0$ as well as $\boldsymbol{n} \times \mathbf{E}|_\Gamma = \mathbf{0}$, $\boldsymbol{n} \times \mathbf{H}|_\Gamma = \mathbf{0}$, $\boldsymbol{n} \times \mathbf{H}^c|_\Gamma = \mathbf{0}$. The transmission conditions and Theorem 6.8.1 gives $\mathbf{E} = \mathbf{0}$ and $\mathbf{H} = \mathbf{0}$ in Ω. This completes the proof of Theorem 6.8.7. ■

6.9 Representation Formulae

To be self contained we begin with the Stratton–Chu formulae [409], Stratton [408, p.466] and Schechter [377]. From the Maxwell equations (6.4.1)–(6.4.4) **in the bounded domain** Ω one obtains

$$\begin{aligned} \mathbf{curl}\,\mathbf{curl}\,\mathbf{E} &= k^2\mathbf{E} + i\omega\mu\mathbf{J}_0, \\ \mathbf{curl}\,\mathbf{curl}\,\mathbf{H} &= k^2\mathbf{H} + \mathbf{curl}\,\mathbf{J}_0 \end{aligned} \tag{6.9.1}$$

with k^2 given in (6.4.8).

Green's second formula for a pair $(\mathbf{P}, \mathbf{Q})^\top$ of continuously diffentiable vector fields reads as

$$\int\limits_\Omega (\mathbf{Q}\cdot\mathbf{curl}\,\mathbf{curl}\,\mathbf{P} - \mathbf{P}\cdot\mathbf{curl}\,\mathbf{curl}\,\mathbf{Q})dy = \int\limits_\Gamma \{\boldsymbol{P}\times\mathbf{curl}\,\mathbf{Q} - \mathbf{Q}\times\mathbf{curl}\,\mathbf{P}\}\cdot\boldsymbol{n}\,ds. \tag{6.9.2}$$

Let $\mathbf{Q} = G_k\boldsymbol{a}$ with a constant vector $\boldsymbol{a}$, and where

$$G_k(\boldsymbol{x} - \boldsymbol{y}) = \frac{1}{4\pi}\frac{e^{ik|\boldsymbol{x}-\boldsymbol{y}|}}{|\boldsymbol{x} - \boldsymbol{y}|}$$

is the fundamental solution of the Helmholtz equation (see (2.1.4) and (6.2.4)). We see that for $\boldsymbol{x} \neq \boldsymbol{y}$,

$$\begin{aligned} \mathbf{curl}\,\mathbf{Q} &= \mathbf{grad}\ G_k \times \boldsymbol{a}, \\ \mathbf{curl}\,\mathbf{curl}\,\mathbf{Q} &= \boldsymbol{a}k^2G_k + \mathbf{grad}\ (\boldsymbol{a}\cdot\mathbf{grad}\ G_k). \end{aligned} \tag{6.9.3}$$

Letting

$$\Omega_\delta := \Omega \setminus \{\boldsymbol{y} \in \Omega \wedge |\boldsymbol{y} - \boldsymbol{x}| \le \delta\},\ \delta > 0$$

and substituting $\mathbf{Q}$ and $\boldsymbol{P} = \mathbf{E}$ into Green's formula (6.9.2) in Ω_δ, we obtain

$$\int_{\Omega_\delta} G_k i\omega\mu \mathbf{J}_0 d\boldsymbol{y} \cdot \boldsymbol{a} - \int_{\Omega_\delta} \mathbf{E} \cdot \mathbf{grad}\,(\boldsymbol{a} \cdot \mathbf{grad}\, G_k) d\boldsymbol{y}$$

$$= \int_{\partial\Omega_\delta} \{ i\omega\mu G_k (\boldsymbol{n} \times \mathbf{H}) + (\boldsymbol{n} \times \mathbf{E}) \times \mathbf{grad}\, G_k \} ds_{\boldsymbol{y}} \cdot \boldsymbol{a}.$$

With

$$\int_{\Omega_\delta} \mathbf{E} \cdot \mathbf{grad}\,(\boldsymbol{a} \cdot \mathbf{grad}\, G_k) d\boldsymbol{y}$$

$$= \int_{\Omega_\delta} \mathrm{div}\{(\mathbf{E}(\boldsymbol{a} \cdot \mathbf{grad}\, G_k)\} d\boldsymbol{y} - \int_{\Omega_\delta} (\mathrm{div}\mathbf{E}) \mathbf{grad}\, G_k d\boldsymbol{y} \cdot \boldsymbol{a}$$

$$= \int_{\Omega_\delta} (\boldsymbol{n} \cdot \mathbf{E}) \mathbf{grad}\, G_k ds_y \cdot \boldsymbol{a} - \int_{\partial\Omega_\delta} (\mathrm{div}\mathbf{E}) \mathbf{grad}\, G_k d\boldsymbol{y} \cdot \boldsymbol{a}$$

and with (6.4.3) we finally get

$$\int_{\Omega_\delta} \{ i\omega\mu G_k \mathbf{J}_0 + \frac{\varrho}{\varepsilon} \mathbf{grad}\; G_k \} d\boldsymbol{y} \cdot \boldsymbol{a} =$$

$$\int_{\partial\Omega_\delta} \{ i\omega\mu G_k (\boldsymbol{n} \times \mathbf{H}) + (\boldsymbol{n} \cdot \mathbf{E}) \mathbf{grad}\; G_k + (\boldsymbol{n} \times \mathbf{E}) \times \mathbf{grad}\, G_k \} ds_{\boldsymbol{y}} \cdot \boldsymbol{a}$$

for arbitrary $\boldsymbol{a} \in \mathbb{R}^3$. Now, using tacitly the identity

$$\mathbf{E} = (\boldsymbol{n} \times \mathbf{E}) \times \boldsymbol{n} + (\mathbf{E} \cdot \boldsymbol{n}) \boldsymbol{n},$$

i.e. $\mathbf{E} = \pi_t \mathbf{E} + (\mathbf{E} \cdot \boldsymbol{n}) \boldsymbol{n}$ (see Section 4.5), we notice that if $\mathbf{grad}_{\boldsymbol{y}} G_k(\boldsymbol{x} - \boldsymbol{y})$ is replaced by its normal component $(\mathbf{grad}_{\boldsymbol{y}} G_k(\boldsymbol{x} - \boldsymbol{y}) \cdot \boldsymbol{n}) \boldsymbol{n}$, then the last two terms in the integrand on the right–hand side of the above integral can be simplified as $\mathbf{E}(\boldsymbol{y}) \frac{\partial G_k(\boldsymbol{x} - \boldsymbol{y})}{\partial n_{\boldsymbol{y}}}$. Hence, a simple computation shows that for $\delta \to 0$ one has the **Stratton–Chu representation formula** in the **interior domain** for the **electric field**,

$$\mathbf{E}(\boldsymbol{x}) = \int_{\Omega} [G_k(\boldsymbol{x} - \boldsymbol{y}) i\omega\mu \mathbf{J}_0(\boldsymbol{y}) + \frac{1}{\varepsilon} \varrho(\boldsymbol{y}) \mathbf{grad}_{\boldsymbol{y}} G_k(\boldsymbol{x} - \boldsymbol{y})] d\boldsymbol{y} \tag{6.9.4}$$

$$- \int_{\Gamma} \{ i\omega\mu G_k(\boldsymbol{x} - \boldsymbol{y}) (\boldsymbol{n}(\boldsymbol{y}) \times \mathbf{H}(\boldsymbol{y})) + (\boldsymbol{n} \cdot \mathbf{E}) \mathbf{grad}_{\boldsymbol{y}} G_k(\boldsymbol{x} - \boldsymbol{y})$$
$$+ (\boldsymbol{n}(\boldsymbol{y}) \times \mathbf{E}(\boldsymbol{y})) \times \mathbf{grad}_{\boldsymbol{y}} G_k(\boldsymbol{x} - \boldsymbol{y}) \} ds_{\boldsymbol{y}} \quad \text{for any } \boldsymbol{x} \in \Omega.$$

In a quite analogous way, one derives the corresponding formula of the **Stratton–Chu representation formula** for the **magnetic field**,

$$\mathbf{H}(\boldsymbol{x}) = \int\limits_{\Omega} \mathbf{J}_0(\boldsymbol{y}) \times \mathbf{grad}_{\boldsymbol{y}} G_k(\boldsymbol{x}-\boldsymbol{y}) d\boldsymbol{y} \tag{6.9.5}$$
$$- \int\limits_{\Gamma} \{(-i\omega\varepsilon + \sigma) G_k(\boldsymbol{x}-\boldsymbol{y})\big(\boldsymbol{n}(\boldsymbol{y}) \times \mathbf{E}(\boldsymbol{y})\big) + (\boldsymbol{n}\cdot\mathbf{H})\mathbf{grad}_{\boldsymbol{y}} G_k(\boldsymbol{x}-\boldsymbol{y})$$
$$+ \big(\boldsymbol{n}(\boldsymbol{y}) \times \mathbf{H}(\boldsymbol{y})\big) \times \mathbf{grad}_{\boldsymbol{y}} G_k(\boldsymbol{x}-\boldsymbol{y})\big\} ds_{\boldsymbol{y}} \quad \text{for any } \boldsymbol{x} \in \Omega,$$

where we have employed the formula

$$\int\limits_{\Omega} G_k(\boldsymbol{x}-\boldsymbol{y}) \mathbf{curl}\ \mathbf{J}_0(\boldsymbol{y}) d\boldsymbol{y} = \int\limits_{\Omega} \mathbf{J}_0(\boldsymbol{y}) \times \mathbf{grad}_y G_k(\boldsymbol{x}-\boldsymbol{y}) d\boldsymbol{y}$$
$$+ \int\limits_{\Gamma} \big(\boldsymbol{n} \times \mathbf{J}_0(\boldsymbol{y})\big) G_k(\boldsymbol{x}-\boldsymbol{y}) ds_{\boldsymbol{y}}$$

by using the identity

$$\int\limits_{\Omega} \mathbf{curl}\ \mathbf{F}(\boldsymbol{y}) d\boldsymbol{y} = \int\limits_{\Gamma} \boldsymbol{n} \times \mathbf{F}(\boldsymbol{y}) ds_{\boldsymbol{y}}$$

(see Stratton [408, p.604]) .

In the **exterior domain** $\Omega^c = \mathbb{R}^3 \setminus \overline{\Omega}$ one obtains

$$\mathbf{E}(\boldsymbol{x}) = \int\limits_{\Omega^c} [G_k(\boldsymbol{x}-\boldsymbol{y}) i\omega\mu \mathbf{J}_0(\boldsymbol{y}) + \frac{1}{\varepsilon}\varrho(\boldsymbol{y}) \mathbf{grad}_{\boldsymbol{y}} G_k(\boldsymbol{x}-\boldsymbol{y})] d\boldsymbol{y} \tag{6.9.6}$$
$$+ \int\limits_{\Gamma} \{i\omega\mu G_k(\boldsymbol{x}-\boldsymbol{y})\big(\boldsymbol{n}(\boldsymbol{y}) \times \mathbf{H}(\boldsymbol{y})\big) + (\boldsymbol{n}\cdot\mathbf{E})\mathbf{grad}_{\boldsymbol{y}} G_k(\boldsymbol{x}-\boldsymbol{y})$$
$$+ \big(\boldsymbol{n}(\boldsymbol{y}) \times \mathbf{E}(\boldsymbol{y})\big) \times \mathbf{grad}_{\boldsymbol{y}} G_k(\boldsymbol{x}-\boldsymbol{y})\big\} ds_{\boldsymbol{y}} \quad \text{for } \boldsymbol{x} \in \Omega^c,$$

and

$$\mathbf{H}(\boldsymbol{x}) = \int\limits_{\Omega^c} \mathbf{J}_0(\boldsymbol{x}) \times \mathbf{grad}_{\boldsymbol{y}} G_k(\boldsymbol{x}-\boldsymbol{y}) d\boldsymbol{y} \tag{6.9.7}$$
$$+ \int\limits_{\Gamma} \{(-i\omega\varepsilon + \sigma) G_k(\boldsymbol{x}-\boldsymbol{y})\big(\boldsymbol{n}(\boldsymbol{y}) \times \mathbf{E}(\boldsymbol{y})\big) + (\boldsymbol{n}\cdot\mathbf{H})\mathbf{grad}_{\boldsymbol{y}} G_k(\boldsymbol{x}-\boldsymbol{y})$$
$$+ \big(\boldsymbol{n}(\boldsymbol{y}) \times \mathbf{H}(\boldsymbol{y})\big) \times \mathbf{grad}_{\boldsymbol{y}} G_k(\boldsymbol{x}-\boldsymbol{y})\big\} ds_{\boldsymbol{y}} \quad \text{for any } \boldsymbol{x} \in \Omega^c.$$

Here we used the Silver–Müller radiation conditions (6.7.2) and assume that $\mathbf{J}_0$ has compact support, supp $\mathbf{J}_0 \Subset \Omega^c$.

We remark that in the representation formulae (6.9.4), (6.9.5), (6.9.6) and (6.9.7) $\boldsymbol{n} \times \mathbf{H}$ and $\boldsymbol{n} \times \mathbf{E}$ are the Cauchy data for $\mathbf{E}(\boldsymbol{x})$ and $\mathbf{H}(\boldsymbol{x})$ while the terms $\boldsymbol{n} \cdot \mathbf{E}$ and $\boldsymbol{n} \cdot \mathbf{H}$ can be replaced by what are called the surface divergence of $\boldsymbol{n} \times \mathbf{H}$ and $\boldsymbol{n} \times \mathbf{E}$, respectively. From vectorial calculus, we see that

$$\operatorname{div}(\boldsymbol{n} \times \mathbf{E}) = \mathbf{E} \cdot \mathbf{curl}\ \boldsymbol{n} - \boldsymbol{n} \cdot \mathbf{curl}\,\mathbf{E} = -\boldsymbol{n} \cdot \mathbf{curl}\ \mathbf{E},$$

since the field of normals is a gradient, which implies $\mathbf{curl}\ \boldsymbol{n} = \mathbf{0}$. Then from (6.4.1), we obtain

$$\boldsymbol{n} \cdot \mathbf{H} = \frac{-\operatorname{div}(\boldsymbol{n} \times \mathbf{E})}{i\omega\mu}.$$

Similarly,

$$\operatorname{div}(\boldsymbol{n} \times \mathbf{H}) = -\boldsymbol{n} \cdot \mathbf{curl}\,\mathbf{H}$$

and from (6.4.2), we obtain

$$\begin{aligned}\boldsymbol{n} \cdot \mathbf{E} &= (\sigma - i\omega\varepsilon)^{-1}\{\boldsymbol{n} \cdot \mathbf{curl}\ \mathbf{H} - \boldsymbol{n} \cdot \mathbf{J}_0\} \\ &= -(\sigma - i\omega\varepsilon)^{-1}\{\operatorname{div}(\boldsymbol{n} \times \mathbf{H}) + \boldsymbol{n} \cdot \mathbf{J}_0)\}.\end{aligned}$$

In fact, the surface divergence of a tangent vector field $\boldsymbol{v}$ is defined by its associated lifting $\widetilde{\boldsymbol{v}}$ as $\operatorname{div}_\Gamma \boldsymbol{v} = \operatorname{div}\widetilde{\boldsymbol{v}}|_\Gamma$. In this sense,

$$\operatorname{div}(\boldsymbol{n} \times \mathbf{E})|_\Gamma = \operatorname{div}_\Gamma(\boldsymbol{n} \times \mathbf{E}) \quad \text{and} \quad \operatorname{div}(\boldsymbol{n} \times \mathbf{H})|_\Gamma = \operatorname{div}_\Gamma(\boldsymbol{n} \times \mathbf{H}).$$

6.10 Boundary Integral Equations for Electromagnetic fields

We begin with the Stratton–Chu representation formulae Stratton and Chu [409, (19),(20) Chap.8] (6.9.4), (6.9.5) for the solution of the Maxwell equations (6.4.1)–(6.4.5) within a given domain Ω. If $\boldsymbol{x} \in \Omega$ tends to the boundary $\partial\Omega = \Gamma$, then from standard jump relations of potential theory (see (6.10.8)–(6.10.11)), we arrive at

$$\begin{aligned}\boldsymbol{n}(\boldsymbol{x}) \times \mathbf{E}(\boldsymbol{x}) = \big(\mathbf{T}_\mathbf{E}(\boldsymbol{n} \times \mathbf{H})\big)(\boldsymbol{x}) + \frac{1}{2}\big(\boldsymbol{n}(\boldsymbol{x}) \times \mathbf{E}(\boldsymbol{x})\big) - \big(\mathbf{R}(\boldsymbol{n} \times \mathbf{E})\big)(\boldsymbol{x}) \\ + \boldsymbol{n}(\boldsymbol{x}) \times \mathbf{E}_p(\boldsymbol{x})\end{aligned} \tag{6.10.1}$$

for $\boldsymbol{x} \in \Gamma$. where

$$\begin{aligned}\mathbf{E}_p(\boldsymbol{x}) = (\sigma - i\omega\varepsilon)^{-1} \int\limits_\Gamma (\mathbf{grad}_{\boldsymbol{y}} G_k(\boldsymbol{x} - \boldsymbol{y}))\boldsymbol{n}(\boldsymbol{y}) \cdot \mathbf{J}_0(\boldsymbol{y}) ds_{\boldsymbol{y}} \\ + \int\limits_\Omega \{G_k(\boldsymbol{x} - \boldsymbol{y}) i\omega\mu \mathbf{J}_0(\boldsymbol{y}) + \frac{1}{\varepsilon}\varrho(\boldsymbol{y})\mathbf{grad}_{\boldsymbol{y}} G_k(\boldsymbol{x} - \boldsymbol{y})\} d\boldsymbol{y}\end{aligned}$$

is known for given $\mathbf{J}_0(\boldsymbol{y})$, and the boundary integral operators are defined by

$$\mathbf{T_E}(\boldsymbol{n} \times \mathbf{H})(\boldsymbol{x}) = -i\omega\mu \boldsymbol{n}(\boldsymbol{x}) \times \int_\Gamma G_k(\boldsymbol{x}-\boldsymbol{y})\big(\boldsymbol{n}(\boldsymbol{y}) \times \mathbf{H}(\boldsymbol{y})\big) ds_{\boldsymbol{y}} \tag{6.10.2}$$

$$+ (-i\omega\varepsilon + \sigma)^{-1} \boldsymbol{n}(\boldsymbol{x}) \times \int_\Gamma \mathbf{grad}_{\boldsymbol{y}} G_k(\boldsymbol{x}-\boldsymbol{y})\big(\mathrm{div}_\Gamma(\boldsymbol{n}(\boldsymbol{y}) \times \mathbf{H}(\boldsymbol{y})\big) ds_y$$

$$= -\big\{i\omega\mu \boldsymbol{n} \times \mathbf{V}_k(\boldsymbol{n} \times \mathbf{H}) + (-i\omega\varepsilon + \sigma)^{-1} \boldsymbol{n} \times \mathbf{grad}_{\boldsymbol{x}} V_k(\mathrm{div}_\Gamma(\boldsymbol{n} \times \mathbf{H}))\big\}$$

and the singular Calderon–Zygmund integral operator $\mathbf{R}$ is given by

$$\mathbf{R}(\boldsymbol{n} \times \mathbf{E})(\boldsymbol{x}) := \text{p.v.}\, \boldsymbol{n}(\boldsymbol{x}) \times \int_{\Gamma \setminus \{\boldsymbol{x}\}} \{(\boldsymbol{n}(\boldsymbol{y}) \times \mathbf{E}(\boldsymbol{y})) \times \mathbf{grad}_{\boldsymbol{y}} G_k(\boldsymbol{x}-\boldsymbol{y})\} ds_y$$

$$= \boldsymbol{n}(\boldsymbol{x}) \times \mathbf{curl}_{\boldsymbol{x}} \int_{\Gamma \setminus \{\boldsymbol{x}\}} G_k(\boldsymbol{x}-\boldsymbol{y})(\boldsymbol{n} \times \mathbf{E}) ds_{\boldsymbol{y}} \quad \text{for } \boldsymbol{x} \in \Gamma. \tag{6.10.3}$$

Similarly we have from (6.9.5) for $\Omega \ni \boldsymbol{x} \to \Gamma$,

$$\boldsymbol{n}(\boldsymbol{x}) \times \mathbf{H}(\boldsymbol{x}) = \big(\mathbf{T_H}(\boldsymbol{n} \times \mathbf{E})\big)(\boldsymbol{x}) + \{(\frac{1}{2}\mathbf{I} - \mathbf{R})(\boldsymbol{n} \times \mathbf{H})\}(\boldsymbol{x}) + \boldsymbol{n}(\boldsymbol{x}) \times \mathbf{H}_p(\boldsymbol{x}), \tag{6.10.4}$$

for $\boldsymbol{x} \in \Gamma$ where

$$\mathbf{H}_p(\boldsymbol{x}) := \int_\Omega \mathbf{J}_0(\boldsymbol{y}) \times \mathbf{grad}_{\boldsymbol{y}} G_k(\boldsymbol{x}-\boldsymbol{y}) d\boldsymbol{y}$$

is known for given $\mathbf{J}_0(\boldsymbol{y})$. Moreover,

$$\mathbf{T_H}(\boldsymbol{n} \times \mathbf{E})(\boldsymbol{x}) := -(-i\omega\varepsilon + \sigma)\boldsymbol{n}(\boldsymbol{x}) \times \int_\Gamma G_k(\boldsymbol{x}-\boldsymbol{y})\big(\boldsymbol{n}(\boldsymbol{y}) \times \mathbf{E}(\boldsymbol{y})\big) ds_{\boldsymbol{y}}$$

$$+ \frac{1}{i\omega\mu} \boldsymbol{n}(\boldsymbol{x}) \times \int_\Gamma \mathbf{grad}_{\boldsymbol{y}} G_k(\boldsymbol{x}-\boldsymbol{y}) \mathrm{div}_\Gamma\big(\boldsymbol{n}(\boldsymbol{y}) \times \mathbf{E}(\boldsymbol{y})\big) ds_{\boldsymbol{y}} \tag{6.10.5}$$

$$= -\big\{(-i\omega\varepsilon + \sigma)\boldsymbol{n} \times \mathbf{V}_k(\boldsymbol{n} \times \mathbf{E}) + \frac{1}{i\omega\mu} \boldsymbol{n} \times \mathbf{grad}_{\boldsymbol{x}} V_k\big(\mathrm{div}_\Gamma(\boldsymbol{n} \times \mathbf{E})\big)\big\}.$$

Now let us consider the Maxwell equations (6.4.1)–(6.4.5) in the **exterior domain** $\Omega^c = \mathbb{R}^3 \setminus \overline{\Omega}$ for solutions $(\mathbf{E}, \mathbf{H})^\top$ satisfying the Silver–Müller radiation conditions (6.7.2) where $\mathbf{J}_0$ has compact support as well as ϱ. For $\Omega^c \ni \boldsymbol{x} \to \Gamma$, with the standard jump relations we obtain from (6.9.6) the equation on Γ:

$$\boldsymbol{n}(\boldsymbol{x}) \times \mathbf{E}(\boldsymbol{x}) \tag{6.10.6}$$

$$= -\big(\mathbf{T_E}(\boldsymbol{n} \times \mathbf{H})\big)(\boldsymbol{x}) + \frac{1}{2}\big(\boldsymbol{n}(\boldsymbol{x}) \times \mathbf{E}(\boldsymbol{x})\big) + \mathbf{R}(\boldsymbol{n} \times \mathbf{E})(\boldsymbol{x}) + \boldsymbol{n}(\boldsymbol{x}) \times \mathbf{E}_p^c(\boldsymbol{x})$$

for $\boldsymbol{x} \in \Gamma$ where

$$\mathbf{E}_p^c(\boldsymbol{x}) := -(\sigma - i\omega\varepsilon)^{-1} \int\limits_{\Gamma} (\mathbf{grad}_{\,\boldsymbol{y}} G_k(\boldsymbol{x}-\boldsymbol{y}))\boldsymbol{n}(\boldsymbol{y}) \cdot \mathbf{J}_0(\boldsymbol{y}) ds_{\boldsymbol{y}} + \int\limits_{\Omega^c} \{G_k(\boldsymbol{x}-\boldsymbol{y}) i\omega\mu \mathbf{J}_0(\boldsymbol{y}) + \frac{1}{\varepsilon}\varrho(\boldsymbol{y})\mathbf{grad}_{\,\boldsymbol{y}} G_k(\boldsymbol{x}-\boldsymbol{y})\} d\boldsymbol{y}$$

Similarly, from (6.9.7) we find

$$\begin{aligned} \boldsymbol{n}(\boldsymbol{x}) \times \mathbf{H}(\boldsymbol{x}) = &-\mathbf{T}_{\mathbf{H}}(\boldsymbol{n} \times \mathbf{E}) + \frac{1}{2}(\boldsymbol{n}(\boldsymbol{x}) \times \mathbf{H}(\boldsymbol{x})) \\ &+ \mathbf{R}(\boldsymbol{n} \times \mathbf{H})(\boldsymbol{x}) + \boldsymbol{n}(\boldsymbol{x}) \times \mathbf{H}_p^c(\boldsymbol{x}) \end{aligned} \tag{6.10.7}$$

for $\boldsymbol{x} \in \Gamma$ where

$$\mathbf{H}_p^c(\boldsymbol{x}) := \int\limits_{\Omega^c} \mathbf{J}_0(\boldsymbol{y}) \times \mathbf{grad}_{\,\boldsymbol{y}} G_k(\boldsymbol{x}-\boldsymbol{y}) d\boldsymbol{y}.$$

We now include derivations of three typical jump relations of the single–layer potential in electromagnetics (see, e.g. Hsiao and Kleinman [194], Nedelec [331, p.240]). We follow the standard convention and assume here that Γ is a Lyapounov surface. The subscripts plus and minus denote limits from $\Omega^c = \mathbb{R}^3 \setminus \overline{\Omega}$ and Ω, respectively. In the following derivations for simplicity we denote by $\mathbf{J}(\boldsymbol{y}) := \boldsymbol{n} \times \mathbf{H}(\boldsymbol{y})$, which can be replaced by $\mathbf{M}(\boldsymbol{y}) = \boldsymbol{n} \times \mathbf{E}(\boldsymbol{y})$ as well.

$$\lim_{\boldsymbol{x}\to\Gamma^{\pm}} \boldsymbol{n}(\boldsymbol{x}) \times \mathbf{curl}_{\,\boldsymbol{x}} \int\limits_{\Gamma} G_k(\boldsymbol{x}-\boldsymbol{y})\mathbf{J}(\boldsymbol{y}) ds_{\boldsymbol{y}} = \pm\frac{1}{2}\mathbf{J}(\boldsymbol{x}) + \boldsymbol{n}(\boldsymbol{x}) \times \mathbf{curl}_{\,\boldsymbol{x}} \int\limits_{\Gamma} G_k(\boldsymbol{x}-\boldsymbol{y})\mathbf{J}(\boldsymbol{y}) ds_{\boldsymbol{y}}. \tag{6.10.8}$$

The derivation is based on the vector identity **I** (1) in the Appendix B:

$$\begin{aligned} \boldsymbol{n}(\boldsymbol{x}) \times \mathbf{curl}_{\boldsymbol{x}} \int\limits_{\Gamma} G_k(\boldsymbol{x}-\boldsymbol{y})\mathbf{J}(\boldsymbol{y}) ds_{\boldsymbol{y}} &= \boldsymbol{n}(\boldsymbol{x}) \times \int\limits_{\Gamma} \mathbf{grad}_{\,\boldsymbol{x}} G_k(\boldsymbol{x}-\boldsymbol{y}) \times \mathbf{J}(\boldsymbol{y}) ds_{\boldsymbol{y}} \\ &= \int\limits_{\Gamma} (\boldsymbol{n}(\boldsymbol{x})\cdot\mathbf{J}(\boldsymbol{y}))\,(\mathbf{grad}_{\,\boldsymbol{x}} G_k(\boldsymbol{x}-\boldsymbol{y})) ds_{\boldsymbol{y}} - \int\limits_{\Gamma} (\boldsymbol{n}(\boldsymbol{x})\cdot\mathbf{grad}_{\,\boldsymbol{x}} G_k(\boldsymbol{x}-\boldsymbol{y}))\,\mathbf{J}(\boldsymbol{y}) ds_{\boldsymbol{y}} \\ &= \int\limits_{\Gamma} ((\boldsymbol{n}(\boldsymbol{x}) - \boldsymbol{n}(\boldsymbol{y})) \cdot \mathbf{J}(\boldsymbol{y}))(\mathbf{grad}_{\,\boldsymbol{x}} G_k(\boldsymbol{x}-\boldsymbol{y})) ds_{\boldsymbol{y}} \\ &\quad - \int\limits_{\Gamma} (\boldsymbol{n}(\boldsymbol{x}) \cdot \mathbf{grad}_{\,\boldsymbol{x}} G_k(\boldsymbol{x}-\boldsymbol{y}))\mathbf{J}(\boldsymbol{y}) ds_{\boldsymbol{y}}, \end{aligned}$$

where the limit of the last term on the right side is equal to

$$\pm\frac{1}{2}\mathbf{J}(\boldsymbol{x}) - \int\limits_{\Gamma\setminus\{\boldsymbol{x}\}} \frac{\partial}{\partial n_{\boldsymbol{x}}} G_k(\boldsymbol{x}-\boldsymbol{y})\mathbf{J}(\boldsymbol{y})ds_{\boldsymbol{y}}, \quad \text{as } \boldsymbol{x}\to\Gamma^{\pm}.$$

The second jump relation reads

$$\begin{aligned}\lim_{\boldsymbol{x}\to\Gamma^{\pm}} \boldsymbol{n}(\boldsymbol{x})\cdot\mathbf{curl}_{\boldsymbol{x}}\int\limits_{\Gamma} G_k(\boldsymbol{x}-\boldsymbol{y})\mathbf{J}(\boldsymbol{y})ds_{\boldsymbol{y}} \\ = \int\limits_{\Gamma}\big((\boldsymbol{n}(\boldsymbol{x})-\boldsymbol{n}(\boldsymbol{y}))\times\mathbf{grad}_{\boldsymbol{x}}G_k(\boldsymbol{x}-\boldsymbol{y})\big)\big)\cdot\mathbf{J}(\boldsymbol{y})ds_{\boldsymbol{y}} \\ + \int\limits_{\Gamma} G_k(\boldsymbol{x}-\boldsymbol{y})\,\mathrm{curl}_{\Gamma}\mathbf{J}(\boldsymbol{y})ds_{\boldsymbol{y}}\end{aligned} \tag{6.10.9}$$

The derivation uses the vector identities $\mathbf{I}(3)$, $\mathbf{III}(5)$ and $\mathbf{V}(6)$ in the Appendix B:

$$\begin{aligned}\boldsymbol{n}(\boldsymbol{x})\cdot\mathbf{curl}_{\boldsymbol{x}}\int\limits_{\Gamma} G_k(\boldsymbol{x}-\boldsymbol{y})\mathbf{J}(\boldsymbol{y})ds_{\boldsymbol{y}} &= \boldsymbol{n}(\boldsymbol{x})\cdot\int\limits_{\Gamma}\mathbf{grad}_{\boldsymbol{x}}G_k(\boldsymbol{x}-\boldsymbol{y})\times\mathbf{J}(\boldsymbol{y})ds_{\boldsymbol{y}} \\ &= \int\limits_{\Gamma}\mathbf{J}(\boldsymbol{y})\cdot\big(\boldsymbol{n}(\boldsymbol{x})\times\mathbf{grad}_{\boldsymbol{x}}G_k(\boldsymbol{x}-\boldsymbol{y})\big)ds_{\boldsymbol{y}} \\ &= \int\limits_{\Gamma}\mathbf{J}(\boldsymbol{y})\cdot\big((\boldsymbol{n}(\boldsymbol{x})-\boldsymbol{n}(\boldsymbol{y}))\times\mathbf{grad}_{\boldsymbol{x}}G_k(\boldsymbol{x}-\boldsymbol{y})\big)ds_{\boldsymbol{y}} \\ &+ \int\limits_{\Gamma}\mathbf{J}(\boldsymbol{y})\cdot\big(\boldsymbol{n}(\boldsymbol{y})\times\mathbf{grad}_{\boldsymbol{x}}G_k(\boldsymbol{x}-\boldsymbol{y})\big)ds_{\boldsymbol{y}},\end{aligned}$$

where the last term on the right hand side is equal to

$$\begin{aligned}&-\int\limits_{\Gamma}\mathbf{J}(\boldsymbol{y})\cdot\Big(\boldsymbol{n}(\boldsymbol{y})\times\mathbf{grad}_{\Gamma,\boldsymbol{y}}G_k(\boldsymbol{x}-\boldsymbol{y})\Big)ds_{\boldsymbol{y}} \\ &= \int\limits_{\Gamma}\mathbf{J}(\boldsymbol{y})\cdot\mathbf{curl}_{\Gamma,\boldsymbol{y}}G_k(\boldsymbol{x}-\boldsymbol{y})ds_{\boldsymbol{y}} = \int\limits_{\Gamma}G_k(\boldsymbol{x}-\boldsymbol{y})\,\mathrm{curl}_{\Gamma}\,\mathbf{J}(\boldsymbol{y})ds_{\boldsymbol{y}}\end{aligned}$$

due to III (5) and VI (6) in the Appendix B. This completes the derivation of (6.10.9).

The third jump relation is

$$\lim_{\boldsymbol{x}\to\Gamma^\pm} \boldsymbol{n}(\boldsymbol{x}) \times \mathbf{grad}_{\boldsymbol{x}} \int_\Gamma G_k(\boldsymbol{x}-\boldsymbol{y})\mathrm{div}_\Gamma \mathbf{J}(\boldsymbol{y})ds_{\boldsymbol{y}}$$

$$= \boldsymbol{n}(\boldsymbol{x}) \times \mathbf{grad}_{\boldsymbol{x}} \int_\Gamma G_k(\boldsymbol{x}-\boldsymbol{y})\mathrm{div}_\Gamma \mathbf{J}(\boldsymbol{y})ds_{\boldsymbol{y}}, \tag{6.10.10}$$

$$= \int_\Gamma (\boldsymbol{n}(\boldsymbol{x}) - \boldsymbol{n}(\boldsymbol{y})) \times (\mathbf{grad}_{\boldsymbol{x}} G_k(\boldsymbol{x}-\boldsymbol{y}))\mathrm{div}_\Gamma \mathbf{J}(\boldsymbol{y})ds_{\boldsymbol{y}}$$

$$- \int_\Gamma G_k(\boldsymbol{x}-\boldsymbol{y})\,(\mathbf{curl}_{\Gamma,\boldsymbol{y}}\,\mathrm{div}_\Gamma \mathbf{J}(\boldsymbol{y}))\,ds_{\boldsymbol{y}} \tag{6.10.11}$$

The derivation makes use of formulae **III**(1), **IV**(2) in the Appendix B:

$$\boldsymbol{n}(\boldsymbol{x}) \times \mathbf{grad}_{\boldsymbol{x}} \int_\Gamma G_k(\boldsymbol{x}-\boldsymbol{y})\mathrm{div}_\Gamma \mathbf{J}(\boldsymbol{y})ds_{\boldsymbol{y}}$$

$$= \boldsymbol{n}(\boldsymbol{x}) \times \int_\Gamma \mathbf{grad}_{\boldsymbol{x}} G_k(\boldsymbol{x}-\boldsymbol{y})\mathrm{div}_\Gamma \mathbf{J}(\boldsymbol{y})ds_{\boldsymbol{y}}$$

$$= \boldsymbol{n}(\boldsymbol{x}) \times \int_\Gamma \left(\mathbf{grad}_{\Gamma,\boldsymbol{x}} G_k(\boldsymbol{x}-\boldsymbol{y}) + \boldsymbol{n}_{\boldsymbol{x}} \frac{\partial}{\partial n_x} G_k(\boldsymbol{x}-\boldsymbol{y})\right) \mathrm{div}_\Gamma \mathbf{J}(\boldsymbol{y})ds_{\boldsymbol{y}},$$

which has the limit (6.10.10):

$$\mp\frac{1}{2}\boldsymbol{n}(\boldsymbol{x}) \times (\boldsymbol{n}(\boldsymbol{x})\mathrm{div}_\Gamma \mathbf{J}(\boldsymbol{x})) + \boldsymbol{n}(\boldsymbol{x}) \times \mathbf{grad}_{\boldsymbol{x}} \int_\Gamma G_k(\boldsymbol{x}-\boldsymbol{y})\mathrm{div}_\Gamma \mathbf{J}(\boldsymbol{y})ds_{\boldsymbol{y}}$$

$$= \boldsymbol{n}(\boldsymbol{x}) \times \mathbf{grad}_{\boldsymbol{x}} \int_\Gamma G_k(\boldsymbol{x}-\boldsymbol{y})\mathrm{div}_\Gamma \mathbf{J}(\boldsymbol{y})ds_{\boldsymbol{y}},$$

since $\boldsymbol{n}(\boldsymbol{x}) \times \boldsymbol{n}(\boldsymbol{x}) = \mathbf{0}$.

On the other hand, the form of (6.10.11) can be derived as follows:

$$
\begin{aligned}
\boldsymbol{n}(\boldsymbol{x}) \times \mathbf{grad}_{\boldsymbol{x}} \int_\Gamma & G_k(\boldsymbol{x}-\boldsymbol{y}) \mathrm{div}_\Gamma \mathbf{J}(\boldsymbol{y}) ds_{\boldsymbol{y}} \\
&= \int_\Gamma (\boldsymbol{n}(\boldsymbol{x}) - \boldsymbol{n}(\boldsymbol{y})) \times \mathbf{grad}_{\boldsymbol{x}} G_k(\boldsymbol{x}-\boldsymbol{y}) \mathrm{div}_\Gamma \mathbf{J}(\boldsymbol{y}) ds_{\boldsymbol{y}} \\
&\quad + \int_\Gamma \boldsymbol{n}(\boldsymbol{y}) \times \mathbf{grad}_{\boldsymbol{x}} G_k(\boldsymbol{x}-\boldsymbol{y}) \mathrm{div}_\Gamma \mathbf{J}(\boldsymbol{y}) ds_{\boldsymbol{y}} \\
&= \int_\Gamma (\boldsymbol{n}(\boldsymbol{x}) - \boldsymbol{n}(\boldsymbol{y})) \times \mathbf{grad}_{\boldsymbol{x}} G_k(\boldsymbol{x}-\boldsymbol{y}) \mathrm{div}_\Gamma \mathbf{J}(\boldsymbol{y}) ds_{\boldsymbol{y}} \\
&\quad - \int_\Gamma (\boldsymbol{n}(\boldsymbol{y}) \times \mathbf{grad}_{\boldsymbol{y}} G_k(\boldsymbol{x}-\boldsymbol{y})) \mathrm{div}_\Gamma \mathbf{J}(\boldsymbol{y}) ds_{\boldsymbol{y}} \\
&= \int_\Gamma \big(\boldsymbol{n}(\boldsymbol{x}) - \boldsymbol{n}(\boldsymbol{y})\big) \times \mathbf{grad}_{\boldsymbol{x}} G_k(\boldsymbol{x}-\boldsymbol{y}) \mathrm{div}_\Gamma \mathbf{J}(\boldsymbol{y}) ds_{\boldsymbol{y}} \\
&\quad + \int_\Gamma \mathbf{curl}_{\Gamma,\boldsymbol{y}} \big(g_k(\boldsymbol{x}-\boldsymbol{y}) \mathrm{div}_\Gamma \mathbf{J}(\boldsymbol{y})\big) ds_{\boldsymbol{y}} \\
&\quad - \int_\Gamma G_k(\boldsymbol{x}-\boldsymbol{y}) \mathbf{curl}_{\Gamma,\boldsymbol{y}} \mathrm{div}_\Gamma \mathbf{J}(\boldsymbol{y}) ds_{\boldsymbol{y}} \\
&= \int_\Gamma \big((\boldsymbol{n}(\boldsymbol{x}) - \boldsymbol{n}(\boldsymbol{y})\big) \times \mathbf{grad}_{\boldsymbol{x}} G_k(\boldsymbol{x}-\boldsymbol{y}) \mathrm{div}_\Gamma \mathbf{J}(\boldsymbol{y}) ds_{\boldsymbol{y}} \\
&\quad - \int_\Gamma G_k(\boldsymbol{x}-\boldsymbol{y}) \mathbf{curl}_{\Gamma,\boldsymbol{y}} \big(\mathrm{div}_\Gamma \mathbf{J}(\boldsymbol{y})\big) ds_{\boldsymbol{y}}
\end{aligned}
$$

due to *VI* (2) in the Appendix B. As will be seen, these three formulae will be useful concerning boundary integral equations in electromagnetics. In particular, (6.10.8)–(6.10.10) have been employed repeatedly in the derivation of the equations (6.10.1)–(6.10.7).

6.10.1 The Calderon projector and the capacity operators

From the homogeneous boundary integral equations (6.10.1) and (6.10.4) on Γ where $\mathbf{J}_0 = \mathbf{0}$ one obtains for all solutions of the homogeneous Maxwell system (6.4.1)–(6.4.5) the identity

$$
\begin{pmatrix} \boldsymbol{n} \times \mathbf{E} \\ \boldsymbol{n} \times \mathbf{H} \end{pmatrix} = \begin{pmatrix} \frac{1}{2}\mathbf{I} - \mathbf{R}, & \mathbf{T_E} \\ \mathbf{T_H}, & \frac{1}{2}\mathbf{I} - \mathbf{R} \end{pmatrix} \begin{pmatrix} \boldsymbol{n} \times \mathbf{E} \\ \boldsymbol{n} \times \mathbf{H} \end{pmatrix}
$$

where the boundary integral operator

$$
\mathcal{C}_\Omega = \begin{pmatrix} \frac{1}{2}\mathbf{I} - \mathbf{R}, & \mathbf{T_E} \\ \mathbf{T_H}, & \frac{1}{2}\mathbf{I} - \mathbf{R} \end{pmatrix} \tag{6.10.12}
$$

is known as the **Calderon projector** from the space $C^\alpha(\Gamma)^3 \times C^\alpha(\Gamma)^3$ into $C^\alpha(\Gamma)^3 \times C^\alpha(\Gamma)^3$ with the porperty

$$\mathcal{C}_\Omega \circ \mathcal{C}_\Omega = \mathcal{C}_\Omega \tag{6.10.13}$$

which implies the operator relations:

$$\begin{aligned} \mathbf{T_E} \circ \mathbf{T_H} &= \tfrac{1}{4}\mathbf{I} - \mathbf{R}^2, & \mathbf{R} \circ \mathbf{T_E} &= -\mathbf{T_E} \circ \mathbf{R}, \\ \mathbf{T_H} \circ \mathbf{R} &= -\mathbf{R} \circ \mathbf{T_H}, & \mathbf{T_H} \circ \mathbf{T_E} &= \tfrac{1}{4}\mathbf{I} - \mathbf{R}^2. \end{aligned} \tag{6.10.14}$$

If we define the operator $\mathcal{A}_\Omega$ by

$$\mathcal{C}_\Omega = \frac{1}{2}\mathbf{I} + \mathcal{A}_\Omega\,, \; \mathcal{A}_\Omega = \begin{pmatrix} -\mathbf{R} & \mathbf{T_E} \\ \mathbf{T_H} & -\mathbf{R} \end{pmatrix}, \tag{6.10.15}$$

hence

$$\mathcal{A}_\Omega \begin{pmatrix} \boldsymbol{n} \times \mathbf{E} \\ \boldsymbol{n} \times \mathbf{H} \end{pmatrix} = \frac{1}{2} \begin{pmatrix} \boldsymbol{n} \times \mathbf{E} \\ \boldsymbol{n} \times \mathbf{H} \end{pmatrix}.$$

for $(\mathbf{E}, \mathbf{H})^\top$, the solutions of the homogeneous Maxwell equations in Ω. With

$$\mathcal{B}_\Omega = \begin{pmatrix} \mathbf{T_E} & -\mathbf{R} \\ -\mathbf{R} & \mathbf{T_H} \end{pmatrix} \tag{6.10.16}$$

we arrive with the identities (6.10.13) at the equations

$$4\mathcal{A}_\Omega^2 = \mathbf{I} \tag{6.10.17}$$

and

$$4\mathcal{A}_\Omega \circ \mathcal{B}_\Omega = \begin{pmatrix} \mathbf{0} & \mathbf{I} \\ \mathbf{I} & \mathbf{0} \end{pmatrix}. \tag{6.10.18}$$

This implies

$$\begin{aligned} \mathcal{B}_\Omega \begin{pmatrix} \mathbf{0} & \mathbf{I} \\ \mathbf{I} & \mathbf{0} \end{pmatrix} &= \begin{pmatrix} \mathbf{0} & \mathbf{I} \\ \mathbf{I} & \mathbf{0} \end{pmatrix} \mathcal{B}_\Omega = \mathcal{A}_\Omega, \\ \mathcal{B}_\Omega^2 &= \mathcal{A}_\Omega^2 = \frac{1}{4}\mathbf{I} \end{aligned} \tag{6.10.19}$$

and the celebrated relation

$$2\mathcal{B}_\Omega \begin{pmatrix} \boldsymbol{n} \times \mathbf{H} \\ \boldsymbol{n} \times \mathbf{E} \end{pmatrix} = 2\mathcal{A}_\Omega \begin{pmatrix} \boldsymbol{n} \times \mathbf{E} \\ \boldsymbol{n} \times \mathbf{H} \end{pmatrix} = \begin{pmatrix} \boldsymbol{n} \times \mathbf{E} \\ \boldsymbol{n} \times \mathbf{H} \end{pmatrix} \tag{6.10.20}$$

for all solutions $(\mathbf{E}, \mathbf{H})$ of the homogeneous Maxwell equations in Ω. The operator

$$2\mathcal{B}_\Omega = 2 \begin{pmatrix} \mathbf{T_E}, & -\mathbf{R} \\ -\mathbf{R}, & \mathbf{T_H} \end{pmatrix} \tag{6.10.21}$$

is called the **capacity operator** since it maps the (Neumann, Dirichlet) data to the (Dirichlet, Neumann) data of any solution pair $(\mathbf{E}, \mathbf{H})^\top$ (see also Cessenat [63, p.65], Nedelec [331, p.116]).

Now let us consider the Maxwell equations (6.4.1)–(6.4.5) in the **exterior domain** $\Omega^c - \mathbb{R}^3 \setminus \overline{\Omega}$ for solutions $(\mathbf{E}, \mathbf{H})$ satisfying the Silver–Müller radiation conditions (6.7.2). Then for $\mathbf{J}_0 = \mathbf{0}$ and $\varrho = 0$, we find the corresponding Calderon projector

$$\mathcal{C}_{\Omega^c} \begin{pmatrix} \boldsymbol{n} \times \mathbf{E}_s \\ \boldsymbol{n} \times \mathbf{H}_s \end{pmatrix} := \begin{pmatrix} \frac{1}{2}\mathbf{I} + \mathbf{R} & -\mathbf{T_E} \\ -\mathbf{T_H} & \frac{1}{2}(\mathbf{I} + \mathbf{R}) \end{pmatrix} \begin{pmatrix} \boldsymbol{n} \times \mathbf{E}_s \\ \boldsymbol{n} \times \mathbf{H}_s \end{pmatrix} = \begin{pmatrix} \boldsymbol{n} \times \mathbf{E}_s \\ \boldsymbol{n} \times \mathbf{H}_s \end{pmatrix} \tag{6.10.22}$$

for all solutions $(\mathbf{E}, \mathbf{H})^\top$ of the homogeneous Maxwell equations (6.4.1)–(6.4.5) which satisfy the Silver–Müller radiation conditions (6.7.2). Note, that

$$\mathcal{C}_{\Omega^c} + \mathcal{C}_{\Omega} = \mathbf{I}. \tag{6.10.23}$$

If we now define $\mathcal{A}_{\Omega^c}$ by

$$\mathcal{C}_{\Omega^c} = \frac{1}{2}\mathbf{I} + \mathcal{A}_{\Omega^c}$$

then

$$\mathcal{A}_{\Omega^c} = -\mathcal{A}_{\Omega} \quad \text{and} \quad \mathcal{B}_{\Omega^c} = -\mathcal{B}_{\Omega} \tag{6.10.24}$$

is the **Calderon or capacity operator** for the exterior domain Ω^c.

Similar to the boundary value problems in the previous chapters, the Calderon projector provides us two boundary integral equations for unknowns either $\boldsymbol{n} \times \mathbf{E}$ or $\boldsymbol{n} \times \mathbf{H}$, on the boundary. This is called the **direct method** for defining boundary integral equations for solving boundary value problems of the Maxwell's system (6.4.1)–(6.4.4) in Ω or Ω^c.

6.10.2 Weak solutions for a fundamental problem

One of the difficulties in treating the solutions of boundary value problems in electromagnetics by using boundary integral equations is that the imbedding of the solution space $\mathbf{H}(\mathbf{curl}, \Omega) \subset \boldsymbol{L}_2(\Omega)$ is not compact in contrast to $\mathbf{H}^1(\Omega)$ solutions of acoustic problems. To circumvent this difficulty one can employ the Helmholtz (or Hodge) decomposition of the solution of the Maxwell equations in Ω as well as the solution space of the corresponding integral equation on Γ. In this section we confine the analysis to simple model problems.

6.10.2.1 Interior Dirichlet problem in Ω.

Let $(\mathbf{E}, \mathbf{H})^\top \in \mathbf{H}(\mathbf{curl}, \Omega)$ be the solution of the Dirichlet problem for the homogeneous Maxwell equations (6.4.1)–(6.4.4), namely

$$\begin{aligned} \mathbf{curl}\,\mathbf{E}(\boldsymbol{x}) &= i\omega\mu\mathbf{H}(\boldsymbol{x}) & & \\ \mathbf{curl}\,\mathbf{H}(\boldsymbol{x}) &= (-i\omega\varepsilon+\sigma)\mathbf{E}(\boldsymbol{x}) & \text{in } \Omega, & \\ \boldsymbol{n}(\boldsymbol{x})\times\mathbf{E}(\boldsymbol{x})|_\Gamma &= \boldsymbol{m}(\boldsymbol{x}) & \text{on } \Gamma, & \end{aligned} \tag{6.10.25}$$

where $\boldsymbol{m}$ on Γ is given. This problem is also called the **cavity problem**. For its weak solution it suffices to consider only the electric field $\mathbf{E}\in\mathbf{H}(\mathbf{curl}\,,\Omega)$ of the equation

$$\mathbf{curl}\,\mathbf{curl}\,\mathbf{E}-k^2\mathbf{E}=\mathbf{0} \tag{6.10.26}$$

where k^2 is given by (6.4.8). Then, according to Green's first identity (App **B**(8)) the weak formulation of (6.10.26) reads:

Given $\boldsymbol{m}\in\mathbf{H}^{-1/2}(\mathrm{div}_\Gamma,\Gamma)$ find $\mathbf{E}\in\mathbf{H}(\mathbf{curl}\,,\Omega)$ such that $\boldsymbol{m}-\boldsymbol{n}\times\mathbf{E}|_\Gamma=\mathbf{0}$ and

$$\begin{aligned} \mathfrak{A}(\mathbf{E},\mathbf{E}^t) &:= \int\limits_\Omega \{(\mathbf{curl}\,\mathbf{E})\cdot\overline{(\mathbf{curl}\,\mathbf{E}^t)}-k^2\mathbf{E}\cdot\overline{\mathbf{E}^t}\}dy \\ &= \int\limits_\Gamma (\mathbf{curl}\,\mathbf{E})\cdot(\boldsymbol{n}\times\overline{\mathbf{E}^t})ds_y = 0 \end{aligned} \tag{6.10.27}$$

for all test functions $\mathbf{E}^t\in\mathbf{H}_0(\mathbf{curl}\,,\Omega)$.

First let $\widetilde{\mathbf{E}}\in\mathbf{H}(\mathbf{curl}\,,\Omega)$ be an extension of the boundary values $\boldsymbol{m}=\boldsymbol{n}\times\widetilde{\mathbf{E}}|_\Gamma$ into Ω and set $\mathbf{E}_0:=\mathbf{E}-\widetilde{\mathbf{E}}$. Then $\mathbf{E}_0\in\mathbf{H}_0(\mathbf{curl}\,,\Omega)$ satisfies the variational equation

$$\mathfrak{A}(\mathbf{E}_0,\mathbf{E}^t)=-\mathfrak{A}(\widetilde{\mathbf{E}},\mathbf{E}^t)\quad\text{for all}\quad\mathbf{E}^t\in\mathbf{H}_0(\mathbf{curl}\,,\Omega) \tag{6.10.28}$$

where $\mathfrak{A}(\widetilde{\mathbf{E}},\mathbf{E}^t)$ is a given linear functional operating on $\mathbf{E}^t$. To consider the existence of a solution (6.10.28), we have to use the **Helmholtz decomposition** (see e.g. Ammari and Nedelec [9], Cessenat [63, p.52], Dautray and Lions [96]):

$$\mathbf{H}(\mathbf{curl}\,,\Omega)=\mathbf{M}(\Omega)\oplus\mathbf{P}(\Omega) \tag{6.10.29}$$

where

$$\mathbf{M}(\Omega):=\{\boldsymbol{u}\in\mathbf{H}_0(\mathbf{curl}\,,\Omega)\quad\text{with}\quad\int\limits_\Omega \boldsymbol{u}\cdot\mathbf{grad}\,p\,d\boldsymbol{x}=0\,\forall\,p\in H_0^1(\Omega)\},$$

$$\mathbf{P}(\Omega):=\{\boldsymbol{v}=\mathbf{grad}\,p\quad\text{with}\quad p\in H_0^1(\Omega)\}.$$

$\mathbf{M}(\Omega)$ and $\mathbf{P}(\Omega)$ are orthogonal with respect to the scalar product

$$((\boldsymbol{u},\boldsymbol{v})):=\int\limits_\Omega \mathbf{curl}\,\boldsymbol{u}\cdot\overline{\mathbf{curl}\,\boldsymbol{v}}\,dy+\int\limits_\Omega \boldsymbol{u}\cdot\overline{\boldsymbol{v}}\,dy, \tag{6.10.30}$$

since for $\boldsymbol{u}\in\mathbf{M}(\Omega)$ and $\boldsymbol{v}\in\mathbf{P}(\Omega)$ there holds

$$((\boldsymbol{u},\boldsymbol{v})) := \int\limits_\Omega (\mathbf{curl}\,\boldsymbol{u})\cdot(\mathbf{curl}\,\mathbf{grad}\,p)dy + \int\limits_\Omega \boldsymbol{u}\cdot\mathbf{grad}\,pdy = 0.$$

Moreover, $p|_\Gamma = 0$ implies the tangential derivative of p, i.e. $\boldsymbol{n}\times\mathbf{grad}\,p|_\Gamma = \mathbf{0}$ as well.

The bilinear equation (6.10.27) also reads

$$\mathfrak{A}(\mathbf{E},\mathbf{E}^t) = \mathfrak{A}(\mathbf{E}-\tilde{\mathbf{E}},\mathbf{E}^t) + \mathfrak{A}(\tilde{\mathbf{E}},\mathbf{E}^t) = 0$$

where $\tilde{\mathbf{E}} \in \mathbf{H}(\mathbf{curl}\,,\Omega)$ satisfies $\boldsymbol{n}\times\tilde{\mathbf{E}} = \boldsymbol{m}$ on Γ, whereas $\mathbf{E}-\tilde{\mathbf{E}} = \mathbf{E}_0$ is the solution of the cavity problem and satisfies the boundary condition $\boldsymbol{n}\times\mathbf{E}_0|_\Gamma = \mathbf{0}$ and is determined by the system of equations for $\mathbf{E}_0 = \boldsymbol{u} + \mathbf{grad}\;p$ with $\boldsymbol{u}\in\mathbf{M}(\Omega)$ and for $p\in H^1_0(\Omega)$.

According to the Helmholtz decomposition (6.10.29) we test (6.10.28) first with $\boldsymbol{u}^t\in\mathbf{M}(\Omega)$ and obtain for $\boldsymbol{u}\in\mathbf{M}(\Omega)$ the variational equation

$$\int\limits_\Omega \{(\mathbf{curl}\,\boldsymbol{u}\cdot\overline{\mathbf{curl}\,\boldsymbol{u}^t dy}\} + \int\limits_\Omega \boldsymbol{u}\cdot\overline{\boldsymbol{u}^t}dy - (1+k^2)\int\limits_\Omega \boldsymbol{u}\cdot\overline{\boldsymbol{u}^t}dy = -\mathfrak{A}(\tilde{\mathbf{E}},\boldsymbol{u}^t). \tag{6.10.31}$$

for all $\boldsymbol{u}^t\in\mathbf{M}(\Omega)$.

Testing with the second component of the decomposition, with $\mathbf{grad}\,p^t$ we obtain

$$\int\limits_\Omega \mathbf{grad}\,p\cdot\overline{\mathbf{grad}\,p^t dy} = \frac{1}{k^2}\mathfrak{A}(\tilde{\mathbf{E}},\mathbf{grad}\,p^t)dy \quad \text{for all} \quad \mathbf{grad}\,p^t\in\mathbf{P}(\Omega). \tag{6.10.32}$$

Note that

$$\operatorname{div}\boldsymbol{u} = 0 \quad \text{in} \quad \Omega \tag{6.10.33}$$

for $\boldsymbol{u}\in\mathbf{M}(\Omega)$ since

$$0 = \int\limits_\Omega \boldsymbol{u}\cdot\mathbf{grad}\,pdy = -\int\limits_\Omega p\operatorname{div}\boldsymbol{u}dy \quad \text{for all} \quad p\in H^1_0(\Omega)$$

due to $p|_\Gamma = 0$.

Theorem 6.10.1. *Let Ω be a bounded Lipschitz domain in $\mathbb{R}^3$. Then $\mathbf{M}(\Omega)\subset\mathbf{H}^1(\Omega)$ and, hence, $\mathbf{M}(\Omega)$ is compactly $\mathbb{R}^3$ imbedded into $\boldsymbol{L}_2(\Omega)$.*

As a consequence we have Gårding's inequality

$$\operatorname{Re}\mathfrak{A}(\boldsymbol{u}+\mathbf{grad}\,p,\boldsymbol{u}+\mathbf{grad}\,p) \geq \|\boldsymbol{u}+\mathbf{grad}\,p\|^2_\Omega - (1+|k^2|)\|\boldsymbol{u}+\mathbf{grad}\,p\|^2_{L_2(\Omega)} \tag{6.10.34}$$

where

$$\|\boldsymbol{u}+\mathbf{grad}\,p\|_\Omega := ((\boldsymbol{u}+\mathbf{grad}\,p,\boldsymbol{u}+\mathbf{grad}\,p))^{\frac{1}{2}} \tag{6.10.35}$$

is the Hilbert space norm associated with the inner product$((\cdot,\cdot))$ in (6.10.30).

Proof of Theorem 6.10.1: For the proof we follow Girauld and Raviart [146, Theorem 3.2]. Since $\boldsymbol{u} \in \big(L_2(\Omega)\big)^3$ satisfying (6.10.33) we can extend $\boldsymbol{u}$ to all of $\mathbb{R}^3$ with $\boldsymbol{u}_{\text{ext}} \in \big(L_2(\mathbb{R}^3)\big)^3$ having compact support and satisfying $\text{div}\boldsymbol{u}_{\text{ext}} = 0$ in $\mathbb{R}^3$.

The Fourier transformed components of $\boldsymbol{u}_{\text{ext}}$,

$$\widehat{u}_j(\boldsymbol{\xi}) := (2\pi)^{-3/2} \int\limits_{\mathbb{R}^3} e^{-i\boldsymbol{x}\cdot\boldsymbol{\xi}} u_{\text{ext,j}}(\boldsymbol{x}) d\boldsymbol{x}$$

are holomorphic in $\mathbb{R}^3$ and $\text{div}\,\boldsymbol{u}_{\text{ext}} = 0$ in $\mathbb{R}^3$ yields

$$\sum_{j=1}^{3} \xi_j \widehat{u}_j = 0. \tag{6.10.36}$$

Now, we must find a function $\widehat{\varphi} \in (\big(L_2(\mathbb{R}^3)\big)^3$ such that $\textbf{curl}\,\boldsymbol{\varphi} = \boldsymbol{u}_{\text{ext}}$, i.e.

$$\widehat{u}_1 = 2\pi i(\xi_2\widehat{\varphi}_3 - \xi_3\widehat{\varphi}_2)\,,\ \widehat{u}_2 = 2\pi i(\xi_3\widehat{\varphi}_1 - \xi_1\widehat{\varphi}_3)\,,\ \widehat{u}_3 = 2\pi i(\xi_1\widehat{\varphi}_2 - \xi_2\widehat{\varphi}_1).$$

Note, the third equation is a consequence of the first two and (6.10.36). Hence, in order to fix $\boldsymbol{\varphi}$ we add the Coulomb gauge condition

$$\sum_{j=1}^{3} \xi_j \widehat{\varphi}_j = 0. \tag{6.10.37}$$

This implies

$$\begin{aligned}
\widehat{\varphi}_1 &= (i|\boldsymbol{\xi}|^2)^{-1}(\xi_3\widehat{u}_2 - \xi_2\widehat{u}_3),\\
\widehat{\varphi}_2 &= (i|\boldsymbol{\xi}|^2)^{-1}(\xi_1\widehat{u}_3 - \xi_3\widehat{u}_1),\\
\widehat{\varphi}_3 &= (i|\boldsymbol{\xi}|^2)^{-1}(\xi_2\widehat{u}_1 - \xi_1\widehat{u}_2).
\end{aligned}$$

Since $\xi_j\widehat{\varphi}_j \in L_2(\mathbb{R}^3)$ and $|\widehat{\varphi}_i| \le (2\pi|\boldsymbol{\xi}|)^{-1}(|\widehat{\boldsymbol{u}}_j| + |\widehat{\boldsymbol{u}}_k|)$ there holds $\boldsymbol{\varphi} \in \big(H^1(\Omega)\big)^3$ if $\widehat{\boldsymbol{\varphi}}$ is bounded at $\boldsymbol{\xi} = \boldsymbol{0}$. Note, that $\widehat{\boldsymbol{u}}(\boldsymbol{\xi})$ is holomorphic and $\widehat{\boldsymbol{u}}(\boldsymbol{0}) = \boldsymbol{0}$ due to (6.10.36) and

$$\widehat{u}_j(\boldsymbol{\xi}) = \sum_{\ell=1}^{3} \xi_\ell \frac{\partial \widehat{u}_j}{\partial \xi_\ell}(\boldsymbol{0}) + \mathcal{O}(|\boldsymbol{\xi}|^2)$$

near $\boldsymbol{\xi} = \boldsymbol{0}$. This implies that $\widehat{\boldsymbol{\varphi}}(\boldsymbol{\xi})$ is bounded near $\boldsymbol{\xi} = \boldsymbol{0}$ and the inverse Fourier transform (3.1.14) gives

$$\boldsymbol{\varphi}(\boldsymbol{x}) = (2\pi)^{-3/2} \int\limits_{\mathbb{R}^3} e^{i\boldsymbol{x}\cdot\boldsymbol{\xi}} \widehat{\boldsymbol{\varphi}}(\boldsymbol{\xi}) d\boldsymbol{\xi}. \tag{6.10.38}$$

Hence,

$$\mathbf{curl}\,\varphi = \boldsymbol{u} \quad \text{and} \quad \operatorname{div}\varphi = 0$$

due to (6.10.37) and $\boldsymbol{u} \in \left(\mathbf{H}^1(\mathbb{R}^3)\right)^3 = \left(\mathbf{H}^1(\mathbb{R}^3)\right)^3|_\Omega$. As a consequence,

$$\mathbf{curl}\,(\mathbf{curl}\,\varphi) = -\Delta\varphi + \mathbf{0}, \quad \text{hence } \mathbf{curl}\,\boldsymbol{u} \in (L_2(\mathbb{R}^3))^3$$

which implies $\varphi \in \left(H^2(\Omega)\right)^3 = \left(H^2(\mathbb{R}^3)\right)^3|_\Omega$, therefore, $\boldsymbol{u} \in \left(H^1(\Omega)\right)^3$ and $\mathbf{M}(\Omega) \subset \left(H^1(\Omega)\right)^3$ which is compactly imbedded into $\left(L_2(\Omega)\right)^3$ as we have proposed. ∎

Note that $\mathfrak{A}$ (see (6.10.34)) is continuous with respect to the Hilbert space norm $|||\cdot|||$.

As a consequence of the Gårding inequality (6.10.34), the Fredholm alternative is valid (see Chapter 5.3.3), uniqueness would imply the existence of a unique solution to (6.10.28) which satisfies the Maxwell equations (6.10.25) where $\mathbf{curl}\,\mathbf{E}(\boldsymbol{x}) = i\omega\mu\mathbf{H}(\boldsymbol{x})$ defines the magnetic field $\mathbf{H}(\boldsymbol{x})$.

6.10.2.2 A reduction to boundary integral equations.

The Dirichlet problem (6.10.25) or cavity problem, can be reduced to boundary integral equations based on the Stratton–Chu formulae (6.9.4) and (6.9.5).

To be consistent with the previous chapters we first introduce the single layer and double layer potentials for the Maxwell equations as follows:

$$\begin{aligned}(\mathfrak{V}\boldsymbol{\lambda})(\boldsymbol{x}) := & -i\omega\mu\int\limits_\Gamma G_k(\boldsymbol{x}-\boldsymbol{y})\boldsymbol{\lambda}(\boldsymbol{y})ds_{\boldsymbol{y}} \\ & + \frac{1}{(-i\omega\varepsilon+\sigma)}\int\limits_\Gamma (\mathbf{grad}_{\,\boldsymbol{y}}G_k(\boldsymbol{x}-\boldsymbol{y}))(\operatorname{div}_\Gamma\boldsymbol{\lambda}(\boldsymbol{y}))ds_{\boldsymbol{y}},\end{aligned} \tag{6.10.39}$$

$$\begin{aligned}(\mathfrak{W}\varphi)(\boldsymbol{x}) := & \int\limits_\Gamma \varphi(\boldsymbol{y})\times\mathbf{grad}_{\,\boldsymbol{y}}G_k(\boldsymbol{x}-\boldsymbol{y})ds_{\boldsymbol{y}} \\ = & \;\mathbf{curl}\int\limits_\Gamma G_k(\boldsymbol{x}-\boldsymbol{y})\varphi(\boldsymbol{y})ds_{\boldsymbol{y}} \quad \text{for } \boldsymbol{x}\in\Omega.\end{aligned}$$

These boundary potentials have the mapping properties

$$\begin{aligned}\mathfrak{V} &: \mathbf{H}^{-1/2}(\operatorname{div}_\Gamma,\Gamma) \to \mathbf{H}(\mathbf{curl}\,,\Omega), \\ \mathfrak{W} &: \mathbf{H}^{-1/2}(\operatorname{div}_\Gamma,\Gamma) \to \mathbf{H}(\mathbf{curl}\,,\Omega) \quad \text{continuously}\;.\end{aligned} \tag{6.10.40}$$

For the Dirichlet problem we seek the solution by the layer ansatz

$$\mathbf{E}(\boldsymbol{x}) := (\mathfrak{V}\boldsymbol{\lambda})(\boldsymbol{x})\,,\; \boldsymbol{x}\in\Omega. \tag{6.10.41}$$

The boundary condition $\boldsymbol{n} \times \mathbf{E}_{|\Gamma} = \boldsymbol{m} \in \mathbf{H}_{||}^{-1/2}(\mathrm{div}_\Gamma, \Gamma)$ then leads to the boundary integral equation

$$\boldsymbol{n}(\boldsymbol{x}) \times \mathbf{E}(\boldsymbol{x}) = (\mathbf{T_E}\boldsymbol{\lambda})(\boldsymbol{x}) = \boldsymbol{m}(\boldsymbol{x}) \quad \text{for } \boldsymbol{x} \in \Gamma \tag{6.10.42}$$

and $\mathbf{T_E}$ is given by (6.10.2), i.e.,

$$\begin{aligned}(\mathbf{T_E})\boldsymbol{\lambda}(\boldsymbol{x}) = {} & -i\omega\mu\boldsymbol{n}(\boldsymbol{x}) \times (\mathbf{V}_k\boldsymbol{\lambda})(\boldsymbol{x}) \\ & - (-i\omega\varepsilon + \sigma)^{-1}\boldsymbol{n}(\boldsymbol{x}) \times \big(\mathbf{grad}_\Gamma V_k(\mathrm{div}_\Gamma\boldsymbol{\lambda})\big)(\boldsymbol{x})\,,\ \boldsymbol{x} \in \Gamma,\end{aligned} \tag{6.10.43}$$

$$\mathbf{V}_k\boldsymbol{\lambda} := \int\limits_\Gamma G_k(\boldsymbol{x}-\boldsymbol{y})\boldsymbol{\lambda}(\boldsymbol{y})ds_{\boldsymbol{y}} \tag{6.10.44}$$

(see (2.1.6) and Section 10.1.1).

Given $\boldsymbol{m} \in \mathbf{H}^{-\frac{1}{2}}(\mathrm{div}_\Gamma, \Gamma)$, we seek $\boldsymbol{\lambda} \in \mathbf{H}^{-\frac{1}{2}}(\mathrm{div}_\Gamma, \Gamma)$ by using the variational equation for (6.10.43) by testing (6.10.43) with $\boldsymbol{n}(\boldsymbol{x}) \times \boldsymbol{\lambda}^t(\boldsymbol{x})$ to obtain

$$\begin{aligned} & - i\omega\mu \int\limits_\Gamma \big(\boldsymbol{n}(\boldsymbol{x}) \times \mathbf{V}_k\boldsymbol{\lambda}(\boldsymbol{x})\big) \cdot \big(\boldsymbol{n}(\boldsymbol{x}) \times \overline{\boldsymbol{\lambda}^t}(\boldsymbol{x})\big)ds_{\boldsymbol{x}} \\ & = -i\omega\mu \int\limits_\Gamma (\mathbf{V}_k\boldsymbol{\lambda})(\boldsymbol{x}) \cdot \pi_t\overline{\boldsymbol{\lambda}^t}(\boldsymbol{x})ds_{\boldsymbol{x}} = -i\omega\mu \int\limits_\Gamma (V_k\boldsymbol{\lambda})(\boldsymbol{x}) \cdot \overline{\boldsymbol{\lambda}}^t(\boldsymbol{x})ds_{\boldsymbol{x}}.\end{aligned} \tag{6.10.45}$$

Similarly,

$$\begin{aligned} & \int\limits_\Gamma \big(\mathbf{grad}\,(V_k\mathrm{div}_\Gamma\boldsymbol{\lambda}) \times \boldsymbol{n}(\boldsymbol{x})\big) \cdot \big(\boldsymbol{n}(\boldsymbol{x}) \times \overline{\boldsymbol{\lambda}^t}(\boldsymbol{x})\big)ds_{\boldsymbol{x}} \\ & \qquad = - \int\limits_\Gamma \pi_t\mathbf{grad}\ (V_k\mathrm{div}_\Gamma\boldsymbol{\lambda}) \cdot \overline{\boldsymbol{\lambda}}^t(\boldsymbol{x})ds_{\boldsymbol{x}} \\ & \qquad = - \int\limits_\Gamma \mathbf{grad}_\Gamma(V_k\mathrm{div}_\Gamma\boldsymbol{\lambda}) \cdot \overline{\boldsymbol{\lambda}^t}(\boldsymbol{x})ds_{\boldsymbol{x}} \\ & \qquad = \int\limits_\Gamma (V_k\mathrm{div}_\Gamma\boldsymbol{\lambda}) \cdot \mathrm{div}_\Gamma\overline{\boldsymbol{\lambda}^t}ds_{\boldsymbol{x}}.\end{aligned} \tag{6.10.46}$$

Collecting these relations for (6.10.43) gives the weak variational equation

$$\begin{aligned} & \int\limits_\Gamma \boldsymbol{m}(\boldsymbol{x}) \cdot (\boldsymbol{n}(\boldsymbol{x}) \times \overline{\boldsymbol{\lambda}^t}(\boldsymbol{x}))ds_{\boldsymbol{x}} = \int\limits_\Gamma \big(\boldsymbol{m}(\boldsymbol{x}) \times \boldsymbol{n}(\boldsymbol{x})\big) \cdot \overline{\boldsymbol{\lambda}^t}(\boldsymbol{x})ds_{\boldsymbol{x}} \\ & = -i\omega\mu \int\limits_\Gamma (\mathbf{V}_k\boldsymbol{\lambda})(\boldsymbol{x}) \cdot \overline{\boldsymbol{\lambda}^t}(\boldsymbol{x})ds_{\boldsymbol{x}} + (-i\omega\varepsilon_0 + \sigma)^{-1} \int\limits_\Gamma (V_k\mathrm{div}_\Gamma\boldsymbol{\lambda})\mathrm{div}_\Gamma\overline{\boldsymbol{\lambda}^t}(\boldsymbol{x})ds_{\boldsymbol{x}}\end{aligned}$$

or, in short,

$$\langle \mathbf{V}_k \boldsymbol{\lambda}, \overline{\boldsymbol{\lambda}^t} \rangle_\Gamma - \frac{1}{k^2} \langle V_k \mathrm{div}_\Gamma \boldsymbol{\lambda}, \mathrm{div}_\Gamma \overline{\boldsymbol{\lambda}^t} \rangle_\Gamma = \frac{-i}{\omega\mu} \langle \boldsymbol{n} \times \boldsymbol{m}, \overline{\boldsymbol{\lambda}^t} \rangle_\Gamma \tag{6.10.47}$$

for all $\boldsymbol{\lambda}^t \in \mathbf{H}_{||}^{-1/2}(\mathrm{div}_\Gamma, \Gamma)$.

For obtaining a Gårding inequality for the boundary integral equation (6.10.47), we again use a Helmholtz (Hodge) decomposition of the solution on the boundary in the following form (which in fact has no direct relation with the Helmholtz decomposition of the corresponding solution in the domain Ω):

$$\mathbf{H}_{||}^{-1/2}(\mathrm{div}_\Gamma, \Gamma) = \mathbf{grad}_\Gamma H^{3/2}(\Gamma)/\mathbb{C} \oplus \mathbf{curl}_\Gamma H^{1/2}(\Gamma)/\mathbb{C}. \tag{6.10.48}$$

The spaces on the right are orthogonal with respect to the $L_2(\Gamma)$–duality. In fact, we note that $\mathbf{curl}_\Gamma H^{1/2}(\Gamma) \subset \mathbf{H}^{-1/2}(\Gamma)$ whereas $\mathbf{grad}_\Gamma H^{3/2}(\Gamma) \subset \mathbf{H}^{1/2}(\Gamma) \subset \mathbf{H}^{-1/2}(\Gamma)$ compactly. The space on the left is purposely chosen in order to obtain $\varphi \in H^{3/2}(\Gamma)$ and $\psi \in H^{1/2}(\Gamma)$ so that the $L_t^2(\Gamma)$–duality $\langle \mathbf{grad}_\Gamma \varphi, \mathbf{curl}_\Gamma \overline{\psi} \rangle$ is well defined. Therefore, in terms of the decomposition, we write $\boldsymbol{\lambda} = \mathbf{grad}_\Gamma \varphi + \mathbf{curl}_\Gamma \psi$ and $\boldsymbol{\lambda}^t = \mathbf{grad}_\Gamma \varphi^t + \mathbf{curl}_\Gamma \psi^t$. Then the equation (6.10.47), i.e.

$$\begin{aligned} \langle \mathbf{V}_k(\mathbf{grad}_\Gamma \varphi + \mathbf{curl}_\Gamma \psi), \overline{\boldsymbol{\lambda}^t} \rangle - \frac{1}{k^2} \langle V_k \mathrm{div}_\Gamma (\mathbf{grad}_\Gamma \varphi + \mathbf{curl}_\Gamma \psi), \mathrm{div}_\Gamma \overline{\boldsymbol{\lambda}^t} \rangle_\Gamma \\ = -\frac{i}{\omega\mu} \langle \boldsymbol{n} \times \boldsymbol{m}, \overline{\boldsymbol{\lambda}^t} \rangle \end{aligned} \tag{6.10.49}$$

will first be tested by $\mathbf{grad}_\Gamma \varphi^t$ to obtain

$$\begin{aligned} &\langle \mathbf{V}_k \mathbf{grad}_\Gamma \varphi, \mathbf{grad}_\Gamma \overline{\varphi^t} \rangle + \langle \mathbf{V}_k \mathbf{curl}_\Gamma \psi, \mathbf{grad}_\Gamma \overline{\varphi^t} \rangle_\Gamma \\ &- \frac{1}{k^2} \langle V_k \Delta_\Gamma \varphi, \Delta_\Gamma \overline{\varphi^t} \rangle_\Gamma = -\frac{i}{\omega\mu} \langle \boldsymbol{n} \times \boldsymbol{m}, \mathbf{grad}_\Gamma \overline{\varphi}^t \rangle_\Gamma \end{aligned} \tag{6.10.50}$$

and then tested by $\mathbf{curl}_\Gamma \psi^t$ to yield

$$\begin{aligned} &\langle \mathbf{V}_k(\mathbf{grad}_\Gamma \varphi), \mathbf{curl}_\Gamma \overline{\psi^t} \rangle + \langle \mathbf{V}_k \mathbf{curl}_\Gamma \psi, \mathbf{curl}_\Gamma \overline{\psi^t} \rangle_\Gamma \\ &= -\frac{i}{\omega\mu} \langle \boldsymbol{n} \times \boldsymbol{m}, \mathbf{curl}_\Gamma \overline{\psi^t}. \rangle \end{aligned} \tag{6.10.51}$$

in view of $\mathrm{div}_\Gamma \mathbf{curl}_\Gamma \psi^t = 0$.

Multiplying the Equation (6.10.50) by $-k^2$ and adding to the Equation (6.10.51), the **weak variational equation** reads:

Find $(\varphi, \psi) \in H^{3/2}(\Gamma) \times H^{1/2}(\Gamma)$ **such that**

$$\begin{aligned} \mathcal{A}_\Gamma(\varphi, \psi; \varphi^t, \psi^t) := &\langle V_k \Delta_\Gamma \varphi, \Delta_\Gamma \overline{\varphi^t} \rangle_\Gamma + \langle \mathbf{V}_k \mathbf{curl}_\Gamma \psi, \mathbf{curl}_\Gamma \overline{\psi^t} \rangle_\Gamma \\ &+ \langle \mathbf{V}_k \mathbf{grad}_\Gamma \varphi, \mathbf{curl}_\Gamma \overline{\psi^t}, \rangle_\Gamma \\ &- k^2 \langle \mathbf{V}_k \mathbf{grad}_\Gamma \varphi, \mathbf{grad}_\Gamma \overline{\varphi^t} \rangle_\Gamma - k^2 \langle \mathbf{V}_k \mathbf{grad}_\Gamma \varphi, \mathbf{curl}_\Gamma \overline{\psi^t} \rangle_\Gamma \\ &= \frac{i}{\omega\mu} \langle \boldsymbol{n} \times \boldsymbol{m}, (k^2 \mathbf{grad}_\Gamma \overline{\varphi^t} - \mathbf{curl}_\Gamma \overline{\psi^t}) \rangle_\Gamma \end{aligned} \tag{6.10.52}$$

for all $(\varphi^t, \psi^t) \in H^{3/2}(\Gamma) \times H^{1/2}(\Gamma)$.

This sesquilinear variational equation is called the **Rumsey principle** (see Nedelec [331, p.245]). From (2.1.9) we have the pseudohomogeneous expansion of the fundamental solution

$$G_k(\boldsymbol{x}-\boldsymbol{y}) = G_0(\boldsymbol{x}-\boldsymbol{y}) + \frac{ik}{4\pi} + S_k(\boldsymbol{x}-\boldsymbol{y}) \tag{6.10.53}$$

with the kernel function S_k of the remainder, given by

$$S_k(\boldsymbol{x}-\boldsymbol{y}) = \frac{1}{4\pi|\boldsymbol{x}-\boldsymbol{y}|}\sum_{m=2}^{\infty}\frac{(ik|\boldsymbol{x}-\boldsymbol{y}|)^m}{m!}. \tag{6.10.54}$$

Lemma 6.10.2. *Based on the mapping properties of* $\mathbf{V}_k$ *in Theorems 5.6.2 and 5.6.8*

$$\mathbf{V}_k : \mathbf{H}^{-1/2}(\Gamma) \to \mathbf{H}^1(\Omega),$$
$$\gamma_0\mathbf{V}_k : \mathbf{H}^{-1/2}(\Gamma) \to \mathbf{H}^{1/2}(\Gamma),$$

there holds the following Gårding inequality

$$\operatorname{Re}\{\langle \mathbf{V}_k\boldsymbol{\lambda}, \boldsymbol{\lambda}\rangle + \langle \mathbf{C}_k\boldsymbol{\lambda}, \boldsymbol{\lambda}\rangle\} \geq \alpha\|\boldsymbol{\lambda}\|^2_{\mathbf{H}^{-1/2}(\Gamma)} \tag{6.10.55}$$

with $\alpha > 0$ *and some compact linear operator*

$$\mathbf{C}_k : \mathbf{H}^{-1/2}(\Gamma) \to \mathbf{H}^{1/2}(\Gamma).$$

In the weak varational formulation (6.10.52) the following compactness properties hold.

Lemma 6.10.3. *For* $\varphi \in H^{3/2}(\Gamma)$ *and* $\psi \in H^{1/2}(\Gamma)$, *the following linear operators are compact:*

$$\mathbf{V}_k\mathbf{grad}_\Gamma : H^{3/2}(\Gamma) \to \mathbf{curl}_\Gamma H^{1/2}(\Gamma) \Subset \mathbf{H}^{-1/2}(\Gamma), \tag{6.10.56}$$
$$\mathbf{V}_k\mathbf{curl}_\Gamma : H^{1/2}(\Gamma) \to \mathbf{grad}_\Gamma H^{1/2}(\Gamma) \Subset \mathbf{H}^{-1/2}(\Gamma). \tag{6.10.57}$$

In view of Lemata 6.10.2 and 6.10.3 we obtain:

Theorem 6.10.4. *The sesquilinear form* $\mathcal{A}_\Gamma$ *in* (6.10.52) *for Re* $k^2 > 0$ *is strongly elliptic on* $H^{3/2}(\Gamma) \times H^{1/2}(\Gamma)$ *satisfying the Gårding inequality*

$$\begin{aligned}\operatorname{Re}\mathcal{A}_\Gamma(\varphi, \psi; \varphi, \psi) \geq\; & c_0\{\|\varphi\|^2_{H^{3/2}} + \|\psi\|^2_{H^{1/2}(\Gamma)}\}\\ & - c_1(\|\varphi\|^2_{L_2(\Gamma)} + \|\psi\|^2_{L_2(\Gamma)})\end{aligned} \tag{6.10.58}$$

where $c_0 > 0$ *and* $c_1 \geq 0$ *are constants independent of* (φ, ψ).

The Gårding inequality (6.10.58) implies that the linear operator equation associated with the weak variational formula (6.10.52) is a Fredholm mapping with index zero. As a consequence, since $\sigma > 0$ and, hence, Im$k > 0$, uniqueness implies that the sesquilinear variational equation (6.10.52) has a unique solution $(\varphi, \psi) \in H^{3/2}(\Gamma) \times H^{1/2}(\Gamma)$ with $\boldsymbol{\lambda} = \mathbf{grad}_\Gamma\varphi + \mathbf{curl}_\Gamma\psi$ which generates via (6.10.41) a unique solution of the cavity problem of the Maxwell equations.

6.11 Application of the Electromagnetic Potentials to Eddy Current Problems

In this section we consider a classical formulation of eddy current problems in terms of three fields $\boldsymbol{A}, \varphi$ and ψ which was introduced in Leon and Rodger [264, 90–93] and Biró and Preis [31, 3145–3159].

6.11.1 The '$(\boldsymbol{A}, \varphi) - (\boldsymbol{A}) - (\psi)$' formulation in the bounded domain

In what follows, we consider all the functions involved to be sufficiently smooth. Later on, the precise function spaces will be specified.

As in the figure below, consider three bounded domains Ω_c, Ω_A and Ω in $\mathbb{R}^3$ such that $\overline{\Omega}_c \subset \Omega_A, \overline{\Omega}_A \subset \Omega$ and $\overline{\Omega} \subset \mathbb{R}^3$, with the boundaries $\partial\Omega_c = \Gamma_c, \partial\Omega_A = \Gamma_A$ and $\partial\Omega = \Gamma$. Here Ω_c is the conducting domain, and a source current density $\mathbf{J}_0$ is given in $\Omega_A \setminus \overline{\Omega}_c$. We denote by $\Omega_D := \Omega \setminus \overline{\Omega}_c$ the subdomain of Ω occupied by dielectric material with $\varrho = 0$ and constant μ_A and ε_A in $\Omega_A \setminus \overline{\Omega}_c$ and constant μ_0, ε_0 in the free space $\Omega \setminus \overline{\Omega}_A$. The approach is based on the $(\boldsymbol{A}, \varphi) - (\boldsymbol{A}) - (\psi)$ formulation in the conductor Ω_c and dielectric material $\Omega_D = \Omega \setminus \overline{\Omega}_c$, respectively. Both $\boldsymbol{A}$'s are magnetic vector potentials, while φ and ψ are respectively the electric and magnetic scalar potentials.

For given $\mathbf{J}_0$ in $\Omega_A \setminus \overline{\Omega}_c$ and the homogeneous boundary condition $\boldsymbol{n} \times \mathbf{H} = \mathbf{0}$ on Γ, the time harmonic Maxwell equations (6.4.1)–(6.4.5) for the eddy current problem now read:

$$\mathbf{curl}\,\mathbf{E} = \begin{cases} i\omega\mu_c\mathbf{H} & \text{in } \Omega_c, \\ i\omega\mu_A\mathbf{H} & \text{in } \Omega_A \setminus \overline{\Omega}_c, \\ i\omega\mu_0\mathbf{H} & \text{in } \Omega_\psi = \Omega \setminus \overline{\Omega}_A; \end{cases} \tag{6.11.1}$$

$$\mathbf{curl}\,\mathbf{H} = \begin{cases} \sigma\mathbf{E} & \text{in } \Omega_c \\ \mathbf{J}_0 & \text{in } \Omega_A \setminus \overline{\Omega}_c \\ \mathbf{0} & \text{in } \Omega_\psi. \end{cases} \tag{6.11.2}$$

$\mathbf{J}_0$ is the given electric current distribution with $\operatorname{div} \mathbf{J}_0 = 0$ and

$$\operatorname{div} \boldsymbol{D} = 0 \quad \text{in } \Omega_D, \tag{6.11.3}$$

where

$$\boldsymbol{D} = \begin{cases} \varepsilon_c\mathbf{E} & \text{in } \Omega_c \\ \varepsilon_A\mathbf{E} & \text{in } \Omega_A \setminus \overline{\Omega}_c, \\ \varepsilon_0\mathbf{E} & \text{in } \Omega_\psi. \end{cases} \tag{6.11.4}$$

Moreover,

$$\operatorname{div} \mathbf{B} = 0 \quad \text{in } \Omega, \tag{6.11.5}$$

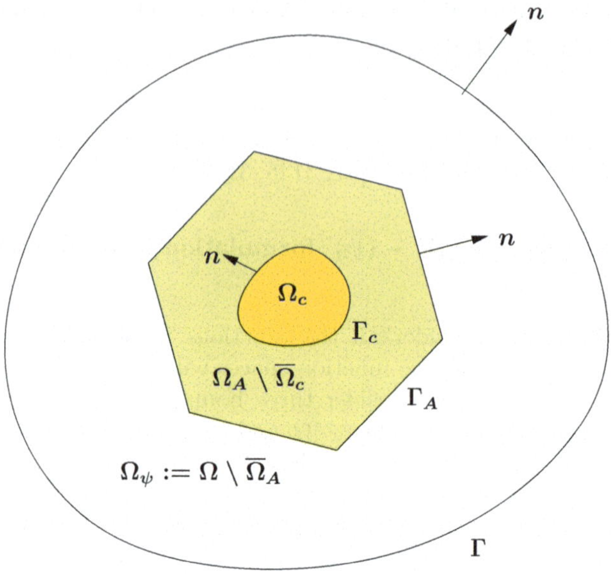

Fig. 6.2. An eddy current configuration

where

$$\mathbf{B} = \begin{cases} \mu_c \mathbf{H} & \text{in } \Omega_c, \\ \mu_A \mathbf{H} & \text{in } \Omega_A \setminus \overline{\Omega}_c, \\ \mu_0 \mathbf{H} & \text{in } \Omega_\psi. \end{cases} \tag{6.11.6}$$

From Equation (6.11.5) in Ω_c we introduce a vector potential $\boldsymbol{A}$ such that

$$\mathbf{B} = \mathbf{curl}\, \boldsymbol{A} \quad \text{in } \Omega_c.$$

Then from Equation (6.11.1) we have

$$\mathbf{curl}\, (\mathbf{E} - i\omega \boldsymbol{A}) = \mathbf{0} \quad \text{in } \Omega_c.$$

This implies that there exists a scalar potential φ such that

$$\mathbf{E} = i\omega \boldsymbol{A} + i\omega\, \mathbf{grad}\, \varphi \quad \text{in } \Omega_c.$$

Then from Equation (6.11.2) we have

$$\mathbf{curl}\, \frac{1}{\mu_c} \mathbf{curl}\, \boldsymbol{A} - \sigma i\omega (\boldsymbol{A} + \mathbf{grad}\, \varphi) = \mathbf{0} \quad \text{in } \Omega_c, \tag{6.11.7}$$

and, consequently,

$$\operatorname{div} \{\sigma i\omega (\boldsymbol{A} + \mathbf{grad}\, \varphi)\} = 0 \quad \text{in } \Omega_c. \tag{6.11.8}$$

In $\Omega_A \setminus \overline{\Omega}_c$, again from Equation (6.11.5) there exists a vector potential $\boldsymbol{A}$ with

$$\mathbf{curl}\,\boldsymbol{A} = \mathbf{B}.$$

Hence, from Equation (6.11.2) we arrive at

$$\mathbf{curl}\,\frac{1}{\mu_A}\,(\mathbf{curl}\,\boldsymbol{A}) = \mathbf{J}_0 \quad \text{in} \;\; \Omega_A \setminus \overline{\Omega}_c. \tag{6.11.9}$$

Finally, in $\Omega_\psi = \Omega \setminus \overline{\Omega}_A$, again from Equation (6.11.2), there exists a scalar potential ψ such that

$$\mathbf{H} = \omega\,\mathbf{grad}\,\psi$$

and from Equation (6.11.5) we obtain from div $\mathbf{B} = 0$ finally

$$\mathrm{div}(\mu_0\,\mathbf{grad}\,\psi) = 0 \quad \text{in} \;\; \Omega_\psi = \Omega \setminus \overline{\Omega}_A. \tag{6.11.10}$$

In order to ensure uniqueness of $\boldsymbol{A}$ in Ω_A, we choose the simple **Coulomb's gauge function**, namely

$$\mathrm{div}\boldsymbol{A} = 0 \quad \text{in} \;\; \Omega_A \setminus \Gamma_c. \tag{6.11.11}$$

In summery, our '$(\boldsymbol{A},\varphi) - (\boldsymbol{A}) - (\psi)$' formulation consists of the following system of Maxwell equations:

$$\begin{cases} \mathbf{curl}\,\frac{1}{\mu_c}\;\mathbf{curl}\,\boldsymbol{A} - \sigma i\omega(\boldsymbol{A} + \mathbf{grad}\,\varphi) = \mathbf{0}, \\ \qquad \mathrm{div}\{\sigma i\omega(\boldsymbol{A} + \mathbf{grad}\,\varphi)\} = 0 \quad \text{in} \;\; \Omega_c; \\ \qquad\qquad \mathbf{curl}\,\frac{1}{\mu_A}\mathbf{curl}\,\boldsymbol{A} = \mathbf{J}_0 \quad \text{in} \;\; \Omega_A \setminus \overline{\Omega}_c; \\ \qquad\qquad \mathrm{div}(\mu_0\,\mathbf{grad}\;\psi) = 0 \quad \text{in} \;\; \Omega \setminus \overline{\Omega}_A; \end{cases} \tag{6.11.12}$$

together with the Coulomb gauge condition

$$\mathrm{div}\boldsymbol{A} = 0 \quad \text{in} \;\; \Omega_A \setminus \Gamma_c. \tag{6.11.13}$$

For the transmission conditions on Γ_c, we have

$$\begin{cases} \boldsymbol{n} \times \frac{1}{\mu_c}\mathbf{curl}\,\mathbf{A}^- = \boldsymbol{n} \times \frac{1}{\mu_A}\mathbf{curl}\,\mathbf{A}^+, \\ \boldsymbol{n} \cdot \sigma i\omega(\mathbf{A}^- + \mathbf{grad}\,\varphi^-) = 0, \\ \boldsymbol{n} \cdot \mathbf{A}^- - \boldsymbol{n} \cdot \mathbf{A}^+ = 0. \end{cases} \tag{6.11.14}$$

On the interface Γ_A, we have

$$\begin{cases} \boldsymbol{n} \times \frac{1}{\mu_A}\mathbf{curl}\,\mathbf{A}^- = \boldsymbol{n} \times \omega\,\mathbf{grad}\,\psi^+, \\ \boldsymbol{n} \cdot \mathbf{curl}\,\mathbf{A}^- = \boldsymbol{n} \cdot \omega\mu_0\,\mathbf{grad}\,\psi^+, \\ \boldsymbol{n} \cdot \mathbf{A}^- = 0. \end{cases} \tag{6.11.15}$$

Finally, we impose the boundary condition on Γ:

$$\boldsymbol{n} \cdot \mathbf{H}|_{\Gamma} = h \in H^{-1/2}(\Gamma) \tag{6.11.16}$$

We now add an existence theorem for the weak solution of the eddy current problem in the bounded domain Ω.

Let us first specify the function spaces in the domains and on the Lipschitz boundaries:

$$\begin{aligned}
\mathcal{Z} &= \mathbf{H}_0(\mathrm{div}, \Omega_A) \cap \mathbf{H}(\mathbf{curl}, \Omega_A) \\
&:= \{\boldsymbol{Z} : \boldsymbol{Z} \cdot \boldsymbol{n}_A = 0 \quad \text{on} \quad \Gamma_A \quad \text{and} \quad \mathrm{div}\, \boldsymbol{Z} \in L^2(\Omega_A)\,, \; \mathbf{curl}\; \boldsymbol{Z} \in \mathbf{L}^2(\Omega_A)\} \\
\mathcal{M} &= \mathbf{H}^1(\Omega_c)/\mathbb{C}; \\
H^1_\Gamma(\Omega_\psi) &:= \{\psi \in H^1(\Omega_\psi) \wedge \mathbf{grad}_{\,\Gamma}\psi := \boldsymbol{n} \times \mathbf{grad}\, \psi \times \boldsymbol{n} = \mathbf{0} \quad \text{on} \quad \Gamma\}.
\end{aligned}$$

Moreover, we also need the trace spaces of $\mathbf{H}(\mathbf{curl}, \Omega_A)$. Recall from Section 4.5 that

$$\gamma_t(\boldsymbol{u}) := \boldsymbol{n} \times \boldsymbol{u}|_{\Gamma_A} \quad \text{and} \quad \pi_t(\boldsymbol{u}) := -\boldsymbol{n} \times (\boldsymbol{n} \times \boldsymbol{u}|_{\Gamma_A})$$

denote respectively the tangential trace and projection operators. Then the operators

$$\begin{aligned}
\gamma_t \;&:\; \mathbf{H}(\mathbf{curl}, \Omega_A) \to \mathbf{H}_{||}^{-1/2}(\mathrm{div}_\Gamma, \Gamma_A) \\
\text{and } \pi_t \;&:\; \mathbf{H}(\mathbf{curl}, \Omega_A) \to \mathbf{H}_{\perp}^{-1/2}(\mathbf{curl}_{\,\Gamma}, \Gamma_A)
\end{aligned}$$

are linear, continuous, surjective and possess continuous right–inverses, where

$$\begin{aligned}
\mathbf{H}_{||}^{-1/2}(\mathrm{div}_\Gamma, \Gamma_A) &:= \{\boldsymbol{\lambda} \in \mathbf{H}_{||}^{-1/2}(\Gamma_A) \,|\, \mathrm{div}_\Gamma(\boldsymbol{\lambda}) \in H^{-1/2}(\Gamma_A)\}, \\
\mathbf{H}_{\perp}^{-1/2}(\mathbf{curl}_{\,\Gamma}, \Gamma_A) &:= \{\boldsymbol{\lambda} \in \mathbf{H}_{\perp}^{-1/2}(\underset{\Gamma}{\mathrm{curl}}(\boldsymbol{\lambda}) \in H^{-1/2}(\Gamma_A)\},
\end{aligned}$$

are the trace spaces for $\mathbf{H}(\mathbf{curl}, \Omega)$ equipped with graph norms (see Theorem 4.5.3). The duality pairing between $\mathbf{H}_{||}^{-1/2}(\mathrm{div}_\Gamma, \Gamma_A)$ and $\mathbf{H}_{\perp}^{-1/2}(\mathbf{curl}_{\,\Gamma}, \Gamma_A)$ is well defined by

$$\int_{\Gamma_A} \gamma_t(\boldsymbol{u}) \cdot \pi_t(\boldsymbol{v}) ds = \int_{\Omega_A} (\mathbf{curl}\; \boldsymbol{u} \cdot \boldsymbol{v} - \boldsymbol{u} \cdot \mathbf{curl}\; \boldsymbol{v}) d\boldsymbol{y} \quad \text{for} \quad \boldsymbol{u}, \boldsymbol{v} \in \mathbf{H}(\mathbf{curl}, \Omega_A). \tag{6.11.17}$$

Following Rodriguez [365], we now state and prove the following

Lemma 6.11.1. (See Rodriguez [365, Theorem 6, p.104–107])

Let $\mathbf{J}_0 \in \widetilde{\mathbf{H}}^{-1}(\Omega_A \setminus \overline{\Omega^c})$ *and* $h \in H^{-1/2}(\Gamma)$ *be given and* $\mathbf{E}$ *and* $\mathbf{H}$ *be a solution of the Maxwell equations* (6.11.1), (6.11.2) *defining* $(\mathbf{A}, \varphi, \psi) \in \mathcal{Z} \times \mathcal{M} \times H^1_\Gamma(\Omega_\psi)$ *by* $\mathbf{E} = i\omega \boldsymbol{A} + i\omega\, \mathbf{grad}\, \varphi$ *and* (6.11.7) *in* Ω_c, (6.11.9) *in* $\Omega_A \setminus \overline{\Omega}_c$ *and* ψ *with* $\omega\, \mathbf{grad}\, \psi = \mathbf{H}$ *satisfying* (6.11.11), (6.11.12) *and*

(6.11.13) in $\Omega \setminus \Gamma_c$ together with the transmission and boundary conditions (6.11.14), (6.11.15) and $\mathbf{H} \cdot \boldsymbol{n} = h$ on Γ.

Then $(\boldsymbol{A}, \varphi, \psi)$ fulfills weakly the variational equation

$$\mathcal{A}\big((\boldsymbol{A}, \varphi, \psi),\, (\boldsymbol{Z}, u, \eta)\big) = \int\limits_{\Omega_A \setminus \overline{\Omega}_c} \mathbf{J}_0 \cdot \overline{\boldsymbol{Z}} d\boldsymbol{y} + \omega \int\limits_{\Gamma} \overline{\eta} \mu_0 h ds \qquad (6.11.18)$$

for all $(\boldsymbol{Z}, u, \eta) \in \mathcal{Z}_0 \times \mathcal{M} \times \overset{\circ}{H}{}^1_{\Gamma}(\Omega_\psi)$ with

$$\mathcal{Z}_0 := \{\boldsymbol{Z} \in \mathcal{Z} \quad \text{satisfying } \mathbf{curl} \frac{1}{\mu_A} \operatorname{curl} \boldsymbol{Z} = \mathbf{0} \quad \text{in } \Omega \setminus \overline{\Omega}_c\},$$
$$\overset{\circ}{H}{}^1_{\Gamma}(\Omega_\psi) := \{\psi \in H^1_{\Gamma}(\Omega_\psi) \quad \text{with } \mathbf{grad}\, \psi \cdot \boldsymbol{n} = 0 \quad \text{on } \Gamma\}$$

and where the bilinear form $\mathcal{A}$ is given by

$$\begin{aligned}
&\mathcal{A}\big((\boldsymbol{A}, \varphi, \psi),\, (\boldsymbol{Z}, u, \eta)\big) \\
&:= \int\limits_{\Omega_A} \frac{1}{\mu} \{\mathbf{curl}\ \boldsymbol{A} \cdot \mathbf{curl}\ \overline{\boldsymbol{Z}} + (\operatorname{div} \boldsymbol{A}) \cdot (\operatorname{div} \overline{\boldsymbol{Z}})\} d\boldsymbol{y} \\
&\quad + \omega^2 \int\limits_{\Omega_\psi} \mu_0 \mathbf{grad}\ \psi \cdot \mathbf{grad}\, \overline{\eta}\, d\boldsymbol{y} \\
&\quad - i\omega \int\limits_{\Omega_c} \sigma(\boldsymbol{A} + \mathbf{grad}\ \varphi) \cdot (\overline{\boldsymbol{Z}} + \mathbf{grad}\ \overline{u})\, d\boldsymbol{y} \qquad (6.11.19) \\
&\quad - \omega \int\limits_{\Gamma_A} (\mathbf{grad}\ \psi \times \boldsymbol{n}) \cdot \pi_t(\overline{\boldsymbol{Z}}) ds + \omega \int\limits_{\Gamma_A} (\mathbf{grad}\ \overline{\eta} \times \boldsymbol{n}) \cdot \pi_t(\boldsymbol{A}) ds,
\end{aligned}$$

where μ is defined by (6.11.24) below.

Proof: For the proof let us begin with the variational equations in Ω_c by testing (6.11.8) with $\overline{u}$, i.e.

$$\int\limits_{\Omega_c} \operatorname{div}\{\sigma i\omega(\boldsymbol{A} + (\mathbf{grad}\ \varphi)\} \overline{u} d\boldsymbol{y} = 0.$$

Integration by parts gives

$$\begin{aligned}
&\int\limits_{\Gamma_c} (\sigma i\omega \boldsymbol{A}) \overline{u} \cdot \boldsymbol{n}_c ds - \int\limits_{\Omega_c} i\omega\sigma \boldsymbol{A} \cdot (\mathbf{grad}\ \overline{u}) d\boldsymbol{y} \\
&+ \int\limits_{\Gamma_c} \sigma i\omega\, \mathbf{grad}\ \varphi \overline{u} \cdot \boldsymbol{n}_c ds - \int\limits_{\Omega_c} i\omega\sigma \mathbf{grad}\ \varphi \cdot (\mathbf{grad}\ \overline{u}) d\boldsymbol{y} = 0;
\end{aligned}$$

and (6.11.14) in Ω_c yields

$$\int_{\Omega_c} i\omega\sigma(\boldsymbol{A}+\mathbf{grad}\,\varphi)\cdot(\mathbf{grad}\,\overline{u})d\boldsymbol{y}=0 \quad \text{for all } \ u\in H^1(\Omega_c). \tag{6.11.20}$$

In Ω_c we also have the first of the equations in (6.11.12) which in weak form with test fields $\overline{\boldsymbol{Z}}\in\mathcal{Z}:=\mathbf{H}_0(\mathrm{div},\Omega_A)\cap\mathbf{H}(\mathbf{curl},\Omega_A)$ reads as

$$\int_{\Omega_c}\{(\mathbf{grad}\,\frac{1}{\mu_c})\times\mathbf{curl}\,\boldsymbol{A}\}\cdot\overline{\boldsymbol{Z}}\,d\boldsymbol{y}+\int_{\Omega_c}\frac{1}{\mu_c}(\mathbf{curl}\,\mathbf{curl}\,\boldsymbol{A})\cdot\overline{\boldsymbol{Z}}\,d\boldsymbol{y}$$
$$-\int_{\Omega_c} i\omega\sigma(\boldsymbol{A}+\mathbf{grad}\,\varphi)\cdot\overline{\boldsymbol{Z}}\,d\boldsymbol{y}=0.$$

With Green's first identity we find

$$\int_{\Omega_c}\{(\mathbf{grad}\,\frac{1}{\mu_c})\times\mathbf{curl}\,\boldsymbol{A}\}\cdot\overline{\boldsymbol{Z}}\,d\boldsymbol{y}+\int_{\Omega_c}\{(\mathbf{grad}\,\frac{1}{\mu_c})\times\overline{\boldsymbol{Z}}\}\cdot\mathbf{curl}\,\boldsymbol{A}\,d\boldsymbol{y}$$
$$+\int_{\Omega_c}\frac{1}{\mu_c}\mathbf{curl}\,\boldsymbol{A}\cdot\mathbf{curl}\,\overline{\boldsymbol{Z}}\,d\boldsymbol{y}-\int_{\Omega_c} i\omega\sigma(\boldsymbol{A}+\mathbf{grad}\,\varphi)\cdot\overline{\boldsymbol{Z}}\,d\boldsymbol{y}$$
$$+\int_{\partial\Omega_c}\boldsymbol{n}\cdot\{\frac{1}{\mu_c}(\mathbf{curl}\,\boldsymbol{A})\times\overline{\boldsymbol{Z}}\}ds=0.$$

The commutator rule for the scalar triple product implies cancelling of the first two integrals and for the last surface integral with Equations (6.11.13) and (6.11.12) we have the relation

$$\int_{\Omega_c}\frac{1}{\mu_c}\{\mathbf{curl}\,\boldsymbol{A}\cdot\mathbf{curl}\,\overline{\boldsymbol{Z}}+(\mathrm{div}\,\boldsymbol{A})\cdot(\mathrm{div}\,\overline{\boldsymbol{Z}})\}d\boldsymbol{y} \tag{6.11.21}$$
$$-\int_{\Omega_c} i\omega\sigma(\boldsymbol{A}+\mathbf{grad}\,\varphi)\cdot\overline{\boldsymbol{Z}}\,d\boldsymbol{y}+\int_{\Gamma_c}(\boldsymbol{n}\times\frac{1}{\mu_c}\mathbf{curl}\,\boldsymbol{A})\cdot\overline{\boldsymbol{Z}}ds=0.$$

Next, in $\Omega_A\setminus\overline{\Omega}_c$ we have from (6.11.2) and (6.11.7) the relation

$$\mathbf{curl}\,(\frac{1}{\mu_A}\mathbf{curl}\,\boldsymbol{A})=\mathbf{J}_0$$

which we test also with $\overline{\boldsymbol{Z}}\in\mathcal{Z}$:

$$\int_{\Omega_A\setminus\overline{\Omega}_c}\mathbf{curl}\,(\frac{1}{\mu_{\boldsymbol{A}}}\mathbf{curl}\,\boldsymbol{A})\cdot\overline{\boldsymbol{Z}}\,d\boldsymbol{y}=\int_{\Omega_A\setminus\overline{\Omega}_c}\mathbf{J}_0\cdot\overline{\boldsymbol{Z}}\,d\boldsymbol{y}.$$

Again with Green's first identity and the chain rule together with the commutator property of the scalar triple product we obtain

$$\int_{\Omega_A\setminus\overline{\Omega}_c} \frac{1}{\mu_{\boldsymbol{A}}}\mathbf{curl}\,\boldsymbol{A}\cdot\mathbf{curl}\,\overline{\boldsymbol{Z}}\,d\boldsymbol{y} + \int_{\partial(\Omega_A\setminus\overline{\Omega}_c)} ((\frac{1}{\mu_{\boldsymbol{A}}}\mathbf{curl}\,\boldsymbol{A})\times\overline{\boldsymbol{Z}})\cdot\boldsymbol{n}\,ds$$
$$= \int_{\Omega_A\setminus\overline{\Omega}_c} \mathbf{J}_0\cdot\overline{\boldsymbol{Z}}\,d\boldsymbol{y} \quad \text{for all } \boldsymbol{Z}\in\mathcal{Z}. \tag{6.11.22}$$

On Γ_A, (6.11.14) gives

$$\int_{\Gamma_A} \boldsymbol{n}\cdot\frac{1}{\mu_{\boldsymbol{A}}}((\mathbf{curl}\,\boldsymbol{A})\times\overline{\boldsymbol{Z}})ds = \int_{\Gamma_A}(\boldsymbol{n}\times\frac{1}{\mu_{\boldsymbol{A}}}\mathbf{curl}\,\boldsymbol{A})\cdot\overline{\boldsymbol{Z}}\,ds$$
$$= \int_{\Gamma_A}(\boldsymbol{n}\times\omega\mathbf{grad}\,\psi^+)\cdot\overline{\boldsymbol{Z}}\,ds$$

and on Γ_c:

$$-\int_{\Gamma_c} \boldsymbol{n}\cdot\frac{1}{\mu_{\boldsymbol{A}}}((\mathbf{curl}\,\boldsymbol{A})\times\overline{\boldsymbol{Z}})\,ds = -\int_{\Gamma_c}(\boldsymbol{n}\times\frac{1}{\mu_{\boldsymbol{A}}}\mathbf{curl}\,\boldsymbol{A})\cdot\overline{\boldsymbol{Z}}\,ds$$
$$= -\int_{\Gamma_c}(\boldsymbol{n}\times\frac{1}{\mu_c}\mathbf{curl}\,\boldsymbol{A})\cdot\overline{\boldsymbol{Z}}\,ds.$$

Adding (6.11.20), (6.11.21) and (6.11.22) with (6.11.11) and (6.11.15) results in

$$\int_{\Omega_A}\frac{1}{\mu}\{\mathbf{curl}\,\boldsymbol{A}\cdot\mathbf{curl}\,\overline{\boldsymbol{Z}} + \operatorname{div}\boldsymbol{A}\cdot\operatorname{div}\overline{\boldsymbol{Z}}\}d\boldsymbol{y}$$
$$-i\omega\int_{\Omega_c}\sigma(\boldsymbol{A}+\mathbf{grad}\,\varphi)\cdot(\overline{\boldsymbol{Z}}+\mathbf{grad}\,\overline{u})\,d\boldsymbol{y}$$
$$+\int_{\Gamma_A}(\boldsymbol{n}\times\omega\,\mathbf{grad}\,\psi^+)\cdot\overline{\boldsymbol{Z}}\,ds = \int_{\Omega_A\setminus\overline{\Omega}_c}\mathbf{J}_0\cdot\overline{\boldsymbol{Z}}\,d\boldsymbol{y} \tag{6.11.23}$$

for all $u\in H^1(\Omega_c)$, $\boldsymbol{Z}\in\mathcal{Z}(\Omega_A)$ and with

$$\mu = \begin{cases}\mu_c & \text{in } \Omega_c,\\ \mu_A & \text{in } \Omega_A\setminus\Omega_c\end{cases}. \tag{6.11.24}$$

Finally in Ω_ψ, (6.11.10) yields with integration by parts and (6.11.15):

$$\omega^2 \int\limits_{\Omega_\psi} \mu_0 \,\mathbf{grad}\, \psi \cdot \mathbf{grad}\, \overline{\eta}\, d\boldsymbol{y} = \omega^2 \int\limits_{\partial\Omega_\psi} \overline{\eta}\mu_0 \,\mathbf{grad}\, \psi \cdot \boldsymbol{n}\, ds$$

$$= \omega^2 \int\limits_{\Gamma} \overline{\eta}\, \mu_0 \,\mathbf{grad}\, \psi \cdot \boldsymbol{n}\, ds \tag{6.11.25}$$

$$- \omega \int\limits_{\Gamma_A} \overline{\eta}\, \mathbf{curl}\, \boldsymbol{A} \cdot \boldsymbol{n}\, ds \quad \text{for all} \quad \eta \in H^1_\Gamma(\Omega_\psi).$$

For the last integral in (6.11.25) we apply (6.11.17) and obtain:

$$\int\limits_{\Gamma_A} \overline{\eta}\, \mathbf{curl}\, \boldsymbol{A} \cdot \boldsymbol{n} ds = \int\limits_{\Omega_A} \mathrm{div}\, (\overline{\eta}\, \mathbf{curl}\, \boldsymbol{A})\, d\boldsymbol{y}$$

$$= \int\limits_{\Omega_A} \mathbf{grad}\, \overline{\eta} \cdot \mathbf{curl}\, \boldsymbol{A} d\boldsymbol{y} = \int\limits_{\Omega_A} \mathrm{div}(\boldsymbol{A} \times \mathbf{grad}\, \overline{\eta}) d\boldsymbol{y}$$

$$= - \int\limits_{\Gamma_A} \gamma_t(\,\mathbf{grad}\, \overline{\eta}) \cdot \pi_t(\boldsymbol{A}) ds.$$

Hence, (6.11.25) becomes

$$\omega^2 \int\limits_{\Omega\psi} \mu_0 \mathbf{grad}\, \psi \cdot \mathbf{grad}\, \overline{\eta}\, d\boldsymbol{y} - \omega \int\limits_{\Gamma_A} \gamma_t(\mathbf{grad}\, \overline{\eta}) \cdot \pi_t \boldsymbol{A} ds = \omega \int\limits_{\Gamma} \overline{\eta}\mu_0 \mathbf{H} \cdot \boldsymbol{n} ds. \tag{6.11.26}$$

Finally by adding (6.11.23) and (6.11.26) we find the desired variational equation (6.11.18) for $(\boldsymbol{A}, \varphi, \psi) \in \mathcal{Z} \times \mathcal{M} \times H^1_\Gamma(\Omega_\psi)$. and all test functions $(\boldsymbol{Z}, u, \eta) \in \mathcal{Z}_0 \times \mathcal{M} \times \overset{\circ}{H}{}^1_\Gamma(\Omega_\psi)$. ■

In order to prove continuity and coerciveness of $\mathcal{A}$ we shall need the following elementary lemma:

Lemma 6.11.2. *Let $c > 0$ and $\kappa \in (0, \frac{1}{2})$. Then*

$$(c + 2d)^2 > \kappa c^2 - 8\kappa d^2 \quad \textit{for all} \quad d \in \mathbb{R}. \tag{6.11.27}$$

Proof: Set $\frac{d}{c} = \frac{1}{2}(t - 1)$ or $t = (1 + 2\frac{d}{c})$. Then the inequality (6.11.27) is equivalent to

$$\Phi(t) := (2\kappa + 1)t^2 - 4\kappa t + \kappa > 0 \quad \text{for all} \quad t \in \mathbb{R}.$$
$$\Phi'(t_0) := 0 = 2(2\kappa + 1)t_0 - 4\kappa = 0$$

implies that with $t_0 = 2\kappa(2\kappa + 1)^{-1}$

$$\Phi(t) \geq \Phi(t_0) = \kappa(2\kappa + 1)^{-1} > 0 \quad \text{for all} \quad t \in \mathbb{R}$$

which is equivalent with (6.11.27). ■

Theorem 6.11.3. Rodriguez ([365, Theorem 6])
The sesquilinear form (6.11.9) *is continuous on* $\mathcal{Z} \times \mathcal{M} \times H^1_\Gamma(\Omega_\psi)$ *satisfying*

$$\begin{aligned}&|\mathcal{A}((\boldsymbol{A}, \varphi, \psi)\,;\,(\boldsymbol{Z}, u, \eta))| \qquad (6.11.28)\\ &\leq c(\|\boldsymbol{A}\|_{\mathcal{Z}} + \|\varphi\|_{\mathcal{M}} + \|\psi\|_{H^1_\Gamma(\Omega_\psi)}) \cdot (\|\boldsymbol{Z}\|_{\mathcal{Z}} + \|u\|_{\mathcal{M}} + \|\eta\|_{H^1_\Gamma(\Omega_\psi)}).\end{aligned}$$

Moreover, for $(\boldsymbol{Z}, u, \eta) \in \mathcal{Z}_0 \times \mathcal{M} \times \overset{o}{H}{}^1_\Gamma(\Omega_\psi)$, *we have the coerciveness estimate*

$$\mathcal{A}((\boldsymbol{Z}, u, \eta), (\boldsymbol{Z}, u, \eta)) \geq \kappa_0(\|\boldsymbol{Z}\|_{\mathcal{Z}} + \|u\|_{\mathcal{M}} + \|\eta\|_{H^1_\Gamma(\Omega_\psi)})^2. \qquad (6.11.29)$$

Proof: We begin with the inequality (6.11.28) and consider

$$\begin{aligned}&|\mathcal{A}((\boldsymbol{A}, \varphi.\psi), (\boldsymbol{Z}, u, \eta))|^2\\ &= |\int\limits_{\Omega_A} \frac{1}{\mu}\{\textbf{curl}\ \boldsymbol{A} \cdot \textbf{curl}\ \overline{\boldsymbol{Z}} + (\text{div}\,\boldsymbol{A}) \cdot (\text{div}\,\overline{\boldsymbol{Z}})\}d\boldsymbol{y} + \omega^2 \int\limits_{\Omega_\psi} \mu_0 \textbf{grad}\ \psi \cdot \textbf{grad}\ \overline{\eta} d\boldsymbol{y}\\ &\quad - i\omega \int\limits_{\Omega_c} \sigma(\boldsymbol{A} + \textbf{grad}\ \varphi) \cdot (\overline{\boldsymbol{Z}} + \textbf{grad}\ \overline{u})d\boldsymbol{y}\\ &\quad - \omega \int\limits_{\Gamma_A} \{(\textbf{grad}\ \psi \times \boldsymbol{n}) \cdot \pi_t(\overline{\boldsymbol{Z}}) + (\textbf{grad}\ \overline{\eta} \times \boldsymbol{n}) \cdot \pi_t(\boldsymbol{A})\}ds|^2\end{aligned}$$

$$\begin{aligned}&\leq \{\|\frac{1}{\mu}\|_{L^\infty(\Omega_A)}\|(\textbf{curl}\ \boldsymbol{A}\|_{L^2(\Omega_A)}\|\textbf{curl}\,\boldsymbol{Z}\|_{L^2(\Omega_A)} + \|div\,\boldsymbol{A}\|_{L^2(\Omega_A)}\|div\boldsymbol{Z}\|_{L^2(\Omega_A)})\\ &\quad + \omega^2\|\mu_0\|_{L^\infty(\Omega_\psi)}\|\textbf{grad}\ \psi\|_{L^2(\Omega_\psi)}\|\textbf{grad}\ \eta\|_{L^2(\Omega_\psi)}\\ &\quad + \omega\|\sigma\|_{L^\infty(\Omega_c)}\|\boldsymbol{A} + \textbf{grad}\ \varphi\|_{L^2(\Omega_c)}\|\boldsymbol{Z} + \textbf{grad}\ u\|_{L^2(\Omega_c)}\\ &\quad + \omega\|\textbf{grad}\,\psi\|_{H^{-\frac{1}{2}}(\Gamma_A)}\|\pi_t(\overline{\boldsymbol{Z}})\|_{H^{\frac{1}{2}}(\Gamma_A)} + \omega\|\textbf{grad}\,\eta\|_{H^{-\frac{1}{2}}(\Gamma_A)}\|\pi_t(\boldsymbol{A})\|_{H^{\frac{1}{2}}(\Gamma_A)}\}^2\\ &\leq c\|\frac{1}{\mu}\|^2_{L^\infty(\Omega_A)}\{\|\boldsymbol{A}\|^2_{\mathcal{Z}} + \|\boldsymbol{Z}\|^2_{\mathcal{Z}}\}^2\\ &\quad + \omega^4\|\mu_0\|^2_{l^\infty(\Omega_\psi)}\{\|\psi\|^2_{H^1_\Gamma(\Omega_\psi)} + \|\eta\|^2_{H^1_\Gamma(\Omega_\psi)}\}^2\\ &\quad + \omega^2\|\sigma\|^2_{L^\infty(\Omega_c)}\{\|\boldsymbol{A}\|^2_{L^2(\Omega_c)} + \|\varphi\|^2_{\mathcal{M}} + \|\boldsymbol{Z}\|^2_{L^2(\Omega_c)} + \|u\|^2_{\mathcal{M}}\}^2\\ &\quad + \omega^2\{\|\psi\|^2_{H^1_\Gamma(\Omega_\psi)} + \|\eta\|^2_{H^1_\Gamma(\Omega_\psi)} + \|\boldsymbol{A}\|^2_{\mathcal{Z}} + \|\boldsymbol{Z}\|^2_{\mathcal{Z}}\}^2\\ &\leq c'(\|\boldsymbol{A}\|_{\mathcal{Z}} + \|\varphi\|_{\mathcal{M}} + \|\psi\|_{H^1_\Gamma(\Omega_\psi)})^2(\|\boldsymbol{Z}\|_{\mathcal{Z}} + \|u\|_{\mathcal{M}} + \|\eta\|_{H^1_\Gamma(\Omega_\psi)})^2,\end{aligned}$$

which implies (6.11.28). The constant c' depends on ω, μ, μ_0 and σ.

For showing the coerciveness estimate (6.11.29) we shall employ Lemma 6.11.2.

Let

$$a := \int\limits_{\Omega_A} \frac{1}{\mu}(|\mathbf{curl}\ \boldsymbol{Z}|^2 + |\mathrm{div}\boldsymbol{Z}|^2)d\boldsymbol{y},$$

$$b := \int\limits_{\Omega_\psi} \mu_0|\mathbf{grad}\ \eta|^2 d\boldsymbol{y}\,,\ c := \int\limits_{\Omega_c} (|\boldsymbol{Z}|^2 + |\mathbf{grad}\ u|^2)d\boldsymbol{y},$$

$$e := \int\limits_{\Omega_c} \sigma\ \mathrm{Re}\,(\mathbf{grad}\ u\cdot \overline{\boldsymbol{Z}})d\boldsymbol{y}\,,\ f = \int\limits_{\Gamma_A} \mathrm{Im}(\mathbf{grad}\ \overline{\eta}\times \boldsymbol{n}_A\cdot \pi_\tau(\overline{\boldsymbol{Z}})ds$$

and $d := e + f$ be the constants in the Lemma 6.11.2. Then a simple computation shows that

$$\begin{aligned}\big|\mathcal{A}((\boldsymbol{Z},u,\eta),(\boldsymbol{Z},u,\eta)))\big|^2 &= (a+\omega^2 b)^2 + \omega^2(c+2d)^2\\ &\geq (a^2+\omega^4 b^2) + \omega^2(\kappa c^2 - 8\kappa d^2).\end{aligned}$$

Since $a \geq \frac{K}{\mu_{\max}}\|\boldsymbol{Z}\|^2_{\mathcal{Z}}$ and $b \geq \mu_{0\,\min}\|\mathbf{grad}\ \eta\|^2_{L^2(\Omega_\psi)}$,

$$\begin{aligned}\big|\mathcal{A}((\boldsymbol{Z},u,\eta)\,,\ (\boldsymbol{Z},u,\eta))\big|^2 \geq{}& \frac{K^2}{\mu^2_{\max}}\|\boldsymbol{Z}\|^4_{\mathcal{Z}} + \omega^4\mu^2_{0\,\min}\|\mathbf{grad}\ \eta\|^2_{L^2(\Omega_\psi)}\\ &+\omega^2\kappa\|\sigma^{\frac{1}{2}}\mathbf{grad}\ u\|^4_{L^2(\Omega_c)} - 16\omega^2\kappa(e^2+f^2).\end{aligned}$$

Moreover, we see that for all $\varepsilon > 0$,

$$e^2 = \Big(\int\limits_{\Omega_c} |\sigma\mathbf{grad}\ u\cdot\overline{\boldsymbol{Z}}|d\boldsymbol{y}\Big)^2 \leq \frac{\varepsilon}{2}\Big(\int\limits_{\Omega_c}\sigma|\mathbf{grad}\ u|^2 d\boldsymbol{y}\Big)^2 + \frac{1}{2\varepsilon}\Big(\int\limits_{\Omega_c}\sigma|\boldsymbol{Z}|^2 d\boldsymbol{y}\Big)^2$$

and with the trace theorem

$$\begin{aligned}f^2 &\leq \|\mathbf{grad}\ \eta\times\boldsymbol{n}_A\|^2_{H_{||}^{-\frac{1}{2}}(\mathrm{div}_\Gamma,\Gamma_A)}\cdot\|\pi_\tau(\overline{\boldsymbol{Z}})\|^2_{H_\perp^{-\frac{1}{2}}(\mathbf{curl}_\Gamma,\Gamma_A)}\\ &\leq c''\big(\|\mathbf{grad}\ \eta\|^4_{L^2(\Omega_\psi)} + \|\boldsymbol{Z}\|^4_{\mathcal{Z}}\big)\end{aligned}$$

with some constants c'' (independent of η and $\boldsymbol{Z}$), we get

$$\begin{aligned}&|\mathcal{A}((\boldsymbol{Z},u,\eta),(\boldsymbol{Z},u,\eta))|^2 \geq \Big(\frac{K^2}{\mu^2_{\max}} - 16\omega^2\frac{\kappa}{2\varepsilon} - C''\omega^2\kappa\Big)\|\boldsymbol{Z}\|^4_{\mathcal{Z}}\\ &+\Big(\omega^2\kappa - 16\omega^2\kappa\frac{\varepsilon}{2}\Big)\|\sigma^{\frac{1}{2}}\mathbf{grad}\ u\|^4_{\boldsymbol{L}^2(\Omega_c)} + (\omega^2\mu_{0\,\min} - 16c''\omega^2\kappa)\|\mathbf{grad}\ \eta\|^4_{\boldsymbol{L}^2(\Omega_\psi)}\\ &\geq \frac{K^2}{2\mu^2_{\max}}\|\boldsymbol{Z}\|^4_{\mathcal{Z}} + \frac{1}{2}\omega^2\kappa\|\sigma^{\frac{1}{2}}\mathbf{grad}\ u\|^4_{\boldsymbol{L}^2(\Omega_c)} + \frac{1}{2}\omega^2\mu^2_{0\,\min}\|\mathbf{grad}\ \eta\|^4_{\boldsymbol{L}^2(\Omega_\psi)}\end{aligned}$$

for $\varepsilon = \frac{1}{16}$ and $\kappa = \min\big\{K^2(32c''\omega\mu_{\max})^{-1}, (32c'')^{-1}\big\}$.
If $c_0^2 := \min\{\frac{K^2}{2\mu_{\max}}, \frac{1}{2}\omega^2\kappa, \frac{1}{2}\omega^2\mu_{0\,\min}\}$ then

$$|\mathcal{A}((\boldsymbol{Z},u,\eta),(\boldsymbol{Z},u,\eta))|^2 \geq c_0^2\{\|\boldsymbol{Z}\|_{\mathcal{Z}}^4 + \|u\|_{\mathcal{M}}^4 + \|\eta\|_{H^1(\Omega_\psi)}^4\}$$
$$\geq c_0^2 3^{-3}\big(\|\boldsymbol{Z}\|_{\mathcal{Z}} + \|u\|_{\mathcal{M}} + \|\eta\|_{H^1(\Omega_\psi)}\big)^4$$

which yields (6.11.29) with $\kappa_0 := c_0 3^{-\frac{3}{2}}$. This completes the proof of Theorem 6.11.3. ∎

Theorem 6.11.4. *Let* $\mathbf{J}_0 \in \widetilde{\mathbf{H}}^{-1}(\Omega_A \setminus \overline{\Omega}_c)$ *and* $h \in H^{-\frac{1}{2}}(\Gamma)$ *be given. Then there exists exactly one solution* $(\boldsymbol{A},\varphi,\psi) \in \mathcal{Z} \times \mathcal{M} \times H^1_\Gamma(\Omega_\psi)$ *of the variational equation* (6.11.18) *where the test function* $(\boldsymbol{Z},u,\eta)$ *traces the space* $\mathcal{Z}_0 \times \mathcal{M} \times H^1_\Gamma(\Omega_\psi)$. *Note that here* $\omega > 0$ *is a necessary assumption.*

Proof: The sesquilinear form $\mathcal{A}\big((\boldsymbol{A},\varphi,\psi),(\boldsymbol{Z},u,\eta)\big)$ is continuous on the Hilbert space $\mathcal{Z} \times \mathcal{M} \times H^1_\Gamma(\Omega_\psi)$ and coercive on the closed subspace $\mathcal{Z}_0 \times \mathcal{M} \times \overset{0}{H^1}{}_\Gamma(\Omega_\psi)$ due to Theorem 6.11.3. Then the Lax–Milgram theorem yields the existence of a unique solution to the variational equation as proposed. ∎

Remark 6.11.1: Since $\varphi \in \mathcal{M} = H^1(\Omega_c)/\mathbb{C}$, φ is uniquely determined only up to an additive complex constant, whereas $\mathbf{E} = i\omega\boldsymbol{A} + i\omega\,\mathbf{grad}\,\varphi$ is uniquely determined in Ω_c.

In terms of $(\boldsymbol{A},\varphi,\psi)$, the magnetic field $\mathbf{H}$ of the eddy current problem is given by

$$\mathbf{H} = \begin{cases} \frac{1}{\mu_c}\,\mathbf{curl}\,\boldsymbol{A} & \text{in } \Omega_c, \\ \frac{1}{\mu_A}\,\mathbf{curl}\,\boldsymbol{A} & \text{in } \Omega_A \setminus \overline{\Omega}_c, \\ \omega\;\;\mathbf{grad}\,\psi & \text{in } \Omega_\psi, \end{cases}$$

whereas the electrical field is only determined in Ω_c as

$$\mathbf{E} = i\omega\boldsymbol{A} + i\omega\,\mathbf{grad}\,\varphi \quad \text{in } \Omega_c.$$

In order to find $\mathbf{E}$ in Ω_D, one needs to solve a transmission problem in terms of $\mathbf{E}_c$ on Γ_c.

We will return to this topic when treating the exterior eddy current problem in the next section.

6.11.2 The $'(\boldsymbol{A},\varphi) - (\psi)'$ formulation in an unbounded domain

In this section, we consider an eddy current problem in an unbounded domain. For simplicity, let Ω_c be a bounded conducting domain in $\mathbb{R}^3$ with Lipschitz boundary Γ_c and $\Omega_D := \mathbb{R}^3 \setminus \overline{\Omega_c}$ is occupied with a lossless dielectric medium (see Figure 6.11.2). We assume that the electric conductivity σ and the magnetic permeability μ are in $L^\infty(\Omega_c)$ such that

$$\begin{cases} 0 < \mu_{min} \leq \mu_c \leq \mu'_{max} \\ 0 < \sigma_{min} \leq \sigma \leq \sigma_{max}, \end{cases} \quad \text{for } x \in \Omega_c.$$

Moreover, the applied current density $\mathbf{J}_0$ is assumed to be in $\boldsymbol{L}^2(\Omega_c)$. The magnetic permeability and the electric permittivity are assumed to be positive constants μ_0 and ε_0, respectively in Ω_D.

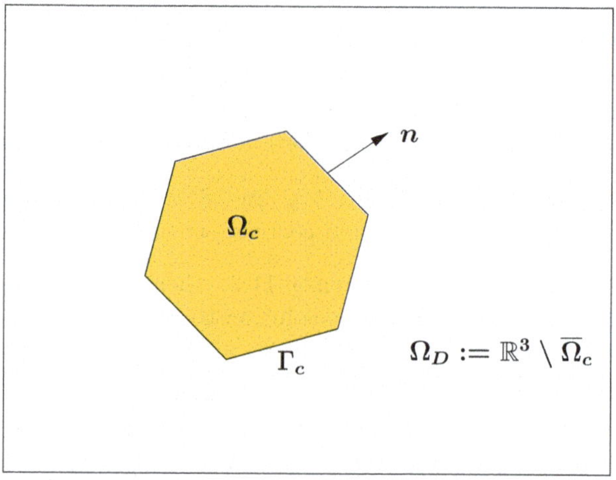

Fig. 6.3. An eddy current problem in unbounded domain.

For an exterior eddy current problem of this kind, it is most conveniently handled by a coupling of boundary and field equation methods (see, e.g., Gatica and Hsiao [139]). Our presentation follows the $'(\boldsymbol{A}, \varphi) - (\psi)'$ formulation from Rodriguez and Valli [366]. That is, we use the field equations in Ω_c for the magnetic vector potential $\boldsymbol{A}$, and for an electric scalar potential φ. In Ω_D, a symmetric boundary integral equation formulation for the scalar magnetic potential ψ is employed.

For given $\mathbf{J}_0$ in Ω_c, the time harmonic Maxwell equations of the magnetic field $\mathbf{H}$ and the electric field $\mathbf{E}$ for the eddy current problem read:

$$\mathbf{curl}\,\mathbf{E} = \begin{cases} i\,\omega\mu_c\mathbf{H} & \text{in} \quad \Omega_c, \\ i\,\omega\mu_0\mathbf{H} & \text{in} \quad \Omega_D, \end{cases} \tag{6.11.30}$$

$$\mathbf{curl}\,\mathbf{H} = \begin{cases} \sigma\mathbf{E} + \mathbf{J}_0 & \text{in} \quad \Omega_c, \\ \mathbf{0} & \text{in} \quad \Omega_D, \end{cases} \tag{6.11.31}$$

$$\operatorname{div}\mathbf{B} = 0, \quad \text{in} \quad \mathbb{R}^3, \tag{6.11.32}$$

$$\operatorname{div}\boldsymbol{D} = 0, \quad \text{in} \quad \Omega_D, \tag{6.11.33}$$

where Eq.(6.11.32) is the consequence of Eq.(6.11.30) and $\boldsymbol{D} = \varepsilon_0\,\mathbf{E}$ in Ω_D. For the boundary-field equation methods, we need also the transmission conditions:

$$\left.\begin{aligned} \boldsymbol{n}\cdot\mathbf{B}^- &= \boldsymbol{n}\cdot\mathbf{B}^+ \quad \text{on} \quad \Gamma_c, \\ \boldsymbol{n}\times\mathbf{H}^- &= \boldsymbol{n}\times\mathbf{H}^+ \quad \text{on} \quad \Gamma_c, \\ \boldsymbol{n}\times\mathbf{E}^- &= \boldsymbol{n}\times\mathbf{E}^+ \quad \text{on} \quad \Gamma_c \end{aligned}\right\} \tag{6.11.34}$$

together with the condition at infinity:

$$\mathbf{H}(\boldsymbol{x}) = \mathcal{O}(|\boldsymbol{x}|^{-1}), \quad \text{and} \;\; \mathbf{E}(\boldsymbol{x}) = \mathcal{O}(|\boldsymbol{x}|^{-1}) \tag{6.11.35}$$

For the $'(\boldsymbol{A},\varphi)-(\psi)'$ formulation, we see from Equation (6.11.32) in Ω_c, we may introduce a magnetic vector potential $\boldsymbol{A}$ such that

$$\mathbf{B} = \mathbf{curl}\;\boldsymbol{A} \quad \text{in} \quad \Omega_c.$$

Then from Equation (6.11.30) in Ω_c we have

$$\mathbf{curl}\,(\mathbf{E} - i\omega\boldsymbol{A}) = \mathbf{0}.$$

This implies that there exists an electric scalar potential φ such that

$$\mathbf{E} = i\omega(\boldsymbol{A} + \mathbf{grad}\,\varphi). \tag{6.11.36}$$

As a consequence of (6.11.31) in Ω_c, we have

$$\mathbf{curl}\;\frac{1}{\mu_c}\mathbf{curl}\;\boldsymbol{A} - i\sigma\omega(\boldsymbol{A} + \mathbf{grad}\,\varphi) = \mathbf{J}_0 \quad \text{in} \quad \Omega_c. \tag{6.11.37}$$

subject to the constraint

$$\text{div}\{-i\sigma\omega(\boldsymbol{A} + \mathbf{grad}\,\varphi)\} = \text{div}\,\mathbf{J}_0 \qquad \text{in} \qquad \Omega_c. \tag{6.11.38}$$

In order to have unique vector potential $\boldsymbol{A}$, as usual we choose the Coulomb gauge condition

$$\text{div}\;\boldsymbol{A} = 0 \quad \text{in} \quad \Omega_c.$$

Next, from Equation (6.11.31) in Ω_D, there exits a magnetic scalar potential ψ such that

$$\mathbf{H} = \mathbf{grad}\;\psi$$

and from Equation (6.11.31) we obtain

$$-\text{div}\;(\mu_0\;\mathbf{grad}\;\psi) = 0 \quad \text{or} \quad -\Delta\psi = 0 \quad \text{in} \quad \Omega_D, \tag{6.11.39}$$

since μ_0 is constant.

In summary, the $'(\boldsymbol{A},\varphi)-(\psi)'$ formulation for the eddy current problem consists of the following system of equations for the unknowns $(\boldsymbol{A},\varphi,\psi)$:

$$\left\{\begin{aligned} &\mathbf{curl}\,\frac{1}{\mu_c}\;\mathbf{curl}\;\boldsymbol{A} - i\sigma\omega(\boldsymbol{A} + \mathbf{grad}\;\varphi) = \mathbf{J}_0 \quad \text{in} \quad \Omega_c, \\ &\qquad \text{div}\,\{-i\sigma\omega(\boldsymbol{A} + \;\mathbf{grad}\;\varphi)\} = \text{div}\,\mathbf{J}_0 \quad \text{in} \quad \Omega_c, \\ &\qquad\qquad -\Delta\psi = 0 \quad \text{in} \quad \Omega_D \end{aligned}\right. \tag{6.11.40}$$

with the Coulomb gauge condition

$$\operatorname{div} \boldsymbol{A} = 0 \quad \text{in} \;\; \Omega_c, \tag{6.11.41}$$

where φ is unique up to an additive constant. The corresponding transmission conditions on Γ_c now read

$$\begin{cases} \boldsymbol{n} \times \dfrac{1}{\mu_c} \,\mathbf{curl}\, \boldsymbol{A}^- = \boldsymbol{n} \times \mathbf{grad}\, \psi^+, \\ \boldsymbol{n} \cdot \dfrac{1}{\mu_0} \,\mathbf{curl}\, \boldsymbol{A}^- = \dfrac{\partial}{\partial n} \psi^+, \\ \boldsymbol{n} \cdot \{-i\sigma\omega(\boldsymbol{A}^- + \mathbf{grad}\, \varphi^-)\} = \boldsymbol{n} \cdot \mathbf{J}_0^-, \\ \boldsymbol{n} \cdot \boldsymbol{A}^- = 0 \end{cases} \tag{6.11.42}$$

together with conditions at infinity

$$|\psi(\boldsymbol{x})| + |\mathbf{grad}\, \psi(\boldsymbol{x})| = \mathcal{O}(|\boldsymbol{x}|^{-1}) \quad \text{as} \quad |\boldsymbol{x}| \to \infty. \tag{6.11.43}$$

We are interested here in the weak solutions of the eddy problem consisting of Equations (6.11.40), (6.11.41), (6.11.42) and (6.11.43).

First, notice that we may remove (6.11.41) and consider the modified system of equations

$$\begin{cases} \mathbf{curl} \dfrac{1}{\mu_c} \,\mathbf{curl}\, \boldsymbol{A} - \dfrac{1}{\mu_{max}} \,\mathbf{grad}\, \operatorname{div} \boldsymbol{A} \\ \qquad -i\sigma\omega(\boldsymbol{A} + \mathbf{grad}\, \varphi) = \mathbf{J}_0 \quad \text{in} \quad \Omega_c, \\ \operatorname{div} \{-i\sigma\,\omega(\boldsymbol{A} + \mathbf{grad}\, \varphi)\} = \operatorname{div} \mathbf{J}_0 \quad \text{in} \quad \Omega_c, \\ -\Delta\psi = 0 \quad \text{in} \quad \Omega_D \\ \boldsymbol{n} \times \dfrac{1}{\mu_c} \,\mathbf{curl}\, \boldsymbol{A}^- = \boldsymbol{n} \times \mathbf{grad}\, \psi^+ \quad \text{on} \quad \Gamma_c \\ \boldsymbol{n} \cdot \dfrac{1}{\mu_0} \mathbf{curl}\, \boldsymbol{A}^- = \dfrac{\partial}{\partial n} \psi^+ \quad \text{on} \quad \Gamma_c \\ \boldsymbol{n} \cdot \{-i\sigma\omega(\boldsymbol{A}^- + \mathbf{grad}\, \varphi^-)\} = \boldsymbol{n} \cdot \mathbf{J}_0^- \quad \text{on} \quad \Gamma_c \\ \boldsymbol{n} \cdot \boldsymbol{A}^- = 0 \quad \text{on} \quad \Gamma_c \\ |\psi(\boldsymbol{x})| + |\,\mathbf{grad}\, \psi(\boldsymbol{x})| = \mathcal{O}(|\boldsymbol{x}|^{-1}) \quad \text{as} \quad |\boldsymbol{x}| \to \infty. \end{cases} \tag{6.11.44}$$

It is easy to see that this modified system and the original system of equations (6.11.40), (6.11.41), (6.11.42) and (6.11.43) are equivalent. In particular, any solution to (6.11.44) will satisfy (6.11.41). This can be seen as follows: By taking the divergence of the first equation in (6.11.44), denoted by $(6.11.44)_1$, we obtain that

$$\Delta\,(\operatorname{div} \boldsymbol{A}) = 0 \quad \text{in} \;\; \Omega_c$$

in view of $(6.11.44)_2$. Then taking the scalar product of $(6.11.44)_1$ with $\boldsymbol{n}$ and making use of the condition $(6.11.44)_6$, we see that

$$\frac{1}{\mu_{max}}\frac{\partial}{\partial n}\operatorname{div}\boldsymbol{A}|_{\Gamma_c} = \boldsymbol{n}\cdot\mathbf{curl}\,(\frac{1}{\mu_c}\mathbf{curl}\,\boldsymbol{A})|_{\Gamma_c} = -\operatorname{div}_{\Gamma_c}\{\boldsymbol{n}\times\frac{1}{\mu_c}\,\mathbf{curl}\,\boldsymbol{A}^-\}$$
$$= -\operatorname{div}_{\Gamma_c}\{\boldsymbol{n}\times\mathbf{grad}\,\psi^+\} = 0 \quad \text{on } \Gamma_c.$$

This implies div $\boldsymbol{A}$ is a solution of the Laplace equation in Ω_c and satisfies the homogeneous Neumann boundary condition on Γ_c. Hence div $\boldsymbol{A}$ is a constant and the constant is zero from the condition $(6.11.44)_7$.

To this end, it is worth mentioning that in order to determine solutions $\boldsymbol{A}$ (and φ) in Ω_c, in addition to the differential equations for $\boldsymbol{A}$ and φ in Ω_c, the natural boundary condition needed for $\boldsymbol{A}$ on Γ_c from variational point of view is the condition $(6.11.44)_4$ of (6.11.44), namely

$$\boldsymbol{n}\times\frac{1}{\mu_c}\mathbf{curl}\,\boldsymbol{A}^- = \boldsymbol{n}\times\mathbf{grad}\,\psi^+ = -\mathbf{curl}_{\Gamma_c}\,\psi^+, \quad \text{on } \Gamma_c,$$

which depends on $\psi^+ = \gamma(\psi)$, the trace of ψ on Γ_c. On the other hand, ψ is the solution of the Laplace equation in the unbounded domain Ω_D with a natural boundary condition in terms of the trace of $\boldsymbol{A}$, namely

$$\frac{\partial}{\partial n}\psi^+ = \boldsymbol{n}\cdot\frac{1}{\mu_0}\,\mathbf{curl}\,\boldsymbol{A}^- \quad \text{on } \Gamma_c,$$

which is known, if the solution $\boldsymbol{A}$ is available. Thus, the eddy current problem in the unbounded domain defined by (6.11.44) should be reduced to a nonlocal boundary value problem in the bounded domain Ω_c for the unknowns $\boldsymbol{A},\varphi$ satisfying PDEs in the domain Ω_c ,and the unknown $\psi^+ = \gamma(\psi)$ satisfying a suitable BIE on the boundary Γ_c, the nonlocal boundary condition.

If $\psi \in H_{loc}(\Omega_D)$ (or the weighted space $H^1(\Omega_D;\lambda)$, see Section 4.5) satisfies the Laplace equation $(6.11.44)_3$ of (6.11.44), we have the Green representation formula of the solution in terms of single– and double–layer boundary potentials, $\mathfrak{V}$ and $\mathfrak{W}$, respectively:

$$\psi(\boldsymbol{x}) = \int\limits_{\Gamma_c}\frac{\partial}{\partial n_{\boldsymbol{y}}}E(\boldsymbol{x},\boldsymbol{y})\psi^+(\boldsymbol{y})ds_{\boldsymbol{y}} - \int_{\Gamma_c}E(\boldsymbol{x},\boldsymbol{y})\frac{\partial}{\partial n}\psi^+(\boldsymbol{y})ds_{\boldsymbol{y}}$$
$$=: \mathfrak{W}(\psi^+)(\boldsymbol{x}) - \mathfrak{V}(\frac{\partial\psi^+}{\partial n})(\boldsymbol{x})\,,\ \boldsymbol{x}\in\Omega_D, \tag{6.11.45}$$

where $E(\boldsymbol{x},\boldsymbol{y}) = 1/(4\pi|\boldsymbol{x}-\boldsymbol{y}|)$ is the fundamental solution of $-\Delta$. The Cauchy data $(\psi^+, \partial\psi^+/\partial n)^T$ of the solution ψ satisfy the relations:

$$\begin{pmatrix}\psi^+\\ \\ \frac{\partial\psi^+}{\partial n}\end{pmatrix} = \begin{pmatrix}\frac{1}{2}I+K & -V\\ \\ -D & \frac{1}{2}I-K'\end{pmatrix}\begin{pmatrix}\psi^+\\ \\ \frac{\partial\psi^+}{\partial n}\end{pmatrix} \quad \text{on } \Gamma_c, \tag{6.11.46}$$

where the boundary integral operators K, V, D, K' are related to the limits of the boundary potentials $\mathfrak{V}$ and $\mathfrak{W}$ from Ω_D which are defined explicitly in

Section 1.4.1. Their mapping properties are given in Theorems 1.2.3 and 5.6.2. In terms of the transmission conditions $(6.11.44)_5$, we have from (6.11.46)

$$\psi^+(\boldsymbol{x}) = (\frac{1}{2}I + K)\psi^+(\boldsymbol{x}) - V(\boldsymbol{n}\cdot\frac{1}{\mu_0}\,\mathbf{curl}\,\boldsymbol{A}^-)(\boldsymbol{x})\,,\ \boldsymbol{x}\in\Gamma_c, \tag{6.11.47}$$

$$0 = D(\psi^+)(\boldsymbol{x}) + (\frac{1}{2}I + K')(\boldsymbol{n}\cdot\frac{1}{\mu_0}\mathbf{curl}\,\boldsymbol{A}^-)(\boldsymbol{x})\,,\ \boldsymbol{x}\in\Gamma_c. \tag{6.11.48}$$

As will be seen, in the transmission condition $(6.11.44)_4$, ψ^+ is replaced by the right hand of Equation (6.11.47), while Equation (6.11.48) serves as a nonlocal boundary condition and it is a boundary integral equation of the first kind for the unknown ψ^+ provided $\boldsymbol{n}\cdot\mu_0^{-1}\mathbf{curl}\,\boldsymbol{A}^+$ is given.

We are now in a position to define the nonlocal boundary value problem which are equivalent to the eddy current problem in the unbounded domain formulated in (6.11.44). Let us first specify the function spaces in the domain Ω_c and on the Lipschitz boundary Γ_c for the unknowns

$$(\boldsymbol{A},\varphi,\psi^+)\in\boldsymbol{\mathcal{Z}}\times H^1(\Omega_c)/\mathbb{C}\times H^{1/2}(\Gamma_c)/\mathbb{C},$$

where the function space $\boldsymbol{\mathcal{Z}}$ is defined as in Section 6.11.1 by

$$\boldsymbol{\mathcal{Z}} = \mathbf{H}(\mathbf{curl},\Omega_c)\cap\mathbf{H}_0(\mathrm{div},\Omega_c)$$

$$:= \{\boldsymbol{Z}\in\boldsymbol{L}^2(\Omega_c) : \boldsymbol{Z}\in\boldsymbol{L}^2(\Omega_c), \mathrm{div}\,\boldsymbol{Z}\in L^2(\Omega_c) \quad\text{and}\quad \boldsymbol{Z}\cdot\boldsymbol{n}=0 \quad\text{on}\quad \Gamma_c\}$$

equipped with the norm

$$||\boldsymbol{Z}||^2_{\boldsymbol{\mathcal{Z}}} := ||\boldsymbol{Z}||^2_{0,\Omega_c} + ||\mathrm{div}\boldsymbol{Z}||^2_{0,\Omega_c} + ||\mathbf{curl}\ \boldsymbol{Z}||^2_{0,\Omega_c}.$$

The nonlocal boundary value problem for the eddy current problem in the unbounded domain now reads: Given $\mathbf{J}_0\in\boldsymbol{L}^2(\Omega_c)$, find $(\boldsymbol{A},\varphi,\psi^+)\in\boldsymbol{\mathcal{E}}\times H^1(\Omega_c)/\mathbb{C}\times H^{1/2}(\Gamma_c)/\mathbb{C}$ such that

$$\left\{\begin{aligned}
\mathbf{curl}\frac{1}{\mu_c}\ \mathbf{curl}\ \boldsymbol{A} - \frac{1}{\mu_{max}}\mathbf{grad}\ \mathrm{div}\,\boldsymbol{A}&\\
-i\sigma\,\omega(\boldsymbol{A}+\mathbf{grad}\ \varphi) = \mathbf{J}_0 &\quad\text{in}\ \Omega_c,\\
\mathrm{div}\,\{-i\sigma\,\omega(\boldsymbol{A}+\mathbf{grad}\ \varphi)\} = \mathrm{div}\,\mathbf{J}_0 &\quad\text{in}\quad \Omega_c,\\
\mu_0\,D(\psi^+) + (\frac{1}{2}I+K')(\boldsymbol{n}\cdot\ \mathbf{curl}\ \boldsymbol{A}^-) = 0 &\quad\text{on}\ \Gamma_c,\\
\boldsymbol{n}\times\frac{1}{\mu_c}\mathbf{curl}\ \boldsymbol{A}^- = -\,\mathbf{curl}_{\Gamma_c}\ \psi^+ &\quad\text{on}\ \Gamma_c\quad\text{with}\\
\psi^+ = (\frac{1}{2}I+K)\psi^+ - V(\boldsymbol{n}\cdot\frac{1}{\mu_0}\mathbf{curl}\ \boldsymbol{A}^-) &\quad\text{on}\ \Gamma_c,\\
\boldsymbol{n}\cdot\{-i\sigma\omega(\boldsymbol{A}^- + \mathbf{grad}\ \varphi^-)\} = \boldsymbol{n}\cdot\mathbf{J}_0^- &\quad\text{on}\ \Gamma_c,\\
\boldsymbol{n}\cdot\boldsymbol{A}^- = 0 &\quad\text{on}\ \Gamma_c.
\end{aligned}\right. \tag{6.11.49}$$

We now consider the existence theorem for the weak solution of the nonlocal boundary value problem.

To obtain a variational formulation of the nonlocal boundary value problem (6.11.49), notice that from $(6.11.49)_1$, and $(6.11.49)_4$, we have that $\frac{1}{\mu_c}\mathbf{curl}\, \boldsymbol{A} \in \mathbf{H}(\mathbf{curl}, \Omega_c)$. Multiplying $(6.11.49)_1$ by the test function $\boldsymbol{Z} \in \boldsymbol{\mathcal{Z}}$ and integrating by parts lead to

$$\int_{\Omega_c} \mathbf{J}_0 \cdot \overline{\boldsymbol{Z}}\, d\boldsymbol{x} = \int_{\Omega_c} (\frac{1}{\mu_c}\,\mathbf{curl}\ \boldsymbol{A}) \cdot\ (\mathbf{curl}\ \overline{\boldsymbol{Z}})\, d\boldsymbol{x} + \frac{1}{\mu_{max}} \int_{\Omega_c} \operatorname{div} \boldsymbol{A} \operatorname{div} \overline{\boldsymbol{Z}}\, d\boldsymbol{x}$$
$$- i\omega \int_{\Omega_c} \sigma(\boldsymbol{A} + \mathbf{grad}\, \varphi) \cdot \overline{\boldsymbol{Z}}\, d\boldsymbol{x} + \int_{\Gamma_c} (\boldsymbol{n} \times \frac{1}{\mu_c}\, curl_\Gamma\, \boldsymbol{A}) \cdot \overline{\boldsymbol{Z}}\, ds$$
$$- \frac{1}{\mu_{max}} \int_{\Gamma_c} \operatorname{div} \boldsymbol{A}(\boldsymbol{n} \cdot \overline{\boldsymbol{Z}}) ds. \tag{6.11.50}$$

Since $\boldsymbol{Z} \in \boldsymbol{\mathcal{Z}}$, the last term on the right hand side vanishes, and from $(6.11.49)_4$, we see that

$$\int_{\Gamma_c} (\boldsymbol{n} \times \frac{1}{\mu_c}\mathbf{curl}\, \boldsymbol{A}) \cdot \overline{\boldsymbol{Z}}\, ds = - \int_{\Gamma_c} \mathbf{curl}_{\Gamma_c}\, \psi^+ \cdot \overline{\boldsymbol{Z}}\, ds = - \int_{\Gamma_c} \psi^+\, curl_{\Gamma_c} \overline{\boldsymbol{Z}}\, ds$$
$$= - \int_{\Gamma_c} \Big(\frac{1}{2}I + K)\psi^+ - \frac{1}{\mu_0} V(\boldsymbol{n} \cdot\ \mathbf{curl}\ \boldsymbol{A}^-\Big)\ \mathbf{curl}\ \overline{\boldsymbol{Z}} \cdot \boldsymbol{n} d\boldsymbol{x} \tag{6.11.51}$$
$$= -\Big\langle \boldsymbol{n} \cdot\ \mathbf{curl}\, \overline{\boldsymbol{Z}},\ \Big(\frac{1}{2}I + K)\psi^+ - \frac{1}{\mu_0} V(\boldsymbol{n} \cdot \mathbf{curl}\, \boldsymbol{A}^-)\Big)\Big\rangle_{\Gamma_c}.$$

Recall the notation $\langle\ ,\ \rangle_{\Gamma_c}$ denotes the duality pairing between $H^{-1/2}(\Gamma_c)$ and $H^{1/2}(\Gamma_c)$ (cf. Section 4.5).

Similarly, multiplying $(6.11.49)_2$ by the test function $u \in H(\Omega_c)/\mathbb{C}$ and integrating by parts, we obtain

$$\int_{\Gamma_c} \operatorname{div} \mathbf{J}_0\, \overline{u}\, d\boldsymbol{x} = \int_{\Gamma_c} \operatorname{div} \{-i\sigma\, \omega(\boldsymbol{A} + \mathbf{grad}\ \varphi)\} \overline{u}\, d\boldsymbol{x}$$
$$= i\,\omega \int_{\Omega_c} \sigma(\boldsymbol{A} +\ \mathbf{grad}\ \varphi)\, \mathbf{grad}\ \overline{u}\, dx - i\,\omega \int_{\Gamma_c} \sigma \boldsymbol{n} \cdot (\boldsymbol{A} +\ \mathbf{grad}\ \varphi) \overline{u}\, ds.$$

Hence, using $(6.11.49)_5$, we obtain

$$-i\,\omega \int_{\Omega_c} \sigma(\boldsymbol{A} +\ \mathbf{grad}\ \varphi)\, \mathbf{grad}\ \overline{u}\, d\boldsymbol{x} = -i\,\omega \int_{\Gamma_c} \sigma \boldsymbol{n} \cdot (\boldsymbol{A} +\ \mathbf{grad}\ \varphi) \overline{u}\, ds$$
$$- \int_{\Omega_c} \operatorname{div} \mathbf{J}_0 \overline{u} d\boldsymbol{x} = \int_{\Omega_c} \mathbf{J}_0 \cdot\ \mathbf{grad}\ \overline{u}\, d\boldsymbol{x}. \tag{6.11.52}$$

Finally, multiplying $(6.11.49)_3$ by the test function $\eta \in H^{1/2}(\Gamma_c)/\mathbb{C}$, and integrating over the boundary Γ_c yield

$$\Big\langle \mu_0\, D\psi^+ + \Big(\frac{1}{2}I + K'\Big)\Big(\boldsymbol{n}\cdot \mathbf{curl}\ \boldsymbol{A}\Big), \overline{\eta}\Big\rangle_{\Gamma_c} = 0. \tag{6.11.53}$$

Collecting (6.11.50) , (6.11.51), (6.11.52) and (6.11.53), we have the weak formulation of (6.11.49):

For given $\mathbf{J}_0 \in \boldsymbol{L}^2(\Omega_c)$, *find* $(\boldsymbol{A}, \varphi, \psi^+) \in \boldsymbol{\mathcal{Z}} \times H^1(\Omega_c)/\mathbb{C} \times H^{1/2}(\Gamma_c)/\mathbb{C}$ *such that*

$$\begin{cases} \displaystyle\int_{\Omega_c} (\frac{1}{\mu_c}\mathbf{curl}\ \boldsymbol{A}) \cdot (\mathbf{curl}\ \overline{\boldsymbol{Z}})\, d\boldsymbol{x} + \frac{1}{\mu_{max}} \int_{\Omega_c} (\operatorname{div} \boldsymbol{A})\,(\operatorname{div} \overline{\boldsymbol{Z}})\, d\boldsymbol{x} \\ \displaystyle\quad - i\omega \int_{\Omega_c} \sigma, (\boldsymbol{A} + \mathbf{grad}\ \varphi)\cdot \overline{\boldsymbol{Z}}\, d\boldsymbol{x} \\ \displaystyle + \Big\langle \boldsymbol{n}\cdot \mathbf{curl}\ \overline{\boldsymbol{Z}},\ -(\frac{1}{2}I + K\Big)\psi^+ + \frac{1}{\mu_0} V\Big(\boldsymbol{n}\cdot \mathbf{curl}\, \boldsymbol{A}\Big)\Big\rangle_{\Gamma_c} = \int_{\Omega_c} \mathbf{J}_0 \cdot \overline{\boldsymbol{Z}}\, d\boldsymbol{x}, \\ \displaystyle\quad - i\omega \int_{\Omega_c} \sigma (\boldsymbol{A} + \mathbf{grad}\ \varphi)\cdot (\mathbf{grad}\ \overline{u})\, d\boldsymbol{x} = \int_{\Omega_c} \mathbf{J}_0 \cdot \mathbf{grad}\ \overline{u}\, d\boldsymbol{x}, \\ \displaystyle\quad \Big\langle \mu_0 D\psi^+ + \Big(\frac{1}{2}I + K'\Big)\Big(\boldsymbol{n}\cdot \mathbf{curl}\ \boldsymbol{A}\Big), \overline{\eta}\Big\rangle_{\Gamma_c} = 0 \end{cases} \tag{6.11.54}$$

for all $(\boldsymbol{Z}, u, \eta) \in \boldsymbol{\mathcal{H}} := \boldsymbol{\mathcal{Z}} \times H^1(\Omega_c)/\mathbb{C} \times H^{1/2}(\Gamma_c)/\mathbb{C}$. In order to discuss existence and uniqueness of the weak solution, we rewrite (6.11.54) in a more compact form:

$$\boldsymbol{\mathcal{A}}((\boldsymbol{A}, \varphi, \psi), \boldsymbol{Z}, u, \eta) = \boldsymbol{\ell}((\boldsymbol{Z}, u, \eta)) \quad \text{for all} \quad (\boldsymbol{Z}, u, \eta) \in \boldsymbol{\mathcal{H}}, \tag{6.11.55}$$

where

$$\boldsymbol{\ell}((\boldsymbol{Z}, u, \eta)) := \int_{\Omega_c} \mathbf{J}_0 \cdot (\overline{\boldsymbol{Z}} + \mathbf{grad}\ \overline{u})\, d\boldsymbol{x}$$

is the linear functional and $\boldsymbol{\mathcal{A}}$ is the sesquilinear form defined on $\boldsymbol{\mathcal{H}}$ by

$$
\begin{aligned}
&\mathcal{A}((\boldsymbol{A},\varphi,\psi),(\boldsymbol{Z},u,\eta)) \\
&:= \int_{\Omega_c} \Big\{ (\frac{1}{\mu_c}\mathbf{curl}\,\boldsymbol{A}) \cdot (\mathbf{curl}\,\overline{\boldsymbol{Z}}) + \frac{1}{\mu_{max}}(\mathrm{div}\,\boldsymbol{A})(\mathrm{div}\,\overline{\boldsymbol{Z}}) \Big\} d\boldsymbol{x} \\
&- i\omega \int_{\Omega_c} \sigma(\boldsymbol{A} + \mathbf{grad}\,\varphi) \cdot (\overline{\boldsymbol{Z}} + \mathbf{grad}\,\overline{u}) d\boldsymbol{x} \\
&- \Big\langle \boldsymbol{n}\cdot\mathbf{curl}\,\overline{\boldsymbol{Z}}, \Big(\frac{1}{2}I + K\Big)\psi^{+} \Big\rangle_{\Gamma_0} + \Big\langle \Big(\frac{1}{2}I + K'\Big)(\boldsymbol{n}\cdot\mathbf{curl}\,\boldsymbol{A}),\, \overline{\eta} \Big\rangle_{\Gamma_0} \\
&+ \Big\langle \boldsymbol{n}\cdot\,\mathbf{curl}\,\overline{\boldsymbol{Z}}, \frac{1}{\mu_0} V(\boldsymbol{n}\cdot\,\mathbf{curl}\,\boldsymbol{A}) \Big\rangle_{\Gamma_c} + \mu_0 \Big\langle D\psi^{+}, \overline{\eta} \Big\rangle_{\Gamma_c}.
\end{aligned}
\tag{6.11.56}
$$

We now state and prove the main result in this section.

Theorem 6.11.5. (Rodriguez and Valli [366, Theorem 7.5]) *The sesquilinear form $\mathcal{A}(\cdot,\cdot)$ in (6.11.56) is continuous on $\mathcal{H}_{\Gamma_c} = \Big\{(\boldsymbol{Z},u,\eta) \in \mathcal{H}$ with $\int_{\Omega_c} u d\boldsymbol{x} = \boldsymbol{0}$ and $\int_{\Gamma_c} \eta ds = 0\Big\}$ satisfying*

$$
|\mathcal{A}((\boldsymbol{A},\varphi,\psi),(\boldsymbol{Z},u,\eta))| \le M \|(\boldsymbol{A},\varphi,\psi)\|_{\mathcal{H}_{\Gamma_c}} \|(\boldsymbol{Z},u,\eta)\|_{\mathcal{H}_{\Gamma_c}}. \tag{6.11.57}
$$

Moreover, for $(\boldsymbol{Z},u,\eta) \in \mathcal{H}_{\boldsymbol{\Gamma_c}}$, we have the coerciveness estimate

$$
|\mathcal{A}((\boldsymbol{Z},u,\eta),(\boldsymbol{Z},u,\eta))| \ge \alpha \|(\boldsymbol{Z},u,\eta)\|^2_{\mathcal{H}_{\boldsymbol{\Gamma_c}}} \quad \text{for all} \quad (\boldsymbol{Z},u,\eta) \in \mathcal{H}_{\boldsymbol{\Gamma_c}} \tag{6.11.58}
$$

with $\alpha > 0$. Consequently, there is a unique weak solution of (6.11.55) *in the function space $\mathcal{H}_{\boldsymbol{\Gamma_c}}$.*

Proof: In order to apply the Lax-Milgram lemma to establish the existence and uniqueness of the weak solution of (6.11.55), we need to check the continuity of the sesquilinear form $\mathcal{A}(\cdot\cdot)$ and the boundedness of the functional $\boldsymbol{\ell}(\cdot)$. These are clear and in the following, we only concentrate on the coercivity of $\mathcal{A}(\cdot\cdot)$.

We begin with the examination of the real part of $\mathcal{A}((\boldsymbol{Z},u,\eta),(\boldsymbol{Z},u,\eta))$,

$\mathrm{Re}\,\mathcal{A}((\boldsymbol{Z},u,\eta),(\boldsymbol{Z},u,\eta))$ containing the following terms:

$$
\begin{aligned}
&\int_{\Omega_c} (\frac{1}{\mu_c}\mathbf{curl}\,\boldsymbol{Z}) \cdot (\mathbf{curl}\,\overline{\boldsymbol{Z}})\, d\boldsymbol{x}, + \frac{1}{\mu_{max}} \int_{\Omega_c} (\mathrm{div}\,\boldsymbol{Z})\,(\mathrm{div}\,\overline{\boldsymbol{Z}})\, d\boldsymbol{x} \ge \frac{1}{\mu_{max}} k_1 \,\|\boldsymbol{Z}\|^2_{\boldsymbol{\mathcal{Z}}}, \\
&\frac{1}{\mu_0}\Big\langle V(\boldsymbol{n}\cdot\mathbf{curl}\;\boldsymbol{Z}), (\boldsymbol{n}\cdot\mathbf{curl}\;\boldsymbol{Z}) \Big\rangle_{\Gamma_c} \ge \frac{1}{\mu_0}\, k_2 \|\boldsymbol{n}\cdot\mathbf{curl}\,\boldsymbol{Z}\|^2_{H^{-1/2}(\Gamma_c)}, \\
&\mu_0 \Big\langle D\eta, \eta \Big\rangle_{\Gamma_c} \ge \mu_0\, k_3 \,\|\eta\|^2_{H^{1/2}(\Gamma_c)}.
\end{aligned}
\tag{6.11.59}
$$

For the inequality of $(6.11.59)_1$, see for instance Girault and Raviart [147, Lemma I.3.6], while for inequalities $(6.11.59)_1$ and $(6.11.59)_2$ see Section 5.5. Thus, we have the estimates:

$$
\begin{aligned}
&| \text{Re } \boldsymbol{\mathcal{A}}((\boldsymbol{Z}, u, \eta), (\boldsymbol{Z}, u, \eta))| \\
&\quad \geq \frac{1}{\mu_{max}} k_1 \, \|\boldsymbol{Z}\|^2_{\boldsymbol{\mathcal{Z}}} + k_2 \, \|\boldsymbol{n} \cdot \mathbf{curl} \; \boldsymbol{Z}\|^2_{H^{-1/2}(\Gamma_c)} + k_3 \, \|\eta\|^2_{H^{1/2}(\Gamma_c)} \\
&\quad \geq k_4 \Big(\|\boldsymbol{Z}\|^2_{\boldsymbol{\mathcal{Z}}} + \|\boldsymbol{n} \cdot \mathbf{curl} \; \boldsymbol{Z}\|^2_{H^{-1/2}(\Gamma_c)} + \|\eta\|^2_{H^{1/2}(\Gamma_c)} \Big),
\end{aligned}
\tag{6.11.60}
$$

where $k_4 := min\{k_1/\mu^{-1}_{max}, k_2, k_3\}$.

For the imaginary part of $\boldsymbol{\mathcal{A}}((\boldsymbol{Z}, u, \eta), (\boldsymbol{Z}, u, \eta))$, it is a little more involved. In view of the negative sign $-i\omega$ in (6.11.56), we begin with

$$
\begin{aligned}
&- \text{Im } \boldsymbol{\mathcal{A}}((\boldsymbol{Z}, u, \eta), (\boldsymbol{Z}, u, \eta)) \\
&= \omega \int_{\Omega_c} \sigma \, (\boldsymbol{Z} + \mathbf{grad} \; u) \cdot (\overline{\boldsymbol{Z}} + \mathbf{grad} \; \overline{u}) \; d\boldsymbol{x} \\
&+ 2 \, \text{Im} \Big\langle \boldsymbol{n} \cdot \mathbf{curl} \; \overline{\boldsymbol{Z}}, \Big(\frac{1}{2} I + K\Big)\eta \Big\rangle_{\Gamma_0} \\
&\geq \omega \, \sigma_{min} \int_{\Omega_c} \Big\{ (|\mathbf{grad} \; u|^2 + |\boldsymbol{Z}|^2) + 2 \quad \text{Re} \quad (\boldsymbol{Z} \cdot \mathbf{grad} \; \overline{u}) \Big\} d\boldsymbol{x} \\
&+ 2 \quad \text{Im} \quad \Big\langle \boldsymbol{n} \cdot \mathbf{curl} \; \overline{\boldsymbol{Z}}, \Big(\frac{1}{2} I + K\Big)\eta \Big\rangle_{\Gamma_0}.
\end{aligned}
\tag{6.11.61}
$$

We now examine the real and imaginary terms in (6.11.61) separately as follows.

$$
\Big| \int_{\Omega_c} 2 \; \text{Re} \; (\boldsymbol{Z} \cdot \mathbf{grad} \, \overline{u}) \; d\boldsymbol{x} \Big| \leq \frac{1}{2} \; \|\mathbf{grad} \; u\|^2_{0,\Omega_c} + 2 \, \|\boldsymbol{Z}\|^2_{0,\Omega_c}, \tag{6.11.62}
$$

while for the imaginary term there exists a constant $k_5 > 0$ such that

$$
\begin{aligned}
2 \Big| \; \text{Im} \Big\langle \boldsymbol{n} \cdot \mathbf{curl} \; \overline{\boldsymbol{Z}}, \Big(\frac{1}{2} I + K\Big)\eta \Big\rangle_{\Gamma_0} \Big| &\leq k_5 \; \|\eta\|_{H^{1/2}(\Gamma_c)} \; \|\boldsymbol{n} \cdot \mathbf{curl} \, \boldsymbol{Z}\|_{H^{-1/2}(\Gamma_c)} \\
&\leq \frac{k_5}{2} \|\eta\|^2_{H^{1/2}(\Gamma_c)} + \frac{k_5}{2} \|\boldsymbol{n} \cdot \mathbf{curl} \, \boldsymbol{Z}\|^2_{H^{-1/2}(\Gamma_c)}.
\end{aligned}
\tag{6.11.63}
$$

Substituting (6.11.62) and (6.11.63) into (6.11.61) yields the estimate:

$$
\begin{aligned}
&- \text{Im } \boldsymbol{\mathcal{A}}((\boldsymbol{Z}, u, \eta), (\boldsymbol{Z}, u, \eta)) \\
&\geq \frac{1}{2} \omega \sigma_{min} \|\mathbf{grad} \; u\|^2_{0,\Omega_c} - \omega \sigma_{min} \|\boldsymbol{Z}\|^2_{0,\Omega_c} \\
&- \frac{k_5}{2} \|\eta\|^2_{H^{1/2}(\Gamma_c)} - \frac{k_5}{2} \|\boldsymbol{n} \cdot \mathbf{curl} \, \boldsymbol{Z}\|^2_{H^{-1/2}(\Gamma_c)}.
\end{aligned}
\tag{6.11.64}
$$

From the Poincaré inequality, there is a constant $k_6 > 0$ such that

$$
\|\mathbf{grad} \, u\|^2_{0,\Omega_c} \geq k_6 \|u\|^2_{H^1(\Omega_c)}. \tag{6.11.65}
$$

These together with (6.11.64) imply that

$$\begin{aligned} &\left|- \text{Im}\, \boldsymbol{\mathcal{A}}((\boldsymbol{Z},u,\eta),(\boldsymbol{Z},u,\eta))\right| \\ &\geq \frac{1}{2}\omega\sigma_{min}k_6\|u\|^2_{H^1(\Omega_c)} - \omega\sigma_{min}\|\boldsymbol{Z}\|^2_{0,\Omega_c} \\ &- \frac{k_5}{2}\|\eta\|^2_{H^{1/2}(\Gamma_c)} - \frac{k_5}{2}\|\boldsymbol{n}\cdot\textbf{curl}\ \boldsymbol{Z}\|^2_{H^{-1/2}(\Gamma_c)}. \end{aligned} \tag{6.11.66}$$

Then for any positive parameter q, we see that

$$\begin{aligned} &(1+q)|\boldsymbol{\mathcal{A}}(\boldsymbol{Z},u,\eta),(\boldsymbol{Z},u,\eta))| \\ &\geq |\ \text{Re}\ \boldsymbol{\mathcal{A}}((\boldsymbol{Z},u,\eta),(\boldsymbol{Z}u,\eta))| + q|-\ \text{Im}\ \boldsymbol{\mathcal{A}}((\boldsymbol{Z},u,\eta),(\boldsymbol{Z},u,\eta))| \\ &\geq (k_4 - \omega\,\sigma_{min}\,q)\ \|\boldsymbol{Z}\|^2_{\boldsymbol{\mathcal{Z}}} + \frac{k_6}{2}\omega\sigma_{min}q\|u\|^2_{H^1(\Omega_c} \\ &+ (k_4 - \frac{k_5}{2}q)\|\eta\|^2_{H^{1/2}(\Gamma_c)} + (k_4 - \frac{k_5}{2}q)\|\boldsymbol{n}\cdot\textbf{curl}\,\boldsymbol{Z}\|^2_{H^{-1/2}(\Gamma_c)}. \end{aligned} \tag{6.11.67}$$

Let $0 < q < 1$ be sufficiently small so that for fixed k_4, k_5, and k_6, we require that

$$(k_4 - \omega\,\sigma_{min}\,q) > 0 \quad \text{and} \quad (k_4 - \frac{k_5}{2}q) > 0.$$

Hence if we choose any q such that

$$0 < q < min\{\frac{k_4}{\omega\sigma_{min}}, \frac{2k_4}{k_5}\}$$

and define

$$\alpha := \frac{1}{1+q}\ min\{k_4 - \omega\sigma_{min}q, \frac{k_6}{2}\omega\sigma_{min}\,q, k_4 - \frac{k_5}{2}\},$$

we arrive at the desired result (6.11.58). This completes the proof. ■

Remark 6.11.2: We should remark that our time-harmonic Maxwell equations formulation are in terms of the factor $e^{-i\omega\,t}$, and hence the time-harmonic Faraday equation (see for instance $(6.11.68)_1$) has a different sign from conventional engineering literature.

6.11.3 Electric field in the dielectric domain Ω_D.

It ist noteworthy to mention that in the conductor domain Ω, once the magnetic vector potential $\boldsymbol{A}$ and electrical potential φ are found, the electric field $\mathbf{E}$ is known and given by formula (6.11.35), namely

$$\mathbf{E}_c = -i\omega(\boldsymbol{A} + \textbf{grad}\,\varphi) \quad \text{in} \quad \Omega_c$$

However, for the eddy current problem in the dielectric domain Ω_D, $\mathbf{E}$ is still unknown, and we need to find $\mathbf{E}$ in Ω_D by solving the following system:

$$\begin{cases} \mathbf{curl}\,\mathbf{E} - i\,\omega\,\mu_0\mathbf{H} = & \text{in} \quad \Omega_D, \\ \operatorname{div}(\varepsilon_0\,\mathbf{E}) = 0 & \text{in} \quad \Omega_D, \\ \boldsymbol{n}\times\mathbf{E} = \boldsymbol{n}\times\mathbf{E}_c & \text{on} \quad \Gamma_c, \\ \int\limits_{\Gamma_c} \boldsymbol{n}\cdot(\varepsilon_0\,\mathbf{E})\,d\boldsymbol{x} = \mathbf{0}, & \\ \mathbf{E}(\boldsymbol{x}) = \mathcal{O}(|\boldsymbol{x}|^{-1}) & \text{as} \quad |\boldsymbol{x}|\to\infty, \\ \mathbf{curl}\;\mathbf{E}(\boldsymbol{x}) = \mathcal{O}(|\boldsymbol{x}|^{-1}) & \text{as} \quad |\boldsymbol{x}|\to\infty. \end{cases} \tag{6.11.68}$$

This is an exterior Dirichlet problem for the electric field $\mathbf{E}$. It seems that it is best to reduce the problem to a boundary integral equation on Γ_c. To do so, first set

$$\mathbf{E}_p(\boldsymbol{x}) := i\,\omega\mu_0\,\mathbf{curl} \int\limits_{\Omega_D} \frac{\mathbf{H}(\boldsymbol{y})}{4\pi|\boldsymbol{x}-\boldsymbol{y}|}d\boldsymbol{y},$$

and define $\mathbf{E}_0 := \mathbf{E} - \mathbf{E}_p$. Then we see that

$$\mathbf{curl}\;\mathbf{curl}\;\mathbf{E}_0 = \mathbf{0}, \quad \text{and} \quad \operatorname{div}\mathbf{E}_0 = 0.$$

This simplifies the problem for $\mathbf{E}_0$ by solving a modified exterior problem in the exterior domain Ω_D:

$$\begin{cases} \mathbf{curl}\;\mathbf{curl}\;\mathbf{E}_0 & = \mathbf{0} \;\;\text{and}\;\; \operatorname{div}\,\mathbf{E}_0 = 0 & \text{in } \Omega_D, \\ \pi_t\mathbf{E}_0 & = \pi_t(\mathbf{E}_c - \mathbf{E}_p) := \boldsymbol{f} & \text{on } \Gamma_c \\ \mathbf{E}_0 & = \mathcal{O}(|\boldsymbol{x}|^{-1}) & \text{as } |\boldsymbol{x}|\to\infty. \\ \mathbf{curl}\;\mathbf{E}(\boldsymbol{x}) & = \mathcal{O}(|\boldsymbol{x}|^{-1}) & \text{as } |\boldsymbol{x}|\to\infty. \end{cases} \tag{6.11.69}$$

In order to solve (6.11.69) by boundary integral equations, it is useful, if we have some information concerning the subject of vector potentials for the Laplacian

$$\Delta\mathbf{E}_0 := \mathbf{grad}\;(\operatorname{div}\mathbf{E}_0) - \mathbf{curl}\;\mathbf{curl}\;\mathbf{E}_0. \tag{6.11.70}$$

in three–dimensional Lipschitz-domains.

A short digression to vector potentials. Let Γ be a closed Lipschitz boundary in $\mathbb{R}^3$. We denote its interior and exterior domains by Ω^- and Ω^+, respectively. Suppose that $\boldsymbol{u}$ is a vector field satisfying

$$\begin{aligned} &\Delta\boldsymbol{u} = \mathbf{0}, \quad \text{and} \;\; \operatorname{div}\boldsymbol{u} = 0 \;\; \text{in} \;\; \mathbb{R}^3\setminus\Gamma \\ &\boldsymbol{u} = \mathcal{O}(|\boldsymbol{x}|^{-1}) \;\; \text{and} \;\; \mathbf{curl}\;u = \mathcal{O}(|\boldsymbol{x}|^{-1}) \;\; \text{as} \;\; |\boldsymbol{x}| \longrightarrow \infty. \end{aligned} \tag{6.11.71}$$

Based on the formula (6.11.70) in an analogue to the Stratton-Chu formulas for the Maxwell's equations, we have the representation formula for the solution of (6.11.71) in the form

$$\begin{aligned}\boldsymbol{u}(\boldsymbol{x}) = \quad & \mathbf{curl} \int_\Gamma G(\boldsymbol{x}-\boldsymbol{y})[\boldsymbol{n}\times\boldsymbol{u}]_\Gamma ds_y + \int_\Gamma G(\boldsymbol{x}-\boldsymbol{y})[\boldsymbol{n}\times\mathbf{curl}\ \boldsymbol{u}]_\Gamma\, ds_{\boldsymbol{y}} \\ & - \mathbf{grad} \int_\Gamma G(\boldsymbol{x}-\boldsymbol{y})[\boldsymbol{n}\cdot\boldsymbol{u}]_\Gamma\, ds_{\boldsymbol{y}} \quad \text{for } \boldsymbol{x}\neq\Gamma, \end{aligned} \tag{6.11.72}$$

where the fundamental solution for the - Laplacian in $\mathbb{R}^3$ is denoted by

$$G(\boldsymbol{x}-\boldsymbol{y}) = \frac{1}{4\pi}\frac{1}{|\boldsymbol{x}-\boldsymbol{y}|}.$$

In the representation formula (6.11.72), the notation $[\boldsymbol{\lambda}]_\Gamma = \boldsymbol{\lambda}^+ - \boldsymbol{\lambda}^-$ stands for the jump of $\boldsymbol{\lambda}$ across Γ from Ω^+ and Ω^-, respectively. Similar to the scalar case, we now introduce the potentials for $x \neq \Gamma$:

$$\begin{cases} \mathfrak{V}\boldsymbol{\lambda}(\boldsymbol{x}) := \int_\Gamma G(\boldsymbol{x}-\boldsymbol{y})\boldsymbol{\lambda}(\boldsymbol{y})\, ds_{\boldsymbol{y}}\,, & \text{vectorial single layer potential}\ , \\ \mathfrak{W}\boldsymbol{\mu}(\boldsymbol{x}) := \mathbf{curl} \int_\Gamma G(\boldsymbol{x}-\boldsymbol{y})\gamma_t(\boldsymbol{\mu})(\boldsymbol{y})\, ds_y, & \text{vectorial double layer potential}\ , \\ \mathbf{Q}\varphi(\boldsymbol{x}) := \mathbf{grad} \int_\Gamma G(\boldsymbol{x}-\boldsymbol{y})\varphi(y) ds_y. & \text{nonstandard potential}\ , \end{cases} \tag{6.11.73}$$

together with the vectorial Newton potential:

$$\mathfrak{N}\boldsymbol{f}(\boldsymbol{x}) := \int_{\mathbb{R}^3} G(\boldsymbol{x}-\boldsymbol{y})\,\boldsymbol{f}(\boldsymbol{y})d\boldsymbol{y}, \tag{6.11.74}$$

which defines a continuous mapping

$$\mathfrak{N} : \mathbf{H}^{-1}_{comp}(\mathbb{R}^3) \longrightarrow \mathbf{H}^1_{loc}(\mathbb{R}^3).$$

As we know from (4.2.30), the simple layer potentials and the Newton potentials are related for special distributions, namely, $\mathfrak{V}(\boldsymbol{\lambda}) = \mathfrak{N}\gamma_0^*\boldsymbol{\lambda} = \mathfrak{N}(\boldsymbol{\lambda}\otimes\delta_\Gamma)$. For smooth test function $\boldsymbol{\varphi} \in \boldsymbol{\mathcal{D}}(\mathbb{R}^3)$, it follows that

$$(\mathfrak{V}(\boldsymbol{\lambda},\boldsymbol{\varphi})_{\boldsymbol{L}^2_{\mathbb{R}^3}} = (\mathfrak{N}\boldsymbol{\lambda}\otimes\delta_\Gamma)\,,\,\boldsymbol{\varphi})_{\mathbb{R}^3} = \langle\boldsymbol{\lambda},\pi_t(\mathfrak{N}\boldsymbol{\varphi})\rangle_\Gamma = (\mathfrak{N}\pi_t^*(\boldsymbol{\lambda}),\boldsymbol{\varphi})_{\mathbb{R}^3}.$$

Hence, we obtain that

$$\mathfrak{V}^* = \pi_t\circ\mathfrak{N} \quad \text{and} \quad \mathfrak{V} = \mathfrak{N}\circ\pi_t^* \tag{6.11.75}$$

by a density argument, and moreover, the mapping

$$\mathfrak{V} = \mathfrak{N} \circ \pi_t^* : \mathbf{H}_{||}^{-1/2}(\Gamma) \longrightarrow \mathbf{H}_{loc}^1(\mathbb{R}^3)$$

is continuous, since the L^2-adjoint operator $\pi_t^* : \mathbf{H}_{||}^{-1/2}(\Gamma) \longrightarrow \mathbf{H}_{loc}^1(\mathbb{R}^3)$ is continuous (see Figure 6.4 below).

We note that for $\boldsymbol{\lambda} \in \mathbf{H}_{||}^{-1/2}(\Gamma)$, if we require that $\mathrm{div}_\Gamma \boldsymbol{\lambda} = 0$, then $\mathrm{div}\, \mathfrak{V}(\boldsymbol{\lambda}) = 0$ in $\mathbb{R}^3 \setminus \Gamma$. This follows from the following result.

Lemma 6.11.6. *If* $\boldsymbol{\lambda} \in \mathbf{H}_{||}^{-1/2}(\mathrm{div}_\Gamma, \Gamma)$, *then*

$$\mathrm{div}\, \mathfrak{V}\boldsymbol{\lambda} = \int\limits_\Gamma G(\boldsymbol{x} - \boldsymbol{y})\, \mathrm{div}_\Gamma \boldsymbol{\lambda}(\boldsymbol{y})\, ds_y \quad \textit{for } \boldsymbol{x} \in \mathbb{R}^3 \setminus \Gamma.$$

Proof: The proof of this lemma is based on the Stokes formula,

$$\int\limits_\Gamma \mathbf{grad}_\Gamma G(\boldsymbol{x} - \boldsymbol{y}) \cdot \boldsymbol{\lambda} ds_y + \int\limits_\Gamma G(\boldsymbol{x} - \boldsymbol{y})\, \mathrm{div}_\Gamma \boldsymbol{\lambda} ds_y = 0,$$

since $\int\limits_\Gamma \mathrm{div}_\Gamma \boldsymbol{w} ds_{\boldsymbol{y}} = \mathbf{0}$ for any tangential field $\boldsymbol{w}$. (See, e.g. Colton and Kress [75, Theorem 2.29] and MacCamy and Stephan [278, Lemma 2.3]). ∎

We may now restrict $\boldsymbol{\lambda}$ to the space $\mathbf{H}_{||}^{-1/2}(\mathrm{div}_\Gamma, \Gamma)$ with $\mathrm{div}_\Gamma \boldsymbol{\lambda} = 0$, which will be denoted by $\mathbf{H}_{||}^{-1/2}(\mathrm{div}_\Gamma 0, \Gamma)$. Since $\Delta \mathfrak{V}\boldsymbol{\lambda}(\boldsymbol{x}) = \mathbf{0}$ for $\boldsymbol{x} \neq \Gamma$, it follows that $\mathbf{curl}\,\mathbf{curl}\, \mathfrak{V}(\boldsymbol{\lambda})(\boldsymbol{x}) = \mathbf{0}$ as well for $\boldsymbol{x} \neq \Gamma$. In view of $\mathbf{curl}\,\mathbf{curl}\, \mathfrak{V}(\boldsymbol{\lambda})(\boldsymbol{x}) = \mathbf{0}$, we apply Green' s formula to $\mathfrak{V}(\boldsymbol{\lambda})$ in $\Omega^\mp$ with the test function $\boldsymbol{\varphi} \in \left(C_0^\infty(\mathbb{R}^2)\right)^3$ to obtain

$$\begin{aligned}\int\limits_{\mathbb{R}^3} \mathfrak{V}(\boldsymbol{\lambda}) \cdot \mathbf{curl}\,\mathbf{curl}\, \boldsymbol{\varphi} dx &= \Big\langle [\gamma_t \mathfrak{V}(\boldsymbol{\lambda})]_\Gamma, \pi_t \mathbf{curl}\, \boldsymbol{\varphi} \Big\rangle_t - \Big\langle [\gamma_t\, \mathbf{curl}\, \mathfrak{V}(\boldsymbol{\lambda})]_\Gamma, \pi_t \boldsymbol{\varphi} \Big\rangle_t \\ &= -\Big\langle [\gamma_t \mathbf{curl}\, \mathfrak{V}(\boldsymbol{\lambda})]_\Gamma, \pi_t \boldsymbol{\varphi} \Big\rangle_t, \end{aligned} \tag{6.11.76}$$

and by making use of the condition $\mathrm{div}\, \mathfrak{V}(\boldsymbol{\lambda}) = 0$, the left–hand side of (6.11.76) is also equal to

$$-\int\limits_{\mathbb{R}^3} \mathfrak{V}(\boldsymbol{\lambda}) \cdot \Delta \boldsymbol{\varphi}\, dx = -\Big\langle \boldsymbol{\lambda}, (\mathfrak{V}^* \circ \Delta)\, \boldsymbol{\varphi} \Big\rangle_t.$$

This implies that

$$\Big\langle [\gamma_t(\mathbf{curl}\, \mathfrak{V}(\boldsymbol{\lambda}))]_\Gamma, \pi_t \boldsymbol{\varphi} \Big\rangle_t = \Big\langle\, \boldsymbol{\lambda}, (\mathfrak{V}^* \circ \Delta) \boldsymbol{\varphi} \Big\rangle_t = -\Big\langle \boldsymbol{\lambda}, \pi_t \boldsymbol{\varphi} \Big\rangle_t,$$

the last step following from the definition of $\mathfrak{V}^*$ in (6.11.75). Hence,

$$[\gamma_t(\mathbf{curl}\ \mathfrak{V}(\boldsymbol{\lambda}))]_\Gamma = -\boldsymbol{\lambda} \quad \text{in} \quad \mathbf{H}_{||}^{-1/2}(\text{div}_\Gamma, \Gamma), \tag{6.11.77}$$

since $\left(C_0^\infty(\mathbb{R}^3)\right)^3$ is dense in $\mathbf{H}(\mathbf{curl}\,, \Omega^\mp)$ and $\pi_t \boldsymbol{u} \in \mathbf{H}_\perp^{-1/2}(\mathbf{curl}_\Gamma, \Gamma)$ for $\boldsymbol{u} \in \mathbf{H}(\mathbf{curl}\,, \Omega^\mp)$.

As will be seen, the property of the vectorial double–layer potential will depend upon properties of the vectorial simple–layer potential $\mathfrak{V}$ as well as the nonstandard potential $\mathbf{Q}$. So, let us first examine the property of $\mathbf{Q}$. This non–standard potential has the form

$$\mathbf{Q}\,(\varphi) = \mathbf{grad}\,\mathcal{V}(\varphi),$$

where $\mathcal{V}(\varphi)$ is the scalar simple layer potential and from the property of $\mathcal{V}$ in Chapter 5 we can conclude that

$$\mathbf{Q} : H^{-1/2}(\Gamma) \longrightarrow \mathbf{H}(\mathbf{curl}\,, \mathbb{R}^3)$$

is a continuous operator.

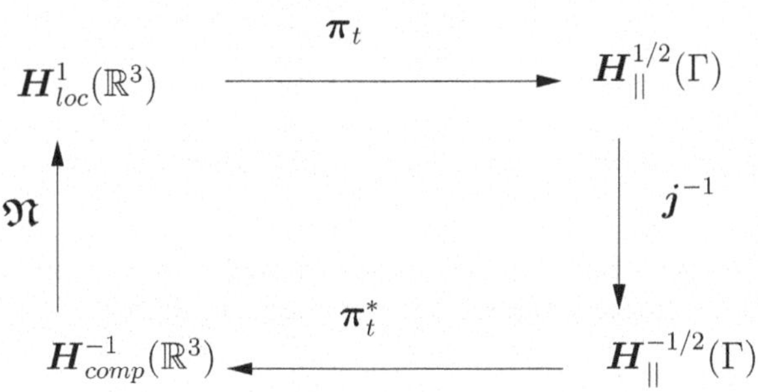

Fig. 6.4. Mapping property of the simple–layer potential

It remains now to consider the vectorial double-layer potential. The vectorial double layer potential for a tangential vectorial field $\boldsymbol{\mu}$ is defined by

$$\mathfrak{W}(\boldsymbol{\mu}) := \mathbf{curl}\ \mathfrak{V}\,(\gamma_t\boldsymbol{\mu})). \tag{6.11.78}$$

It can be shown that if $\boldsymbol{u} = \mathfrak{W}(\boldsymbol{\mu})$, then $\boldsymbol{\mu} = [\pi_t(\boldsymbol{u})] \in \mathbf{H}_\perp^{-1/2}(\Gamma)$ for $\boldsymbol{u} \in \mathbf{H}(\mathbf{curl}\,, \Omega^-) \cup \mathbf{H}_{loc}(\mathbf{curl}\,, \mathbb{R}^3)$. From the mapping property of the vectorial simple-layer operator $\mathfrak{V}$, this implies that

$$\mathfrak{W} : \mathbf{H}_\perp^{-1/2}(\Gamma) \longrightarrow \mathbf{H}(\text{div}, \mathbb{R}^3) \tag{6.11.79}$$

is a continuous operator. For obtaining further properties of the double layer potential, similar to the scalar case in Chapter 5, we need some information concerning the solution of the aforementioned **Dirichlet problem** of the transmission problem consisting of (6.11.71) together with the given Dirichlet boundary condition:

$$\pi_t(\boldsymbol{u}) = \boldsymbol{f} \in \mathbf{H}_{\perp}^{-1/2}(\mathbf{curl}_{\Gamma}, \Gamma) \quad \text{on } \Gamma, \tag{6.11.80}$$

To this end, let $\mathfrak{S}$ be the solution operator of the Dirichlet problem on both sides of Γ defined by

$$\begin{aligned}\mathfrak{S} : \mathbf{H}_{\perp}^{-1/2}(\mathbf{curl}_{\Gamma}, \Gamma) \mapsto \boldsymbol{W}(\mathbf{curl}^2, \Omega^- \cup \Omega^+) \cap \mathbf{H}(\text{div}\ 0; \Omega^- \cup \Omega^+) : \\ \mathfrak{S} : \boldsymbol{f} \mapsto \boldsymbol{u} = \mathfrak{S}(\boldsymbol{f}),\end{aligned} \tag{6.11.81}$$

where $\boldsymbol{W}(\mathbf{curl}, \Omega)$ and $\boldsymbol{W}(\mathbf{curl}^2, \Omega)$ are weighted Sobolev spaces defined by

$$\begin{aligned}\boldsymbol{W}(\mathbf{curl}, \Omega) &:= \Big\{\boldsymbol{v} \in \mathcal{D}(\Omega)' | \frac{\boldsymbol{v}}{\sqrt{1+|\boldsymbol{x}|^2}} \in \boldsymbol{L}^2(\Omega),\ \mathbf{curl}\ \boldsymbol{v} \in \boldsymbol{L}^2(\Omega)\Big\} \\ \boldsymbol{W}(\mathbf{curl}^2, \Omega) &:= \Big\{\boldsymbol{v} \in \boldsymbol{W}(\mathbf{curl}, \Omega) | \mathbf{curl}\ \mathbf{curl}\ \boldsymbol{v} \in \boldsymbol{L}^2(\Omega)\Big\}\end{aligned} \tag{6.11.82}$$

(cf. Section 4.4 for the weighted Sobolev space $\mathbf{H}^m(\mathbb{R}^n; \boldsymbol{\lambda})$ with $n = 3, m = 1$. $\lambda = 0, \kappa = -1$). Then from (6.11.72), we have the representation of the solutions in the form:

$$\mathfrak{S}(\boldsymbol{f}) = \pm\mathfrak{W}(\boldsymbol{f}) + \mathfrak{V}\Big([\gamma_t(\ \mathbf{curl}\ \mathfrak{S}(\boldsymbol{f})\ )]_{\Gamma}\Big) - \mathbf{grad}\ \mathcal{V}\Big([\gamma_n(\mathfrak{S}(\boldsymbol{f}))]_{\Gamma}\Big) \quad \text{in } \Omega^{\pm}, \tag{6.11.83}$$

from which, the vectorial double-layer potential $\mathfrak{W}$ can be explicitly expressed in terms of the solution operator $\mathfrak{S}$. For instance, in Ω^+,

$$\mathfrak{W} = \Big\{\boldsymbol{I} - \mathfrak{V} \circ \gamma_t \circ \mathbf{curl}\ - \mathbf{grad}\ \circ \mathbf{V} \circ \gamma_n\Big\} \circ \mathfrak{S}.$$

This indicates that the properties of $\mathfrak{W}$ can be deduced from those of $\mathfrak{S}$. A careful analysis with the help of continuity properties of potentials $\mathfrak{V}$ and $\mathbf{Q}$ and of the trace operators, one can show that

$$\mathfrak{W} : \mathbf{H}_{\perp}^{-1/2}(\mathbf{curl}_{\Gamma}; \Gamma) \longrightarrow \mathbf{H}(\,\text{div}\,0;\ \Omega^{\pm}) \cup \mathbf{H}(\mathbf{curl}\ ; \Omega^{\pm})$$

is continuous.

Now we are in a position to define the boundary integral operators from the potentials defined in (6.11.73). We consider here only the simple, double-layer and the nonstandard boundary integral operators, $\pi_t(\mathfrak{V}), \pi_t^{\pm}(\mathfrak{W})$ and $\pi_t^{\pm}(\mathbf{Q})$. We summarize the results in the following theorem.

Theorem 6.11.7. *The following boundary integral operators are continuous:*

$$\left.\begin{aligned}\pi_t(\mathfrak{V}) : \mathbf{H}_{||}^{-1/2}(\text{div}_{\Gamma} 0, \Gamma) &\longrightarrow \mathbf{H}_{||}^{1/2}(\text{div}_{\Gamma}\, 0, \Gamma) \\ \pi_t^{\pm}(\mathfrak{W}) : \mathbf{H}_{\perp}^{-1/2}(\mathbf{curl}_{\Gamma}\,,\ \Gamma) &\longrightarrow \mathbf{H}_{\perp}^{-1/2}(\mathbf{curl}_{\Gamma}\,,\ \Gamma) \\ \pi_t^{\pm}(\mathbf{Q}) : H^{-1/2}(\Gamma) &\longrightarrow \mathbf{H}_{\perp}^{-1/2}(\Gamma).\end{aligned}\right\} \tag{6.11.84}$$

Proof: The continuity properties for the vectorial simple–layer and double–layer boundary integral operators, $\pi_t(\mathfrak{V})$ and $\pi_t^{\pm}(\mathfrak{W})$ should be clear. We shall consider here only the nonstandard boundary integral operator $\pi_t^{\pm}(\mathbf{Q})$:

$$\pi_t^{\pm}\mathbf{Q}(\varphi)(\boldsymbol{x}) = \pi_t^{\pm}\,\mathbf{grad}\,(\mathfrak{V}\varphi)(\boldsymbol{x}) = \mathbf{grad}_{\Gamma}(V\varphi)(\boldsymbol{x}), \quad \text{for } \boldsymbol{x}\in\Gamma,$$

where $(V\varphi)(\boldsymbol{x}) = \int\limits_{\Gamma} G(\boldsymbol{x}-\boldsymbol{y})\varphi(\boldsymbol{y})ds_{\boldsymbol{y}}$ is the scalar simple–layer boundary integral operator:

$$V : H^{-1/2}(\Gamma) \mapsto H^{1/2}(\Gamma).$$

Then, as a mapping,

$$\pi_t^{\pm}(\mathbf{Q}) : \varphi \in H^{-1/2}(\Gamma) \longrightarrow \mathbf{grad}_{\Gamma}V\varphi \in \mathbf{H}_{\perp}^{-1/2}(\mathbf{curl}_{\Gamma}\,,\,\Gamma),$$

is continuous, and since $\mathbf{grad}_{\Gamma}V\varphi \in \mathbf{H}_{\perp}^{-1/2}(\Gamma)$, this completes the proof of the desired continuity property of the nonstandard boundary integral operator $\pi_t^{\pm}\mathbf{Q}(\varphi)$ (cf. Section 4.5). ■

It should be mentioned that as in the scalar case, there are two other vectorial boundary integral operators, $\gamma_t^{\pm}(\mathbf{curl}\ \mathfrak{V})(\boldsymbol{\lambda})$ and $\gamma_t^{\pm}(\mathbf{curl}\ \mathfrak{W})(\boldsymbol{\mu})$, the vectorial adjoint and hypersingular boundary integral operators. These boundary integral operators will be needed for the Neumann problems, which will not be included in this short digression; however, their development should be followed in the same manner as in the scalar case. For readers interested in these operators, we refer them to the comprehensive paper by Claeys and Hiptmair [72].

In closing this short digression, we now state and prove the following useful theorem for the Dirichlet problem to be considered in (6.11.68).

Theorem 6.11.8. *The sesquilinear form $\langle\boldsymbol{\lambda}, \pi_t\mathfrak{V}\boldsymbol{\lambda}\rangle_t$, induced by the boundary integral operator $\pi_t\mathfrak{V}$ is $\mathbf{H}_{||}^{-1/2}(\mathrm{div}0,\Gamma)$ – elliptic, that is, there exists a positive constant α such that*

$$\Re e\Big\langle\boldsymbol{\lambda}, \pi_t\mathfrak{V}\boldsymbol{\lambda}\Big\rangle_t \geq \alpha\|\boldsymbol{\lambda}\|^2_{\mathbf{H}_{||}^{-1/2}(\mathrm{div}0,\Gamma)} \quad \text{for all } \boldsymbol{\lambda}\in\mathbf{H}_{||}^{-1/2}(\mathrm{div}\,0,\Gamma). \quad (6.11.85)$$

Proof: For given $\boldsymbol{\lambda} \in \mathbf{H}_{||}^{-1/2}(\mathrm{div}\,0,\Gamma)$, let $\boldsymbol{u} = \mathfrak{V}\boldsymbol{\lambda}$ in $\Omega^-\cup\Omega^+$. Then $\boldsymbol{u}\in\boldsymbol{W}(\mathbf{curl}^2,\Omega^-\cup\Omega^+)$ is a solution of the transmission problem:

$$\begin{aligned} \mathbf{curl\ curl}\,\boldsymbol{u} = \mathbf{0},\ \mathrm{div}\,\boldsymbol{u} = 0 \quad &\text{in } \Omega^-\cup\Omega^+, \\ [\pi_t\boldsymbol{u}]_\Gamma = 0 \quad \text{and} \quad &[\gamma_t(\mathbf{curl}\,\boldsymbol{u})]_\Gamma = \boldsymbol{\lambda} \\ \boldsymbol{u} = \mathcal{O}(|\boldsymbol{x}|^{-1}),\ \mathbf{curl}\,\boldsymbol{u} = &\mathcal{O}(|\boldsymbol{x}|^{-1}) \quad \text{as } |\boldsymbol{x}|\longrightarrow\infty. \end{aligned}$$

An application of Green's formula yields

$$\Big\langle\boldsymbol{\lambda}, \pi_t\mathfrak{V}\boldsymbol{\lambda}\Big\rangle = \Big\langle[\gamma_t\boldsymbol{u}], \pi_t\boldsymbol{u}\Big\rangle = \|\mathbf{curl}\ u\|^2_{\boldsymbol{L}^2(\Omega^-\cup\Omega^+)}. \quad (6.11.86)$$

To complete the proof, we need a similar generalized first Green's formula (e.g. Lemma 5.1.1) to dominate the norm of $\boldsymbol{\lambda}$. In particular, for $\boldsymbol{v} \in \mathbf{H}_\perp^{-1/2}(\mathrm{curl}_\Gamma, \Gamma)$, Green's formula implies that

$$\begin{aligned}
|\langle \gamma_t(\mathbf{curl}\ \boldsymbol{u}), \pi_t \mathcal{Z}\boldsymbol{v}\rangle_t| &= |(\mathbf{curl}\ \boldsymbol{u}, \mathbf{curl}\ \mathcal{Z}\boldsymbol{v})_{\boldsymbol{L}^2(\Omega)}| \\
&\le c\|\mathbf{curl}\ \boldsymbol{u}\|_{\boldsymbol{L}^2(\Omega)}\|\boldsymbol{v}\|_{\mathbf{H}_\perp^{-1/2}(\mathbf{curl}\,_\Gamma,\Gamma)}.
\end{aligned}$$

Thus,

$$\begin{aligned}
\|\gamma_t(\mathbf{curl}\ \boldsymbol{u})\|_{\mathbf{H}_{||}^{-1/2}(\mathrm{div}_\Gamma\,,\,\Gamma)} &= \sup_{\|\boldsymbol{v}\|_{\mathbf{H}_\perp^{-1/2}(\mathbf{curl}\,_\Gamma,\Gamma)}=1} |\langle \gamma_t(\mathbf{curl}\ \boldsymbol{u}), \pi_t \mathcal{Z}\boldsymbol{v}\rangle_t| \\
&\le c\|\mathbf{curl}\ \boldsymbol{u}\|_{\boldsymbol{L}^2(\Omega)}
\end{aligned}$$

for $\Omega = \Omega^\pm$, where

$$\mathcal{Z} : \mathbf{H}_\perp^{-1/2}(\mathbf{curl}\,_\Gamma\,,\ \Gamma) \mapsto \mathbf{H}(\mathbf{curl}\,, \Omega)$$

is a continuous right inverse of the operator π_t. Consequently,

$$\|\boldsymbol{\lambda}\|^2_{\mathbf{H}_{||}^{-1/2}(\mathrm{div}_\Gamma\,,\,\Gamma)} = \|[\gamma_t(\mathbf{curl}\ \boldsymbol{u})]\|^2_{\mathbf{H}_{||}^{-1/2}(\mathrm{div}_\Gamma\,,\,\Gamma)} \le c\|\mathbf{curl}\ u\|^2_{\boldsymbol{L}^2(\Omega)}$$

and we obtain the result with $\alpha = c^{-1}$. ■

We now return to the problem for **the electric field in the dielectric domain** Ω_D defined by (6.11.69). Let $\boldsymbol{f} = \pi_t(\mathbf{E}_c - \mathbf{E}_p)$ be given. We are going to solve the problem by using the direct method in terms of the boundary integral equation of the first kind. For given Dirichlet boundary data $\pi_t(\mathbf{E}_0) = \boldsymbol{f} \in \mathbf{H}_\perp^{-1/2}(\mathbf{curl}\,_\Gamma, \Gamma)$, the solution assumes a representation in the form

$$\begin{aligned}
\mathbf{E}_0(\boldsymbol{x}) = &\int_\Gamma G(\boldsymbol{x}-\boldsymbol{y})\boldsymbol{\lambda}(\boldsymbol{y})ds_{\boldsymbol{y}} - \mathbf{grad}_{\boldsymbol{x}} \int_\Gamma G(\boldsymbol{x}-\boldsymbol{y})(\boldsymbol{n}\cdot\mathbf{E}_0)(\boldsymbol{y})ds_{\boldsymbol{y}} \\
&+ \mathbf{curl}_{\boldsymbol{x}} \int_\Gamma G(\boldsymbol{x}-\boldsymbol{y})(\boldsymbol{n}\times\boldsymbol{f})ds_{\boldsymbol{y}}, \qquad (6.11.87)
\end{aligned}$$

where $\boldsymbol{\lambda} = \gamma_t(\mathbf{curl}\ \mathbf{E}_0) \in \mathbf{H}_{||}^{-1/2}(\Gamma)$ and $\varphi = \boldsymbol{n}\cdot\boldsymbol{u} \in H^{-1/2}(\Gamma)$ are the unknowns. The corresponding boundary integral equation is a first kind boundary integral equation for $\boldsymbol{\lambda}$ and φ in the form:

$$\pi_t(\mathfrak{V}\boldsymbol{\lambda})(\boldsymbol{x}) - \mathbf{grad}\,_\Gamma V(\varphi)(\boldsymbol{x}) = \boldsymbol{q}(\boldsymbol{x})\,, \quad \text{for } \boldsymbol{x}\in\Gamma, \qquad (6.11.88)$$

where $\boldsymbol{q}$ is given by

$$\boldsymbol{q}(\boldsymbol{x}) := -\frac{1}{2}\boldsymbol{f} + \pi_t\,\mathbf{curl} \int_\Gamma G(\boldsymbol{x}-\boldsymbol{y})(\boldsymbol{n}\times\boldsymbol{f})ds_{\boldsymbol{y}} \in \mathbf{H}_\perp^{-1/2}(\mathbf{curl}\,_\Gamma, \Gamma).$$

For the weak solution of (6.11.88), multiplying with test functions $\xi \in \mathbf{H}_{||}^{-1/2}(\mathrm{div}_\Gamma, \Gamma)$ yields the variational equation:

$$\langle \boldsymbol{\xi}, \pi_t \mathfrak{V} \boldsymbol{\lambda} \rangle_t - \langle \mathbf{grad}_\Gamma V\varphi, \boldsymbol{\xi} \rangle_t = \langle \boldsymbol{q}, \boldsymbol{\xi} \rangle_t \,, \quad \forall\, \boldsymbol{\xi} \in \mathbf{H}_{||}^{-1/2}(\mathrm{div}_\Gamma, \Gamma) \qquad (6.11.89)$$

It is worth mentioning that (6.11.89) can not be solved without an additional condition, since in contrast to the EFIE in electromagnetics, it contains more than one unknown (while the EFIE has $\boldsymbol{n} \times \mathbf{H}$ and $\boldsymbol{n}\cdot\mathbf{E}$ as the unknowns, but the latter can be replaced by $\mathrm{div}_\Gamma(\boldsymbol{n} \times \mathbf{H})$) . One way to circumvent this difficulty is to solve the equation (6.11.89) by a two–step procedure. Note that from the Stokes formula, we obtain

$$\langle \mathbf{grad}_\Gamma V\varphi, \boldsymbol{\xi} \rangle_t = -\langle V\varphi, \mathrm{div}_\Gamma \boldsymbol{\xi} \rangle_t,$$

where $\mathrm{div}_\Gamma \boldsymbol{\xi} \in H^{-1/2}(\Gamma)$. Here are the two steps: first testing (6.11.89) by $\boldsymbol{\xi} \in \mathbf{H}_{||}^{-1/2}(\mathrm{div}_\Gamma 0, \Gamma)$ and next testing the equation by $\boldsymbol{\xi} \in \mathbf{H}_{||}^{-1/2}(\mathrm{div}_\Gamma, \Gamma)$, we obtain two equations as follows:

$$\begin{aligned} \langle \boldsymbol{\xi}, \pi_t \mathfrak{V} \boldsymbol{\lambda} \rangle_t &= \langle \boldsymbol{q}, \boldsymbol{\xi} \rangle_t, \quad \forall\, \boldsymbol{\xi} \in \mathbf{H}_{||}^{-1/2}(\mathrm{div}_\Gamma 0, \Gamma) \\ \langle V\varphi, \mathrm{div}_\Gamma \boldsymbol{\xi} \rangle_t &= \langle \boldsymbol{q}, \boldsymbol{\xi} \rangle - \langle \boldsymbol{\xi}, \pi_t \mathfrak{V} \boldsymbol{\lambda} \rangle_t, \quad \forall\, \boldsymbol{\xi} \in \mathbf{H}_{||}^{-1/2}(\mathrm{div}_\Gamma, \Gamma) \end{aligned} \qquad (6.11.90)$$

The first equation in (6.11.90) is uniquely solvable for given $\boldsymbol{q} \in \mathbf{H}_{\perp}^{-1/2}(\mathbf{curl}_\Gamma, \Gamma)$ in view of Theorem 6.11.8. After solving for $\boldsymbol{\lambda} \in \mathbf{H}_{||}^{-1/2}(\mathrm{div}_\Gamma, \Gamma)$, the second equation in (6.11.90) is uniquely solvable for φ as well for given $\boldsymbol{\lambda} \in \mathbf{H}_{||}^{-1/2}(\mathrm{div}_\Gamma, \Gamma)$, and for given $\boldsymbol{q} \in \mathbf{H}_{\perp}^{-1/2}(\mathbf{curl}_\Gamma, \Gamma)$. Since V is $H^{-1/2}(\Gamma)$–elliptic. In this way, we obtain the complete electric field $\mathbf{E} = \mathbf{E}_c + \mathbf{E}_p$ in the domain Ω_D.

Coupling procedures for eddy current problems in unbounded domains have been developed by many individuals directly in terms of magnetic fields $\mathbf{H}$ (see for instance Hiptmair [181], Meddahi and Selgas [292], Girault and Raviart [147] and Malgrange [280]).

6.11.4 Vector potentials — revisited

We remark that the electric filed $\mathbf{E}_0$ in the dielectric field satisfying a special form of the three–dimensional vector potential equation or the Hodge–Laplace equation in $\mathbb{R}^3$. The two–step method for solving the electric field in the dielectric domain Ω_D is in fact equivalent to solve a system of three simple–integral equations of the first kind for the three unknowns $\gamma_t(\mathbf{curl}\, \mathbf{E}_0)$ and $\gamma_n(\mathbf{E}_0)$ from the saddle–point of view in Claeys and Hiptmair [72]. This motivates us following Claeys and Hiptmair [72] to re-examining the Hodge–Laplace equation as follows:

Again let Γ be the closed Lipschitz boundary in $\mathbb{R}^3$. We denote its inside and outside domains by Ω^- and Ω^+, respectively. Suppose $\boldsymbol{u}$ is a vector field satisfying the Hodge–Laplace equation

$$\Delta_{HL}\boldsymbol{u} := \textbf{grad}\ \operatorname{div}\boldsymbol{u} - \textbf{curl}\ \textbf{curl}\ \boldsymbol{u} = \mathbf{0} \quad \text{in} \quad \Omega^-(\text{ or } \Omega^+). \tag{6.11.91}$$

We begin with the solution $\boldsymbol{u}$ in Ω^-. For $\boldsymbol{u} \in \mathbf{H}(\textbf{curl}, \Omega^-) \cap \mathbf{H}(\operatorname{div}, \Omega^-)$, from the Green formula, $\boldsymbol{u}$ admits the representation:

$$\begin{aligned}\boldsymbol{u}(\boldsymbol{x}) &= -\int_\Gamma G(\boldsymbol{x}-\boldsymbol{y})(\boldsymbol{n}\times\textbf{curl}\ \boldsymbol{u})\,ds_{\boldsymbol{y}} + \textbf{grad}_{\boldsymbol{x}}\int_\Gamma G(\boldsymbol{x}-\boldsymbol{y})(\boldsymbol{n}\cdot\boldsymbol{u})\,ds_{\boldsymbol{y}}\\ &\quad - \textbf{curl}_{\boldsymbol{x}}\int_\Gamma G(\boldsymbol{x}-\boldsymbol{y})\,(\boldsymbol{n}\times\pi_t\boldsymbol{u})\,ds_{\boldsymbol{y}} + \int_\Gamma G(\boldsymbol{x}-\boldsymbol{y})\boldsymbol{n}(\boldsymbol{y})\operatorname{div}\boldsymbol{u}(\boldsymbol{y})\,ds_{\boldsymbol{y}}\\ &= \mathfrak{V}_{HL}(\gamma_t\textbf{curl}\ \boldsymbol{u}, \gamma_n\boldsymbol{u})(\boldsymbol{x}) - \mathfrak{W}_{HL}(\pi_t\boldsymbol{u}, \operatorname{div}\boldsymbol{u})(\boldsymbol{x}), \quad \boldsymbol{x}\in\Omega^-,\end{aligned} \tag{6.11.92}$$

where $\mathfrak{V}_{HL}$ and $\mathfrak{W}_{HL}$ are the corresponding single– and double–layer potentials for the Hodge–Laplacian defined by

$$\begin{cases}\mathfrak{V}_{HL}(\gamma_t\textbf{curl}\ \boldsymbol{u}, \gamma_n\boldsymbol{u})(\boldsymbol{x}) := -\displaystyle\int_\Gamma G(\boldsymbol{x}-\boldsymbol{y})(\boldsymbol{n}\times\textbf{curl}\ \boldsymbol{u})\,ds_{\boldsymbol{y}}\\ \qquad\qquad + \textbf{grad}_{\boldsymbol{x}}\displaystyle\int_\Gamma G(\boldsymbol{x}-\boldsymbol{y})(\boldsymbol{n}\cdot\boldsymbol{u})\,ds_{\boldsymbol{y}}\\ \mathfrak{W}_{HL}(\pi_t\boldsymbol{u}, \operatorname{div}\boldsymbol{u})(\boldsymbol{x}) := \textbf{curl}_{\boldsymbol{x}}\displaystyle\int_\Gamma G(\boldsymbol{x}-\boldsymbol{y})\,(\boldsymbol{n}\times\pi_t\boldsymbol{u})\,ds_{\boldsymbol{y}}\\ \qquad\qquad - \displaystyle\int_\Gamma G(\boldsymbol{x}-\boldsymbol{y})\boldsymbol{n}\operatorname{div}\boldsymbol{u}\,ds_{\boldsymbol{y}}.\end{cases} \tag{6.11.93}$$

Here in contrast to the scalar Laplacian, we have *four* Cauchy data : $\Big(\pi_t\boldsymbol{u}, \operatorname{div}\boldsymbol{u}, \gamma_t\textbf{curl}\ \boldsymbol{u}, \gamma_n\boldsymbol{u}\Big) \in \mathbf{H}_\perp^{-1/2}(\textbf{curl}_\Gamma, \Gamma)\times H^{1/2}(\Gamma)\times\mathbf{H}_\parallel^{-1/2}(\operatorname{div}_\Gamma, \Gamma)\times H^{-1/2}(\Gamma)$.

We now define four basic boundary integral operators as in the case of the scalar Laplacian:

$$\begin{aligned}\mathbf{V}\begin{pmatrix}\gamma_t\textbf{curl}\ \boldsymbol{u}\\ \gamma_n\boldsymbol{u}\end{pmatrix}(\boldsymbol{x}) &:= \Big\{\begin{pmatrix}\pi_t\\ \operatorname{div}\end{pmatrix}\mathfrak{V}_{HL}(\gamma_t\textbf{curl}\ \boldsymbol{u}, \gamma_n\boldsymbol{u})\Big\}^-\\ &= \begin{pmatrix}-\pi_t\int_\Gamma G(\boldsymbol{x}-\boldsymbol{y})(\boldsymbol{n}\times\textbf{curl}\ \boldsymbol{u})\,ds_{\boldsymbol{y}} + \textbf{grad}_\Gamma\int_\Gamma G(\boldsymbol{x}-\boldsymbol{y})(\boldsymbol{n}\cdot\boldsymbol{u})\,ds_y\\ -\int_\Gamma G(\boldsymbol{x}-\boldsymbol{y})\operatorname{div}_\Gamma(\boldsymbol{n}\times\textbf{curl}\ \boldsymbol{u})\,ds_{\boldsymbol{y}}\end{pmatrix}\\ &:= \Big\{\begin{pmatrix}\pi_t\\ \operatorname{div}\end{pmatrix}\mathfrak{V}_{HL}(\gamma_t\textbf{curl}\ \boldsymbol{u}, \gamma_n\boldsymbol{u})\Big\}^+,\end{aligned}$$

$$\boldsymbol{K}\begin{pmatrix}\pi_t\boldsymbol{u}\\ \operatorname{div}\boldsymbol{u}\end{pmatrix}(x) := \left\{\left\{\begin{pmatrix}\pi_t\\ \operatorname{div}\end{pmatrix}\mathfrak{W}_{HL}(\pi_t\boldsymbol{u}, \operatorname{div}\boldsymbol{u})\right\}\right\}$$
$$= \begin{pmatrix} \pi_t\mathbf{curl}_{\boldsymbol{x}} \int\limits_{\Gamma\setminus\{\boldsymbol{x}\}} G(\boldsymbol{x}-\boldsymbol{y})(\boldsymbol{n}\times\pi_t\boldsymbol{u})\,ds_{\boldsymbol{y}} - \pi_t\int\limits_{\Gamma} G(\boldsymbol{x}-\boldsymbol{y})\,\boldsymbol{n}\operatorname{div}\boldsymbol{u}\,ds_{\boldsymbol{y}}\\ -\operatorname{div}_{\boldsymbol{x}}\int\limits_{\Gamma\setminus\{\boldsymbol{x}\}} G(\boldsymbol{x}-\boldsymbol{y})\boldsymbol{n}\operatorname{div}\boldsymbol{u}\,ds_{\boldsymbol{y}}\end{pmatrix} \tag{6.11.94}$$

$$\boldsymbol{K}'\begin{pmatrix}\gamma_t\,\mathbf{curl}\,\boldsymbol{u}\\ \gamma_n\boldsymbol{u}\end{pmatrix}(\boldsymbol{x}) := \left\{\left\{\begin{pmatrix}\gamma_t\mathbf{curl}\\ \gamma_n\end{pmatrix}\mathfrak{V}_{HL}(\gamma_t\,\mathbf{curl}\,\boldsymbol{u}, \gamma_n\boldsymbol{u})\right\}\right\}$$
$$= \begin{pmatrix} -\gamma_t\,\mathbf{curl}\int\limits_{\Gamma\setminus\{\boldsymbol{x}\}} G(\boldsymbol{x}-\boldsymbol{y})(\boldsymbol{n}\times\mathbf{curl}\,\boldsymbol{u})\,ds_{\boldsymbol{y}}\\ -\boldsymbol{n}\cdot\int\limits_{\Gamma} G(\boldsymbol{x}-\boldsymbol{y})(\boldsymbol{n}\times\mathbf{curl}\,\boldsymbol{u})ds_{\boldsymbol{y}} + \int\limits_{\Gamma\setminus\{\boldsymbol{x}\}}(\partial G(\boldsymbol{x}-\boldsymbol{y})/\partial n_{\boldsymbol{x}})(\boldsymbol{u}\cdot\boldsymbol{n})ds_{\boldsymbol{y}}\end{pmatrix}$$

$$\boldsymbol{D}\begin{pmatrix}\pi_t\boldsymbol{u}\\ \operatorname{div}\boldsymbol{u}\end{pmatrix}(\boldsymbol{x}) := -\left\{\begin{pmatrix}\gamma_t\mathbf{curl}\\ \gamma_n\end{pmatrix}\mathfrak{W}_{HL}(\pi_t\boldsymbol{u}, \operatorname{div}\boldsymbol{u})\right\}^-$$
$$= \begin{pmatrix} \mathbf{curl}_\Gamma\int\limits_\Gamma G(\boldsymbol{x}-\boldsymbol{y})\operatorname{div}_\Gamma(\boldsymbol{n}\times\pi_t\boldsymbol{u})ds_{\boldsymbol{y}} + \boldsymbol{n}\times\int\limits_\Gamma G(\boldsymbol{x}-\boldsymbol{y})\,\mathbf{curl}_\Gamma(\operatorname{div}\boldsymbol{u})ds_{\boldsymbol{y}}\\ \operatorname{div}_\Gamma\left\{\boldsymbol{n}\times\int\limits_\Gamma G(\boldsymbol{x}-\boldsymbol{y})(\boldsymbol{n})\times(\pi_t\boldsymbol{u})(\boldsymbol{y})ds_{\boldsymbol{y}}\right\} + \boldsymbol{n}\cdot\int\limits_\Gamma G(\boldsymbol{x}-\boldsymbol{y})\boldsymbol{n}\operatorname{div}\boldsymbol{u}ds_{\boldsymbol{y}}\end{pmatrix}$$
$$= -\left\{\begin{pmatrix}\gamma_t\mathbf{curl}\\ \gamma_n\end{pmatrix}\mathfrak{W}_{HL}(\pi_t\boldsymbol{u}, \operatorname{div}\boldsymbol{u})\right\}^+$$

In the formulations above, we have adopted the notations:

$$\{\boldsymbol{v}\}^{\mp} := \gamma^{\mp}\boldsymbol{v} \quad\text{and}\quad \{\{\boldsymbol{v}\}\} := \frac{1}{2}\left(\{\boldsymbol{v}\}^- + \{\boldsymbol{v}\}^+\right).$$

A simple computation shows that the Cauchy data from Ω^- on Γ lead to the equations

$$\begin{pmatrix}\pi_t\boldsymbol{u}\\ \operatorname{div}\boldsymbol{u}\\ \gamma_t\,\mathbf{curl}\,\boldsymbol{u}\\ \gamma_n\boldsymbol{u}\end{pmatrix} = \begin{pmatrix}\frac{1}{2}\boldsymbol{I}-\boldsymbol{K}, & \mathbf{V}\\ \boldsymbol{D}, & \frac{1}{2}\boldsymbol{I}+\boldsymbol{K}'\end{pmatrix}\begin{pmatrix}\pi_t\boldsymbol{u}\\ \operatorname{div}\boldsymbol{u}\\ \gamma_t\,\mathbf{curl}\,\boldsymbol{u}\\ \gamma_n\boldsymbol{u}\end{pmatrix} \quad\text{on } \Gamma. \tag{6.11.95}$$

For any solution of (6.11.91) in Ω^-, the Cauchy data on Γ are reproduced by operators on the right–hand side of (6.11.95), namely by

$$\boldsymbol{C}_{\Omega^-} := \begin{pmatrix}\frac{1}{2}\boldsymbol{I}-\boldsymbol{K}, & \mathbf{V}\\ \boldsymbol{D}, & \frac{1}{2}\boldsymbol{I}+\boldsymbol{K}'\end{pmatrix}, \tag{6.11.96}$$

which therefore defines the *Calderon projector* $\boldsymbol{C}_{\Omega^-}$ for the Hodge–Laplacian with respect to Ω^-.

Similarly, for the solution $\boldsymbol{u}$ of (6.11.91) in the exterior domain Ω^+ , if $\boldsymbol{u}\in\mathbf{H}_{loc}(\mathbf{curl}\,, \Omega^+)\cap\mathbf{H}_{loc}(\operatorname{div}, \Omega^+)$ satisfying the conditions at infinity

$$\boldsymbol{u} = \mathcal{O}|\boldsymbol{x}|^{-1}) \quad\text{and}\quad \mathbf{curl}\,\boldsymbol{u} = \mathcal{O}(|\boldsymbol{x}|^{-1}) \quad\text{as } |\boldsymbol{x}|\longrightarrow\infty, \tag{6.11.97}$$

then $\mathbf{u}$ can be represented in the form

$$\begin{aligned}\boldsymbol{u}(\boldsymbol{x}) &= \Big\{\mathbf{curl}_{\boldsymbol{x}} \int_\Gamma G(\boldsymbol{x}-\boldsymbol{y})(\boldsymbol{n}\times\pi_t\boldsymbol{u})ds_{\boldsymbol{y}} - \int_\Gamma G(\boldsymbol{x}-\boldsymbol{y})\boldsymbol{n}(\boldsymbol{y})\operatorname{div}\boldsymbol{u}(\boldsymbol{y})\,ds_{\boldsymbol{y}}\Big\} \\ &\quad - \Big\{-\int_\Gamma G(\boldsymbol{x}-\boldsymbol{y})(\boldsymbol{n}\times\mathbf{curl}\ \boldsymbol{u})ds_{\boldsymbol{y}} + \mathbf{grad}_{\boldsymbol{x}}\int_\Gamma G(\boldsymbol{x}-\boldsymbol{y})(\boldsymbol{n}\cdot\boldsymbol{u})ds_{\boldsymbol{y}}\Big\} \\ &= \mathfrak{W}_{HL}(\pi_t\boldsymbol{u},\operatorname{div}\boldsymbol{u})(\boldsymbol{x}) - \mathfrak{V}_{HL}(\gamma_t\mathbf{curl}\ \boldsymbol{u},\gamma_n\boldsymbol{u})(\boldsymbol{x}), \quad \text{for } \boldsymbol{x}\in\Omega^+. \end{aligned} \tag{6.11.98}$$

In the same manner we may consider the Cauchy data from Ω^+ on Γ, which leads to the equations

$$\begin{pmatrix}\pi_t\boldsymbol{u}\\ \operatorname{div}\boldsymbol{u}\\ \gamma_t\mathbf{curl}\ \boldsymbol{u}\\ \gamma_n\boldsymbol{u}\end{pmatrix} = \begin{pmatrix}\frac12\boldsymbol{I}+\boldsymbol{K}, & -\mathbf{V}\\ -\boldsymbol{D}, & \frac12\boldsymbol{I}-\boldsymbol{K}'\end{pmatrix}\begin{pmatrix}\pi_t\boldsymbol{u}\\ \operatorname{div}\boldsymbol{u}\\ \gamma_t\mathbf{curl}\ \boldsymbol{u})\\ \gamma_n\boldsymbol{u}\end{pmatrix} \quad \text{on } \Gamma. \tag{6.11.99}$$

Clearly,

$$\boldsymbol{C}_{\Omega^+} = \boldsymbol{I} - \boldsymbol{C}_{\Omega^-}.$$

As in the case for the scalar Laplacian, we may consider two basic boundary value problems for the Hodge–Laplacian in Ω^+ consisting of either boundary conditions of the **Dirichlet type** or boundary conditions of the **Neumann type**. In both cases, from (6.11.99), the boundary value problems can be reduced to a system of boundary integral equations of the first kind.

Since our purpose is to find the electric field in the dielectric by solving the exterior problem (6.11.69), which belongs to the Dirichlet type of boundary value problem for the Hodge–Laplacian, in the following we shall only consider the Dirichlet type boundary value problem. For the Neumann type boundary value problem, we refer to the papers of Claeys and Hiptmair [72], and for natural boundary–value problems see Taylor [415, p.361]. For a comprehensive treatment of the general Hodge Laplacian, we refer to the monograph of Mitrea et al. [311].

The Dirichlet type boundary value problem states: Given

$$\boldsymbol{f}\in\mathbf{H}_\perp^{-1/2}(\mathbf{curl}_\Gamma,\Gamma) \quad \text{and} \quad g\in H^{1/2}(\Gamma)$$

find $\boldsymbol{u}\in\mathbf{H}_{loc}(\mathbf{curl},\Omega^+)\cap\mathbf{H}_{loc}(\operatorname{div},\Omega^+)$ satisfying (6.11.91) in Ω^+, and the boundary conditions

$$\pi_t\boldsymbol{u} = \boldsymbol{f} \quad \text{and} \quad \operatorname{div}\boldsymbol{u} = g \quad \text{on } \Gamma$$

together with the conditions at infinity (6.11.97). We seek a solution in the form (6.11.98). This reduces to the boundary integral equation system of the first kind for the unknowns $\gamma_t\,\mathbf{curl}\ \boldsymbol{u}$ and $\boldsymbol{u}\cdot\boldsymbol{n}$ from (6.11.99):

$$\mathbf{V}\begin{pmatrix}\gamma_t\mathbf{curl}\ \boldsymbol{u}\\ \gamma_n\boldsymbol{u}\end{pmatrix} = -\Big(\frac12\boldsymbol{I}-\boldsymbol{K}\Big)\begin{pmatrix}\pi_t\boldsymbol{u}\\ \operatorname{div}\boldsymbol{u}\end{pmatrix} = -(\frac12\boldsymbol{I}-\boldsymbol{K})\begin{pmatrix}\boldsymbol{f}\\ g\end{pmatrix} \tag{6.11.100}$$

since $\pi_t \boldsymbol{u} = \boldsymbol{f}$ and $\mathrm{div}\boldsymbol{u} = g$ on Γ. This is a direct boundary integral formulation and hence the right–hand side of (6.11.100) is always in the range of $\mathbf{V}$. The weak formulation of (6.11.100) now reads:

Given $(\boldsymbol{f}, g) \in \mathbf{H}_{\perp}^{-1/2}(\mathbf{curl}_{\Gamma}, \Gamma) \times H^{1/2}(\Gamma)$, *find* $(\gamma_t\, \mathbf{curl}\, \boldsymbol{u}, \boldsymbol{u}\cdot\boldsymbol{n})$ $\in \mathbf{H}_{\|}^{-1/2}(\mathrm{div}_{\Gamma}, \Gamma) \times H^{-1/2}(\Gamma)$ such that

$$\left\langle \mathbf{V}\begin{pmatrix} \gamma_t \mathbf{curl}\, \boldsymbol{u} \\ \gamma_n \boldsymbol{u} \end{pmatrix}, \begin{pmatrix} \gamma_t \mathbf{curl}\, \boldsymbol{v} \\ \gamma_n \boldsymbol{v} \end{pmatrix} \right\rangle = -\left\langle \left(\frac{1}{2}\boldsymbol{I} - \boldsymbol{K}\right)\begin{pmatrix} \boldsymbol{f} \\ g \end{pmatrix}, \begin{pmatrix} \gamma_t \mathbf{curl}\, \boldsymbol{v} \\ \gamma_n \boldsymbol{v} \end{pmatrix} \right\rangle \tag{6.11.101}$$

for all $(\gamma_t \mathbf{curl}\, \boldsymbol{v}, \boldsymbol{v}\cdot\boldsymbol{n}) \in \mathbf{H}_{\|}^{-1/2}(\mathrm{div}_{\Gamma}, \Gamma) \times H^{-1/2}(\Gamma)$. Or componentwise, we have explicitly

$$\begin{cases} \Big\langle \gamma_t \mathbf{curl}\, \boldsymbol{v}, \mathbf{V}\,(\gamma_t\, \mathbf{curl}\, \boldsymbol{u})\Big\rangle + \Big\langle \mathrm{div}_{\,\Gamma}\,(\gamma_t \mathbf{curl}\, \boldsymbol{v}), V(\boldsymbol{u}\cdot\boldsymbol{n})\Big\rangle = \ell_1(\gamma_t \mathbf{curl}\, \boldsymbol{v}) \\ \qquad\qquad\qquad\qquad \Big\langle (\boldsymbol{v}\cdot\boldsymbol{n}), V\,(\mathrm{div}_{\Gamma}(\gamma_t\, \mathbf{curl}\, \boldsymbol{u}))\Big\rangle = \ell_2(\boldsymbol{v}\cdot\boldsymbol{n}) \end{cases} \tag{6.11.102}$$

for all $(\gamma_t \mathbf{curl}\, \boldsymbol{v}, \boldsymbol{v}\cdot\boldsymbol{n}) \in \mathbf{H}_{\|}^{-1/2}(\mathrm{div}_{\Gamma}, \Gamma) \times H^{-1/2}(\Gamma)$ where $V(\varphi) := \int_{\Gamma} G(\boldsymbol{x}-\boldsymbol{y})\varphi(\boldsymbol{y})\, ds_{\boldsymbol{y}}$, whereas $\mathbf{V} = V\boldsymbol{I}$, and ℓ_1 and ℓ_2 are functionals defined by

$$\begin{pmatrix} \ell_1(\gamma_t\, \mathbf{curl}\, \boldsymbol{v}) \\ \ell_2(\boldsymbol{v}\cdot\boldsymbol{n}) \end{pmatrix} := \left\langle \left(\frac{1}{2}\boldsymbol{I} - \boldsymbol{K}\right)\begin{pmatrix} \boldsymbol{f} \\ g \end{pmatrix}, \begin{pmatrix} \gamma_t\, \mathbf{curl}\, \boldsymbol{v} \\ \gamma_n \boldsymbol{v} \end{pmatrix} \right\rangle \tag{6.11.103}$$

Note that this is a system of three equations for three unknowns since $\gamma_t\, \mathbf{curl}$ is two–dimensional. As previously mentioned the variational formulation (6.11.101) has a saddle–point structure which can be analyzed by the standard method for the mixed formulations by following Gatica [138] and Claeys and Hiptmair [72, Lemma 3.5 and Corollary 3.6]. Let us denote the unknowns $(\gamma_t\, \mathbf{curl}\, \boldsymbol{u}, \boldsymbol{u}\cdot\boldsymbol{n})$ by $(\boldsymbol{\sigma}, p)$ and the corresponding test functions by $(\boldsymbol{\tau}, q)$. Then we may rewrite the system in a more familiar form :

Find $(\gamma_t\, \mathbf{curl}\, \boldsymbol{u}, \boldsymbol{u}\cdot\boldsymbol{n}) \in \mathbf{H}_{\|}^{-1/2}(\mathrm{div}_{\Gamma}, \Gamma) \times H^{-1/2}(\Gamma)$ such that

$$\begin{cases} a\,(\boldsymbol{\sigma}, \boldsymbol{\tau}) + b(\boldsymbol{\tau}, p) = \ell_1(\boldsymbol{\tau}), \quad \text{for all } \ \boldsymbol{\tau} \in \mathbf{H}_{\|}^{-1/2}(\mathrm{div}_{\Gamma}, \Gamma) =: \mathbf{H} \\ \qquad\qquad b\,(\boldsymbol{\sigma}, q) = \ell_2(q), \quad \text{for all } \ q \in H^{-1/2}(\Gamma) =: Q, \end{cases} \tag{6.11.104}$$

where $a(\bullet, \bullet)$ on $\mathbf{H} \times \mathbf{H}$ and $b(\bullet, \bullet)$ on $\mathbf{H} \times Q$ are two continuous bilinear forms defined by

$$a(\boldsymbol{\sigma}, \boldsymbol{\tau}) := \langle \boldsymbol{\tau}, \mathbf{V}(\boldsymbol{\sigma})\rangle, \tag{6.11.105}$$

$$b(\boldsymbol{\tau}, q) := \langle q,\, V(\mathrm{div}_{\,\Gamma}\, \boldsymbol{\tau})\rangle, \tag{6.11.106}$$

Associated with the bilinear form $a(\bullet, \bullet)$, two linear continuous operators $A : \mathbf{H} \to \mathbf{H}'$ and $A^t : \mathbf{H} \to \mathbf{H}'$ are defined by

$$\langle A\boldsymbol{\sigma}, \boldsymbol{\tau}\rangle_{\mathbf{H}'\times\mathbf{H}} = \langle \boldsymbol{\sigma}, A^t\boldsymbol{\tau}\rangle_{\mathbf{H}'\times\mathbf{H}} = a(\boldsymbol{\sigma}, \boldsymbol{\tau}), \quad \forall\, \boldsymbol{\tau},\, \boldsymbol{\sigma} \in \mathbf{H}. \tag{6.11.107}$$

Similarly, associated with the bilinear form $b(\bullet,\bullet)$, two linear operators $B : \mathbf{H} \to Q'$ and $B^t : Q \to \mathbf{H}'$ are defined by

$$\left\langle B\boldsymbol{\tau}, q\right\rangle_{Q'\times Q} = \left\langle \boldsymbol{\tau}, B^t q\right\rangle_{\mathbf{H}\times\mathbf{H}'} = b(\boldsymbol{\tau}, q), \quad \forall\, \boldsymbol{\tau} \in \mathbf{H},\, \forall q \in Q. \tag{6.11.108}$$

The problem (6.11.104) can also be written as

$$\begin{cases} A\boldsymbol{\sigma} + B^t p = \ell_1 & \text{in } \mathbf{H}', \\ B\boldsymbol{\sigma} = \ell_2 & \text{in } Q'. \end{cases} \tag{6.11.109}$$

We shall consider the formulations (6.11.104) and (6.11.109) to be the same, referring to one or the other according to the convenience.

For the existence theorem for the solutions of this problem, we begin with the kernel of the bilinear form $b(\bullet,\bullet)$ (or Ker B) in the space $\mathbf{H}$:

$$\begin{aligned} \mathbf{H}_0 &:= \left\{\boldsymbol{\tau} \in \mathbf{H} \,|\, b(\boldsymbol{\tau}, q) = 0, \quad \text{for all } \; q \in Q\right\} \\ &= \left\{\boldsymbol{\tau} \in \mathbf{H} \,|\, \operatorname{div}_\Gamma \boldsymbol{\tau} = 0\right\}. \end{aligned} \tag{6.11.110}$$

The bilinear form $a(\bullet,\bullet)$ is coercive on $\mathbf{H}_0$, that is, there exists a constant $\alpha_0 > 0$ such that

$$a(\boldsymbol{\tau}, \boldsymbol{\tau}) \geq \alpha_0 \|\boldsymbol{\tau}\|^2_{\mathbf{H}_{\|}^{-1/2}(\Gamma)} = \alpha_0 \|\boldsymbol{\tau}\|^2_{\mathbf{H}} \quad \text{for } \boldsymbol{\tau} \in \mathbf{H}_0. \tag{6.11.111}$$

Then let A_0 be the restriction of A to $\mathbf{H}_0$ (or Ker B) which shows that A_0 is an isomorphism from $\mathbf{H}_0$ to $(\mathbf{H}_0)'$, which is equivalent to saying that there is an $\alpha_1 > 0$ such that

$$\inf_{\boldsymbol{\tau}_0 \in \mathbf{H}_0} \sup_{\boldsymbol{\sigma}_0 \in \mathbf{H}_0} \frac{a(\boldsymbol{\sigma}_0, \boldsymbol{\tau}_0)}{\|\boldsymbol{\sigma}_0\|_{\mathbf{H}}} \|\boldsymbol{\tau}_0\|_{\mathbf{H}} \geq \alpha_1 \tag{6.11.112}$$

which implies Ker $A_0^t = 0$ that is, A_0 is surjective, and

$$\inf_{\boldsymbol{\sigma}_0 \in \mathbf{H}_0} \sup_{\boldsymbol{\tau}_0 \in \mathbf{H}_0} \frac{a(\boldsymbol{\sigma}_0, \boldsymbol{\tau}_0)}{\|\boldsymbol{\sigma}_0\|_{\mathbf{H}} \|\boldsymbol{\tau}_0\|_{\mathbf{H}}} \geq \alpha_1 \tag{6.11.113}$$

which implies Ker $A_0 = \mathbf{0}$ that is, A_0 is **injective**.

Before we consider the inf–sup condition for the bilinear form $b(\boldsymbol{\sigma}, q)$, we note that as in the potential theory for the scalar Laplacian, if $(\boldsymbol{\sigma}_0, p_0) \in \mathbf{H} \times Q$ is a solution of the corresponding homogeneous equations of (6.11.104), consider

$$\boldsymbol{u}(\boldsymbol{\sigma}) = -\mathfrak{V}_{HL}(\boldsymbol{\sigma}_0, p_0)(\boldsymbol{x}) \quad \text{for } \boldsymbol{x} \in \mathbb{R}^3 \setminus \Gamma. \tag{6.11.114}$$

Then $\boldsymbol{u}(\boldsymbol{x})$ satisfies a transmission problem for $\Delta_{HL}\boldsymbol{u}(x) = \mathbf{0}$ in $\mathbb{R}^3 \setminus \Gamma$ with jumping transmission conditions:

$$\left[\begin{pmatrix} \pi_t \boldsymbol{u} \\ \operatorname{div} \boldsymbol{u} \end{pmatrix}\right]\Big|_{\Gamma} = \mathbf{0} \quad \text{and} \quad \left[\begin{pmatrix} \gamma_t \, \mathbf{curl}\, \boldsymbol{u} \\ \gamma_n \boldsymbol{u} \end{pmatrix}\right]\Big|_{\Gamma} = \begin{pmatrix} \boldsymbol{\sigma}_0 \\ p_0 \end{pmatrix}. \tag{6.11.115}$$

As a result, it can be shown that

$$\boldsymbol{\sigma}_0 = \mathbf{0} \quad \text{and} \quad p_0 \in Q_0, \tag{6.11.116}$$

where $Q_0 = \operatorname{Ker} B^t$ is the kernel of the bilinear form $b(\bullet, \bullet)$ in the space Q:

$$Q_0 := \big\{ p* \in Q \,|\, b(\boldsymbol{\tau}, p*) = \big\langle \operatorname{div}_{\Gamma} \boldsymbol{\tau}, p* \big\rangle = 0, \forall \boldsymbol{\tau} \in \mathbf{H} \big\}, \tag{6.11.117}$$

which is a finite–dimensional subspace of Q whose dimension equals the second Betti number of Γ, together with the Euler–Poincaré characteristic (see Kress [244, 245, 246], Martensen [283] for $\kappa = 0$ Cessenat [63, p.335] and Rodriguez and Valli [366], Claeys and Hiptmair [72, Section 7.1] for $\kappa \neq 0$). This means, from Fredholm's alternative, the right–hand side of (6.11.100) needs to satisfy the orthogonality condition

$$\left\langle \left(\frac{1}{2}\boldsymbol{I} - \boldsymbol{K}\right) \begin{pmatrix} \boldsymbol{f} \\ g \end{pmatrix}, \begin{pmatrix} \mathbf{0} \\ q \end{pmatrix} \right\rangle_{\Gamma} = \begin{pmatrix} \mathbf{0} \\ 0 \end{pmatrix} \quad \text{for all } q \in Q_0 = \operatorname{Ker} B^t.$$

This leads to the condition

$$\langle (\frac{1}{2}I - K)g,\, q\rangle_{\Gamma} = \langle g,\, (\frac{1}{2}I - K')\, q\rangle_{\Gamma} = 0 \quad \text{for all } q \in Q_0 = \operatorname{Ker} B^t, \tag{6.11.118}$$

where

$$K'(q)(\boldsymbol{x}) = \int\limits_{\Gamma \setminus \{x\}} \frac{\partial}{\partial n_{\boldsymbol{x}}} G(\boldsymbol{x} - \boldsymbol{y})\, q(\boldsymbol{y})\, ds_{\boldsymbol{y}}, \quad \text{for } \boldsymbol{x} \in \Gamma$$

is the transposed double–layer boundary integral operator for the three–dimensional scalar Laplacian (see, e.g. Equation (1.2.9) in Chapter 1).

Now let $\widetilde{Q}$ be the orthogonal complement of Q_0. Since $\widetilde{Q}$ equals $\operatorname{div}_{\Gamma} \mathbf{H}_{\|}^{-1/2}(\operatorname{div}_{\Gamma}, \Gamma)$, it is a closed subspace of Q and, hence, the range of B is closed in $\widetilde{Q}'$. For $\ell_2 \in$ range (B), we get $\widetilde{Q}' =$ range (B). According to Boffi, Brezzi and Fortin [34, §4.24], the operator B^t is invertible and there is a constant $\beta > 0$ such that

$$\|B^t q\|_{\mathbf{H}'} \geq \beta \|q\|_{\widetilde{Q}} = \beta \|q\|_{Q/Q_0} \quad \text{for all } q \in Q.$$

Thus, we have the inf–sup condition

$$\inf_{0 \neq q \in Q} \sup_{0 \neq \boldsymbol{\tau} \in \mathbf{H}} \frac{b(\boldsymbol{\tau}, q)}{\|\boldsymbol{\tau}\|_{\mathbf{H}} \, \|q\|_{Q/Q_0}} \geq \beta > 0 \tag{6.11.119}$$

According to Boffi, Brezzi and Fortin [34, Theorem 4.2.4], as consequences of (6.11.112), (6.11.113) and (6.11.119) we obtain the following existence result.

Theorem 6.11.9. *Under the condition* (6.11.118), *the problem* (6.11.104) *has a solution* $(\boldsymbol{\sigma}, p)$ *where* $\boldsymbol{\sigma}$ *is uniquely determined, and* p *is determined up to an element of the finite–dimansional* Ker B^t.

We now return to the problem (6.11.69) for the electric field $\mathbf{E}_0$ in the dielectric domain Ω_D (i.e. Ω^+). In terms of the notations in the problem (6.11.104), we have $(\boldsymbol{\sigma}, p) = (\gamma_t \mathbf{curl}\, \mathbf{E}_0, \mathbf{E}_0 \cdot \boldsymbol{n})$ for the unknowns. The given data are $\pi_t \mathbf{E}_0 = \boldsymbol{f}$ and $\operatorname{div} \mathbf{E}_0 = g = 0$ on the boundary Γ_c. So this is a special case where $\ell_2 = 0$ in the second equation of (6.11.104). From Green's formula, it follows

$$\operatorname{div} \mathbf{E}_0 = 0 \quad \text{on } \Gamma_c \quad \text{implies} \quad \operatorname{div} \mathbf{E}_0 = 0 \quad \text{in } \Omega_D.$$

Hence, if we seek a solution $\mathbf{E}_0$ of (6.11.69) in the form

$$\begin{aligned} \mathbf{E}_0(\boldsymbol{x}) &= \mathfrak{W}_{HL}(\boldsymbol{f}, 0)(\boldsymbol{x}) - \mathfrak{V}_{HL}(\boldsymbol{\sigma}, p)(\boldsymbol{x}) \\ &= \mathbf{curl}_{\boldsymbol{x}} \int_{\Gamma} G(\boldsymbol{x}-\boldsymbol{y})\, (\boldsymbol{n} \times \boldsymbol{f})(\boldsymbol{y})\, ds_{\boldsymbol{y}} \qquad (6.11.120) \\ &\quad - \Big\{ - \int_{\Gamma} G(\boldsymbol{x}-\boldsymbol{y}) \boldsymbol{\sigma}(\boldsymbol{y})\, ds_{\boldsymbol{y}} + \mathbf{grad}_{\boldsymbol{x}} \int_{\Gamma} G(\boldsymbol{x}-\boldsymbol{y})\, p(\boldsymbol{y}) ds_{\boldsymbol{y}} \Big\} \text{ for } \boldsymbol{x} \in \Omega_D \end{aligned}$$

then $(\boldsymbol{\sigma}, p)$ satisfies the system of integral equations

$$\begin{cases} a\,(\boldsymbol{\sigma}, \boldsymbol{\tau}) + b(\boldsymbol{\tau}, p) = \ell_f(\boldsymbol{\tau}), & \text{for all } \boldsymbol{\tau} \in \mathbf{H} \\ \qquad\quad b\,(\boldsymbol{\sigma}, q) = 0, & \text{for all } q \in Q, \end{cases} \qquad (6.11.121)$$

where

$$\ell_f(\boldsymbol{\tau}) := \langle (\frac{1}{2}\boldsymbol{I} - \boldsymbol{K})(\boldsymbol{f}), \boldsymbol{\tau} \rangle$$

$$\text{with } \boldsymbol{K}(\boldsymbol{f}) := \pi_t\, \mathbf{curl}_{\boldsymbol{x}} \int_{\Gamma_c \setminus \{\boldsymbol{x}\}} G(\boldsymbol{x}-\boldsymbol{y})(\boldsymbol{n} \times \boldsymbol{f})(\boldsymbol{y})\, ds_{\boldsymbol{y}}.$$

Or in terms of the operators A, B and B^t, the system of boundary integral equations (6.11.109) can be written as

$$\begin{cases} A\boldsymbol{\sigma} + B^t p = (\frac{1}{2}\boldsymbol{I} - \boldsymbol{K})\,\boldsymbol{f} & \text{in } \mathbf{H}', \\ \qquad B\boldsymbol{\sigma} = \mathbf{0} & \text{in } Q'. \end{cases} \qquad (6.11.122)$$

We end this section by including an existence and uniqueness result for the solutions of (6.11.121) or (6.11.122) as a corollary of Theorem 6.11.9 in the following

Corollary 6.11.10. *The equations* (6.11.121) *have a solution* $(\boldsymbol{\sigma}, p)$ *where* $\boldsymbol{\sigma}$ *is uniquely determined, and* p *is determined up to an element of the* $\operatorname{Ker}(B^t)$. *Moreover, setting*

$$\widetilde{\beta} := \inf_{0\neq q\in Q} \sup_{0\neq \boldsymbol{\tau}\in \mathbf{H}} \frac{b(\boldsymbol{\tau}, q)}{\|\boldsymbol{\tau}\|_{\mathbf{H}} \|q\|_{Q/Q_0}} > 0, \tag{6.11.123}$$

we have

$$\|\boldsymbol{\sigma}\|_{\mathbf{H}} \leq \frac{1}{\alpha_1}\|\ell_f\|_{\mathbf{H}'} \quad \textit{and} \quad \|p\|_{Q/Q_0} \leq \frac{2M_a}{\alpha_1\widetilde{\beta}}\|\ell_f\|_{\mathbf{H}'}, \tag{6.11.124}$$

with M_a being the continuity constant such that

$$|a(\boldsymbol{\sigma},\boldsymbol{\tau})| \leq M_a \|\boldsymbol{\sigma}\|_{\mathbf{H}} \|\boldsymbol{\tau}\|_{\mathbf{H}}, \quad \textit{for all } \boldsymbol{\sigma},\boldsymbol{\tau}\in\mathbf{H}.$$

The estimates (6.11.124) are straight forward and their derivations will be omitted. We remark that in this special case, the orthogonal condition (6.11.118) is automatically fulfilled.

6.12 Applications of boundary integral equations to scattering problems

(See also the Monograph by D. Colton and R. Kress [75].)

In all the basic scattering problems, we are given $\omega, \varepsilon^+\mu^+$, incident fields $\mathbf{E}^i, \mathbf{H}^i$ and Γ, and we must determine the total fields

$$\mathbf{E} = \mathbf{E}^i + \mathbf{E}^s\,,\ \mathbf{H} = \mathbf{H}^i + \mathbf{H}^s \tag{6.12.1}$$

so that the scattered fields $\mathbf{E}^s, \mathbf{H}^s$ satisfy Maxwell's equations in Ω^c:

$$\begin{aligned}\mathbf{curl}\,\mathbf{E}^s &= i\omega\mu^+\mathbf{H}^s\\ \mathbf{curl}\,\mathbf{H}^s &= -i\omega\varepsilon^+\mathbf{E}^s\end{aligned} \tag{6.12.2}$$

and the Silver–Müller conditions (6.7.2) together with appropriate boundary conditions (see Agranovich [6, p.28]). We consider as an incident field any pair $(\mathbf{E}^i, \mathbf{H}^i)$ of either entire solutions of the homogeneous exterior Maxwell equations, or fields that solve the exterior Maxwell equations everywhere in Ω and also in $\Omega^c \setminus \overline{\Omega}_0$, where $\overline{\Omega}_0$ is the support of the sources of the incident waves. We assume that all the sources originate in Ω^c. In particular, for plane wave incidence in the direction of $\widehat{\boldsymbol{k}}$ we define the normalized fields:

$$\begin{aligned}\mathbf{E}^i(\boldsymbol{x}) &= \widehat{\boldsymbol{a}}\exp\{ik\widehat{\boldsymbol{k}}\cdot\boldsymbol{x}\},\\ \mathbf{H}^i(\boldsymbol{x}) &= Y^+\widehat{\boldsymbol{b}}\exp\{ik\widehat{\boldsymbol{k}}\cdot\boldsymbol{x}\}\,,\ Y^+ = \sqrt{\frac{\varepsilon^+}{\mu^+}},\end{aligned}$$

where the polarization vectors $\widehat{\boldsymbol{a}}$ and $\widehat{\boldsymbol{b}}$ and the direction of propagation $\widehat{\boldsymbol{k}}$ form a right–handed orthonormal system, i.e. $\widehat{\boldsymbol{a}}\times\widehat{\boldsymbol{b}} = \widehat{\boldsymbol{k}}$. Note that the plane waves do not satisfy the Silver–Müller radiation conditions (6.7.2). The incident fields $(\mathbf{E}^i, \mathbf{H}^i)$ will represent the total field produced by the sources in absence of any scatterers. They satisfy similar Stratton–Chu representation formulae, but in particular

$$\mathbf{0} = i\omega\mu^+ \int_\Gamma G_k^+(\boldsymbol{x}-\boldsymbol{y})\boldsymbol{n} \times \mathbf{H}^i(\boldsymbol{y})ds_{\boldsymbol{y}} + \int_\Gamma \boldsymbol{n} \cdot \mathbf{E}^i(\boldsymbol{y})\mathbf{grad}_y G_k^+(\boldsymbol{x}-\boldsymbol{y})ds_{\boldsymbol{y}}$$
$$+ \int_\Gamma (\boldsymbol{n} \times \mathbf{E}^i(\boldsymbol{y})) \times \mathbf{grad}_{\boldsymbol{y}} G_k^+(\boldsymbol{x}-\boldsymbol{y})ds_{\boldsymbol{y}},$$
$$\mathbf{0} = -i\omega\varepsilon^+ \int_\Gamma G_k^+(\boldsymbol{x}-\boldsymbol{y})\boldsymbol{n} \times \mathbf{E}^i(\boldsymbol{y})ds_{\boldsymbol{y}} + \int_\Gamma \boldsymbol{n} \cdot \mathbf{H}^i(\boldsymbol{y})\mathbf{grad}_y G_k^+(\boldsymbol{x}-\boldsymbol{y})ds_{\boldsymbol{y}}$$
$$+ \int_\Gamma (\boldsymbol{n} \times \mathbf{H}^i(\boldsymbol{y})) \times \mathbf{grad}_{\boldsymbol{y}} G_k^+(\boldsymbol{x}-\boldsymbol{y})ds_{\boldsymbol{y}} \quad \text{for } \boldsymbol{x} \in \Omega^c,$$

since the incident fields satisfy the Maxwell equations in the region occupied by Ω in the absence of the scatterer (see e.g. Dassios and Kleinman [94, p.54] and Hsiao and Kleinman [194]).

From Equation (6.9.6) and (6.9.7) with $\mathbf{J}_0 = \mathbf{0}$ and $\varrho = 0$, we arrive at

$$\mathbf{E}(\boldsymbol{x}) = \mathbf{E}^i(\boldsymbol{x}) + \int_\Gamma i\omega\mu^+ G_k(\boldsymbol{x}-\boldsymbol{y})\big(\boldsymbol{n}(\boldsymbol{y}) \times \mathbf{H}(\boldsymbol{y})\big)ds_{\boldsymbol{y}} \tag{6.12.3}$$
$$+ \int_\Gamma \big(\boldsymbol{n}(\boldsymbol{y}) \cdot \mathbf{E}(\boldsymbol{y})\big)\mathbf{grad}_{\boldsymbol{y}} G_k(\boldsymbol{x}-\boldsymbol{y})ds_{\boldsymbol{y}}$$
$$+ \int_\Gamma \big(\boldsymbol{n}(\boldsymbol{y}) \times \mathbf{E}(\boldsymbol{y})\big) \times \mathbf{grad}_{\boldsymbol{y}} G_k(\boldsymbol{x}-\boldsymbol{y})ds_{\boldsymbol{y}},$$
$$\mathbf{H}(\boldsymbol{x}) = \mathbf{H}^i(\boldsymbol{x}) - \int_\Gamma i\omega\varepsilon^+ G_k(\boldsymbol{x}-\boldsymbol{y})\big(\boldsymbol{n}(\boldsymbol{y}) \times \mathbf{E}(\boldsymbol{y})\big)ds_{\boldsymbol{y}} \tag{6.12.4}$$
$$+ \int_\Gamma \big(\boldsymbol{n}(\boldsymbol{y}) \cdot \mathbf{H}(\boldsymbol{y})\big)\mathbf{grad}_{\boldsymbol{y}} G_k(\boldsymbol{x}-\boldsymbol{y})ds_{\boldsymbol{y}}$$
$$+ \int_\Gamma \big(\boldsymbol{n}(\boldsymbol{y}) \times \mathbf{H}(\boldsymbol{y})\big) \times \mathbf{grad}_{\boldsymbol{y}} G_k(\boldsymbol{x}-\boldsymbol{y})ds_{\boldsymbol{y}} \quad \text{for } \boldsymbol{x} \in \Omega^c.$$

If the scatterer is a perfect electric conductor then (6.12.3) and (6.12.4) with (6.6.4) lead to the representation formulae

$$\mathbf{E}(\boldsymbol{x}) = \mathbf{E}^i(\boldsymbol{x}) + \int_\Gamma i\omega\mu^+ G_k(\boldsymbol{x}-\boldsymbol{y})\big(\boldsymbol{n}(\boldsymbol{y}) \times \mathbf{H}(\boldsymbol{y})\big)ds_{\boldsymbol{y}} \tag{6.12.5}$$

$$+ \int_\Gamma \big(\boldsymbol{n}(\boldsymbol{y}) \cdot \mathbf{E}(\boldsymbol{y})\big)\mathbf{grad}_{\boldsymbol{y}} G_k(\boldsymbol{x}-\boldsymbol{y})ds_{\boldsymbol{y}}\,,\ \boldsymbol{x} \in \Omega^c,$$

$$\mathbf{H}(\boldsymbol{x}) = \mathbf{H}^i(\boldsymbol{x}) + \int_\Gamma \big(\boldsymbol{n}(\boldsymbol{y}) \times \mathbf{H}(\boldsymbol{y})\big) \times \mathbf{grad}_{\boldsymbol{y}} G_k(\boldsymbol{x}-\boldsymbol{y})ds_{\boldsymbol{y}}\,,\ \boldsymbol{x} \in \Omega^c. \tag{6.12.6}$$

Now we consider the application of boundary integral equations to scattering problems.

6.12.1 Scattering by a perfect electric conductor, EFIE and MFIE

(see also Mitrea, D., Mitrea, M. and Pipher [308], Buffa, Costabel and Sheen [52] and Buffa, Hiptmair, von Petersdorff and Schwab [53].)

Let a perfect electric conductor occupy the bounded region $\overline{\Omega} \subset \mathbb{R}^3$ with the Lipschitz boundary $\Gamma = \partial\Omega$ which is immersed into an isotropic homogeneous medium with constant ε^+ and μ^+. When illuminated by the incident field, $\mathbf{E}^i(\boldsymbol{x})$ and $\mathbf{H}^i(\boldsymbol{x})$, we consider the following boundary value problem together with the homogeneous boundary conditions (6.6.4), i.e.

$$\boldsymbol{n} \cdot \mathbf{H}|_\Gamma = 0\,,\ \boldsymbol{n} \times \mathbf{E}|_\Gamma = \mathbf{0}.$$

This leads to the boundary integral equations **EFIE**

$$i\omega\mu^+ \boldsymbol{n}(\boldsymbol{x}) \times \int_\Gamma G_k(\boldsymbol{x}-\boldsymbol{y})\big(\boldsymbol{n}(\boldsymbol{y}) \times \mathbf{H}(\boldsymbol{y})\big)ds_{\boldsymbol{y}}$$

$$+\boldsymbol{n}(\boldsymbol{x}) \times \int_\Gamma \boldsymbol{n}(\boldsymbol{y}) \cdot \mathbf{E}(\boldsymbol{y})\mathbf{grad}_{\boldsymbol{y}} G_k(\boldsymbol{x}-\boldsymbol{y})ds_{\boldsymbol{y}} = -\boldsymbol{n} \times \mathbf{E}^i, \quad \text{for } \boldsymbol{x} \in \Gamma, \tag{6.12.7}$$

and **MFIE**

$$\frac{1}{2}(\boldsymbol{n} \times \mathbf{H})(\boldsymbol{x}) - \boldsymbol{n}(\boldsymbol{x}) \times \int_\Gamma \big(\boldsymbol{n}(\boldsymbol{y}) \times \mathbf{H}(\boldsymbol{y})\big) \times \mathbf{grad}_{\boldsymbol{y}} G_k(\boldsymbol{x}-\boldsymbol{y})ds_{\boldsymbol{y}}$$

$$= \boldsymbol{n} \times \mathbf{H}^i(\boldsymbol{x})\,, \quad \text{for } \boldsymbol{x} \in \Gamma. \tag{6.12.8}$$

Equation (6.12.7) is known as the electric field integral equation (**EFIE**) whereas (6.12.8) is the magnetic field integral equation (**MFIE**), Both are integral equations for $\boldsymbol{n} \times \mathbf{H}$ on Γ. We note in (6.12.7), that the term $\boldsymbol{n} \cdot \mathbf{E}$ can be replaced by

$$\boldsymbol{n} \cdot \mathbf{E} = \frac{1}{i\omega\varepsilon^+}\mathrm{div}_\Gamma(\boldsymbol{n} \times \mathbf{H}) \quad \text{on } \Gamma, \tag{6.12.9}$$

which is the equation of continuity on the boundary Γ (see Jones [220, (6.84)]). Then **EFIE** becomes an integral equation on Γ for $\boldsymbol{n} \times \mathbf{H}$ on Γ only. The integral equation **MFIE** also determines the only unknown $\boldsymbol{n} \times \mathbf{H}$ on Γ since $\boldsymbol{n} \times \mathbf{E}|_\Gamma = \mathbf{0}$ and $\boldsymbol{n} \cdot \mathbf{H}|_\Gamma = 0$ on Γ.

Remark 6.12.1: The **EFIE** (6.12.7) and **MFIE** (6.12.8) result from a method based on the Green representation theorem of solutions of Maxwell equations, known as the **direct method**. We only need to seek the solution either from (6.12.7) or from (6.12.8), but not both.

On the other hand, one may also obtain boundary integral equations by the indirect method or layer ansatz as used by Colton and Kress [75], Hsiao and Kleinman [194]. For instance, we may seek the scattered electric field $\mathbf{E}^{\hat{s}}(\boldsymbol{x})$ in the form of either simple- or double–layer potentials in order to arrive at two boundary integral equations for the unknown moment or density. Again one needs only to solve one of the two boundary integral equations. Then the scattered magnetic field $\mathbf{H}^s(\boldsymbol{x})$ can be recovered from the corresponding Maxwell equations. In what follows, for simplicity, we will suppress + signs for μ^+ and ε^+.

Remark 6.12.2: Alternatively, the Stratton–Chu representation formula (6.12.5) can be rewritten in the form

$$\mathbf{E}(\boldsymbol{x}) = \mathbf{E}^i(\boldsymbol{x}) - \frac{1}{i\omega\varepsilon}(\mathbf{grad}\ \mathrm{div} + k^2)\boldsymbol{A}(\boldsymbol{x})\,,\ x \in \Omega^c, \tag{6.12.10}$$

where

$$\boldsymbol{A}(\boldsymbol{x}) := \int_\Gamma G_k(\boldsymbol{x} - \boldsymbol{y})(\boldsymbol{n} \times \mathbf{H})(\boldsymbol{y})ds_{\boldsymbol{y}}. \tag{6.12.11}$$

As will be seen, this formula is particularly useful for deriving boundary integral equations for scattering by a perfectly electric conducting thin wire or antenna.

We include here a derivation of Equation (6.12.10). We concentrate on the term:

$$\int_\Gamma (\boldsymbol{n} \cdot \mathbf{E}(\boldsymbol{y}))\mathbf{grad}_{\boldsymbol{y}} G_k(\boldsymbol{x} - \boldsymbol{y})ds_{\boldsymbol{y}}$$

in the Stratton–Chu formula (6.12.5). We note that from (6.12.9) we get

$$\begin{aligned}\int_\Gamma (\boldsymbol{n}\cdot\mathbf{E}(\boldsymbol{y}))\mathbf{grad}_{\boldsymbol{y}} G_k(\boldsymbol{x}-\boldsymbol{y})ds_{\boldsymbol{y}} &= \frac{1}{i\omega\varepsilon}\int_\Gamma \mathbf{grad}_{\boldsymbol{y}} G_k(\boldsymbol{x}-\boldsymbol{y})\mathrm{div}_\Gamma\big(\boldsymbol{n}\times\mathbf{H}(\boldsymbol{y})\big)ds_{\boldsymbol{y}} \\ &= \frac{-1}{i\omega\varepsilon}\mathbf{grad}_{\boldsymbol{x}}\int_\Gamma G_k(\boldsymbol{x}-\boldsymbol{y})\mathrm{div}_\Gamma\big(\boldsymbol{n}\times\mathbf{H}(\boldsymbol{y})\big)ds_{\boldsymbol{y}} \\ &= \frac{-1}{i\omega\varepsilon}\mathbf{grad}_{\boldsymbol{x}}\mathrm{div}_{\boldsymbol{x}}\boldsymbol{A}(\boldsymbol{x}),\end{aligned}$$

where we have employed formula

$$\begin{aligned}\mathrm{div}\boldsymbol{A}(\boldsymbol{x}) &= \mathrm{div}\int_\Gamma G_k(\boldsymbol{x}-\boldsymbol{y})\big(\boldsymbol{n}\times\mathbf{H}(\boldsymbol{y})\big)ds_{\boldsymbol{y}} \\ &= \int_\Gamma G_k(\boldsymbol{x}-\boldsymbol{y})\,\mathrm{div}_\Gamma\big(\boldsymbol{n}\times\mathbf{H}(\boldsymbol{y})\big)ds_{\boldsymbol{y}}\end{aligned} \tag{6.12.12}$$

(see Colton and Kress [75, Theorem 2.2]).

The formula (6.12.10) now follows immediately from the Stratton–Chu formula (6.12.3) and (6.12.12).

Remark 6.12.3: (**Pocklington's integral equation**). In the special case of a perfectly conducting wire or thin–wire antenna, equation (6.12.10) leads to a much simplified **EFIE**. For simplicity, we assume that the wire is a circular cylinder of radius a which is much smaller than its length l. Let us assume that an incident wave $\mathbf{E}^i$ impinges on the surface of a conducting wire. It induces a surface current density

$$\mathbf{J}_\Gamma = \boldsymbol{n}\times\mathbf{H}$$

that in turn is used to find a scattered electric field

$$\mathbf{E}^s(\boldsymbol{x}) = -\frac{1}{i\omega\varepsilon}(\mathbf{grad}\ \mathrm{div} + k^2)\boldsymbol{A}(\boldsymbol{x})$$

with

$$\boldsymbol{A}(\boldsymbol{x}) := \int_\Gamma G_k(\boldsymbol{x}-\boldsymbol{y})\mathbf{J}_\Gamma(\boldsymbol{y})ds_{\boldsymbol{y}},$$

Since the wire is a circular cylinder, it is convenient to use the cylindrical coordinate system where $\boldsymbol{x} = (\varrho, \Phi, z)$ denotes the observation points and $\boldsymbol{y} = (a, \Phi_\Gamma, z_\Gamma)$ denotes the source points on the surface of the wire.

A very reasonable assumption is that the surface current density $\mathbf{J}_\Gamma(\boldsymbol{y})$ is entirely axially directed with azimuthal symmetry for $a \ll \ell$. This assumption leads to the consideration of a linear z–directed current density, $\mathbf{J}_{\Gamma_z}(\boldsymbol{y})$, and

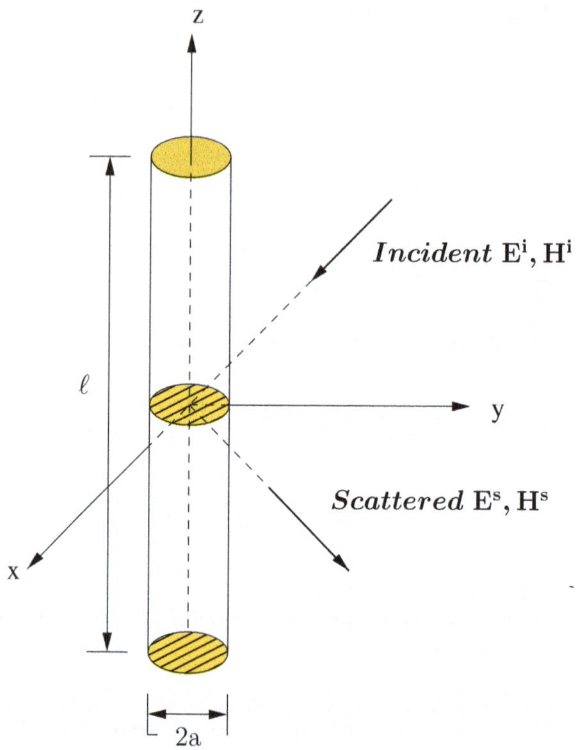

Fig. 6.5. Plane incident wave on a conducting wire.

for the observation point of the wire surface only the z–component of $\mathbf{E}^s(\boldsymbol{x})$ is needed. That is

$$\mathbf{E}^s_z(\boldsymbol{x}) = -\frac{1}{i\omega\varepsilon}(\mathbf{grad}\ \mathrm{div} + k^2)\boldsymbol{A}_z(\boldsymbol{x})\,,\ \boldsymbol{x} \in \Gamma,$$

by neglecting the edge effects, where

$$\boldsymbol{A}_z(\boldsymbol{x}) = \int\limits_{\Gamma} G_k(\boldsymbol{x}-\boldsymbol{y})\mathbf{J}_{\Gamma_z}(\boldsymbol{y})dz_\Gamma = \int\limits_{-\ell/2}^{\ell/2} \Big\{ \int\limits_0^{2\pi} \mathbf{J}_{\Gamma_z}(\boldsymbol{y})G_k(\boldsymbol{x}-\boldsymbol{y})ad\Phi_\Gamma \Big\} dz_\Gamma.$$

Moreover, if the wire is very thin, then the surface current density $\mathbf{J}_{\Gamma_z}$ is not a function of the azimuthal angle Φ_Γ, and we can write $\mathbf{J}_{\Gamma_z} = \mathbf{J}_{\Gamma_z}(z_\Gamma)$ and obtain

$$\boldsymbol{A}_z(\boldsymbol{x}) = \int\limits_{-\ell/2}^{\ell/2} \mathbf{J}_{\Gamma_z}(z_\Gamma)\Big\{ \int\limits_0^{2\pi} \frac{e^{ikR}}{4\pi R} d\Phi_\Gamma \Big\} adz_\Gamma,$$

where

$$R = |\boldsymbol{x} - \boldsymbol{y}| = \{\varrho^2 + a^2 - 2\varrho a \cos(\Phi - \Phi_\Gamma) + (z - z_\Gamma)^2\}^{1/2}.$$

This shows that

$$\mathbf{E}_z^s(\boldsymbol{x}) = -\frac{1}{i\omega\varepsilon}(\mathbf{grad}\ \mathrm{div} + k^2) \int_{-\ell/2}^{\ell/2} \mathbf{J}_{\Gamma_z}(z_\Gamma)\Big\{\int_0^{2\pi} \frac{e^{ikR}}{4\pi R} d\Phi_\Gamma\Big\} a dz_\Gamma.$$

In order to derive an integral equation for $\mathbf{J}_{\Gamma_z}$, we require the boundary condition on the tangent component of the total field over the surface of the wire:

$$\mathbf{E}_z^i + \mathbf{E}_z^s = \mathbf{0} \quad \text{for} \quad \varrho = a, \tag{6.12.13}$$

where $\mathbf{E}^i$ is the impregnate field assumed to be azimuthally symmetric. Also, we introduce the total current flowing is the z_Γ–direction as

$$\mathbf{J}(z_\Gamma) = 2\pi a \mathbf{J}_{\Gamma_z}(z_\Gamma)$$

and obtain

$$i\omega\varepsilon\mathbf{E}_z^i(a, \Phi, z) = (\mathbf{grad}\ \mathrm{div} + k^2) \int_{-\ell/2}^{\ell/2} \frac{\mathbf{J}(z_\Gamma)}{2\pi a}\Big\{\int_0^{2\pi} \frac{e^{ik\widetilde{R}}}{4\pi \widetilde{R}} d\Phi_\Gamma\Big\} a dz_\Gamma, \tag{6.12.14}$$

where

$$\widetilde{R} := \sqrt{(z - z_\Gamma)^2 + [2a \sin\frac{(\Phi - \Phi_\Gamma)}{2}]^2}.$$

Because of the symmetry of the scatterer, the integration over Φ_Γ is independent of the reference point Φ, we may set $\Phi = 0$ and obtain finally the integral equation

$$\Big(\frac{d^2}{dz^2} + k^2\Big) \int_{-\ell/2}^{\ell/2} \mathbf{J}(z_\Gamma) g_\Gamma(z - z_\Gamma) dz_\Gamma = i\omega\varepsilon\mathbf{E}_z^i(z) \tag{6.12.15}$$

where as in (6.12.13) and (6.12.14),

$$\mathbf{E}_z^i(z) := \mathbf{E}_z^i(a, 0, z)\,,\ g_\Gamma(z - z_\Gamma) := \frac{1}{2\pi}\int_0^{2\pi} \frac{e^{ikR_0}}{4\pi R_0} d\Phi_\Gamma,$$

$$R_0 := \widetilde{R}|_{\Phi=0} = \sqrt{(z - z_\Gamma)^2 + 4a^2 \sin^2\big(\frac{\Phi_\Gamma}{2}\big)}.$$

We note that in (6.12.15) the second order partial differential operator **grad** div has been reduced to an ordinary differential operator with respect to z, since the integral is a function of z only. A further simplification results, if we impose the assumption $ka \ll 1$. Thus

$$\widetilde{R}_0 \cong \sqrt{(z - z_\Gamma)^2 + a^2}$$

and the integral equation (6.12.15) becomes

$$\left(\frac{d^2}{dz^2} + k^2\right) \int_{-\ell/2}^{\ell/2} \mathbf{J}(z_\Gamma) \frac{e^{ik\widetilde{R}_o}}{4\pi \widetilde{R}_0} dz_p = i\omega\varepsilon \mathbf{E}_z^i(z) \tag{6.12.16}$$

with a nonsingular kernel.

The integral equation (6.12.16), is known as the **Pocklington's integral equation** developed for the thin wire or antenna (see e.g. Balanis [21, (8.22)] and Barelishli [24, p.301]).

An existence theorem for EFIE

To end this section, we now include an existence theorem for the solution for EFIE in the scattering by a perfect electric conductor. We begin with the representation of the electric field $\mathbf{E}(\boldsymbol{x}), \boldsymbol{x} \in \Omega^c$ by using III(2) in Appendix B:

$$\begin{aligned}
\mathbf{E}(\boldsymbol{x}) &= \mathbf{E}^i(\boldsymbol{x}) + i\omega\mu \int_\Gamma G_k(\boldsymbol{x} - \boldsymbol{y})(\boldsymbol{n}(\boldsymbol{y}) \times \mathbf{H}(\boldsymbol{y}))\, ds_{\boldsymbol{y}} \qquad (6.12.17)\\
&\quad + \frac{1}{i\omega\epsilon} \int_\Gamma \mathbf{grad}_{\boldsymbol{y}}\, G_k(\boldsymbol{x} - \boldsymbol{y})\, \mathrm{div}_\Gamma(\boldsymbol{n}(\boldsymbol{y}) \times \mathbf{H}(\boldsymbol{y}))\, ds_{\boldsymbol{y}}\\
&= \mathbf{E}^i(\boldsymbol{x}) + i\omega\mu \left(\mathfrak{V}(\pi_t^* \mathbf{J})\right)(\boldsymbol{x}) - \frac{1}{i\omega\epsilon} \mathbf{grad}_{\boldsymbol{x}}\, \mathrm{div}_{\boldsymbol{x}} \left(\mathfrak{V}(\pi_t^* \mathbf{J})\right)(\boldsymbol{x}),\\
&= \mathbf{E}^i(\boldsymbol{x}) + i\omega\mu \left(\mathfrak{V}\mathbf{J}\right)(\boldsymbol{x}) - \frac{1}{i\omega\epsilon} \mathbf{grad}_{\boldsymbol{x}}\, \mathrm{div}_x \left(\mathfrak{V}\mathbf{J}\right)(\boldsymbol{x}),\\
&= \mathbf{E}^i(\boldsymbol{x}) + i\omega\mu \left(\mathfrak{V}\mathbf{J}\right)(\boldsymbol{x}) - \frac{1}{i\omega\epsilon} \mathbf{grad}_x \left(\mathfrak{V}\mathrm{div}_\Gamma \mathbf{J}\right)(\boldsymbol{x}),
\end{aligned}$$

where $\pi_t^* \mathbf{J} = \boldsymbol{n}(\boldsymbol{y}) \times \mathbf{H}(y)$, and where $\pi_t^* : \left(\boldsymbol{L}_t^2(\Gamma)\right)^2 \longrightarrow \left(\boldsymbol{L}^2(\Gamma)\right)^2$ is the adjoint operator of π_t and $\mathfrak{V}$ is the simple–layer potential defined by

$$\mathfrak{V}\boldsymbol{\lambda}(\boldsymbol{x}) := \int_\Gamma \frac{1}{4\pi} \frac{e^{ik|\boldsymbol{x}-\boldsymbol{y}|}}{|\boldsymbol{x} - \boldsymbol{y}|} \boldsymbol{\lambda}(\boldsymbol{y}) ds_{\boldsymbol{y}}\,,\ \boldsymbol{x} \notin \Gamma$$

Remark 6.12.4: Some explanations should be in order. It has been shown in Buffa, Costabel and Schwab [51] that for $\boldsymbol{\lambda} \in \mathbf{H}_{||}^{-1/2}(\Gamma), \mathfrak{V}(\boldsymbol{\lambda}) = \mathfrak{V}(\pi_t^* \boldsymbol{\lambda})$ (Note: $\gamma\mathfrak{V}(\boldsymbol{\lambda}) = \gamma\mathfrak{V}(\pi_t^* \boldsymbol{\lambda}) \in \mathbf{H}^{1/2}(\Gamma)$). Then one may replace $\boldsymbol{n} \times \mathbf{H}$ by the corresponding $\mathbf{J}_\gamma \in \mathbf{H}_{||}^{-\frac{1}{2}}(\mathrm{div}_\Gamma, \Gamma)$ in $\mathfrak{V}$, since $\mathfrak{V}(\boldsymbol{n} \times \mathbf{H}) = \mathfrak{V}(\pi_t^* \mathbf{J}_\gamma) = \mathfrak{V}(\mathbf{J}_\gamma)$. For simplicity, we suppress the subindex γ.

It is understood that here π_t is really $\pi_t \circ \mathcal{Z}$ where $\mathcal{Z}$ is a right inverse of the trace operator γ, and the adjoint operator π_t^* is the adjoint operator of $\pi_t \circ \mathcal{Z}$ according to the notation in Buffa, Costabel and Schwab [51] as discussed in Remark 4.2.1 (see Section 4.5).

In terms of the single–layer potential operator $\mathfrak{V}$, the EFIE from the representation formula (6.12.17) for the unknown function $\mathbf{J} \in \mathbf{H}_{||}^{-1/2}(\operatorname{div}_\Gamma, \Gamma)$ is given by

$$\pi_t\Big(\mathbf{V}\,(\mathbf{J})\Big) + \frac{1}{k^2}\,\mathbf{grad}_{\,\Gamma}\Big(V(\operatorname{div}_\Gamma \mathbf{J})\Big) = \boldsymbol{g}\,,\ \boldsymbol{x} \in \Gamma, \tag{6.12.18}$$

where $\mathbf{V} := \gamma\mathfrak{V}$ is the single–layer boundary integral operator and $\boldsymbol{g} = -\pi_t \mathbf{E}^i / i\omega\mu \in \mathbf{H}_\perp^{1/2}(\mathbf{curl}_{\,\Gamma}, \Gamma)$ is given in terms of the incident field $\mathbf{E}^i(x)$. The weak formulation of (6.12.18) now reads:
Given $\boldsymbol{g} \in \mathbf{H}_\perp^{-1/2}(\mathbf{curl}_{\,\Gamma}, \Gamma)$, find $\mathbf{J} \in \mathbf{H}_{||}^{-1/2}(\operatorname{div}_\Gamma, \Gamma)$ such that

$$\Big\langle \overline{\mathbf{J}^t}, \pi_t \mathbf{V}(\mathbf{J})\Big\rangle - \frac{1}{k^2}\Big\langle \operatorname{div}_\Gamma \overline{\mathbf{J}^t}, V(\operatorname{div}_\Gamma \mathbf{J})\Big\rangle = \Big\langle \overline{\mathbf{J}^t}, \boldsymbol{g}\Big\rangle \tag{6.12.19}$$

for all test fields $\mathbf{J}^t \in \mathbf{H}_{||}^{-1/2}(\operatorname{div}_\Gamma, \Gamma)$ (by making use of the Stokes formula). The variational formulation (6.12.19) is exactly the same as the one derived from the **Rumsey Variational Principle** (see Rumsey [368], Nedelec [331] and De La Bourdonnage [98]).

In order to discuss the existence of the variational solution $\mathbf{J}$, following Buffa, Costabel and Schwab [51], we need the Hodge decomposition of the tangential vector space $\mathbf{H}_{||}^{-1/2}(\operatorname{div}_\Gamma, \Gamma)$ introduced in Buffa and Ciarlet [50],

$$\mathbf{H}_{||}^{-1/2}(\operatorname{div}_\Gamma, \Gamma) = \mathbf{grad}_{\,\Gamma}\mathcal{H} \oplus \mathbf{curl}_{\,\Gamma}\, H^{1/2}(\Gamma) \tag{6.12.20}$$
$$\text{and}\ \ \mathcal{H}(\Gamma) := \big\{\alpha \in H^1(\Gamma)\ \ \text{with}\ \ \Delta_\Gamma \alpha \in H^{-1/2}(\Gamma), \langle \alpha, 1\rangle_\Gamma = 0\big\}.$$

Set

$$\mathbf{J} := \mathbf{grad}_{\,\Gamma}\,\varphi + \mathbf{curl}_{\,\Gamma}\,\psi\,,\ \mathbf{J}^t := \mathbf{grad}_{\,\Gamma}\,\varphi^t + \mathbf{curl}_{\,\Gamma}\,\psi^t \tag{6.12.21}$$

and $\Delta_\Gamma \varphi \in H^{-1/2}(\Gamma)$ (see De la Bourdonnaye [98] and Nedelec [331]).

Lemma 6.12.1. *Since* $\mathbf{J} = \mathbf{grad}_{\,\Gamma}\varphi + \mathbf{curl}_{\,\Gamma}\psi$ *with* $\varphi \in \mathcal{H}(\Gamma)$ *and* $\psi \in H^{1/2}(\Gamma)$*, there holds (cf. Appendix B, V(3))*

$$\operatorname{div}_\Gamma \mathbf{J} = \Delta_\Gamma \varphi \in H^{-1/2}(\Gamma).$$

Proof:

$$\begin{aligned}\operatorname{div}_\Gamma \mathbf{J} &= \operatorname{div}_\Gamma \mathbf{grad}_{\,\Gamma}\varphi + \operatorname{div}_\Gamma \mathbf{curl}_{\,\Gamma}\psi \\ &= \Delta_\Gamma \varphi - \operatorname{div}_\Gamma(\gamma_t \mathbf{grad}_{\,\Gamma}\psi) \\ &= \Delta_\Gamma \varphi - \big(\operatorname{div}(\boldsymbol{n} \times \mathbf{grad}_{\,\Gamma}\psi)\big)|_\Gamma\end{aligned}$$

due to Appendix B,III(5). Then

$$\begin{aligned}(\mathrm{div}(\boldsymbol{n}\times \mathbf{grad}_{\Gamma}\psi)) &= \mathbf{grad}_{\Gamma}\psi \cdot \mathbf{curl}_{\Gamma}\boldsymbol{n} - \boldsymbol{n}\cdot \mathbf{curl}_{\Gamma}\mathbf{grad}_{\Gamma}\psi \\ &= \mathbf{grad}_{\Gamma}\psi \cdot \mathbf{curl}_{\Gamma}\boldsymbol{n}\end{aligned}$$

since $\mathbf{curl}_{\Gamma}\mathbf{grad}_{\Gamma}\psi = 0$ due to Appendix B, V(4).

For $\Gamma \in C^1$ the exterior normal vector $\boldsymbol{n}$ can be written as $\boldsymbol{n} = \mathbf{grad}\,\Phi$ with a local implicit representation $\Phi = \text{const}$ of Γ, and $\mathbf{curl}_{\Gamma}\boldsymbol{n} = \nabla\times\nabla\Phi = \mathbf{0}$. This completes the proof. ■

Remark 6.12.5: For $\mathbf{J}$ we have

$$\begin{aligned}\|\mathbf{J}\|^2_{\mathbf{H}_{\|}^{-1/2}(\mathrm{div}_{\Gamma},\Gamma)} &= \|\mathbf{J}\|^2_{\mathbf{H}_{\|}^{-1/2}(\Gamma)} + \|\mathrm{div}_{\Gamma}\mathbf{J}\|^2_{H^{-1/2}(\Gamma)} \\ &= \|\mathbf{grad}_{\Gamma}\varphi\|^2_{\mathbf{H}_{\|}^{-1/2}(\Gamma)} + \|\mathbf{curl}_{\Gamma}\psi\|^2_{\mathbf{H}_{\|}^{-1/2}(\Gamma)} + \|\Delta_{\Gamma}\varphi\|^2_{H^{-1/2}(\Gamma)}\end{aligned}$$

and

$$\|\mathbf{grad}_{\Gamma}\varphi\|^2_{\mathbf{H}_{\|}^{-1/2}(\Gamma)} \le \|\mathbf{grad}_{\Gamma}\varphi\|^2_{\boldsymbol{L}_2(\Gamma)} \le c\|\Delta_{\Gamma}\varphi\|^2_{\mathbf{H}^{-1/2}(\Gamma)}$$

which implies that the norm

$$\|\mathbf{curl}_{\Gamma}\varphi\|_{\mathbf{H}_{\|}^{-1/2}(\Gamma)} + \|\Delta\varphi\|_{\mathbf{H}^{-1/2}(\Gamma)}$$

is equivalent to the norm $\|\mathbf{J}\|_{\mathbf{H}_{\|}^{-1/2}(\mathrm{div}_{\Gamma},\Gamma)}$.

From (6.12.21), in terms of φ, ψ and the test functions φ^t, ψ^t, the variational equation (6.12.19) becomes

$$\begin{aligned}&\Big\langle \pi_t\mathbf{V}\Big(\mathbf{grad}_{\Gamma}\varphi + \mathbf{curl}_{\Gamma}\psi\Big), \mathbf{grad}_{\Gamma}\overline{\varphi^t} + \mathbf{curl}_{\Gamma}\overline{\psi^t}\Big\rangle - \frac{1}{k^2}\Big\langle V(\Delta_{\Gamma}\varphi)\,,\, \Delta_{\Gamma}\overline{\varphi}^t\Big\rangle \\ &= \Big\langle \boldsymbol{g}\,,\, \mathbf{grad}_{\Gamma}\overline{\varphi^t} + \mathbf{curl}_{\Gamma}\overline{\psi^t}\Big\rangle, \qquad (6.12.22)\end{aligned}$$

from which we define the sesquilinear form

$$\begin{aligned}\boldsymbol{\mathcal{A}}(\varphi,\psi;\overline{\varphi}^t,\overline{\psi}^t) &:= \Big\langle \pi_t\mathbf{V}_k\Big(\mathbf{grad}_{\Gamma}\varphi + \mathbf{curl}_{\Gamma}\psi\Big), \mathbf{grad}_{\Gamma}\overline{\varphi^t} + \mathbf{curl}_{\Gamma}\overline{\psi^t}\Big\rangle \\ &\quad - \frac{1}{k^2}\Big\langle \boldsymbol{C}_k(\Delta_{\Gamma}\varphi)\,,\, \Delta_{\Gamma}\overline{\varphi}^t\Big\rangle \\ &=: \mathbf{B}(\mathbf{grad}_{\Gamma}\varphi + \mathbf{curl}_{\Gamma}\psi, \mathbf{grad}_{\Gamma}\overline{\varphi^t} + \mathbf{curl}_{\Gamma}\overline{\psi^t}) \\ &\quad - \frac{1}{k^2}\boldsymbol{C}_k(\Delta_{\Gamma}\varphi, \Delta_{\Gamma}\overline{\varphi}^t) \\ &= \mathbf{B}_0(\mathbf{grad}_{\Gamma}\varphi + \mathbf{curl}_{\Gamma}\psi, \mathbf{grad}_{\Gamma}\overline{\varphi}^t + \mathbf{curl}_{\Gamma}\overline{\psi}^t) \\ &\quad - \frac{1}{k^2}\boldsymbol{C}_0(\Delta_{\Gamma}\varphi, \Delta_{\Gamma}\overline{\varphi}^t) + \mathbf{R}(\varphi,\psi;\overline{\varphi^t},\overline{\psi^t}) \qquad (6.12.23)\end{aligned}$$

where

$$\mathbf{B}_0(\bullet,\bullet) = \langle \pi_t \mathbf{V}_0(\mathbf{grad}_{\Gamma}\varphi + \mathbf{curl}_{\Gamma}\psi)\,,\ (\mathbf{grad}_{\Gamma}\overline{\varphi}^t + \mathbf{curl}_{\Gamma}\overline{\psi}^t)\rangle$$
$$\boldsymbol{C}_0(\bullet,\bullet) = \langle V_0(\Delta_{\Gamma}\varphi)\,,\ \Delta_{\Gamma}\overline{\varphi}^t\rangle$$

and the remaining term $\mathbf{R}$ is a compact perturbation of the principal parts. Note that $\mathcal{A}$ is the sesquilinear form of a system of three boundary integral equations for two functions φ, ψ depending on the variables on Γ. First, we show continuity.

Lemma 6.12.2. *The sesquilinear form* $\boldsymbol{\mathcal{A}}(\varphi,\psi;\overline{\varphi^t},\overline{\psi^t})$ *is continuous on* $\mathbf{H}\times\mathbf{H}$ *where* $\mathbf{H} = H^{-1/2}(\Gamma)\times\mathbf{H}_{\|}^{-1/2}(\Gamma)\simeq\mathbf{H}_{\|}^{-1/2}(\mathrm{div}_{\Gamma},\Gamma)$ *(see* (6.12.20)*).*

Proof: Since $\mathbf{V}_k\mathbf{H}_{\|}^{-1/2}(\Gamma) = \mathbf{H}_{\|}^{1/2}(\Gamma) = \big(\mathbf{H}_{\|}^{-1/2}(\Gamma)\big)'$ and $\mathbf{V}_k\mathbf{H}_{\perp}^{-1/2}(\Gamma) = \mathbf{H}_{\perp}^{1/2}(\Gamma) = \big(\mathbf{H}_{\perp}^{-1/2}(\Gamma)\big)'$, we have

$$|\boldsymbol{\mathcal{A}}(\varphi,\psi;\overline{\varphi^t},\overline{\psi^t})|$$
$$\leq |\langle\pi_t\mathbf{V}_k(\mathbf{grad}_{\Gamma}\varphi+\mathbf{curl}_{\Gamma}\psi),(\mathbf{grad}_{\Gamma}\overline{\varphi}^t+\mathbf{curl}_{\Gamma}\overline{\psi}^t)\rangle|+\frac{1}{k^2}|\langle V_k\Delta_{\Gamma}\varphi,\Delta_{\Gamma}\overline{\varphi}^t\rangle|$$
$$\leq \Big(\|(\pi_t\mathbf{V}_k(\mathbf{grad}_{\Gamma}\varphi)\|_{\mathbf{H}_{\|}^{\frac{1}{2}}(\Gamma)} + \|\pi_t\mathbf{V}_k\mathbf{curl}_{\Gamma}\psi)\|_{\mathbf{H}_{\|}^{\frac{1}{2}}(\Gamma)}\Big)\times$$
$$(\|\mathbf{grad}_{\Gamma}\overline{\varphi}^t\|_{\mathbf{H}_{\|}^{-\frac{1}{2}}(\Gamma)}+\|(\mathbf{curl}_{\Gamma}\overline{\psi}^t\|_{\mathbf{H}_{\|}^{-\frac{1}{2}}(\Gamma)})+\frac{1}{k^2}\|V_k\Delta_{\Gamma}\varphi\|_{H^{\frac{1}{2}}(\Gamma)}\|\Delta_{\Gamma}\overline{\varphi}^t\|_{H^{-\frac{1}{2}}(\Gamma)}$$

the mapping properties of the single layer potential operator of the Helmholtz equation (see Theorem 5.6.2 and Theorem 5.8.1) assure the continuity estimate for any $k>0$:

$$\begin{aligned}|\boldsymbol{\mathcal{A}}(\varphi,\psi;\overline{\varphi}^t,\overline{\psi}^t)| \leq c_{V,k}\{&(\|\mathbf{grad}_{\Gamma}\varphi\|_{\mathbf{H}_{\|}^{-1/2}(\Gamma)} + \|\mathbf{grad}_{\Gamma}\overline{\varphi}^t\|_{\mathbf{H}_{\|}^{-1/2}(\Gamma)})\\ &\times(\|\mathbf{curl}_{\Gamma}\psi\|_{\mathbf{H}_{\|}^{-1/2}(\Gamma)} + \|\mathbf{curl}_{\Gamma}\overline{\psi}^t\|_{\mathbf{H}_{\|}^{-1/2}(\Gamma)})\}\\ &+\frac{1}{k^2}\,c_{V,k}\{\|\Delta\varphi\|_{H^{-1/2}(\Gamma)}\cdot\|\Delta\overline{\varphi}^t\|_{H^{-1/2}(\Gamma)}\}\end{aligned}$$

for all (φ,ψ) and $(\overline{\varphi}^t,\overline{\psi}^t)\in\mathcal{H}(\Gamma)\times H^{-1/2}(\Gamma)$ and with a positive constant $c_{V,k}$ depending only on k and Γ.

Note that continuity of the principal parts $\mathbf{B}_0, \boldsymbol{C}_0$ and the remainder $\boldsymbol{R}$ follow as well. ∎

Lemma 6.12.3. *The sesquilinear forms*

$$\langle\pi_t\mathbf{V}_k\mathbf{grad}_{\Gamma}\varphi,\mathbf{curl}_{\Gamma}\overline{\psi}^t\rangle\,,\ \langle\pi_t\mathbf{V}_k\mathbf{curl}_{\Gamma}\psi,\mathbf{grad}_{\Gamma}\overline{\varphi}^t\rangle$$
$$\text{and}\ \ \langle\pi_t\mathbf{V}_k\mathbf{grad}_{\Gamma}\varphi,\mathbf{grad}_{\Gamma}\overline{\varphi}^t\rangle$$

are compact for $\varphi,\varphi^t\in\mathcal{H}(\Gamma)$*, since* $\mathbf{grad}_{\Gamma}\varphi$ *and* $\mathbf{grad}_{\Gamma}\varphi^t\in\boldsymbol{L}_2(\Gamma)$ *and* $\mathbf{curl}_{\Gamma}\psi$ *and* $\mathbf{curl}_{\Gamma}\psi^t\in\mathbf{H}_{\|}^{-1/2}(\Gamma)$.

Proof: Since $\mathbf{V}_k$ maps $\mathbf{H}^{-1/2}(\Gamma)$ into $\mathbf{H}^{1/2}(\Gamma)$, one obtains

$$\begin{aligned}|\langle \pi_t \mathbf{V}_k \mathbf{grad}\,_\Gamma \varphi, \mathbf{curl}\,_\Gamma \overline{\psi}^t |\rangle &\le c_1 \|\pi_t \mathbf{V}_k \mathbf{grad}\,_\Gamma \varphi\|_{\mathbf{H}_\|^{1/2}(\Gamma)} \cdot \|\mathbf{curl}\,_\Gamma \overline{\psi}^t\|_{\mathbf{H}_\|^{-1/2}(\Gamma)},\\ &\le c_2 \|\mathbf{grad}\,_\Gamma \varphi\|_{\mathbf{H}_\|^{-1/2}(\Gamma)} \|\mathbf{curl}\,_\Gamma \overline{\psi}^t\|_{\mathbf{H}_\|^{-1/2}(\Gamma)},\\ &\le c_2 \|\mathbf{grad}\,_\Gamma \varphi\|_{\boldsymbol{L}_2(\Gamma)} \|\mathbf{curl}\,_\Gamma \overline{\psi}^t\|_{\mathbf{H}_\|^{-1/2}(\Gamma)}.\end{aligned}$$

Since the imbedding $\boldsymbol{L}_2(\Gamma) = \mathbf{H}^0(\Gamma) \hookrightarrow \mathbf{H}^{-1/2}(\Gamma)$ is compact due to the Rellich Lemma Theorem 4.1.6, the sesquilinear form is compact.

For the second sesquilinear form we have

$$\begin{aligned}|\langle \pi_t \mathbf{V}_k \mathbf{curl}\,_\Gamma \psi, \mathbf{grad}\,_\Gamma \overline{\varphi}^t\rangle| &\le c_2 \|\pi_t \mathbf{V} \mathbf{curl}\,_\Gamma \psi\|_{\mathbf{H}_\|^{1/2}(\Gamma)} \|\mathbf{grad}\,_\Gamma \overline{\varphi}^t\|_{\mathbf{H}_\|^{-1/2}(\Gamma)}\\ &\le c_3 \|\mathbf{curl}\,_\Gamma \psi\|_{\mathbf{H}_\|^{-1/2}(\Gamma)} \|\mathbf{grad}\,_\Gamma \overline{\varphi}^t\|_{\boldsymbol{L}_2(\Gamma)}\end{aligned}$$

and again compactness of $\boldsymbol{L}_2(\Gamma) \hookrightarrow \mathbf{H}^{-1/2}(\Gamma)$ implies that the sesquilinear form is compact.

The third sesquilinear form implies

$$\begin{aligned}|\langle \pi_t \mathbf{V}_k \mathbf{grad}\,_\Gamma \varphi, \mathbf{grad}\,_\Gamma \overline{\varphi}^t\rangle| &\le \|\pi_t \mathbf{V}_k \mathbf{grad}\,_\Gamma \varphi\|_{\mathbf{H}_\|^{1/2}(\Gamma)} \|\mathbf{grad}\,_\Gamma \overline{\varphi}^t\|_{\mathbf{H}_\|^{-1/2}(\Gamma)}\\ &\le c_v \|\mathbf{grad}\,_\Gamma \varphi\|_{\mathbf{H}_\|^{-1/2}(\Gamma)} \|\mathbf{grad}\,_\Gamma \overline{\varphi}^t\|_{\mathbf{H}_\|^{-1/2}(\Gamma)}\\ &\le c_v \|\mathbf{grad}\,_\Gamma \varphi\|_{\boldsymbol{L}_2(\Gamma)} \|\mathbf{grad}\,_\Gamma \overline{\varphi}^t\|_{\boldsymbol{L}_2(\Gamma)}.\end{aligned}$$

and the compactness of $\boldsymbol{L}_2(\Gamma) \hookrightarrow \mathbf{H}^{-1/2}(\Gamma)$ again implies the compactness of the sesquilinear form. ∎

Since in (6.12.19)

$$\langle \overline{\mathbf{J}}^t, \pi_t \mathbf{V}(\mathbf{J})\rangle = \langle \mathbf{J}, \pi_t \mathbf{V}(\overline{\mathbf{J}}^t)\rangle$$

we may rewrite (6.12.22) in the form

$$\begin{aligned}\langle \mathbf{grad}\,_\Gamma \varphi + \mathbf{curl}\,_\Gamma \psi, \pi_t \mathbf{V}(\mathbf{grad}\,_\Gamma \overline{\varphi}^t + \mathbf{curl}\,_\Gamma \overline{\psi}^t)\rangle + \langle -\Delta_\Gamma \varphi, k^{-2} V(\Delta_\Gamma \overline{\varphi}^t)\rangle\\ = \langle \boldsymbol{g}, \mathbf{grad}\,_\Gamma \overline{\varphi}^t + \mathbf{curl}\,_\Gamma \overline{\psi}^t\rangle\end{aligned}$$

which motivates us to introduce a modified form in Theorem 6.12.4 below.

Let $\boldsymbol{e}_1, \boldsymbol{e}_2, \boldsymbol{n}$ be an orthonormal basis of local coordinates at $\boldsymbol{x} \in \Gamma$. Then $\mathbf{grad}\,_\Gamma \varphi = \partial_1 \varphi \boldsymbol{e}_1 + \partial_2 \varphi \boldsymbol{e}_2$ and $\mathbf{curl}\,_\Gamma \psi = \partial_2 \varphi \boldsymbol{e}_1 - \partial_1 \varphi \boldsymbol{e}_2$.

Theorem 6.12.4. *The following Gårding inequality holds for the principal part of* $\boldsymbol{\mathcal{A}}$, *that is, there exist a compact linear operator* $\boldsymbol{\mathcal{C}}$ *and a positive coerciveness constant* c_0 *depending on* k^2 *and* Γ *such that*

$$\begin{aligned}
&\mathrm{Re}\left\langle (-\Delta_\Gamma\varphi, \mathbf{grad}_\Gamma\varphi\,,\ \mathbf{curl}_\Gamma\psi), \boldsymbol{\Theta}\begin{pmatrix} k^{-2}V_0\Delta_\Gamma\overline{\varphi} \\ \pi_t\mathbf{V}_0\,\mathbf{grad}_\Gamma\,\overline{\varphi} \\ \pi_t\mathbf{V}_0\,\mathbf{curl}_\Gamma\,\overline{\psi}\end{pmatrix}\right\rangle \\
&+\langle(-\Delta_\Gamma\varphi, \mathbf{grad}_\Gamma\varphi\,,\ \mathbf{curl}_\Gamma\psi), \boldsymbol{\mathcal{C}}(-\Delta_\Gamma\overline{\varphi}, \mathbf{grad}_\Gamma\overline{\varphi}\,,\ \mathbf{curl}_\Gamma\overline{\psi})\rangle \qquad (6.12.24)\\
&\geq c_0\big\{\|\Delta_\Gamma\varphi\|^2_{H^{-1/2}(\Gamma)} + \|\mathbf{grad}_\Gamma\varphi\|^2_{\mathbf{H}_\|^{-1/2}(\Gamma)} + \|\mathbf{curl}_\Gamma\psi\|^2_{\mathbf{H}_\|^{-1/2}(\Gamma)}\big\},
\end{aligned}$$

where

$$\boldsymbol{\Theta} = \begin{pmatrix} -1 & 0 & 0 \\ 0 & 1 & 0 \\ 0 & 0 & 1\end{pmatrix}.$$

The proof will be given in Section 12.9 and follows from strong ellipticity of $\boldsymbol{\mathcal{A}}$.

Since the sesquilinear form $\langle\pi_t\mathbf{V}_k\mathbf{grad}_\Gamma\varphi, \mathbf{grad}_\Gamma\overline{\varphi}^t\rangle$ is compact on $\mathbf{H}\times\mathbf{H}$ due to Lemma 6.12.3 this sesquilinear form could be discarded in (6.12.24) for getting strong ellipticity of the remaining 2×2–system. However, Gårding's inequality for the extended system (6.12.24) supports the convergence and error analysis for corresponding Galerkin boundary element approximation.

From Gårding's inequality (6.12.24) we conclude that the system of equations (6.12.19) defines a Fredholm mapping of index zero. However, there exist examples where the dimension of the mapping's kernel is non zero for a whole family of integral equations (see Hellwig [175, Chap. 5.4.5], [176], [177]). Hence, in order the Fredholm alternative to hold for the system (6.12.22) we need in the Fredholm resolvent set for $k^2 > 0$ at least one point in $\mathbb{C}$ where we have uniqueness.

Lemma 6.12.5. *For* $0 < k < k_1 = \lambda_1^{-1/2}$ *where* λ_1 *is given in Theorem 6.8.2, the solution of EFIE is unique.*

Proof: Since EFIE is equivalent to (6.12.18) non uniquenes would imply that to all $k \in (0, k_1)$ there exists $\mathbf{J}_0 = \mathbf{grad}_\Gamma\varphi_0 + \mathbf{curl}_\Gamma\psi_0 \neq \mathbf{0}$ satisfying (6.12.19) with $\mathbf{J} = \mathbf{J}^t = \mathbf{J}_0$ and $\boldsymbol{g} = \mathbf{0}$:

$$\langle\overline{\mathbf{J}}_0, \pi_t\mathbf{V}\mathbf{J}_0\rangle - \frac{1}{k^2}\langle\Delta_\Gamma\overline{\varphi}_0, V\Delta_\Gamma\varphi_0\rangle = 0.$$

Due to the invertibility of V and its coerciveness this implies

$$c_V\|\Delta_\Gamma\varphi_0\|^2_{H^{-1/2}(\Gamma)} \leq \langle\Delta_\Gamma\overline{\varphi}_0, V\Delta_\Gamma\varphi_0\rangle = k^2\langle\overline{\mathbf{J}}_0, \pi_t\mathbf{V}\mathbf{J}_0\rangle$$

where c_V does not depend on k.

Without loss of generality, $\|\mathbf{J}_0\|_{\mathbf{H}_\|^{-1/2}(\mathrm{div}_\Gamma,\Gamma)} = 1$ for all $k \in (0, k_1)$. Then for $k \to 0$ and $\langle\overline{\mathbf{J}}_0, \pi_t\mathbf{V}\mathbf{J}_0\rangle \leq c_V\|\mathbf{J}_0\|^2_{\mathbf{H}_\|^{-1/2}(\mathrm{div}_\Gamma,\Gamma)} = c_V$ and $0 < k$ sufficiently small, $\|\Delta_\Gamma\varphi_0\|_{H^{-1/2}(\Gamma)}$ becomes arbitrarily small. For $k = 0$, $\Delta_\Gamma\varphi_0 = 0$, $\mathbf{J}_0 = \mathbf{0}$ in contrast to $\|\mathbf{J}_0\|_{\mathbf{H}_\|^{-1/2}(\mathrm{div}_\Gamma,\Gamma)} = 1$. But the resolvent set is open and contains $0 < k \in \mathbb{C}$ where we still have uniqueness. Hence, the solution $\mathbf{J}$ in (6.12.18) for given $\boldsymbol{g}$ is unique. ∎

Theorem 6.12.6. *If $k^2 > 0$ is not an eigenvalue of the Dirichlet problem for the interior Maxwell equations, for any given $\boldsymbol{g} \in \mathbf{H}_{\perp}^{-\frac{1}{2}}(\mathbf{curl}_{\Gamma} 0, \Gamma)$ there exists a unique solution $\mathbf{J} \in \mathbf{H}_{||}^{-\frac{1}{2}}(\mathrm{div}_{\Gamma}, \Gamma)$ of the equation* (6.12.18).

Proof: The Gårding inequality (6.12.24) implies that for equation (6.12.18) the Fredholm alternative is valid. Since then uniqueness implies existence if k^2 is not an eigenvalue of the integral equation for small $k^2 > 0$, the equation (6.12.18) is uniquely solvable and therefore the Fredholm index, $\kappa = \alpha - \beta = 0$ and $\alpha = \beta = 0$. This implies unique solvability (6.12.22) for all k^2 not an eigenvalue of (6.12.18), which is the case for k^2 not an eigenvalue of the interior Dirichlet problem of Maxwell's equation (see e.g. Maz'ya [289], Buffa, Costabel and Sheen [52]) and one has the a priori estimate

$$\|\mathbf{J}\|_{\mathbf{H}_{||}^{-\frac{1}{2}}(\mathrm{div}_{\Gamma}, \Gamma)} \leq c \|\boldsymbol{g}\|_{\mathbf{H}_{\perp}^{-\frac{1}{2}}(\mathrm{div}_{\Gamma}, \Gamma)}$$

with a constant $c(k^2, \Gamma) > 0$. ∎

A saddlepoint formulation for the construction of the solution of the EFIE

Buffa, Costabel and Schwab [51] present also a saddle point formulation for solving the **EFIE** with a boundary element method where the incident field on Γ is supposed to be given in the form

$$\pi_t \mathbf{E}^i = \mathbf{grad}_{\Gamma} \alpha + \mathbf{curl}_{\Gamma} \beta \quad \text{with} \quad \alpha \in H^{1/2}(\Gamma), \ \beta \in \mathcal{H}(\Gamma).$$

In addition to the unknown

$$\mathbf{J} = \mathbf{grad}_{\Gamma} \varphi + \mathbf{curl}_{\Gamma} \psi, \ \varphi \in \mathcal{H}(\Gamma), \ \psi \in H^{1/2}(\Gamma)$$

and test fields

$$\mathbf{J}^t = \mathbf{grad}_{\Gamma} \varphi^t + \mathbf{curl}_{\Gamma} \psi^t, \ \varphi^t \in \mathcal{H}(\Gamma), \ \psi^t \in H^{1/2}(\Gamma)$$

an additional unknown $m = -\Delta_{\Gamma} \varphi$ and $m^t = -\Delta_{\Gamma} \varphi^t$ are introduced as well as a Lagrangian λ and λ^t in $H_*^{-1/2}(\Gamma)$. The saddle point problem reads:

Find $(\varphi, \psi, m, \lambda) \in \mathbb{X}$ satisfying the EFIE saddle point equations

$$\begin{aligned}
& i\omega\mu\langle \mathbf{grad}_{\Gamma} \varphi + \mathbf{curl}_{\Gamma} \psi, \pi_t \mathbf{V}_k(\mathbf{grad}_{\Gamma} \overline{\varphi}^t + \mathbf{curl}_{\Gamma} \overline{\psi}^t)\rangle \\
& \qquad - \langle \mathbf{grad}_{\Gamma} \lambda, \mathbf{grad}_{\Gamma} \overline{\varphi}^t \rangle = -\langle \mathbf{grad}_{\Gamma} \alpha, \mathbf{grad}_{\Gamma} \overline{\varphi}^t \rangle - \langle \Delta_{\Gamma} \beta, \overline{\psi}^t \rangle, \\
& -\frac{i}{\varepsilon\omega} \langle m, \pi_t V_k \overline{m}^t \rangle + \langle \lambda, \overline{m}^t \rangle = 0, \qquad (6.12.25) \\
& -\langle \mathbf{grad}_{\Gamma} \varphi, \mathbf{grad}_{\Gamma} \overline{\lambda}^t \rangle + \langle m, \overline{\lambda}^t \rangle = 0
\end{aligned}$$

for all $(\varphi^t, \psi^t, m^t, \lambda^t) \in \mathbb{X}$ with the Hilbert space

$$\mathbb{X} = H^1(\Gamma)/\mathbb{C} \times H^{1/2}(\Gamma)/\mathbb{C} \times H_*^{-1/2}(\Gamma) \times H^1(\Gamma)/\mathbb{C}.$$

With similar arguments as in Lemma 6.12.3 it follows that the sequilinear forms

$$\omega\mu\langle \mathbf{grad}_\Gamma\varphi + \mathbf{curl}_\Gamma\psi, -\pi_t \mathbf{V}_k \mathbf{grad}_\Gamma\overline{\lambda}^t\rangle, \omega\mu\langle \mathbf{grad}_\Gamma\varphi, \pi_t\mathbf{V}_k\mathbf{curl}_\Gamma\overline{\psi t}\rangle,$$
$$\omega\mu\langle \mathbf{curl}_\Gamma\psi, \pi_t\mathbf{V}_k\mathbf{grad}_\Gamma\overline{\varphi}^t\rangle, \langle\lambda, \overline{m}^t\rangle \text{ and } \langle m, \overline{\varphi}^t\rangle$$

are compact on $\mathbb{X}$. Hence, these terms in the System (6.12.25) are moved to the compact part of the sequilinear form, and the system now reads:

$$\begin{aligned}
&\mathfrak{B}\big((\varphi,\psi,m,\lambda)\,,\,(\overline{\varphi}^t,\overline{\psi}^t,\overline{m}^t,\overline{\lambda}^t)\big) := &(6.12.26)\\
&\quad i\omega\mu\langle \mathbf{grad}_\Gamma\varphi, \pi_t\mathbf{V}_k\mathbf{grad}_\Gamma\overline{\varphi}^t\rangle + i\omega\mu\langle \mathbf{curl}_\Gamma\psi, \pi_t\mathbf{V}_k\mathbf{curl}_\Gamma\overline{\psi}^t\rangle\\
&-\langle \mathbf{grad}_\Gamma\lambda, \mathbf{grad}_\Gamma\overline{\varphi}^t\rangle + \boldsymbol{\mathcal{C}}(\varphi,\psi,\varphi^t,\psi^t) = -\langle \mathbf{grad}_\Gamma\alpha, \mathbf{grad}_\Gamma\varphi\rangle - \langle \Delta_\Gamma\beta, \psi\rangle,\\
&\quad -\frac{i}{\varepsilon\omega}\langle m, \pi_t V_k\overline{m}^t\rangle + \langle\lambda, \overline{m}^t\rangle = 0,\\
&\quad -\langle \mathbf{grad}_\Gamma\varphi, \mathbf{grad}_\Gamma\overline{\lambda}^t\rangle + \langle m, \overline{\lambda}^t\rangle = 0.
\end{aligned}$$

In order to obtain strong ellipticity, the authors Buffa, Costabel and Schwab [51] find

$$\boldsymbol{\Theta} = \begin{pmatrix} 0 & 0 & 0 & -1\\ 0 & 1 & 0 & 0\\ 0 & 0 & -1 & 0\\ -1 & 0 & 0 & 0 \end{pmatrix}$$

and the principal part modulo compact sesquilinear forms reads

$$\begin{aligned}
&\mathfrak{B}_0\big((\varphi,\psi,m,\lambda)\,,\,\boldsymbol{\Theta}(\overline{\varphi},\overline{\psi},\overline{m},\overline{\lambda})\big)\\
&= \left\langle (\mathbf{grad}_\Gamma\varphi, \mathbf{curl}_\Gamma\psi, m, \mathbf{grad}_\Gamma\lambda), \boldsymbol{\Theta}\begin{pmatrix} i\omega\mu\pi_t\mathbf{V}_0(\mathbf{grad}_\Gamma\overline{\varphi})\\ i\omega\mu\pi_t\mathbf{V}_0(\mathbf{curl}_\Gamma\overline{\psi})\\ -(i/\omega\varepsilon)V_0(\overline{m})\\ \mathbf{grad}_\Gamma\overline{\lambda}\end{pmatrix}\right\rangle \quad (6.12.27)\\
&= \frac{i}{\varepsilon\omega}\langle m, \pi_t V_0\overline{m}\rangle + i\omega\mu\langle \mathbf{grad}_\Gamma\varphi, \pi_t\mathbf{V}_0\overline{\mathbf{grad}_\Gamma\varphi}\rangle\\
&+i\omega\mu\langle \mathbf{curl}_\Gamma\psi, \pi_t\mathbf{V}_0\overline{\mathbf{curl}_\Gamma\psi}\rangle + \langle \mathbf{grad}_\Gamma\lambda, \overline{\mathbf{grad}_\Gamma\lambda}\rangle + \langle \mathbf{grad}_\Gamma\varphi, \overline{\mathbf{grad}_\Gamma\varphi}\rangle.
\end{aligned}$$

Hence, $\mathfrak{B}_0$ satisfies the coerciveness inequalities

$$\begin{aligned}
&\operatorname{Im}\mathfrak{B}_0\big((\varphi,\psi,m,\lambda), \boldsymbol{\Theta}(\overline{\varphi},\overline{\psi},\overline{m},\overline{\lambda})\big) \geq &(6.12.28)\\
&\frac{c_1}{\varepsilon\omega}\|m\|^2_{H^{-1/2}(\Gamma)} + \omega\mu c_2\|\mathbf{grad}_\Gamma\varphi\|^2_{H^{-1/2}(\Gamma)} + \omega\mu c_2\|\mathbf{curl}_\Gamma\psi\|^2_{\mathbf{H}^{-1/2}(\Gamma)},\\
&\operatorname{Re}\mathfrak{B}_0\big((\varphi,\psi,m,\lambda), \boldsymbol{\Theta}(\overline{\varphi},\overline{\psi},\overline{m},\overline{\lambda})\big) \geq \|\mathbf{grad}_\Gamma\lambda\|^2_{\mathbf{L}^2_t(\Gamma)} + \|\mathbf{grad}_\Gamma\varphi\|^2_{\mathbf{L}^2_t(\Gamma)}
\end{aligned}$$

and the sesquilinear form $\mathfrak{B}$ satisfies a corresponding Gårding inequality. As a consequence, the Galerkin approximation of the equations (6.12.25) is stable in $\mathbb{X}$ and asymtotically qusioptimal convergent as analysed also by Buffa, Costabel and Schwab in [51].

Since $\Delta_\Gamma \varphi = m \in H^{-1/2}(\Gamma)$ is on Γ explicitly available if the system (6.12.26) is numerically apprimated with boundary element Galerkin methods, also φ may be approximated on Γ by using surface Crouzeix –Raviard elements as investigated by Hailong Guo in [164].

Note that in terms of the solution $\mathbf{J} \in \mathbf{H}_{||}^{-1/2}(\mathrm{div}_\Gamma, \Gamma)$ of the **EFIE**, the electric field $\mathbf{E}$ is given from the representation formula (6.12.5) in Ω^c, namely:

$$\begin{aligned}\mathbf{E}(\boldsymbol{x}) = \mathbf{E}^i(\boldsymbol{x}) &+ i\omega\mu \int\limits_\Gamma G_k(\boldsymbol{x}-\boldsymbol{y})\mathbf{J}(\boldsymbol{y})ds_{\boldsymbol{y}} \\ &+ \frac{1}{i\omega\varepsilon} \int\limits_\Gamma \mathbf{grad}_{\,y} G_k(\boldsymbol{x}-\boldsymbol{y})\mathrm{div}_\Gamma \mathbf{J}(\boldsymbol{y})ds_{\boldsymbol{y}},\end{aligned}$$

and the magnetic field $\mathbf{H}$ can be obtained in terms of $\mathbf{E}(\boldsymbol{x})$ from the Maxwell system in the form:

$$\begin{aligned}\mathbf{H}(\boldsymbol{x}) &= \mathbf{curl}\ \mathbf{E}(\boldsymbol{x})/i\omega\mu \\ &= \frac{\mathbf{curl}\,\mathbf{E}^i(\boldsymbol{x})}{i\omega\mu} + \mathbf{curl} \int\limits_\Gamma G_k(\boldsymbol{x}-\boldsymbol{y})\mathbf{J}(\boldsymbol{y})ds_{\boldsymbol{y}} \\ &= \mathbf{H}^i(\boldsymbol{x}) + \int\limits_\Gamma \mathbf{J}(\boldsymbol{y}) \times \mathbf{grad}_{\,y} G_k(\boldsymbol{x}-\boldsymbol{y})ds_{\boldsymbol{y}}\,,\ \boldsymbol{x} \in \Omega^c,\end{aligned}$$

which is the same as in (6.12.6), where

$$\mathbf{H}^i(\boldsymbol{x}) := \mathbf{curl}\ \mathbf{E}^i(\boldsymbol{x})/i\omega\mu.$$

On the other hand, we may alternatively begin with the representation formula of $\mathbf{H}(\boldsymbol{x})$ in Ω^c given by (6.12.6):

$$\mathbf{H}(\boldsymbol{x}) = \mathbf{H}^i(\boldsymbol{x}) + \int\limits_\Gamma \mathbf{J}(\boldsymbol{y}) \times \mathbf{grad}_{\,y} G_k(\boldsymbol{x}-\boldsymbol{y})ds_{\boldsymbol{y}}$$

with $\mathbf{J} \in \mathbf{H}_{||}^{-1/2}(\mathrm{div}_\Gamma, \Gamma)$ being the solution of the **MFIE** given by (6.12.8):

$$\frac{1}{2}\mathbf{J}(\boldsymbol{y}) - \boldsymbol{n}(\boldsymbol{x}) \times \int\limits_\Gamma \mathbf{J}(\boldsymbol{y}) \times \mathbf{grad}_{\,y} G_k(\boldsymbol{x}-\boldsymbol{y})ds_{\boldsymbol{y}} = \gamma_t \mathbf{H}^i(\boldsymbol{x})\,,\ \boldsymbol{x} \in \Gamma. \quad (6.12.29)$$

In terms of the boundary integral operator

$$\begin{aligned}\boldsymbol{R}(\mathbf{J})(\boldsymbol{x}) := \boldsymbol{n}(\boldsymbol{x}) &\times \int\limits_\Gamma \mathbf{J}(\boldsymbol{y}) \times \mathbf{grad}_{\,y} G_k(\boldsymbol{x}-\boldsymbol{y})ds_{\boldsymbol{y}} \\ &= \boldsymbol{n}(\boldsymbol{x}) \times \mathbf{curl} \int\limits_\Gamma G_k(\boldsymbol{x}-\boldsymbol{y})ds_{\boldsymbol{y}}\,,\ \boldsymbol{x} \in \Gamma\end{aligned}$$

from (6.12.7), the **MFIE** can be rewritten in the form:

$$(\frac{1}{2}\boldsymbol{I} - \boldsymbol{R})\mathbf{J}(\boldsymbol{x}) = \gamma_t \mathbf{H}^i(\boldsymbol{x}), \ \boldsymbol{x} \in \Gamma. \tag{6.12.30}$$

Before we discuss the solvability of (6.12.30), we see that in terms of the solution $\mathbf{J}(\boldsymbol{x})$ of (6.12.30), the magnetic field $\mathbf{H}(\boldsymbol{x})$ is given from the representation formula (6.12.6) and the electric field $\mathbf{E}(\boldsymbol{x})$ in Ω^c can be recovered from the Maxwell system:

$$\mathbf{E}(\boldsymbol{x}) = \frac{\mathbf{curl}\ \mathbf{H}^i(\boldsymbol{x})}{-i\omega\varepsilon} + \frac{\mathbf{curl}}{-i\omega\varepsilon}\int\limits_\Gamma \mathbf{J}(\boldsymbol{y}) \times \mathbf{grad}_{\boldsymbol{y}} G_k(\boldsymbol{x}-\boldsymbol{y}) ds_{\boldsymbol{y}}.$$

A simple computation shows that if $\mathbf{E}^i(\boldsymbol{x}) = \frac{\mathbf{curl}\ \mathbf{H}^i(\boldsymbol{x})}{-i\omega\varepsilon}$, then

$$\begin{aligned}\mathbf{E}(\boldsymbol{x}) &= \mathbf{E}^i(\boldsymbol{x}) + \frac{\mathbf{curl}\ \mathbf{curl}}{-i\omega\varepsilon}\int\limits_\Gamma G_k(\boldsymbol{x}-\boldsymbol{y})\mathbf{J}(\boldsymbol{y}) ds_{\boldsymbol{y}} \\ &= \mathbf{E}^i(\boldsymbol{x}) - \frac{1}{i\omega\varepsilon}\{-\Delta_{\boldsymbol{x}} + \mathbf{grad}_{\boldsymbol{x}}\mathrm{div}_{\boldsymbol{x}}\}\int\limits_\Gamma G_k(\boldsymbol{x}-\boldsymbol{y})\mathbf{J}(\boldsymbol{y}) ds_{\boldsymbol{y}}\end{aligned}$$

$$\begin{aligned}\mathbf{E}(\boldsymbol{x}) = \mathbf{E}^i(\boldsymbol{x}) &- \frac{1}{i\omega\varepsilon}\{k^2\int\limits_\Gamma G_k(\boldsymbol{x}-\boldsymbol{y})\mathbf{J}(\boldsymbol{y}) ds_{\boldsymbol{y}} \\ &- \int\limits_\Gamma \mathbf{grad}_{\boldsymbol{y}} G_k(\boldsymbol{x}-\boldsymbol{y}))\mathrm{div}_\Gamma \mathbf{J}(\boldsymbol{y}) ds_{\boldsymbol{y}}\} \\ = \mathbf{E}^i(\boldsymbol{x}) &+ i\omega\mu\int\limits_\Gamma G_k(\boldsymbol{x}-\boldsymbol{y})\mathbf{J}(\boldsymbol{y}) ds_{\boldsymbol{y}} \\ &+ \frac{1}{i\omega\varepsilon}\int\limits_\Gamma \mathbf{grad}_{\boldsymbol{y}} G_k(\boldsymbol{x}-\boldsymbol{y})\mathrm{div}_\Gamma \mathbf{J}(\boldsymbol{y}) ds_{\boldsymbol{y}},\end{aligned}$$

which is exactly the same as (6.12.5), since

$$\frac{1}{i\omega\varepsilon}\mathrm{div}_\Gamma \mathbf{J}(\boldsymbol{y}) = \boldsymbol{n}(\boldsymbol{y}) \cdot \mathbf{E}(\boldsymbol{y}).$$

Now let us consider the solution of (6.12.30).

Solvability of MFIE

For smooth Γ (say $\Gamma \in C^{1+\alpha}$), the kernel of $\boldsymbol{R}$ is **weakly singular** and hence $\boldsymbol{R} \in \psi dO_{-1}$. This can be seen as follows:

$$\begin{aligned}
\boldsymbol{R}(\mathbf{J})(\boldsymbol{x}) &= \boldsymbol{n}(\boldsymbol{x}) \times \mathbf{curl} \int\limits_{\Gamma} G_k(\boldsymbol{x}-\boldsymbol{y})\mathbf{J}(\boldsymbol{y})ds_{\boldsymbol{y}} \quad \text{for } \boldsymbol{x} \in \Gamma, \\
&= \int\limits_{\Gamma} \boldsymbol{n}(\boldsymbol{x}) \times \big(\mathbf{grad}_{\boldsymbol{x}} G_k(\boldsymbol{x}-\boldsymbol{y}) \times \mathbf{J}(\boldsymbol{y})\big)ds_{\boldsymbol{y}} \\
&= \int\limits_{\Gamma} \Big\{\big(\boldsymbol{n}(\boldsymbol{x}) \cdot \mathbf{J}(\boldsymbol{y})\big)\mathbf{grad}_{\boldsymbol{x}} G_k(\boldsymbol{x}-\boldsymbol{y}) - \frac{\partial G_k}{\partial n_{\boldsymbol{x}}}(\boldsymbol{x}-\boldsymbol{y})\mathbf{J}(\boldsymbol{y})\Big\}ds_{\boldsymbol{y}} \\
&= \int\limits_{\Gamma} \Big[\big(\boldsymbol{n}(\boldsymbol{x}) - \boldsymbol{n}(\boldsymbol{y})\big) \cdot \mathbf{J}(\boldsymbol{y})\Big]\mathbf{grad}_{\boldsymbol{x}} G_k(\boldsymbol{x}-\boldsymbol{y})ds_{\boldsymbol{y}} \\
&\quad - \int\limits_{\Gamma} \Big[\frac{\partial G_k}{\partial n_{\boldsymbol{x}}}(\boldsymbol{x}-\boldsymbol{y})\mathbf{J}(\boldsymbol{y})\Big]ds_{\boldsymbol{y}}
\end{aligned}$$

Since $\boldsymbol{n}(\boldsymbol{y}) \cdot \mathbf{J}(\boldsymbol{y}) = 0$ for $\mathbf{J} \in \mathbf{H}_{||}^{-1/2}(\mathrm{div}_\Gamma, \Gamma)$. Thus, for smooth Γ,

$$\boldsymbol{R} : \mathbf{H}^{-1/2}(\Gamma) \to \mathbf{H}^{-1/2}(\Gamma) \overset{\text{comp}}{\hookrightarrow} \mathbf{H}^{-1/2}(\Gamma)$$

is a compact operator and the **MFIE** is a Fredholm integral equation of the second kind.

Hence, uniqueness implies existence. However, it is well-known that solutions of **MFIE** as well as **EFIE** for perfectly electric conducting scatterers are non–unique for irregular frequencies which correspond to interior Maxwell resonances of the scatterer. It is not our intent to address this question here. Various methods of eliminating these spurious resonances have been proposed (see, e.g. Colton and Kress [75], Mautz and Harrington [285], Yaghjian [441], and Jost [221]). For smooth boundary Γ, variational solutions of (6.12.30) in standard Sobolev spaces may be treated similar to boundary integral equations in Section 5.1.6. for classical solutions of (6.12.30), we refer to the work of Colton and Kress [75] (see also Hsiao, Kleinman and Wang [195]).

It is noteworthy to mention that in order to apply the Fredholm alternative to an integral equation of the second kind such as the **MFIE** with non–unique solutions, we need to investigate the corresponding homogeneous adjoint (or rather transposed) equations. For $(\boldsymbol{\lambda}, \boldsymbol{\varphi}) \in \boldsymbol{L}_t^2(\Gamma)$, we see that the transposed operator $\boldsymbol{R}'$ of $\boldsymbol{R}$ with respect to the duality product

$$\langle \boldsymbol{\lambda}, \boldsymbol{\varphi} \rangle := \int\limits_{\Gamma} \boldsymbol{\lambda} \cdot \boldsymbol{\varphi}\, ds,$$

is given by $\boldsymbol{R}' = \gamma_t \boldsymbol{R} \gamma_t$ and hence the L_2 - adjoint operator $\boldsymbol{R}^* \boldsymbol{\lambda} = \overline{(\boldsymbol{R}' \overline{\boldsymbol{\lambda}})}$ (see Colton and Kress [75, 2.82]). This can be seen as follows:

$$
\begin{aligned}
\langle \boldsymbol{\lambda}, \boldsymbol{R}(\boldsymbol{\varphi}) \rangle &= \int_\Gamma \boldsymbol{\lambda}(\boldsymbol{x}) \cdot (\boldsymbol{n}(\boldsymbol{x}) \times \mathbf{curl}_{\boldsymbol{x}}(\mathbf{V}(\boldsymbol{\varphi}))ds_{\boldsymbol{x}} \\
&= \int_\Gamma \Big\{ \mathbf{curl}_{\boldsymbol{x}} \int_\Gamma G_k(\boldsymbol{x}-\boldsymbol{y})\boldsymbol{\varphi}(\boldsymbol{y})ds_{\boldsymbol{y}} \Big\} \cdot (\boldsymbol{\lambda}(\boldsymbol{x}) \times \boldsymbol{n}(\boldsymbol{x}))ds_{\boldsymbol{x}} \\
&= \int_\Gamma \Big\{ \int_\Gamma (\mathbf{grad}_{\boldsymbol{x}} G_k(\boldsymbol{x}-\boldsymbol{y})) \times \boldsymbol{\varphi}(\boldsymbol{y})ds_{\boldsymbol{y}} \Big\} \cdot (\boldsymbol{\lambda}(\boldsymbol{x}) \times \boldsymbol{n}(\boldsymbol{x}))ds_{\boldsymbol{x}} \\
&= \int_\Gamma \boldsymbol{\varphi}(\boldsymbol{y}) \cdot \Big[\int_\Gamma (\boldsymbol{\lambda}(\boldsymbol{x}) \times \boldsymbol{n}(\boldsymbol{x})) \times \mathbf{grad}_{\boldsymbol{x}} G_k(\boldsymbol{x}-\boldsymbol{y})ds_{\boldsymbol{x}} \Big] ds_{\boldsymbol{y}} \\
&= \int_\Gamma \boldsymbol{\varphi}(\boldsymbol{y}) \cdot \Big\{ \int_\Gamma (\mathbf{grad}_{\boldsymbol{x}} G_k(\boldsymbol{x}-\boldsymbol{y})) \times (\boldsymbol{n}(\boldsymbol{x}) \times \boldsymbol{\lambda}(\boldsymbol{x}))ds_{\boldsymbol{x}} \Big\} ds_{\boldsymbol{y}} \\
&= \int_\Gamma \boldsymbol{\varphi}(\boldsymbol{y}) \cdot \Big\{ -\mathbf{curl}_{\boldsymbol{y}} \int_\Gamma G_k(\boldsymbol{x}-\boldsymbol{y})(\boldsymbol{n}(\boldsymbol{x}) \times \boldsymbol{\lambda}(\boldsymbol{x}))ds_{\boldsymbol{x}} \Big\} ds_{\boldsymbol{y}} \\
&= \int_\Gamma \Big\{ \boldsymbol{n}(\boldsymbol{y}) \times (\boldsymbol{n}(\boldsymbol{y}) \times \boldsymbol{\varphi}(\boldsymbol{y}) \Big\} \bullet \\
&\qquad \bullet \Big\{ \mathbf{curl}_{\boldsymbol{y}} \int_\Gamma G_k(\boldsymbol{x}-\boldsymbol{y})(\boldsymbol{n}(\boldsymbol{x}) \times \boldsymbol{\lambda}(\boldsymbol{x}))ds_{\boldsymbol{x}} \Big\} ds_{\boldsymbol{y}}
\end{aligned}
$$

$$
\begin{aligned}
&= \int_\Gamma \Big(\boldsymbol{n}(\boldsymbol{y}) \times \boldsymbol{\varphi}(\boldsymbol{y}) \Big) \bullet \\
&\qquad \bullet \Big(\mathbf{curl}_{\boldsymbol{y}} \int_\Gamma G_k(\boldsymbol{x}-\boldsymbol{y})(\boldsymbol{n}(\boldsymbol{x}) \times \boldsymbol{\lambda}(\boldsymbol{x}))ds_{\boldsymbol{x}} \Big) \times \boldsymbol{n}(\boldsymbol{y})ds_{\boldsymbol{y}} \\
&= -\int_\Gamma (\boldsymbol{n}(\boldsymbol{y}) \times \boldsymbol{\varphi}(\boldsymbol{y})) \cdot \{ \boldsymbol{R}(\boldsymbol{n} \times \boldsymbol{\lambda} \}(\boldsymbol{y})ds_{\boldsymbol{y}} \\
&= -\int_\Gamma \boldsymbol{\varphi}(\boldsymbol{y}) \cdot \{ \boldsymbol{R}(\boldsymbol{n} \times \boldsymbol{\lambda}) \times \boldsymbol{n}(\boldsymbol{y}) \} ds_{\boldsymbol{y}} \\
&= \langle \boldsymbol{\varphi}(\boldsymbol{y}), \, \boldsymbol{n} \times \boldsymbol{R}(\boldsymbol{n} \times \lambda) \rangle.
\end{aligned}
$$

Before we discuss the transposed equation of **MFIE** (6.12.30), we note that **MFIE** is a Fredholm equation of the second kind obtained from the direct method, which seems to have inherited some complications. So for simplicity, let us first consider the solution of the magnetic field $\mathbf{H} \in \Omega^c$ by using the indirect method based on a layer potential ansatz. We may seek a solution in the form:

$$
\mathbf{H}(\boldsymbol{x}) = \mathbf{curl}_{\boldsymbol{x}} \int_\Gamma G_k(\boldsymbol{x}-\boldsymbol{y})\boldsymbol{\varphi}(\boldsymbol{y})ds_{\boldsymbol{y}} \quad \text{in } \Omega^c
$$

in terms of an unknown density boundary field $\boldsymbol{\varphi}$. This reduces to the boundary integral equation of the second kind in the form

$$(\frac{1}{2}\boldsymbol{I} + \boldsymbol{R})\boldsymbol{\varphi} = \gamma_t \mathbf{H}^{\text{inc}}(\boldsymbol{x}) \quad \text{for } \boldsymbol{x} \in \Gamma. \tag{6.12.31}$$

However (6.12.31) appears startingly different from the corresponding equation (6.12.30) in that the equation involves the operator $\frac{1}{2}\boldsymbol{I} + \boldsymbol{R}$ rather than $\frac{1}{2}\boldsymbol{I} - \boldsymbol{R}$ (see Hsiao and Kleinman [194]).

If (6.12.31) can be solved, we then define the associated electric field $\mathbf{E}$ by

$$\mathbf{E}(\boldsymbol{x}) = \frac{1}{-i\omega\varepsilon}\mathbf{curl}\ \mathbf{H}(\boldsymbol{x}) \in \Omega^c.$$

Then $\mathbf{H}$ and $\mathbf{E}$ satisfy the Maxwell systems in Ω^c automatically. In particular, we see that

$$\begin{aligned}
\mathbf{curl}\ \mathbf{E}(\boldsymbol{x}) &= \frac{-1}{i\omega\varepsilon}\mathbf{curl}\,\mathbf{curl}\ \mathbf{H}(\boldsymbol{x}) \\
&= \frac{-1}{i\omega\varepsilon}(-\Delta + \mathbf{grad}\ \mathrm{div})\mathbf{H}(\boldsymbol{x}) \\
&= \frac{1}{i\omega\varepsilon}(\Delta\mathbf{H}(\boldsymbol{x})) \\
&= \frac{1}{i\omega\varepsilon}\mathbf{curl}\int_\Gamma \Delta_{\boldsymbol{x}}(G_k(\boldsymbol{x}-\boldsymbol{y})\boldsymbol{\varphi}(\boldsymbol{y}))ds_{\boldsymbol{y}} \\
&= \frac{-k^2}{i\omega\varepsilon}\mathbf{curl}\int_\Gamma G_k(\boldsymbol{x}-\boldsymbol{y})\boldsymbol{\varphi}(\boldsymbol{y})ds_{\boldsymbol{y}} \\
&= i\omega\mu\mathbf{H}(\boldsymbol{x}).
\end{aligned}$$

For the transposed equation of (6.12.30) let

$$\mathbf{H}(\boldsymbol{x}) = \mathbf{H}^i(\boldsymbol{x}) - \mathbf{curl}\int_\Gamma G_k(\boldsymbol{x}-\boldsymbol{y})\boldsymbol{\psi}(\boldsymbol{y})ds_{\boldsymbol{y}} \quad \text{in } \Omega,$$

be the magnetic field in Ω, where $\boldsymbol{\psi} = \boldsymbol{n} \times \boldsymbol{\lambda}$ is a tangential vector on Γ to be determined. Then if $\pi_t\mathbf{H} = \mathbf{0}$, we see that $\boldsymbol{\lambda}$ is a solution of the transposed equation of (6.12.31) in the form

$$(\frac{1}{2}\boldsymbol{I} + \boldsymbol{R}')\boldsymbol{\lambda} = \pi_t\mathbf{H}^i \quad \text{on } \Gamma.$$

Then the corresponding homogeneous transposed equation has only the trivial solution, provided k^2 is not an eigenvalue for the homogeneous interior Dirichlet problem:

$$\left.\begin{aligned}
\mathbf{curl}\,(\boldsymbol{\varepsilon}^{1/2}\mathbf{E}) &= ik(\mu^{1/2}\mathbf{H}), \\
\mathbf{curl}\,(\mu^{1/2}\mathbf{H}) &= -ik(\varepsilon^{1/2}\mathbf{E})
\end{aligned}\right\} \quad \text{in } \Omega$$

together with the corresponding boundary condition

$$\pi_t \mathbf{H} = \mathbf{0} \quad \text{on } \Gamma.$$

Thus, (6.12.31) has at most one solution if k^2 is not an eigenvalue.

Conversely, if k^2 is an eigenvalue for the homogeneous interior Dirichlet problem, then the corresponding homogeneous transposed equation will have nontrivial solutions $\boldsymbol{\lambda}_0$. In contrast to the direct method, the given datum $\gamma_t \mathbf{H}^i$ in (6.12.31) is required to satisfy the orthogonality condition:

$$\int_\Gamma \gamma_t \mathbf{H}^i \cdot \boldsymbol{\lambda}_0 \, ds = 0$$

which is a necessary condition for the existence of solutions of (6.12.31).

Next, we return to the **MFIE** (6.12.30). It is quite surprising that the homogeneous transposed equation

$$(\frac{1}{2}\boldsymbol{I} - \boldsymbol{R}')\boldsymbol{\lambda} = \mathbf{0} \quad \text{on } \Gamma \tag{6.12.32}$$

is not the correct homogeneous corresponding transposed equation for the equation (6.12.30) in the sense that (6.12.32) can not be identified as an integral equation for the solution of any homogeneous interior Maxwell boundary value problem. In fact, one may verify that the original integral equation (6.12.30) can be alternatively rewritten in terms of the tangential magnetic field strength $\mathbf{H}$ in $\mathbf{J} = \boldsymbol{n} \times \mathbf{H}$, whence the operator $\boldsymbol{R}'$ as the transposed operator to $\boldsymbol{R}$ appears. This can be seen as follows. From (6.12.30), let $\mathbf{J} = (\boldsymbol{n} \times \mathbf{H})$. Then

$$\frac{1}{2}(\boldsymbol{n} \times \mathbf{H}) - \boldsymbol{R}(\boldsymbol{n} \times \mathbf{H}) = \gamma_t \mathbf{H}^i,$$

or,

$$\frac{1}{2}(\boldsymbol{n} \times \mathbf{H}) - \boldsymbol{R}(\boldsymbol{n} \times \pi_t \mathbf{H}) = \gamma_t \mathbf{H}^i$$

since $\boldsymbol{n} \times \mathbf{H} = \boldsymbol{n} \times \pi_t \mathbf{H}$. Then by applying $-\gamma_t$ to the above equation, we obtain

$$(\frac{1}{2}\boldsymbol{I} + \boldsymbol{R}')(\pi_t \mathbf{H}) = \pi_t \mathbf{H}^i, \tag{6.12.33}$$

That is, the equation (6.12.29) for the exterior Maxwell problem is equivalent to (6.12.33). This results from the corresponding homogeneous transposed equation for (6.12.30) (i.e. (6.12.33)) in the form

$$(\frac{1}{2}\boldsymbol{I} + \boldsymbol{R})\mathbf{J}(\boldsymbol{x}) = \mathbf{0}, \ \boldsymbol{x} \in \Gamma,$$

which can be identified as a boundary integral equation reduced from an interior Maxwell problem with appropriate homogeneous boundary conditions.

For details and non–uniqueness of the electromagnetic scattering integral equation solutions in general, we refer the readers to the paper by Langenberg [260].

To end this section, we remark that for Lipschitz continuous boundary Γ, it is known that $\boldsymbol{R}$ is no longer compact. Further investigation is needed. Either of **EFIE** or **MFIE** can be used to solve for the equivalent surface current **J**. However, the particular choice of **EFIE** or **MFIE** will in general be dictated by the geometry of the scatterer. The **MFIE** becomes useless when Γ shrinks to an infinitely thin scatterer, while **EFIE** is ideally suited for the thin cylinder.

6.12.2 Scattering by a dielectric body

In this section, we concider the scattering from a dielectric body Ω with Lipschitz boundary surface Γ which is immersed into an isotropic homogeneous medium $\Omega^c := \mathbb{R}^3 \setminus \overline{\Omega}$.

We denote the physical parameters and quantities as follows:

$$\varepsilon, \mu, k = \omega\sqrt{\mu\varepsilon}\,,\ \mathbf{H}(\boldsymbol{x}), \mathbf{E}(\boldsymbol{x}) \quad \text{in } \Omega$$
$$\varepsilon^+, \mu^+, k^+ = \omega\sqrt{\mu^+\varepsilon^+}\,,\ \mathbf{H}^+(\boldsymbol{x}), \mathbf{E}^+(\boldsymbol{x}) \quad \text{in } \Omega^c,$$

where all the physical parameters are assumed to be positive constants. The incidence electromagnetic fields are denoted by $\mathbf{E}^i$ or $\mathbf{H}^i$. It is understood that the electric and magnetic fields as $\mathbf{E}(\boldsymbol{x}), \mathbf{H}(\boldsymbol{x})$ and $\mathbf{E}^+(\boldsymbol{x}), \mathbf{H}^+(\boldsymbol{x})$ satisfy corresponding Maxwell equations in Ω and Ω^c, respectively. On the interface Γ, we require the **transmission conditions**:

$$\boldsymbol{n} \times (\mathbf{E}^+ - \mathbf{E}) = \mathbf{0}\,,\ \boldsymbol{n} \times (\mathbf{H}^+ - \mathbf{H}) = \mathbf{0} \quad \text{on } \Gamma$$

together with the Silver–Müller radiation conditions (6.7.2) for the scattered field as $|\boldsymbol{x}| \to \infty$.

We consider two approaches:

(**a**) The volume integral equation approach on Ω based on the equivalence principle in electromagnetics (see, e.g. Balanis [21, p.327], Barkeshli [24] and Harrington [172], to name a few), and

(**b**) The boundary integral equation approach on Γ based on the Green representation theorem.

For simplicity, we begin with

(a) The volume integral equation approach

Equivalence principles are extremely useful concepts in electromagnetics. Many source distributions outside a region may produce the same fields inside the region. Moreover, equivalence principles are based on the uniqueness theorem for solutions of the Maxwell equations. In general there are two classes of equivalence principles: Volumetric and Surface principles.

We will illustrate the volumetric equivalence principle here and refer to the references above for the surface equivalence principle.

It is presumed that the dielectric body occupies the region Ω, and that μ and ε are the given permeability and permittivity as mentioned before; in the exterior the corresponding quantities are taken as μ^+ and ε^+, respectively. An incident wave $\mathbf{E}^i, \mathbf{H}^i$ strikes the dielectric body from $\Omega^c := \mathbb{R}^3 \setminus \overline{\Omega}$. As a result, the obstacle generates in Ω^c a scattered field $\mathbf{E}^s, \mathbf{H}^s$ which is required to satisfy the Silver–Müller radiation conditions at infinity. In addition, a total field $\mathbf{E}, \mathbf{H}$ is transmitted into Ω. The transmission conditions are required for the total field as follows:

$$\boldsymbol{n} \times (\mathbf{E}^i + \mathbf{E}^s)_+ = \boldsymbol{n} \times (\mathbf{E})_- \,, \; \boldsymbol{n} \times (\mathbf{H}^i + \mathbf{H}^s)_+ = \boldsymbol{n} \times (\mathbf{H})_- . \qquad (6.12.34)$$

Inside the electric body, $\mathbf{E}$ and $\mathbf{H}$ satisfy the Maxwell equations which can be rewritten in the form

$$\mathbf{curl}\,\mathbf{E} - i\omega\mu^+\mathbf{H} = -i\omega(\mu^+ - \mu)\mathbf{H}\,, \quad \text{in } \Omega, \qquad (6.12.35)$$

$$\mathbf{curl}\,\mathbf{H} + i\omega\varepsilon^+\mathbf{E} = -i\omega(\varepsilon - \varepsilon^+)\mathbf{E}\,, \quad \text{in } \Omega. \qquad (6.12.36)$$

With this modification made in the fields of these equations, all fields can now be interpreted as existing in the interior homogeneous region Ω bounded by Γ of constitutive parameters ε^+ and μ^+ together with dependent electric are magnetic current densities

$$\mathbf{J}_e = -i\omega(\varepsilon - \varepsilon^+)\mathbf{E}, \qquad (6.12.37)$$

$$\mathbf{M}_e = -i\omega(\mu^+ - \mu)\mathbf{H}, \qquad (6.12.38)$$

respectively. In this context, the influence of the dielectric body is represented as attributable to volume distributions of $\mathbf{J}_\ell$ and $\mathbf{H}_\ell$ in the same medium as in the exterior Ω^c.

The total field $\mathbf{E}, \mathbf{H}$ can be expressed in terms of electromagnetic potentials introduced in Section 6.5. To do so, we first decompose the fields and obtain the corresponding scattered fields. By using superposition, one obtains the total scattered fields.

For simplicity, let us first consider the case when $\mu = \mu^+$ but $\varepsilon \neq \varepsilon^+$. Then from the magnetic vector potential, we see that

$$\mathbf{E}^s(\boldsymbol{x}) = \frac{-1}{i\omega\varepsilon^+}(\mathbf{grad}\ \mathrm{div} + (k^+)^2) \int_\Omega G_k^+(\boldsymbol{x} - \boldsymbol{y})\mathbf{J}_e(\boldsymbol{y})ds_{\boldsymbol{y}} \qquad (6.12.39)$$

$$\mathbf{H}^s(\boldsymbol{x}) = \mathbf{curl} \int_\Omega G_k^+(\boldsymbol{x} - \boldsymbol{y})\mathbf{J}_e(\boldsymbol{y})ds_{\boldsymbol{y}}, \qquad (6.12.40)$$

where

$$k^+ := \omega\sqrt{\mu^+\varepsilon^+}\,, \; G_k^+ = \frac{1}{4\pi}\frac{e^{ik^+|\boldsymbol{x}-\boldsymbol{y}|}}{|\boldsymbol{x} - \boldsymbol{y}|}.$$

In the same manner, in the case where $\mu \neq \mu^+$, but $\varepsilon = \varepsilon^+$, we obtain

$$\mathbf{E}^s(\boldsymbol{x}) = \mathbf{curl} \int_\Omega G_k^+(\boldsymbol{x}-\boldsymbol{y})\mathbf{M}_e(\boldsymbol{y})d\boldsymbol{y}, \tag{6.12.41}$$

$$\mathbf{H}^s(\boldsymbol{x}) = \frac{1}{i\omega\mu^+}(\mathbf{grad}\ \mathrm{div} + (k^+)^2)\int_\Omega G_k^+(\boldsymbol{x}-\boldsymbol{y})\mathbf{M}_e(\boldsymbol{y})d\boldsymbol{y}. \tag{6.12.42}$$

Finally, by superposition, we obtain

$$\mathbf{E}(\boldsymbol{x}) = \mathbf{E}^i(\boldsymbol{x}) + \big(\mathbf{grad}\ \mathrm{div} + (k^+)^2\big)\int_\Omega G_k^+(\boldsymbol{x}-\boldsymbol{y})(\frac{\varepsilon}{\varepsilon^+}-1)\mathbf{E}(\boldsymbol{y})d\boldsymbol{y}$$

$$+ i\omega\mathbf{curl}\int_\Omega (\mu-\mu^+)\mathbf{H}(\boldsymbol{y})G_k^+(\boldsymbol{x}-\boldsymbol{y})d\boldsymbol{y}, \tag{6.12.43}$$

$$\mathbf{H}(\boldsymbol{x}) = \mathbf{H}^i(\boldsymbol{x}) + \big(\mathbf{grad}\ \mathrm{div} + (k^+)^2\big)\int_\Omega (\frac{\mu}{\mu^+}-1)G_k^+(\boldsymbol{x}-\boldsymbol{y})\mathbf{E}(\boldsymbol{y})d\boldsymbol{y}$$

$$- i\omega\mathbf{curl}\int_\Omega (\varepsilon-\varepsilon^+)G_k^+(\boldsymbol{x}-\boldsymbol{y})\mathbf{E}(\boldsymbol{y})d\boldsymbol{y} \tag{6.12.44}$$

(see Jones [220, p.382]).

We notice equations (6.12.43) and (6.12.44) form a system of volume integral equations for determining **E** and **H** inside the dielectric body Ω, and provide a representation of the fields outside the dielectric body when $x \in \Omega^c$. It can be shown that the representation has continuous tangent components and therefore the transmission conditions (6.12.34) will be satisfied.

Remark 6.12.6: One advantage of these volume integral equations is that μ, ε will not necessarily be constant, and we may concider an inhomogeneous dielectric body as well.

Remark 6.12.7: In (6.12.37) and (6.12.38) the terms $\mathbf{J}_e$ and $\mathbf{M}_e$ are often referred to as equivalent volumetric currents, and can be rewritten in the form

$$\begin{aligned}\mathbf{J}_e &= -i\omega\varepsilon^+(\frac{\varepsilon}{\varepsilon^+}-1)\mathbf{E} &&= -i\omega\varepsilon^+(\varepsilon_r-1)\mathbf{E},\\ \mathbf{M}_c &= i\omega\mu^+(\frac{\mu}{\mu^+}-1)\mathbf{H} &&= i\omega\mu^+(\mu_r-1)\mathbf{H},\end{aligned}$$

and (ε_r, μ_r) are the relative permittivity and permeability of the medium. Thus, the presence of the dielectric material is actually equivalent to having the equivalent volumetric currents $\mathbf{J}_e$ and $\mathbf{M}_c$ radiating in the unbounded space.

These volumetric currents may be directly used to construct volumetric integral equations for scattering from a dielectric obstacle. The equivalent electric current inside the scatterer is often denoted by

$$\mathbf{J}_{eq} = -i\frac{k^+}{z^+}\chi_e \mathbf{E}, \; k^+ = \omega\sqrt{\varepsilon^+\mu^+}, \; z^+ = \sqrt{\frac{\mu^+}{\varepsilon^+}},$$

where $\chi_e := \varepsilon_r - 1$. In the special case when $\mu = \mu^+$, (6.12.43) reduces to the volumetric integral equation in the form of

$$\mathbf{J}_{eq}(\boldsymbol{x}) - \chi_e\big((k^+)^2 + \mathbf{grad}\ \mathrm{div}\big)\int_{\Omega} G_k^+(\boldsymbol{x}-\boldsymbol{y})\mathbf{J}_{eq}(\boldsymbol{y})d\boldsymbol{y} = -i\frac{k^+\chi_e}{z^+}\mathbf{E}^i(\boldsymbol{x}) \tag{6.12.45}$$

for all $\boldsymbol{x} \in \Omega$. Equation (6.12.45) is the volumetric **EFIE** for scattering from a dielectric scatterer. (See Barelishli [24, p.307].)

Remark 6.12.8: The existence theorem for solutions of (6.12.43), (6.12.44) and (6.12.45) are not currently known (see Jones [220, Section 6.28]).

Next we consider

(b) The boundary integral equation approach

In the following, let $\mathbf{E}_+(\boldsymbol{x})$ and $\mathbf{H}_+(\boldsymbol{x})$ be the total fields in Ω^c. (See Figure 6.6) and

$$G_{k^+}(\boldsymbol{x}-\boldsymbol{y}) := \frac{1}{4\pi}\frac{e^{ik^+|\boldsymbol{x}-\boldsymbol{y}|}}{|\boldsymbol{x}-\boldsymbol{y}|}, \; G_k(\boldsymbol{x}-\boldsymbol{y}) := \frac{1}{4\pi}\frac{e^{ik|\boldsymbol{x}-\boldsymbol{y}|}}{|\boldsymbol{x}-\boldsymbol{y}|}, \quad \text{with}$$
$$k^+ = \omega\sqrt{\mu^+\varepsilon^+}, \; k = \omega\sqrt{\mu\varepsilon}.$$

Then we have the integral equations for the scattering by a dielectric body as follows:

$$\begin{aligned} & i\omega\mu^+\boldsymbol{n}\times\int_{\Gamma}\big(\boldsymbol{n}\times\mathbf{H}_+(\boldsymbol{y})\big)\big(G_{k^+}(\boldsymbol{x}-\boldsymbol{y}) + \frac{\mu}{\mu^+}G_k(\boldsymbol{x}-\boldsymbol{y})\big)ds_{\boldsymbol{y}} \\ & + \frac{1}{i\omega\varepsilon^+}\boldsymbol{n}\times\int_{\Gamma}\mathrm{div}_{\Gamma}\big(\boldsymbol{n}\times\mathbf{H}_+(\boldsymbol{y})\big)\mathbf{grad}_{\,y}\big(G_{k^+}(\boldsymbol{x}-\boldsymbol{y}) + \frac{\varepsilon^+}{\varepsilon}G_k(\boldsymbol{x}-\boldsymbol{y})\big)ds_{\boldsymbol{y}} \\ & + \boldsymbol{n}\times\int_{\Gamma}\big(\boldsymbol{n}\times\mathbf{E}_+(\boldsymbol{y})\big)\times\mathbf{grad}_{\,y}\big(G_{k^+}(\boldsymbol{x}-\boldsymbol{y}) + G_k(\boldsymbol{x}-\boldsymbol{y})\big)ds_{\boldsymbol{y}}, \\ & = -\boldsymbol{n}\times\mathbf{E}^i(\boldsymbol{x}) \quad \text{for } \boldsymbol{x}\in\Gamma, \end{aligned} \tag{6.12.46}$$

and

$$
\begin{aligned}
&- i\omega\varepsilon^+ \boldsymbol{n} \times \int_\Gamma \Big(G_{k^+}(\boldsymbol{x}-\boldsymbol{y}) + \frac{\varepsilon}{\varepsilon^+} G_k(\boldsymbol{x}-\boldsymbol{y})\Big)\big(\boldsymbol{n} \times \mathbf{E}_+(\boldsymbol{y})\big) ds_{\boldsymbol{y}} \\
&= \frac{1}{i\omega\mu^+} \boldsymbol{n} \times \int_\Gamma \operatorname{div}_\Gamma\big(\boldsymbol{n} \times \mathbf{E}_+(\boldsymbol{y})\big) \mathbf{grad}_{\boldsymbol{y}}\Big(G_{k^+}(\boldsymbol{x}-\boldsymbol{y}) + \frac{\mu^+}{\mu} G_k(\boldsymbol{x}-\boldsymbol{y})\Big) ds_{\boldsymbol{y}} \\
&\quad + \boldsymbol{n} \times \int_\Gamma \big(\boldsymbol{n} \times \mathbf{H}_+(\boldsymbol{y})\big) \times \mathbf{grad}_y\big(G_{k^+}(\boldsymbol{x}-\boldsymbol{y}) + G_k(\boldsymbol{x}-\boldsymbol{y})\big) ds_{\boldsymbol{y}} \\
&= -\boldsymbol{n} \times \mathbf{H}^i(\boldsymbol{y}) \quad \text{for} \quad \boldsymbol{x} \in \Gamma.
\end{aligned} \tag{6.12.47}
$$

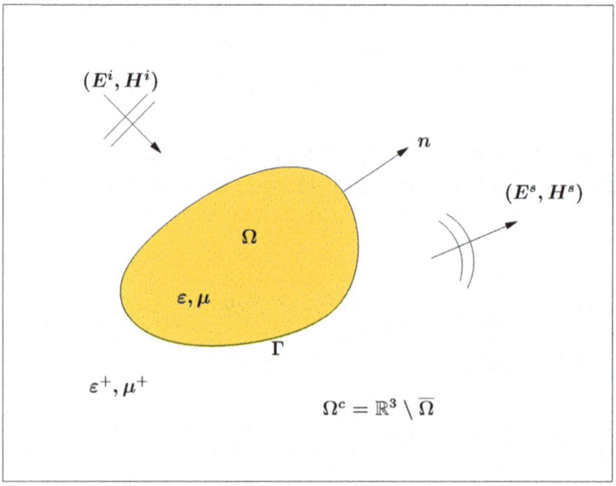

Fig. 6.6. Scattering by a dielectric body

We have used the notation: $\mathbf{E}_+(\boldsymbol{y})$ and $\mathbf{H}_+(\boldsymbol{y})$ for the total fields in Ω^c The boundary integral equations are in terms of $\mathbf{E}_+(x), \mathbf{H}_+(x)$ in Ω^c.

To derive (6.12.46), (6.12.47), as usual, we start from the Stratton–Chu representation formulae for the fields in Ω and Ω^c (c.f. (6.9.4), (6.9.5), (6.12.5) and (6.12.6) with $\mathbf{J}_0 = \mathbf{0}$ and $\varrho = 0\,,\ \sigma = 0$). In the representation of the fields in Ω, we first replace $\mathbf{E}, \mathbf{H}$ in terms of $\mathbf{E}_+, \mathbf{H}_+$ by making use of the transmission conditions.

$$
\boldsymbol{n} \times \mathbf{E} = \boldsymbol{n} \times \mathbf{E}_+\,,\ \boldsymbol{n} \times \mathbf{H} = \boldsymbol{n} \times \mathbf{H}_+ \quad \text{on} \quad \Gamma,
$$

which imply that

$$
\boldsymbol{n} \cdot \mu\mathbf{H} = \boldsymbol{n} \cdot \mu^+\mathbf{H}_+ \quad \text{and} \quad \boldsymbol{n} \cdot \varepsilon\mathbf{E} = \boldsymbol{n} \cdot \varepsilon^+\mathbf{E}_+ \quad \text{on} \quad \Gamma.
$$

Then combining the integral equations, we finally obtain the boundary integral equations for only $\mathbf{E}_+$ and $\mathbf{H}_+$.

For definiteness, we begin with the Stratton–Chu formulae in Ω and obtain the boundary integral equations:

$$\begin{aligned}
-\frac{1}{2}\boldsymbol{n} \times \mathbf{E}(\boldsymbol{x}) &= (i\omega\mu)\boldsymbol{n} \times \int_{\Gamma} G_k(\boldsymbol{x}-\boldsymbol{y})\big(\boldsymbol{n} \times \mathbf{H}(\boldsymbol{y})\big)ds_{\boldsymbol{y}} \\
&\quad + \boldsymbol{n} \times \int_{\Gamma} (\boldsymbol{n} \cdot \mathbf{E}(\boldsymbol{y}))\mathbf{grad}_{\boldsymbol{y}} G_k(\boldsymbol{x}-\boldsymbol{y})ds_{\boldsymbol{y}} \\
&\quad + \boldsymbol{n} \times \int_{\Gamma} (\boldsymbol{n} \times \mathbf{E}(\boldsymbol{y})) \times \mathbf{grad}_{\boldsymbol{y}} G_k(\boldsymbol{x}-\boldsymbol{y})ds_{\boldsymbol{y}} \\
&= (i\omega\mu^+)\boldsymbol{n} \times \int_{\Gamma} \frac{\mu}{\mu^+} G_k(\boldsymbol{x}-\boldsymbol{y})\big(\boldsymbol{n} \times \mathbf{H}_+(\boldsymbol{y})\big)ds_{\boldsymbol{y}} \\
&\quad + \boldsymbol{n} \times \int_{\Gamma} (\boldsymbol{n} \cdot \mathbf{E}_+(\boldsymbol{y}))\frac{\varepsilon^+}{\varepsilon}\mathbf{grad}_{\boldsymbol{y}} G_k(\boldsymbol{x}-\boldsymbol{y})ds_{\boldsymbol{y}} \\
&\quad + \boldsymbol{n} \times \int_{\Gamma} (\boldsymbol{n} \times \mathbf{E}_+(\boldsymbol{y})) \times \mathbf{grad}_{\boldsymbol{y}} G_k(\boldsymbol{x}-\boldsymbol{y})ds_{\boldsymbol{y}}\,, \quad \boldsymbol{x} \in \Gamma,
\end{aligned} \tag{6.12.48}$$

and

$$\begin{aligned}
-\frac{1}{2}\boldsymbol{n} \times \mathbf{H}(\boldsymbol{x}) &= -(i\omega\varepsilon)\boldsymbol{n} \times \int_{\Gamma} G_k(\boldsymbol{x}-\boldsymbol{y})\big(\boldsymbol{n} \times \mathbf{E}(\boldsymbol{y})\big)ds_{\boldsymbol{y}} \\
&\quad + \boldsymbol{n} \times \int_{\Gamma} (\boldsymbol{n} \cdot \mathbf{H}(\boldsymbol{y}))\mathbf{grad}_{\boldsymbol{y}} G_k(\boldsymbol{x}-\boldsymbol{y})ds_{\boldsymbol{y}} \\
&\quad + \boldsymbol{n} \times \int_{\Gamma} (\boldsymbol{n} \times \mathbf{H}(\boldsymbol{y})) \times \mathbf{grad}_{\boldsymbol{y}} G_k(\boldsymbol{x}-\boldsymbol{y})ds_{\boldsymbol{y}} \\
&= -(i\omega\varepsilon^+)\boldsymbol{n} \times \int_{\Gamma} \frac{\varepsilon}{\varepsilon^+} G_k(\boldsymbol{x}-\boldsymbol{y})\big(\boldsymbol{n} \times \mathbf{E}_+(\boldsymbol{y})\big)ds_{\boldsymbol{y}} \\
&\quad + \boldsymbol{n} \times \int_{\Gamma} (\boldsymbol{n} \cdot \mathbf{H}_+(\boldsymbol{y}))\frac{\mu^+}{\mu}\mathbf{grad}_{\boldsymbol{y}} G_k(\boldsymbol{x}-\boldsymbol{y})ds_{\boldsymbol{y}} \\
&\quad + \boldsymbol{n} \times \int_{\Gamma} (\boldsymbol{n} \times \mathbf{H}_+(\boldsymbol{y})) \times \mathbf{grad}_{\boldsymbol{y}} G_k(\boldsymbol{x}-\boldsymbol{y})ds_{\boldsymbol{y}}\,, \quad \boldsymbol{x} \in \Gamma.
\end{aligned} \tag{6.12.49}$$

On the other hand, the Stratton–Chu formulae in Ω^c lead to the boundary integral equations:

$$\frac{1}{2}\big(\boldsymbol{n}\times\mathbf{E}_+(\boldsymbol{x})\big)=\boldsymbol{n}\times\mathbf{E}^i(\boldsymbol{x})+(i\omega\mu^+)\boldsymbol{n}(\boldsymbol{x})\times\int_\Gamma G_{k^+}(\boldsymbol{x}-\boldsymbol{y})\big(\boldsymbol{n}\times\mathbf{H}_+(\boldsymbol{y})\big)ds_{\boldsymbol{y}}$$
$$+\boldsymbol{n}\times\int_\Gamma\big(\boldsymbol{n}\cdot\mathbf{E}_+(\boldsymbol{y})\big)\mathbf{grad}_{\boldsymbol{y}}G_{k^+}(\boldsymbol{x}-\boldsymbol{y})ds_{\boldsymbol{y}}$$
$$+\boldsymbol{n}\times\int_\Gamma\big(\boldsymbol{n}\times\mathbf{E}_+(\boldsymbol{y})\big)\times\mathbf{grad}_{\boldsymbol{y}}G_{k^+}(\boldsymbol{x}-\boldsymbol{y})ds_{\boldsymbol{y}}\,,\ \boldsymbol{x}\in\Gamma, \tag{6.12.50}$$

$$\frac{1}{2}\big(\boldsymbol{n}\times\mathbf{H}_+(\boldsymbol{y})\big)=\boldsymbol{n}\times\mathbf{H}^i(\boldsymbol{x})-(i\omega\varepsilon^+)\boldsymbol{n}\times\int_\Gamma G_{k^+}(\boldsymbol{x}-\boldsymbol{y})\big(\boldsymbol{n}\times\mathbf{E}_+(\boldsymbol{y})\big)ds_{\boldsymbol{y}}$$
$$+\boldsymbol{n}\times\int_\Gamma\big(\boldsymbol{n}\cdot\mathbf{H}_+(\boldsymbol{y})\big)\mathbf{grad}_{\boldsymbol{y}}G_{k^+}(\boldsymbol{x}-\boldsymbol{y})ds_{\boldsymbol{y}}$$
$$+\boldsymbol{n}\times\int_\Gamma\big(\boldsymbol{n}\times\mathbf{H}_+(\boldsymbol{y})\big)\times\mathbf{grad}_{\boldsymbol{y}}G_{k^+}(\boldsymbol{x}-\boldsymbol{y})ds_{\boldsymbol{y}}\,,\ \boldsymbol{x}\in\Gamma. \tag{6.12.51}$$

By adding (6.12.48) and (6.12.50), we obtain

$$(i\omega\mu^+)\boldsymbol{n}\times\int_\Gamma\Big(G_{k^+}(\boldsymbol{x}-\boldsymbol{y})+\frac{\mu}{\mu^+}G_k(\boldsymbol{x}-\boldsymbol{y})\Big)\big(\boldsymbol{n}\times\mathbf{H}_+(\boldsymbol{y})\big)ds_{\boldsymbol{y}}$$
$$+\boldsymbol{n}\times\int_\Gamma\big(\boldsymbol{n}\cdot\mathbf{E}_+(\boldsymbol{y})\big)\mathbf{grad}_y\Big(G_{k^+}(\boldsymbol{x}-\boldsymbol{y})+\frac{\varepsilon^+}{\varepsilon}G_k(\boldsymbol{x}-\boldsymbol{y})\Big)ds_{\boldsymbol{y}}$$
$$+\boldsymbol{n}\times\int_\Gamma\big(\boldsymbol{n}\times\mathbf{E}_+(\boldsymbol{y})\big)\times\mathbf{grad}_{\boldsymbol{y}}\big(G_{k^+}(\boldsymbol{x}-\boldsymbol{y})+G_k(\boldsymbol{x}-\boldsymbol{y})\big)ds_{\boldsymbol{y}}$$
$$=-\boldsymbol{n}\times\mathbf{E}_+(\boldsymbol{x})\,,\ \boldsymbol{x}\in\Gamma. \tag{6.12.52}$$

Note

$$\boldsymbol{n}\cdot\mathbf{E}_+(\boldsymbol{y})=\frac{1}{i\omega\varepsilon^+}\mathrm{div}_\Gamma\big(\boldsymbol{n}\times\mathbf{H}_+(\boldsymbol{y})\big)\quad\text{on }\ \Gamma.$$

Similarly, by adding (6.12.49) and (6.12.51), we obtain

$$-(i\omega\varepsilon^+)\boldsymbol{n}\times\int_\Gamma\Big(G_{k^+}(\boldsymbol{x}-\boldsymbol{y})+\frac{\varepsilon}{\varepsilon^+}G_k(\boldsymbol{x}-\boldsymbol{y})\Big)\big(\boldsymbol{n}\times\mathbf{E}_+(\boldsymbol{y})\big)ds_{\boldsymbol{y}}$$
$$+\boldsymbol{n}\times\int_\Gamma\big(\boldsymbol{n}\cdot\mathbf{H}_+(\boldsymbol{y})\big)\mathbf{grad}_{\boldsymbol{y}}\Big(G_{k^+}(\boldsymbol{x}-\boldsymbol{y})+\frac{\mu^+}{\mu}G_k(\boldsymbol{x}-\boldsymbol{y})\Big)ds_{\boldsymbol{y}}$$
$$+\boldsymbol{n}\times\int_\Gamma\big(\boldsymbol{n}\times\mathbf{H}_+(\boldsymbol{y})\big)\times\mathbf{grad}_{\boldsymbol{y}}\big(G_{k^+}(\boldsymbol{x}-\boldsymbol{y})+G_k(\boldsymbol{x}-\boldsymbol{y})\big)ds_{\boldsymbol{y}}$$
$$=-\boldsymbol{n}\times\mathbf{H}^i(\boldsymbol{x})\,,\ \boldsymbol{x}\in\Gamma \tag{6.12.53}$$

with

$$\boldsymbol{n} \cdot \mathbf{H}_+(\boldsymbol{y}) = \frac{-1}{i\omega\mu^+} \mathrm{div}_\Gamma(\boldsymbol{n} \times \mathbf{E}_+(\boldsymbol{y})) \quad \text{on } \Gamma.$$

This completes the derivations of (6.12.52) and (6.12.53).

6.12.3 Scattering by objects with impedance boundary conditions

When propagating a wave through an actual conductor, the electromagnetic wave amplitude falls off exponentially to $e^{-d\delta}$, where d is the distance into the conductor. Thus, in most cases, the field can be considered as not penetrating into good conductors. However, when it is necessary to account for dissipation losses in the conductor, we may proceed as follows: The electromagnetic field is found by assuming perfect conductivity with the boundary condition: of the conductor:

$$\boldsymbol{n} \times \mathbf{E} = \mathbf{0},$$

from which we obtain the approximate surface current density:

$$\mathbf{J} = \boldsymbol{n} \times \mathbf{H}.$$

Next, the perfect conductivity boundary condition is modified by assuming an inhomogenous boundary condition:

$$\boldsymbol{n} \times \mathbf{E} = Z_s \boldsymbol{n} \times \mathbf{J},$$

where Z_s is the surface impedance of the conductor given by

$$Z_s = \frac{1-i}{\sigma\delta}$$

with δ being the skin depth and σ the conductivity. This eventually leads to the approximated boundary condition:

$$-\boldsymbol{n} \times (\boldsymbol{n} \times \mathbf{E}) = Z_s \boldsymbol{n} \times \mathbf{H}$$

or equivalently,

$$\boldsymbol{n} \times \mathbf{E} = Z_s \boldsymbol{n} \times (\boldsymbol{n} \times \mathbf{H}) \quad \text{on } \Gamma. \tag{6.12.54}$$

The condition (6.12.54) is called the **impedance boundary condition**.

Remark 6.12.9: The surface of a conductor will not be perfectly conducting but the field will not penetrate deeply into the scatterer, particularly when it is coated. Then it might be enough to account for the properties by imposing an impedance boundary condition on the boundary instead of considering a transmission boundary value problem (see Jost [221]).

For the scattering problem with a prescribed impedance boundary condition, the basic boundary integral equations are:

$$
\begin{aligned}
\boldsymbol{n} \times \boldsymbol{n} \times \int_{\Gamma} & \{ i\omega\mu(\boldsymbol{n} \times \mathbf{H}(\boldsymbol{y}))G_k(\boldsymbol{x}-\boldsymbol{y}) \\
& + Z_s\big(\boldsymbol{n} \times (\boldsymbol{n} \times \mathbf{H}(\boldsymbol{y}))\big) \times \mathbf{grad}_{\boldsymbol{y}} G_k(\boldsymbol{x}-\boldsymbol{y}) \\
& + \frac{1}{i\omega\varepsilon} \mathrm{div}_\Gamma\big(\boldsymbol{n} \times \mathbf{H}(\boldsymbol{y})\big) \mathbf{grad}_{\boldsymbol{y}} G_k(\boldsymbol{x}-\boldsymbol{y}) \} ds_{\boldsymbol{y}} \\
+ Z_s \boldsymbol{n} \times \int_{\Gamma} & \{ \big(Z_s \boldsymbol{n} \times \mathbf{H}(\boldsymbol{y})\big) \times \mathbf{grad}_{\boldsymbol{y}} G_k(\boldsymbol{x}-\boldsymbol{y}) \\
& - \frac{1}{i\omega\mu} \mathrm{div}_\Gamma\big(Z_s \boldsymbol{n} \times \mathbf{H}(\boldsymbol{y})\big) \mathbf{grad}_{\boldsymbol{y}} G_k(\boldsymbol{x}-\boldsymbol{y}) \\
& - (i\omega\varepsilon)\big(Z_s \boldsymbol{n} \times (\boldsymbol{n} \times \mathbf{H}(\boldsymbol{y}))\big) \mathbf{grad}_{\boldsymbol{y}} G_k(\boldsymbol{x}-\boldsymbol{y}) \Big\} ds_{\boldsymbol{y}} \\
= -\boldsymbol{n} \times \big(\boldsymbol{n} & \times \mathbf{E}^i(x)\big) - Z_s\big(\boldsymbol{n} \times \mathbf{H}^i(\boldsymbol{x})\big), \ \boldsymbol{x} \in \Gamma.
\end{aligned} \tag{6.12.55}
$$

and

$$
\begin{aligned}
\frac{1}{2}(\boldsymbol{n} \times \mathbf{H}) - \boldsymbol{n} \times \{ & \int_{\Gamma} (\boldsymbol{n} \times \mathbf{H}) \times \mathbf{grad}_{\boldsymbol{y}} G_k(\boldsymbol{x}-\boldsymbol{y}) ds_{\boldsymbol{y}} \\
- i\omega\varepsilon & \int_{\Gamma} Z_s\big(\boldsymbol{n} \times (\boldsymbol{n} \times \mathbf{H})\big) G_k(\boldsymbol{x}-\boldsymbol{y}) ds_{\boldsymbol{y}} \\
- \frac{1}{i\omega\mu} & \int_{\Gamma} \mathrm{div}_\Gamma\big(Z_s \boldsymbol{n} \times (\boldsymbol{n} \times \mathbf{H}\big) \mathbf{grad}_{\boldsymbol{y}} G_k(\boldsymbol{x}-\boldsymbol{y}) ds_{\boldsymbol{y}} \} \\
= \boldsymbol{n} \times \mathbf{H}^i & \quad \text{for} \ \ \boldsymbol{x} \in \Gamma.
\end{aligned} \tag{6.12.56}
$$

In the following,we provide the derivations of the boundary integral equations (6.12.55) and (6.12.56). We begin with (6.12.56) by using the Stratton–Chu formula for the magnetic field $\mathbf{H}$ in Ω^c:

$$
\begin{aligned}
\mathbf{H}(\boldsymbol{x}) = \mathbf{H}^i(\boldsymbol{x}) - i\omega\varepsilon & \int_{\Gamma} (\boldsymbol{n} \times \mathbf{E}(\boldsymbol{y})) G_k(\boldsymbol{x}-\boldsymbol{y}) ds_{\boldsymbol{y}} \\
& + \int_{\Gamma} (\boldsymbol{n} \cdot \mathbf{H}(\boldsymbol{y})) \mathbf{grad}_{\boldsymbol{y}} G_k(\boldsymbol{x}-\boldsymbol{y}) ds_{\boldsymbol{y}} \\
& + \int_{\Gamma} (\boldsymbol{n} \times \mathbf{H}(\boldsymbol{y})) \times \mathbf{grad}_{\boldsymbol{y}} G_k(\boldsymbol{x}-\boldsymbol{y}) ds_{\boldsymbol{y}}, \ \boldsymbol{x} \in \Omega^c.
\end{aligned}
$$

Now let $\boldsymbol{x} \to \Gamma$, and apply the impedance boundary condition (6.12.54). This results in the following relation:

$$\frac{1}{2}(\boldsymbol{n}\times\mathbf{H}(\boldsymbol{x})) - \boldsymbol{n}\times\Big\{\int_\Gamma (\boldsymbol{n}\times\mathbf{H}(\boldsymbol{y}))\times\mathbf{grad}_{\boldsymbol{y}}G_k(\boldsymbol{x}-\boldsymbol{y})ds_{\boldsymbol{y}}$$
$$- i\omega\varepsilon\int_\Gamma Z_s\big(\boldsymbol{n}\times\big(\boldsymbol{n}\times\mathbf{H}(\boldsymbol{y})\big)\big)\mathbf{grad}_{\boldsymbol{y}}G_k(\boldsymbol{x}-\boldsymbol{y})ds_{\boldsymbol{y}}$$
$$-\frac{1}{i\omega\mu}\int_\Gamma \mathrm{div}_\Gamma\big(Z_s\boldsymbol{n}\times\big(\boldsymbol{n}\times\mathbf{H}(\boldsymbol{y})\big)\big)\mathbf{grad}_{\boldsymbol{y}}G_k(\boldsymbol{x}-\boldsymbol{y})ds_{\boldsymbol{y}}\Big\}$$
$$= \boldsymbol{n}\times\mathbf{H}^i(\boldsymbol{x}) \quad \text{for } \boldsymbol{x}\in\Omega^c.$$

Next, the derivation of (6.12.55) is more involved. Let us first show that

$$\frac{1}{2}Z_s(\boldsymbol{n}\times\mathbf{H}) + \boldsymbol{n}\times\boldsymbol{n}\times\Big\{\int_\Gamma Z_s(\boldsymbol{n}\times(\boldsymbol{n}\times\mathbf{H}))\times\mathbf{grad}_{\boldsymbol{y}}G_k(\boldsymbol{x}-\boldsymbol{y})ds_{\boldsymbol{y}}$$
$$+\int_\Gamma i\omega\mu(\boldsymbol{n}\times\mathbf{H})G_k(\boldsymbol{x}-\boldsymbol{y}))ds_{\boldsymbol{y}}$$
$$+\int_\Gamma \frac{1}{i\omega\varepsilon}\mathrm{div}_\Gamma(\boldsymbol{n}\times\mathbf{H})\mathbf{grad}_{\boldsymbol{y}}G_k(\boldsymbol{x}-\boldsymbol{y}))ds_{\boldsymbol{y}}\Big\}$$
$$= -\boldsymbol{n}\times(\boldsymbol{n}\times\mathbf{E}^i) \quad \text{for } \boldsymbol{x}\in\Gamma. \tag{6.12.57}$$

The Stratton–Chu formula for the $\mathbf{E}$ field in Ω^c reads:

$$\mathbf{E}(\boldsymbol{x}) = \mathbf{E}^i(\boldsymbol{x}) + i\omega\mu\int_\Gamma G_k(\boldsymbol{x}-\boldsymbol{y})\big(\boldsymbol{n}\times\mathbf{H}(\boldsymbol{y})ds_{\boldsymbol{y}}$$
$$+\int_\Gamma (\boldsymbol{n}\cdot\mathbf{E}(\boldsymbol{y})\times\mathbf{grad}_{\boldsymbol{y}}G_k(\boldsymbol{x}-\boldsymbol{y})ds_{\boldsymbol{y}}$$
$$+\int_\Gamma (\boldsymbol{n}\times\mathbf{E}(\boldsymbol{y}))\times\mathbf{grad}_{\boldsymbol{y}}G_k(\boldsymbol{x}-\boldsymbol{y})ds_{\boldsymbol{y}}\,,\ \boldsymbol{x}\in\Omega^c.$$

Again, let $\boldsymbol{x}\to\Gamma$ and apply the impedance boundary condition (6.12.54). We obtain

$$\frac{1}{2}Z_s\boldsymbol{n}\times(\boldsymbol{n}\times\mathbf{H}(x)) = \boldsymbol{n}\times\mathbf{E}^i(x) + i\omega\mu\boldsymbol{n}\times\int_\Gamma G_k(\boldsymbol{x}-\boldsymbol{y})\big(\boldsymbol{n}\times\mathbf{H}(\boldsymbol{y})\big)ds_{\boldsymbol{y}}$$
$$+\frac{1}{i\omega\varepsilon}\boldsymbol{n}\times\int_\Gamma \big(\mathrm{div}\big(\boldsymbol{n}\times\mathbf{H}(\boldsymbol{y})\big)\mathbf{grad}_{\boldsymbol{y}}G_k(\boldsymbol{x}-\boldsymbol{y})ds_{\boldsymbol{y}}$$
$$+\boldsymbol{n}\times\int_\Gamma Z_s\big(\boldsymbol{n}\times\big(\boldsymbol{n}\times\mathbf{H}(\boldsymbol{y})\big)\big)\times\mathbf{grad}_{\boldsymbol{y}}G_k(\boldsymbol{x}-\boldsymbol{y})ds_{\boldsymbol{y}}\,,\ \boldsymbol{x}\in\Gamma.$$

By applying the cross–product $\boldsymbol{n}\times$ to the above boundary integral equation once more, we obtain the desired equation:

$$\frac{1}{2}Z_s(\boldsymbol{n}\times\mathbf{H})(\boldsymbol{x}) + \boldsymbol{n}\times\boldsymbol{n}\times\Big\{\int\limits_\Gamma (Z_s\boldsymbol{n}\times(\boldsymbol{n}\times\mathbf{H}(\boldsymbol{y})))\times\mathbf{grad}_{\boldsymbol{y}}G_k(\boldsymbol{x}-\boldsymbol{y}))ds_{\boldsymbol{y}}$$
$$+ i\omega\mu\int\limits_\Gamma G_k(\boldsymbol{x}-\boldsymbol{y})(\boldsymbol{n}\times\mathbf{H}(\boldsymbol{y})))ds_{\boldsymbol{y}}$$
$$+\frac{1}{i\omega\varepsilon}\int\limits_\Gamma \mathrm{div}_\Gamma(\boldsymbol{n}\times\mathbf{H}(\boldsymbol{y}))\mathbf{grad}_{\boldsymbol{y}}G_k(\boldsymbol{x}-\boldsymbol{y})ds_{\boldsymbol{y}}\Big\}$$
$$= -\boldsymbol{n}\times(\boldsymbol{n}\times\mathbf{E}^i)(\boldsymbol{x})\,,\ \boldsymbol{x}\in\Gamma,$$

since

$$\boldsymbol{n}\times\big(Z_s\boldsymbol{n}\times\big(\boldsymbol{n}\times\mathbf{H}(\boldsymbol{x})\big)\big) = \boldsymbol{n}\times\big(\boldsymbol{n}\times\mathbf{E}(\boldsymbol{x})\big) = -Z_s\big(\boldsymbol{n}\times\mathbf{H}(\boldsymbol{x})\big).$$

To complete the derivation of (6.12.55) we first multiply (6.12.56) by Z_s:

$$\frac{1}{2}Z_s\big(\boldsymbol{n}\times\mathbf{H}(\boldsymbol{x})\big) = Z_s\boldsymbol{n}\times\Big\{\int\limits_\Gamma \big(\boldsymbol{n}\times\mathbf{H}(\boldsymbol{y})\big)\times\mathbf{grad}_{\boldsymbol{y}}G_k(\boldsymbol{x}-\boldsymbol{y})ds_{\boldsymbol{y}}$$
$$-\frac{1}{i\omega\mu}\int\limits_\Gamma \mathrm{div}_\Gamma\big(Z_s\boldsymbol{n}\times\big(\boldsymbol{n}\times\mathbf{H}(\boldsymbol{y})\big)\big)\mathbf{grad}_{\boldsymbol{y}}G_k(\boldsymbol{x}-\boldsymbol{y})ds_{\boldsymbol{y}}$$
$$-i\omega\varepsilon\int\limits_\Gamma Z_s\big(\boldsymbol{n}\times\big(\boldsymbol{n}\times\mathbf{H}(\boldsymbol{y})\big)\big)G_k(\boldsymbol{x}-\boldsymbol{y})ds_{\boldsymbol{y}}\Big\}$$
$$+Z_s(\boldsymbol{n}\times\mathbf{H}^i)(\boldsymbol{x})\,,\ \boldsymbol{x}\in\Gamma. \tag{6.12.58}$$

Substituting $\frac{1}{2}Z_s(\boldsymbol{n}\times\mathbf{H})(\boldsymbol{x})$ from (6.12.58) into (6.12.57) yields (6.12.55). This completes the derivations of (6.12.55) and (6.12.56).

Remark 6.12.10: Equation (6.12.55) is the analogue of the **EFIE** and reduces to it when $Z_s = 0$ (i.e. $\sigma\to\infty$), while (6.12.56) reduces to **MFIE**. (See Jones [220, (6.134),(6.135)].)

In closing the section, we note that the integral equations here are more complicated than those for perfect conductivity as well as for the transmission problems. Some uniqueness theorems are available (see Jones [220] and Colton and Kress [75]).In Picard and Mc Ghee [347], a unified Hilbert space approach for treating extended Maxwell systems is included. We refer to Colton and Kress [77] for inverse acoustic and electromagnetic scattering theory, and to Angell, Hsiao and Wen [15] for two–dimensional inverse scattering problems in electromagnetics. For optimization methods in electromagnetics, we refer to the monograph by Angell and Kirsch [14], while for computational methods see Christiansen [67], Jin [218], Monk [314], and monographs edited by Ciarlet et al. [113], and de Castro and Valli [78], respectively. Today, for treating exterior and transmission problems in electromagnetics, the coupling of boundary and finite element methods becomes a very effective and popular solution procedure (see e.g. Bendali [27, 28], Hiptmair [182], Hsiao [193], Hsiao, Monk and Nigam [200] and Langer, Pauly and Repin [200]. In Sterz and Schwab [407], an eddy current problem with impedance boundary condition has been investigated. The problem was reduced to a s calar hypersingular boundary integral equation on the surface Γ of the conductor. Strong ellipticity of the associated nonsymmetric hypersingular operator was established.

7. Introduction to Pseudodifferential Operators

The pseudodifferential operators provide a unified treatment of differential and integral operators. They are based on the intensive use of the Fourier transformation $\mathcal{F}$ (3.1.12) and its inverse $\mathcal{F}^{-1} = \mathcal{F}^*$ (3.1.14). The linear pseudodifferential operators can be characterized by generalized Fourier multipliers, called symbols. The class of pseudodifferential operators form an algebra, and the operations of composition, transposition and adjoining of operators can be analyzed by algebraic calculations of the corresponding symbols.

Moreover, this class of pseudodifferential operators is invariant under diffeomorphic coordinate transformations. As linear mappings between distributions, the pseudodifferential operators can also be represented as Hadamard's finite part integral operators, whose Schwartz kernels can be computed explicitly from their symbols or as integro–differential operators and vice versa. For elliptic pseudodifferential operators, we construct parametrices. For elliptic differential operators, in addition, we construct Levi functions and also fundamental solutions if they exist. Since the latter provide the most convenient basis for boundary integral equation formulations, we also present a short survey on fundamental solutions.

7.1 Basic Theory of Pseudodifferential Operators

In this introductory section for pseudodifferential operators we collect various basic results of pseudodifferential operators without giving all of the proofs. The primary sources for most of the assertions are due to Hörmander's pioneering work in [184, 185, 186, 188].

We begin with the definition of the classical symbols for operators on functions and distributions defined in some domain $\Omega \subset \mathbb{R}^n$ (see [186]).

Definition 7.1.1. *For $m \in \mathbb{R}$, we define the symbol class $\boldsymbol{S}^m(\Omega \times \mathbb{R}^n)$ of order m to consist of the set of functions $a \in C^\infty(\Omega \times \mathbb{R}^n)$ with the property that, for any compact set $K \Subset \Omega$ and multi–indices α, β there exist positive constants $c(K, \alpha, \beta)$ such that*[1]

[1] For our readers, we here purposely do not use the notation D^α since in pseudodifferential analysis D^α is reserved for $\left(-i\frac{\partial}{\partial x}\right)^\alpha$.

G. C. Hsiao und W. L. Wendland, *Boundary Integral Equations*,
Applied Mathematical Sciences 164,
https://doi.org/10.1007/978-3-030-71127-6_7

$$\left| \left(\frac{\partial}{\partial x} \right)^{\beta} \left(\frac{\partial}{\partial \xi} \right)^{\alpha} a(x,\xi) \right| \le c(K,\alpha,\beta)\langle \xi \rangle^{m-|\alpha|} \tag{7.1.1}$$

for all $x \in K$ *and* $\xi \in \mathbb{R}^n$, *where*

$$\langle \xi \rangle := (1 + |\xi|^2)^{\frac{1}{2}} . \tag{7.1.2}$$

Remark 7.1.1: In Hörmander [188], the classical symbols need to satisfy (7.1.1) for all $x \in \mathbb{R}^n$, $\xi \in \mathbb{R}^n$. This approach eliminates some of the difficulties with operators defined only on Ω. Since we are dealing with problems where Ω is not necessarily $\mathbb{R}^n$ we prefer to present the local version as in [186].

A simple example is the polynomial

$$a(x,\xi) = \sum_{|\alpha| \le m} a_\alpha(x)\xi^\alpha . \tag{7.1.3}$$

A function $a_m^0 \in C^\infty(\Omega \times (\mathbb{R}^n \setminus \{0\}))$ is said to be *positively homogeneous of degree* m (with respect to ξ) if it satisfies

$$a_m^0(x, t\xi) = t^m a_m^0(x,\xi) \quad \text{for every } t > 0 \quad \text{and all } \xi \in \mathbb{R}^n \setminus \{0\} . \tag{7.1.4}$$

If a_m^0 is positively homogeneous of degree m then

$$a_m(x,\xi) = \chi(\xi) a_m^0(x,\xi) \in \boldsymbol{S}^m(\Omega \times \mathbb{R}^n) \tag{7.1.5}$$

will be a symbol provided χ is a C^∞–cut–off function with $\chi(\eta) = 0$ in the vicinity of $\eta = 0$ and $\chi(\eta) \equiv 1$ for $|\eta| \ge 1$.

In connection with the symbols $a \in \boldsymbol{S}^m(\Omega \times \mathbb{R}^n)$, we define the associated *standard pseudodifferential operator* A *of order* m.

Definition 7.1.2.

$$A(x, -iD)u := (2\pi)^{-n/2} \int\limits_{\mathbb{R}^n} e^{ix\cdot\xi} a(x,\xi)\widehat{u}(\xi) d\xi \quad \textit{for } u \in C_0^\infty(\Omega) \quad \textit{and } x \in \Omega . \tag{7.1.6}$$

Here $\widehat{u}(\xi) = \mathcal{F}_{x \mapsto \xi} u(x)$ denotes the Fourier transform of u (see (3.1.12)). The operator in (7.1.6) can also be written in the form

$$A(x,-iD)u = \mathcal{F}^{-1}_{\xi \mapsto x}\big(a(x,\xi)\mathcal{F}_{y\mapsto\xi} u(y)\big) \tag{7.1.7}$$

or as an iterated integral,

$$A(x,-iD)u = (2\pi)^{-n} \int\limits_{\mathbb{R}^n} \int\limits_{\Omega} e^{i(x-y)\cdot\xi} a(x,\xi) u(y) dy d\xi . \tag{7.1.8}$$

The set of all standard pseudodifferential operators $A(x,-iD)$ of order m will be denoted by $OPS^m(\Omega \times \mathbb{R}^n)$ and forms a linear vector space together with the usual linear operations. Note that the differential operator of order m,

$$A(x,-iD) = \sum_{|\alpha|\le m} a_\alpha(x)\left(-i\frac{\partial}{\partial x}\right)^\alpha\,, \tag{7.1.9}$$

with C^∞–coefficients on Ω belongs to $OPS^m(\Omega \times \mathbb{R}^n)$ with the symbol (7.1.3).

For a standard pseudodifferential operator one has the following mapping properties.

Theorem 7.1.1. (Hörmander [188, Theorem 18.1.6], Egorov and Shubin [109, p.7 and Theorem 1.4, p.18], Petersen [346, p.169 Theorem 2.4]) *The operator $A \in OPS^m(\Omega \times \mathbb{R}^n)$ defined by (7.1.6) is a continuous operator*

$$A : C_0^\infty(\Omega) \to C^\infty(\Omega)\,. \tag{7.1.10}$$

The operator A can be extended to a continuous linear mapping from $\widetilde{H}^s(K)$ into $H^{s-m}_{\rm loc}(\Omega)$ for any compact subset $K \Subset \Omega$. Furthermore, in the framework of distributions, A can also be extended to a continuous linear operator

$$A : \mathcal{E}'(\Omega) \to \mathcal{D}'(\Omega)\,.$$

The proof is available in textbooks as e.g. in Egorov and Shubin [109], Petersen [346]. The main tool for the proof is the well known Paley–Wiener–Schwartz theorem, Theorem 3.1.3.

For any linear continuous operator $A : C_0^\infty \to C^\infty(\Omega)$ there exists a distribution $K_A \in \mathcal{D}'(\Omega \times \Omega)$ such that

$$Au(x) = \int_\Omega K_A(x,y)u(y)dy \quad \text{for } \; u \in C_0^\infty(\Omega)$$

where the integration is understood in the distributional sense. Due to the Schwartz kernel theorem, the *Schwartz kernel* K_A is uniquely determined by the operator A (see Hörmander [188], Schwartz [382] or Taira [412, Theorem 4.5.1]).

For an operator $A \in OPS^m(\Omega \times \mathbb{R}^n)$, the Schwartz kernel $K_A(x,y)$ has the following smoothness property.

Theorem 7.1.2. (Egorov and Shubin [109, p.7]) *The Schwartz kernel K_A of $A(x,D) \in OPS^m(\Omega \times \mathbb{R}^n)$ is in $C^\infty(\Omega \times \Omega \setminus \{(x,x)\,|\,x \in \mathbb{R}^n\})$.*

Moreover, by use of the identity

$$e^{iz\cdot\xi} = |z|^{-2N}(-\Delta_\xi)^N e^{iz\cdot\xi} \quad \textit{with} \;\; N \in \mathbb{N}\,, \tag{7.1.11}$$

K_A has the representation in terms of the symbol,

$$K_A(x,y) = k(x,x-y) = |y-x|^{-2N}(2\pi)^{-n} \int\limits_{\mathbb{R}^n} \big((-\Delta_\xi)^N a(x,\xi)\big)\, e^{i(x-y)\cdot\xi} d\xi$$
$$\text{for } x \neq y \tag{7.1.12}$$

where $N \geq \left[\frac{m+n}{2}\right] + 1$ *for* $m+n \geq 0$ *and* $N = 0$ *for* $m+n < 0$. *In the latter case,* K_A *is continuous in* $\Omega \times \Omega$ *and for* $u \in C_0^\infty(\Omega)$ *and* $x \in \Omega$ *we have the representation*

$$Au(x) = \int\limits_\Omega k(x,x-y)u(y)dy\,. \tag{7.1.13}$$

If $n+m \geq 0$ *let* $\psi \in C_0^\infty(\Omega)$ *be a cut–off function with* $\psi(x) = 1$ *for all* $x \in \operatorname{supp} u$. *Then*

$$Au(x) = \sum_{|\alpha|\leq 2N} c_\alpha(x) D^\alpha u(x) \tag{7.1.14}$$
$$+ \int\limits_\Omega k(x,x-y)\Big\{u(y) - \sum_{|\alpha|\leq 2N} \frac{1}{\alpha!}(y-x)^\alpha D^\alpha u(x)\Big\}\psi(y)dy$$

where

$$c_\alpha(x) = (2\pi)^{-n} \int\limits_{\mathbb{R}^n}\int\limits_\Omega \frac{1}{\alpha!}(y-x)^\alpha \psi(y) e^{i(x-y)\cdot\xi} dy\, a(x,\xi) d\xi\,.$$

Alternatively,

$$Au(x) = (A_1u)(x) + (A_2u)(x)\,;$$

here

$$(A_1u)(x) = \int\limits_\Omega k_1(x,x-y)u(y)dy \tag{7.1.15}$$

with

$$k_1(x,x-y) = (2\pi)^{-n} \int\limits_{\mathbb{R}^n} (1-\chi(\xi))a(x,\xi)e^{i(x-y)\cdot\xi} d\xi\,,$$

and the integro–differential operator

$$(A_2u)(x) = \int\limits_\Omega k_2(x,x-y)(-\Delta_y)^N u(y)dy \tag{7.1.16}$$

where

$$k_2(x,x-y) = (2\pi)^{-n} \int\limits_{\mathbb{R}^n} |\xi|^{-2N}\chi(\xi)a(x,\xi)e^{i(x-y)\cdot\xi} d\xi$$

and $\chi(\xi)$ *is the cut–off function with* $\chi(\xi) = 0$ *for* $|\xi| \leq \varepsilon$ *and* $\chi(\xi) = 1$ *for* $|\xi| \geq R$.

Remark 7.1.2: The representation (7.1.14) will be also expressed in terms of Hadamard's finite part integrals later on.

Remark 7.1.3: In the theorem we use the identity (7.1.11) to ensure that the integral (7.1.12) decays at infinity with sufficiently high order, whereas for $\xi = 0$, we employ the regularization given by the Hadamard's finite part integrals. In fact, the trick (7.1.11) used in (7.1.12) leads to the definition of the *oscillatory integrals* which allows to define pseudodifferential operators of order $m \geq -n$ (see e.g. Hadamard [168], Hörmander [188, I p.238], Wloka et al [440]).

Proof:

i) $\boldsymbol{n+m<0}$ **:** In this case, the integral has a weakly singular kernel, hence,

$$Au(x) = (2\pi)^{-n} \int\limits_{\mathbb{R}^n} u(y) \int\limits_{\mathbb{R}^n} a(x,\xi) e^{i(x-y)\cdot\xi} d\xi dy$$

and

$$k(x, x-y) = (2\pi)^{-n} \int\limits_{\mathbb{R}^n} a(x,\xi) e^{i(x-y)\cdot\xi} d\xi\,.$$

For the regularity, we begin with the identity (7.1.11) setting $z = x - y$. Since $(-\Delta_\xi)^N a(x,\xi) \in \boldsymbol{S}^{m-2N}(\Omega \times \mathbb{R}^n)$ integration by parts yields

$$k(x, x-y) = |x-y|^{-2N} (2\pi)^{-n} \int\limits_{\mathbb{R}^n} \left\{(-\Delta_\xi)^N a(x,\xi)\right\} e^{i(x-y)\cdot\xi} d\xi\,.$$

This representation can be differentiated $2N$ times with respect to x and y for $x \neq y$. Since $N \in \mathbb{N}$ is arbitrary, k is C^∞ for $x \neq y$ and continuous for $x = y$.

ii) $\boldsymbol{n+m \geq 0}$ **:** Here, with

$$v(y) := \Big\{u(y) - \sum_{|\alpha|\leq 2N} \frac{1}{\alpha!}(x-y)^\alpha D^\alpha u(x)\Big\}\,,$$

(7.1.14) has the form

$$\begin{aligned} Au(x) \;&=\; \sum_{|\alpha|\leq 2N} c_\alpha(x) D^\alpha u(x) \\ &\quad + (2\pi)^{-n} \int\limits_{\mathbb{R}^n} \int\limits_{\Omega} v(y)\psi(y) e^{i(x-y)\cdot\xi} dy a(x,\xi) d\xi\,. \end{aligned}$$

Note that the latter integral exists since the inner integral defines a C^∞ function which decays faster than any power of ξ due to the Palay–Wiener–Schwartz theorem 3.1.3. With (7.1.11) we obtain

$$\begin{aligned} I &:= (2\pi)^{-n} \int\limits_{\mathbb{R}^n} \int\limits_{\text{supp}(\psi)} v(y)\psi(y)e^{i(x-y)\cdot\xi} dy \; a(x,\xi) d\xi \\ &= (2\pi)^{-n} \int\limits_{\mathbb{R}^n} \int\limits_{\text{supp}(\psi)} v(y)\psi(y)|y-x|^{-2}\{(-\Delta_\xi)e^{i(x-y)\cdot\xi}\} dy \; a(x,\xi) d\xi \end{aligned}$$

and with integration by parts

$$\begin{aligned} I = \lim_{R\to\infty} \Big[(2\pi)^{-n} & \int\limits_{|\xi|\le R} \int\limits_{\text{supp}(\psi)} \big(v(y)\psi(y)|y-x|^{-2}\big) e^{i(x-y)\cdot\xi} dy \\ & \qquad\qquad \big((-\Delta_\xi)a(x,\xi)\big)d\xi \\ -(2\pi)^{-n} & \int\limits_{|\xi|=R} \int\limits_{\text{supp}(\psi)} \big(v(y)\psi(y)|y-x|^{-2}\big) i(x-y)e^{i(x-y)\cdot\xi} dy \cdot \tfrac{\xi}{|\xi|} a(x,\xi) dS_\xi \\ +(2\pi)^{-n} & \int\limits_{|\xi|=R} \int\limits_{\text{supp}(\psi)} \big(v(y)\psi(y)|y-x|^{-2}\big) e^{i(x-y)\cdot\xi} dy \big(\nabla_\xi a(x,\xi)\big) \cdot \tfrac{\xi}{|\xi|} dS_\xi \Big] \\ = \lim_{R\to\infty} & [I_1(R) + I_2(R) + I_3(R)] \end{aligned}$$

Now we examine each of the Integrals I_j for $j = 1, 2, 3$ separately.

Behaviour of $I_1(R)$: The inner integral of I_1 reads

$$\Phi_1(x,\xi) := \int\limits_{\text{supp}(\psi)\subset\mathbb{R}^n} \big(v(y)\psi(y)|y-x|^{-2}\big) e^{i(x-y)\cdot\xi} dy$$

and is a C^∞–function of ξ which decays faster than any power of $|\xi|$; in particular of an order higher than $n+m-2$, as can be shown with integration by parts and employing (7.1.11) with z and ξ exchanged. Hence, the limit

$$\lim_{R\to\infty} I_1(R) = (2\pi)^{-n} \int\limits_{\mathbb{R}^n} \int\limits_{\Omega} v(y)\psi(y)|y-x|^{-2} e^{i(x-y)\cdot\xi} dy \big((-\Delta_\xi)a(x,\xi)\big) d\xi$$

exists.

Behaviour of $I_2(R)$: Similarly, the inner integral of I_2,

$$\Phi_2(x,\xi) := i \int\limits_{\Omega} \big(v(y)\psi(y)|y-x|^{-2}\big)(x-y)e^{i(x-y)\cdot\xi} dy\,,$$

is a C^∞–function of ξ which decays faster than any power of $|\xi| = R$; in particular of an order higher than $m+n-2$. Therefore,

$$I_2(R) = O(R^{m+n-1-\nu}) \to 0 \quad \text{for } R\to\infty \quad \text{if } \nu > m+n-1\,.$$

Behaviour of $I_3(R)$: The inner integral of I_3 is $\Phi_1(x,\xi)$. Now, choose $\nu > m+n-2$, then $\lim\limits_{R\to\infty} I_3(R) = 0$.

To complete the proof, we repeat the process by applying this technique N times. Finally, we obtain

$$A(v\psi) = (2\pi)^{-n} \int\limits_{\mathbb{R}^n} \int\limits_{\mathrm{supp}(\psi)} \big(v(y)\psi(y)|y-x|^{-2N}\big)e^{i(x-y)\cdot\xi}dy \times \\ \times\big((-\Delta_\xi)^N a(x,\xi)\big)d\xi\,.$$

Replacing $v(y)$ by its definition, we find (7.1.14) after interchanging the order of integration.

In the same manner as for $m+n<0$ one finds that

$$\int\limits_{\mathbb{R}^n} (-\Delta_\xi)^N a(x,\xi)e^{i(x-y)\cdot\xi}d\sigma$$

is in C^∞ for $x \neq y$. The alternative formulae (7.1.15) and (7.1.16) are direct consequences of the Fourier transform of $\big((-\Delta_\xi)^N a(x,\xi)\big)$. ■

For the symbol calculus in $\boldsymbol{S}^m(\Omega\times\mathbb{R}^n)$ generated by the algebra of pseudodifferential operators it is desirable to introduce the notion of asymptotic expansions of symbols and use families of symbol classes.

We note that the symbol classes $\boldsymbol{S}^m(\Omega\times\mathbb{R}^n)$ have the following properties:

a) For $m \leq m'$ we have the inclusions $\boldsymbol{S}^{-\infty}(\Omega\times\mathbb{R}^n) \subset \boldsymbol{S}^m(\Omega\times\mathbb{R}^n) \subset \boldsymbol{S}^{m'}(\Omega\times\mathbb{R}^n)$ where $\boldsymbol{S}^{-\infty}(\Omega\times\mathbb{R}^n) := \bigcap\limits_{m\in\mathbb{R}} \boldsymbol{S}^m(\Omega\times\mathbb{R}^n)$.

b) If $a \in \boldsymbol{S}^m(\Omega\times\mathbb{R}^n)$ then $\left(\frac{\partial}{\partial x}\right)^\beta \left(\frac{\partial}{\partial\xi}\right)^\alpha a \in \boldsymbol{S}^{m-|\alpha|}(\Omega\times\mathbb{R}^n)$.

c) For $a \in \boldsymbol{S}^m(\Omega\times\mathbb{R}^n)$ and $b \in \boldsymbol{S}^{m'}(\Omega\times\mathbb{R}^n)$ we have $ab \in \boldsymbol{S}^{m+m'}(\Omega\times\mathbb{R}^n)$.

Definition 7.1.3. *Given $a \in \boldsymbol{S}^{m_0}(\Omega\times\mathbb{R}^n)$ and a sequence of symbols $a_{m_j} \in \boldsymbol{S}^{m_j}(\Omega\times\mathbb{R}^n)$ with $m_j \in \mathbb{R}$ and $m_j > m_{j+1} \to -\infty$. We call the formal sum $\sum\limits_{j=0}^{\infty} a_{m_j}$ an* asymptotic expansion *of a if for every $k>0$ there holds*

$$a - \sum_{j=0}^{k-1} a_{m_j} \in \boldsymbol{S}^{m_k}(\Omega\times\mathbb{R}^n) \quad \textit{and we write} \quad a \sim \sum_{j=0}^{\infty} a_{m_j}\,.$$

The leading term $a_{m_0} \in \boldsymbol{S}^{m_0}(\Omega\times\mathbb{R}^n)$ is called the principal symbol.

In fact, if a sequence a_{m_j} is given, the following theorem holds.

Theorem 7.1.3. *Let $a_{m_j} \in \boldsymbol{S}^{m_j}(\Omega\times\mathbb{R}^n)$ with $m_j > m_{j+1} \to -\infty$ for $j\to\infty$. Then there exists a symbol $a \in \boldsymbol{S}^{m_0}(\Omega\times\mathbb{R}^n)$, unique modulo $\boldsymbol{S}^{-\infty}(\Omega\times\mathbb{R}^n)$, such that for all $k>0$ we have*

$$a - \sum_{j=0}^{k-1} a_{m_j} \in \boldsymbol{S}^{m_k}(\Omega \times \mathbb{R}^n)\,. \tag{7.1.17}$$

For the proof of this theorem one may set

$$a(x,\xi) := \sum_{j=0}^{\infty} \chi\Big(\frac{\xi}{t_j}\Big) a_{m_j}(x,\xi) \tag{7.1.18}$$

by using a C^∞ cut–off function $\chi(\eta)$ with $\chi(\eta) = 1$ for $|\eta| \geq 1$ and $\chi(\eta) = 0$ for $|\eta| \leq \frac{1}{2}$ and by using a sequence of real scaling factors with $t_j \to \infty$ for $j \to \infty$ sufficiently fast. For the details of the proof see Hörmander [186, 1.1.9], Taylor [416, Chap. II, Theorem 3.1].

With the help of the asymptotic expansions it is useful to identify in $\boldsymbol{S}^m(\Omega \times \mathbb{R}^n)$ the subclass $\boldsymbol{S}^m_{c\ell}(\Omega \times \mathbb{R}^n)$ called classical (also polyhomogeneous) symbols.

A symbol $a \in \boldsymbol{S}^m(\Omega \times \mathbb{R}^n)$ is called a *classical symbol* if there exist functions $a_{m-j}(x,\xi)$ with $a_{m-j} \in \boldsymbol{S}^{m-j}(\Omega \times \mathbb{R}^n)$, $j \in \mathbb{N}_0$ such that $a \sim \sum_{j=0}^\infty a_{m-j}$ where each a_{m-j} is of the form (7.1.5) with (7.1.4) and is of homogeneous degree $m_j = m - j$; i.e., satisfying

$$a_{m-j}(x,t\xi) = t^{m-j} a_{m-j}(x,\xi) \text{ for } t \geq 1 \quad \text{and} \quad |\xi| \geq 1\,. \tag{7.1.19}$$

The set of all classical symbols of order m will be denoted by $\boldsymbol{S}^m_{c\ell}(\Omega \times \mathbb{R}^n)$.

We remark that for $a \in \boldsymbol{S}^m_{c\ell}(\Omega \times \mathbb{R}^n)$, the homogeneous functions $a_{m-j}(x,\xi)$ for $|\xi| \geq 1$ are uniquely determined. Moreover, note that the asymptotic expansion means that for all $|\alpha|, |\beta| \geq 0$ and every compact $K \Subset \Omega$ there exist constants $c(K,\alpha,\beta)$ such that

$$\Big|\Big(\frac{\partial}{\partial x}\Big)^\beta \Big(\frac{\partial}{\partial \xi}\Big)^\alpha \Big(a(x,\xi) - \sum_{j=0}^N a_{m-j}(x,\xi)\Big)\Big| \leq c(K,\alpha,\beta)\langle\xi\rangle^{m-N-|\alpha|-1} \tag{7.1.20}$$

holds for every $N \in \mathbb{N}_0$.

For a given symbol or for a given asymptotic expansion, the associated operator A is given via the definition (7.1.8) or with Theorem 7.1.3, in addition. On the other hand, if the operator A is given, an equally important question is how to find the corresponding symbol since by examining the symbol one may deduce further properties of A. In order to reduce A to the form $A(x,D)$ in (7.1.6), we need the concept of properly supported operators with respect to their Schwartz kernels K_A.

Definition 7.1.4. *A distribution $K_A \in \mathcal{D}'(\Omega \times \Omega)$ is called properly supported if* $\operatorname{supp} K_A \cap (K \times \Omega)$ *and* $\operatorname{supp} K_A \cap (\Omega \times K)$ *are both compact in $\Omega \times \Omega$ for every compact subset $K \Subset \Omega$* (Treves [420, Vol.I p.25]).

We recall that the support of K_A is defined by the relation $z \in \operatorname{supp} K_A \subset \mathbb{R}^{2n}$ if and only if for every neighbourhood $\mathcal{U}(z)$ there exists $\varphi \in C_0^\infty(\mathcal{U}(z))$

such that $\langle K_A, \varphi \rangle \neq 0$. In accordance with Definition 7.1.4, any continuous linear *operator* $A : C_0^\infty \to C^\infty(\Omega)$ is called *properly supported* if and only if its Schwartz kernel $K_A \in \mathcal{D}'(\Omega \times \Omega)$ is a properly supported distribution.

In order to characterize properly supported operators we recall the following lemma.

Lemma 7.1.4. *(see* Folland [126, Proposition 8.12]*) A linear continuous operator* $A : C_0^\infty(\Omega) \to C^\infty(\Omega)$ *is properly supported if and only if the following two conditions hold:*

i) For any compact subset $K_y \Subset \Omega$ *there exists a compact* $K_x \Subset \Omega$ *such that* $\operatorname{supp} v \subset K_y$ *implies* $\operatorname{supp}(Av) \subset K_x$.
ii) For any compact subset $K_x \Subset \Omega$ *there exists a compact* $K_y \Subset \Omega$ *such that* $\operatorname{supp} v \cap K_y = \emptyset$ *implies* $\operatorname{supp}(Av) \cap K_x = \emptyset$.

Proof:

We first show the necessity of *i)* and *ii)* for any given properly supported operator A. To show *i)*, let K_y be any fixed compact subset of Ω. Then define

$$K_x := \{x \in \Omega \mid \quad \text{there exists} \quad y \in K_y \quad \text{with} \quad (x,y) \in \operatorname{supp} K_A \cap (\Omega \times K_y)\}$$

where K_A is the Schwartz kernel of A. Clearly, K_x is a compact subset of Ω since $\operatorname{supp} K_A \cap (\Omega \times K_y)$ is compact because K_A is properly supported. In order to show i), i.e. $\operatorname{supp}(Av) \subset K_x$, we consider any $v \in C_0^\infty(\Omega)$ with $\operatorname{supp} v \subset K_y$ and $\psi \in C_0^\infty(\Omega \setminus K_x)$. Then

$$\begin{aligned}\langle Av, \psi \rangle &= \int\limits_\Omega \int\limits_\Omega K_A(x,y)v(y)dy\psi(x)dx \\ &= \int\limits_{K_y} \int\limits_\Omega K_A(x,y)\psi(x)v(y)dxdy = 0\end{aligned}$$

since for $x \in \operatorname{supp} \psi \subset \Omega \setminus K_x$ and all $y \in \operatorname{supp} v$ we have $(x,y) \notin \operatorname{supp} K_A \cap (\Omega \times K_y)$. Hence, $\langle Av, \psi \rangle = 0$ for all $\psi \in C_0^\infty(\Omega \setminus K_x)$ and we find $\operatorname{supp}(Av) \subset K_x$.

For *ii)*, let K_x be any compact fixed subset of Ω. Then define

$$K_y := \{y \in \Omega \mid \quad \text{there exists} \quad x \in K_x \quad \text{with} \quad (x,y) \in \operatorname{supp} K_A \cap (K_x \times \Omega)\},$$

which is a compact subset of Ω. For $v \in C_0^\infty(\Omega \setminus K_y)$ and $\psi \in C_0^\infty(\operatorname{supp}(Av) \cap K_x)$ we have

$$\langle Av, \psi \rangle = \int\limits_\Omega \int\limits_\Omega K_A(x,y)v(y)\psi(x)dydx = 0$$

since for every $y \in \operatorname{supp} v$ we have $y \notin K_y$ and $(x,y) \notin \operatorname{supp} K_A \cap (K_x \times \Omega)$ for all $x \in \operatorname{supp} \psi$. Hence, $\langle Av, \psi \rangle = 0$ for all $\psi \in C_0^\infty(\operatorname{supp}(Av) \cap K_x)$ which implies $\operatorname{supp}(Av) \cap K_x = \emptyset$.

Next, we show the sufficiency of *i)* and *ii)*. Let K_x be any compact subset of Ω and K_y the corresponding subset in *ii)* such that $\operatorname{supp} v \cap K_y = \emptyset$ implies $\operatorname{supp}(Av) \cap K_x = \emptyset$. We want to show that $\operatorname{supp} K_A \cap (K_x \times \Omega)$ is compact. To this end, we consider $\psi \in C_0^\infty(\Omega)$ with $\operatorname{supp} \psi \subset K_x$. Then

$$\begin{aligned}\langle Av, \psi \rangle &= \int\limits_\Omega \int\limits_\Omega K_A(x,y)v(y)\psi(x)dydx \\ &= \int\limits_\Omega \int\limits_{\Omega \setminus K_y} K_A(x,y)v(y)dy\psi(x)dx = 0\end{aligned}$$

for every $v \in C_0^\infty(\Omega)$ with $\operatorname{supp} v \cap K_y = \emptyset$ due to *ii)*. This implies $\operatorname{supp} K_A \cap \big(K_x \times (\Omega \setminus K_y)\big) = \emptyset$; so,

$$\operatorname{supp} K_A \cap (K_x \times \Omega) = \operatorname{supp} K_A \cap (K_x \times K_y) \subset K_x \times K_y \,.$$

Hence, $\operatorname{supp} K_A \cap (K_x \times \Omega)$ is compact because $K_x \times K_y$ is compact. To show that for any chosen compact $K_y \Subset \Omega$, the set $\operatorname{supp} K_A \cap (\Omega \times K_y)$ is compact we invoke *i)* and take K_x to be the corresponding compact set in Ω. Then for any $v \in C_0^\infty(\Omega)$ with $\operatorname{supp} v \subset K_y$, we have $\operatorname{supp}(Av) \subset K_x$. Hence, for every $\psi \in C_0^\infty(\Omega \setminus K_x)$,

$$\int\limits_\Omega \int\limits_\Omega K_A(x,y)v(y)dy\psi(x)dx = \langle Av, \psi \rangle = 0 \,.$$

As a consequence, we find $\operatorname{supp} K_A \cap (\Omega \times K_y) = \operatorname{supp} K_A \cap (K_x \times K_y) \subset K_x \times K_y$; so, $\operatorname{supp} K_A \cap (\Omega \times K_y)$ is compact. Thus, K_A is properly supported. ∎

Lemma 7.1.4 implies the following corollary.

Corollary 7.1.5. *The operator $A : C_0^\infty(\Omega) \to C^\infty$ is properly supported if and only if the following two conditions hold:*

i) For any compact subset $K_y \Subset \Omega$ there exists a compact $K_x \Subset \Omega$ such that

$$A : C_0^\infty(K_y) \to C_0^\infty(K_x) \quad \textit{is continuous.} \tag{7.1.21}$$

ii) For any compact subset $K_x \Subset \Omega$ there exists a compact $K_y \Subset \Omega$ such that

$$A^\top : C_0^\infty(K_x) \to C_0^\infty(K_y) \quad \textit{is continuous.} \tag{7.1.22}$$

Proof: Since $A : C_0^\infty(\Omega) \to C^\infty(\Omega)$ is continuous, *i)* in Lemma 7.1.4 is equivalent to (7.1.21).

Now, if A is properly supported then $K_A^\top(x,y) = K_{A^\top}(y,x)$ is the Schwartz kernel of $A^\top$ and, hence, also properly supported. Therefore *i)* in Lemma 7.1.4 is valid for $K_{A^\top}(x,y) = K_A^\top(y,x)$ which, for this case, is equivalent to (7.1.22). Consequently, (7.1.21) and (7.1.22) are satisfied if A is properly supported.

Conversely, if (7.1.21) and (7.1.22) hold, then *i)* in Lemma 7.1.4 is already satisfied. In order to show *ii)* let us choose any compact subset K_x of Ω and let K_y be the corresponding compact subset in (7.1.22). Let $v \in C_0^\infty(\Omega \setminus K_y)$. Then $Av \in C_0^\infty(\Omega)$ due to (7.1.21). Now, let $\psi \in C_0^\infty(\Omega)$ be any function with $\operatorname{supp}\psi \subset K_x$. Then $\operatorname{supp} A^\top \psi \subset K_y$ because of (7.1.22). Hence,

$$\langle Av, \psi\rangle = \langle v, A^\top \psi\rangle = 0$$

for any such ψ. Therefore, $\operatorname{supp}(Av) \cap K_x = \emptyset$ which is the second proposition ii) of Lemma 7.1.4. ■

As an obvious consequence of Corollary 7.1.5, the following proposition is valid.

Proposition 7.1.6. *If A and $B : C_0^\infty(\Omega) \to C^\infty(\Omega)$ are properly supported, then the composition $A \circ B$ is properly supported, too.*

Theorem 7.1.7. *If $A \in OPS^m(\Omega \times \mathbb{R}^n)$ is properly supported then*

$$a(x,\xi) = e^{-ix\cdot\xi}(Ae^{i\xi\bullet})(x) \tag{7.1.23}$$

is the symbol (see Taylor [416, Chap. II, Theorem 3.8]*).*

To express the transposed operator $A^\top$ of $A \in OPS^m(\Omega \times \mathbb{R}^n)$ with the given symbol $a(x,\xi)$, $a \in \boldsymbol{S}^m(\Omega \times \mathbb{R}^n)$ we use the distributional relation

$$\langle Au, v\rangle = \langle u, A^\top v\rangle = \int\limits_\Omega Au(x)v(x)dx = \int\limits_\Omega u(x)A^\top v(x)dx \text{ for } u,v \in C_0^\infty(\Omega)\,.$$

Then from the definition (7.1.8) of A, we find

$$A^\top v(x) = (2\pi)^{-n} \int\limits_{\mathbb{R}^n}\int\limits_\Omega e^{i(x-y)\cdot\xi} a(y,-\xi)v(y)dyd\xi\,. \tag{7.1.24}$$

From this representation it is not transparent that $A^\top$ is a standard pseudodifferential operator since it does not have the standard form (7.1.8). In fact, the following theorem is valid.

Theorem 7.1.8. (Taylor [416, Chap. II, Theorem 4.2]) *If $A \in OPS^m(\Omega \times \mathbb{R}^n)$ is properly supported, then $A^\top \in OPS^m(\Omega \times \mathbb{R}^n)$.*

If, however, $A \in OPS^m(\Omega \times \mathbb{R}^n)$ is **not** properly supported, then $A^\top$ belongs to a slightly larger class of operators. Note that if we let

$${}^\top a(x,y,\xi) := a(y,-\xi)$$

then we may rewrite $A^\top v$ in the form

$$A^\top v(x) = (2\pi)^{-n} \int\limits_{\mathbb{R}^n} \int\limits_{\Omega} {}^\top a(x,y,\xi) e^{i(x-y)\cdot\xi} v(y) dy d\xi \,. \tag{7.1.25}$$

Therefore, Hörmander in [186] introduced the more general class of Fourier integral operators of the form

$$Au(x) = (2\pi)^{-n} \int\limits_{\mathbb{R}^n} \int\limits_{\Omega} e^{i(x-y)\cdot\xi} a(x,y,\xi) u(y) dy d\xi \tag{7.1.26}$$

with the *amplitude function* $a \in \boldsymbol{S}^m(\Omega \times \Omega \times \mathbb{R}^n)$ and with the special phase function $\varphi(x,y,\xi) = (x-y)\cdot\xi$. The integral in (7.1.26) is understood in the sense of *oscillatory integrals* by employing the same procedure as in (7.1.11), (7.1.12). (See e.g. Hörmander [186], Treves [420, Vol. II p.315]). This class of operators will be denoted by $\mathcal{L}^m(\Omega)$.

Theorem 7.1.9. (Hörmander [186, Theorem 2.1.1], Taira [412, Theorem 6.5.2]) *Every operator* $A \in \mathcal{L}^m(\Omega)$ *can be written as*

$$A = A_0(x, -iD) + R$$

where $A_0(x,-iD) \in OPS^m(\Omega \times \mathbb{R}^n)$ *is properly supported and* $R \in \mathcal{L}^{-\infty}(\Omega)$ *where*

$$\mathcal{L}^{-\infty}(\Omega) := \bigcap_{m\in\mathbb{R}} \mathcal{L}^m(\Omega) = OPS^{-\infty}(\Omega \times \Omega \times \mathbb{R}^n)\,.$$

Proof: A simple proof by using a proper mapping is available in [412] which is not constructive. Here we present a constructive proof based on the presentation in Petersen [346] and Hörmander [188, Prop.18.1.22]. It is based on a partition of the unity over Ω with corresponding functions $\{\phi_\ell\}$, $\phi_\ell \in C_0^\infty(\Omega)$, $\sum_\ell \phi_\ell(x) = 1$. For every $x_0 \in \Omega$ and balls $B_\varepsilon(x_0) \subset \Omega$ with fixed $\varepsilon > 0$, the number of ϕ_ℓ with $\operatorname{supp}\phi_\ell \cap B_\varepsilon(x_0) \neq \emptyset$ is finite. Let us denote by $\mathcal{I}_1 = \{(j,\ell) \mid \operatorname{supp}\phi_j \cap \operatorname{supp}\phi_\ell \neq \emptyset\}$ and by $\mathcal{I}_2 = \{(j,\ell) \mid \operatorname{supp}\phi_j \cap \operatorname{supp}\phi_\ell = \emptyset\}$ the corresponding index sets. Then

$$\begin{aligned} Au(x) &= \sum_{j,\ell} \phi_j(x)(A\phi_\ell u)(x) \\ &= \sum_{\mathcal{I}_1} \phi_j(x) A\phi_\ell u + \sum_{\mathcal{I}_2} \phi_j(x) A\phi_\ell u \,. \end{aligned}$$

For every $(j,\ell) \in \mathcal{I}_2$ we see that the corresponding Schwartz kernel is given by

$$\phi_j(x) k(x, x-y) \phi_\ell(y)$$

which is $C_0^\infty(\Omega \times \Omega)$. Moreover, for any pair $(x_0, y_0) \in \Omega \times \Omega$, the sum

$$\sum_{\mathcal{I}_2} \phi_j(x) k(x, x-y) \phi_\ell(y) \quad \text{for } (x,y) \in B_\varepsilon(x_0) \times B_\varepsilon(y_0)$$

for $\varepsilon > 0$ chosen appropriately small, has only finitely many nontrivial terms. Hence,

$$R := \sum_{\mathcal{I}_2} \phi_j A \phi_\ell \in OPS^{-\infty}(\Omega \times \mathbb{R}^n).$$

The operator $Q := \sum_{\mathcal{I}_1} \phi_j A \phi_\ell$, however, is properly supported since it has the Schwartz kernel $\sum_{\mathcal{I}_1} \phi_j(x) k(x, x-y) \phi_\ell(y)$. Each term has compact support for fixed x with respect to y and with fixed y with respect to x.

Now, if $x \in K \Subset \Omega$ then, due to the compactness of K, we have $\operatorname{supp} \phi_j \cap K \neq \emptyset$ only for finitely many indices j whose collection we call $\mathcal{J}(K)$. For every $j \in \mathcal{J}(K)$, there are only finitely many ℓ with $\operatorname{supp} \phi_j \cap \operatorname{supp} \phi_\ell \neq \emptyset$; we denote the collection of these ℓ by $\mathcal{J}(K, j)$. Hence, if x traces K, then $\phi_j(x) k(x, x-y) \phi_\ell(y) \neq 0$ only for finitely many indices (j, ℓ) with $j \in \mathcal{J}(K)$ and $\ell \in \mathcal{J}(K, j)$; and we see that

$$\operatorname{supp} \sum_{\substack{j \in \mathcal{J}(K) \\ \ell \in \mathcal{J}(K,j)}} \big(\phi_j(x) K(x, x-y) \phi_\ell(y)\big) \cap (K \times \Omega)$$

is compact since there are only finitely many terms in the sum. If $y \in K \Subset \Omega$, we find with the same arguments that

$$(K \cap \Omega) \cap \operatorname{supp} \sum_{\substack{\ell \in \mathcal{J}(K) \\ j \in \mathcal{J}(K,\ell)}} \big(\phi_j(x) k(x, x-y) \phi_\ell(y)\big)$$

is compact. Hence, Q is properly supported. ∎

We emphasize that $\mathcal{L}^m(\Omega) \neq OPS^m(\Omega \times \mathbb{R}^n) \subset \mathcal{L}^m(\Omega)$. The operators $R \in \mathcal{L}^{-\infty}$ are called *smoothing operators*. (In the book by Taira [412] the smoothing operators are called regularizers.)

Definition 7.1.5. *A continuous linear operator $A : C_0^\infty(\Omega) \to \mathcal{D}'(\Omega)$ is called a smoothing operator if it extends to a continuous linear operator from $\mathcal{E}'(\Omega)$ into $C^\infty(\Omega)$.*

The following theorem characterizes the smoothing operators in terms of our operator classes.

Theorem 7.1.10. (Taira [412, Theorem 6.5.1, Theorem 4.5.2])
The following four conditions are equivalent:

(i) *A is a smoothing operator,*
(ii) $A \in \mathcal{L}^{-\infty}(\Omega) = \bigcap_{m \in \mathbb{R}} \mathcal{L}^m(\Omega)$,
(iii) *A is of the form* (7.1.26) *with some $a \in \boldsymbol{S}^{-\infty}(\Omega \times \Omega \times \mathbb{R}^n)$,*
(iv) *A has a $C^\infty(\Omega \times \Omega)$ Schwartz kernel.*

Theorem 7.1.11. (Hörmander [186, p.103], Taylor [416, Chap. II, Theorem 3.8]) *Let $A \in \mathcal{L}^m(\Omega)$ be properly supported. Then*

$$a(x,\xi) = e^{-ix\cdot\xi}A(e^{i\xi\cdot\bullet})(x) \tag{7.1.27}$$

with $a \in \boldsymbol{S}^m(\Omega \times \mathbb{R}^n)$ *and* $A = A(x, -iD) \in OPS^m(\Omega \times \mathbb{R}^n)$. *Furthermore, if* A *has an amplitude* $a(x,y,\xi)$ *with* $a \in \boldsymbol{S}^m(\Omega \times \Omega \times \mathbb{R}^n)$ *then* $a(x,\xi)$ *has the asymptotic expansion*

$$a(x,\xi) \sim \sum_{\alpha \geq 0} \frac{1}{\alpha!}\Big(\Big(\frac{\partial}{\partial \xi}\Big)^\alpha \Big(-i\frac{\partial}{\partial y}\Big)^\alpha a(x,y,\xi)\Big)_{|y=x} . \tag{7.1.28}$$

Here we omit the proof.

Theorems 7.1.9 and 7.1.11 imply that if $A \in \mathcal{L}^m(\Omega)$ then A can always be decomposed in the form

$$A = A_0 + R$$

with $A_0 \in OPS^m(\Omega \times \mathbb{R}^n)$ having a properly supported Schwartz kernel and a smoothing operator R. Clearly, A_0 is not unique. Hence, to any A and A_0 we can associate with A a symbol

$$a(x,\xi) := e^{-ix\cdot\xi}A_0(e^{iy\cdot\xi})(x) \tag{7.1.29}$$

and the corresponding asymptotic expansion (7.1.28). Now we define

$$\begin{aligned} \sigma_A \ := \ & \text{the equivalence class of all the symbols associated} \\ & \text{with } A \ \text{ defined by (7.1.29) in } \boldsymbol{S}^m(\Omega \times \mathbb{R}^n)/\boldsymbol{S}^{-\infty}(\Omega \times \mathbb{R}^n)\,. \end{aligned} \tag{7.1.30}$$

This equivalence class is called the *complete symbol class* of $A \in \mathcal{L}^m(\Omega)$.

Clearly, the mapping

$$\mathcal{L}^m(\Omega) \ni A \mapsto \sigma_A \in \boldsymbol{S}^m(\Omega \times \mathbb{R}^n)/\boldsymbol{S}^{-\infty}(\Omega \times \mathbb{R}^n)$$

induces an isomorphism

$$\mathcal{L}^m(\Omega)/\mathcal{L}^{-\infty} \to \boldsymbol{S}^m(\Omega \times \mathbb{R}^n)/\boldsymbol{S}^{-\infty}(\Omega \times \mathbb{R}^n)\,.$$

The equivalence class defined by

$$\begin{aligned} \sigma_{mA} := \ & \text{the equivalence class of all the symbols associated} \\ & \text{with } A \ \text{ defined by (7.1.29) in } \boldsymbol{S}^m(\Omega \times \mathbb{R}^n)/\boldsymbol{S}^{m-1}(\Omega \times \mathbb{R}^n) \end{aligned} \tag{7.1.31}$$

is called the *principal symbol class* of A which induces an isomorphism

$$\mathcal{L}^m(\Omega)/\mathcal{L}^{m-1} \to \boldsymbol{S}^m(\Omega \times \mathbb{R}^n)/\boldsymbol{S}^{m-1}(\Omega \times \mathbb{R}^n)\,.$$

As for equivalence classes in general, one often uses just one representative (such as the asymptotic expansion (7.1.28)) of the class σ_A or σ_{mA}, respectively, to identify the whole class in $\boldsymbol{S}^m(\Omega \times \mathbb{R}^n)$.

In view of Theorem 7.1.10, we now collect the mapping properties of the pseudodifferential operators in $\mathcal{L}^m(\Omega)$ in the following theorem without proof.

Theorem 7.1.12. *If $A \in \mathcal{L}^m(\Omega)$ then the following mappings are continuous (see* Treves [420, Chap. I, Corollary 2.1 and Theorem 2.1], Taira [412, Theorem 6.5.9]*):*

$$\begin{array}{lcll} A & : & C_0^\infty(\Omega) & \to & C^\infty(\Omega)\,, \\ A & : & \mathcal{E}'(\Omega) & \to & \mathcal{D}'(\Omega)\,, \\ A & : & H^s_{\text{comp}}(\Omega) & \to & H^{s-m}_{\text{loc}}(\Omega)\,. \end{array} \tag{7.1.32}$$

If in addition, $A \in \mathcal{L}^m(\Omega)$ is properly supported, then the mappings can be extended to continuous mappings as follows [420, Chap. I, Proposition 3.2], [412, Theorem 6.5.9]*:*

$$\begin{array}{lcll} A & : & C_0^\infty(\Omega) & \to & C_0^\infty(\Omega)\,, \\ A & : & C^\infty(\Omega) & \to & C^\infty(\Omega)\,, \\ A & : & \mathcal{E}'(\Omega) & \to & \mathcal{E}'(\Omega)\,, \\ A & : & \mathcal{D}'(\Omega) & \to & \mathcal{D}'(\Omega)\,, \\ A & : & H^s_{\text{comp}}(\Omega) & \to & H^{s-m}_{\text{comp}}(\Omega)\,, \\ A & : & H^s_{\text{loc}}(\Omega) & \to & H^{s-m}_{\text{loc}}(\Omega)\,. \end{array} \tag{7.1.33}$$

Based on Theorem 7.1.9 and the concept of the principal symbol, one may consider the algebraic properties of pseudodifferential operators in $\mathcal{L}^m(\Omega)$. The class $\mathcal{L}^m(\Omega)$ is closed under the operations of taking the transposed and the adjoint of these operators.

Theorem 7.1.13. *If $A \in \mathcal{L}^m(\Omega)$ then its transposed $A^\top \in \mathcal{L}^m(\Omega)$ and its adjoint $A^* \in \mathcal{L}^m(\Omega)$. The corresponding complete symbol classes have the asymptotic expansions*

$$\sigma_{A^\top} \sim \sum_{\alpha\geq 0} \frac{1}{\alpha!}\left(\frac{\partial}{\partial\xi}\right)^\alpha \left(-i\frac{\partial}{\partial x}\right)^\alpha \Big(\sigma_A(x,-\xi)\Big)^\top \tag{7.1.34}$$

and

$$\sigma_{A^*} \sim \sum_{\alpha\geq 0} \frac{1}{\alpha!}\left(\frac{\partial}{\partial\xi}\right)^\alpha \left(-i\frac{\partial}{\partial x}\right)^\alpha \Big(\sigma_A(x,\xi)\Big)^* \tag{7.1.35}$$

where σ_A denotes one of the representatives of the complete symbol class.

For the detailed proof we refer to [420, Chap. I, Theroem 3.1].

With respect to the composition of operators, the class of all $\mathcal{L}^m(\Omega)$ is not closed. However, the properly supported pseudodifferential operators in $\mathcal{L}^m(\Omega)$ form an algebra. More precisely, we have the following theorem [188, 18.1], [416, Chap.II, Theorem 4.4.].

Theorem 7.1.14. *Let $A \in \mathcal{L}^{m_1}(\Omega)\,,\ B \in \mathcal{L}^{m_2}(\Omega)$ and one of them be properly supported. Then the composition*

$$A \circ B \in \mathcal{L}^{m_1+m_2}(\Omega) \tag{7.1.36}$$

and we have the asymptotic expansion for the complete symbol class:

$$\sigma_{A\circ B} \sim \sum_{\alpha\geq 0} \frac{1}{\alpha!}\Big(\Big(\frac{\partial}{\partial\xi}\Big)^{\alpha}\sigma_A(x,\xi)\Big)\Big(\Big(-i\frac{\partial}{\partial x}\Big)^{\alpha}\sigma_B(x,\xi)\Big)\,. \tag{7.1.37}$$

Here, $\sigma_A(x,\xi)$ and $\sigma_B(x,\xi)$ denote one of the respective representatives in the corresponding equivalence classes of the complete symbol classes.

For the proof, we remark that with Theorem 7.1.9 we have either $A = A_0 + R$ and $B = B_0$ or $A = A_0$ and $B = B_0 + Q$. Then

$$A\circ B \;=\; A_0\circ B_0 + R\circ B_0$$

or

$$A\circ B \;=\; A_0\circ B_0 + A_0\circ Q\,.$$

Since A_0 and B_0 are properly supported, $R\circ B_0$ or $A_0\circ Q$ are continuous mappings from $\mathcal{E}'(\Omega)$ to $C^\infty(\Omega)$ and, hence, are smoothing operators. (Note that for A and B both not properly supported, the composition generates the term $R\circ Q$, the composition of two regularizers, which is not defined in general.)

For the remaining products $A_0\circ B_0$, we can use that without loss of generality, $A_0 = A_0(x,D)\in OPS^{m_1}(\Omega)$ and $B_0 = B_0(x,D)\in OPS^{m_2}(\Omega)$. Hence, one obtains for $C := A_0\circ B_0$ a representation in the form (7.1.26) with the amplitude function

$$c(x,z,\eta) = \int\limits_{\mathbb{R}^n}\int\limits_{\Omega} a(x,\xi)e^{i(x-y)\cdot\xi}b(y,\eta)e^{i(y-z)\cdot\eta}dyd\xi$$

for which one needs to show $c\in S^{m_1+m_2}(\Omega\times\Omega\times\mathbb{R}^n)$ and, with the Leibniz rule, to evaluate the asymptotic expansion (7.1.28) for c. For details see e.g. [420, Chap. I, Theorem 3.2].

As an immediate consequence of Theorem 7.1.14, the following corollary holds.

Corollary 7.1.15. *Let $A\in\mathcal{L}^{m_1}(\Omega)$, $B\in\mathcal{L}^{m_2}(\Omega)$ and one of them be properly supported. Then the commutator satisfies*

$$[A,B] := A\circ B - B\circ A \in \mathcal{L}^{m_1+m_2-1}(\Omega)\,. \tag{7.1.38}$$

Proof: For $A\circ B$ and $B\circ A$ we have Theorem 7.1.14 and the commutator's symbol has the asymptotic expansion

$$\begin{aligned}\sigma_{[A,B]} \sim \sum_{|\alpha|\geq 1}\frac{1}{\alpha!}\Big\{&\Big(\Big(\frac{\partial}{\partial\xi}\Big)^{\alpha}\sigma_A(x,\xi)\Big)\Big(\Big(-i\frac{\partial}{\partial x}\Big)^{\alpha}\sigma_B(x,\xi)\Big)\\ &-\Big(\Big(\frac{\partial}{\partial\xi}\Big)^{\alpha}\sigma_B(x,\xi)\Big)\Big(\Big(-i\frac{\partial}{\partial x}\Big)^{\alpha}\sigma_A(x,\xi)\Big)\Big\}\,.\end{aligned}$$

Hence, the order of $[A, B]$ equals $m_1 - 1 + m_2$. ■

To conclude this section, we now return to a subclass, the class of classical pseudodifferential operators, which is very important in connection with elliptic boundary value problems and boundary integral equations.

Definition 7.1.6. *A pseudodifferential operator $A \in \mathcal{L}^m(\Omega)$ is said to be* classical *if its complete symbol σ_A has a representative in the class $\boldsymbol{S}^m_{c\ell}(\Omega \times \mathbb{R}^n)$, cf. (7.1.19). We denote by $\mathcal{L}^m_{c\ell}(\Omega \times \mathbb{R}^n)$ the set of all classical pseudodifferential operators of order m.*

We remark that for $\mathcal{L}^m_{c\ell}(\Omega \times \mathbb{R}^n)$, the mapping

$$\mathcal{L}^m_{c\ell}(\Omega) \ni A \mapsto \sigma_A \in \boldsymbol{S}^m_{c\ell}(\Omega \times \mathbb{R}^n)/\boldsymbol{S}^{-\infty}(\Omega \times \mathbb{R}^n)$$

induces the isomorphism

$$\mathcal{L}^m_{c\ell}(\Omega)/\mathcal{L}^{-\infty}(\Omega) \to \boldsymbol{S}^m_{c\ell}(\Omega \times \mathbb{R}^n)/\boldsymbol{S}^{-\infty}(\Omega \times \mathbb{R}^n)\,. \tag{7.1.39}$$

Moreover, we have

$$\mathcal{L}^{-\infty}(\Omega) = \bigcap_{m \in \mathbb{R}} \mathcal{L}^m_{c\ell}(\Omega)\,. \tag{7.1.40}$$

In this case, for $A \in \mathcal{L}^m_{c\ell}(\Omega)$, the principal symbol σ_{mA} has a representative

$$a^0_m(x,\xi) := |\xi|^m a_m\left(x, \frac{\xi}{|\xi|}\right) \tag{7.1.41}$$

which belongs to $C^\infty(\Omega \times (\mathbb{R}^n \setminus \{0\}))$ and is positively homogeneous of degree m with respect to ξ. The function $a^0_m(x,\xi)$ in (7.1.41) is called the *homogeneous principal symbol* of $A \in \mathcal{L}^m_{c\ell}(\Omega)$. Note, in contrast to the principal symbol of A defined in (7.1.31), the homogeneous principal symbol is only a single function which represents the whole equivalence class in (7.1.31). Correspondingly, if we denote by

$$a^0_{m-j}(x,\xi) := |\xi|^{m-j} a_{m-j}\left(x, \frac{\xi}{|\xi|}\right) \tag{7.1.42}$$

the homogeneous parts of the asymptotic expansion of the classical symbol σ_{mA}, which have the properties

$$\begin{array}{rclllll} a^0_{m-j}(x,\xi) & = & a_{m-j}(x,\xi) & \text{for } |\xi| & \geq & 1 & \text{and} \\ a^0_{m-j}(x,t\xi) & = & t^{m-j}a^0_{m-j}(x,\xi) & \text{for all } t & > & 0 & \text{and } \xi \neq 0\,, \end{array}$$

then σ_{mA} may be represented asymptotically by the formal sum $\sum\limits_{j=0}^{\infty} a^0_{m-j}(x,\xi)$.

Coordinate Changes

Let $\Phi : \Omega \to \Omega'$ be a C^∞–diffeomorphism between the two open subsets Ω and Ω' in $\mathbb{R}^n$, i.e. $x' = \Phi(x)$. The diffeomorphism is bijective and $x = \Phi^{-1}(x')$.

If $v \in \mathcal{D}(\Omega')$ then

$$u(x) := (v \circ \Phi)(x) =: (\Phi^* v)(x) \tag{7.1.43}$$

defines $u \in \mathcal{D}(\Omega)$ and the linear mapping $\Phi^* : \mathcal{D}(\Omega') \to \mathcal{D}(\Omega)$ is called the *pullback* of v by Φ.

Correspondingly, the *pushforward* of u by Φ is defined by the linear mapping $\Phi_* : \mathcal{D}(\Omega) \to \mathcal{D}(\Omega')$ given by

$$v(x') := u \circ \Phi^{-1}(x') =: (\Phi_* u)(x') \,. \tag{7.1.44}$$

Now we consider the behaviour of a pseudodifferential operator $A \in \mathcal{L}^m(\Omega)$ under changes of coordinates given by a C^∞–diffeomorphism Φ.

Theorem 7.1.16. (Hörmander [188, Th.18.1.17]) *Let $A \in \mathcal{L}^m(\Omega)$ then $A_\Phi := \Phi_* A \Phi^* \in \mathcal{L}^m(\Omega')$. Moreover, if $A = A_0 + R$ with A_0 properly supported and R a smoothing operator then in the decomposition*

$$A_\Phi = \Phi_* A_0 \Phi^* + \Phi_* R \Phi^* \,,$$

$\Phi_* A_0 \Phi^*$ is properly supported and $\Phi_* R \Phi^*$ is a smoothing operator on Ω'. The complete symbol class of A_Φ has the following asymptotic expansion

$$a_\Phi(x', \xi') \sim \sum_{0 \le |\alpha|} \frac{1}{\alpha!} \Big(\Big(\frac{\partial}{\partial \xi'}\Big)^\alpha \Big(-i \frac{\partial}{\partial y'}\Big)^\alpha a_\Phi(x', y', \xi') \Big)_{|y'=x'} \tag{7.1.45}$$

where

$$\begin{aligned} a_\Phi(x', y', \xi') &= \Big(\det \frac{\partial \Phi}{\partial y}(y)\Big)^{-1} \Big(\det \Xi^\top(x', y')\Big)^{-1} \psi\Big(\frac{x'-y'}{\varepsilon}\Big) \\ &\quad\; a\Big(x, \big(\Xi^\top(x', y')\big)^{-1} \xi'\Big) \end{aligned} \tag{7.1.46}$$

with $x = \Phi^{-1}(x')$, $y = \Phi^{-1}(y')$ and where the smooth matrix–valued function $\Xi(x', y')$ is defined by

$$\Xi(x', y')(x' - y') = \Phi^{-1}(x') - \Phi^{-1}(y') \tag{7.1.47}$$

which implies

$$\Xi(x', x') = \Big(\frac{\partial \Phi}{\partial x}\Big)^{-1} \,. \tag{7.1.48}$$

The constant $\varepsilon > 0$ is chosen such that $\big(\Xi(x', y')\big)^{-1}$ exists for $|x' - y'| \le 2\varepsilon$.

(7.1.45) *yields for the principal symbols the relation*

$$\sigma_{A_\Phi,m}(x',\xi') = \sigma_{A,m}\Big(\Phi^{(-1)}(x')\,,\ \Big(\frac{\partial\Phi}{\partial x}\Big)^{\top}\xi'\Big)\,. \tag{7.1.49}$$

Here, $\psi(z)$ is a C_0^∞ cut–off function with $\psi(z) = 1$ for all $|z| \leq \frac{1}{2}$ and $\psi(z) = 0$ for all $|z| \geq 1$.

Remark 7.1.4: The relation between A_Φ and A can be seen in Figure 7.1.1

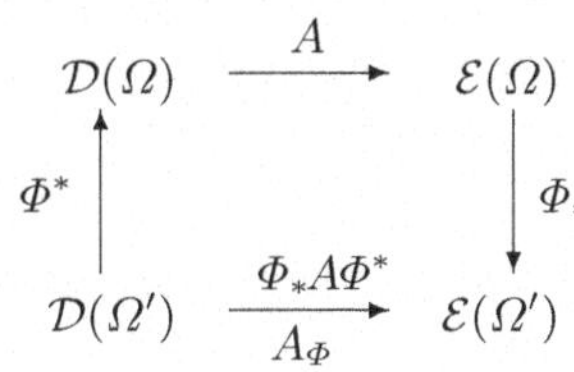

Figure 7.1.1: The commutative diagram of a pseudodifferential operator under change of coordinates

Proof: **i)** If A_0 is properly supported, so is $\Phi_* A_0 \Phi^*$ due to the following. With Lemma 7.1.4, we have that to any compact subset $K_{y'} \Subset \Omega'$ the subset $\Phi^{(-1)}(K_{y'}) = K_y \subset \Omega$ is compact. Then to $K_y \Subset \Omega$ there exists the compact subset $K_x \Subset \Omega$ such that for every $u \in \mathcal{D}(\Omega)$ with $\operatorname{supp} u \subset K_y$ there holds $\operatorname{supp} A_0 u \subset K_x$. Consequently, for any $v \in \mathcal{D}(\Omega')$ with $\operatorname{supp} v \subset K_{y'}$ we have $\operatorname{supp} \Phi^* v \subset \Phi^{(-1)}(K_{y'}) = K_y$ and $\operatorname{supp} A_0 \Phi^* v \subset K_x$. This implies

$$\operatorname{supp} \Phi_* A_0 \Phi^* v = \operatorname{supp} A_{0\Phi} v = \Phi(\operatorname{supp} A_0 \Phi^* v) \subset \Phi(K_x) =: K_{x'}\,.$$

On the other hand, if $K_{x'} \Subset \Omega'$ is a given compact subset then $K_x := \Phi^{(-1)}(K_{x'})$ is a compact subset in Ω to which there exists a compact subset $K_y \Subset \Omega$ with the property that for every $u \in \mathcal{D}(\Omega)$ with $\operatorname{supp} u \cap K_y = \emptyset$ there holds $\operatorname{supp} A_0 u \cap K_x = \emptyset$. Now let $K_{y'} := \Phi(K_y)$ which is a compact subset of Ω'. Thus, if $v \in \mathcal{D}(\Omega')$ with $\operatorname{supp} v \cap K_{y'} = \emptyset$ then

$$\operatorname{supp} \Phi^* v \cap K_y = (\Phi^{(-1)} \operatorname{supp} v) \cap K_y = \Phi^{(-1)}(\operatorname{supp} v \cap K_{y'}) = \emptyset$$

and, since A_0 is properly supported, $\operatorname{supp} A_0 \Phi^* v \cap K_x = \emptyset$. Hence, $\Phi^{(-1)}\big(\Phi(\operatorname{supp} A_0 \Phi^* v) \cap \Phi(K_x)\big) = \emptyset$ which yields

$$\operatorname{supp} \Phi_* A_0 \Phi^* v \cap K_{x'} = \emptyset\,.$$

According to Lemma 7.1.4, this implies that $\Phi_* A_0 \Phi^*$ is properly supported.

Clearly, for $v \in \mathcal{D}(\Omega')$ we have

$$(\Phi_* R \Phi^* v)(x') = \int\limits_{\Omega'} k_R\big(\Phi^{(-1)}(x')\,,\ \Phi^{(-1)}(y')\big) v(y') \Big(\det\Big(\frac{\partial \Phi}{\partial y}\Big)\Big)^{-1} dy'$$

with the Schwartz kernel $k_R \in C^\infty(\Omega \times \Omega)$ of R. The transformed operator $\Phi_* R \Phi^*$ has the Schwartz kernel

$$k_R(\Phi^{(-1)}(x')\,,\,\Phi^{(-1)}(y'))\Big(\det\Big(\frac{\partial\Phi}{\partial y}\Big)\Big)^{-1} \in C^\infty(\Omega' \times \Omega')\,.$$

Hence, $\Phi_* R \Phi^*$ is a smoothing operator.

ii) It remains to show that the properly supported part $\Phi_* A_0 \Phi^* \in OPS^m(\Omega' \times \mathbb{R}^n)$. For any $v \in \mathcal{D}(\Omega')$ we have the representation

$$\begin{aligned}(\Phi_* A_0 \Phi^* v)(x') &= (A_0 \Phi^* v)(\Phi^{(-1)}(x'))\\ &= (2\pi)^{-n} \int\limits_{\mathbb{R}^n}\int\limits_{\Omega} e^{i(x-y)\cdot\xi} a(x,\xi)(v\circ\Phi)(y)dyd\xi\\ &= (2\pi)^{-n} \int\limits_{\mathbb{R}^n}\quad\int\limits_{\Omega' \wedge |x'-y'| \le 2\varepsilon} e^{i(x-y)\cdot(\Xi^\top(x',y')\xi)} a(x,\xi)v(y') \times\\ &\qquad \times\psi\Big(\frac{x'-y'}{\varepsilon}\Big)\Big(\det\frac{\partial\phi}{\partial y}\Big)^{-1} dy'd\xi + Rv\end{aligned}$$

where

$$\begin{aligned}(Rv)(x) &= (2\pi)^{-n}\int\limits_{\mathbb{R}^n}\int\limits_{\Omega'} e^{i(x-y)\cdot\xi} a(x,\xi)v(y')\Big(1-\psi\Big(\frac{x'-y'}{\varepsilon}\Big)\Big) \times\\ &\qquad \times \Big(\det\frac{\partial\phi}{\partial y}\Big)^{-1} dy'd\xi\end{aligned}$$

and $y = \Phi^{(-1)}(y')\,,\; x = \Phi^{(-1)}(x')$.

With the affine transformation of coordinates,

$$\Xi^\top(x',y')\xi = \xi'\,,$$

the first integral reduces to

$$(A_{\Phi,0}v)(x') = (2\pi)^{-n}\int\limits_{\mathbb{R}^n}\int\limits_{\Omega'} e^{i(x'-y')\cdot\xi'} v(y') a_\Phi(x',y',\xi')dy'd\xi'$$

with $a_\Phi(x',y',\xi')$ given by (7.1.46). Since $a(x,\xi)$ belongs to $\boldsymbol{S}^m(\Omega\times,\mathbb{R}^n)$, with the chain rule one obtains the estimates

$$\Big|\Big(\frac{\partial}{\partial x'}\Big)^\beta\Big(\frac{\partial}{\partial y'}\Big)^\gamma\Big(\frac{\partial}{\partial \xi'}\Big)^\alpha a_\Phi(x',y',\xi')\Big| \le c(K,\alpha,\beta,\gamma)(1+|\xi'|)^{m-|\alpha|}$$

for all $(x',y') \in K \times K\,,\; \xi' \in \mathbb{R}^n$ and any compact subset $K \Subset \Omega'$. Hence, $a_\Phi \in \boldsymbol{S}^m(\Omega' \times \Omega' \times \mathbb{R}^n)$.

Since the Schwartz kernel of $A_{\Phi,0}$ satisfies

$$K_{A_{\Phi,0}}(x', x'-y') = K_{\Phi_* A_0 \Phi^*}(x', x'-y')\psi\Big(\frac{x'-y'}{\varepsilon}\Big)$$

and $\Phi_* A_0 \Phi^*$ is properly supported, so is $A_{\Phi,0}$. Now Theorem 7.1.11 can be applied to $A_{\Phi,0}$.

Since the function $v(y')\Big(1-\psi\big(\frac{x'-y'}{\varepsilon}\big)\Big)\Big(\det\frac{\partial\phi}{\partial y}\Big)^{-1}$ vanishes in the vicinity of $x'=y'$ identically, we can proceed in the same manner as in the proof of Theorem 7.1.2 and obtain

$$\begin{aligned} Rv(x') &= (2\pi)^{-n}\int\limits_{\mathbb{R}^n}\int\limits_{\Omega'} e^{i(x-y)\cdot\xi}v(y')\Big(1-\psi\Big(\frac{x'-y'}{\varepsilon}\Big)\Big)\times \\ &\qquad \times\Big(\det\frac{\partial\phi}{\partial y}\Big)^{-1}\Big((-\Delta_\xi)^N a(x,\xi)\Big)dy'd\xi\,. \end{aligned}$$

Here, for $N > m+n+1$ we may interchange the order of integration and find that R is a smoothing integral operator with C^∞–kernel. Therefore, $\Phi_* A_0 \Phi^* \in OPS^m(\Omega'\times\mathbb{R}^n)$ and the asymptotic formula (7.1.45) is a consequence of Theorem 7.1.11. This completes the proof. ∎

The transformed symbol in (7.1.45) is not very practical from the computational point of view since one needs the explicit form of the matrix $\Xi(x',y')$ in (7.1.47) and its inverse. Alternatively, one may use the following asymptotic expansion of the symbol.

Lemma 7.1.17. *Under the same conditions as in Theorem 7.1.16, the complete symbol class of A_Φ also has the asymptotic expansion*

$$a_\Phi(x',\xi') \sim \sum_{0\le\alpha}\frac{1}{\alpha!}\Big(\Big(\frac{\partial}{\partial\xi}\Big)^\alpha a(x,\xi)\Big)_{|_{\xi=\frac{\partial\Phi}{\partial x}^\top\xi'}}\Big(\Big(-i\frac{\partial}{\partial z}\Big)^\alpha e^{i\xi'\cdot r}\Big)_{|_{z=x}} \tag{7.1.50}$$

where $x'=\Phi(x)$ and

$$r = \Phi(z)-\Phi(x)-\frac{\partial\Phi}{\partial x}(x)(z-x)\,. \tag{7.1.51}$$

Moreover,

$$\Big(\Big(-i\frac{\partial}{\partial z}\Big)^\alpha e^{i\xi'\cdot r}\Big)_{|z=x} = \begin{cases} 0 & \textit{for } |\alpha|=1\,,\\ (-i)^{|\alpha|-1}\Big(\Big(\frac{\partial}{\partial x}\Big)^\alpha\Phi(x)\Big)\cdot\xi' & \textit{for } |\alpha|\ge 2\,. \end{cases} \tag{7.1.52}$$

Proof: In view of Theorem 7.1.16 it suffices to consider properly supported operators A_0 and $A_{\Phi,0}$. Therefore the symbol class to $A_{\Phi,0}$ can be found by using formula (7.1.29), i.e.,

$$\begin{aligned} \sigma_{A_{\Phi,0}}(x',\xi') &= e^{-ix'\cdot\xi'}A_{\Phi,0}e^{i\xi'\cdot\bullet}(x') \\ &= e^{-i\Phi(x)\cdot\xi'}A_0 e^{i\xi'\cdot\Phi(\cdot)}(x) \quad \text{for } x=\Phi^{-1}(x') \end{aligned}$$

after coordinate transform. With the substitution

$$\Phi(z) = \Phi(x) + \frac{\partial \Phi}{\partial x}(x)(z-x) + r(x,z)$$

we obtain

$$\begin{aligned}\sigma_{A_{\Phi,0}}(x',\xi') &= e^{-ix\cdot\left(\frac{\partial\Phi}{\partial x}^{\top}\xi'\right)}(2\pi)^{-n}\int\limits_{\mathbb{R}^n}\int\limits_{\Omega} e^{-i(x-z)\cdot\xi}a(x,\xi)\times \\ &\qquad \times e^{iz\cdot\left(\frac{\partial\Phi}{\partial x}^{\top}\xi'\right)}e^{ir(x,z)\cdot\xi'}dzd\xi \\ &= e^{-ix\cdot\zeta}A\circ Be^{i\bullet\cdot\zeta}\end{aligned}$$

where $A = a(x,D)$ and B is the multiplication operator defined by the symbol

$$\sigma_B(z,\xi) = e^{ir(x,z)\cdot\xi'}$$

with x and ξ' fixed as constant parameters. With formula (7.1.29) we then find

$$\sigma_{A_{\Phi,0}}(x',\xi') \sim \sigma_{A\circ B}(x,\zeta) \quad \text{where } x' = \Phi(x) \quad \text{and} \quad \zeta = \frac{\partial\Phi}{\partial x}^{\top}\xi'.$$

The symbol of $A\circ B$ is given by (7.1.37), which now implies

$$\sigma_{A_{\Phi,0}}(x',\xi') \sim \sum_{\alpha\geq 0}\frac{1}{\alpha!}\left(\left(\frac{\partial}{\partial\xi}\right)^{\alpha}a(x,\xi)\right)_{|\xi=\zeta}\left(\left(-i\frac{\partial}{\partial z}\right)^{\alpha}e^{ir(x,z)\cdot\xi'}\right)_{|z=x}$$

with constant $\zeta = \frac{\partial\Phi}{\partial x}^{\top}\xi'$, which is the desired formula (7.1.50).

To show (7.1.52) we employ induction for α with $|\alpha|\geq 1$. In particular we see that

$$\frac{\partial}{\partial z_j}e^{i\xi'\cdot r(x,y)}_{|z=x} = \left\{ie^{i\xi'\cdot r(x,z)}\left(\frac{\partial\Phi}{\partial z_j}(z) - \frac{\partial\Phi}{\partial x_j}(x)\right)\cdot\xi'\right\}_{|z=x} = 0$$

and

$$\begin{aligned}\frac{\partial}{\partial z_k}\frac{\partial}{\partial z_j}e^{i\xi'\cdot r(x,z)}_{|z=x} &= \Big\{i^2e^{i\xi'\cdot r(x,z)}\left(\frac{\partial\Phi}{\partial z_j}(z) - \frac{\partial\Phi}{\partial x_j}(x)\right)\cdot\xi'\times \\ &\qquad \left(\frac{\partial\Phi}{\partial z_k}(z) - \frac{\partial\Phi}{\partial x_k}(x)\right)\cdot\xi' \\ &\qquad + ie^{i\xi'\cdot r(x,z)}\frac{\partial^2\Phi}{\partial z_j\partial z_k}(z)\cdot\xi'\Big\}_{|z=x} \\ &= i\frac{\partial^2\Phi}{\partial x_j\partial x_k}(x)\cdot\xi',\end{aligned}$$

which is (7.1.52) for $1 \le |\alpha| \le 2$. For $|\alpha| > 2$, we observe that every differentiation yields terms which contain at least one factor of the form $\Big(\frac{\partial \Phi}{\partial z_\ell}(z) - \frac{\partial \Phi}{\partial x_\ell}(x)\Big) \cdot \xi'$ which vanishes for $z = x$, except the last term which is of the form

$$i\Big(\frac{\partial}{\partial x}\Big)^\alpha \Phi(x) \cdot \xi' ,$$

which proves (7.1.52). ■

We comment that in this proof we did not use the explicit representation of $A_{\Phi,0}$ in the form

$$(A_{\Phi,0}v)(x) = (2\pi)^{-n} \int\limits_{\mathbb{R}^n} \int\limits_{\Omega'} a\Big(x, \frac{\partial \Phi}{\partial x}^\top \xi'\Big) e^{ih + i(x'-y')\cdot\xi'} \times$$
$$\times v(y)\Big(\det \frac{\partial \Phi}{\partial y}\Big)^{-1} \Big(\det \frac{\partial \Phi}{\partial x}\Big) dy' d\xi'$$

where

$$h(x', y', \xi') = \Big\{\frac{\partial \Phi}{\partial x}(x' - y')\Big\} \cdot \xi' \quad \text{with } x = \Phi^{-1}(x') \quad \text{and} \quad y = \Phi^{-1}(y') .$$

Now it follows from Theorem 7.1.11 that

$$a_\Phi(x', \xi') \quad \sim \quad \sum_{|\alpha| \ge 0} \frac{1}{\alpha!} \Big(\Big(\frac{\partial}{\partial \xi'}\Big)^\alpha \Big(-i\frac{\partial}{\partial y'}\Big)^\alpha a_\Phi(x', y', \xi')\Big)_{|y'=x'}$$

where

$$a_\Phi(x', y', \xi') \quad = \quad a\Big(x, \frac{\partial \Phi}{\partial x}^\top \xi'\Big) e^{ir} \Big(\det \frac{\partial \Phi}{\partial y}\Big)^{-1} \Big(\det \frac{\partial \Phi}{\partial x}\Big) .$$

Since $\Big(\det \frac{\partial \Phi}{\partial x}\Big) a\Big(x, \frac{\partial \Phi}{\partial x}^\top \xi'\Big)$ is independent of y', interchanging differentiation yields the formula

$$a_\Phi(x', \xi') \sim \sum_{0 \le |\alpha|} \frac{1}{\alpha!} \Big(\det \frac{\partial \Phi}{\partial x}\Big) \Big(\frac{\partial}{\partial \xi'}\Big)^\alpha \Big\{ a\Big(x, \Big(\frac{\partial \Phi}{\partial x}\Big)^\top \xi'\Big) \times$$
$$\times \Big(-i\frac{\partial}{\partial y'}\Big)^\alpha \Big(e^{ih(x',y',\xi')} \Big(\det \frac{\partial \Phi}{\partial y}(y')\Big)^{-1}\Big)\Big\}_{|y'=x'} . \qquad (7.1.53)$$

Needless to say that both formulae (7.1.50) and (7.1.53) are equivalent, however, in (7.1.53) one needs to differentiate the Jacobian $\Big(\det \frac{\partial \Phi}{\partial y}\Big)^{-1}$ which makes (7.1.53) less desirable than (7.1.50) from the practical point of view.

7.2 Elliptic Pseudodifferential Operators on $\Omega \subset \mathbb{R}^n$

Elliptic pseudodifferential operators form a special class of $\mathcal{L}^m(\Omega)$ which are essentially invertible. This section we devote to some basic properties of this class of operators on Ω. In particular, we will discuss the connections between the parametrix and the Green's operator for elliptic boundary value problems. The development here will also be useful for the treatment of boundary integral operators later on.

Definition 7.2.1. *A symbol $a(x,\xi)$ in $\boldsymbol{S}^m(\Omega \times \mathbb{R}^n)$ is elliptic of order m if there exists a symbol $b(x,\xi)$ in $\boldsymbol{S}^{-m}(\Omega \times \mathbb{R}^n)$ such that*

$$ab - 1 \in \boldsymbol{S}^{-1}(\Omega \times \mathbb{R}^n)\,. \tag{7.2.1}$$

This definition leads at once to the following criterion.

Lemma 7.2.1. *A symbol $a \in \boldsymbol{S}^m(\Omega \times \mathbb{R}^n)$ is elliptic if and only if for any compact set $K \Subset \Omega$ there are constants $c(K) > 0$ and $R(K) > 0$ such that*

$$|a(x,\xi)| \geq c(K)\langle\xi\rangle^m \quad \textit{for all } x \in K \quad \textit{and} \quad |\xi| \geq R(K)\,. \tag{7.2.2}$$

We leave the proof to the reader.

For a pseudodifferential operator $A \in \mathcal{L}^m(\Omega)$ we say that A is *elliptic of order m*, if one of the representatives of the complete symbol σ_A is elliptic of order m.

For classical pseudodifferential operators $A \in \mathcal{L}^m_{c\ell}(\Omega)$, the ellipticity of order m can easily be characterized by the homogeneous principal symbol a^0_m and the ellipticity condition

$$a^0_m(x,\xi) \neq 0 \quad \text{for all } x \in \Omega \quad \text{and} \quad |\xi| = 1\,. \tag{7.2.3}$$

As an example we consider the second order linear differential operator P as in (5.1.1) for the scalar case $p = 1$ which has the complete symbol

$$a(x,\xi) = \sum_{j,k=1}^{n} a_{jk}(x)\xi_j\xi_k - i\sum_{k=1}^{n}\Big(b_k(x) - \sum_{j=1}^{n}\frac{\partial a_{jk}}{\partial x_j}(x)\Big)\xi_k + c(x)\,. \tag{7.2.4}$$

In particular, Condition (7.2.3) is fulfilled for a strongly elliptic P satisfying (5.4.1); then

$$a^0_2(x,\xi) := \sum_{j,k=1}^{n} a_{jk}(x)\xi_j\xi_k \neq 0 \quad \text{for } |\xi| \neq 0 \quad \text{and} \quad x \in \Omega\,, \tag{7.2.5}$$

and P is elliptic of order $m = 2$.

Definition 7.2.2. *A parametrix Q_0 for the operator $A \in \mathcal{L}^m(\Omega)$ is a properly supported operator which is a two-sided inverse for A modulo $\mathcal{L}^{-\infty}(\Omega)$:*

$$\begin{aligned} A \circ Q_0 - I &= C_1 \in \mathcal{L}^{-\infty}(\Omega) \\ \textit{and} \qquad & \\ Q_0 \circ A - I &= C_2 \in \mathcal{L}^{-\infty}(\Omega) \end{aligned} \tag{7.2.6}$$

Theorem 7.2.2. (Hörmander[186], Taylor [416, Chap.III, Theorem 1.3]) *$A \in \mathcal{L}^m(\Omega)$ is elliptic of order m if and only if there exists a parametrix $Q_0 \in \mathcal{L}^{-m}(\Omega)$ satisfying* (7.2.6).

Proof: **(i)** Let A be a given operator in $\mathcal{L}^m(\Omega)$ which is elliptic of order m. Then to the complete symbol class of A there is a representative $a(x,\xi)$ satisfying (7.2.2). If Q would be known then (7.1.37) with (7.2.6) would imply

$$\sigma_{Q\circ A} = 1 \sim \sum_{0\leq j} q_{-m-j}(x,\xi)a(x,\xi) + \sum_{1\leq|\alpha|} \frac{1}{\alpha!}\Big(\frac{\partial}{\partial\xi}\Big)^\alpha \Big(\sum_{0\leq\ell} q_{-m-\ell}(x,\xi)\Big)\times$$
$$\times\Big(-i\frac{\partial}{\partial x}\Big)^\alpha a(x,\xi)\,. \quad (7.2.7)$$

So, to a we choose a function $\chi \in C^\infty(\Omega\times\mathbb{R}^n)$ with $\chi(x,\xi) = 1$ for all $x \in \Omega$ and $|\xi| \geq C \geq 1$ where C is chosen appropriately large, and $\chi = 0$ in some neighbourhood of the zeros of a. Then define

$$q_{-m}(x,\xi) := \chi(x,\xi)a(x,\xi)^{-1} \quad (7.2.8)$$

and, in view of (7.2.7), recursively for $j = 1,2,\ldots$,

$$q_{-m-j}(x,\xi) := \quad (7.2.9)$$
$$-\sum_{1\leq|\alpha|\leq j}\frac{1}{\alpha!}\Big(\Big(\frac{\partial}{\partial\xi}\Big)^\alpha q_{-m-j+|\alpha|}(x,\xi)\Big)\Big(\Big(-i\frac{\partial}{\partial x}\Big)^\alpha a(x,\xi)\Big)q_{-m}(x,\xi)\,.$$

From this construction one can easily show that $q_{-m-j} \in \boldsymbol{S}^{-m-j}(\Omega\times\mathbb{R}^n)$. The sequence $\{q_{-m-j}\}$ can be used to define an operator $Q \in OPS^{-m}(\Omega\times\mathbb{R}^n)$ with $\sum\limits_{j=0}^{\infty} q_{-m-j}$ as the asymptotic expansion of σ_Q, see Theorem 7.1.3. With Q available, we use the decomposition of $Q = Q_0 + R$, Theorem 7.1.9, where $Q_0 \in OPS^{-m}(\Omega\times\mathbb{R}^n)$ is properly supported and still has the same symbol as Q. Then we have with Theorem 7.1.14 that $Q_0\circ A \in \mathcal{L}^0(\Omega)$ and by construction in view of (7.1.37) $\sigma_{Q_0\circ A} \sim 1$. Hence, $Q_0\circ A - I = R_2 \in \mathcal{L}^{-\infty}(\Omega)$.

In the next step it remains to show that $A\circ Q_0 - I \in \mathcal{L}^{-\infty}(\Omega)$, too.

The choice of q_{-m} also shows that Q_0 is an elliptic operator of order $-m$. Therefore, to Q_0 there exists an operator $Q_0^{(-1)} \in OPS^m$ which is also properly supported and satisfies $Q_0^{(-1)}\circ Q_0 - I = R_1 \in \mathcal{L}^{-\infty}(\Omega)$, where R_1 is properly supported, due to Proposition 7.1.6. Then

$$\begin{aligned} Q_0^{(-1)}\circ Q_0\circ A\circ Q_0 - I &= Q_0^{(-1)}\circ(I+R_2)\circ Q_0 - I\\ &= Q_0^{(-1)}\circ Q_0 - I + Q_0^{(-1)}\circ R_2\circ Q_0\\ &= R_1 + Q_0^{(-1)}\circ R_2\circ Q_0\,.\end{aligned}$$

Since both, Q_0 and $Q_0^{(-1)}$ are properly supported, $Q_0^{(-1)}\circ R_2\circ Q_0$ again is a smoothing operator due to Theorem 7.1.14. Hence, $Q_0^{(-1)}\circ Q_0\circ A\circ Q_0 - I =: S_1 \in \mathcal{L}^{-\infty}(\Omega)$.

On the other hand,

$$\begin{aligned} A \circ Q_0 - I &= Q_0^{(-1)} \circ Q_0 \circ A \circ Q_0 + (I - Q_0^{(-1)} \circ Q_0) \circ A \circ Q_0 - I \\ &= S_1 - R_1 \circ A \circ Q_0 \end{aligned}$$

and $R_1 \circ A \circ Q_0$ is a smoothing operator since Q_0 and R_1 are both properly supported and $R_1 \in \mathcal{L}^{-\infty}(\Omega)$. This shows with the previous steps that $A \circ Q_0 - I \in \mathcal{L}^{-\infty}(\Omega)$. Hence, Q_0 satisfies (7.2.6) and is a parametrix.

(ii) If Q is a parametrix satisfying (7.2.6), then with (7.1.37) we find

$$\sigma_{A \circ Q} - aq \sim \sum_{|\alpha| \geq 1} \frac{1}{\alpha!}\Big(\Big(\frac{\partial}{\partial \xi}\Big)^\alpha \sigma_A(x,\xi)\Big)\Big(\Big(-i\frac{\partial}{\partial x}\Big)^\alpha q(x,\xi)\Big).$$

Hence, $\sigma_{A\circ Q} - aq \in \boldsymbol{S}^{-1}(\Omega \times \mathbb{R}^n)$, i.e. (7.2.1) with $b = q$. Moreover, $\sigma_{A\circ Q} - 1 \in \boldsymbol{S}^{-\infty}(\Omega \times \mathbb{R}^n)$ because of (7.2.6). ■

7.2.1 Systems of Pseudodifferential Operators

The previous approach for a scalar elliptic operator can be extended to general elliptic systems. Let us consider the $p \times p$ system of pseudodifferential operators

$$A = ((A_{jk}))_{p\times p}$$

with symbols $a^{jk}(x;\xi) \in \boldsymbol{S}^{s_j+t_k}(\Omega \times \mathbb{R}^n)$ where we assume that there exist two p–tuples of numbers $s_j\,,\ t_k \in \mathbb{R}\,;\ j,k = 1,\dots,p$. As a special example we consider the *Agmon–Douglis–Nirenberg elliptic system of partial differential equations*

$$\sum_{k=1}^{p} \sum_{|\beta|=0}^{s_j+t_k} a_\beta^{jk}(x) D^\beta u_k(x) = f_j(x) \quad \text{for } j = 1,\dots,p. \tag{7.2.10}$$

Without loss of generality, assume $s_j \leq 0$. The symbol matrix associated with (7.2.10) is given by

$$\sum_{|\beta|=0}^{s_j+t_k} a_\beta^{jk}(x) i^{|\beta|} \xi^\beta = a^{jk}(x;\xi) \tag{7.2.11}$$

and its principal part is now defined by

$$a_{s_j+t_k}^{jk}(x;\xi) := \sum_{|\beta|=s_j+t_k} a_\beta^{jk}(x) i^{|\beta|} \xi^\beta \tag{7.2.12}$$

where $a_{s_j+t_k}^{jk}(x;\xi)$ is set equal to zero if the order of $a^{jk}(x;\xi)$ is less than $s_j + t_k$. Then

$$a^{jk0}_{s_j+t_k}(x;\xi) = |\xi|^{s_j+t_k} a^{jk}_{s_j+t_k}\Big(x;\frac{\xi}{|\xi|}\Big) \tag{7.2.13}$$

is a representative of the corresponding homogeneous principal symbol.

For the more general system of operators $A_{jk} \in \mathcal{L}^{s_j+t_k}(\Omega)$ we can associate in the same manner as for the differential operators in (7.2.10) representatives (7.2.11) by first using the principal symbol class (7.1.31) of A_{jk} and then taking the homogeneous representatives (7.2.13) and neglecting those $a^{jk}_{s_j+t_k}$ from (7.1.31) of order less than $s_j + t_k$.

Definition 7.2.3. *Let the characteristic determinant $H(x,\xi)$ for the system $((A_{jk}))$ be defined by*

$$H(x,\xi) := \det\, \big((a^{jk0}_{s_j+t_k}(x;\xi))\big)_{p\times p}\,. \tag{7.2.14}$$

Then the system is elliptic in the sense of Agmon–Douglis–Nirenberg[2] *if*

$$H(x,\xi) \neq 0 \quad \textit{for all} \;\; x \in \Omega \quad \textit{and} \;\; \xi \in \mathbb{R}^n \setminus \{0\}\,. \tag{7.2.15}$$

With this definition of ellipticity, Definition 7.2.2 of a parametrix $Q_0 = ((Q_{jk}))$ for the operator $A = ((A_{jk}))_{p\times p}$ with $A_{jk} \in \mathcal{L}^{s_j+t_k}(\Omega)$ as well as Theorem 7.2.2 with $Q_{jk} \in \mathcal{L}^{s_j+t_k}(\Omega)$ remain valid.

The Stokes system (2.3.1) is a simple example of an elliptic system in the sense of Agmon–Douglis–Nirenberg. If we identify the pressure in (2.3.1) with u_{n+1} for $n = 2$ or 3 then the Stokes system takes the form (7.2.10) with $p = n+1$, for $n = 3$:

$$\sum_{k=1}^{4}\sum_{|\beta|=0}^{s_j+t_j} a^{jk}_\beta D^\beta u_k = \begin{pmatrix} -\mu\Delta & 0 & 0 & \frac{\partial}{\partial x_1} \\ 0 & -\mu\Delta & 0 & \frac{\partial}{\partial x_2} \\ 0 & 0 & -\mu\Delta & \frac{\partial}{\partial x_3} \\ \frac{\partial}{\partial x_1} & \frac{\partial}{\partial x_2} & \frac{\partial}{\partial x_3} & 0 \end{pmatrix} \begin{pmatrix} u_1 \\ u_2 \\ u_3 \\ u_4 \end{pmatrix} = \boldsymbol{f}$$

where $s_j = 0$ for $j = \overline{1,n}$, $s_{n+1} = -1$ and $t_k = 2$ for $k = 1,n$, $t_{n+1} = 1$. The coefficients are given by $a^{jk}_\beta = -\mu\delta^{jk}$ for $|\beta| = 2$ and $j,k = \overline{1,n}$;

$$\begin{array}{llllllll} a^{14}_{(1,0,0)} = a^{24}_{(0,1,0)} = a^{34}_{(0,0,1)} = a^{41}_{(1,0,0)} = a^{42}_{(0,1,0)} = a^{43}_{(0,0,1)} = 1 & \text{for } n = 3\,; \\ a^{13}_{(1,0)} = a^{23}_{(0.1)} = a^{31}_{(1,0)} = a^{32}_{(0,1)} = 1 & \text{for } n = 2 \end{array}$$

and all other coefficients are zero. Then the symbol matrix (7.2.12) for (2.3.1) reads

$$a(x,\xi) = \begin{pmatrix} \mu|\xi|^2 & 0 & 0 & i\xi_1 \\ 0 & \mu|\xi|^2 & 0 & i\xi_2 \\ 0 & 0 & \mu|\xi|^2 & i\xi_3 \\ i\xi_1 & i\xi_2 & i\xi_3 & 0 \end{pmatrix}$$

[2] Originally called Douglis–Nirenberg elliptic.

For $n = 2$, the third column and third row are to be discarded. The characteristic determinant becames

$$H(x,\xi) = \det\ a(x,\xi) = \mu^{n-1}|\xi|^{2n} \quad \text{for} \quad n = 2,3\,.$$

Hence, the Stokes system is elliptic in the sense of Agmon–Douglis–Nirenberg.

Theorem 7.2.3. *(see also* Chazarain and Piriou [64, Chap.4, Theorem 7.7]*)* $A = ((A_{jk}))_{p\times p}$ *is elliptic in the sense of Agmon–Douglis–Nirenberg if and only if there exists a properly supported parametrix* $Q_0 = ((Q_{jk}))_{p\times p}$ *with* $Q_{jk} \in \mathcal{L}^{-t_j-s_k}(\Omega)$.

Proof: Assume that A is elliptic. By the definition of ellipticity, the homogeneous principal symbol is then of the form

$$a^{jk0}_{s_j+t_k}(x;\xi) = ((|\xi|^{s_j}\delta_{j\ell}))((a^{\ell r}_{s_j+t_k}(x;\widehat{\Theta})\,))((|\xi|^{t_r}\delta_{rk}))$$

where $\widehat{\Theta} = \frac{\xi}{|\xi|}$ and Einstein's rule of summation is used, and where

$$H(x;\widehat{\Theta}) = \det\ ((a^{jk}_{s_j+t_k}(x;\widehat{\Theta})\,)) \quad \text{for every} \;\; x\in\Omega \;\; \text{and} \;\; \widehat{\Theta}\in \mathbb{R}^n\,,\; |\widehat{\Theta}| = 1\,.$$

This implies that for every fixed $x \in \Omega$ there exists $R(x) > 0$ such that $\det((a^{jk}(x;\xi)\,)) \neq 0$ for all $|\xi| \geq R(x)$. By setting $a(x;\xi) := ((a^{jk}(x;\xi)\,))_{p\times p}$ and using an appropriate cut–off function $\chi \in C^\infty(\Omega\times\mathbb{R}^n)$ with $\chi(x,\xi) = 1$ for all $x\in\Omega$ and $|\xi| \geq 2R(x)$ and with $\chi(x,\xi) = 0$ in some neighbourhood of the zeros of $\det a(x,\xi)$, we define $q_{-m}(x,\xi)$ by (7.2.8) as a matrix–valued function. Recursively, then $q_{-m-j}(x,\xi)$ can be obtained by (7.2.9) providing us with asymptotic symbol expansions of all the matrix elements of the operator Q which we can find due to Theorem 7.1.3. The properly supported parts of Q define the desired Q_0 satisfying the second equation in (7.2.6), i.e., Q_0 is a left parametrix. To show that Q_0 is also a right parametrix, i.e. to satisfy the first equation in (7.2.6), the arguments are the same as in the scalar case in the proof of Theorem 7.2.2.

Conversely, if Q is a given parametrix for A with $Q_{jk} \in \mathcal{L}^{-t_j-s_k}(\Omega)$ being properly supported then its homogeneous principal symbol has the form

$$\begin{aligned}((q^{jk0}_{-t_j-s_k}(x;\xi)\,)) &= ((|\xi|^{-t_j-s_k}q^{jk0}_{-t_j-s_k}(x;\widehat{\Theta})\,))\\ &= ((|\xi|^{-t_j}\delta_{j\ell}))q^{\ell r0}_{-t_j-s_k}(x;\widehat{\Theta})\,))((|\xi|^{-s_k}\delta_{rk}))\,.\end{aligned}$$

This implies with (7.2.6)

$$a^{jk0}_{s_j+t_k}(x;\xi) = |\xi|^{s_j+t_k}((q^{-1}_{t_j+s_k}(x;\widehat{\Theta})\,))_{jk}$$

where $q^{-1}_{t_j+s_k}$ denotes the inverse of $((q^{jk0}_{-t_j-s_k}))$. The latter exists since

$$\det\ ((q^{jk0}_{-t_j-s_k}(x;\widehat{\Theta})\,))\det\ ((a^{jk0}_0(x;\widehat{\Theta})\,)) = 1\,.$$

Hence, A is elliptic in the sense of Agmon–Douglis–Nirenberg. Because of (7.2.6) we also have $A_{jk} \in \mathcal{L}^{s_j+t_k}(\Omega)$. ∎

In Fulling and Kennedy [130] one finds the construction of the parametrix of elliptic differential operators on manifolds.

7.2.2 Parametrix and Fundamental Solution

To illustrate the idea of the parametrix we consider again the linear second order scalar elliptic differential equation (5.1.1) for $p = 1$, whose differential operator can be seen as a classical elliptic pseudodifferential operator of order $m = 2$.

Let us first assume that P has constant coefficients and, moreover, that the symbol satisfies

$$a(\xi) = \sum_{j,k=1}^{n} a^{jk}\xi_j\xi_k - i\sum_{k=1}^{n} b^k\xi_k + c \neq 0 \quad \text{for all} \quad \xi \in \mathbb{R}^n\,. \tag{7.2.16}$$

Then the solution of (5.1.1) for $f \in C_0^\infty(\Omega)$ can be written as

$$u(x) = Nf(x) := \mathcal{F}^{-1}_{\xi\mapsto x}\left(\frac{1}{a(\xi)}\mathcal{F}_{y\mapsto\xi}f\right)(x)\,. \tag{7.2.17}$$

The operator $N \in \mathcal{L}^{-2}(\Omega)$ defines the *volume potential* for the operator P, the generalized Newton potential or *free space Green's operator*. When $f(y)$ is replaced by $\delta(y - x)$ then we obtain the fundamental solution $E(x, y)$ for P explicitly:

$$E(x, y) = (2\pi)^{-n}\int\limits_{\mathbb{R}^n} \frac{1}{a(\xi)}e^{i(x-y)\cdot\xi}d\xi\,. \tag{7.2.18}$$

If condition (7.2.16) is replaced by

$$a(\xi) \neq 0 \quad \text{for} \quad \xi \neq 0$$

(as for $c = 0$ in (7.2.16)) then, as in (7.2.8), we take

$$q_{-2}(\xi) = \chi(\xi)\frac{1}{a(\xi)}$$

where the cut–off function $\chi \in C^\infty(\mathbb{R}^n)$ satisfies

$$\chi(\xi) = 1 \quad \text{for} \quad |\xi| \geq 1 \quad \text{and} \quad \chi(\xi) = 0 \quad \text{for} \quad |\xi| \leq \frac{1}{2}\,. \tag{7.2.19}$$

We define the *parametrix* by

$$Qf := \mathcal{F}^{-1}_{\xi\mapsto x}(q_{-2}(\xi)\mathcal{F}_{y\mapsto\xi}f)$$

with $Q \in \mathcal{L}^{-2}(\mathbb{R}^n)$; and we have

$$u(x) = Qf(x) + Rf(x)$$

with the remainder

$$Rf(x) = \mathcal{F}^{-1}_{\xi \mapsto x}\Big((1 - \chi(\xi))\frac{1}{a(\xi)}\widehat{f}(\xi)\Big) . \tag{7.2.20}$$

Since $f \in C_0^\infty(\Omega)$ we have $\widehat{f} \in C^\infty$ from the Paley–Wiener–Schwartz theorem 3.1.3 (iii). Hence, with $\big(1 - \chi(\xi)\big) = 0$ for $|\xi| \geq 1$ and $\frac{1}{a(\xi)}$ having a singularity only at $\xi = 0$ and having quadratic growth, due to Lemma 3.2.1, $\big(1 - \chi(\xi)\big)\frac{1}{a(\xi)}\widehat{f}(\xi)$ defines a distribution in $\mathcal{E}'$. Hence, $Rf \in C^\infty$ and $R \in \mathcal{L}^{-\infty}(\Omega)$ by applying Theorem 3.1.3 (ii). (Note that $R \notin OPS^{-\infty}(\Omega \times \mathbb{R}^n)$).

Thus, $Q + R \in \mathcal{L}^{-2}(\Omega)$ defining again $N := Q + R$, the Newton potential operator, which can now still be written as

$$Nf(x) = \mathcal{F}^{-1}_{\xi \mapsto x}\left(\frac{1}{a(\xi)}\mathcal{F}_{y \mapsto \xi} f(y)\right)(x) . \tag{7.2.21}$$

As in the previous case, we still can define the fundamental solution

$$E(x, y) = N\delta(x - y) \tag{7.2.22}$$

where $\delta(x - y)$ is the Dirac functional with singularity at y. Note that $\delta \in \mathcal{E}'(\mathbb{R}^n)$ and $N : \mathcal{E}'(\mathbb{R}^n) \to \mathcal{D}'(\mathbb{R}^n)$ is well defined. $E(x, y)$ will satisfy (3.6.4).

We now return to the more general scalar elliptic equation (5.1.1) with C^∞–coefficients (but $p = 1$). Here, the symbol $a(x, \xi)$ in (7.2.4) depends on both, x and ξ. The construction of the symbols (7.2.8) and (7.2.9) leads to an asymptotic expansion and via (7.1.18), i.e., by

$$q(x, \xi) := \sum_{j=0}^{\infty} \Xi\left(\frac{\xi}{t_j}\right) q_{-2-j}(x, \xi)$$

we define the symbol q where $\Xi \in C^\infty$ is a cut–off function satisfying $\Xi(y) = 0$ for $|y| \leq \frac{1}{2}$ and $\Xi(y) = 1$ for $|y| \geq 1$. Then $q \in \boldsymbol{S}^{-2}(\Omega \times \mathbb{R}^n)$ and

$$Q(x, -iD)f(x) = \frac{1}{(2\pi)^n} \int\limits_{\mathbb{R}^n} \int\limits_{\Omega} e^{i(x-y)\xi} q(x, \xi) f(y) dy d\xi \tag{7.2.23}$$

is a (perhaps not properly supported) parametrix $Q \in \mathcal{L}^{-2}(\Omega)$.

With $Q = Q_0 + R$, the equations (7.2.6) yield

$$\begin{array}{rclclcl} P \circ Q &=& P \circ (Q_0 + R) &=& I + C_1 + P \circ R &=:& I + R_1 , \\ Q \circ P &=& (Q_0 + R) \circ P &=& I + C_2 + R \circ P &=:& I + R_2 . \end{array} \tag{7.2.24}$$

Since P is properly supported, $P \circ R$ and $R \circ P$ are still smoothing operators; so are R_1 and R_2. The equations (7.2.24) read for the differential equation:

$$P \circ (Qf)(x) = f(x) + \int_\Omega r_1(x,y) f(y) dy \qquad (7.2.25)$$

and

$$Q(x, -iD) \circ Pu(x) = u(x) + \int_\Omega r_2(x,y) u(y) dy \qquad (7.2.26)$$

where $r_1(x,y)$ and $r_2(x,y)$ are the $C^\infty(\mathbb{R}^n)$–Schwartz kernels of the smoothing operators R_1 and R_2, respectively. We note that both smoothing operators R_1 and R_2 can be constructed since in (7.2.24) the left-hand sides are known.

Theorem 7.2.4. *If R_2 extends to a continuous operator $C^\infty(\Omega) \to C^\infty(\Omega)$ and* $\ker(Q \circ P) = \{0\}$ *in $C^\infty(\Omega)$ then the Newton potential operator N is given by*

$$u(x) = (Nf) := (I + R_2)^{-1} \circ Q(x, -iD) f(x) \,. \qquad (7.2.27)$$

where $f \in C_0^\infty(\Omega)$. Moreover, in this case, the fundamental solution exists and is given by

$$E(x,y) = (N\delta_y)(x) = \big((I + R_2)^{-1} \circ Q(x, -iD)\delta_y\big)(x) \qquad (7.2.28)$$

where $\delta_y(\cdot) := \delta(\cdot - y)$.

Our procedure for constructing the fundamental solution, in principle, can be extended to general elliptic systems. Similar to the second order case we will arrive at an expression as (7.2.28). The difficulty is to show the invertibility as well as the continuous extendibility of $I + R_2$. We comment that this is, in general, not necessarily possible. Hence, for general elliptic linear differential operators with variable coefficients it may not always be possible to find a fundamental solution. We shall provide a list of references concerning the construction of fundamental solutions at the end of the section. In any case, even if the fundamental solution does not exist, for elliptic operators one can always construct a parametrix.

From the practical point of view, however, it is more desirable to construct only a finite number of terms in (7.2.9) rather than the complete symbol. This leads us to the idea of Levi functions.

7.2.3 Levi Functions for Scalar Elliptic Equations

From the homogeneous constant coefficient case we have an explicit formula to construct the fundamental solution via the Fourier transformation. However, in the case of variable coefficients, fundamental solutions can not be constructed explicitly, in general. This leads us to the idea of freezing coefficients in the principal part of the differential operator and, by following the idea of the parametrix construction, one constructs the *Levi functions* which dates back to the work of E.E. Levi [266] and Hilbert [179]. These can be used as approximations of the fundamental solution for the variable coefficient equations. In general, one can show that Levi functions always exist even if there is no fundamental solution.

Pomp in [351] developed an iterative scheme for constructing Levi functions of arbitrary order for general elliptic systems, from the distributional point of view. Here we combine his approach with the concept of pseudodifferential operators.

To illustrate the idea, we use the scalar elliptic equation of the form (3.6.1) with $p = 1$ and the order 2κ (instead of $2m$). For fixed $y \in \Omega$, we define

$$P_0(y)v(x) = \sum_{|\alpha|=2\kappa} a_\alpha(y) D_x^\alpha v(x) \tag{7.2.29}$$

and write the original equation in the form

$$Pu(x) = P_0(y)u(x) - T(y)u(x) = f(x)\,, \tag{7.2.30}$$

where

$$\begin{aligned} T(y)u(x) &= \big(P_0(y) - P\big)u(x) \qquad (7.2.31)\\ &= \sum_{|\alpha|=2\kappa} \big(a_\alpha(y) - a_\alpha(x)\big) D_x^\alpha u(x) - \sum_{|\alpha|<2\kappa} a_\alpha(x) D_x^\alpha u(x)\,. \end{aligned}$$

Note that $P_0(y)$ is the principal part of P with coefficients frozen at y. Its fundamental solution exists and has the form

$$E_0(x, z; y) = (2\pi)^{-n} \int\limits_{\mathbb{R}^n} \frac{e^{i(x-z)\cdot\xi}}{a_{2\kappa}^0(y, \xi)} d\xi \tag{7.2.32}$$

with a parameter y, where

$$a_{2\kappa}^0(y, \xi) = (-1)^\kappa \sum_{|\alpha|=2\kappa} a_\alpha(y)\xi^\alpha \tag{7.2.33}$$

is the (principal) symbol of $P_0(y)$. If one identifies $z = y$, then $L_0(x, y) = E_0(x, y; y)$ is called the *Levi function* of order 0 for P in (7.2.30) and has the singular behaviour of the form

$$L_0(x, y) = |x - y|^{2\kappa - n} \ell_{00}(x, |x - y|, \widehat{\Theta}) + \log|x - y| \ell_{01}(x, |x - y|, \widehat{\Theta}) \tag{7.2.34}$$

with $\widehat{\Theta} = (x - y)/|x - y|$ where the functions ℓ_{00} and ℓ_{01} are $C^\infty(\Omega \times \mathbb{R}_+ \times \{\widehat{\Theta} \in \mathbb{R}^n \,|\, |\widehat{\Theta}| = 1\})$. The integral operator defined by the kernel $L_0(x, y)$ will be denoted by $\underset{\sim}{N}_0$;

$$(\underset{\sim}{N}_0 v)(x) := \int\limits_{\Omega} L_0(x,y)v(y)dy = \frac{1}{(2\pi)^n} \int\limits_{\mathbb{R}^n}\int\limits_{\Omega} \frac{e^{i(x-y)\cdot\xi}}{a^0_{2\kappa}(y,\xi)} v(y)dyd\xi\,. \quad (7.2.35)$$

This is a pseudodifferential operator of the form (7.1.26) with the amplitude function

$$a(x,y,\xi) = a^0_{2\kappa}(y,\xi)^{-1}\,;$$

and $L_0(x,y)$ is the Schwartz kernel of $\underset{\sim}{N}_0$.

We remark that the only singularity of $a^0_{2\kappa}(y,\xi)^{-1}$ at $\xi = 0$ can be handled, by introducing the cut–off function $\chi(\xi)$, in the same manner as in Remark 3.2.2. Hence, $\chi(\xi)a^0_{2\kappa}(y,\xi)^{-1} \in \boldsymbol{S}^{-2\kappa}(\Omega\times\Omega\times\mathbb{R}^n)$ and $\underset{\sim}{N}_0 \in \mathcal{L}^{-2\kappa}(\Omega)$.

To derive higher order Levi functions, we return to (7.2.30). Define the pencil of pseudodifferential operators by

$$\begin{aligned} \big(A_0(x,-iD;y)f\big)(x) &= \mathcal{F}^{-1}_{\xi\mapsto x}\big(a^0_{2\kappa}(y,\xi)\big)^{-1}\mathcal{F}_{z\mapsto\xi}f(z) \\ &= \int\limits_{\Omega} E_0(x,z;y)f(z)dz\,. \end{aligned} \quad (7.2.36)$$

Then $A_0(x,D;y) \in \mathcal{L}^{-2\kappa}(\Omega)$ for every $y\in\Omega$. Moreover,

$$P_0(y)A_0(x,-iD;y) = I \quad\text{and}\quad A_0(x,-iD;y)P_0(y) = I\,. \quad (7.2.37)$$

In accordance with the Neumann series, we introduce the recursive sequence of operators

$$\begin{aligned} A_j(x,-iD;y) &:= \sum_{\ell=0}^{j}\big(A_0(x,-iD;y)T(y)\big)^{\ell}A_0(x,-iD;y) \quad (7.2.38)\\ &= A_0(x,-iD;y) + A_{j-1}(x,-iD;y)T(y)A_0(x,-iD;y)\,, \end{aligned}$$

$j = 1,2,\ldots$. We note that

$$\begin{aligned} PA_j(x,-iD;y) &= \big(P_0(y)-T(y)\big)\sum_{\ell=0}^{j}\big(A_0(\cdot,-iD;y)T(y)\big)^{\ell}A_0(x,-iD;y) \\ &= I - \big(T(y)A_0(\cdot,-iD;y)\big)^{j+1} \end{aligned} \quad (7.2.39)$$

for every $y\in\Omega$ and every $j\in\mathbb{N}_0$.

We are now in the position to define Levi functions of various orders recursively as follows.

$$\begin{aligned} L_0(x,y) &:= A_0(x,-iD;y)\delta(\cdot-y)\,,\\ L_j(x,y) &:= L_0(x,y) + A_0(x,-iD;y)T(y)L_{j-1}(x,y)\,.\\ &= A_0(x,-iD;y)\sum_{\ell=0}^{j}\big(T(y)A_0(\cdot,-iD;y)\big)^{\ell}\delta(\cdot-y)\,,\\ &= A_j(x,-iD;y)\delta(\cdot-y)\,. \end{aligned} \quad (7.2.40)$$

$L_j(x,y)$ is the Levi function of order j for the operator P.

By applying P to $L_j(x,y)$ we obtain

$$\begin{aligned}
PL_j(x,y) &= \big(P_0(y) - T(y)\big)L_j(x,y) \\
&= \big(P_0(y) - T(y)\big)A_0(x,-iD;y)\sum_{\ell=0}^{j}\big(T(y)A_0(\cdot,-iD;y)\big)^{\ell}\delta(\cdot - y) \\
&= \Big\{\sum_{\ell=0}^{j}\big(T(y)A_0(\cdot,-iD;y)\big)^{\ell} - \sum_{\ell=1}^{j+1}\big(T(y)A_0(\cdot,-iD;y)\big)^{\ell}\Big\}\delta(\cdot - y) \\
&= \delta(x-y) - \big(T(y)A_0(\cdot,-iD;y)\big)^{j+1}\delta(\cdot - y)\,.
\end{aligned}$$

By introducing $N_0(x,y) := L_0(x,y)$ and

$$N_{j+1}(x,y) := A_0(x,-iD;y)T(y)N_j(\cdot,y) \tag{7.2.41}$$

we can write

$$PL_j = \delta(x-y) - T(y)N_{j+1}(x,y)\,. \tag{7.2.42}$$

In the following, we want to show that every $N_j(x,y)$ is a Schwartz kernel of a pseudodifferential operator $\underset{\sim}{N}_j \in \mathcal{L}^{-j-2\kappa}(\Omega)\,,\; j = 0,1,\ldots$ and, hence, the kernel $N_j(x,y)$ has the form

$$\begin{aligned}
N_j(x,y) &= |x-y|^{-n+j+2\kappa}\Big(f_j(x,|x-y|,\widehat{\Theta}) + \big(\log|x-y|\big)g_j(x,|x-y|,\widehat{\Theta})\Big)\,, \\
&\qquad f_j, g_j \in C^\infty(\Omega\times \mathbb{R}\times\{\widehat{\Theta}\in\mathbb{R}^n \mid |\widehat{\Theta}| = 1\})\,.
\end{aligned} \tag{7.2.43}$$

(The latter will be shown in Theorem 8.1.8).

We further consider the kernels

$$T_j(x,y) := T(y)N_j(x,y) \tag{7.2.44}$$

and will show that $T_j(x,y)$ is also a Schwartz kernel of a pseudodifferential operator $\underset{\sim}{T}_j \in \mathcal{L}^{-j-1}(\Omega)$. Hence, $T_j(x,y)$ also has the form (7.2.43) with $j+2\kappa$ replaced by $j+1$.

Before we proceed to prove these assertions, we remark that $N_j(x,y)$ can be computed explicitly by using the fundamental solution $E_0(x,z;y)$ in (7.2.32); namely by the recursion procedure

$$\begin{aligned}
N_0(x,y) &:= E_0(x,y;y) = L_0(x,y)\,, \\
N_\ell(x,y) &= \int_\Omega E_0(x,z;y)T_z(y)N_{\ell-1}(z,y)dz \\
&\qquad\qquad \text{for } \ell = 1,2,\ldots,j+1\,.
\end{aligned} \tag{7.2.45}$$

We remark that the kernels in (7.2.45) are all of weakly singular type as in (7.2.43) (with $j \geq -1$). The composite integrals of this kind can also be analyzed according to estimates obtained by Sobolev [393] and Mikhlin [303] (see Mikhlin and Prößdorf [305, p. 214]).

So, the Levi function $L_j(x, y)$ of order j has the form

$$L_j(x, y) = \sum_{\ell=0}^{j} N_\ell(x, y) . \tag{7.2.46}$$

With the Levi function available, we may seek a solution of the partial differential equation (7.2.30) in the form

$$u(x) = \int_\Omega L_j(x, y)\Phi(y)dy . \tag{7.2.47}$$

By applying the distributions in (7.2.42) to $u(x)$, we obtain the equation

$$f(x) = \Phi(x) - \int_\Omega T(y)N_{j+1}\Phi(y)dy . \tag{7.2.48}$$

This is a Fredholm integral equation of the second kind for the unknown density $\Phi(x)$. The integral operator $\underset{\sim}{T}_{j+1}$ in (7.2.48) has the kernel

$$T_{j+1}(x, y) = T(y)N_{j+1}(x, y)$$

belonging to $C^{j+1-n}(\Omega \times \Omega)$ for $j + 1 - n \geq 0$. If the kernel $T_{j+1}(x, y)$ is properly supported, then the integral operator defines a continuous mapping $C^\infty(\Omega) \to C^\infty(\Omega)$ and for (7.2.48), the classical Fredholm theory is available.

We now return to the proof of the above made assertions for $N_j(x, y)$ and $T_j(x, y)$.

Lemma 7.2.5. *Let $N_j(x, y)$ be the Schwartz kernel of a pseudodifferential operator $\underset{\sim}{N}_j \in \mathcal{L}^{-j-2\kappa}$, $j \in \mathbb{N}_0$. Then*

$$T(y)N_j(x, y) =: T_j(x, y) \tag{7.2.49}$$

is the Schwartz kernel of a pseudodifferential operator $\underset{\sim}{T}_j \in \mathcal{L}^{-j-1}$ and

$$\underset{\sim}{T}_j = \underset{\sim}{T} \circ \underset{\sim}{N}_j$$

where $\underset{\sim}{T} \in \mathcal{L}^{2\kappa-1}$ is generated by the operation $T(y)N_j(x, y)$ on the Schwartz kernel N_j of $\underset{\sim}{N}_j$.

Note that the operator $\underset{\sim}{T}$ is defined here implicitly and does not coincide with the differential operator $T(y)$.

Proof: From the definition of $N_j(x, y)$ with $\underset{\sim}{N}_j$ we have

$$\underset{\sim j}{N}\delta(\cdot - y) = N_j(x,y) = \frac{1}{(2\pi)^n}\int\limits_{\mathbb{R}^n} \sigma_{\underset{\sim j}{N}}(x,\xi)e^{i(x-y)\cdot\xi}d\xi + N_{j\infty}(x,y)$$

where $N_{j\infty}(x,y)$ is the C^∞–kernel of a smoothing operator. Here, the symbol $\sigma_{\underset{\sim j}{N}} \in S^{-j-2\kappa}(\Omega \times \mathbb{R}^n)$ is a representative of the complete symbol of $\underset{\sim j}{N}$. Applying $T(y)$ to the kernel, we first consider the new kernel

$$\begin{aligned}&\frac{\partial}{\partial x_k}N_j(x,y)\\ &= \frac{1}{(2\pi)^n}\int\limits_{\mathbb{R}^n}\Big\{\frac{\partial\sigma_{\underset{\sim j}{N}}}{\partial x_k}(x,\xi) - i\xi_k\sigma_{N_j}(x,\xi)\Big\}e^{i(x-y)\cdot\xi}d\xi + \frac{\partial}{\partial x_k}N_{j\infty}(x,y)\,.\end{aligned}$$

Hence, the differentiated kernel defines a new operator with the new symbol

$$\Big\{\frac{\partial\sigma_{\underset{\sim j}{N}}}{\partial x_k}(x,\xi) - i\xi_k\sigma_{\underset{\sim j}{N}}(x,\xi)\Big\} \in S^{-j-2\kappa+1}(\Omega \times \mathbb{R}^n)\,. \tag{7.2.50}$$

Repeating this argument, we find that the Schwartz kernels

$$\sum_{|\alpha|<2\kappa} a_\alpha(x)D_x^\alpha N_j(x,y) \quad\text{and}\quad D_x^\alpha N_j(x,y) \quad\text{for}\quad |\alpha| = 2\kappa$$

define pseudodifferential operators in $S^{-j-2\kappa+|\alpha|}(\Omega \times \mathbb{R}^n)$ for $|\alpha| < 2\kappa$ and $S^{-j}(\Omega \times \mathbb{R}^n)$, respectively. Therefore,

$$\int\limits_\Omega D_x^\alpha N_j(x,y) \bullet dy = \int\limits_\Omega k_{\alpha j}(x,y) \bullet dy \quad\text{for}\quad |\alpha| = 2\kappa$$

defines a Schwartz kernel $k_{\alpha j}(x,y)$ for a pseudodifferential operator in $\mathcal{L}^{-j}(\Omega)$. If $k_p(x,y)$ is the properly supported part of the Schwartz kernel $k_{\alpha j}(x,y)$ due to Theorems 7.1.9 and 7.1.11 then its symbol can be computed by

$$\sigma_K(x,\xi) = \int\limits_\Omega k_p(x,y)e^{i(y-x)\cdot\xi}d\xi\,. \tag{7.2.51}$$

The factors $(a_\alpha(y) - a_\alpha(x))$ in (7.2.31) induce new kernels of the form

$$\big(a_\alpha(y) - a_\alpha(x)\big)D_x^\alpha N_j(x,y) = \big(a_\alpha(y) - a_\alpha(x)\big)k_{\alpha j}(x,y)$$

and can be written as asymptotic sums

$$\sum_{|\beta|\geq 1} c_{\alpha j\beta}(x)(x-y)^\beta k_{\alpha j}(x,y)$$

by using the Taylor expansion for a_α. The new Schwartz kernels

$$(x-y)^\alpha k_p(x,y) + (x-y)^\alpha k_\infty(x,y)$$

generate properly supported pseudodifferential operators with symbols

$$\sigma(x,\xi) = \int\limits_\Omega (x-y)^\beta k_p(x,y) e^{i(y-x)\cdot\xi} dy\,.$$

The latter can be rewritten as

$$\begin{aligned}\sigma(x,\xi) &= \left(-i\frac{\partial}{\partial\xi}\right)^\beta \int\limits_\Omega k_p(x,y)e^{i(y-x)\cdot\xi}dy \\ &= \left(-i\frac{\partial}{\partial\xi}\right)^\beta \sigma_K(x,\xi) \in \boldsymbol{S}^{-j-|\beta|}(\Omega\times\mathbb{R}^n) \quad\text{with } |\beta|\geq 1\,.\end{aligned}$$

Consequently, $T(y)N_j(x,y) = T_j(x,y)$ defines the Schwartz kernel of a pseudodifferential operator $\underset{\sim j}{T} \in \mathcal{L}^{-j-1}(\Omega)$ and, also, defines the operator $\underset{\sim}{T}$ with $\underset{\sim}{T} \in \mathcal{L}^{2\kappa-1}(\Omega)$. ∎

We now justify the assumption for $\underset{\sim j}{N}$ made in Lemma 7.2.5. For $\underset{\sim j}{N}$ given by the Schwartz kernel N_j in (7.2.43), we have already shown $\underset{\sim 0}{N} \in \mathcal{L}^{-2\kappa}(\Omega)$. For $j\in\mathbb{N}$, the following lemma justifies the desired property of $\underset{\sim j}{N} \in \mathcal{L}^{j-2\kappa}$.

Lemma 7.2.6. *Let $T_j(x,y)$ be the Schwartz kernel of a pseudodifferential operator $\underset{\sim j}{T} \in \mathcal{L}^{-j-1}$ with $j\in\mathbb{N}_0$. Then*

$$A_0(x,D;y)T_j(\cdot,y) =: N_{j+1}(x,y) \tag{7.2.52}$$

defines the Schwartz kernel of a pseudodifferential operator $\underset{\sim j+1}{N} \in \mathcal{L}^{-(j+2\kappa+1)}(\Omega)$.

Note that the special choice of T_j in (7.2.44) and N_j in (7.2.41) leads to the assumption in Lemma 7.2.6.

Proof: From the definition (7.2.52) of the kernel function N_{j+1} one has the representation

$$N_{j+1}(x,y) = \frac{1}{(2\pi)^n}\int\limits_{\mathbb{R}^n}\int\limits_\Omega \frac{1}{a^0_{2\kappa}(y,\xi)} e^{i(x-z)\cdot\xi} T_j(z,y)dzd\xi\,.$$

Hence, this kernel defines the operator

$$(\underset{\sim j+1}{N} f)(x) = \frac{1}{(2\pi)^n}\int\limits_{\mathbb{R}^n}\int\limits_\Omega\int\limits_\Omega \frac{e^{i(x-z)\cdot\xi}}{a^0_{2\kappa}(y,\xi)} T_j(z,y)f(y)dzdyd\xi\,.$$

Since $\underset{\sim j}{T} \in \mathcal{L}^{-j-1}$, we also have $\underset{\sim j}{T}^\top \in \mathcal{L}^{-j-1}$ due to Theorem 7.1.8 and, for the properly supported part T_{jp} we may use (7.1.27) which yields

$$\int\limits_\Omega e^{i(y-z)\cdot\xi} T_{jp}(z,y)dz = \sigma_{T_{jp}^\top}(y,-\xi)\,.$$

Hence, performing the integration with respect to z yields

$$\begin{aligned}(\underset{\sim j+1}{N} f)(x) &= \frac{1}{(2\pi)^n} \int\limits_{\mathbb{R}^n}\int\limits_\Omega \frac{e^{i(x-y)\cdot\xi}}{a^0_{2\kappa}(y,\xi)} \sigma_{T_{jp}^\top}(y,-\xi) f(y)dyd\xi \\ &+ \frac{1}{(2\pi)^n} \int\limits_{\mathbb{R}^n}\int\limits_\Omega\int\limits_\Omega \frac{e^{i(x-z)\cdot\xi}}{a^0_{2\kappa}(y,\xi)} T_{j\infty}(z,y) f(y)dzdyd\xi\,.\end{aligned} \tag{7.2.53}$$

The first term on the right–hand side defines an operator of the form (7.1.26) with the amplitude function

$$a(x,y,\xi) = \sigma_{T_{jp}^\top}(y,-\xi)\frac{\chi(\xi)}{a^0_{2\kappa}(y,\xi)} \in \boldsymbol{S}^{-(j+2\kappa+1)}(\Omega\times\Omega\times\mathbb{R}^n)\,. \tag{7.2.54}$$

Here, $\chi(\xi)$ is again the cut–off function with $\chi(\xi)=1$ for $|\xi|\geq 1$ and $\chi(\xi)=0$ for $|\xi|\leq \frac{1}{2}$. The remaining $(1-\chi(\xi))a(x,y,\xi)$ is a distribution with compact support, and, hence, defines a smoothing operator.

The second term in (7.2.53) has the Schwartz kernel

$$k(x,y) := \int\limits_{\mathbb{R}^n}\int\limits_\Omega \frac{e^{i(x-z)\cdot\xi}}{a^0_{2\kappa}(y,\xi)} T_{j\infty}(z,y)dzd\xi$$

which is $C^\infty(\Omega\times\Omega)$. Hence, it defines a smoothing operator. This implies $\underset{\sim j+1}{N} \in \mathcal{L}^{-(j+2\kappa+1)}(\Omega)$, completing the proof. ∎

7.2.4 Levi Functions for Elliptic Systems

In view of the construction of Levi functions for scalar elliptic differential operators of order 2κ we now are able to extend the approach to systems (7.2.10), i.e.

$$\sum_{k=1}^p \sum_{|\beta|=0}^{s_j+t_k} a^{jk}_\beta(x) D^\beta u_k(x) = f_j(x)\,,\ j=1,\dots,p$$

where $s_j\leq 0$ and $t_k\geq 0$, which are elliptic in the sense of Agmon–Douglis–Nirenberg, i.e.

$$H_{2\kappa}(x,\xi) := \det\Big(\Big(\sum_{|\beta|=s_j+t_k} a^{jk}_\beta i^{|\beta|}\xi^\beta\Big)\Big)_{p\times p} \neq 0$$

for $\xi \in \mathbb{R}^n \setminus \{0\}$ where a_β^{jk} is set to be zero for terms with order less than $s_j + t_k$. We note that the determinant itself defines a scalar elliptic differential operator

$$H_{2\kappa}(x;D) := \det \Big(\Big(\sum_{|\beta|=s_j+t_k} a_\beta^{jk}(x) D^\beta \Big)\Big)$$

of order $2\kappa = \left(\sum_{j=1}^p s_j + \sum_{k=1}^p t_k\right)$ whose homogeneous symbol is $H_{2\kappa}(x;\xi)$. Let $B(x,\xi)$ denote the cofactor matrix of $a_0(x,\xi) = ((a^{jk}_{s_j+t_k}(x;\xi)))_{p\times p}$ defined by

$$a_0(x,\xi)B(x,\xi) = B(x,\xi)a_0(x,\xi) = H_{2\kappa}(x,\xi)((\delta_{jk}))_{p\times p}\,. \tag{7.2.55}$$

It is not difficult to see that the cofactor matrix $((B_{k\ell}(x,\xi)))_{p\times p}$ given implicitly by (7.2.55) defines a system of differential operators $B_{k\ell}(x,iD)$ of orders $2\kappa - t_k - s_\ell$. Then we can define

$$a_0^{-1}(y;D) \;:=\; B(y,iD)H^{-1}_{-2\kappa}(y,iD)\,,$$

where

$$\big(H^{-1}_{-2\kappa}(y,iD)f\big)(x) \;:=\; \frac{1}{(2\pi)^n}\int\limits_{\mathbb{R}^n}\int\limits_{\Omega} \frac{e^{i(x-z)\cdot\xi}}{H_{2\kappa}(y,\xi)} f(z)dzd\xi\,. \tag{7.2.56}$$

The integral in (7.2.56) is defined in the sense of Hadamard's finite part integral. Hence, the integral operator $H^{(-1)}_{-2\kappa}(y;iD)$ is a pseudodifferential operator of order -2κ having the homogeneous symbol $1/H_{2\kappa}(y,\xi)$ with y as a frozen parameter. Consequently, the operator $a_0^{(-1)}$ is a right inverse to $a_0(y;D)$, the principal part of the differential equations with constant coefficients frozen at y. The differential operator $a(x;D)$ can be decomposed by

$$a(x;D) = a_0(y;D) - T(y;D)$$

where $T(y;D)$ denotes the difference of the differential operators, i.e.,

$$T(y;D) = a_0(y,D) - a(x,D)\,.$$

Now, the inverse of $a(x,D)$ can formally be written in the form of Neumann's series

$$a^{-1}(x,D) = \sum_{\ell\ge 0} \big(a_0^{(-1)}(y;D)T(y;D)\big)^\ell a_0^{(-1)}(y;D)\,.$$

This leads us to the following successively defined sequence of kernel functions

$$\begin{aligned} N^{(0)}(x,y) &:= a_0^{(-1)}(y;D)((\delta(x-y)\delta_{jk}))_{p\times p}\,, \qquad (7.2.57)\\ N^{(\ell+1)}(x,y) &:= a_0^{(-1)}(y;D)T(y;D)N^{(\ell)}(x,y) \quad \text{for } \ell = 0,1,\ldots\,. \end{aligned}$$

The Levi function of order μ to the system of differential equations (7.2.10) assumes the form

$$L^{(\mu)}(x,y) := \sum_{\ell=0}^{\mu} N^{(\ell)}(x,y)\,. \tag{7.2.58}$$

In the same way as for the scalar case we may seek the solution $u(x) = \big(u_1(x), \ldots, u_p(x)\big)^\top$ to the system (7.2.10) in the form

$$u_k(x) = \sum_{\ell=1}^{p} \int_{\Omega} L_{k\ell}^{(\mu)}(x,y)\phi_\ell(y)dy \quad \text{for } x \in \Omega \tag{7.2.59}$$

where $\phi(x) = \big(\phi_1(x), \ldots, \phi_p(x)\big)^\top$ denotes an unknown vector–valued density. By applying the differential operator A to (7.2.59) we obtain a system of domain integral equations of the second kind,

$$f_j(x) = \phi_j(x) - \sum_{k=1}^{p} \int_{\Omega} T_{jk}(y,D) N_{km}^{(\mu+1)}(x,y)\phi_m(y)dy \tag{7.2.60}$$

by using the relation

$$AL^{(\mu)}(x,y) = \delta(x-y) - T(x,D)N^{(\mu+1)}(x,y)\,. \tag{7.2.61}$$

The integral operator in (7.2.60) has the kernel

$$T(y,D)N^{(\mu+1)}(x,y)$$

which belongs to the class $C^\lambda(\Omega\times\Omega)$ with $\lambda = \mu+1-n-\max_{k=1,\ldots,p}|s_k|$. For ℓ sufficiently large and compact Ω, the integral equation (7.2.60) is a classical Fredholm integral equation of the second kind with continuous kernel in Ω.

7.2.5 Strong Ellipticity and Gårding's Inequality

One of the advantages of considering integral operators as pseudodifferential operators is that the mapping properties of the boundary integral operators can be deduced by examining the symbols of the pseudodifferential operators. On the other hand, Gårding's inequality for the integral operators plays a fundamental role in the variational formulation of the integral equations. The latter follows from the definition of uniform strong ellipticity of pseudodifferential operators.

Definition 7.2.4. (see Stephan et al [405]) *We call a system of pseudodifferential operators*
$A_{jk} \in \mathcal{L}_{c\ell}^{s_j+t_k}(\Omega)$ *uniformly strongly elliptic if to the principal part matrix*

$a^0(x;\xi) = \big((a^{jk0}_{s_j+t_k}(x;\xi))\big)_{p\times p}$ *there exist a* C^∞*–matrix valued function* $\Theta(x) = \big((\Theta_{jk}(x))\big)_{p\times p}$ *and a constant* $\gamma_0 > 0$ *such that*

$$\operatorname{Re}\zeta^\top\Theta(x)a^0(x,\xi)\overline{\zeta} \geq \gamma_0|\zeta|^2 \tag{7.2.62}$$

for all $x \in \Omega$, $\zeta \in \mathbb{C}^p$ *and* $\xi \in \mathbb{R}^n$ *with* $|\xi| = 1$.

Remark 7.2.1: In order to show Gårding's inequality let Λ^α be the *Bessel potential* of order $\alpha \in \mathbb{R}$ which is the pseudodifferential operator $(-\Delta+1)^{\alpha/2}$ with the symbol $(|\xi|^2+1)^{\alpha/2}$ defining isomorphisms $H^\sigma(\mathbb{R}^n) \to H^{\sigma-\alpha}(\mathbb{R}^n)$ for every $\sigma \in \mathbb{R}$.

Theorem 7.2.7. *Let* $\Omega \subset \mathbb{R}^n$ *be a bounded domain and let* A *be a strongly elliptic system of pseudodifferential operators and let* $K \Subset \Omega$ *be a compact subregion. Then there exist constants* $\gamma_1 > 0$ *and* $\gamma_2 \geq 0$ *such that Gårding's inequality holds in the form*

$$\begin{aligned}&\operatorname{Re}(w, \Lambda^\sigma\Theta\Lambda^{-\sigma}Aw)_{\prod_{\ell=1}^p H^{(t_\ell-s_\ell)/2}(\Omega)}\\ &\geq \gamma_1\|w\|^2_{\prod_{\ell=1}^p H^{t_\ell}(\Omega)} - \gamma_2\|w\|^2_{\prod_{\ell=1}^p H^{t_\ell-1}(\Omega)}\end{aligned} \tag{7.2.63}$$

for $w \in \prod_{\ell=1}^p H^{t_\ell}_{\text{comp}}(\Omega)$ *with supp* $w \subset K$ *where* $\Lambda^\sigma = \big((\Lambda^{s_j}\delta_{j\ell})\big)_{p\times p}$ *(see* (4.1.44)*) and where* $\gamma_1 > 0$ *and* $\gamma_2 \geq 0$ *depend on* Θ, A, Ω, *the compact* $K \Subset \Omega$ *and on the indices* t_j *and* s_k.

The last term in (7.2.63) *defines a linear compact operator* $C : \prod_{\ell=1}^p H^{t_\ell}_{\text{comp}}(\Omega) \to \prod_{\ell=1}^p H^{-s_\ell}_{\text{comp}}(\Omega)$ *which is given by*

$$(v, Cw)_{\prod_{\ell=1}^p H^{(t_\ell-s_\ell)/2}(\Omega)} = \gamma_2(v,w)_{\prod_{\ell=1}^p H^{t_\ell-1}(\Omega)}\,.$$

With this operator C, *the Gårding inequality* (7.2.63) *becomes*

$$\operatorname{Re}\big(w, (\Lambda^\sigma\Theta\Lambda^{-\sigma}A + C)w\big)_{\prod_{\ell=1}^p H^{(t_\ell-s_\ell)/2}(\Omega)} \geq \gamma_1\|w\|^2_{\prod_{\ell=1}^p H^{t_\ell}(\Omega)}\,. \tag{7.2.64}$$

Proof: [3] For the proof let us consider the operator

$$A_1 := \Theta\Lambda^{-\sigma}A\Lambda^{-\tau}$$

where $\tau = (t_1,\dots,t_p)$ and $\sigma = (s_1,\dots,s_p)$, which is now a pseudodifferential operator of order zero and let us prove (7.2.63) for A_1 first. The strong ellipticity of A implies that the operator

$$\operatorname{Re}A_1 = \tfrac{1}{2}(A_1 + A_1^*)$$

[3] The authors are grateful to one of the reviewers who suggested the following proof.

has a positive definite hermitian principal symbol matrix $a^0_{\operatorname{Re} A_1}(x,\xi)$ whose positive definite hermitian square root $\big(a^0_{\operatorname{Re} A_1}(x,\xi)\big)^{\frac{1}{2}} \in \boldsymbol{S}^0(\Omega\times\mathbb{R}^n)$ defines via

$$Bu(x) := (2\pi)^{n/2}\int\limits_{\mathbb{R}^n} e^{ix\cdot\xi}\big(a^0_{\operatorname{Re} A_1}(x,\xi)\big)^{\frac{1}{2}}\widehat{u}(\xi)d\xi = B_0u(x) + Ru(x)$$

a strongly elliptic pseudodifferential operator of order zero, where B_0 denotes its properly supported part and $R \in \mathcal{L}^{-\infty}(\Omega)$ due to Theorem 7.1.9. Hence, B admits a properly supported parametrix Q_0 due to Theorem 7.2.3. For $u \in L^2(\Omega)$ with $\operatorname{supp} u \subset K$ extended by zero to $\widetilde{u} \in L^2(\mathbb{R}^n)$, formula (7.2.6) yields

$$\widetilde{u} = Q_0B\widetilde{u} + C_2\widetilde{u} = Q_0B_0\widetilde{u} + (Q_0R + C_2)\widetilde{u}\,, \tag{7.2.65}$$

where $C_2 \in \mathcal{L}^{-\infty}$, and $Q_0R \in \mathcal{L}^{-\infty}$ because of Theorem 7.1.14. Since $\widetilde{u}$ has compact support and Q_0B_0 is properly supported due to Proposition 7.1.6, there exists a compact set $K_1 \Subset \Omega$ and $\operatorname{supp} Q_0B_0\widetilde{u} \subset K_1$. Hence, $\operatorname{supp}\,(Q_0R + C_2) \subset K \cup K_1 \Subset \Omega$. Now, (7.2.65) yields

$$\|\widetilde{u}\|_{L^2(\mathbb{R}^n)} = \|u\|_{L^2(K)} \le c_1\|B_0\widetilde{u}\|_{L^2(\mathbb{R}^n)} + c_2\|\widetilde{u}\|_{H^{-1}(\mathbb{R}^n)}$$

with some $c_1 > 0$ since $B_0\widetilde{u}$ has compact support in Ω. Hence, there exists $\gamma_0 > 0$ such that

$$(B_0^*B_0\widetilde{u},\widetilde{u})_{L^2(\mathbb{R}^n)} \ge \gamma_0\|\widetilde{u}\|^2_{L^2(\mathbb{R}^n)} - c_2'\|\widetilde{u}\|^2_{H^{-1}(\mathbb{R}^n)},\,.$$

Since $\operatorname{Re} A_1 = B^*B + C = B_0^*B_0 + B_0^*R + R^*B_0 + R^*R + C$ with some $C \in \mathcal{L}^{-1}_{c\ell}(\Omega)$ due to Theorem 7.1.14, we get

$$\begin{aligned}\operatorname{Re}(A_1u,u) &= \operatorname{Re}(A_1\widetilde{u},\widetilde{u})\\ &= (B_0\widetilde{u},B_0\widetilde{u}) + (R\widetilde{u},B_0\widetilde{u}) + (B_0\widetilde{u},R\widetilde{u}) + (R\widetilde{u},R\widetilde{u}) + (C\widetilde{u},\widetilde{u})\end{aligned}$$

and

$$\begin{aligned}\operatorname{Re}(A_1u,u)_{L^2(\Omega)} &\ge \gamma_0\|u\|^2_{L^2(\Omega)} - c_3\|\widetilde{u}\|^2_{H^{-1}(\mathbb{R}^n)} - c_4\|\widetilde{u}\|_{L^2(\Omega)}\|\widetilde{u}\|_{H^{-1}(\mathbb{R}^n)}\\ &\ge \frac{\gamma_0}{2}\|u\|^2_{L^2(\Omega)} - c_5\|\widetilde{u}\|^2_{H^{-1}(\Omega)}\end{aligned} \tag{7.2.66}$$

with an appropriate constant c_5. This, in fact, is already the proposed Gårding inequality for the special case $\alpha_{jk} = 0$.

Now, for the general case, we employ the isometries Λ^α from $H^s(\mathbb{R}^n)$ to $H^{s-\alpha}(\mathbb{R}^n)$ and replace u by $\eta\Lambda^\tau w$, where $\eta \in C_0^\infty(\overset{\circ}{K}_2)$ and $K_2 \Subset \Omega$ with $K \Subset \overset{\circ}{K}_2$, $\eta|_K \equiv 1$ and $0 \le \eta \le 1$. For $w \in H^\tau(\Omega)$ with $\operatorname{supp} w \subset K \Subset \Omega$ now apply (7.2.66) and obtain

$$\begin{aligned}
\mathrm{Re}\,(\eta\Lambda^\tau w, \Theta\Lambda^{-\sigma}A\Lambda^{-\tau}\eta\Lambda^\tau w)_{L^2(\mathbb{R}^n)} & \\
&\geq \gamma_0'\|\eta\Lambda^\tau w\|^2_{L^2(\Omega)} - c_3'\|\eta\Lambda^\tau w\|^2_{H^{-1}(\Omega)} \\
&\geq \gamma_0'\|\Lambda^\tau w\|^2_{L^2(\Omega)} - c_3''\|w\|^2_{H^{\tau-1}(\Omega)} - 4\gamma_0'\|[\eta,\Lambda^\tau]w\|^2_{L^2(\Omega)} \\
&\geq \gamma_0''\|w\|^2_{H^\tau(\Omega)} - c_6\|w\|^2_{H^{\tau-1}(\Omega)}
\end{aligned} \tag{7.2.67}$$

with $\gamma_0'' > 0$ since $[\eta, \lambda^\tau] \in \mathcal{L}^{\tau-1}_{c\ell}$ due to Corollary 7.1.15. The left–hand side can be reformulated as

$$\begin{aligned}
(\eta\Lambda^\tau w, A_1\eta\Lambda^\tau w) = \;& \mathrm{Re}\,(w, \Lambda^\tau\Theta\Lambda^{-\sigma}Aw)_{L^2} \\
&+ \big([\eta,\Lambda^\tau]w, A_1\eta\Lambda^\tau w\big) + \big(\Lambda^\tau w, A_1[\eta,\Lambda^\tau]w\big)\,.
\end{aligned}$$

So, with Corollary 7.1.15,

$$\begin{aligned}
\mathrm{Re}\,(w, \Lambda^\sigma\Theta\Lambda^{-\sigma}Aw)_{\prod\limits_{j=1}^{p} H^{(t_j-s_j)/2}(\Omega)} &= \mathrm{Re}\,(w, \Lambda^\tau\Theta\Lambda^{-\sigma}Aw)_{L^2(\Omega)} \\
\geq \gamma_0''\|\Lambda^\tau w\|^2_{L^2(\Omega)} - c_6\|w\|^2_{H^{\tau-1}(\Omega)} &- c_7\|\Lambda^\tau w\|_{L^2(\Omega)}\,\|\Lambda^{\tau-1}w\|_{L^2(\Omega)} \\
\geq \gamma_0''\|w\|^2_{H^\tau(\Omega)} - c_8\|w\|^2_{H^{\tau-1}(\Omega)} &
\end{aligned}$$

as proposed. ■

Remark 7.2.2: In the special case when $\alpha_{jk} = 2\alpha$ is constant, where 2α is the same order for all A_{jk}, we may choose $\alpha = s_j = t_k$, and if the system is strongly elliptic then Gårding's inequality (7.2.64) reduces to the familiar form

$$\mathrm{Re}\,((\Theta A + C)w, w)_{L^2(\Omega)} \geq \gamma_1\|w\|^2_{H^\alpha(\Omega)} \text{ with } \gamma_1 > 0\,, \text{ for all } w \in H^\alpha_{\mathrm{comp}}(\Omega)\,.$$

(For differential operators A_{jk} see Miranda [307, p.252]).

7.3 Review on Fundamental Solutions

As is well known, the fundamental solution plays a decisive role in the method of boundary integral equations. It would be desirable to have a general constructive method to find an explicit calculation of fundamental solutions. However, in general it is not easy to obtain the fundamental solution explicitly whereas Levi functions can always be constructed. The latter will allow us to construct at least *local fundamental solutions*.

The existence of a fundamental solution is closely related to the existence of solutions of

$$Pu = f \quad \text{in } \Omega\,. \tag{7.3.1}$$

More precisely, we have the following theorem (Komech [235], Miranda [307]).

Theorem 7.3.1. *A necessary and sufficient condition for the existence of a fundamental solution for P is that the equation* (7.3.1) *admits at least one solution $u \in \mathcal{E}'$ for every $f \in \mathcal{E}'$.*

Consequently, in all the cases when variational solutions exist, we also have a fundamental solution.

To our knowledge it seems that the most comprehensive survey on fundamental solutions of elliptic equations is due to Miranda in [307]. Here we quote his useful remarks:

"The first research on fundamental matrices looks at particular systems and is due to C. Somigliana [396], E.E. Levi [265], G. Giraud [144, 145]. For elliptic systems in the sense of Lopatinskii this study then was taken up in general by this author [273], first for equations with constant coefficients and then, with Levi's method in a sufficiently restricted domain, for equations with variable coefficients. For elliptic systems in the sense of Petrowskii the case of constant coefficients was treated by F. John [219] as a particular case of systems with analytic coefficients and by C.B. Morrey [315, 316] in the case $r = 1$. Y.B. Lopatinskii [272] proved, with Levi's method, the existence in the small of fundamental matrices for systems with variable coefficients, while the existence in certain cases of *principal fundamental matrices* was established by A. Avantaggiati [18]. For elliptic systems in the sense of Douglis and Nirenberg the construction of the fundamental matrix, in the case of constant coefficients, was done by these authors [103] with a procedure due to F. Bureau [57]. Also for these systems, V.V. Grusin [161] and A. Avantaggiati [18] proved the existence of principal fundamental matrices in certain cases. For the case of variable coefficients, some indication relating to the existence in the small is given in §10.6 of the volume [187] by Hörmander. Finally, for certain systems of particular type see D. Greco [154] and L.L. Parasjuk [343].

As for the case of a single equation, the problem of existence of a fundamental matrix can be related on one hand with that of the existence for arbitrary f of at least one solution of the equation $\mathcal{M}u = f$ and on the other hand with that of the validity of the unique continuation property[4]. Each one of these questions has separately been an object for study, but the relationship between the results obtained had to be deepened later.

Regarding the first question we note that from theorems of existence in the small of a fundamental matrix it follows that in every sufficiently restricted domain an elliptic system always admits at least one solution."

In the recent publication [351], Pomp also discusses currently available existence results for fundamental solutions.

7.3.1 Local Fundamental Solutions

In this section we present Levi's method for constructing local fundamental solutions.

[4] $\mathcal{M}u = Pu$ in our notation as in (7.3.1).

Let us assume that a fundamental solution $E(x,y)$ exists. Then with the Levi function $L_N(x,y)$ constructed in (7.2.40) for a scalar operator P or $L^{(N)}(x,y) = L_N(x,y)$ by (7.2.58) for an Agmon–Douglis–Nirenberg system we consider

$$W(x,y) := E(x,y) - L_N(x,y) \quad \text{for } x,y, \in \Omega\,. \tag{7.3.2}$$

Then for fixed $y \in \Omega$,

$$P_x W(x,y) = \delta(x-y) - \big(\delta(x-y) - T(y)N_{N+1}(x,y)\big) \tag{7.3.3}$$

due to (7.2.42) or (7.2.61). The function $T_{N+1}(x,y)$ is the Schwartz kernel of a pseudodifferential operator in $\mathcal{L}^{-N-2}(\Omega)$ and, moreover, $T_{N+1} \in C^{N+1-n}(\Omega \times \Omega)$. For $N \geq n-1$, T_{N+1} is at least continuous. Now, $W(x,y)$ can be represented in the form (7.2.47) or (7.2.59), namely

$$W(x,y) = \int\limits_{\Omega} L_N(x,z)\Phi(z,y)dz \tag{7.3.4}$$

with $\Phi(x,y)$ the solution of the integral equation

$$T_{N+1}(x,y) = \Phi(x,y) - \int\limits_{\Omega} T_{N+1}(x,z)\Phi(z,y)dz\,. \tag{7.3.5}$$

If the existence of $E(x,y)$ is assumed then (7.3.5) has a solution $\Phi(x,y)$ and $E(x,y)$ is given by

$$E(x,y) = L_N(x,y) + \int\limits_{\Omega} L_N(x,z)\Phi(z,y)dz\,. \tag{7.3.6}$$

Conversely, if (7.3.5) has a solution $\Phi(x,y)$ which is at least continuous for $(x,y) \in \Omega \times \Omega$ then $E(x,y)$ in (7.3.6) is the desired fundamental solution.

Note that Fredholm's alternative for the integral equation (7.3.5) can be applied if the integral operator in (7.3.5) defines a compact mapping on the solution space for $\Phi(x,y)$, e.g., on the space of continuous functions. In view of the properties of T_{N+1}, this can be guaranteed if Ω is replaced by a compact region $\Omega' \Subset \Omega$ with a sufficiently smooth boundary $\partial\Omega'$. In this case, the whole construction procedure of the Levi function should be carried out on Ω'. If then for (7.3.5) uniqueness of the solution holds then the corresponding uniquely determined function $\Phi(x,y)$ generates a local fundamental solution by (7.3.6) associated with Ω'.

If $|\Omega'|$ is sufficiently small then the integral operator's norm in (7.3.5) is less than 1 and Banach's fixed point theorem in the form of Neumann's series will provide the local existence of $\Phi(x,y)$ and, hence, that of $E(x,y)$.

If (7.3.5) has eigensolutions, the representation of $E(x,y)$ in (7.3.6) and the integral equation are to be appropriately modified. We refer to the details in Miranda [307] and in Ljubič [271]. For bounded regions in $\mathbb{R}^2$ and elliptic systems in the sense of Petrovski, Fichera also gave a complete constructive existence proof in [118].

For real analytic coefficients in Agmon–Douglis–Nirenberg elliptic systems, John presents the construction of a local fundamental solution in [219]. In [425] Vekua constructs, for second order systems with the Laplacian Δ as principal part and analytic coefficients and also for higher order scalar equations in $\mathbb{R}^2$, the fundamental solution. For a scalar elliptic differential operator having C^∞–coefficients, Theorem 13.3.3 by Hörmander [188, Vol II] provides a local fundamental solution.

7.3.2 Fundamental Solutions in $\mathbb{R}^n$ for Operators with Constant Coefficients

In the previous chapters we have introduced fundamental solutions for particular elliptic partial differential equations which are defined in the whole space $\mathbb{R}^n$. Fundamental solutions in the whole space $\mathbb{R}^n$ are often referred to as *principal fundamental solutions.* In particular, we have principal fundamental solutions for the Laplacian (1.1.2), the Helmholtz equation (2.1.4), the Lamé system (2.3.10) and the biharmonic equation (2.4.7).

Notice that all these partial differential equations have constant coefficients.

Scalar equations with constant coefficients

For scalar differential equations with constant coefficients, in particular the elliptic ones, the principal fundamental solution always exists due to Malgrange and Ehrenpreis (see also Folland [125], Wagner [429]).

Theorem 7.3.2. (Ehrenpreis [111], Malgrange [280]) *For the scalar partial differential operator P with symbol $\sum\limits_{|\alpha|\leq m} i^{|\alpha|}c_\alpha \xi^\alpha$, the fundamental solution exists and is given by the complex contour integral*

$$E(x,y) = \frac{(2\pi)^{-n/2}}{\overline{a_m(\xi)}} \int\limits_{\lambda\in\mathbb{C}\wedge|\lambda|=1} \lambda^m e^{i\lambda\xi\cdot(x-y)} \Big\{\mathcal{F}^{-1}_{\eta\mapsto(x-y)} \frac{\overline{a(\eta+\lambda\xi)}}{a(\eta+\lambda\xi)}\Big\} \frac{d\lambda}{2\pi i\lambda}\,. \tag{7.3.7}$$

where $a_m(\xi) = \sum\limits_{|\alpha|=m} i^m c_\alpha \xi^\alpha$ is the principal symbol and $\xi \in \mathbb{C}$ is fixed such that $a_m(\xi) \neq 0$ (see Wagner [429]).

A different formula by König for this case can also be found in [231]. Treves constructs in [418] fundamental solutions for several differential polynomials.

Wagner in [428] illustrates how to apply the method of Hörmander's stairs to construct fundamental solutions in this case. In [341, 342], Ortner presents a collection of some fundamental solutions for constant coefficient operators. For second order systems in $\mathbb{R}^2$ with constant coefficients see also Clements [73].

John in [219, (3.53)] obtains more explicit expressions for fundamental solutions depending on n odd or even.

For $\boldsymbol{n}$ **odd** he obtains the general representation formula

$$E(x,y) = \frac{i}{4(2\pi i)^n}(\Delta_y)^{(n-1)/2} \int\limits_{|\xi|=1} \operatorname{sign}[\xi \cdot (x-y)] \oint\limits_{|\lambda|=M} \frac{e^{i\lambda(x-y)\cdot\xi}}{a(\lambda\xi)} d\lambda d\omega_\xi \tag{7.3.8}$$

where M is chosen large enough such that all zeros of $a(\lambda\xi)$ are contained in the circle $|\lambda| = M$. He also simplifies this formula for subcases of odd n.

For $\boldsymbol{n}$ **even** and $n < m$, he obtains the explicit expression [219, (3.64)]

$$E(x,y) = \frac{i}{(2\pi)^{n+1}} \int\limits_{|\xi|=1, \xi \in \mathbb{R}^n} \log|(x-y)\cdot\xi| \oint\limits_{|\lambda|=M} \frac{e^{i\lambda(x-y)\cdot\xi}}{a(\lambda\xi)} \lambda^{n-1} d\lambda d\omega_\xi \,. \tag{7.3.9}$$

For even $n \geq m$, he presents $E(x,y)$ only for the case of the homogeneous equation with $a(\xi) = a_m(\xi)$:

$$E(x,y) = -\frac{1}{m!(2\pi i)^n}(\Delta_y)^{n/2} \int\limits_{|\xi|=1, \xi \in \mathbb{R}^n} \big((x-y)\cdot\xi\big)^m \log|(x-y)\cdot\xi| \frac{1}{a_m(-i\xi)} d\omega_\xi \,. \tag{7.3.10}$$

In special cases, these integrations can be carried out explicitly. For instance, if P is the iterated Laplacian in $\mathbb{R}^n$,

$$Pu = \Delta^q u \quad \text{with} \quad q \in \mathbb{N} \tag{7.3.11}$$

then a fundamental solution is given in [219, (3.5)],

$$E(x,y) = \frac{(-1)^q \Gamma(\frac{n}{2} - q)}{2^{2q}\pi^{n/2}\Gamma(q)} |x-y|^{2q-n} \quad \text{for } n \quad \text{odd and for even } n > 2q\,; \tag{7.3.12}$$

and for even $n \leq 2q$, one has [219, (3.6)],

$$E(x,y) = \frac{-(-1)^{\frac{n}{2}}}{2^{2q-1}\pi^{n/2}(q-1)!(q-\frac{n}{2})!} |x-y|^{2q-n} \log|x-y| \,. \tag{7.3.13}$$

For the Helmholtz operator in $\mathbb{R}^n$,

$$Pu = \Delta u + k^2 u$$

one has

$$E(x,y) = \frac{1}{4}\Big(\frac{k}{2\pi|x-y|}\Big)^{\frac{n-2}{2}} H_{\frac{n-2}{2}}(k|x-y|) \tag{7.3.14}$$

where H_ν is the Bessel function of the second kind.

Petrovski elliptic systems with constant coefficients

Fichera in [121] has carried out Fritz John's construction of the fundamental solution for even order, $m = 2m'$, homogeneous elliptic systems in the sense of Petrovski which have the form

$$Pu = \sum_{|\alpha|=|\beta|=m'} a_{\alpha\beta} D^{\alpha+\beta} u \quad \text{in } \mathbb{R}^n \tag{7.3.15}$$

where $u = (u_1, \dots, u_p)^\top$ and $a_{\alpha,\beta} = a^*_{\beta,\alpha}$. The symbol matrix P is then given by

$$\sigma_{\alpha\beta}(\xi) = (-1)^{m'} a_{\alpha\beta} \xi^{\alpha+\beta}; \tag{7.3.16}$$

and ellipticity in the sense of Petrovski means:

$$Q(\xi) := \det\big((-1)^{m'} a_{\alpha\beta}\xi^{\alpha+\beta}\big) \neq 0 \quad \text{for all } \xi \in \mathbb{R}^n \quad \text{with } \xi \neq 0\,. \tag{7.3.17}$$

Let $\widetilde{L}(\xi)$ be defined by the inverse of the symbol matrix times Q, i.e.,

$$\widetilde{L}(\xi) = Q(\xi)\big(\big((-1)^{m'} a_{\alpha\beta}\xi^{\alpha+\beta}\big)\big)^{-1} \tag{7.3.18}$$

which consists of the cofactors to the symbols $\sigma_{\alpha\beta}(\xi)$ and, hence, of homogeneous polynomials. Due to Fichera [121], the fundamental solution then is given explicitly by

$$E(x,y) = \widetilde{L}(x,D)S(x,y) \tag{7.3.19}$$

where for n **odd**,

$$S(x,y) = \frac{1}{4(2\pi i)^{n-1}(m-1)!}(\Delta_y)^{\frac{n-1}{2}} \int\limits_{|\xi|=1} \frac{|(x-y)\cdot\xi|^{m-1}}{Q(\xi)} d\omega_\xi \tag{7.3.20}$$

and for n **even**

$$S(x,y) = \frac{-1}{4(2\pi i)^n m!}(\Delta_y)^{\frac{n}{2}} \int\limits_{|\xi|=1} \frac{|(x-y)\cdot\xi|^m \log|(x-y)\cdot\xi|}{Q(\xi)} d\omega_\xi\,. \tag{7.3.21}$$

Agmon–Douglis–Nirenberg elliptic systems with constant coefficients

In [219] Fritz John represents explicitly the fundamental solution for the principal part of an elliptic system (7.2.10), i.e.,

$$P_j u = \sum_{k=1}^{p} \sum_{|\beta|=s_j+t_k} a^{jk}_\beta D^\beta u_k \quad \text{for } j = 1, \dots, p, \tag{7.3.22}$$

having the symbol matrix (7.2.13), i.e.,

$$\sigma_{jk}(\xi) := a^{jk0}_{s_j+t_k}(\xi) = \sum_{|\beta|=s_j+t_k} a^{jk0}_\beta i^{|\beta|}\xi^\beta . \qquad (7.3.23)$$

Let $m_k := \max_{j=1,\dots,p}(s_j+t_k) \geq 0$ and let $P^{q\ell}(\xi)$ denote the components of the inverse matrix to $\big(\!\big(\sigma_{jk}(\xi)\big)\!\big)$. Then the fundamental solution is given by the matrix components for **n odd**:

$$E^\ell_q(x,y) = \frac{1}{4(2\pi i)^{n-1}(m_q-1)!}\times \qquad (7.3.24)$$
$$\times(\Delta_y)^{\frac{n-1}{2}} \int\limits_{|\eta|=1,\eta\in\mathbb{R}^n} \big((x-y)\cdot\eta\big)^{m_q-1}\operatorname{sign}\big((x-y)\cdot\eta\big)P^{q\ell}(-i\eta)d\omega_\eta\,,$$

and for **n even**,

$$E^\ell_q(x,y) = \frac{-1}{m_q!(2\pi i)^n}\times \qquad (7.3.25)$$
$$\times(\Delta_y)^{\frac{n}{2}} \int\limits_{|\eta|=1,\eta\in\mathbb{R}^n} \big((x-y)\cdot\eta\big)^{m_q}\log|(x-y)\cdot\eta|P^{q\ell}(-i\eta)d\omega_\eta\,.$$

7.3.3 Existing Fundamental Solutions in Applications

For various existing problems in applications, explicit fundamental solutions are available. Here we collect a list of a few relevant references which may be useful but is by no means complete.

For harmonic elastodynamics, fundamental solutions were given in [191]. For the Yukawa equation, for $\Delta^m - s\Delta^{m-1}$ and the Oseen equation one finds fundamental solutions in [198]. In the book by Natroshvili [321] one finds the construction of the fundamental solution of anisotropic elasticity in $\mathbb{R}^3$ (see also Natroshvili et al [322, 323]). The book by Kythe [255] contains several examples ranging from classical potential theory, elastostatics, elastodynamics, Stokes flows, piezoelectricity to some nonlinear equations such as the p–Laplacian and the Einstein–Yang–Mills equations. The proceedings by Benitez (ed.) [29] contain the topics: methodology for obtaining fundamental solutions, numerical evaluation of fundamental solutions and some applications for not–well–known fundamental solutions.

In the engineering community one may find many further interesting examples of fundamental solutions. To name a few, we list some of these publications: Balaš et al [23], Bonnet [37], Brebbia et al [44], Gaul et al [140], Hartmann [173], Kohr and Pop [233], Kupradze et al [253] and Pozrikidis [353] — without claiming completeness.

8. Pseudodifferential Operators as Integral Operators

All of the integral operators with nonintegrable kernels are given in terms of computable Hadamard's partie finie, i.e. finite part integrals [168], which can be applied to problems in applications (Guiggiani [163], Schwab et al [380]).

In this chapter, we discuss the interpretation of pseudodifferential operators as integral operators. In particular, we show that every classical pseudodifferential operator is an integral operator with integrable or nonintegrable kernel plus a differential operator of the same order as that of the pseudodifferential operator in case of a nonnegative integer order. In addition, we also give necessary and sufficient conditions for integral operators to be classical pseudodifferential operators in the domain. Symbols and admissible kernels are closely related based on the asymptotic expansions of the symbols and corresponding pseudohomogeneous expansions of the kernels as examined by Seeley [385]. The main theorems in this context are Theorems 8.1.1, 8.1.6 and 8.1.7 below.

8.1 Pseudohomogeneous Kernels

In order to introduce appropriate kernel functions for integral operators we need the concept of pseudohomogeneous functions as defined in Definition 3.2.1, (3.2.1). In this context, we first recapture the definition of pseudohomogeneous functions.

Definition 8.1.1. *A function $k_q(x,z)$ is a $C^\infty(\Omega\times\mathbb{R}^n\setminus\{0\})$ pseudohomogeneous function (w.r.t. to z) of degree $q\in\mathbb{R}$:*

$$\begin{array}{rcll} k_q(x,tz) &=& t^q k_q(x,z) \ \ \textit{for every} \ \ t>0 \ \ \textit{and} \ \ z\neq 0 & \textit{if}\, q\notin\mathbb{N}_0\,; \\ k_q(x,z) &=& f_q(x,z)+\log|z|p_q(x,z) & \textit{if}\, q\in\mathbb{N}_0\,, \end{array} \tag{8.1.1}$$

where $p_q(x,z)$ is a homogeneous polynomial in z of degree q having C^∞-coefficients and where the function $f_q(x,z)$ satisfies

$$f_q(x,tz)=t^q f_q(x,z) \ \ \textit{for every} \ \ t>0 \ \ \textit{and} \ \ z\neq 0\,.$$

In short, we denote the class of pseudohomogeneous functions of degree $q\in\mathbb{R}$ by Ψhf_q. Note that for $q\notin\mathbb{N}_0$, in this case k_q is positively homogeneous of degree q with respect to z as well as f_q in the case $q\in\mathbb{N}_0$.

G. C. Hsiao und W. L. Wendland, *Boundary Integral Equations*,
Applied Mathematical Sciences 164,
https://doi.org/10.1007/978-3-030-71127-6_8

A *kernel function* $k(x, x-y)$ with $(x, y) \in \Omega \times \Omega$, $x \neq y$, is said to have a *pseudohomogeneous expansion* of degree q if there exist $k_{q+j} \in \Psi hf_{q+j}$ for $j \in \mathbb{N}_0$ such that

$$k(x, x-y) - \sum_{j=0}^{J} k_{q+j}(x, x-y) \in C^{q+J-\delta}(\Omega \times \Omega) \qquad (8.1.2)$$

for some δ with $0 < \delta < 1$. In what follows, for simplicity, we call this class of kernel functions *pseudohomogeneous kernels of degree* q.

The class of all kernel functions with pseudohomogeneous expansions of degree $q \in \mathbb{R}$ will be denoted by $\Psi hk_q(\Omega)$.

As examples, the fundamental solutions of the Helmholtz equation are of this form: $E_k(x, y) \in \Psi hk_0$ in (2.1.18) for $n = 2$ and $E_k(x, y) \in \Psi hk_{-1}$ for $n = 3$ in (2.1.19).

We note that for $q > -n$, the integral operator

$$\int_{\Omega} k_q(x, x-y)u(y)dy \qquad (8.1.3)$$

for $u \in C_0^\infty(\Omega)$ is well defined as an improper integral. For $q \leq -n$, however, $k_q(x, x-y)$ is non integrable except that $u(y)$ and its derivatives up to the order $\kappa := [-n-q]$ vanish at $y = x$. For fixed $x \in \Omega$, hence, (8.1.3) defines a homogeneous distribution on $C_0^\infty(\Omega \setminus \{x\})$. In order to extend this distribution to all of $C_0^\infty(\Omega)$, we use the Hadamard finite part concept which is the most natural one for integral operators.

Definition 8.1.2. *Let Ω_0 with $\overline{\Omega}_0 \Subset \Omega$ be a convex domain with $x \in \Omega_0$ and $R(\widehat{\Theta})$ the radial distance from x to $y \in \partial\Omega_0$ depending on the angular coordinate $\widehat{\Theta} = \frac{x-y}{|x-y|}$ on the unit sphere.*

The Hadamard finite part integral operator is defined by

$$\text{p.f.} \int_{\Omega} k_q(x, x-y)u(y)dy := \qquad (8.1.4)$$

$$\int_{\Omega} k_q(x, x-y)\Big\{u(y) - \sum_{|\alpha| \leq \kappa} \frac{1}{\alpha!}(y-x)^\alpha D^\alpha u(x)\Big\}dy + \sum_{|\alpha| \leq \kappa} d_\alpha(x) D^\alpha u(x)$$

where $\kappa := [-n-q]$ and $\varrho := -n-q-\kappa$. For $0 < \varrho < 1$,

$$\begin{aligned} d_\alpha(x) &= -\frac{1}{(j+\varrho)}\frac{1}{\alpha!} \int_{|\widehat{\Theta}|=1} R^{-j-\varrho}(\widehat{\Theta})\widehat{\Theta}^\alpha k_q(x, -\widehat{\Theta})d\omega \\ &\quad + \frac{1}{\alpha!} \int_{\Omega \setminus \Omega_0} k_q(x, x-y)(y-x)^\alpha dy \end{aligned} \qquad (8.1.5)$$

for all multi–indices α with $|\alpha| = \kappa - j$ and $j = 0, \ldots, \kappa$.

If $\varrho = 0$, then

$$d_\alpha(x) = \frac{1}{\alpha!} \int\limits_{|\widehat{\Theta}|=1} \log R(\widehat{\Theta})\widehat{\Theta}^\alpha k_q(x, -\widehat{\Theta})d\omega + \frac{1}{\alpha!} \int\limits_{\Omega\setminus\Omega_0} k_q(x, x-y)(y-x)^\alpha dy \tag{8.1.6}$$

for $|\alpha| = \kappa$, i.e. $j = 0$; and the $d_\alpha(x)$ for $j = 1, \ldots, \kappa$ are given by (8.1.5) *where $\varrho = 0$.*

Note that in the one–dimensional case $n = 1$, integrals over $|\widehat{\Theta}| = 1$ will always be understood as

$$\int\limits_{|\widehat{\Theta}|=1} \Psi(x, \widehat{\Theta})d\omega = \Psi(x, 1) + \Psi(x, -1) \tag{8.1.7}$$

according to Definition 3.2.2.

Here, Ω_0 with $x \in \Omega_0$ and $\overline{\Omega}_0 \subset \Omega$ is any star–shaped domain with respect to x, and $\widehat{\Theta} = (y-x)/|y-x|$, where $d\omega(\widehat{\Theta})$ is the surface element on the unit sphere and $R(\widehat{\Theta})$ describes the boundary of Ω_0 in polar coordinates. Note that this definition resembles the Definition 3.2.2 of Hadamard's finite part integral.

In the following we are going to show that every classical pseudodifferential operator can be written as an integral operator in the form of Hadamard's finite part integral operators with pseudohomogeneous kernels together with a differential operator and vice versa. Our presentation follows closely the article by Seeley [385] (see also [384]), the book by Pomp [351] and Schwab et al [380]. In view of the definition of $\mathcal{L}^m_{c\ell}(\Omega)$ and Theorem 7.1.2, this means that for $A \in \mathcal{L}^m_{c\ell}(\Omega)$ we have the general representation

$$(Au)(x) = \sum_{|\alpha|\le 2N} c_\alpha(x)D^\alpha u(x) + \text{p.f.} \int\limits_\Omega k(x, x-y)u(y)dy \tag{8.1.8}$$

where

$$\begin{aligned} c_\alpha(x) &= (2\pi)^{-n}\frac{1}{\alpha!}\Big\{ \int\limits_{\mathbb{R}^n}\int\limits_\Omega (y-x)^\alpha \psi(y)e^{i(x-y)\cdot\xi}dya(x,\xi)d\xi \\ &\quad - \text{p.f.} \int\limits_\Omega k(x, x-y)(y-x)^\alpha\psi(y)dy\Big\} \end{aligned} \tag{8.1.9}$$

with $k(x, x-y)$ given by (7.1.12).

Note that for $n + m \notin \mathbb{N}_0$ and $m + n \ge 0$, the coefficients $a_\alpha(x)$ must vanish since the finite part integral is a classical pseudodifferential operator of order m whereas the orders of $a_\alpha(x)D^\alpha(x)$ are $|\alpha| \in \mathbb{N}_0$.

8.1.1 Integral Operators as Pseudodifferential Operators of Negative Order

We begin with the case $A \in \mathcal{L}^m_{c\ell}(\Omega)$ for $m < 0$ since, in this case, the corresponding integral operators do not involve finite part integrals.

Theorem 8.1.1. (Seeley [385, p. 209]) *Let $m < 0$. Then $A \in \mathcal{L}^m_{c\ell}(\Omega)$ if and only if*

$$(Au)(x) = \int\limits_{\Omega} k(x, x-y)u(y)dy \quad \textit{for all } u \in C_0^\infty(\Omega) \tag{8.1.10}$$

with the Schwartz kernel k satisfying $k \in \Psi hk_{-m-n}(\Omega)$.

Remark 8.1.1: Note that for the elliptic differential equations, the parametrix Q_0 (7.2.6) as well as the free space Green's operator (7.2.11) are of the form (8.1.10) with $m < 0$.

The proof needs some detailed relations concerning the properties of pseudohomogeneous kernels. We, therefore, first present these details and return to the proof of Theorem 8.1.1 later on.

Lemma 8.1.2. *Let $k_{-n} \in \Psi hf_{-n}$, i.e. $k_{-n}(x,z)$ be a homogeneous function of degree $-n$ and let $\psi(r)$ be a C^∞–cut–off function, $0 \le \psi(r) \le 1$ with $\psi(r) = 0$ for $r \ge 2$ and $\psi(r) = 1$ for $0 \le r \le 1$. Then*

$$\begin{aligned} b_\psi(x,\xi) &:= \text{p.f.} \int\limits_{\mathbb{R}^n} k_{-n}(x,z)\psi(|z|)e^{-iz\cdot\xi}dz \\ &= (2\pi)^{n/2}\mathcal{F}_{z\mapsto\xi}\big(k_{-n}(x,z)\psi(|z|)\big) \end{aligned}$$

has the following properties:

$$b_\psi(x,\xi) = a_\psi(x,\xi) + c(x)\{a_0 + a_1 \log|\xi|\} \tag{8.1.11}$$

with $a_\psi \in \boldsymbol{S}^0(\Omega \times \mathbb{R}^n)$ and $c(x) := \frac{1}{\omega_n} \int\limits_{|\widehat{\Theta}|=1} k_{-n}(x, \widehat{\Theta})d\omega$, for $n \ge 2$ where, for $n \ge 2$, $a_0 = (2^{-1}\pi^{-n/2})c_0^{(n)}$, $a_1 = (2^{-1}\pi^{-n/2})c_{-1}^{(n)}$ with the constants $c_0^{(n)}$, $c_{-1}^{(n)}$ given in Gelfand and Shilov [143, p.364], *i.e.,*

$$\begin{aligned} c_0^{(n)} &= -c_E \quad \textit{with the Euler–Mascheroni constant } c_E. \\ c_{-1}^{(n)} &= 2. \end{aligned}$$

Moreover,

$$\lim_{t\to\infty} a_\psi\Big(x, t\frac{\xi}{|\xi|}\Big) = a^0_\psi(x,\xi_0) \quad \textit{with } \xi_0 := \frac{\xi}{|\xi|} \tag{8.1.12}$$

and

$$a^0_\psi(x,\xi_0) = \int\limits_{|\widehat{\Theta}|=1} \big(k_{-n}(x,\widehat{\Theta}) - c(x)\big)\Big\{\log\frac{1}{|\widehat{\Theta}\cdot\xi_0|} - \frac{i\pi}{2}\frac{\widehat{\Theta}\cdot\xi_0}{|\widehat{\Theta}\cdot\xi_0|}\Big\}d\omega \quad (8.1.13)$$

In the case $n=1$*, one has* $c(x) = \frac{1}{2}\big(k_{-1}(x,1)+k_{-1}(x,-1)\big)$*,*
$a_0 = 2\alpha_0$*,* $a_1 = -2$ *where* $\alpha_0 = \int\limits_0^1 (\cos t - 1)\frac{dt}{t} + \int\limits_1^\infty \cos t\frac{dt}{t}$*; and*

$$a^0(x,\xi_0) = -\tfrac{1}{2}\big(k_{-1}(x,1) - k_{-1}(x,-1)\big)i\pi\,\mathrm{sign}\,\xi_0\,. \quad (8.1.14)$$

Proof: From the definition of b_ψ it follows with the Paley–Wiener Theorem 3.1.3 that $b_\psi \in C^\infty(\Omega\times{\rm I\!R}^n)$. Hence, we only need to consider the behaviour of $b_\psi(x,\xi)$ for $|\xi|\geq q$. There, with the Fourier transform of distributions, we write

$$\begin{aligned} b_\psi(x,\xi) &= (2\pi)^{n/2}\mathcal{F}_{z\mapsto\xi}k_{-n}(x,z) \\ &\quad -(2\pi)^{n/2}\mathcal{F}_{z\mapsto\xi}\{k_{-n}(x,z)\big(1-\psi(|z|)\big)\} \\ &= I_1(x,\xi) - I_2(x,\xi)\,. \end{aligned}$$

We first consider $I_2(x,\xi)$ for $|\xi|\geq 1$ and the multiindex γ with $|\gamma|\geq 1$:

$$\xi^\gamma I_2(x,\xi) = \int\limits_{|z|\geq 1} e^{-iz\cdot\xi}\Big(-i\frac{\partial}{\partial z}\Big)^\gamma\{k_{-n}(x,z)\big(1-\psi(|z|)\big)\}dz\,.$$

Since

$$\Big|\Big(-i\frac{\partial}{\partial z}\Big)^\gamma\{k_{-n}(x,z)\big(1-\psi(|z|)\big)\}\Big| \leq c_\gamma(x)|z|^{-n-|\gamma|}\,,$$

we have

$$|\xi^\gamma I_2(x,\xi)| \leq c_\gamma(x) \quad \text{for } x\in\Omega \text{ and any } \gamma \text{ with } |\gamma|\geq 1\,.$$

Hence,

$$I_2 \in S^{-\infty}(\Omega\times{\rm I\!R}^n) \quad\text{and}\quad \Big|I_2\Big(x,t\frac{\xi}{|\xi|}\Big)\Big| \leq t^{-|\gamma|}c_\gamma(x) \quad \text{for } t\geq 1\,. \quad (8.1.15)$$

Now we consider the case $n\geq 2$. For $I_1(x,\xi)$ we first modify the homogeneous kernel function as

$$\overset{\circ}{k}_{-n}(x,z) := k_{-n}(x,z) - c(x)|z|^{-n}\,, \quad (8.1.16)$$

which satisfies the *Tricomi condition*

$$\int\limits_{|\widehat{\Theta}|=1} \overset{\circ}{k}_{-n}(x,\widehat{\Theta})d\omega(\widehat{\Theta}) = 0\,. \quad (8.1.17)$$

Then

$$\begin{aligned} I_1(x,\xi) &= (2\pi)^{n/2}\mathcal{F}_{z\mapsto\xi}\overset{\circ}{k}_{-n}(x,z) + c(x)(2\pi)^{n/2}\mathcal{F}_{z\mapsto\xi}|z|^{-n} \\ &= a^0_\psi(x,\xi) + c(x)\{a_0 + a_1\log|\xi|\} \end{aligned}$$

where, after introducing polar coordinates, we arrive at

$$\begin{aligned} a^0_\psi(x,\xi) &= \text{p.f.} \int\limits_{|\widehat{\theta}|=1}\int\limits_0^\infty \overset{\circ}{k}_{-n}(x,\widehat{\Theta})e^{-i\widehat{\Theta}\cdot\xi_0 r}r^{-1}dr d\omega(\widehat{\Theta}) \\ &= \int\limits_{|\widehat{\theta}|=1}\overset{\circ}{k}_{-n}(x,\widehat{\Theta})\Big\{\int\limits_0^1 (e^{-i\widehat{\Theta}\cdot\xi_0 r}-1)r^{-1}dr \\ &\qquad\qquad + \int\limits_1^\infty e^{-i\widehat{\Theta}\cdot\xi_0 r}r^{-1}dr\Big\}d\omega(\widehat{\Theta}) \end{aligned}$$

with (8.1.17). From the Calderón–Zygmund formula (see also Mikhlin and Prössdorf[305, p. 249])

$$\Big\{\int\limits_0^\infty (e^{-i\widehat{\Theta}\cdot\xi_0 r}-1)\frac{dr}{r} + \int\limits_1^\infty e^{-i\widehat{\Theta}\cdot\xi_0 r}\frac{dr}{r}\Big\} = -\log|\widehat{\Theta}\cdot\xi_0| - \frac{i\pi}{2}\frac{\widehat{\Theta}\cdot\xi_0}{|\widehat{\Theta}\cdot\xi_0|} + \alpha_0 \tag{8.1.18}$$

With some constant α_0 given explicitly in [305] we find

$$a^0_\psi(x,\xi) = -\int\limits_{|\widehat{\Theta}|=1}\overset{\circ}{k}_{-n}(x,\widehat{\Theta})\Big\{\log|\widehat{\Theta}\cdot\xi_0| + \frac{i\pi}{2}\frac{\widehat{\Theta}\cdot\xi_0}{|\widehat{\Theta}\cdot\xi_0|} + \alpha_0\Big\}d\omega\,.$$

For a given vector $\xi = |\xi|\xi_0 \in \mathbb{R}^n$, the integrand is weakly singular along the circle $\widehat{\Theta}\cdot\xi_0 = 0$ on the unit sphere $|\widehat{\Theta}| = 1$. Hence, $a^0_\psi(x,\xi)$ is uniformly bounded for every $x \in K \Subset \Omega$ and all $\xi \in \mathbb{R}^n\setminus\{0\}$ and is homogeneous of degree zero.

For $n = 1$, the proof follows in the same manner, except the Tricomi condition (8.1.17) now is replaced by

$$k^0_{-1}(x,1) + k^0_{-1}(x,-1) = 0\,. \tag{8.1.19}$$

Here, instead of the Calderón–Zygmund formula (8.1.18) we use its one–dimensional version, i.e.,

$$\text{p.f.}\int\limits_0^\infty e^{-i\xi z}\frac{dz}{z} = -\log|\xi| - \frac{i\pi}{2}\frac{\xi}{|\xi|} + \alpha_0\,. \tag{8.1.20}$$

Collecting the above results, we find by using (8.1.17),

$$b_\psi(x,\xi) = a_\psi^0(x,\xi) - I_2(x,\xi) + c(x)\{a_0 + a_1 \log|\xi|\}\,.$$

Since

$$\lim_{t\to\infty} a_\psi(x,t\xi_0) = a_\psi^0(x,\xi_0) - \lim_{t\to\infty} I_2(x,t\xi_0) = a_\psi^0(x,\xi_0)$$

from (8.1.15) we obtain the desired property (8.1.12). ■

The next lemma, due to Seeley [385], is concerned with the differentiation and primitives of pseudohomogeneous functions.

Lemma 8.1.3. *Let $m \in \mathbb{R}$, $k \in \mathbb{N}_0$ and $m-k > -n$.*

(i) *If $f(z) \in \Psi hf_m$ then $\left(-i\frac{\partial}{\partial z}\right)^\alpha f(z) \in \Psi hf_{m-k}$ for all $|\alpha| = k$. If $m \in \mathbb{N}$ and if f_m, the homogeneous part of f, satisfies the compatibility conditions*

$$\int\limits_{|z|=1} f_m(\widehat{\Theta})\widehat{\Theta}^\alpha d\omega = 0 \quad \textit{for all } |\alpha| = m \quad \textit{for } n \geq 2 \tag{8.1.21}$$

where $\widehat{\Theta} = \frac{z}{|z|}$, and

$$f(1) + (-1)^m f_m(-1) = 0 \quad \textit{for } n = 1\,,$$

then $\frac{\partial f_m}{\partial z_j}$ will satisfy

$$\int\limits_{|z|=1} \frac{\partial f_m}{\partial z_j}\widehat{\Theta}^\beta d\omega = 0 \quad \textit{for all } |\beta| = m-1 \quad \textit{if } n \geq 2 \tag{8.1.22}$$

and

$$f_m'(1) + (-1)^{m-1} f_m'(-1) = 0 \quad \textit{for } n = 1\,.$$

(ii) *Let $n \geq 2$ and $h^{(\alpha)}(z) \in \Psi hf_{m-k}$ for all α with $|\alpha| = k$ with the additional property that there exists some function $F \in C^\infty(\mathbb{R}^n \setminus \{0\})$ and*

$$h^{(\alpha)}(z) = \left(-i\frac{\partial}{\partial z}\right)^\alpha F(z) \quad \textit{for all } \alpha \quad \textit{with } |\alpha| = k\,.$$

Then there exists a function $f \in \Psi hf_m$ with the property that

$$F(z) - f(z) = p_{k-1}(z)$$

is a polynomial of degree less than k.
If $m \neq 0$ then $f \in \Psi hf_m$ is uniquely determined.
In the one–dimensional case $n = 1$, the space $\mathbb{R} \setminus \{0\}$ is not connected. Therefore we obtain only the weaker result

$$F(z) - f(z) = p^{\pm}_{k-1}(z)$$

with two polynomials for $z > 0$ and $z < 0$, respectively, which in general might be different, provided $m \notin \mathbb{N}_0$ or $m \in \mathbb{N}_0$ and $m - k \geq 0$. In the case $m \in \mathbb{N}_0$ and $m - k < 0$, the function $h^{(k)}(z)$ must also satisfy the additional assumption

$$h^{(k)}(1) - (-1)^k h^{(k)}(-1) = 0 .$$

Proof:

(i) The function $f \in \Psi h f_m$ is of the form $f = f_m(z)$ with $f_m(tz) = t^m f_m(z)$ for $t > 0$ and $z \neq 0$, provided $m \notin \mathbb{N}_0$. Then differentiation with $|\alpha| = 1$ yields for $h^{(\alpha)}_{m-1}(z) := \frac{\partial^\alpha f_m}{\partial z^\alpha}(z)$ the relation

$$t^m h^{(\alpha)}_{m-1}(z) = t^m \frac{\partial^\alpha f_m}{\partial z^\alpha}(z) = t \frac{\partial^\alpha f_m}{\partial z^\alpha}(tz) = t h^{(\alpha)}_{m-1}(tz) .$$

Hence, $h^{(\alpha)}_{m-1} \in \Psi h f_{m-1}$ and also $m - 1 \notin \mathbb{N}_0$.
For $m \in \mathbb{N}_0$, the function f is of the form

$$f(z) = f_m(z) + p_m(z) \log |z|$$

where $f_m(z)$ is a positively homogeneous function and $p_m(z)$ is a homogeneous polynomial of degree m. Then

$$\frac{\partial}{\partial z_k} f(z) = \frac{\partial}{\partial z_k} f_m(z) + z_k p_m(z) \cdot \frac{1}{|z|^2} + \log |z| \frac{\partial}{\partial z_k} p_m(z) .$$

The first two terms now are positively homogeneous of degree $m - 1$ whereas $\frac{\partial}{\partial z_k} p_m(z)$ is a homogeneous polynomial of degree $m - 1$. By repeating the same argument k times, the assertion follows.
If, in addition, (8.1.21) is fulfilled then we distinguish the cases $n \geq 2$ and $n = 1$. We begin with the case $n \geq 2$ and consider the integral

$$I(r) := \int \left(\frac{\partial f_m}{\partial z_j} z^\beta \right) \Big|_{|z|=r} d\omega = r^{2m-2} \int\limits_{|\widehat{\Theta}|=1} \frac{\partial f_m}{\partial z_j}(\widehat{\Theta}) \widehat{\Theta}^\beta d\omega$$

where β is an arbitrarily chosen multiindex with $|\beta| = m - 1$ and $d\omega$ denotes the surface element of the unit sphere. Integrating $r^2 I(r) = \left(\sum_{\ell=1}^n z_\ell^2\right) I(r)$ from $r = 0$ to ϱ, we obtain with the divergence theorem:

$$\int_0^{\varrho} r^{n+1} I(r) dr = \int_{V(\varrho)} \frac{\partial f_m}{\partial z_j} z^\beta \left(\sum_{\ell=1}^n z_\ell^2 \right) dv$$

$$= - \int_{V(\varrho)} f_m(z) \frac{\partial}{\partial z_j} \left(z^\beta \sum_{\ell=1}^n z_\ell^2 \right) dv + \int_{|z|=\varrho} f_m(z) z^\beta \left(\sum_{\ell=1}^n z_\ell^2 \right) z_j \varrho^{n-2} d\omega$$

$$= - \int_{V(\varrho)} f_m(z) \frac{\partial}{\partial z_j} \left(z^\beta \sum_{\ell=1}^n z_\ell^2 \right) dv + \varrho^{2m+n} \int_{|\widehat{\Theta}|=1} f_m(\widehat{\Theta}) \widehat{\Theta}^\beta \widehat{\Theta}_j d\omega \,.$$

$V(\varrho)$ denotes the volume of the ball with radius ϱ. The last integrals on the right–hand side vanish due to assumption (8.1.21). Then taking $\frac{d}{d\varrho}$ on both sides at $\varrho = 1$, we find

$$I(1) = - \int_{|\widehat{\Theta}|=1} f_m(\widehat{\Theta}) \left\{ \sum_{\ell=1}^n \left(\frac{\partial z^\beta}{\partial z_j} (\widehat{\Theta}) \right) \widehat{\Theta}_\ell^2 + \widehat{\Theta}^\beta \widehat{\Theta}_j \right\} d\omega \,.$$

Each of the polynomials $\frac{\partial^\beta}{\partial z_j}(\widehat{\Theta})\widehat{\Theta}_\ell^2$ and $\widehat{\Theta}^\beta \widehat{\Theta}_j$ is homogeneous of order $|\beta| + 1 = m$. Again, with (8.1.21) for each of these terms, we obtain

$$I(1) = \int_{|\widehat{\Theta}|=1} \frac{\partial f}{\partial z_j} \widehat{\Theta}^\beta d\omega = 0 \quad \text{for any } \; |\beta| = m-1 \,.$$

Now we consider the case $n = 1$. Then

$$f_m'(1) + f_m'(-1)(-1)^{m-1} = \int_0^1 \{ f'(1) + f_m'(-1)(-1)^{m-1} \} dr$$

$$= \int_0^1 f_m'(r) r^{-m+1} dr + \int_0^1 f_m'(-r) r^{-m+1} (-1)^{m-1} dr \,,$$

since $f_m'(r)$ is positively homogeneous of degree $(m-1)$.
Integration by parts gives with (8.1.21):

$$f_m'(1) + f_m'(-1)(-1)^{m-1} = - \int_0^1 f_m(r)(1-m) r^{-m} dr$$

$$+ (-1)^{m-1} \int_0^1 f_m(-r)(1-m) r^{-m} dr$$

$$= - \{ f_m(1) + f_m(-1)(-1)^m \} = 0 \,,$$

as proposed.

(ii) For this part of the theorem we confine ourselves only to the case $k = 1$. Then the general case can be treated by standard induction.
We will consider the three different subcases $m < 0$, $m > 0$ and $m = 0$, separately. For $m < 0$, let

$$\frac{\partial F}{\partial z_j} = ih_{m-1}^{(j)}(z) \in \Psi hf_{m-1} \quad \text{for } j = 1, \ldots, n\,.$$

Since we assume that $F(z)$ is a given primitive of the $h_{m-1}^{(j)}$ the indefinite line integral

$$F(z) = i \int\limits_{\infty}^{z} \sum_{j=1}^{n} h_{m-1}^{(j)}(\zeta) d\zeta_j + c_0 \quad \text{for } z \in \mathrm{I\!R}^n \setminus \{0\}\,,$$

for $n \geq 2$ is independent of the path of integration avoiding the origin. All that remains to show is the positive homogeneity of

$$f(z) := \int\limits_{\infty}^{z} \sum_{j=1}^{n} h_{m-1}^{(j)}(\zeta) d\zeta_j$$

which is well defined because of

$$\lim_{\zeta \to \infty} h_{m-1}^{(j)}(\zeta) = h_{m-1}^{(j)}(\widehat{\Theta}) \lim_{\zeta \to \infty} |\zeta|^{m-1} = 0$$

uniformly and $m - 1 < -1$. A simple change of variables $\zeta_j = t\widetilde{\zeta_j}$ with any $t > 0$ shows with the homogeneity of $h_{m-1}^{(j)}$ for $z \neq 0$ that

$$f(z) = \frac{1}{m} \sum_{j=1}^{n} h_{m-1}^{(j)}(z) z_j$$

which is obviously homogeneous of degree m. This is Euler's formula for positively homogeneous functions of degree m.

For $n = 1$, the integration paths are from $(\text{sign}\, z)\infty$ to z and are different for $z > 0$ and $z < 0$. Hence, in general one obtains two different constants $c_0^{\pm}$ accordingly.

For $m > 0$ and $m \notin \mathrm{I\!N}$, we proceed in the same manner as before by defining

$$f(z) := \int\limits_{0}^{z} \sum_{j=1}^{n} h_{m-1}^{(j)}(\zeta) d\zeta_j = \frac{1}{m} \sum_{j=1}^{n} h_{m-1}^{(j)}(z) z_j\,.$$

This representation also shows that f is uniquely determined.

Again, in the case $n = 1$, one may obtain different contants $c_0^{\pm}$ corresponding to $z > 0$ or $z < 0$, respectively.

Now, let us consider the case $m \in \mathbb{N}$ and $n \geq 2$. Here,

$$h^{(j)}(z) = h_{m-1}^{(j)}(z) + p_{m-1}^{(j)}(z) \log|z| = -i \frac{\partial F}{\partial z_j}(z)$$

is pseudohomogeneous of degree $m - 1$. Hence,

$$\frac{\partial F}{\partial z_j}(tz) = i t^{m-1} h_{m-1}^{(j)}(z) + i t^{m-1} p_{m-1}^{(j)}(z) \{\log|z| + \log t\}$$

for every $t > 0$ and $z \neq 0$. For $F(z)$, we find the path–independent line integral

$$F(z) = \sum_{j=1}^{n} \int_{0}^{z} \frac{\partial F}{\partial z_j}(\zeta) d\zeta_j + c$$

where c is any constant. By changing the variable of integration $\zeta = \tau z$ and integrating along the straight lines where $\tau \in [0, 1]$, we find

$$\begin{aligned} F(z) &= i \sum_{j=1}^{n} \int_{\tau=0}^{1} \tau^{m-1} \left(h_{m-1}^{(j)}(z) + p_{m-1}^{(j)}(z) \log|z| \right) z_j d\tau \\ &\quad + i \sum_{j=1}^{n} \int_{0}^{1} \tau^{m-1} \log \tau d\tau p_{m-1}^{(j)}(z) z_j + c, \end{aligned}$$

i.e., the explicit representation

$$F(z) = \frac{i}{m} \sum_{j=1}^{n} \left(h_{m-1}^{(j)}(z) + p_{m-1}^{(j)}(z) \log|z| - \frac{1}{m} p_{m-1}^{(j)}(z) \right) z_j + c$$

with any constant c. Therefore,

$$\begin{aligned} f_m(z) &= \frac{i}{m} \sum_{j=1}^{n} \left(h_{m-1}^{(j)}(z) - \frac{1}{m} p_{m-1}^{(j)}(z) \right) z_j, \\ p_m(z) &= \frac{i}{m} \sum_{j=1}^{n} p_{m-1}^{(j)}(z) z_j. \end{aligned}$$

Both functions, $f_m(z)$ and the polynomial $p_m(z)$ are homogeneous of degree m and $F = f_m + p_m \log|z| + c$.

Finally, for $n \geq 2$, let us consider the case $m = 0$. Here the desired form of f is

$$f(z) = c_0 + g_0(z) + p_0 \log |z|$$

where g_0 and p_0 both are to be found. The yet unknown function $g_0(z)$ is positively homogeneous of degree zero, hence, the desired function

$$\sum_{j=1}^{n} \frac{\partial g_0}{\partial z_j}(z) z_j \quad \text{is positively homogeneous}$$

of degree zero again, due to the first part of our theorem. Now we require that the condition

$$\int\limits_{|z|=1} \sum_{j=1}^{n} \frac{\partial g_0}{\partial z_j} \Theta_j d\omega = 0 \tag{8.1.23}$$

is satisfied. This implies with

$$\frac{\partial F}{\partial z_j} = \frac{\partial g_0}{\partial z_j} + p_0 \frac{z_j}{|z|^2} = h^{(j)}(z) \tag{8.1.24}$$

that p_0 must be chosen as to satisfy

$$p_0 \int\limits_{|z|=1} \sum_{j=1}^{n} \frac{z_j}{|z|^2} \Theta_j d\omega = \sum_{j=1}^{n} \int\limits_{|z|=1} h^{(j)}(z) \Theta_j d\omega \,. \tag{8.1.25}$$

With the constant p_0, we now define the function

$$g_0(z) = \sum_{j=1}^{n} \int\limits_{\overline{\Theta}}^{z} \frac{\partial g_0}{\partial z_j} d\zeta_j := \sum_{j=1}^{n} \int\limits_{\overline{\Theta}}^{z} \left(h^{(j)}(\zeta) - p_0 \frac{\zeta_j}{|\zeta|^2} \right) d\zeta_j \tag{8.1.26}$$

where $\overline{\Theta}$ is a point on the unit sphere which fulfills

$$\sum_{j=1}^{n} \left(h^{(j)}(\overline{\Theta}) - p_0 \overline{\Theta}_j \right) \overline{\Theta}_j = 0 \,.$$

This is always possible because of (8.1.24). Since $h^{(j)} = \frac{\partial F}{\partial z_j}$ and $\frac{z_j}{|z|^2} = \frac{\partial \log |z|}{\partial z_j}$, the value of the line integral in (8.1.26) is independent of the path as long as $\mathbb{R}^n \setminus \{0\}$ is simply connected. It will be independent of the path in every case if for a closed curve $\mathcal{C}$ on the unit sphere, $\mathcal{C} : z = \Theta(\tau)$ for $\tau \in [0, T]$ with $\Theta(0) = \Theta(T) = 0$, the integral

$$\sum_{j=1}^{n} \oint\limits_{\mathcal{C}} \left(h^{(j)}(\Theta(\tau)) - p_0 \Theta(\tau) \right) \frac{d\Theta_j}{d\tau} d\tau$$

vanishes. Since $h^{(j)}(z) = \frac{\partial F}{\partial z_j}$ with $F \in C^\infty(\mathbb{R}^n \setminus \{0\})$ and

$$\frac{d}{d\tau}\sum_{j=1}^{n}\Theta_j^2(\tau)=\sum_{j=1}^{n}\Theta_j(\tau)\frac{\partial\Theta_j}{\partial\tau}=0$$

for any curve on $|\Theta(\tau)|=1$, we find, indeed,

$$\sum_{j=1}^{n}\oint_{C}\frac{\partial g_0}{\partial z_j}d\zeta_j=0\,.$$

To show that g_0 is positively homogeneous of degree zero, we first show with (8.1.24) and (8.1.25)

$$\begin{aligned}\int_{\overline{\Theta}}^{t\overline{\Theta}}\sum_{j=1}^{n}\frac{\partial g_0}{\partial z_j}(\zeta)d\zeta_j &= \sum_{j=1}^{n}\int_{\overline{\Theta}}^{t\overline{\Theta}}\Big\{h^{(j)}(\tau\overline{\Theta})-p_0\frac{\tau\overline{\Theta}_j}{\tau^2}\Big\}\overline{\Theta}_j d\tau\\ &= \log t\Big\{\sum_{j=1}^{n}h^{(j)}(\overline{\Theta})\overline{\Theta}_j-p_0\Big\}=0\end{aligned}$$

for every $t>0$. Then, with $\widetilde{\zeta}_j=\zeta_j/|z|$,

$$\begin{aligned}g_0(z) &= \sum_{j=1}^{n}\int_{\overline{\Theta}}^{|z|\overline{\Theta}}\frac{\partial g_0}{\partial z_j}d\zeta_j+\sum_{j=1}^{n}\int_{|z|\overline{\Theta}}^{z}\frac{\partial g_0}{\partial z_j}d\zeta_j\\ &= 0+\sum_{j=1}^{n}\int_{\overline{\Theta}}^{z/|z|}\frac{\partial g_0}{\partial z_j}(|z|\widetilde{\zeta})|z|d\widetilde{\zeta}_j\\ &= \sum_{j=1}^{n}\int_{\overline{\Theta}}^{z/|z|}\frac{\partial g_0}{\partial z_j}(\widetilde{\zeta})d\widetilde{\zeta}_j=g_0\left(\frac{z}{|z|}\right)\end{aligned}$$

where we have used the homogeneity of $\frac{\partial g_0}{\partial z_j}$ due to (8.1.24).

We note that we may add to g_0 an arbitrarily chosen constant which defines a homogeneous function of degree zero. Hence, g_0 is not unique.

In the remaining case $n=1$ and $m=0$, for $k=1$ we have for $h^{(1)}\in\Psi hf_{-1}$ the explicit representation

$$h^{(1)}(z)=h^{(1)}(\operatorname{sign}z)\cdot\frac{1}{|z|}\,.$$

Hence,

$$f(z)=(\operatorname{sign}z)\cdot h^{(1)}(\operatorname{sign}z)\log|z|+c_0^{\pm}\,,$$

and only with

$$h^{(1)}(1) - (-1)h^{(1)}(-1) = 0$$

we find

$$f(z) = h^{(1)}(1)\log|z| + c_0^{\pm} \in \Psi h f_0$$

as proposed. ■

In the following lemma we show how one can derive the homogeneous symbol from the integral operator with a kernel given as a pseudohomogeneous function.

Lemma 8.1.4. *Let $k_\kappa(x,z) \in \Psi h f_\kappa$ with $\kappa > -n$. Then the integral operator K_m defined by*

$$(K_m u)(x) = \int_\Omega k_\kappa(x, x-y)u(y)dy \quad \textit{for } x \in \Omega\,,\ u \in C_0^\infty(\Omega) \tag{8.1.27}$$

belongs to $\mathcal{L}^m_{c\ell}(\Omega)$ where $m = -\kappa - n$.

In addition, if $\psi(z)$ is a C^∞ cut–off function with $\psi(z) = 1$ for $|z| \le \varepsilon$, $\varepsilon > 0$, and $\psi(z) = 0$ for $|z| \ge R$ then

$$a_\psi(x,\xi) := e^{-ix\cdot\xi} \int_\Omega k_\kappa(x, x-y)\psi(x-y)e^{iy\cdot\xi}dy \tag{8.1.28}$$

is a symbol in $S^m_{c\ell}(\Omega \times \mathbb{R}^n)$ and the limit

$$a_m(x,\xi) := \lim_{t\to+\infty} t^{-m} a_\psi(x, t\xi) \tag{8.1.29}$$

exists for $x \in \Omega$ and $\xi \ne 0$ and defines the positively homogeneous representative of degree m of the complete symbol class $\sigma_{K_m} \sim a_m(x,\xi)$.

Proof: With $\psi(z)$, first write

$$(K_m u)(x) = \int_\Omega k_\kappa(x,x-y)\psi(x-y)u(y)dy + \int_\Omega k_\kappa(x,x-y)\big(1-\psi(x-y)\big)u(y)dy\,.$$

Since the kernel $\{k_\kappa(x, x-y)\big(1-\psi(x-y)\big)\}$ of the second integral operator is a $C^\infty(\Omega \times \Omega)$ function, the latter defines a smoothing operator due to Theorem 7.1.10 (iv). Hence, it suffices to consider only the first operator

$$(K_\psi u)(x) := \int_\Omega k_\kappa(x, x-y)\psi(x-y)u(y)dy\,.$$

Note that $k_\kappa(x,z)\psi(z)$ is weakly singular and has compact support with respect to z, hence, the symbol $a_\psi(x,\xi)$ given by (8.1.28) is just its Fourier transformation and, therefore, due to the Paley–Wiener–Schwartz theorem

(Theorem 3.1.3) $a_\psi \in C^\infty(\Omega \times \mathbb{R}^n)$. It remains to show that $a_\psi \in S^m_{c\ell}(\Omega \times \mathbb{R}^n)$, that $a_m(x,\xi)$ is well defined by (8.1.29) and, finally, that $a_\psi \sim a_m$ (for $|\xi| \geq 1$).

We first show the existence of the limit a_m. To do so, choose a multi–index γ with $|\gamma| > -m$ and consider

$$\begin{aligned}
&\xi^\gamma\big(a_\psi(x,\xi) - t^{-m}a_\psi(x,t\xi)\big)\\
&= \xi^\gamma \int\limits_{\mathbb{R}^n} e^{-iz\cdot\xi}k_\kappa(x,z)\Big\{\psi(z) - \psi\Big(\frac{z}{t}\Big)\Big\}dz\\
&\quad + \xi^\gamma \int\limits_{\mathbb{R}^n} e^{-iz\cdot\xi}p_\kappa(x,z)\psi\Big(\frac{z}{t}\Big)dz\log t\\
&= \int\limits_{\varepsilon\leq|z|} e^{-iz\cdot\xi}\Big(-i\frac{\partial}{\partial z}\Big)^\gamma\Big(k_\kappa(x,z)\Big\{\psi(z) - \psi\Big(\frac{z}{t}\Big)\Big\}\Big)dz\\
&\quad + \int\limits_{\mathbb{R}^n} e^{-iz\cdot\xi}\sum_{\substack{0\leq\beta\leq\gamma\\|\beta|\leq\kappa}}\int\limits_{\varepsilon\leq|z|\leq R} e^{-iz\cdot\xi}\Big(\Big(-i\frac{\partial}{\partial z}\Big)^\beta p_\kappa(x,z)\Big)\times\\
&\qquad\qquad \times c_{\beta\gamma}\Big(-i\frac{\partial}{\partial z}\Big)^{\gamma-\beta}\psi\Big(\frac{z}{t}\Big)dz\log t
\end{aligned}$$

provided $t \geq R$, by using the Leibniz rule. Then we obtain

$$\xi^\gamma\big(a_\psi(x,\xi) - t^{-m}a_\psi(x,t\xi)\big) = I_1(t) + \sum_{0\leq\lambda<\gamma} I_{2\lambda\gamma}(t) + \sum_{\substack{0\leq\beta\leq\gamma\\|\beta|\leq\kappa}} I_{3\beta\gamma}(t)$$

where

$$\begin{aligned}
I_1(t) &:= \int\limits_{|z|\geq\varepsilon} e^{-iz\cdot\xi}k^{(\gamma)}_{\kappa-|\gamma|}(x,z)\Big\{\psi(z) - \psi\Big(\frac{z}{t}\Big)\Big\}dz\\
&\quad\text{with } k^{(\gamma)}_{\kappa-|\gamma|}(x,z) := \Big(-i\frac{\partial}{\partial z}\Big)^\gamma k_\kappa(x,z);\\
I_{2\lambda\gamma}(t) &:= c_{\lambda\gamma}\int\limits_{\varepsilon\leq|z|} e^{-iz\cdot\xi}k^{(\lambda)}_{\kappa-|\lambda|}(x,z)\Big\{\psi_{\gamma-\lambda}(z) - t^{|\lambda|-|\gamma|}\psi_{\gamma-\lambda}\Big(\frac{z}{t}\Big)\Big\}dz\\
&\quad\text{with } \psi_{\gamma-\lambda}(z) := \Big(-i\frac{\partial}{\partial z}\Big)^{\gamma-\lambda}\psi(z) \neq 0 \quad\text{for } \varepsilon < |z| < R;\\
I_{3\beta\gamma}(t) &= c_{\beta\gamma}\int\limits_{|z|\leq tR} e^{-iz\cdot\xi}p^{(\beta)}_\kappa(x,z)\psi_{\gamma-\beta}\Big(\frac{z}{t}\Big)dz\,t^{-|\gamma-\beta|}\log t\,.
\end{aligned}$$

We note from Lemma 8.1.3 part (i) that the kernels $k^{(\lambda)}_{\kappa-|\lambda|}$ belong to $\Psi hf_{\kappa-|\lambda|}$.

Since $|\gamma| > -m$, for fixed $x \in \Omega$, the integral $I_1(t)$ converges uniformly as $t \to \infty$ for all $|z| \geq \varepsilon$:

$$\lim_{t\to\infty} I_1(t) = \int\limits_{|z|\geq\varepsilon} e^{-iz\cdot\xi} k^{(\gamma)}_{\kappa-|\gamma|}(x,z)\,\{\psi(z)-1\}\,dz\,.$$

A simple manipulation shows that the second term in $I_{2\lambda\gamma}(t)$ can be estimated by

$$\begin{aligned}
&\Big| c_{\lambda\gamma} \int\limits_{\varepsilon\leq|z|\leq tR} k^{(\lambda)}_{\kappa-|\lambda|}(x,z) t^{|\lambda|-|\gamma|} \psi_{\gamma-\lambda}\Big(\frac{z}{t}\Big)\,dz\Big| \\
&\leq\ c(x,\lambda,\gamma,\varepsilon,R) \int\limits_{r=\varepsilon}^{tR} r^{\kappa-|\lambda|} t^{|\lambda|-|\gamma|} r^{n-1}(1+|\log r|)dr = O(t^{-1}\log t) \\
&\to\ 0 \quad \text{as } t\to\infty \quad \text{since } |\gamma|>\kappa+n\,.
\end{aligned}$$

For the third term we find similarly

$$|I_{3\beta\gamma}(t)| = O(t^{-1}\log t)\to 0 \quad \text{as } t\to\infty\,.$$

Hence, the limit exists; and for $\xi\neq 0$ we find

$$\begin{aligned}
a_m(x,\xi) \ &:=\ \lim_{t\to\infty} t^{-m} a_\psi(x,t\xi) \qquad (8.1.30)\\
&=\ a_\psi(x,\xi) - \xi^{-\gamma} \int\limits_{|z|\geq\varepsilon} e^{-iz\cdot\xi}\, k^{(\gamma)}_{\kappa-|\gamma|}(x,z)\,\{\psi(z)-1\}\,dz \\
&\quad -\xi^{-\gamma} \sum_{0\leq\lambda<\gamma} c_{\lambda\gamma} \int\limits_{\varepsilon\leq|z|\leq R} e^{-iz\cdot\xi}\, k^{(\lambda)}_{\kappa-|\lambda|}(x,z)\,\psi_{\gamma-\lambda}(z)dz\,.
\end{aligned}$$

Clearly, (8.1.30) implies with the Paley–Wiener–Schwartz theorem 3.1.3 that $a_m(x,\xi)\in C^\infty\big(\Omega\times(\mathbb{R}^n\setminus\{0\})\big)$. From the definition (8.1.29) it is also clear, that $a_m(x,\xi)$ is positively *homogeneous of degree* m:

$$\begin{aligned}
a_m(x,\tau\xi) \ &=\ \tau^m \lim_{t\to+\infty} \tau^{-m} t^{-m} a_\Psi(x,\tau t\xi) \\
&=\ \tau^m a_m(x,\xi) \quad \text{for } \tau>0 \quad \text{and} \quad \xi\neq 0\,.
\end{aligned}$$

Finally, we show that $a_\psi(x,\xi)-\chi(\xi)a_m(x,\xi)\in S^{-\infty}(\Omega\times\mathbb{R}^n)$ for any C^∞–cut–off function χ with $\chi(\xi)=0$ for $|\xi|\leq\frac{1}{2}$ and $\chi(\xi)=1$ for all $|\xi|\geq 1$.

Let $\widetilde{\Omega}\Subset\Omega$ be any compact subset; consider any multiindices α,β,γ. Then, for $|\xi|\geq 1$ from (8.1.30),

$$
\begin{aligned}
&\Big(\frac{\partial}{\partial x}\Big)^{\beta}\xi^{\gamma}\Big(\frac{\partial}{\partial \xi}\Big)^{\alpha}\big(a_{\psi}(x,y)-a_m(x,\xi)\big)\\
&=\Big(\frac{\partial}{\partial \xi}\Big)^{\alpha}\int\limits_{|z|\ge\varepsilon} e^{-iz\cdot\xi}\Big(\Big(\frac{\partial}{\partial x}\Big)^{\beta}k^{(\gamma)}_{\kappa-|\gamma|}(x,z)\Big)\{\psi(z)-1\}dz\\
&+\sum_{0\le\lambda<\gamma}c_{\lambda\gamma}\Big(\frac{\partial}{\partial \xi}\Big)^{\alpha}\int\limits_{\varepsilon\le|z|\le R} e^{-iz\cdot\xi}\Big(\Big(\frac{\partial}{\partial x}\Big)^{\beta}k^{(\lambda)}_{\kappa-|\lambda|}(x,z)\Big)\psi_{\gamma-\lambda}(z)dz\\
&=\int\limits_{|z|\ge\varepsilon} e^{-iz\cdot\xi}(-iz)^{\alpha}\Big(\Big(\frac{\partial}{\partial x}\Big)^{\beta}k^{(\gamma)}_{\kappa-|\gamma|}(x,z)\Big)\{\psi(z)-1\}dz\\
&+\sum_{0\le\lambda<\gamma}c_{\lambda\gamma}\int\limits_{\varepsilon\le|z|\le R} e^{-iz\cdot\xi}(-iz)^{\alpha}\Big(\Big(\frac{\partial}{\partial x}\Big)^{\beta}k^{(\lambda)}_{\kappa-|\lambda|}(x,z)\Big)\psi_{\gamma-\lambda}(z)dz\,.
\end{aligned}
$$

For any α and β and for every γ with $|\gamma| > \kappa + n + |\alpha|$, the first integral on the right–hand side is uniformly bounded for $x \in \widetilde{\Omega}$ and all $\xi \in \mathbb{R}^n$. The remaining integrals on the right–hand side are the Fourier transforms of C^∞–functions with compact support in $\varepsilon \le |z| \le R$ and are infinitely differentiable for $x \in \widetilde{\Omega}$. Consequently, with Theorem 3.1.3, we have for any α, β and for every γ with $|\gamma| > \kappa + n + |\alpha|$ the estimates

$$
\sup_{x\in\widetilde{\Omega},|\xi|\ge 1}\Big|\Big(\frac{\partial}{\partial x}\Big)^{\beta}\Big(\frac{\partial}{\partial \xi}\Big)^{\alpha}\big(a_{\psi}(x,\xi)-a_m(x,\xi)\big)\Big||\xi^{\gamma}| \le C(\widetilde{\Omega},\alpha,\beta,\gamma)\,.
$$

Hence, $\big(a_\psi(x,\xi)-\chi(\xi)a_m(x,\xi)\big) \in S^{-\infty}(\Omega\times\mathbb{R}^n)$ and $a_\psi(x,\xi) \in S^m_{c\ell}(\Omega\times\mathbb{R}^n)$. ∎

Remark 8.1.2: For $\kappa \in \mathbb{N}_0$, consider the whole class of integral operators with the kernels of the form

$$
k_\kappa(x,z) = f_\kappa(x,z) + p_\kappa(x,z)\log|z| + q_\kappa(x,z) \qquad (8.1.31)
$$

where $q_\kappa(x,z) = \sum_{|\gamma|=\kappa} c_\gamma(x)z^\gamma$ is a homogeneous polynomial of degree κ. Then all k_κ of the form (8.1.31) belong to Ψhf_κ. For given f_κ and p_κ, and for any choice of $q_\kappa(x,z)$ in (8.1.31), k_κ has the **same** positively homogeneous principal symbol

$$
a^0_m(x,\xi) = a_m\left(x,\frac{\xi}{|\xi|}\right)|\xi|^m
$$

since the integral operator defined by the kernel function q_κ is a smoothing operator.

For fixed x and $\kappa \notin \mathbb{N}_0$, the symbol of $k_\kappa(x,z) = f_\kappa(x,z)$ is a positively homogeneous function given by the Fourier transform of f_κ which also defines a homogeneous distribution of degree $m = -\kappa - n$ with a singularity at $\xi = 0$ for $m < 0$.

Let us recall that a distribution u is said to be *homogeneous of degree* $\lambda \in \mathbb{C}$ if

$$t^{\lambda+n}\langle u, \varphi(xt)\rangle = \langle u, \varphi\rangle \quad \text{for all} \quad \varphi \in C_0^\infty(\mathbb{R}^n) \quad \text{and} \quad t > 0\,. \tag{8.1.32}$$

For the case $\kappa \in \mathbb{N}_0$, and pseudohomogeneous kernels of the form (8.1.31), we again consider the Fourier transform of the kernel. The latter defines the density of a distribution whose Fourier transform can be written as

$$\widehat{k_\kappa}(x,\xi) = (p_\kappa \widehat{(x,\cdot)\log|\cdot|})(x,\xi) + \mathcal{F}_{z\mapsto\xi} f_\kappa(x,z)\,.$$

We first consider

$$\begin{aligned}\widehat{f_\kappa(x,\bullet)}(\xi) &= (2\pi)^{-n/2} \int\limits_{|z|\le R} f_\kappa(x,z)\psi(z)e^{-iz\cdot\xi}dz \\ &\quad + |\xi|^{-2k}(2\pi)^{-n/2} \int\limits_{|z|\ge\varepsilon} e^{-iz\cdot\xi}(-\Delta_z^k)\Big(f_\kappa(x,z)(1-\psi(z))\Big)dz\end{aligned} \tag{8.1.33}$$

which exists for $\xi \neq 0$ since both integrals exist, provided that for the second one $2k > \kappa + n$. Similar to the previous analysis, one can easily show that $\widehat{f}_\kappa(x,\xi) = O(|\xi|^m)$ as $|\xi| \to 0$. Moreover, with the identity

$$\widehat{\varphi\Big(\frac{\bullet}{t}\Big)}(z) = t^n \widehat{\varphi}(tz) \quad \text{for} \quad t > 0$$

we obtain

$$\begin{aligned}\int\limits_{\mathbb{R}^n} \widehat{f}_\kappa(x,\xi)\varphi\Big(\frac{\xi}{t}\Big)d\xi &= t^n \int\limits_{\mathbb{R}^n} f_\kappa(x,z)\widehat{\varphi}(tz)dz \\ = t^{n+m}\int\limits_{\mathbb{R}^n} f_\kappa(x,z)\widehat{\varphi}(z)dz &= t^{m+n}\int\limits_{\mathbb{R}^n} \widehat{f}_\kappa(x,\xi)\varphi(\xi)d\xi\end{aligned} \tag{8.1.34}$$

for all $\varphi \in C_0^\infty(\mathbb{R}^n \setminus \{0\})$ and $t > 0$, $m = -\kappa - n$. According to the definition of homogeneity of generalized functions in Hörmander [188, Vol.I, Definition 3.2.2], Equation (8.1.34) implies that $\widehat{f}_\kappa(x,\xi)$ is a positively homogeneous function of degree m for $\xi \neq 0$. To extend $\widehat{f}_\kappa(x,\xi)$ as a distribution on $C_0^\infty(\mathbb{R}^n)$, we use the canonical extension through the finite part integral

$$\text{p.f.} \int_{\mathbb{R}^n} \widehat{f}_\kappa(x,\xi)\varphi(\xi)d\xi \tag{8.1.35}$$

$$= \int_{|\xi|<R} \widehat{f}_\kappa(x,\xi)\Big\{\varphi(\xi) - \sum_{|\alpha|\le\kappa} \frac{1}{\alpha!}\xi^\alpha \frac{\partial^\alpha\varphi}{\partial\xi^\alpha}(0)\Big\}d\xi$$

$$- \sum_{|\alpha|<\kappa} \frac{1}{\kappa-|\alpha|}\frac{1}{\alpha!} \int_{|\theta|=1} R^{-\kappa+|\alpha|}\,\theta^\alpha \widehat{f}_\kappa(x,\theta)d\omega(\theta)\frac{\partial^\alpha\varphi}{\partial\xi^\alpha}(0)$$

$$+ \sum_{|\alpha|=\kappa} \frac{1}{\alpha!} \int_{|\theta|=1} \log R\, \theta^\alpha \widehat{f}_\kappa(x,\theta)d\omega(\theta)\frac{\partial^\alpha\varphi}{\partial\xi^\alpha}(0) + \int_{|\xi|\ge R} \widehat{f}_\kappa(x,\xi)\varphi(\xi)d\xi$$

for $\varphi \in C_0^\infty(\mathbb{R}^n)$ and any $R > 0$.

If the logarithmic term is present, the canonical extension is not homogeneous anymore, but only pseudohomogeneous.

For the term $p_\kappa(x,z)\log|z|$, the Fourier transform can be obtained explicitly (see Schwartz [382, VII. 7.16]),

$$\widehat{p_\kappa(x,\cdot)\log|\cdot|} = \sum_{|\gamma|=\kappa} c_\gamma(x)\left(i\frac{\partial}{\partial\xi}\right)^\gamma \left\{C_n\frac{1}{|\xi|^n} + C_n'\delta_0(\xi)\right\}$$

where $p_\kappa(x,z) = \sum_{|\gamma|=\kappa} c_\gamma(x)z^\gamma$, where $\delta_0(\xi)$ is the Dirac functional and where C_n and C_n' are universal constants. Here, the first terms $\left(i\frac{\partial}{\partial\xi}\right)^\gamma \frac{1}{|\xi|^n}$ are homogeneous functions and, again, we use the canonical extension

$$\int_{\mathbb{R}^n} p_\kappa(x,z)\log|z|\widehat{\varphi}(z)dz \tag{8.1.36}$$

$$= \text{p.f.} \int_{\mathbb{R}^n} \sum_{|\gamma|=\kappa} c_\gamma(x)\left(\left(\frac{\partial}{\partial\xi}\right)^\gamma \frac{C_n}{|\xi|^n}\right)\varphi(\xi)d\xi + i^\kappa C_n' \sum_{|\gamma|=\kappa} c_\gamma(x)\frac{\partial\varphi^\gamma}{\partial\xi^\gamma}(0)$$

where the finite part integral takes a form corresponding to (8.1.35).

Summing (8.1.35) and (8.1.36) yields the representation of the Fourier transform of the kernel $k_\kappa(x,z)$ in the sense of distributions, i. e., $a_m(x,\xi) = \widehat{k}_\kappa(x,\xi)$.

To test the homogeneity of the distribution given by $\widehat{k}_\kappa(x,\xi)$, we choose $R = t$ in (8.1.35) and (8.1.36) to compute p.f. $\int_{\mathbb{R}^n} \widehat{k}_\kappa(x,\xi)\varphi\left(\frac{\xi}{t}\right)d\xi$. Then, by setting $t = 1$ in the latter we get p.f. $\int_{\mathbb{R}^n} \widehat{k}_\kappa(x,\xi)\varphi(\xi)d\xi$. Hence, we obtain

$$\begin{aligned}\langle \widehat{k}_\kappa(x,\cdot),\varphi\rangle &= \text{p.f.}\int\limits_{\mathbb{R}^n} \widehat{k}_\kappa(x,\xi)\varphi(\xi)d\xi + i^\kappa C_n' \sum_{|\gamma|=\kappa} c_\gamma(x)\frac{\partial\varphi^\gamma}{\partial\xi^\gamma}(0) \qquad (8.1.37)\\
&= t^{-m-n}\Big\{\text{p.f.}\int\limits_{\mathbb{R}^n} \widehat{k}_\kappa(x,\xi)\varphi\Big(\frac{\xi}{t}\Big)d\xi\\
&\quad + i^\kappa C_n' \sum_{|\gamma|=\kappa} c_\gamma(x)\Big(\Big(\frac{\partial}{\partial\xi}\Big)^\gamma\varphi\Big(\Big(\frac{\xi}{t}\Big)\Big)|_{\xi=0}\Big\}\\
&\quad - \log t \sum_{|\gamma|=\kappa}\frac{1}{\alpha!}\int\limits_{|\theta|=1}\theta^\alpha\Big(\widehat{f}_\kappa(x,\theta)\\
&\quad + \sum_{|\gamma|=\kappa} c_\gamma(x)\Big(\Big(i\frac{\partial}{\partial\xi}\Big)^\gamma\frac{C_n}{|\xi|^n}\Big)|_{\xi=0}\Big)d\omega(\theta)\frac{\partial^\alpha\varphi}{\partial\xi^\alpha}(0)\,.\end{aligned}$$

From Equation (8.1.37) we see, that the Fourier transform $\widehat{k}_\kappa(x,\xi)$ defines a pseudohomogeneous distribution which becomes homogeneous only if the following compatibility condition is fulfilled:

$$\int\limits_{|\theta|=1}\theta^\alpha\Big(\widehat{f}_\kappa(x,\theta) + \sum_{|\gamma|=\kappa} c_\gamma(x)\Big(\Big(i\frac{\partial}{\partial\xi}\Big)^\gamma\frac{C_n}{|\xi|^n}\Big)|_{\xi=\theta}\Big)d\omega(\theta) = 0\,. \qquad (8.1.38)$$

On the other hand, as we shall see, if we have $k_\kappa \in \psi h f_\kappa$, the corresponding Fourier transformed $\widehat{k}_\kappa(x,\xi) = a(x,\xi)$ always defines the homogeneous symbol of order m. We will return to this remark later on in Theorem 8.1.6.

Lemma 8.1.5. *Let $a_m(x,\xi) \in C^\infty\big(\Omega\times(\mathbb{R}^n\setminus\{0\})\big)$ be a positively homogeneous function of degree $m < -n$; i. e.,*

$$a_m(x,t\xi) = t^m a_m(x,\xi) \quad \textit{for all } x\in\Omega\,,\ t>0\,,\ \xi\neq 0\,.$$

Then

$$k_{-m-n}(x,z) := (2\pi)^{-n}\ \text{p.f.}\int\limits_{\mathbb{R}^n} e^{iz\cdot\xi}a_m(x,\xi)d\xi \qquad (8.1.39)$$

exists and defines a positively pseudohomogeneous kernel function of degree $-m-n$.

Proof: Let $\chi(\xi)$ be a C^∞ cut–off function with $\chi(\xi) = 1$ for $|\xi| \geq 1$ and $\chi(\xi) = 0$ for $|\xi| \leq \frac{1}{2}$. Then each of the following integrals,

$$\begin{aligned}k(x,z) &= (2\pi)^{-n}\int_{\mathbb{R}^n} e^{iz\cdot\xi}a_m(x,\xi)\chi(\xi)d\xi\\
&\quad + (2\pi)^{-n}\ \text{p.f.}\int_{\mathbb{R}^n} e^{iz\cdot\xi}\big(1-\chi(\xi)\big)a_m(x,\xi)d\xi\,,\end{aligned}$$

exists since the first has integrable integrand because of $m < -n$ and the second one is defined by the Fourier transform of a distribution with compact support, therefore it is a $C^\infty(\Omega \times \mathbb{R}^n)$–function according to the Paley–Wiener–Schwartz theorem, Theorem 3.1.3. Hence, for x fixed, $k(x,z)$ is a Fourier transformed positively homogeneous distribution.

We now show that $k \in C^\infty(\Omega \times \mathbb{R}^n \setminus \{0\})$. For $z \neq 0$ one finds with (7.1.11)

$$\Big(-i\frac{\partial}{\partial z}\Big)^\alpha k(x,z) = (2\pi)^{-n}|z|^{-2k} \int\limits_{\mathbb{R}^n \cap \{|\xi| \geq \frac{1}{2}\}} e^{iz\cdot\xi}(\Delta_\xi)^k\big(\xi^\alpha a_m(x,\xi)\chi(\xi)\big)d\xi$$
$$+(2\pi)^{-n}\Big(-i\frac{\partial}{\partial z}\Big)^\alpha \int\limits_{\mathbb{R}^n \cap \{|\xi| \leq 1\}} e^{iz\cdot\xi}\big(1-\chi(\xi)\big)a_m(x,\xi)d\xi\,. \tag{8.1.40}$$

The first integral on the right–hand side is continuous if $|\alpha| < -n - m + 2k$ for any chosen $k \in \mathbb{N}_0$ whereas the second integral is $C^\infty(\Omega \times \mathbb{R}^n)$ for any $\alpha \geq 0$. Hence, $k(x,z) \in C^\infty(\Omega \times \mathbb{R}^n \setminus \{0\})$.

Since, in general, the Fourier transform of a positively homogeneous distribution is not necessarily positively homogeneous, we consider first the derivatives

$$k^{(\alpha)}(x,z) := \Big(-i\frac{\partial}{\partial z}\Big)^\alpha k(x,z)$$

given by (8.1.40) with $|\alpha| > -m - n$ and k chosen with $2k > |\alpha| + n + m$. For these α one will see that $k^{(\alpha)}$ is positively homogeneous of degree $-m-n-|\alpha|$. For this purpose we use (7.1.11) and begin with the limit

$$\lim_{t\to 0+} t^{m+n+|\alpha|}k^{(\alpha)}(x,tz)$$
$$= (2\pi)^{-n}\lim_{t\to 0+} t^{m+n+|\alpha|}\Big\{|tz|^{-2k}\int\limits_{\mathbb{R}^n \cap \{|\xi| \geq \frac{1}{2}\}} e^{itz\cdot\xi}(\Delta_\xi)^k\big(\xi^\alpha a_m(x,\xi)\chi(\xi)\big)d\xi$$
$$+\int\limits_{\mathbb{R}^n \cap \{|\xi| \leq 1\}} e^{itz\cdot\xi}\xi^\alpha\big(1-\chi(\xi)\big)a_m(x,\xi)d\xi\Big\}$$
$$= (2\pi)^{-n}\lim_{t\to 0+}\Big\{|z|^{-2k}\int\limits_{|\xi'| \geq t\frac{1}{2}} e^{iz\cdot\xi'}(\Delta_{\xi'})^k\Big(\xi'^\alpha a_m(x,\xi')\chi\Big(\frac{\xi'}{t}\Big)\,d\xi'$$
$$+\int\limits_{|\xi'| \leq t} e^{iz\cdot\xi'}\xi'^\alpha\Big(1-\chi\Big(\frac{\xi'}{t}\Big)\Big)a_m(x,\xi')d\xi'\Big\},$$

where the transformation $\xi' = t\xi$ and the positive homogeneity of $a_m(x,\xi)$ have been employed. Since $|\alpha| > -m-n$, the integrand of the second integral on the right–hand side is continuous, uniformly bounded and tends to zero as $t \to 0+$. The first term on the right–hand side can be rewritten as follows:

$$\begin{aligned}
&\lim_{t\to 0+} t^{m+n+|\alpha|}k^{(\alpha)}(x,tz)\\
&= (2\pi)^{-n}|z|^{-2k}\int\limits_{\frac{1}{2}\le|\xi'|} e^{iz\cdot\xi'}(\Delta_{\xi'})^k(\xi'^{\alpha}a_m(x,\xi')\chi(\xi'))d\xi'\\
&+(2\pi)^{-n}\lim_{t\to 0+}|z|^{-2k}\int\limits_{\frac{1}{2}t\le|\xi'|\le 1} e^{iz\cdot\xi'}(\Delta_{\xi'})^k\Big(\xi^{\alpha'}a_m(x,\xi')\Big(\chi\Big(\frac{\xi'}{t}\Big)\Big)\chi(\xi')\Big)d\xi'\\
&= (2\pi)^{-n}|z|^{-2k}\int\limits_{\frac{1}{2}\le|\xi|} e^{iz\cdot\xi}(\Delta_{\xi})^k(\xi^{\alpha}a(x,\xi)\chi(\xi)d\xi\\
&\quad +(2\pi)^{-n}\int\limits_{|\xi|\le 1} e^{iz\cdot\xi}\xi^{\alpha}a(x,\xi)(1-\chi(\xi))d\xi\\
&= k^{(\alpha)}(x,z)\,,
\end{aligned}$$

since the second term has a continuous, uniformly bounded integrand which tends to the corresponding integrand in (8.1.40). Thus,

$$\lim_{t\to 0+} t^{m+n+|\alpha|}k^{(\alpha)}(x,tz) = k^{(\alpha)}(x,z)\,. \tag{8.1.41}$$

Equation (8.1.41) implies that $k^{(\alpha)}(x,z)\in \Psi hf_{\kappa-|\alpha|}$ is positively homogeneous with $\kappa=-m-n$. Hence, for $n\ge 2$ it follows from Lemma 8.1.3 part (ii) that $k(x,z)-p_{|\alpha|-1}(z)\in\Psi hf_{-m-n}$ where $p_{|\alpha|-1}(x,z)$ is some polynomial of degree less than $|\alpha|$.

In what follows, for $n\ge 2$ we shall now present the explicit form of the kernel

$$k(x,z) = f_{-m-n}(x,z) + p_{-m-n}(x,z)\log|z|$$

which shows that $p_{|\alpha|-1}=0$. From Definition 8.1.2 applied to (8.1.39) we find with $\Omega_0=\{\xi\in\mathbb{R}^n\,|\,|\xi|<R\}$ for $\kappa=-m-n>0$,

$$\begin{aligned}
k(x,z) = (2\pi)^{-n}\Big\{&\int\limits_{|\xi|<R} a_m(x,\xi)\Big\{e^{i\xi\cdot z}-\sum_{|\alpha|\le\kappa}\frac{1}{\alpha!}(iz)^{\alpha}\xi^{\alpha}\Big\}d\xi \\
&-\sum_{|\alpha|<\kappa}\frac{1}{\kappa-|\alpha|}\frac{1}{\alpha!}\int\limits_{|\theta|=1} R^{-\kappa+|\alpha|}\theta^{\alpha}a_m(x,\theta)d\omega(\theta)(iz)^{\alpha}\\
&+\sum_{|\alpha|=\kappa}\frac{1}{\alpha!}\int\limits_{|\theta|=1}(\log R)\,\theta^{\alpha}a_m(x,\theta)d\omega(\theta)(iz)^{\alpha}+\int\limits_{|\xi|\ge R} e^{i\xi\cdot z}a_m(x,\xi)d\xi\Big\}\,.
\end{aligned} \tag{8.1.42}$$

Here $R>0$ can be chosen arbitrarily. The dilatation of the kernel with $t>0$ reads after proper change of variables,

$$\begin{aligned}
k(x,tz) &= (2\pi)^{-n}\Bigg\{t^\kappa \int\limits_{|\xi|<tR} a_m(x,\xi)\Big\{e^{i\xi\cdot z} - \sum_{|\alpha|\le\kappa}\frac{1}{\alpha!}(iz)^\alpha\xi^\alpha\Big\}d\xi \\
&\quad - \sum_{|\alpha|<\kappa}\frac{1}{\kappa-|\alpha|}\frac{1}{\alpha!}R^{-\kappa+|\alpha|}\,t^{|\alpha|}\int\limits_{|\theta|=1}\theta^\alpha a_m(x,\theta)d\omega(\theta)(iz)^\alpha \\
&\quad + \sum_{|\alpha|=\kappa}\frac{1}{\alpha!}t^\kappa\log R\int\limits_{|\theta|=1}\theta^\alpha a_m(x,\theta)d\omega(\theta)(iz)^\alpha \\
&\quad + t^\kappa\int\limits_{|\xi|\ge tR} e^{i\xi\cdot z}a_m(x,\xi)d\xi\Bigg\}\,. \qquad (8.1.43)
\end{aligned}$$

Comparing (8.1.43) with (8.1.42) suggests the choice of $R=\frac{1}{t}$ which yields

$$k(x,tz) = t^{-m-n}k(x,z) + t^{-m-n}\log t\,p_\kappa(x,z) \qquad (8.1.44)$$

where

$$p_\kappa(x,z) := -\sum_{|\alpha|=\kappa}\frac{(2\pi)^{-n}}{\alpha!}\int\limits_{|\theta|=1}\theta^\alpha a_m(x,\theta)d\omega(\theta)(iz)^\alpha\,. \qquad (8.1.45)$$

Then we obtain

$$k(x,z) = t^{m+n}k(x,tz) - \log t\,p_\kappa(x,z)$$

for any $t>0$. Hence, the choice $t=1/|z|$ gives

$$k(x,z) = |z|^{-m-n}k\left(x,\frac{z}{|z|}\right) + \log|z|\cdot p_{-m-n}(x,z)\,. \qquad (8.1.46)$$

This is a pseudohomogeneous function with the positively homogeneous part

$$\begin{aligned}
&f_{-m-n}(x,z) := |z|^{-m-n}(2\pi)^{-n} \qquad (8.1.47)\\
&\Big\{\int\limits_{|\xi|<1} a_m(x,\xi)\Big\{e^{i\Theta\cdot\xi} - \sum_{|\alpha|\le\kappa}\frac{1}{\alpha!}(i\Theta)^\alpha\xi^\alpha\Big\}d\xi \\
&-\sum_{|\alpha|<\kappa}\frac{1}{\kappa-|\alpha|}\frac{1}{\alpha!}\int\limits_{|\vartheta|=1}\vartheta^\alpha a_m(x,\vartheta)d\omega(\vartheta)(i\Theta)^\alpha + \int\limits_{|\xi|\ge1}e^{i\xi\cdot\Theta}a_m(x,\xi)d\xi\Big\}
\end{aligned}$$

where $\Theta = \dfrac{z}{|z|}$, and the homogeneous polynomial is given by (8.1.45) with $\kappa=-m-n$.

If $n=1$, then Lemma 8.1.3 part (ii) implies that $k(x,z)-p^{\pm}_{|\alpha|-1}(z)\in \Psi hf_{m-1}$. On the other hand,

$$k(x,z) = (2\pi)^{-1} \int\limits_{\mathbb{R}^1} e^{iz\xi} a_m^0(x,\xi)d\xi$$

$$= (2\pi)^{-1}\Big\{ \int\limits_0^\infty e^{iz\xi} a_m^0(x,1)\xi^m d\xi + \int\limits_0^\infty e^{-iz\xi} a_m^0(x,-1)\xi^m d\xi \Big\}$$

and the one–dimensional Fourier transform of ξ_+^m in the book by Gelfand and Shilov [143, p. 360] provides us with the explicit formula

$$k(x,z) = |z|^{-m-1} i\Gamma(m+1)\big\{ e^{im\frac{\pi}{2}} a_m^0(x, \operatorname{sign} z) - e^{-im\frac{\pi}{2}} a_m^0(x, -\operatorname{sign} z)\big\}\,.$$

Hence, the polynomials $p^{\pm}_{|\alpha|-1}(x,z) \equiv 0$ as in the cases for $n \geq 2$. This completes the proof of Lemma 8.1.5. ∎

Now we are in the position to prove Theorem 8.1.1.

Proof of Theorem 8.1.1: (i) We begin with $A \in \mathcal{L}^m_{c\ell}(\Omega)$ given, provided $m < -n$. Then there exists a properly supported operator A_m of order m such that $A - A_m$ has a Schwartz kernel in $C^\infty(\Omega \times \mathbb{R}^n)$; and the symbol $a(x,\xi)$ of A_m is given by

$$a(x,\xi) = e^{-ix\cdot\xi} A_m e^{i\xi\cdot\bullet} \in \boldsymbol{S}^m_{c\ell}(\Omega \times \mathbb{R}^n)\,.$$

With this symbol, we define

$$k(x,z) := (2\pi)^{-n} \text{ p.f.} \int\limits_{\mathbb{R}^n} e^{iz\cdot\xi} a(x,\xi) d\xi\,. \tag{8.1.48}$$

In order to find the pseudohomogeneous expansion of k, we use the asymptotic expansion of the symbol

$$a(x,\xi) \sim \sum_{j\geq 0} a_{m-j}(x,\xi)$$

where each $a_{m-j} \in C^\infty(\Omega \times \mathbb{R}^n \setminus \{0\})$ is a positively homogeneous function of degree $m - j < -n$. To each a_{m-j}, we define the kernel

$$k_{-m-n+j}(x,z) := (2\pi)^{-n} \text{ p.f.} \int\limits_{\mathbb{R}^n} e^{i\xi\cdot z} a_{m-j}(x,\xi) d\xi \tag{8.1.49}$$

which is positively homogeneous of degree $-m-n+j$ due to Lemma 8.1.5. Now we show that the remainder

$$k_J(x,z) := k(x,z) - \sum_{j<J} k_{-m-n+j}(x,z) \in C^\ell(\Omega \times \mathbb{R}^n) \tag{8.1.50}$$

for $\ell < J - m - n$. By definition, we then have

$$
\begin{aligned}
k_J(x,z) &= (2\pi)^{-n}\,\text{p.f.}\int\limits_{\mathbb{R}^n} e^{i\xi\cdot z}\Big(a(x,\xi)-\sum_{j<J}a_{m-j}(x,\xi)\Big)d\xi \\
&= (2\pi)^{-n}\,\text{p.f.}\int\limits_{|\xi|\le 1} e^{i\xi\cdot z}\Big(1-\chi(\xi)\Big)\Big(a(x,\xi)-\sum_{j<J}a_{m-j}(x,\xi)\Big)d\xi \\
&\quad + (2\pi)^{-n}\,\text{p.f.}\int\limits_{|\xi|\ge \frac12} e^{i\xi\cdot z}\Big(a(x,\xi)-\sum_{j<J}a_{m-j}(x,\xi)\Big)\chi(\xi)d\xi
\end{aligned}
$$

where $\chi(\xi)$ is a C^∞ cut–off–function with $0 \le \chi(\xi) \le 1$, $\chi(\xi) = 0$ for $|\xi| \le \frac12$ and $\chi(\xi) = 1$ for $|\xi| \ge 1$. The first integral on the right–hand side is a distribution with compact support defining a function $k_{J1}(x,z) \in C^\infty(\Omega\times\mathbb{R}^n)$ due to Theorem 3.1.3. The second integral $k_{J2}(x,z)$ has an integrable integrand since it decays as $|\xi|^{m-J}$. Moreover, we may differentiate and obtain

$$
\Big(-i\frac{\partial}{\partial z}\Big)^\alpha k_{J2}(x,z) = (2\pi)^{-n}\int\limits_{|\xi|\ge\frac12} e^{i\xi\cdot z}\xi^\alpha\Big(a(x,\xi)-\sum_{j<J}a_{m-j}(x,\xi)\Big)\chi(\xi)d\xi\,,
$$

which implies $\big(-i\frac{\partial}{\partial z}\big)^\alpha k_{J2}(x,z) \in C^0(\Omega\times\mathbb{R}^n)$ for $m-J+|\alpha| < -n$, i. e. (8.1.50). This shows that $k(x,z)\in\Psi hk_\kappa$ with $\kappa = -m-n$.

It remains to show that

$$
(A_m u)(x) = \int\limits_\Omega k(x,x-y)u(y)dy \quad \text{for} \quad u\in C_0^\infty(\Omega) \quad \text{and} \quad x\in\Omega
$$

with the kernel $k(x,z)$ defined by (8.1.48). Collecting the terms in the above expansion, from (8.1.50) and Lemma 8.1.5 we have

$$
\begin{aligned}
(A_m u)(x) &= (2\pi)^{-n}\int\limits_{\mathbb{R}^n}\int\limits_\Omega e^{i(x-y)\cdot\xi}a(x,\xi)u(y)dyd\xi \\
&= (2\pi)^{-n}\int\limits_{\mathbb{R}^n}\int\limits_\Omega e^{i(x-y)\cdot\xi}\Big(a(x,\xi)-\sum_{j<J}a_{m-j}(x,\xi)\Big)u(y)dyd\xi \\
&\quad +\sum_{j<J}\int\limits_\Omega k_{\kappa+j}(x,x-y)u(y)dy \\
&= \int\limits_\Omega\Big(k_J(x,x-y)+\sum_{j<J}k_{\kappa+j}(x,x-y)\Big)u(y)dy \\
&= \int\limits_\Omega k(x,x-y)u(y)dy\,,
\end{aligned}
$$

i. e., the proposed representation:

$$(Au)(x) = \int\limits_{\Omega} k(x, x-y)u(y)dy + \int\limits_{\Omega} r(x,y)u(y)dy$$

where $k \in \Psi hk_\kappa$ and $r \in C^\infty(\Omega \times \Omega)$.

(ii) For $-n \leq m < 0$ we choose $\ell \in \mathbb{N}$ with $2\ell > m+n$ and define $k_b(x,z)$ for the modified symbol $a(x,\xi)|\xi|^{-2\ell} =: b(x,\xi)$. Then one has

$$k_b(x,z) = (2\pi)^{-n} \int\limits_{\mathbb{R}^n} e^{i\xi\cdot z} a(x,\xi)|\xi|^{-2\ell} d\xi$$

with $k_b \in \Psi hk_{-m-n+2\ell}$ from the previous case $m - 2\ell < -n$. With the distributional derivatives, this implies

$$k(x,z) = (-\Delta_z)^\ell k_b(x,z) = (2\pi)^{-n} \int\limits_{\mathbb{R}^n} e^{i\xi\cdot z} a(x,\xi) d\xi \tag{8.1.51}$$

with $k \in \Psi hk_{-m-n}$ as proposed. The rest of the proof follows as in case **(i)**. Hence, we have shown, if $A \in \mathcal{L}^m(\Omega)$, then A admits the representation (8.1.10).

Conversely, let $k \in \psi hk_\kappa(\Omega)$ with $\kappa = -m-n$ and $m < 0$ be given. Then k admits an asymptotic expansion into pseudohomogeneous functions $k_{\kappa-j} \in \Psi kf_\kappa$ for $j \in \mathbb{N}_0$, see (8.1.2). We define

$$a_\psi(x,\xi) := \int\limits_{|z|\leq R} e^{-iz\cdot\xi} k(x,z)\psi(z)dz\,,$$

where the cut–off function ψ is defined as in Lemma 8.1.4; here we choose $R = 1$ and $\varepsilon = \frac{1}{2}$. Now apply Lemma 8.1.4 to each $k_{\kappa+j} \in \Psi hf_{\kappa+j}$. Then the operators $K_{\kappa+j}$ defined by

$$(K_{\kappa+j}u)(x) = \int\limits_{\Omega} k_{\kappa+j}(x, x-y)u(y)dy$$

belong to $\mathcal{L}^{m-j}_{c\ell}(\Omega)$ and each of them has the positively homogeneous symbol $a_{m-j}(x,\xi)$ defined by (8.1.29). Correspondingly, the operators defined by

$$\begin{aligned}(\widetilde{K}_{\kappa+j}u)(x) &= \int\limits_{\Omega} k_{\kappa+j}(x,x-y)\psi(x-y)u(y)dy \\ &= (K_{\kappa+j}u)(x) + \int\limits_{\Omega} k_{\kappa+j}(x,x-y)\big(\psi(x-y)-1\big)dy\end{aligned}$$

also belong to $\mathcal{L}^{m-j}_{c\ell}(\Omega)$ with the same positively homogeneous representative $a_{m-j}(x,\xi)$ of the complete symbol class $\widetilde{a}_{m-j}$, since the second integral operator on the right–hand side has a $C^\infty(\Omega\times\Omega)$ kernel. Now,

$$
\begin{aligned}
R_a(x,\xi) &:= \xi^\gamma \left(\frac{\partial}{\partial x}\right)^\beta \left(\frac{\partial}{\partial \xi}\right)^\alpha \Big(a_\psi(x,\xi) - \sum_{j<J} \widetilde{a}_{m-j}(x,\xi)\Big) \\
&= \xi^\gamma \int\limits_{|z|\le 1} e^{-iz\cdot\xi}(-iz)^\alpha \Big(\frac{\partial^\beta k(x,z)}{\partial x^\beta} - \sum_{j<J} \frac{\partial^\beta k_{\kappa+j}(x,z)}{\partial x^\beta}\Big)\psi(z)dz \\
&= \int\limits_{|z|\le 1} e^{-iz\cdot\xi} \left(-i\frac{\partial}{\partial z}\right)^\gamma \Big\{(-iz)^\alpha \frac{\partial^\beta}{\partial x^\beta}\Big(k_{\kappa+J}(x,z) + \Re_{\kappa+J+1}(x,z)\Big)\psi(z)\Big\}dz
\end{aligned}
$$

where $k_{\kappa+J}$ is the next pseudohomogeneous function and $\Re_{\kappa+J+1}$ the corresponding remainder in the asymptotic kernel expansion in (8.1.2). Due to the Paley–Wiener–Schwartz theorem 3.1.3, the function $R_a(x,\xi)$ is in $C^\infty(\Omega \times \mathrm{I\!R}^n)$. Since $|\gamma - \alpha| = J - m$, we have

$$
R_a(x,\xi) = (2\pi)^{n/2}\mathcal{F}_{z\mapsto\xi} k_{-n}^{(\alpha,\beta,\gamma)}(x,z) + \int\limits_{|z|\le 1} e^{-iz\cdot\xi}O(|z|^{-n+1})dz + c_a(x,\xi)
$$

where

$$
k_{-n}^{(\alpha,\beta,\gamma)}(x,z) = (-i)^{|\gamma|+|\alpha|}\left(\frac{\partial^\beta}{\partial x^\beta}\right)^\beta \left(\frac{\partial}{\partial z}\right)^\gamma z^\alpha k_{\kappa+y}(x,z)
$$

is a homogeneous function of degree $-n$ and, hence, of the form (8.1.1). The term $c_a(x,\xi)$ can be estimated in the same manner as in the proof of Lemma 8.1.2, namely

$$
|\xi^\delta c_a(x,\xi)| \le c_{K,\delta} \quad \text{for any } |\delta| > 1 \quad \text{for } x \in K \Subset \Omega
$$

with any compact K. The second term in R_a is uniformly bounded since it has a weakly singular kernel. Lemma 8.1.2 implies the estimates

$$
|R_a(x,\xi)| \le c_{1K} + c_{2K}\log|\xi| \quad \text{for } |\xi| \ge 1 \quad \text{and } x \in K
$$

with any compact set $K \Subset \Omega$. Consequently, we obtain estimates

$$
\begin{aligned}
&\Big|\left(\frac{\partial}{\partial x}\right)^\beta \left(\frac{\partial}{\partial \xi}\right)^\alpha \Big(a_\psi(x,\xi) - \sum_{j<J}\widetilde{a}_{m-j}(x,\xi)\Big)\Big| \\
&\le \Big|\left(\frac{\partial}{\partial x}\right)^\beta \left(\frac{\partial}{\partial \xi}\right)^\alpha \widetilde{a}_{m-J}(x,\xi)\Big| \\
&\le \big(c_1(\alpha,\beta,K) + c_2(\alpha,\beta,K)\log|\xi|\big)\langle\xi\rangle^{m-J-|\alpha|-1} + c_3(\alpha,\beta,K)\langle\xi\rangle^{m-J-|\alpha|} \\
&\le c_4(\alpha,\beta,K)\langle\xi\rangle^{m-J-|\alpha|} \quad \text{for all } |\xi| \ge 1\,,\ x \in K\,.
\end{aligned}
$$

Hence,

$$
a_\psi(x,\xi) \sim \sum \widetilde{a}_{m-j}(x,\xi) \sim \sum a_{m-j}(x,\xi)
$$

as claimed. This completes the proof of Theorem 8.1.1. ■

8.1.2 Non–Negative Order Pseudodifferential Operators as Hadamard Finite Part Integral Operators

We now consider the case $m \geq 0$. We begin with $0 < m \notin \mathbb{N}$ and consider a positively homogeneous kernel $f_\kappa(x,z)$ of degree $\kappa = -m-n$.

Theorem 8.1.6. *Let $m > 0$, $m \notin \mathbb{N}$ and $\kappa = -m-n$. Then every positively homogeneous function $f_\kappa(x,z)$ of degree κ, i.e. $f_\kappa \in \Psi h f_\kappa$ defines via*

$$(A_m u) := \text{p.f.} \int\limits_\Omega f_\kappa(x, x-y)u(y)dy \quad \textit{for } u \in C_0^\infty(\Omega) \tag{8.1.52}$$

a classical pseudodifferential operator $A_m \in \mathcal{L}^m_{c\ell}(\Omega)$. The complete symbol class σ_{A_m} is given by the positively homogeneous representative $a_m(x,\xi)$ of order m defined by

$$a_m(x,\xi) = \text{p.f.} \int\limits_{\mathbb{R}^n} f_\kappa(x,z)e^{-iz\cdot\xi}dz\,. \tag{8.1.53}$$

Proof: The function $a_m(x,\xi)$ defined by (8.1.53) is well defined for all $\xi \in \mathbb{R}^n$ due to (8.1.4) with $u(z) = e^{-iz\cdot\xi}$. The positive homogeneity of $a_m(x,t\xi) = t^m a_m(x,\xi)$ for $t > 0$ and $\xi \neq 0$ follows immediately also from (8.1.4):

$$\begin{aligned} a_m(x,t\xi) &= \Big\{ \int\limits_{|z|\leq R} f_\kappa(x,z)\Big\{e^{-iz\cdot t\xi} - \sum_{|\alpha|<m} \frac{1}{\alpha!} z^\alpha(-it\xi)^\alpha\Big\}dz \\ &\quad + \sum_{|\alpha|<m} \frac{1}{m-|\alpha|}\frac{1}{\alpha!} \int\limits_{|\vartheta|=1} R^{-m+|\alpha|}\vartheta^\alpha f_\kappa(x,\vartheta)d\omega(\vartheta)(-it\xi)^\alpha \\ &\quad + \int\limits_{|z|\geq R} e^{-iz\cdot t\xi} f_\kappa(x,z)dz\Big\} \end{aligned} \tag{8.1.54}$$

where new coordinates $z' = zt$ and the choice of $R = t^{-1}$ provides us with the desired identity.

Now, write

$$\begin{aligned} (A_m u)(x) &= \text{p.f.} \int\limits_\Omega f_\kappa(x,x-y)\psi(x-y)u(y)dy \\ &\quad + \int\limits_\Omega f_\kappa(x,x-y)\big(1-\psi(x-y)\big)u(y)dy\,. \end{aligned}$$

Since the kernel of the second integral operator is in $C^\infty(\Omega\times\Omega)$, it defines a smoothing operator due to Theorem 7.1.10 (iv). The kernel of the first operator $f_\kappa(x,z)\psi(z)$, for fixed $x \in \Omega$, has compact support. Therefore, the symbol

$$a(x,\xi) := \text{ p.f.} \int\limits_{\mathbb{R}^n} e^{-iz\cdot\xi} f_\kappa(x,z)\psi(z)dz$$

is in $C^\infty(\Omega \times \mathbb{R}^n)$ due to the Paley–Wiener–Schwartz theorem 3.1.3. Moreover, for $|\xi| \geq 1$ we find

$$\begin{aligned}
&\Big|\xi^\gamma \Big(\frac{\partial}{\partial x}\Big)^\beta \Big(\frac{\partial}{\partial \xi}\Big)^\alpha \big(a(x,\xi) - a_m(x,\xi)\big)\Big| \\
&= |\xi|^{-2k} \Big| \int\limits_{|z|\geq 1} e^{-iz\cdot\xi}(-\Delta_z)^k \Big(-i\frac{\partial}{\partial z}\Big)^\gamma (-iz)^\alpha \Big(\frac{\partial}{\partial x}\Big)^\beta \{f_\kappa(x,z)\big(\psi(z)-1\big)\}dz\Big| \\
&\leq c(\alpha,\beta,\gamma) \qquad (8.1.55)
\end{aligned}$$

for every γ and α,β with the choice of k such that $2k \geq |\alpha| - |\gamma|$ and the choice $\varepsilon = 1$, $R = 2$ for ψ (Lemma 8.1.4).

For $|\xi| \geq 1$, this estimate implies $a \sim a_m$ and the complete symbol class of A_m is given just by the homogeneous symbol (8.1.53). ■

Similarly, for $m \in \mathbb{N}_0$ and $\kappa = -m - n$ we have the following relation between f_κ and A_m.

Theorem 8.1.7. *Let $m \in \mathbb{N}_0$ and $\kappa = -m-n$. A pseudohomogeneous function $f_\kappa(x,z) \in \Psi hf_\kappa$ defines the Schwartz kernel of a pseudodifferential operator in $\mathcal{L}^m_{c\ell}(\Omega)$ if and only if the Tricomi compatibility conditions*

$$\int\limits_{|\Theta|=1} \Theta^\alpha f_\kappa(x,\Theta)d\omega(\Theta) = 0 \quad \textit{for all } |\alpha| = m \qquad (8.1.56)$$

are satisfied.

Proof: As in the case of $m \notin \mathbb{N}_0$, we write with a C^∞ cut–off function ψ as in Lemma 8.1.4,

$$\begin{aligned}
(A_m u)(x) &= \text{ p.f.} \int\limits_\Omega f_\kappa(x,x-y)\psi(x-y)u(y)dy \\
&\quad + \int\limits_\Omega f_\kappa(x,x-y)\big(1-\psi(x-y)\big)u(y)dy\,.
\end{aligned}$$

The second term defines again a smoothing operator due to Theorem 7.1.10 (iv).

The case $m = 0$ now follows from Lemma 8.1.2 where $c(x) = 0$ is just the Tricomi condition for $m = 0$. For $m \in \mathbb{N}$, as in the case $m \notin \mathbb{N}_0$, we define

$$\widehat{f}_\kappa(x,\xi) := (2\pi)^{-n/2}\,\text{p.f.}\int\limits_{\mathbb{R}^n} e^{-iz\cdot\xi} f_\kappa(x,z)dz\,.$$

Similarly, for the properly supported operator

$$A_{m0}u := \text{ p.f.}\int\limits_{\Omega} f_\kappa(x,x-y)\psi(x-y)u(y)dy$$

operating on $u \in C_0^\infty(\Omega)$, we define the symbol

$$a(x,\xi) := \text{ p.f.}\int\limits_{\mathbb{R}^n} e^{-iz\cdot\xi} f_\kappa(x,z)\psi(z)dz\,, \tag{8.1.57}$$

which is a well–defined $C^\infty(\Omega\times\mathbb{R}^n)$ function due to the Paley–Wiener–Schwartz theorem 3.1.3. Hence,

$$A_{m0}u(x) := (2\pi)^{-n/2}\,\text{p.f.}\int\limits_{\mathbb{R}^n} e^{iz\cdot\xi} a(x,\xi)\widehat{u}(\xi)d\xi\,.$$

Moreover, for every α,β,γ we find the estimate

$$\Big|\xi^\gamma\Big(\frac{\partial}{\partial x}\Big)^\beta\Big(\frac{\partial}{\partial\xi}\Big)^\alpha\big(a(x,\xi)-(2\pi)^{n/2}\widehat{f}_\kappa(x,\xi)\big)\Big| \le c(\alpha,\beta,\gamma)$$

for $|\xi|\ge 1$ in the same manner as in (8.1.55) in the case $m\notin\mathbb{N}_0$. Hence, $\widehat{f}_\kappa(x,\xi)$ represents the complete asymptotic expansion of $a(x,\xi)$. However, in order to have positive homogeneity for $\widehat{f}_\kappa$, we use the same procedure as in (8.1.42), (8.1.43) and obtain for $m\ge 1$ finally

$$\begin{aligned}\widehat{f}_\kappa(x,\xi) &= |\xi|^m \widehat{f}_\kappa\Big(x,\frac{\xi}{|\xi|}\Big) \\ &\quad -(2\pi)^{-n/2}\log|\xi|\sum_{|\alpha|=m}\frac{1}{\alpha!}\int\limits_{|\Theta|=1}\Theta^\alpha f_\kappa(x,\Theta)d\omega(\Theta)(-i\xi)^\alpha\,.\end{aligned} \tag{8.1.58}$$

This relation shows that the Tricomi condition is necessary for $\widehat{f}_\kappa(x,\xi)$ to define a homogeneous symbol. Consequently, $a(x,\xi)\in \boldsymbol{S}^m(\Omega\times\mathbb{R}^n)$ only if the Tricomi condition (8.1.56) is fulfilled. ■

Theorem 8.1.8. i) *Let $0<m\notin\mathbb{N}$ and $\kappa=-m-n$. Then $A\in\mathcal{L}^m_{c\ell}(\Omega)$ if and only if*

$$(Au)(x) = \text{ p.f.}\int\limits_{\Omega} k(x,x-y)u(y)dy \quad \textit{for all } u\in C_0^\infty(\Omega) \tag{8.1.59}$$

with the Schwartz kernel satisfying $k\in\Psi hk_\kappa(\Omega)$. The kernel function k admits the pseudohomogeneous asymptotic expansion

$$k(x, x-y) \sim \sum_{j\geq 0} k_{\kappa+j}(x, x-y)$$

with $k_{\kappa+j} \in \Psi hf_{\kappa+j}$ *for* $j \in \mathbb{N}_0$.

ii) *If* $m \in \mathbb{N}_0$ *then* $A \in \mathcal{L}^m_{c\ell}(\Omega)$ *if and only if it has the representation*

$$(Au)(x) = \sum_{|\alpha|\leq m} a_\alpha(x) D^\alpha u(x) + \text{ p.f.} \int_\Omega k(x, x-y)u(y)dy \qquad (8.1.60)$$

for all $u \in C_0^\infty(\Omega)$, *the Schwartz kernel satisfies* $k \in \Psi hk_\kappa(\Omega)$; *and, in addition, the Tricomi conditions*

$$\int_{|\widehat{\Theta}|=1} \widehat{\Theta}^\alpha k_{\kappa+j}(x, \widehat{\Theta}) d\omega(\widehat{\Theta}) = 0 \quad \textit{for all } \alpha \quad \textit{with } |\alpha| = m - j \qquad (8.1.61)$$

and all $j \in \mathbb{N}_0$ *with* $j \leq m$ *are fulfilled. The coefficients* $a_\alpha(x)$ *are given by*

$$\left\{ \begin{array}{l} a_\alpha(x) = \frac{1}{\alpha!}\Big(A((\bullet - x)^\alpha \psi(|\bullet - x|)|_x \\ \qquad - \text{ p.f.} \displaystyle\int_\Omega k(x, x-y)(y-x)^\alpha \psi(|y-x|)dy \Big) , \\ \textit{where } \psi \in C_0^\infty \textit{ is a cut–off function with } \psi(z) = 1 \textit{ for} \\ |z| \leq \varepsilon, \varepsilon > 0 \textit{ and } \psi(z) = 0 \textit{ for } |z| \geq 2\varepsilon\,. \end{array} \right. \qquad (8.1.62)$$

Proof: i) Let $A \in \mathcal{L}^m_{c\ell}(\Omega)$ be given where $m \geq 0$. Then to A there exists a properly supported operator A_m of order m such that $A - A_m$ has a $C^\infty(\Omega \times \mathbb{R}^n)$ Schwartz kernel. $A_m \in OPS^m(\Omega \times \mathbb{R}^n)$ has the symbol

$$a(x,\xi) := e^{-ix\cdot\xi}(A_m e^{i\xi\cdot\bullet})(x) \in \boldsymbol{S}^m_{c\ell}(\Omega \times \mathbb{R}^n)\,.$$

Then $a(x,\xi)$ admits a complete symbol class with positively homogeneous representatives $a^0_{m-j}(x,\xi) \in C^\infty(\Omega \times \mathbb{R}^n \setminus \{0\})$ and

$$a(x,\xi) \sim \sum_{j\geq 0} a^0_{m-j}(x,\xi)\,.$$

The kernel $k(x, x-y)$ of the operator A_m then satisfies

$$\begin{aligned} k(x,z) &= (2\pi)^{-n} \int_{\mathbb{R}^n} e^{i\xi\cdot z} a(x,\xi)\big(1 - \chi(\xi)\big) d\xi \\ &\quad + |z|^{-2k}(2\pi)^{-n} \int_{\mathbb{R}^n} e^{i\xi\cdot z}(-\Delta_\xi)^k \{a(x,\xi)\chi(\xi)\} d\xi \end{aligned}$$

provided $2k > m + n$, where $\chi(\xi)$ with $0 \leq \chi(\xi) \leq 1$ is a C^∞ cut–off function with $\chi(\xi) = 0$ for $|\xi| \leq \frac{1}{2}$ and $\chi(\xi) = 1$ for $|\xi| \geq 1$. Corresponding to the asymptotic expansion of the symbol $a(x,\xi)$, we define

$$k_{\kappa+j}(x,z) := (2\pi)^{-n/2}\mathcal{F}^{-1}_{\xi\mapsto z}a^0_{m-j}(x,\xi) - \sum_{|\alpha|\le 2k} a_{\alpha,\kappa+j}(x)(-D)^\alpha\delta(z)\,. \tag{8.1.63}$$

In the case $j \le m+n$, the kernel in (8.1.63) is given by

$$\begin{aligned} k_{\kappa+j}(x,z) &:= (2\pi)^{-n}\int\limits_{\mathbb{R}^n} e^{i\xi\cdot z}a^0_{m-j}(x,\xi)\big(1-\chi(\xi)\big)d\xi \\ &+|z|^{-2k}(2\pi)^{-n}\int\limits_{\mathbb{R}^n} e^{i\xi\cdot z}(-\Delta_\xi)^k\{a^0_{m-j}(x,\xi)\chi(\xi)\}d\xi \end{aligned} \tag{8.1.64}$$

for $2k > m+n-j$, where both integrals exist.

For the homogeneity of $k_{m-j}(x,z)$ we see that with $\xi' = t\xi$, we have

$$\begin{aligned} &k_{\kappa+j}(x,tz) \\ &= (2\pi)^{-n}\int\limits_{\mathbb{R}^n} e^{i\xi'\cdot z}a^0_{m-j}(x,\xi')t^{-m+j}\Big(1-\chi\Big(\frac{\xi'}{t}\Big)\Big)t^{-n}d\xi' \\ &\quad +|zt|^{-2k}(2\pi)^{-n}\int\limits_{\mathbb{R}^n} e^{i\xi'\cdot z}(-\Delta_{\xi'})^k t^{2k-m+j-n}\Big\{a^0_{m-j}(x,\xi')\chi\Big(\frac{\xi'}{t}\Big)\Big\}d\xi' \\ &= t^{\kappa+j}k_{\kappa+j}(x,z) \end{aligned}$$

since the cut–off function χ is arbitrary. Hence, $k_{\kappa+j} \in \Psi hf_{\kappa+j}$ for $\kappa + j < 0$, i.e. $k_{\kappa+j}$ is even positively homogeneous of degree $\kappa+j$. In the case $m \in \mathbb{N}_0$, the homogeneous functions $k_{\kappa+j}(x,x-y)$ are the kernels of the pseudodifferential operators $a^0_{m-j}(x,-iD) \in OPS^{m-j}(\Omega\times\mathbb{R}^n)$ which are of nonnegative orders if $m-j\ge 0$. Hence, for $j=0,\dots,m$, Theorem 8.1.7 implies that $k_{\kappa+j}(x,z)$ must satisfy the Tricomi conditions (8.1.61).

For $j \ge m+n$, we distinguish the subcases $m \notin \mathbb{N}_0$ and $m \in \mathbb{N}_0$. We begin with $m \notin \mathbb{N}_0$. Then $j > m+n$ and (8.1.63) reads as

$$k_{\kappa+j}(x,z) = (2\pi)^{-n}\ \text{p.f.}\int\limits_{\mathbb{R}^n} e^{i\xi\cdot z}a^0_{m-j}(x,\xi)d\xi \tag{8.1.65}$$

which clearly exists. In particular, with the homogeneity of $a^0_{m-j}(x,\xi)$, we find

$$\begin{aligned} k_{\kappa+j}(x,tz) = (2\pi)^{-n}\Big\{ &\int\limits_{|\xi|\le R} a^0_{m-j}(x,\xi)\{e^{i\xi\cdot tz} - \sum_{|\alpha|<\kappa+j}\frac{1}{\alpha!}\xi^\alpha(itz)^\alpha\}d\xi \\ &- \sum_{|\alpha|<\kappa+j}\frac{1}{\kappa+j-|\alpha|}\frac{1}{\alpha!}\int\limits_{|\Theta|=1} R^{-\kappa-j+|\alpha|}\Theta^\alpha a^0_{m-j}(x,\Theta)d\omega(\Theta)(itz)^\alpha \\ &+ \int\limits_{|\xi|\ge R} e^{i\xi\cdot tz}a^0_{m-j}(x,\xi)d\xi\Big\} \end{aligned}$$

Now set $t = 1$ and $R = 1$, which gives the form of $k_{\kappa+j}(x,z)$. Next, set $R = t^{-1}$ and change the coordinates $\xi' = t\xi$ which yields the identity

$$k_{\kappa+j}(x,tz) = t^{\kappa+j} k_{\kappa+j}(x,z)\,.$$

Again, $k_{\kappa+j} \in \psi h f_{\kappa+j}$.

For $m \in \mathbb{N}_0$, let us consider first the case $\kappa + j = 0$. Then (8.1.63) reads

$$k_0(x,z) = (2\pi)^{-n/2} \mathcal{F}^{-1}_{\xi\mapsto z} a^0_{-n}(x,\xi) - \sum_{|\alpha|\le 2k} a_{\alpha,0}(x)(-D)^\alpha \delta(z)$$

where $a^0_{-n}(x,\xi) = |\xi|^{-n} a^0_{-n}(x,\widehat{\Theta})$. In this case Lemma 8.1.2 implies

$$\begin{aligned} k_0(x,z) &= c(x)\{a_0 + a_1 \log|z|\} \\ &+ (2\pi)^{-n} \int\limits_{|\widehat{\Theta}|=1} \big(a^0_{-n}(x,\widehat{\Theta}) - c(x)\big) \left\{ \log \frac{1}{|\widehat{\Theta}\cdot z_0|} + \frac{i\pi}{2} \frac{\widehat{\Theta}\cdot z_0}{|\widehat{\Theta}\cdot z_0|} \right\} d\omega(\widehat{\Theta}) \end{aligned} \tag{8.1.66}$$

where $z_0 = z/|z|$ and $c(x) = \frac{1}{\omega_n} \int\limits_{|\widehat{\Theta}|=1} a^0_{-n}(x,\widehat{\Theta}) d\omega(\widehat{\Theta})$. The constants a_0 and a_1 are explicitly given by Lemma 8.1.2. This shows that $k_0(x,z) \in \psi h f_0(\Omega \times \mathbb{R}^n)$ and for $x \neq y$, the function $k_0(x, x-y)$ is the Schwartz kernel of $a^0_{-n}(x,-iD)$.

For the remaining case $\kappa + j \in \mathbb{N}$, i.e. $j > m + n$, we have again the representation (8.1.65) for the Schwartz kernel of $a_{m-j}(x,-iD)$. For the homogeneity of $k_{\kappa+j}$, we see that

$$\begin{aligned} k_{\kappa+j}(x,tz) &= (2\pi)^{-n} \Big\{ \int\limits_{|\xi|\le R} a^0_{m-j}(x,\xi)\Big\{e^{i\xi\cdot tz} - \sum_{|\alpha|\le\kappa+j} \frac{1}{\alpha!}\xi^\alpha (itz)^\alpha\Big\} d\xi \\ &- \sum_{|\alpha|<\kappa+j} \frac{1}{\kappa+j-|\alpha|} \frac{1}{\alpha!} \int\limits_{|\Theta|=1} R^{-\kappa-j+|\alpha|} \widehat{\Theta}^\alpha a^0_{m-j}(x,\widehat{\Theta}) d\omega(\widehat{\Theta})(itz)^\alpha \\ &+ \sum_{|\alpha|=\kappa+j} \frac{1}{\alpha!} \int\limits_{|\Theta|=1} (\log R) \widehat{\Theta}^\alpha a^0_{m-j}(x,\widehat{\Theta}) d\omega(\widehat{\Theta})(itz)^\alpha + \int\limits_{R\le|\xi|} e^{i\xi\cdot tz} a^0_{m-j}(x,\xi) d\xi \Big\} \end{aligned} \tag{8.1.67}$$

By choosing first $R = 1$ one finds representations of $k_{\kappa+j}(x,tz)$ and of $k_{\kappa+j}(x,z)$ where $t = 1$. Transforming $\xi' = t\xi$ and afterwards choosing $R = t^{-1}$, we finally obtain

$$k(x,tz) = t^{\kappa+j} \Big\{ k_{\kappa+j}(x,z) - \log t \sum_{|\alpha|=\kappa+j} \frac{1}{\alpha!} \int\limits_{|\Theta|=1} \widehat{\Theta}^\alpha a^0_{m-j}(x,\widehat{\Theta}) d\omega(\widehat{\Theta})(iz)^\alpha \Big\}.$$

Hence, $k_{\kappa+j} \in \Psi h f_{\kappa+j}(\Omega \times \mathbb{R}^n)$ can be written explicitly as

$$k_{\kappa+j}(x,z) = |z|^{\kappa+j}(2\pi)^{-n}\,\text{p.f.}\int\limits_{\mathbb{R}^n} e^{i\xi\cdot z_0} a^0_{m-j}(x,\xi)d\xi - (\log|z|)p_{\kappa+j}(x,z)$$

with

$$p_{\kappa+j}(x,z) = \sum_{|\alpha|=\kappa+j}\frac{1}{\alpha!}\int\limits_{|\Theta|=1}\widehat{\Theta}^\alpha a^0_{m-j}(x,\widehat{\Theta})d\omega(\widehat{\Theta})(iz)^\alpha\,.$$

It remains to establish that

$$k(x,z) = \sum_{0\le j<J} k_{\kappa+j} \quad \text{is in } C^\ell(\Omega\times\mathbb{R}^n) \quad \text{for } 0\le\ell<\kappa+J.$$

From the definitions of k and $k_{\kappa+j}$ we obtain

$$\begin{aligned}
&\Big(-i\frac{\partial}{\partial z}\Big)^\alpha\Big(\frac{\partial}{\partial x}\Big)^\beta\Big(k(x,z) - \sum_{0\le j<J} k_{\kappa+j}(x,z)\Big)\\
&= (2\pi)^{-n}\int\limits_{\mathbb{R}^n} e^{i\xi\cdot z}\xi^\alpha\Big(\frac{\partial}{\partial x}\Big)^\beta\Big(a(x,\xi) - \sum_{j<J} a_{m-j}(x,\xi)\Big)d\xi\\
&= (2\pi)^{-n}\int\limits_{|\xi|\le 1}(1-\chi(\xi))e^{i\xi\cdot z}\xi^\alpha\Big(\frac{\partial}{\partial x}\Big)^\beta\Big(a - \sum_{j<J} a_{m-j}\Big)d\xi\\
&\quad + (2\pi)^{-n}\int\limits_{|\xi|\ge\frac{1}{2}}\chi(\xi)e^{i\xi\cdot z}\xi^\alpha\Big(\frac{\partial}{\partial x}\Big)^\beta\Big(a - \sum_{j<J} a_{m-j}\Big)d\xi
\end{aligned}$$

where $|\alpha|\le\ell<\kappa+J$. The first of the integrals on the right–hand side is in $C^\infty(\Omega\times\mathbb{R}^n)$ due to the Paley–Wiener–Schwartz theorem 3.1.3. The integrand of the second integral can be estimated by $c(\alpha,\,\beta,\,m,\,J)|\xi|^{|\alpha|+m-J}$ where $|\alpha|+m-J<\kappa+J+m-J=-n$. Therefore, this integral is continuous for all these α and any β.

(ii) Now let $k\in\psi hk_\kappa$ with $\kappa=-m-n$ for $m\ge0$ be given. Then k admits the pseudohomogeneous expansion $\sum\limits_{j\ge0} k_{\kappa+j}$ with

$$k(x,z) - \sum_{0\le j<J} k_{\kappa+j}(x,z)\in C^\ell(\Omega\times\mathbb{R}^n) \quad \text{for } 0\le\ell<\kappa+J \tag{8.1.68}$$

where $k_{\kappa+j}\in\psi hf_{\kappa+j}$. If $m\in\mathbb{N}_0$ then, for $0\le j\le m$ we require $k_{\kappa+j}$ to satisfy the Tricomi conditions (8.1.56) with $\kappa+j$ instead of κ.

Note that condition (8.1.68) implies $k\in C^\infty(\Omega\times\mathbb{R}^n\setminus\{0\})$. With a cut-off function $\psi(z)$, the integral operator

$$Au(x) := \text{p.f.}\int\limits_\Omega k(x,x-y)u(y)dy$$

operating on $u \in C_0^\infty(\Omega)$ can be written as

$$Au = A_m u + Ru = \text{p.f.} \int\limits_\Omega k(x, x-y)\psi(x-y)u(y)dy + \int\limits_\Omega \big(1-\psi(x-y)\big)k(x,x-y)u(y)dy$$

where R is a smoothing operator since it has a C^∞–kernel. For A_m we have the representation

$$A_m u(x) = \mathcal{F}^{-1}_{\xi \mapsto x}\big(a(x,\xi)\widehat{u}(\xi)\big)$$

where

$$a(x,\xi) := (2\pi)^{n/2}\mathcal{F}_{z\mapsto\xi}\big(k(x,z)\psi(z)\big) \tag{8.1.69}$$

is in $C^\infty(\Omega \times \mathbb{R}^n)$ due to the Paley–Wiener–Schwartz theorem 3.1.3.

For the pseudohomogeneous functions $k_{\kappa+j}$ we know from Theorem 8.1.7, that the operators

$$A_{m-j}u(x) = \int\limits_\Omega k_{\kappa+j}(x, x-y)u(y)dy \quad \text{for} \quad u \in C_0^\infty(\Omega)$$

are pseudodifferential operators $A_{m-j} \in \mathcal{L}^{m-j}_{c\ell}(\Omega)$ where for $0 \le j \le m$ with $m \in \mathbb{N}_0$ the Tricomi conditions are assumed to be satisfied. Moreover, their homogeneous symbols are given by

$$a^0_{m-j}(x,\xi) = (2\pi)^{n/2}\mathcal{F}_{z\mapsto\xi}k_{\kappa+j}(x,z)\,.$$

So, in order to show $A_m \in \mathcal{L}^m_{c\ell}(\Omega)$ it suffices to show

$$a(x,\xi) - \sum_{j<L} a^0_{m-j}(x,\xi)\chi(\xi) \in \boldsymbol{S}^{m-L}(\Omega\times\mathbb{R}^n) \tag{8.1.70}$$

where $\chi(\xi)$is the cut–off function with $\chi(\xi) = 0$ for $|\xi| \le \frac{1}{2}$ and $\chi(\xi) = 1$ for $|\xi| \ge 1$. We note that (8.1.70) is equivalent to the estimate

$$\Big|\Big(\frac{\partial}{\partial\xi}\Big)^\alpha\Big(\frac{\partial}{\partial x}\Big)^\beta\Big(a(x,\xi) - \sum_{j<J} a^0_{m-j}(x,\xi)\Big)\Big| \le c(\alpha,\beta,J)|\xi|^{m-|\alpha|-J} \tag{8.1.71}$$

for $|\xi| \ge 1$ and any multiindices α and β. For showing (8.1.71), let $L \in \mathbb{N}$ and the multiindex γ be chosen arbitrarily and consider the estimate

$$\Big|\xi^\gamma\Big(\frac{\partial}{\partial\xi}\Big)^\alpha\Big(\frac{\partial}{\partial x}\Big)^\beta\Big(a(x,\xi)-\sum_{j<L}a^0_{m-j}(x,\xi)\Big)\tag{8.1.72}$$

$$\leq \Big|\int_{|z|\leq 1} e^{-iz\cdot\xi}\Big(\frac{\partial}{\partial z}\Big)^\gamma\Big\{(-iz)^\alpha\Big(\frac{\partial^\beta k(x,z)}{\partial x^\beta}-\sum_{j<L}\frac{\partial^\beta k_{\kappa+j}(x,z)}{\partial x^\beta}\Big)\psi(z)\Big\}dz\Big|$$

$$+\sum_{j<L}\Big|\int_{|z|\geq\frac12} e^{-iz\cdot\xi}\Big(-i\frac{\partial}{\partial z}\Big)^\gamma\Big\{(-iz)^\alpha\frac{\partial^\beta k_{\kappa+j}(x,z)}{\partial x^\beta}(1-\psi(z))\Big\}dz\Big|$$

where (8.1.69) is employed. From (8.1.68) we conclude that the integrand of the first integral on the right–hand side in (8.1.72) is in $C^q(\Omega\times\mathbb{R}^n)$ with $q=\kappa+L-1+|\alpha|-|\gamma|$. Hence, this integral is bounded for $q=0$ which corresponds to the choice of γ so that $|\gamma|=L-1+|\alpha|-m-n$. For the remaining terms we notice that, for any integer k we have the estimates

$$|\xi|^{2k}\Big|\int_{|z|\geq\frac12} e^{-iz\cdot\xi}\Big(-i\frac{\partial}{\partial z}\Big)^\gamma\Big\{(-iz)^\alpha\frac{\partial^\beta k_{\kappa+j}}{\partial x^\beta}(x,z)(1-\psi(z))\Big\}dz\Big|$$

$$= \Big|\int_{|z|\geq\frac12} e^{-iz\cdot\xi}(\Delta_z)^k\Big(-i\frac{\partial}{\partial z}\Big)^\gamma\{\cdots\}dz\Big|$$

$$\leq c\int_{|z|\geq\frac12}|z|^{-2k-|\gamma|+\kappa+j+|\alpha|}dz<\infty$$

provided $-2k-|\gamma|+\kappa+j+|\alpha|<-n$. So, we choose $2k\geq -m-|\gamma|+L+|\alpha|$. Then (8.1.72) implies the estimates

$$\Big|\Big(\frac{\partial}{\partial\xi}\Big)^\alpha\Big(\frac{\partial}{\partial x}\Big)^\beta\Big(a(x,\xi)-\sum_{j<L}a^0_{m-j}(x,\xi)\Big)\Big|$$

$$\leq c(\alpha,\beta,m,L)|\xi|^{-|\gamma|}\tag{8.1.73}$$

$$\leq c(\alpha,\beta,m,L)|\xi|^{m-|\alpha|-L+n+1}\quad\text{for }|\xi|\geq 1\,.$$

To obtain the desired estimates (8.1.71) we now choose $L=J+n+1$ and exploit the homogeneity of $\Big(\frac{\partial}{\partial x}\Big)^\beta a^0_{m-j}(x,\xi)$ of degree $m-j$. Then it follows with (8.1.73) and the triangle inequality that

$$\Big|\Big(\frac{\partial}{\partial\xi}\Big)^\alpha\Big(\frac{\partial}{\partial x}\Big)^\beta\Big(a(x,\xi)-\sum_{j<L}a^0_{m-j}(x,\xi)\Big)\Big|$$

$$\leq \sum_{J\leq j\leq J+n}\Big|\Big(\frac{\partial}{\partial\xi}\Big)^\alpha\Big(\frac{\partial}{\partial x}\Big)^\beta a^0_{m-j}(x,\xi)\Big|+c(\alpha,\beta,m,J+n+1)|\xi|^{m-|\alpha|-J}$$

$$\leq c_1(\alpha,\beta,m,J)|\xi|^{m-|\alpha|-J}\quad\text{for }|\xi|\geq 1\,,$$

which are the desired estimates (8.1.71).

Consequently, $A_m \in \mathcal{L}^m_{c\ell}(\Omega)$ and $A \in \mathcal{L}^m_{c\ell}(\Omega)$ as proposed. ■

8.1.3 Parity Conditions

Definition 8.1.3. (Hörmander [188, Vol.I, Section 3.2])

The pseudohomogeneous function $k_q \in \Psi hf_q$ for $q \in \mathbb{Z}$ is of parity σ if it satisfies the condition

$$k_q(x, -z) = (-1)^\sigma \, k_q(x, z) \quad \textit{for} \;\; z \neq 0\,. \tag{8.1.74}$$

We now state the following crucial result concerning the transformation of the finite part integral operators.

Lemma 8.1.9. *The parity condition (8.1.74) for $k_q \in \Psi hf_q$ with $q \in \mathbb{Z}$, is satisfied if and only if the corresponding homogeneous symbol $a^0_{-n-q}(x, \xi)$ satisfies the parity condition*

$$a^0_{-n-q}(x, -\xi) = (-1)^\sigma a^0_{-n-q}(x, \xi) \quad \textit{for} \;\; \xi \neq 0\,. \tag{8.1.75}$$

Proof: **i)** Let $k_q \in \Psi hf_q$ satisfy the parity condition (8.1.74). Then with Lemma 8.1.4, the symbol

$$a_\psi(x, \xi) := e^{-ix\cdot\xi} \int\limits_{\mathbb{R}^n} k_q(x, x-y)\psi(x-y) e^{iy\cdot\xi} dy$$

associated with a suitable cut–off function $\psi \in C^\infty_0$ with $\psi(z) = 1$ for $|z| \leq \varepsilon$, $\varepsilon > 0$, and $\psi(z) = 0$ for $|z| \geq 2\varepsilon$ and $\psi(-z) = \psi(z)$ satisfies the parity condition, i.e.

$$\begin{aligned}
a_\psi(x, -\xi) &= \int\limits_{\mathbb{R}^n} e^{iz\cdot\xi} k_q(x, z)\psi(z) dz \\
&= \int\limits_{\mathbb{R}^n} e^{-iz\cdot\xi} k_q(x, -z)\psi(z) dz \\
&= \int\limits_{\mathbb{R}^n} e^{-iz\cdot\xi} (-1)^\sigma k_q(x, z)\psi(z) dz \\
&= (-1)^\sigma e^{-i\cdot\xi} \int\limits_{\mathbb{R}^n} k_q(x, x-y) e^{iy\cdot\xi} \psi(x-y) dy \\
&= (-1)^\sigma a_\psi(x, \xi)\,.
\end{aligned}$$

Hence, by taking the limit $0 < t \to \infty$, we have

$$\begin{aligned}
a^0_{-n-q}(x, -\xi) &= \lim_{t\to\infty} t^{n+q} a_\psi(x, -t\xi) \\
&= (-1)^\sigma \lim_{t\to\infty} t^{n-q} a_\psi(x, t\xi) \\
&= (-1)^\sigma a^0_{-n-q}(x, \xi)\,.
\end{aligned}$$

ii) Conversely, the kernel $k_q(x, x-y)$ can be expressed in the form (7.1.12) , i.e.

$$k_q(x, x-y) = |y-x|^{-2N}(2\pi)^{-n} \int\limits_{\mathbb{R}^n} (-\Delta_\xi)^N a^0_{-n-q}(x,\xi) e^{i(x-y)\cdot\xi} d\xi\,.$$

Then

$$k_q(x, x-y) = (-1)^\sigma k_q(x, x-y)$$

follows immediately from $a^0_{-n-q}(x,-\xi) = (-1)^\sigma a^0_{-n-q}(x,\xi)$. ■

As a consequence of Lemma 8.1.9, the parity conditions will provide us with a criterion when the local differential operator in the representation (8.1.60) will vanish.

Theorem 8.1.10. *Let $A \in \mathcal{L}^m_{c\ell}(\Omega)$, $m \in \mathbb{N}_0$ and suppose the parity conditions*

$$a^0_{m-j}(x,-\xi) = (-1)^{m-j+1} a^0_{m-j}(x,\xi) \quad \textit{for } 0 \le j \le m \tag{8.1.76}$$

for the homogeneous terms in the symbol expansion of A. Then

$$(Au)(x) = \text{p.f.} \int\limits_\Omega k(x, x-y) u(y) dy\,. \tag{8.1.77}$$

Proof: From the representation (8.1.60) we observe that

$$A\big((\bullet - x)^\beta \psi(|\bullet - x|)\big)|_x = 0 + \int\limits_\Omega (k(x, x-y)(y-x)^\beta \psi(|y-x|) dy$$

for any $|\beta| > m$ and $A \in \mathcal{L}^m_{c\ell}(\Omega)$. This implies

$$\begin{aligned}
&\sum_{j=0}^m a_{m-j}(x,D)\big((\bullet - x)^\beta \psi(|\bullet - x|)\big) + A_R\big((\bullet - x)^\beta \psi(|\bullet - x|)\big)(x) \\
&\quad = \sum_{j=0}^m \Big\{ \sum_{|\alpha| = m-j} a_\alpha(x) D^\alpha_y \big((y-x)^\beta \psi(|y-x|)\big)|_{y=x} \\
&\qquad\qquad + \text{p.f.} \int\limits_\Omega k_{\kappa+j}(x, x-y)(y-x)^\beta \psi(|y-x|) dy \Big\} \\
&\qquad\qquad + \int k_R(x, x-y)(y-x)^\beta \psi(|y-x|) dy
\end{aligned} \tag{8.1.78}$$

for any $|\beta| \ge 0$. Now, choose $|\beta| = m$. Then, with the previous remark we have

$$\begin{aligned}
\beta! a_\beta(x) &= a_m(x,D)\big((\bullet - x)^\beta \psi(|\bullet - x|)\big) \\
&\quad - \text{p.f.} \int\limits_\Omega k_\kappa(x, x-y)(y-x)^\beta \psi(|y-x|)dy \\
&= (2\pi)^{-n/2} \int\limits_{\mathbb{R}^n}\int\limits_{\mathbb{R}^n} (y-x)^\beta e^{i(x-y)\cdot\xi}\psi(|y-x|)a_m(x,\xi)dyd\xi \\
&\quad - \text{p.f.} \int\limits_{\mathbb{R}^n} k_\kappa(x, x-y)(y-x)^\beta \psi(|y-x|)dy\,.
\end{aligned}$$

We want to show $a_\beta(x) = 0$ under assumption (8.1.76). First we note that Lemma 8.1.9 implies the parity conditions

$$k_{\kappa+j}(x, x-y) = (-1)^{m-j+1} k_{\kappa+j}(x, x-y) \tag{8.1.79}$$

for the homogeneous term of the asymptotic kernel expansion. A straightforward computation shows for $j = 0$ by substituting $z = y - x$ and replacing ξ by $-\xi$, that

$$\begin{aligned}
a_\beta(x) &= \tfrac{1}{\beta!}\Big((2\pi)^{-n/2} \int\limits_{\mathbb{R}^n}\int\limits_{\mathbb{R}^n} z^\beta e^{iz\cdot\xi} a_m(x,-\xi)\psi(|z|)dzd\xi \\
&\quad - \text{p.f.} \int\limits_{\mathbb{R}^n} k_\kappa(x,-z)z^\beta \psi(|z|)dz\Big)\,.
\end{aligned}$$

Now, use the parity conditions for a_m as well as for k_κ after transforming z to $-z$ and obtain

$$a_\beta(x) = -a_\beta(x)\,, \quad \text{i.e. } a_\beta(x) = 0 \quad \text{for all } |\beta| = m\,.$$

Substituting this into (8.1.78) and choosing $|\beta| = m - 1$, we obtain

$$\begin{aligned}
\beta! a_\beta(x) &= a_1(x,D)\big((\bullet - x)^\beta \psi(|\bullet - x|)\big) \\
&\quad - \text{p.f.} \int\limits_\Omega k_{\kappa+1}(x, x-y)(y-x)^\beta \psi(|\bullet - x|)dy \\
&= (2\pi)^{-n/2} \int\limits_{\mathbb{R}^n}\int\limits_{\mathbb{R}^n} (y-x)^\beta e^{i(x-y)\cdot\xi}\psi(|\bullet - x|)a_1(x,\xi)dyd\xi \\
&\quad - \text{p.f.} \int\limits_\Omega k_{\kappa+1}(x, x-y)(y-x)^\beta \psi(|\bullet - x|)dy
\end{aligned}$$

with the application of our theorem to the particular operator $a_0(x,D)$. Applying again the parity conditions (8.1.76) and (8.1.79), we find in the same manner as for $|\beta| = m$, now

$$a_\beta(x) = 0 \quad \text{for all } |\beta| = m - 1\,.$$

Repeating this process we finally obtain $a_\beta(x) = 0$ for all $|\beta| \le m$ which assures the representation (8.1.77) of our theorem. ■

8.1.4 A Summary of the Relations between Kernels and Symbols

As we have seen in this section, a pseudodifferential operator $A \in \mathcal{L}^m_{c\ell}(\Omega)$ can be expressed in terms of either a given Schwartz kernel or in terms of its symbol. The relations between these two representations have been given explicitly so far. However, for ease of reading, we now summarize these relations in the following.

Kernel to symbol

Let the operator $A \in \mathcal{L}^m_{c\ell}(\Omega)$ be given in the form (8.1.8) or, more precisely, in the form (8.1.60) where the Schwartz kernel of the finite part integral operator in (8.1.60) is a given function in the class $\Psi hk_\kappa(\Omega)$ with $\kappa = -m-n$ for fixed $m \in \mathbb{R}$ (see Definition 8.1.1). Then this kernel has an asymptotic expansion in the form (8.1.2), i.e.,

$$k(x, x-y) \sim \sum_{j=0}^{\infty} k_{\kappa+j}(x, x-y) \quad \text{where} \quad k_{\kappa+j} \in \Psi hf_{\kappa+j}\,.$$

In the case that $m - j \in \mathbb{N}_0$, we assume that the corresponding terms in the asymptotic expansion satisfy the Tricomi conditions (8.1.56), i.e.,

$$\int\limits_{|\Theta|=1} \Theta^\alpha k_{\kappa+j}(x,\Theta)d\omega(\Theta) = 0 \quad \text{for all} \quad |\alpha| = m-j\,,$$

(see Theorem 8.1.7). For $n=1$, again this formula is interpreted as in (3.2.14). With a properly supported part A_0 of A, the symbol corresponding to A_0 can be computed according to the formula (7.1.23), i.e.,

$$a(x,\xi) := e^{-ix\cdot\xi} A_0(e^{i\xi\bullet})(x)\,.$$

The symbol $a(x,\xi)$ admits for $A \in \mathcal{L}^m_{c\ell}(\Omega)$ a classical asymptotic expansion

$$a(x,\xi) \sim \sum_{j=0}^{\infty} a^0_{m-j}(x,\xi) \tag{8.1.80}$$

(see (7.1.42)). Each term in the expansion (8.1.80) can be calculated from the corresponding pseudohomogeneous terms of the kernel expansion explicitly depending on $m-j$.

For $m - j < 0$ we have:

$$a^0_{m-j}(x,\xi) = \lim_{t\to+\infty} \int\limits_{\mathbb{R}^n} k_{\kappa+j}(x,z)\psi(\tfrac{z}{t})e^{-i\xi\cdot z}dz \quad \text{for} \quad x \in \Omega \tag{8.1.81}$$

where the cut–off function $\psi(z) = 1$ for $|z| \le \frac{1}{2}$ and $\psi(z) = 0$ for $|z| > 1$ (see Lemma 8.1.4, equations (8.1.28) and (8.1.29)).

For $m \notin \mathbb{N}_0$ and $m - j > 0$ we have

$$a^0_{m-j}(x,\xi) = \text{p.f.} \int\limits_{\mathbb{R}^n} k_{\kappa+j}(x,z)e^{-i\xi\cdot z}dz \tag{8.1.82}$$

(see Theorem 8.1.6, formula (8.1.53)).

For $m \in \mathbb{N}_0$ and $m - j \geq 0$ and if the Tricomi conditions (8.1.56) are satisfied, then we have

$$a^0_{m-j}(x,\xi) = \sum_{|\alpha|=m-j} c_\alpha(x)(i\xi)^\alpha + \text{p.f.} \int\limits_{\mathbb{R}^n} k_{\kappa+j}(x,z)e^{-i\xi\cdot z}dz \tag{8.1.83}$$

(see Theorem 8.1.7, formula (8.1.57)).

Symbol to kernel

For the operator $A \in \mathcal{L}^m_{c\ell}(\Omega)$, let the classical symbol $a(x,\xi)$ be given by its homogeneous classical asymptotic symbol expansion (8.1.80). Then the corresponding pseudohomogeneous kernel expansion can be calculated explicitly as follows.

For $m - j < 0$ we have

$$k_{\kappa+j}(x,z) = (2\pi)^{-n}\,\text{p.f.} \int\limits_{\mathbb{R}^n} e^{iz\cdot\xi}a^0_{m-j}(x,\xi)d\xi \quad \text{for } x \in \Omega \quad \text{and} \quad z \in \mathbb{R}^n \tag{8.1.84}$$

(see in the proof of Theorem 8.1.1, formulae (8.1.48), (8.1.51)).

For $m - j \geq 0$ we have

$$\begin{aligned} k_{\kappa+j}(x,z) &= (2\pi)^{-n} \int\limits_{\mathbb{R}^n} e^{i\xi\cdot z}a^0_{m-j}(x,\xi)\psi(\xi)d\xi \\ &+|z|^{-2\ell}(2\pi)^{-n} \int\limits_{\mathbb{R}^n} e^{i\xi\cdot z}(-\Delta_\xi)^\ell\big\{a^0_{m-j}(x,\xi)\big(1-\psi(\xi)\big)\big\}d\xi \end{aligned} \tag{8.1.85}$$

where $\ell \in \mathbb{N}$ satisfying $2\ell > m + n - j$ (see in the proof of Theorem 8.1.8 **i)**, formula (8.1.64)). This formula is valid for arbitrary $m \in \mathbb{R}$.

If $m \in \mathbb{N}_0$ then the Schwartz kernel is given by (8.1.85) for $j = 0, \dots, m$ whereas the operator $a^0_{m-j}(x, -iD)$ contains a differential operator of the form $\sum\limits_{|\alpha|=m-j} c_\alpha(x)D^\alpha$. Then the coefficients $c_\alpha(x)$ can be recovered by formula (8.1.9), i.e.

$$\begin{aligned} c_\alpha(x) &= (2\pi)^{-n}\frac{1}{\alpha!}\Big\{\int\limits_{\mathbb{R}^n}\int\limits_{\Omega}(y-x)^\alpha\psi(y)e^{i(x-y)\cdot\xi}dy\,a^0_{m-j}(x,\xi)d\xi \\ &- \text{p.f.}\int\limits_{\Omega} k_{\kappa+j}(x,x-y)(y-x)^\alpha\psi(y)dy\Big\} \quad \text{for } |\alpha| = m-j\,. \end{aligned} \tag{8.1.86}$$

If the decomposition

$$a^0_{m-j}(x,\xi) = \sum_{|\alpha|=m-j} c_\alpha(x)(i\xi)^\alpha + a^{00}_{m-j}(x,\xi)$$

is known then in formula (8.1.86) one can simply replace a^0_{m-j} by a^{00}_{m-j}.

8.2 Coordinate Changes and Pseudohomogeneous Kernels

As we have seen in Theorems 8.1.1 and 8.1.8, all the pseudodifferential operators $A \in \mathcal{L}^m_{c\ell}(\Omega)$ have the general form (8.1.8) with a kernel function $k \in \Psi hk_\kappa(\Omega)$. In particular, for $m < 0$ the integral operator is weakly singular and the change of coordinates $x' = \Phi(x)$ with Φ a diffeomorphism and $x = \Phi^{(-1)}(x') = \Psi(x')$ results in the traditional transformation formula

$$\int_\Omega k(x,x-y)u(y)dy = \int_{\Omega'} k\big(\Psi(x')\,,\,\Psi(x') - \Psi(y')\big)u\big(\Psi(y')\big)J(y')dy' \quad (8.2.1)$$

with the Jacobian $J(y') = \Big(\det \dfrac{\partial\Psi}{\partial y'}\Big)$. The new kernel $\widetilde{k}(x',x'-y')$ is still weakly singular in Ω'.

In the case $m \geq 0$, however, k is strongly singular and the traditional transformation rule (8.2.1) needs to be modified.

In what follows, we first examine the properties of the pseudohomogeneous kernel under the change of coordinates.

Lemma 8.2.1. *Let*

$$k(x,x-y) \sim \sum_{j\geq 0} k_{\kappa+j}(x,x-y) \quad (8.2.2)$$

(see (8.1.2)*) be the pseudohomogeneous asymptotic expansion of* $k \in \Psi hk_\kappa$. *Then the transformed kernel*

$$\widetilde{k}(x',x'-y') := k\big(\Psi(x')\,,\,\Psi(x') - \Psi(y')\big) \sim \sum_{p\geq 0} \widetilde{k}_{\kappa+p}(x',x'-y') \quad (8.2.3)$$

is in Ψhk_κ, *too.*

Proof: Since with Φ, also $\Psi(y')$ is a smooth diffeomorphism, the Taylor expansion of $z = x - y$ about x' can be written as the asymptotic expansion

$$x - y = \Psi(x') - \Psi(y') \sim \sum_{|\alpha|\geq 1} \pi_\alpha(x',x'-y')\,. \quad (8.2.4)$$

The homogeneous polynomials π_α are given explicitly as

$$\pi_\alpha(x', x'-y') = \frac{(-1)^{\langle\alpha\rangle}}{\alpha!}\frac{\partial^\alpha \Psi}{\partial x'^\alpha}(x')(x'-y')^\alpha\,. \tag{8.2.5}$$

By using the homogeneity of π_α and setting $\Theta' = \dfrac{y'-x'}{|y'-x'|}$, the relation (8.2.4) yields

$$|x-y| = |x'-y'|\,\Big|\sum_{|\alpha|\geq 1}|x'-y'|^{|\alpha|-1}\pi_\alpha(x', -\Theta')\Big|\,. \tag{8.2.6}$$

Hence,

$$\Theta = \frac{y-x}{|y-x|} = -\frac{\sum_{|\alpha|=1}\pi_\alpha(x',-\Theta') + \sum_{|\alpha|\geq 2}|x'-y'|^{|\alpha|-1}\pi_\alpha(x',-\Theta')}{\big|\sum_{|\alpha|=1}\pi_\alpha(x',-\Theta') + \sum_{|\alpha|\geq 2}|x'-y'|^{|\alpha|-1}\pi_\alpha(x',-\Theta')\big|}$$

holds for every $|x'-y'| > 0$. This implies with $|x'-y'| \to 0$,

$$\Theta = \Theta(\Theta') = -\frac{\sum_{|\alpha|=1}\pi_\alpha(x',-\Theta')}{\big|\sum_{|\alpha|=1}\pi_\alpha(x',-\Theta')\big|}\,. \tag{8.2.7}$$

The denominator does not vanish since $\frac{\partial\Psi}{\partial x'}$ is invertible. In fact, $\Theta(\Theta')$ defines a diffeomorphic mapping of the unit sphere onto itself.

Now we consider first the positively homogeneous terms of the asymptotic expansion of $k(x, x-y)$, namely

$$f(x, x-y) \sim \sum_{j\geq 0} f_{\kappa+j}(x, x-y)\,. \tag{8.2.8}$$

In terms of the transformation and homogeneity of $f_{\kappa+j}$, this reads

$$\begin{aligned}\sum_{j\geq 0} f_{\kappa+j}(x, x-y) &= \sum_{j\geq 0}|x-y|^{\kappa+j} f_{\kappa+j}\big(\Psi(x'), \Theta(\Theta')\big)\\ &= \sum_{j\geq 0}|x'-y'|^{\kappa+j}\Big|\sum_{|\alpha|\geq 1}|x'-y'|^{|\alpha|-1}\pi_\alpha(x',-\Theta')\Big|^{\kappa+j} f_{\kappa+j}\big(\Psi(x'), \Theta(\Theta')\big)\,.\end{aligned}$$

The second factor can be rewritten in the form

$$\begin{aligned}&\Big|\sum_{|\alpha|=1}\pi_\alpha(x',-\Theta') + \sum_{|\alpha|\geq 2}|x'-y'|^{|\alpha|-1}\pi_\alpha(x',-\Theta')\Big|^{\kappa+j}\\ &= \Big|\sum_{|\alpha|=1}\pi_\alpha(x',-\Theta')\Big|^{\kappa+j}\Big\{1 + \sum_{\ell\geq 1}|x'-y'|^\ell q_{\ell,\kappa+j}(x',\Theta')\Big\}\end{aligned}$$

where the functions $q_{\ell,\kappa+j}(x',\Theta')$ defined on $(\Omega' \times \{|\Theta'| = 1\})$ are obtained by using a power series representation. Collecting terms, we obtain

$$\sum_{j\ge 0} f_{\kappa+j}(x,x-y) \sim \sum_{p\ge 0} |x'-y'|^{\kappa+p} \widetilde{f}_{\kappa+p}(x',\Theta') \tag{8.2.9}$$

where

$$\widetilde{f}_{\kappa+p}(x',\Theta') = \sum_{\substack{j+\ell=p\\ j,\ell\ge 0}} \Big|\sum_{|\alpha|=1} \pi_\alpha(x',-\Theta')\Big|^{\kappa+j} q_{\ell,\kappa+j}(x',\Theta') f_{\kappa+j}\big(\Psi(x'),\Theta(\Theta')\big)\,.$$

If $\kappa \in \mathbb{Z}$ then

$$k(x,x-y) \sim \sum_{j\ge 0} f_{\kappa+j}(x,x-y) + \sum_{\substack{j\ge 0\\ \kappa+j\ge 0}} \log|x-y| p_{\kappa+j}(x,x-y)$$

where $p_{\kappa+j}(x,z)$ are homogeneous polynomials of degree $\kappa+j$. According to our previous analysis it suffices to consider only the terms

$$Q(x,x-y) := \sum_{\substack{j\ge 0\\ \kappa+j\ge 0}} \log|x-y| p_{\kappa+j}(x,x-y)\,.$$

In view of (8.2.6) we may rewrite

$$\begin{aligned} Q(x,x-y) \quad \sim \quad & \sum_{\substack{j\ge 0\\ \kappa+j\ge 0}} \log|x'-y'| p_{\kappa+j}(x,x-y) \\ & + \log\Big|\sum_{|\alpha|=1} \pi_\alpha(x',-\Theta')\Big| p_{\kappa+j}(x,x-y) \qquad (8.2.10)\\ & + \log\Big(\Big|1+\sum_{\ell\ge 1} |x'-y'|^{\ell} q_{\ell,1}(x',\Theta')\Big|\Big) p_{\kappa+j}(x,x-y)\,. \end{aligned}$$

For the polynomials $p_{\kappa+j}$ we use (8.2.4) to obtain

$$\begin{aligned} p_{\kappa+j}(x,x-y) \quad &\sim \quad p_{\kappa+j}\big(\Psi(x')\,,\sum_{|\alpha|\ge 1} \pi_\alpha(x',x'-y')\big) \\ &\sim \sum_{\ell\ge j+\kappa} \widetilde{p}_{\kappa+\ell,j}(x',x'-y') \qquad (8.2.11) \end{aligned}$$

with the homogeneous polynomials $\widetilde{p}_{\kappa+\ell}$ of degrees $\kappa+\ell$.

By substituting (8.2.11) into (8.2.10) we see that the first sum has the form

$$\sum_{\ell\ge 1} \log|x'-y'| \Big\{ \sum_{\substack{-\kappa\le j\le \ell,\\ 0\le j}} \widetilde{p}_{\kappa+\ell,j}(x',x'-y') \Big\} =: \sum_{\ell\ge 0} \log|x'-y'| p'_{\ell+\kappa}(x',x'-y')\,. \tag{8.2.12}$$

The other two terms have an asymptotic expansion of the same form as (8.2.4), namely

$$\sum_{p\geq 0} |x' - y'|^{\kappa+p} f'_{\kappa+p}(x', \Theta') \tag{8.2.13}$$

which follows in the same manner as for the homogeneous expansion. Collecting (8.2.11) and (8.2.13) yields

$$\begin{aligned} k(x, x-y) \quad \sim \quad & \sum_{p\geq 0} |x' - y'|^{\kappa+p} \big(\widetilde{f}_{\kappa+p}(x', \Theta') + f'_{\kappa+p}(x', \Theta')\big) \\ & + \sum_{\substack{\ell \geq 0, \\ \ell+\kappa\geq 0}} \log |x' - y'| p'_{\ell+\kappa}(x', x' - y') \,. \end{aligned}$$

as proposed. ■

8.2.1 The Transformation of General Hadamard Finite Part Integral Operators under Change of Coordinates

Now we return to the transformation properties of the operator $A \in \mathcal{L}^m_{c\ell}(\Omega)$ in the form (8.1.8),

$$(Au)(x) = \sum_{|\alpha|\leq m} a_\alpha(x) D^\alpha u(x) + \text{ p.f.} \int_\Omega k(x, x-y) u(y) dy \,. \tag{8.2.14}$$

If $m \notin \mathbb{N}_0$ then $a_\alpha(x) = 0$. For $m < 0$, the integral operator is weakly singular and its transformation was already discussed. For $m \geq 0$, we regularize the finite part integral in (8.2.14) and write

$$\begin{aligned} (Au)(x) \quad = \quad & \sum_{|\alpha|\leq m} \big(a_\alpha(x) + d_\alpha(x)\big) D^\alpha u(x) \qquad (8.2.15) \\ & + \int_\Omega k(x, x-y) \Big\{ u(y) - \sum_{|\alpha|\leq m} \frac{1}{\alpha!} (y-x)^\alpha D^\alpha u(x) \Big\} dy \end{aligned}$$

where

$$d_\alpha(x) = \frac{1}{\alpha!} \text{ p.f.} \int_\Omega k(x, x-y)(y-x)^\alpha dy \tag{8.2.16}$$

(see also (8.1.5) and (8.1.6)). For the transformation of the derivatives due to $x = \Psi(x')$, we need the identity (3.4.1), i.e.

$$\frac{\partial}{\partial x_j} = \sum_{m,\ell=1}^{n} \frac{\partial \Psi_j}{\partial x'_m}(x') g^{m\ell}(x') \frac{\partial}{\partial x'_\ell} =: \widetilde{D}_j \,, \tag{8.2.17}$$

where the Riemanian metric is given by the tensor (3.4.2), i.e.

$$g_{ik} = \left(\frac{\partial \Psi}{\partial x'_j} \cdot \frac{\partial \Psi}{\partial x'_k} \right)$$

and the $g^{m\ell}(x')$ are given by its inverse, see (3.4.4).

Both, the differential operators and the regularized integral in (8.2.15) can now be transformed in the usual way. With $\widetilde{u}(x') := u\big(\Psi(x')\big)$, we obtain

$$(Au)(x) = (\widetilde{A}\widetilde{u})(x') = \sum_{|\alpha|\le m} \Big(a_\alpha\big(\Psi(x')\big) + d_\alpha\big(\Psi(x')\big)\Big)\widetilde{D}^\alpha \widetilde{u}(x')$$
$$+ \int_{\Omega'} \widetilde{k}(x', x'-y')J(y')\Big\{\widetilde{u}(y') - \sum_{|\alpha|\le m} \frac{1}{\alpha!}\big(\Psi(y') - \Psi(x')\big)^\alpha \widetilde{D}^\alpha \widetilde{u}(x')\Big\}dy' . \tag{8.2.18}$$

From Lemma 8.2.1 we know that the transformed kernel $\widetilde{k} \in \Psi hk_\kappa$ with $\kappa = -n-m$. Therefore, we may write (8.2.18) in terms of a finite part integral and obtain

$$\begin{aligned}(\widetilde{A}\widetilde{u})(x') &= \sum_{|\alpha|\le m} \Big\{ a_\alpha\big(\Psi(x')\big) + d_\alpha\big(\Psi(x')\big) \\ &\quad - \frac{1}{\alpha!}\ \text{p.f.} \int_{\Omega'} \widetilde{k}(x', x'-y')J(y')\big(\Psi(y') - \Psi(x')\big)^\alpha dy' \Big\} \widetilde{D}^\alpha \widetilde{u}(x') \\ &\quad + \text{p.f.} \int_{\Omega'} \widetilde{k}(x', x'-y')J(y')\widetilde{u}(y')dy' \\ &= (Au)\big(\Psi(x')\big) . \end{aligned} \tag{8.2.19}$$

We summarize these results in the following theorem.

Theorem 8.2.2. *Let $m \in \mathbb{R}$. The operator $A \in \mathcal{L}^m_{c\ell}(\Omega)$ in the form (8.2.14) under change of coordinates $x = \Psi(x')$ becomes $\widetilde{A} \in \mathcal{L}^m_{c\ell}(\Omega')$ and has the form*

$$\begin{aligned}(Au)\big(\Psi(x')\big) = (\widetilde{A}\widetilde{u})(x') &= \sum_{|\alpha|\le m} a_\alpha\big(\Psi(x')\big)\widetilde{D}^\alpha \widetilde{u}(x') + \sum_{|\alpha|\le m} b_\alpha(x')\widetilde{D}^\alpha \widetilde{u}(x') \\ &\quad + \text{p.f.} \int_{\Omega'} \widetilde{k}(x', x'-y')J(y')\widetilde{u}(y')dy' \end{aligned} \tag{8.2.20}$$

where

$$\widetilde{k}(x', x'-y') := k\big(\Psi(x'), \Psi(x') - \Psi(y')\big) \tag{8.2.21}$$

and

$$\begin{aligned} b_\alpha(x') &:= \frac{1}{\alpha!}\Big\{ \text{p.f.} \int_{\Omega} k(x, x-y)(y-x)^\alpha dy|_{x=\Psi(x')} \\ &\quad - \text{p.f.} \int_{\Omega'} \widetilde{k}(x', x'-y')J(y')\big(\Psi(y') - \Psi(x')\big)^\alpha dy' \Big\} . \end{aligned} \tag{8.2.22}$$

For $m < 0$, the differential operator terms with the coefficients a_α and b_α do not appear and

$$\widetilde{A}\widetilde{u}(x') = \int_{\Omega'} \widetilde{k}(x', y' - x')J(y')\widetilde{u}(y')dy' . \tag{8.2.23}$$

In comparison with A in (8.2.14), we note that for $m \geq 0$ the coefficients of the differential operator part contain the *extra terms* $b_\alpha(x')$, which for $m \in \mathbb{N}_0$, in general, do **not** vanish. In the special case when $m = 0$, see Mikhlin and Prössdorf [305, Formulae (8) p.223 and (11) p.226]. This is a very important difference from the coordinate transformation for operators with weakly singular or regular kernel functions as in (8.2.23). Fore more general m see also Sellier [387].

In the case $0 < m \notin \mathbb{N}_0$ we shall show that the $b_\alpha(x')$ in (8.2.22) will vanish. For this purpose we first need the following lemma concerning the coordinate transformation Ψ.

Lemma 8.2.3. *Let $\varrho_\varepsilon(\omega)$ denote the distance between $x = \Psi(x')$ and $y = \Psi(y')$ where $|y' - x'| = \varepsilon > 0$ and $y = x + \varrho_\varepsilon(\omega)\omega$ is the image of the sphere in terms of polar coordinates $(\varrho_\varepsilon, \omega)$ about x. Then, for sufficiently small $\varepsilon > 0$ and Ψ a C^∞–diffeomorphism, one admits the asymptotic representation*

$$\varrho_\varepsilon(\omega) = \sum_{k=1}^{N} c_k(\omega)\varepsilon^k + O(\varepsilon^{N+1}) \tag{8.2.24}$$

for any $N \in \mathbb{N}$. The coefficients $c_k(\omega)$ satisfy the parity conditions

$$c_k(-\omega) = (-1)^{k+1}c_k(\omega) . \tag{8.2.25}$$

Proof: With the C^∞–diffeomorphism $\Phi(x) = x'$, $\Phi(y) = y'$, inverse to Ψ, the Taylor expansion about x' reads

$$y' - x' = \sum_{|\alpha| \geq 1} \frac{1}{\alpha!} D^\alpha \Phi(x) \omega^\alpha \varrho^{|\alpha|} .$$

Hence,

$$\varepsilon^2 = |y' - x'|^2 = \sum_{|\alpha|\geq 1, |\beta| \geq 1} \big(D^\alpha\Phi(x)\omega^\alpha\big) \cdot \big(D^\beta\Phi(x)\omega^\beta\big)\varrho_\varepsilon^{|\alpha|+|\beta|} \tag{8.2.26}$$

which implies, if we define

$$F(\varepsilon, \varrho, \omega) := -\varepsilon + \varrho\sqrt{\sum_{|\alpha|\geq 1, |\beta|\geq 1} \big(D^\alpha\Phi(x)\omega^\alpha\big) \bullet \big(D^\beta\Phi(x)\omega^\beta\big)\varrho^{|\alpha|-|\beta|-2}}$$

for $|\varrho| \leq \varrho_0$ and $|\varepsilon| \leq \varepsilon_0$, $|\omega| = 1$, that the equation

$$F(\varepsilon, \varrho, \omega) = 0$$

admits a unique C^∞–solution

$$\varrho = f(\varepsilon, \omega), \quad \text{i.e.} \quad F\big(\varepsilon, f(\varepsilon, \omega), \omega\big) = 0$$

since

$$\frac{\partial F}{\partial \varrho}|_{\varrho=0} = \sqrt{\sum_{\substack{|\alpha|=1\\|\beta|=1}} \big(D^\alpha \Phi(x)\omega^\alpha\big) \cdot \big(D^\beta \Phi(x)\omega^\beta\big)} > 0$$

for all ω on the unit sphere, because Φ is diffeomorphic. Moreover, $f(\varepsilon, \omega)$ is also defined for $\varepsilon < 0$ where $f(\varepsilon, \omega) < 0$. In particular, we see that

$$-f(-\varepsilon, -\omega) = f(\varepsilon, \omega) \tag{8.2.27}$$

since

$$F(-\varepsilon, -\varrho, \omega) = -F(\varepsilon, \varrho, \omega) = 0\,.$$

Since F is C^∞, so is f the asymptotic expansion (8.2.24) for any $N \in \mathbb{N}$. Hence, the parity condition (8.2.25) then is an immediate consequence of (8.2.27). ■

Lemma 8.2.4. *For $0 < m \notin \mathbb{N}_0$, the coefficients $b_\alpha(x') = 0$ in (8.2.20).*

Proof: Since $A \in \mathcal{L}^m_{c\ell}(\Omega)$, it follows from Theorem 8.1.8 that the kernel k belongs to $\Psi hk_\kappa(\Omega)$ with $\kappa = -n - m$. Hence,

$$k(x, x-y) = \sum_{j=0}^{L} k_{\kappa+j}(x, x-y) + k_R(x, x-y)$$

where $k_R \in C^{\kappa+L-\delta}(\Omega \times \Omega)$ with some $\delta \in (0,1)$ and $k_{\kappa+j} \in \Psi hf_{\kappa+j}$. For L sufficiently large,

$$\begin{aligned}
&\text{p.f.} \int\limits_\Omega k_R(x, x-y)(y-x)^\alpha dy|_{x=\Psi(x')} \\
&= \int\limits_\Omega k_R(x, x-y)(y-x)^\alpha dy|_{x=\Psi(x')} \\
&= \int\limits_{\Omega'} \widetilde{k}_R(x', x'-y')J(y')\big(\Psi(y') - \Psi(x')\big)^\alpha dy'
\end{aligned}$$

since the integrand is continuous. To show that $b_\alpha(x') = 0$ it therefore suffices to prove that

$$\text{p.f.} \int\limits_{\Omega'} k_q(x, x-y)(y-x)^\alpha \chi(|y-x|) J(y') dy'$$
$$= \text{p.f.} \int\limits_{\Omega} k_q(x, x-y)(y-x)^\alpha \chi(|y-x|) dy \tag{8.2.28}$$

for $q = \kappa + j \notin \mathbb{N}_0$ and $j = 0, \ldots, L$, in view of the asymptotic expansion of k. In the integral on the left–hand side set $x = \Psi(x')$ and $y = \Psi(y')$. The C^∞ cut–off function $\chi(\varrho)$ has the properties $\chi(\varrho) = 1$ for $0 \le \varrho \le \varrho_0$ and $\chi(\varrho) = 0$ for $2\varrho_0 \le \varrho$ with some fixed $\varrho_0 > 0$. For $\varepsilon > 0$, the integral on the right–hand side is given by

$$\begin{aligned}
&\text{p.f.} \int_{\Omega} k_q(x, x-y)(y-x)^\alpha \chi(|y-x|) dy \\
&= \text{p.f.}_{\varepsilon \to 0} \int\limits_{|\omega|=1} \int\limits_{r=\varepsilon}^{2\varrho_0} r^{q+|\alpha|+n-1} \chi(r) dr\, k_q(x, -\omega)\omega^\alpha d\omega \\
&= \text{p.f.}_{\varepsilon \to 0} \Big\{ \frac{\varrho_0^{q+|\alpha|+n}}{q+|\alpha|+n} + \int\limits_{\varrho_0}^{2\varrho_0} r^{q+|\alpha|+n-1}\chi(r) dr - \frac{\varepsilon^{q+|\alpha|+n}}{q+|\alpha|+n} \Big\} \times \\
&\qquad \times \int\limits_{|\omega|=1} k_q(x, -\omega)\omega^\alpha d\omega \\
&= c_\chi \int\limits_{|\omega|=1} k_q(x, -\omega)\omega^\alpha d\omega
\end{aligned}$$

where

$$c_\chi = \frac{\varrho^{q+|\alpha|+n}}{q+|\alpha|+n} + \int\limits_{\varrho_0}^{2\varrho_0} r^{q+|\alpha|+n-1}\chi(r) dr$$

is a constant.

In the same manner, for the integral on the left–hand side in (8.2.28), and by employing the transformation, we obtain

$$\begin{aligned} I'(\varepsilon) &:= \int\limits_{\Omega'\setminus\{|y'-x'|<\varepsilon\}} k_q(\Psi(x'),\Psi(x')-\Psi(y'))(\Psi(y')-\Psi(x'))^\alpha \times \\ &\qquad\qquad \times\chi(|\Psi(y')-\Psi(x')|)J(y')dy' \\ &= \int\limits_{\Omega\setminus\{|y-x|<\varrho_\varepsilon(\omega)\}} k_q(x,x-y)(y-x)^\alpha\chi(|y-x|)dy \\ &= \int\limits_{|\omega|=1}\int\limits_{\varrho=\varrho_\varepsilon(\omega)}^{2\varrho_0} \varrho^{q+n-1+|\alpha|}\chi(\varrho)d\varrho k_q(x,-\omega)\omega^\alpha d\omega \\ &= \int\limits_{|\omega|=1}\left\{\frac{\varrho_0^{q+|\alpha|+n}}{q+|\alpha|+n} + \int\limits_{\varrho_0}^{2\varrho_0} r^{q+|\alpha|+n-1}\chi(r)dr - \frac{\varrho_\varepsilon(\omega)^{q+|\alpha|+n}}{q+|\alpha|+n}\right\}\times \\ &\qquad \times k_q(x,-\omega)\omega^\alpha d\omega\,. \end{aligned}$$

Hence, with the expansion of $\varrho_\varepsilon(\omega)$ in Lemma 8.2.3,

$$\begin{aligned} &\operatorname*{p.f.}_{\varepsilon\to 0} I'(\varepsilon) \\ &= c_\chi \int\limits_{|\omega|=1} k_q(x,-\omega)\omega^\alpha d\alpha - \operatorname*{p.f.}_{\varepsilon\to 0}\int\limits_{|\omega|=1}\frac{1}{q+|\alpha|+n}P(\varepsilon,\omega)k_q(x,-\omega)\omega^\alpha d\omega \end{aligned}$$

with

$$P(\varepsilon,\omega) := \Big\{\sum_{k=1}^N c_k(\omega)\varepsilon^k + O(\varepsilon^{N+1})\Big\}^{q+|\alpha|+n}.$$

By using the geometric series, $P(\varepsilon,\omega)$ can be written in the form

$$\begin{aligned} P(\varepsilon,\omega) &= -\varepsilon^{q+n+|\alpha|}\big(c_1(\omega)\big)^{q+n+|\alpha|}\Big\{1+\sum_{k=1}^N\frac{c_k}{c_1}(\omega)\varepsilon^{k-1}+O(\varepsilon^N)\Big\}^{q+n+|\alpha|} \\ &= -\varepsilon^{q+n+|\alpha|}\big(c_1(\omega)\big)^{q+n+|\alpha|}\times \\ &\qquad \times\sum_{\ell=0}^\infty\binom{q+n+|\alpha|}{\ell}\Big(\sum_{k=2}^N\frac{c_k}{c_1}(\omega)\varepsilon^{k-1}+O(\varepsilon^N)\Big)^\ell \\ &= -\sum_{\ell\ge 0}\varepsilon^{q+n+|\alpha|+\ell}\widetilde{c}_\ell(\omega)+O(\varepsilon^{q+n+|\alpha|+N}) \end{aligned}$$

where $q+n+|\alpha|+\ell\neq 0$ for all $\ell\in\mathbb{N}_0$. Hence,

$$\begin{aligned} I'(\varepsilon) &= -\sum_{\ell\ge 0}\varepsilon^{q+n+|\alpha|+\ell}\int\limits_{|\omega|=1}\widetilde{c}_\ell(\omega)k_q(x,-\omega)\omega^\alpha d\omega \\ &\quad + \int\limits_{|\omega|=1} c_\chi(\omega)k_q(x,-\omega)\omega^\alpha d\omega + O(\varepsilon^{q+n+|\alpha|+N})\,. \end{aligned}$$

Consequently, for N sufficiently large, we have

$$\underset{\varepsilon\to 0}{\text{p.f.}}\, I'(\varepsilon) = \int\limits_{|\omega|=1} c_\chi(\omega) k_q(x,-\omega)\omega^\alpha d\omega\,. \tag{8.2.29}$$

Hence,

$$\underset{\varepsilon\to 0}{\text{p.f.}} \int\limits_{|\omega|=1} P(\varepsilon,\omega)\frac{1}{q+|\alpha|+n} k_q(x,-\omega)\omega^\alpha = 0$$

and, consequently,

$$\underset{\varepsilon\to 0}{\text{p.f.}}\, I'(\varepsilon) = c_\chi \int\limits_{|\omega|=1} k_q(x,-\omega)\omega^\alpha d\omega$$

which proves (8.2.28).

Note that $c_\chi(\omega)$ is independent of the coordinate transformation, therefore this integral is independent of the special choice of Ψ, and, hence, is invariant with respect to the change of coordinates. ∎

In general, for $m \in \mathbb{N}_0$, the extra differential operator $\sum\limits_{|\alpha|\le m} b_\alpha(x')\widetilde{D}^\alpha \widetilde{u}(x')$ is unpleasant for $m > 2$ in view of (8.2.17). However, if the transformation is a linear diffeomorphism $\Psi_L(x')$ then this extra differential operator can be represented in a more simplified manner (see Kieser [225, Theorem 2.2.9]).

Theorem 8.2.5. *If $k \in \Psi hk_\kappa$ with $\kappa = -n-m$, $m \in \mathbb{N}_0$, and $\Psi_L(x')$ is a bijective linear transformation then the coefficients $b_\alpha(x')$ can be written explictly in the form*

$$b_\alpha(x') = \frac{1}{\alpha!}\int\limits_{|\omega|=1} k_{-n-|\alpha|}(x,-\omega)\omega^\alpha \log R(\omega) d\omega \quad \textit{for } |\alpha| \le m\,,\ x = \Psi_L(x')\,. \tag{8.2.30}$$

Here, $k = \sum_{j=0}^J k_{\kappa+j} + k_R$ where $k_R \in C^{\kappa+J-\delta}(\Omega\times\Omega)$ with some $\delta \in (0,1)$ and $k_{\kappa+j} \in \Psi hf_{\kappa+j}(\Omega)$. Moreover, $R(\omega) := |L^{-1}\omega|^{-1}$ where L is the matrix representation for the linear transformation $\Psi_L : y - x = L(y'-x')$.

Proof: From (8.2.22) it is clear, that the difference defining $b_\alpha(x')$ is determined by the singular behavior of the integrals near to $y = x$ only. In view of (3.2.18) for the finite part integrals, we now consider

$$\int\limits_{\Omega'\setminus\{|x'-y'|<\varepsilon\}} \widetilde{k}_{\kappa+j}(x',x'-y')J(y')\big(\Psi_L(y')-\Psi_L(x')\big)^\alpha dy'$$

$$= \int\limits_{\Omega\setminus\{|x-y|<\varepsilon R(\omega)\}} k_{\kappa+j}(x,x-y)(y-x)^\alpha dy\,.$$

Hence,

$$b_\alpha(x') = \frac{1}{\alpha!}\,\text{p.f.}\Big\{ \int\limits_{\Omega\setminus\{|x-y|<\varepsilon\}} k_{\kappa+j}(x,x-y)(y-x)^\alpha dy - \int\limits_{\Omega\setminus\{|x-y|<\varepsilon R(\omega)\}} k_{\kappa+j}(x,x-y)(y-x)^\alpha dy\Big\}.$$

Since $\kappa + j = -m - n + j \notin \mathbb{N}_0$, the kernel function $k_{\kappa+j}$ is positively homogeneous without log–terms. Then, in terms of polar coordinates $y - x = \varrho\omega$, we have

$$b_\alpha(x') = \frac{1}{\alpha!}\,\text{p.f.}\Big\{ \int\limits_{|\omega|=1} k_{\kappa+j}(x,-\omega)\omega^\alpha \int\limits_{r=\varepsilon}^{\varepsilon R(\omega)} r^{\kappa-m+|\alpha|+j-1} dr d\omega\Big\}.$$

Since $\kappa = -m - n$, we have to distinguish two cases:
$|\alpha| \neq m - j$:

$$\begin{aligned} b_\alpha(x') &= \frac{1}{\alpha!}\,\underset{\varepsilon\to 0}{\text{p.f.}}\,\varepsilon^{|\alpha|-m+j}\frac{1}{m-|\alpha|+j}\times \\ &\times \int\limits_{|\omega|=1} \big(R(\omega)^{|\alpha|-m+j} - 1\big) k_{\kappa+j}(x,-\omega)\omega^\alpha d\omega = 0 \end{aligned}$$

since $|\alpha| - m + j \neq 0$.

$|\alpha| = m - j$:

$$b_\alpha(x') = \frac{1}{\alpha!} \int\limits_{|\omega|=1} k_{\kappa+j}(x,-\omega)\omega^\alpha \log R(\omega) d\omega\,.$$

This gives with $j = m - |\alpha|$ and $\kappa = -n - m$ the desired formula (8.2.30). ■

8.2.2 The Class of Invariant Hadamard Finite Part Integral Operators under Change of Coordinates

For $m \in \mathbb{N}_0$, we shall see that the extra terms b_α in Theorem 8.2.2 also vanish for a large class of operators whose kernel functions satisfy the *parity conditions*.

We now state the following crucial result concerning the transformation of finite part integral operators.

Theorem 8.2.6. (Kieser [225, Theorem 2.2.12])

For $A \in \mathcal{L}^m_{c\ell}(\Omega)$ *with* $m \in \mathbb{N}_0$ *let the Schwartz kernel of* A *be* $k(x, x-y) \sim \sum\limits_{j\geq 0} k_{\kappa+j}(x, x-y)$ *and let* $k_{\kappa+j}$ *satisfy the parity conditions* (8.1.74) *with*

$$\sigma_j = j - m + 1 \quad \textit{for } 0 \leq j \leq m\,. \tag{8.2.31}$$

Then the finite part integral is invariant under change of coordinates; namely

$$\text{p.f.} \int\limits_{\Omega} k(x, x-y)u(y)dy = \text{p.f.} \int\limits_{\Omega'} \widetilde{k}(x', x'-y')\widetilde{u}(y')J(y')dy' \tag{8.2.32}$$

where $\widetilde{k}$ *is given by* (8.2.21).

Remark 8.2.1: This class of operators includes all of the boundary integral operators generated by the reduction to the boundary of regular elliptic boundary value problems based on Green's formula. The proof of this theorem is delicate and will be presented after we establish some preliminary results.

Let $\mathcal{P}$ denote the class of all polynomials in ε of the form

$$\wp(\omega, \varepsilon) = \sum_{j\geq 0} a_j(\omega)\varepsilon^j \quad \text{with} \quad a_j(-\omega) = (-1)^j a_j(\omega)$$

where $\omega \in \mathbb{R}^n$ with $|\omega| = 1$.

Lemma 8.2.7. *If* $\wp_1, \wp_2 \in \mathcal{P}$ *then* $\wp_1 + \wp_2 \in \mathcal{P}$ *and* $\wp_1\wp_2 \in \mathcal{P}$.

Proof: Let $\wp_1 = \sum\limits_{j\geq 0} a_j(\omega)\varepsilon^j$ and $\wp_2 = \sum\limits_{k\geq 0} b_k(\omega)\varepsilon^k$. Then

$$\wp_1 + \wp_2 = \sum_{\ell\geq 0} c_\ell(\omega)\varepsilon^\ell \quad \text{with} \quad c_\ell = a_\ell + b_\ell\,.$$

Clearly,

$$c_\ell(-\omega) = a_\ell(-\omega) + b_\ell(-\omega) = (-1)^\ell\big(a_\ell(\omega) - b_\ell(\omega)\big) = (-1)^\ell c_\ell(\omega)\,.$$

Similarly, by the use of the Cauchy product,

$$\wp_1\wp_2 = \sum_{\ell\geq 0} c_\ell(\omega)\varepsilon^\ell \quad \text{where} \quad c_\ell(\omega) = \sum_{j+k=\ell} a_j(\omega)b_k(\omega)\,.$$

Hence,

$$c_\ell(-\omega) = \sum_{j+k=\ell} a_j(-\omega)b_k(-\omega) = (-1)^\ell \sum_{j+k=\ell} a_j(\omega)b_k(\omega) = (-1)^\ell c_\ell(\omega)\,.$$

■

Lemma 8.2.8. *Let $k_q \in \Psi hf_q(\Omega)$. In the case when $q \le -n$ is an integer then we require in addition that k_q satisfies the parity condition*

$$k_q(x, y-x) = (-1)^{-q-n+1} k_q(x, x-y) \quad \text{for } y \ne x\,. \tag{8.2.33}$$

Then the following invariance properties hold:

$$\begin{aligned} &\text{p.f.} \int\limits_{\Omega'} k_q(x, x-y)(y-x)^\alpha \chi(|y-x|) J(y') dy' \\ &= \quad \text{p.f.} \int\limits_{\Omega} k_q(x, x-y)(y-x)^\alpha \chi(|y-x|) dy\,. \end{aligned} \tag{8.2.34}$$

In the integral on the left–hand side it is understood that $x = \Psi(x')$ and $y = \Psi(y')$. The C^∞ cut–off function $\chi(\varrho)$ has the properties $\chi(\varrho) = 1$ for $0 \le \varrho \le \varrho_0$, $\chi(\varrho) = 0$ for $2\varrho_0 \le \varrho$ for some fixed $\varrho_0 > 0$. In formula (8.2.34) we assume that $B_{2\varrho_0} \subset \Omega$ and $\Psi\big(B_\varepsilon(x')\big) \subset B_{\varrho_0}(x)$ for all sufficiently small $\varepsilon > 0$.

Proof: With the polar coordinates $y-x = r\omega$, the right–hand side of (8.2.34) is given by

$$I_r = \underset{\varepsilon\to 0}{\text{p.f.}} \int\limits_{|\omega|=1} \int\limits_{r=\varepsilon}^{2\varrho_0} k_q(x, -\omega)\omega^\alpha \chi(r) r^{q+|\alpha|+n-1} dr d\omega\,.$$

For the left–hand side of (8.2.34), since for $\varepsilon > 0$ the integrand is regular, we have similarly

$$I_\ell = \underset{\varepsilon\to 0}{\text{p.f.}} \int\limits_{|\omega|=1} \int\limits_{r=\varepsilon(\omega)}^{2\varrho_0} k_q(x, -\omega)\omega^\alpha \chi(r) r^{q+|\alpha|+n-1} dr d\omega\,.$$

Now we first consider the case $q + n + |\alpha| = 0$. Then,

$$\begin{aligned} I_r &= \underset{\varepsilon\to 0}{\text{p.f.}} \Big\{ \log \varrho_0 + \int\limits_{\varrho_0}^{2\varrho_0} r^{-1}\chi(r) dr - \log\varepsilon \Big\} \int\limits_{|\omega|=1} k_q(x, -\omega)\omega^\alpha d\omega \\ &= c_\chi(0) \int\limits_{|\omega|=1} k_q(x, -\omega)\omega^\alpha d\omega \end{aligned}$$

with the constant $c_\chi(0) := \log \varrho_0 + \int\limits_{\varrho_0}^{2\varrho_0} r^{-1}\chi(r) dr$.

For the left–hand side of (8.2.34) we have

$$I_\ell = \underset{\varepsilon\to 0}{\text{p.f.}} \Big\{ \log \varrho_0 + \int\limits_{\varrho_0}^{2\varrho_0} r^{-1}\chi(r)dr - \log \varrho_\varepsilon(\omega)\Big\} \int\limits_{|\omega|=1} k_q(x,-\omega)\omega^\alpha d\omega\,.$$

In terms of the expansion (8.2.24) of $\varrho_\varepsilon(\omega)$ in Lemma 8.2.3, we have

$$\begin{aligned}
-\log \varrho_\varepsilon(\omega) &= -\log\Big\{\sum_{k=1}^N c_k(\omega)\varepsilon^k + O(\varepsilon^{N+1})\Big\}\\
&= -\log\varepsilon - \log c_1(\omega) - \log\Big(1 + \sum_{k=2}^N \frac{c_k(\omega)}{c_1(\omega)}\varepsilon^{k-1} + O(\varepsilon^N)\Big)
\end{aligned}$$

from which we obtain

$$\begin{aligned}
&\underset{\varepsilon\to 0}{\text{p.f.}}\big(-\log\varrho_\varepsilon(\omega)\big)\\
&= \underset{\varepsilon\to 0}{\text{p.f.}}\Big\{-\log c_1(\omega) - \log\varepsilon + \sum_{\ell=1}^\infty \frac{1}{\ell}\Big(\sum_{k=2}^N \frac{c_k}{c_1}\varepsilon^{k-1} + O(\varepsilon^N)\Big)^\ell\Big\}\\
&= -\log c_1(\omega)\,.
\end{aligned}$$

Hence,

$$I_\ell = c_\chi(0)\int\limits_{|\omega|=1} k_q(x,-\omega)\omega^\alpha d\omega - \int\limits_{|\omega|=1}\big(\log c_1(\omega)\big)k_q(x,\omega)\omega^\alpha d\omega\,.$$

Since the integration is taken over the unit sphere, we have

$$\begin{aligned}
&-\int\limits_{|\omega|=1} \log\big(c_1(\omega)\big)k_q(x,-\omega)\omega^\alpha d\omega\\
&= -\int\limits_{|\omega|=1} \log c_1(-\omega)k_q(x,+\omega)(-\omega)^\alpha d(-\omega)\\
&= -\int\limits_{|\omega|=1} \log c_1(\omega)(-1)^{-q-n-1+|\alpha|}k_q(x,-\omega)\omega^\alpha d\omega\\
&= \int\limits_{|\omega|=1} \log c_1(\omega)k_q(x,-\omega)\omega^\alpha d\omega
\end{aligned}$$

which implies

$$\int\limits_{|\omega|=1} \log c_1(\omega)k_q(x,-\omega)\omega^\alpha d\omega = 0\,.$$

Consequently,

$$I_\ell = c_\chi(0) \int\limits_{|\omega|=1} k_q(x,-\omega)\omega^\alpha d\omega = I_r \,,$$

which implies the proposed result (8.2.34) in this case.

Next, we consider the case $-p := q + n + |\alpha| < 0$. Then the right–hand side in (8.2.34) takes the form

$$\begin{aligned} I_r &= \underset{\varepsilon\to 0}{\text{p.f.}} \Big\{ -\frac{\varrho_0^{-p}}{p} + \int\limits_{\varrho_0}^{2\varrho_0} r^{-p-1}\chi(r)dr + \frac{\varepsilon^{-p}}{p} \Big\} \int\limits_{|\omega|=1} k_q(x,-\omega)\omega^\alpha d\omega \\ &= c_\chi(p) \int\limits_{|\omega|=1} k_q(x,-\omega)\omega^\alpha d\omega \end{aligned}$$

where

$$c_\chi(p) = \int\limits_{\varrho_0}^{2\varrho_0} r^{-p-1}\chi(r)dr - \frac{\varrho_0^{-p}}{p}\,.$$

Similarly,

$$I_\ell = c_\chi(p) \int\limits_{|\omega|=1} k_q(x,-\omega)\omega^\alpha d\omega + \frac{1}{p}\underset{\varepsilon\to 0}{\text{p.f.}} \int\limits_{|\omega|=1} P(\varepsilon,\omega) k_q(x,-\omega)\omega^\alpha d\omega$$

with

$$P(\varepsilon,\omega) = \big(\varrho_\varepsilon(\omega)\big)^{-p}\,.$$

By inserting the expansion (8.2.24) of $\varrho_\varepsilon(\omega)$, we obtain

$$\begin{aligned} P(\varepsilon,\omega) &= \Big(\sum_{k=1}^N c_k(\omega)\varepsilon^k + O(\varepsilon^{N+1})\Big)^{-p} \\ &= c_1(\omega)^{-p}\varepsilon^{-p}\Big\{1 + \sum_{k=2}^N \frac{c_k(\omega)}{c_1(\omega)}\varepsilon^{k-1} + O(\varepsilon^N)\Big\}^{-p} \\ &= c_1(\omega)^{-p}\varepsilon^{-p}\Big\{\sum_{\ell=0}^\infty (-1)^\ell\Big(\sum_{k=2}^N \frac{c_k(\omega)}{c_1(\omega)}\varepsilon^{k-1} + O(\varepsilon^N)\Big)^\ell\Big\}^p\,. \end{aligned}$$

We note that by setting $\wp_1 := \sum\limits_{k=2}^N \frac{c_k(\omega)}{c_1(\omega)}\varepsilon^{k-1}$ that $\wp_1 \in \mathcal{P}$ since (8.2.25) holds. Now, apply $\ell \cdot p$ times Lemma 8.2.7; then it follows that

$$P(\varepsilon,\omega) = c_1(\omega)^{-p}\varepsilon^{-p}\big(\wp + O(\varepsilon^N)\big)$$

with $\wp \in \mathcal{P}$. If $N = p+1$, then

$$\underset{\varepsilon\to 0}{\text{p.f.}}\, P(\varepsilon,\omega) = \underset{\varepsilon\to 0}{\text{p.f.}}\, c_1(\omega)^{-p}\varepsilon^{-p}\wp(\varepsilon,\omega)\,.$$

Therefore, employing $\wp(\varepsilon,\omega) = \sum_{j=0}^{p+1} a_j(\omega)\varepsilon^j$ with $a_j(-\omega) = (-1)^j a_j(\omega)$, we find

$$\begin{aligned}
&\frac{1}{p}\underset{\varepsilon\to 0}{\text{p.f.}} \int\limits_{|\omega|=1} P(\varepsilon,\omega)k_q(x,-\omega)\omega^\alpha d\alpha \\
&\quad = \frac{1}{p}\underset{\varepsilon\to 0}{\text{p.f.}} \int\limits_{|\omega|=1} \sum_{j=0}^{p+1} a_j(\omega)\varepsilon^{j-p}k_q(x,-\omega)\omega^\alpha d\omega \\
&\quad = \frac{1}{p}\int\limits_{|\omega|=1} a_p(\omega)k_q(x,-\omega)\omega^\alpha d\omega\,.
\end{aligned}$$

The latter integral as an integral over the unit sphere satisfies

$$\begin{aligned}
&\frac{1}{p}\int\limits_{|\omega|=1} a_p(\omega)k_q(x,-\omega)\omega^\alpha d\omega \\
&\quad = \frac{1}{p}\int\limits_{|\omega|=1} a_p(-\omega)k_q(x,\omega)(-\omega)^\alpha d(-\omega) \\
&\quad = \frac{1}{p}(-1)^{p-q-n-1+|\alpha|}\int\limits_{|\omega|=1} a_p(\omega)k_q(x,-\omega)\omega^\alpha d\omega \\
&\quad = -\frac{1}{p}\int\limits_{|\omega|=1} a_p(\omega)k_q(x,-\omega)\omega^\alpha d\omega
\end{aligned}$$

and, hence, vanishes. This implies again

$$I_\ell = c_\chi(p)\int\limits_{|\omega|=1} a_p(\omega)k_q(x,-\omega)\omega^\alpha d\omega = I_r\,.$$

This completes the proof of Lemma 8.2.8. ■

We are now able to prove Theorem 8.2.6.

Proof of Theorem 8.2.6: With the asymptotic pseudohomogeneous expansion of the Schwartz kernel of A,

$$k(x,x-y) = \sum_{j=0}^{L} k_{\kappa+j}(x,x-y) + k_R(x,x-y)\,,$$

to prove (8.2.32), it suffices to consider the coordinate transformation for a typical term in the expansion since k_R is not singular any more. To this end consider $k_q = k_{\kappa+j}$ in Lemma 8.2.8 and we have the parity condition

$$k_q(x, x-y) = k_{\kappa+j}(x, x-y) = (-1)^{\sigma_j} k_{\kappa+j}(x, y-x)$$

with $\sigma_j = j - m + 1$. Hence,

$$\begin{aligned} k_q(x, x-y) &= (-1)^{j-m+1} k_q(x, y-x) \\ &= (-1)^{q+n+1} k_q(x, y-x) \\ &= (-1)^{-q-n-1} k_q(x, y-x) \end{aligned}$$

as assumed in (8.2.33) and $k_q \in \Psi h f_q$. Therefore, with L sufficiently large,

$$\begin{aligned} \text{p.f.} \int_\Omega k(x, x-y) u(y) dy &= \text{p.f.} \sum_{j=0}^{L} \int_\Omega k_{\kappa+j}(x, x-y) u(y) dy \\ &\quad + \int_\Omega k_R(x, x-y) u(y) dy \\ &= \text{p.f.} \sum_{j=0}^{L} \int_{\Omega'} \widetilde{k}_{\kappa+j}(x', x'-y') \widetilde{u}(y') J(y') dy' \\ &\quad + \int_{\Omega'} \widetilde{k}_R(x', x'-y') \widetilde{u}(y') J(y') dy' \\ &= \text{p.f.} \int_{\Omega'} \widetilde{k}(x', x'-y') \widetilde{u}(y') J(y') dy' , \end{aligned}$$

which completes the proof of Theorem 8.2.6. ■

In fact, **the parity conditions required in Theorem 8.2.6 are invariant under the change of coordinates** since that is a special case of the following theorem.

Theorem 8.2.9. *Let $A \in \mathcal{L}^m_{c\ell}(\Omega)$ with $m \in \mathbb{Z}$ having the pseudohomogeneous kernel expansion of the Schwartz kernel*

$$k_A(x, x-y) \sim \sum_{j \geq 0} k_{\kappa+j}(x, x-y) .$$

Suppose that the parity conditions

$$k_{\kappa+j}(x, x-y) = (-1)^{m-j+\sigma_0} k_{\kappa+j}(x, x-y) \quad \textit{for } 0 \leq j \leq L \tag{8.2.35}$$

with a fixed $\sigma_0 \in \mathbb{N}_0$ are satisfied. Let Φ be a diffeomorphism defining by $x' = \Phi(x)$, a change of coordinates, and let

$$\widetilde{k}_{A_\Phi}(x', x'-y') \sim \sum_{j\geq 0} \widetilde{k}_{\kappa+j}(x', x'-y')$$

be the pseudohomogeneous kernel expansion of the transformed Schwartz kernel to A_Φ *in terms of the new coordinates. Then the parity conditions in* (8.2.35) *are invariant under the change of coordinates, i.e.*

$$\widetilde{k}_{\kappa+j}(x', x'-y') = (-1)^{m-j+\sigma_0}\widetilde{k}_{\kappa+j}(x', x'-y') \quad \textit{for } 0 \leq j \leq L\,. \tag{8.2.36}$$

Proof: For the proof we employ Lemma 8.1.9. Then (8.2.35) is equivalent to

$$a^0_{m-j}(x, -\xi) = (-1)^{m-k-\sigma_0} a^0_{m-j}(x, \xi) \quad \text{for } \xi \neq 0 \text{ and } 0 \leq j \leq L\,, \tag{8.2.37}$$

where a^0_{m-j} are the homogeneous symbols of the complete symbol expansion of A. Since the symbol $a_\Phi(x', \xi')$ is given by (7.1.45) with (7.1.46), by rearranging terms, the corresponding asymptotic homogeneous expansion of the symbol reads

$$\begin{aligned}
&\sum_{j\geq 0} a^0_{\Phi, m-j}(x', \xi') \\
&= \sum_{j\geq 0} \Big\{ \sum_{\ell=0}^{j} \sum_{|\alpha|=\ell} \frac{1}{\alpha!}\Big(-i\frac{\partial}{\partial y'}\Big)^\alpha \Big(\det\frac{\partial\Phi}{\partial y}\Big)^{-1} \big(\Xi^\top(x', y')\big)^{-1} \Psi\Big(\frac{x'-y'}{\varepsilon}\Big) \times \\
&\qquad \times \Big(\frac{\partial}{\partial \xi'}\Big)^\alpha a^0_{m-j+\ell}\Big(x, \big(\Xi^\top(x', y')\big)^{-1} \xi'\Big)\Big\}\big|_{y'=x'}\,.
\end{aligned}$$

We observe that (8.2.37) implies

$$\begin{aligned}
&\Big(\frac{\partial}{\partial \xi'}\Big)^\alpha a^0_{m-j+\ell}\Big(x, -\big(\Xi^\top(x', y')\big)^{-1} \xi'\Big) \\
&\qquad = (-1)^{m-j+\sigma_0} \Big(\frac{\partial}{\partial \xi'}\Big)^\alpha a^0_{m-j+\ell}\Big(x, \big(\Xi^\top(x', y')\big)^{-1} \xi'\Big)\,.
\end{aligned}$$

Then it follows from the above representation that

$$a^0_{\Phi, m-j}(x', -\xi') = (-1)^{m-j+\sigma_0} a^0_{\Phi, m-j}(x', -\xi')$$

which implies (8.2.37) by the application of Lemma 8.1.9 again. ■

To conclude this section, we observe for the Tricomi conditions, that as a consequence of Theorem 7.1.13 and $A \in \mathcal{L}^m_{c\ell}(\Omega)$, we also have $\widetilde{A} \in \mathcal{L}^m_{c\ell}(\Omega')$. Hence, one may apply Theorem 8.1.7 to the kernels $k_{\kappa+j}$ and $\widetilde{k}_{\kappa+j}$ of A and $\widetilde{A}$, respectively. This yields the following lemma which shows that **the Tricomi conditions are invariant under the change of coordinates**.

Lemma 8.2.10. *Let* $A \in \mathcal{L}^m_{c\ell}(\Omega)$ *and* $m \in \mathbb{N}_0$, $k(x, x-y) \sim \sum\limits_{j\geq 0} k_{\kappa+j}(x, x-y)$ *and* $\widetilde{k}(x', x'-y') \sim \sum\limits_{j\geq 0} \widetilde{k}_{\kappa+j}(x', x'-y')$. *Then the Tricomi conditions*

$$\int\limits_{|\Theta|=1} \Theta^\alpha k_{\kappa+j}(x, \Theta) d\omega(\Theta) = 0 = \int\limits_{|\Theta|=1} \Theta^\alpha \widetilde{k}_{\kappa+j}(x', \Theta) d\omega(\Theta) \tag{8.2.38}$$

are satisfied for all $|\alpha| = m - j$ *and* $0 \leq j \leq m$.

We remark that for $0 \leq j \leq m$, one has $\kappa + j = -n - (m - j) < 0$. Hence, from Definition 8.1.1, we conclude that $k_{\kappa+j}$ and $\widetilde{k}_{\kappa+j}$ are both positively homogeneous of degree $\kappa + j$.

9. Pseudodifferential and Boundary Integral Operators

This chapter concerns the relation between the boundary integral operators and classical pseudodifferential operators. A large class of boundary integral operators including those presented in the previous chapters belong to the special class of classical pseudodifferential operators on compact manifolds. We are particularly interested in strongly elliptic systems of pseudodifferential operators providing Gårding's inequality, see Theorem 9.1.4. The particular class of operators in the domain having symbols of rational type enjoys many special properties such as their relation to Newton potentials, which define genuine pseudodifferential operators in $\mathbb{R}^n$ and which satisfy in particular the transmission conditions covered in the work of Boutet de Monvel for a more general class of pseudodifferential operators. The traces of their composition with tensor product distributions involving $\delta_{\Gamma}^{(k)}$ (i.e. the trace of Poisson operators by Boutet de Monvel [39, 40, 41] and Grubb [159]), generate, in a natural way, boundary integral operators as pseudodifferential operators on the boundary manifold.

In fact, it should be mentioned that Boutet de Monvel found a complete calculus of Green operators (see Chapter 10) where all compositions of these operators belong to the same class. For the special class of pseudodifferential operators with symbols of rational type, corresponding results are presented in the Theorems 9.5.5 and 9.5.8.

To obtain these results, we present a detailed analysis of the boundary potentials (or Poisson operators) in the vicinity of the boundary which is based on properties of the pseudohomogeneous expansions of the Schwartz kernel (see the extension properties Theorems 9.3.2 and 9.3.8). It should be noted that in contrast to the requirements of C^{∞}–extensions as for the transmission conditions, here we only need finitely many conditions and corresponding finite regularity of Γ. Our conditions can be obtained directly from the Schwartz kernels which is costumarily employed in practical applications. Once the resulting boundary operator given by the original Schwartz kernel reduced to Γ satisfies the Tricomi conditions it automatically becomes a pseudodifferential operator on the boundary.

The relations between various conditions such as the extension conditions, Tricomi conditions, parity conditions as well as transmission conditions are summarized in Table 9.3.1. Moreover, the invariance properties and trans-

G. C. Hsiao und W. L. Wendland, *Boundary Integral Equations*,
Applied Mathematical Sciences 164,
https://doi.org/10.1007/978-3-030-71127-6_9

mission conditions under change of local coordinates of Γ lead to the result of Theorem 9.4.6 which implies that all these boundary integral operators are invariant under the change of coordinates. In particular, the Hadamard finite part integral operators in this class transform into Hadamard's finite part integral operators without producing local differential operator contributions.

We also collect in this chapter the relevant mapping properties and jump relations for the boundary potentials (or Poisson operators) as well as for the volume potentials, which are covered by Boutet de Monvel's work for even more general pseudodifferential operators satisfying the transmission conditions including those of operators with symbols of rational type.

The last section is devoted to the concept of strong ellipticity and Fredholm properties of boundary pseudodifferential operators.

The presentation of this chapter is partially based on the book by Chazarain and Piriou [64], the PHD dissertation by Kieser [225] and Hörmander's book [188, Vol.III] and covers only a small part of Boutet de Monvel's analysis (see [39, 40, 41], Grubb [159] and Schulze [379]).

9.1 Pseudodifferential Operators on Boundary Manifolds

In Section 3.3 we introduced the parametric representation of the boundary $\Gamma = \partial\Omega$ (cf.(3.3.6),(A.1)). This defines Γ as an $(n-1)$–dimensional parameterized surface. We may also consider Γ as a manifold immersed into $\mathbb{R}^n$ in the sense of differential geometry and associate Γ with an *atlas* $\mathfrak{A}$ which is a family of *local charts* $\{(O_r\,,\,U_r\,,\,\chi_r)\,|\,r\in I\}$. Each of the local charts is a triplet with $U_r\subset\mathbb{R}^{n-1}$ an open subset of the *parametric space* $\mathbb{R}^{n-1}$; and where the representation $x=T_r(\varrho')=\chi_r^{(-1)}(\varrho')$ for $\varrho'\in U_r$ defines a parameterized patch $O_r:=T_r(U_r)$ of the surface Γ, or, reversely, $U_r=\chi_r(O_r)$. The mappings T_r and χ_r are both bijective and bicontinuous, hence, $T_r=\chi_r^{(-1)}$ is a homeomorphism, see Figure 9.1.1. For an atlas we require $\Gamma=\bigcup\limits_{r\in I}O_r$. Moreover, if $O_{rt}:=O_r\cap O_t\neq\emptyset$ then the mapping

$$\Phi_{rt}:=\chi_t\circ T_r=\chi_t\circ\chi_r^{(-1)}\;:\;\chi_r(O_{rt})\to\chi_t(O_{rt}) \tag{9.1.1}$$

is supposed to be a sufficiently smooth diffeomorphism. (For details see also Section 3.3.) Note that any union of two atlases is again an atlas on Γ.

Definition 9.1.1. *Let $A:\mathcal{D}(\Gamma)\to\mathcal{E}(\Gamma)$ be a continuous linear operator. Then A is said to be in the class $\mathcal{L}^m(\Gamma)$ of pseudodifferential operators if for every chart $(O_r\,,\,U_r\,,\,\chi_r)$ the associated local operator*

$$A_{\chi_r}:=\chi_{r*}A\chi_r^*\;:\;\mathcal{D}(U_r)\to\mathcal{E}(U_r) \tag{9.1.2}$$

belongs to $\mathcal{L}^m(U_r)$. Here,

$$\begin{aligned}
\chi_{r*}v &:= v\circ T_r \quad \textit{maps } v \textit{ given on } O_r \textit{ to } v\circ\chi_r^{(-1)}\,,\\
&\qquad\qquad \textit{a function on } U_r\,;\\
\chi_r^* u &:= u\circ\chi_r \quad \textit{maps } u \textit{ given on } U_r \textit{ to } u\circ\chi_r\,,\\
&\qquad\qquad \textit{a function on } O_r\subset\Gamma\,,
\end{aligned}\tag{9.1.3}$$

are defining the pushforward χ_{r*} *and* pullback χ_r^*, *respectively.*

A function v on Γ is said to be in the class $C^k(\Gamma)$ if for every chart the pushforward has the property $v\circ T_r = \chi_{r*}v\in C^k(U_r)$. Hence,

$$\begin{aligned}
\chi_{r*} &: C^\infty(O_r) \to C^\infty(U_r)\,,\\
\chi_r^* &: C^\infty(U_r) \to C^\infty(O_r)
\end{aligned}\tag{9.1.4}$$

and

$$\begin{aligned}
\chi_{r*} &: C_0^\infty(O_r) \to C_0^\infty(U_r)\,,\\
\chi_r^* &: C_0^\infty(U_r) \to C_0^\infty(O_r)\,.
\end{aligned}\tag{9.1.5}$$

In addition, we require the following smoothing property *to be satisfied: If $\varphi,\ \psi\in C_0^\infty(\Gamma)$ with $\operatorname{supp}\varphi\cap\operatorname{supp}\psi=\emptyset$ then the composition $\varphi A\psi\bullet$ extends to a smoothing operator on Γ, i.e., for every $u\in C^\infty(\Gamma)$ one has $\varphi A\psi u\in C^\infty(\Gamma)$.*

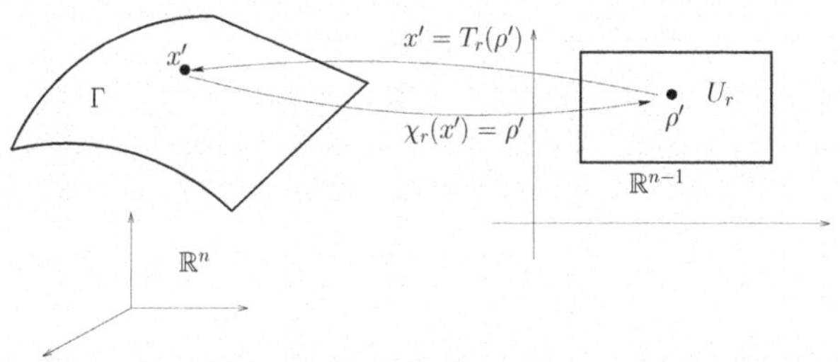

Figure 9.1.1: The local surface representation.

An immediate consequence of these definitions is the following theorem.

Theorem 9.1.1. *Let $A\in\mathcal{L}^m(\Gamma)$ and let $\mathfrak{A}=\{(O_r\,,\ U_r\,,\ \chi_r)\,|\,r\in I\}$ be an atlas on Γ. Then for every pair of charts (O_r,U_r,χ_r), (O_t,U_t,χ_t) and the induced mapping Φ_{rt} in (9.1.1), the induced local pseudodifferential operators satisfy the compatibility relations*

$$\chi_t^* A_{\chi_t}\chi_{t*} = \chi_r^* A_{\chi_r}\chi_{r*} = A \quad \textit{on } \mathcal{D}(O_{rt}) \tag{9.1.6}$$

and

$$A_{\chi_t} = \chi_{t*}\chi_r^* A_{\chi_r}\chi_{r*}\chi_t^* = \Phi_{rt*}A_{\chi_r}\Phi_{rt}^* = (A_{\chi_r})_{\Phi_{rt}} \quad \textit{on } \mathcal{D}\big(\chi_t(O_{rt})\big)\,; \tag{9.1.7}$$

(*see* Remark 7.1.4).

For the principal symbols of the local operators A_{χ_r} *one has*

$$\sigma_{A_{\chi_r},m}\big(\chi_r(x),\xi\big) = \sigma_{A_{\chi_t},m}\big(\chi_t(x),\xi'\big) \tag{9.1.8}$$

where $x = T_r(\varrho') = T_t(\tau') \in O_{rt} \subset \Gamma$ *and the variables* ξ *and* ξ' *in* $\mathbb{R}^{n-1}\setminus\{0\}$ *are related by the equations*

$$\sum_{\iota=1}^{n-1} \frac{\partial T_t}{\partial \tau'_\iota}(\tau')\xi_\iota = \sum_{\iota=1}^{n-1} \frac{\partial T_r}{\partial \varrho'_\iota}(\varrho')\xi'_\iota \,. \tag{9.1.9}$$

Conversely, if a family of local pseudodifferential operators $A_{\chi_r} \in \mathcal{L}^m(U_r)$ *is given satisfying* (9.1.7) *for the whole atlas* $\mathfrak{A}$ *and satisfying the smoothing property in Definition* 9.1.1, *then* (9.1.6) *defines a pseudodifferential operator*$A \in \mathcal{L}^m(\Gamma)$.

Proof: The equations (9.1.6) and (9.1.7) are immediate consequences of the previous definitions.

For the transformation proposed in (9.1.8) we employ the coordinate transformation $\Phi_{rt} = \chi_t \circ \chi_r^{(-1)}$ given in (9.1.1), apply (7.1.49) to (9.1.7) and obtain

$$\sigma_{A_{\chi_t},m}(\tau',\xi') = \sigma_{A_{\chi_r},m}\Big(\Phi_{rt}^{(-1)}(\tau')\,,\,\Big(\frac{\partial \Phi_{rt}}{\partial \varrho'}\Big)^{\top}\xi'\Big)\,.$$

The argument in the right–hand side can be expressed component–wise as

$$\xi_\lambda := \Big(\Big(\frac{\partial \Phi_{rt}}{\partial \varrho'}\Big)^{\top}\xi'\Big)_\lambda = \sum_{\iota=1}^{n-1} \frac{\partial \tau'_\lambda}{\partial \varrho'_\iota}\xi'_\iota = \sum_{\iota=1}^{n-1}\sum_{\ell=1}^{n} \frac{\partial \tau'_\lambda}{\partial x_\ell}\frac{\partial x_\ell}{\partial \varrho'_\iota}\xi'_\iota\,,\ \lambda = 1,\dots,n-1\,.$$

Inserting (3.4.27) with $g_t^{n\lambda} = 0$ and the inverse $\gamma_t^{\mu\lambda}$ to $\gamma_{t\mu\lambda} = \sum_{\ell=1}^{n} \frac{\partial x_\ell}{\partial \tau'_\mu}\frac{\partial x_\ell}{\partial \tau'_\lambda}$ on Γ (c.f. (3.4.2), (3.4.4)) we find with (3.4.24), (3.4.28):

$$\xi_\lambda = \sum_{\iota=1}^{n-1}\sum_{\mu=1}^{n-1}\sum_{\ell=1}^{n} \frac{\partial x_\ell}{\partial \tau'_\mu}\frac{\partial x_\ell}{\partial \varrho'_\iota}\gamma_t^{\mu\lambda}\xi'_\iota \quad \text{for } \lambda = 1,\dots,n-1\,. \tag{9.1.10}$$

On the other hand, if (9.1.9) is satisfied, the scalar multiplication of both sides of (9.1.9) by $\frac{\partial T_t}{\partial \tau'_\mu}$ gives

$$\sum_{\iota=1}^{n-1} \gamma_{t\mu\iota}\xi_\iota = \sum_{\iota=1}^{n-1}\sum_{\ell=1}^{n} \frac{\partial x_\ell}{\partial \varrho'_\iota}\frac{\partial x_\ell}{\partial \tau'_\mu}\xi'_\iota\,.$$

Multiplication by $\gamma_t^{\mu\lambda}$ yields

$$\xi_\lambda = \sum_{\mu=1}^{n-1}\sum_{\iota=1}^{n-1}\sum_{\ell=1}^{n} \frac{\partial x_\ell}{\partial \varrho'_\iota}\frac{\partial x_\ell}{\partial \tau'_\mu}\gamma_t^{\mu\lambda}\xi'_\iota ,$$

which coincides with (9.1.10). Hence, (9.1.9) is equivalent to (9.1.10) and (9.1.8) together with (9.1.9) is justified. ■

This theorem suggests to us how to define the principal symbol of $A \in \mathcal{L}^m(\Gamma)$. To this end we introduce the tangent and the cotangent bundles at $x \in \Gamma$. For any fixed $x \in \Gamma$, we denote by the *tangent space* $\mathcal{T}_x(\Gamma)$ the $(n-1)$–dimensional vector space of all tangent vectors to Γ at x. If $x \in O_r$ for some chart (O_r, U_r, χ_r) then the vectors $\frac{\partial x}{\partial \varrho'_\iota} = \frac{\partial T_r(\varrho')}{\partial \varrho'_\iota}$, $\iota = 1, \ldots, (n-1)$, form a basis of $\mathcal{T}_x(\Gamma)$. By $\mathcal{T}^*_x(\Gamma)$, the *cotangent space*, we denote the space of all linear functionals ζ operating on the tangent space $\mathcal{T}_x$. For the above chart, the values

$$\xi_\iota = \zeta\left(\frac{\partial x}{\partial \varrho'_\iota}\right) \tag{9.1.11}$$

define the contravariant coordinates of ζ with respect to the basis $\frac{\partial x}{\partial \varrho'_\iota}$ of $\mathcal{T}_x$. The collection of all the tangent spaces for x tracing Γ is called the *tangent bundle* of $\mathcal{T}(\Gamma)$ and, correspondingly, the collection of all the cotangent spaces is called the *cotangent bundle* $\mathcal{T}^*(\Gamma)$ of Γ.

In terms of this terminology the identity (9.1.9) is just the transformation between the contravariant coordinates ξ and ξ' for the same linear functional $\zeta \in \mathcal{T}^*_x$ where

$$\xi'_\iota = \zeta\left(\frac{\partial x}{\partial \tau'_\iota}\right),$$

under the change of variables (9.1.1). Now we are in the position to define the *principal symbol* $\sigma_{A,m}(x,\zeta)$ of $A \in \mathcal{L}^m(\Gamma)$ for $x \in \Gamma$ and $\zeta \in \mathcal{T}^*_x(\Gamma)\setminus\{0\}$ as follows: For $x \in \Gamma$ we choose a chart (O_r, U_r, χ_r) with $x \in O_r$. Then A_{χ_r} is well defined and to any chosen $\zeta \in \mathcal{T}^*_x(\Gamma)\setminus\{0\}$ there belongs a vector $\xi \in \mathbb{R}^{n-1}$ given by (9.1.11). To $A_{\chi_r} \in \mathcal{L}^m(U_r)$ and $\xi \in \mathbb{R}^{n-1}$ then we find the complete symbol class (7.1.30) and also via (7.1.31) the corresponding principal symbol

$$\sigma_{A_{\chi_r},m}(\varrho',\xi) =: \sigma_{A,m}(x,\zeta) \tag{9.1.12}$$

in the class $\boldsymbol{S}^m(U_r \times \mathbb{R}^{n-1})/\boldsymbol{S}^{m-1}(U_r \times \mathbb{R}^{n-1})$.

Theorem 9.1.1 now implies that for fixed $x \in \Gamma$ and fixed $\zeta \in \mathcal{T}^*_x$ the value of the principal symbol is invariant with respect to changing charts. Consequently, $\sigma_{A,m}(x,\zeta)$ is well defined on the manifold Γ and the cotangent bundle $\mathcal{T}^*(\Gamma)$ and we write

$$\sigma_{A,m} \in \boldsymbol{S}^m\big(\Gamma \times \mathcal{T}^*(\Gamma)\big)\,.$$

In order to extend the domain of the definition of $A \in \mathcal{L}^m(\Gamma)$ from local functions to functions $u \in C^\infty(\Gamma)$ we need the concept of the *partition of* *unity subordinate to the open covering* $\{O_r\}_{r\in I}$ of an atlas $\mathfrak{A}$. Since we always assume that Γ is a compact boundary manifold, Heine–Borel's theorem implies that we may consider only finite atlases, i.e. I is a finite index set. Then there exists a C^∞–partition of unity subordinate to the finite covering $\{O_r\}_{r\in I}$. This means that there are nonnegative functions $\varphi_r \in C_0^\infty(O_r)$, i.e. $\chi_{r*}\varphi_r \in C_0^\infty(U_r)$, such that

$$\sum_{r\in I} \varphi_r(x) = 1 \quad \text{for all} \quad x \in \Gamma\,. \tag{9.1.13}$$

For its construction we refer to Schechter [377, Section 9-4]. In addition, to the partition of unity let $\{\psi_r\}_{r\in I}$ be another system of functions $\psi_r \in C_0^\infty(O_r)$, i.e. $\chi_{r*}\psi_r \in C_0^\infty(U_r)$, with the property

$$\psi_r(x) = 1 \quad \text{for all} \quad x \in \operatorname{supp}\varphi_r\,.$$

Then A can be written as

$$\begin{aligned} A &= \sum_{r\in I} A_{r1} + R_1 \quad \text{with} \quad A_{r1} := \varphi_r A \psi_r\,,\\ A &= \sum_{r\in I} A_{r2} + R_2 \quad \text{with} \quad A_{r2} := \psi_r A \varphi_r\,, \end{aligned} \tag{9.1.14}$$

and R_1 and R_2 both are smoothing operators mapping $C^\infty(\Gamma)$ into $C^\infty(\Gamma)$ due to the smoothing property required in Definition 9.1.1.

In terms of the local operators A_{χ_r} defined in (9.1.2), A can be written as

$$A = \sum_{r\in I} \varphi_r \chi_r^* A_{\chi_r} \chi_{r*}\psi_r + R_1 = \sum_{r\in I} \psi_r \chi_r^* A_{\chi_r} \chi_{r*}\varphi_r + R_2\,. \tag{9.1.15}$$

As a consequence of the mapping properties of pseudodifferential operators in the parametric domains formulated in Theorem 7.1.12, we have the following mapping properties for pseudodifferential operators on Γ.

Theorem 9.1.2. *If* $A \in \mathcal{L}^m(\Gamma)$ *then the following mappings are continuous:*

$$\begin{aligned} A &: C^\infty(\Gamma) \to C^\infty(\Gamma)\,,\\ A &: \mathcal{E}'(\Gamma) \to \mathcal{E}'(\Gamma)\,,\\ A &: H^s(\Gamma) \to H^{s-m}(\Gamma)\,. \end{aligned} \tag{9.1.16}$$

Here $H^s(\Gamma)$ and $H^{s-m}(\Gamma)$ are the standard trace spaces with norm and topology defined in (4.2.27) for every $s \in \mathbb{R}$.

9.1.1 Ellipticity on Boundary Manifolds

Since the pseudodifferential operators on the manifold Γ are characterized by their representations with respect to an atlas of Γ and its local charts $\mathfrak{A} = \{(O_r, U_r, \chi_r)\,|\, r \in I\}$, the concept of ellipticity in the domain given in Definition 7.2.1 carries over to the pseudodifferential operators on Γ.

Definition 9.1.2. *Let $A \in \mathcal{L}^m_{c\ell}(\Gamma)$ and let $\mathfrak{A} = \{(O_r, U_r, \chi_r) \,|\, r \in I\}$ be an atlas on Γ and let $\sigma_{A_{\chi_r,m}}(\chi_r(x), \xi)$ be the corresponding family of principal symbols. Then A is called elliptic on Γ if there exists an operator $B \in \mathcal{L}^{-m}(\Gamma)$ such that*

$$\sigma_{A\chi_r,m}(\chi_r(x), \xi)\sigma_{B\chi_r,-m}(\chi_r(x), \xi) - 1 \in \boldsymbol{S}^{-1}(U_r \times \mathrm{I\!R}^n) \tag{9.1.17}$$

Clearly, Lemma 7.2.1 remains valid for an elliptic pseudodifferential operator on Γ. If $A = ((A_{jk}))_{p\times p}$ is given as a system of pseudodifferential operators on Γ then also Definition 7.2.3 of ellipticity in the sense of Agmon–Douglis–Nirenberg can be carried over in the same manner to $((A_{jk}))_{p\times p}$ on Γ.

Since for every elliptic operator $A \in \mathcal{L}^m_{c\ell}(\Gamma)$ there exists a parametrix $Q_0 \in \mathcal{L}^{-m}_{c\ell}(\Gamma)$, as a consequence we have the following theorem.

Theorem 9.1.3. *Every elliptic operator $A \in \mathcal{L}^m_{c\ell}(\Gamma)$ is a Fredholm operator, $A : \mathcal{H}_1 = H^{s+m}(\Gamma) \to \mathcal{H}_2 := H^s(\Gamma)$. The parametrix $Q_0 \in \mathcal{L}^{-m}_{c\ell}$ is also elliptic and a Fredholm operator $Q_0 : \mathcal{H}_2 \to \mathcal{H}_1$. Moreover,*

$$\mathrm{index}(A) = -\mathrm{index}(Q_0)\,.$$

Proof: Since Γ is compact, the operators C_1 and C_2 in (7.2.6) now are compact linear operators in $H^{s+m}(\Gamma)$ and $H^s(\Gamma)$, respectively, and the results follow from Theorem 7.2.4 with $Q_1 = Q_2 = Q_0$. ∎

Remark 9.1.1: As a consequence, the equation

$$Au = f \quad \text{on } \Gamma \tag{9.1.18}$$

with $A \in \mathcal{L}^m_{c\ell}(\Gamma)$ is solvable if and only if $f \in H^s(\Gamma)$ satisfies the compatibility conditions (5.3.27) with the adjoint pseudodifferential operator A^*. In

accordance with Stephan et a [405] we call the system $((A_{jk}))_{p\times p}$ on Γ*strongly elliptic* if to the principal symbol matrices $a^0(x,\xi) = \big((a^{jk0}_{s_j+t_k}(\chi_r(x),\xi))\big)_{p\times p}$ on the charts (O_r, U_r, χ_r) of the atlas $\mathfrak{A}$ there exists a C^∞ matrix–valued function $\Theta(x) = ((\Theta_{j\ell}))_{p\times p}$ on Γ, and a constant $\beta_0 > 0$ such that for all $x \in \Gamma$, all $\zeta \in \mathbb{C}^p$ and all $\xi' \in \mathrm{I\!R}^{n-1}$ with $|\xi'| = 1$

$$\mathrm{Re}\,\zeta^\top \Theta(x) a^0(x,\xi')\overline{\zeta} \geq \beta_0|\zeta|^2 \tag{9.1.19}$$

is satisfied. In terms of the Bessel potential on Γ defined by $\Lambda^\alpha_\Gamma = (-\Delta_\Gamma + 1)^{\alpha/2}$ where Δ_Γ is the Laplace–Beltrami operator (3.4.64) for the Laplacian on Γ, Theorem 7.2.7 implies the following Gårding inequality on the whole of Γ since Γ here is a compact manifold.

Theorem 9.1.4. *If $A = ((A_{jk}))_{p\times p}$ is a strongly elliptic system of pseudodifferential operators on Γ then there exist constants $\beta_0 > 0$ and $\beta_1 \geq 0$ such that Gårding's inequality holds in the form*

$$\operatorname{Re}\,(w, \Lambda_\Gamma^\sigma \Theta \Lambda_\Gamma^{-\sigma} A w)_{\prod_{\ell=1}^p H^{(t_\ell - s_\ell)/2}(\Gamma)} \\ \geq \beta_0 \|w\|^2_{\prod_{\ell=1}^p H^{t_\ell}(\Gamma)} - \beta_1 \|w\|^2_{\prod_{\ell=1}^p H^{t_\ell - 1}(\Gamma)} \qquad (9.1.20)$$

for all $w \in \prod_{\ell=1}^p H^{t_\ell}(\Gamma)$, *where* $\Lambda_\Gamma^\sigma = ((\Lambda_\Gamma^{s_j}\delta_{j\ell}))_{p\times p}$. *The last lower order term in* (9.1.20) *defines a linear compact operator* $C : \prod_{\ell=1}^p H^{t_\ell}(\Gamma) \to \prod_{\ell=1}^p H^{-s_\ell}(\Gamma)$ *which is given by*

$$(v, Cw)_{\prod_{\ell=1}^p H^{(t_\ell - s_\ell)/2}(\Gamma)} = \beta_1 (v, w)_{\prod_{\ell=1}^p H^{t_\ell - 1}(\Gamma)} \,.$$

With this compact operator C, *the Gårding inequality* (9.1.20) *takes the form*

$$\operatorname{Re}\big(w, (\Lambda_\Gamma^\sigma \Theta \Lambda_\Gamma^{-\sigma} A + C) w\big)_{\prod_{\ell=1}^p H^{(t_\ell - s_\ell)/2}(\Gamma)} \geq \beta_0 \|w\|^2_{\prod_{\ell=1}^p H^{t_\ell}(\Gamma)} \,. \qquad (9.1.21)$$

Remark 9.1.2: As a consequence, any strongly elliptic pseudodifferential operator or any strongly elliptic system of pseudodifferential operators defines a Fredholm operator of index zero since for the corresponding bilinear form

$$a(v, w) := (v, \Lambda_\Gamma^\sigma \Theta \Lambda_\Gamma^{-\sigma} A w)_{\prod_{\ell=1}^p H^{(t_\ell - s_\ell)/2}(\Gamma)}$$

one may apply Theorem 5.3.10 which means that the classical Fredholm alternative holds implying dim $\mathcal{N}(A) =$ dim $\mathcal{N}(A^*)$, hence, index $(A) = 0$.

9.1.2 Schwartz Kernels on Boundary Manifolds

For the representation of $A \in \mathcal{L}^m(\Gamma)$ in terms of the Schwartz kernel we consider the local operator $A_{\chi_r} \in \mathcal{L}^m(U_r)$ which has the representation

$$\begin{aligned}(A_{\chi_r}\chi_{r*}v)(\varrho') = & \sum_{|\alpha|\leq m} a_\alpha^{(r)}(\varrho')(D_{\varrho'}^\alpha \chi_{r*}v)(\varrho') \\ & + \text{p.f.} \int_{U_r} k^{(r)}(\varrho', \varrho' - \tau')(\chi_{r*}v)(\tau')d\tau' \,.\end{aligned} \qquad (9.1.22)$$

with the Schwartz kernel $k^{(r)}$ given in Theorems 7.1.2, 8.1.1 and 8.1.8.

For two charts (O_r, U_r, χ_r), (O_t, U_t, χ_t) with $O_{rt} = O_r \cap O_t \neq \emptyset$ let $\Phi_{rt} = \chi_t \circ \chi_r^{(-1)}$ be the associated diffeomorphism from $\chi_r(O_{rt})$ to $\chi_t(O_{rt})$. Then the representation (9.1.22) is transformed into

$$\begin{aligned}(A_{\chi_t}\chi_{t*}v)(\tau') = & \sum_{|\alpha|\leq m} a_\alpha^{(t)}(\tau')(D_{\tau'}^\alpha \chi_{t*}v)(\tau') \\ & + \text{p.f.} \int_{U_t} k^{(t)}(\tau', \tau' - \lambda')(\chi_{t*}v)(\lambda')d\lambda'\end{aligned}$$

where the kernel $k^{(t)}(\tau', \tau' - \lambda') = \widetilde{k}^{(r)}(\tau', \tau' - \lambda')J(\lambda')$ and the coefficients $a_\alpha^{(r)}(\varrho')$ and $a_\alpha^{(t)}(\tau')$ transform according to formulae (8.2.3), (8.2.12), (8.2.19).

As a consequence of (9.1.22), the operator $A \in \mathcal{L}^m(\Gamma)$ now has the representations (9.1.15) by substituting (9.1.22) into (9.1.15). The smoothing operators R_1 and R_2, respectively, have the representation

$$R_j v(x) = \sum_{r \in I} \int\limits_{U_r} K_{R_j}\big(x, T_r(\varrho')\big)(\chi_{r*}\varphi_r)(\chi_{r*}v)ds_r \quad \text{for } j = 1, 2 \tag{9.1.23}$$

where ds_r denotes the surface element of $O_r \subset \Gamma$ in terms of the parametric representation in U_r under T_r.

9.2 Boundary Operators Generated by Domain Pseudodifferential Operators

We are interested in boundary operators on functions given on Γ, in some n–dimensional domain that contains Γ in its interior. This amounts to studying the trace of a pseudodifferential operator A given on some tubular neighbourhood of Γ.

In Chapter 3 we introduced the local coordinates (3.3.2) which can be used to define a special atlas $\widetilde{\mathfrak{A}}$ of some spatial tubular neighbourhood of Γ in $\mathbb{R}^n$. Let (O_r, U_r, χ_r) be any local chart of a finite atlas for Γ with the parametrization $T_r = \chi_r^{(-1)}$. Then for every point $x \in \widetilde{O}_r \subset \mathbb{R}^n$ where $\widetilde{O}_r$ is an open set containing O_r, we define the mapping

$$x = \Psi_r(\varrho) := T_r(\varrho') + \varrho_n n(\varrho'), \ \varrho = (\varrho', \varrho_n) \tag{9.2.1}$$

for $\varrho' \in U_r \subset \mathbb{R}^{n-1}$ and $\varrho_n \in (-\varepsilon, \varepsilon)$ with $\varepsilon > 0$. Then, as in (3.3.7), $\widetilde{O}_r = \Psi_r\big(U_r \times (-\varepsilon, \varepsilon)\big) \subset \mathbb{R}^n$. Note that for a smooth surface Γ and appropriate $\varepsilon > 0$, the inverse mapping $\Phi_r = \Psi_r^{(-1)}$ exists,

$$\varrho = \Phi_r(x) \quad \text{for } x \in \widetilde{O}_r, \tag{9.2.2}$$

which maps $\Omega \cap \widetilde{O}_r$ onto $(\varrho', \varrho_n) \in U_r \times (-\varepsilon, 0)$. The boundary patch O_r is mapped to $U_r \times \{0\}$, i.e. $\varrho_n = 0$ (see Figure 9.2.1).

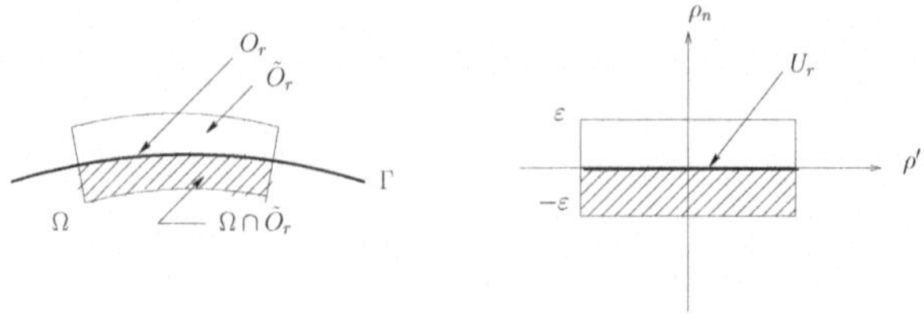

Figure 9.2.1: The tubular neighbourhood of O_r

In addition, we call $\widetilde{\Omega} := \bigcup\limits_{r\in I} \widetilde{O}_r \subset \mathbb{R}^n$ a *tubular neighbourhood* of Γ. In Chapter 3, for such special mappings, we already discussed the following relations:

$$\begin{aligned} g_{jk} &= g_{kj} = \frac{\partial \Psi}{\partial \varrho_j} \cdot \frac{\partial \Psi}{\partial \varrho_k}\,, \; j,k = 1,\dots,n\,, \\ g_{\nu\mu} &= (1-\varrho_n^2 K)\gamma_{\nu\mu} - 2\varrho_n(1-\varrho_n H)L_{\nu\mu}\,, \\ g_{\nu n} &= \delta_{\nu n} \quad \text{for } \nu,\mu = 1,\dots,n-1\,, \end{aligned} \tag{9.2.3}$$

see (3.4.2), (3.4.21) and (3.4.22). Moreover, we have

$$ds(\varrho') = \sqrt{\gamma} d\varrho'\,, \tag{9.2.4}$$

see (3.4.16). From (3.4.31), with integration by parts, one obtains

$$\Big(\frac{\partial^\top}{\partial n}\Big)^k \varphi = (-1)^k \Phi^* \frac{1}{\sqrt{g}} \Big(\frac{\partial}{\partial \varrho_n}\Big)^k (\sqrt{g}\,\Phi_* \varphi) = (-1)^k \frac{1}{\sqrt{g}} \Big(\frac{\partial}{\partial n}\Big)^k (\sqrt{g}\varphi)$$

for $\varphi \in C^\infty(\widetilde{O}_r)$ and any $k \in \mathbb{N}_0$.

For the special distribution of the form $u \otimes \delta_\Gamma$ we also have (see (3.7.1))

$$\Phi_{r*}(u \otimes \delta_\Gamma) = (\chi_{r*} u)(\varrho') \otimes \delta(\varrho_n)\,. \tag{9.2.5}$$

If $\{\varphi_r\}_{r\in I}$ is the partition of unity (9.1.13) subordinate to the finite atlas $\mathfrak{A}$ then it is obvious how to construct a partition of unity $\{\phi_r\}_{r\in I}$ subordinate to the atlas $(\widetilde{O}_r, U_r \times (-\varepsilon,\varepsilon), \Phi_r)_{r\in I}$ for $\widetilde{\Omega} \cap \{|\varrho_n| \le \frac{1}{2}\}$ with $\phi_r \in C_0^\infty(\widetilde{O}_r)$ and

$$\sum_{r\in I} \phi_r(x) = 1 \quad \text{for all } x \in \widetilde{\Omega} \quad \text{with } |\varrho_n| \le \frac{1}{2}\,.$$

Correspondingly, $\psi_r \in C_0^\infty(\widetilde{O}_r)$ are functions with $\psi_r|_{\operatorname{supp}\phi_r} = 1$. Then for $A \in \mathcal{L}^m_{c\ell}(\widetilde{\Omega})$ we have

$$Aw(x) = \sum_{r\in I} \psi_r(x) \Phi_r^* A_{\Phi_r,0} \Phi_{r*}(\phi_r w) + R_2 w \tag{9.2.6}$$

and for $w = u \otimes \delta_\Gamma$ we get

$$\begin{aligned} A(u \otimes \delta_\Gamma) &= \sum_{r\in I} \psi_r(x)\Phi_r^* A_{\Phi_r,0}\big((\chi_{r*}\varphi_r u) \otimes \delta(\eta_n)\big) \\ &\quad + \sum_{r\in I} \int_{U_r} K_{R_2}\big(x, x - T_r(\eta')\big)(\chi_{r*}\varphi_r u)(\eta')ds(\eta') \\ &= \sum_{r\in I} \Phi_r^*(\Phi_{r*}\psi_r)A_{\Phi_r,0}\big((\chi_{r*}\varphi_r u) \otimes \delta(\eta_n)\big) \\ &\quad + \sum_{r\in I} \int_{U_r} K_R\big(x, x - T_r(\eta')\big)(\chi_{r*}\varphi_r u)(\eta')ds(\eta') \end{aligned} \tag{9.2.7}$$

where the symbol of $A_{\Phi_r,0}$ is $a\big(\varrho;(\xi',\xi_n)\big)$ with $\xi = (\xi',\xi_n)$ and $\xi' \in \mathbb{R}^{n-1}$. Hence, we need to characterize the composition of $A_{\Phi_r,0} \in OPS^m \big(U_r \times (-\varepsilon,\varepsilon), \mathbb{R}^n\big)$ with distributions of the form $v(\varrho') \otimes \delta(\varrho_n)$.

In order to discuss surface potentials supported by Γ we begin with the special case when $\Gamma = \mathbb{R}^{n-1}$ in the parametric domain. The surface potential operators are also called Poisson operators in Boutet de Monvel's theory [39, 40, 41] and Grubb [159].

9.3 Surface Potentials on the Plane $\mathbb{R}^{n-1}$

According to the representation (9.2.7), a typical term of the pseudodifferential operator in the half space in terms of local coordinates in the domain $\widetilde{O}_r$ belonging to the chart (O_r, U_r, χ_r) where $U_r \subset \mathbb{R}^{n-1}$corresponds to Γ, is given by

$$\big(A_{\Phi_r,0} v \otimes \delta(\varrho_n)\big)(\varrho) = \sum_{|\alpha'|\leq m} a_{\alpha'}(\varrho)D^{\alpha'}v(\varrho') + \int_{\mathbb{R}^{n-1}} k\big(\varrho, \varrho - (\eta',0)\big)v(\eta')d\eta'$$
$$\text{for } \varrho = (\varrho',\varrho_n) \text{ with } \varrho_n \neq 0\,. \tag{9.3.1}$$

Here $\alpha = (\alpha',0) \in \mathbb{N}_0^n$ are the multi–indices corresponding to the tangential derivatives, $v = \chi_{r*}\varphi_r u \in C_0^\infty(U_r)$, and $k \in \Psi h_\kappa(\widetilde{U}_r)$ is the Schwartz kernel of the pseudodifferential operator $A_{\Phi_r,0} \in \mathcal{L}^m_{c\ell}(\widetilde{U}_r)$ where $\widetilde{U}_r = U_r \times (-\varepsilon,\varepsilon) = \Psi_r(\widetilde{O}_r)$. In (9.3.1), the operator defined by

$$\widetilde{Q}v(\varrho) = \int_{\mathbb{R}^{n-1}} k\big(\varrho, \varrho - (\eta',0)\big)v(\eta')d\eta' \tag{9.3.2}$$

gives a surface potential supported by the plane $\varrho_n = 0$ which is also called *Poisson operator* due to Boutet de Monvel [41].

As a consequence of Theorems 8.1.1 and 8.1.8, the Schwartz kernel of $A_{\Phi_r,0}$ admits an asymptotic expansion

$$k(\varrho, \varrho - \eta) \sim \sum_{j \geq 0} k_{\kappa+j}(\varrho, \varrho - \eta),$$

and hence, due to (8.1.2),

$$\begin{aligned} \widetilde{Q}v(\varrho) &= \sum_{0 \leq j \leq J} \int_{\mathbb{R}^{n-1}} k_{\kappa+j}(\varrho, \varrho - (\eta', 0)) v(\eta') d\eta' \\ &\quad + \int_{\mathbb{R}^{n-1}} k_R(\varrho, \varrho - (\eta', 0)) v(\eta') d\eta', \end{aligned} \tag{9.3.3}$$

where $k_R \in C^{\kappa+j+J-\delta}(\widetilde{U}_r \times \widetilde{U}_r)$. Clearly, $\widetilde{Q}v \in C^\infty(\widetilde{U}_r \setminus \{\varrho_n = 0\})$. However, for $\varrho_n \to 0$, the limit of $\widetilde{Q}v(\varrho', \varrho_n)$ might not exist, in general. In fact, if the limit exists and even defines a C^∞ function on $U_r \times [0, \varepsilon)$ and on $U_r \times (-\varepsilon, 0]$ then $A_{\Phi_r, 0}$ is said to satisfy the *transmission condition* (see Boutet de Monvel [39, 40, 41] and Hörmander [188, Definition 18.2.13]). In the following we now shall investigate these limits in terms of the pseudohomogeneous kernel expansion since these are explicitly given by the boundary integral operators in applications.

We first consider the limits of the typical terms in (9.3.3), i.e.,

$$\int_{\mathbb{R}^{n-1}} k_{\kappa+j}(\varrho, \varrho - (\eta', 0)) v(\varrho') d\eta',$$

since the remaining integral in (9.3.3) with a continuous kernel k_R has always a limit and is continuous across $\varrho_n = 0$. The kernels $k_{\kappa+j}$ satisfy the inequalities

$$\begin{aligned} |k_{\kappa+j}(\varrho, \varrho - \eta)| &\leq \frac{c}{|\varrho - \eta|^{m+n-j}} \\ \text{and} & \\ |D_\varrho^\beta D_z^\alpha k(\varrho, z)| &\leq c_{\alpha\beta} |z|^{j-m-n-|\alpha|} \quad \text{for } \kappa + j < 0 \text{ and for } \kappa + j \notin \mathbb{N}_0 \\ \text{and} & \\ |k_{\kappa+j}(\varrho, \varrho, \eta)| &\leq \frac{c}{|\varrho - \eta|^{m+n-j}} (1 + |\log|x - y||), \\ |D_\varrho^\beta D_z^\alpha k(\varrho, z)| &\leq c_{\alpha,\beta} |z|^{j-m-n-|\alpha|} (1 + |\log|x - y||) \text{ if } \kappa + j \in \mathbb{N}_0. \end{aligned}$$

Therefore, we shall distinguish two cases: $m - j + 1 < 0$ and $m - j + 1 \geq 0$.

We begin with the simple case where $m - j + 1 < 0$. Here, the kernels satisfy

$$|k_{\kappa+j}(\varrho, \varrho - (\eta', 0))| \leq \frac{c}{(\varrho_n^2 + |\varrho' - \eta'|^2)^{(m+n-j)/2}}$$

and depend on its first argument continuously. Therefore, for $\varrho_n = 0$, we obtain surface integrals whose kernels are at most weakly singular. Consequently, in this case we have continuous limits across $\varrho_n = 0$ and

$$\lim_{\varrho_n\to 0}\int\limits_{\mathbb{R}^{n-1}} k_{\kappa+j}\big(\varrho,\varrho-(\eta',0)\big)v(\eta')d\eta' = \int\limits_{\mathbb{R}^{n-1}} k_{\kappa+j}\big((\varrho,0)\,,\,(\varrho'-\eta',0)\big)v(\eta')d\eta'\,. \tag{9.3.4}$$

Therefore, we can choose $J=[m+1]$ in (9.3.3).

For the remaining case $m-j+1\geq 0$, the limits will generally not exist. We begin with the investigation by using a regularization procedure on U_r, namely

$$\begin{aligned}
\int\limits_{\mathbb{R}^{n-1}} k_{\kappa+j}\big(\varrho,\varrho-(\eta',0)\big)v(\eta')d\eta' &= \int\limits_{\mathbb{R}^{n-1}} k_{\kappa+j}(1-\psi)vd\eta' \\
&+ \int\limits_{\mathbb{R}^{n-1}} k_{\kappa+j}\big(\varrho,\varrho-(\eta',0)\big)\Big\{v(\eta')-\sum_{|\alpha|=0}^{[m-j+1]}\tfrac{1}{\alpha!}D^\alpha_{\varrho'}v(\varrho')(\eta'-\varrho')^\alpha\Big\}\psi(|\eta'-\varrho'|)d\eta' \\
&+ \sum_{|\alpha|=0}^{[m-j+1]}\tfrac{1}{\alpha!}D^\alpha_{\varrho'}v(\varrho')\Big(\text{p.f.}\int\limits_{\mathbb{R}^{n-1}} k_{\kappa+j}\big((\varrho',0)\,;\,(\varrho'-\eta',0)\big)(\eta'-\varrho')^\alpha\psi d\eta'\Big) \\
&+ \sum_{|\alpha|=0}^{[m-j+1]}\tfrac{1}{\alpha!}I(\varrho,\kappa+j,\alpha)D^\alpha_{\varrho'}v(\varrho')\,.
\end{aligned}$$

Here, $\alpha=(\alpha_1,\dots,\alpha_{n-1})\in\mathbb{N}_0^{n-1}$ is a multi–index and the coefficients are defined by

$$\begin{aligned}
I(\varrho,\kappa+j,\alpha) \;&:= \text{p.f.}\int\limits_{\mathbb{R}^{n-1}}\Big(k_{\kappa+j}(\varrho,\varrho-(\eta',0)\big)-k_{\kappa+j}\big((\varrho',0)\,,\,(\varrho'-\eta',0)\big)\Big)\times \\
&\qquad\times(\eta'-\varrho')^\alpha\psi(|\eta'-\varrho'|)d\eta'\,,
\end{aligned}$$

where we tacitly have used the C_0^∞ cut–off function $\psi(t)$ to guarantee the compact support of the integrand. Here ψ has the properties $\psi(t)=1$ for $0\leq t\leq c_0$, $\psi(t)=0$ for $2c_0\leq t$ and $0\leq\psi\leq 1$ for all $t\geq 0$.

To investigate the limits of I, we use polar coordinates $\eta'-\varrho'=t\Theta'\in\mathbb{R}^{n-1}$ and $0\leq t\in\mathbb{R}$ with $|\Theta'|=1$. Then

$$\begin{aligned}
I(\varrho,\kappa+j,\alpha) \;&= \int\limits_{c_0}^{2c_0}\int\limits_{|\Theta'|=1}\big\{k_{\kappa+j}(\varrho_n n+\varrho'\,;\,\varrho_n n-t\Theta') \\
&\qquad - k_{\kappa+j}\big((\varrho',0)\,;\,(-t\Theta',0)\big)\big\}\Theta'^{\alpha}d\omega(\Theta')t^{|\alpha|+n-2}\psi(t)dt \\
&\quad+\widetilde{I}(\varrho,\kappa+j,\alpha)\,,
\end{aligned}\tag{9.3.5}$$

where $\widetilde{I}$ is defined by

$$\widetilde{I}((\varrho,\kappa+j,\alpha) := \int\limits_{t=0}^{c_0} \int\limits_{|\Theta'|=1} k_{\kappa+j}(\varrho_n n + \varrho'; \varrho_n n - t\Theta')\Theta'^{\alpha} d\omega(\Theta') t^{|\alpha|+n-2} dt$$

$$- \text{p.f.} \int\limits_{t=0}^{c_0} \int\limits_{|\Theta'|=1} k_{\kappa+j}\big((\varrho',0)\,;\,(-\Theta',0)\big)\Theta'^{\alpha} d\omega(\Theta') t^{|\alpha|+j-m-2} dt$$

where the homogeneity property of $k_{\kappa+j}$ has been employed in the second integral.

In the limit $\varrho_n \to 0$ we see that the first integral in the right–hand side of (9.3.5) vanishes and

$$\begin{aligned}\lim_{\varrho_n\to 0} \widetilde{Q}v(\varrho) &= \text{p.f.} \int\limits_{\mathbb{R}^{n-1}} k\big((\varrho',0)\,;\,(\varrho'-\eta',0)\big)v(\eta')d\eta' \\ &\quad + \lim_{\varrho_n\to 0} \sum_{j=0}^{[m+1]} \sum_{|\alpha|=0}^{[m-j+1]} \tfrac{1}{\alpha!} \widetilde{I}(\varrho,\kappa+j,\alpha) D^{\alpha}_{\varrho'} v(\varrho')\,.\end{aligned}$$

Hence, this limit exists for every $v \in \mathcal{D}(U_r)$ if and only if the limits

$$\lim_{\varrho_n\to 0} \sum_{j=0}^{[m+1]} \widetilde{I}(\varrho,\kappa+j,\alpha) \quad \text{for every} \;\; \alpha \in \mathbb{N}_0^{n-1} \;\; \text{with} \;\; 0 \le |\alpha| \le [m+1] \tag{9.3.6}$$

exist.

Lemma 9.3.1. i) *For* $0 < m - j + 1 \notin \mathbb{N}$ *the limits in* (9.3.6) *exist if and only if for every multi–index* $\alpha \in \mathbb{N}_0^{n-1}$ *with* $0 \le |\alpha| \le [m+1]$ *the following extension conditions are satisfied:*

For each $\nu = 0, \ldots, [m+1]$:

$$\sum_{j=0}^{[m+1]-|\alpha|-\nu} \frac{(-1)^{\ell}}{\ell!} d^{\pm}(\varrho',\kappa+j,\alpha,\ell) = 0\,, \tag{9.3.7}$$

where $\ell = [m+1] - |\alpha| - \nu - j \ge 0$. *If the conditions* (9.3.7) *are satisfied then*

$$\lim_{\varrho_n\to 0\pm} \sum_{j=0}^{[m+1]} \widetilde{I}(\varrho,\kappa+j,\alpha) = 0\,.$$

ii) *For* $m - j + 1 \in \mathbb{N}_0$, *the limits in* (9.3.6) *exist if and only if for every* α *with* $0 \le |\alpha| < m - j + 1$ *the extension conditions* (9.3.7) *for each* $\nu = 1, \ldots, m+1$ *are satisfied together with the Tricomi conditions*

$$\int\limits_{|\Theta'|=1} k_{\kappa+j}\big((\varrho',0)\,;\,(-\Theta',0)\big)\Theta'^{\alpha} d\omega(\Theta') = 0 \quad \textit{for} \;\; |\alpha| = m-j+1\,. \tag{9.3.8}$$

If the conditions (9.3.7) *and* (9.3.8) *are fulfilled then*

$$\lim_{\varrho_n \to 0\pm} \sum_{j=0}^{[m+1]} \widetilde{I}(\varrho, \kappa + j, \alpha) = \sum_{j=0}^{m+1-|\alpha|} \frac{(-1)^\ell}{\ell!} d^\pm(\varrho', \kappa + j, \alpha, \ell) , \tag{9.3.9}$$

where $\ell = m + 1 - |\alpha| - j$.
Here, in i) *and* ii), *for all* $|\alpha| \leq [m - j + 1]$ *and* $0 \leq \ell \leq [m - j + 1] - |\alpha|$, *the coefficients* $d^\pm$ *are defined by*

$$d^\pm(\varrho', \kappa + j, \alpha, \ell) := \tag{9.3.10}$$
$$\int_0^1 \sqrt{1+\tau^2}^{\kappa+j} \Big\{ \tau^{|\alpha|+n-2} \int_{|\Theta'|=1} k^{(\ell)}_{\kappa+j}\Big((\varrho', 0) ; \frac{\pm n - \tau\Theta'}{\sqrt{1+\tau^2}}\Big) \Theta'^\alpha d\omega(\Theta')$$
$$+ \tau^{m-j-|\alpha|} \int_{|\Theta'|=1} k^{(\ell)}_{\kappa+j}\Big((\varrho', 0) ; \frac{\pm \tau n - \Theta'}{\sqrt{1+\tau^2}}\Big) \Theta'^\alpha d\omega(\Theta') \Big\} d\tau$$

where
$$k^{(\ell)}_{\kappa+j}(\varrho, z) := \big(n(\varrho') \cdot \nabla_\varrho\big)^\ell k_{\kappa+j}(\varrho; z) .$$

Remarks 9.3.1: The $\pm$ signs indicate the one–sided limits from $\varrho_n > 0$ and $\varrho_n < 0$ corresponding to the limits from Ω^c and Ω, respectively. In general, the limits from both sides may not necessarily exist simultaneously. Note that for $m + 1 \in \mathbb{N}_0$ and $|\alpha| = m - j + 1$, the second integral in (9.3.10) exists if and only if the Tricomi condition (9.3.8) holds whereas the coefficients $d^\pm$ are well defined otherwise.

Proof: From the definition of $\widetilde{I}$ in (9.3.4), Taylor expansion of $k_{\kappa+j}$ about $\varrho_n = 0$ in the first argument and the definition of finite part integrals, we see that

$$\widetilde{I}(\varrho, \kappa + j, \alpha) =$$
$$\sum_{\ell=0}^{[m-j+1]-|\alpha|} \tfrac{1}{\ell!} \varrho_n^\ell \int_0^{c_0} \int_{|\Theta'|=1} k^{(\ell)}_{\kappa+j}\big((\varrho', 0) ; \varrho_n n - t\Theta'\big) \Theta'^\alpha d\omega(\Theta') t^{|\alpha|+n-2} dt$$
$$+ \tfrac{1}{(R+1)!} \varrho_n^{R+1} \int_0^{c_0} \int_{|\Theta'|=1} k^{(R)}_{\kappa+j}(\varrho', \varrho_n ; \varrho_n n - t\Theta') \Theta'^\alpha d\omega(\Theta') t^{|\alpha|+n-2} dt$$
$$- P_f(\varrho', \kappa + j, m - j + 1, \alpha) ,$$

where $k^{(R)}_{\kappa+j}$ denotes the remainder term, $R = [m - j + 2] - |\alpha|$ and

$$P_f(\varrho', \kappa+j, m-j+1, \alpha) = \int\limits_{|\Theta'|=1} k_{\kappa+j}\big((\varrho', 0);\, (-\Theta', 0)\big)\Theta'^{\alpha} d\omega(\Theta') \times$$
$$\times \begin{cases} \frac{1}{(|\alpha|+j-m-1)} c_0^{|\alpha|+j-m-1} & \text{for } m-j+1-|\alpha| \neq 0\,, \\ \ln c_0 & \text{for } |\alpha| = m-j+1\,. \end{cases}$$

Now, by using the homogeneity of $k^{(\ell)}_{\kappa+j}$ and the new variable $\frac{t}{|\varrho_n|} = t'$ we obtain

$$\begin{aligned} \widetilde{I} \ = \ & \sum_{\ell=0}^{[m-j+1]-|\alpha|} (-1)^{\ell} |\varrho_n|^{\ell+j-m+|\alpha|-1} \times \\ & \times \frac{1}{\ell!}\Big\{ \int\limits_{t'=0}^{c_0/|\varrho_n|} \int\limits_{|\Theta'|=1} k^{(\ell)}_{\kappa+j}\big((\varrho', 0);\, \pm n - t'\Theta'\big)\Theta'^{\alpha} d\omega(\Theta') t'^{|\alpha|+n-2} dt' \Big\} \\ & + (-1)^{R+1} |\varrho_n|^{[m+1]-m} \times \\ & \times \frac{1}{(R+1)!} \int\limits_{t'=0}^{c_0/|\varrho_n|} \int\limits_{|\Theta'|=1} k^{(R)}_{\kappa+j}(\varrho', \varrho_n; \pm n - t'\Theta')\Theta'^{\alpha} d\omega(\Theta') t'^{|\alpha|+n-2} dt' \\ & - P_f(\varrho', \kappa+j, m-j+1, \alpha)\,. \end{aligned}$$

Next, for $|\varrho_n| < c_0$, we split the interval of integration into $0 \leq t' \leq 1$ and $1 \leq t' =: \frac{1}{\tau} \leq \frac{c_0}{|\varrho_n|}$. This leads to

$$\begin{aligned} \widetilde{I} = & \sum_{\ell=0}^{[m-j+1]-|\alpha|} (-1)^{\ell} |\varrho_n|^{\ell+j-m+|\alpha|-1} \times \qquad (9.3.11) \\ & \times \frac{1}{\ell!}\Big\{ \int\limits_{\tau=0}^{1} \int\limits_{|\Theta'|=1} k^{(\ell)}_{\kappa+j}\Big((\varrho', 0);\, \frac{\pm n - \tau\Theta'}{\sqrt{1+\tau^2}}\Big) \sqrt{1+\tau^2}^{\,\kappa+j} \Theta'^{\alpha} d\omega(\Theta') \tau^{|\alpha|+n-2} d\tau \\ & + \int\limits_{\frac{|\varrho_n|}{c_0}}^{1} \int\limits_{|\Theta'|=1} k^{(\ell)}_{\kappa+j}\Big((\varrho', 0);\, \frac{\pm \tau n - \Theta'}{\sqrt{1+\tau^2}}\Big) \sqrt{1+\tau^2}^{\,\kappa+j} \Theta'^{\alpha} d\omega(\Theta') \tau^{m-j-|\alpha|} d\tau \Big\} \\ & - P_f(\varrho', \kappa+j, m-j+1, \alpha) + |\varrho_n|^{[m+1]-m} C^{(R)}\,, \end{aligned}$$

where

$$C^{(R)} = -\frac{(-1)^R}{(R+1)!}\Big\{ \int\limits_{\tau=0}^{1} \int\limits_{|\Theta'|=1} k^{(R)}_{\kappa+j}\Big(\varrho', \varrho_n\,;\ \frac{\pm n - \tau\Theta'}{\sqrt{1+\tau^2}}\Big) \times$$

$$\times \sqrt{1+\tau^2}^{\kappa+j}\Theta'^{\alpha} d\omega(\Theta')\tau^{|\alpha|+n-2} d\tau$$

$$+ \int\limits_{\frac{|\varrho_n|}{c_0}}^{1} \int\limits_{|\Theta'|=1} k^{(R)}_{\kappa+j}\Big(\varrho', \varrho_n\,;\ \frac{\pm\tau n - \Theta'}{\sqrt{1+\tau^2}}\Big)\sqrt{1+\tau^2}^{\kappa+j}\Theta'^{\alpha} d\omega(\Theta')\tau^{m-j-|\alpha|}d\tau\Big\}$$

with $R = [m-j+2] - |\alpha|$.

Let us first complete the proof for the special case i) where $0 < m-j+1 \notin \mathbb{N}$. From (9.3.11) we see that

$$\begin{aligned}
\widetilde{I} \ &= \ |\varrho_n|^{-(m-j+1-|\alpha|)}\Big\{ d^{\pm}(\varrho', \kappa+j, \alpha, 0) \\
&\quad - \int\limits_{0}^{\frac{|\varrho_n|}{c_0}} \int\limits_{|\Theta'|=1} k_{\kappa+j}\big((\varrho',0)\,;\ \pm\tau n - \Theta'\big)\Theta'^{\alpha} d\omega(\Theta')\tau^{m-j-|\alpha|}d\tau\Big\} \\
&\quad - \frac{c_0^{|\alpha|+j-m-1}}{|\alpha|+j-m-1} \int\limits_{|\Theta'|=1} k_{\kappa+j}\big((\varrho',0)\,;\ (-\Theta',0)\big)\Theta'^{\alpha} d\omega(\Theta') \\
&\quad + \sum_{1\le\ell}^{[m-j+1]-|\alpha|} (-1)^{\ell}|\varrho_n|^{\ell-(m-j+1-|\alpha|)}\tfrac{1}{\ell!}\Big\{ d^{\pm}(\varrho', \kappa+j, \alpha, \ell) \\
&\quad - \int\limits_{0}^{\frac{|\varrho_n|}{c_0}} \int\limits_{|\Theta'|=1} k^{(\ell)}_{\kappa+j}\big((\varrho',0)\,;\ \pm\tau n - \Theta'\big)\Theta'^{\alpha} d\omega(\Theta')\tau^{m-j-|\alpha|}d\tau\Big\} \\
&\quad + O(|\varrho_n|^{[m+1]-m}),
\end{aligned} \tag{9.3.12}$$

since in the remainder integrals $n \ge 2$ and $|\alpha| < m+1-j$, hence $C^{(R)}$ is bounded.

In the first integral on the right–hand side we apply the Taylor formula to $k_{\kappa+j}\big((\varrho',0)\,;\ \pm\tau n - \Theta'\big)$ with respect to τ about 0 which yields

$$
\begin{aligned}
&-\int\limits_{0}^{\frac{|\varrho_n|}{c_0}}\int\limits_{|\Theta'|=1} k_{\kappa+j}\big((\varrho',0);\ \pm\tau n-\Theta'\big)\Theta'^{\alpha}d\omega(\Theta')\tau^{-|\alpha|+m-j}d\tau\\
&=-\int\limits_{0}^{\frac{|\varrho_n|}{c_0}}\int\limits_{|\Theta'|=1} k_{\kappa+j}\big((\varrho',0);\ (-\Theta',0)\big)\Theta'^{\alpha}d\omega(\Theta')\tau^{-|\alpha|+m-j}d\tau\\
&\qquad+O(|\varrho_n|^{m-j+2-|\alpha|})\\
&=\frac{c_0^{|\alpha|+j-m-1}}{|\alpha|+j-m-1}\int\limits_{|\Theta'|=1} k_{\kappa+j}\big((\varrho',0);\ (-\Theta',0)\big)\Theta'^{\alpha}d\omega(\Theta')|\varrho_n|^{m-j+1-|\alpha|}\\
&\qquad+O(|\varrho_n|^{m-j+2-|\alpha|})\,.
\end{aligned}
$$

Moreover, for the last integrals on the right–hand side of (9.3.12) we obtain

$$
\int\limits_{0}^{\frac{|\varrho_n|}{c_0}}\int\limits_{|\Theta'|=1} k^{(\ell)}_{\kappa+j}\big((\varrho',0);\ \pm\tau n-\Theta'\big)\Theta'^{\alpha}d\omega(\Theta')\tau^{-|\alpha|+m-j}d\tau = O(|\varrho_n|^{m-j-|\alpha|+1})\,.
$$

So,

$$
\widetilde{I}=\sum_{\ell=0}^{[m-j+1]-|\alpha|}\tfrac{(-1)^{\ell}}{\ell!}|\varrho_n|^{\ell-(m-j+1-|\alpha|)}d^{\pm}(\varrho',\kappa+j,\alpha,\ell)\ +O(|\varrho_n|^{[m+1]-m})\,. \tag{9.3.13}
$$

Hence,

$$
\begin{aligned}
&\sum_{j=0}^{[m+1]}\widetilde{I}(\varrho,\kappa+j,\alpha)\\
&\quad=\sum_{j=0}^{[m+1]}\sum_{\ell=0}^{[m-j+1]-|\alpha|}\tfrac{(-1)^{\ell}}{\ell!}|\varrho_n|^{\ell-(m+1-j-|\alpha|)}d^{\pm}(\varrho',\kappa+j,\alpha,\ell)\\
&\qquad+O(|\varrho_n|^{[m+1]-m})\\
&\quad=\sum_{\nu=0}^{[m+1]-|\alpha|}|\varrho_n|^{-\nu+[m+1]-(m+1)}\sum_{j=0}^{[m+1]-|\alpha|-\nu}\tfrac{(-1)^{\ell}}{\ell!}d^{\pm}(\varrho',\kappa+j,\alpha,\ell)\\
&\qquad+O(|\varrho_n|^{[m+1]-m})
\end{aligned}
$$

where $\ell=[m+1]-|\alpha|-\nu-j$.

For any given $\alpha\in\mathbb{N}_0^{n-1}$ with $0\le|\alpha|\le[m+1]$, the limit of $\sum_{j=0}^{[m+1]}\widetilde{I}(\varrho,\kappa+j,\alpha)$ exists as $|\varrho_n|\to0$ if and only if for each $\nu=0,\dots,[m+1]$, the coefficient of $|\varrho_n|^{-\nu-(m+1)+[m+1]}$ vanishes.

In the case ii) where $m-j+1 \in \mathbb{N}_0$, we can treat all the terms of $\widetilde{I}$ in (9.3.11) in the same manner as in the case i) except for $\ell = m-j-|\alpha|+1$. If $|\alpha| \le m-j$, we again find

$$\begin{aligned}\widetilde{I}(\varrho,\kappa+j,\alpha) &= \sum_{\ell=0}^{m-j-|\alpha|} \frac{(-1)^\ell}{\ell!}|\varrho_n|^{\ell-(m-j+1-|\alpha|)} d^{\pm}(\varrho',\kappa+j,\alpha,\ell)\\ &+ \frac{(-1)^{m-j+1-|\alpha|}}{(m-j+1-|\alpha|)!} d^{\pm}(\varrho',\kappa+j,\alpha,m-j+1-|\alpha|)\\ &+ O(|\varrho_n \ln|\varrho_n||)\,. \end{aligned} \tag{9.3.14}$$

If $|\alpha| = m-j+1$ then (9.3.11) reduces to one term in the sum, i.e.,

$$\begin{aligned}&\widetilde{I}(\varrho,\kappa+j,\alpha)\\ &= \int\limits_{\tau=0}^{1}\int\limits_{|\Theta'|=1} k_{\kappa+j}\Big((\varrho',0)\,;\,\frac{\pm n - \tau\Theta'}{\sqrt{1+\tau^2}}\Big)\sqrt{1+\tau^2}^{\kappa+j}\Theta'^{\alpha}\,d\omega(\Theta')\tau^{|\alpha|+n-2}d\tau\\ &\quad+ \int\limits_{\frac{|\varrho_n|}{c_0}}^{1}\int\limits_{|\Theta'|=1} k_{\kappa+j}\Big((\varrho',0)\,;\,\frac{\pm\tau n - \Theta'}{\sqrt{1+\tau^2}}\Big)\sqrt{1+\tau^2}^{\kappa+j}\Theta'^{\alpha}\,d\omega(\Theta')\tau^{-1}d\tau\\ &\quad+O(|\varrho_n \ln|\varrho_n||) - P_f(\varrho',\kappa+j,|\alpha|,\alpha)\,.\end{aligned}$$

Now we use a one term Taylor expansion of $k_{\kappa+j}\big((\varrho',0)\,;\,\pm\tau n - \Theta'\big)$ about $\tau = 0$ and obtain

$$\begin{aligned}&\widetilde{I}(\varrho,\kappa+j,\alpha)\\ &= \int\limits_{0}^{1}\int\limits_{|\Theta'|=1} k_{\kappa+j}\Big((\varrho',0)\,;\,\frac{\pm n - \tau\Theta'}{\sqrt{1+\tau^2}}\Big)\sqrt{1+\tau^2}^{\kappa+j}\Theta'^{\alpha}d\omega(\Theta')\tau^{|\alpha|+n-2}d\tau\\ &\quad+\Big(\ln\frac{c_0}{|\varrho_n|}\Big)\int\limits_{|\Theta'|=1} k_{\kappa+j}\big((\varrho',0)\,;\,(-\Theta',0)\big)\Theta'^{\alpha}d\omega(\Theta')\\ &\quad+\int\limits_{0}^{1}\int\limits_{|\Theta'|=1}\Big\{\frac{\partial}{\partial t}k_{\kappa+j}\big((\varrho',0)\,;\,\pm tn - \Theta'\big)\Big\}\Big|_{t=\vartheta\tau}\Theta'^{\alpha}d\omega(\Theta')d\tau\\ &\quad-P_f(\varrho',\kappa+j,|\alpha|,\alpha) + O(|\varrho_n \ln|\varrho_n||)\end{aligned}$$

where $0 < \vartheta < 1$.

Clearly, the limit $\lim\limits_{\varrho_n \to 0} \widetilde{I}(\varrho,\kappa+j,\alpha)$ exists if and only if the Tricomi condition

$$\int\limits_{\Theta'|=1} k_{\kappa+j}\big((\varrho',0)\,;\,(-\Theta',0)\big)\Theta'^{\alpha}d\omega(\Theta') = 0$$

is satisfied. If this is the case then $P_f(\varrho', \kappa+j, |\alpha|, \alpha) = 0$ for $|\alpha| = m-j+1$. Moreover, then the integrals defining $d^{\pm}(\varrho', \kappa+j, \alpha, 0)$ given by (9.3.10) exist in this case and

$$\widetilde{I}(\varrho, \kappa+j, \alpha) = d^{\pm}(\varrho', \kappa+j, \alpha, 0) + O(|\varrho_n \ln |\varrho_n||)\,.$$

■

As a consequence of Lemma 9.3.1, we have the following main result for the surface potentials on the plane.

Theorem 9.3.2. *Let $A_{\Phi r,0} \in \mathcal{L}^m_{c\ell}(\widetilde{U}_r)$ and $\widetilde{Q}$ be the associated integral operator defined by (9.3.2). Then, for $m < -1$, the operator $\widetilde{Q}$ has a continuous extension to $\varrho_n \le 0$. For $m > -1$ and $m+1 \notin \mathbb{N}_0$, $\widetilde{Q}v(\varrho)$ has a continuous extension to $\varrho_n \le 0$ if and only if for all $j \in \mathbb{N}_0$ with $0 \le j \le m+1$ and every multi–index $\alpha \in \mathbb{N}_0^{n-1}$ with $0 \le |\alpha| < m+1$, the extension conditions*

$$\sum_{j=0}^{[m+1]-|\alpha|-j-\nu} \frac{(-1)^{\ell}}{\ell!} d^{-}(\varrho', \kappa+j, \alpha, \ell) = 0 \quad \textit{for each} \quad \nu = 0, \ldots, [m+1] \tag{9.3.15}$$

(cf. (9.3.7)) hold, where $\ell = [m+1] - |\alpha| - j - \nu$. If the limit exists, then

$$\lim_{0>\varrho_n\to 0} \widetilde{Q}v(\varrho) = Qv(\varrho') = \text{p.f.} \int_{\mathbb{R}^{n-1}} k((\varrho', 0)\,;\, (\varrho'-\eta', 0))v(\eta')d\eta' \tag{9.3.16}$$

and it defines on $U_r \subset \mathbb{R}^{n-1}$ a pseudodifferential operator $Q \in \mathcal{L}^{m+1}_{c\ell}(U_r)$ with the Schwartz kernel of $A_{\Phi r,0}$,

$$k\big((\varrho', 0)\,;\, (\varrho'-\eta', 0)\big) \in \Psi hk_{m+1}$$

having the pseudohomogeneous expansion

$$k\big((\varrho', 0)\,;\, (\varrho'-\eta', 0)\big) \sim \sum_{j\ge 0} k_{\kappa+j}\big((\varrho', 0)\,;\, (\varrho'-\eta', 0)\big)\,. \tag{9.3.17}$$

Moreover, the finite part operator in (9.3.16) is invariant under the change of coordinates in U_r.

If $m \ge -1$ and $m+1 \in \mathbb{N}_0$, then $\widetilde{Q}v(\varrho)$ has a continuous extension to $\varrho_n \le 0$ if and only if in addition to the extension conditions (9.3.15) for each $\nu = 1, \ldots, m+1$ also the Tricomi conditions

$$\int_{|\Theta'|=1} k_{\kappa+j}\big((\varrho', 0)\,;\, (-\Theta', 0)\big)\Theta'^{\alpha} d\omega(\Theta') = 0 \quad \textit{for all } \alpha \textit{ with } |\alpha| = m-j+1 \tag{9.3.18}$$

are satisfied where $\Theta' \in \mathbb{R}^{n-1}$. In the latter case, the limit is given by

$$Qv(\varrho') = \text{p.f.} \int\limits_{\mathbb{R}^{n-1}} k\big((\varrho',0)\,;\,(\varrho'-\eta',0)\big)v(\eta')d\eta' - \mathcal{T}v(\varrho')\,, \tag{9.3.19}$$

where $\mathcal{T}$ is the tangential differential operator

$$\mathcal{T}v(\varrho') = \sum_{0\le|\alpha|\le m+1} \sum_{j=0}^{m+1-|\alpha|} \frac{(-1)^{m-j-|\alpha|}}{(m-j+1-|\alpha|)!} \times$$
$$\times\, d^-\big(\varrho',\kappa+j,\alpha,m+1-j-|\alpha|\big)\tfrac{1}{\alpha!}D^\alpha_{\varrho'}v(\varrho') \tag{9.3.20}$$

and, again, $Q \in \mathcal{L}^{m+1}_{c\ell}(U_r)$.

Proof: It only remains to show that for $m+1 \notin \mathbb{N}_0$, the finite part integral operator Q is invariant under the change of coordinates on $U_r \subset \mathbb{R}^{n-1}$. This fact follows from an application of Theorem 8.2.2 together with Lemma 8.2.4 to this operator. In view of Lemma 9.3.1 we then only need to show that the limit $Q \in \mathcal{L}^{m+1}_{c\ell}(U_r)$. The latter is a consequence of the pseudohomogeneous expansion (9.3.17), where

$$k_{-n-m+j}\big((\varrho',0)\,;\,(\varrho'-\eta',0)\big) = k_{-(n-1)-(m+1)+j}\big((\varrho',0)\,;\,(\varrho'-\eta',0)\big)\,.$$

For $m+1 \in \mathbb{N}_0$, the Tricomi conditions (9.3.18) are satisfied due to Theorem 8.1.8 (with (8.1.61) and n replaced by $(n-1)$), and again the fact that the Schwartz kernel has the pseudohomogeneous asymptotic expansion (9.3.17). Consequently, $Q \in \mathcal{L}^{m+1}_{c\ell}(U_r)$ also in this case. ■

We remark that Theorem 9.3.2 holds for $\varrho_n \ge 0$ correspondingly if d^- is replaced by d^+ in (9.3.15) and (9.3.20) whereas (9.3.19) changes slightly. In this case, we denote by Q_c the corresponding pseudodifferential operator;

$$Q_cv(\varrho') = \lim_{0<\varrho_n\to 0} \widetilde{Q}v(\varrho) = \text{p.f.} \int\limits_{\mathbb{R}^{n-1}} k\big((\varrho',0)\,;\,(\varrho'-\eta',0)\big)v(\eta')d\eta' + \mathcal{T}_cv(\varrho')\,, \tag{9.3.21}$$

where $\mathcal{T}_c$ is the tangential operator

$$\mathcal{T}_cv(\varrho') := \sum_{0\le|\alpha|\le m+1} \sum_{j=1}^{m+1-|\alpha|} \frac{(-1)^{m-j-|\alpha|+1}}{(m-j-|\alpha|+1)!} \times$$
$$\times\, d^+\big(\varrho',\kappa+j,\alpha,m+1-j-|\alpha|\big)\frac{1}{\alpha!}D^\alpha_{\varrho'}v(\varrho')\,.$$

We comment that for $m+1 \in \mathbb{N}_0$, the finite part integral operator in (9.3.19) and in (9.3.21) are in general not invariant under the change of coordinates in U_r.

Note that we claim in Theorem 9.3.2 only the continuous extension of $\widetilde{Q}$ up to $\varrho_n = 0$ whereas it is not yet clear when this extension is in C^∞ up to $\varrho_n = 0$ as is required for the transmission condition.

For the derivatives with respect to ϱ', however, we shall only need more extension conditions.

Corollary 9.3.3. *Let $\widetilde{Q}$ possess a continuous limit for $\varrho_n \to 0-$. Then also the limits*

$$\lim_{\varrho_n \to 0-} \Big(\frac{\partial}{\partial \varrho_j}\widetilde{Q}v\Big)(\varrho) = D_{\varrho_j} \circ Qv(\varrho') \quad \textit{for } j = 1,\dots,n-1 \tag{9.3.22}$$

exist and equal the pseudodifferential operators $D_{\varrho_j} \circ Q \in \mathcal{L}^{m+2}_{c\ell}(\Gamma)$ on the plane $\varrho_n = 0$. They have the representations

$$D_{\varrho_j} \circ Qv(\varrho') = \text{p.f.} \int\limits_{\mathbb{R}^{n-1}} \Big\{\frac{\partial}{\partial \varrho_j} k\big((\varrho',0)\,;\,(\varrho'-\eta'),0\big)\Big\} v(\eta')d\eta' - D_{\varrho_j}\mathcal{T}v(\varrho'), \tag{9.3.23}$$

where for $m+1 \notin \mathbb{N}_0$ we set $\mathcal{T} = 0$.

Clearly, this Corollary holds also for $\varrho_n \to 0+$ and Q_c, correspondingly.

Proof: Let $\varrho_n < 0$. Then, for $k = 1,\dots,n-1$ one has with integration by parts

$$\begin{aligned}
&\sum_{j=0}^{[m+1]} \frac{\partial}{\partial \varrho_k} \int\limits_{\mathbb{R}^{n-1}} k_{\kappa+j}\big(\varrho; (\varrho'-\eta', \varrho_n)\big)v(\eta')d\eta' \\
&\qquad + \int\limits_{\mathbb{R}^{n-1}} \frac{\partial}{\partial \varrho_k} k_R\big(\varrho; (\varrho'-\eta', \varrho_n)\big)v(\eta')d\eta' \\
&\quad = \sum_{j=0}^{m+1} \int\limits_{\mathbb{R}^{n-1}} \Big\{\frac{\partial k_{\kappa+j}}{\partial \varrho_k}\big(\varrho; (z', \varrho_n)\big)\Big\}_{|z'=\varrho'-\eta'} v(\eta')d\eta' \\
&\qquad + \int\limits_{\mathbb{R}^{n-1}} \frac{\partial}{\partial \varrho_k} k_R\big(\varrho; (z', \varrho_n)\big)_{|z'=\varrho'-\eta'} v(\eta')d\eta' + \widetilde{Q}\Big(\frac{\partial v}{\partial \eta_k}\Big)(\varrho)\,.
\end{aligned}$$

The limit of the second term exists and becomes the operator $Q \circ D_\eta \in \mathcal{L}^{m+2}_{c\ell}(U_r)$ since $\frac{\partial v}{\partial \eta} \in C_0^\infty$. The limit of the first term,

$$\lim_{\varrho_n \to 0} \sum_{j=0}^{m+1} \int\limits_{\mathbb{R}^{n-1}} \Big\{\frac{\partial k_{\kappa+j}}{\partial \varrho_k}\big((\varrho', \varrho_n)\,;\,(z', \varrho_n)\big)\Big\}_{|z'=\varrho'-\eta'} v(\eta')d\eta'$$

exists since, for $\varrho' \in U_r$,

$$\frac{\partial}{\partial \varrho_k}\Big\{\sum_{j=0}^{[m+1]-|\alpha|-\nu} \frac{(-1)^\ell}{\ell!} d^-(\varrho', \kappa+j, \alpha, \ell)\Big\} = 0$$

by differentiation of (9.3.15). For $m+1 \in \mathbb{N}_0$, the corresponding Tricomi conditions are also satisfied by differentiation of (9.3.18) with respect to ϱ'_k if $0 < k < n$. Hence, for the limit we obtain by again using integration by parts,

$$\lim_{\varrho_n \to 0-} \Big(\frac{\partial}{\partial \varrho_k} \widetilde{Q}v(\varrho)\Big) = \text{p.f.} \int\limits_{\mathbb{R}^{n-1}} \Big\{\frac{\partial}{\partial \varrho_k} k\big((\varrho',0)\,;\,(z',0)\big)\Big\}|_{z'=\varrho'-\eta'} v(\eta')d\eta'$$

$$+ \text{ p.f.} \int\limits_{\mathbb{R}^{n-1}} k\big((\varrho',0)\,;\,(\varrho'-\eta',0)\big)\frac{\partial v}{\partial \eta_k}(\eta')d\eta'$$

$$- \sum_{0\le|\alpha|\le m+1} \sum_{j=0}^{m+1-|\alpha|} \frac{(-1)^{m-j-|\alpha|}}{(m-j-|\alpha|+1)!}$$

$$\Big(\frac{\partial}{\partial \varrho_k} d^-(\varrho', \kappa+j, \alpha, m+1-j-|\alpha|)\Big) D^\alpha_{\varrho'} v(\varrho') - \mathcal{T}\frac{\partial}{\partial \varrho_k} v(\varrho')$$

$$= \text{p.f.} \int\limits_{\mathbb{R}^{n-1}} \Big\{\frac{\partial}{\partial \varrho_k} k\big((\varrho',0)\,;\,(\varrho'-\eta',0)\big)\Big\} v(\eta')d\eta' - \frac{\partial}{\partial \varrho_k}\mathcal{T}v(\varrho')$$

which shows (9.3.20). ∎

Now let us consider the derivative with respect to ϱ_n. In this case for $\varrho_n < 0$, we have

$$\begin{aligned}
\Big(\frac{\partial}{\partial \varrho_n}\widetilde{Q}v\Big)(\varrho) &= \sum_{j=0}^{[m+1]+1} \int\limits_{\mathbb{R}^{n-1}} \frac{\partial}{\partial \varrho_n} k_{\kappa+j}\big(\varrho; \varrho-(\eta',0)\big)v(\eta')d\eta' \\
&\quad + \int\limits_{\mathbb{R}^{n-1}} \frac{\partial}{\partial \varrho_n} k_R\big(\varrho; \varrho-(\eta',0)\big)v(\eta')d\eta' \\
&= \sum_{j=0}^{[m+1]+1} \int\limits_{\mathbb{R}^{n-1}} \big(n(\varrho')\cdot \nabla_\varrho k_{\kappa+j}(\varrho, z)\big)|_{z=\varrho-(\eta',0)} v(\eta')d\eta' \\
&\quad + \sum_{j=0}^{[m+1]+1} \int\limits_{\mathbb{R}^{n-1}} \big(n(\varrho')\cdot \nabla_z k_{\kappa+j}(\varrho, z)\big)|_{z=\varrho-(\eta',0)} v(\eta')d\eta' \\
&\quad + \int\limits_{\mathbb{R}^{n-1}} \frac{\partial}{\partial \varrho_n} k_R\big(\varrho; \varrho-(\eta',0)\big)v(\eta')d\eta' \\
&=: \widetilde{Q}_1 v(\varrho) + \widetilde{Q}_2 v(\varrho) + \widetilde{Q}_3 v(\varrho)\,.
\end{aligned} \tag{9.3.24}$$

The operator $\widetilde{Q}_1$ with the Schwartz kernel $\sum_{j=0}^{[m+1]+1} n(\varrho')\cdot\nabla\varrho k_{\kappa+j}(\varrho; z)$ is a sum of pseudodifferential operators in $\mathcal{L}^{m-j}_{c\ell}(\widetilde{U}_r)$, while the operator $\widetilde{Q}_2$ is a sum of pseudodifferential operators in $\mathcal{L}^{m-j+1}_{c\ell}(\widetilde{U}_r)$ and has the kernel

$\sum_{j=0}^{[m+1]+1} n(\varrho')\cdot\nabla_z k_{\kappa+j}(\varrho;z)$. For the limit as $\varrho_n \to 0^-$, we perform the same calculations as in the proof of Lemma 9.3.1 and obtain

$$\begin{aligned}
&\lim_{\varrho_n\to 0^-}\Big(\frac{\partial}{\partial\varrho_n}\widetilde{Q}v\Big)(\varrho)\\
&= \text{p.f.}\sum_{j=0}^{[m+1]+1}\int\limits_{\mathbb{R}^{n-1}} k^{(1)}_{\kappa+j}\big((\varrho',0)\,;\,(\varrho'-\eta',0)\big)v(\eta')d\eta'\\
&\quad + \text{p.f.}\sum_{j=0}^{[m+1]+1}\int\limits_{\mathbb{R}^{n-1}} k_{2,\kappa+j-1}\big((\varrho',0)\,;\,(\varrho'-\eta',0)\big)v(\eta')d\eta'\\
&\quad + \lim_{\varrho_n\to 0^-}\sum_{j=0}^{[m+1]+1}\Big[\sum_{|\alpha|=0}^{[m-j+1]}\frac{1}{\alpha!}D^\alpha_{\varrho'}v(\varrho')\times\\
&\quad \times\Big\{|\varrho_n|^{-(m-j+1-|\alpha|)}\sum_{\ell=0}^{[m-j+1]-|\alpha|}\frac{(-1)^\ell}{\ell!}d^-(\varrho',\kappa+j,\alpha,\ell+1)\Big\}\\
&\quad + \sum_{|\alpha|=0}^{[m-j+2]}\frac{1}{\alpha!}D^\alpha_{\varrho'}v(\varrho')\times\\
&\quad \times\Big\{|\varrho_n|^{-(m-j+2-|\alpha|)}\sum_{\ell=0}^{[m-j+2]-|\alpha|}\frac{(-1)^\ell}{\ell!}d^{(2)-}(\varrho',\kappa+j-1,\alpha,\ell)\Big\}\Big]
\end{aligned}$$

for all $v\in C_0^\infty(U_r)$. Here we have denoted by $d^{(2)}$ the coefficients as in (9.3.10) for the operator $\widetilde{Q}_2$.

From this calculation, we see that as $\varrho_n\to 0$ the limit of $\frac{\partial}{\partial\varrho_n}\widetilde{Q}v$ exists for every v if and only if the limits of both operators $\widetilde{Q}_1$ and $\widetilde{Q}_2$ exist independently, since the coefficients of $D^\alpha_{\varrho'}v(\varrho')$ are of different orders, namely $|\varrho_n|^{-(m-j+1-|\alpha|)}$ and $|\varrho_n|^{-(m-j+2-|\alpha|)}$.

Corollary 9.3.4. *Under the assumptions of Theorem* 9.3.2, *for* $m>-1$ *and* $m+1\notin\mathbb{N}_0$, *the limit*

$$\begin{aligned}
\lim_{\varrho_n\to 0^-}\big(\tfrac{\partial}{\partial\varrho_n}\widetilde{Q}v\big)(\varrho) &= \lim_{\varrho_n\to 0^-}(\widetilde{Q}_1v)(\varrho)+\lim_{\varrho_n\to 0^-}(\widetilde{Q}_2v)(\varrho)\\
&\quad + \sum_{\mathbb{R}^{n-1}}\Big(\frac{\partial}{\partial\varrho_n}k_R\Big)\big((\varrho',0)\,;\,(\varrho'-\eta',0)\big)v(\eta')d\eta'
\end{aligned}$$

exists if and only if the first two limits on the right–hand side exist separately, where $\widetilde{Q}_1$ *and* $\widetilde{Q}_2$ *are defined in* (9.3.24). *Moreover, the extension conditions for* $\widetilde{Q}_1$ *now read:*

For each $\nu = 0, \dots, [m+1]$:

$$\sum_{j=0}^{[m+1]-|\alpha|-\nu} \frac{(-1)^\ell}{\ell!} d^-(\varrho', \kappa + j, \alpha, \ell + 1) = 0\,, \quad (9.3.25)$$

where $\ell = [m+1] - |\alpha| - j - \nu$ *for all* $\alpha \in \mathbb{N}_0^{n-1}$ *with* $0 \le |\alpha| < m+1$.

For $\widetilde{Q}_2$ *which has the pseudohomogeneous asymptotic expansion*

$$\sum_{j\ge 0} k_{2,\kappa+j-1}(\varrho; z) := \sum_{j\ge 0} n(\varrho') \cdot \nabla_z k_{\kappa+j}(\varrho; z)\,,$$

the extension conditions are:

For each $\nu = 0, \dots, [m+2]$:

$$\sum_{j=0}^{[m+2]-|\alpha|-\nu} \frac{(-1)^\ell}{\ell!} d_2^-(\varrho', \kappa + j - 1, \alpha, \ell) = 0\,, \quad (9.3.26)$$

where $\ell = [m+2] - |\alpha| - j - \nu$, *for all* $\alpha \in \mathbb{N}_0^{n-1}$ *with* $0 \le |\alpha| < m - j + 2$, *where the coefficients* d_2^- *are given by* (9.3.10), *correspondingly.*

Proof: From the previous discussion, we now consider $\widetilde{Q}_1$ and $\widetilde{Q}_2$ separately. The operator $\widetilde{Q}_1$ has the Schwartz kernel $n(\varrho') \cdot \nabla_\varrho k(\varrho; z)$ with the pseudohomogeneous expansion

$$\sum_{j\ge 0} \big(n(\varrho') \cdot \nabla_\varrho\big) k_{\kappa+j}(\varrho; z) = \sum_{j\ge 0} k^{(1)}_{\kappa+j}(\varrho; z)\,.$$

Applying Theorem 9.3.2 to $\widetilde{Q}_1$, we have the desired result for $\widetilde{Q}_1$. For $\widetilde{Q}_2$, which has the Schwartz kernel $n(\varrho') \cdot \nabla_z k(\varrho; z)$, we again apply Theorem 9.3.2 to obtain the desired result. ∎

In the case $m + 2 \in \mathbb{N}_0$, the extension conditions (9.3.25) must be satisfied for $\nu = 1, \dots, m+2$. In addition, one needs the Tricomi conditions

$$\int\limits_{|\Theta'|=1} k^{(1)}_{\kappa+j}\big((\varrho', 0)\,;\,(-\Theta', 0)\big)\Theta'^\alpha d\omega(\Theta') \;=\; 0 \;\text{ for } |\alpha| = m - j + 1$$

and

$$\int\limits_{|\Theta'|=1} k_{2,\kappa+j-1}\big((\varrho', 0)\,;\,(-\Theta', 0)\big)\Theta'^\alpha d\omega(\Theta') \;=\; 0 \;\text{ for } |\alpha| = m - j + 2\,.$$

We again remark that for $\varrho_n \to 0^+$, one can see that the same arguments carry over and, consequently, the Corollary 9.3.4 holds for this case as well.

For higher order derivatives with respect to ϱ_n, it is clear that we need to apply Corollary 9.3.4 repeatedly. This means, in each step we will have to apply a relation like (9.3.24) to the corresponding higher–order derivatives, and more extension conditions will be needed depending on the order of the derivatives.

We recall that for $m+1 \notin \mathbb{N}_0$, in Theorem 9.3.2 the finite part integral operators Q and Q_c are invariant under the change of coordinates on $U_r \subset \mathbb{R}^{n-1}$. This fact follows from an application of Theorem 8.2.2 together with Lemma 8.2.4 to these operators. In order to have the same invariance properties also for the cases $m+1 \in \mathbb{N}_0$, we consider now the subclass of operators $A_{\Phi r,0} \in \mathcal{L}^m_{c\ell}(\widetilde{U}_r)$ whose asymptotic kernel expansions satisfy the parity conditions (8.2.35) with $\sigma_0 = 0$, i.e.,

$$k_{\kappa+j}(\varrho;\eta-\varrho) = (-1)^{m-j}k_{\kappa+j}(\varrho;\varrho-\eta) \quad \text{for } 0 \le j \le m+1\,. \tag{9.3.27}$$

In view of Theorem 8.2.6, this condition implies that the finite part integral (9.3.16) is now also invariant under the change of coordinates since n is to be replaced by $(n-1)$ and m by $m+1$ in (8.2.31).

Lemma 9.3.5. *Let $A_{\Phi r,0} \in \mathcal{L}^m_{c\ell}(\widetilde{U}_r)$ with $m+1 \in \mathbb{N}_0$. Let the asymptotic kernel expansion of $A_{\Phi r,0}$ satisfy the special parity conditions* (9.3.27)*. Then the Tricomi conditions* (9.3.18) *are satisfied.*

Proof: The parity conditions (9.3.27) imply with $\widetilde{\Theta}' = -\Theta'$,

$$\begin{aligned}
&\int\limits_{|\Theta'|=1} k_{\kappa+j}\big(\varrho;(-\Theta',0)\big)\Theta'^{\alpha}d\omega \\
&\qquad = \int\limits_{|\Theta'|=1} k_{\kappa+j}\big(\varrho;(\Theta',0)\big)\Theta'^{\alpha}d\omega(-1)^{m-j} \\
&\qquad = \int\limits_{|\widetilde{\Theta}'|=1} k_{\kappa+j}\big(\varrho;(-\widetilde{\Theta}',0)\big)\widetilde{\Theta}'^{\alpha}d\omega(-1)^{m-j+|\alpha|} \\
&\qquad = -\int\limits_{|\widetilde{\Theta}'|=1} k_{\kappa+j}\big(\varrho;(-\widetilde{\Theta}',0)\big)\widetilde{\Theta}'^{\alpha}d\omega \quad \text{for } |\alpha| = m-j+1\,.
\end{aligned}$$

Hence,

$$\int\limits_{|\Theta'|=1} k_{\kappa+j}\big(\varrho;(-\Theta',0)\big)\Theta'^{\alpha}\,d\omega = 0 \quad \text{for } |\alpha| = m-j+1\,.$$

∎

Theorem 9.3.6. *Let $A_{\Phi r,0} \in \mathcal{L}^m_{c\ell}(\widetilde{U}_r)$ with $m+1 \in \mathbb{N}_0$ satisfy the special parity conditions* (9.3.27) *in addition to the extension conditions* (9.3.15) *for $\nu = 1,\dots,m+1$. Let $\Lambda : U'_r \to U_r$ be a coordinate transform in $\mathbb{R}^{n-1}$.*

Then the finite part integral operator in (9.3.19) *has the following invariance property:*

$$\text{p.f.} \int\limits_{\mathbb{R}^{n-1}} k\big((\varrho',0)\,;\,(\varrho'-\eta',0)\big)v(\eta')d\eta'$$
$$= \text{p.f.} \int\limits_{\mathbb{R}^{n-1}} \widetilde{k}(\varrho''\,;\,\varrho''-\eta'')\widetilde{v}(\eta'')J(\eta'')d\eta'' \qquad (9.3.28)$$

where $\widetilde{k}(\varrho'',\varrho''-\eta'') = k\big((\Lambda(\varrho''),0)\,;\,(\Lambda(\varrho'')-\Lambda(\eta''),0)\big)$, $\widetilde{v}(\eta'') = v\big(\Lambda(\eta'')\big)$ *and* $J(\eta'') = \det\left(\frac{\partial \Lambda}{\partial \eta''}\right)$.

Remark 9.3.2: This theorem is of particular importance for boundary integral operators from the computational point of view since these finite part integrals in (9.3.28) can be calculated by using appropriate transformations of boundary elements to canonical representations. We shall return to this topic when we discuss a special class of operators having symbols of the rational type.

Proof: Here we use the representation of the finite part integral in terms of the pseudohomogeneous expansion of the Schwartz kernel of $A_{\Phi r,0}$ which is also pseudohomogeneous with respect to $(\varrho'-\eta')$ in

$$\begin{aligned}
&\text{p.f.} \int\limits_{\mathbb{R}^{n-1}} k\big((\varrho',0)\,;\,(\varrho'-\eta',0)\big)v(\eta')d\eta' \\
&= \sum_{j=0}^{m+1} \int\limits_{\mathbb{R}^{n-1}} k_{\kappa+j}\big((\varrho',0)\,;\,(\varrho'-\eta',0)\big)v(\eta')d\eta' \\
&\quad + \int\limits_{\mathbb{R}^{n-1}} k_R\big((\varrho',0)\,;\,(\varrho'-\eta',0)\big)v(\eta')d\eta'\,.
\end{aligned}$$

Then we apply Theorem 8.2.6 to the $(n-1)$–dimensional finite part integral operators

$$\text{p.f.} \int\limits_{\mathbb{R}^{n-1}} k_{\kappa+j}\big((\varrho',0)\,;\,(\varrho'-\eta',0)\big)v(\eta')d\eta'$$

for $0 \le j \le m+1$ whose kernels satisfy the parity conditions (8.1.74), i.e.,

$$k_{\kappa+j}\big((\varrho',0)\,;\,(\varrho'-\eta',0)\big) = (-1)^{j-(m+1)+1}k_{\kappa+j}\big((\varrho',0)\,;\,(\eta'-\varrho',0)\big)$$

as required in Theorem 8.2.6 if n is replaced by $n-1$ and m by $m+1$. Hence, these finite part integrals are invariant under the change of coordinates and the remainder integrals are weakly singular and transform in the standard way. ■

Up to now, the extension conditions are given only for a fixed, particular manifold which is the plane $\varrho_n = 0$ in $\widetilde{U}_r = U_r \times (-\varepsilon, \varepsilon)$. Since $A \in \mathcal{L}^m_{c\ell}(\widetilde{U}_r)$ is given in the whole region $\widetilde{U}_r$, there is no reason to define Q solely on one single fixed plane $\varrho_n = 0$. In fact, as will be seen, the concept of our approach for obtaining Q from $A_{\Phi r,0}$ can be made more universal in the sense that there is a family of operators $Q(\tau)$ generated by $A_{\Phi r,0}$ for the family of planes characterized by $\varrho_n = \text{const} = \tau$. More precisely, we now consider for each fixed $\tau \in I_\tau$, where I_τ is some small neighbourhood of 0, the surface integral

$$\widetilde{Q}(\tau)v(\varrho) = A_{\Phi r,0}\big(v \otimes \delta(\eta_n - \tau)\big)(\varrho) \tag{9.3.29}$$

at the plane $\varrho_n = \tau$. This amounts to considering the family of operators $A_{\Phi r,0}(\tau)$ given by the family of Schwartz kernels $k(\varrho + \tau n; z)$ where we have replaced ϱ_n by the new coordinate $\varrho_n + \tau$; and the associated limits as $\varrho_{\to} 0\pm$ of the family of surface potentials

$$\begin{aligned}
&\lim_{\varrho_n \to 0-} \big(\widetilde{Q}(\tau)v\big)(\varrho + \tau n) \\
&= \lim_{\varrho_n \to 0-} \int\limits_{\mathbb{R}^{n-1}} k\big(\varrho + \tau n\,;\, (\varrho' - \eta', \varrho_n)\big) v(\eta') d\eta' =: Q(\tau)v(\varrho')\,.
\end{aligned} \tag{9.3.30}$$

For $\varrho_n \to 0+$, $Q_c(\tau)v(\varrho')$ is defined correspondingly.

In order to ensure the existence of these limits as in Theorem 9.3.2 for the whole family of planes with $\tau \in I_\tau$, we need more restrictive extension and Tricomi conditions than were previously required. In particular, in the coefficients $d^\pm(\varrho', \kappa + j, \alpha, \ell)$ in (9.3.10) and in the Tricomi conditions (9.3.8) we replace the kernels $k_{\kappa+j}\big((\varrho', 0); \dots\big)$ by $k_{\kappa+j}(\varrho; \dots)$ and write $d^\pm(\varrho, \kappa + j, \alpha, \ell)$, correspondingly. Then ϱ varies in $\widetilde{U}_r$.

These more restrictive *canonical extension conditions* are defined by

$$\sum_{j=0}^{[m+1]-|\alpha|-\nu} \frac{(-1)^\ell}{\ell!} d^\pm(\varrho, \kappa + j, \alpha, \ell) = 0 \quad \text{for } \varrho \in \widetilde{U}_r = U_r \times (-\varepsilon, \varepsilon) \tag{9.3.31}$$

and for $0 \le |\alpha| \le m + 1$ and $\nu = 0, \dots, [m + 1]$ if $m + 1 \notin \mathbb{N}_0$,
and for $\nu = 1, \dots, m + 1$ if $m + 1 \in \mathbb{N}_0$, where $\ell = [m + 1] - |\alpha| - \nu - j \ge 0$;
the *canonical Tricomi conditions* are defined by

$$\int\limits_{|\Theta'|=1} k_{\kappa+j}\big(\varrho; (-\Theta', 0)\big)\Theta'^\alpha d\omega(\Theta') = 0 \quad \text{for } \varrho \in \widetilde{U}_r = U_r \times (-\varepsilon, \varepsilon) \tag{9.3.32}$$

and $|\alpha| = m - j + 1$, where $0 \le j \le m + 1$ and $m \in \mathbb{N}_0$.

Lemma 9.3.7. *Under the canonical extension conditions* (9.3.31) *we have*

$$d^{\pm}(\varrho, \kappa + j, \alpha, \ell) = 0 \quad \text{for} \quad \varrho \in \widetilde{U}_r \quad \text{and all} \quad \ell \in \mathbb{N}_0 \tag{9.3.33}$$

in the case $m + 1 \notin \mathbb{N}_0$ *for* $0 \leq |\alpha| \leq [m+1]$ *and* $0 \leq j \leq [m+1] - |\alpha|$ *and in the case* $m + 1 \in \mathbb{N}_0$ *for* $0 \leq |\alpha| \leq m - j$ *and* $0 \leq j \leq m$.

Proof: Let us begin with the case $m + 1 \notin \mathbb{N}_0$.

Our proof will be based on induction with respect to $p := [m+1] - |\alpha| - \nu$ with $0 \leq p \leq [m+1]$ which corresponds to $\nu = 0, \ldots, [m+1]$. Then $0 \leq j \leq p$ in (9.3.31).

For $p = 0$, in (9.3.31) we have

$$d^{\pm}(\varrho, \kappa, \alpha, 0) = 0\,, \quad \text{where} \ \ |\alpha| \leq [m+1]\,.$$

Then differentiation yields

$$\big(n(\varrho')\cdot\nabla_{\varrho}\big)^{\ell} d^{\pm}(\varrho, \kappa, \alpha, 0) = d^{\pm}(\varrho, \kappa, \alpha, \ell) = 0 \quad \text{for} \ \ \varrho \in \widetilde{U}_r \quad \text{and every} \ \ \ell \in \mathbb{N}_0\,.$$

Now assume that (9.3.33) is fulfilled for all $0 \leq j \leq p$, $|\alpha| \leq [m+1] - j$ and $\ell \in \mathbb{N}_0$. Then (9.3.31) with $p+1$ instead of p gives

$$\sum_{j=0}^{p} \frac{(-1)^{p+1-j}}{(p+1-j)!} d^{\pm}(\varrho, \kappa + j, \alpha, p + 1 - j) + d^{\pm}(\varrho, \kappa + p + 1, \alpha, 0) = 0\,.$$

By induction assumption, $d^{\pm}(\varrho, \kappa + j, \alpha, p + 1 - j) = 0$ for $j \leq p$ and $|\alpha| \leq [m+1] - j$. Hence,

$$d^{\pm}(\varrho, \kappa + p + 1, \alpha, 0) = 0 \quad \text{for all} \ \ |\alpha| \leq [m+1] - p - 1$$

and differentiation with respect to ϱ implies

$$d^{\pm}(\varrho, \kappa + p + 1, \alpha, \ell) = 0 \quad \text{for all} \ \ \ell \in \mathbb{N}_0\,.$$

Hence, induction up to $p = [m+1]$ completes the proof for $m + 1 \notin \mathbb{N}_0$.

In the case $m + 1 \in \mathbb{N}_0$, the proof is exactly the same as above, now for $0 \leq p \leq m$. ■

As a consequence of Lemma 9.3.7, Theorem 9.3.2 remains valid for the whole family of planes Γ_τ with $\tau \in I_\tau$, the family parameter. Without loss of generality we suppose that all the members of this family have local parametric representations defined on the same parametric domain $\widetilde{U}_r = U_r \times (-\varepsilon, \varepsilon)$.

Theorem 9.3.8. *Let* $A_{\Phi\tau,0} \in \mathcal{L}^m_{c\ell}(\widetilde{U}_r)$ *denote the family of operators generated by the family of planes* Γ_τ. *Let* $\widetilde{Q}(\tau)$ *be the corresponding family of integral operators defined by* (9.3.29) *for* $\varrho_n \neq 0$. *For* $m < -1$ *all* $\widetilde{Q}(\tau)v(\varrho)$ *have continuous extensions up to* $\varrho_n = 0$. *For* $m > -1$ *and* $m + 1 \notin \mathbb{N}$, *all*

$\widetilde{Q}(\tau)v(\varrho)$ have continuous extensions up to $\varrho_n \leq 0$ ($\varrho \geq 0$ respectively), if and only if the conditions (9.3.31) are satisfied with d^- (d^+ respectively) for $\varrho \in \widetilde{U}_r$. If these limits exist then

$$\lim_{0>\varrho_n\to 0} \widetilde{Q}(\tau)v(\varrho) = \lim_{0<\varrho_n\to 0} \widetilde{Q}(\tau)v(\varrho) = Q(\tau)v(\varrho') \tag{9.3.34}$$
$$= \text{p.f.} \int_{\mathbb{R}^{n-1}} k\big((\varrho',\tau);\,(\varrho'-\eta',0)\big)v(\eta')d\eta'$$

for the whole family. Moreover, $Q(\tau) \in \mathcal{L}^{m+1}_{c\ell}(U_r)$ are pseudodifferential operators on U_r with the kernel functions $k\big((\varrho',\tau);\,(\varrho'-\eta',0)\big) \in \psi hk_{m+1}$ having the pseudohomogeneous expansions (9.3.17) where $(\varrho',0)$ must be replaced by (ϱ',τ). Moreover, the finite part operator in (9.3.34) is invariant under the change of coordinates in U_r.

If $m \geq -1$ and $m+1 \in \mathbb{N}_0$ then the whole family has continuous extensions up to $\varrho_n \leq 0$ ($\varrho_n \geq 0$ respectively), if and only if the canonical extension conditions (9.3.31) and canonical Tricomi conditions (9.3.32) both are satisfied with d^- (d^+ respectively) for $\varrho \in \widetilde{U}_r$.

In the latter case, the limits are given by

$$\lim_{0>\varrho_n\to 0} \widetilde{Q}(\tau)v(\varrho) = Q(\tau)v(\varrho') \tag{9.3.35}$$
$$= \text{p.f.} \int_{\mathbb{R}^{n-1}} k\big((\varrho',\tau);\,(\varrho'-\eta',0)\big)v(\eta')d\eta' - \mathcal{T}v(\varrho')$$

and

$$\lim_{0<\varrho_n\to 0} \widetilde{Q}(\tau)v(\varrho) = Q_c(\tau)v(\varrho')$$
$$= \text{p.f.} \int_{\mathbb{R}^{n-1}} k\big((\varrho',\tau);\,(\varrho'-\eta',0)\big)v(\eta')d\eta' + \mathcal{T}_c v(\varrho')\,,$$

respectively, where $\mathcal{T}$ and $\mathcal{T}_c$ are tangential differential operators given by

$$\left.\begin{matrix} \mathcal{T}v(\varrho') \\ \mathcal{T}_c v(\varrho') \end{matrix}\right\} = \mp \sum_{|\alpha|\leq m+1} d^{\mp}\big((\varrho',\tau),\,\kappa+m+1-|\alpha|,0\big)\tfrac{1}{\alpha!}D^{\alpha}_{\varrho'}v(\varrho')\,, \tag{9.3.36}$$

respectively.

Proof: Since the proof is identical with that for Theorem 9.3.2 we only have to justify the formula (9.3.35). Since now the canonical extension conditions (9.3.31) and the canonical Tricomi conditions (9.3.32) are more restrictive, Lemma 9.3.7 is valid and all the terms in (9.3.20) with $0 \leq j \leq m - |\alpha|$ are zero because of (9.3.33). ■

We observe that the tangential operators given by (9.3.36) are much simpler than those given in (9.3.20) where the extension conditions were required only at $\varrho_n = 0$.

Clearly, with the canonical extension conditions, the corollaries 9.3.3 and 9.3.4 concerning the derivatives of the potentials can be also modified in the same way as for Theorem 9.3.8.

If, in addition, we require the special parity conditions (9.3.27), then the tangential differential operators $\mathcal{T}$ and $\mathcal{T}_c$ in (9.3.36) even coincide and the finite part integral operators will be invariant under change of coordinates.

Theorem 9.3.9. *If the operator $A_{\Phi_r,0} \in \mathcal{L}^m_{c\ell}(\widetilde{U}_r)$ for $m+1 \in \mathbb{N}_0$ satisfies the canonical extension conditions* (9.3.31) *and the special parity conditions* (9.3.27) *then in Theorem* 9.3.8 *we have*

$$\mathcal{T}_c = \mathcal{T} = \tfrac{1}{2}[Q_c - Q]\,. \tag{9.3.37}$$

Moreover, the finite part integral in (9.3.35) *is invariant under the change of coordinates.*

Proof: Lemma 9.3.5 implies that the canonical Tricomi conditions (9.3.32) are satisfied. Then

$$Q_c - Q = \mathcal{T}_c + \mathcal{T}\,. \tag{9.3.38}$$

On the other hand, from the definition of the coefficients $d^{\pm}$ in (9.3.10) we see as in the proof of Lemma 9.3.1 that the integrals

$$\begin{aligned} d^+(\varrho,\kappa+j,\alpha,0) = & \int\limits_0^1 \tau^{|\alpha|+n-2} \int\limits_{|\Theta'|=1} k_{\kappa+j}\big(\varrho;(n-\tau\Theta')\big)\Theta'^{\alpha} d\omega(\Theta')d\tau \\ & + \int\limits_0^1 \tau^{m-j-|\alpha|} \int\limits_{|\Theta'|=1} k_{\kappa+j}\big(\varrho;(\tau n-\Theta')\big)\Theta'^{\alpha} d\omega(\Theta')d\tau \end{aligned}$$

exist for $j = m+1-|\alpha|$ since the canonical Tricomi conditions (9.3.32) are satisfied. By the change of coordinates $\tau\Theta' = \eta'$ in the first integral and $\tau^{-1}\Theta' = \eta'$ in the second integral,

$$d^+(\varrho,\kappa+j,\alpha,0) = \int\limits_{\mathbb{R}^{n-1}} k_{\kappa+j}(\varrho, n-\eta')\eta'^{\alpha} d\eta'\,.$$

Correspondingly,

$$d^-(\varrho,\kappa+j,\alpha,0) = \int\limits_{\mathbb{R}^{n-1}} k_{\kappa+j}(\varrho, -n-\eta')\eta'^{\alpha} d\eta'\,.$$

Hence, by applying the parity conditions (9.3.27), we obtain

$$\begin{aligned} d^-(\varrho, \kappa + j, \alpha, 0) &= (-1)^{m-j} \int\limits_{\mathbb{R}^{n-1}} k_{\kappa+j}(\varrho, n + \eta')\eta'^{\alpha} d\eta' \\ &= (-1)^{m-j+|\alpha|} \int\limits_{\mathbb{R}^{n-1}} k_{\kappa+j}(\varrho, n - \widetilde{\eta}')\widetilde{\eta}'^{\alpha} d\widetilde{\eta}' \\ &= (-1)^{m-j+|\alpha|} d^+(\varrho, \kappa + j, \alpha, 0)\,. \end{aligned}$$

Then $j = m + 1 - |\alpha|$ yields

$$d^-(\varrho, \kappa + m + 1 - |\alpha|, \alpha, 0) = -d^+(\varrho, \kappa + m + 1 - |\alpha|, \alpha, 0)\,.$$

substituting these relations into (9.3.36), the desired result follows from (9.3.38). ■

As an immediate consequence of Theorem 9.3.9 together with 9.3.8 we have the following representation of Q and Q_c, respectively, in terms of the jump relations (9.3.37) which extend the classical Plemelj–Sochotzki conditions of complex function theory to our situation.

Corollary 9.3.10. *Let $A \in \mathcal{L}^m_{c\ell}(\widetilde{\Omega})$ satisfy all the conditions of Theorem 9.3.9. Then with the tangential differential operator $\mathcal{T}$ in* (9.3.36) *the operator Q has the representation*

$$Qv(\varrho') = \text{ p.f.} \int\limits_{\mathbb{R}^{n-1}} k((\varrho', 0)\,;\, (\varrho' - \eta', 0))v(\eta')d\eta' - \mathcal{T}v(\varrho') \qquad (9.3.39)$$

for every $v \in C^\infty_0(U_r)$- The operator Q_c has the same representation with $\mathcal{T}$ replaced by $-\mathcal{T}$.

Remark 9.3.3: If one is interested having a C^∞–extension up to $\varrho_n = 0$, then a countable number of infinitely many extension conditions are required. In performance of these calculations, clearly a chain rule for the Schwartz kernel is needed and it will become rather tedious and cumbersome. However, if the calculations are based on the Fourier transform, then the differentiation is transformed to algebraic calculations and the extension conditions can be formulated in terms of the asymptotic symbols. These conditions were derived by Boutet de Monvel [41, 2.3], Grubb [159, Sec.12 and 2.2] and Hörmander [188]) and are called the *transmission conditions.*

Theorem 9.3.11. (Hörmander [188, Vol.III, p.108]) *Let $A_{\Phi r,0} \in \mathcal{L}^m_{c\ell}(\widetilde{U}_r)$. Then the transmission conditions*

$$\begin{aligned} &\Big(\frac{\partial}{\partial \varrho}\Big)^\beta \Big(\frac{\partial}{\partial \xi}\Big)^\alpha a^0_{m-j}(\varrho', 0; 0, +1) \\ &\qquad = e^{i\pi(m-j-|\alpha|)} \Big(\frac{\partial}{\partial \varrho}\Big)^\beta \Big(\frac{\partial}{\partial \xi}\Big)^\alpha a^0_{m-j}(\varrho', 0; 0, -1) \end{aligned} \qquad (9.3.40)$$

for all $|\alpha| \geq 0$, $|\beta| \geq 0$ *and* $j \in \mathbb{N}_0$ *are necessary and sufficient for* $\widetilde{Q}v$ *to have a* C^∞*–extension for* $\varrho_n < 0$ *up to* $\varrho_n \to O-$ *, i.e. from* Ω*, and*

$$\Big(\frac{\partial}{\partial\varrho}\Big)^\beta \Big(\frac{\partial}{\partial\xi}\Big)^\alpha a^0_{m-j}(\varrho',0;\,0,-1)$$
$$= e^{i\pi(m-j-|\alpha|)}\Big(\frac{\partial}{\partial\varrho}\Big)^\beta \Big(\frac{\partial}{\partial\xi}\Big)^\beta a^0_{m-j}(\varrho',0;\,0,+1) \quad (9.3.41)$$

for $\varrho_n > 0$ *up to* $\varrho_n \to O^+$*, i.e., from the exterior. Here* $a^0_{m-j}(\varrho,\xi)$ *are the homogeneous terms of the expansion of the symbol* $a(\varrho,\xi)$ *for* $A_{\Phi,r,0}$ *(see Definition* 7.1.6 *and* (7.1.42)*). For* $m \in \mathbb{Z}$*, the two conditions coincide.*

In conclusion, we remark that the operators $A_{\Phi r,0} \in \mathcal{L}^m_{c\ell}(\widetilde{U}_r)$ which satisfy the special parity conditions (9.3.27) for all $j \geq 0$ and all $\varrho \in \widetilde{U}_r$ which implies $m \in \mathbb{Z}$, will also satisfy the transmission conditions (9.3.40) in view of Lemma 8.1.9. This shows that for $m \in \mathbb{Z}$, the transmission conditions (9.3.40) are invariant under change of coordinates due to Lemma 8.2.8. For the invariance of the transmission property under change of coordinates preserving Γ (see Boutet de Monvel [41]).

Table 9.3.1. Extension, Tricomi, parity and transmission conditions

Properties of $A_{\Phi r,0} \in \mathcal{L}^m_{c\ell}(\widetilde{U}_r)$	Main results	Theorems
extension conditions and Tricomi conditions (9.3.15), (9.3.18)	Q and $Q_c \in \mathcal{L}^{m+1}_{c\ell}(U_r)$	Theorem 9.3.2
extension conditions and parity conditions* (9.3.15), (9.3.27)	Q and Q_c and their invariance	Theorem 9.3.6
canonical extension and Tricomi conditions (9.3.31), (9.3.32)	Q and Q_c and simplified $\mathcal{T}$ and $\mathcal{T}_c$	Theorem 9.3.8
canonical extension conditions and parity conditions* (9.3.31), (9.3.27)	Q and Q_c, their invariance and simplified $\mathcal{T} = \mathcal{T}_c = \frac{1}{2}[Q]$	Theorem 9.3.9
transmission conditions (9.3.40)	$\widetilde{Q}v$ and $\widetilde{Q}_cv$ have C^∞–extensions and invariance	Theorem 9.3.11

* The parity conditions imply Tricomi conditions (Lemma 9.3.5) and also the transmission conditions (Lemma 8.1.9).

The results of this section can be summarized as follows.

If $m+1 \notin \mathbb{N}_0$ then the extension conditions (9.3.15) are already sufficient to guarantee the existence of Q and Q_c, and the invariance of the finite part representations under change of coordinates (see Theorem 9.3.2).

The case $m+1 \in \mathbb{N}_0$ is more involved. In order to give an overview of the development, we summarize the results in Table 9.3.1.

9.4 Pseudodifferential Operators with Symbols of Rational Type

Theorem 9.3.2 and its corollaries as well as Theorems 9.3.8 and 9.3.9 show that the extension conditions seem to be rather restrictive and are cumbersome to verify. Nevertheless, in most applications, including classical boundary potentials, we deal with a class of pseudodifferential operators which will satisfy all of these conditions. This class of operators is characterized by the*symbols of rational type* and will be considered in this section.

We begin with the definition for the symbol $a \in \boldsymbol{S}^m_{c\ell}(\widetilde{\Omega} \times \mathbb{R}^n)$ which also gives the definition when $\widetilde{\Omega}$ is restricted to $\widetilde{U}_r = U_r \times (-\varepsilon, \varepsilon)$.

Definition 9.4.1. *The symbol $a \in \boldsymbol{S}^m_{c\ell}(\widetilde{\Omega} \times \mathbb{R}^n)$ is called of rational type if in the complete asymptotic expansion $a \sim \sum_{j\geq 0} a_{m-j}$, each symbol $a_{m-j}(x,\xi)$, which is homogeneous of degree $m-j$ for $|\xi| \geq 1$, is a rational function of ξ, i.e. the homogeneous symbol a^0_{m-j} is given by the equation*

$$a^0_{m-j}(x,\xi) = \Big\{ \sum_{|\alpha|=m-j+d(j)} c_\alpha(x)\xi^\alpha \Big\} \Big/ \Big\{ \sum_{|\beta|=d(j)} b_\beta(x)\xi^\beta \Big\} \tag{9.4.1}$$

for $\xi \in \mathbb{R}^n \setminus \{0\}$ with some $d(j) \in \mathbb{N}_0$ satisfying $m-j+d(j) \in \mathbb{N}_0$.

Note that $a \in \boldsymbol{S}^m(\widetilde{\Omega} \times \mathbb{R}^n)$ can only be of rational type if $m \in \mathbb{Z}$. Moreover, Equations (7.1.45) and (7.1.46) imply that the class of symbols of rational type is invariant under diffeomorphic coordinate changes.

The latter implies that the symbol of A_{Φ_r} in (9.4.1) is of rational type iff $A \in \mathcal{L}^m_{c\ell}(\Omega)$ is of rational type. We further note that our assumptions imply that a_{m-j} is C^∞ and has compact support with respect to $\varrho \in \widetilde{U}_r$. In addition, the symbols of rational type satisfy the following parity conditions.

A more general class of pseudodifferential symbols, which includes those of rational type was considered by Boutet Monvel [41] and Grubb [159]. These symbols all satisfy the transmission conditions but in general not necessarily the following parity conditions.

Lemma 9.4.1. *Let* $a(x,\xi) \sim \sum_{j\geq 0} a_{m-j}(x,\xi)$ *be a symbol* $a \in \boldsymbol{S}^m_{c\ell}(\widetilde{\Omega} \times \mathbb{R}^n)$ *of rational type. Then the corresponding homogeneous rational symbols* $a^0_{m-j}(x,\xi)$ *satisfy the special parity conditions (cf.* (8.1.74), (8.2.35)*)*

$$a^0_{m-j}(x,-\xi) = (-1)^{m-j} a^0_{m-j}(x,\xi) \quad \textit{for every } j \in \mathbb{N}_0 \tag{9.4.2}$$

as well as (9.3.27),

$$k_{\kappa+j}(x;y-x) = (-1)^{m-j} k_{\kappa+j}(x;x-y) \quad \textit{for every } j \in \mathbb{N}_0\,, \tag{9.4.3}$$

and also the transmission conditions (cf. (9.3.40)*)*

$$\Big(\Big(\frac{\partial}{\partial\xi'}\Big)^{\gamma} a^0_{m-j}\Big)(x,(0,-1)) = (-1)^{m-j-|\gamma|}\Big(\Big(\frac{\partial}{\partial\xi'}\Big)^{\gamma} a^0_{m-j}\Big)(x,(0,1))\,. \tag{9.4.4}$$

Therefore, also the Tricomi conditions (9.3.18) *are satisfied in view of Lemma* 9.3.5.

We remark that these transmission conditions (9.4.4) will imply (9.3.40) after differentiation of (9.4.4) with respect to $x = \varrho$ at $\varrho_n = 0$.

Proof: The proof follows from the definition of the homogeneous symbols a^0_{m-j} in (9.4.1) immediately; i.e.

$$\begin{aligned} a^0_{m-j}(x,-\xi) &= \Big\{ \sum_{|\alpha|=m-j+d(j)} c_\alpha(x)\xi^\alpha(-1)^{|\alpha|}\Big\} \Big/ \Big\{ \sum_{|\beta|=d(j)} b_\beta \xi^\beta (-1)^{|\beta|}\Big\} \\ &= (-1)^{m-j} a^0_{m-j}(x,\xi)\,. \end{aligned}$$

Lemma 8.1.9 then implies the conditions (9.4.3).

For the transmission condition we note that a straightforward computation yields

$$\Big(\frac{\partial}{\partial\xi_\ell} a^0_{m-j}\Big)(x,(0,-1)) = (-1)^{m-j-1}\Big(\frac{\partial}{\partial\xi_\ell} a^0_{m-j}\Big)(x,(0,1))$$

for any $\ell = 1,\dots,n-1$.

Since the derived symbols on both sides are rational again, induction implies the desired conditions (9.4.4). ∎

This special class of symbols of rational type is invariant with respect to the following operations, such as diffeomorphism, transposition and composition.

Lemma 9.4.2. *Let* $a(x,\xi)$ *be a symbol of rational type belonging to* $A \in \mathcal{L}^m_{c\ell}(\Omega)$. *Then the symbols of* $A^\top, A^*$ *and* A_Φ *are of rational type, too. If* A *and* B *have symbols of rational type then* $A \circ B$ *also has a symbol of rational type.*

Proof: From formula (7.1.34) for the symbol expansion of $A^\top$, it can be seen that the corresponding homogeneous term $a^{(T)}_{m-k}(x,\xi)$ of the asymptotic expansion of $\sigma_{A^\top}$ is given in terms of a linear combination of at most $k+1$ derivatives of a^0_{m-j} in (9.4.1) with $j+|\alpha|=k$. Hence, each of the resulting functions $a^{(T)}_{m-k}(x,\xi)$ is rational.

With the same arguments for (7.1.35) and (7.1.37) it follows that A^* and $A\circ B$ have symbols of rational type.

Concerning a diffeomorphic coordinate transformation, we use formula (7.1.45) for the asymptotic expansion of the symbol a_Φ. Again, only a finite number of derivatives of finitely many rational functions

$$a^0_{m-j}\big(\Psi(x'), \Xi^\top(x',y')^{-1}\xi'\big)$$

will contribute to one typical homogeneous term $a^0_{\Phi,m-k}(x',\xi')$ in the resulting asymptotic expansion of the transformed symbol. Hence, $a^0_{\Phi,m-k}(x',\xi')$ is a rational function. ∎

As we observe, the pseudodifferential operators $A\in\mathcal{L}^m_{c\ell}(\widetilde{\Omega})$ with symbols of rational type enjoy all the essential properties such as the special parity conditions and canonical extensions so that their surface potentials $\widetilde{Q}v$ have continuous limits Q and Q_c, respectively with $Q,Q_c\in\mathcal{L}^{m+1}_{c\ell}(U_r)$.In particular, Theorem 9.3.9 holds for A which can be formulated in a more specific form. In contrast to Theorem 9.3.8, we now present a proof different from the one for Theorem 9.3.8, which takes advantage of the special features of rational type symbols. In particular, we are able to show that the surface potential $\widetilde{Q}v$ is C^∞ up to the surface $\varrho_n=0$. In addition, we shall give precise mapping properties of $\widetilde{Q}$, Q and Q_c in Sobolev spaces which will be crucial for boundary integral equation methods.

All of these methods are covered by Boutet de Monvel's analysis for a more general class of operators including those with symbols of rational type, where the surface potential operators $\widetilde{Q}$ are called *Poisson operators* (Boutet de Monvel [39, 40, 41] and Grubb [159]).

Theorem 9.4.3. (Seeley [384]) *Let $A\in\mathcal{L}^m_{c\ell}(\widetilde{\Omega})$ for $m\in\mathbb{Z}$ have a symbol of rational type. Let $A_{\Phi_r,0}\in OPS^m(\widetilde{U}_r\times\mathbb{R}^n)$ be the properly supported part of the transformed operator in* (9.2.7) *under transformation* (9.2.1) *having the complete symbol $a(\varrho,\xi)\sim\sum_{j\geq 0}a_{m-j}(\varrho,\xi)$ with $a_{m-j}\in\boldsymbol{S}^{m-j}_{c\ell}(\widetilde{U}_r\times\mathbb{R}^n)$ where $a^0_{m-j}(\varrho,\xi)$ is of the form* (9.4.1) *(with x replaced by ϱ) for every $j\in\mathbb{N}_0$ and where $\widetilde{U}_r=U_r\times(-\varepsilon,\varepsilon)$. Then the Poisson operator $\widetilde{Q}$ defined by*

$$\widetilde{Q}v(\varrho):=A_{\Phi_r,0}\big(v\otimes\delta(\eta_n)\big)(\varrho)\quad\textit{for}\quad\varrho_n<0\quad\textit{and}\quad v\in\mathcal{D}(U_r)\tag{9.4.5}$$

where $\varrho=(\varrho',\varrho_n)$, $\varrho'\in U_r$, has a continuous extension up to $\varrho_n\leq 0$, and maps $\mathcal{D}(U_r)$ to $C^\infty(U_r\times\{0\geq\varrho_n>-\varepsilon\})$.

Moreover, the limit

$$Qv(\varrho') = \lim_{0>\varrho_n\to 0} A_{\Phi_r,0}(v\otimes\delta)(\varrho',\varrho_n)\,, \tag{9.4.6}$$

exists and defines a pseudodifferential operator $Q \in \mathcal{L}^{m+1}_{c\ell}(U_r)$ *which has the complete symbol* $q(\varrho',\xi') \in \boldsymbol{S}^{m+1}_{c\ell}(U_r\times\mathbb{R}^{n-1})$. *Each term in its asymptotic expansion* $q \sim \sum\limits_{j\geq 0} q_{m+1-j}(\varrho',\xi')$ *is explicitly given by the contour integral*

$$q_{m+1-j}(\varrho',\xi') = (2\pi)^{-1}\int\limits_{\mathfrak{c}} a_{m-j}\big((\varrho',0),(\xi',z)\big)dz \text{ for } |\xi'|\leq 1 \tag{9.4.7}$$

and

$$q_{m+1-j}(\varrho',\xi') = |\xi'|^{m+1-j} q_{m+1-j}(\varrho',\xi'_0) \quad \text{for } |\xi'|\geq 1 \tag{9.4.8}$$

with $\xi'_0 := \xi'/|\xi'|$ *and* $q_{m+1-j}(\varrho',\xi'_0)$ *given by* (9.4.7). *The contour* $\mathfrak{c}\subset\mathbb{C}$ *consists of the points*

$$\mathfrak{c} = \big\{z\in[-c_0,c_0]\cup\{z=c_0e^{i\vartheta}\ :\ 0\geq\vartheta\geq-\pi\}\big\} \tag{9.4.9}$$

in the lower–half plane, and is clockwise oriented where $c_0 > 0$ *is chosen sufficiently large so that all the poles of* $a^0_{m-j}\big((\varrho',0),(\xi'_0,z)\big)$ *in the lower half–plane are enclosed in the interior of* $\mathfrak{c}\subset\mathbb{C}$.

Proof: In order to approximate the Dirac functional in the distribution $w = v(\varrho')\otimes\delta(\varrho_n)$ by functions in $C_0^\infty\big(U_r\times(-\varepsilon,\varepsilon)\big)$ we take a cut–off function $g\in C_0^\infty(-1,1)$ with $0\leq g(t)\leq 1$ and $\int\limits_{-1}^{1} g(t)dt = \sqrt{2\pi}$ and define

$$w_\eta := v(\varrho')\frac{1}{\eta}g\Big(\frac{\varrho_n}{\eta}\Big) \in \mathcal{D}(\mathbb{R}^n) \quad \text{with} \quad \eta>0\,.$$

Then $w_\eta\to w$ in $\mathcal{D}'(\mathbb{R}^n)$ for $\eta\to 0$. Moreover,

$$e^{i\varrho_n z}\widehat{g}(\eta z) = \frac{1}{\sqrt{2\pi}}e^{i\varrho_n z}\int\limits_{\xi_n=-1}^{1} e^{-i\eta\xi_n z}g(\xi_n)d\xi_n \tag{9.4.10}$$

satisfies the estimate

$$|e^{i\varrho_n z}\widehat{g}(\eta z)| \leq c(N)e^{\eta|\operatorname{Im} z|-\varrho_n\operatorname{Im} z}(1+|z|)^{-N} \tag{9.4.11}$$

for every $N\in\mathbb{N}$ and for $z\in\mathbb{C}$ due to the Paley–Wiener–Schwartz Theorem 3.1.3. The constant $c(N)$ depends on g and N but neither on η nor on z.

If $|z|\leq C$, $\operatorname{Im} z\leq 0$ and $0\geq\varrho_n\geq -1$ then

$$\lim_{0<\eta\to 0} e^{i\varrho_n z}\widehat{g}(\eta z) = e^{i\varrho_n z} \tag{9.4.12}$$

uniformly for any C fixed.

We also need information concerning the poles of the meromorphic function $a^0_{m-j}((\varrho',\varrho_n),(\xi',z))$ depending on ϱ_n and ξ', i.e. the zeros of

$$\wp_{d(j)}(\varrho,\xi',z) := \sum_{|\beta|=d(j)} b_\beta(\varrho',\varrho_n)\xi'^{\beta'} z^{\beta_n} \tag{9.4.13}$$

where $\beta = (\beta',\beta_n) = (\beta_1,\dots,\beta_{n-1},\beta_n)$.

Since $\varrho_n \in [-1,0]$ and $\varrho' \in K_r \Subset U_r$ with a fixed compact subset K_r, for $|\xi'| = 1$, all zeros of $\wp_{d(j)}$ in the lower half–plane are contained in $\{z \in \mathbb{C} \mid \operatorname{Im} z \le 0 \wedge |z| \le c_0\}$ with some appropriately chosen constant $c_0 > 1$. Since $\wp_{d(j)}$ is a homogeneous polynomial of degree $d(j)$ with respect to (ξ',z), there holds

$$\wp_{d(j)}(\varrho,\xi',z) = |\xi'|^d \wp_{d(j)}(\varrho,\xi'_0,\tfrac{z}{|\xi'|}) = 0$$

for $0 < |\xi'| \ne 1$ if and only if $z = |\xi'|z_0$ where z_0 satisfies $\wp_{d(j)}(\varrho,\xi'_0,z_0) = 0$. Hence, all poles z_0 of $a^0_{m-j}((\varrho',\varrho_n),(\xi',z))$ with negative imaginary part are contained in the interior of the set

$$\{z \in \mathbb{C} \mid \operatorname{Im} z \le 0 \wedge |z| \le c_0|\xi'|\}\,.$$

Now let us consider the family of contour integrals

$$\begin{aligned}\widetilde{q}_{m+1-j}(\varrho,\xi') &:= (2\pi)^{-1}\oint_{\mathfrak{c}} e^{i\varrho_n z} a_{m-j}(\varrho,(\xi',z))dz \quad \text{for } |\xi'| \le 1\,,\\ &:= (2\pi)^{-1}\oint_{|\xi'|\mathfrak{c}} e^{i\varrho_n z} a^0_{m-j}(\varrho,(\xi',z))dz \quad \text{for } |\xi'| \ge 1\end{aligned} \tag{9.4.14}$$

where

$$|\xi'|\mathfrak{c} := \{z \in \mathbb{C} \mid z \in \mathbb{R} : -|\xi'|c_0 \le z \le |\xi'|c_0\} \cup \{z = c_0|\xi'|e^{i\vartheta}\,,\ 0 \ge \vartheta \ge -\pi\}$$

which is oriented clock–wise. For $|\xi'| \ge 1$, one has $|(\xi',z)| \ge 1$ and $a_{m-j} = a^0_{m-j}$. Then

$$\lim_{0>\varrho_n\to 0} \widetilde{q}_{m+1-j}(\varrho,\xi') = \widetilde{q}_{m+1-j}((\varrho',0),\xi') = q_{m+1-j}(\varrho',\xi') \tag{9.4.15}$$

and

$$\widetilde{q}_{m+1-j}((\varrho',0),\xi') = |\xi'|^{m-j+1} q_{m+1-j}(\varrho',\xi'_0) \quad \text{for } |\xi'| \ge 1 \tag{9.4.16}$$

where $\xi'_0 = \xi'/|\xi'|$ and q_{m+1-j} is given by (9.4.7).

To show (9.4.6) for $\widetilde{Q}v$, we write

$$\begin{aligned}
&a_{m-j}(\varrho, D)w_\eta(\varrho)\\
&= (2\pi)^{-n} \int\limits_{|\xi'|\le 1} e^{i\varrho'\cdot\xi'}\widehat{v}(\xi') \int\limits_{|\xi_n|\le c_0} e^{i\varrho_n\xi_n} a_{m-j}\big((\varrho',\varrho_n),(\xi',\xi_n)\big)\widehat{g}(\eta\xi_n)d\xi_n d\xi'\\
&\quad +(2\pi)^{-n} \int\limits_{|\xi'|\le 1}\int\limits_{|\xi_n|\ge c_0} e^{i\varrho'\cdot\xi'}\widehat{v}(\xi')a^0_{m-j}\big((\varrho',\varrho_n),(\xi',\xi_n)\big)e^{i\varrho_n\xi_n}\widehat{g}(\eta\xi_n)d\xi_n d\xi'\\
&\quad +(2\pi)^{-n} \int\limits_{|\xi'|\ge 1}\int\limits_{\mathbb{R}} e^{i\varrho'\cdot\xi'}\widehat{v}(\xi')a^0_{m-j}\big((\varrho',\varrho_n),(\xi',\xi_n)\big)e^{i\varrho_n\xi_n}\widehat{g}(\eta\xi_n)d\xi_n d\xi'\\
&=: \quad I_1 + I_2 + I_3\,.
\end{aligned}$$

Since in I_1 the domain of integration is compact and the integrand is C^∞, we get

$$\lim_{0<\eta\to 0} I_1 = (2\pi)^{-n} \int\limits_{|\xi'|\le 1} e^{i\varrho'\cdot\xi'}\widehat{v}(\xi') \int\limits_{|\xi_n|\le c_0} e^{i\varrho_n\xi_n} a_{m-j}\big((\varrho',\varrho_n),(\xi',\xi_n)\big)d\xi_n d\xi'\,. \tag{9.4.17}$$

In I_2 we change the path of integration of the inner integral and obtain

$$\begin{aligned}
&\int\limits_{|\xi_n|\ge c_0} a^0_{m-j}\big((\varrho',\varrho_n),(\xi',\xi_n)\big)e^{i\varrho_n\xi_n}\widehat{g}(\eta\xi_n)d\xi_n\\
&\qquad\qquad = \int\limits_{\mathfrak{c}\setminus[-c_0,c_0]} a^0_{m-j}\big((\varrho',\varrho_n),(\xi',z)\big)e^{i\varrho_n z}\widehat{g}(\eta z)dz\,,
\end{aligned}$$

since the estimate (9.4.11) allows us to choose $N > m-j+1$ and the integrand is holomorphic in $\operatorname{Im} z \le 0 \wedge |z| \ge c_0$ provided $0 \le \eta < -\varrho_n$. Then we write

$$I_2 = (2\pi)^{-n} \int\limits_{|\xi'|\le 1} e^{i\varrho'\cdot\xi'}\widehat{v}(\xi') \int\limits_{\mathfrak{c}\setminus[-c_0,c_0]} a^0_{m-j}\big((\varrho',\varrho_n),(\xi',z)\big)e^{i\varrho_n z}\widehat{g}(\eta z)dz d\xi'$$

and the integrand depends continuously on η on the compact domain of integration. Taking the limit yields

$$\begin{aligned}
&\lim_{0<\eta\to 0} I_2\\
&= (2\pi)^{-n+1} \int\limits_{|\xi'|\le 1} e^{i\varrho'\cdot\xi'}\widehat{v}(\xi')(2\pi)^{-1} \int\limits_{\mathfrak{c}\setminus[-c_0,c_0]} e^{i\varrho_n z}a^0_{m-j}\big((\varrho',0),(\xi',z)\big)dz d\xi'\,.
\end{aligned}$$

This together with (9.4.17) gives

$$\lim_{0<\eta\to 0}(I_1+I_2) = (2\pi)^{-n+1} \int\limits_{|\xi'|\le 1} e^{i\varrho'\cdot\xi'}\widehat{v}(\xi')\widetilde{q}_{m+1-j}(\varrho,\xi')d\xi' \tag{9.4.18}$$

since, on $z \in \mathfrak{c} \setminus [-c_0, c_0]$ one has $a_{m-j}(\varrho, (\xi', z)) = a^0_{m-j}(\varrho, (\xi', z))$.

It remains now to consider I_3. We can write the inner integral as

$$
\begin{aligned}
&(2\pi)^{-1} \int\limits_{\mathbb{R}} a^0_{m-j}\big((\varrho', \varrho_n), (\xi', \xi_n)\big) e^{i\varrho_n \xi_n} \widehat{g}(\eta\xi_n) d\xi_n \\
&= (2\pi)^{-1} \int\limits_{|\xi'|\mathfrak{c}} a^0_{m-j}\big((\varrho', \varrho_n), (\xi', z)\big) e^{i\varrho_n z} \widehat{g}(\eta z) dz \\
&= (2\pi)^{-1} |\xi'|^{m-j+1} \int\limits_{\mathfrak{c}} a^0_{m-j}\big((\varrho', \varrho_n), (\xi'_0, z')\big) e^{i\varrho_n |\xi'| z'} \widehat{g}(\eta|\xi'|z') dz' .
\end{aligned}
$$

This is due to the holomorphy of the integrand for $\operatorname{Im} z \leq 0,\ 1 \leq c_0 \leq |\xi'| c_0 \leq |z|$, the estimate (9.4.11) for $\eta \leq -\varrho_n$ and the homogeneity of a^0_{m-j} for $|\xi'| \geq 1$. Now we have, for every $\xi' \in \mathbb{R}^{n-1}$ with $|\xi'| \geq 1$,

$$
\begin{aligned}
&\lim_{0<\eta\to 0} (2\pi)^{-1} \int\limits_{\mathbb{R}} a^0_{m-j}\big((\varrho', \varrho_n), (\xi', \xi_n)\big) e^{i\varrho_n \xi_n} \widehat{g}(\eta\xi_n) d\xi_n \\
&= \lim_{0<\eta\to 0} (2\pi)^{-1} |\xi'|^{m-j+1} \int\limits_{\mathfrak{c}} a^0_{m-j}\big((\varrho', \varrho_n), (\xi', z')\big) e^{i\varrho_n |\xi'| z'} \widehat{g}(\eta|\xi'|z') dz' \\
&= |\xi'|^{m-j+1} (2\pi)^{-1} \int\limits_{\mathfrak{c}} a^0_{m-j}\big(\varrho,\ (\xi'_0, z')\big) e^{i\varrho_n |\xi'| z'} dz' \\
&= \widetilde{q}_{m+1-j}(\varrho, \xi') .
\end{aligned}
$$

Hence, the integrand of

$$
\begin{aligned}
I_3 = (2\pi)^{-n+1} \int\limits_{1\leq|\xi'|} & e^{i\varrho' \cdot \xi'} \widehat{v}(\xi') (2\pi)^{-1} |\xi'|^{m-j+1} \times \\
& \times \int\limits_{\mathfrak{c}} a^0_{m-j}\big((\varrho', \varrho_n), (\xi', z')\big) e^{i\varrho_n |\xi'| z'} \widehat{g}(\eta|\xi'|z') dz' d\xi'
\end{aligned}
$$

converges point–wise as $\eta \to 0$. Consequently,

$$
(2\pi)^{-n+1} \int\limits_{1\leq|\xi'|} e^{i\varrho' \cdot \xi'} \widehat{v}(\xi') \widetilde{q}_{m+1-j}(\varrho, \xi') d\xi' = \lim_{\eta\to 0} I_3 \tag{9.4.19}
$$

due to the Lebesgue theorem. Collecting (9.4.18) and (9.4.19) yields

$$
\widetilde{Q}_{m+1-j} v(\varrho) := (2\pi)^{-n+1} \int\limits_{\mathbb{R}^{n-1}} e^{i\varrho' \cdot \xi'} \widehat{v}(\xi') \widetilde{q}_{m+1-j}(\varrho, \xi') d\xi' . \tag{9.4.20}
$$

Clearly, $\widetilde{Q}_{m+1-j} v(\varrho)$ exists and is a C^∞–function for $\varrho_n < 0$. It follows from (9.4.15) that the integrand of (9.4.20) converges pointwise to

$$e^{i\varrho'\cdot\xi'}\widehat{v}(\xi')q_{m+1-j}(\varrho',\xi')$$

if $0 > \varrho_n \to 0$. Therefore, the Lebesgue theorem implies

$$\lim_{0>\varrho_n\to 0}\widetilde{Q}_{m+1-j}v(\varrho) = q_{m+j-1}(\varrho', D)v \quad \text{for } j = 0, 1, \dots .$$

Here, $q_{m+1-j} \in \boldsymbol{S}^{m-j+1}_{c\ell}(U_r \times \mathbb{R}^{n-1})$ is given explicitly by (9.4.7). To this asymptotic sequence, there exists a symbol $q \in \boldsymbol{S}^{m+1}_{c\ell}(U_r \times \mathbb{R}^{n-1})$ with $q \sim \sum\limits_{j\geq 0} q_{m+1-j}$ due to Theorem 7.1.3. The symbol then defines a pseudodifferential operator $q(\varrho', D') \in \mathcal{L}^{m+1}_{c\ell}(U_r)$ according to Definition 7.1.2. Hence, there exists a smoothing operator R such that $Q = q(\varrho', D') + R$ with Q given by (9.4.6).

To see this, note that $\Big(q - \sum\limits_{j=0}^{L} q_{m+1-j}\Big) \in \boldsymbol{S}^{m+1-L}(U_r \times \mathbb{R}^{n-1})$ and, hence, has a Schwartz kernel in $C^{L-m-n-\delta}(U_r \times U_r)$ due to Theorem 8.1.1 and (8.1.2) for some δ with $0 < \delta < 1$ provided $L > m + n + 1$. On the other hand, $A_R = \big(A_{\Phi_r,0} - \sum\limits_{j=0}^{L} A_{m-j}\big) \in \mathcal{L}^{m-L}_{c\ell}(\widetilde{\Omega} \times \mathbb{R}^n)$ has a Schwartz kernel $k_{A_R}(\varrho, \varrho - \eta)$ in $C^{-m-n+L-\delta'}(\widetilde{\Omega} \times \widetilde{\Omega})$ for some δ' with $0 < \delta' < 1$. Therefore the restriction of the kernel, $k_{A_R}\big((\varrho', 0)\,,\,((\varrho' - \eta'), 0)\big)$ belongs to the same class and is the Schwartz kernel of the operator $Q - \sum\limits_{j=0}^{L} Q_{m+1-j}$. Consequently,

$$R := Q - q(\varrho', D') = -\Big(q(\varrho', D') - \sum_{j=0}^{L} q_{m+1-j}(\varrho', D')\Big) + \Big(Q - \sum_{j=0}^{L} Q_{m+1-j}\Big) \tag{9.4.21}$$

has a Schwartz kernel in the class $C^{L-m-n-1}(U_r \times U_r)$ for every $L > m+n+1$, i.e., a C^∞–Schwartz kernel. This completes the proof of Theorem 9.4.3. ■

Corollary 9.4.4. *Let $A \in \mathcal{L}^m_{c\ell}(\widetilde{\Omega})$ satisfy all the conditions of Theorem 9.4.3. Then*

$$\widetilde{Q}_c v(\varrho) := A_{\Phi_r,0}(v \otimes \delta)(\varrho', \varrho_n) \quad \textit{for } \varrho' \in U_r\,,\ \varrho_n > 0 \quad \textit{and } v \in \mathcal{D}(U_r) \tag{9.4.22}$$

has a continuous extension to $\varrho_n \geq 0$ and $\widetilde{Q}_c$ maps $C^\infty_0(U_r)$ to $C^\infty\big(U_r \times (0, \varepsilon)\big)$. The trace

$$Q_c v(\varrho') = \lim_{0<\varrho_n\to 0} A_{\Phi_r,0}(v \otimes \delta)(\varrho', \varrho_n) \tag{9.4.23}$$

defines an operator $Q_c \in \mathcal{L}^{m+1}_{c\ell}(U_r)$ having the complete symbol $q_c \sim \sum\limits_{j\geq 0} q_{m+1-j,c}(\varrho', \xi')$ where $q_{m+1-j,c}(\varrho', \xi')$ is explicitly given by

$$q_{m+1-j,c}(\varrho',\xi') = (2\pi)^{-1}\int\limits_{\mathfrak{c}_c} a_{m-j}\big((\varrho',0),(\xi',z)\big)dz \quad for \ |\xi'|\leq 1 \tag{9.4.24}$$

and

$$q_{m+1-j,c}(\varrho',\xi') = |\xi'|^{m+1-j} q_{m+1-j,c}(\varrho',\xi_0') \quad for \ |\xi'|\geq 1$$

where $\xi_0' = \xi'/|\xi'|$. *Here, the contour*

$$\mathfrak{c}_c := \big\{z\in\mathbb{R} : -c_0\leq z\leq c_0\big\}\cup\big\{z = c_0 e^{i\vartheta} \quad with \ 0\leq\vartheta\leq\pi\big\}$$

in the upper half–plane is counterclockwise oriented and $c_0 > 0$ *is chosen sufficiently large so that all the poles of* $a_{m-j}\big((\varrho',0),(\xi',z)\big)$ *in the upper half–plane are enclosed in the interior of* $\mathfrak{c}_c\subset\mathbb{C}$.

The proof is identical to that of Theorem 9.4.3 except $\varrho_n\geq\eta>0$ and $\operatorname{Im} z\geq 0$.

Lemma 9.4.5. *Let* Γ_τ *be the family of planes* $\varrho_n = \tau = \text{const}$ *in* $\widetilde{U}_r$ *with* $\tau\in I_\tau$. *Then the limits of* $\widetilde{Q}(\tau)v(\varrho)$ *given by* (9.4.22) *in* (9.4.5) *and* (9.4.22) *for* $\varrho_n\to\tau\pm$, *respectively, exist for every* $\tau\in I_\tau$ *and depend continuously on* τ.

Proof: The poles of $a^0_{m-j}\big((\varrho',\tau+\varrho_n);\ (\xi',z)\big)$, i.e., the zeros of $\wp_{d(j)}\big((\varrho+\tau n);\ (\xi',z)\big)$ in (9.4.13) depend continuously on τ. Therefore, the contour integrals defined in (9.4.14) also depend on τ continuously. Hence, each of the potentials $(\widetilde{Q}_{m+1-j}v)(\tau)(\varrho)$ in the family given by (9.4.20) converges to the corresponding limit $q_{m+j-1}(\tau,\varrho',D)v$ for each Γ_τ. Each of these limits then depends continuously on τ.

As in (9.4.21) the remaining operator R has a kernel that now depends continuously on the parameter τ, it follows that the limit operators $Q(\tau)$ and $Q_c(\tau)$ exist and depend on τ continuously. ∎

Remark 9.4.1: As a consequence of Lemma 9.4.5, the existence of the limits Q and Q_c implies that the extension conditions are fulfilled for each $\tau\in I_\tau$. The latter implies that therefore the canonical extension conditions (9.3.31) are satisfied. This means that together with the special parity conditions (9.4.3) all the assumptions of Theorem 9.3.8 are satisfied for our class of operators A having symbols of rational type. Hence, together with Lemma 9.4.1, Theorem 9.3.9 holds as well.

To sum up these consequences we now have the following theorem which, again, is the Plemelj–Sochozki jump relation for the class of operators having symbols of rational type.

Theorem 9.4.6. *Let* $A\in\mathcal{L}^m_{c\ell}(\widetilde{\Omega})$ *have a symbol of rational type. Let* $A_{\Phi r,0}\in\mathcal{L}^m_{c\ell}(\widetilde{U}_r)$ *be the properly supported part of* $A_{\Phi r}$ *in the parametric domain. Then the jump of* $A_{\Phi r,0}\big(v\otimes\delta(\varrho_n)\big)$ *across* $\varrho_n = 0$ *is given by*

$$\lim_{0<\varrho_n\to 0}\left\{A_{\Phi_r,0}(v\otimes\delta)(\varrho',\varrho_n)-A_{\Phi_r,0}(v\otimes\delta)(\varrho',-\varrho_n)\right\}=[Q]v(\varrho') \quad (9.4.25)$$

and $\frac{1}{2}[Q]$ *is the tangential differential operator*

$$\mathcal{T}=\mathcal{T}_c=\frac{1}{2}[Q](\varrho',D') \quad (9.4.26)$$

$$=\frac{1}{2}\sum_{0\le j\le m+1}\ \sum_{|\alpha|=m-j+1}(-i)^{|\alpha|+1}\frac{1}{\alpha!}\frac{\partial^\alpha a_{m-j}}{(\partial\xi')^\alpha}\big((\varrho',0),(0,1)\big)D^\alpha_{\varrho'}\,.$$

Moreover, the representation (9.3.37) *holds and the invariance* (9.3.28) *under change of coordinates remains valid.*

Proof: Subtracting (9.4.7) from (9.4.24) we obtain for the jump across $\varrho_n=0$, the relation

$$[Q_{m+1-j}]v(\varrho')=[q_{m+1-j}(\varrho',D')]v(\varrho')$$

with the symbol

$$[q_{m+1-j}](\varrho',\xi')=(2\pi)^{-1}\oint\limits_{|z|=R}a^0_{m-j}\big((\varrho',0),(\xi',z)\big)dz \quad (9.4.27)$$

provided $R\ge c_0\ge 1$ since then $|\xi'|^2+|z|^2\ge 1$, where $a_{m-j}=a^0_{m-j}$ and the integrand is holomorphic for $|z|\ge c_0$.

Since Theorem 9.3.8 is valid here, $[q_{m+1-j}]$ is a differential operator. In order to show the particular form of $\mathcal{T}$ in (9.4.26), we combine the homogeneity of a^0_{m-j} with the Taylor formula about $(0,\xi_n)$ and obtain

$$a_{m-j}\big((\varrho',0),(\xi',\xi_n)\big)=\sum_{0\le|\gamma|\le N}\xi_n^{m-j-|\gamma|}\frac{\partial^\gamma a_{m-j}}{(\partial\xi')^\gamma}\big((\varrho',0),(0,1)\big)\frac{1}{\gamma!}\xi'^\gamma$$
$$+a^+_{m-jR}\big((\varrho',0),(\xi',\xi_n)\big)\ \text{ provided }\ \xi_n\ge 1\,. \quad (9.4.28)$$

In the same manner we have

$$a_{m-j}\big((\varrho',0),(\xi',\xi_n)\big)$$
$$=\sum_{0\le|\gamma|\le N}\xi_n^{m-j-|\gamma|}(-1)^{m-j-|\gamma|}\frac{\partial^\gamma a_{m-j}}{(\partial\xi')^\gamma}\big((\varrho',0),(0,-1)\big)\frac{1}{\gamma!}\xi'^\gamma$$
$$+a^-_{m-jR}\big((\varrho',0),(\xi',\xi_n)\big)\ \text{ for }\ \xi_n\le -1\,. \quad (9.4.29)$$

Clearly, the remainders satisfy the estimate

$$|a^\pm_{m-jR}|\le c_{jN}|\xi_n|^{m-j-N-1}|\xi'|^{N+1}\ \text{ for }\ \begin{cases}\xi_n\ge 1 & \text{and } +\\ \xi_n\le -1 & \text{and } -\end{cases} \quad (9.4.30)$$

uniformly in $\xi' \in \mathbb{R}^{n-1}$.

Because of the transmission conditions (9.4.4), we have

$$a_{m-j}\big((\varrho',0),(\xi',\xi_n)\big) = a_{m-jH}\big((\varrho',0),(\xi',\xi_n)\big) + a^{\pm}_{m-jR}$$

for $\xi_n \ge 1$ or $\xi_n \le -1$, respectively, where

$$a_{m-jH} = \sum_{0\le|\gamma|\le N} \xi_n^{m-j-|\gamma|} \frac{\partial^\gamma a_{m-j}}{(\partial\xi')^\gamma}\big((\varrho',0),(0,1)\big)\frac{1}{\gamma!}\xi'^\gamma . \tag{9.4.31}$$

The main idea for showing that (9.4.27) is of the form (9.4.26), is to replace a^0_{m-j} in (9.4.27) by a_{m-jH}:

$$\begin{aligned}
[q_{m+1-j}](\varrho',\xi') &= (2\pi)^{-1} \oint\limits_{|z|=R} a_{m-jH}(\varrho',\xi',z)dz \\
&+ (2\pi)^{-1} \oint\limits_{|z|=R} \big(a^0_{m-j}\big((\varrho',0),(\xi',z)\big) - a_{m-jH}(\varrho',\xi',z)\big)dz\,.
\end{aligned}$$

The first term can be calculated explicitly by the residue theorem,

$$(2\pi)^{-1} \oint\limits_{|z|=R} a_{m-jH}(\varrho',\xi',z)dz = i \sum_{|\gamma|=m-j+1} \frac{\partial^\gamma a_{m-j}}{(\partial\xi')^\gamma}\big((\varrho',0),(0,1)\big)\frac{1}{\gamma!}\xi'^\gamma \tag{9.4.32}$$

if $N > m-j+1$ in (9.4.31).

To estimate the remaining integral we want to show

$$\begin{aligned}
(2\pi)^{-1} &\oint\limits_{|z|=R} \big(a^0_{m-j} - a_{m-jH}\big)(\varrho',0),(\xi',z)\big)dz \\
&= (2\pi)^{-1} \int\limits_{\xi_n\le -R} a^-_{m-jR}\big((\varrho',0),(\xi',\xi_n)\big)d\xi_n \\
&\quad -(2\pi)^{-1} \int\limits_{R\le\xi_n} a^+_{m-jR}\big((\varrho',0)\,,\ (\xi',\xi_n)\big)d\xi_n\,.
\end{aligned}$$

The latter then can be estimated by the use of (9.4.30). To this end, employ (9.4.10), introduce polar coordinates about $z=0$ and write

$$\begin{aligned}
&\int\limits_{\mathfrak{c}^+_R} \big(a^0_{m-j} - a_{m-jH}\big)\big((\varrho',0),(\xi',z)\big)dz \\
&= \lim_{\varrho_n\to 0+}\Big\{\lim_{\eta\to 0+} \int\limits_{\mathfrak{c}^+_R} \big(a^0_{m-j} - a_{m-jH}\big)\big((\varrho',\varrho_n),(\xi',z)\big)e^{i\varrho_n z}\widehat{g}(\eta z)dz\Big\}
\end{aligned}$$

with $0 < \eta < \varrho_n$ where $\widehat{g}(\eta z)$ is defined by (9.4.10). Here, $\mathfrak{c}_R^+$ is the contour defined by $\mathfrak{c}_R^+ = \{Re^{i\vartheta} \text{ with } 0 \le \vartheta < \pi\}$. Because of the estimate (9.4.11), for every $\eta > 0$, with an application of the Cauchy integral theorem applied to the integrand which is a holomorphic function in the upper half–plane $\operatorname{Im} z \ge 0$ for $|z| \ge R \ge c_0$, we obtain

$$\begin{aligned}
&\int\limits_{\mathfrak{c}_R^+} (a_{m-j}^0 - a_{m-jH})\big((\varrho',0),(\xi',z)\big)dz \\
&= - \lim_{\varrho_n\to 0+}\Big\{\lim_{\eta\to 0+} \int\limits_{|\xi_n|\ge R} (a_{m-j}^0 - a_{m-jH})\big((\varrho',\varrho_n),\xi\big)e^{i\varrho_n\xi_n}\widehat{g}(\eta\xi_n)d\xi_n\Big\} \\
&= \int\limits_{\xi_n\le -R} a_{m-jR}^-\big((\varrho',o)(\xi',\xi_n)\big)d\xi_n - \int\limits_{\xi_n\ge R} a_{m-jR}^+\big((\varrho',0),(\xi',\xi_n)\big)d\xi_n\,.
\end{aligned}$$

Similarly, with $\mathfrak{c}_R^- := \{z = \mathrm{Re}^{i\delta} \text{ with } \pi \le \delta \le 2\pi\}$, we obtain

$$\begin{aligned}
&\int\limits_{\mathfrak{c}_R^-} (a_{m-j}^0 - a_{m-jH})\big((\varrho',0),(\xi',z)\big)dz \\
&= \int\limits_{\xi_n\le -R} a_{m-jR}^-\big((\varrho',0),(\xi',\xi_n)\big)d\xi_n - \int\limits_{R\le\xi_n} a_{m-jR}^+\big((\varrho',0)\,,\,(\xi',\xi_n)\big)d\xi_n\,.
\end{aligned}$$

Collecting terms, we find

$$\begin{aligned}
[q_{m+1-j}](\varrho',\xi') &= i\sum_{|\gamma|=m-j+1}\frac{1}{\gamma!}\frac{\partial^\gamma a_{m-j}}{(\partial\xi')^\gamma}\big((\varrho',0),(0,1)\big)\xi'^\gamma \\
&\quad + (2\pi)^{-1}\int\limits_{\xi_n\le -R} a_{m-jR}^- d\xi_n - (2\pi)^{-1}\int\limits_{R\le\xi_n} a_{m-jR}^+ d\xi_n
\end{aligned}$$

for every $R \ge c_0$ and every fixed $\varrho' \in U_r$ and $\xi' \in \mathbb{R}^{n-1}\setminus\{0\}$. By taking the limit as $R \to \infty$, in view of the estimate (9.4.30) we obtain the desired result. ∎

Remark 9.4.2: In Theorem 9.4.6, as we mentioned already in Remark 9.4.1, the invariance property of the finite part integral is crucial for the behaviour of potential operators. In particular this type of potential operators includes all the classical boundary integral operators as investigated by R. Kieser in [225]. This implies that numerical computations for this class of operators can be performed by the use of classical coordinate transformations.

We now turn to investigating the mapping properties of the Poisson operator $\widetilde{Q}$ defined by (9.4.5).

Theorem 9.4.7. *The linear operator* $\widetilde{Q}$ *defined by* (9.4.5) *satisfies the estimate*

$$\|\varphi\widetilde{Q}v\|_{H^{s-m-\frac{1}{2}}(\mathbb{R}^n\cap\{\varrho_n\leq 0\})} \leq c\|v\|_{H^s(K)} \tag{9.4.33}$$

for any $s\in\mathbb{R}$ *where* $\varphi\in C_0^\infty(\widetilde{U}_r)$, *and any compact set* $K \Subset U_r$ *and every* $v\in C_0^\infty(K)$. *The constant* c *depends on* K,φ *and* s.

The corresponding estimate for $\widetilde{Q}_c$ *reads*

$$\|\varphi\widetilde{Q}_c v\|_{H^{s-m-\frac{1}{2}}(\mathbb{R}^n\cap\{\varrho_n\geq 0\})} \leq c\|v\|_{H^s(K)}\,. \tag{9.4.34}$$

To prove this theorem we first collect some preliminary results.

Lemma 9.4.8. *Let* $K\Subset\mathbb{R}^{n-1}$ *be a compact set. Then the distributions of the form*

$$u(\varrho) = v(\varrho')\otimes\delta(\varrho_n) \quad \textit{with} \quad \varrho = (\varrho',\varrho_n) \quad \textit{and} \quad v\in C_0^\infty(K)$$

will satisfy the estimate

$$\|u\|_{H^{s-\frac{1}{2}}(K\times\mathbb{R})} \leq c\|v\|_{H^s(K)} \tag{9.4.35}$$

provided $s<0$.

Proof: Using duality, the trace theorem and the trace operator (4.2.32), we find for $\varphi\in C_0^\infty(\mathbb{R}^n)$:

$$\begin{aligned}\|u\|_{H^{s-\frac{1}{2}}(K\times\mathbb{R})} &= \sup_{\|\varphi\|_{\frac{1}{2}-s}\leq 1} |\langle u,\varphi\rangle_{\mathbb{R}^n}| = \sup_{\|\varphi\|_{\frac{1}{2}-s}\leq 1}\Big|\int\limits_K v(\varrho')(\gamma_0\varphi)d\varrho'\Big| \\ &\leq c'\cdot\|v\|_{H^s(K)}\cdot \sup_{\|\varphi\|_{\frac{1}{2}-s}\leq 1}\|\gamma_0\varphi\|_{H^{-s}(K)} \leq c\cdot\|v\|_{H^s(K)}\end{aligned}$$

for any $s\in\mathbb{R}$ with $s<0$. ∎

We remark that, as a consequence of Lemma 9.4.8 we already obtained the desired estimate (9.4.33) provided $s<0$. In order to extend this estimate to arbitrary $s\geq 0$, we first analyze the function $\widetilde{q}_{m+1-j}$ in more detail.

Lemma 9.4.9. *Let* $\alpha,\beta\in\mathbb{N}_0$ *and* $K\Subset U_r$ *be any compact subset. Then* $\widetilde{q}_{m+1-j}(\varrho,\xi')$ *given in* (9.4.14) *satisfies the estimates*

$$|\varrho_n^\beta D_{\varrho_n}^\alpha\widetilde{q}_{m+1-j}(\varrho,\xi')| \leq c\langle\xi'\rangle^{m+1+\alpha-\beta-j} \tag{9.4.36}$$

for $\varrho_n\in[-\varepsilon,0)$, $\varrho'\in K$ *and* $\xi'\in\mathbb{R}^{n-1}$ *where* $c=c(j,\alpha,\beta,K)$ *is a constant.*

Proof: From (9.4.14) we obtain, by using the transformation $z=z'|\xi'|, z'\in \mathfrak{c}$, $|\xi'|\geq 1$ and with integration by parts with respect to z',

$$\begin{aligned}
&\varrho_n^\beta D_{\varrho_n}^\alpha \widetilde{q}_{m+1-j}(\varrho,\xi') \\
&= \frac{1}{2\pi}\int\limits_{|\xi'|\mathfrak{c}} \varrho_n^\beta \sum_{0\le\gamma\le\alpha}\binom{\alpha}{\gamma}(iz)^\gamma e^{i\varrho_n z} D_{\varrho_n}^{\alpha-\gamma} a_{m-j}^0\big(\varrho,(\xi',z)\big)dz \\
&= \frac{1}{2\pi}\int\limits_{|\xi'|\mathfrak{c}} \sum_{0\le\gamma\le\alpha}\binom{\alpha}{\gamma}\varrho_n^\beta\left(\frac{i}{\varrho_n}\right)^\beta e^{i\varrho_n z} D_z^\beta\left\{(iz)^\gamma D_{\varrho_n}^{\alpha-\gamma} a_{m-j}^0\big(\varrho,(\xi',z)\big)\right\}dz \\
&= \frac{1}{2\pi}\sum_{0\le\gamma\le\alpha}|\xi'|^{-\beta+\gamma+m-j+1}\binom{\alpha}{\gamma}\times \\
&\qquad\times\int\limits_{\mathfrak{c}}(i)^\beta e^{i\varrho_n|\xi'|z'} D_{z'}^\beta\left\{(iz')^\gamma D_{\varrho_n}^{\alpha-\gamma} a_{m-j}^0\big(\varrho,(\xi_0',z')\big)\right\}dz'\,.
\end{aligned}$$

Then we obtain the estimates

$$\left|\varrho_n^\beta D_{\varrho_n}^\alpha \widetilde{q}_{m+1-j}(\varrho,\xi')\right| \le c|\xi'|^{m+1-j+\alpha-\beta} \quad \text{for } |\xi'|\ge 1$$

whereas for $|\xi'|\le 1$ the left–hand side is bounded since it is smooth on $|\xi'|\le 1$ and has compact support with respect to ϱ and $\varrho_n\le 0$. This yields (9.4.36). ■

Lemma 9.4.9 indicates that $\widetilde{q}(\varrho,\xi')$ for $\varrho_n\le 0$ has properties almost like a symbol although the inequality (9.4.36) is not strong enough for $\varrho_n\to 0$. However, for applying the techniques of Fourier transform we need a C^∞ extension of $\widetilde{q}_{m+1-j}$ to $\varrho_n\ge 0$. To this end we use the *Seeley extension theorem* proved in [383] by putting

$$p_{m+1-j}(\varrho,\xi') := \begin{cases} \widetilde{q}_{m+1-j}\big((\varrho',\varrho_n),\xi'\big) & \text{for } \varrho_n\le 0 \text{ and} \\ \sum\limits_{p=1}^{\infty}\lambda_p\widetilde{q}_{m+1-j}\big((\varrho',-2^p\varrho_n),\xi'\big) & \text{for } \varrho_n>0 \end{cases} \tag{9.4.37}$$

where the sequence of real weights

$$\lambda_p = \prod_{\ell\ne p}^{1,\infty}\frac{1-2^{-\ell}}{1-2^{p-\ell}} \quad \text{for } p\in\mathbb{N} \tag{9.4.38}$$

satisfies the infinitely many identities

$$\sum_{p=1}^{\infty}\lambda_p 2^{p\ell} = (-1)^\ell \quad \text{for all } \ell\in\mathbb{N} \tag{9.4.39}$$

and these series converge absolutely. With the relations (9.4.39) one can easily show that all derivatives of $\widetilde{q}_{m+1-j}(\varrho,\xi)$ are continuously extended across

$\varrho_n = 0$ since the term–wise differentiated series (9.4.37) still converges absolutely and uniformly. Since $\widetilde{q}_{m+1-j}$ has compact support with respect to ϱ and $\varrho_n \le 0$, also p_{m+1-j} has compact support in ϱ.

The extended functions p_{m+1-j} define operators

$$P_{m+1-j}v(\varrho) \;:=\; (2\pi)^{-n+1} \int\limits_{\mathbb{R}^{n-1}} e^{i\varrho'\cdot\xi'} p_{m+1-j}(\varrho,\xi')\widehat{v}(\xi')d\xi' \quad (9.4.40)$$

for $\varrho \in \big(U_r \times (-\varepsilon,\varepsilon)\big)$ and

$$P_{m+1-j}v(\varrho) \;=\; \widetilde{Q}_{m+1-j}v(\varrho) \quad \text{for} \quad \varrho_n \le 0 \quad (9.4.41)$$

where $\widetilde{Q}_{m+1-j}$ is given by (9.4.20) via (9.4.16). In order to analyze the mapping properties of P_{m+1-j}, we now consider the Fourier transformed $\widehat{p}_{m+1-j}$ of p_j, i.e.

$$\widehat{p}_{m+1-j}(\zeta,\xi') := (2\pi)^{-n/2} \int\limits_{\mathbb{R}^n} e^{-i\zeta\cdot\varrho} p_{m+1-j}(\varrho,\xi')d\varrho\,. \quad (9.4.42)$$

Lemma 9.4.10. *For any $q, r \in \mathbb{N}$, there exist constants $c(q,r,j)$ such that $\widehat{p}_{m+1-j}$ satisfies the estimates*

$$|\widehat{p}_{m+1-j}(\zeta,\xi')| \le c(1+|\xi'|)^{m-j}(1+|\zeta'|)^{-r}\Big(1+\frac{|\zeta_n|}{1+|\xi'|}\Big)^{-q}. \quad (9.4.43)$$

Proof: We begin with the case $|\xi'| \ge 1$. Then from the definition of p_{m+1-j}, we have

$$\widehat{p}_{m+1-j}(\zeta,\xi') = (2\pi)^{-n/2} \int\limits_{\varrho'\in\mathbb{R}^{n-1}} e^{-i\varrho'\cdot\zeta'} \times$$

$$\times\Big\{ \int\limits_{\varrho_n=-\infty}^{0} e^{-i\varrho_n\zeta_n}(2\pi)^{-1} \int\limits_{z\in|\xi'|\mathfrak{c}} e^{i\varrho_n z} a^0_{m-j}\big((\varrho',\varrho_n),(\xi',z)\big)dz d\varrho_n$$

$$+ \int\limits_{\varrho_n=0}^{\infty} e^{-i\varrho_n\zeta_n}(2\pi)^{-1} \int\limits_{z\in|\xi'|\mathfrak{c}} \sum_{p=1}^{\infty} \lambda_p e^{i2^p\varrho_n z} a^0_{m-j}\big((\varrho', -2^p\varrho_n),(\xi',z)\big)dz d\varrho_n \Big\} d\varrho'.$$

By the homogeneity of a^0_{m-j} and $\xi' = |\xi'|\xi'_0$ and by scaling $\widetilde{\varrho}_n = |\xi'|\varrho_n$, $\widetilde{\zeta}_n = |\xi'|^{-1}\zeta_n$ and $z = |\xi'|z'$, one obtains

$$\widehat{p}_{m+1-j}(\zeta,\xi') = |\xi'|^{m-j}(2\pi)^{-n/2-1} \int\limits_{\varrho'\in\mathbb{R}^{n-1}} e^{-i\varrho'\cdot\zeta'}\{\dots\}d\varrho',$$

where

$$\{\dots\} = \Big\{ \int\limits_{\widetilde{\varrho}_n=-\infty}^{0} e^{-i\widetilde{\varrho}_n\widetilde{\zeta}_n} \oint\limits_{z'\in\mathfrak{c}} e^{i\widetilde{\varrho}_n z'} a^0_{m-j}\Big((\varrho', \frac{\widetilde{\varrho}_n}{|\xi'|}, (\xi'_0, z')\Big) dz' d\widetilde{\varrho}_n$$

$$+ \int\limits_{\widetilde{\varrho}_n=0}^{\infty} e^{-i\widetilde{\varrho}_n\widetilde{\zeta}_n} \sum_{p=1}^{\infty} \lambda_p \oint\limits_{z'\in\mathfrak{c}} e^{-i\widetilde{\varrho}_n 2^p z'} a^0_{m-j}\Big((\varrho', -2^p\frac{\widetilde{\varrho}_n}{|\xi'|}), (\xi'_0, z')\Big) dz' d\widetilde{\varrho}_n \Big\}.$$

In the integrands above, we see that the second integral contains the function

$$\sum_{p=1}^{\infty} \lambda_p f\Big(\varrho', -\frac{2^p\widetilde{\varrho}_n}{|\xi'|}, \xi'_0\Big),$$

which is the C^∞ Seeley extension of

$$f\Big(\varrho', \frac{\widetilde{\varrho}_n}{|\xi'|}, \xi'_0\Big) := (2\pi)^{-1} \oint\limits_{z'\in\mathfrak{c}} e^{i\widetilde{\varrho}_n z'} a^0_{m-j}\Big(\Big(\varrho', \frac{\widetilde{\varrho}_n}{|\xi'|}\Big), (\xi'_0, z')\Big) dz'$$

from $\widetilde{\varrho}_n \le 0$ to $\widetilde{\varrho}_n > 0$. The function f has compact support with respect to $(\varrho', \widetilde{\varrho}_n)$ for every fixed ξ' and, moreover, is uniformly bounded. Therefore, its Seeley extension $F\big(\varrho', \frac{\widetilde{\varrho}_n}{|\xi'|}\big)$ has the same properties and its function series converges absolutely and uniformly. Since

$$\widehat{p}_{m+1-j}(\zeta, \xi') = |\xi'|^{m-j}(2\pi)^{-n/2} \int\limits_{\widetilde{\varrho}\in\mathbb{R}^n} e^{-i\widetilde{\varrho}\cdot\widetilde{\zeta}} F\Big((\varrho', \frac{\widetilde{\varrho}_n}{|\xi'|}), \xi'_0\Big) d\widetilde{\varrho}$$

where $\widetilde{\varrho} = (\varrho', \widetilde{\varrho}_n)$ and $\widetilde{\zeta} = (\zeta', \widetilde{\zeta}_n)$, the Paley–Wiener–Schwartz theorem 3.1.3 yields the estimates

$$|\widehat{p}_{m+1-j}(\zeta, \xi')| \le c_N(1 + |\widetilde{\zeta}|)^{-N}|\xi'|^{m-j} \text{ for } \widetilde{\zeta} \in \mathbb{R}^n \text{ and } \xi' \in \mathbb{R}^{n-1},\ |\xi'| \ge 1.$$

Here, since $N \in \mathbb{N}$ is arbitrary, the latter implies with $N = r + q$,

$$|\widehat{p}_{m+1-j}(\zeta, \xi')| \le c_N(1 + |\zeta'|)^{-r}\left(1 + \frac{|\zeta_n|}{1 + |\xi'|}\right)^{-q}(1 + |\xi'|)^{m-j},$$

the proposed estimate (9.4.43) for $|\xi'| \ge 1$.

For $|\xi'| \le 1$ one finds the uniform estimates

$$|\widehat{p}_{m+1-j}(\zeta, \xi')| \le c_N(1 + |\zeta'|)^{-r}(1 + |\zeta_n|)^{-q}$$

by using the Paley–Wiener–Schwartz theorem 3.1.3 again. Then $1 \le 1+|\xi'| \le 2$ implies the desired estimate (9.4.43) for $|\xi'| \le 1$ as well. This completes the proof. ■

Lemma 9.4.10, in contrast to Lemma 9.4.9, will give us the following crucial mapping properties.

Lemma 9.4.11. *The operator $\widetilde{Q}_{m+1-j}$ defined by* (9.4.20) *satisfies the estimates*

$$\begin{aligned}\|\widetilde{Q}_{m+1-j}v\|_{H^{s-m+j-\frac{1}{2}}(\mathbb{R}^n\cap\{\varrho_n\le 0\})} & \\ \le\; & \|P_{m+1-j}v\|_{H^{s-m+j-\frac{1}{2}}(\mathbb{R}^n)} \qquad (9.4.44)\\ \le\; & c\|v\|_{H^s(\mathbb{R}^{n-1})} \quad \textit{for } v\in C_0^\infty(U_r)\,.\end{aligned}$$

Proof: Our proof essentially follows the presentation in the book by Chazarain and Piriou [64, Chap.5, Sec.2]. Since

$$\|\psi\|_{H^t(\mathbb{R}^n_-)} = \inf\big\{\|\Psi\|_{H^t(\mathbb{R}^n)} \;\text{ with }\; \Psi|_{\overline{\mathbb{R}^n_-}} = \psi\big\}$$

where $\overline{\mathbb{R}^n_-} = \mathbb{R}^n\cap\{\varrho_n\le 0\}$ and

$$(\widetilde{Q}_{m+1-j}v)(\varrho) = (P_{m+1-j}v)(\varrho) \quad\text{for}\quad \varrho_n<0\,,$$

the first estimate in (9.4.44) is evident. In order to estimate the norm of $P_{m+1-j}v$ we use the Fourier transform of $P_{m+1-j}v$. The Parseval identity implies

$$\begin{aligned}&\langle P_{m+1-j}v,\varphi\rangle\\ &:= \int\limits_{\mathbb{R}^n} P_{m+1-j}v(\varrho)\overline{\varphi(\varrho)}d\varrho = \langle\widehat{P_{m+1-j}v},\widehat{\varphi}\rangle\\ &= (2\pi)^{-n+1}\int\limits_{\zeta\in\mathbb{R}^n}\int\limits_{\xi'\in\mathbb{R}^{n-1}} \widehat{p}_{m+1-j}(\zeta-\widetilde{\xi},\xi')(1+|\zeta|)^{s-m+j-\frac{1}{2}}\times\\ &\qquad\times\widehat{v}(\xi')(1+|\zeta|)^{-s+m-j+\frac{1}{2}}\widehat{\varphi}(-\zeta)d\xi'd\zeta\end{aligned}$$

where $\widetilde{\xi}=(\xi',0)\in\mathbb{R}^n$ for any $\varphi\in C_0^\infty(\mathbb{R}^n)$. Consequently,

$$|\langle P_{m+1-j}v,\varphi\rangle| \le \|\varphi\|_{H^{-s+m-j+\frac{1}{2}}(\mathbb{R}^n)}\|V_{m+1-j}v\|_{L^2(\mathbb{R}^n)}$$

where

$$V_{m+1-j}v(\zeta) := (1+|\zeta|)^{s-m+j-\frac{1}{2}}\widehat{P_{m+1-j}v}(\zeta)\,.$$

The estimate (9.4.43) of $\widehat{p}_{m+1-j}$ in Lemma 9.4.10 gives us

$$\begin{aligned}\|V_{m+1-j}v\|_{L^2(\mathbb{R}^n)} \le\; & c\Big(\int\limits_{\mathbb{R}^n}\Big(\int\limits_{\mathbb{R}^{n-1}}(1+|\xi'|)^s\,|\widehat{v}(\xi')|\Big\{(1+|\zeta|)^{s-m+j-\frac{1}{2}}\times\\ &\times(1+|\xi'|)^{m-j-s}(1+|\zeta'-\xi'|)^{-r}\Big(1+\frac{|\zeta_n|}{1+|\xi'|}\Big)^{-q}\Big\}d\xi'\Big)^2d\zeta\Big)^{\frac{1}{2}}\\ =:\; & \|W_{m+1-j}v\|_{L^2(\mathbb{R}^n)}\\ =\; & \|\int\limits_{\mathbb{R}^{n-1}}(1+|\xi'|)^s\,|\widehat{v}(\xi')|\chi_{m+1-j}(\cdot,\xi')d\xi'\|_{L^2(\mathbb{R}^n)}\end{aligned}$$

where $\chi_{m+1-j}(\zeta,\xi')$ is defined by the kernel function $\{\dots\}$ above. We note that $\chi_{m+1-j}(\zeta,\xi')$ contains one term of convolutional type $|\zeta'-\xi'|$ in $\mathbb{R}^{n-1}$ which motivates us to estimate χ_{m+1-j} by a convolutional kernel with the help of *Peetre's inequality*, namely

$$\left(\frac{1+|\xi'|}{1+|\zeta'|}\right)^{\ell} \le (1+|\xi'-\zeta'|)^{|\ell|} \quad \text{for any } \ell\in\mathbb{R}. \tag{9.4.45}$$

In $\chi_{m+1-j}(\zeta,\xi')$ we use (9.4.45) with $\ell = m-j-s$ and, also, with $\ell = 1$ and obtain

$$\chi_{m+1-j}(\zeta,\xi') \le (1+|\zeta|)^{s-m+j-\frac12}(1+|\zeta'|)^{m-j-s}\times$$
$$\times\,(1+|\xi'-\zeta'|)^{|m-j-s|-r}\left(1+\frac{|\zeta_n|}{(1+|\zeta'|)(1+|\xi'-\zeta'|)}\right)^{-q}.$$

The last factor in χ_{m+1-j} contains a term of the form $|\zeta_n|/(1+|\zeta'|)$ which motivates us to introduce the new variable $\widetilde{\zeta}_n = \zeta_n/(1+|\zeta'|)$. Then

$$\begin{aligned}\|W_{m+1-j}v\|^2_{L^2} &= \int\limits_{\mathbb{R}}\int\limits_{\mathbb{R}^{n-1}} |(W_{m+1-j}v(\zeta',(1+|\zeta'|)\widetilde{\zeta}_n))|^2(1+|\zeta'|)d\zeta' d\widetilde{\zeta}_n\\ &\le \int\limits_{\mathbb{R}}\int\limits_{\mathbb{R}^{n-1}}\Big\{\int\limits_{\mathbb{R}^{n-1}} (1+|\xi'|)^s\,|\widehat{v}(\xi')|\,\times\\ &\times\ \chi_{m+1-j}\big((\zeta',\widetilde{\zeta}_n(1+|\zeta'|),\xi')\big)d\xi'(1+|\zeta'|)^{\frac12}\Big\}^2 d\zeta' d\widetilde{\zeta}_n\end{aligned}$$

and

$$\begin{aligned}&\chi_{m+1-j}\big((\zeta',\widetilde{\zeta}_n(1+|\zeta'|),\xi')\big)(1+|\zeta'|)^{\frac12}\\ &\le\ (1+|\zeta'|)^{m-j-s+\frac12}\left(1+\sqrt{|\zeta'|^2+(1+|\zeta'|)^2|\widetilde{\zeta}_n|^2}\right)^{s-m+j-\frac12}\times\\ &\qquad\times(1+|\zeta'-\xi'|)^{|m-j-s|-r}\left(1+\frac{|\widetilde{\zeta}_n|}{1+|\zeta'-\xi'|}\right)^{-q}.\end{aligned}$$

Since

$$\tfrac{1}{\sqrt{2}}(1+|\zeta'|)(1+|\widetilde{\zeta}_n|) \le (1+\sqrt{\dots}) \le (1+|\zeta'|)(1+|\widetilde{\zeta}_n|),$$

we finally find

$$\begin{aligned}&\chi_{m+1-j}\big((\zeta',\widetilde{\zeta}_n(1+|\zeta'|),\xi')\big)(1+|\zeta'|)^{\frac12}\\ &\le\ c(1+|\zeta'-\xi'|)^{|m-j-s|-r}\left(1+\frac{|\widetilde{\zeta}_n|}{1+|\zeta'-\xi'|}\right)^{-q}(1+|\widetilde{\zeta}_n|)^{s-m+j-\frac12}.\end{aligned}$$

Hence,

$$\|W_{m+1-j}v\|^2_{L^2(\mathbb{R}^n)} \leq c\int\limits_{\mathbb{R}}\Big\|\int\limits_{\mathbb{R}^{n-1}}(1+|\xi'|)^s|\widehat{v}(\xi')|(1+|\bullet-\xi'|)^{|m-j-s|-r}\times$$

$$\times\Big(1+\frac{|\widetilde{\zeta}_n|}{1+|\bullet-\xi'|}\Big)^{-q}d\xi'\Big\|^2_{L^2(\mathbb{R}^{n-1})}\times(1+|\widetilde{\zeta}_n|)^{2(s-m+j)-1}d\widetilde{\zeta}_n\,. \quad (9.4.46)$$

For every fixed $\widetilde{\zeta}_n \in \mathbb{R}$, the inner integral is a convolution integral with respect to ξ' whose L^2–norm can be estimated as:

$$\begin{aligned}\|\dots\|_{L^2(\mathbb{R}^{n-1})} &\leq c\|(1+|\bullet|)^s|\widehat{v}(\bullet)|\,\|_{L^2(\mathbb{R}^{n-1})}\times \\ &\quad\times\int\limits_{\mathbb{R}^{n-1}}(1+|\eta'|)^{|m-j-s|-r}\left(1+\frac{|\widetilde{\zeta}_n|}{1+|\eta'|}\right)^{-q}d\eta' \\ &\leq c\|v\|_{H^s(\mathbb{R}^{n-1})}(1+|\widetilde{\zeta}_n|)^{-q}\int\limits_{\mathbb{R}^{n-1}}(1+|\eta'|)^{|m-j-s|-r+q}d\eta'\,.\end{aligned} \quad (9.4.47)$$

This estimate holds for any choice of r and $q \in \mathbb{N}$ provided the integral

$$\int\limits_{\mathbb{R}^{n-1}}(1+|\eta'|)^{|m-j-s|-r+q}d\eta' = c_{rq} < \infty\,,$$

which is true for $|m-j-s|-r+q < -n+1$.

Inserting (9.4.47) into (9.4.46) implies

$$\|W_{m+1-j}v\|^2_{L^2(\mathbb{R}^n)} \leq cc^2_{rq}\|v\|^2_{H^s(\mathbb{R}^{n-1})}\int\limits_{\mathbb{R}}(1+|\widetilde{\zeta}_n|)^{2(s-m+j)-1-2q}d\widetilde{\zeta}_n\,.$$

Choose $q > \max\{s-m+j-1,\ 1\}$ and then $r > |m-j-s|+n-1+q$, which guarantees the existence of the integrals in (9.4.46). In turn, we obtain the desired estimate (9.4.44). ∎

Proof of Theorem 9.4.7**:** By using the pseudohomogeneous expansion of the Schwartz kernel to $A_{\Phi_r,0}$, we write

$$A_{\Phi_r,0} = \sum_{j=0}^{L}A_{m-j\Phi_r,0} + A_{\Phi_r R} \quad (9.4.48)$$

where $A_{m-j\Phi_r,0} \in \mathcal{L}^{m-j}_{c\ell}(\widetilde{\Omega})$, and the Schwartz kernel k_R of $A_{\Phi_r R}$ is in the class $C^{-m-n+L-\delta'}(\widetilde{\Omega}\times\widetilde{\Omega})$ with some $\delta' \in (0,1)$. Correspondingly, we have

$$\begin{aligned}\widetilde{Q}_{m+1-j}v(\varrho) &= (2\pi)^{-n+1}\int\limits_{\mathbb{R}^{n-1}}e^{i\varrho'\cdot\xi'}\widehat{v}(\xi')\widetilde{q}_{m+1-j}(\varrho,\xi')d\xi' \\ &= A_{m-j\Phi_r,0}(v\otimes\delta)(\varrho) \quad\text{for } \varrho_n \neq 0\end{aligned} \quad (9.4.49)$$

and

$$A_{\Phi_r R}(v \otimes \delta)(\varrho) = \int\limits_{\mathbb{R}^{n-1}} k_R\big(\varrho, \varrho - (\eta', 0)\big) v(\eta') d\eta'$$

for $L \geq m+1$ and $\varrho \in \widetilde{\Omega}$. Hence, we may decompose $\widetilde{Q}$ in the form

$$\widetilde{Q}v(\varrho) = \sum_{j=0}^{L} \widetilde{Q}_{m+1-j} v(\varrho) + A_{\Phi_r R}(\varrho \otimes \delta)(\varrho)\,,$$

where $A_{\Phi_r,R} \in \mathcal{L}_{c\ell}^{m-L-1}(\widetilde{\Omega})$. Therefore, with Lemma 9.4.8 we can estimate the last term on the right–hand side as

$$\begin{aligned} \|\varphi A_{\Phi_r,R}(v \otimes \delta)\|_{H^{s-m-\frac{1}{2}}(\mathbb{R}^n)} & \\ \leq c\|v \otimes \delta\|_{H^{m-L+\frac{1}{2}}(\mathbb{R}^{n-1}\times\mathbb{R})} &\leq c\|v\|_{H^{m-L+1}(K)} \end{aligned}$$

provided $L > s+1$.

Here, $c = c(s, L, K, \varphi)$. We note that we need to consider $\varphi A_{\Phi_r,R}$ instead of $A_{\Phi_r R}$ since the middle estimate is only valid if ϱ varies in a compact subset of $\widetilde{Q}$. For the estimates of $\varphi\widetilde{Q}v$, we use Lemma 9.4.9 and obtain

$$\begin{aligned} \|\varphi\widetilde{Q}v\|\|_{H^{s-m-\frac{1}{2}}(\mathbb{R}^n \cap\{\varrho_n \leq 0\})} &\leq \sum_{j=0}^{L} c_j \|v\|_{H^{s-j}(\mathbb{R}^{n-1})} + c\|v\|_{H^{s-L+1}(K)} \\ &\leq c'\|v\|_{H^s(K)} \end{aligned}$$

if we choose $L > \max\{s+1, 1\}$. The proof of estimate (9.4.34) follows in exactly the same manner. This completes the proof of Theorem 9.4.7. ∎

Corollary 9.4.12. *The linear operator $\widetilde{Q}$ defined by* (9.4.5) *is a continuous mapping from $C_0^\infty(U_r)$ into $C^\infty(\bar{\mathbb{R}}^n_- \cap \widetilde{U}_r)$ and also into $C^\infty(\bar{\mathbb{R}}^n_+ \cap \widetilde{U}_r)$ where $\bar{\mathbb{R}}^n_-$ and $\bar{\mathbb{R}}^n_+$ denote the lower and upper closed half space, respectively.*

The proof is an immediate consequence of Theorem 9.4.7 since s can be chosen arbitrarily large in (9.4.33) and (9.4.34).

In order to extend Theorem 9.4.7 to sectional traces, let us introduce the following definition.

Definition 9.4.2. *For $k \in \mathbb{N}_0$, let $V \in C^k\big((-\varepsilon, \varepsilon)\,,\, \mathcal{D}'(U_r)\big)\,,\ \varrho_n \mapsto V(\varrho_n)$ $\in \mathcal{D}'(U_r)$ be a family of distributions which are k times continuously differentiable on $\varrho_n \in (-\varepsilon, \varepsilon)$. Then for the distribution $v \in \mathcal{D}'\big((U_r \times (-\varepsilon, \varepsilon)\big)$ defined as*

$$\langle v, \varphi\rangle = \int\limits_{\mathbb{R}} \langle V(\varrho_n)\,,\ \varphi(\bullet, \varrho_n)\rangle d\varrho_n \quad \textit{for}\ \ \varphi \in \mathcal{D}\big(U_r \times (-\varepsilon, \varepsilon)\big)\,,$$

the distributions

$$\gamma_j v := \Big(\Big(\frac{d}{d\varrho_n}^j V\Big)(0) \in \mathcal{D}'(U_r) \quad for \;\; 0 \le j \le k$$

are called the s e c t i o n a l t r a c e s o f o r d e r s *j on the plane $\varrho_n = 0$.*

Theorem 9.4.13. (Chazarain and Piriou [64, Theorem 5.2.4. i)])
Let $A \in \mathcal{L}^m_{c\ell}(\Omega \times \widetilde{\Omega})$ have a symbol of rational type and $v \in \mathcal{D}'(\Gamma)$. Then

$$\widetilde{Q}v(x) := \big(A(v \otimes \delta_\Gamma)\big)(x) \quad for \;\; x \in \Omega$$

and

$$\widetilde{Q}_c v(x) := \big(A(v \otimes \delta_\Gamma)\big)(x) \quad for \;\; x \in \widetilde{\Omega} \setminus \overline{\Omega}$$

have, respectively, the sectional traces $\gamma_k \widetilde{Q}v$ and $\gamma_{kc}\widetilde{Q}_c v$, of any order $k \in \mathbb{N}_0$ on Γ. Moreover, for $\varphi \in C^\infty_0(\widetilde{U}_r)$ and for any $s \in \mathbb{R}$ and any $k \in \mathbb{N}_0$, we have the estimate

$$\|\varphi\Big(\frac{\partial}{\partial\varrho_n}\Big)^k \widetilde{Q}v\|_{H^{s-m-\frac{1}{2}}(\mathbb{R}^n \cap \{\varrho_n \le 0\})} \le c\|v\|_{H^s(K)} \tag{9.4.50}$$

for every $v \in H^s(K)$, where $K \Subset U_r$ is any compact set. The constant c depends on K, φ, s and k. The corresponding estimate for $\Big(\frac{\partial}{\partial\varrho_n}\Big)^k \widetilde{Q}_c$ holds as well, and we have

$$\|\varphi\Big(\frac{\partial}{\partial\varrho_n}\Big)^k \widetilde{Q}_c v\|_{H^{s-m-\frac{1}{2}}(\mathbb{R}^n \cap \{\varrho_n \ge 0\})} \le c\|v\|_{H^s(K)}\,. \tag{9.4.51}$$

Proof: As before, we fix one of the local charts $(O_r, U_r\,,\, \chi_r)$ in the tubular neighbourhood $\widetilde{\Omega}$ of Γ and consider the representation of A in the parametric domain $\widetilde{U}_r = U_r \times (-\varepsilon, \varepsilon)$ in the form (9.4.48). Then one of the terms as in (9.4.49), we decompose,

$$A_{m-j,\Phi r,0}(v \otimes \delta)(\varrho) = \widetilde{B}_{m+1-j}v(\varrho) + \widetilde{R}v(\varrho) \quad \text{for} \;\; \varrho_n \in (-\varepsilon, \varepsilon]\,,$$

where

$$\widetilde{B}_{m+1-j}v(\varrho) := \int\limits_{|\xi'| \ge 1} e^{i\varrho'\cdot\xi'}\widehat{v}(\xi')p_{m+1-j}(\varrho, \xi')d\xi' \tag{9.4.52}$$

and

$$\widetilde{R}_{m+1-j}v(\varrho) := \int\limits_{|\xi'| \le 1} e^{i\varrho'\cdot\xi'}\widehat{v}(\xi')p_{m+1-j}(\varrho, \xi')d\xi'\,,$$

where $p_{m+1-j}(\varrho, \xi')$ is given in (9.4.37) and $\widetilde{R}$ is a smoothing operator with a C^∞ Schwartz kernel. Since $p_{m+1-j}(\varrho, \xi')$ is in $C^\infty(U_r \times \mathbb{R}^{n-1})$, the right–hand side in (9.4.52) defines a distribution in $C^\infty((-\varepsilon, \varepsilon), \mathcal{D}'(U_r))$, and the sectional traces

$$\gamma_k \widetilde{B}_{m+1-j} v = \lim_{0<\varrho_n \to 0} \int\limits_{|\xi'|\geq 1} e^{i\varrho' \cdot \xi'} \widehat{v}(\xi') \Big(\frac{\partial}{\partial \varrho_n}\Big)^k p_{m+1-j}(\varrho, \xi')\Big) d\xi'$$

$$= \int\limits_{|\xi'|\geq 1} e^{i\varrho' \cdot \xi'} \widehat{v}(\xi') \Big(\Big(\frac{\partial^k p_{m+1-j}}{\partial \varrho_n^k}\Big)(\varrho', 0; \xi') d\xi'$$

are well defined distributions on $\mathcal{D}(U_r)$ for any $k \in \mathbb{N}_0$. Hence, the sectional traces of order k exist as proposed.

As for the estimate (9.4.50), consider now $\frac{\partial^k p_{m+1-j}}{\partial \varrho_n^k}(\varrho, \xi')$ in (9.4.40) instead of $p_{m+1-j}(\varrho, \xi')$. Then all of the estimates remain the same for $k \in \mathbb{N}_0$ fixed. Hence, Lemmata 9.4.9 and 9.4.10 remain valid for $\frac{\partial^k p_{m+1-j}}{\partial \varrho_n^k}(\varrho, \xi')$ and Lemma 9.4.11 holds for $\Big(\frac{\partial^k}{\partial \varrho_n^k} \widetilde{Q}_{m+1-j} v\Big)$ as well, and (9.4.50) follows. For $\varrho \geq 0$, the proof is the same, where $q_{m+1-j,c}$ is defined by (9.4.24) and $p_{m+1-j,c}$ in (9.4.37), where the cases $\varrho_n \leq 0$ and $\varrho_n > 0$ are to be exchanged with $\varrho_n \geq 0$ and $\varrho_n < 0$. ■

9.5 Surface Potentials on the Boundary Manifold Γ

In the previous section we discussed surface potentials and their limits for $\varrho_n \to 0$ only for the half–space. There we introduced the extension conditions (9.3.7), the canonical extension and Tricomi conditions (9.3.31), (9.3.32), the special parity conditions (9.3.27) and the transmission conditions according to Hörmander (9.3.40) as well as the operators with symbols of rational type. Up to now, our discussion was confined to the half–space situation only. To extend these ideas to surface potentials for a general boundary manifold Γ, we need to generalize these concepts to manifolds. Let us recall that for $A \in \mathcal{L}^m_{c\ell}(\widetilde{\Omega})$ and any local chart (O_r, U_r, χ_r) of an atlas $\mathfrak{A}(\Gamma)$ and the corresponding local transformation $\Phi_r : \widetilde{O}_r \to (U_r \times (-\varepsilon, \varepsilon)) =: \widetilde{U}_r$ with $(\varrho', \varrho_n) = \Phi_r(x' + \varrho_n n(x'))$ and $(\varrho_n', 0) = \Phi_r(x')$ and $\varrho' = \chi_r(x')$ for $x' \in \Gamma$, and its inverse $x = \Psi_r(\varrho)$ (see (9.2.1) and (9.2.2)), the surface potential in the half space is defined by

$$\widetilde{Q}_r v(\varrho) = A_{\Phi r,0}\big(v \otimes \delta(\eta_n)\big)(\varrho) \tag{9.5.1}$$

for any $v \in C_0^\infty(U_r)$.

Hence, it is natural to consider the mapping defined by

$$\widetilde{Q}_\Gamma u(x) := A(u \otimes \delta_\Gamma)(x) \tag{9.5.2}$$

for $x = x' + \varrho_n n(x')$ with $x \in \widetilde{O}_r \setminus \Gamma$ and $v = \chi_{r*}u$ where $u \in C_0^\infty(O_r)$. Then

$$\widetilde{Q}_\Gamma u = \Phi_r^* \widetilde{Q}_r \chi_{r*} u = \Phi_r^* A_{\Phi r,0} \Phi_{r*}(u \otimes \delta_\Gamma) = \Phi_r^* A_{\Phi r,0}\big(\chi_{r*}u \otimes \delta(\varrho_n)\big)\,. \quad (9.5.3)$$

To consider the limits of $\widetilde{Q}_\Gamma u(x)$ as $x \to x' \in O_r \subset \Gamma$, we need the following definition of the extension conditions.

Definition 9.5.1. *The operator $\widetilde{Q}_\Gamma$ satisfies the extension conditions for $\varrho_n \to 0^-$ ($\varrho_n \to 0^+$) if and only if for every local chart of an atlas $\mathfrak{A}(\Gamma)$ the associated operators*

$$\widetilde{Q}_r = \Phi_{r*} \widetilde{Q}_\Gamma \chi_r^*$$

satisfy the extension conditions (9.3.7) *for $m+1 \notin \mathbb{N}_0$ and $\nu = 0, \dots, [m+1]$; and, for $m+1 \in \mathbb{N}_0$, the extension conditions* (9.3.7) *for $\nu = 1, \dots, m+1$.*

For $m+1 \in \mathbb{N}_0$, the operator $\widetilde{Q}_\Gamma$ is said to satisfy the Tricomi conditions iff all the operators $\widetilde{Q}_r$ satisfy the Tricomi conditions (9.3.8).

Now we are using a partition of unity $\{\varphi_r\}_{r\in I}$ to an atlas $\mathfrak{A}(\Gamma)$ as in Section 9.1 and the family $\{\psi_r\}_{r\in I}$ with $\psi_r \in C_0^\infty(O_r)$ and $\psi_r(x') = 1$ for $x' \in \operatorname{supp}\varphi_r$. Then we are in the position to establish the following theorem for the surface potential.

Theorem 9.5.1. *Let $A \in \mathcal{L}^m_{c\ell}(\widetilde{\Omega} \cup \Omega)$ with $m \in \mathbb{R}$. Then the limit*

$$Q_\Gamma u(x') := \lim_{0>\varrho_n\to 0} \widetilde{Q}_\Gamma u(x) \quad \textit{where } x = x' + \varrho_n n(x') \in \widetilde{\Omega} \qquad (9.5.4)$$

always exists if $m < -1$. For $m > -1$ it exists in the case $m+1 \notin \mathbb{N}_0$ iff $\widetilde{Q}_\Gamma$ satisfies the extension conditions. For $m+1 \in \mathbb{N}_0$, the limit exists iff $\widetilde{Q}_\Gamma$ satisfies the extension conditions and the Tricomi conditions. If the limit exists then $Q_\Gamma \in \mathcal{L}^{m+1}_{c\ell}(\Gamma)$ is a pseudodifferential operator of order $m+1$ on Γ. These results hold also for $Q_{\Gamma,c}u(x') = \lim_{0<\varrho_n\to 0} \widetilde{Q}_\Gamma u(x)$, correspondingly.

Proof: For every chart (O_r, U_r, χ_r) of the atlas $\mathfrak{A}(\Gamma)$ and $u \in C_0^\infty(O_r)$, the limit

$$\lim_{0>\varrho_n\to 0} \widetilde{Q}_r(\chi_{r*}u)(\varrho) = Q_r(\chi_{r*}u)(\varrho')$$

exists iff the extension conditions (9.3.7) for $m+1 \notin \mathbb{N}_0$ and (9.3.7) and (9.3.8) for $m+1 \in \mathbb{N}_0$ are satisfied due to Theorem 9.3.2. Furthermore, Q_r is a pseudodifferential operator $Q_r \in \mathcal{L}^{m+1}_{c\ell}(U_r)$.

For $x \in \widetilde{O}_r \cap \widetilde{O}_t \cap \Omega$ we have

$$\begin{aligned}
\widetilde{Q}_\Gamma u(x) &= A(u \otimes \delta_\Gamma)(x) = \big(\Phi_r^* \Phi_{r*} A \Phi_r^* \Phi_{r*}(u \otimes \delta_\Gamma)\big)(x)\\
&= \Phi_r^* A_{\Phi r}\big((\chi_{r*}u) \otimes \delta(\eta_n)\big)(x)\\
&= \big(\Phi_r^* \widetilde{Q}_r(\chi_{r*}u)\big)(x)\\
&= \big(\Phi_t^* \widetilde{Q}_t(\chi_{t*}u)\big)(x)\,.
\end{aligned}$$

For $\varrho_n \to 0$, i.e., $x \to x' \in O_r$, the limits exist due to the extension conditions and, hence, are equal. This means

$$(Q_\Gamma u)(x') = (\chi_r^* Q_r \chi_{r*} u)(x') = (\chi_t^* Q_t \chi_{t*} u)(x')$$

for arbitrary charts with $x' \in O_r \cap O_t = O_{rt} \neq \emptyset$ and $u \in C_0^\infty(O_{rt})$. Consequently, in terms of the partition of unity, by

$$Q_\Gamma u := \sum_{r\in I} \psi_r(x')\chi_r^* Q_r \chi_{r*}\varphi_r u + \sum_{r\in I} (1 - \psi_r(x'))\chi_r^* Q_r \chi_{r*}\varphi_r u\,, \qquad (9.5.5)$$

the pseudodifferential operator $Q_\Gamma \in \mathcal{L}_{c\ell}^{m+1}(\Gamma)$ is well defined.

For $Q_{\Gamma,c}$, the proof follows in the same manner. ∎

In order to express Q_Γ in (9.5.5) in terms of finite part integral operators, we observe that from the definition of $\widetilde{Q}_\Gamma$ in (9.5.2) it follows that Equation (9.5.3) can be written in terms of the Schwartz kernels of A and $A_{\Phi r,0}$, respectively, i.e.,

$$\begin{aligned}
\widetilde{Q}_\Gamma u(x) &= \int\limits_{y'\in\Gamma} k_A(x, x - y')u(y')ds_{y'} \\
&= \int\limits_{\eta'\in U_r} k_A\big(\Psi_r(\varrho)\,,\, \Psi_r(\varrho) - T_r(\eta')\big)u\big(T_r(\eta')\big)\sqrt{\gamma_r(\eta')}d\eta' \\
&= \int\limits_{\eta'\in U_r} k_{A_{\Phi r,0}}\big(\varrho, \varrho - (\eta',0)\big)u\big(T_r(\eta')\big)d\eta' \quad \text{for } x \in \widetilde{O}_r \setminus \Gamma
\end{aligned} \qquad (9.5.6)$$

and any $u \in C_0^\infty(O_r)$, where $x = \psi_r(\varrho)\,,\ y' = T_r = \psi_r(\eta',0)$ and γ_r is given by (3.4.15). Consequently,

$$k_{A_{\Phi r,0}}\big(\varrho, \varrho - (\eta',0)\big) = k_A\big(\Psi_r(\varrho)\,,\, \Psi_r(\varrho) - \Psi_r(\eta',0)\big)\sqrt{\gamma_r(\eta')} \qquad (9.5.7)$$

for $\varrho_n \neq 0$ and, by continuity, for all $\varrho \in \widetilde{U}_r \setminus (\eta',0)$. Since $k_A \in \psi hk_{-m-n}$, Lemma 8.2.1 implies $k_{A_{\Phi r,0}} \in \psi hk_{-m-n}$ in $\widetilde{U}_r$ and for $\varrho_n = 0$, the kernel function $k_{A_{\Phi r,0}}\big((\eta',0)\,;\,(\varrho - \eta',0)\big)$ belongs to the class $\psi hk_{-(m+1)-(n-1)}$ on U_r. Therefore the corresponding finite part integral operators define on U_r pseudodifferential operators in $\mathcal{L}_{c\ell}^{m+1}(U_r)$.

For these operators we can show that the following lemma holds.

Lemma 9.5.2. *Let $A \in \mathcal{L}_{c\ell}^m(\Omega \cup \widetilde{\Omega})$ and suppose that the finite part integral operator*

$$Q_{f,r}v(\varrho') = \text{p.f.} \int\limits_{U_r} k_A\big(T_r(\varrho')\,;\, T_r(\varrho') - T_r(\eta')\big)v(\eta')\sqrt{\gamma_r(\eta')}d\eta' \qquad (9.5.8)$$

for $\varrho' \in U_r$ is invariant under change of coordinates. Then the finite part integral on Γ given by

$$\begin{aligned}
\text{p.f.} \int_\Gamma & k_A(x', x'-y')u(y')ds_{y'} \\
:= & \sum_{r\in I} \psi_r(x') \,\text{p.f.} \int_{U_r} k_A\big(x'; x' - T_r(\eta')\big)(\chi_{r*}\varphi_r u)(\eta')\sqrt{\gamma_r(\eta')}d\eta' \\
& + \sum_{r\in I} (1-\psi_r(x')) \int_{O_r} k_A(x', x'-y')\varphi_r u(y')ds_{y'} \qquad (9.5.9)
\end{aligned}$$

defines a pseudodifferential operator on Γ belonging to the class $\mathcal{L}^{m+1}_{c\ell}(\Gamma)$.

Let $\Lambda := \chi_r \circ T_t : U_t \to U_r$ denote the coordinate transform generated by the local charts (O_{rt}, U_r, χ_r) and (O_{rt}, U_t, χ_t) and let $\eta'' \in U_t$, $\eta' = \Lambda(\eta'') \in U_r$.

We shall now show the relations

$$\chi_r^* Q_{f,r} \chi_{r*} = \chi_t^* Q_{f,t} \chi_{t*} \quad \text{on } \mathcal{D}(O_{rt}). \qquad (9.5.10)$$

With $v_r(\varrho') := \chi_{r*} u$ we have $v_t(\varrho'') = v_r\big(\Lambda(\varrho'')\big)$ since $\chi_r^* v_r = u = \chi_t^* v_t$. From (9.2.4) we have for the surface element

$$ds(\eta') = \sqrt{\gamma_r}d\eta' \quad \text{and} \quad ds(\eta'') = \sqrt{\gamma_t}d\eta''.$$

With

$$\frac{\partial \varrho'_\ell}{\partial \varrho''_j} = \sum_{k,m} \frac{\partial x_k}{\partial \varrho'_m} g'^{m\ell} \frac{\partial x_k}{\partial \varrho''_j}, \; \varrho'_n = \varrho''_n = \varrho_n,$$

from (3.4.5) together with

$$\begin{aligned}
\sum_j \frac{\partial \varrho'_\ell}{\partial \varrho''_j} g''^{ji} &= \sum_{k,m,j} \frac{\partial x_k}{\partial \varrho'_m} g'^{m\ell} \frac{\partial x_k}{\partial \varrho''_j} g''^{ji} \\
&= \sum_{k,m} \frac{\partial x_k}{\partial \varrho'_m}\frac{\partial \varrho''_i}{\partial x_k} g'^{m\ell} = \sum_m \frac{\partial \varrho''_i}{\partial \varrho'_m} g'^{m\ell}
\end{aligned}$$

we obtain

$$\det\Big(\Big(\frac{\partial \varrho'_e}{\partial \varrho''_j}\Big)\Big)(g'')^{-1} = \det\Big(\Big(\frac{\partial \varrho''_i}{\partial \varrho'_m}\Big)\Big)(g')^{-1},$$

hence,

$$\Big(\det\Big(\Big(\frac{\partial \varrho'_\ell}{\partial \varrho''_j}\Big)\Big)\Big)^2 = \frac{g''}{g'}.$$

Then, for $\varrho_n = 0$, with (3.4.24) and $\varrho'_n = \varrho''_n = \varrho_n$, we obtain

$$\Big(\det \frac{\partial \Lambda}{\partial \eta''}\Big)^2 = \frac{\gamma_t}{\gamma_r}$$

and, consequently, since the finite part integral operator $Q_{f,r}$ in (9.5.8) is invariant under change of coordinates by assumption, we find

$$\begin{aligned}
&(Q_{f,r}v_r)\big(\Lambda(\varrho'')\big)\\
&= \text{p.f.} \int\limits_{U_t} k_A\big(T_r(\Lambda(\varrho''))\,,\ (T_r(\Lambda(\varrho'')) - T_r(\Lambda(\eta'')))\big)\sqrt{\gamma_r}\Big(\det \frac{\partial \Lambda}{\partial \eta''}\Big) \times\\
&\qquad \times v_r\big(\Lambda(\eta'')\big)d\eta''\\
&= \text{p.f.} \int\limits_{U_t} k_A\big((T_t(\varrho''))\,,\ (T_t(\varrho'') - T_t(\eta''))\big)v_t(\eta'')\sqrt{\gamma_t}d\eta''\\
&= \quad Q_{f,t}v_t(\varrho'')\,. \qquad (9.5.11)
\end{aligned}$$

This shows that $Q_{f,t}$ and $Q_{f,r}$ satisfy the compatibility condition (9.5.10) and Theorem 9.1.1 implies that the family $\chi_r^* Q_{f,r}\chi_{r*}$ defines by (9.5.9) a pseudodifferential operator in $\mathcal{L}^{m+1}_{c\ell}(\Gamma)$. ■

Now we return to the operator Q_Γ defined by (9.5.5). In terms of the locals charts, as an immediate consequence of Theorem 9.3.2 and Theorem 9.5.1, the representation (9.5.5) of Q_Γ leads to the following lemma.

Lemma 9.5.3. *Let $\widetilde{Q}_\Gamma$ satisfy the extension conditions (see Definition 9.5.1). If $m+1 \in \mathbb{N}_0$ let $\widetilde{Q}_\Gamma$ also satisfy the Tricomi conditions. Then Q_Γ admits the representation in the form*

$$\begin{aligned}
&Q_\Gamma u(x')\\
&= \sum_{r\in I} \psi_r(x')\chi_r^*\Big\{ \text{p.f.} \int\limits_{U_r} k_{A_{\Phi r,0}}\big((\varrho',0)\,;\ (\varrho'-\eta',0)\big)(\chi_{r*}\varphi_r u)(\eta')d\eta'\\
&\qquad - \mathcal{T}_r(\chi_{r*}\varphi_r u)\Big\}(x')\\
&\qquad + \sum_{r\in I}(1-\psi_r(x'))\chi_r^* \int\limits_{U_r} k_{A_{\Phi r,0}}\big((\varrho',0)\,;\ (\varrho'-\eta',0)\big)\chi_{r*}(\varphi_r u)(\eta')d\eta'\,,\\
&\qquad x' = T_r(\varrho') \in \Gamma \qquad (9.5.12)
\end{aligned}$$

where $\mathcal{T}_r$ is the tangential differential operator in (9.3.20) associated with any chart (O_r, U_r, χ_r) in the atlas $\mathfrak{A}(\Gamma)$, and where $k_{A_{\Phi r,0}}$ is given by (9.5.7).

If $m+1 \notin \mathbb{N}_0$ then (9.5.12) holds with $\mathcal{T} = 0$ and, moreover, is invariantly defined by

$$Q_\Gamma u(x') = \text{p.f.} \int\limits_\Gamma k_A(x', x'-y')u(y')ds_{y'}\,. \qquad (9.5.13)$$

For the exterior limits with $0 < \varrho_n \to 0$, the same results hold for $A_{\Gamma,c}$ if $-\mathcal{T}_r$ in (9.5.12) is replaced by $+\mathcal{T}_r$.

We note that the finite part integral in (9.5.12) for $m+1 \in \mathbb{N}_0$ is not generally invariant under the coordinate transformations generated by changes of the local charts. However, with the appropriate parity conditions for the Schwartz kernel of A, the finite part integrals in (9.5.12) will become invariant.

Definition 9.5.2. *The operator $A \in \mathcal{L}^m_{c\ell}(\Omega \cup \widetilde{\Omega})$ with $m+1 \in \mathbb{N}_0$ is said to satisfy the special parity conditions in $\Omega \cup \widetilde{\Omega}$ iff the members of the pseudohomogeneous kernel expansion of A satisfy*

$$k_{\kappa+j}(x, y-x) = (-1)^{m-j} k_{\kappa+j}(x, x-y) \quad \textit{for } 0 \le j \le m+1 \qquad (9.5.14)$$

and $x \in \Omega \cup \widetilde{\Omega}$, $y \in \mathbb{R}^n$.

We remark that due to Theorem 8.2.9 with $\sigma_0 = 0$, the special parity conditions (9.5.14) are invariant under changes of coordinates. Hence, this definition implies the special parity conditions (9.3.27) for every $\widetilde{Q}_r$ of the form (9.5.1) and $\varrho \in \widetilde{U}_r$ in any chart of the atlas. Therefore, by using Theorem 9.3.6, we have the following theorem.

Theorem 9.5.4. *Let $A \in \mathcal{L}^m_{c\ell}(\widetilde{\Omega})$ with $m+1 \in \mathbb{N}_0$ satisfy the special parity conditions* (9.5.14) *and let $\widetilde{Q}_\Gamma$ satisfy the extension conditions at Γ. Then Q_Γ as in* (9.5.12) *can also be defined invariantly by*

$$Q_\Gamma u(x') \;=\; \text{p.f.} \int_\Gamma k_A(x', x'-y') u(y') ds_{y'} - \mathcal{T} u(x') , \qquad (9.5.15)$$

so is $Q_{\Gamma,c}$, i.e.,

$$Q_{\Gamma,c} u(x') \;=\; \text{p.f.} \int_\Gamma k_A(x', x'-y') u(y') ds_{y'} + \mathcal{T}_c u(x') \qquad (9.5.16)$$

for $x' \in \Gamma$.

Proof: The invariance of the finite part integral operator is now a consequence of Theorem 9.3.6 since the special parity conditions guarantee that the finite part integrals in (9.5.12) are invariant under the change of coordinates generated by two compatible charts of the atlas. Hence, Lemma 9.5.3 implies that Q_Γ and Q_{Γ_c} in (9.5.15) and (9.5.3) are defined invariantly, respectively. ∎

In order to introduce the concept of canonical extension conditions for Γ, let Γ_τ with $\tau \in I_\tau$ denote the family of parallel surfaces associated with the local chart (O_r, U_r, χ_r) by

$$\Gamma_\tau \,:\, x = \Psi_r(\varrho', \tau) = x' + \tau n(x') \quad \text{where } x' = T_r(\varrho') \in \Gamma \quad \text{and } \varrho' \in U_r .$$

Then we may define

$$\widetilde{Q}_{\Gamma_\tau} u(x) := A(u \otimes \delta_{\Gamma_\tau})(x) = \int\limits_{y' \in \Gamma_\tau} k_A\big(x, (x' + \tau n(x') - y', 0)\big) u(y') ds_{y'} \,. \tag{9.5.17}$$

For $u \in C_0^\infty(U_r)$ and $x = x' = (\tau - \varrho_n) n(x') \in \widetilde{O}_r$ and $\varrho \neq 0$ we have

$$\widetilde{Q}_{\Gamma_\tau} u(x) = \Phi_r^* \widetilde{Q}_r(\tau) \chi_{r*} u \,.$$

where $\widetilde{Q}_r(\tau)$ is defined by (9.5.1) in $\widetilde{U}_r$. Hence, in order to obtain the limits of $\widetilde{Q}_{\Gamma_\tau} u(x)$ in (9.5.17) as $\varrho_n \to 0$; i.e., $x \to x' + \tau u(x') \in \Gamma_\tau$ for each $\tau \in I_\tau$, we now carry over the definition of the canonical extension conditions from the family of planes in the parametric domain to the family of parallel surfaces Γ_τ in $\widetilde{\Omega}$.

Definition 9.5.3. *We say that the* canonical extension conditions *are satisfied for A at Γ iff for each chart (O_r, U_r, χ_r) of the atlas $\mathfrak{A}(\Gamma)$ the Schwartz kernel of the associated operator $A_{\Phi r,0} \in \mathcal{L}^m_{c\ell}(\widetilde{U}_r)$ satisfies the canonical extension conditions* (9.3.31) *where $\varrho \in \widetilde{U}_r$.*

Clearly, for operators $A \in \mathcal{L}^m_{c\ell}(\Omega \cup \widetilde{\Omega})$ with $m + 1 \in \mathbb{N}_0$ satisfying the canonical extension conditions in the vicinity of Γ and, in addition, the canonical Tricomi conditions (9.3.32) in every local chart, Theorem 9.5.4 holds with more simplified tangential differential operators $\mathcal{T}$ and $\mathcal{T}_c$, respectively, in view of Theorem 9.3.8.

Now we are in the position to formulate the following fundamental theorem for the surface potential on Γ in the case $m+1 \in \mathbb{N}_0$. For $m+1 \notin \mathbb{N}_0$, the corresponding result is already formulated in Theorem 9.5.1, where the extension conditions are sufficient without further restrictions.

Theorem 9.5.5. *Let $A \in \mathcal{L}^m_{c\ell}(\widetilde{\Omega} \cup \Omega)$ with $m+1 \in \mathbb{N}_0$ satisfy the canonical extension conditions (cf. Definition* 9.5.3*) and the special parity conditions* (9.5.14)*. Then the limits of $\widetilde{Q}_\Gamma u$ from the interior Ω as well as from the exterior $\mathbb{R}^n \setminus \overline{\Omega}$ exist. We have*

$$\begin{aligned} \lim_{0 > \varrho_n \to 0} \widetilde{Q}_\Gamma u(x' + \varrho_n n(x')) &= Q_\Gamma u(x') \\ &= \text{p.f.} \int\limits_{y' \in \Gamma} k_A(x', x' - y') u(y') ds'_y - \mathcal{T} u(x') \quad \textit{for } x' \in \Gamma \end{aligned}$$

and

$$\begin{aligned} \lim_{0 < \varrho_n \to 0} \widetilde{Q}_\Gamma u(x' + \varrho_n n(x')) &= Q_{\Gamma,c} u(x') \\ &= \text{p.f.} \int\limits_{y' \in \Gamma} k_A(x', x' - y') u(y') ds_{y'} + \mathcal{T} u(x') \quad \textit{for } x' \in \Gamma \,, \end{aligned}$$

where the finite part integrals are invariantly defined on Γ.

Here Q_Γ and $Q_{\Gamma,c}$ are pseudodifferential operators in $\mathcal{L}^{m+1}_{c\ell}(\Gamma)$ which satisfy the generalized Plemelj–Sochozki jump relations

$$\mathcal{T} = \mathcal{T}_c = \tfrac{1}{2}(Q_{\Gamma,c} - Q_\Gamma) = \tfrac{1}{2}[Q]_\Gamma \,.$$

Remark 9.5.1: Here the finite part integral operators are invariant under the change of the local representation of the boundary Γ in the sense of Theorem 9.3.6. This result is due to R. Kieser [225, Satz 4.3.9].

Under our assumptions in the theorem it is clear that the limits of $\widetilde{Q}_{\Gamma_\tau} u\big(x' + \tau n(x') + \varrho_n n(x')\big)$ for the parallel surfaces Γ_τ also exist for each $\tau \in I_\tau$ and define corresponding pseudodifferential operators Q_{Γ_τ}, $Q_{\Gamma_\tau,c}$ on each of the parallel surfaces on Γ_τ.

On the other hand, if all these limits exist for every $\tau \in I_\tau$, then A satisfies the canonical extension conditions.

Corollary 9.5.6. *Let $S = \sum\limits_{q=1}^{n} c_q(x)\mathcal{D}_q$ be a tangential operator on the surface parallel to Γ with C^∞ coefficients $c_q(x)$ and $\mathcal{D}_q$ defined as in (3.4.33), and let $A \in \mathcal{L}^m_{c\ell}(\Omega \cup \widetilde{\Omega})$ as well as $S \circ A \in \mathcal{L}^{m+1}_{c\ell}(\Omega \cup \widetilde{\Omega})$ with $m+1 \in \mathbb{N}_0$ satisfy all the conditions of Theorem 9.5.5. Then Theorem 9.5.5 remains valid for $S \circ A$. In particular, we have*

$$S \circ \mathcal{T} = S \circ \mathcal{T}_c = \tfrac{1}{2}[S \circ Q]_\Gamma = \tfrac{1}{2} S \circ [Q_{\gamma,c} - Q_\Gamma] \,. \tag{9.5.18}$$

Proof:

$$\begin{aligned} &\lim_{0 > \varrho_n \to 0} S \circ A(u \otimes \delta_\Gamma)\big(x' + \varrho_n n(x')\big) \\ &\quad = \text{p.f.} \int\limits_{y' \in \Gamma} (S \circ k_A)(x', x' - y') u(y') ds_{y'} - \widetilde{\mathcal{T}} u(x') \quad \text{for } x' \in \Gamma \end{aligned}$$

due to Theorem 9.5.5 where

$$\widetilde{\mathcal{T}} u = \widetilde{\mathcal{T}}_c u = \tfrac{1}{2}(\gamma_c - \gamma)\big(S \circ A(u \otimes \delta_\Gamma)\big) \,.$$

However, since S is a tangential operator. we have

$$\gamma S = S|_\Gamma \gamma \quad \text{and} \quad \gamma_c S = S|_\Gamma \gamma_c \,.$$

Consequently, with Theorem 9.5.5 applied to A,

$$\widetilde{\mathcal{T}} u = \widetilde{\mathcal{T}}_c u = S|_\Gamma \tfrac{1}{2}(\gamma_c - \gamma) A(u \otimes \delta_\Gamma) = S|_\Gamma \circ \mathcal{T} u = S|_\Gamma \circ \mathcal{T}_c u \,.$$

∎

So far $\widetilde{Q}_\Gamma$ has continuous limits under the extension conditions. For surface potentials $\widetilde{Q}v(\varrho)$ on the plane $\mathbb{R}^{n-1}$ we have shown that for operators with symbols of rational type, the potentials $\widetilde{Q}v(\varrho)$ for $\varrho_n \neq 0$ and $v \in C_0^\infty(U_r)$, can be extended to C^∞–functions up to $\varrho_n \to 0^\pm$, respectively, and they define continuous mappings on Sobolev spaces. To obtain corresponding results for $\widetilde{Q}_\Gamma$ when Γ is a general surface, we now confine ourselves to this subclass of operators $A \in \mathcal{L}^m_{c\ell}(\widetilde{\Omega} \cup \Omega)$ having symbols of rational type. Then Lemma 9.4.2 implies that all the operators $A_{\Phi r,0}$ generated by local charts of an atlas $\mathfrak{A}(\Gamma)$ have symbols of rational type, too. As a consequence, the following theorem can be established and resembles all the previous properties.

Theorem 9.5.7. *Let $A \in \mathcal{L}^m_{c\ell}(\widetilde{\Omega} \cup \Omega)$ with $m \in \mathbb{Z}$ have a symbol $a(x,\xi)$ whose j–th asymptotic terms $a_{m-j}(x,\xi)$ are of rational type for $j = 0,\dots,m$ with respect to any local chart. Then all the assumptions of Theorem* 9.5.5 *are fulfilled and Theorem* 9.5.5 *applies to this class of operators. By the same arguments, Corollary* 9.5.6 *remains valid also for this class of operators, provided a_{j-m} are of rational type for $j = 0,\dots,m+1$.*

For this special class of operators, the mapping properties of surface potentials on the plane $\mathbb{R}^{n-1}$ in Sobolev spaces have been given in Theorem 9.4.7 and Corollary 9.4.12 implies that these surface potentials define continuous mappings from $C_0^\infty(U_r)$ to $C^\infty(\widetilde{U} \cap \overline{\mathbb{R}^n_-})$ and to $C^\infty(\widetilde{U} \cap \overline{\mathbb{R}^n_+})$, respectively. For operators other than of rational type, the latter mapping properties are not true, in general. For this reason, we now only consider the mapping conditions for surface potentials on Γ for the class of operators with symbols of rational type.

Theorem 9.5.8. *Let $A \in \mathcal{L}^m_{c\ell}(\widetilde{\Omega} \cup \Omega)$ with $m \in \mathbb{Z}$ have a symbol of rational type. Then the operators $\widetilde{Q}_\Gamma$ given by* (9.5.2) *with $x \in \Omega$ or $x \in \Omega^c$, respectively, define the following linear mappings which are continuous:*

$$\widetilde{Q}_\Gamma \ : \ \begin{array}{lcl} H^s(\Gamma) & \to & H^{s-m-\frac{1}{2}}(\Omega) \ \text{ and} \\ H^s(\Gamma) & \to & H^{s-m-\frac{1}{2}}(\widetilde{\Omega} \cap \Omega^c) \ \text{ for } s \in \mathbb{R}\,, \end{array} \tag{9.5.19}$$

and, hence,

$$\widetilde{Q}_\Gamma \ : \ \begin{array}{lcl} C^\infty(\Gamma) & \to & C^\infty(\overline{\Omega}) \ \text{ and} \\ C^\infty(\Gamma) & \to & C^\infty(\widetilde{\Omega} \setminus \Omega)\,; \end{array} \tag{9.5.20}$$

$$\widetilde{Q}_\Gamma \ : \ \begin{array}{lcl} \mathcal{E}'(\Gamma) & \to & \mathcal{E}'(\overline{\Omega}) \ \text{ and} \\ \mathcal{E}'(\Gamma) & \to & \mathcal{E}'(\widetilde{\Omega} \setminus \Omega)\,. \end{array} \tag{9.5.21}$$

Proof: With the partition of unity $\{\varphi_r\}_{r\in I}$ to an atlas $\mathfrak{A}(\Gamma)$ and the corresponding family of functions $\psi_r \in C_0^\infty(O_r)$ with $\psi_r(x) = 1$ for $x \in \operatorname{supp}\varphi_r$, we can write $\widetilde{Q}_\Gamma u(x)$ for $u \in C^\infty$ in the form

$$\widetilde{Q}_\Gamma u(x) = \sum_{r\in I} \psi_r(x)(\Phi_r^*\widetilde{Q}_r\chi_{r*}\varphi_r u)(x) + \sum_{r\in I}\int\limits_{y'\in\Gamma} (1-\psi_r(x))k_A(x,x-y')\varphi_r(y')u(y')ds_{y'}\,.$$

Since in the last sum the kernels $(1-\psi_r(x))k_A(x,x-y')\varphi_r(y')$ are C^∞–functions of $x \in \widetilde{\Omega}$ and $y' \in \Gamma$, the mapping results follow from those of $\widetilde{Q}_r$ in Theorem 9.4.7 and its Corollary 9.4.12. ∎

In closing this section, we remark that the operators Q_Γ and $Q_{\Gamma,c}$ whenever they exist as the limits of $\widetilde{Q}_\Gamma$, belong to $\mathcal{L}^{m+1}_{c\ell}(\Gamma)$. This implies in particular that Q_Γ and $Q_{\Gamma,c}$ are continuous mappings on the Sobolev spaces,

$$Q_\Gamma\,,\ Q_{\Gamma,c}\,:\,H^s(\Gamma)\to H^{s-m-1}(\Gamma)\quad\text{for every}\quad s\in\mathbb{R}\,.$$

and, consequently, are also continuous mappings on $C^\infty(\Gamma)$.

9.6 Volume Potentials

With the mapping properties available for the pseudodifferential operators we first consider the volume potentials Af and their mapping properties.

For $f \in \widetilde{H}^s(\Omega\cup\widetilde{\Omega})$ and $A \in \mathcal{L}^m_{c\ell}(\Omega\cup\widetilde{\Omega})$, we denote by $Af(x)$ for $x \in \Omega\cup\widetilde{\Omega}$ the *volume potential*. Then clearly $Af \in H^{s-m}_{\mathrm{loc}}(\Omega\cup\widetilde{\Omega})$. Moreover, then (4.2.32) implies that

$$\|\gamma_0 Af\|_{H^{s-m-\frac{1}{2}}(\Gamma)} + \|\gamma_{0c} Af\|_{H^{s-m-\frac{1}{2}}(\Gamma)} \le c\|f\|_{\widetilde{H}^s(\Omega\cup\widetilde{\Omega})}\quad\text{for}\quad s > m+\frac{1}{2}\,,\tag{9.6.1}$$

where c is independent of f. For the remaining indices s and in what follows, we now confine ourselves to the subclass of operators with symbols of rational type.

Theorem 9.6.1. (Boutet de Monvel [39])
Let $A \in \mathcal{L}^m_{c\ell}(\Omega\cup\widetilde{\Omega})$. Then the following mappings are continuous:

$$A\,:\,\widetilde{H}^s(\Omega)\longrightarrow H^{s-m}(\Omega)\quad\textit{for every}\quad s\in\mathbb{R}\,,\tag{9.6.2}$$

$$A\,:\,H^s(\Omega)\longrightarrow H^{s-m}(\Omega)\quad\textit{for}\quad s>-\frac{1}{2}\,.\tag{9.6.3}$$

The corresponding results hold in $(\widetilde{\Omega}\setminus\overline{\Omega})$ if Ω is replaced by $\overset{\circ}{K}$ where K is any compact set with $K\subset\widetilde{\Omega}\setminus\Omega$.

Note that for $s \to \infty$ the estimate (9.6.3) implies that A maps $C^\infty(\overline{\Omega})$ continuously into $C^\infty(\overline{\Omega})$.

Proof:

i) Let $K \Subset \Omega \cup \widetilde{\Omega}$ be a compact set with $\overline{\Omega} \Subset \overset{\circ}{K}$ where $\overset{\circ}{K}$ is the open interior of K, and let $f \in \widetilde{H}^s(\Omega)$. Then $Af \in H^{s-m}_{\text{loc}}(\Omega \cup \widetilde{\Omega})$ and

$$\|Af\|_{H^{s-m}(K)} \le c\|f\|_{\widetilde{H}^s(\Omega)}$$

with some constant c depending on K, A and s. For any $w \in C_0^\infty(\Omega)$ we then find

$$|(Af, w)_{L^2(\Omega)}| = |(Af, w)_{L^2(K)}| \le c'\|f\|_{\widetilde{H}^s(\Omega)}\|w\|_{\widetilde{H}^{m-s}(\Omega)} .$$

This yields by duality

$$\begin{aligned} &\|(Af)\|_{H^{s-m}(\Omega)} \\ &\quad = \sup\left\{|(Af, w)_{L^2(\Omega)}| \quad \text{for} \quad w \in C_0^\infty(\Omega) \quad \text{with} \quad \|w\|_{\widetilde{H}^{m-s}(\Omega)} \le 1\right\} \end{aligned}$$

and

$$\|Af\|_{H^{s-m}(\Omega)} \le c'\|f\|_{\widetilde{H}^s(\Omega)}$$

which implies (9.6.2) since $C_0^\infty(\Omega)$ is dense in $\widetilde{H}^s(\Omega)$.

ii) Now let $f \in C_0^\infty(K)$ and define

$$f^0(x) := \begin{cases} f(x) & \text{for } x \in \overline{\Omega}, \\ 0 & \text{for } x \notin \overline{\Omega}, \end{cases}$$

where K and w are as above. Then

$$(Af^0, w)_{L^2(K)} = (f^0, A^*w)_{L^2(K)} = (f^0, A^*w)_{L^2(\overline{\Omega})}$$

and the restriction

$$r_\Omega A^*w(x) := \begin{cases} A^*w(x) & \text{for } x \in \overline{\Omega}, \\ 0 & \text{for } x \notin \overline{\Omega} \end{cases}$$

satisfies $r_\Omega A^*w \in \widetilde{H}^{-s}(\Omega)$, provided $-s < \frac{1}{2}$. Hence,

$$\begin{aligned} |(Af^0, w)_{L^2(\overline{\Omega})}| &= |(Af^0, w)_{L^2(K)}| \\ &= |(f^0, r_\Omega A^*w)_{L^2(K)}| \\ &\le c\|f^0\|_{H^s(\Omega)}\|A^*w\|_{\widetilde{H}^{-s}(\Omega)} \\ &\le c'\|f^0\|_{H^\sigma(\Omega)}\|w\|_{\widetilde{H}^{m-s}(\Omega)} . \end{aligned}$$

Again, taking the supremum for $\|w\|_{\widetilde{H}^{m-s}(\Omega)} \le 1$ we find by duality the estimate

$$\|Af\|_{H^{m-s}(\Omega)} \le c\|f\|_{H^s(\Omega)}\,, \quad \text{if } s > -\frac{1}{2}\,,$$

i.e., (9.6.3) since $C^\infty(\overline{\Omega})$ is dense in $H^s(\Omega)$. ■

For the subclass of operators having symbols of rational type, the traces of the volume potentials on the surface Γ exist for some scale of Sobolev spaces as in the standard trace theorem. It should be noted that the operators $\gamma_0 Af$ belong to the class of Boutet de Monvel's trace operators. He also developed the complete calculus for treating compositions of trace and Poisson operators (Boutet de Monvel [39, 40, 41], Grubb [159], Schulze [379] and Chazarain and Piriou [64, p.287]).

Theorem 9.6.2. *Let $A \in \mathcal{L}^m_{c\ell}(\Omega \cup \widetilde{\Omega})$ have a symbol of rational type. Then for $f \in \widetilde{H}^s(\Omega)$ and $s > m + \frac{1}{2}$ we have the estimate*

$$\|\gamma_0 Af\|_{H^{s-m-\frac{1}{2}}(\Gamma)} \le c\|f\|_{\widetilde{H}^s(\Omega)}\,. \tag{9.6.4}$$

If $f \in H^s(\Omega)$ then

$$\|\gamma_0 Af\|_{H^{s-m-\frac{1}{2}}(\Gamma)} \le c\|f\|_{H^s(\Omega)} \tag{9.6.5}$$

provided $s > \max\{-\frac{1}{2}\,,\ m + \frac{1}{2}\}$. The constant c is independent of f.

The corresponding results hold in $(\widetilde{\Omega} \setminus \overline{\Omega})$ if Ω is replaced by $\overset{\circ}{K}$, where K is any compact set with $K \subset \widetilde{\Omega} \setminus \Omega$ and $\overline{\Omega} \subset \overset{\circ}{K}$.

Proof: If $f \in \widetilde{H}^s(\Omega)$ then $Af \in H^{s-m}(\Omega)$ due to (9.6.2), and the Trace Theorem 4.2.1 implies (9.6.4) for $s - m > \frac{1}{2}$.

For $f \in H^s(\Omega)$ we obtain $Af \in H^{s-m}(\Omega)$ from (9.6.3) provided aditionally $s > -\frac{1}{2}$. ■

If, however, $f \in \widetilde{H}^s(\widetilde{\Omega} \setminus \overline{\Omega})$, where $f|_\Omega = 0$, then one has more generally

Theorem 9.6.3. (Chazarain and Piriou [64, Theorem 5.2.2])
Let $A \in \mathcal{L}^m_{c\ell}(\Omega \cup \widetilde{\Omega})$ have a symbol of rational type. Then, for $f \in \widetilde{H}^s(\widetilde{\Omega} \setminus \overline{\Omega})$ and $\sigma \in \mathbb{R}$, the sectional trace $\gamma_0 Af$ exists and

$$\|\gamma_0 Af\|_{H^{s-m-\frac{1}{2}}(\Gamma)} \le c\|f\|_{\widetilde{H}^s(\widetilde{\Omega} \setminus \overline{\Omega})}\,. \tag{9.6.6}$$

For $f \in \widetilde{H}^s(\Omega)$, we have

$$\|\gamma_{0c} Af\|_{H^{s-m-\frac{1}{2}}(\Gamma)} \le c\|f\|_{\widetilde{H}^s(\Omega)}\,, \tag{9.6.7}$$

where the constants c are independent of f.

Proof: For the proof of (9.6.6), let $v \in C^\infty(\Gamma)$ and $f \in C_0^\infty(\widetilde{\Omega} \setminus \overline{\Omega})$ with $K := \operatorname{supp} f$. Then

$$\int\limits_\Gamma v\gamma_0 Af ds = (v \otimes \delta_\Gamma\,,\, Af)_{L^2(\Omega\cup\widetilde{\Omega})} = \big(A^*(v \otimes \delta_\Gamma), f\big)_{L^2(\Omega\cup\widetilde{\Omega})}\,,$$

where $A^* \in \mathcal{L}^m_{c\ell}(\Omega \cup \widetilde{\Omega})$ has a symbol of rational type as well, and $A^*(v\otimes\delta_\Gamma)$ is a Poisson operator or surface potential operating on v. Since $f = 0$ in Ω, we obtain

$$\begin{aligned}\big|\big(A^*(v \otimes \delta_\Gamma), f\big)_{L^2(\widetilde{\Omega}\setminus\overline{\Omega})}\big| &= |(\widetilde{Q}v, f)_{L^2(\widetilde{\Omega}\setminus\overline{\Omega})}| \\ &\le c\|f\|_{\widetilde{H}^s(\widetilde{\Omega}\setminus\overline{\Omega})}\|\widetilde{Q}_c v\|_{H^{-s}(K)}\,,\end{aligned}$$

where $\widetilde{Q}_c$ is the operator defined by (9.4.5) with respect to A^* and for $0 < \varrho_n \to 0$. From Theorem 9.4.7 with s replaced by $m + \frac{1}{2} - s$, we obtain

$$\Big|\int\limits_\Gamma v\gamma_0 Af ds\Big| \le c\|f\|_{\widetilde{H}^s(\widetilde{\Omega}\setminus\overline{\Omega})}\|v\|_{H^{m+\frac{1}{2}-s}(\Gamma)}$$

and

$$\|\gamma_0 Af\|_{H^{s-m-\frac{1}{2}}(\Gamma)} = \sup_{\|v\|_{H^{m+\frac{1}{2}-s}(\Gamma)}\le 1} \Big|\int\limits_\Gamma v\gamma_0 Af ds\Big| \le c\|f\|_{\widetilde{H}^s(\widetilde{\Omega}\setminus\overline{\Omega})}$$

as proposed.
The proof of (9.6.7) follows in the same manner if $\widetilde{\Omega}\setminus\overline{\Omega}$ is replaced by Ω. ∎

9.7 Strong Ellipticity and Fredholm Properties

For completeness, in this section we now give an overview and collect some basic results for boundary integral equations with respect to their mapping and solvability properties, the concept of strong ellipticity, variational formulation and Fredholmness.

In Section 7.2.5, the concept of strong ellipticity for systems of pseudodifferential operators in $\mathbb{R}^n$ has been introduced. Theorem 7.2.7 is the consequence of strong ellipticity in the form of Gårding's inequalities associated with corresponding bilinear forms on compact subdomains. However, inequality (7.2.63) is restricted to Sobolev spaces on compact subregions. On the other hand, if Gårding's inequality is satisfied for operators defined on a compact boundary manifold Γ, it holds on the whole Sobolev spaces on Γ. In this case, the boundary integral equations can be reformulated in the form of variational equations on Γ in terms of duality pairings on boundary Sobolev spaces associated with the scalar products in Gårding's inequality.

With the formulation in variational form together with Gårding's inequality Theorem 5.3.10 may be applied and provides the Fredholm alternative for the solvability of these variational formulations of the corresponding integral equations.

In Chapter 5 we already have shown that Gårding's inequality is satisfied for boundary integral equations which are intimately associated with strongly elliptic boundary value problems for partial differential equations satisfying Gårding's inequality in the domain. There Gårding's inequalities for these boundary integral equations are consequences of the Gårding inequalities for the corresponding partial differential equations. We emphasize that the Gårding inequalities for the corresponding boundary integral equations are formulated in terms of the corresponding energy spaces.

For more general pseudodifferential operators on Γ, we now require strong ellipticity on Γ as in Section 9.1.1 which will lead to Gårding inequalities for a whole family of variational formulations in various Sobolev spaces including the corresponding energy spaces. For the benefit of the reader let us recall the definition of strong ellipticity for a system $((A_{jk}))_{p\times p}$ of pseudodifferential operators on Γ with $A_{jk} \in \mathcal{L}^{s_j+t_k}_{c\ell}(\Gamma)$.

We say that A is strongly elliptic, if there exists a constant $\beta_0 > 0$ and a C^∞–matrix valued function $\Theta(x) = \big((\Theta_{\ell j}(x))\big)_{p\times p}$ such that

$$\mathrm{Re}\,\zeta^\top \Theta(x) a^0(x,\xi')\overline{\zeta} \geq \beta_0|\zeta|^2 \tag{9.7.1}$$

for all $x \in \Gamma$ and $\xi' \in \mathbb{R}^{-1}$ with $|\xi'| = 1$ and for all $\zeta \in \mathbb{C}^p$. Here, $a^0(x,\xi') = \big((a^0_{jk}(x,\xi'))\big)$ is the principal part symbol matrix of A. For strongly elliptic operators, Theorem 9.1.4 is valid. The corresponding variational formulation reads:

For given $f \in \prod_{\ell=1}^p H^{-s_\ell}(\Gamma)$ find $u \in \prod_{\ell=1}^p H^{t_\ell}(\Gamma)$ such that

$$\begin{aligned} a(v,u) &:= \langle v, \Lambda^\sigma_\Gamma \Theta \Lambda^{-\sigma}_\Gamma A u\rangle_{\prod_{\ell=1}^p H^{(t_\ell - s_\ell)/2}(\Gamma)} \\ &= \langle v, \Lambda^\sigma_\Gamma \Theta \Lambda^{-\sigma}_\Gamma f\rangle_{\prod_{\ell=1}^p H^{(t_\ell - s_\ell)/2}(\Gamma)} \end{aligned} \tag{9.7.2}$$

for all $v \in \prod_{\ell=1}^p H^{t_\ell}(\Gamma)$ where $\Lambda^\sigma_\Gamma = ((\Lambda^{s_j}_\Gamma \delta_{j\ell}))_{p\times p}$ and $\Lambda^{s_j}_\Gamma = (-\Delta_\Gamma + 1)^{s_j/2}$ with the Laplace–Beltrami operator Δ_Γ.

As we notice, in Equation (9.7.2) the bilinear form is formulated with respect to the Sobolev space $\prod_{\ell=1}^p H^{(t_\ell - s_\ell)/2}(\Gamma)$–inner product. In practice, however, one prefers the L_2–scalar product instead. For this purpose, we employ $\Lambda^{\tau-\sigma}_\Gamma$ with $\tau = (t_1,\dots,t_p)$ as a preconditioner. Then the bilinear form becomes

$$\langle v, \Lambda^\tau_\Gamma \Theta \Lambda^{-\sigma}_\Gamma A u\rangle_{L^2(\Gamma)} = \langle v, \Lambda^\tau_\Gamma \Theta \Lambda^{-\sigma}_\Gamma f\rangle_{L^2(\Gamma)}\,. \tag{9.7.3}$$

Alternatively, we may rewrite (9.7.3) in the form

$$a(v,u) = \langle Bv, Au\rangle_{L^2(\Gamma)} = \langle Bv, f\rangle_{L^2(\Gamma)} \quad \text{with } Bv := \Lambda^{-\sigma}\Theta^*\Lambda^\tau v\,. \tag{9.7.4}$$

Special examples of this formulation have been introduced in Section 5.6; in particular the equations (5.6.27) and (5.6.28) for the variational formulations of boundary integral equations of the second kind where $\Theta = ((\delta_{jk}))_{p\times p}$.

In (5.6.27), we may identify Λ_Γ with V^{-1} and in (5.6.28) Λ_Γ with D since their principal symbols are the same.

The bilinear form in (9.7.3) or (9.7.4) satisfies a Gårding inequality in the form

$$\mathrm{Re}\,\{a(w,w) + \langle w, \Lambda_\Gamma^{\tau-\sigma} Cw\rangle_{L^2(\Gamma)}\} \geq \beta_0' \|w\|^2_{\prod_{\ell=1}^p H^{t_\ell}(\Gamma)} \tag{9.7.5}$$

with some constant $\beta_0' > 0$ for all $w \in \prod_{\ell=1}^p H^{t_\ell}(\Gamma)$ where

$$\Lambda_\Gamma^{\tau-\sigma} C = ((\Lambda_\Gamma^{t_\ell - s_\ell} C_{\ell j}))_{p\times p}$$

contains the compact operators $\Lambda_\Gamma^{t_\ell - s_\ell} C_{\ell j} : H^{t_j}(\Gamma) \to H^{-t_\ell}(\Gamma)$ due to Theorem 9.1.4. Here, from Gårding's inequality (9.7.5) we may consider $\prod_{\ell=1}^p H^{t_\ell}(\Gamma)$ as the energy space for the boundary integral operator A and the variational equations (9.7.3) or (9.7.4).

However, in order to analyze these equations in other Sobolev spaces as well, we need a variational formulation for the mapping $A : \prod_{\ell=1}^p H^{t_\ell+\varepsilon}(\Gamma) \to \prod_{k=1}^p H^{-s_k+\varepsilon}(\Gamma)$ for any $\varepsilon \in \mathbb{R}$. To this end, we introduce the bilinear variational equations of the same boundary equation $Au = f$ for $u \in \prod_{\ell=1}^p H^{t_\ell+\varepsilon}(\Gamma)$, i.e.

$$a_\varepsilon(v,u) := \langle v, \Lambda^{\tau+\varepsilon}\Theta\Lambda^{-\sigma+\varepsilon} Au\rangle_{L^2(\Gamma)} = \langle v, f_\varepsilon\rangle_{L^2(\Gamma)} \tag{9.7.6}$$

where $f_\varepsilon = \Lambda^{\tau+\varepsilon}\Theta\Lambda^{-\sigma+\varepsilon} f$ with $f \in \prod_{\ell=1}^p H^{-s_\ell+\varepsilon}(\Gamma)$; or equivalently,

$$a_\varepsilon(v,u) = \langle B_\varepsilon v, Au\rangle_{L^2(\Gamma)} = \langle B_\varepsilon v, f\rangle_{L^2(\Gamma)} \tag{9.7.7}$$

where

$$B_\varepsilon v = \Lambda^{-\sigma+\varepsilon}\Theta^*\Lambda^{\tau+\varepsilon} v\,.$$

For the solvability of (9.7.6) or (9.7.7), we now consider corresponding Gårding inequalities for all the bilinear forms $a_\varepsilon(u,v)$, where $\varepsilon \in \mathbb{R}$.

Lemma 9.7.1. *If $A = ((A_{jk}))_{p\times p}$ is a strongly elliptic system of pseudodifferential operators $A_{jk} \in \mathcal{L}_{c\ell}^{s_j+t_k}(\Gamma)$ on Γ, then for every $\varepsilon \in \mathbb{R}$ there exist constants $\gamma_\varepsilon > 0$, $\gamma_{1\varepsilon} \geq 0$ such that Gårding's inequality holds in the form*

$$\mathrm{Re}\, a_\varepsilon(w,w) + \beta_{1\varepsilon}\|w\|^2_{\prod_{\ell=1}^p H^{t_\ell - 1+\varepsilon}(\Gamma)} \geq \beta_\varepsilon \|w\|^2_{\prod_{\ell=1}^p H^{t_\ell+\varepsilon}(\Gamma)} \tag{9.7.8}$$

for all $w \in \prod_{\ell=1}^p H^{t_\ell+\varepsilon}(\Gamma)$.

Proof: Let $A_{jk}^\varepsilon := \Lambda^\varepsilon A_{jk}\Lambda^{-\varepsilon}$, then $A_{jk}^\varepsilon \in \mathcal{L}_{c\ell}^{s_j+t_k}(\Gamma)$ and the principal symbol matrices A_{jk}^ε and A_{jk} are the same. Hence, $A^\varepsilon = ((A_{jk}^\varepsilon))_{p\times p}$ is also strongly elliptic for any $\varepsilon \in \mathbb{R}$ and Theorem 9.1.4 may be applied to A^ε providing Gårding's inequality (9.1.20) with corresponding constants γ_0 and γ_1

depending on ε. By setting $v = \Lambda^\varepsilon w$ with $w \in \prod_{\ell=1}^p H^{t_\ell+\varepsilon}(\Gamma)$ and employing the fact that Λ^ε is an isomorphism we find

$$\begin{aligned}
\operatorname{Re} a_\varepsilon(w,w) &= \operatorname{Re}\langle w, \Lambda^{\tau+\varepsilon}\Theta\Lambda^{-\sigma+\varepsilon}Aw\rangle_{L^2(\Gamma)} \\
&= \operatorname{Re}\langle v, \Lambda^{\tau}\Theta\Lambda^{-\sigma}A^\varepsilon v\rangle_{L^2(\Gamma)} \\
&\geq \beta_0\|v\|^2_{\prod_{\ell=1}^p H^{t_\ell}(\Gamma)} - \beta_1\|v\|^2_{\prod_{\ell=1}^p H^{t_\ell-1}(\Gamma)} \\
&\geq \beta_\varepsilon\|w\|^2_{\prod_{\ell=1}^p H^{t_\ell+\varepsilon}(\Gamma)} - \beta_{1\varepsilon}\|w\|^2_{\prod_{\ell=1}^p H^{t_\ell+\varepsilon-1}(\Gamma)},
\end{aligned}$$

which corresponds to (9.7.8). ■

Note that the compact operators C_ε are now defined by the relations

$$\langle v, C_\varepsilon w\rangle_{\prod_{\ell=1}^p H^{(t_\ell-s_\ell)/2}(\Gamma)} = \langle v, \Lambda^{\tau-\sigma+2\varepsilon}C_\varepsilon w\rangle_{L^2(\Gamma)} = \langle v, w\rangle_{\prod_{\ell=1}^p H^{t_\ell-1+\varepsilon}(\Gamma)}. \tag{9.7.9}$$

As a consequence of Gårding's inequality (9.7.8), for fixed $\varepsilon \in \mathbb{R}$ the boundary integral operators A^ε are Fredholm operators of index zero. More precisely, the following Fredholm theorem holds.

Theorem 9.7.2. *Let $\mathcal{H} = \prod_{\ell=1}^p H^{t_\ell+\varepsilon}(\Gamma)$ with fixed $\varepsilon \in \mathbb{R}$. Then for the variational equation* (9.7.7), *i.e.*

$$a_\varepsilon(v,u) = \langle v, f_\varepsilon\rangle_{L^2(\Gamma)} \quad \text{for all } v \in \mathcal{H} \tag{9.7.10}$$

there holds Fredholm's alternative:

i) *either* (9.7.10) *has exactly one solution for any $f_\varepsilon \in \prod_{\ell=1}^p H^{-t_\ell-\varepsilon}(\Gamma)$, i.e., any $f \in \prod_{\ell=1}^p H^{-s_\ell+\varepsilon}$ or*

ii) *the homogeneous problem*

$$a_\varepsilon(v,u_0) = 0 \quad \text{for all } v \in \mathcal{H} \tag{9.7.11}$$

has a finite–dimensional eigenspace of dimension k. In this case, the inhomogeneous variational equation (9.7.10) *has solutions if and only if the orthogonality conditions*

$$\ell_{f_\varepsilon}(v_{0(j)}) = \langle v_{0(j)}, f_\varepsilon\rangle_{L^2(\Gamma)} = 0 \tag{9.7.12}$$

are satisfied where $v_{0(1)}, \dots, v_{0(k)}$ are a basis of the eigenspace to the adjoint homogeneous variational equation

$$a_\varepsilon(v_0, w) = 0 \quad \text{for all } w \in \mathcal{H}.$$

The dimension of this eigenspace is the same as that of (9.7.11).

Clearly, this Theorem is identical to Theorem 5.3.10 which was established already in Section 5.3.3 since a_ε satisfies the Gårding inequality (9.7.8).

Now, for the family of variational equations (9.7.7) corresponding to the parameter $\varepsilon \in \mathbb{R}$ we have the following version of the *shift–theorem.*

Theorem 9.7.3. i) **Uniqueness:** *If for any* $\varepsilon_0 \in \mathbb{R}$, *the corresponding variational equation* (9.7.10) *with* a_{ε_0} *is uniquely solvable then each of these equations with* $\varepsilon \in \mathbb{R}$ *is also uniquely solvable.*

ii) **Solvability:** *If* $u_0 \in H^p_{\ell=1} H^{t_\ell+\varepsilon_0}(\Gamma)$ *is an eigensolution of the homogeneous equation* (9.7.11) *with* a_{ε_0} *for some* ε_0 *then* $u_0 \in \bigcap_{\varepsilon\in\mathbb{R}} \prod_{\ell=1}^p H^{t_\ell+\varepsilon}(\Gamma)$ $= C^\infty(\Gamma)$. *Hence, the eigenspaces of all the equations* (9.7.11) *with* $\varepsilon \in \mathbb{R}$ *are identical.*

iii) **Regularity:** *If* $u \in \prod_{\ell=1}^p H^{t_\ell+\varepsilon_0}(\Gamma)$ *is a solution of the variational equation*

$$a_{\varepsilon_0}(v,u) = \ell_{f_{\varepsilon_0}}(v) \quad \textit{for all} \quad v \in \prod_{\ell=1}^p H^{t_\ell+\varepsilon}(\Gamma)$$

with $f \in \prod_{\ell=1}^p H^{-s_\ell+\varepsilon}(\Gamma)$ *and* $\varepsilon \geq \varepsilon_0$ *then* $u \in \prod_{\ell=1}^p H^{t_\ell+\varepsilon}(\Gamma)$.

Proof: We begin by showing ii) first, i.e. $u_0 \in \bigcap_{\varepsilon\in\mathbb{R}} \prod_{\ell=1}^p H^{t_\ell+\varepsilon_0}(\Gamma) = C^\infty(\Gamma)$ for every eigensolution $u_0 \in \mathcal{H}$ of (9.7.11) with any fixed $\varepsilon = \varepsilon_0$. Then, by the definition of a_{ε_0} we have for $u_0 \in \prod_{\ell=1}^p H^{t_\ell+\varepsilon_0}(\Gamma)$:

$$0 = a_{\varepsilon_0}(v,u_0) = \langle v, \Lambda^{\tau+\varepsilon_0}\Theta\Lambda^{-\sigma+\varepsilon_0}Au_0\rangle_{L^2(\Gamma)} \quad \text{for every} \quad v \in C^\infty(\Gamma)\,.$$

This implies

$$Au_0 = 0 \quad \text{in} \quad \prod_{\ell=1}^p H^{-s_\ell+\varepsilon_0}(\Gamma)$$

since $\Lambda^{\tau+\varepsilon_0}$, Θ and $\Lambda^{-\sigma+\varepsilon_0}$ are isomorphisms. Then, by duality, we also have

$$\langle v, \Lambda^{\tau+\varepsilon_0+1}\Theta\Lambda^{-\sigma+\varepsilon_0+1}Au_0\rangle_{L^2(\Gamma)} = 0 \quad \text{for} \quad v \in C^\infty\,.$$

Hence,

$$a_{\varepsilon_0+1}(v,u_0) = 0 \quad \text{for all} \quad v \in C^\infty\,,$$

which implies with Gårding's inequality

$$\|u_0\|^2_{\prod_{\ell=1}^p H^{t_\ell+\varepsilon_0+1}} \leq \frac{1}{\beta_{\varepsilon_0+1}}\beta_{1,\varepsilon_0+1}\|u_0\|^2_{\prod_{\ell=1}^p H^{t_\ell+\varepsilon_0}}\,.$$

By induction we find $u_0 \in \bigcap_{\varepsilon\geq\varepsilon_0} \prod_{\ell=1}^p H^{t_\ell+\varepsilon}(\Gamma)$ and with the embedding $H^\alpha(\Gamma) \subset H^\beta(\Gamma)$ for $\alpha \geq \beta$ we obtain $u_0 \in \bigcap_{\varepsilon\geq\mathbb{R}} \prod_{\ell=1}^p H^{t_\ell+\varepsilon}(\Gamma) = C^\infty(\Gamma)$. Consequently, if u_0 is an eigenfunction of a_{ε_0}, it is in $C^\infty(\Gamma)$ and satisfies $Au_0 = 0$ which implies $a_\varepsilon(v,u_0) = 0$ for every $\varepsilon \in \mathbb{R}$ and $v \in C^\infty(\Gamma)$. This establishes the assertion ii).

For the uniqueness in all the spaces let us assume that equation (9.7.10) with a_{ε_0} has a unique solution $u \in \prod_{\ell=1}^p H^{t_\ell+\varepsilon_0}(\Gamma)$ and that there is non-uniqueness for the equation with a_{ε_1} for some $\varepsilon_1 \neq \varepsilon_0$. Then the dimension of the eigenspace to a_{ε_1} satisfies $k \geq 1$; the eigensolutions are in $C^\infty(\Gamma)$ and are eigensolutions to a_ε for every $\varepsilon \in \mathbb{R}$, in particular for $\varepsilon = \varepsilon_0$, which contradicts our assumption of uniqueness. This establishes that uniqueness for one fixed ε_0 implies uniqueness for all $\varepsilon \in \mathbb{R}$, i.e. assertion i).

To show assertion iii), let $u_{\varepsilon_0} \in \prod_{\ell=1}^{p} H^{t_\ell+\varepsilon_0}(\Gamma)$ be a solution of

$$a_{\varepsilon_0}(v, u_{\varepsilon_0}) = \langle v, f_{\varepsilon_0}\rangle_{L^2(\Gamma)} \quad \text{for all } v \in C^\infty(\Gamma) \tag{9.7.13}$$

with $f_{\varepsilon_0} = \Lambda^{\tau+\varepsilon_0}\Theta\Lambda^{-\sigma+\varepsilon_0} f$ and $f \in \prod_{\ell=1}^{p} H^{-s_\ell+\varepsilon}(\Gamma)$ where $\varepsilon > \varepsilon_0$. Hence,

$$Au_{\varepsilon_0} = f \in \prod_{\ell=1}^{p} H^{-s_\ell+\varepsilon}(\Gamma) \tag{9.7.14}$$

due to the definition of the variational equation (9.7.6) and since $\Lambda^{\sigma+\varepsilon_0}\Theta\Lambda^{-\sigma+\varepsilon_0}$ is an isomorphism. From equation (9.7.14) and by duality, we then find that u_{ε_0} is also a solution of

$$\begin{aligned} a_{\varepsilon_0+1}(v, u_{\varepsilon_0}) &= \langle v, \Lambda^{\tau+\varepsilon_0+1}\Theta\Lambda^{-\sigma+\varepsilon_0+1}Au_{\varepsilon_0}\rangle_{L^2(\Gamma)} \\ &= \langle \Lambda^{-\sigma+\varepsilon_0+1}\Theta^*\Lambda^{\tau+\varepsilon_0+1}v, f\rangle_{L^2(\Gamma)} = \langle v, f_{\varepsilon_0+1}\rangle_{L^2(\Gamma)} \end{aligned} \tag{9.7.15}$$

for all $v \in C^\infty(\Gamma)$ provided $\varepsilon - \varepsilon_0 \geq 1$. If temporarily we assume that $u_{\varepsilon_0} \in \prod_{\ell=1}^{p} H^{t_\ell+\varepsilon_0+1}(\Gamma)$ then, by using Gårding's inequality (9.7.8) for a_{ε_0+1}, we find

$$\begin{aligned} &\|u_{\varepsilon_0}\|^2_{\prod_{\ell=1}^{p} H^{t_\ell+\varepsilon_0+1}(\Gamma)} \\ &\leq \frac{1}{\beta_{\varepsilon_0+1}}\big\{\beta_{1\varepsilon_0+1}\|u_{\varepsilon_0}\|^2_{\prod_{\ell=1}^{p} H^{t_\ell+\varepsilon_0}(\Gamma)} + \operatorname{Re}\langle \Lambda^{\tau+\varepsilon_0+1}u_{\varepsilon_0}, \Theta\Lambda^{-\sigma+\varepsilon_0+1}\rangle_{L^2(\Gamma)}\big\} \\ &\leq \frac{1}{\beta_{\varepsilon_0+1}}\big\{\beta_{1\varepsilon_0+1}\|u_{\varepsilon_0}\|^2_{\prod_{\ell=1}^{p} H^{t_\ell+\varepsilon_0}(\Gamma)} \\ &\qquad + c_{\varepsilon_0}\|f\|_{\prod_{\ell=1}^{p} H^{-s_\ell+\varepsilon_0+1}(\Gamma)}\|u_{\varepsilon_0}\|_{\prod_{\ell=1}^{p} H^{t_\ell+\varepsilon_0+1}(\Gamma)}\big\}. \end{aligned}$$

This inequality implies the a priori estimate

$$\|u_{\varepsilon_0}\|_{\prod_{\ell=1}^{p} H^{t_\ell+\varepsilon_0+1}(\Gamma)} \leq c(\varepsilon_0)\big\{\|f\|_{\prod_{\ell=1}^{p} H^{-s_\ell+\varepsilon}(\Gamma)} + \|u_{\varepsilon_0}\|_{\prod_{\ell=1}^{p} H^{t_\ell+\varepsilon_0}(\Gamma)}\big\} \tag{9.7.16}$$

with $\varepsilon \geq \varepsilon_0 + 1$. In order to relax the temporary regularity, assumption on u_{ε_0} we now apply a standard density argument. More precisely, let $f_j \in C^\infty(\Gamma)$ be a sequence of functions approximating f in $\prod_{\ell=1}^{p} H^{-\sigma_\ell+\varepsilon}(\Gamma)$ from (9.7.14). Then we find $u_{\varepsilon_0 j} \in C^\infty(\Gamma)$ as corresponding solutions of (9.7.13), also satisfying (9.7.14) with f_j instead of f, and $u_{\varepsilon_0 j} \to u_{\varepsilon_0}$ in $\prod_{\ell=1}^{p} H^{t_\ell+\varepsilon_0}(\Gamma)$. Because of $u_{\varepsilon_0 j} \in C^\infty(\Gamma)$, these functions automatically satisfy (9.7.16) with $u_{\varepsilon_0 j}$ instead of u_{ε_0} and f_j instead of f. Then, for $j \to \infty$, the right–hand side of (9.7.16) converges to the corresponding expression with f and u_{ε_0}. Hence, from the left–hand side we see that $u_{\varepsilon_0 j}$ is a Cauchy sequence in the space $\prod_{\ell=1}^{p} H^{t_\ell+\varepsilon_0+1}(\Gamma)$ which converges to the limit u_{ε_0} in this space, hence we obtain $u_{\varepsilon_0} \in \prod_{\ell=1}^{p} H^{t_\ell+\varepsilon_0+1}(\Gamma)$. If $\varepsilon - \varepsilon_0 \in \mathbb{N}$ then by induction and after finally many steps we obtain the proposed regularity.

If $\varepsilon - \varepsilon_0 \notin \mathbb{N}$, first prove the estimate for the integer case and then use interpolation to obtain the same conclusion. ∎

9.8 Strong Ellipticity of Boundary Value Problems and Associated Boundary Integral Equations

In Section 5.6.6 we considered a class of direct boundary integral equations whose variational sesquilinear forms satisfy Gårding's inequalities since the corresponding boundary value problems are strongly elliptic. Grubb presented in [158] a general approach for general, normal elliptic boundary problems satisfying coerciveness or Gårding's inequalities, and constructed associated pseudodifferential operators on the boundary which form there a strongly elliptic system as well. Here we present a short review of the work by Costabel et al in [90], where a more special case of elliptic even order boundary value problems were considered.

9.8.1 The Boundary Value and Transmission Problems

Let us consider the boundary value problem (3.9.1) and (3.9.2), i.e.,

$$Pu = 0 \quad \text{in } \Omega\,, \tag{9.8.1}$$

$$R\gamma u = \varphi \quad \text{on } \Gamma \tag{9.8.2}$$

where P is an elliptic operator (3.9.1) of order $2m$ with $m \in \mathbb{N}$, and R is a $m \times 2m$ matrix $R = ((R_{jk}))_{\substack{j=1,\dots,m,\\ k=1,\dots,2m}}$ and every R_{jk} is a $p \times p$ matrix of tangential differential operators with C^∞ coefficients and of equal orders

$$\text{ord } R_{jk} = \mu_j - k + 1\,,\ 0 \le \mu_j \le 2m-1\,,\ j = 1,\dots,m\,. \tag{9.8.3}$$

By R_j we denote the j–th row of (3.9.2). Thus,

$$R_j\gamma u := \sum_{k=1}^{2m} R_{jk}\gamma_{k-1}u = \sum_{k=1}^{2m} R_{jk}\Big(\frac{\partial^{k-1}}{\partial n^{k-1}}u\Big)\big|_\Gamma\,,\ j = 1,\dots,m \tag{9.8.4}$$

is the restriction to Γ of a differential operator of order μ_{j-1} defined in $\widetilde{\Omega}$. For the complementing boundary operators $S = ((S_{jk}))_{\substack{j=1,\dots,m,\\ k=1,\dots,2m}}$ we assume the orders

$$\text{ord } S_{jk} = 2m - \mu_j - k \tag{9.8.5}$$

and require the fundamental assumption (3.9.4) for $\mathcal{M} = \binom{R}{S}$, i.e., $\mathcal{N} = \mathcal{M}^{-1}$ is a matrix of tangential differential operators. For the boundary value problem (9.8.1) and (9.8.2), we further assume that the Lopatinski–Shapiro conditions (3.9.6) and (3.9.7) are satisfied. Related with the boundary value problem we will consider a transmission problem where $\widetilde{\Omega} \subset \mathbb{R}^n$ is a bounded domain that contains $\overline{\Omega} \subset \widetilde{\Omega}$ and where $\Omega^e := \Omega^c \cap \widetilde{\Omega}$ denotes a bounded exterior. $\Gamma_e := \partial\Omega^e \setminus \Gamma$ is the exterior boundary. The traces $\gamma_k u = \big(\frac{\partial}{\partial n}\big)^k u$ on Γ are defined by sectional traces for any $u \in H^t(\Omega)$ satisfying the differential equation (9.8.1) in Ω.

Lemma 9.8.1. *Let $u \in H^t(\Omega)$ satisfy the differential equation* (9.8.1) *in Ω, where $t \geq 0$. Then u has sectional traces $\gamma_k u$ on Γ of any order $k \in \mathbb{N}_0$ and*

$$\|\gamma_k u\|_{H^{t-k-\frac{1}{2}}(\Gamma)} \leq c\|u\|_{H^t(\Omega)} . \tag{9.8.6}$$

Proof: If $u \in H^t(\Omega)$ then with Theorem 4.1.5 we can extend u to $\widetilde{u} \in \widetilde{H}^t(\Omega \cup \widetilde{\Omega})$ and the mapping $u \mapsto \widetilde{u}$ is continuous from $H^t(\Omega)$ to $\widetilde{H}^t(\Omega \cup \widetilde{\Omega})$. Hence,

$$f := P\widetilde{u} \in \widetilde{H}^{t-2m}(\widetilde{\Omega} \setminus \overline{\Omega})$$

since $P\widetilde{u} = 0$ in $\overline{\Omega}$. To the elliptic operator P there exist a properly supported parametrix $Q \in \mathcal{L}^{-2m}_{c\ell}(\Omega \cup \widetilde{\Omega})$ due to Theorem 7.2.2 which has a symbol of rational type. Relation (7.2.6) implies

$$\widetilde{u} = Qf + R \circ P\widetilde{u}$$

with a smoothing operator $R \in \mathcal{L}^{-\infty}(\Omega \cup \widetilde{\Omega})$. Then

$$\Big(\frac{\partial}{\partial n}\Big)^k \circ Q \in \mathcal{L}^{-2m+k}_{c\ell}(\Omega \cup \widetilde{\Omega}) ,$$

and (9.6.6) in Theorem 9.6.3 implies with m replaced by $-2m+k$ and $\sigma = t - 2m$ the desired estimate:

$$\begin{aligned}
\|\gamma_k u\|_{H^{t-k-\frac{1}{2}}(\Gamma)} &\leq \|\gamma_0\Big(\frac{\partial}{\partial n}\Big)^k \circ Qf\|_{H^{t-k-\frac{1}{2}}(\Gamma)} \\
&\quad + \|\gamma_0\Big(\frac{\partial}{\partial n}\Big)^k R \circ P\widetilde{u}\|_{H^{t-k-\frac{1}{2}}(\Gamma)} \\
&\leq c_1\|f\|_{\widetilde{H}^{t-2m}(\widetilde{\Omega}\setminus\overline{\Omega})} + c_2\|\widetilde{u}\|_{\widetilde{H}^t(\Omega\cup\widetilde{\Omega})} \\
&\leq c_3\|\widetilde{u}\|_{\widetilde{H}^t(\Omega\cup\widetilde{\Omega})} \leq c_4\|u\|_{H^t(\Omega)} .
\end{aligned}$$

■

Clearly, Lemma 9.8.1 also holds for any $u \in H^t(\Omega^e)$ satisfying $Pu = 0$ in Ω^e for the sectional traces $\gamma_{kc}u$. Then Γ defines a transmission boundary between Ω and Ω^e. (See also Section 5.6.3.)

Let C^∞_e denote the restrictions of $C^\infty_0(\widetilde{\Omega})$ functions to $\overline{\Omega^e}$, i.e.,

$$C^\infty_e = \{u = \varphi|_{\overline{\Omega^e}} \text{ where } \varphi \in C^\infty_0(\widetilde{\Omega})\} . \tag{9.8.7}$$

Now we impose the following Assumptions A.1–A.3:

A1: *There exist two sesquilinear forms $a_\Omega(u,v)$ and $a_{\Omega^e}(u_e,v_e)$ on $C^\infty(\overline{\Omega})$ and C^∞_e, respectively, such that*

$$\begin{aligned}
&\mathrm{Re}\,\{a_\Omega(u,u) + a_{\Omega^e}(u_e,u_e)\} \qquad\qquad (9.8.8)\\
&= \mathrm{Re}\sum_{j=1}^m \int_\Gamma \big((R_j\gamma u)^\top \overline{(S_j\gamma u)} - (R_j\gamma u_e)^\top \overline{(S_j\gamma u_e)}ds \\
&\qquad + \sum_{j=1}^m \int_{\Gamma_e} (R_j\gamma u)^\top \overline{(S_j\gamma u_e)}\big)ds
\end{aligned}$$

is fulfilled for every $u \in C^\infty(\overline{\Omega})$, $u_e \in C_e^\infty$ *satisfying* $Pu = 0$ *in* Ω *and* $Pu_e = 0$ *in* Ω^e.

Assumption A1 corresponds to the validity of the first Green's formula on Ω and on Ω^e. For the sesquilinear forms a_Ω and a_{Ω^e} we require continuity and Gårding's inequality as follows.

A2: *There exits a positive constant c such that*

$$\begin{aligned} &|a_\Omega(u,v)| + |a_{\Omega^e}(u_e,v_e)| \\ &\quad \le c\big\{(\|u\|_{H^m(\Omega)} + \|u_e\|_{H^m(\Omega^e)})(\|v\|_{H^m(\Omega)} + (\|v_e\|_{H^m(\Omega^e)})\big\} \end{aligned} \tag{9.8.9}$$

for all $u, v \in C^\infty(\overline{\Omega})$ *and* $u_e, v_e \in C_e^\infty$.

A3: *Gårding's inequality: There exist positive constants* β_0 *and* ε, *and a constant c such that*

$$\begin{aligned} &\operatorname{Re}\big\{a_\Omega(u,u) + a_{\Omega^e}(u_e,u_e)\big\} \\ &\ge \beta_0(\|u\|^2_{H^m(\Omega)} + \|u_e\|^2_{H^m(\Omega^e)}) - c(\|u\|^2_{H^{m-\varepsilon}(\Omega)} + \|u_e\|^2_{H^{m-\varepsilon}(\Omega^e)}) \end{aligned} \tag{9.8.10}$$

for all $(u.u_e) \in C^\infty(\overline{\Omega}) \times C_e^\infty$ *satisfying the transmission condition*

$$R\gamma u = R\gamma u_e \quad \text{on } \Gamma\,. \tag{9.8.11}$$

Note that under these assumptions the sesquilinear forms

$$\int\limits_\Omega (Pu)^\top \overline{v} dx \quad \text{and} \quad \int\limits_{\Omega^e} (Pu_e)^\top \overline{v}_e dx \tag{9.8.12}$$

satisfy Gårding's inequalities on the subspaces of $H^m(\Omega)$ and of $H^m(\Omega^e)$ with elements fulfilling the boundary conditions

$$R\gamma u = 0 \quad \text{on } \Gamma \quad \text{and } R\gamma u_e = 0 \quad \text{on } \Gamma;\ R\gamma u_e = 0 \quad \text{on } \Gamma_e := \partial\Omega^e \cap \partial\widetilde{\Omega}\,. \tag{9.8.13}$$

9.8.2 The Associated Boundary Integral Equations of the First Kind

As exemplified in Section 3.9, the solution to the boundary value problem (9.8.1) and (9.8.2) can in Ω be represented in terms of Poisson operators or multilayer potentials via (3.9.18) and (3.9.18), i.e.,

$$u(x) = -\sum_{\ell=0}^{2m-1}\sum_{t=0}^{2m-\ell-1} \mathcal{K}_t\big(\mathcal{P}_{t+\ell+1}\big\{\sum_{j=1}^m \mathcal{N}_{\ell+1,j}\varphi_j + \sum_{j=1}^m \mathcal{N}_{\ell+1,m+j}\lambda_j\big\} \otimes \delta_\Gamma\big) \tag{9.8.14}$$

for $x \in \Omega$, where $\lambda = (\lambda_1,\dots,\lambda_m)^\top = S\gamma u$ denotes the yet unknown Cauchy data of u on Γ. Invoking Theorem 9.4.3 for the traces of Poisson operators we obtain on Γ, by taking traces of (9.8.14), the relations

$$\gamma u = -\sum_{j=1}^{m}\sum_{\ell=0}^{2m-1}\sum_{t=0}^{2m-\ell-1}\gamma_0\sum_{k=0}^{2m-1}\Big(\frac{\partial}{\partial n}\Big)^{k-1}\mathcal{K}_t((\mathcal{P}_{t+\ell+1}\mathcal{N}_{\ell+1,m+j}\lambda_j)\otimes\delta_\Gamma)$$
$$-\sum_{j=1}^{m}\sum_{\ell=0}^{2m-1}\sum_{t=0}^{2m-\ell-1}\gamma_0\sum_{k=0}^{2m-1}\Big(\frac{\partial}{\partial n}\Big)^{k-1}\mathcal{K}_t((\mathcal{P}_{t+\ell+1}\mathcal{N}_{\ell+1,j}\varphi_j)\otimes\delta_\Gamma)\,. \tag{9.8.15}$$

Applying the tangential differential operator R to these relations one obtains boundary integral equations of the first kind for λ on Γ, namely

$$\sum_{j=1}^{m}\sum_{\ell=0}^{2m-1}\sum_{t=0}^{2m-\ell-1} -R_q\gamma\mathcal{K}_t\mathcal{P}_{t+\ell+1}\mathcal{N}_{\ell+1,m+j}\lambda_j$$
$$= \varphi_q + \sum_{j=1}^{m}\sum_{\ell=0}^{2m-1}\sum_{t=0}^{2m-\ell-1} R_q\gamma\mathcal{K}_t\mathcal{P}_{t+\ell+1}\mathcal{N}_{\ell+1,j}\varphi_j\,, \tag{9.8.16}$$

in short

$$\sum_{j=1}^{m} A_{qj}\lambda_j = \psi_q\,,\ q = 1,\ldots,m\,, \tag{9.8.17}$$

where the operators A_{qj} and right–hand sides ψ_q are defined by (9.8.16).

In view of Theorem 9.4.3, the operators A_{qj} are classical pseudodifferential operators $A_{qj} \in \mathcal{L}_{c\ell}^{\mu_q+\mu_j+1-2m}(\Gamma)$ which also can be represented via (8.1.84) and (8.1.85) in terms of boundary integral operators with Hadamard's finite part integral operators composed with tangential differential operators.

As shown by Costabel et al in [90], we have the following theorem.

Theorem 9.8.2. ([90, Theorem 3.9]) *Under the assumptions A.1–A.3 there exist positive constants β_0' and δ, and a constant c_0' such that*

$$\mathrm{Re}\sum_{j,q=1}^{m}\int_\Gamma \overline{\lambda}_q A_{qj}\lambda_j ds \ge \beta_0'\sum_{j=1}^{m}\|\lambda_j\|^2_{H^{\mu_j-m+\frac12}(\Gamma)} - c_0'\sum_{j=1}^{m}\|\lambda_j\|^2_{H^{\mu_j-m+\frac12-\delta}(\Gamma)} \tag{9.8.18}$$

for all $\lambda = (\lambda_1,\ldots,\lambda_m)^\top \in \prod_{j01}^{m} H^{-m+\mu_j-m+\frac12}(\Gamma)$.

For showing Gårding's inequality (9.8.18) let us consider the following transmission problem (similar to Sections 3.9.2 and 5.5.3).

9.8.3 The Transmission Problem and Gårding's inequality

For any given $\lambda \in C^\infty(\Gamma)$ let us define the boundary potentials

$$\left.\begin{matrix} u(x) \\ u_e(x) \end{matrix}\right\} := -\sum_{j=1}^{m}\sum_{\ell=1}^{2m-1}\sum_{t=0}^{2m-\ell-1}\mathcal{K}_t\{(\mathcal{P}_{t+\ell+1}\mathcal{N}_{\ell+1,m+j}\lambda_j)\otimes\delta_\Gamma\}$$

$$\text{for } x\in\begin{cases}\Omega\\ \Omega^e\end{cases}. \tag{9.8.19}$$

Then these functions are defined by Poisson operators based on the pseudodifferential operators $\mathcal{K}_t$ with the Schwartz kernels $\left(\left(\frac{\partial'}{\partial n_y}\right)' E(x,y)\right)^\top$ in $\Omega\times\widetilde{\Omega}$ having symbols of rational type, see (3.8.1) and (3.8.2). Then $u(x)$ and $u_e(x)$ can be extended to $C^\infty(\overline{\Omega})$ and $C^\infty(\overline{\Omega^e})$, respectively, due to Theorem 9.5.8. Moreover, u satisfies (9.8.1) in Ω and $Pu_e=0$ in Ω^e.

With the jumps of their traces across Γ,

$$[\gamma u] = \gamma_c u_e - \gamma u \quad \text{on } \Gamma$$

we find for u and u_e given by (9.8.19),

$$[\gamma u] = -\sum_{j=1}^{m}\sum_{\ell=1}^{2m-1}\sum_{t=0}^{2m-\ell-1}[\gamma\mathcal{K}_t]\mathcal{P}_{t+\ell+1}\mathcal{N}_{\ell+1,m+j}\lambda_j\,; \tag{9.8.20}$$

and with (3.9.35) and (3.9.36) we obtain

$$R[\gamma u] = 0 \quad \text{on } \Gamma\,.$$

On the other hand, (9.8.15) gives

$$A\lambda = -R\gamma u = -R\gamma_c u_e \quad \text{on } \Gamma\,,$$

and

$$\sum_{q,j=1}^{m}\int_\Gamma \overline{\lambda}_q A_{qj}\lambda_j ds = -\int_\Gamma (R\gamma u)^\top\,\overline{\lambda} ds\,.$$

Now, we invoke Assumption A1 in the form of (9.8.8) and obtain

$$\begin{aligned} &\mathrm{Re}\sum_{q,j=1}^{m}\int_\Gamma \overline{\lambda}_q A_{qj}\lambda_j ds \\ &= \mathrm{Re}\left\{a_\Omega(u,u)+a_{\Omega^e}(u_e,u_e)\right\} - \mathrm{Re}\sum_{j=1}^{m}\int_{\Gamma_e}(R_j\gamma u_e)^\top(\overline{S_j\gamma u_e})ds \end{aligned} \tag{9.8.21}$$

where on Γ_e the representation (9.8.19) is to be inserted. Since the integration in (9.8.19) is performed over Γ whereas the traces in (9.8.21) are taken on Γ_e, and the Schwartz kernels of the operators in (9.8.19) are in $C^\infty(\Gamma\times\Gamma_e)$, the mappings

$$\lambda \in \prod_{j=1}^{m} H^{-m+\mu_j+\frac{1}{2}-\varepsilon}(\Gamma) \to (S_j\gamma u_e|_{\Gamma_e}\,,\; R_j\gamma u_e|_{\Gamma_e}) \in L^2(\Gamma_e)\times L^2(\Gamma_e)$$

are continuous. So,

$$\mathrm{Re}\sum_{q,j=1}^{m}\int_\Gamma \lambda_q A_{qj}\overline{\lambda}_j ds \geq \mathrm{Re}\,\{a_\Omega(u,u)+a_{\Omega^e}(u_e,u_e)\}-c\|\lambda\|^2_{\prod\limits_{j=1}^{m} H^{-m+\mu_j+\frac{1}{2}-\varepsilon}(\Gamma)}.$$

Now, (9.8.10) of Assumption A3 implies

$$\begin{aligned}\mathrm{Re}\sum_{q,j=1}^{m}\int_\Gamma \overline{\lambda}_q A_{qj}\lambda_j ds &\geq \beta_0\{\|u\|^2_{H^m(\Omega)}+\|u_e\|^2_{H^m(\Omega^e)}\}\\ &-c\{\|u\|^2_{H^{m-\varepsilon}(\Omega)}+\|u_e\|^2_{H^{m-\varepsilon}(\Omega^e)}+\|\lambda\|^2_{\prod\limits_{j=1}^{m} H^{-m+\mu_j+\frac{1}{2}-\varepsilon}(\Gamma)}\}.\end{aligned}\tag{9.8.22}$$

Whereas in (9.8.20) we were using the mapping $\lambda \mapsto [\gamma u]|_\Gamma$, we now need for the first terms on the right–hand side in (9.8.22) the mapping $[\gamma u]|_\Gamma \mapsto \lambda$. To this end, in addition to (9.8.15), where the traces are obtained for $\Omega \ni x \to \Gamma$, we also use the representation of u_e for $x \in \Omega_e$:

$$\begin{aligned}u_e(x) = \sum_{j=1}^{m}\sum_{\ell=0}^{2m-1}\sum_{t=0}^{2m-\ell-1}&\int_\Gamma \Big\{\Big(\frac{\partial'}{\partial n_y}\Big)^t E(x,y)^\top\Big\}\times\\ &\times \mathcal{P}_{t+\ell+1}\Big\{\mathcal{N}_{\ell+1,j}\varphi^e_j+\mathcal{N}_{\ell+1,m+j}\lambda^e_j\Big\}ds\\ -\sum_{\ell=0}^{2m-1}\sum_{t=0}^{2m-\ell-1}&\int_{\Gamma_e}\Big\{\Big(\frac{\partial'}{\partial n_y}\Big)^t E(x,y)^\top\Big\}\mathcal{P}_{t+\ell+1}\gamma_{e\ell}u ds\,.\end{aligned}\tag{9.8.23}$$

As we can see, all the operators on the right–hand side of (9.8.23) are Poisson operators. Then we apply the trace γ_c for $\Omega^e \ni x \to \Gamma$, subtract (9.8.15) and apply the tangential differential operator S to obtain from (3.9.29) the relation

$$\lambda = S[\gamma u] = B[\gamma u] - C\lambda \tag{9.8.24}$$

where $B = ((B_{jk}))_{\substack{j=1,\dots,m,\\ k=1,\dots,2m}}$ is defined by the tangential operators of S and the traces from Ω^e and Ω, respectively, having symbols of rational type. Hence, with Theorem 9.4.3 and Corollary 9.4.4, the operators $B_{jk} \in \mathcal{L}^{2m-\mu_j-k}_{c\ell}(\Gamma)$ are classical pseudodifferential operators on the manifold Γ.

The operator C in (9.8.22) is obtained from applying $S\gamma_c$ on Γ to the last terms in (9.8.21), which are surface potentials but with charges $\gamma_e u_e$ on Γ_e, which are given by applying γ_e on Γ to the potential u_e in Ω^e in (9.8.19), where λ is given on Γ. Since for $x \in \Gamma$ and $y \in \Gamma_e$ we have $E(\cdot,\cdot) \in C^\infty(\Gamma \times \Gamma_e)$, we conclude that

$$C : \prod_{j=1}^{m} H^{\mu_j - m + \frac{1}{2} - \delta}(\Gamma) \to \prod_{j=1}^{m} H^{\mu_j - m + \frac{1}{2}}(\Gamma)$$

is continuous for every $\delta > 0$. We choose $\delta = \varepsilon$. Thus, (9.8.24) yields an estimate

$$\begin{aligned} \|\lambda\|^2_{\prod\limits_{j=1}^{m} H^{\mu_j - m + \frac{1}{2}}(\Gamma)} &\leq c_1 \|[\gamma u]\|^2_{\prod_{j=1}^{m} H^{\mu_j - k + \frac{1}{2}}(\Gamma)} + c_2 \|\lambda\|^2_{\prod\limits_{j=1}^{m} H^{\mu_j - m + \frac{1}{2} - \varepsilon}(\Gamma)} \\ &\leq 2c_1 \big\{ \|\gamma u\|^2_{\prod_{k=1}^{2m} H^{\mu_j - k + \frac{1}{2}}(\Gamma)} + \|\gamma_c u_e\|^2_{\prod_{k=1}^{2m} H^{\mu_j - k + \frac{1}{2}}(\Gamma)} \big\} \\ &\quad + c_2 \|\lambda\|^2_{\prod\limits_{j=1}^{m} H^{\mu_j - m + \frac{1}{2} - \varepsilon}(\Gamma)} . \end{aligned} \tag{9.8.25}$$

Since u and u_e are solutions of $Pu = 0$ in Ω and $Pu_e = 0$ in Ω^e, respectively, we can apply for their sectional traces Lemma 9.8.1, and finally get

$$\|\lambda\|^2_{\prod\limits_{j=1}^{m} H^{\mu_j - m + \frac{1}{2}}(\Gamma)} \leq c_3 \big\{ \|u\|^2_{H^m(\Omega)} + \|u_e\|^2_{H^m(\Omega^e)} \big\} + c_2 \|\lambda\|^2_{\prod\limits_{j=1}^{m} H^{\mu_j - m + \frac{1}{2} - \varepsilon}(\Gamma)} \tag{9.8.26}$$

with $c_3 > 0$.

For $\|u\|_{H^{m-\varepsilon}(\Omega)}$ and $\|u_e\|_{H^{m-\varepsilon}(\Omega)}$ in (9.8.22) we employ Theorem 9.4.7 for the Poisson operators in (9.8.19) where $s = \mu_j - m + \frac{1}{2} - \varepsilon$ and obtain

$$\|u\|^2_{H^{m-\varepsilon}(\Omega)} + \|u_e\|^2_{H^{m-\varepsilon}(\Omega^e)} \leq \|\lambda\|^2_{\prod\limits_{j=1}^{m} H^{\mu_j - m + \frac{1}{2} - \varepsilon}(\Gamma)} . \tag{9.8.27}$$

Inserting (9.8.26) and (9.8.27) into (9.8.22) we finally obtain

$$\operatorname{Re} \sum_{q,j=1}^{m} \int_{\Gamma} \overline{\lambda_q} A_{qj} \lambda_j ds \geq \beta_1 \|\lambda\|^2_{\prod\limits_{j=1}^{m} H^{\mu_j - m + \frac{1}{2}}(\Gamma)} - c \|\lambda\|^2_{\prod\limits_{j=1}^{m} H^{\mu_j - m + \frac{1}{2} - \varepsilon}(\Gamma)} ,$$

the proposed Gårding inequality (9.8.18).

9.9 Remarks

The treatment of elliptic boundary value problems via their reformulation in terms of integral equations in the domain and on the boundary has a long history going back to G. Green [155], C.F. Gauss [141, 142], C. Neumann [335] and H. Poincaré [350]. More recently, this approach has lead to the general theory of elliptic boundary value problems in terms of systems of pseudodifferential equations in the domain and on the boundary manifold. So, our approach may be seen as a rather particular and explicit example within the general theory. The general approach extending the theory of pseudodifferential operators see, is due to Boutet de Monvel. He developed a complete calculus where all the compositions of pseudodifferential operators satisfying the transmission conditions with trace operators, Poisson operators and Newton potential operators are included. He established a class of operators encompassing the elliptic boundary value problems as well as their solution operators. Moreover, his class of operators is an algebra closed under composition (see e. g., Boutet de Monvel [39, 40, 41], Grubb [159], Eskin [112], Schrohe and Schulze [378, 379], Vishik and Eskin [427]). M. Bach in [19] and Bach, Nazarov and Wendland in [20] used a system of pseudodifferential operators to model and analyse the propagation of cracks. For the general theory of pseudodifferential operators there is a vast amount of literature available, e.g., Agranovich [6], Calderón and Zygmund [61], Chazarain and Piriou [64], Dieudonné [100], Egorov and Shubin [109], Folland [125, 126], Giraud [144], Hörmander [187], Petersen [346], Seeley [384], Taira [412], Taylor [416], Treves [420], Wloka et al [440], Park [344] to name a few.

10. Integral Equations on $\Gamma \subset \mathbb{R}^3$ Recast as Pseudodifferential Equations

The treatment of boundary value problems in Chapter 5 was based on variational principles and lead us to corresponding boundary integral equations in weak formulations. The mapping properties of the boundary integral operators were derived from the variational solutions of the corresponding partial differential equations. This approach is restricted to only those boundary integral operators associated with boundary value problems which can be formulated in terms of general variational principles based on Gårding's inequality.

On the other hand, the boundary integral operators can also be considered as special classes of pseudodifferential operators. In the previous chapters 7 to 9, we have presented some basic properties of pseudodifferential operators. The purpose of this chapter is to apply the basic tools from previous pseudodifferential operator theory to concrete examples of this class of boundary integral equations for elliptic boundary value problems in applications. In particular, the three–dimensional boundary value problems for the Helmholtz equation in scattering theory, the Lamé equations of linear elasticity and the Stokes system will serve as model problems. Two–dimensional problems will be pursued in the next chapter.

For the specific examples in Chapters 2 and 3 our reduction of boundary value problems to boundary integral equations is based on the availability of the explicit fundamental solution E. As is well known, in applications often the fundamental solution cannot be constructed explicitly (see, e.g., Section 7.3 and for anisotropic elasticity in $\mathbb{R}^3$ Natroshvili [321]). However, one may still be able to modify the setting of the reduction from boundary value problems to integral equations either based on the fundamental solution of the principal part of the differential equations if available (see Mikhailov [295]) or on the Levi functions including some parametrix as in Section 7.2 (see Pomp [351]). We will, in the following, refer to the latter as "generalized fundamental solutions". These always generate classical pseudodifferential operators on $\Omega \cup \widetilde{\Omega}$ as will be verified below. These pseudodifferential operators belong to a subclass of the Boutet de Monvel algebra which has its origin in works of Vishik and Eskin [427] and Boutet de Monvel [39, 40, 41] in the 60's for regular elliptic pseudodifferential boundary value problems (see also Grubb [159]). Specifically, following the presentation in [159] let us consider a simple

G. C. Hsiao und W. L. Wendland, *Boundary Integral Equations*,
Applied Mathematical Sciences 164,
https://doi.org/10.1007/978-3-030-71127-6_10

model of this algebra,

$$\begin{aligned} Pu &:= -\Delta u + qu = f \quad \text{in } \Omega \subset \mathbb{R}^3\,, \\ \gamma_0 u &= \varphi \quad \text{on } \Gamma \end{aligned} \tag{10.0.1}$$

with the smooth coefficient $q = q(x) \geq q_0 > 0$ in $\overline{\Omega}$, and a positive constant q_0. Then the unique solution u can be represented in the form

$$u = V\lambda - W\varphi + Nf - Nqu \quad \text{in } \Omega\,, \tag{10.0.2}$$

where V and W are the simple and double layer potentials of the Laplacian, i.e. (1.2.1), (1.2.2), respectively, and N is the Newton potential

$$Nf(x) = \int\limits_{\Omega} E(x,y)f(y)dy \quad \text{for } x \in \Omega$$

in terms of the fundamental solution $E(x,y)$ of the Laplacian given by (1.1.2) for $n = 3$, and λ, φ are the Cauchy data of u in (10.0.1). Then we may rewrite (10.0.2) in the form

$$Au - V\lambda = Nf - W\varphi \quad \text{in } \Omega \tag{10.0.3}$$

by setting

$$Au := Iu + Nqu\,. \tag{10.0.4}$$

Taking the trace of (10.0.3), we obtain the boundary integral equation

$$\gamma_0 Nqu - \gamma_0 V\lambda = \gamma_0 Nf - (I + \gamma_0 W)\varphi \quad \text{on } \Gamma\,. \tag{10.0.5}$$

Hence, the boundary value problem is now reduced to the following coupled system of domain and boundary integral equations,

$$\mathcal{A}\begin{pmatrix} u \\ \lambda \end{pmatrix} := \begin{pmatrix} A\,, & -V \\ \gamma_0 Nq\,, & -\gamma_0 V \end{pmatrix}\begin{pmatrix} u \\ \lambda \end{pmatrix} = \begin{pmatrix} N\,, & -W \\ \gamma_0 N\,, & -(\gamma_0 W + I) \end{pmatrix}\begin{pmatrix} f \\ \varphi \end{pmatrix} \tag{10.0.6}$$

for the unknowns u in Ω and λ on Γ.

The matrix $\mathcal{A}$ of operators on the left–hand side of (10.0.6) is a special simple case of operators belonging to the Boutet de Monvel operator algebra. In terms of his terminology, A in (10.0.4) is a *pseudodifferential operator in Ω* with the symbol $\{|\xi|^2 + q(x)\}^{-1}$ for $|\xi| \geq 1$ which is of rational type and, hence, satisfies the transmission conditions (9.3.40). The simple layer potential V is a *Poisson operator*, $\Upsilon = \gamma_0 Nq$ is called a *trace operator* and $-\gamma_0 V$ is a *pseudodifferential operator* on Γ. The matrix of operators in (10.0.6) is called a general *Green's operator* on Ω. The operator $\mathcal{A}$ in (10.0.6) is invertible and we have

$$\mathcal{A}^{-1} = \begin{pmatrix} B^{-1} & , -B^{-1}V(\gamma_0 V)^{-1} \\ (\gamma_0 V)^{-1}\gamma_0 N_q B^{-1} & , -(\gamma_0 V)^{-1}(I + \gamma_0 N_q B^{-1} V(\gamma_0 V)^{-1}) \end{pmatrix} \tag{10.0.7}$$

where

$$B = \big(A - V(\gamma_0 V)^{-1}\gamma_0 N_q\big) \quad \text{and} \quad N_q g := N(qg)\,; \tag{10.0.8}$$

also belongs to the algebra.

By following Grubb's example in [159, 160], we may also formulate the general Green operator associated with the boundary value problem (10.0.1):

$$\mathcal{A} = \begin{pmatrix} P_\Omega + G & ,\mathcal{K} \\ \Upsilon & ,Q \end{pmatrix} \tag{10.0.9}$$

where we have

$$P_\Omega u := (-\Delta \widetilde{u} + q\widetilde{u})|_\Omega \tag{10.0.10}$$

with $\widetilde{u}$ the extension of u by zero on $\mathbb{R}^3 \setminus \Omega$; $\Upsilon u = \gamma_0 u$ on Γ; the Poisson operator is given by

$$\mathcal{K} = B^{-1}\{V(\gamma_0 V)^{-1}(I + \gamma_0 W) - W\}\,, \tag{10.0.11}$$

and Q is a pseudodifferential operator on the boundary, defined by $\Upsilon\mathcal{K} = \gamma_0\mathcal{K} = I$, the identity in this example. The singular Green's operator G on Ω is here defined by

$$G = B^{-1}\{ - V(\gamma_0 V)^{-1}\gamma_0 N + N\} - N\,. \tag{10.0.12}$$

The above simple example shows that the general Green operators in (10.0.6) and (10.0.9) exhibit all of the essential features in the Boutet de Monvel theory which are formed in terms of compositions of boundary and domain integral operators, their inverses and traces. These also provide the basis of the boundary integral equation approach.

Similarly, in general, in terms of generalized fundamental solutions, the reduction to boundary integral equations can still be achieved (see e.g. Mikhailov [295]).To be more specific, let $F(x,y)$ be a generalized fundamental solution for the differential equation (3.9.1) in $\Omega \cup \widetilde{\Omega}$ together with the boundary conditions (3.9.2). If the distribution φ in (3.7.10) is replaced by F instead of E then the representation formula (3.9.18) is replaced by the generalized representation formula for the solution of the partial differential equation under consideration, i.e.

$$\begin{aligned} u(x) - \int\limits_{\Omega\setminus\{x\}} T(x,y)u(y)dy &= \int\limits_\Omega F(x,y)f(y)dy \\ - \sum_{\ell=0}^{2m-1}\sum_{p=0}^{2m-\ell-1} \mathcal{K}_p\mathcal{P}_{p+\ell+1}\Big\{\sum_{j=1}^m \mathcal{N}_{\ell+1,j}\varphi_j &+ \sum_{j=1}^m \mathcal{N}_{\ell+1,m+j}\lambda_j\Big\}(x)\,. \end{aligned} \tag{10.0.13}$$

Here we tacitly employed the property

$$P^\top_{(y)} F(x,y) = \delta(x-y) - T(x,y)$$

for generalized fundamental solutions. In particular, for $F = E$ one has $T(x,y) = 0$ and recovers the representation formula (3.9.29), whereas for a parametrix F, the kernel $T(x,y)$ is C^∞; for a Levi function of degree $j \geq 0$, we have $T(x,y) = T_{j+1}(x,y)$ with $T \in \Psi hk_{j-1}$ and the operator $\underset{\sim}{T}_{j+1} \in \mathcal{L}^{-j-2}(\Omega \cup \widetilde{\Omega})$ (see (7.2.42) and (7.2.43)).

The operators $\mathcal{K}_p$ now are defined in (3.8.2) with E replaced by F. For ease of reading, we abbreviate Equation (10.0.13) as follows:

$$u(x) - \underset{\sim}{T} u(x) = \underset{\sim}{N} f(x) + \mathfrak{V}\lambda(x) - \mathfrak{W}\varphi(x) \quad \text{for } x \in \Omega\,. \tag{10.0.14}$$

In the same manner as in Section 3.9, we apply the boundary operators $R\gamma$ and $S\gamma$ to equation (10.0.14) and obtain (in addition to (10.0.14)) the two sets of boundary integral equations corresponding to (3.9.30):

$$\varphi - R\gamma\underset{\sim}{T} u = R\gamma\underset{\sim}{N} f + R\gamma\mathfrak{V}\lambda - R\gamma\mathfrak{W}\varphi\,, \tag{10.0.15}$$

$$\lambda - S\gamma\underset{\sim}{T} u = S\gamma\underset{\sim}{N} f + S\gamma\mathfrak{V}\lambda - S\gamma\mathfrak{W}\varphi\,. \tag{10.0.16}$$

For solving the boundary value problem (3.9.1), (3.9.2) with given f in Ω and φ on Γ, we now have at least two choices.

Coupled domain and boundary integral equations

Equations (10.0.14) together with (10.0.15) define a coupled system of domain and boundary integral equations,

$$\begin{pmatrix} I - \underset{\sim}{T} & , & -\mathfrak{V} \\ R\gamma\underset{\sim}{T} & , & R\gamma\mathfrak{V} \end{pmatrix} \begin{pmatrix} u \\ \lambda \end{pmatrix} = \begin{pmatrix} \underset{\sim}{N} & , & -\mathfrak{W} \\ -R\gamma\underset{\sim}{N} & , & I + R\gamma\mathfrak{W} \end{pmatrix} \begin{pmatrix} f \\ \varphi \end{pmatrix}, \tag{10.0.17}$$

whereas from (10.0.14) and (10.0.16) we arrive at the system

$$\begin{pmatrix} I - \underset{\sim}{T}, & -\mathfrak{V} \\ -S\gamma\underset{\sim}{T}, & I - S\gamma\mathfrak{V} \end{pmatrix} \begin{pmatrix} u \\ \lambda \end{pmatrix} = \begin{pmatrix} \underset{\sim}{N}, & -\mathfrak{W} \\ S\gamma\underset{\sim}{N}, & -S\gamma\mathfrak{W} \end{pmatrix} \begin{pmatrix} f \\ \varphi \end{pmatrix}. \tag{10.0.18}$$

We seek the solution $u \in H^{s+m}(\Omega)$ with the given and unknown Cauchy data φ and λ in the spaces

$$R\gamma u = \varphi \in \prod_{j=1}^{m} H^{m+s-\mu_j-\frac{1}{2}}(\Gamma) \quad \text{and}$$

$$S\gamma u = \lambda \in \prod_{j=1}^{m} H^{-m+s+\mu_j+\frac{1}{2}}(\Gamma)\,,$$

respectively. R and S are given rectangular matrices of tangential differential operators along Γ and extended to $\widetilde{\Omega}$ having the orders $\mathrm{order}(R_{jk}) = \mu_j - k + 1$ where $0 \le \mu_j \le 2m-1$ and $\mathrm{order}(S_{jk}) = 2m - \mu_j - k$ and satisfying the fundamental assumptions, see (3.9.4) and (3.9.5).

In order to eliminate u in the domain, the Fredholm integral operator $I - \underset{\sim}{T}$ of the second kind must be invertible, which can always be guaranteed by an appropriate choice of F; and since $\underset{\sim}{T}$ becomes a compact integral operator on various solution spaces on Ω.

On the other hand, the boundary integral equations (10.0.15) and (10.0.16) for the unknown Cauchy datum λ on Γ are boundary integral equations of the first and second kind, respectively, as in Section 3.9.

To set up a variational formulation for the system (10.0.17) or (10.0.18) , one needs the mapping properties of the corresponding operators which can be obtained by the use of the mapping properties of the corresponding pseudodifferential operators in the domain, their composition with differential operators and with traces on the boundary. We may classify these operators in the following categories:

Mapping properties

1) The pseudodifferential operators $\underset{\sim}{N} \in \mathcal{L}^{-2m}_{c\ell}(\Omega \cup \widetilde{\Omega})$ and $\underset{\sim}{T} \in \mathcal{L}^{-j-2}_{c\ell}(\Omega \cup \widetilde{\Omega})$ map distributions defined on Ω into itself. They have the mapping properties

$$\begin{array}{llll} \underset{\sim}{N} & : H^{-m+s}(\Omega) & \to H^{m+s}(\Omega) & \text{if } s > m - \tfrac{1}{2}, \\ \underset{\sim}{T} & : H^{m+s}(\Omega) & \to H^{m+s+j+2}(\Omega) \hookrightarrow H^{m+s}(\Omega) & \text{if } s > -m - \tfrac{1}{2}, \end{array} \tag{10.0.19}$$

(see Theorem 9.6.1) where the last imbedding is compact since $j \ge 0$.

2) The operators $\mathfrak{V}$ and $\mathfrak{W}$ are surface potentials defining Poisson operators which map distributions with support on Γ into distributions defined on $\Omega \cup \widetilde{\Omega}$. They have the following mapping properties (see Theorem 9.5.8):

$$\mathfrak{V} \; : \; \prod_{j=1}^{m} H^{-m+s+\mu_j+\frac{1}{2}}(\Gamma) \to H^{m+s}(\Omega)\,, \tag{10.0.20}$$

$$\mathfrak{W} \; : \; \prod_{j=1}^{m} H^{m+s-\mu_j-\frac{1}{2}}(\Gamma) \to H^{m+s}(\Omega)\,. \tag{10.0.21}$$

3) The trace operators defined on Γ by traces of pseudodifferential operators on $\Omega \cup \widetilde{\Omega}$ (see (3.9.3) and Theorem 9.6.2):
Let $s_0 := \mu_m + \frac{1}{2} - m$ and $\widetilde{s}_0 := \max\{s_0, m - \frac{1}{2}\}$. Then

$$
\begin{aligned}
R\gamma \underset{\sim}{N} &: \widetilde{H}^{-m+s}(\Omega) \to \prod_{j=1}^{m} H^{m+s-\mu_j-\frac{1}{2}}(\Gamma) \quad \text{for } s > s_0\,,\\
R\gamma \underset{\sim}{N} &: H^{-m+s}(\Omega) \to \prod_{j=1}^{m} H^{m+s-\mu_j-\frac{1}{2}}(\Gamma) \quad \text{for } s > \widetilde{s}_0\,,
\end{aligned}
\tag{10.0.22}
$$

Correspondingly, with $s_1 := \nu_1 + \frac{1}{2} - m$ and $\widetilde{s}_1 := max\{s_1, m - \frac{1}{2}\}$ we have

$$
\begin{aligned}
S\gamma \underset{\sim}{N} &: \widetilde{H}^{-m+s}(\Omega) \to \prod_{j=1}^{m} H^{-m+s+\mu_j+\frac{1}{2}}(\Gamma) \quad \text{for } s > s_1\,,\\
S\gamma \underset{\sim}{N} &: H^{-m+s}(\Omega) \to \prod_{j=1}^{m} H^{-m+s+\mu_j+\frac{1}{2}}(\Gamma) \quad \text{for } s > \widetilde{s}_1\,;
\end{aligned}
\tag{10.0.23}
$$

$$
\begin{aligned}
R\gamma \underset{\sim}{T} &: \widetilde{H}^{m+s}(\Omega) \to \prod_{j=1}^{m} H^{m+s-\mu_j-\frac{1}{2}}(\Gamma) \quad \text{compactly for } s > s_1\,,\\
R\gamma \underset{\sim}{T} &: H^{m+s}(\Omega) \to \prod_{j=1}^{m} H^{m+s-\mu_j-\frac{1}{2}}(\Gamma) \quad \text{compactly for } s > \widetilde{s}_1\,;
\end{aligned}
\tag{10.0.24}
$$

$$
\begin{aligned}
S\gamma \underset{\sim}{T} &: \widetilde{H}^{m+s}(\Omega) \to \prod_{j=1}^{m} H^{-m+s+\mu_j+\frac{1}{2}}(\Gamma) \quad \text{compactly for } s > s_1\,,\\
S\gamma \underset{\sim}{T} &: H^{m+s}(\Omega) \to \prod_{j=1}^{m} H^{-m+s+\mu_j+\frac{1}{2}}(\Gamma) \quad \text{compactly for } s > \widetilde{s}_1\,.
\end{aligned}
\tag{10.0.25}
$$

Note that all of these mapping properties remain valid if Ω is replaced by $\Omega^c \cap K$ with any compact $K \Subset \mathbb{R}^n$.

4) Pseudodifferential operators on Γ involving traces of Poisson operators, namely:

$$\gamma\mathfrak{V} \in \mathcal{L}_{c\ell}^{-2m+2\mu_j+1}(\Gamma) \,:\, \prod_{j=1}^{m} H^{-m+s+\mu_j+\frac{1}{2}}(\Gamma) \to \prod_{j=1}^{m} H^{m+s-\mu_j-\frac{1}{2}}(\Gamma)\,; \tag{10.0.26}$$

$$S\gamma\mathfrak{V} \in \mathcal{L}_{c\ell}^{0}(\Gamma) \,:\, \prod_{j=1}^{m} H^{-m+s+\mu_j+\frac{1}{2}}(\Gamma) \to \prod_{j=1}^{m} H^{m+s+\mu_j+\frac{1}{2}}(\Gamma)\,; \tag{10.0.27}$$

$$R\gamma\mathfrak{W} \in \mathcal{L}_{c\ell}^{0}(\Gamma) \,:\, \prod_{j=1}^{m} H^{m+s-\mu_j-\frac{1}{2}}(\Gamma) \to \prod_{j=1}^{m} H^{m+s-\mu_j-\frac{1}{2}}(\Gamma)\,; \tag{10.0.28}$$

$$RS\gamma\mathfrak{W} \in \mathcal{L}_{c\ell}^{2m-2\mu_j-1}(\Gamma) \,:\, \prod_{j=1}^{m} H^{m+s-\mu_j-\frac{1}{2}}(\Gamma) \to \prod_{j=1}^{m} H^{-m+s+\mu_j+\frac{1}{2}}(\Gamma)\,, \tag{10.0.29}$$

for $s \in \mathbb{R}$ (see Theorems 9.5.5 and 9.5.7).

All the operators in equations (10.0.17) and (10.0.18) are intimately related to the properties of the generalized Newton potentials having symbols of rational type. Their actions on distributions supported by the boundary surface as in Theorems 9.4.3 and 9.5.1 generate all of the boundary integral operators needed for the solution of elliptic boundary problems as in (10.0.17) or (10.0.18). R. Duduchava considered these systems in [104] in more general Sobolev and Besov spaces.

The rest of the chapter is devoted to the concrete examples of boundary integral operators in applications. We begin with the Newton potential defined by fundamental solutions, Levi functions or parametrices. These integral operators with Schwartz kernels are pseudodifferential operators in $\mathbb{R}^3$. Next, we consider the traces of Newton potentials in relation to suitable function spaces for the weak solution of boundary value problems. The boundary potentials can be considered as Newton potentials applied to distributions supported on the boundary manifold. Hence, the jump relations may be computed classically as well as by using pseudodifferential calculus.

Moreover, for the boundary integral operators generated by boundary value problems, we find their symbols and invariance properties (Kieser's theorem). Finally, we present the computation of the solution's derivatives for the corresponding boundary integral equations.

10.1 Newton Potential Operators for Elliptic Partial Differential Equations and Systems

The latter approach can be presented schematically. For illustration let us consider the elliptic partial differential operator P of order $2m$ given by

$$Pu(x) = (-1)^m \sum_{|\alpha|\le 2m} a_\alpha(x) D^\alpha u(x) = f(x) \quad \text{in } \Omega \cup \widetilde{\Omega} \subset \mathbb{R}^n \tag{10.1.1}$$

where $\widetilde{\Omega}$ is the tubular neighbourhood of $\Gamma = \partial\Omega$ as defined in Section 9.2. The ellipticity condition reads here

$$\sum_{|\alpha|=2m} a_\alpha(x)\xi^\alpha \ne 0 \quad \text{for all } \xi \in \mathbb{R}^n \quad \text{with } |\xi| = 1 \quad \text{and } x \in \Omega \cup \widetilde{\Omega}\,.$$

The operator P is obviously a pseudodifferential operator of order $2m$ with the polynomial symbol

$$\sigma_P(x,\xi) = (-1)^m \sum_{|\alpha|\le 2m} a_\alpha(x)(i\xi)^\alpha\,.$$

Clearly, the inverse $\big(\sigma_P(x,\xi)\big)^{-1}$ is of rational type. In particular, if the coefficients a_α are constants, then $\big(\sigma_P(\xi)\big)^{-1}$ is the symbol of the parametrix which also corresponds to the Newton potential operator defined by a fundamental solution. As for the Levi function of order $j \in \mathbb{N}_0$, the corresponding Newton potential is given by (7.2.47), i.e.

$$\underset{\sim}{L} f(x) := \int_{\Omega\cup\widetilde{\Omega}} L_j(x,y) f(y)dy = \sum_{\ell=0}^{j} \int_{\Omega\cup\widetilde{\Omega}} N_\ell(x,y)f(y)dy = \sum_{\ell=0}^{j} \underset{\sim}{N}_\ell f(x) \tag{10.1.2}$$

where $\underset{\sim}{N}_\ell \in \mathcal{L}_{c\ell}^{-\ell-2m}(\Omega\cup\widetilde{\Omega})$ are pseudodifferential operators due to Lemma 7.2.6. Our goal is to show that all of the Newton potentials mentioned above are pseudodifferential operators with symbols of rational type. More generally, we have the following theorem.

Theorem 10.1.1. *The pseudodifferential operator given by a generalized Newton potential in terms of either a parametrix or Levi function of an elliptic partial differential operator as well as an elliptic system in the sense of Agmon, Douglis and Nirenberg has a symbol of rational type.*

Remark 10.1.1: Since a fundamental solution is a special parametrix, Theorem 10.1.1 is particularly valid for the Newton potential in terms of a fundamental solution.

Proof: We begin with the proof for the case of a parametrix and first consider the scalar elliptic case. From the proof of Theorem 7.2.2 we notice that $q_{-2m}(x,\xi)$ defined by (7.2.8) is of rational type since $a(x,\xi)$ is polynomial. Since all the homogeneous terms in the symbol expansion of the parametrix given by the recursion formula (7.2.9) are finite compositions of differentiation and rational type symbols, Lemma 9.4.2 implies that each term $q_{-2m-j}(x,\xi)$ is also of rational type, by induction.

Now, for an elliptic system in the sense of Agmon–Douglis–Nirenberg (7.2.10), the characteristic determinant (7.2.14) is polynomial. Therefore, all the elements of the matrix–valued symbol $q_{-2m}(x,\xi)$ given by (7.2.8) are of rational type. Hence, the arguments for the rational type of each term in the matrices defined by the recursion formula (7.2.9) remain the same as in the scalar case.

ii) For the Levi functions in the case of scalar elliptic equations we begin with $\underset{\sim}{N}_0$ given by (7.2.35) where Ω is replaced by $\Omega \cup \widetilde{\Omega}$ with the kernel L_0 whose amplitude is

$$a(x,y;\xi) = \big(a^0_{2m}(y,\xi)\big)^{-1}.$$

The latter, clearly is a rational function of ξ depending also on y. Hence the symbol of $\underset{\sim}{N}_0$ is given by the asymptotic expansion (7.1.28) where for each homogeneous term only a finite number of differentiations appears. Then Lemma 9.4.2 implies that $\underset{\sim}{N}_0$ has a symbol of rational type. In order to show that $\underset{\sim}{N}_{\ell+1} \in \mathcal{L}_{c\ell}^{-(\ell+2m+1)}(\Omega \cup \widetilde{\Omega})$ given by (7.2.52) in Lemma 7.2.6 has a symbol of rational type, we use the argument of induction assuming that $\underset{\sim}{N}_\ell$ has a symbol of rational type. Since $\underset{\sim}{T}$ given by (7.2.31) has the polynomial amplitude

$$(-1)^m\Big\{\sum_{|\alpha|=2m}\big(a_\alpha(y)-a_\alpha(x)\big)(i\xi)^\alpha - \sum_{|\alpha|<2m} a_\alpha(x)(i\xi)^\alpha\Big\},$$

$\underset{\sim}{T}$ has a polynomial symbol of order $2m-1$ in view of (7.1.28) in Theorem 7.1.11. Then the operator $\underset{\sim}{T}_m = \underset{\sim}{T} \circ \underset{\sim}{N}_m \in \mathcal{L}_{c\ell}^{-\ell-1}(\Omega \cup \widetilde{\Omega})$ has a symbol of rational type because of Lemma 9.4.2. By the representation of $\underset{\sim}{N}_{\ell+1}$ in (7.2.53) it has the amplitude function given in (7.2.54) which is now of rational type with respect to ξ. Applying Theorem 7.1.11 together with Lemma 9.4.2 we obtain that $\underset{\sim}{N}_{\ell+1}$ has a symbol of rational type.

In view of the definition of the Newton potential in (10.1.2), also $\underset{\sim}{L}_j$ has a symbol of rational type.

To complete the proof it remains to show that the Newton potentials in terms of Levi functions for elliptic systems in the sense of Agmon–Douglis–Nirenberg again have a symbol of rational type. Now, the characteristic determinant $H_{2m}(x,\xi)$ in Section 7.2.4 is a polynomial in ξ of degree $2m$ whereas the cofactor matrix differential operator symbols $B_{k\ell}(x,\xi)$ are polynomials of orders $2m - t_k - s_\ell$. Hence, the matrix–valued amplitude $a_0^{-1}(y;\xi)$ is of rational type and defines a matrix pseudodifferential operator $\underset{\sim}{a}_0^{(-1)}$ with a matrix symbol of rational type elements as can be seen by repeating the arguments as for the scalar case. Consequently, the matrix–valued pseudodifferential operator $\underset{\sim}{N}_0$ given by the Schwartz kernel $N^{(0)}(x,y)$ in (7.2.57) has a symbol matrix with elements of rational type. Again, the pseudodifferential operator $\underset{\sim}{T}$ corresponding to

$$T(y;-iD) = a_0(y;-iD) - a(x;-iD)$$

has an amplitude matrix with polynomial elements and, hence, has a symbol matrix with polynomial elements. By similar arguments as in the scalar case, the pseudodifferential operator $\underset{\sim}{N}^{(\ell+1)}$ defined by the recursion relation

$$\underset{\sim}{N}^{(\ell+1)} := \underset{\sim 0}{a}^{(-1)} \circ \underset{\sim}{T} \circ \underset{\sim}{N}^{(\ell)}$$

now has a symbol matrix with elements of rational type provided $\underset{\sim}{N}^{(\ell)}$ has a symbol matrix of rational type. Thus, by induction, the proposed assertion can be established which completes the proof. ■

Continuity of the trace of Newton potentials

As a consequence of Theorem 10.1.1, since all the generalized Newton potential operators $A \in \mathcal{L}_{c\ell}^{-2m}(\Omega \cup \widetilde{\Omega})$ are classical pseudodifferential operators with symbols of rational type, in addition to Theorem 7.1.12 one may apply Theorems 9.6.1 and 9.6.2 to A and obtain in the scalar case the mapping properties as follows:

$$\|\gamma_0 A f\|_{H^{2m-\frac{1}{2}+\sigma}(\Gamma)} \leq c\|f\|_{\widetilde{H}^\sigma(\Omega)} \quad \text{for } \sigma > -2m + \tfrac{1}{2}\,, \tag{10.1.3}$$

$$\|\gamma_0 A f\|_{H^{2m-\frac{1}{2}+\sigma}(\Gamma)} \leq c\|f\|_{H^\sigma(\Omega)} \quad \text{for } \sigma > \sigma_0 \tag{10.1.4}$$

$$\text{where } \sigma_0 = \max\{-\tfrac{1}{2}, -2m + \tfrac{1}{2}\}\,.$$

In the case of an elliptic system in the sense of Agmon–Douglis–Nirenberg, the appropriate mapping properties for each of the elements of the matrix Newton potential operator can be obtained by again using Theorems 9.6.1 and 9.6.2 in addition to Theorem 7.1.12.

10.1.1 Generalized Newton Potentials for the Helmholtz Equation

To illustrate the different generalized Newton potentials in terms of the fundamental solution, parametrices and Levi functions we use the inhomogeneous Helmholtz equation (see Section 2.1),

$$Pu := -(\Delta + \mathsf{k}^2)u = f \quad \text{in } \Omega \cup \widetilde{\Omega} \subset \mathbb{R}^3 \tag{10.1.5}$$

where the fundamental solution which satisfies the radiation condition (2.1.2) is given explicitly in (2.1.4) as

$$E_{\mathsf{k}}(x,y) = \frac{e^{i\mathsf{k}|z|}}{4\pi|z|} \quad \text{with } z = x - y\,.$$

So, the classical Newton potential is of the form

$$Nf(x) = \int\limits_{\mathrm{I\!R}^3} E_{\mathsf{k}}(x,y)f(y)dy \tag{10.1.6}$$

where f has compact support in $\Omega \cup \widetilde{\Omega} \subset \mathrm{I\!R}^3$.

Symbol and kernel expansions

Since the Helmholtz operator P is a scalar strongly second order operator with constant coefficients with the symbol $\sigma_P = |\xi|^2 - \mathsf{k}^2$ for $\xi \in \mathrm{I\!R}^3$, the Newton potential N in (10.1.6) is a classical pseudodifferential operator $N \in \mathcal{L}^{-2}_{c\ell}(\mathrm{I\!R}^3)$. In this case, we also have the complete symbol

$$\sigma_N = \frac{1}{|\xi|^2 - \mathsf{k}^2} = \sum_{p=0}^{\infty} \frac{\mathsf{k}^{2p}}{|\xi|^{2+2p}} = \sum_{p=0}^{\infty} a^0_{-2-2p}(\xi) \quad \text{for } |\xi| > \mathsf{k}\,, \tag{10.1.7}$$

which defines the classical symbol of order -2 as in (7.1.19) and, obviously, is of rational type. As in Section 7.2, we use the cut–off function χ in (7.2.31) to define a parametrix $H \in \mathcal{L}^{-2}_{c\ell}(\mathrm{I\!R}^2)$ by

$$Hf := \mathcal{F}^{-1}_{\xi \mapsto x}\Big(\Big(\chi\Big(\frac{1}{|\xi|^2 - \mathsf{k}^2}\Big)\mathcal{F}_{y\mapsto\xi} f(y)\Big)\,. \tag{10.1.8}$$

We remark that in this case, as we shall see, the difference $N - H$ between the Newton potential and the parametrix is a smoothing operator. From the homogeneous expansion σ_N in (10.1.7), the relation (8.1.49) leads to the homogeneous kernel

$$k_{-1+2p}(z) = (2\pi)^{-2}\ \text{p.f.} \int\limits_{\mathrm{I\!R}^3} e^{i\xi\cdot z} a^0_{-2-2p}(\xi)d\xi\,. \tag{10.1.9}$$

By using the explicit Fourier transform in Gelfand and Shilov [143, p. 363] we obtain

$$k_{-1+2p}(z) = \mathsf{k}^{2p}(2\pi)^{-3}2^{-2p-\frac{1}{2}}(2\pi)^{3/2}\frac{\Gamma(\frac{1}{2}-p)}{\Gamma(p+1)}|z|^{2p-1}$$

and, with the properties of the gamma–function Γ,

$$\begin{aligned} \frac{\Gamma(\frac{1}{2}-p)}{\Gamma(p+1)} &= (-1)^p\sqrt{\pi}\frac{1}{(2p)!}2^{2p}\,, \\ k_{-1+2p}(z) &= \mathsf{k}^{2p}(-1)^p\frac{1}{4\pi}\frac{1}{(2p)!}|z|^{2p-1}\,. \end{aligned} \tag{10.1.10}$$

This defines the asymptotic pseudohomogeneous kernel expansion

$$\sum_{p=0}^{\infty} k_{-1+2p}(z) = \frac{1}{4\pi} \sum_{p=0}^{\infty} \frac{(-\mathsf{k}^2)^p}{(2p)!} |z|^{2p-1} \tag{10.1.11}$$

where $z = x - y$. If we use only a finite number of terms in the expansion (10.1.11) we obtain the Levi functions

$$L_{2q}(x,y) = \frac{1}{4\pi} \sum_{p=0}^{q} \frac{(-\mathsf{k}^2)^p}{(2p)!} |x-y|^{2p-1} \tag{10.1.12}$$

corresponding to (7.2.46) with $j = 2q$. The associated volume potentials in terms of Levi functions of order $2q$ are given by

$$\underset{\sim 2q}{L} f(x) = \frac{1}{4\pi} \sum_{p=0}^{q} \int_{\mathbb{R}^3} \frac{(-\mathsf{k}^2)^p}{(2p)!} |x-y|^{2p-1} f(y) dy\,. \tag{10.1.13}$$

We observe that the asymptotic expansion in (10.1.11) even converges for all $z \in \mathbb{R}^3 \setminus \{0\}$ and defines a Schwartz kernel

$$k(z) := \sum_{p=0}^{\infty} \frac{1}{4\pi} \frac{(-\mathsf{k}^2)^p}{(2p)!} |z|^{2p-1}$$

with $k \in \Psi hk_{-1}(\mathbb{R}^3)$. Then Theorem 8.1.1 implies that the integral operator

$$\underset{\sim}{L} f(x) := \int_{\mathbb{R}^3} k(x-y) f(y) dy$$

is a classical pseudodifferential operator of order -2.

On the other hand, the fundamental solution $E_{\mathsf{k}}(x,y)$ has a series expansion in the form

$$\begin{aligned} E_{\mathsf{k}}(x,y) &= \frac{1}{4\pi} \sum_{p=0}^{\infty} \frac{(-\mathsf{k}^2)^p}{(2p)!} |z|^{2p-1} + \frac{i\mathsf{k}}{4\pi} \sum_{p=0}^{\infty} \frac{(-\mathsf{k}^2)^p}{(2p+1)!} |z|^{2p} \\ &= \frac{1}{4\pi} \frac{1}{|z|} \cos(\mathsf{k}|z|) + \frac{i\mathsf{k}}{4\pi} \frac{\sin(\mathsf{k}|z|)}{\mathsf{k}|z|}\,. \end{aligned} \tag{10.1.14}$$

This shows that the difference

$$E_{\mathsf{k}}(x,y) - k(x-y) = \frac{i\mathsf{k}}{4\pi} \mathrm{sinc}(\mathsf{k}|x-y|)$$

defines a C^∞–kernel function, namely $\mathrm{sinc}(\mathsf{k}|x-y|) = \dfrac{\sin(\mathsf{k}|x-y|)}{\mathsf{k}|x-y|}$.

Clearly, the real part $k(x-y) = \operatorname{Re} E_{\mathsf{k}}(x,y)$ also is a fundamental solution which, however, does not satisfy the Sommerfeld radiation condition (2.1.2). Therefore, by using the cut–off function $\psi(z)$ as in Theorem 7.1.16 with

$\psi(z) = 1$ for $|z| \le \frac{1}{2}$ and $\psi(z) = 0$ for $|z| \ge 1$, the Newton potential can be decomposed in the form

$$\begin{aligned} Nf(x) &= \int\limits_{\mathbb{R}^3} k(x-y)f(y)dy + \frac{i\mathsf{k}}{4\pi}\int\limits_{\mathbb{R}^3} \operatorname{sinc}(\mathsf{k}|x-y|)f(y)dy \\ &= \int\limits_{\mathbb{R}^3} k(x-y)\psi(k|x-y|)f(y)dy \\ &\quad + \int\limits_{\mathbb{R}^3} \{k(x-y)\big(1-\psi(|x-y|) + \frac{i\mathsf{k}}{4\pi}\operatorname{sinc}(\mathsf{k}|x-y|)\}f(y)dy \\ &= N_0 f(x) + Rf(x) \end{aligned}$$

where the integral operator N_0 is properly supported and R is a smoothing operator corresponding to Theorem 7.1.9. By applying Theorem 7.1.7 to $N_0 \in OPS^{-2}(\mathbb{R}^3 \times \mathbb{R}^3)$ we find the symbol $a(x,\xi)$ of N_0 in the form

$$\begin{aligned} a(x,\xi) &= e^{-ix\cdot\xi}(N_0 e^{i\xi\bullet})(x) \\ &= \sum_{p=0}^{\infty} \frac{1}{4\pi} e^{-ix\cdot\xi} \frac{(-\mathsf{k}^2)^p}{(2p)!} \int\limits_{\mathbb{R}^3} |x-y|^{2p-1}\psi(|x-y|)e^{i\xi\cdot y}dy\,. \end{aligned}$$

Since N_0, N and H all coincide modulo smoothing operators, the asymptotic expansions of their symbols belong to the same equivalence class, i.e., the complete symbol class is characterized by the homogeneous symbol expansion

$$\sum_{p=0}^{\infty} a^0_{-2-2p}(\xi) = \sum_{p=0}^{\infty} \frac{\mathsf{k}^{2p}}{|\xi|^{2+2p}} \sim a(x,\xi) \in \boldsymbol{S}^{-2}_{c\ell}(\mathbb{R}^3 \times \mathbb{R}^3) \tag{10.1.15}$$

as in (10.1.7).

Clearly, N_0, N, H, $\underset{\sim}{L}_{2q}$ and $\underset{\sim}{L}$ all have the mapping properties (10.1.3) and (10.1.4).

10.1.2 The Newton Potential for the Lamé System

We now consider the simplest nontrivial elliptic system of equations in $\mathbb{R}^3$ in applications, namely the Lamé system (2.2.1),

$$Pu = -\Delta^* u = -\mu\Delta u - (\lambda+\mu)\text{graddiv}u = f \quad \text{in } \Omega \cup \widetilde{\Omega} \subset \mathbb{R}^3\,. \tag{10.1.16}$$

Its quadratic symbol matrix can be calculated from (5.4.12) as

$$\sigma_P(\xi) = -\mathcal{G}^\top(i\xi)\widetilde{\mathcal{C}}\mathcal{G}(i\xi) = \big((\mu|\xi|^2\delta_{jk} + (\mu+\lambda)\xi_j\xi_k)\big)_{3\times 3}\,. \tag{10.1.17}$$

Hence, the characteristic determinant is given by

$$\det \sigma_P(\xi) = \mu^2(\lambda + 2\mu)|\xi|^6 , \tag{10.1.18}$$

P is strongly elliptic and the inverse to $\sigma_P(\xi)$ defines the symbol of the Newton potential, i.e.

$$\big(\sigma_P(\xi)\big)^{-1} = \frac{1}{\mu|\xi|^4}\Big(\Big(|\xi|^2\delta_{jk} - \frac{(\lambda+\mu)}{(\lambda+2\mu)}\,\xi_j\xi_k\Big)\Big)_{3\times 3}. \tag{10.1.19}$$

The Fourier inverse of $(\sigma_P)^{-1}$ defines the fundamental matrix $E(x,y)$ for P, see (7.2.18), i.e.

$$\begin{aligned} E(x,y) &= (2\pi)^{-3}\,\text{p.f.}\int\limits_{\mathbb{R}^3} \big(\sigma_P(\xi)\big)^{-1} e^{i(x-y)\cdot\xi} d\xi \\ &= \frac{1}{\mu(2\pi)^3}\Big(\Big(\,\text{p.f.}\int\limits_{\mathbb{R}^3} \frac{1}{|\xi|^2} e^{i(x-y)\cdot\xi} d\xi\delta_{jk} \\ &\qquad + \frac{\lambda+\mu}{\lambda+2\mu}\,\frac{\partial^2}{\partial x_j \partial x_k}\,\text{p.f.}\int\limits_{\mathbb{R}^3} \frac{1}{|\xi|^4} e^{i(x-y)\cdot\xi} d\xi\Big)\Big)_{3\times 3} \\ &= \frac{1}{8\pi\mu}\,\frac{(3\mu+\lambda)}{(\lambda+2\mu)}\Big(\Big(\frac{\delta_{jk}}{|x-y|} + \frac{\lambda+\mu}{3\mu+\lambda}\,\frac{(x_j-y_j)(x_k-y_k)}{|x-y|^3}\Big)\Big)_{3\times 3} \end{aligned} \tag{10.1.20}$$

as in (2.2.2). Since the symbol in (10.1.19) is homogeneous, the fundamental solution defines the Schwartz kernel of the Newton potential. Although one may still define a parametrix as in (10.1.8) by multiplying the homogeneous symbol in (10.1.19) by the cut–off function $\chi(\xi)$ one would not gain anything. As in the scalar case, by using the cut–off function Ψ from Theorem 7.1.16, the Newton potential can be decomposed in the form

$$N\boldsymbol{f} = N_0\boldsymbol{f} + R\boldsymbol{f} =$$

$$\int\limits_{\mathbb{R}^3} E(x,y)\psi(|x-y|)\boldsymbol{f}(y)dy + \int\limits_{\mathbb{R}^3} E(x,y)\big(1-\psi(|x-y|)\big)\boldsymbol{f}(y)dy$$

where $N_0 \in OPS^{-2}(\mathbb{R}^3 \times \mathbb{R}^3)$ is properly supported and R is smoothing.

10.1.3 The Newton Potential for the Stokes System

As we have seen in Section 2.3.3, the fundamental solution of the Stokes system can be obtained from that of the Lamé system by taking the limit $\lambda \to +\infty$. Hence, we recover (2.3.10) from (10.1.20) and also the corresponding symbols from (10.1.19) and (2.3.10) as

$$\sigma_{E_{St}}(\xi) = \frac{1}{\mu|\xi|^4}\Big(\Big(|\xi|^2\delta_{jk} - \frac{1}{2}\xi_j\xi_k\Big)\Big)_{3\times 3} \tag{10.1.21}$$

and

$$\sigma_Q(\xi) = \frac{2i}{|\xi|^2}\xi_j \,. \tag{10.1.22}$$

The corresponding Newton potentials then read

$$\begin{aligned} N_{St}\boldsymbol{f} &= N_{St0}\boldsymbol{f} + N_{St1}\boldsymbol{f} \\ &= \int\limits_{\mathbb{R}^3} \psi(|x-y|)E_{St}(x,y)\boldsymbol{f}(y)dy + \int\limits_{\mathbb{R}^3} (1-\psi(|x-y|)\; E_{St}(x,y)\boldsymbol{f}(y)dy \end{aligned}$$

and

$$Q\boldsymbol{f} = Q_0\boldsymbol{f} + Q_1\boldsymbol{f}$$

with

$$Q_0\boldsymbol{f} = \int\limits_{\mathbb{R}^3} \psi(|x-y|)Q(x,y)\cdot\boldsymbol{f}(y)dy \,.$$

Clearly, from (10.1.21) and (10.1.22) we conclude that $N_{St0} \in OPS^{-2}(\mathbb{R}^3\times\mathbb{R}^3)$ and $Q_0 \in OPS^{-1}(\mathbb{R}^3\times\mathbb{R}^3)$.

10.2 Surface Potentials for Second Order Equations

For the special case of second order systems, $m = 1$, we find two kinds of boundary conditions (3.9.10) and (3.9.12). The representation formula (3.7.10) with the fundamental solution E is of the form

$$\begin{aligned} u(x) &= \int\limits_{\Omega} E(x,y)f(y)dy - \int\limits_{\Gamma} E(x,y)\mathcal{P}_1\gamma_0(y)ds_y \\ &\quad - \int\limits_{\Gamma} E(x,y)\mathcal{P}_2(y)\gamma_1 u(y)ds_y \\ &\quad - \int\limits_{\Gamma} \Big(\frac{\partial}{\partial n_y}^{\top} E(x,y)\Big)^{\top} \mathcal{P}_2(y)\gamma_0 u(y)ds_y \,. \end{aligned} \tag{10.2.1}$$

Here $\dfrac{\partial}{\partial n_y}^{\top} = -\Big(\dfrac{\partial}{\partial n_y} - 2H\Big)$ with $H = -\frac{1}{2}\nabla\cdot n$; the mean curvature of Γ (see (3.5.5)). $\mathcal{P}_1$ and $\mathcal{P}_2$ are given by (3.4.62). Note that, in general, $\mathcal{P}_1$ contains first order tangential differentiation. By using the definition of $\frac{\partial}{\partial n_y}^{\top}$, we find

$$\begin{aligned} u(x) &= \int\limits_{\Omega} E(x,y)f(y)dy \\ &\quad - \int\limits_{\Gamma} E(x,y)\{\mathcal{P}_1\gamma_0 u + 2H\mathcal{P}_2\gamma_0 u + \mathcal{P}_2\gamma_1 u\}ds_y \qquad (10.2.2) \\ &\quad + \int\limits_{\gamma} \Big(\frac{\partial}{\partial n_y}E(x,y)\Big)^\top \mathcal{P}_2\gamma_0 u(y)ds_y\,, \end{aligned}$$

i.e., a volume potential, a simple layer surface potential and a double layer potential.

Hence, the simplest case of $\mathfrak{V}$ and $\mathfrak{W}$ in equations (10.0.20) and (10.0.21) now corresponds to

$$Pu(x) = -\sum_{j,k=1}^{n} \frac{\partial}{\partial x_j}\Big(a_{jk}(x)\frac{\partial u}{\partial x_k}\Big) + \sum_{j=1}^{n} a_j(x)\frac{\partial u}{\partial x_j} + c(x)u$$

with the Dirichlet conditions $R\gamma u = u|_\Gamma$ and the conormal derivative as the complementary boundary operator

$$S\gamma = \sum_{j,k=1}^{n} n_j(x)a_{jk}(x)\frac{\partial}{\partial x_k}|_\Gamma\,.$$

Then the generalized Newton potentials of Section 10.1 have symbols of rational type and generate corresponding simple and double layer potentials supported by Γ.

Simple layer potentials

In this case, the simple layer potential has the form

$$\mathfrak{V}\lambda(x) = N(\lambda\otimes\delta_\Gamma) = \widetilde{Q}_\Gamma\lambda(x) = \int\limits_{y\in\Gamma} N(x, x-y')\lambda(y')ds_\Gamma(y') \qquad (10.2.3)$$

where N is the Schwartz kernel of the generalized Newton potential. As a continuous linear mapping, the surface potential $\widetilde{Q}_\Gamma$ in (10.2.3) has all the mapping properties given in Theorem 9.5.8 for an operator of order -2 in (9.5.19). Moreover, this surface potential $\mathfrak{V}$ has limits for $x \to \Gamma$ from $x \in \Omega$ as well as $x \in \widetilde{\Omega} \cap \Omega^c$ because of Theorem 9.5.1:

$$\begin{aligned} \lim_{x\to x'\in\Gamma} \widetilde{Q}_\Gamma\lambda(x) &= \int\limits_{y'\in\Gamma} N(x', x'-y')\lambda(y')ds_\Gamma(y') \quad \text{for } x' \in \Gamma \\ &=: Q_\Gamma\lambda(x') = V\lambda(x')\,. \qquad (10.2.4) \end{aligned}$$

This boundary integral operator is a pseudodifferential operator $V \in \mathcal{L}_{c\ell}^{-2m+1}(\Gamma)$ with $m = 1$ which is the simple layer boundary integral operator introduced in Chapter 2.

Since the homogeneous principal symbol of the generalized Newton potential is given here by the matrix valued function

$$a^0_{-2}(x,\xi) = \Big\{ \sum_{j,k=1}^{n} a_{jk}(x)\xi_j\xi_k \Big\}^{-1}, \tag{10.2.5}$$

Formula (9.4.7) provides us the principal symbol

$$q^0_{-1}(x',\xi') = \frac{1}{2\pi}\int_{\mathfrak{c}} \{z^2 + bz + a\}^{-1} dz a^{-1}_{nn} \tag{10.2.6}$$

where

$$a = a^{-1}_{nn}\sum_{j,k=1}^{n-1} a_{jk}\xi_j\xi_k\,,\; b = a^{-1}_{nn}\sum_{j=1}^{n-1}(a_{jn}+a_{nj})\xi_j$$

and the curve $\mathfrak{c} \in \mathbb{C}$ is given in Theorem 9.4.3 and oriented clockwise.

Double layer potentials

Since in (10.2.3) we have already considered the simple layer potential, it remains to analyze the double layer potential

$$\begin{aligned}\mathfrak{W}u(x) &= \int_\Gamma \Big(\frac{\partial}{\partial n_y}E(x,y)\Big)^\top \mathcal{P}_2(y)u(y)ds_y \\ &= \int_{\widetilde{\Omega}} \Big(\frac{\partial}{\partial n_y}E(x,y)\Big)^\top \mathcal{P}_2(y)\big(u(y')\otimes\delta_\Gamma\big)dy\end{aligned}$$

for $x \in \widetilde{\Omega}\setminus\Gamma$ in terms of Newton potentials. Here the latter is defined by the Schwartz kernel $\big(\frac{\partial}{\partial n_y}E(x,y)\big)^\top \mathcal{P}_2(y)$ which generates a pseudodifferential operator $A \in \mathcal{L}^{-1}_{c\ell}(\widetilde{\Omega})$. Hence, only its principal part will contribute to the tangential differential operator $\mathcal{T}$ in Theorem 9.5.5 which here is of order 0.

To compute q_0 we apply Theorem 9.4.3 in the tubular coordinates $\varrho \in \widetilde{U}_r$ for which only the principal symbol of A is needed. Now we use the representation of the second order differential operator P in the form (3.4.56), i.e.

$$Pu = \mathcal{P}_0 u + \mathcal{P}_1\frac{\partial u}{\partial n} + \mathcal{P}_2\frac{\partial^2 u}{\partial n^2}$$

where the tangential differential operators $\mathcal{P}_0$ and $\mathcal{P}_1$ of orders 2, respectively, and 1 and the coefficient $\mathcal{P}_2$ are given (3.4.57)–(3.4.59). Hence, the principal symbol of A has the form

$$\sigma_{A_{\Phi_r},0}(x,\xi) = -i\xi_n\{p_0(x,\xi') + p_1(x,\xi')\xi_n + \xi_n^2\}^{-1} \tag{10.2.7}$$

where $p_0(x,\xi')$ is the principal part of $-\mathcal{P}_0(x,i\xi')/\mathcal{P}_2(x)$ and $p_1(x,\xi')$ the principal part of $i\mathcal{P}_1(x,i\xi')/\mathcal{P}_2(x)$. Furthermore, $x = \Psi(\varrho)$ with $\varrho = (\varrho',\varrho_k) \in \widetilde{U}_r$. Hence, q_0 is given by (9.4.7), i.e.

$$q_0(\varrho',\xi') = -\frac{i}{(2\pi)}\int\limits_{\mathfrak{c}} \{p_0(x,\xi') + p_1(x,\xi')z + z^2\}^{-1} z\,dz \tag{10.2.8}$$

where $x = T_r(\varrho')$ and $\mathfrak{c}$ is the contour defined by (9.4.9) circumventing the pole in the lower half–plane clockwise.

10.2.1 Strongly Elliptic Differential Equations

The simple layer potential for the scalar equation

If (10.1.2) is a scalar, real elliptic equation of second order, then it is even a strongly elliptic equation where we have

$$4a - b^2 \geq \gamma_0|\xi'|^2 \quad \text{with } \gamma_0 > 0\,. \tag{10.2.9}$$

Then the residue theorem implies

$$q^0_{-1} = a_{nn}(x')\Big\{4a_{nn}(x')\sum_{j,k=1}^{n-1} a_{jk}(x')\xi_j\xi_k - \Big(\sum_{j=1}^{n-1}(a_{jn}+a_{nj}\xi_j)\Big)^2\Big\}^{-\frac{1}{2}}. \tag{10.2.10}$$

More generally, if Γ is a given surface, one may prefer to use tubular coordinates (3.3.7). However, since here we only used the homogeneous principal symbol, the conclusion remains the same and the principal symbol in the tubular coordinates can be obtained by the transformation formula (7.1.49). We shall return to the tubular coordinates in the next section.

As a consequence of (10.2.9) we conclude the following lemma.

Lemma 10.2.1. *For a scalar strongly elliptic second order partial differential operator* (10.1.1)*, the simple layer boundary integral operator $V \in \mathcal{L}^{-1}_{c\ell}(\Gamma)$ is strongly elliptic (see* (7.2.62)*) in the sense that there exists a function $\Theta(x') \neq 0$ on Γ and a positive constant $\gamma_0 > 0$ such that*

$$\mathrm{Re}\,\Theta(x')\sigma^0_V(x',\xi') \geq \gamma_0 \quad \textit{for all } \xi' \in \mathbb{R}^{n-1} \quad \textit{with } |\xi'| = 1\,.$$

The simple layer potential for strongly elliptic second order systems

In order to guarantee strong ellipticity for the Newton potential, i.e. for the principal symbol a^0_{-2} of a second order system in (10.2.5), we establish the following lemma.

Lemma 10.2.2. *Let the adjoint differential operator P^**

$$P^*v(x) = -\sum_{j=1}^{n}\frac{\partial}{\partial x_j}\Big(a^*_{jk}(x)\frac{\partial v}{\partial x_k}\Big) - \sum_{j=1}^{n}\frac{\partial}{\partial x_j}(a^*_j v) + c^* v \tag{10.2.11}$$

be strongly elliptic. Then the Newton potential with principal symbol given in (10.2.5) *is strongly elliptic.*

Proof: With $A := ((a_{jk}))_{np\times np}$, which is invertible since A^* is invertible, for any given $\zeta \in \mathbb{C}$ we choose $\eta = A^\top \zeta$ and find with $\Theta_{-1}(x) := A^{-1}\Theta$ where Θ belongs to A^* in (7.2.62),

$$\begin{aligned}\mathrm{Re}\,\eta^\top \Theta_{-1} A^{-1}\overline{\eta} &= \zeta^\top A(A^{-1}\Theta)A^*\overline{\zeta}\\ &= \mathrm{Re}\,\zeta^\top \Theta A^* \overline{\zeta} \geq \gamma_0|\zeta|^2 \geq \gamma_0'|\eta|^2 \quad \text{for all } \eta \in \mathbb{C}^n\end{aligned}$$

since $|\eta|^2 \leq c|\zeta|^2$. Hence, A^{-1} is strongly elliptic and $\gamma_0' = \frac{\gamma_0}{c} > 0$. ∎

We are now in the position to show strong ellipticity of the simple layer pseudodifferential operator Q_Γ with the principal symbol given by (10.2.5).

Theorem 10.2.3. *For a strongly elliptic adjoint differential operator* (10.2.11), *the simple layer potential operator Q_Γ corresponding to P with the principal symbol given by* (10.2.6) *is strongly elliptic on Γ.*

Proof: In view of the transformation formula (7.1.49) for principal symbols under change of coordinates it suffices to prove the theorem for Γ identified with the tangent hyperplane $x_n = 0$, i.e. for q^0_{-1} given by (10.2.6),

$$q^0_{-1}(x',\xi') = \frac{1}{2\pi}\int\limits_{-R}^{R} a^0_{-2}\big((x',0);\,(\xi',\xi_n)\big)d\xi_n$$

$$+\frac{i}{2\pi}\int\limits_{\vartheta=0}^{-\pi} a^0_{-2}\big((x',0);\,(\xi',\,\mathrm{Re}\,e^{i\vartheta})\big)\,\mathrm{Re}\,e^{i\vartheta}d\vartheta$$

with $R \geq R_0$ sufficiently large so that all the poles of $a^0_{-2}\big((x',0);\,(\xi',z)\big)$ for $|\xi'| = 1$ are contained in $|z| \leq \frac{1}{2}R_0$, $\mathrm{Im}\,z \leq 0$. Since a^0_{-2} in (10.2.5) is homogeneous of order -2, the second integral on the right–hand side is of order R^{-1}. Hence, we obtain

$$q^0_{-1}(x',\xi') = \frac{1}{2\pi}\int\limits_{-\infty}^{\infty} a^0_{-2}\big((x',0);\,(\xi',\xi_n)\big)d\xi_n\,.$$

Now we first consider the case that $\Theta = ((\delta_{\ell m}))_{p\times p}$ in the definition (7.2.62) of strong ellipticity for P^*. Then for $\eta \in \mathbb{C}^p$ we have

$$\begin{aligned}\eta^\top q^0_{-1}(x',\xi')\overline{\eta} &= \frac{1}{2\pi}\int\limits_{-\infty}^{\infty} \eta^\top a^0_{-2}\big((x',0);\,(\xi',\xi_n)\big)\overline{\eta}\;d\xi_n\\ &= \frac{1}{2\pi}\int\limits_{-\infty}^{\infty} \zeta^\top \sigma^0_{P^*}\big((x',0),\,(\xi',\xi_n)\big)\overline{\zeta}\;d\xi_n\end{aligned}$$

with the substitution

$$\zeta^\top(\xi', \xi_n) = \eta^\top a^0_{-2}\big((x',0)\,;\;(\xi',\xi_n)\big)\,. \tag{10.2.12}$$

Strong ellipticity for P^* with $\Theta = ((\delta_{\ell m}))_{p\times p}$ implies that

$$\begin{aligned}
\operatorname{Re}\eta^\top q^0_{-1}(x',\xi')\overline{\eta} &= \frac{1}{2\pi}\int\limits_{-\infty}^{\infty} \operatorname{Re}\big(\zeta^\top \sigma^0_{P^*}\big((x',0)\,,\;(\xi',\xi_n)\big)\overline{\zeta}\big)d\xi_n \\
&\geq \gamma_0 \frac{1}{2\pi}\int\limits_{-\infty}^{\infty} |\zeta(\xi',\xi_n)|^2(1+|\xi_n|^2)d\xi_n \\
&\geq \gamma_0 \frac{1}{2\pi}\int\limits_{-1}^{1} |\zeta(\xi',\xi_n)|^2 d\xi_n \quad \text{for } |\xi'| = 1\,.
\end{aligned}$$

On the other hand, since

$$\overline{\eta} = \sigma^0_{P^*}\big((x',0)\,;\;(\xi',\xi_n)\big)\overline{\zeta}\,,$$

we have the estimate

$$|\eta|^2 \leq |\zeta|^2 \max_{|\xi'|=1,\xi_n\in[-R_0,R_0]} |\sigma^0_{P^*}\big((x',0)\,;\;(\xi',\xi_n)\big)| = c|\zeta|^2\,.$$

This implies the strong ellipticity of q^0_{-1}:

$$\operatorname{Re}\eta^\top q^0_{-1}(x',\xi')\,\overline{\eta} \geq \gamma_1|\eta|^2 \quad \text{for all } |\xi'| = 1 \quad \text{and} \quad \eta\in\mathbb{C}^p \tag{10.2.13}$$

with $\gamma_1 > 0$.

For general $\Theta(x)$ consider the modified differential operator $\widetilde{P}^* = \Theta P^*$ instead of P^*. Then we find the relation

$$\begin{aligned}
\widetilde{q}^0_{-1}(x',\xi') &= \frac{1}{2\pi}\int\limits_{-\infty}^{+\infty} \Big\{\sum_{j,k=1}^{3} a_{jk}\Theta^*\xi_j\xi_k\Big\}^{-1} d\xi_n \\
&= \Theta^{*-1}(x',0)\frac{1}{2\pi}\int\limits_{-\infty}^{+\infty} \Big\{\sum_{j,k=1}^{3} a_{jk}\xi_j\xi_k\Big\}^{-1} d\xi_n \\
&= \Theta^{*-1}(x',0)q^0_{-1}(x,\xi')\,.
\end{aligned}$$

By using the strong ellipticity (10.2.13) we obtain

$$\operatorname{Re}\eta^\top \Theta^{*-1}(x',0)q^0_{-1}(x',\xi')\,\overline{\eta} = \operatorname{Re}\eta^\top \widetilde{q}^0_{-1}(x',\xi')\,\overline{\eta} \geq \gamma_1^2|\eta|^2$$

with $\gamma_1 > 0$ for all $\eta\in\mathbb{C}^p$ and $|\xi'| = 1$. ∎

The double layer potential for the scalar equation

The poles of the integrand in (10.2.8) are given by

$$z_{1|2} = -\frac{1}{2}p_1 \pm \frac{i}{2}\sqrt{4p_0 - p_1^2}$$

due to the strong ellipticity of the scalar, real differential operator P. Hence, by the residue theorem we find

$$q_0(\varrho', \xi') = -\frac{1}{2} + \frac{i}{2}\frac{p_1(x,\xi')}{\sqrt{4p_0(x,\xi') - p_1^2(x,\xi')}}$$

and

$$q_{0c}(\varrho', \xi') = \frac{1}{2} + \frac{i}{2}\frac{p_1(x,\xi')}{\sqrt{4p_0(x,\xi') - p_1^2(x,\xi')}}\,.$$

Thus, the tangential differential operator is given here by

$$\mathcal{T}u(x') = \frac{1}{2}u(x')\,.$$

Moreover, for the double layer potential we obtain the jump relation

$$\begin{aligned}&\lim_{0>\varrho_n\to 0} \mathfrak{W}u\big(T_r(\varrho') + \varrho_n n(\varrho')\big)\\ &= -\frac{1}{2}u(x') + \text{p.f.}\int\limits_{\Gamma\setminus\{x'\}} \Big(\frac{\partial}{\partial n_y}E(x',y)\Big)\mathcal{P}_2(y)u(y)ds_y \end{aligned} \tag{10.2.14}$$

where, in general, the finite part integral operator defines a Cauchy–Mikhlin singular integral operator which is a pseudodifferential operator of order 0 on the boundary Γ.

However, since $p_1(\xi')$ is the principal part of $i\mathcal{P}_1(x, i\xi')/\mathcal{P}_2(x)$ as in (10.2.7) and (3.4.58), $p_1(\xi')$ is real for a real elliptic scalar differential operator P of second order (3.4.53). Hence,we have

$$\operatorname{Re}\Theta q_0(\varrho', \xi') = 1 \quad \text{for} \quad \Theta = -2\,.$$

Consequently, we have the following lemma.

Lemma 10.2.4. *For a scalar, real elliptic second order differential operator* (10.1.1), *the corresponding operator*

$$-\tfrac{1}{2}I + \mathcal{K}$$

defined by (10.2.14) *with the double layer potential* $\mathcal{K} \in \mathcal{L}^0_{c\ell}(\Gamma)$ *is a strongly elliptic pseudodifferential operator of order zero.*

In the special case as for the Laplacian (also the Helmholtz operator), see Remark 3.4.1, we have $p_1(x,\xi') = 2H(x)$ and therefore the finite part integral operator $\mathcal{K}$ in (10.2.14) even becomes weakly singular.

10.2.2 Surface Potentials for the Helmholtz Equation

As explicit examples for the surface potentials we return to the Helmholtz equation in $\mathbb{R}^3$. With the Newton potential given by (10.1.5) we define the

simple layer surface potential

$$\widetilde{Q}_\Gamma v(x) = \int\limits_{\mathbb{R}^3} E_\mathsf{k}(x,y)\big(v(y')\otimes\delta_\Gamma\big)dy = \int\limits_\Gamma E_\mathsf{k}(x,y)v(y')ds_{y'}\,.$$

For $x \to \Gamma$ we obtain the simple layer boundary integral operator

$$\begin{aligned} V_\mathsf{k}(x) &= \lim_{x\to\Gamma} \widetilde{Q}_\Gamma v(x) = \int\limits_\Gamma E_\mathsf{k}(x,y')v(y')ds_y \\ &= \frac{1}{4\pi}\int\limits_\Gamma \frac{e^{i\mathsf{k}|x-y'|}}{|x-y'|}v(y')ds_{y'} \quad \text{for } x\in\Gamma \end{aligned} \tag{10.2.15}$$

which is a pseudodifferential operator $V_\mathsf{k} \in \mathcal{L}^{-1}_{c\ell}(\Gamma)$.

If Γ is given by $x_n = 0$ then the complete symbol of V_k has a homogeneous asymptotic expansion $\Sigma q_{-1-j}(x',\xi')$ and, from Formula (9.4.7), we find

$$q_{-2p} \quad = \quad 0 \tag{10.2.16}$$

and

$$\begin{aligned} q_{-1-2p}(x',\xi') &= (2\pi)^{-1}\int\limits_{\mathbb{C}} a^0_{-2-2p}\big((x'0)\,,\,(\xi',z)\big)dz \\ &= (2\pi)^{-1}\int\limits_{\mathbb{C}} \mathsf{k}^{2p}\frac{dz}{|\xi'|^{2p+2}+z^2} \\ &= \frac{1}{2}|\xi'|^{-1-2p}\mathsf{k}^{2p} \quad \text{for } |\xi'|\geq 1 \quad \text{and } p\in\mathbb{N}_0 \end{aligned} \tag{10.2.17}$$

by the use of the residue theorem since the only pole in the lower half plane of $\mathbb{C}$ is at $-i|\xi'|^{p+1}$.

For a general surface Γ given by $x' = T(\varrho')$ one first has to employ canonical coordinates and represent the Newton potentials in terms of these coordinates, and then compute the symbol of the boundary potential operator V_k from $V_{\mathsf{k}\Phi_r,0}$ the transformed operator's symbol in Theorem 9.4.3. To illustrate the idea let

$$\begin{aligned} \Phi(x) = \varrho : \widetilde{\Omega} \to \widetilde{U} \quad &\text{and} \\ x = \Psi(\varrho) = T(\varrho') + \varrho_n n(\varrho') \end{aligned} \tag{10.2.18}$$

denote the diffeomorphism (3.3.7) of the canonical coordinates. Then the principal symbol of the Newton potential with respect to the canonical coordinates is transformed according to (7.1.49), i.e.

$$\sigma^0_{N_k\Phi,-2}(\varrho,\zeta) = \sigma^0_{N_k,-2}\Big(\Psi(\varrho)\,,\,\Big(\frac{\partial\Phi}{\partial x}\Big)^\top\zeta\Big)\,,$$

where ζ denotes the Fourier variables corresponding to $\varrho \in \widetilde{U}$. The Jacobian matrix for the canonical coordinates can be computed explicitly by using (3.4.28) and (3.4.29) in the form

$$\Big(\frac{\partial\Phi}{\partial x}\Big)^\top = \Big(g^{11}\frac{\partial\Psi}{\partial\varrho_1} + g^{12}\frac{\partial\Psi}{\partial\varrho_2}\,,\ g^{12}\frac{\partial\Psi}{\partial\varrho_1} + g^{22}\frac{\partial\Psi}{\partial\varrho_2}\,,\ \boldsymbol{n}\Big)\,. \tag{10.2.19}$$

In what follows, we shall adhere to the Einstein summation convention where for repeated Greek indices $\lambda,\nu,\alpha,\beta,\gamma\dots$ the summation index runs from 1 to 2, whereas for Roman indices $j,k,\ell,q\dots$ it runs from 1 to 3.

The relation (10.2.19) implies

$$\begin{aligned}
|\xi|^2 &= \Big|\Big(\frac{\partial\Phi}{\partial x}\Big)^\top\zeta\Big|^2 = \zeta^\top\frac{\partial\Phi}{\partial x}\Big(\frac{\partial\Phi}{\partial x}\Big)^\top\zeta = \zeta_\ell\frac{\partial\Phi_\ell}{\partial x_k}\frac{\partial\Phi_j}{\partial x_k}\zeta_j \qquad (10.2.20)\\
&= \zeta_\ell\frac{\partial x_k}{\partial\varrho_m}g^{m\ell}\frac{\partial x_k}{\partial\varrho_q}g^{qj}\zeta_j\\
&= \zeta^\top\begin{pmatrix} g^{11} & g^{12} & 0\\ g^{12} & g^{22} & 0\\ 0 & 0 & 1\end{pmatrix}\zeta = \{\zeta_1^2g^{11} + 2\zeta_1\zeta_2g^{12} + \zeta_2^2g^{22} + \zeta_3^2\}\,.
\end{aligned}$$

This implies for the principal symbol

$$\sigma_{N_k\Phi,-2}(\varrho,\zeta) = \Big|\Big(\frac{\partial\Phi}{\partial x}\Big)^\top\zeta\Big|^{-2} = \{\zeta_1^2g^{11} + 2\zeta_1\zeta_2g^{12} + \zeta_2^2g^{22} + \zeta_3^2\}^{-1}$$

since

$$g_{mq} = \frac{\partial x_k}{\partial\varrho_m}\frac{\partial x_k}{\partial\varrho_q} \quad\text{and}\quad g_{mq}g^{qj} = \delta^j_m\,,$$

see (3.4.2) and (3.4.4).

For the principal symbol (10.2.20) of N_k we apply Theorem 9.4.3 again for $\varrho_n = 0$ and obtain

$$q^0_{-1}(\varrho',\zeta') = \tfrac{1}{2}\{\zeta_\alpha\gamma^{\alpha\beta}(g')\zeta_\beta\}^{-\frac{1}{2}} \tag{10.2.21}$$

with $\zeta' = (\zeta_1,\zeta_2) \in \mathbb{R}^2 \setminus \{0\}$ and $\gamma^{\alpha\beta}$ the contravariant fundamental tensor of Γ, (3.4.14), (3.4.15).

Obviously, in view of (10.2.21), the simple layer potential V_k is a strongly elliptic pseudodifferential operator $V_k \in \mathcal{L}^{-1}_{c\ell}(\Gamma)$ since

$$q^0_{-1}(\varrho',\zeta') = \tfrac{1}{2}\{\zeta_\alpha\gamma^{\alpha\beta}\zeta_\beta\}^{-\frac{1}{2}} \geq \gamma_0|\zeta'|^{-1} \quad \text{for all } \zeta' \in \mathbb{R}^2 \tag{10.2.22}$$

with a positive constant γ_0 depending on the local fundamental tensor $\gamma^{\alpha\beta}$. Of course, the strong ellipticity also follows from Lemma 10.2.1.

The double layer potential

The double layer potential W_k of the Helmholtz equation in (2.1.5) will now be considered as a surface potential generated by an appropriate Newton potential applied to the boundary distribution $v(x') \otimes \delta_\Gamma$. To facilitate the computations, it is more convenient to begin with the adjoint operator

$$\begin{aligned} W_k' u(x) :=& \frac{1}{4\pi} \int_\Omega \Big(n(x) \cdot \nabla_x \frac{e^{ik|x-y|}}{|x-y|} \Big) \psi(|x-y|) u(y) dy \\ &+ \frac{1}{4\pi} \int_\Omega \Big(n(x) \cdot \nabla_x \frac{e^{ik|x-y|}}{|x-y|} \Big) \big(1 - \psi(|x-y|)\big) u(y) dy \end{aligned} \tag{10.2.23}$$

where ψ is the same cut–off function as in Theorem 7.1.16. Since the first integral on the right–hand side defines a properly supported pseudodifferential operator in Ω, its symbol can be computed as

$$\sigma_{W_k'}(x,\xi) = -n(x) \cdot \int_{\mathbb{R}^3} e^{-ik\xi\cdot(x-y)} \big(\nabla_y E_k(x,y)\big) \psi(y) dy$$

based on Theorem 7.1.11 after employing

$$\nabla_x E_k(x,y) = -\nabla_y E_k(x,y)$$

for the fundamental solution $E_k = \dfrac{1}{4\pi} \dfrac{e^{ik|x-y|}}{|x-y|}$ (see (2.1.4)). Then integration by parts yields

$$\sigma_{W_k'}(x,\xi) = i(n(x)\cdot\xi) \int_{\mathbb{R}^3} e^{-i\xi\cdot(x-y)} E_k \psi dy = n(x) \cdot \int_{\mathbb{R}^3} e^{-i\xi\cdot(x-y)} (\nabla_y \psi) E_k dy\,.$$

The second integral on the right–hand side decays, due to the Paley–Wiener Theorem 3.1.3, since $\nabla_y \psi(|x-y|)$ has a compact support not containing $x = y$. From the decomposition of E_k in (10.1.14) it is clear that only $\frac{1}{4\pi}\frac{1}{z}\cos(k|z|)$ contributes to the symbol $\sigma_{W'}$ and, again, we obtain the complete symbol expansion of rational type

$$\frac{1}{4\pi} \int_{\mathbb{R}^3} e^{-i\xi\cdot z} \frac{1}{|z|} \cos(k|z|) \psi(|z|) dz = \sum_{p=0}^{\infty} \frac{k^{2p}}{|\xi|^{2+2p}} + R_\infty(\xi)$$

where $R_\infty \in \boldsymbol{S}^{-\infty}(\mathbb{R}^3)$. Hence,

$$\sigma_{W_k'}(x,\xi) = i(n(x)\cdot\xi) \sum_{p=0}^{\infty} \frac{k^{2p}}{|\xi|^{2+2p}}\,. \tag{10.2.24}$$

For $p = 0$, we find the principal symbol

$$\sigma_{W'_{\mathsf{k}},-1}(x,\xi) = i(n(x)\cdot\xi)\frac{1}{|\xi|^2}\,. \tag{10.2.25}$$

Therefore, by using Formula (7.1.34) in Theorem 7.1.13, we find the symbol of W_{k} as

$$\sigma_{W_{\mathsf{k}}} \sim \sum_{|\alpha|\geq 0} \frac{1}{\alpha!}\Big(\frac{\partial}{\partial\xi}\Big)^{\alpha}\Big(-i\frac{\partial}{\partial x}\Big)^{\alpha}\Big\{-i(n(x)\cdot\xi)\sum_{p=0}\frac{\mathsf{k}^{2p}}{|\xi|^{2p+2}}\Big\}$$

and its principal symbol

$$\sigma^0_{W_{\mathsf{k}},-1}(x,\xi) = -i(n\cdot\xi)\frac{1}{|\xi|^2}\,. \tag{10.2.26}$$

In order to obtain the boundary integral operators on Γ, we again apply Theorem 9.4.3 to W_{k} and W'_{k}, respectively. Then we have for Γ identified with $x_n = 0$,

$$q^0_{W_{\mathsf{k}},0}(x',\xi') = -\frac{1}{2\pi}\int\limits_{\mathfrak{c}} \frac{zdz}{z^2+|\xi'|^2} = -\frac{1}{2} \tag{10.2.27}$$

since the only pole in the lower half plane of $\mathbb{C}$ is $z = -i|\xi'|$. In the same way we obtain

$$q^0_{W'_{\mathsf{k}},0}(x',\xi') = \frac{1}{2}\,.$$

These are the principal symbols of Q_Γ and Q'_Γ defined by the limits of W_{k} and of W'_{k} for $x_n \to 0^-$, respectively.

We notice that

$$\begin{aligned} Qv(x') &= \lim_{x_n\to 0^-} W_{\mathsf{k}}v(x) = -\frac{1}{2}v(x') + \text{p.f.}\int\limits_\Gamma \Big(\frac{\partial}{\partial n_y}E_{\mathsf{k}}(x,y)\Big)v(y)ds_y \\ &\text{and} \\ Q_cv(x') &= \lim_{x_n\to 0^+} W_{\mathsf{k}}v(x) = \frac{1}{2}v(x') + \text{p.f.}\int\limits_\Gamma \Big(\frac{\partial}{\partial n_y}E_{\mathsf{k}}(x,y)\Big)v(y)ds_y\,. \end{aligned} \tag{10.2.28}$$

Correspondingly, for the transposed operators, one obtains

$$\begin{aligned} Q'v(x') &= \lim_{x_n\to 0^-} W'_{\mathsf{k}}v(x) = \frac{1}{2}v(x') + \text{p.f.}\int\limits_\Gamma \Big(\frac{\partial}{\partial n_x}E_{\mathsf{k}}(x,y)\Big)v(y)ds_y \\ &\text{and} \\ Q'_cv(x') &= \lim_{x_n\to 0^+} W'_{\mathsf{k}}v(x) = -\frac{1}{2}v(x') + \text{p.f.}\int\limits_\Gamma \Big(\frac{\partial}{\partial n_x}E_{\mathsf{k}}(x,y)\Big)v(y)ds_y\,, \end{aligned} \tag{10.2.29}$$

resembling the well–known jump relations for the classical layer potential W_{k} and its adjoint W'_{k}. In particular, we notice

$$\frac{1}{2}[\mathcal{Q}_c - \mathcal{Q}] = \mathcal{T} = \mathcal{T}_c = \frac{1}{2}I$$

as in (9.3.37). From our symbol computations for $q^0_{W_\mathsf{k},0}$ and $q^0_{W'_\mathsf{k},0}$ it follows that the acoustic double layer boundary integral operator

$$K_\mathsf{k}v(y) = \text{p.f.} \int\limits_\Gamma \Big(\frac{\partial}{\partial n_y}E_\mathsf{k}(x,y)\Big)v(y)ds_y$$

and its adjoint

$$K'_\mathsf{k}v(y) = \text{p.f.} \int\limits_\Gamma \Big(\frac{\partial}{\partial n_x}E_\mathsf{k}(x,y)\Big)v(y)ds_y$$

both must be pseudodifferential operators on Γ of order at most -1 which implies that their Schwartz kernels $\frac{\partial}{\partial n_y}E_\mathsf{k}(x,y)$ and $\frac{\partial}{\partial n_x}E_\mathsf{k}(x,y)$ are in $\Psi hk_{-1}(\Gamma)$, i.e., they are weakly singular.

For a curved surface Γ, we need to employ the canonical, tubular coordinates for W_k and $W_{\mathsf{k}'}$ and then compute the corresponding symbols for $x \to 0$. From our general results in Section 10.2.2 it is clear that the principal symbol of $\mathcal{Q}_\Gamma$ remains the same. Here, $\mathcal{T}_\Gamma = -\frac{1}{2}I$ also on Γ.

Now we consider the principal symbols of the operators K_k and $K'_\mathsf{k} \in \mathcal{L}^{-1}_{c\ell}(\Gamma)$ for whose computation we shall need $a^0_{W_\mathsf{k},0}(x,\xi) + a^0_{W_\mathsf{k},-2}(x,\xi)$ in the tubular coordinates (10.2.18). These can be obtained by Formula (7.1.34) for the symbol of the transposed operator, i.e., from

$$\sigma_{W'_\mathsf{k}}(x,\xi) = i\big(n(x)\cdot\xi\big)\Big\{\frac{1}{|\xi|^2} + O(|\xi|^{-4})\Big\}$$

which yields

$$\begin{aligned}
\sigma_{W_\mathsf{k}}(x,\xi) &= \sum_{|\alpha|\geq 0} \frac{1}{\alpha!}\Big(\frac{\partial}{\partial\xi}\Big)^\alpha\Big(-i\frac{\partial}{\partial x}\Big)^\alpha \sigma_{W'_\mathsf{k}}(x,-\xi)\\
&= -i\big(n(x)\cdot\xi\big)\frac{1}{|\xi|^2} + \frac{\partial}{\partial\xi_\ell}\Big(-i\frac{\partial}{\partial x_\ell}\Big)\Big(-i\frac{n(x)\cdot\xi}{|\xi|^2}\Big) + O(|\xi|^{-3})\\
&= -i\big(n(x)\cdot\xi\big)\frac{1}{|\xi|^2} - \frac{\nabla\cdot n(x)}{|\xi|^2} + 2\Big(\frac{\partial n_k}{\partial x_\ell}\frac{\xi_k\xi_\ell}{|\xi|^4}\Big) + O(|\xi|^{-3})\,.
\end{aligned}$$

In the tubular coordinates we use (3.4.19) and, moreover, (3.4.6), (3.4.17) and (3.4.28) to obtain

$$\frac{\partial n_k}{\partial x_\ell} = \frac{\partial x_\ell}{\partial \varrho_m}g^{mj}\frac{\partial n_k}{\partial \varrho_j} = \frac{\partial x_\ell}{\partial \varrho_m}g^{m\lambda}\frac{\partial n_k}{\partial \varrho_\lambda} = -L^\nu_\lambda\frac{\partial y_k}{\partial \varrho_\nu}g^{\nu\lambda}\frac{\partial x_\ell}{\partial \varrho_\mu}$$

and

$$\begin{aligned}
\sigma_{W_\mathsf{k}}(x,\xi) = -i\big(n(x)\cdot\xi\big)\frac{1}{|\xi|^2} + \frac{2H(x)}{|\xi|^2}\\
-2L^\nu_\lambda g^{\mu\lambda}\xi_k\frac{\partial y_k}{\partial \varrho_\nu}\xi_\ell\frac{\partial x_\ell}{\partial \varrho_\mu}\frac{1}{|\xi|^4} + O(|\xi|^{-3})\,.
\end{aligned} \tag{10.2.30}$$

The transformation of the symbol in terms of the tubular coordinates ϱ and $\zeta = \frac{\partial\Psi}{\partial\varrho}^{\top}\xi$ now reads by the use of Lemma 7.1.17:

$$a_{W_{k,\Phi}}(\varrho,\zeta) = \sum_{|\alpha|\geq 0} \frac{1}{\alpha!}\Big(\big(\frac{\partial}{\partial\xi}\big)^{\alpha}\sigma_W(x,\xi)\Big)\Big|_{\xi=\frac{\partial\Phi}{\partial x}^{\top}\zeta} \times \Big(\big(-i\frac{\partial}{\partial z}\big)^{\alpha}e^{i\zeta\cdot r}\Big)\Big|_{z=x=\Psi(\varrho)}$$

with $r = \Phi(z) - \Phi(x) - \frac{\partial\Phi}{\partial x}(x)(z-x)$. Because of (7.1.52) and since we only need terms up to the order -2, this yields for (10.2.30) with $\zeta_\lambda = \xi_\ell\frac{\partial x_\ell}{\partial\varrho_\lambda}$ for $\lambda = 1,2$; $\zeta_3 = n(x)\cdot\xi$ and (10.2.20), finally

$$a_{W_{k,\Phi}}(\varrho,\zeta) = \big(-i\zeta_3+2H(x)\big)\frac{1}{|\xi|^2} - 2\Big(g^{\mu\lambda}L^{\nu}_{\lambda}\zeta_\mu\zeta_\nu|_\Gamma\frac{1}{|\xi|^4}\Big) + O(|\zeta|^{-3}) \quad (10.2.31)$$

where $|\xi|^2 = \zeta_3^2 + d^2(\zeta)$ and $d^2(\zeta) = \zeta_\lambda g^{\lambda\nu}\zeta_\nu$.

With this relation (10.2.31), Theorem 9.4.3 can be applied for $\varrho_n \to 0^-$ and provides us with the symbols

$$\begin{aligned} q^0_0(\varrho',\zeta') &= -\frac{1}{2\pi}\int_c \frac{izdz}{z^2+d_0^2} = -\frac{1}{2}, \\ q^0_{-1}(\varrho',\zeta') &= \frac{1}{2\pi}\int_c 2\Big\{\frac{H(x)(z^2+d_0^2) - \sum_{\nu,\lambda,\nu=1}^{2} g^{\mu\nu}L^{\nu}_{\lambda}\zeta_\mu\zeta_\nu|_\Gamma}{(z^2+d_0^2)^2}\Big\}dz \\ &= \frac{H(x)}{d_0(\zeta)} - \frac{\zeta_\nu L^{\nu\mu}\zeta_\mu}{2d_0^3(\zeta)} \end{aligned} \quad (10.2.32)$$

with

$$d_0(\varrho',\zeta') := \{\zeta_\lambda\gamma^{\lambda\nu}\zeta_\nu\}^{\frac{1}{2}}. \quad (10.2.33)$$

The symbol $q^0_{-1}(\varrho',\zeta')$ is now the principal symbol of the acoustic double layer boundary integral operator $K_k \in \mathcal{L}^{-1}_{c\ell}(\Gamma)$.

10.2.3 Surface Potentials for the Lamé System

We begin with the

Simple layer potential

For the Lamé system (10.1.16) we define the simple layer potential

$$\mathfrak{V}\lambda(x) = N(\lambda\times\delta_\Gamma)(x) = \widetilde{Q}_\Gamma\lambda(x) = \int_{y'\in\Gamma} E(x,y')\lambda(y')ds_\Gamma(y')$$

with E given by (2.2.2). For $x \to \Gamma$, the simple layer boundary integral operator

$$V\lambda(x') = \lim_{\Omega\ni x\to\Gamma}\widetilde{Q}_\Gamma\lambda(x) = \int_\Gamma E(x',y')\lambda(y')ds_\Gamma(y') \quad \text{for } x'\in\Gamma$$

defines a pseudodifferential operator $V \in \mathcal{L}^{-1}_{c\ell}(\Gamma)$.

For Γ identified with $x_3 = 0$, we get with Formula (9.4.7) for $\big(\sigma_P(\xi)\big)^{-1}$ given by (10.1.19) the symbol

$$\begin{aligned} q_{-1}(x',\xi') &= \frac{1}{2\pi}\int\limits_{c}\big(\sigma_p^{-1}(\xi',z)\big)dz = \frac{1}{2\pi}\int\limits_{c}\frac{\delta_{jk}}{\mu\{z^2+|\xi'|^2\}}dz \\ &- \frac{1}{2\pi}\int\limits_{c}\frac{\lambda+\mu}{\mu(\lambda+2\mu)}\frac{1}{\{z^2+|\xi'|^2\}^2}\begin{pmatrix}\xi_1^2 & , & \xi_1\xi_2 & , & \xi_1 z\\ \xi_1\xi_2 & , & \xi_2^2 & , & \xi_2 z\\ \xi_1 z & , & \xi_2 z & , & z^2\end{pmatrix}dz\,. \end{aligned}$$

By using the residue formula, we have

$$\begin{aligned} \frac{1}{2\pi}\int\limits_{c}\frac{dz}{d_0^2+z^2} &= \frac{1}{2}d_0^{-1}, & \frac{1}{2\pi}\int\limits_{c}\frac{z^2dz}{z^2+d_0^2} &= -\frac{1}{2}d_0\,,\\ \frac{1}{2\pi}\int\limits_{c}\frac{zdz}{z^2+d_0^2} &= -\frac{i}{2}, & & \\ \frac{1}{2\pi}\int\limits_{c}\frac{dz}{(d_0^2+z^2)^2} &= \frac{1}{4}d_0^{-3}, & \frac{1}{2\pi}\int\limits_{c}\frac{z^2dz}{(d_0^2+z^2)^2} &= \frac{1}{4}d_0^{-1}\,,\\ \frac{1}{2\pi}\int\limits_{c}\frac{z^3dz}{(z^2+d_0)^2} &= -\frac{i}{2}, & \frac{1}{2\pi}\int\limits_{c}\frac{zdz}{(z^2+d_0)^2} &= 0\,. \end{aligned} \tag{10.2.34}$$

This yields the complete homogeneous symbol matrix of V on Γ,

$$\begin{aligned} q_{-1}(\xi') &= \frac{1}{2\mu|\xi'|}\delta_{jk} - \frac{\lambda+\mu}{4\mu(\lambda+2\mu)|\xi'|^3}\begin{pmatrix}\xi_1^2 & , & \xi_1\xi_2 & , & 0\\ \xi_1\xi_2 & , & \xi_2^2 & , & 0\\ 0 & , & 0 & , & |\xi'|^2\end{pmatrix}\\ &= \frac{(\lambda+3\mu)}{4|\xi'|^3\mu(\lambda+2\mu)}\begin{pmatrix}|\xi'|^2+\kappa\xi_2^2 & , & -\kappa\xi_1\xi_2 & , & 0\\ -\kappa\xi_2\xi_1 & , & |\xi'|^2+\kappa\xi_1^2 & , & 0\\ 0 & , & 0 & , & |\xi'|^2\end{pmatrix} \end{aligned} \tag{10.2.35}$$

where $\kappa = \frac{\lambda+\mu}{\lambda+3\mu}$.

Obviously, since $0<\kappa<1$, this matrix is positive definite, therefore, q_{-1} is a strongly elliptic symbol satisfying:

$$\mathrm{Re}\,\zeta^\top q_{-1}(\xi')\overline{\zeta} \geq \gamma_0|\xi'|^{-1}|\zeta|^2 \quad\text{for all } 0\neq\xi'\in\mathbb{R}^2 \text{ and } \zeta\in\mathbb{C}^3 \tag{10.2.36}$$

with $\gamma_0 = \frac{1}{2(\lambda+2\mu)} > 0$.

We remark that the strong ellipticity (10.2.36) also follows from Lemma 10.2.2.

For a general surface Γ we have to use the canonical coordinates (10.2.18) and for $\xi = \frac{\partial\Phi}{\partial x}^\top\zeta$ the relation (10.2.19). Then the principal symbol (10.1.19) reads

$$\sigma^0_{N_\Phi,-2}(\varrho,\zeta) = \mu\{\zeta_3^2 + g^{\nu\lambda}\zeta_\nu\zeta_\lambda\}^{-2}\Big(\Big(\{\zeta_3^2+\zeta_\nu g^{\nu\lambda}\zeta_\lambda\}\delta_{jk} - \frac{\lambda+\mu}{\lambda+2\mu}\Big(n_j\zeta_3 + g^{\nu\lambda}\frac{\partial x_j}{\partial\varrho_\nu}\zeta_\lambda\Big)\cdot\Big(n_k\zeta_3 + g^{\nu\lambda}\frac{\partial x_k}{\partial\varrho_\nu}\zeta_\lambda\Big)\Big)\Big)_{3\times 3}. \tag{10.2.37}$$

For Γ, we now compute the principal symbol of the simple layer boundary integral operator V on Γ according to Theorem 9.4.3:

$$q^0_{-1}(\varrho',\zeta') = \frac{1}{2\pi}\int\limits_{c}\Big(\Big(\frac{1}{\mu\{z^2+d_0^2\}}\delta_{jk} - \frac{\lambda+\mu}{\mu(\lambda+2\mu)}\frac{1}{\{z^2+d_0^2\}^2}\times \times\Big\{n_kn_jz^2 + n_jz\gamma^{\lambda\nu}\frac{\partial x_k}{\partial\varrho_\nu}\zeta_\lambda + n_kz\gamma^{\lambda\nu}\frac{\partial x_j}{\partial\varrho_\nu}\zeta_\lambda + (\gamma^{\lambda\nu}\frac{\partial x_j}{\partial\varrho_\nu}\zeta_\lambda)(\gamma^{\alpha\beta}\frac{\partial x_k}{\partial\varrho_\beta}\zeta_\alpha)\Big\}\Big)\Big)_{3\times 3}dz\,.$$

With (10.2.34) we obtain the explicit symbol matrix of V:

$$q^0_{-1}(\varrho',\zeta') = \Big(\Big(\frac{1}{2\mu d_0}\delta_{jk} - \frac{\lambda+\mu}{4\mu(\lambda+2\mu)}\frac{1}{d_0^3}\Big(d_0^2 n_jn_k + \Big(\gamma^{\iota\nu}\frac{\partial x_j}{\partial\varrho_\nu}\zeta_\iota\Big)\Big(\gamma^{\alpha\beta}\frac{\partial x_k}{\partial\varrho_\alpha}\zeta_\beta\Big)\Big)\Big)\Big)_{3\times 3}. \tag{10.2.38}$$

The double layer potential for the Lamé system

From the Betti–Somigliana representation formula (2.2.4) we now consider the double layer potential given by (2.2.8), i.e.

$$W\varphi(x) = \int\limits_\Gamma \big(T_y(x,y)E(x,y)\big)^\top\varphi(y)ds_y \quad\text{for } x\notin\Gamma\,,$$

where T_y denotes the boundary traction operator given by (2.2.5). Similar to the case of the Helmholtz operator in Section 10.2.2, the composition with the fundamental solution matrix given by (2.2.2),

$$\big(T_yE(x,y)\big)^\top = N_W(x,y)$$

defines in the tubular neighbourhood $\widetilde{\Omega}\subset\mathbb{R}^3$ of Γ a pseudohomogeneous Schwartz kernel generating a pseudodifferential operator A_W of order -1 with symbol of rational type since both, T_y and the elastic volume potential operators are of rational type. To facilitate the computation, again we begin with the transposed operator $A_W^\top$ which has the kernel $T_xE(x,y)$.

With the symbol of T_x,

$$\sigma_{T_x}(x,\xi) = i\big(\big(\lambda n_j(x)\xi_k + \mu n_k(x)\xi_j + \mu\big(n(x)\cdot\xi\big)\delta_{jk}\big)\big)_{3\times 3} \tag{10.2.39}$$

and according to (7.1.37), the complete symbol of $A_W^\top$ is given by

$$\begin{aligned}
\sigma_{T_x}(x,\xi) \circ \sigma_{N_{p-1}}(\xi) &= \sigma_{A_W^\top}(x,\xi) \\
&= i\big((\lambda n_j(x)\xi_k + \mu n_k(x)\xi_j + \mu(\xi \cdot n(x))\delta_{jk})\big)_{3\times 3} \\
&\qquad \times \frac{1}{\mu|\xi|^4}\Big(\Big(|\xi|^2\delta_{k\ell} - \frac{(\lambda+\mu)}{(\lambda+2\mu)}\xi_k\xi_\ell\Big)\Big)_{3\times 3} \\
&= -\frac{i}{\mu|\xi|^4}\frac{\lambda+\mu}{\lambda+2\mu}\big((\lambda|\xi|^2 n_j(x)\xi_\ell + 2\mu(n(x)\cdot\xi)\xi_j\xi_\ell)\big)_{3\times 3} \\
&\quad + \frac{i}{\mu|\xi|^2}\big((\lambda n_j(x)\xi_\ell + \mu\xi_j n_\ell(x) + \mu(n(x)\cdot\xi)\delta_{j\ell})\big)_{3\times 3}\,.
\end{aligned} \tag{10.2.40}$$

For A_W, the transposed operator, we then get from (7.1.34) the principal symbol

$$\sigma^0_{A_W}(x,\xi) = \sigma^0_{A_W^\top}(x,-\xi)^\top\,. \tag{10.2.41}$$

Since

$$W\varphi(x) \;=\; A_W(\varphi \otimes \delta_\Gamma)$$

then

$$Q_\Gamma\varphi(x) \;=\; \lim_{\Omega \ni x \to \Gamma} W\varphi(x)$$

is a pseudodifferential operator $Q_\Gamma \in \mathcal{L}^0_{c\ell}(\Gamma)$ due to Theorem 9.5.5. To compute the principal symbol of Q_Γ we apply Theorem 9.4.3. In particular, if $x_n = 0$ corresponds to the surface Γ we obtain

$$q^0_0(x',\xi') = \frac{1}{2\pi}\int\limits_{c} \sigma^0_{A_W}(x',0;\,\xi',z)dz\,.$$

By using (10.2.40) with (10.2.41), we first compute the entries of q_0 for $j \neq 3$ and $\ell \neq 3$ by using $n_1(x) = n_2(x) = 0\,,\; n_3(x) = 1$:

$$\begin{aligned}
q_0^{j\ell 0}(x',\xi') &= -\frac{i}{2\pi}\int\limits_{c}\frac{zdz}{z^2+|\xi'|^2}\delta_{j\ell} + \frac{i}{2\pi}\int\limits_{c}\frac{z^2dz}{(z^2+|\xi'|^2)^2}\frac{2(\lambda+\mu)}{\lambda+2\mu}\xi_j\xi_\ell \\
&= -\frac{1}{2}\delta_{j\ell} \quad \text{for } j,\ell = 1,2
\end{aligned} \tag{10.2.42}$$

in view of (10.2.34). If $j = \ell = 3$, then with (10.2.34)

$$\begin{aligned}
q_0^{330}(x',\xi') &= \frac{i}{2\pi}\int\limits_{c}\frac{zdz}{z^2+|\xi'|^2}\Big(\frac{\lambda}{\mu}\frac{(\lambda+\mu)}{(\lambda+2\mu)} - \frac{(\lambda+2\mu)}{\mu}\Big) \\
&\quad + \frac{i}{2\pi}\int\limits_{c}\frac{z^3dz}{(z^2+|\xi'|^2)^2}2\frac{(\lambda+\mu)}{(\lambda+2\mu)} \\
&= \frac{1}{2}\frac{1}{\mu}\frac{1}{\lambda+2\mu}\{\lambda(\lambda+\mu) - (\lambda+2\mu)^2 + 2(\lambda+\mu)\mu\} \\
&= -\frac{1}{2}\,.
\end{aligned}$$

For $\ell = 3$ and $j = 1, 2$ we find, again with (10.2.34),

$$q_0^{j30}(x', \xi') = -\frac{2i}{2\pi}\frac{(\lambda+\mu)}{(\lambda+2\mu)}\int\limits_{\mathfrak{C}}\frac{z^2 dz}{(z^2+|\xi'|^2)^2}\xi_j + \frac{i}{2\pi}\int\limits_{\mathfrak{C}}\frac{dz}{z^2+|\xi'|^2}\xi_j$$

$$= \frac{i}{2}\frac{\mu}{\lambda+2\mu}\frac{\xi_j}{|\xi'|}.$$

Similarly, for $j = 3$ and $\ell = 1, 2$ we obtain

$$q_0^{3\ell 0}(x', \xi') = -\frac{i}{2}\frac{\mu}{\lambda+2\mu}\frac{\xi_\ell}{|\xi'|}.$$

To summarize, we have the principal symbol matrix for a flat surface,

$$q_0^0(x', \xi') = -\frac{1}{2}I + \frac{1}{2}\begin{pmatrix} 0 & , & 0 & , & i\varepsilon\frac{\xi_1}{|\xi'|} \\ 0 & , & 0 & , & i\varepsilon\frac{\xi_2}{|\xi'|} \\ -i\varepsilon\frac{\xi_1}{|\xi'|} & , & -i\varepsilon\frac{\xi_2}{|\xi'|} & , & 0 \end{pmatrix}, \tag{10.2.43}$$

where $\varepsilon = \frac{\mu}{\lambda+2\mu}$.

In view of Theorem 9.4.6, the tangential differential operator $\mathcal{T}$ in this case is of order zero, i.e.

$$\mathcal{T} = \frac{1}{2}I,$$

whereas the skew–symmetric symbolic matrix belongs to the classical singular integral operator of Cauchy–Mikhlin type, namely

$$K\varphi(x') = \text{p.v.}\int\limits_{\Gamma}\big(T_y E(x', y)\big)^\top \varphi(y) ds_y,$$

which is the double layer surface potential operator of linear elasticity (see also (2.2.19)).

From this representation we conclude that q_0 is a strongly elliptic system according to our definition (7.2.4) since inequality (7.2.16) can easily be derived for $\Theta(x) = -((\delta_{jk}))_{3\times 3}$ and with the help of $0 < \varepsilon < 1$.

In the same manner as for the simple layer potential (1.2.28), we can consider the double layer potential for a curved surface Γ in terms of the canonical coordinates and obtain

$$\begin{aligned}
\sigma^0_{A_{W,\Phi}}(\varrho, \zeta) &= -\sigma_{A^\top_{W,\Phi}}(\varrho, \zeta)^\top = \sigma_{A^\top_W}\Big(\Phi^{-1}(\varrho), \frac{\partial \Phi}{\partial x}^\top \zeta\Big)^\top \\
&= \Big(\Big(\frac{i}{\mu}\frac{1}{(\zeta_3^2+d^2)^2}\Big(\frac{\lambda+\mu}{\lambda+2\mu}\Big)\Big\{\lambda(\zeta_3^2+d^2)n_\ell\Big(g^{\lambda\nu}\frac{\partial x_j}{\partial \varrho_\nu}\zeta_\lambda + n_j\zeta_3\Big) \\
&\qquad + 2\mu\zeta_3\Big(g^{\lambda\nu}\frac{\partial x_\ell}{\partial \varrho_\nu}\zeta_\lambda + n_\ell\zeta_3\Big)\Big(g^{\lambda\nu}\frac{\partial x_j}{\partial \varrho_\nu}\zeta_\nu + n_j\zeta_3\Big)\Big\} \\
&\quad - \frac{i}{\mu}\frac{1}{(\zeta_3^2+d^2)}\Big\{\lambda n_\ell\Big(g^{\lambda\nu}\frac{\partial x_j}{\partial \varrho_\nu}\zeta_\nu + n_j\zeta_3\Big) \\
&\qquad + \mu n_j\Big(g^{\lambda\nu}\frac{\partial x_\ell}{\partial \varrho_\nu}\zeta_\lambda + n_\ell\zeta_3\Big) + \mu\zeta_3\delta_{j\ell}\Big\}\Big)\Big)_{3\times 3},
\end{aligned} \tag{10.2.44}$$

where $d = (\zeta_\lambda g^{\lambda\nu} \zeta_\nu)^{\frac{1}{2}}$. Replacing ζ_3 by z and performing the contour integration of formula (9.4.7) we finally again obtain the symbol matrix of $q_0(\varrho', \zeta')$ on the surface $\varrho_n = 0$,

$$q_0^0(\varrho', \zeta') = \Big(\Big(-\frac{1}{2}\delta_{j\ell} - \frac{i}{2}\varepsilon\Big(n_j \gamma^{\lambda\nu} \frac{\partial x_\ell}{\partial \varrho_\nu} \frac{\zeta_\lambda}{d_0} - n_\ell \gamma^{\lambda\nu} \frac{\partial x_j}{\partial \varrho_\nu} \frac{\zeta_\lambda}{d_0}\Big)\Big)\Big)_{3\times 3}$$

with $d_0 = \{\gamma^{\lambda\nu}\zeta_\lambda\zeta_\nu\}^{\frac{1}{2}}$. Obviously, q_0 remains strongly elliptic if Γ is any smooth surface.

10.2.4 Surface Potentials for the Stokes System

The principal symbols of the surface potentials of the Stokes system can be read off explicitly for the simple layer potential by taking $\lambda \to +\infty$ in (10.2.38) and for the double layer potential in (2.3.16) by taling $\varepsilon \to 0$ in (10.2.43).

For the pressure potential in (2.3.10), however, we insert

$$\xi_j = n_j \zeta_3 + \gamma^{\nu\iota} \frac{\partial x_j}{\partial \varrho_\nu} \zeta_\iota$$

into (10.1.22) and find the principal symbol

$$2i(2\pi)^{-1} \int\limits_{c} \frac{n_j z + \gamma^{\nu\iota} \frac{\partial x_j}{\partial \varrho_\nu} \zeta_\iota}{z^2 + d_0^2} dz = i n_j + \frac{1}{2} d_0^{-1} \gamma^{\nu\iota} \frac{\partial x_j}{\partial \varrho_\nu} \zeta_\iota \,, \; j = 1,2,3 \quad (10.2.45)$$

where we employed (10.2.27) and (10.2.34) and where d_0^2 is given in (10.2.33). So, the boundary integral operator of the pressure potential belongs to $\mathcal{L}^0(\Gamma)$.

10.3 Invariance of Boundary Pseudodifferential Operators

In the previous chapters we considered boundary integral operators as pseudodifferential operators on Γ generated by generalized Newton potentials and their compositions with differential operators which are pseudodifferential operators on the domain $\Omega \cup \widetilde{\Omega}$ by applying these to distributions of the special type $v \otimes \delta_\Gamma$. The boundary integral operators then correspond to the traces of these compositions A in the form

$$Q_\Gamma v(x) = \mathcal{T} v(x) + \text{ p.f. } \int\limits_\Gamma k_A(x, x-y) v(y) ds_\Gamma(y) \quad \text{for } x \in \Gamma \,. \quad (10.3.1)$$

The generalized Newton potential operators have symbols of rational type as established in Theorem 10.1.1, and their composition A with differential operators then have symbols of rational type as well, see Lemma 9.4.2. Hence, Theorem 9.5.7 implies the following invariance property formulated as *Kieser's theorem.*

Theorem 10.3.1. *The boundary pseudodifferential operators of the form* (10.3.1) *generated by generalized Newton potentials and their compositions with differential operators are invariant under change of coordinates on* Γ, *i.e., the tangential differential operators* $\mathcal{T}v$ *are transformed in the usual way by applying the standard chain rule whereas the finite part integral operator transforms according to the classical rule of substitution.*

We remark that these invariance properties are due to the special properties of operators with symbols of rational type. They satisfy the canonical extension conditions, Definition 9.5.3, the special parity conditions (9.5.14) and the jump relations as in Theorem 9.5.5. In general, as we noticed in Theorem 8.2.2, the transformation of finite part integral operators produces extra terms under the change of coordinates if the special parity conditions are *not* fulfilled.

Theorem 10.3.1 is of particular importance from the computational point of view. In particular, the double layer potentials in applications belong to this class and the corresponding boundary integral operators enjoy the invariance properties. These include the corresponding singular integral operators of Cauchy–Mikhlin type.

Clearly, this class of operators also satisfies the Tricomi conditions (9.3.18).

10.3.1 The Hypersingular Boundary Integral Operators for the Helmholtz Equation

As a further illustration of Theorem 9.5.5 we now consider the hypersingular surface potential defined by

$$\begin{aligned} A(v\otimes\delta_\Gamma)(x) &:= \int\limits_\Gamma \frac{\partial}{\partial n_x}\Big(\frac{\partial}{\partial n_y}E_{\mathsf{k}}(x,y)\Big)v(y)ds_\Gamma(y) \quad \text{for } x\notin\Gamma \\ &= \int\limits_{\mathbb{R}^3} k_A(x,x-y)\big(v(y')\otimes\delta_\Gamma(y)\big)dy \end{aligned} \tag{10.3.2}$$

where the Schwartz kernel k_A can be computed explicitly,

$$\begin{aligned} k_A(x,x-y) = \frac{1}{2\pi}\frac{e^{ikr}}{r^5}\big\{n(x)\cdot(x-y)n(y)\cdot(y-x)\big((i\mathsf{k}r-1)-1-(i\mathsf{k}r-1)^2\big) \\ + r^2 n(x)\cdot n(y)(i\mathsf{k}r-1)\big\} \end{aligned}$$

where $r=|x-y|$. In order to find the corresponding symbol we use (10.2.26) for $\sigma_{W_{\mathsf{k}}}(x,\xi)$; then with (7.1.37) the complete symbol of $A\in\mathcal{L}^0_{c\ell}(\widetilde{\Omega})$ has the expansion

$$\sigma_A(x,\xi) \sim (-i)^2 \sum_{|\beta|\geq 0} \frac{1}{\beta!}\Big(\Big(\frac{\partial}{\partial\xi}\Big)^\beta (n(x)\cdot\xi)\Big)\times$$
$$\times\Big(-i\frac{\partial}{\partial x}\Big)^\beta \sum_{|\alpha|\geq 0}\frac{1}{\alpha!}\Big(\frac{\partial}{\partial\xi}\Big)^\alpha\Big(-i\frac{\partial}{\partial x}\Big)^\alpha\Big\{(n(x)\cdot\xi)\sum_{p=0}^{\infty}\frac{\mathrm{k}^{2p}}{|\xi|^{2+2p}}\Big\}. \tag{10.3.3}$$

Since $A \in \mathcal{L}^0_{c\ell}(\widetilde{\Omega})$ has a symbol of rational type, Theorems 9.6.1 and 9.6.2 with $m = 0$ provide the mapping properties of the hypersingular surface potential (10.3.2). In order to verify Theorem 9.5.5 for the operator A, particularly the jump relations and the derivation of the symbol of Q_Γ, we need the first terms of the expansion (10.3.3) up to the order $O(|\xi|^{-1})$:

$$\sigma_A(x,\xi) = -\frac{(n(x)\cdot\xi)^2}{|\xi|^2} + i\Big\{(n(x)\cdot\xi)\frac{\nabla\cdot n}{|\xi|^2}2\Big(\xi_j\frac{\partial n}{\partial x_j}\cdot\xi\Big)\frac{(n(x)\cdot\xi)}{|\xi|^4}$$
$$+\big((n(x)\cdot\nabla_x)n(x)\big)\cdot\xi\frac{1}{|\xi|^2}\Big\} + O(|\xi|^{-2}).$$

If we identify Γ with $x_n = 0$ then $\frac{\partial n}{\partial x_j} = 0$ and we find

$$q(x',\xi') = -\frac{1}{2\pi}\int\limits_{\mathcal{C}}\frac{z^2}{z^2+|\xi|^2}dz + O(|\xi'|^{-1}) = \frac{1}{2}\,|\xi'| + O(|\xi'|^{-1}).$$

In this special case we conclude that for the tangential operator we have $\mathcal{T} = 0$ in Theorem 9.5.5 and, moreover,

$$\lim_{x_n\to 0} A(v\otimes\delta_\Gamma)(x)$$
$$= Q_\Gamma v(x') = \text{ p.f.}\int\limits_\Gamma k_A(x',x'-y')v(y')ds_\Gamma(y') \quad\text{for } x'\in\Gamma.$$

Here $Q_\Gamma = D \in \mathcal{L}^1_{c\ell}(\Gamma)$ is the hypersingular boundary integral operator of the Helmholtz equation with the principal symbol

$$q_1^0(x',\xi') = \frac{1}{2}|\xi'| \quad\text{and with}\quad q_0^0(x',\xi') = 0\,.$$

For a general surface Γ, again we transform A into the canonical coordinates (10.2.18) by the use of the transformation formula (7.1.50) together with Formula (7.1.52), which yields

$$\sigma_{A,\Phi}(\varrho,\zeta) = -\frac{(n\cdot\xi)^2}{|\xi|^2} \tag{10.3.4}$$
$$+i\Big\{-(n\cdot\xi)\frac{2H}{|\xi|^2} - 2\Big(\xi_j\frac{\partial n}{\partial x_j}\cdot\xi\Big)\frac{(n\cdot\xi)}{|\xi|^4} + (n\cdot\nabla_x)(n(x)\cdot\xi)\frac{1}{|\xi|^2}\Big\}$$
$$+\frac{1}{2}\Big(\frac{\partial^2}{\partial\xi_j\partial\xi_k}\Big(-\frac{(n(x)\cdot\xi)^2}{|\xi|^2}\Big)\Big)\Big(-i\frac{\partial^2\Phi(x)}{\partial x_j\partial x_k}\Big)\cdot\zeta + O(|\xi|^{-2})$$

where

$$\xi_j = g^{\lambda\nu} \frac{\partial x_j}{\partial \varrho_\nu} \zeta_\lambda + n_j \zeta_3 \tag{10.3.5}$$

according to $\xi = \frac{\partial \Phi}{\partial x}^\top \zeta$ and (10.2.19).

By applying Theorem 9.4.3, we compute the first term of the symbol of Q_Γ by employing (9.4.7). Then we obtain the following result.

Lemma 10.3.2. *The hypersingular boundary integral operator $D_k$in (2.1.17), namely*

$$D_k v(x') = -\text{ p.f.} \int_\Gamma \frac{\partial}{\partial n_x} \Big(\frac{\partial}{\partial n_y} E_k(x', y') \Big)^\top v(y') ds_\Gamma(y') \quad \textit{for } x' \in \Gamma$$

has a symbol of the form

$$\begin{aligned} q(x', \zeta') &= \frac{1}{2} d_0 - \frac{i}{4\sqrt{\gamma}} \frac{\partial}{\partial \varrho_\nu} (\sqrt{\gamma}\, \gamma^{\nu\lambda}) \frac{\zeta_\lambda}{d_0} - \frac{i}{4} g^{\lambda\nu} g^{\alpha\beta} G^\gamma_{\nu\beta} \frac{\zeta_\lambda \zeta_\alpha \zeta_\gamma}{d_0^3} + O(d_0^{-1}) \\ \textit{where } d_0 &= \{\zeta_\lambda \gamma^{\lambda\nu} \zeta_\nu\}^{\frac{1}{2}} . \end{aligned} \tag{10.3.6}$$

Proof: Only the first term on the right–hand side of (10.3.4) will contribute to q_1^0, i.e., by employing (10.2.34) we get

$$q_1^0(\varrho', \zeta') = -\frac{1}{2\pi} \int_c \frac{z^2}{z^2 + d_0^2} dz = \frac{1}{2} d_0 . \tag{10.3.7}$$

For the zeroth order terms we compute the contour integrals separately and will need the residues

$$\frac{1}{2\pi} \int_c \frac{z^2 dz}{(z^2 + d_0^2)^3} = \frac{1}{16} d_0^{-3} \quad \text{and} \quad \frac{1}{2\pi} \int_c \frac{z^3 dz}{(z^2 + d_0^2)^3} = 0 \tag{10.3.8}$$

in addition to those in (10.2.34). Hence,

$$-i2H(x) \frac{1}{2\pi} \int_c \frac{z}{z^2 + d_0^2} dz = -H(x) . \tag{10.3.9}$$

Moreover, from (3.4.6), (3.4.17) and (3.4.28) we obtain

$$\frac{\partial n_k}{\partial x_j} = -\frac{\partial x_j}{\partial \varrho_\nu} L^{\nu\lambda} \frac{\partial x_k}{\partial \varrho_\lambda} \tag{10.3.10}$$

and

$$\begin{aligned} & -2i \frac{1}{2\pi} \int_c \xi_j \frac{\partial n_k}{\partial x_j} \xi_k \frac{z}{(z^2 + d_0^2)^2} dz \\ &= 2i \frac{1}{2\pi} \int_c \xi_j \frac{\partial x_j}{\partial \varrho_\nu} L^{\nu\lambda} \xi_k \frac{\partial x_k}{\partial \varrho_\lambda} \frac{z}{(z^2 + d_0^2)^2} dz \\ &= 2i \zeta_\nu L^{\nu\lambda} \zeta_\lambda \frac{1}{2\pi} \int_c \frac{z}{(z^2 + d_0^2)^2} dz = 0 . \end{aligned} \tag{10.3.11}$$

Furthermore,

$$(n \cdot \nabla_x)\big(n(x) \cdot \xi\big) = \frac{\partial}{\partial \varrho_3}\big(n(\varrho') \cdot \xi\big) = 0\,. \tag{10.3.12}$$

Up to now we have computed all of the contributions to q_0^0 from (7.1.50) for $|\alpha| \leq 1$. It remains to compute the contributions from terms in (7.1.50) for $|\alpha| = 2$. There we first need the explicit expression

$$\begin{aligned}\frac{\partial}{\partial \xi_j}\frac{\partial}{\partial \xi_k}\Big\{-\frac{(n(x)\cdot\xi)^2}{|\xi|^2}\Big\} &= -2\frac{n_j(x)n_k(x)}{|\xi|^2} + 2\frac{n_j(x)(n(x)\cdot\xi)}{|\xi|^4}\cdot 2\xi_k \\ +2\delta_{jk}\frac{(n(x)\cdot\xi)^2}{|\xi|^4} &+ 4\frac{\xi_j n_k(x)(n(x)\cdot\xi)}{|\xi|^4} - 4\frac{\xi_j(n(x)\cdot\xi)^2}{|\xi|^6}2\xi_k\end{aligned} \tag{10.3.13}$$

for the last term in (10.3.4).

Again we collect the terms of the second sum in (10.3.4) and obtain with the first term of (10.3.13) and the chain rule

$$i\frac{n_j n_k}{|\xi|^2}\frac{\partial^2 \Phi}{\partial x_j \partial x_k}\cdot\zeta = i\Big(\Big(\frac{\partial}{\partial n}\Big)^2\Phi\Big)\cdot\zeta\frac{1}{|\xi|^2} = 0 \tag{10.3.14}$$

since $\dfrac{\partial \Phi}{\partial n} = \dfrac{\partial \Phi}{\partial \varrho_3} = \begin{pmatrix}0\\0\\1\end{pmatrix}$ due to (3.4.30).

For the next product we find, again with the chain rule,

$$\begin{aligned}\frac{2n_j\zeta_3}{|\xi|^4}2\xi_k\Big(-\frac{i}{2}\frac{\partial^2\Phi}{\partial x_j\partial x_k}\Big)\cdot\zeta &= -2i\frac{\zeta_3}{|\xi|^4}\frac{\partial}{\partial x_k}\Big(\frac{\partial\Phi}{\partial n}\Big)\cdot\zeta \\ &= -2i\frac{\zeta_3}{|\xi|^4}\Big(\frac{\partial}{\partial x_k}\begin{pmatrix}0\\0\\1\end{pmatrix}\Big)\cdot\zeta = 0\,.\end{aligned} \tag{10.3.15}$$

The third product reads

$$\begin{aligned}-i\frac{\zeta_3^2}{|\xi|^4}\delta_{jk}\frac{\partial^2\Phi}{\partial x_j\partial x_k}\cdot\zeta &= -i\frac{\zeta_3^2}{|\xi|^2}\Delta_x\Phi\cdot\zeta \\ &= -i\frac{\zeta_3^2}{|\xi|^4}\Big(\mathcal{P}_0\Phi - 2H\frac{\partial\Phi}{\partial n} + \frac{\partial^2\Phi}{\partial n^2}\Big)\cdot\zeta \\ &= -i\frac{\zeta_3^2}{(z^2+d_0^2)^2}\Big\{\frac{1}{\sqrt{\gamma}}\frac{\partial}{\partial\varrho_\kappa}\Big(\sqrt{\gamma}\gamma^{\kappa\lambda}\frac{\partial\Phi}{\partial\varrho_\lambda}\Big)\cdot\zeta - 2H\zeta_3 + 0\Big\} \\ &= -i\frac{\zeta_3^2}{(z^2+d_0^2)^2}\frac{1}{\sqrt{\gamma}}\Big(\frac{\partial}{\partial\varrho_\kappa}\sqrt{\gamma}\gamma^{\kappa\lambda}\Big)\zeta_\lambda + 2iH\frac{\zeta_3^3}{(z^2+d_0^2)^2}\,.\end{aligned} \tag{10.3.16}$$

For (10.3.16), the integration over $\mathfrak{c}$ then gives the contribution corresponding to this third product:

$$
\begin{aligned}
&-i\frac{1}{2\pi}\int\limits_{c}\frac{z^2}{(z^2+d_0^2)^2}\delta_{jk}\frac{\partial^2\Phi}{\partial x_j\partial x_k}\cdot\begin{pmatrix}\zeta_1\\ \zeta_2\\ z\end{pmatrix}dz\\
&= -i\frac{1}{\sqrt{\gamma}}\Big(\frac{\partial}{\partial\varrho_\kappa}\sqrt{\gamma}\gamma^{\kappa\lambda}\Big)\zeta_\lambda\frac{1}{2\pi}\int\limits_{c}\frac{z^2}{(z^2+d_0^2)^2}dz+2iH\frac{1}{2\pi}\int\limits_{c}\frac{z^3}{(z^2+d_0^2)^2}dz\\
&= -i\frac{1}{4\sqrt{\gamma}}\Big(\frac{\partial}{\partial\varrho_\kappa}\sqrt{\gamma}\gamma^{\kappa\lambda}\Big)\frac{\zeta_\lambda}{d_0}+H\,. \qquad (10.3.17)
\end{aligned}
$$

For the fourth product from (10.3.13) we obtain

$$
\frac{4\xi_j n_k(x)\zeta_3}{|\xi|^4}\Big(-\frac{i}{2}\frac{\partial^2\Phi}{\partial x_j\partial x_k}\Big)\cdot\zeta=-2i\frac{\xi_j}{|\xi|^4}\zeta_3\frac{\partial}{\partial x_j}\Big(\frac{\partial\Phi}{\partial n}\Big)=0 \qquad (10.3.18)
$$

as in (10.3.15).

Finally, for the last product from (10.3.13), we have

$$
\begin{aligned}
&4i\frac{\zeta_3^2}{|\xi|^6}\xi_j\xi_k\frac{\partial^2\Phi}{\partial x_j\partial x_k}\cdot\zeta\\
&= 4i\frac{\zeta_3^2}{|\xi|^6}\Big(g^{\lambda\nu}\frac{\partial x_j}{\partial\varrho_\nu}\zeta_\lambda+n_j\zeta_3\Big)\Big(g^{\alpha\beta}\frac{\partial x_k}{\partial\varrho_\beta}\zeta_\alpha+n_k\zeta_3\Big)\frac{\partial^2\Phi_q}{\partial x_j\partial x_k}\zeta_q\\
&= 4i\frac{\zeta_3^2}{(\zeta_3^2+d_0^2)^3}\Big\{g^{\lambda\nu}g^{\alpha\beta}\frac{\partial x_j}{\partial\varrho_\nu}\frac{\partial x_k}{\partial\varrho_\beta}\frac{\partial^2\Phi_q}{\partial x_j\partial x_k}\zeta_\lambda\zeta_\alpha\zeta_q\\
&\quad +2g^{\lambda\nu}\frac{\partial x_j}{\partial\varrho_\nu}n_k\frac{\partial^2\Phi_q}{\partial x_j\partial x_k}\zeta_\lambda\zeta_q\zeta_3+n_j\cdot n_k\zeta_3^2\frac{\partial^2\Phi_q}{\partial x_j\partial x_k}\zeta_q\Big\}\\
&=: \Sigma_1+\Sigma_2+\Sigma_3\,. \qquad (10.3.19)
\end{aligned}
$$

Now we examine each term in (10.3.19) separately. We begin with Σ_1. There we have

$$
\begin{aligned}
&g^{\lambda\nu}g^{\alpha\beta}\frac{\partial x_j}{\partial\varrho_\nu}\frac{\partial x_k}{\partial\varrho_\beta}\frac{\partial^2\Phi_q}{\partial x_j\partial x_k}\\
&= g^{\lambda\nu}g^{\alpha\beta}\Big(\frac{\partial}{\partial\varrho_\nu}\Big(\frac{\partial\Phi_q}{\partial x_k}\Big)\Big)\frac{\partial x_k}{\partial\varrho_\beta}\\
&= g^{\lambda\nu}g^{\alpha\beta}\frac{\partial}{\partial\varrho_\nu}\Big(\frac{\partial\Phi_q}{\partial x_k}\frac{\partial x_k}{\partial\varrho_\beta}\Big)-g^{\lambda\nu}g^{\alpha\beta}\frac{\partial\Phi_q}{\partial x_k}\frac{\partial^2 x_k}{\partial\varrho_\nu\partial\varrho_\beta}\\
&= g^{\lambda\nu}g^{\alpha\beta}\frac{\partial}{\partial\varrho_\nu}\Big(\frac{\partial\varrho_q}{\partial\varrho_\beta}\Big)-g^{\lambda\nu}g^{\alpha\beta}\frac{\partial\Phi_q}{\partial x_k}G^r_{\nu\beta}\frac{\partial x_k}{\partial\varrho_r}
\end{aligned}
$$

from (3.4.8) with the Christoffel symbols $G^r_{\nu\beta}$.

Hence, with $\frac{\partial\varrho_q}{\partial\varrho_\beta}=\delta^q_\beta$ and differentiation,

$$
\begin{aligned}
g^{\lambda\nu}g^{\alpha\beta}\frac{\partial x_j}{\partial\varrho_\nu}\frac{\partial x_k}{\partial\varrho_\beta}\frac{\partial^2\Phi_q}{\partial x_k\partial x_k} &= -g^{\lambda\nu}g^{\alpha\beta}G^r_{\nu\beta}\frac{\partial\Phi_q}{\partial\varrho_r}=-g^{\lambda\nu}g^{\alpha\beta}G^r_{\nu\beta}\delta^q_r\\
&= -g^{\lambda\nu}g^{\alpha\beta}G^q_{\nu\beta}\,.
\end{aligned}
$$

By applying the contour integration, the first term in (10.3.19) becomes with the residues in (10.3.8):

$$\begin{aligned}\frac{1}{2\pi}\int\limits_{\mathfrak{c}} \Sigma_1(\varrho';\zeta',z)dz &= \\ -4ig^{\lambda\nu}g^{\alpha\beta}\zeta_\lambda\zeta_\alpha\Big\{\frac{1}{2\pi}\int\limits_{\mathfrak{c}}\frac{z^3}{(z^2+d_0^2)^3}dzG^3_{\nu\beta} + \frac{1}{2\pi}\int\limits_{\mathfrak{c}}\frac{z^2}{(z^2+d_0^2)^3}dzG^\gamma_{\nu\beta}\zeta_\gamma\Big\} \\ &= -\frac{i}{4}g^{\lambda\nu}g^{\alpha\beta}G^\gamma_{\nu\beta}\frac{\zeta_\lambda\zeta_\alpha\zeta_\gamma}{d_0^3}\,. \qquad (10.3.20)\end{aligned}$$

Next, we analyze Σ_2 in (10.3.19) by using (3.4.25) and (3.4.26). We begin with

$$\begin{aligned}&2g^{\lambda\nu}\frac{\partial x_j}{\partial\varrho_\nu}\frac{\partial}{\partial n}\Big(\frac{\partial\Phi_q}{\partial x_j}\Big)\zeta_\lambda\zeta_q\zeta_3 \\ &= \Big(2g^{\lambda\nu}\frac{\partial}{\partial\varrho_3}\Big(\frac{\partial\varrho_q}{\partial\varrho_\nu}\Big) - 2g^{\lambda\nu}\frac{\partial\Phi_q}{\partial x_j}\frac{\partial^2 x_j}{\partial\varrho_\nu\partial\varrho_3}\Big)\zeta_\lambda\zeta_q\zeta_3 \\ &= 0 - 2g^{\lambda\nu}G^r_{\nu 3}\frac{\partial x_j}{\partial\varrho_r}\frac{\partial\Phi_q}{\partial x_j}\zeta_\lambda\zeta_q\zeta_3 \\ &= -2g^{\lambda\nu}G^r_{\nu 3}\frac{\partial\varrho_q}{\partial\varrho_r}\zeta_\lambda\zeta_q\zeta_3 = -2g^{\lambda\nu}G^r_{\nu 3}\delta^q_r\zeta_\lambda\zeta_q\zeta_3 \\ &= -2g^{\lambda\nu}G^\beta_{\nu 3}\zeta_\lambda\zeta_\beta\zeta_3 = (2g^{\lambda\nu}L^\beta_\nu - \varrho_3 2g^{\lambda\nu}G^\varrho_{\nu 3}L^\beta_\varrho)\zeta_\lambda\zeta_\beta\zeta_3 \\ &= 2\gamma^{\lambda\nu}L^\beta_\nu\zeta_\lambda\zeta_\beta\zeta_3 \quad \text{for } \varrho_3 = 0\,.\end{aligned}$$

Hence,

$$\frac{1}{2\pi}\int\limits_{\mathfrak{c}}\Sigma_2(\varrho';\zeta',z)dz = \frac{8i}{2\pi}\int\limits_{\mathfrak{c}}\frac{z^3dz}{(z^2+d_0^2)^3}\gamma^{\lambda\nu}L^\beta_\nu\zeta_\lambda\zeta_\beta = 0\,. \qquad (10.3.21)$$

For the last term Σ_3 of (10.3.19) we get with $\frac{\partial}{\partial\varrho_3}n_k = 0$:

$$\begin{aligned}n_jn_k\frac{\partial^2\Phi_q}{\partial x_j\partial x_k}\zeta_q\zeta_3^2 &= \frac{\partial}{\partial\varrho_3}\Big(\frac{\partial\Phi_q}{\partial x_k}\Big)n_k\zeta_q\zeta_3^2 &&= \frac{\partial}{\partial\varrho_3}\Big(\frac{\partial\Phi_q}{\partial x_k}n_k\Big)\zeta_q\zeta_3^2 \\ &= \frac{\partial}{\partial\varrho_3}\Big(\frac{\partial\varrho_q}{\partial\varrho_3}\Big)\zeta_q\zeta_3^2 &&= \frac{\partial}{\partial\varrho_3}(\delta^q_3)\zeta_q\zeta_3^2 = 0\,,\end{aligned} \qquad (10.3.22)$$

which yields $\Sigma_3 = 0$.

By collecting all terms, in particular, (10.3.7)–(10.3.9), (10.3.11)–(10.3.22), the proposed relation (10.3.6) follows. ∎

Now we show that in this case the tangential differential operator $\mathcal{T} = 0$.

Lemma 10.3.3. *For the hypersingular boundary potential operator as given in (10.3.2), the tangential operator $\mathcal{T}$ in the jump relation (9.4.25) vanishes:*

$$\mathcal{T} = \mathcal{T}_{\mathfrak{c}} = 0\,.$$

Proof: Here we can apply Theorem 9.4.6, i.e. (9.4.26); but we can also compute the jumps of the symbols by (9.4.27) which means nothing but replacing all the contour integrations over $\mathfrak{c}$ in the proof of Theorem 9.4.3 by corresponding counter clockwise oriented contour integrals over the complete circle $|z| = R > 2d_0$. It is easy to see that these integrals corresponding to (10.3.7), (10.3.11), (10.3.20) and (10.3.21) all vanish. In addition, we see that (10.3.9) gives

$$-i2H(x)\frac{1}{2\pi}\oint\limits_{|z|=R}\frac{z}{z^2+d_0^2}dz = 2H(x)$$

whereas for the corresponding integrals in (10.3.17) we find

$$-iH(x)\frac{1}{2\pi}\oint\limits_{|z|=R}\frac{z^3}{(z^2+d_0^2)^2}dz = -2H(x) \quad\text{and}\quad \frac{1}{2\pi}\oint\limits_{|z|=R}\frac{z^2}{(z^2+d_0^2)^2}dz = 0\,.$$

So, for the hypersingular operator of the Helmholtz equation we find

$$[q_1^0 + q_0^0](\varrho', \xi') = 2\mathcal{T} = 0\,, \tag{10.3.23}$$

as expected. ∎

10.3.2 The Hypersingular Operator for the Lamé System

Similar to the case of the Helmholtz equation, we now consider the surface potentials of the hypersingular operator D as in (2.2.31) and define

$$\begin{aligned} A(v\otimes\delta_\Gamma)(x) &:= -\int\limits_\Gamma T_x\big(T_y E(x,y)\big)^\top v(y)ds_\Gamma(y) \\ &= \int\limits_{\mathbb{R}^3} k_A(x,x-y)\big(v(y')\otimes\delta_\Gamma(y)\big)dy \quad\text{for } x\notin\Gamma\,. \end{aligned} \tag{10.3.24}$$

Here the Schwartz kernel is given by

$$\begin{aligned} k_A(x,x-y) = \frac{1}{4\pi}\frac{\mu}{\lambda+2\mu}\frac{1}{|z|^5}\Big(&(3z\cdot n(y)\{2\mu n_j(x)z_k \\ &+\lambda n_k(x)z_j + \lambda z\cdot n(x)\delta_{jk} - \frac{5}{|z|^2}z\cdot n(x)z_jz_k\} \\ &+3\lambda\{n(x)\cdot n(y)z_jz_k + z\cdot n(x)n_j(y)z_k\} \\ &+2\mu\big\{3z\cdot n(x)z_jn_k(y) + |z|^2n_j(y)n_k(x) + |z|^2n(x)\cdot n(y)\delta_{jk}\big\} \\ &-2(\mu-\lambda)n_j(x)n_k(y)|z|^2\big)\Big)_{3\times 3} \end{aligned} \tag{10.3.25}$$

where $z = x - y$ (see Brebbia et al [44, p.191]).

To compute the symbol, we use the complete symbol of A_W by applying (7.1.34) to the symbol of the transposed to the elastic double layer potential (10.2.40) and the symbol of the traction operator T_x to find the symbol of the hypersingular operator by employing (7.1.37) in Theorem 7.1.14:

$$
\begin{aligned}
&\sigma_{-T_x \circ A_W}(x,\xi) = \\
&\quad -\sum_{|\beta|\geq 0}\Big(\frac{1}{\beta!}\Big(\frac{\partial}{\partial\xi}\Big)^{\beta} i\big((\lambda n_j(x)\xi_k + \mu\xi_j n_k(x) + \mu(\xi\cdot n(x))\delta_{jk})\big)_{3\times 3}\Big)\times \\
&\qquad \times\Big(-i\frac{\partial}{\partial x}\Big)^{\beta}\sum_{|\alpha|\geq 0}\frac{1}{\alpha!}\Big(\frac{\partial}{\partial\xi}\Big)^{\alpha}\Big(-i\frac{\partial}{\partial x}\Big)^{\alpha}\times \\
&\quad \times\Big\{\frac{i}{\mu|\xi|^4}\frac{\lambda+\mu}{\lambda+2\mu}\big((\lambda|\xi|^2 n_\ell(x)\xi_m + 2\mu(n(x)\cdot\xi)\xi_\ell\xi_m)\big)_{3\times 3} \\
&\quad -\frac{i}{\mu|\xi|^2}\big((\lambda n_\ell(x)\xi_m + \mu\xi_\ell n_m(x) + \mu(n(x)\cdot\xi)\delta_{\ell m})\big)_{3\times 3}\Big\}^{\top}. \qquad (10.3.26)
\end{aligned}
$$

With the complete symbol available, it is clear, that the jump conditions can be investigated in the same manner as for the Helmholtz equation. Nevertheless, this is very tedious and cumbersome and we therefore shall return to the jump relations by using a different approach later on, based on a different representation formula for the double layer and the hypersingular boundary potentials (see Han [170] and Kupradze et al [253]).

For simplicity, we now confine ourselves to the computation of the principal symbol of $-T_x \circ A_W$ and of D on Γ by choosing $\alpha = \beta = 0$ in (10.3.26). A straightforward computation yields

$$
\begin{aligned}
&\sigma^0_{-T_x\circ A_W}(x,\xi) = \frac{\lambda+\mu}{\mu(\lambda+2\mu)}\frac{1}{|\xi|^4}\big((\lambda^2|\xi|^4 n_j(x)n_k(x) \\
&+2\lambda\mu|\xi|^2(n(x)\cdot\xi)(\xi_j n_k(x) + n_j(x)\xi_k) + 4\mu^2(n(x)\cdot\xi)^2\xi_j\xi_k)\big)_{3\times 3} \\
&-\frac{1}{\mu|\xi|^2}\big((\lambda^2|\xi|^2 n_j(x)n_k(x) + \mu(2\lambda+\mu)(n(x)\cdot\xi)(\xi_j n_k(x) + n_j(x)\xi_k) \\
&+\mu^2\xi_j\xi_k + \mu^2(n(x)\cdot\xi)^2\delta_{jk})\big)_{3\times 3}. \qquad (10.3.27)
\end{aligned}
$$

By setting $n_1(x) = n_2(x) = 0$, $n_3(x) = 1$ and $\xi_3 = z$, we obtain

$$
\begin{aligned}
\sigma^0_{\alpha\beta} &= -\mu\frac{z^2}{z^2+d_0^2}\delta_{\alpha\beta} + \Big(\frac{4\mu(\lambda+\mu)}{\lambda+2\mu}\frac{z^2}{(z^2+d_0^2)^2} - \mu\frac{1}{z^2+d_0^2}\Big)\xi_\alpha\xi_\beta\,, \\
\sigma^0_{\alpha 3} &= \sigma^0_{3\alpha} = 4\mu\frac{\lambda+\mu}{\lambda+2\mu}\Big(\frac{z^3}{(z^2+d_0^2)^2} - \frac{z}{z^2+d_0^2}\Big)\xi_\alpha \quad \text{for } \alpha,\beta = 1,2\,; \\
\sigma^0_{33} &= \frac{4\mu(\lambda+\mu)}{\lambda+2\mu}\Big(\frac{z^4}{(z^2+d_0^2)^2} - 2\frac{z^2}{z^2+d_0^2}\Big) - \frac{\lambda^2}{\lambda+2\mu}\,. \qquad (10.3.28)
\end{aligned}
$$

Now we employ Formula (9.4.7) and use (10.2.34), (10.3.8) and

$$\frac{1}{2\pi} \int\limits_{\mathfrak{c}} \frac{z^4}{(z^2+d_0^2)^2} dz = -\frac{3}{4} d_0$$

to obtain the principal symbol of the hypersingular boundary integral operator D:

$$q^0_{D,1} = \frac{\mu}{2|\xi'|} \begin{pmatrix} |\xi'|^2 + \varepsilon_1 \xi_1^2 & , & \varepsilon_1 \xi_1 \xi_2 & , & 0 \\ \varepsilon_1 \xi_2 \xi_1 & , & |\xi'|^2 + \varepsilon_1 \xi_2^2 & , & 0 \\ 0 & , & 0 & , & (1+\varepsilon_1)|\xi'|^2 \end{pmatrix} \qquad (10.3.29)$$

where $\varepsilon_1 = \frac{\lambda}{\lambda + 2\mu}$ and, hence, $|\varepsilon_1| < 1$.

Obviously, for any $\xi' \in \mathbb{R}^2$ with $|\xi'| = 1$, the matrix $q^0_{D,1}$ is symmetric and positive definite. Consequently, inequality (9.1.19) is satisfied with $\Theta = ((\delta_{jk}))_{3\times 3}$ and $\gamma_0 = (1 - |\varepsilon_1|) > 0$. Therefore the hypersingular boundary integral operator matrix D of linear elasticity is strongly elliptic and satisfies a Gårding inequality (9.1.21) on Γ where $t_\ell = s_\ell = \frac{1}{2}$.

For the jump relation, i.e. for obtaining the operator $\mathcal{T}$ in (9.4.26) we could follow the calculations as for the Helmholtz equation which will be cumbersome, as mentioned before. Alternatively, we use the representation (2.2.34) of the hypersingular operator, i.e.

$$\begin{aligned} Dv(x) = -T_x W v(x) &= -\frac{\mu}{4\pi} \text{ p.f.} \int\limits_\Gamma \frac{\partial^2}{\partial n_x \partial n_y} \Big(\frac{1}{|x-y|} \Big) v(y) ds_y \\ &- \text{p.f.} \int\limits_\Gamma \mathcal{M}\big(x, n(x)\big) \Big\{ 4\mu^2 E(x,y) - \frac{\mu}{2\pi} \frac{1}{|x-y|} I \Big\} \mathcal{M}\big(\partial_y, n(y)\big) v(y) ds_y \\ &+ \frac{\mu}{4\pi} \text{ p.f.} \int\limits_\Gamma \Big(\mathcal{M}\big(\partial_x, n(x)\big) \frac{1}{|x-y|} I \mathcal{M}\big(\partial_y, n(y)\big)\Big)^\top v(y) ds_y \,. \end{aligned} \qquad (10.3.30)$$

We notice that the first term on the right–hand side of (10.3.30) corresponds to the hypersingular boundary potential (10.3.2) of the Laplacian, i.e. of the Helmholtz equation for $\mathsf{k} = 0$. For the latter, $\mathcal{T} = \mathcal{T}_c = 0$ due to Lemma 10.3.3, and from Lemma 10.3.2 we obtain its symbol in the form $\mu q(x', \xi')((\delta_{jk}))_{3\times 3}$ with $q(x', \xi')$ given by (10.3.6). For the next two terms we use the property that the elements of the operator matrix

$$\mathcal{M}\big(\partial_x, n(x)\big) = \big((m_{jk}(\partial_x, n(x)))\big)_{3\times 3}$$

where

$$m_{jk}\big(\partial_y, n(y)\big) = n_k(y) \frac{\partial}{\partial y_j} - n_j(y) \frac{\partial}{\partial y_k} = -m_{kj}\big(\partial_y, n(y)\big) \qquad (10.3.31)$$

are tangential differential operators and apply Corollary 9.5.6 and Theorem 9.5.7. To facilitate the presentation we first write the second term in the form

$$-\mu(\partial_x, n(x)) \int\limits_{y' \in \Gamma} \Big\{4\mu E(x,y') - \frac{1}{2\pi}\frac{1}{|x-y'|} I\Big\} \mathcal{M}(\partial_{y'}, n(y'))v(y')ds_{y'}$$
$$= -\mu\mathcal{M} \circ \widetilde{N}_1 \circ \mathcal{M}(v \otimes \delta_\Gamma)(x) \quad (10.3.32)$$

where

$$\widetilde{N}_1 f(x) = \int\limits_{\Omega\cup\widetilde{\Omega}} \Big\{4\mu E(x,y) - \frac{1}{2\pi}\frac{1}{|x-y|} I\Big\} f(y)dy$$

is a Newton potential $\widetilde{N}_1 \in \mathcal{L}^{-2}_{c\ell}(\mathbb{R}^2)$ with weakly singular kernel whose corresponding boundary potential with

$$f(y) = w(y') \otimes \delta_\Gamma(y)$$

has a continuous extension to Γ without jump (see (10.2.4)). Consequently, Corollary 9.5.6 and Theorem 9.5.7 imply, that

$$\lim_{x\to x'\in\Gamma} \big(-\mu\mathcal{M} \circ \widetilde{N}_1 \circ \mathcal{M}(v \otimes \delta_\Gamma)(x)\big)$$
$$= -\mathcal{M}(\partial_{x'}, n(x')) \int\limits_{\Gamma} \Big\{4\mu E(x',y') - \frac{1}{2\pi}\frac{1}{|x'-y'|}\Big\} \mathcal{M}(\partial_{y'}, n(y'))v(y')ds_{y'}$$

where for the corresponding tangential operator we have $\mathcal{T} = 0$.

For the last term we write this operator matrix in terms of its components which are compositions of tangential operators with the Newton potential $\widetilde{V}$ of the Laplacian, i.e.

$$\frac{\mu}{4\pi}\sum_{\ell=1}^{3} \int\limits_{\Omega\cup\widetilde{\Omega}} m_{k\ell}(\partial_x, n(x)) \frac{1}{|x-y|} m_{\ell j}(\partial_y, n(y))\big(v_k(y') \otimes \delta_\Gamma\big)dy$$
$$= \mu \sum_{\ell=1}^{3} (\partial_x, n(x)) \circ \widetilde{V} \circ m_{\ell j}(v_k \otimes \delta_\Gamma) \quad (10.3.33)$$

where we have employed the Corollary 9.5.6 for interchanging the tangential differentiation and integration. In the same manner as for $\widetilde{N}_1$ we have

$$\lim_{x\to x'\in\Gamma} \mu m_{k\ell}(\partial_x, n(x)) \circ \big(\widetilde{V} \circ m_{\ell j}(v_k \otimes \delta_\Gamma)\big)(x)$$
$$= \frac{\mu}{4\pi} m_{k\ell}(\partial_{x'}, n(x')) \int\limits_{\Gamma} \frac{1}{|x'-y'|} m_{\ell j}(\partial_{y'}, n(y'))v_k(y')ds_{y'}\,;$$

the corresponding tangential operator $\mathcal{T}$ again vanishes.

To conclude this section we summarize one of the main results for the hypersingular boundary potentials in elasticity in the following lemma.

Lemma 10.3.4. *The hypersingular boundary potential Dv satisfies*

$$\begin{aligned} Dv(x') &:= -T_{x'}\,\text{p.f.}\int_\Gamma \big(T_y E(x',y)\big)^\top v(y)ds_y \qquad (10.3.34)\\ &= -\,\text{p.f.}\int_\Gamma T_{x'}\big(T_y E(x',y)\big)^\top v(y)ds_y \\ &= -\frac{\mu}{4\pi}\,\text{p.f.}\int_\Gamma \Big(\frac{\partial^2}{\partial n_{x'}\partial n_y}\,\frac{1}{|x'-y|}\Big)v(y)ds_y \end{aligned}$$

$$-\ \text{p.f.}\int_\Gamma \mathcal{M}(\partial_{x'},n(x'))\Big\{4\mu^2E(x',y') - \frac{\mu}{2\pi|x'-y'|}I\Big\}\mathcal{M}(\partial_{y'},n(y'))v(y')ds_{y'}$$

$$+\ \text{p.f.}\,\frac{\mu}{4\pi}\int_\Gamma \Big(\mathcal{M}(\partial_{x'},n(x'))\frac{1}{|x'-y'|}\mathcal{M}(\partial_{y'},n(y'))\Big)^\top v(y')ds_{y'}$$

for $x' \in \Gamma$.

Moreover, D has a strongly elliptic principal symbol given by (10.3.29).

We remark that the last two finite part integrals are Cauchy principal value integrals due to the corresponding Tricomi conditions (9.3.8) with $m = -1$ and $j = |\alpha| = 0$ from (10.3.32) and (10.3.33), respectively, since all the operators involved here have symbols of rational type (see Lemma 9.4.1).

10.3.3 The Hypersingular Operator for the Stokes System

In view of Section 2.3.3, in particular relation (2.3.48) between the hypersingular operator of the Lamé and that of the Stokes system it is clear that the principal symbol of the Stokes system, in the tubular Hadamard coordinates is given by (10.3.29) with $\varepsilon_1 = 1 = \lim_{\lambda\to+\infty}\lambda/(\lambda+2\mu)$, i.e.,

$$q^0_{D_{st},1} = \frac{\mu}{2|\xi'|}\begin{pmatrix} |\xi'|^2+\xi_1^2 & , & \xi_1\xi_2 & , & 0\\ \xi_1\xi_2 & , & |\xi'|^2+\xi_2^2 & , & 0\\ 0 & , & 0 & , & 2|\xi'|^2\end{pmatrix} \qquad (10.3.35)$$

which is a strongly elliptic symbol matrix. Correspondingly, the hypersingular operator of the Stokes system can be obtained from (10.3.34) by sending $\lambda \to +\infty$, i.e., sending $c \to 0$ in (2.3.47) and (2.3.48). Therefore, one obtains the last relation in (10.3.34) where now $E(x,y)$ is to be chosen as in (2.3.10) according to the Stokes system.

10.4 Derivatives of Boundary Potentials

In practice one often needs to compute not only the Cauchy data of the solution of boundary value problems and their potentials but also their gradients or higher order derivatives in Ω (or Ω^c) as well as near and up to the boundary Γ. As is well known, these derivatives can be considered as linear combinations of normal and tangential derivatives in $\widetilde{\Omega}$, the vicinity of Γ. In the framework of pseudodifferential operators, for derivatives of potentials, the following lemma provides us with the jump relations for the derivatives of boundary potentials based on Theorem 9.4.6.

Lemma 10.4.1. *Let $A \in \mathcal{L}^m_{c\ell}(\widetilde{\Omega})$ have a symbol of rational type. Further let*

$$m_{jk}(x, \partial_x) = n_k(x')\frac{\partial}{\partial x_j} - n_j(x')\frac{\partial}{\partial x_k} \tag{10.4.1}$$

denote the extension of the Günter derivative form Γ into its neighbourhood. Then

$$\begin{aligned} &\lim_{0>\varrho_n\to 0} m_{jk}(x,\partial_x)\circ A(u\otimes\delta_\Gamma)(x'+\varrho_n n(x')) \\ &= \text{p.f.}\int\limits_{y'\in\Gamma} \big(m_{jk}(x',\partial_{x'})k_A(x',x'-y')\big)u(y')ds_{y'} - m_{jk}\circ\mathcal{T}u(x') \end{aligned} \tag{10.4.2}$$

where $x' \in \Gamma$ and $\mathcal{T}$ is the tangential differential operator given in Theorem 9.4.6.

For the normal derivative we find

$$\begin{aligned} &\lim_{0>\varrho_n\to 0} \frac{\partial}{\partial n}\circ A(u\otimes\delta_\Gamma)(x'+\varrho_n n(x')) \\ &= \text{p.f.}\int\limits_{y'\in\Gamma} \Big(\frac{\partial}{\partial n_x}k_A(x,x-y')\Big)\Big|_{x=x'}u(y')ds_{y'} - \mathcal{T}_n u(x') \quad \text{for } x'\in\Gamma\,. \end{aligned} \tag{10.4.3}$$

Here $\mathcal{T}_n$ is the tangential differential operator corresponding to the new pseudodifferential operator $\frac{\partial}{\partial n}\circ A$ with the Schwartz kernel $\frac{\partial}{\partial n_x}k_A(x,x-y)$. In terms of the symbol of A_{Φ_r} and $\frac{\partial}{\partial n} = \frac{\partial}{\partial \varrho_n}$, the operator $\mathcal{T}_n$ reads

$$\begin{aligned} \mathcal{T}_n = &\frac{1}{2}\sum_{0\le j\le m+2}\ \sum_{|\alpha|=m+2-j}(-i)^{|\alpha|}\Big(\Big(\frac{\partial}{\partial\xi'}\Big)^\alpha a_{m-j}((\varrho',0),\xi)\Big)\Big|_{\xi=(0,1)}D^\alpha_{\varrho'} \\ &+\frac{1}{2}\sum_{0\le j\le m+1}\ \sum_{|\alpha|=m+1-j}(-i)^{|\alpha|+1}\Big(\Big(\frac{\partial}{\partial\xi'}\Big)^\alpha\frac{\partial}{\partial\varrho_n}a_{m-j}(\varrho,\xi)\Big)\Big|_{\substack{\xi=(0,1)\\ \varrho=(\varrho',0)}} D^\alpha_{\varrho'}\,. \end{aligned} \tag{10.4.4}$$

Clearly, the corresponding limits for $x \in \Omega^c$ also satisfy the equations (10.4.2) *and* (10.4.3) *with + instead of − in front of the tangential operators.*

Proof: The first limit, (10.4.2) is a special case of Corollary 9.5.6. For the second limit in (10.4.3), we apply Theorem 9.4.6 to the composed operator $\frac{\partial}{\partial\varrho_n}A_{\Phi_r}$ in the tubular coordinates, which means in the parametric half space $\varrho_n < 0$. With the symbol of $\frac{\partial}{\partial\varrho_n}\otimes A_{\Phi_r}$ obtained from (7.1.37),

$$\sigma_{\frac{\partial}{\partial \varrho_n} \circ A} = i\xi_n \sum_{j \geq 0} a_{m-j}(\varrho, \xi) \Big(\frac{\partial}{\partial \xi_n}(\xi_n) \Big) \Big(-i \frac{\partial}{\partial \varrho_n} \sum_{j \geq 0} a_{m-j}(\varrho, \xi) \Big) .$$

A straightforward application of (9.4.26) to this symbol gives the desired formula (10.4.4). ■

If $m_{jk}(x, \partial_x)u(x)$ and $\frac{\partial u}{\partial n}(x)$ are available, then the relation

$$\begin{aligned} \frac{\partial v}{\partial x_j}(x) &= \sum_{k=1}^{n} n_k(x) m_{jk}(x, \partial_x) v(x) + n_j(x) \frac{\partial v}{\partial n}(x) \\ &= \mathcal{D}_j v + n_j(x) \frac{\partial v}{\partial n}(x) \end{aligned} \tag{10.4.5}$$

(see (3.4.32) and (3.4.33)) allows one to compute arbitrary derivatives

$$\lim_{0 > \varrho_n \to 0} \Big(\frac{\partial}{\partial x_j} A(u \otimes \delta_\Gamma) \Big) \big(x' + \varrho_n(x') \big) .$$

The general procedure for computing derivatives of the solution of the boundary value problems in terms of boundary potentials and the Cauchy data can be summarized as follows: First one computes the Cauchy data by using any kind of boundary integral equations as desired. Next, from the Cauchy data we compute $\mathcal{M}u$. Together with $\frac{\partial u}{\partial n}$ which either is given or can be obtained from the Cauchy data, we then compute $\frac{\partial u}{\partial x_j}$ by using (10.4.5). This approach can be repeated and yields an appropriate bootstrapping procedure for finding higher order derivatives (see Schwab et al [381]).

For computing $\mathcal{M}u$ we employ the commutator

$$[\mathcal{M}, A](u \otimes \delta_\Gamma)(x) = (\mathcal{M} \circ A - A \circ \mathcal{M})(u \otimes \delta_\Gamma)(x) \tag{10.4.6}$$

which has the same order as $A \in \mathcal{L}^m_{c\ell}(\widetilde{\Omega})$ due to Corollary 7.1.15 since $\mathcal{M} \in \mathcal{L}^1_{c\ell}(\widetilde{\Omega})$.

Correspondingly, on the manifold Γ, we have the commutator

$$[\mathcal{M}_\Gamma, Q_\Gamma] \in \mathcal{L}^{m+1}_{c\ell}(\Gamma)$$

where $Q_\Gamma \in \mathcal{L}^{m+1}_{c\ell}(\Gamma)$ is defined by (9.5.4), (9.5.15).

Theorem 10.4.2. (Schwab et al [381]) *Let $A \in \mathcal{L}^m_{c\ell}(\widetilde{\Omega})$ with symbol of rational type. Then the commutators in* (10.4.6) *in terms of the Schwartz kernel $k_A(x, x-y)$ can be expressed explicitly as follows:*

$$[m_{j\ell}, A](u \otimes \delta_\Gamma)(x)$$
$$= \int\limits_{y\in\Gamma} \{m_{j\ell}(x, \partial_x)k_A(x, x-y) - k_A(x, x-y)m_{j\ell}(y, \partial_y)\}u(y)ds_y$$
$$= \int\limits_{\Gamma} \Big\{m_{j\ell}(x, \partial_x)k_A(x, z)$$
$$+ \Big(\big(n_\ell(x) - n_\ell(y)\big)\frac{\partial}{\partial z_j} - \big(n_j(x) - n_j(y)\big)\frac{\partial}{\partial z_\ell}\Big)k_A(x, z)\Big\}\Big|_{z=x-y}u(y)ds_y \tag{10.4.7}$$

for $x \notin \Gamma$.

Moreover, for $\Omega \ni x \to \Gamma$ and Q_Γ defined by (9.5.4) and (9.5.15), we have

$$[m_{j\ell}, Q_\Gamma]u(x) = -[m_{j\ell}, \mathcal{T}]u(x') + \text{p.f.} \int\limits_{\Gamma} \Big\{m_{j\ell}(x', \partial_{x'})k_A(x', z)$$
$$+\Big(\big(n_\ell(x')-n_\ell(y')\big)\frac{\partial}{\partial z_j}-\big(n_j(x')-n_j(y')\big)\frac{\partial}{\partial z_\ell}\Big)k_A(x', z)\Big\}\Big|_{z=x'-y'}u(y')ds_{y'} \tag{10.4.8}$$

for $x' \in \Gamma$. In local coordinates for $x' \in \Gamma$, $\varrho' = \chi_r(x')$ and for $A_{\Phi_r,0}$, the corresponding formula reads

$$\Big[\frac{\partial}{\partial\varrho_\nu}, Q_{\Gamma,\Phi_r}\Big]u(\varrho') = -\Big[\frac{\partial}{\partial\varrho_\nu}, \mathcal{T}_{\Phi_r}\Big]u(\varrho') \tag{10.4.9}$$
$$+ \text{p.f.} \int\limits_{U_r} \Big(\frac{\partial}{\partial\varrho_\nu}k_{A_{\Phi_r,0}}\big((\varrho', 0), z\big)\Big)\Big|_{z=(\varrho'-\eta',0)}u(T_r(\eta'))d\eta'\,,\ \nu = 1, \ldots, n-1\,.$$

Here $k_{A_{\Phi_r,0}}$ is given by (9.5.7) in terms of the parametric representation of the boundary manifold Γ.

Proof: Since in (10.4.7) $x \notin \Gamma$, differentiation and integration are interchangeable and integration by parts is allowed. To prove (10.4.8) we first apply Lemma 10.4.1 and obtain with Theorem 9.5.4

$$m_{j\ell}(x', \partial_{x'})Q_\Gamma u(x') = -m_{j\ell}(x', \partial_{x'})\mathcal{T}u(x')$$
$$+ \text{p.f.} \int\limits_{\Gamma} \big(m_{j\ell}(x', \partial_{x'})k_A(x', x'-y')\big)u(y')ds_{y'}\,. \tag{10.4.10}$$

Next, for $x \notin \Gamma$ we consider the boundary potentials

$$\widetilde{Q}_\Gamma(m_{j\ell}u)(x) = \int\limits_{\Gamma} k_A(x, x-y')\big(m_{j\ell}(y', \partial_{y'})u(y')\big)ds_{y'}$$

and

$$\widetilde{Q}'_\Gamma u(x) \;:=\; -\int\limits_\Gamma \big(m_{j\ell}(y', \partial_{y'})k_A(x, x-y')\big)u(y')ds_{y'}\,.$$

For the latter we note that its kernel is a Schwartz kernel of a pseudodifferential operator in $\mathcal{L}^{m+1}_{c\ell}(\widetilde{\Omega})$ with symbol of rational type. Moreover,

$$\widetilde{Q}_\Gamma(m_{j\ell}u)(x) - \widetilde{Q}'_\Gamma u(x) = 0 \quad \text{for all } \; x \notin \Gamma\,.$$

By applying Theorem 9.5.5 and subsequently Theorem 9.5.7, we then obtain

$$\mathcal{T}(m_{j\ell}u)(x') - \mathcal{T}'u(x') = \tfrac{1}{2}[\widetilde{Q}_\Gamma(m_{j\ell}u) - \widetilde{Q}'_\Gamma u]_\Gamma = 0$$

which implies

$$\begin{aligned}
&\lim_{\Omega\ni x\to\Gamma} \widetilde{Q}_\Gamma(m_{j\ell}u)(x)\\
&= -\mathcal{T}(m_{j\ell}u) + \text{ p.f.} \int\limits_{y'\in\Gamma} k_A(x', x'-y')m_{j\ell}(y', \partial_{y'})u(y')ds_{y'}\\
&= \lim_{\Omega\ni x\to\Gamma} \widetilde{Q}'_\Gamma u(x)\\
&= -\mathcal{T}'u(x') - \text{ p.f.} \int\limits_{y'\in\Gamma} \big(m_{j\ell}(y', \partial_{y'})k_A(x', x'-y')\big)u(y')ds_{y'}\,.
\end{aligned}$$

Since $\mathcal{T}'u = \mathcal{T}m_{j\ell}u$, we find

$$\begin{aligned}
&Q_\Gamma m_{j\ell}u(x') = -\mathcal{T}m_{j\ell}u(x')\\
&- \text{ p.f.} \int\limits_{y'\in\Gamma,} \big(m_{j\ell}(y', \partial_{y'})\big)k_A(x', x'-y')u(y')ds_{y'} \quad \text{for } \; x' \in \Gamma\,. \qquad (10.4.11)
\end{aligned}$$

Hence, the commutator has the representation

$$\begin{aligned}
&[m_{j\ell}, Q_\Gamma]u(x') = -[m_{j\ell}, \mathcal{T}]u(x')\\
&\qquad + \text{ p.f.} \int\limits_\Gamma \big\{\big(m_{j\ell}(x', \partial_{x'}) + m_{j\ell}(y', \partial_{y'})\big)k_A(x', x'-y')\big\}u(y')ds_{y'}\,.
\end{aligned}$$

By using the definition of $m_{jk}(x, \partial_x)$ in (10.4.1) and $z = x-y$, the formula (10.4.8) follows with the standard chain rule.

For the formula (10.4.9), we note that in the tubular coordinates the transformed pseudodifferential operator $A_{\Phi_r,0}$ defines a boundary potential $\widetilde{Q}_{\Gamma,\Phi_r}$ in the half space where $n(x') = n(y')$. Hence (10.4.9) follows from (10.4.8) since the last terms involving $\frac{\partial}{\partial z}k_A(x', z)$ vanish. ■

To illustrate how to compute the derivatives $\frac{\partial}{\partial x_\ell}$ on Γ, let us consider one of the general equations (10.0.15) or (10.0.16), respectively, e.g. Equation

(10.0.15) where φ is given and λ is obtained from solving the system (10.0.15) for u in Ω and for λ on Γ. Then we apply $m_{jk}(x, \partial_x)$ to the equation (10.0.15) and obtain

$$m_{jk}\varphi - m_{jk}R\gamma \underset{\sim}{T}u = m_{jk}R\gamma \underset{\sim}{N}f + R\gamma\mathfrak{V}(m_{jk}\lambda) + [m_{jk}\,,\ R\gamma\mathfrak{V}]\lambda - m_{jk}R\gamma\mathfrak{W}\varphi \quad (10.4.12)$$

which is an integral equation on Γ for $m_{jk}\lambda$ if we compute $m_{jk}\varphi$ directly from the given data φ on Γ. Once we have $m_{jk}\varphi$ and $m_{jk}\lambda$ available, with the matrix $\mathcal{N}$ of the tangential differential operators given in (3.9.5), we obtain

$$\Big(m_{jk}u\,,\ m_{jk}\frac{\partial u}{\partial n}, \dots, m_{jk}\frac{\partial^{m-1}}{\partial n^{m-1}}u\Big)^\top = \mathcal{N}\begin{pmatrix} m_{jk}\varphi \\ m_{jk}\lambda\end{pmatrix} + [m_{jk}\,,\ \mathcal{N}]\begin{pmatrix}\varphi\\ \lambda\end{pmatrix}. \quad (10.4.13)$$

With (10.4.5), we finally recover all the derivatives of the first order on Γ,

$$\frac{\partial u}{\partial x_j}|_\Gamma = \sum_{k=1}^n n_k(x)m_{jk}(x, \partial_x)u(x) + n_j(x)\frac{\partial u}{\partial n}(x)\,.$$

For calculating the second or higher order derivatives, we employ a bootstrapping procedure by repeating the above procedure with φ and λ replaced by $\mathcal{D}^t\varphi$ and $\mathcal{D}^t\lambda$ (see (3.4.40)) for $t = 1, \dots, m-1$ consecutively and obtain the derivatives

$$D^\alpha u|_\Gamma = \prod_{\ell=1}^n \Big(\sum_{k=1}^n n_k(x)m_{\ell k}(x, \partial_x) + n_\ell(x)\frac{\partial}{\partial n}\Big)^{\alpha_\ell} u|_\Gamma \quad (10.4.14)$$

for $|\alpha| \le 2m-1$ where $2m$ is the order of P.

If one needs the $2m$–th order derivatives on Γ, then for $|\alpha| = 2m$, in (10.4.14), $\big(\frac{\partial}{\partial n}\big)^{2m} u$ cannot be computed from the boundary integral operators alone. Since u is a solution of the partial differential equation (10.1.1), $Pu = f$, we may use this equation in the tubular coordinates along Γ in the form (3.4.49) (or (A.0.20)). By inserting (10.4.14) into one of these relations, we finally obtain with (3.4.51)

$$\Big(\frac{\partial}{\partial n}\Big)^{2m} u\big|_\Gamma = (\underset{\sim}{\mathcal{P}}_{2m})^{-1}\Big\{f - \sum_{k=0}^{2m-1}\underset{\sim}{\mathcal{P}}_k\frac{\partial^k u}{\partial n^k}\Big\}\big|_\Gamma \quad (10.4.15)$$

where $\underset{\sim}{\mathcal{P}}_k$ are tangential operators of orders at most $2m-k$. The derivatives $\underset{\sim}{\mathcal{P}}_k\frac{\partial^k u}{\partial n^k}$ for $k = 0, \dots, m-1$ have been obtained by the bootstrapping procedure described above.

Of course, instead of using equation (10.0.15), one may use (10.0.16) instead with the same procedure.

In the following we return to the examples of the Helmholtz equation and the Lamé system.

10.4.1 Derivatives of the Solution to the Helmholtz Equation

We begin with the representation of the solution of the Helmholtz equation,

$$u(x) = \int\limits_\Gamma E_k(x,y)\frac{\partial u}{\partial n}(y)ds_y - \int\limits_\Gamma \Big(\frac{\partial}{\partial n_y}E_k(x,y)\Big)u(y)ds_y \quad \text{for } x \in \Omega\,. \tag{10.4.16}$$

Here the Cauchy data are $\varphi = u|_\Gamma$ and $\sigma = \frac{\partial u}{\partial n}|_\Gamma$. If $\sigma = \frac{\partial u}{\partial n}|_\Gamma$ is given in $H^{s-1}(\Gamma)$ then we first solve for φ the integral equation of the second kind (2.1.11) (or of the first kind with the hypersingular operator, respectively, see Table 2.1.1). Equation (2.1.11) reads

$$\frac{1}{2}\varphi(x) + \int\limits_\Gamma \Big(\frac{\partial}{\partial n_y}E_k(x,y)\Big)\varphi(y)ds_y = \int\limits_\Gamma E_k(x,y)\sigma(y)ds_y\,; \tag{10.4.17}$$

and $\varphi \in H^s(\Gamma)$ due to the mapping properties of the boundary integral operators and to the shift theorem 9.7.3. Then we may apply the operator $m_{j\ell}(x,\partial_x)$ on $u(x)$ in (10.4.16),

$$\begin{aligned}\varphi_{j\ell}(x) := m_{j\ell}(x,\partial_x)u(x) &= \int\limits_\Gamma \big(m_{j\ell}(x,\partial_x)E_k(x,y)\big)\sigma(y)ds_y \\ &\quad - \int\limits_\Gamma \big(m_{j\ell}(x,\partial_x)\frac{\partial}{\partial n_y}E_k(x,y)\big)u(y)ds_y \quad \text{for } x\in\Omega\,.\end{aligned} \tag{10.4.18}$$

Taking the limit $x \to \Gamma$, we obtain from (10.4.18) a resulting boundary integral equation of the second kind for the desired tangential derivative $\varphi_{j\ell}(x) = m_{j\ell}(x,\partial_x)u(x)$:

$$\begin{aligned}&\frac{1}{2}\varphi_{j\ell}(x) + \int\limits_\Gamma \Big(\frac{\partial}{\partial n_y}E_k(x,y)\Big)\varphi_{j\ell}\sigma(y)ds_y \\ &\quad = \text{p.f.}\int\limits_\Gamma \big(m_{j\ell}(x,\partial_x)E_k(x,y)\big)\sigma(y)ds_y - ([m_{j\ell}\,,\,K_k]\varphi)(x)\end{aligned} \tag{10.4.19}$$

where φ is now the solution of the boundary integral equation (10.4.17) and σ is given. Here $[m_{j\ell}\,,\,K_k]$ is the commutator given by

$$\begin{aligned}([m_{j\ell}\,,\,K_k]\varphi)(x) &= (m_{j\ell}\circ K_k - K_k\circ m_{j\ell})\varphi(x) \qquad (10.4.20)\\ &= \text{p.f.}\int\limits_\Gamma \Big(m_{j\ell}(x,\partial_x)\frac{\partial}{\partial n_y}E_k(x,y) - \frac{\partial}{\partial n_y}E_k(x,y)m_{j\ell}(y,\partial_y)\Big)\varphi(y)ds_y \\ &= \text{p.f.}\int\limits_\Gamma \Big\{m_{j\ell}(x,\partial_x)\frac{\partial}{\partial n_y}E_k(x,y) + m_{j\ell}(y,\partial_y)\frac{\partial}{\partial n_y}E_k(x,y)\Big\}\varphi(y)ds_y\end{aligned}$$

which is a weakly singular integral operator on φ due to Theorem 10.4.2. Now, $\varphi_{j\ell}(x)$ can be obtained by solving the integral equation (10.4.19) of the second kind and all derivatives $\frac{\partial}{\partial x_j}$ on Γ can be obtained from (10.4.5). In fact, in the following we show directly that the kernel of the commutator in (10.4.20) is of order $O(r^{-1})$ where $r = |x-y|$. To this end, let us write

$$\frac{\partial E_k(x,y)}{\partial n_y} = (y-x)\cdot n(y) f(r)$$

where $f(r) = \frac{1}{4\pi r^3} e^{ikr}(ikr-1) = O(r^{-3})$ and $r = |x-y|$. Then

$$\begin{aligned}\big(m_{j\ell}(x,\partial_x) + m_{j\ell}(y,\partial_y)\big)\frac{\partial E_k}{\partial n_y} &= (y-x)\cdot n(y)\big(m_{j\ell}(x,\partial_x) + m_{j\ell}(y\partial_y)\big) f(r)\\ &\quad + f(r)\big(m_{j\ell}(x,\partial_x) + m_{j\ell}(y,\partial_y)\big)(y-x)\cdot n(y)\,. \qquad (10.4.21)\end{aligned}$$

For the first term it is easy to show that it is weakly singular for smooth Γ since

$$\begin{aligned} &(y-x)\cdot n(y)\big(m_{j\ell}(x,\partial_x) + m_{j\ell}(y,\partial_y)\big) f(r)\\ &= (y-x)\cdot n(y)\Big\{\big(n_j(x) - n_j(y)\big)\frac{\partial}{\partial y_\ell} - \big(n_\ell(x) - n_\ell(y)\big)\frac{\partial}{\partial y_j}\Big\} f(r)\\ &= O(r^2)\cdot O(r^{-3}) = O(r^{-1})\,. \qquad (10.4.22)\end{aligned}$$

For the second term we use the chain rule and Taylor expansion about y to obtain

$$\begin{aligned} &\big(m_{jk}(x,\partial_x) + m_{jk}(y,\partial_y)\big)\big(n(y)\cdot(y-x)\big) \qquad (10.4.23)\\ &= -n_k(x)n_j(y) + n_j(x)n_k(y) + \Big\{n_k(y)\frac{\partial n}{\partial y_j}(y) - n_j(y)\frac{\partial n}{\partial y_k}(y)\Big\}\cdot(y-x)\\ &= \sum_{\ell=1}^{n}\Big\{n_j\frac{\partial n_k}{\partial y_\ell} - n_k\frac{\partial n_j}{\partial y_\ell} + n_k\frac{\partial n_\ell}{\partial y_j} - n_j\frac{\partial n_\ell}{\partial y_k}\Big\}\Big|_y (y_\ell - x_\ell) + O(r^2)\,.\end{aligned}$$

From (3.4.6) with $\frac{\partial n_k}{\partial \varrho_n} = 0$ and (3.4.28) we obtain

$$\begin{aligned}\frac{\partial n_\ell}{\partial y_k} &= \sum_{\mu,\lambda=1}^{n-1}\frac{\partial y_k}{\partial \varrho_\mu} g^{\mu\lambda}\frac{\partial n_\ell}{\partial \varrho_\lambda} = -\sum_{\mu,\lambda,\nu=1}^{n-1}\frac{\partial y_k}{\partial \varrho_\mu} g^{\mu\lambda} L_\lambda^\nu \frac{\partial y_\ell}{\partial \varrho_\nu}\\ &= -\sum_{\mu,\nu=1}^{n-1}\frac{\partial y_k}{\partial \varrho_\mu} L^{\mu\nu}\frac{\partial y_\ell}{\partial \varrho_\nu} = \frac{\partial n_k}{\partial y_\ell} \qquad (10.4.24)\end{aligned}$$

since $L^{\mu\nu} = L^{\nu\mu}$, where we have employed (3.4.17). Inserting (10.4.24) into (10.4.23), we find

$$\begin{aligned}
&\big(m_{jk}(x,\partial_x)+m_{jk}(y,\partial_y)\big)\big(n(y)\cdot(y-x)\big)\\
&= \sum_{\ell=1}^{n}\sum_{\mu,\nu=1}^{n-1}\Big\{-n_j(y)\frac{\partial y_\ell}{\partial\varrho_\mu}L^{\mu\nu}\frac{\partial y_k}{\partial\varrho_\nu}+n_k(y)\frac{\partial y_\ell}{\partial\varrho_\mu}L^{\mu\nu}\frac{\partial y_j}{\partial\varrho_\nu}\\
&\qquad -n_k(y)\frac{\partial y_j}{\partial\varrho_\mu}L^{\mu\nu}\frac{\partial y_\ell}{\partial\varrho_\nu}+n_j(y)\frac{\partial y_k}{\partial\varrho_\mu}L^{\mu\nu}\frac{\partial y_\ell}{\partial\varrho_\nu}\Big\}(y_\ell-x_\ell)+O(r^2)\\
&= O(r^2)\,. \qquad (10.4.25)
\end{aligned}$$

With $f(r)=O(r^{-3})$ and collecting (10.4.25) and (10.4.22) we finally obtain that (10.4.21), i.e. the kernel of (10.4.20), is of order $O(r^{-1})$.

Since φ and σ on Γ are known, boundary integral equation (10.4.19) is the same as (10.4.17), only the right–hand side has been modified.

Clearly, if we apply the operator m_{rs} again, now to equation (10.4.19), we may apply the same arguments as before to compute higher order tangential derivatives of φ. For higher order normal derivatives we use the Helmholtz equation in the form (10.4.15) and later on apply m_{rs} to the given data $\sigma=\frac{\partial u}{\partial n}|_\Gamma$.

10.4.2 Computation of Stress and Strain on the Boundary for the Lamé System

In applications one is not only interested in the boundary traction and boundary displacement of the elastic field but also in the complete stress and strain on Γ. We now show how the procedure for computing derivatives of potentials in this section can be applied to this situation.

As in the case of the Helmholtz equation let us begin with the representation formula (2.2.4) with $\boldsymbol{f}=\boldsymbol{0}$,

$$\boldsymbol{v}(x)=\int_\Gamma E(x,y)\boldsymbol{t}(y)ds_y-\int_\Gamma \big(T_yE(x,y)\big)^\top\boldsymbol{\varphi}(y)ds_y \quad \text{for } x\in\Omega \qquad (10.4.26)$$

where $\boldsymbol{t}(y)=T_y\boldsymbol{v}(y)$ now denotes the boundary traction and $\boldsymbol{\varphi}=\boldsymbol{v}|_\Gamma$ the boundary displacement. We assume that both are known from solving (2.2.16) or (2.2.17) for the relevant missing Cauchy data with $\boldsymbol{\sigma}=\boldsymbol{t}$. To obtain the complete stress or strain up to the boundary, we need all of the first derivatives $\frac{\partial v_j}{\partial x_i}$ on Γ. For this we need to compute $\varphi_{\ell jk}:=m_{\ell j}\varphi_k$ on Γ which can be obtained from one of the boundary integral equations, e.g. from (2.2.16), i.e.

$$\begin{aligned}
\boldsymbol{\varphi}_{\ell j}(x) &= (\tfrac{1}{2}I-K)\boldsymbol{\varphi}_{\ell j} \qquad (10.4.27)\\
&\quad -[m_{\ell j},K]\boldsymbol{\varphi}+\text{p.v.}\int_\Gamma m_{\ell j}(x,\partial_x)E(x,y)\boldsymbol{t}(y)ds_y \quad \text{for } x\in\Gamma\,.
\end{aligned}$$

Here $\boldsymbol{\varphi}_{\ell j}=(\varphi_{\ell j},\dots,\varphi_{\ell jn})^\top$ is the desired new unknown. Note that this is the same system of singular integral equations for all the $\boldsymbol{\varphi}_{\ell j}$ as in (2.2.16) for φ, with only modified right–hand sides. In particular, we note that the commutator

$$[m_{j\ell}, K]\varphi(x) = \text{p.v.} \int\limits_{\Gamma} \big((m_{j\ell}(x,\partial_x) + m_{j\ell}(y,\partial_y)\big)\big(T_y E(x,y)\big)^\top \varphi(y) ds_y \tag{10.4.28}$$

is of the same type singularity as K, hence, it is an operator of the class $\mathcal{L}^0_{c\ell}(\Gamma)$. We shall further discuss the commutator in more detail at the end of this section. After solving (10.4.27) for $\varphi_{\ell jk}$ with $\ell, j, k = 1, \dots, n$, we now may compute the stress tensor σ_{ij} and the strain tensor ε_{ij} on Γ as follows:

If $\varphi_{\ell jk}$ on Γ are available, we introduce the surface divergence operator

$$\begin{aligned} \mathrm{Div}_\Gamma \boldsymbol{\varphi} &= n_p \varphi_{rpr} = n_p \Big(n_p \frac{\partial}{\partial x_r} - n_r \frac{\partial}{\partial x_p} \Big) \varphi_r \\ &= \Big(\mathrm{div} \boldsymbol{v} - n \cdot \frac{\partial \boldsymbol{v}}{\partial n} \Big) \Big|_\Gamma \quad \text{on } \Gamma . \end{aligned} \tag{10.4.29}$$

Here and in what follows we use again the Einstein summation convention. Then the boundary traction (2.2.5) can be rewritten in the form

$$\begin{aligned} t_\ell &= (\lambda + \mu)(\mathrm{div}\boldsymbol{v}) n_\ell + \mu \frac{\partial v_\ell}{\partial n} + \mu \varphi_{\ell rr} \\ &= (\lambda + \mu)\Big(\mathrm{Div}_\Gamma \boldsymbol{\varphi} + n \cdot \frac{\partial \boldsymbol{v}}{\partial n} \Big) n_\ell + \mu \varphi_{\ell rr} . \end{aligned} \tag{10.4.30}$$

Multiplication by n_ℓ and summation lead to the identity

$$(\lambda + 2\mu) n \cdot \frac{\partial \boldsymbol{v}}{\partial n} |_\Gamma = n \cdot \boldsymbol{t} - \lambda \mathrm{Div}_\Gamma \boldsymbol{\varphi} \tag{10.4.31}$$

from which with (10.4.30) we obtain

$$\mu \frac{\partial v_\ell}{\partial n} |_\Gamma = t_\ell - \frac{\lambda + \mu}{\lambda + 2\mu} n \cdot \boldsymbol{t} n_\ell - 2\mu \frac{\lambda + \mu}{\lambda + 2\mu} \mathrm{Div}_\Gamma \boldsymbol{\varphi} n_\ell - \mu \varphi_{\ell rr} . \tag{10.4.32}$$

Now the stress tensor on Γ can be written in the form

$$\begin{aligned} \sigma_{ij} &= \lambda \mathrm{div} \boldsymbol{v} \delta_{ij} + \mu \Big(\frac{\partial v_i}{\partial x_j} + \frac{\partial v_j}{\partial x_i} \Big) = \lambda \Big(\mathrm{Div}_\Gamma \boldsymbol{\varphi} + \boldsymbol{n} \cdot \frac{\partial \boldsymbol{v}}{\partial n} \Big) \delta_{ij} \\ &\quad + \mu \Big(\Big(\frac{\partial v_i}{\partial x_j} - n_j \frac{\partial v_i}{\partial n} \Big) + \Big(\frac{\partial v_j}{\partial x_i} - n_i \frac{\partial v_j}{\partial n} \Big) + n_j \frac{\partial v_i}{\partial n} + n_i \frac{\partial v_j}{\partial n} \Big) \\ &= \lambda (\mathrm{Div}_\Gamma \boldsymbol{\varphi}) \delta_{ij} + \frac{\lambda}{\lambda + 2\mu} (n \cdot \boldsymbol{t} - \lambda \mathrm{Div}_\Gamma \boldsymbol{\varphi}) \delta_{ij} \\ &\quad + \mu (n_q \varphi_{jqi} + n_q \varphi_{iqj} + n_j \varphi_{rir} + n_i \varphi_{rjr}) \\ &\quad + n_j \Big\{ t_i - n_i \frac{\lambda + \mu}{\lambda + 2\mu} n \cdot \boldsymbol{t} - n_i \frac{2\mu(\lambda + \mu)}{\lambda + 2\mu} \mathrm{Div}_\Gamma \boldsymbol{\varphi} \Big\} \\ &\quad + n_i \Big\{ t_j - n_j \frac{\lambda + \mu}{\lambda + 2\mu} n \cdot \boldsymbol{t} - n_j \frac{2\mu(\lambda + \mu)}{\lambda + 2\mu} \mathrm{Div}_\Gamma \boldsymbol{\varphi} \Big\} ; \end{aligned}$$

or explicitly in terms of the computed $\varphi_{\ell jk}$ as

$$\begin{aligned}\sigma_{ij_{|\Gamma}} &= \frac{1}{\lambda+2\mu}(n\cdot \boldsymbol{t} + 2\mu \mathrm{Div}_\Gamma \boldsymbol{\varphi})\big(\lambda\delta_{ij} - 2(\lambda+\mu)n_i n_j\big) \\ &\quad + n_j t_i + n_i t_j + \mu(n_r\varphi_{jri} + n_r\varphi_{irj} + n_j\varphi_{rir} + n_i\varphi_{rjr})\,. \end{aligned} \tag{10.4.33}$$

Then from Hooke's law we also obtain the strain tensor on Γ, i.e.

$$\begin{aligned} 2\mu\varepsilon_{ij_{|\Gamma}} &= \mu\Big(\frac{\partial v_j}{\partial x_i} + \frac{\partial v_i}{\partial x_j}\Big)_{|\Gamma} \\ &= \mu(n_k\varphi_{ikj} + n_k\varphi_{jki} + n_j\varphi_{rir} + n_i\varphi_{rjr}) \\ &\quad + n_i t_j + n_j t_i - 2\frac{\lambda+\mu}{\lambda+2\mu}(n\cdot\boldsymbol{t} + 2\mu\mathrm{Div}_\Gamma\boldsymbol{\varphi})n_i n_j\,. \end{aligned} \tag{10.4.34}$$

To this end, for three–dimensional problems with $n = 3$, let us first rewrite the kernel of K according to our formula (2.2.30):

$$\begin{aligned} k_K(x, x-y) &= \big(T_y E(x,y)\big)^\top \\ &= \frac{\partial}{\partial n_y}\Big(\frac{1}{4\pi}\frac{1}{r}\Big)I + \Big(\Big(m_{st}(y,\partial_y)\Big(\frac{1}{4\pi}\frac{1}{r}\Big)\Big)\Big)_{3\times 3} \\ &\quad - 2\mu\big(\big(m_{sr}(y,\partial_y)E_{rt}(x,y)\big)\big)_{3\times 3} \end{aligned} \tag{10.4.35}$$

where $E(x,y) = \big((E_{rt}(x,y))\big)_{3\times 3}$. Then the elements of the commutator

$$\begin{aligned} &[m_{j\ell}, K_{st}]\varphi_t(x) = \\ &\int\limits_\Gamma \big(m_{j\ell}(x,\partial_x) + m_{j\ell}(y,\partial_y)\big)\Big(\frac{\partial}{\partial n_y}\frac{1}{4\pi}\frac{1}{r}\Big)\varphi_s(y)ds_y \\ &+ \text{p.v.}\int\limits_\Gamma \big(m_{j\ell}(x,\partial_x) + m_{j\ell}(y,\partial_y)\big)\Big(m_{st}(y,\partial_y)\Big(\frac{1}{4\pi}\frac{1}{r}\Big)\Big)\varphi_t(y)ds_y \\ &- 2\mu\text{p.v.}\int\limits_\Gamma \big(m_{j\ell}(x,\partial_x) + m_{j\ell}(y,\partial_y)\big)\big(m_{sr}(y,\partial_y)E_{rt}(x,y)\big)\varphi_t(y)ds_y \end{aligned}$$

for $s, j, \ell = 1, 2, 3$.

Now we examine each term on the right–hand side. The first term corresponds to the commutator (10.4.20) for $k = 0$ and is a weakly singular kernel defining on Γ a pseudodifferential operator in the class $\mathcal{L}^{-1}_{c\ell}(\Gamma)$. The next two terms, however, define kernels of classical singular integral operators which are in the class $\mathcal{L}^0_{c\ell}$. For the kernel of the second term we have

$$\begin{aligned}
&\frac{1}{4\pi}\big(m_{j\ell}(x,\partial_x)+m_{j\ell}(y,\partial_y)\big)m_{st}(y,\partial_y)\frac{1}{r} \\
&\quad = \frac{1}{4\pi}m_{st}(y,\partial_y)\big(m_{j\ell}(x,\partial_x)+m_{j\ell}(y,\partial_y)\big)\frac{1}{r} \\
&\qquad + \frac{1}{4\pi}\big(m_{j\ell}(y,\partial_y)m_{st}(y,\partial_y)+m_{st}(y,\partial_y)m_{j\ell}(y,\partial_y)\big)\frac{1}{r} \\
&\quad = \frac{1}{4\pi}m_{st}(y,\partial_y)\Big\{\big(n_\ell(y)-n_\ell(x)\big)\frac{\partial}{\partial y_j}+\big(n_j(y)-n_j(x)\big)\frac{\partial}{\partial y_\ell}\Big\}\frac{1}{r} \\
&\qquad + \frac{1}{4\pi}\big[m_{j\ell}(y,\partial_y)\,,\,m_{st}(y,\partial_y)\big]\frac{1}{r}
\end{aligned}$$

with the commutator $[m_{j\ell},m_{st}]_{|\Gamma}$. If we employ (10.4.24) on Γ then

$$\begin{aligned}
&[m_{j\ell}(y,\partial_y)\,,\,m_{st}(y,\partial_y)]_{|\Gamma} \\
&\quad = \frac{\partial n_t}{\partial y_j}m_{k\ell}(y,\partial_y)+\frac{\partial n_k}{\partial y_\ell}m_{tj}(y,\partial_y)+\frac{\partial n_\ell}{\partial y_t}m_{jk}(y,\partial_y)+\frac{\partial n_j}{\partial y_k}m_{\ell t}(y,\partial_y)
\end{aligned}$$

defines a tangential differential operator of the first order on Γ. Collecting terms, we conclude that for the second term we have a singular integral operator in the form

$$\begin{aligned}
&\frac{1}{4\pi}\text{ p.v.}\int_\Gamma\Big\{\big(m_{jk}(x,\partial_x)+m_{jk}(y,\partial_y)\big)m_{st}(y,\partial_y)\frac{1}{r}\Big\}\varphi_t(y)ds_y \\
&\quad = \frac{1}{4\pi}\text{ p.v.}\int_\Gamma\Big\{m_{st}(y,\partial_y)\frac{1}{r}\Big(\big(n_\ell(y)-n_\ell(x)\big)(y_j-x_j) \\
&\qquad\qquad + \big(n_j(y)-n_j(x)\big)(y_\ell-x_\ell)\Big)\Big\}\varphi_t(y)ds_y \\
&\qquad + \frac{1}{4\pi}\text{ p.v.}\int_\Gamma\Big\{[m_{j\ell}(y,\partial_y)\,,\,m_{st}(y,\partial_y)]\frac{1}{r}\Big\}\varphi_t(y)ds_y\,.
\end{aligned}$$

The last term of the commutator can be treated in the same manner resulting in the singular integral operator

$$\begin{aligned}
&-2\mu\text{ p.v.}\int_\Gamma\big(m_{j\ell}(x,\partial_x)+m_{j\ell}(y,\partial_y)\big)\big(m_{sr}(y,\partial_y)E_{rt}(x,y)\big)\varphi_t(y)ds_y \\
&\quad = -2\mu\text{ p.v.}\int_\Gamma\Big\{m_{st}(y,\partial_y)\Big\{\big(n_\ell(y)-n_\ell(x)\big)\frac{\partial E_{rt}(x,y)}{\partial y_j} \\
&\qquad\qquad + \big(n_j(y)-n_j(x)\big)\frac{\partial E_{rt}(x,y)}{\partial y_\ell}\Big\}\Big\}\varphi_t(y)ds_y \\
&\qquad - 2\mu\text{ p.v.}\int_\Gamma\{[m_{j\ell}(y,\partial_y)\,,\,m_{st}(y,\partial_y)]E_{rt}(x,y)\}\varphi_t(y)ds_y\,.
\end{aligned}$$

10.5 Remarks

For higher order tangential derivatives of φ we may repeatedly apply $m_{j\ell}(y, \partial_y)$ to (10.4.27) again. Of course, the same bootstrapping procedure can be used for computing tangential derivatives of the boundary tractions $\boldsymbol{t}$ via one of the boundary integral operators (2.2.16) or (2.2.17) between the Cauchy data. For computing higher order normal derivatives, we need to employ the corresponding formula (10.4.15) for the Lamé system in local coordinates together with their normal derivatives as desired.

In concluding this section, we see how we can compute the derivatives of the boundary potentials on the boundary by solving the **same boundary integral equations** for the tangential derivatives as for the Cauchy data by appropriate modifications of the right–hand sides but without additional effort. This means that available computer packages can be adopted with only slight modifications. Moreover, from the mathematical point of view, the boundary integral equations for the tangential derivatives in the variational formulation satisfy the same Gårding inequalities as the bilinear forms of the original boundary integral equations.

11. Boundary Integral Equations on Curves in $\mathbb{R}^2$

In Chapter 10 we presented the essence of boundary integral equations recast as pseudodifferential operators on boundary manifolds $\Gamma \subset \mathbb{R}^n$ for $n = 3$. In this chapter we present the two–dimensional theory of classical pseudodifferential and boundary integral operators based on Fourier analysis. In general, the representations of boundary potentials are based on the local charts and local coordinates (3.3.3)–(3.3.5). For $n = 2$, however, every closed Jordan curve Γ_j as a part of Γ admits a global parametric representation (4.2.45). Moreover, any function defined on a closed curve can be identified with a 1–periodic function on $\mathbb{R}$. These global representations allow one to use Fourier series expansions in the theory of pseudodifferential operators. The latter lead to explicit expressions in terms of corresponding Fourier coefficients. Moreover, the Sobolev spaces on Γ can be characterized in terms of the function's Fourier coefficients as in the Lemmata 4.2.4, 4.2.5 and Corollary 4.2.6.

If the boundary Γ consists of p distinct, simply closed, smooth curves Γ_j , $j = 1, \ldots, p$, then for each curve Γ_j we may consider the periodic extension of a function defined on Γ_j and identify the original function on Γ with a p–vector valued periodic function.

For simplicity, we first consider Ω a bounded, simply connected domain having one simply closed smooth boundary curve Γ which admits a global regular parametric 1–periodic representation (3.3.7) or (4.2.45), i.e.

$$\Gamma : x = T(t) \quad \text{for } t \in \mathbb{R} \quad \text{and} \quad T(t+1) = T(t) \tag{11.0.1}$$

satisfying $\left|\frac{dT}{dt}\right| \geq c_0 > 0$ for all $t \in \mathbb{R}$. Hence, we need just one global chart for the representation of Γ, and the local representation in Section 9.1 can now be identified with the global representation and t stands for $\underset{\sim}{\varrho}'$. The tubular neighbourhood $\widetilde{\Omega}$ of Γ now also has one global chart defined by

$$x = \Psi(t, \varrho_n) = T(t) + \varrho_n n(t) \tag{11.0.2}$$

for $t \in \mathbb{R}$, $\varrho_n \in [-\varepsilon, \varepsilon]$ and $x \in \widetilde{\Omega}$. Clearly, all of the formulations in Chapter 9 in terms of local charts (O_r, U_r, χ_r) can now be reduced to global formulations based on (11.0.2), i.e. $O_r = \widetilde{\Omega}$, $U_r = \mathbb{R} \times (-\varepsilon, \varepsilon)$ and χ_r the inverse to T.

G. C. Hsiao und W. L. Wendland, *Boundary Integral Equations*,
Applied Mathematical Sciences 164,
https://doi.org/10.1007/978-3-030-71127-6_11

11.1 Representation of the basic operators for the 2D–Laplacian in terms of Fourier series

Before we give the general formulations concerning the pseudodifferential operators on Γ in terms of Fourier series expansions let us illustrate the main ideas by analyzing the four basic boundary integral operators V, K, K' and D of the two–dimensional Laplacian as introduced in Section 1.2. Further examples can be found in [197] and will also be treated below. In particular, we want to compute their symbols in terms of Fourier coefficients as well as the action of these operators on Fourier series.

For the simple layer potential V on the closed curve Γ, parameterized globally according to (11.0.1), we write

$$\begin{aligned} V\sigma(x) &= \frac{1}{2\pi}\int\limits_\Gamma \log\frac{1}{|x-y|}\sigma(y)ds_y \qquad (11.1.1)\\ &= -\frac{1}{4\pi}\int\limits_0^1 \log\left(4\sin^2\pi(t-\tau)\right)\widetilde{\sigma}(\tau)d\tau + \frac{1}{4\pi}\int\limits_\Gamma k_\infty(t,\tau)\widetilde{\sigma}(\tau)d\tau \end{aligned}$$

where $x=T(t)$, $y=T(\tau)$, $\widetilde{\sigma}(\tau)=\sigma\big(T(\tau)\big)\frac{ds_y}{d\tau}$, and

$$k_\infty(t,\tau) := 2\log\Big(\frac{|2\sin\pi(t-\tau)|}{|T(t)-T(\tau)|}\Big) \qquad (11.1.2)$$

by using the identity $|e^{i2\pi t}-e^{i2\pi\tau}| = 2|\sin\pi(t-\tau)|$ in the representation (11.1.1). Moreover, we let τL be the arc length parameterization on Γ where L is the length of Γ. For $\Gamma\in C^\infty$, the kernel $k_\infty\in C^\infty(\Gamma\times\Gamma)$ is biperiodic and defines a smoothing operator in $\mathcal{L}^{-\infty}(\Gamma)$.

If the periodic function $\widetilde{\sigma}$ is represented by its Fourier series expansion

$$\widetilde{\sigma}(\tau) = \sum_{\ell=\mathbb{Z}}\widehat{\sigma}_\ell e^{2\pi i\ell\tau} \quad\text{with } \widehat{\sigma}_\ell = L\int\limits_0^1 e^{-2\pi i\ell\tau}\sigma(\tau)d\tau \qquad (11.1.3)$$

(see (4.2.46) and (4.2.47)) then we obtain the Fourier series representation of the simple layer potential operator V,

$$\begin{aligned} V\sigma((T(t)) &= -\frac{1}{4\pi}\int\limits_0^1 \log\left(4\sin^2\pi(t-\tau)\right)\sum_{\ell\in\mathbb{Z}}\widehat{\sigma}_\ell e^{2\pi i\ell\tau}d\tau\\ &\quad + \frac{1}{4\pi}\sum_{\ell\in\mathbb{Z}}\int\limits_\Gamma k_\infty(t,\tau)e^{2\pi i\ell\tau}d\tau\widehat{\sigma}_\ell\\ &= \sum_{\ell\neq 0}\frac{1}{4\pi|\ell|}\widehat{\sigma}_\ell e^{2\pi i\ell t} + (K_\infty\sigma)(t)\,. \qquad (11.1.4)\end{aligned}$$

By using the Fourier series form of the Sobolev space norms from Lemma 4.2.4 we see from (11.1.4) that for $\widetilde{\sigma} \in H^s(\Gamma)$,

$$\begin{aligned}\|V\widetilde{\sigma} - K_\infty\sigma\|^2_{H^{s+1}(\Gamma)} &= \sum_{\ell\neq 0}\Big|\frac{1}{4\pi|\ell|}\widehat{\sigma}_\ell\Big|^2|\ell|^{2s+2} \\ &\leq \Big(\frac{1}{4\pi}\Big)^2\sum_{\ell\in\mathbb{Z}}|\widehat{\sigma}_\ell|^2(\ell+\delta_{\ell 0})^{2s} = \Big(\frac{L}{4\pi}\Big)^2\|\sigma\|^2_{H^s(\Gamma)}\,.\end{aligned}$$

This implies the mapping property

$$\|V\sigma\|_{H^{s+1}(\Gamma)} \leq c_1\|\sigma\|_{H^s(\Gamma)} + c_2(s,s_0)\|\sigma\|_{H^{s_0}(\Gamma)} \tag{11.1.5}$$

for any $s_0 \leq s$; and the order of V is -1 as this explicit computation shows. Consequently, in view of Parseval's equality, (11.1.4) also implies

$$\begin{aligned}L^2\frac{1}{4\pi}\|\sigma\|^2_{H^{-\frac{1}{2}}(\Gamma)} &= \frac{1}{4\pi}\sum_{\substack{\ell\neq 0\\ \ell\in\mathbb{Z}}}\frac{1}{|\ell|}|\widehat{\sigma}_\ell|^2 + \frac{1}{4\pi}|\widehat{\sigma}_0|^2 \\ &= \langle V,\sigma,\sigma\rangle_{L^2(\Gamma)} + \langle C\sigma,\sigma\rangle_{L^2(\Gamma)} \qquad (11.1.6)\end{aligned}$$

with the compact operator C defined by

$$C\sigma = -K_\infty\sigma + \frac{L}{4\pi}\int_0^1 \sigma d\tau\,.$$

Clearly, (11.1.6) is Gårding's inequality for V and V is strongly elliptic.

The symbol of V can be computed from Theorem 7.1.11 since V is in $\mathcal{L}^{-1}_{c\ell}(\Gamma)$ because it stems from the trace of the two–dimensional operator Δ^{-1} having symbol of rational type in $\mathbb{R}^2$ and, moreover, is properly supported since Γ is compact. Formula (7.1.27) then yields, for

$$V\sigma(x) = \frac{1}{2\pi}\int_\Gamma \log\frac{1}{|x-y|}\psi(|x-y|)\sigma(y)ds_y + R_\Gamma\sigma \tag{11.1.7}$$

with an appropriate cut–off function ψ and R_Γ having a C^∞ kernel, the symbol in the global parameterization can be computed in the form

$$\begin{aligned}a_\psi(t,\xi) &= e^{-it\xi}\int_0^1\Big(-\frac{1}{2\pi}\log|2\sin\pi(t-\tau)|\Big)\psi(|t-\tau|)e^{i\xi\tau}d\tau \\ &= -\frac{1}{2\pi}\int_{\mathbb{R}}\big(\psi(|t-\tau|)e^{-i\xi(t-\tau)}\log|t-\tau|\big)d\tau \\ &\quad -\frac{1}{2\pi}\int_{\mathbb{R}} e^{-i\xi(t-\tau)}\log\Big|\frac{2\sin\pi(t-\tau)}{t-\tau}\Big|\psi(|t-\tau|)d\tau\,.\end{aligned}$$

Since the last term on the right–hand side is a C^∞-function with compact support, its Fourier transform is C^∞ and decays faster than any order of $|\xi|^{-N}$ for $N \in \mathbb{N}$ according to the Paley–Wiener–Schwartz Theorem 3.1.3. Hence, for the symbol we only need to consider the first term, which is given by

$$a_\psi(t,\xi) = -\frac{1}{2\pi}\int\limits_{\mathbb{R}} e^{i\xi z}\log|z|\psi(|z|)dz\,. \tag{11.1.8}$$

According to Definition 8.1.1, the kernel $k_0(x,z) = \log|z|$ in (11.1.8) is a pseudohomogeneous function of degree 0. Therefore, for finding the complete homogeneous symbol we employ formula (8.1.29) and find

$$a_{-1}(t,\xi) = \lim_{\varrho\to+\infty} \varrho a_\psi(t,\varrho\xi)\,. \tag{11.1.9}$$

However, it is not clear how to compute this limit without detailed information on the oscillating integral in (11.1.8). We therefore integrate by parts and obtain

$$\begin{aligned}
a_\psi(t,\xi) &= \frac{1}{i\xi}\,\text{p.f.}\,\frac{1}{2\pi}\int\limits_{\mathbb{R}} e^{i\xi z}\Big(\frac{1}{z}\psi(|z|) + \log|z|\psi'\Big)dz\\
&= \frac{1}{2\pi i\xi}\,\text{p.f.}\int\limits_{\mathbb{R}} e^{i\xi z}\frac{1}{z}\psi(|z|)dz\\
&\quad + \frac{1}{2\pi\xi^2}\int\limits_{\frac12\le|z|\le a}\Big(\frac{1}{z}\psi' + \log|z|\psi''\Big)dz\,.
\end{aligned}$$

Now the limit in (11.1.9) can be performed and yields with the Fourier transform of z^{-1} (see Gelfand and Shilov [143, p.360])

$$\begin{aligned}
a_{-1}(t,\xi) &= \lim_{\varrho\to+\infty}\Big\{\frac{\varrho}{2\pi i\varrho\xi}\,\text{p.f.}\int\limits_{\mathbb{R}} e^{i\varrho z\xi}\frac{1}{z}\psi(|z|)dz\\
&\quad + \frac{1}{2\pi}\frac{1}{\varrho\xi^2}\int\limits_{\frac12\le|z|\le1} e^{i\varrho\xi z}\Big(\frac{1}{z}\psi' + \log|z|\psi''\Big)dz\Big\}\\
&= \lim_{\varrho\to\infty}\frac{1}{2\pi i\xi}\,\text{p.f.}\int\limits_{\mathbb{R}} e^{iz\xi}\frac{1}{z}\psi\Big(\frac{|z|}{\varrho}\Big)dz\\
&= \frac{1}{2\pi i\xi}\,\text{p.f.}\int\limits_{\mathbb{R}} e^{iz\xi}\frac{1}{z}dz = \frac{1}{2\pi i\xi}i\pi\,\text{sign}\,\xi = \frac{1}{2|\xi|}\,.
\end{aligned}$$

Hence,

$$a_{-1}(t,\xi) = \frac{1}{2|\xi|} \tag{11.1.10}$$

is the complete symbol of V and is homogeneous of order -1. Consequently, we may write V according to Formula (7.1.7), i.e.,

$$V\sigma\big(T(t)\big) = L\mathcal{F}^{-1}_{\xi\mapsto t}a(t,\xi)\mathcal{F}_{\tau\mapsto\xi}\sigma(\tau) + R_\infty\sigma \tag{11.1.11}$$

where $a(t,\sigma) = \frac{1}{2|\xi|}\chi(\xi)$ with a cut–off function χ as in (7.1.5). With the Fourier series representation of $\widetilde{\sigma}$, we have

$$\begin{aligned} V\sigma\big(T(t)\big) &= \mathcal{F}^{-1}_{\xi\mapsto t}\Big\{\frac{1}{2|\xi|}\chi(\xi)\mathcal{F}_{\tau\mapsto\xi}\sum_{\ell\in\mathbb{Z}}\widehat{\sigma}_\ell e^{i2\pi\ell\tau}\Big\} + R_\infty\sigma \\ &= \sum_{\ell\in\mathbb{Z}}\widehat{\sigma}_\ell\mathcal{F}^{-1}_{\xi\mapsto t}\Big\{\frac{1}{2|\xi|}\chi(\xi)\mathcal{F}_{\tau\mapsto\xi}e^{i2\pi\ell\tau}\Big\} + R_\infty\sigma\,. \end{aligned} \tag{11.1.12}$$

By using the identity

$$\mathcal{F}_{\tau\mapsto\xi}e^{i2\pi\ell\tau} = \sqrt{2\pi}\delta(\xi - 2\pi\ell) \quad \text{for } \ell\in\mathbb{Z}\,, \tag{11.1.13}$$

(see Gelfand and Shilov [143, p.359]) we obtain

$$\begin{aligned} V\sigma\big(T(t)\big) &= \sum_{\ell\in\mathbb{Z}}\widehat{\sigma}_\ell\mathcal{F}^{-1}_{\xi\mapsto t}\Big\{\sqrt{2\pi}\frac{\chi(\xi)}{2|\xi|}\delta(\xi-2\pi\ell)\Big\} + R_\infty\sigma \qquad (11.1.14) \\ &= \sum_{\ell\in\mathbb{Z}}\widehat{\sigma}_\ell a(t,2\pi\ell)e^{i2\pi\ell t} + \sum_{\ell\in\mathbb{Z}}\widehat{\sigma}_\ell L\int_{\tau=0}^{1} k_\infty(t,\tau)e^{2\pi\ell\tau}d\tau \end{aligned}$$

where $a(t,\xi) = \frac{\chi(\xi)}{2|\xi|}$ is a symbol of V.

In comparison with (11.1.4), we see that the two representations are identical.

This suggests that for periodic functions σ and $A\in\mathcal{L}^m_{c\ell}(\Gamma)$ we may have a representation in the form

$$A\sigma\big(T(t)\big) = \sum_{\ell\in\mathbb{Z}}a(t,2\pi\ell)\widehat{\sigma}_\ell e^{i2\pi\ell t} + R_\infty\sigma \tag{11.1.15}$$

where $a\in\boldsymbol{S}^m_{c\ell}(\Gamma\times\mathbb{R})$ is a representative of the classical symbol of A. We shall return to this point in Section 11.2

For the double layer potential operators K and K' in $\mathbb{R}^2$, we first represent Γ in terms of its arc length parameterizations with $s = Lt$. Then $x = T(t)$, $y = T(\tau)$ and

$$\begin{aligned} T(t) - T(\tau) &= (t-\tau)\frac{dT}{dt}(t) \qquad (11.1.16) \\ &\quad -\frac{(t-\tau)^2}{2}\frac{d^2T}{dt^2}(t) + \frac{(t-\tau)^3}{3!}\frac{d^3T}{dt^3}(t) - \frac{(t-\tau)^4}{4!}\frac{d^4T}{dt^4}(t) + T_R(t,\tau) \end{aligned}$$

where the remainder $T_R \in C^\infty(\Gamma \times \Gamma)$ and is of the form $T_R = \mathcal{O}_5 = (t-\tau)^5 f_\infty(t,\tau)$, where $f_\infty(t,\tau)$ is a smooth function in C^∞ in both variables. We recall (3.4.20) in $\mathbb{R}^2$ which yields $L^{11} = \kappa$ with the curve's curvature $K = \kappa$; and with (3.4.17) and (3.4.18) we obtain Frenet's formulae

$$\frac{d\boldsymbol{t}}{dt} = L\kappa\boldsymbol{n}\,, \quad \frac{d\boldsymbol{n}}{dt} = -L\kappa\boldsymbol{t} \quad \text{and} \quad \frac{d^2\boldsymbol{t}}{dt^2} = L\frac{d\kappa}{dt}\boldsymbol{n} - L^2\kappa^2\boldsymbol{t} \tag{11.1.17}$$

with $\boldsymbol{t}$ the unit tangent and $\boldsymbol{n}$ the unit normal vectors of Γ.

For ease of reading, we introduce the generic notation

$$\mathcal{O}_\ell = (t-\tau)^\ell f_\infty(t,\tau)\,,\ \ell \in \mathbb{N}_0$$

where $f_\infty(t,\tau)$ is a smooth function in C^∞ of both variables. Then we obtain for (11.1.16) the canonical representation of Γ,

$$\begin{aligned}\boldsymbol{x} - \boldsymbol{y} = &\Big((t-\tau)L - \tfrac{1}{6}(t-\tau)^3L^3\kappa^2(t) + \tfrac{1}{8}(t-\tau)^4L^3\kappa\frac{d\kappa}{dt}\Big)\boldsymbol{t}(t)\\ -\Big(\tfrac{1}{2}L^2(t-\tau)^2\kappa - \tfrac{1}{6}(t-\tau)^3L^2\frac{d\kappa}{dt} &+ \tfrac{1}{24}(t-\tau)^4\Big(L^2\frac{d^2\kappa}{dt^2} - L^4\kappa^3\Big)\Big)\boldsymbol{n}(t) + \mathcal{O}_5\,.\end{aligned} \tag{11.1.18}$$

Then, as a consequence we obtain

$$r^2 = |\boldsymbol{y}-\boldsymbol{x}|^2 = L^2(t-\tau)^2\Big[1 - \tfrac{1}{12}(t-\tau)^2L^2\kappa^2(t) + \tfrac{1}{12}(t-\tau)^3L^2\kappa\frac{d\kappa}{dt}(t) + \mathcal{O}_4\Big]\,, \tag{11.1.19}$$

and, together with the Frenet formulae (11.1.17), we have

$$\begin{aligned}\boldsymbol{n}(y) = &\Big\{1 - \tfrac{1}{2}L^2\kappa^2(t)(t-\tau)^2 + \tfrac{1}{2}(t-\tau)^3L^2\kappa\frac{d\kappa}{dt}(t)\Big\}\boldsymbol{n}(t) \qquad (11.1.20)\\ &+\Big\{L\kappa(t)(t-\tau) - L\frac{d\kappa}{dt}(t)\tfrac{1}{2}(t-\tau)^2\\ &+\tfrac{1}{6}(t-\tau)^3\Big(L\frac{d^2\kappa}{dt^2}(t) - L^3\kappa^3(t)\Big)\Big\}\boldsymbol{t}(t) + \mathcal{O}_4\,,\\ \boldsymbol{t}(y) = &\Big\{1 - L^2\kappa^2\tfrac{1}{2}(t-\tau)^2 + \tfrac{1}{2}(t-\tau)^3L^2\kappa\frac{d\kappa}{dt}(t)\Big\}\boldsymbol{t}(t) \qquad (11.1.21)\\ &-\Big\{L\kappa(t-\tau) - L\frac{d\kappa}{dt}(t)\tfrac{1}{2}(t-\tau)^2\\ &+\tfrac{1}{6}(t-\tau)^3\Big(L\frac{d^2\kappa}{dt^2}(t) - L^3\kappa^3(t)\Big)\Big\}\boldsymbol{n}(t) + \mathcal{O}_4\,.\end{aligned}$$

Moreover,

$$\boldsymbol{n}(y)\cdot(\boldsymbol{y}-x) = -\frac{1}{2}\kappa(t)L^2(t-\tau)^2 + \frac{1}{3}(t-\tau)^3L^2\frac{d\kappa}{dt}(t) + \mathcal{O}_4\,. \tag{11.1.22}$$

Hence, with (11.1.19) and (11.1.22) the kernel of the double layer potential reads

$$
\begin{aligned}
&-\frac{1}{2\pi}\frac{\partial}{\partial n_y}\log|x-y| \\
&= \frac{1}{2\pi r^2}\Big(\frac{L^2}{2}(t-\tau)^2\kappa(t) - \frac{L^2}{3}(t-\tau)^3\frac{d\kappa}{dt}(t) + \mathcal{O}_4\Big) \qquad (11.1.23)\\
&= k_\infty(t,\tau) = \frac{1}{4\pi}\kappa(t) - \frac{1}{6\pi}(t-\tau)\frac{d\kappa}{dt}(t) + k_{\infty,1}(t_0,t)
\end{aligned}
$$

with a doubly periodic kernel function $k_\infty \in C^\infty(\Gamma\times\Gamma)$. Consequently, $K\varphi$ has the representation

$$K\varphi(t) = L\int\limits_0^1 k_\infty(t,\tau)\varphi(T(\tau))d\tau = \sum_{\ell\in\mathbb{Z}}\widehat{\varphi}_\ell L\int\limits_0^1 k_\infty(t,\tau)e^{2\pi i\ell\tau}d\tau \qquad (11.1.24)$$

where K is a smoothing operator for $\Gamma\in C^\infty$.

Clearly, the adjoint operator K' has the corresponding representation

$$K'\sigma(t) = L\int\limits_0^1 k_\infty(\tau,t)\sigma(T(\tau))d\tau \qquad (11.1.25)$$

with the same smooth kernel k_∞. The symbol of both operators, therefore, is equal to zero.

For the hypersingular operator D of the Laplacian, as in (1.2.14), we use the Fourier series representation (11.1.4) of V and obtain

$$D\varphi(T(t)) = -\frac{1}{L^2}\frac{d}{dt}V\Big(\frac{d\varphi}{d\tau}\Big) = L^{-1}\pi\sum_{\ell\in\mathbb{Z}}\widehat{\varphi}_\ell|\ell|e^{2\pi i\ell t} + (K_{1\infty}\varphi)(t) \quad (11.1.26)$$

where

$$K_{1\infty}\varphi(t) = L^{-2}\frac{d}{dt}\Big(K_\infty\frac{d\varphi}{d\tau}\Big)(t)\,.$$

In the same manner as for the simple layer potential V, we compute

$$
\begin{aligned}
\|(D\varphi - K_{1\infty}\varphi)\|^2_{H^{s-1}(\Gamma)} &= L^{-2}\pi^2\sum_{\ell\in\mathbb{Z}}|\widehat{\varphi}_\ell|^2|\ell|^{2s}\\
&= L^{-2}\pi^2(\|\varphi\|^2_{H^s(\Gamma)} - |\widehat{\varphi}_0|^2) \le L^{-2}\pi^2\|\varphi\|^2_{H^s(\Gamma)}\,. \quad (11.1.27)
\end{aligned}
$$

This inequality implies the mapping property of D, namely

$$\|D\varphi\|_{H^{s-1}(\Gamma)} \le c_1\|\varphi\|_{H^s(\Gamma)} + c_2(s,s_0)\|\varphi\|_{H^{s_0}(\Gamma)} \qquad (11.1.28)$$

for any $s_0\le s$. Hence, the order of D is $+1$. For the Gårding inequality we also obtain from (11.1.27) with Parseval's equality,

$$
\begin{aligned}
\langle D\varphi, \varphi\rangle_{L^2(\Gamma)} &= L\int\limits_0^1 (D\varphi)(t)\overline{\varphi(t)}dt \\
&= L\pi \sum_{\ell\in\mathbb{Z}} |\widehat{\varphi}_\ell|^2 |\ell + \delta_{0\ell}| + \langle C\varphi, \varphi\rangle_{L^2(\Gamma)} \\
&= L\pi \|\varphi\|^2_{H^{\frac{1}{2}}(\Gamma)} + \langle C\varphi, \varphi\rangle_{L^2(\Gamma)}\,,
\end{aligned}
\tag{11.1.29}
$$

where the smoothing operator is defined by

$$
C\varphi = K_\infty \varphi - L^2\pi \int\limits_0^1 \varphi(T(t))dt\,.
$$

This equality (11.1.29) verifies the strong ellipticity for the hypersingular operator D. On the other hand, from (1.2.14) we have with (11.1.12)

$$
\begin{aligned}
D\varphi(T(t)) &= -\frac{d}{ds_x} V\Big(\frac{d\varphi}{ds}\Big)(T(t)) \\
&= -\frac{1}{L^2}\frac{d}{dt} V\Big(\frac{d\varphi}{d\tau}\Big)(t) \\
&= -\frac{1}{L^2}\frac{d}{dt}\Big\{ \mathcal{F}^{-1}_{\xi\mapsto t}\frac{\chi(\xi)}{2|\xi|}\mathcal{F}_{\tau\mapsto\xi}\frac{d\varphi}{d\tau} + R_\infty \frac{d\varphi}{d\tau}\Big\}(t) \\
&= -\frac{1}{L^2}\mathcal{F}^{-1}_{\xi\mapsto t}\Big(\frac{\chi(\xi)}{2|\xi|}(i\xi)^2 \mathcal{F}_{\tau\mapsto\xi}\varphi(\tau)\Big) - \frac{1}{L^2}\frac{d}{dt}\Big(R_\infty\frac{d\varphi}{d\tau}\Big)(t)\,.
\end{aligned}
$$

Hence, the symbol of D is given by

$$
\sigma_D(t,\xi) = \frac{1}{L}\frac{\chi(\xi)}{2}|\xi|\,. \tag{11.1.30}
$$

Substituting into (11.1.26), we have

$$
D\varphi(T(t)) = \sum_{\ell\in\mathbb{Z}} \widehat{\varphi}_\ell \sigma_D(t, 2\pi\ell)e^{i2\pi\ell t} + R_{1\infty}\varphi(t)\,, \tag{11.1.31}
$$

which again is of the form (11.1.15).

11.2 The Fourier Series Representation of Periodic Operators $A \in \mathcal{L}^m_{c\ell}(\Gamma)$

We return to the general formula (11.1.15) and will show that this is indeed the general representation form for any periodic classical pseudodifferential operator $A \in \mathcal{L}^m_{c\ell}(\Gamma)$. According to Theorem 8.1.8, any given $A \in \mathcal{L}^m_{c\ell}(\Gamma)$ has the representation

$$Av(x) = \text{p.f.} \int_{\Gamma} k(x, x-y)v(y)ds_y + \sum_{j=0}^{m} a_j(x)\Big(-i\frac{\partial}{\partial s_x}\Big)^j v(x) \qquad (11.2.1)$$

where it is understood that the last term only appears when $m \in \mathbb{N}_0$. The Schwartz kernel belongs to the class $\Psi hk_\kappa(\Gamma)$ with $\kappa = -m-1$. In terms of the periodic representation (11.0.1), with $\widetilde{v}(t) := v\big(T(t)\big)$, the operator takes the form

$$A_\Phi \widetilde{v}(t) = \text{p.f.} \int_{\tau=0}^{1} k_\Phi(t, t-\tau)\widetilde{v}(\tau)d\tau + \sum_{j=0}^{m} \widetilde{a}_j(t)\Big(\frac{-id}{dt}\Big)^j \widetilde{v}(t)\,, \qquad (11.2.2)$$

where $\widetilde{v}, \widetilde{a}_j$ and the Schwartz kernel k_Φ are 1–periodic. Again, the last term only appears for $m \in \mathbb{N}_0$ and, moreover, the kernel k_Φ and the coefficients $\widetilde{a}_j(t)$ are obtained from the local transformation from s_x to t.

A symbol $a_\Phi \in \boldsymbol{S}^m_{c\ell}(\Gamma \times \mathbb{R})$ (see (7.1.39)) of A_Φ is given by formula (7.1.27) in Theorem 7.1.11, i.e.,

$$\begin{aligned} a_\Phi(t,\xi) &= e^{-it\xi}(A_\Phi e^{i\xi\bullet})(t) \\ &= e^{-i\xi}\ \text{p.f.} \int_{\tau=0}^{1} k_\Phi(t, t-\tau)e^{i\xi\tau}d\tau + \sum_{j=0}^{m} \widetilde{a}_j(t)\xi^j \qquad (11.2.3) \end{aligned}$$

since A is properly supported because of the compact curve Γ. With this symbol, A can also be written in the form

$$\begin{aligned} A_\Phi \widetilde{v}(t) &= \frac{1}{2\pi}\int_{\mathbb{R}} \int_{\tau=0}^{1} e^{i(t-\tau)\xi} a_\Phi(t,\xi)\widetilde{v}(\tau)d\tau d\xi + R_\infty \widetilde{v}(t) \\ &= \mathcal{F}^{-1}_{\xi \mapsto t}\ a_\Phi(t,\xi)\mathcal{F}_{\tau\mapsto\xi}\ \widetilde{v} + R_\infty \widetilde{v}(t) \qquad (11.2.4) \end{aligned}$$

for any $\widetilde{v} \in C^\infty_0([0,1]) \subset C^\infty_0(\mathbb{R})$ where R_∞ is an appropriate smoothing operator.

Theorem 11.2.1. *Every operator $A \in \mathcal{L}^m_{c\ell}(\Gamma)$ has a Fourier series representation in the form*

$$Av\big(T(t)\big) = \sum_{\ell \in \mathbb{Z}} a_\Phi(t, 2\pi\ell)\widehat{v}_\ell e^{i2\pi\ell t} + R_\infty v \qquad (11.2.5)$$

where $a_\Phi \in \boldsymbol{S}^m_{c\ell}(T \times \mathbb{R})$ is a classical symbol of A_Φ given by (11.2.3)*, and*

$$\widehat{v}_\ell := L\int_0^1 e^{-2\pi i\ell\tau}v\big(T(\tau)\big)d\tau \quad \textit{for } \ell \in \mathbb{Z} \qquad (11.2.6)$$

are the Fourier coefficients of $\widetilde{v}(t)$ (see Agranovich [4],Saranen et al [370]*).*

Remark 11.2.1: The Fourier series representation (11.2.4) of $A \in \mathcal{L}^m_{c\ell}(\Gamma)$ coincides with the definition of periodic pseudodifferential operators defined in the book by Saranen and Vainikko [369, Definition 7.2.1].

Proof: By substituting the Fourier series expansion of $\widetilde{v}$, i.e.

$$\widetilde{v}(\tau) = \sum_{\ell \in \mathbb{Z}} \widehat{v}_\ell e^{2\pi i \ell \tau}$$

into (11.2.4) and exchanging the order of summation and integration — since $\widetilde{v} \in C_0^\infty([0,1])$ and the Fourier series converges in every Sobolev space — we have

$$A_\Phi \widetilde{v}(t) = \sum_{\ell \in \mathbb{Z}} \widehat{v}_\ell \mathcal{F}^{-1}_{\xi \mapsto t} a_\Phi(t,\xi) \mathcal{F}_{\tau \mapsto \xi} e^{2\pi i \ell \tau} + R_\infty \widetilde{v}(t) .$$

By using the identity (11.1.13), the result follows. ∎

The explicit representation (11.2.5) in terms of Fourier series relies on the symbol $a_\Phi(t,\xi)$. If the operator $A \in \mathcal{L}^m_{c\ell}(\Gamma)$ is given as in (11.2.2) then the Schwartz kernel $k_\Phi(t, t-\tau)$ admits the pseudohomogeneous asymptotic expansion

$$k_\Phi(t, t-\tau) = \sum_{j=0}^{L} k_{j-m-1}(t, t-\tau) + R_L(t,\tau) \tag{11.2.7}$$

with $k_{j-m-1} \in \Psi h f_{j-m-1}$, and one may employ the one–dimensional Fourier transform in (11.2.4) to obtain the following explicit formulae for corresponding homogeneous expansion terms of the symbol.

Theorem 11.2.2. *Let $k_{j-m-1} \in \Psi h f_{j-m-1}$ be given where $j \in \mathbb{N}_0$. In the special case $m \in \mathbb{N}_0$ and $0 \le j \le m$, let k_{j-m-1} also satisfy the Tricomi condition* (8.1.56) *(where $n = 1$). Then the homogeneous symbol a^0_{m-j} corresponding to k_{j-m-1} is explicitly given by:*
<u>If $m \notin \mathbb{Z}$:</u>

$$\begin{aligned} a^0_{m-j}(t,\xi) = |\xi|^{m-j}\Gamma(j-m)\big\{ & i^j e^{-i\frac{m\pi}{2}} k_{j-m-1}(t, -\operatorname{sign}\xi) \\ & + (-i)^j e^{i\frac{m\pi}{2}} k_{j-m-1}(t, \operatorname{sign}\xi)\big\} . \end{aligned} \tag{11.2.8}$$

<u>If $m \in \mathbb{Z}$ and $m+1 \le j \in \mathbb{N}_0$:</u>

$$\begin{aligned} a^0_{m-j}(t,\xi) \;=\; & (j-m-1)!\xi^{m-j} i^{j-m} \big\{ f_{j-m-1}(t,-1) + (-1)^{j-m} f_{j-m-1}(t,1) \\ & + i\pi p_{j-m-1}(t,1)(\operatorname{sign}\xi)\big\} \end{aligned} \tag{11.2.9}$$

where

$$k_{j-m-1}(t,z) = f_{j-m-1}(t,z) + p_{j-m-1}(t,1) z^{j-m-1} \log|z| . \tag{11.2.10}$$

If $m \in \mathbb{N}_0$ and $0 \le j \le m$:

$$a^0_{m-j}(t,\xi) = \xi^{m-j}(\operatorname{sign}\xi)\frac{i^{m+1-j}}{(m-j)!}\pi k_{j-m-1}(t,-1)\,. \tag{11.2.11}$$

Proof: Let us begin with the case $m \notin \mathbb{Z}$. Then (8.1.28) for $j > m$ and (8.1.53) for $j < m$ can be written in the compressed form

$$\begin{aligned} a_{\psi,m-j}(t,\xi) = {} & \text{p.f.} \int_{\mathbb{R}} k_{j-m-1}(t,t-\tau)e^{i\xi(\tau-t)}\psi(\tau-t)d\tau \\ & = k_{j-m-1}(t,-1)\ \text{p.f.} \int_0^\infty z^{j-m-1}e^{i\xi z}\psi(z)dz \\ & \quad + k_{j-m-1}(t,1)\ \text{p.f.} \int_0^\infty z^{j-m-1}e^{-i\xi z}\psi(z)dz\,, \end{aligned}$$

and (8.1.29) yields

$$\begin{aligned} a^0_{m-1}(t,\xi) = {} & \lim_{\varrho\to+\infty} \varrho^{j-m}\{k_{j-m-1}(t,-1)\ \text{p.f.} \int_0^\infty z^{j-m-1}e^{i\varrho\xi z}\psi(z)dz \\ & + k_{j-m-1}(t,1)\ \text{p.f.} \int_0^\infty z^{j-m-1}e^{-i\varrho\xi z}\psi(z)dz\} \\ = {} & k_{j-m-1}(t,-1) \lim_{\varrho\to+\infty} \text{p.f.} \int_0^\infty z^{j-m-1}e^{i\xi z}\psi(\tfrac{z}{\varrho})dz \\ & + k_{j-m-1}(t,1) \lim_{\varrho\to\infty} \text{p.f.} \int_0^\infty z^{j-m-1}e^{-i\xi z}\psi(\tfrac{z}{\varrho})dz\,. \end{aligned}$$

Hence, we need the one–dimensional Fourier transform of z^{j-m-1} which can explicitly be found in Gelfand and Shilov [143, p. 360], i.e.,

$$\begin{aligned} a^0_{m-j}(t,\xi) = |\xi|^{m-j} i\Gamma(j-m)\{ & k_{j-m-1}(t,-1)(\operatorname{sign}\xi)i^{(\operatorname{sign}\xi)(j-1)} \cdot e^{-(\operatorname{sign}\xi)\frac{i\pi}{2}m} \\ & -k_{j-m-1}(t,1)(\operatorname{sign}\xi)i^{-(\operatorname{sign}\xi)(j-1)} \cdot e^{(\operatorname{sign}\xi)\frac{i\pi}{2}m}\}\,. \end{aligned}$$

Writing down this result for $\xi > 0$ and $\xi < 0$ separately gives the desired equation (11.2.8).

Next, we consider the case $m \in \mathbb{Z}$ and $m+1 \le j \in \mathbb{N}_0$. Then $k_{j-m-1} \in \psi h f_{j-m-1}$ has the form (11.2.10). With (8.1.28) we find for the homogeneous term

$$\begin{aligned} a_{h\psi,m-j}(t,\xi) &= f_{j-m-1}(t,1)\int\limits_0^\infty z^{j-m-1}e^{-i\xi z}\psi(z)dz \\ &\quad + f_{j-m-1}(t,-1)\int\limits_0^\infty z^{j-m-1}e^{i\xi z}\psi(z)dz\,. \end{aligned} \tag{11.2.12}$$

Hence, for $\xi \neq 0$ we need the limits of

$$I_k(\xi,\varrho) := \varrho^{j-m}\int\limits_0^\infty z^k e^{i\varrho\xi z}\psi(z)dz \quad \text{with } k = j-m-1 \in \mathbb{N}_0$$

for $\varrho \to +\infty$. Integration by parts for $k \geq 1$ yields

$$I_k(\xi,\varrho) = \frac{i}{\xi}\varrho^{j-m-1}k\int\limits_0^\infty z^{k-1}e^{i\varrho\xi z}\psi(z)dz + \frac{i}{\xi}\varrho^{j-m-1}\int\limits_{\frac12}^1 z^k\psi'(z)e^{i\varrho\xi z}dz\,.$$

The second integral is the Fourier transform of a $C_0^\infty(\frac12,1)$–function and, due to the Palay–Wiener–Schwartz theorem (Theorem 3.1.3), therefore, decays faster than $\varrho^{-(j-m)}$. Hence, for $\xi \neq 0$ fixed, we find the recursion relation

$$I_k = \frac{i}{\xi}kI_{k-1}(\xi,\varrho) + O(\varrho^{-1})\,,$$

which gives

$$I_k(\xi,\varrho) = \frac{i^k k!}{\xi^k}I_0(\xi,\varrho) + O(\varrho^{-1})\,,$$

where

$$I_0(\xi,\varrho) = \varrho\int\limits_0^\infty e^{iz\varrho\xi}\psi(z)dz = \frac{i}{\xi} + \frac{i}{\xi}\int\limits_{\frac12}^1 e^{iz\varrho\xi}\psi'(z)dz = \frac{i}{\xi} + O(\varrho^{-1})\,.$$

Hence,

$$\lim_{\varrho\to+\infty} I_{j-m-1}(\xi,\varrho) = \frac{i^{j-m}}{\xi^{j-m}}(j-m-1)!\,. \tag{11.2.13}$$

Inserting (11.2.13) in (11.2.12) gives

$$\begin{aligned} a^0_{h,m-j}(t,\xi) = \frac{(j-m-1)!}{\xi^{j-m}}\{&(-i)^{j-m}f_{j-m-1}(t,1) \\ &+ i^{j-m}f_{j-m-1}(t,-1)\}\,. \end{aligned} \tag{11.2.14}$$

For the logarithmic part in (11.2.10) we introduce

$$J_k(\xi,\varrho) := \varrho^{k+1} \int\limits_{\mathbb{R}} z^k \log|z| \psi(z) e^{i\varrho z\xi} dz$$

and integrate by parts finding

$$J_k(\xi,\varrho) = \frac{ik}{\xi} \varrho^k \int\limits_{\mathbb{R}} z^{k-1} \log|z| \psi(z) e^{i\varrho z\xi} dz$$
$$+ \frac{i}{\xi} \varrho^k \int\limits_{\mathbb{R}} z^{k-1} \psi(z) e^{i\varrho\xi z} dz + \frac{i}{\xi} \varrho^k \int\limits_{\frac{1}{2}\le|z|\le 1} z^k \log|z| \psi'(z) e^{i\varrho\xi z} dz\,,$$
$$J_k(\xi,\varrho) = \frac{ik}{\xi} J_{k-1} + O(\varrho^{-1}) \quad \text{for } k \ge 1$$

and $\xi \neq 0$ fixed since for the last two integrals again the Paley–Wiener–Schwartz theorem (Theorem 3.1.3) can be applied. Thus,

$$J_k(\xi,\varrho) = \frac{i^k k!}{\xi^k} J_0(\xi,\varrho) + O(\varrho^{-1})\,.$$

For J_0 one finds

$$J_0(\xi,\varrho) = \int\limits_{\mathbb{R}} \log|z| e^{iz\xi} \psi\big(\tfrac{z}{\varrho}\big) dz - \varrho \log\varrho \int\limits_{\mathbb{R}} \psi(z) e^{i\varrho z\xi} dz$$
$$= \int\limits_{\mathbb{R}} \log|z| e^{iz\xi} \psi\big(\tfrac{z}{\varrho}\big) dz + O(\varrho^{-1})\,.$$

With $\varrho \to +\infty$ and $\xi \neq 0$ fixed $k = j - m - 1 \ge 0$, and the Fourier transformed of $\log|z|$, we obtain

$$\lim_{\varrho\to+\infty} J_k(\xi,\varrho) = -\frac{\pi}{|\xi|} \frac{(j-m-1)!\; i^{j-m-1}}{\xi^{j-m-1}}\,.$$

Together with (11.2.14), this yields (11.2.9).

In the remaining case $k = m + 1 - j \in \mathbb{N}$ we have from (8.1.53) with the homogeneous kernel k_{-m-1+j}:

$$a^0_{m-j}(t,\xi) = \text{ p.f.} \int\limits_0^\infty k_{-m-1+j}(t,1) z^{-m-1+j} e^{-iz\xi} dz$$
$$+ \text{ p.f.} \int\limits_0^\infty k_{-m-1+j}(t,-1) z^{-m-1+j} e^{iz\xi} dz$$

where we employ the Tricomi condition (8.1.56) for $n = 1$:

$$k_{-m-1+j}(t,1) = (-1)^{m-j+1} k_{-m-1+j}(t,-1)$$

obtaining

$$\begin{aligned} a^0_{m-j}(t,\xi) &= k_{-m-1+j}(t,-1)\{\text{ p.f. }\int_0^\infty z^{-k}e^{iz\xi}dz + (-1)^k \text{ p.f. }\int_0^\infty z^{-k}e^{-iz\xi}dz\} \\ &= k_{-m-1+j}(t,-1)\text{ p.f. }\int_{\mathbb{R}} z^{-k}e^{iz\xi}dz\,. \end{aligned}$$

If we integrate

$$L_k := \text{ p.f. }\int_{\mathbb{R}} z^{-k}e^{iz\xi}dz$$

by parts then we obtain for $k \geq 2$:

$$L_k = \frac{i\xi}{k-1}L_{k-1}\,.$$

Hence, with the Fourier transformed of z^{-1}, we find

$$L_k = \frac{(i\xi)^{k-1}}{(k-1)!}\,L_1 = \frac{i^{m-j}}{(m-j)!}\pi\xi^{m-j}\,\mathrm{sign}\,\xi\,,$$

i.e. (11.2.11). ∎

11.3 Ellipticity Conditions for Periodic Operators on Γ

The representation (11.2.5) of periodic operators $A \in \mathcal{L}^m_{c\ell}(\Gamma)$ gives rise to simple characterization of ellipticity concepts such as ellipticity, strong ellipticity as well as the odd ellipticity. To this end, we employ the asymptotic expansion of the symbol,

$$\begin{aligned} a_\Phi(t,\xi) &= a^0_{\Phi,m}(t,\xi) + a^0_{\Phi,m-1}(t,\xi) + \dots \\ &= a^0_{\Phi,m}(t,\xi) + \big(a_\Phi(t,\xi) - a^0_{\Phi,m}(t,\xi)\big) \\ &= a^0_{\Phi,m}(t,\xi) + \widetilde{a}_\Phi(t,\xi) \end{aligned} \tag{11.3.1}$$

where $\widetilde{a}_\Phi \in \boldsymbol{S}^{m-1}_{c\ell}(\Gamma\times\mathbb{R})$. The principal symbol $a^0_{\Phi,m}(t,\xi)$ is positively homogeneous of degree m and periodic with respect to t. In view of (11.3.1), we may now write

$$Av\big(T(t)\big) = \sum_{0\neq\ell\in\mathbb{Z}} a^0_{\Phi,m}(t,2\pi\ell)\widehat{v}_\ell e^{i2\pi\ell t} + \widetilde{R}v(t) \tag{11.3.2}$$

where $\widetilde{R} \in \mathcal{L}^{m-1}_{c\ell}(\Gamma)$; and by decomposition:

$$
\begin{aligned}
Av\big(T(t)\big) &= \sum_{\ell\geq 1} a^0_{\Phi,m}(t,1)(2\pi\ell)^m \widehat{v}_\ell e^{i2\pi\ell t} \\
&\quad + \sum_{\ell\leq -1} a^0_{\Phi,m}(t,-1)(-2\pi\ell)^m \widehat{v}_\ell e^{i2\pi\ell t} \qquad (11.3.3)\\
&\quad + \sum_{\ell\in\mathbb{Z}} \widehat{v}_\ell \text{ p.f.} \int_0^1 \widetilde{R}(t,\tau)e^{i2\pi\ell\tau}d\tau \,.
\end{aligned}
$$

11.3.1 Scalar Equations

In accordance with (7.2.3), the operator A is **elliptic**, if the two conditions

$$
a^0_{\Phi,m}(t,1) \neq 0 \qquad \text{and} \qquad a^0_{\Phi,m}(t,-1) \neq 0 \qquad (11.3.4)
$$

are both satisfied for all t. Since for an elliptic operator there exists a parametrix and Γ is compact, Theorem 5.3.11 can be applied and A is, hence, a Fredholm operator.

Lemma 11.3.1. *For the elliptic operator $A \in \mathcal{L}^m_{c\ell}(\Gamma)$, the Fredholm index is given explicitly by*

$$
\operatorname{index}(A) = -\frac{1}{2\pi}\int_\Gamma d\arg\Big\{\frac{a^0_{\Phi,m}(t,1)}{a^0_{\Phi,m}(t,-1)}\Big\}\,. \qquad (11.3.5)
$$

Here $\Gamma = \partial\Omega$ may consist of a finite number of simple, plane, closed and smooth Jordan curves which are oriented in such a way that Ω always lies to the left of Γ.

Before we begin with the proof, let us first rewrite the representation (11.3.3) in the form

$$
\begin{aligned}
Av\big(T(t)\big) &= b_+(t)\Big\{\sum_{\ell\in\mathbb{Z}\setminus\{0\}} |2\pi\ell|^m \widehat{v}_\ell e^{2\pi i\ell t} + \widehat{v}_0\Big\} \qquad (11.3.6)\\
&\quad + b_-(t)\Big\{\sum_{\ell\in\mathbb{Z}\setminus\{0\}} |2\pi\ell|^m (\operatorname{sign}\ell)\widehat{v}_\ell e^{2\pi i\ell t} + \widehat{v}_0\Big\} + \widetilde{R}v
\end{aligned}
$$

where the functions $b_\pm(t)$ are defined by

$$
b_\pm(t) := \frac{1}{2}\{a^0_{\Phi,m}(t,1) \pm a^0_{\Phi,m}(t,-1)\}\,. \qquad (11.3.7)
$$

Further let us use the Fourier representation of the Bessel potential operator $\Lambda^m : H^s(\Gamma) \to H^{s-m}(\Gamma)$ for periodic functions, namely

$$
\Lambda^m v(t) := \sum_{\ell\in\mathbb{Z}\setminus\{0\}} |2\pi\ell|^m \widehat{v}_\ell e^{2\pi i\ell t} + \widehat{v}_0\,. \qquad (11.3.8)
$$

Then

$$\Lambda^{-m} \circ \Lambda^m v = \Lambda^m \circ \Lambda^{-m} v = \sum_{\ell \in \mathbb{Z}} \widehat{v}_\ell e^{2\pi i \ell t} = v\,. \tag{11.3.9}$$

We also shall use the Fourier representation of the Hilbert transform $H : H^s(\Gamma) \to H^s(\Gamma)$ for periodic functions, i.e.,

$$Hv(t) = \sum_{\ell \geq 0} \widehat{v}_\ell e^{2\pi i \ell t} - \sum_{\ell < 0} \widehat{v}_\ell e^{2\pi i \ell t}\,. \tag{11.3.10}$$

Clearly,

$$H \circ Hv(t) = \sum_{\ell \in \mathbb{Z}} \widehat{v}_\ell e^{2\pi i \ell t} = v(t)\,. \tag{11.3.11}$$

Moreover,

$$H \circ \Lambda^m = \Lambda^m \circ H \tag{11.3.12}$$

and $\Lambda^m : H^{s+m}(\Gamma) \to H^s(\Gamma)$ as well as $H : H^s(\Gamma) \to H^s(\Gamma)$ are isomorphisms for every m and s in $\mathbb{R}$.

With these operators, (11.3.6) can also be written as

$$Av\big(T(t)\big) = b_+(t)\Lambda^m v(t) + b_-(t) H \circ \Lambda^m v(t) + \widetilde{R}v\,. \tag{11.3.13}$$

Now we are in the position to prove Lemma 11.3.1 by reduction to the case $m = 0$.

Proof of Lemma 11.3.1: If we write (11.3.13) in the form of a classical Cauchy singular integral equation

$$Av = A \circ \Lambda^{-m} w = b_+(t) w + b_-(t) Hw + \widetilde{R} \circ \Lambda^{-m} w$$

for $w = \Lambda^m v$, then from (5.3.23) in Theorem 5.3.11 we obtain

$$\text{index}(A) = \text{index}(A \circ \Lambda^{-m}) = \text{index}(b_+ I + b_- H) \tag{11.3.14}$$

since Λ^{-m} is an isomorphism, hence index $(\Lambda^{-m}) = 0$, and $\widetilde{R} \circ \Lambda^{-m}$ is compact. The operator $b_+ I + b_- H$ is a classical Cauchy singular integral operator on Γ whose Fredholm index is given by

$$\text{index}(b_+ I + b_- H) = -\frac{1}{2\pi} \int_\Gamma d \arg \Big\{ \frac{b_+(t) + b_-(t)}{b_+(t) - b_-(t)} \Big\}\,,$$

even if Γ consists of finitely separated components (see Muskhelishvili [320, formula (45.4)]). With (11.3.7), this gives the proposed formula (11.3.5). ■

According to Definition 7.2.4, the strong ellipticity condition (7.2.62) now reads:

There exists a function $\Theta \in C^\infty(\Gamma)$ and a constant $\beta_0 > 0$ such that

$$\inf_t \min\{ \operatorname{Re}\Theta(t)a^0_{\Phi,m}(t,1)\,,\ \operatorname{Re}\Theta(t)a^0_{\Phi,m}(t,-1)\} \geq \beta_0\,. \tag{11.3.15}$$

Clearly, for a strongly elliptic operator, the zero Fredholm index follows with a direct computation from the index formula (11.3.5) which is consistent with Remark 9.1.2.

We emphasize that the strongly elliptic equation

$$Au = f \quad \text{on } \Gamma$$

with $f \in H^{\ell-\frac{m}{2}}(\Gamma)$ for any fixed $\ell \in \mathbb{R}$ can be treated by the variational method in terms of the corresponding bilinear forms as in Chapter 5. Specifically we may define the bilinear form as

$$a(u,v) = (\Theta Au, v)_{H^\ell(\Gamma)} = (\Theta f, v)_{H^\ell(\Gamma)} \tag{11.3.16}$$

where $\mathcal{H}_1 = H^{\ell+\frac{m}{2}}(\Gamma)$ and $\mathcal{H}_2 = H^{\ell-\frac{m}{2}}(\Gamma)$ and $u,v \in \mathcal{H}_1$.

This variational formulation provides the foundation of corresponding Galerkin methods.

One special feature of periodic pseudodifferential operators on closed curves is the additional concept of the odd ellipticity as introduced in Sloan [390].

Definition 11.3.1. *The operator $A \in \mathcal{L}^m_{c\ell}(\Gamma)$ is called oddly elliptic if there exists a smooth function $\Theta \in C^\infty(\Gamma)$ and a constant $\beta_0 > 0$ such that*

$$\inf_t \min\{ \operatorname{Re}\big(\Theta(t)a^0_{\Phi,m}(t,1)\big)\,,\ -\operatorname{Re}\big(\Theta(t)a^0_{\Phi,m}(t,-1)\big)\} \geq \beta_0 > 0\,. \tag{11.3.17}$$

We remark that the inequality (11.3.17) cannot be generalized to higher dimensions of Γ for a continuous symbol $a^0_{\Phi,m}\left(t,\frac{\xi}{|\xi|}\right)$ since then the unit sphere in $\mathbb{R}^{n-1}$ is not disconnected anymore.

Both, strong as well as odd ellipticity can also be characterized in terms of the functions $b_+(t)$ and $b_-(t)$ defined in (11.3.7).

Theorem 11.3.2.

i) *$A \in \mathcal{L}^m_{c\ell}(\Gamma)$ is strongly elliptic if and only if the condition*

$$b_+(t) + \lambda b_-(t) \neq 0 \quad \textit{for all } t \quad \textit{and all } \lambda \in [-1,1] \tag{11.3.18}$$

is satisfied.

ii) *$A \in \mathcal{L}^m_{c\ell}(\Gamma)$ is oddly elliptic if and only if the condition*

$$b_-(t) + \lambda b_+(t) \neq 0 \quad \textit{for all } t \quad \textit{and all } \lambda \in [-1,1] \tag{11.3.19}$$

is satisfied.

Remark 11.3.1: Both conditions (11.3.18) or (11.3.19) can easily be verified by inspecting the corresponding graphs in the complex plane, see Figure 11.3.1. These conditions play an important rôle in the analysis of singular integral equations and their approximation and numerical treatment (see Gohberg et al [151, 152], Prössdorf and Silbermann [358, 359], Sloan et al [391] and the references therein). In particular, the function $\Theta(t)$ is not needed in (11.3.18) or (11.3.19).

Proof: (See Prössdorf and Schmidt [356, Lemma 4.4].) We first prove the proposition i) for the strongly elliptic case. Suppose that A is strongly elliptic and (11.3.15) is satisfied. Then we show (11.3.18) by contradiction, i.e., for some $\lambda_0 \in [-1,1]$ and t_0 we have

$$b_+(t_0) + \lambda_0 b_-(t_0) = \frac{1}{2}\big(a_+(t_0) + a_-(t_0)\big) + \lambda_0 \frac{1}{2}\big(a_+(t_0) - a_-(t_0)\big) = 0$$

where $a_+(t) := a^0_{\Phi,m}(t,1)$, $a_-(t) := a^0_{\Phi,m}(t,-1)$. This implies after multiplication by $\Theta(t_0)$ and taking the real part,

$$(1+\lambda_0)\,\mathrm{Re}\,\big(\Theta(t_0)a_0(t_0)\big) + (1-\lambda_0)\,\mathrm{Re}\,\big(\Theta(t_0)a_-(t_0)\big) = 0\,.$$

Condition (11.3.15) together with $\lambda_0 \in [-1,1]$ then would imply

$$0 \le (1+\lambda_0)\,\mathrm{Re}\,\big(\Theta(t_0)a_+(t_0)\big) = -(1-\lambda_0)\,\mathrm{Re}\,\big(\Theta(t_0)a_-(t_0)\big) \le 0\,,$$

i.e. $(1+\lambda_0) = 0$ together with $(1-\lambda_0) = 0$ which is impossible.

Conversely, suppose that (11.3.18) is fulfilled. This means that the convex combination satisfies

$$\mu a_+(t) + (1-\mu)a_-(t) \neq 0 \quad \text{for all } t \quad \text{and for all } \mu = \tfrac{1}{2}(1-\lambda) \in [0,1]\,.$$

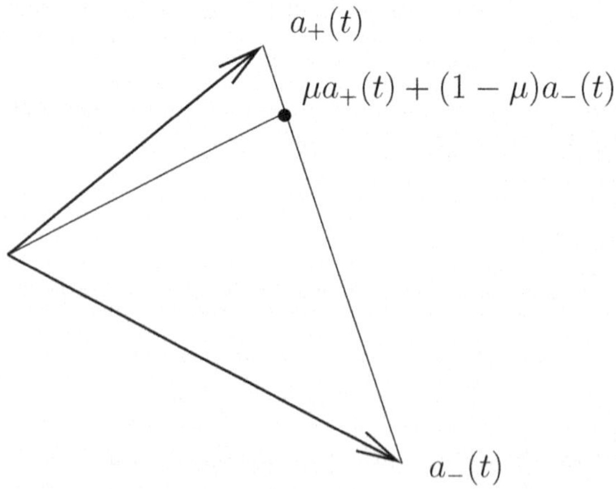

Figure 11.3.1: Strong ellipticity condition.

Therefore, $\left|\arg\frac{a_+}{a_-}\right| < \pi$ and $a_+ \neq 0\,,\ a_- \neq 0$. We may now define

$$\Theta(t) := \exp\big(-\tfrac{i}{2}(\alpha_+ + \alpha_-)\big) \quad \text{where } \ \alpha_\pm := \arg a_\pm\,. \tag{11.3.20}$$

Then

$$\begin{aligned}\Theta(t)a_+(t) &= |a_+(t)|\exp\big(\tfrac{i}{2}(\alpha_+ - \alpha_-)\big)\,,\\ \Theta(t)a_-(t) &= |a_-(t)|\exp\big(-\tfrac{i}{2}(\alpha_+ - \alpha_-)\big)\end{aligned}$$

and

$$\begin{aligned}\mathrm{Re}\,\big(\Theta(t)a_+(t)\big) &= |a_+(t)|\cos\big(\tfrac{1}{2}(\alpha_+ - \alpha_-)\big) \quad > 0\,,\\ \mathrm{Re}\,\big(\Theta(t)a_-(t)\big) &= |a_-(t)|\cos\big(-\tfrac{1}{2}(\alpha_+ - \alpha_-)\big) > 0\,,\end{aligned}$$

which is (11.3.15).

For odd ellipticity, the proof proceeds in the same manner as in the strongly elliptic case by interchanging b_+ and b_- and replacing a_- by $(-a_-)$.

■

A simple example of an oddly elliptic equation is the Cauchy singular integral equation with the Hilbert transform,

$$H_\Gamma u(x) = -\frac{1}{\pi}\int\limits_\Gamma \Big(\frac{d}{ds_y}\log|x-y|\Big)u(y)ds_y + \frac{1}{2\pi}\int\limits_\Gamma u(y)ds_y = f(x)\,.$$

Here the principal symbol is given by

$$a^0_{\Phi,0}(t,\xi) = i\frac{\xi}{|\xi|} \quad \text{for } \ |\xi| \geq 1\,.$$

Hence, $b_+(t) = 0$ and $b_-(t) = i$ and (11.3.19) is satisfied. However, H_Γ is **not** strongly elliptic.

On the other hand, with the Hilbert transform (11.3.10) as a preconditioner, any oddly elliptic equation

$$Au = f \quad \text{on } \ \Gamma$$

can be treated by using the variational methods as in the strongly elliptic case if one defines an associated bilinear equation by

$$a(u,v) := (\Theta HAu, v)_{H^\ell(\Gamma)} = (\Theta Hf, v)_{H^\ell(\Gamma)} \tag{11.3.21}$$

where again $\mathcal{H}_1 = H^{\ell+\frac{m}{2}}(\Gamma)\,,\ \mathcal{H}_2 = H^{\ell-\frac{m}{2}}(\Gamma)$ and $u, v \in \mathcal{H}_1$. Now equation (11.3.21) defines a strongly elliptic bilinear form which fulfills a Gårding inequality and $HA \in \mathcal{L}^m_{c\ell}(\Gamma)$ is strongly elliptic. Clearly, the Fredholm index of the oddly elliptic operator A is zero since H defined by (11.3.10) is an isomorphism with Fredholm index zero and

$$\mathrm{index}(HA) = \mathrm{index}(A)$$

due to (5.3.23) in Theorem 5.3.11.

11.3.2 Systems of Equations

We now extend the concept for one equation to a system of p pseudodifferential equations recalling Section 7.2.1, but now on Γ,

$$\sum_{k=1}^{p} A_{jk} u_k = f_j \quad \text{for } j = 1, \ldots, p \quad \text{on } \Gamma \tag{11.3.22}$$

with operators $A_{jk} \in \mathcal{L}_{c\ell}^{s_j+t_k}(\Gamma)$. Now let us assume that the system (11.3.22) is elliptic in the sense of Agmon–Douglis–Nirenberg, see Definition 7.2.3. Here, $f_j \in H^{s_j}(\Gamma)$ for $j = 1, \ldots, p$ are given and $u_k \in H^{t_k}(\Gamma)$ for $k = 1, \ldots, p$ are the unknowns. By $a^{jk0}_{\Phi,s_j+t_k}(t,\xi)$ we denote the homogeneous principal symbols of A_{jk} on Γ which are 1–periodic in t. By using the representation (11.3.3) for each of the operators in (11.3.22) we obtain

$$\begin{aligned}\sum_{k=1}^{p} A_{jk} u(T(t)) = \sum_{k=1}^{p} \Big\{ \sum_{\ell=1}^{\infty} a^{jk0}_{\Phi,s_j+t_k}(t,1)(2\pi\ell)^{s_j+t_k} \widehat{u}_{k\ell} e^{2\pi i \ell t} \\ + \sum_{\ell \le -1} a^{jk0}_{\Phi,s_j+t_k}(t,-1)(-2\pi\ell)^{s_j+t_k} \widehat{u}_{k\ell} e^{2\pi i \ell t} \Big\} + \sum_{k=1}^{p} R_{jk} u_k\end{aligned} \tag{11.3.23}$$

where the remaining operators $R_{jk} \in \mathcal{L}_{c\ell}^{s_j+t_k-1}(\Gamma)$.

The system (11.3.22) is elliptic in the sense of Agmon–Douglis–Nirenberg if both conditions

$$\begin{aligned} B(t,1) &:= \det\, (\!(a^{jk0}_{\Phi,s_j+t_k}(t,1))\!)_{p\times p} \neq 0 \quad \text{and} \\ B(t,-1) &:= \det\, (\!(a^{jk0}_{\Phi,s_j+t_k}(t,-1))\!)_{p\times p} \neq 0 \quad \text{for every } t \in \mathbb{R} \end{aligned} \tag{11.3.24}$$

are satisfied (see Definition 7.2.3). In this case, a parametrix Q_0 exists according to Theorem 7.2.3 and, hence, $A = (\!(A_{jk})\!)_{p\times p} : \mathcal{H}_1 := \prod_{k=1}^{p} H^{t_k}(\Gamma) \to \mathcal{H}_2 := \prod_{j=1}^{p} H^{-s_j}(\Gamma)$ is a Fredholm operator. The Fredholm index of $(\!(A_{jk})\!)_{p\times p}$ can be computed as follows.

Lemma 11.3.3. *For the elliptic system* (11.3.22) *in the sense of Agmon–Douglis–Nirenberg, the Fredholm index is given by*

$$\operatorname{index}(\!(A_{jk})\!)_{p\times p} = -\frac{1}{2\pi} \int_{\Gamma} d \arg \Big\{ \frac{B(t,1)}{B(t,-1)} \Big\} . \tag{11.3.25}$$

Proof: For the proof we may proceed in the same manner as for the scalar case. In particular, we see that the system may be rewritten in the form

$$\sum_{k=1}^{p} A_{jk}u_k(T(t)) = \sum_{k=1}^{p}\{b_{+,jk}(t)\Lambda^{s_j+t_k}u_k + b_{-,jk}(t)\Lambda^{s_j}H\Lambda^{t_k}u_k\} + \sum_{k=1}^{p}\widetilde{R}_{jk}u_k \qquad (11.3.26)$$

with $\widetilde{R}_{jk} \in \mathcal{L}^{s_j+t_k-1}_{c\ell}(\Gamma)$ which includes R_{jk} and smoothing operators in terms of $\widehat{u}_{k0}$. If we introduce new unknowns $w_k := \Lambda^{t_k}u_k$, with the help of the isomorphic Bessel potential operator (11.3.8), we obtain the system of classical Cauchy singular integral equations

$$\sum_{k=1}^{p} b_{+,jk}(t)w_k(t) + b_{-,jk}(t)Hw_k + \overset{\approx}{R}_{jk}w_k = \Lambda^{-s_j}f_j\,, j = 1,\ldots,p. \qquad (11.3.27)$$

Here

$$\overset{\approx}{R}_{jk} + \Lambda^{-s_j}(b_{+,jk}\Lambda^{s_j} - \Lambda^{s_j}b_{+,jk}) + \Lambda^{-s_j}(b_{-,jk}\Lambda^{s_j} - \Lambda^{s_j}b_{-,jk}) \in \mathcal{L}^{-1}_{c\ell}(\Gamma)\,.$$

The last two terms are the commutators of the pseudodifferential operator $b_{\pm,jk} \in \mathcal{L}^0_{c\ell}(\Gamma)$ with $\Lambda^{s_j} \in \mathcal{L}^{s_j}_{c\ell}(\Gamma)$. By applying Corollary 7.1.15 we see that these commutators are in the class $\mathcal{L}^{s_j-1}_{c\ell}(\Gamma)$ and therefore $\overset{\approx}{R}_{jk} \in \mathcal{L}^{-1}_{c\ell}(\Gamma)$. The Fredholm index of the system (11.3.27) of Cauchy singular equations can be found in the book by Muskhelishvili [320, formula (180.22)] which yields (11.3.25) since the mappings $\Lambda^{s_j+t_k}$ are isomorphisms and $\overset{\approx}{R}_{jk}$ are compact operators. ■

The condition (7.2.62) for strong ellipticity now reads:
There exist a C^∞–matrix–valued 1–periodic function $\Theta(t) = \big((\Theta_{jk}(t))\big)_{p\times p}$ and a constant $\beta_0 > 0$ such that

$$\begin{aligned}\inf_{t\in\mathbb{R}} \min\Big\{ \mathrm{Re} \sum_{\ell,j,k=1}^{p} \zeta_\ell^\top \Theta_{\ell j}(t) a^{jk0}_{\Phi,s_j+t_k}(t,+1)\overline{\zeta}_k\,,\\ \mathrm{Re} \sum_{\ell,j,k=1}^{p} \zeta_\ell^\top \Theta_{\ell j} a^{jk0}_{\Phi,s_j+t_k}(t,-1)\overline{\zeta}_k\Big\}\\ \geq \beta_0|\zeta|^2 \quad \text{for all } \zeta \in \mathbb{C}^p\,. \end{aligned} \qquad (11.3.28)$$

Again, for a strongly elliptic system with $A_{jk} \in \mathcal{L}^{s_j+t_k}_{c\ell}(\Gamma)$, the Fredholm index is zero since any associated bilinear form

$$a(v,u) := (v, \Lambda^\sigma \Theta \Lambda^{-\sigma} Au)_{\prod_{\ell=1}^{p} H^{(t_\ell - s_\ell)/2}(\Gamma)} \qquad (11.3.29)$$

satisfies a Gårding inequality such as (7.2.64) on the boundary curve Γ.

As for one scalar equation, we may define odd ellipticity for the system (11.3.23), namely:

Definition 11.3.2. *The operator $A = ((A_{jk}))_{p\times p}$ is called* oddly elliptic *if there exist a smooth matrix–valued 1–periodic function $\Theta(t)$ and a constant $\beta_0 > 0$ such that*

$$\inf_{t\in\mathbb{R}} \min\Big\{ \mathrm{Re} \sum_{\ell,j,k=1}^{p} \zeta_\ell^\top \Theta_{\ell j}(t) a^{jk,0}_{\Phi,s_j+t_k}(t,1)\overline{\zeta}_k\,, \tag{11.3.30}$$
$$- \mathrm{Re} \sum_{\ell,j,k=1}^{p} \zeta_\ell^\top \Theta_{\ell j}(t) a^{jk,0}_{\Phi,s_j+t_k}(t,-1)\overline{\zeta}_k \Big\} \geq \beta_0|\zeta|^2 \quad \textit{for all } \zeta \in \mathbb{C}^p\,.$$

Again, Theorem 11.3.2 remains valid for the system as follows.

Theorem 11.3.4.

i. The system of operators $((A_{jk}))_{p\times p}$ with $A_{jk} \in \mathcal{L}^{s_j+t_k}_{c\ell}(\Gamma)$ is strongly elliptic if and only if the condition

$$\det\big((b_{+,jk}(t) + \lambda b_{-,jk}(t))\big)_{p\times p} \neq 0 \quad \textit{for all } t \quad \textit{and for all } \lambda \in [-1,1] \tag{11.3.31}$$

is satisfied.

ii. The system is oddly elliptic if and only if the condition

$$\det\big((b_{-,jk}(t) + \lambda b_{+,jk}(t))\big)_{p\times p} \neq 0 \quad \textit{for all } t \quad \textit{and for all } \lambda \in [-1,1] \tag{11.3.32}$$

is satisfied.

Proof: The basic ideas of the proof for the system are similar to the scalar case in Theorem 11.3.2. However, one needs the following results of matrix algebra and functional calculus. The detailed proof for systems was given by Prössdorf and Schmidt in [357]. (See also Prössdorf and Rathsfeld [355] for coefficients with discontinuities.)

i. Let us assume that (11.3.31) is violated, i.e., for some t_0 and $\lambda_0 \in [-1,1]$ we have

$$\det\big((b_{+,jk}(t_0) + \lambda_0 b_{-,jk}(t_0))\big)_{p\times p} = 0\,.$$

Then there exists at least one vector $\zeta_0 \in \mathbb{C}^p$, $\zeta_0 \neq 0$ with

$$\sum_{k=1}^{p} \big(b_{+,jk}(t_0) + \lambda_0 b_{-,jk}(t_0)\big)\overline{\zeta}_{0k} = 0\,, j = 1,\ldots,p\,.$$

Hence,

$$\sum_{k=1}^{p} a^{jk,0}_{\Phi,s_j+t_k}(t_0,1)\overline{\zeta}_{0k} = \sum_{k=1}^{p}\big(b_{+,jk}(t_0)+b_{-,jk}(t_0)\big)\overline{\zeta}_{0k}$$
$$= (1-\lambda_0)\sum_{k=1}^{p} b_{-,jk}(t_0)\overline{\zeta}_{0k}$$

and

$$\sum_{k=1}^{p} a^{jk,0}_{\Phi,s_j+t_k}(t_0,-1)\overline{\zeta}_{0k} = \sum_{k=1}^{p}\big(b_{+,jk}(t_0)-b_{-,jk}(t_0)\big)\overline{\zeta}_{0k}$$
$$= -(1-\lambda_0)\sum_{k=1}^{p} b_{-,jk}(t_0)\overline{\zeta}_{0k}\,.$$

Now, let $\Theta(t_0)$ be any matrix satisfying

$$\mathrm{Re}\sum_{\ell,j,k}\zeta_{0\ell}^{\top}\Theta_{\ell,j}(t_0)a^{jk,0}_{\Phi,s_j+t_k}(t_0,1)\overline{\zeta}_{0k}$$
$$= (1-\lambda_0)\;\mathrm{Re}\sum_{\ell,j,k}\zeta_{0\ell}^{\top}\Theta_{\ell j}(t_0)b_{-,jk}(t_0)\overline{\zeta}_{0k} > 0$$

then for this matrix will hold

$$\mathrm{Re}\sum_{\ell,j,k=1}\zeta_{0\ell}^{\top}\Theta_{\ell j}(t_0)a^{jk,0}_{\Phi,s_j+t_k}(t_0,-1)\overline{\zeta}_{0k}$$
$$= -(1-\lambda_0)\;\mathrm{Re}\sum_{\ell,j,k}\zeta_{0\ell}^{\top}\Theta_{\ell j}(t_0)b_{-,jk}(t_0)\overline{\zeta}_{0k} < 0\,.$$

Therefore, for any choice of $\Theta_{\ell j}(t_0)$, both conditions on (11.3.28) can never be satisfied at the same time. Hence, (11.3.28) implies (11.3.31).

ii. Now, let (11.3.31) be satisfied. Then we need to construct the matrix–valued function $\Theta(t)$, providing (11.3.28). To this end, define the matrix–valued function

$$U(t) := \big(\big(b_{+,jk}(t)\big)\big)^{-1}\big(\big(b_{-,jk}(t)\big)\big)$$

which is well defined since (11.3.31) with $\lambda = 0$ implies the existence of $\big(\big(b_{+,jk}(t)\big)\big)^{-1}$. Then let

$$V(t) := \big\{\big(I+U(t)\big)^{\frac{1}{2}}\big(I-U(t)\big)^{\frac{1}{2}}\big\}^{-1}$$

where the square roots are defined by means of the Dunford–Taylor integral with eigenvalues having positive real parts Kato [223, Section I.5.6.]. With $V(t)$ define

$$N_+(t) := V(t)\big(I+U(t)\big)\,.$$

Then the matrix function

$$H(t) := \int_0^\infty \exp\big(-N_+^*(t)s\big)\exp\big(-N_+(t)s\big)ds$$

can be defined by Bellman [26] and

$$\Theta(t) = H(t)V(t)\big((b_{+,jk}(t))\big)$$

gives the desired matrix providing the two inequalities in (11.3.28).

For the oddly elliptic case, just exchange the roles of $b_{+,jk}$, and $b_{-,jk}$ and of (11.3.28) and (11.3.31). ■

Again, with the Hilbert transform (11.3.10) as preconditioner, the oddly elliptic system of equations (11.3.22) may be treated by variational methods; in this case the associated variational equation with a strongly elliptic bilinear form assumes the form

$$\begin{aligned} a(u,v) &:= (\Lambda^\sigma \Theta \Lambda^{-\sigma} HAu, v)_{\prod_{\ell=1}^p H^{(t_\ell - s_\ell)/2}(\Gamma)} \\ &= (\Lambda^\sigma \Theta \Lambda^{-\sigma} Hf, v)_{\prod_{\ell=1}^p H^{(t_\ell - s_\ell)/2}(\Gamma)}, \end{aligned} \tag{11.3.33}$$

where $\mathcal{H}_1 = \prod_{k=1}^p H^{t_k}(\Gamma)$ and $\mathcal{H}_2 = \prod_{j=1}^p H^{-s_j}(\Gamma)$.

Now $HA : \mathcal{H}_1 \to \mathcal{H}_2$ is a strongly elliptic system; and since H is an isomorphism, we again have for the Fredholm index of the oddly elliptic system operator A:

$$\text{index}(HA) = \text{index}(A) = 0$$

in view of Theorem 5.3.11.

11.3.3 Multiply Connected Domains

Before we treat the general system of pseudodifferential equation on $\Gamma = \partial\Omega$ of a multiply connected domain $\Omega \subset \mathbb{R}^2$ let us begin with the following simple example, the Dirichlet problem

$$\Delta u = 0 \quad \text{in } \Omega \quad \text{and} \quad u|_\Gamma = \varphi \quad \text{on } \Gamma$$

where $\Gamma = \bigcup_{j=1}^q \Gamma_j$ and each of the curves Γ_j is a closed smooth Jordan curve. Moreover, Γ_1 contains all of the other curves in the interior domain bounded by Γ_1 and the curves are mutually disjoint. They all are oriented in such a way that Ω always lies to the left of Γ. On each of the closed curves we use a 1–periodic global parametric representation $y|_{\Gamma_j} = T_j(z)$. By using the direct method, the solution can be expressed by boundary potentials as in

(1.1.7) which leads to the boundary integral equation (1.3.3) for the unknown missing Cauchy datum, the boundary charge $\sigma = \frac{\partial u}{\partial n}|_\Gamma$, i.e.,

$$V\sigma = f(x) := \tfrac{1}{2}\varphi(x) + K\varphi(x) \quad \text{for } x \in \Gamma\,. \tag{11.3.34}$$

In order to write (11.3.34) as a system of periodic integral equations, we introduce the notation

$$\widetilde{\sigma}_\ell(t) := \sigma\big(T_\ell(t)\big)\frac{dT_\ell}{dt}(t) \tag{11.3.35}$$

and write (11.3.34) as

$$\begin{aligned}\sum_{\ell=1}^{q} V_{r\ell}\widetilde{\sigma}_\ell(t) &:= \sum_{\ell=1}^{q}\int\limits_{\tau=0}^{1}\Big(-\frac{1}{2\pi}\log|T_r(t)-T_\ell(\tau)|\Big)\widetilde{\sigma}_\ell(\tau)d\tau \\ &= f_r(t) := f\big(T_r(t)\big) \quad \text{for } r = 1,\ldots,q\,.\end{aligned} \tag{11.3.36}$$

This system is a particular example of a system of pseudodifferential equations as considered in Section 11.3.2. We note that $V_{r\ell}$ for $r \neq \ell$ has a C^∞–kernel. Hence, the symbol matrix has diagonal form, i.e.,

$$a^{r\ell 0}_{\Phi,-1}(t,\xi) = \frac{1}{2|\xi|}\delta_{r\ell} \tag{11.3.37}$$

according to (11.1.10). In terms of the representation (11.3.6) we have the Fourier series representation for the system

$$\begin{aligned}\sum_{\ell=1}^{q}(V_{r\ell}\sigma_\ell)(T(t)) &= \sum_{\ell=1}^{q} b_{+,r\ell}(t)\Big\{\sum_{0\neq j\in\mathbb{Z}}|2\pi j|^{-1}\widehat{\sigma}_j e^{2\pi i j t} + \widehat{\sigma}_0 + \widetilde{R}_{r\ell}\sigma_\ell\Big\} \\ &= \frac{1}{2}\Lambda^{-1}\widetilde{\sigma}_r + \sum_{\ell=1}^{q}\widetilde{R}_{r\ell}\sigma_\ell\end{aligned} \tag{11.3.38}$$

where $b_{+,r\ell} = \frac{1}{2}\delta_{r\ell}$ and, in this case, $b_{-,r\ell} = 0$; $s_\ell = t_\ell = -\frac{1}{2}$ and $\widetilde{R}_{r\ell} \in \mathcal{L}^{-\infty}_{c\ell}(\Gamma)$ for $\ell = 1,\ldots,q$. The last expression corresponds to (11.3.27). Clearly, this is a strongly elliptic system.

For a system of the form (11.3.22) on the boundary Γ of the multiply connected domain Ω we now write

$$u_{k,\ell}(t) \;:=\; u_k|_{\Gamma_\ell}\,,\; f_{j,r} := f_j|_{\Gamma_r} \tag{11.3.39}$$

and

$$A_{jk,r\ell}u_{k,\ell} \;:=\; \big(A_{jk}|_{\Gamma_\ell}u_k|_{\Gamma_\ell}\big)|_{\Gamma_r}\,. \tag{11.3.40}$$

In terms of these notations, the system (11.3.22) for the multiply connected domain now reads

$$\sum_{\ell=1}^{q} \sum_{k=1}^{p} A_{jk,r\ell} u_{k,\ell} = f_{j,r} \quad \text{on } \Gamma_r \quad \text{for } r = 1, \ldots, q \quad \text{and } j = 1, \ldots, p. \tag{11.3.41}$$

For $\ell \neq r$, the integration in the integral operator representation of $A_{jk,r\ell} \in \mathcal{L}_{c\ell}^{s_j+t_k}(\Gamma)$ (see Theorem 8.1.8) acts on Γ_ℓ whereas the observation point $x \in \Gamma_r$, which is separated from Γ_ℓ and $|x - y| \geq \text{dist}(\Gamma_r, \Gamma_\ell) > 0$. Here the Schwartz kernel $A_{jk,r\ell}$ is C^∞–smooth. Hence, for $\ell \neq r$, the symbol matrix elements are zero, i.e.,

$$a_{\Phi,s_j+t_k}^{jk,r\ell 0}(t,\xi) = 0 \quad \text{for } r \neq \ell \quad \text{and } j,k, = 1, \ldots, p.$$

As a consequence, the principal symbol of $A_{jk,r\ell}$ has a block diagonal structure as

$$\begin{pmatrix} a_{\Phi,s_j+t_k}^{jk,110}(t,\xi) & 0 & \cdots & 0 \\ 0 & a_{\Phi,s_j+t_k}^{jk,220}(t,\xi) & & 0 \\ & & & \vdots \\ \vdots & & \ddots & \\ 0 & 0 & \cdots & a_{\Phi,s_j+t_k}^{jk,qq0}(t,\xi) \end{pmatrix} \tag{11.3.42}$$

and $j,k = 1, \ldots, p$. The coupling operators $A_{jk,r\ell}$ for $r \neq \ell$ between different boundary components $\Gamma_\ell \neq \Gamma_r$ are all smoothing operators. Hence, with respect to mapping and Fredholm properties, the complete system can be separated into the single systems with corresponding principal symbols $A_{\Phi,s_j+t_k}^{jk,rr}(t,\xi)$. In particular,

$$\det \left(\left(a_{\Phi,s_j+t_k}^{jk,r\ell}(t,\xi) \right) \right)_{pq \times pq}$$
$$= \prod_{r=1}^{q} \det \left(\left(a_{\Phi,s_j+t_k}^{jk,rr}(t,\xi) \right) \right)_{p \times p} \quad \text{for all } t \quad \text{and } \xi = \pm 1.$$

Therefore, the index formulae (11.3.5) and (11.3.25) also remain valid for the multiply connected case when the integrals are to be taken over all of the components of Γ.

In a similar manner, the concepts of strong and odd ellipticity of the system can now be used for each of the boundary components Γ_r separately.

11.4 Fourier Series Representation of some Particular Operators

To conclude this chapter we now apply Theorem 11.2.2 to a few more basic operators in addition to those of the Laplacian. The general procedure will be to find the symbol's expansion by making use of the boundary integral operators considered and their pseudohomogeneous kernel expansions.

11.4.1 The Helmholtz Equation

We begin with the simple layer potential operator V_{k} in (2.1.6). We need the pseudohomogeneous kernel expansion of $E_{\mathsf{k}}(x,y)$ in (2.1.4). From (2.1.18) we see that

$$\begin{aligned} E_{\mathsf{k}}(x,y) &= E(x,y) - \frac{1}{2\pi}(\log \mathsf{k} + \gamma_0) \qquad (11.4.1)\\ &\quad -\frac{1}{2\pi}\Big\{ \log((\mathsf{k}r)\sum_{m=1}^{\infty} a_m(\mathsf{k}r)^{2m} + \sum_{m=1}^{\infty} b_m(\mathsf{k}r)^{2m}\Big\}. \end{aligned}$$

By using the 1–periodic parameterization $x = T(t)$ and $y = T(\tau)$ we may rewrite $E_{\mathsf{k}}(x,y)$ in the form of a pseudohomogeneous expansion

$$\begin{aligned} E_{\mathsf{k}}\big(T(t),T(\tau)\big) &= -\frac{1}{2\pi}\log|t-\tau|\Big\{1 + a_1\mathsf{k}^2L^2(t-\tau)^2 \\ &\quad - L^4(t-\tau)^4\big(a_1\mathsf{k}^2\frac{1}{12}\kappa^2(t) - a_2 k^4\big) \\ &\quad + a_1\mathsf{k}^2\frac{1}{12}L^4\kappa(t)\frac{d\kappa}{dt}(t)(t-\tau)^5 + \mathcal{O}_6\mathsf{k}^4\Big\} \qquad (11.4.2)\\ &\quad + k_\infty(\mathsf{k},t,\tau) \\ &= \sum_{j=0}^{5} k_{j+1-1}(t,t-\tau) + R_5(t,\tau) \end{aligned}$$

according to (11.2.7) where k_∞ is a C^∞–function of t and τ and

$$k_j(t,t-\tau) = p_j(t,t-\tau)\log|t-\tau|$$

with homogeneous polynomials p_j of degree $j \geq 0$. It is clear that the individual terms of the expansion are no longer 1–periodic in t and τ. Then Theorem 11.2.2 provides us with the first six terms of the homogeneous symbol corresponding to the parametric representation $x = T(t)$ defining the mapping Φ, suppressed in the following notation:

$$
\begin{aligned}
a^0_{-1}(t,\xi) &= -\frac{1}{2\pi}\xi^{-1} i i\pi 1(\operatorname{sign}\xi) = \frac{1}{2|\xi|}\,,\\
a^0_{-2}(t,\xi) &= 0\,,\\
a^0_{-3}(t,\xi) &= -\frac{1}{2\pi}a_1\mathsf{k}^2L^2 2!\xi^{-3}i^3 i\pi \operatorname{sign}\xi = \frac{1}{4}\mathsf{k}^2L^2\frac{1}{|\xi|^3}\,,\\
a^0_{-4}(t,\xi) &= 0\,,\\
a_{-5}(t,\xi) &= \frac{L^4}{2\pi}\big(a_1\mathsf{k}^2\frac{1}{12}\kappa^2(t) - a_2\mathsf{k}^4\big)4!\xi^{-5}i^5 i\pi\operatorname{sign}\xi \qquad (11.4.3)\\
&= \frac{L^4}{4}\big(\mathsf{k}^2\kappa^2(t)+\frac{3}{4}\mathsf{k}^4\big)\frac{1}{|\xi|^5}\,,\\
a^0_{-6}(t,\xi) &= \frac{1}{2\pi}a_1\mathsf{k}^2\frac{1}{12}L^4\kappa(t)\frac{d\kappa}{dt}(t)5!\xi^{-6}i^6 i\pi\operatorname{sign}\xi\\
&= -i\frac{5}{4}\mathsf{k}^2L^4\kappa(t)\frac{d\kappa}{dt}(t)\frac{1}{|\xi|^6}\operatorname{sign}\xi\,,
\end{aligned}
$$

(see Agranovich [6, p. 35 (2.45)]). Consequently, the simple layer potential operator $V_\mathsf{k} \in \mathcal{L}^{-1}_{c\ell}(\Gamma)$ admits a periodic Fourier series representation in the form

$$
V_\mathsf{k}\sigma\big(T(t)\big) = \sum_{\ell\in\mathbb{Z}\setminus\{0\}}\sum_{j=0}^{4} a^0_{-1-j}(t,2\pi\ell)\widehat{\sigma}_\ell e^{i2\pi\ell t} + R_5\sigma \qquad (11.4.4)
$$

where $\widehat{\sigma}_\ell$ is the Fourier coefficient given by

$$
\widehat{\sigma}_\ell = L\int_0^1 e^{-2\pi i\ell\tau}\sigma\big(T(t)\big)d\tau
$$

and

$$
(R_5\sigma)(x) = \int_\Gamma R_5(x,y)\sigma(y)ds_y = \sum_{\ell\in\mathbb{Z}}\widehat{\sigma}_\ell\int_0^1 \big(T(t),T(\tau)\big)e^{2\pi i\ell\tau}d\tau\,.
$$

with $R_5 \in C^5(\Gamma\times\Gamma)$ defining a pseudodifferential operator belonging to $\mathcal{L}^{-7}_{c\ell}(\Gamma)$.

For the double layer potential operator K_k we use the representation with (2.1.20), i.e.

$$
\frac{\partial}{\partial n_y}E_\mathsf{k}(x,y) = \Big(\frac{\partial}{\partial n_y}E(x,y)\Big)\Big\{1+\sum_{m=1}^{\infty}\big(a_m(1+2m\log(\mathsf{k}r))+2mb_m\big)(\mathsf{k}r)^{2m}\Big\}\,.
$$

Then with (11.1.23) we get

$$
\begin{aligned}
\frac{\partial}{\partial n_y}E_\mathsf{k}\big(T(t),T(\tau)\big) &= \frac{1}{2\pi}\Big(\frac{1}{2}\kappa(t) - \frac{1}{3}(t-\tau)\frac{d\kappa}{dt}(t) + k_{\infty,1}(t,\tau)\Big)\times\\
&\qquad\big\{a_1\mathsf{k}^2r^2\log r + (\mathsf{k}r)^4\log r R_{\infty,1}(\mathsf{k}^2;r^2) + R_{\infty,2}(k;r^2)\big\}
\end{aligned}
$$

where the kernels $R_{\infty,1}$ and $R_{\infty,2}$ are C^∞–functions of x and y. Then r^2 will be replaced by (11.1.19) and we obtain the first four terms of the asymptotic kernel expansion:

$$\frac{\partial}{\partial n_y} E_{\mathsf{k}}\big(T(t),T(\tau)\big) = \frac{1}{2\pi} a_1 L^2 \mathsf{k}^2 \Big(\kappa(t)(t-\tau)^2 - \frac{2}{3}(t-\tau)^3 \frac{d\kappa}{dt}(t)\Big) \log|t-\tau| + (t-\tau)^4 \log|t-\tau| k_{\infty,4}(\mathsf{k},t,\tau) + k_{\infty,2}(\mathsf{k},t,\tau)\,.$$

Here, the order of the double layer potential $K_{\mathsf{k}} \in \mathcal{L}^0_{c\ell}(\Gamma)$ should be $m=0$, but for the symbol expansion we find the terms

$$\begin{aligned} a^0_0 &= a^0_{-1} = a^0_{-2} = 0\,,\\ a^0_{-3} &= -\frac{1}{4}\mathsf{k}^2\kappa(t)\frac{L^2}{|\xi|^3}\,,\\ a^0_{-4} &= \frac{i}{2}\mathsf{k}^2\frac{d\kappa}{dt}(t)\frac{L^2}{|\xi|^4}\,\operatorname{sign}\xi\,. \end{aligned} \tag{11.4.5}$$

Consequently, the double layer potential operator K_{k} admits the 1–periodic Fourier series representation

$$K_{\mathsf{k}}\varphi\big(T(t)\big) = L^2 \sum_{\ell\in\mathbb{Z}\setminus\{0\}} \Big(-\frac{1}{4}\mathsf{k}^2\kappa(t)\frac{1}{|2\pi\ell|^3} + \frac{i}{2}\mathsf{k}^2\frac{d\kappa}{dt}(t)\frac{1}{|2\pi\ell|^4}\operatorname{sign}\ell\Big)\widehat{\varphi}_\ell e^{i2\pi\ell t} + R_5\varphi\,.$$

Here, the smooth remaining operator $R_5 \in \mathcal{L}^{-5}_{c\ell}(\Gamma)$ is given by

$$R_5\varphi(x) = \int_\Gamma R_5(x,y)\varphi(y)ds_y = \sum_{\ell\in\mathbb{Z}} \widehat{\varphi}_\ell \int_0^1 \big(T(t),T(\tau)\big)e^{2\pi\ell\tau}d\tau\,,$$

and has a doubly periodical kernel $R_5 \in C^3(\Gamma\times\Gamma)$.

For the operator K'_{k} which is the transposed to K_{k}, one may compute the first four terms of its symbol expansion by employing the formula (7.1.34) directly. However, for illustration, we first compute the pseudohomogeneous kernel expansion of K'_{k} in the same manner as for K_{k}. A simple computation shows that

$$\begin{aligned} \frac{\partial E_{\mathsf{k}}}{\partial n_x} &= \frac{\partial E}{\partial n_x} + R_{\mathsf{k}}(x,y)\\ &= \frac{\partial E}{\partial n_x}\Big\{1 + \sum_{m=1}^\infty \big(a_m(1+2m\log(\mathsf{k}r)) + 2mb_m\big)(\mathsf{k}r)^{2m}\Big\}\,. \end{aligned}$$

With

$$\boldsymbol{n}(x)\cdot(\boldsymbol{x}-\boldsymbol{y}) = -\frac{1}{2}(t-\tau)^2\kappa(t)L^2 + \frac{1}{6}(t-\tau)^3L^2\frac{d\kappa}{dt}(t) + \mathcal{O}_4 \tag{11.4.6}$$

we obtain

$$\frac{\partial E}{\partial n_x}(x,y) = \frac{1}{4\pi}\Big(\kappa(t) - \frac{1}{3}(t-\tau)\frac{d\kappa}{dt}(t)\Big) + \mathcal{O}_2\,.$$

Hence,

$$\frac{\partial}{\partial n_x}E_{\mathsf{k}}(x,y) = \frac{L^2}{2\pi}a_1\mathsf{k}^2\log|t-\tau|\Big\{(t-\tau)^2\kappa(t) - \frac{1}{3}(t-\tau)^3\frac{d\kappa}{dt}(t)\Big\}$$
$$+ (t-\tau)^4\log|t-\tau|k'_{\infty,4}(\mathsf{k},t,\tau) + \mathcal{O}_0\,,$$

where $k'_{\infty,4}$ and $\mathcal{O}_0$ are C^∞–functions of t and τ. With Theorem 11.2.2 we now get the symbols

$$a_0^0 = a_{-1}^0 = a_{-2}^0 = 0 \quad \text{and}$$
$$a_{-3}^0 = -\frac{1}{4}\mathsf{k}^2\kappa(t)\frac{L^2}{|\xi|^3}\,,\; a_{-4}^0 = \frac{i}{4}\mathsf{k}^2\frac{d\kappa}{dt}(t)\frac{L^2}{|\xi|^4}\operatorname{sign}\xi\,,$$

which also follow with (7.1.34) from the symbols of K'_{k} in (11.4.5). Again, for K'_{k} we have the 1–periodic Fourier series representation

$$(K'_{\mathsf{k}}\sigma)(T(t)) = L^2\sum_{\ell\in\mathbb{Z}\setminus\{0\}}\Big\{-\frac{1}{4}\mathsf{k}^2\kappa(t)\frac{1}{|2\pi\ell|^3}$$
$$+\frac{i}{4}\mathsf{k}^2\frac{d\kappa}{dt}(t)\frac{1}{|2\pi\ell|^4}\operatorname{sign}\ell\Big\}\widehat{\sigma}e^{i2\pi\ell t} + R_5^\top\sigma\,,$$

$$R_5^\top\sigma = \int_\Gamma R_5(x,y)\sigma(y)ds_y = \sum_{\ell\in\mathbb{Z}}\widehat{\sigma}_\ell\int_0^1 R_5\big(T(\tau),T(t)\big)e^{2\pi i\ell\tau}d\tau\,.$$

For the hypersingular operator

$$D_{\mathsf{k}}\varphi = -\frac{\partial}{\partial n_x}\int_\Gamma\frac{\partial}{\partial n_y}E_{\mathsf{k}}(x,y)\varphi(y)ds_y \quad \text{with } x\in\Gamma$$

we use the Maue formula [284]

$$D_{\mathsf{k}}\varphi = -\frac{d}{ds_x}V_{\mathsf{k}}\frac{d}{ds_y}\varphi - \mathsf{k}^2V_{\mathsf{k}}\varphi = -\frac{1}{L^2}\frac{d}{dt}V_{\mathsf{k}}\frac{d\varphi}{d\tau} - \mathsf{k}^2V_{\mathsf{k}}\varphi \tag{11.4.7}$$

which can be obtained by a slight modification of the proof of Lemma 1.2.2 for the Laplacian. With the representation (11.4.4) of V_{k} we then find four terms of the 1–periodic Fourier series representation of D_{k}, i.e.

$$D_{\mathsf{k}}\varphi = -\sum_{\ell\in\mathbb{Z}\setminus\{0\}}\sum_{j=0}^{4}\Big\{\frac{1}{L^2}\frac{d}{dt}a_{-1-j}^0(t,2\pi\ell) \tag{11.4.8}$$
$$+\Big(\frac{1}{L^2}(2\pi i\ell)^2 + \mathsf{k}^2\Big)a_{-1-j}^0(t,2\pi\ell)\Big\}\widehat{\varphi}_\ell e^{i2\pi\ell t}$$
$$-\sum_{\ell\in\mathbb{Z}}\widehat{\varphi}_\ell\int_0^1\Big\{\frac{1}{L^2}\frac{d}{dt}R_5\big(T(t),T(\tau)\big)(2\pi\ell) + \mathsf{k}^2R_5\big(T(t),T(\tau)\big)\Big\}e^{2\pi i\ell\tau}d\tau$$

where $a^0_{-1-j}(t,\xi)$ are the symbol terms of V_k given in (11.4.3).

11.4.2 The Lamé System

For the two–dimensional space the fundamental solution of the Lamé system is given in (2.2.2),

$$E(x,y) = \frac{\lambda+3\mu}{4\pi\mu(\lambda+2\mu)}\Big\{ -\log r\delta_{ik} + \frac{\lambda+\mu}{\lambda+3\mu}\frac{1}{r^2}(x_i-y_i)(x_k-y_k)\Big\} \quad (11.4.9)$$

Hence, the simple layer potential can be rewritten as

$$\begin{aligned}(V\sigma)(x) = \frac{\lambda+3\mu}{2\mu(\lambda+2\mu)}\Big\{ &-\frac{1}{2\pi}\int\limits_\Gamma \log r\boldsymbol{\sigma}(y)ds_y \\ &+\frac{\lambda+\mu}{\lambda+3\mu}\frac{1}{2\pi}\int\limits_\Gamma \boldsymbol{L}(x,y)\boldsymbol{\sigma}(y)ds_y\Big\}\end{aligned} \quad (11.4.10)$$

where the tensor $\boldsymbol{L}$ has the form

$$\boldsymbol{L}(x,y) = \frac{1}{r^2}(\boldsymbol{x}-\boldsymbol{y})(\boldsymbol{x}-\boldsymbol{y})^\top .$$

With the parametric representation we have

$$\boldsymbol{L}\big(T(t),T(\tau)\big) = \frac{1}{r^2}\big(T(\tau)-T(t)\big)\big(T(\tau)-T(t)\big)^\top$$

and obtain with (11.1.18) and (11.1.23)

$$\begin{aligned}\boldsymbol{L}\big(T(t),T(\tau)\big) &= [1+(\tau-t)^2c(t,\tau)]^{-1}\times \\ &\times\big(a(t,\tau)\boldsymbol{t}(t)+b(t,\tau)\boldsymbol{n}(t)\big)\big(a(t,\tau)\boldsymbol{t}(t)+b(t,\tau)\boldsymbol{n}(t)\big)^\top + \boldsymbol{L}_\infty(t,\tau)\end{aligned} \quad (11.4.11)$$

where

$$\begin{aligned}a(t,\tau) &= 1-\frac{(\tau-t)^2}{6}L^2\kappa^2(t)\,,\\ b(t,\tau) &= \frac{(\tau-t)}{2}L\kappa(t)+\frac{(\tau-t)^2}{6}L\frac{d\kappa}{dt}(t)\end{aligned} \quad (11.4.12)$$

and $c(t,\tau)$ from (11.1.19) all are C^∞–functions of t and τ.

Consequently, $\boldsymbol{L}\big(T(t),T(\tau)\big)$ is a C^∞–function of t and τ. The symbol of the elastic simple layer potential operator is the same as that of the Laplacian up to a multiplicative constant. More precisely, we have the 1–periodic Fourier series representation

$$(\mathbf{V}\boldsymbol{\sigma})\big(T(t)\big) = \frac{\lambda+3\mu}{2\mu(\lambda+2\mu)}\sum_{\ell\neq0}\frac{1}{4\pi|\ell|}\widehat{\boldsymbol{\sigma}}_\ell e^{2\pi i\ell t} + \boldsymbol{R}_\infty\boldsymbol{\sigma}\big(T(t)\big) \quad (11.4.13)$$

where

$$\boldsymbol{R}_\infty\boldsymbol{\sigma}(x) = \int\limits_\Gamma \boldsymbol{R}_\infty(x,y)\boldsymbol{\sigma}(y)ds_y = \sum_{\ell\in\mathbb{Z}} \widehat{\boldsymbol{\sigma}}_\ell \int\limits_0^1 \boldsymbol{R}_\infty(T(t),T(\tau))e^{2\pi i\ell\tau}d\tau$$

with $\boldsymbol{R}_\infty \in C^\infty(\Gamma\times\Gamma)$, and $\boldsymbol{R}_\infty(T(t),T(\tau))$ is 1–periodic in t and τ.

For the double layer operator of elasticity (2.2.19) we use the expression (2.2.21) with (2.2.23) and obtain the corresponding kernel as

$$\begin{aligned}\big(T_yE(x,y)\big)^\top &= \frac{\mu}{\lambda+2\mu}\Big\{\Big(\boldsymbol{I}+\frac{2(\lambda+\mu)}{\mu}\boldsymbol{L}(x,y)\Big)\frac{\partial}{\partial n_y}\Big(-\frac{1}{2\pi}\log r\Big) \\ &\qquad -\begin{pmatrix}0 & 1\\ -1 & 0\end{pmatrix}\frac{d}{ds_y}\Big(-\frac{1}{2\pi}\log r\Big)\Big\}. \end{aligned} \tag{11.4.14}$$

As we have seen in (11.1.23) and in (11.4.10),

$$\big(T_yE(x,y)\big)^\top = -\frac{\mu}{\lambda+2\mu}\begin{pmatrix}0 & 1\\ -1 & 0\end{pmatrix}\frac{d}{ds_y}\Big(-\frac{1}{2\pi}\log r\Big) + \mathbf{T}_\infty(x,y)$$

where $\mathbf{T}_\infty$ is in $C^\infty(\Gamma\times\Gamma)$. So it remains to compute the symbol of the operator with the kernel given by the first term. For the latter we use integration by parts and (11.1.4) to obtain

$$\begin{aligned}(K\boldsymbol{\sigma})(T(t)) &= -\frac{\mu}{\lambda+2\mu}\begin{pmatrix}0 & 1\\ -1 & 0\end{pmatrix}\Big(-\frac{1}{2\pi}\int\limits_\Gamma \log|x-y|\frac{d\boldsymbol{\sigma}}{ds_y}(y)ds_y\Big) \\ &\qquad + \int\limits_\Gamma \mathbf{T}_\infty(x,y)\boldsymbol{\sigma}(y)ds_y \\ &= \frac{\mu}{\lambda+2\mu}\begin{pmatrix}0 & 1\\ -1 & 0\end{pmatrix}\sum_{\ell\neq 0}\frac{i\,\mathrm{sign}\,\ell}{2L}\widehat{\boldsymbol{\sigma}}_\ell e^{2\pi i\ell t} + \int\limits_\Gamma \boldsymbol{K}_\infty(x,y)\boldsymbol{\sigma}(y)ds_y\end{aligned} \tag{11.4.15}$$

where $\boldsymbol{K}_\infty \in C^\infty(\Gamma\times\Gamma)$ and

$$\int\limits_0^1 \boldsymbol{K}_\infty(T(t)T(\tau))\boldsymbol{\sigma}(\tau)d\tau = \sum_{\ell\in\mathbb{Z}}\int\limits_0^1 \boldsymbol{K}_\infty(T(t),T(\tau))e^{2\pi i\ell\tau}d\tau\,\widehat{\boldsymbol{\sigma}}_\ell\,.$$

The complete symbol of the double layer elastic operator K is therefore given by

$$\sigma_K(t,\xi) = \frac{\mu}{\lambda+2\mu}\begin{pmatrix}0 & 1\\ -1 & 0\end{pmatrix}\frac{i}{2L}\,\mathrm{sign}\,\xi\,. \tag{11.4.16}$$

For the transposed operator K' in (2.2.20) we apply (7.1.34) to find the symbol of K', i.e.

$$\sigma_{K'}(t,\xi) = \sigma_K(t,-\xi) = -\sigma_K(t,\xi)\,. \tag{11.4.17}$$

Hence, the 1–periodic Fourier expansion representation of K' assumes the form

$$\begin{aligned}(K'\varphi)(T(t)) &= \frac{\mu}{\lambda+2\mu}\begin{pmatrix}0 & -1\\ 1 & 0\end{pmatrix}\sum_{\ell\neq 0}\frac{i\,\mathrm{sign}\,\ell}{3L}\widehat{\varphi}_\ell e^{2\pi i\ell t}\\ &\quad+\sum_{\ell\in\mathbb{Z}}\int_0^1 \boldsymbol{K}_\infty(T(\tau),T(t))e^{2\pi i\ell\tau}d\tau\widehat{\varphi}_\ell\,.\end{aligned} \tag{11.4.18}$$

For the elastic hypersingular operator we use the relations (2.2.32) with (2.3.36), i.e.

$$(D\varphi)(x) = -\frac{d}{ds_x}\frac{2\mu(\lambda+\mu)}{(\lambda+2\mu)}\int_\Gamma\Big\{-\frac{1}{2\pi}\log|x-y| + \frac{1}{2\pi}\boldsymbol{L}(x,y)\Big\}\frac{d\varphi}{ds_y}(y)ds_y,$$

$$\begin{aligned}(D\varphi)(T(t)) &= -\frac{1}{L}\frac{2\mu(\lambda+\mu)}{(\lambda+2\mu)}\frac{d}{dt}\sum_{\ell\neq 0}\frac{i}{2L}\,\mathrm{sign}\,\ell\widehat{\varphi}_\ell e^{2\pi i\ell z} + \int_\Gamma \boldsymbol{D}_\infty(x,y)\varphi(y)ds_y\\ &= \frac{\pi}{L^2}\frac{2\mu(\lambda+\mu)}{(\lambda+2\mu)}\sum_{\ell\neq 0}|\ell|\widehat{\varphi}_\ell e^{i2\pi\ell t}\\ &\quad+\sum_{\ell\in\mathbb{Z}}\int_0^1 \boldsymbol{D}_\infty(T(t),T(\tau))e^{i2\pi\ell\tau}d\tau\widehat{\varphi}_\ell\,.\end{aligned} \tag{11.4.19}$$

The complete symbol of D is therefore given by

$$\sigma_D(t,\xi) = \frac{1}{L^2}\frac{\mu(\lambda+\mu)}{(\lambda+2\mu)}|\xi|\boldsymbol{I}\,.$$

11.4.3 The Stokes System

The fundamental solution and the kernel of the pressure operator of the Stokes system in two dimensions are given by (2.3.10), i.e.,

$$E(x,y) = \frac{1}{4\pi\mu}\Big(-(\log r)\boldsymbol{I} + \frac{(x-y)(x-y)^\top}{r^2}\Big), \tag{11.4.20}$$

$$Q(x,y) = -\frac{1}{\pi}(\nabla_y \ln r)^\top = \frac{1}{2\pi r^2}(x-y)^\top \tag{11.4.21}$$

where (11.4.20) can also obtained from (11.4.9) as the limit for $\lambda\to+\infty$. Hence, for the Stokes simple layer potential we obtain from (11.4.10)

$$V_{st}\boldsymbol{\sigma} = \frac{1}{2\mu}\Big\{-\frac{1}{4\pi}\int_\Gamma(\log r^2)\boldsymbol{\sigma}(y)ds_y + \frac{1}{2\pi}\int_\Gamma \boldsymbol{L}(x,y)\boldsymbol{\sigma}(y)ds_y\Big\}$$

with r^2 given by (11.1.19) and $\boldsymbol{L}$ by (11.4.11) and (11.4.12). The corresponding Fourier series representation,

$$\mathbf{V}_{st}\boldsymbol{\sigma}(T(t)) = \frac{1}{2\mu}\sum_{\ell\neq 0}\frac{1}{4\pi|\ell|}\widehat{\boldsymbol{\sigma}}_\ell e^{2\pi i\ell t} + \boldsymbol{R}_{\infty st}\boldsymbol{\sigma}(T(t)) \tag{11.4.22}$$

follows from (11.1.13) with $\lambda \to +\infty$.

For the kernel of the pressure operator Q (11.4.21) in local coordinates we find with (11.1.18) and (11.1.19)

$$\begin{aligned} Q(T(t),T(\tau)) &= \frac{1}{L(t-\tau)}\{1 - \frac{1}{12}(t-\tau)^2L^2\kappa^2(t) \\ &\quad + \frac{1}{24}(t-\tau)^3L^2\kappa(t)\frac{d\kappa}{dt}(t) + \mathcal{O}_4\}\boldsymbol{t}^\top(t) \\ -\Big\{\frac{1}{2}\kappa(t) &+ \frac{1}{24}(t-\tau)^2\frac{d^2\kappa}{dt^2}(t) - \frac{1}{18}(t-\tau)^3L^2\kappa^2\frac{d\kappa}{dt} + \mathcal{O}_4\Big\}\boldsymbol{n}(t)^\top \\ &= \frac{1}{L(t-\tau)}\boldsymbol{t}(t)^\top + k_{1\infty}(t,\tau)\boldsymbol{t}(t)^\top + k_{2\infty}(t,\tau)\boldsymbol{n}(t)^\top . \end{aligned}$$

Hence, with Theorem 11.2.2 we find the corresponding complete symbol as

$$a_Q(t,\xi) = -\frac{i\pi}{L}(\operatorname{sign}\xi)\boldsymbol{t}(t)^\top , \tag{11.4.23}$$

and the Fourier representation

$$\begin{aligned} Q\boldsymbol{\sigma}(T(t)) = -\frac{i}{L}\sum_{\ell\neq 0}\operatorname{sign}(\ell)\widehat{\boldsymbol{\sigma}}_\ell e_\ell^{2\pi i\ell t} \\ + \int_0^1 \big(k_{1\infty}(t,\tau)\boldsymbol{t}(t) + k_{2\infty}(t,\tau)\boldsymbol{n}(t)\big)\cdot\boldsymbol{\sigma}(T(\tau))d\tau . \end{aligned}$$

The double layer potential operator of the Stokes system can be obtained from (11.4.15) by letting $\lambda \to +\infty$. Hence, it is a smoothing operator as well as its adjoint.

Since the hypersingular operator $\boldsymbol{D}_{st}$ of the Stokes system equals the one of elasticity (11.4.19) with $\lambda \to +\infty$ we obtain

$$(\boldsymbol{D}_{st}\varphi)(T(t)) = \frac{2\mu\pi}{L^2}\sum_{\ell\neq 0}|\ell|\widehat{\varphi}_\ell e^{i2\pi\ell t} + \sum_{\ell\neq\mathbb{Z}}\int_0^1 \boldsymbol{D}_{\infty st}(T(t),(T(\tau))e^{2\pi i\ell t}d\tau\widehat{\varphi}_\ell$$

and for its complete symbol

$$a_{D_{st}}(t,\xi) = \frac{1}{L^2}\mu|\xi|\boldsymbol{I} .$$

11.4.4 The Biharmonic Equation

The fundamental solution of the biharmonic operator is given by (2.4.7), i.e.

$$E(x,y) = \tfrac{1}{8\pi} r^2 \log r \quad \text{with } r^2 = |\boldsymbol{x}-\boldsymbol{y}|^2 . \tag{11.4.24}$$

We recall that the boundary integral operators in the Calderón projector C_Ω in (2.4.23) consist of 16 different operators on the boundary curve Γ. In the same manner as in the previous examples we will first present the kernels $k_{\ell j}(x,y)$ of these operators and later on the corresponding symbols based on the pseudohomogeneous kernel expansions, as well as 1–periodic Fourier series representations.

We begin with the first row of the matrix (2.4.23).

$$k_{14}(x,y) = E(x,y) = \tfrac{1}{8\pi} r^2 \log r , \tag{11.4.25}$$

$$k_{13}(x,y) = \frac{\partial}{\partial n_y} E(x,y) = \frac{1}{8\pi} \boldsymbol{n}(y)\cdot(\boldsymbol{y}-\boldsymbol{x})(2\log r + 1) , \tag{11.4.26}$$

$$\begin{aligned} k_{12}(x,y) =& - M_y E(x,y) \\ =& -\frac{1}{4\pi}\Big\{(1+\nu)\log r + \frac{1+3\nu}{2} \\ & \qquad + (1-\nu)\big(\boldsymbol{n}(y)\cdot(\boldsymbol{y}-\boldsymbol{x})\big)^2 \frac{1}{r^2}\Big\} , \end{aligned} \tag{11.4.27}$$

$$\begin{aligned} k_{11}(x,y) =& N_y E(x,y) \\ =& -\frac{1}{4\pi}\Big\{2\boldsymbol{n}(y)\cdot(\boldsymbol{y}-\boldsymbol{x})\frac{1}{r^2} \\ & \qquad + (1-\nu)\frac{d}{ds_y}\Big(\boldsymbol{n}(z)\cdot(\boldsymbol{y}-\boldsymbol{x})\ \boldsymbol{t}(z)\cdot(\boldsymbol{y}-\boldsymbol{x})\frac{1}{r^2}\Big)\Big\}\Big|_{z=y} . \end{aligned} \tag{11.4.28}$$

For the second row, we have the corresponding kernels as:

$$k_{24}(x,y) = \frac{\partial}{\partial n_x} E(x,y) = \frac{1}{8\pi} \boldsymbol{n}(x)\cdot(\boldsymbol{x}-\boldsymbol{y})(2\log r + 1) , \tag{11.4.29}$$

$$\begin{aligned} k_{23}(x,y) =& \frac{\partial}{\partial n_x}\frac{\partial}{\partial n_y} E(x,y) \\ =& \frac{1}{8\pi}\Big\{ -\boldsymbol{n}(x)\cdot\boldsymbol{n}(y)(2\log r + 1) \\ & \qquad + 2\boldsymbol{n}(y)\cdot(\boldsymbol{y}-\boldsymbol{x})\ \boldsymbol{n}(x)\cdot(\boldsymbol{x}-\boldsymbol{y})\frac{1}{r^2}\Big\} , \end{aligned} \tag{11.4.30}$$

$$\begin{aligned} k_{22}(x,y) =& -\frac{\partial}{\partial n_x} M_y E(x,y) \\ =& -\frac{1}{4\pi}\Big\{(1+\nu)\boldsymbol{n}(x)\cdot(\boldsymbol{x}-\boldsymbol{y})\frac{1}{r^2} \\ & \qquad -2(1-\nu)\Big(\boldsymbol{n}(x)\cdot\boldsymbol{n}(y)\ \boldsymbol{n}(y)\cdot(\boldsymbol{y}-\boldsymbol{x})\frac{1}{r^2} \\ & \qquad\qquad + \big(\boldsymbol{n}(y)\cdot(\boldsymbol{y}-\boldsymbol{x})\big)^2 \boldsymbol{n}(x)\cdot(\boldsymbol{x}-\boldsymbol{y})\frac{1}{r^4}\Big)\Big\} , \end{aligned} \tag{11.4.31}$$

$$
\begin{aligned}
k_{21}(x,y) = & -\frac{\partial}{\partial n_x} N_y E(x,y) \qquad (11.4.32)\\
= & -\frac{1}{2\pi}\Big\{\boldsymbol{n}(x)\cdot\boldsymbol{n}(y)\frac{1}{r^2} + 2\boldsymbol{n}(y)\cdot(\boldsymbol{y}-\boldsymbol{x})\,\boldsymbol{n}(x)\cdot(\boldsymbol{x}-\boldsymbol{y})\frac{1}{r^4}\Big\}\\
& -\frac{1-\nu}{4\pi}\Big\{\frac{d}{ds_y}\Big(\boldsymbol{n}(x)\cdot\boldsymbol{t}(z)\,\boldsymbol{n}(z)\cdot(\boldsymbol{y}-\boldsymbol{x})\frac{1}{r^2}\\
& \qquad + \boldsymbol{n}(x)\cdot\boldsymbol{n}(z)\,\boldsymbol{t}(z)\cdot(\boldsymbol{y}-\boldsymbol{x})\frac{1}{r^2}\\
& + 2\boldsymbol{t}(z)\cdot(\boldsymbol{y}-\boldsymbol{x})\,\boldsymbol{n}(z)\cdot(\boldsymbol{y}-\boldsymbol{x})\,\boldsymbol{n}(x)\cdot(\boldsymbol{x}-\boldsymbol{y})\frac{1}{r^4}\Big)\Big\}\Big|_{z=y}\,.
\end{aligned}
$$

The kernels in the third row are given by:

$$
\begin{aligned}
k_{34}(x,y) &= M_x E(x,y) \qquad (11.4.33)\\
&= \frac{1}{4\pi}\Big\{(1+\nu)\log r + \frac{1+3\nu}{2} + (1-\nu)\big(\boldsymbol{n}(x)\cdot(\boldsymbol{x}-\boldsymbol{y})\big)^2\frac{1}{r^2}\Big\},\\
k_{33}(x,y) &= \frac{\partial}{\partial n_y} M_x E(x,y) \qquad (11.4.34)\\
&= \frac{1}{4\pi}\Big\{(1+\nu)\boldsymbol{n}(y)\cdot(\boldsymbol{y}-\boldsymbol{x})\frac{1}{r^2}\\
&\qquad -2(1-\nu)\Big(\boldsymbol{n}(x)\cdot\boldsymbol{n}(y)\,\boldsymbol{n}(x)\cdot(\boldsymbol{x}-\boldsymbol{y})\frac{1}{r^2}\\
&\qquad\qquad +\big(\boldsymbol{n}(x)\cdot(\boldsymbol{x}-\boldsymbol{y})\big)^2\boldsymbol{n}(y)\cdot(\boldsymbol{y}-\boldsymbol{x})\frac{1}{r^4}\Big)\Big\},\\
k_{32}(x,y) &= -\,M_x M_y E(x,y) \qquad (11.4.35)\\
&= -\,\frac{(1-\nu)}{4\pi}\Big\{\Big(1+3\nu+2(1-\nu)\big(\boldsymbol{n}(x)\cdot\boldsymbol{n}(y)\big)^2\Big)\frac{1}{r^2}\\
&\quad +8(1-\nu)\big(\boldsymbol{n}(x)\cdot\boldsymbol{n}(y)\big)\big(\boldsymbol{n}(y)\cdot(\boldsymbol{y}-\boldsymbol{x})\big)\big(\boldsymbol{n}(x)\cdot(\boldsymbol{x}-\boldsymbol{y})\big)\frac{1}{r^4}\\
&\quad +8(1-\nu)\big(\boldsymbol{n}(x)\cdot(\boldsymbol{x}-\boldsymbol{y})\big)^2\big(\boldsymbol{n}(y)\cdot(\boldsymbol{y}-\boldsymbol{x})\big)^2\frac{1}{r^6}\\
&\quad -2(1+\nu)\Big(\big(\boldsymbol{n}(y)\cdot(\boldsymbol{y}-\boldsymbol{x})\big)^2 + \big(\boldsymbol{n}(x)\cdot(\boldsymbol{x}-\boldsymbol{y})\big)^2\Big)\frac{1}{r^4}\Big\},
\end{aligned}
$$

$$
\begin{aligned}
k_{31}(x,y) &= -M_x N_y E(x,y) \qquad (11.4.36)\\
&= \frac{(1-\nu)}{2\pi}\Big\{\big(-\boldsymbol{n}(y)\cdot(\boldsymbol{y}-\boldsymbol{x}) + 2\boldsymbol{n}(x)\cdot\boldsymbol{n}(y)\ \boldsymbol{n}(x)\cdot(\boldsymbol{x}-\boldsymbol{y})\big)\frac{2}{r^4}\\
&+ \frac{d}{ds_y}\Big\{-(1+\nu)\boldsymbol{t}(z)\cdot(\boldsymbol{y}-\boldsymbol{x})\ \boldsymbol{n}(z)\cdot(\boldsymbol{y}-\boldsymbol{x})\frac{1}{r^4}\\
&+ (1-\nu)\boldsymbol{n}(x)\cdot\boldsymbol{t}(z)\ \boldsymbol{n}(x)\cdot\boldsymbol{n}(z)\frac{1}{r^2}\\
&+ 2(1-\nu)\boldsymbol{t}(z)\cdot(\boldsymbol{y}-\boldsymbol{x})\ \boldsymbol{n}(x)\cdot(\boldsymbol{x}-\boldsymbol{y})\ \boldsymbol{n}(x)\cdot\boldsymbol{n}(z)\frac{1}{r^4}\Big\}\\
&+ 8\big(\boldsymbol{n}(x)\cdot(\boldsymbol{x}-\boldsymbol{y})\big)^2\boldsymbol{n}(y)\cdot(\boldsymbol{y}-\boldsymbol{x})\frac{1}{r^6}\\
&+ \frac{d}{ds_y}\Big\{2(1-\nu)\boldsymbol{t}(z)\cdot\boldsymbol{n}(x)\ \boldsymbol{n}(x)\cdot(\boldsymbol{x}-\boldsymbol{y})\ \boldsymbol{n}(z)\cdot(\boldsymbol{y}-\boldsymbol{x})\frac{1}{r^4}\\
&+ 4(1-\nu)\boldsymbol{t}(z)\cdot(\boldsymbol{y}-\boldsymbol{x})\big(\boldsymbol{n}(x)\cdot(\boldsymbol{x}-\boldsymbol{y})\big)^2\boldsymbol{n}(z)\cdot(\boldsymbol{y}-\boldsymbol{x})\frac{1}{r^6}\Big\}\Big\}\Big|_{z=y}\,.
\end{aligned}
$$

Similarly, for the last row we obtain the kernels

$$
\begin{aligned}
k_{44}(x,y) &= N_x E(x,y) \qquad (11.4.37)\\
&= -\frac{1}{4\pi}\Big\{2\big(\boldsymbol{n}(x)\cdot(\boldsymbol{x}-\boldsymbol{y})\frac{1}{r^2}\\
&\quad + (1-\nu)\frac{d}{ds_x}\Big(\boldsymbol{n}(z)\cdot(\boldsymbol{x}-\boldsymbol{y})\ \boldsymbol{t}(z)\cdot(\boldsymbol{x}-\boldsymbol{y})\frac{1}{r^2}\Big)\Big\}\Big|_{z=x}\,,
\end{aligned}
$$

$$
\begin{aligned}
k_{43}(x,y) &= N_x\frac{\partial}{\partial n_y}E(x,y) \qquad (11.4.38)\\
&= \frac{1}{4\pi}\Big\{2\boldsymbol{n}(x)\cdot\boldsymbol{n}(y)\frac{1}{r^2} + (1-\nu)\frac{d}{ds_x}\Big(\boldsymbol{n}(z)\cdot\boldsymbol{n}(y)\ \boldsymbol{t}(z)\cdot(\boldsymbol{x}-\boldsymbol{y})\frac{1}{r^2}\Big)\Big\}\Big|_{z=x}\\
&+ \frac{1}{4\pi}\Big\{4\boldsymbol{n}(x)\cdot(\boldsymbol{x}-\boldsymbol{y})\ \boldsymbol{n}(y)\cdot(\boldsymbol{y}-\boldsymbol{x})\frac{1}{r^4}\\
&+ (1-\nu)\frac{d}{ds_x}\Big[\boldsymbol{n}(z)\cdot(\boldsymbol{x}-\boldsymbol{y})\ \boldsymbol{n}(y)\cdot\boldsymbol{t}(z)\frac{1}{r^2}\\
&+ 2\boldsymbol{n}(z)\cdot(\boldsymbol{x}-\boldsymbol{y})\ \boldsymbol{n}(y)\cdot(\boldsymbol{y}-\boldsymbol{x})\ \boldsymbol{t}(z)\cdot(\boldsymbol{x}-\boldsymbol{y})\frac{1}{r^4}\Big]\Big\}\Big|_{z=x}\,,
\end{aligned}
$$

$$
\begin{aligned}
k_{42}(x,y) &= -N_x M_y E(x,y) \qquad (11.4.39)\\
&= \frac{(1-\nu)}{2\pi}\Big\{\big(-\boldsymbol{n}(x)\cdot(\boldsymbol{x}-\boldsymbol{y}) + 2\boldsymbol{n}(x)\cdot\boldsymbol{n}(y)\;\boldsymbol{n}(y)\cdot(\boldsymbol{y}-\boldsymbol{x})\big)\frac{2}{r^4}\\
&+\frac{d}{ds_x}\Big\{-(1+\nu)\boldsymbol{t}(z)\cdot(\boldsymbol{x}-\boldsymbol{y})\;\boldsymbol{n}(z)\cdot(\boldsymbol{x}-\boldsymbol{y})\frac{1}{r^4}\\
&\qquad +(1-\nu)\boldsymbol{n}(y)\cdot\boldsymbol{t}(z)\;\boldsymbol{n}(z)\cdot\boldsymbol{n}(y)\frac{1}{r^2}\\
&\qquad +2(1-\nu)\boldsymbol{t}(z)\cdot(\boldsymbol{x}-\boldsymbol{y})\;\boldsymbol{n}(y)\cdot(\boldsymbol{y}-\boldsymbol{x})\;\boldsymbol{n}(z)\cdot\boldsymbol{n}(y)\frac{1}{r^4}\Big\}\\
&\qquad +8\big(\boldsymbol{n}(y)\cdot(\boldsymbol{y}-\boldsymbol{x})\big)^2\boldsymbol{n}(x)\cdot(\boldsymbol{x}-\boldsymbol{y})\frac{1}{r^6}\\
&+\frac{d}{ds_x}\Big\{2(1-\nu)\boldsymbol{t}(z)\cdot\boldsymbol{n}(y)\;\boldsymbol{n}(y)\cdot(\boldsymbol{y}-\boldsymbol{x})\;\boldsymbol{n}(z)\cdot(\boldsymbol{x}-\boldsymbol{y})\frac{1}{r^4}\\
&+4(1-\nu)\boldsymbol{t}(z)\cdot(\boldsymbol{x}-\boldsymbol{y})\big(\boldsymbol{n}(y)\cdot(\boldsymbol{y}-\boldsymbol{x})\big)^2\boldsymbol{n}(z)\cdot(\boldsymbol{x}-\boldsymbol{y})\frac{1}{r^6}\Big\}\Big\}\Big|_{z=x},
\end{aligned}
$$

$$
\begin{aligned}
k_{41}(x,y) &= -N_x N_y E(x,y) \qquad (11.4.40)\\
&= -\frac{(1-\nu)^2}{4\pi}\Big\{\frac{d}{ds_x}\frac{d}{ds_y}\Big(\boldsymbol{n}(\zeta)\cdot\boldsymbol{n}(z)\;\boldsymbol{t}(\zeta)\cdot\boldsymbol{t}(z)\frac{1}{r^2}\\
&\qquad +2\boldsymbol{n}(\zeta)\cdot\boldsymbol{n}(z)\;\boldsymbol{t}(\zeta)\cdot(\boldsymbol{y}-\boldsymbol{x})\;\boldsymbol{t}(z)\cdot(\boldsymbol{x}-\boldsymbol{y})\frac{1}{r^4}\Big)\Big\}\Big|_{\substack{z=x\\ \zeta=y}}\\
&-\frac{(1-\nu)}{\pi}\Big\{\frac{d}{ds_x}\Big(\boldsymbol{n}(z)\cdot\boldsymbol{n}(x)\;\boldsymbol{t}(z)\cdot(\boldsymbol{x}-\boldsymbol{y})\frac{1}{r^4}\Big)\Big\}\Big|_{z=x}\\
&-\frac{(1-\nu)}{\pi}\Big\{\frac{d}{ds_y}\Big(\boldsymbol{n}(\zeta)\cdot\boldsymbol{n}(x)\;\boldsymbol{t}(\zeta)\cdot(\boldsymbol{y}-\boldsymbol{x})\frac{1}{r^4}\Big)\Big\}\Big|_{\zeta=y}\\
&-\frac{(1-\nu)}{\pi}\Big\{\frac{d}{ds_x}\Big(\boldsymbol{n}(z)\cdot(\boldsymbol{x}-\boldsymbol{y})\;\boldsymbol{t}(z)\cdot\boldsymbol{n}(y)\frac{1}{r^4}\\
&\qquad +4\boldsymbol{n}(z)\cdot(\boldsymbol{x}-\boldsymbol{y})\;\boldsymbol{t}(z)\cdot(\boldsymbol{x}-\boldsymbol{y})\;\boldsymbol{n}(y)\cdot(\boldsymbol{y}-\boldsymbol{x})\frac{1}{r^6}\Big)\Big\}\Big|_{z=x}\\
&-\frac{(1-\nu)}{\pi}\Big\{\frac{d}{ds_y}\Big(\boldsymbol{n}(\zeta)\cdot(\boldsymbol{y}-\boldsymbol{x})\;\boldsymbol{t}(\zeta)\cdot\boldsymbol{n}(x)\frac{1}{r^4}\\
&\qquad +4\boldsymbol{n}(\zeta)\cdot(\boldsymbol{y}-\boldsymbol{x})\;\boldsymbol{t}(\zeta)\cdot(\boldsymbol{y}-\boldsymbol{x})\;\boldsymbol{n}(x)\cdot(\boldsymbol{x}-\boldsymbol{y})\frac{1}{r^6}\Big)\Big\}\Big|_{\zeta=y}\\
&-\frac{(1-\nu)^2}{4\pi}\Big\{\frac{d}{ds_x}\frac{d}{ds_y}\Big(\boldsymbol{n}(\zeta)\cdot\boldsymbol{t}(z)\;\boldsymbol{n}(z)\cdot\boldsymbol{t}(\zeta)\frac{1}{r^2}\\
&\qquad +2\big(\boldsymbol{n}(\zeta)\cdot(\boldsymbol{y}-\boldsymbol{x})\;\boldsymbol{t}(\zeta)\cdot\boldsymbol{n}(z)\;\boldsymbol{t}(z)\cdot(\boldsymbol{x}-\boldsymbol{y})\\
&\qquad +\boldsymbol{n}(z)\cdot(\boldsymbol{x}-\boldsymbol{y})\;\boldsymbol{t}(z)\cdot\boldsymbol{n}(\zeta)\;\boldsymbol{t}(\zeta)\cdot(\boldsymbol{y}-\boldsymbol{x})\big)\frac{1}{r^4}\\
&\qquad +2\boldsymbol{t}(z)\cdot\boldsymbol{t}(\zeta)\;\boldsymbol{n}(z)\cdot(\boldsymbol{x}-\boldsymbol{y})\;\boldsymbol{n}(\zeta)\cdot(\boldsymbol{y}-\boldsymbol{x})\frac{1}{r^4}\\
&+8\boldsymbol{n}(\zeta)\cdot(\boldsymbol{y}-\boldsymbol{x})\;\boldsymbol{t}(\zeta)\cdot(\boldsymbol{y}-\boldsymbol{x})\;\boldsymbol{n}(z)\cdot(\boldsymbol{x}-\boldsymbol{y})\;\boldsymbol{t}(z)\cdot(\boldsymbol{x}-\boldsymbol{y})\frac{1}{r^6}\Big)\Big\}\Big|_{\substack{z=x\\ \zeta=y}}.
\end{aligned}
$$

In order to calculate the pseudohomogeneous expansions of these kernels we collect some simple relations by using the 1–periodic parametric representation of the boundary Γ based on arclength.

In addition to (11.1.19)–(11.1.22), we shall also need the following expressions, all of them modulo $\mathcal{O}_4$:

$$\boldsymbol{n}(y)\cdot\boldsymbol{n}(x)=\Big(1-L^2\kappa^2\tfrac{1}{2}(t-\tau)^2+\tfrac{1}{2}(t-\tau)^3L^2\kappa\frac{d\kappa}{dt}(t)\Big)\,, \tag{11.4.41}$$

$$\boldsymbol{t}(y)\cdot\boldsymbol{t}(x)=\Big(1-L^2\kappa^2\tfrac{1}{2}(t-\tau)^2+\tfrac{1}{2}(t-\tau)^3L^2\kappa\frac{d\kappa}{dt}(t)\Big)\,, \tag{11.4.42}$$

$$\begin{aligned}\boldsymbol{t}(x)\cdot\boldsymbol{n}(y)=&L(t-\tau)\Big(\kappa(t)-\tfrac{1}{2}(t-\tau)\frac{d\kappa}{dt}(t)\\&-\tfrac{1}{6}(t-\tau)^2(L^2\kappa^3(t)-\frac{d^2\kappa}{dt^2}(t)\Big)\,,\end{aligned} \tag{11.4.43}$$

$$\begin{aligned}\boldsymbol{n}(x)\cdot\boldsymbol{t}(y)=&-L(t-\tau)\Big(\kappa(t)-\tfrac{1}{2}(t-\tau)\frac{d\kappa}{dt}(t)\\&-\tfrac{1}{6}(t-\tau)^2\big(L^2\kappa^3(t)-\frac{d^2\kappa}{dt^2}(t)\big)\Big)\,,\end{aligned} \tag{11.4.44}$$

$$\boldsymbol{n}(y)\cdot(\boldsymbol{y}-\boldsymbol{x})=-\tfrac{1}{2}\kappa(t)L^2(t-\tau)^2+\tfrac{1}{3}(t-\tau)^3L^2\frac{d\kappa}{dt}(t)\,, \tag{11.4.45}$$

$$\boldsymbol{n}(x)\cdot(\boldsymbol{x}-\boldsymbol{y})=-\tfrac{1}{2}\kappa(t)L^2(t-\tau)^2+\tfrac{1}{6}(t-\tau)^3L^2\frac{d\kappa}{dt}(t)\,, \tag{11.4.46}$$

$$\boldsymbol{t}(y)\cdot(\boldsymbol{y}-\boldsymbol{x})=\ -L(t-\tau)\big(1-\tfrac{1}{6}(t-\tau)^2L^2\kappa^2(t)\big)\,, \tag{11.4.47}$$

$$\boldsymbol{t}(x)\cdot(\boldsymbol{x}-\boldsymbol{y})=\ L(t-\tau)\big(1-\tfrac{1}{6}(t-\tau)^2L^2\kappa^2(t)\big)\,. \tag{11.4.48}$$

Now we are in the position to present the pseudohomogeneous expansions of the operator's kernels $k_{j\ell}(t,t-\tau)$ in the Calderón projector C_Ω in (2.4.23); and with Theorem 11.2.2, the corresponding first terms of their symbol expansions. From formulae (11.4.25)–(11.4.28) we obtain

$$\begin{aligned}k_{14}\big(T(t),T(\tau)\big)&=\frac{1}{8\pi}\log|t-\tau|\Big\{L^2(t-\tau)^2-\tfrac{1}{12}L^4(t-\tau)^4\kappa^2(t)\\&\qquad+\tfrac{1}{12}L^3(t-\tau)^5\kappa(t)\frac{d\kappa}{dt}(t)+\mathcal{O}_6\Big\}+\mathcal{O}_0\,,\\ \sigma_{14}(t,\xi)&=L^2\tfrac{1}{4}|\xi|^{-3}+L^4\tfrac{1}{4}\kappa^2(t)|\xi|^{-5}\\&\qquad-iL^4\tfrac{5}{4}\kappa(t)\frac{d\kappa}{dt}(t)|\xi|^{-6}\operatorname{sign}\xi+O(|\xi|^{-7})\,,\end{aligned} \tag{11.4.49}$$

$$\begin{aligned}k_{13}\big(T(t),T(\tau)\big)&=\frac{-1}{4\pi}\log|t-\tau|\times\\&\qquad\times\Big(\tfrac{1}{2}\kappa(t)L^2(t-\tau)^2-\tfrac{1}{3}(t-\tau)^3L^2\frac{d\kappa}{dt}(t)+\mathcal{O}_4\Big)+\mathcal{O}_0\,,\\ \sigma_{13}(t,\xi)&=-\tfrac{1}{4}L^2\kappa(t)|\xi|^{-3}\\&\qquad+i\tfrac{1}{2}L^2\frac{d\kappa}{dt}(t)|\xi|^{-4}\operatorname{sign}\xi+O(|\xi|^{-5})\,,\end{aligned} \tag{11.4.50}$$

$$\begin{aligned} k_{12}\big(T(t),T(\tau)\big) &= -\frac{1}{4\pi}(1+\nu)\log|t-\tau| + \mathcal{O}_0\,,\\ \sigma_{12}(t,\xi) &= (1+\nu)\tfrac{1}{4}|\xi|^{-1}\,, \end{aligned} \tag{11.4.51}$$

$$\begin{aligned} k_{11}\big(T(t),T(\tau)\big) & \qquad \mathcal{O}_0\,,\\ \sigma_{11}(t,\xi) &= 0\,. \end{aligned} \tag{11.4.52}$$

Now the next row of C_Ω is obtained from the representations (11.4.29)–(11.4.32):

$$\begin{aligned} k_{24}\big(T(t),T(\tau)\big) &= -\tfrac{1}{4\pi}\log|t-\tau|\Big(\tfrac{1}{2}\kappa(t)L^2(t-\tau)^2\\ &\qquad -\tfrac{1}{6}(t-\tau)^3L^2\frac{d\kappa}{dt}(t)+\mathcal{O}_4\Big)+\mathcal{O}_0\,,\\ \sigma_{24}(t,\xi) &= -\tfrac{1}{4}L^2\kappa(t)|\xi|^{-3}\\ &\quad + i\tfrac{1}{4}L^2\frac{d\kappa}{dt}(t)|\xi|^{-4}\operatorname{sign}\xi + \mathcal{O}(|\xi|^{-5})\,, \end{aligned} \tag{11.4.53}$$

$$\begin{aligned} k_{23}\big(T(t),(\tau)\big) &= -\tfrac{1}{4\pi}\log|t-\tau|\Big(1-\tfrac{1}{2}L^2\kappa^2(t)(t-\tau)^2\\ &\qquad +\tfrac{1}{2}(t-\tau)^3L^2\kappa\frac{d\kappa}{dt}(t)+\mathcal{O}_4\Big)\,,\\ \sigma_{23}(t,\xi) &= \tfrac{1}{4}|\xi|^{-1}+\tfrac{1}{4}L^2\kappa^2(t)|\xi|^{-3}\\ &\quad - i\tfrac{3}{4}L^2\kappa\frac{d\kappa}{dt}(t)|\xi|^{-4}\operatorname{sign}\xi + O(|\xi|^{-5})\,, \end{aligned} \tag{11.4.54}$$

$$\begin{aligned} k_{22}\big(T(t),T(\tau)\big) &= \mathcal{O}_0\,,\\ \sigma_{22}(t,\xi) &= 0\,, \end{aligned} \tag{11.4.55}$$

$$\begin{aligned} k_{21}\big(T(t),T(\tau)\big) &= -\frac{1}{2\pi}\frac{1}{L^2(t-\tau)^2}+\frac{(1-\nu)}{4\pi}\frac{1}{L}\frac{d}{d\tau}\Big\{\frac{1}{L(t-\tau)}\Big\}+\mathcal{O}_0\\ &= -\frac{1+\nu}{4\pi}\,\frac{1}{L^2(t-\tau)^2}+\mathcal{O}_0\,,\\ \sigma_{21}(t,\xi) &= (1+\nu)\tfrac{1}{4}\,\frac{1}{L^2}|\xi|\,. \end{aligned} \tag{11.4.56}$$

Next, we consider the third row of C_Ω, (11.4.33)–(11.4.36).

$$\begin{aligned} k_{34}\big(T(t),T(\tau)\big) &= \frac{1+\nu}{4\pi}\log|t-\tau|+\mathcal{O}_0\,,\\ \sigma_{34}(t,\xi) &= -(1+\nu)\tfrac{1}{4}|\xi|^{-1}\,, \end{aligned} \tag{11.4.57}$$

$$\begin{aligned} k_{33}\big(T(t),T(\tau)\big) &= \mathcal{O}_0\,,\\ \sigma_{33}(t,\xi) &= 0\,, \end{aligned} \tag{11.4.58}$$

$$\begin{aligned} k_{32}\big(T(t),T(\tau)\big) &= -\frac{1}{4\pi}(1-\nu)(3+\nu)\frac{1}{L^2(t-\tau)^2}+\mathcal{O}_0\,,\\ \sigma_{32}(t,\xi) &= \frac{(1-\nu)(3+\nu)}{4L^2}|\xi|\,, \end{aligned} \tag{11.4.59}$$

$$
\begin{aligned}
k_{31}\big(T(t),T(\tau)\big) &= -\frac{(1-\nu)}{2\pi}\Big\{\frac{\kappa(t)}{L^2(t-\tau)^2} \\
&\qquad + \frac{1}{L^2}\,\frac{1+\nu}{2}\kappa(t)\frac{d}{d\tau}\,\frac{1}{(t-\tau)} + \mathcal{O}_0\Big\} \\
&= -\frac{(1-\nu)(3+\nu)}{4\pi}\kappa(t)\frac{1}{L^2(t-\tau)^2} + \mathcal{O}_0\,, \\
\sigma_{31}(t,\xi) &= \frac{(1-\nu)(3+\nu)}{4L^2}\kappa(t)|\xi|\,. \qquad (11.4.60)
\end{aligned}
$$

Finally, we consider the last row (11.4.37)–(11.4.40).

$$
\begin{aligned}
k_{44}\big(T(t),T(\tau)\big) &= \mathcal{O}_0\,, \\
\sigma_{44}(t,\xi) &= 0\,, \qquad (11.4.61) \\
k_{43}\big(T(t),T(\tau)\big) &= \frac{1}{2\pi}\,\frac{1}{L^2(t-\tau)^2} + \frac{(1-\nu)}{4\pi L^2}\,\frac{d}{dt}\,\frac{1}{(t-\tau)} + \mathcal{O}_0 \\
&= \frac{1}{4\pi}(1+\nu)\frac{1}{L^2(t-\tau)^2} + \mathcal{O}_0\,, \\
\sigma_{43}(t,\xi) &- \frac{(1+\nu)}{4L^2}|\xi|\,, \qquad (11.4.62) \\
k_{42}\big(T(t),T(\tau)\big) &= -\frac{1-\nu}{2\pi}\Big\{\frac{\kappa(t)}{L^2(t-\tau)^2} - \frac{d\kappa}{dt}(t)\frac{1}{L^2(t-\tau)} \\
&\qquad - \frac{(1+\nu)}{2}\,\frac{1}{L^2}\,\frac{d}{dt}\Big(\frac{\kappa(t)}{t-\tau}\Big)\Big\} + \mathcal{O}_0 \\
&= -\frac{1}{4\pi}(1-\nu)(3+\nu)\Big\{\frac{\kappa(t)}{L^2(t-\tau)^2} - \frac{d\kappa}{dt}(t)\frac{1}{L^2(t-\tau)}\Big\} + \mathcal{O}_0\,, \\
\sigma_{42}(t,\xi) &= \frac{(1-\nu)(3+\nu)}{4L^2}\kappa(t)|\xi| \\
&\qquad - i\frac{(1-\nu)(3+\nu)}{4L^2}\,\frac{d\kappa}{dt}(t)\,\mathrm{sign}\,\xi\,, \qquad (11.4.63) \\
k_{41}\big(T(t),T(\tau)\big) &= \frac{(1-\nu)^2}{4\pi L^4}\,\frac{d}{dt}\,\frac{d}{d\tau}\,\frac{1}{(t-\tau)^2} \\
&\quad - \frac{(1-\nu)}{\pi}\,\frac{1}{L}\,\frac{d}{dt}\Big\{\frac{1}{L^3(t-\tau)^3} - \frac{1}{6}\kappa^2(t)\frac{1}{L(t-\tau)}\Big\} \\
&\quad + \frac{(1-\nu)}{\pi}\,\frac{1}{L}\,\frac{d}{d\tau}\Big\{\frac{1}{L^3(t-\tau)^3} - \frac{1}{6}\kappa^2(t)\frac{1}{L(t-\tau)}\Big\} + \mathcal{O}_0 \\
&= \frac{6}{4\pi}(1-\nu)(3+\nu)\frac{1}{L^4(t-\tau)^4} - \frac{1}{3\pi}(1-\nu)\kappa^2(t)\frac{1}{L^2(t-\tau)^2} \\
&\quad + \frac{1}{3\pi}(1-\nu)\kappa\frac{d\kappa}{dt}(t)\frac{1}{L^2(t-\tau)} + \mathcal{O}_0\,,
\end{aligned}
$$

$$\sigma_{41}(t,\xi) = (1-\nu)(3+\nu)\frac{1}{4L^4}|\xi|^3 + (1-\nu)\frac{1}{3L^3}\kappa^2(t)|\xi| \\ - i(1-\nu)\frac{1}{3L^2}\kappa(t)\frac{d\kappa}{dt}(t)\,\mathrm{sign}\,\xi\,. \tag{11.4.64}$$

Since all of the symbol expansions now are available, the action of all these operators on 1–periodic Fourier series have the representation in the form of (11.2.5) where $a_\Phi(t,\xi)$ in (11.2.3) is to be replaced by $\sigma_{jk}(t,\xi)$, as in the other examples.

The first three terms of the asymptotic symbol expansion of the Calderón projector in (2.4.23) read:

$$\sigma_{\mathcal{C}_\Omega}(t,\xi)$$

$$= \begin{pmatrix} \frac{1}{2L} & \frac{1-\nu}{4}|\xi|^{-1} & 0 & \frac{L^2}{4}|\xi|^{-3} \\ \frac{1+\nu}{4L^2}|\xi| & \frac{1}{2L} & \frac{1}{4}|\xi|^{-1} & 0 \\ 0 & \frac{(1-\nu)(3+\nu)}{4L^2}|\xi| & \frac{1}{2L} & -\frac{1+\nu}{4}|\xi|^{-1} \\ \frac{(1-\nu)(3+\nu)}{4L^4}|\xi|^3 & 0 & -\frac{1+\nu}{4L^2}|\xi| & \frac{1}{2L} \end{pmatrix}$$

$$+ \begin{pmatrix} 0 & 0 & -\frac{L^2}{4}\kappa(t)|\xi|^{-3} & 0 \\ 0 & 0 & 0 & -\frac{L^2}{4}\kappa(t)|\xi|^{-3} \\ \frac{(1-\nu)(3+\nu)}{4L^2}\kappa(t)|\xi| & 0 & 0 & 0 \\ 0 & \frac{(1-\nu)(3+\nu)}{4L^2}\kappa(t)|\xi| & 0 & 0 \end{pmatrix}$$

$$+ \begin{pmatrix} 0 & 0 & \frac{iL^2}{2}\frac{d\kappa}{dt}|\xi|^{-4}\,\mathrm{sign}\,\xi & \frac{L^4}{4}\kappa^2(t)|\xi|^{-5} \\ 0 & 0 & \frac{L^2}{4}\kappa^2(t)|\xi|^{-3} & \frac{iL^2}{4}\frac{d\kappa}{dt}|\xi|^{-4}\,\mathrm{sign}\,\xi \\ 0 & 0 & 0 & 0 \\ \frac{1-\nu}{3L^2}\kappa^2|\xi| & -i\frac{(1-\nu)(3+\nu)}{4L^2}\frac{d\kappa}{dt}\,\mathrm{sign}\,\xi & 0 & 0 \end{pmatrix}$$

$$+ \dots$$

To conclude this chapter, we now return to the boundary integral equations associated with the boundary value problems of the biharmonic equation. From the symbols $\sigma_{jk}(t,\xi)$ we see that, in contrast to the previous examples, these systems of equations have different orders.

We begin with the *interior problems* as considered in Section 2.4.2. For the *interior Dirichlet problem*, Theorem 2.4.1, the corresponding system of boundary integral equations (2.4.26), (2.4.27) of the first kind, which are modified to always guarantee unique solvability (independent of so–called Γ–contours), now reads

$$\begin{aligned} V_{23}\sigma_1 + V_{24}\sigma_2 + n_1\omega_1 + n_2\omega_2 &= f_2\,, \\ V_{13}\sigma_1 + V_{14}\sigma_2 + x_1\omega_1 + x_2\omega_2 + \omega_3 &= f_1 \end{aligned} \tag{11.4.65}$$

where the solution $(\sigma_1, \sigma_2) = (Mu, Nu) \in H^{-\frac{1}{2}}(\Gamma) \times H^{-\frac{3}{2}}(\Gamma)$ has to satisfy the additional compatibility conditions

$$\int_\Gamma \Big(\sigma_1 \frac{\partial p_j}{\partial n} + \sigma_2 p_j\Big) ds = 0 \quad \text{with } p_1 = 1\,,\ p_2 = x_1\,,\ p_3 = x_2\,. \tag{11.4.66}$$

As we have seen, the solution $(\omega, \sigma_1, \sigma_2)$ of the the system (11.4.65), (11.4.66) is unique. In terms of pseudodifferential operators, this is a 2×2 system of Agmon–Douglis–Nirenberg type with the orders $s_1 = t_1 = -\frac{1}{2}$ and $s_2 = t_2 = -\frac{3}{2}$; hence the principal symbol matrix (7.2.12) is given by

$$\frac{1}{4}\begin{pmatrix} |\xi|^{-1} & 0 \\ 0 & L^2|\xi|^{-3} \end{pmatrix} \tag{11.4.67}$$

which obviously is strongly elliptic with $\Theta(t) = I$ in (7.2.62). Consequently, due to Theorem 5.3.10, for any given right–hand side $(f_1, f_2) \in H^{\frac{3}{2}+s}(\Gamma) \times H^{\frac{1}{2}+s}(\Gamma)$, the system (11.4.65), (11.4.66) has a unique solution $(\omega, \sigma_1, \sigma_2) \in \mathbb{R}^3 \times H^{-\frac{1}{2}+s}(\Gamma) \times H^{-\frac{3}{2}+s}(\Gamma)$ with any $s \in \mathbb{R}$.

For the *exterior Dirichlet problem* we have the same system of equations, however, the right–hand sides in (11.4.65) and (11.4.66) are different, see (2.4.28).

Similarly, for the interior Neumann problem, the system of boundary integral equations of the first kind is now

$$\begin{aligned} D_{41}\varphi_1 + D_{42}\varphi_2 &= g_2 \\ D_{31}\varphi_1 + D_{32}\varphi_2 &= g_1 \end{aligned} \tag{11.4.68}$$

where the right–hand sides $(g_1, g_2) \in H^{-\frac{1}{2}+s}(\Gamma) \times H^{-\frac{3}{2}+s}(\Gamma)$ are given in Table 2.4.5. These equations always have a three–dimensional kernel given by

$$\mathcal{R} = \Big\{\Big(v, \frac{\partial v}{\partial n}\Big)|_\Gamma \Big| v = a_1x_1 + a_2x_2 + b \quad \text{with } (a_1, a_2, b) \in \mathbb{R}^3\Big\}. \tag{11.4.69}$$

The system (11.4.68) is of Agmon–Douglis–Nirenberg type with $s_1 = t_1 = \frac{3}{2}$ and $s_2 = t_2 = \frac{1}{2}$. The principal symbol is given by

$$(1-\nu)(3+\nu)\frac{1}{4}\begin{pmatrix} \frac{1}{L^4}|\xi|^3 & 0 \\ 0 & \frac{1}{L^2}|\xi| \end{pmatrix} \tag{11.4.70}$$

which obviously is strongly elliptic with $\Theta = I$ since $\nu \in [0, 1)$. Therefore, the Fredholm alternative holds for (11.4.68) and the solution is unique only in $(H^{\frac{3}{2}+s}(\Gamma) \times H^{\frac{1}{2}+s}(\Gamma))/\mathcal{R}$, and (g_1, g_2) have to satisfy orthogonality conditions which are satisfied if the given boundary data $Mu = \psi_1 \in H^{-\frac{1}{2}+s}(\Gamma)$ and $Nu = \psi_2 \in H^{-\frac{3}{2}+s}(\Gamma)$ with $s \in \mathbb{R}$ satisfy the compatibility conditions (2.4.16).

As we discussed in Section 2.4.2, the right–hand sides (g_1, g_2) in (11.4.68) satisfy the orthogonality conditions and the system (11.4.68) then has a solution which is not unique due to the kernel $\mathcal{R}$ of this operator system.

11.5 Remarks

Boundary integral equation methods for two–dimensional boundary value problems and various examples are considered by Yu in [442] based on the work by Feng Kang [117]. Boundary integral equations on periodic functions and corresponding Fourier analysis can be found in the books by Gohberg and Fel'dman [151] and Gohberg and Krupnik [152]. In McLean et al [291] and Lamp et al [259], periodic pseudodifferential operators are treated with Fourier approximations. For a slightly larger class of periodic pseudodifferential equations than those in 11.1–11.4, Saranen and Vainikko present in their book [369] a very extensive analysis employing Fourier series representations.

12. Remarks on Pseudodifferential Operators Related to the Time Harmonic Maxwell Equations

12.1 Introduction

The class of pseudodifferential operators is essentially the smallest algebra of operators which contains all differential operators, all fundamental solutions of elliptic differential operators and all integral operators with pseudohomogeneous kernel expansions. The linear pseudodifferential operators can be characterized by generalized Fourier multipliers, known as *symbols*. The development of the theory of pseudodifferential operators has made it possible to provide a unified treatment of differential and integral operators. These materials are subjects of Chapter 7 to Chapter 11. However, for a quick glance of the materials, we refer to our paper [206] with emphasis on boundary integral equations recast as pseudodifferential equations.

In this chapter, our aim is to follow the approach in Chapter 9 to analyze the Maxwell system from a pseudodifferential equation's point of view.

We begin with the Maxwell equations for time–harmonic electromagnetic fields given by (6.4.1)–(6.4.4), i.e.,

$$\begin{aligned} \mathbf{curl}\,\mathbf{E}(\boldsymbol{x}) &= i\omega\mu\mathbf{H}(\boldsymbol{x}) \\ \mathbf{curl}\,\mathbf{H}(\boldsymbol{x}) &= (\sigma - i\omega\varepsilon)\mathbf{E}(\boldsymbol{x}) + \mathbf{J}_0 \\ \mathrm{div}\,(\varepsilon\mathbf{E}) &= \varrho, \\ \mathrm{div}\,(\mu\mathbf{H}) &= \mathbf{0} \end{aligned} \tag{12.1.1}$$

where $\mathbf{J}_0$ and ϱ are given, satisfying the compatibility condition (6.4.5),

$$\varrho(\boldsymbol{x}) = \mathrm{div}\,\mathbf{J}_0(\boldsymbol{x})/(i\omega - \sigma/\varepsilon). \tag{12.1.2}$$

For simplicity we assume, μ, ε, σ and ω are given as real positive constants. We reduce the system (12.1.1) to a second order partial differential equation for $\mathbf{E}$ in the form

$$\mathbf{PE} := \mathbf{curl}\ \mathbf{curl}\ \mathbf{E} - k^2\mathbf{E} = \boldsymbol{f} := i\omega\mu\,\mathbf{J}_0 \quad \text{in } \mathbb{R}^3 \tag{12.1.3}$$

where $k^2 = (i\mu\omega)(\sigma - i\omega\varepsilon)$. Since

$$\mathbf{curl}\ \mathbf{curl}\ \mathbf{E} = -\Delta\mathbf{E} + \mathbf{grad}\ \mathrm{div}\,\mathbf{E},$$

G. C. Hsiao und W. L. Wendland, *Boundary Integral Equations*,
Applied Mathematical Sciences 164,
https://doi.org/10.1007/978-3-030-71127-6_12

(12.1.3) can now be written as

$$\mathbf{PE} = -(\Delta\mathbf{E} + k^2\mathbf{E}) + \mathbf{grad}\,\mathrm{div}\,\mathbf{E} = \boldsymbol{f} \quad \text{in } \mathbb{R}^3. \tag{12.1.4}$$

In particular, we find the symbol of $\mathbf{P}$, its inverse and the fundamental solution, and define the surface potentials. As will be seen, the standard Sobolev spaces here do not lead to the correct solution space for the Maxwell system.

12.2 Symbols of P and the corresponding Newton potentials

A simple computation shows that the symbol of $\mathbf{P}$ is given by

$$\begin{aligned}\sigma_{\mathbf{P}}(\boldsymbol{\xi}) &= \begin{pmatrix} |\boldsymbol{\xi}|^2 - k^2 - \xi_1^2, & -\xi_2\xi_1, & -\xi_3\xi_1 \\ -\xi_1\xi_2, & |\boldsymbol{\xi}|^2 - k^2 - \xi_2^2, & -\xi_3\xi_2 \\ -\xi_1\xi_3, & -\xi_2\xi_3, & |\boldsymbol{\xi}|^2 - k^2 - \xi_3^2 \end{pmatrix} \\ &= \Big((|\boldsymbol{\xi}|^2 - k^2)\delta_{j\ell} - ((\xi_j\cdot\xi_\ell))\Big)_{j,\ell=1,2,3}.\end{aligned} \tag{12.2.1}$$

For the determinant of $\sigma_{\mathbf{P}}(\boldsymbol{\xi})$ one obtains

$$\mathrm{Det}\;\sigma_{\mathbf{P}}(\boldsymbol{\xi}) = -k^2\big(|\boldsymbol{\xi}|^2 - k^2\big)^2, \tag{12.2.2}$$

and for the inverse of $\sigma_{\mathbf{P}}(\boldsymbol{\xi})$ have

$$\big(\sigma_{\mathbf{P}}(\boldsymbol{\xi})\big)^{-1} = \big(\mathrm{Det}\;\sigma_{\mathbf{P}}(\boldsymbol{\xi})\big)^{-1}(|\boldsymbol{\xi}|^2 - \mathrm{k}^2)\times$$

$$\begin{pmatrix} (|\boldsymbol{\xi}|^2 - k^2) - (\xi_2^2 + \xi_3^2), & \xi_1\xi_2 & \xi_1\xi_3 \\ \xi_1\xi_2 & (|\boldsymbol{\xi}|^2 - k^2) - (\xi_1^2 + \xi_3^2), & \xi_3\xi_2 \\ \xi_1\xi_3 & \xi_2\xi_3 & (|\boldsymbol{\xi}|^2 - k^2) - (\xi_1^2 + \xi_2^2) \end{pmatrix}$$

$$= \frac{-1}{k^2(|\boldsymbol{\xi}|^2 - k^2)}\begin{pmatrix} \xi_1^2 - k^2, & \xi_2\xi_1 & \xi_3\xi_1 \\ \xi_1\xi_2 & \xi_2^2 - k^2 & \xi_3\xi_2 \\ \xi_1\xi_3, & \xi_2\xi_3 & \xi_3^2 - k^2 \end{pmatrix}$$

$$\big(\sigma_{\mathbf{P}}(\boldsymbol{\xi})\big)^{-1} = \Big\{\delta_{j\ell} - \frac{1}{k^2}((\xi_j\xi_\ell))\Big\}\frac{1}{|\boldsymbol{\xi}|^2 - k^2}\,;\; j,\ell \in \{1,2,3\}. \tag{12.2.3}$$

This is also the symbol of the Newton potential associated with $\mathbf{P}$. We notice that this symbol of the Newton potential consists of two parts, in contrast to the elasticity and the Stokes systems as well as the scalar Helmholtz equation:

$$\sigma_{\mathbf{N}}(\boldsymbol{\xi}) = \big(\sigma_{\mathbf{P}}(\boldsymbol{\xi})\big)^{-1} = \mathbf{N}_1(\boldsymbol{\xi}) + \mathbf{N}_2(\boldsymbol{\xi})$$

with

$$\mathbf{N}_1(\boldsymbol{\xi}) = \frac{1}{|\boldsymbol{\xi}|^2 - k^2}\delta_{j\ell}\,,\; \mathbf{N}_2(\boldsymbol{\xi}) = -\frac{1}{k^2}\,((\xi_j\xi_\ell))\frac{1}{|\boldsymbol{\xi}|^2 - k^2}. \tag{12.2.4}$$

Correspondingly, two parametrices are given by

$$\begin{aligned}\mathbf{H}_1 \boldsymbol{f} &:= \mathfrak{F}^{-1}_{\boldsymbol{\xi}\mapsto\boldsymbol{x}}\Big(\chi(|\boldsymbol{\xi}|)\frac{1}{|\boldsymbol{\xi}|^2-k^2}\mathfrak{F}_{\boldsymbol{y}\mapsto\boldsymbol{\xi}}\boldsymbol{f}(\boldsymbol{y})\Big),\\ \mathbf{H}_2 \boldsymbol{f} &:= -\frac{1}{k^2}\mathfrak{F}^{-1}_{\boldsymbol{\xi}\mapsto\boldsymbol{x}}\Big(\chi(|\boldsymbol{\xi}|)((\xi_j\xi_\ell))\frac{1}{|\boldsymbol{\xi}|^2-k^2}\mathfrak{F}_{\boldsymbol{y}\mapsto\boldsymbol{\xi}}\boldsymbol{f}(\boldsymbol{y})\Big)\end{aligned} \tag{12.2.5}$$

with the cut–off function $\chi(\eta)$ is given in (7.1.5), where $\mathbf{H}_1 \in \mathcal{O}PS^{-2}(\mathbb{R}^3 \times \mathbb{R}^3)$ and $\mathbf{H}_2 \in \mathcal{O}PS^0(\mathbb{R}^3 \times \mathbb{R}^3)$ are properly supported. The asymptotic expansions for $\mathbf{N}_1$ and $\mathbf{N}_2$ are:

$$\begin{aligned}\sigma_{\mathbf{N}_1}(\boldsymbol{\xi}) &= \Big\{\frac{1}{|\boldsymbol{\xi}|^2} + \frac{k^2}{|\boldsymbol{\xi}|^2} + \mathcal{O}\big(|\boldsymbol{\xi}|^{-6}k^4\big)\Big\}\delta_{j\ell}\\ \sigma_{\mathbf{N}_2}(\boldsymbol{\xi}) &= -((\xi_j\xi_\ell))\sum_{p=0}^{\infty}\frac{k^{2p-2}}{|\boldsymbol{\xi}|^{2+2p}}\\ &= -((\xi_j\xi_\ell))\Big\{\frac{1}{k^2}\,\frac{1}{|\boldsymbol{\xi}|^2} + \frac{1}{|\boldsymbol{\xi}|^4} + \mathcal{O}\big(|\boldsymbol{\xi}|^{-6}k^2\big)\Big\}.\end{aligned} \tag{12.2.6}$$

Then

$$(\mathbf{N}_1 - \mathbf{H}_1)\boldsymbol{f} = \mathfrak{F}^{-1}_{\boldsymbol{\xi}\mapsto\boldsymbol{x}}\psi(|\boldsymbol{\xi}|)\mathcal{O}(|\boldsymbol{\xi}|^{-6}k^4)\mathfrak{F}_{\boldsymbol{y}\mapsto\boldsymbol{\xi}}\boldsymbol{f}(\boldsymbol{y})$$

is a smoothing operator, where $\psi(\eta) = \big(1-\chi(y)\big)$ is a cut–off function with $\psi(\eta) \equiv 1$ for small $\eta \geq 0$ and $\psi(\eta) \equiv 0$ for $1 \leq \eta$ (see (7.1.5)).

Similarly,

$$(\mathbf{N}_2 - \mathbf{H}_2)\boldsymbol{f} = \mathcal{F}^{-1}_{\boldsymbol{\xi}\mapsto\boldsymbol{x}}\psi(|\boldsymbol{\xi}|)\mathcal{O}(|\boldsymbol{\xi}|)^{-6}k^2)\mathcal{F}^{-1}_{\boldsymbol{y}\mapsto\boldsymbol{\xi}}\boldsymbol{f}(\boldsymbol{y}).$$

Based on (12.2.3), the fundamental solution can be recovered by using the Fourier inverse transform which leads to

$$\begin{aligned}\boldsymbol{\Gamma}(\boldsymbol{x},\boldsymbol{y}) &= \{\boldsymbol{I} + \frac{1}{k^2}\,\mathbf{grad}_{\boldsymbol{x}}\,\mathrm{div}_{\boldsymbol{x}}\}\mathbf{G}_k(\boldsymbol{x}-\boldsymbol{y})\\ \mathbf{G}_k(\boldsymbol{x}-\boldsymbol{y}) &= \frac{1}{4\pi}\,\frac{e^{ik|\boldsymbol{x}-\boldsymbol{y}|}}{|\boldsymbol{x}-\boldsymbol{y}|}\boldsymbol{I}.\end{aligned} \tag{12.2.7}$$

12.3 Representation formulae

In order to find the surface potentials on the boundary $\Gamma = \partial\Omega$ of a bounded domain Ω, we follow Section 10.2 and begin with the representation formula which is obtained by using the second Green's formula (6.9.2) and Appendix B, II(8),

$$\begin{aligned}&\int_\Omega \{\mathbf{Q}\cdot\mathbf{curl}\,\mathbf{curl}\,\mathbf{E} - \mathbf{E}\cdot\mathbf{curl}\,\mathbf{curl}\,\mathbf{Q}\}d\boldsymbol{y}\\ &= \int_\Gamma (\mathbf{E}\times\mathbf{curl}\,\mathbf{Q} - \mathbf{Q}\times\mathbf{curl}\,\mathbf{E})\cdot\boldsymbol{n}ds_y\end{aligned}$$

with $\mathbf{E}$ and $\mathbf{Q} = \boldsymbol{\Gamma}$ from (12.2.7):

$$\begin{aligned}\mathbf{E}(\boldsymbol{x}) &= \int_{\Omega} \boldsymbol{\Gamma}(\boldsymbol{x}-\boldsymbol{y})\boldsymbol{f}(\boldsymbol{y})d\boldsymbol{y} - \langle \boldsymbol{\Gamma}(\boldsymbol{x}-\boldsymbol{y}), (\boldsymbol{n} \times \mathbf{curl}\,\mathbf{E})\rangle_{\Gamma} \\ &\qquad - \langle \mathbf{curl}_{\boldsymbol{y}}\boldsymbol{\Gamma}(\boldsymbol{x}-\boldsymbol{y}), (\boldsymbol{n} \times \mathbf{E}(\boldsymbol{y}))\rangle_{\Gamma}, \\ &= \int_{\Omega} \boldsymbol{\Gamma}(\boldsymbol{x}-\boldsymbol{y})\boldsymbol{f}(\boldsymbol{y})d\boldsymbol{y} - \langle \boldsymbol{I}, \boldsymbol{\Gamma}(\boldsymbol{x}-\boldsymbol{y})(\boldsymbol{n} \times \mathbf{curl}\;\mathbf{E})\rangle_{\Gamma} \\ &\qquad + \langle \boldsymbol{I}, \mathbf{curl}_{\boldsymbol{y}}\boldsymbol{\Gamma}(\boldsymbol{x}-\boldsymbol{y})(\boldsymbol{n} \times \mathbf{E}(\boldsymbol{y}))\rangle_{\Gamma}.\end{aligned}$$

Or, since $\boldsymbol{n} \times \mathbf{E} = \boldsymbol{n} \times \pi_t\mathbf{E}$, we obtain

$$\begin{aligned}\mathbf{E}(\boldsymbol{x}) &= \int_{\Omega} \boldsymbol{\Gamma}(\boldsymbol{x}-\boldsymbol{y})\boldsymbol{f}(\boldsymbol{y})d\boldsymbol{y} - \int_{\Gamma} \boldsymbol{\Gamma}(\boldsymbol{x}-\boldsymbol{y})(\boldsymbol{n} \times \mathbf{curl}\;\mathbf{E})ds_y \\ &\qquad + \int_{\Gamma} \mathbf{curl}_{\boldsymbol{y}}\boldsymbol{\Gamma}(\boldsymbol{x}-\boldsymbol{y})\big(\boldsymbol{n} \times \pi_t\mathbf{E}(\boldsymbol{y})\big)ds_y\end{aligned} \tag{12.3.1}$$

in which we may consider $\boldsymbol{n} \times \mathbf{curl}\;\mathbf{E}$ and $\pi_t\mathbf{E}$ as the Neumann and Dirichlet Cauchy data of $\mathbf{PE}$ in (12.1.3), respectively. Alternatively, Equation (12.3.1) may be written in the form:

$$\begin{aligned}\mathbf{E}(\boldsymbol{x}) &= \int_{\Omega} \boldsymbol{\Gamma}(\boldsymbol{x}-\boldsymbol{y})\boldsymbol{f}(\boldsymbol{y})d\boldsymbol{y} \\ &\qquad - \int_{\Gamma} G_k(\boldsymbol{x}-\boldsymbol{y})\big(\boldsymbol{n}(\boldsymbol{y}) \times \mathbf{curl}\;\mathbf{E}(\boldsymbol{y})\big)ds_{\boldsymbol{y}} \\ &\qquad + \frac{1}{k^2}\int_{\Gamma} \big(\mathbf{grad}_{\boldsymbol{y}}G_k(\boldsymbol{x}-\boldsymbol{y})\big)\mathrm{div}_{\Gamma}\;\big(\boldsymbol{n} \times \mathbf{curl}\;\mathbf{E}(\boldsymbol{y})\big)ds_{\boldsymbol{y}} \\ &\qquad + \int_{\Gamma} \big(\mathbf{grad}_{\boldsymbol{y}}G_k(\boldsymbol{x}-\boldsymbol{y})\big) \times (\boldsymbol{n} \times \mathbf{E})(\boldsymbol{y})ds_{\boldsymbol{y}} \quad \text{for } \boldsymbol{x} \in \Omega,\end{aligned} \tag{12.3.2}$$

where we can show that

$$\begin{aligned}\int_{\Omega} \boldsymbol{\Gamma}(\boldsymbol{x}-\boldsymbol{y})\boldsymbol{f}(\boldsymbol{y})d\boldsymbol{y} &= \int_{\Omega} \{G_k(\boldsymbol{x}-\boldsymbol{y})(i\omega\mu)\mathbf{J}_0(\boldsymbol{y}) + \frac{1}{\varepsilon}\varrho(\boldsymbol{y})\mathbf{grad}_{y}G_k(\boldsymbol{x}-\boldsymbol{y})\}d\boldsymbol{y} \\ &\qquad +(\sigma - i\omega\varepsilon)^{-1}\int_{\Gamma} \{\mathbf{grad}_{\boldsymbol{y}}G_k(\boldsymbol{x}-\boldsymbol{y})\,\boldsymbol{n} \cdot \mathbf{J}_0(\boldsymbol{y})\}ds_{\boldsymbol{y}}.\end{aligned}$$

Hence, Equation (12.3.2) is exactly the equation (6.9.4) with $i\omega\mu(\boldsymbol{n} \times \mathbf{H})$ replaced by $\boldsymbol{n} \times \mathbf{curl}\;\mathbf{E}$ and

$$\boldsymbol{n} \cdot \mathbf{E} = -\big[k^{-2}\mathrm{div}_{\Gamma}(\boldsymbol{n} \times \mathbf{curl}\;\mathbf{E}) + (\sigma - i\omega\varepsilon)^{-1}\boldsymbol{n} \cdot \mathbf{J}_0\big].$$

Thus, for $\boldsymbol{f} = \mathbf{0}$ the electric field $\mathbf{E}(\boldsymbol{x})$ in Ω is completely determined if the corresponding Cauchy data on Γ are known.

12.4 Symbols of the Electromagnetic Boundary Potentials

In the following, for simplicity, we consider only the principal symbols, i.e., the surface Γ is considered as a two–dimensional plane with a local orthonormal positively oriented basis $\boldsymbol{e}_1, \boldsymbol{e}_2, \boldsymbol{n}$.

The simple layer potential has the form

$$\mathfrak{V}\boldsymbol{\lambda}(\boldsymbol{x}) := -\int_{\Gamma} \boldsymbol{\Gamma}(\boldsymbol{x}-\boldsymbol{y})\boldsymbol{\lambda}(\boldsymbol{y})ds_{\boldsymbol{y}}\,,\; x \in \Omega \tag{12.4.1}$$

where $\boldsymbol{\lambda}$ is a tangential field on Γ and can be written as a Newton potential operating on the distribution $\gamma_0^*\boldsymbol{\lambda} = \boldsymbol{\lambda} \otimes \delta_\Gamma$ which has the form

$$\begin{aligned}
\mathfrak{V}\boldsymbol{\lambda}(\boldsymbol{x}) &= -\int_{\mathbb{R}^3} \boldsymbol{\Gamma}(\boldsymbol{x}-\boldsymbol{y})\big(\boldsymbol{\lambda}(\boldsymbol{y}) \otimes \delta_\Gamma\big)d\boldsymbol{y}\,,\; \boldsymbol{x} \in \mathbb{R}^3 \setminus \Gamma \\
&= -\int_{\mathbb{R}^3} \{\boldsymbol{I} + \frac{1}{k^2}\mathbf{grad}_{\boldsymbol{x}}\mathrm{div}_{\boldsymbol{x}}\} G_k(\boldsymbol{x}-\boldsymbol{y})\big(\boldsymbol{\lambda}(\boldsymbol{y}) \otimes \delta_\Gamma\big)d\boldsymbol{y} \\
&= -\int_{\mathbb{R}^3} G_k(\boldsymbol{x}-\boldsymbol{y})\big(\boldsymbol{\lambda}(\boldsymbol{y}) \otimes \delta_\Gamma\big)d\boldsymbol{y} \\
&\quad -\frac{1}{k^2}\int_{\mathbb{R}^3} \mathbf{grad}_{\boldsymbol{x}}\mathrm{div}_{\boldsymbol{x}}\big\{G_k(\boldsymbol{x}-\boldsymbol{y})\big(\boldsymbol{\lambda}(\boldsymbol{y}) \otimes \delta_\Gamma\big)\big\}d\boldsymbol{y}.
\end{aligned} \tag{12.4.2}$$

The first potential considered as a Newton potential

$$\mathfrak{V}_1\boldsymbol{\lambda}(\boldsymbol{x}) = -\int_{\mathbb{R}^3} G_k(\boldsymbol{x}-\boldsymbol{y})\big(\boldsymbol{\lambda}(\boldsymbol{y}) \otimes \delta_\Gamma\big)d\boldsymbol{y} \tag{12.4.3}$$

is a pseudodifferential operator in $\boldsymbol{\mathcal{L}}^{-2}_{c\ell}(\mathbb{R}^3)$, hence,

$$\lim_{\boldsymbol{x}\to\boldsymbol{x}'} \mathfrak{V}_1\boldsymbol{\lambda}(\boldsymbol{x}) = -\int_{\boldsymbol{y}'\in\Gamma} G_k(\boldsymbol{x}'-\boldsymbol{y}')\boldsymbol{\lambda}(\boldsymbol{y}')ds_{\boldsymbol{y}'} \quad \text{for } \boldsymbol{x}' \in \Gamma$$

which defines a pseudodifferential operator of order $m+1=-1$; $\mathbf{V}_k \in \boldsymbol{\mathcal{L}}^{-1}_{c\ell}(\Gamma)$ on Γ where $\mathbf{V}_k$ is the boundary integral operator

$$\mathbf{V}_k\boldsymbol{\lambda}(\boldsymbol{x}') = \int_{\boldsymbol{y}'\in\Gamma} G_k(\boldsymbol{x}'-\boldsymbol{y}')\boldsymbol{\lambda}(\boldsymbol{y}')ds_{\boldsymbol{y}'} \quad \text{for } \boldsymbol{x}' \in \Gamma.$$

The second potential can also be considered as a Newton potential applied to $\gamma_0^*\boldsymbol{\lambda} = \boldsymbol{\lambda}(\boldsymbol{y}') \otimes \delta_\Gamma$:

$$\mathfrak{V}_2\boldsymbol{\lambda}(\boldsymbol{x}) := -\frac{1}{k^2}\int\limits_{\mathbb{R}^3} \mathbf{grad}_{\boldsymbol{x}}\mathrm{div}_{\boldsymbol{x}}\{G_k(\boldsymbol{x}-\boldsymbol{y})(\boldsymbol{\lambda}(\boldsymbol{y})\otimes\delta_\Gamma)\}d\boldsymbol{y} \qquad (12.4.4)$$

for $\boldsymbol{x} \in \mathbb{R}^3 \setminus \Gamma$ which is a pseudodifferential operator in $\boldsymbol{\mathcal{L}}^0_{c\ell}(\mathbb{R}^3)$ with the symbol

$$\begin{aligned}
\sigma_{\mathfrak{V}_2}(\boldsymbol{\xi}) &= -\begin{matrix} \\ \boldsymbol{e}_1 \\ \boldsymbol{e}_2 \\ \boldsymbol{n}\end{matrix}\begin{matrix} \begin{matrix}\boldsymbol{e}_1 & \quad\boldsymbol{e}_2 & \quad\boldsymbol{n}\end{matrix} \\ \begin{pmatrix} \xi_1^2 & \xi_1\xi_2 & \xi_1\xi_3 \\ \xi_1\xi_2 & \xi_2^2 & \xi_2\xi_3 \\ \xi_1\xi_3 & \xi_2\xi_3 & \xi_3^2 \end{pmatrix}\end{matrix} \frac{1}{|\boldsymbol{\xi}|^2}\frac{1}{k^2}\Big\{\sum_{\ell=0}^{\infty}\Big(\frac{k^2}{|\boldsymbol{\xi}|^2}\Big)^\ell\Big\} \\
&= \boldsymbol{a}_0^0(\boldsymbol{\xi}) + \mathcal{O}\Big(|\boldsymbol{\xi}|^{-2}\frac{1}{k^2}\Big),
\end{aligned} \qquad (12.4.5)$$

the principal part of which is given explicitly as

$$\boldsymbol{a}_0^0(\boldsymbol{\xi}) = -\begin{pmatrix} \xi_1^2 & \xi_1\xi_2 & \xi_1\xi_3 \\ \xi_1\xi_2 & \xi_2^2 & \xi_2\xi_3 \\ \xi_1\xi_3 & \xi_2\xi_3 & \xi_3^2 \end{pmatrix}\frac{1}{|\boldsymbol{\xi}|^2}\frac{1}{k^2}. \qquad (12.4.6)$$

According to Theorem 7.1.12, the remainder defines a pseudodifferential operator of order -2 whose limit $\Omega \ni \boldsymbol{x} \to \boldsymbol{x}' \in \Gamma$ exists and defines a boundary integral operator or pseudodifferential operator on Γ in $\mathcal{L}^{-1}_{c\ell}(\Gamma)$.
For the principal part $\boldsymbol{a}_0^0(\boldsymbol{\xi})$ Theorem 9.5.1 can also be applied, if Hörmander's transmission condition is satisfied. Indeed, if we choose the coordinates such that the direction of the normal vector $\boldsymbol{n}$ coincides with the third component. Hörmander's transmission condition (9.3.40) in Theorem 9.3.11 with $m=0\,,\ j=0\,,\ \alpha=0$ reads

$$\begin{pmatrix}0&0&0\\0&0&0\\0&0&1\end{pmatrix} = \Big(\frac{\partial}{\partial\boldsymbol{\xi}}\Big)^\alpha \boldsymbol{a}^0(\boldsymbol{\varrho}',0;0,+1) = \Big(\frac{\partial}{\partial\boldsymbol{\xi}}\Big)^\alpha \boldsymbol{a}^0(\boldsymbol{\varrho}',0;0,-1) = \begin{pmatrix}0&0&0\\0&0&0\\0&0&1\end{pmatrix}.$$

Hence, the transmission and extension conditions are satisfied.

Next, we observe that the parity condition (8.1.75) for the only case $-3-q=0$ for $q \in \mathbb{Z}$ is obviously satisfied for $\sigma = 0$ since all entries in $\boldsymbol{a}_0^0(\boldsymbol{\xi})$ in (12.4.6) satisfy $(-\xi_j)(-\xi_\ell) = \xi_j\xi_\ell$. Lemma 9.4.1 implies that also the Tricomi conditions are satisfied. As a consequence we conclude that the operator $\mathfrak{V}_2$ is a pseudodifferential operator in $\boldsymbol{\mathcal{L}}^0_{c\ell}(\mathbb{R}^3)$ and its restriction to Γ defines a pseudodifferential operator on Γ in $\boldsymbol{\mathcal{L}}^1_{c\ell}(\Gamma)$. However, the boundary integral operator of order 1 on Γ might contain a surface tangential differential operator $\mathcal{T}$ (see Theorem 9.4.6). Hence,

$$
\begin{aligned}
\lim_{\Omega \ni \boldsymbol{x} \to \boldsymbol{x}' \in \Gamma} (-k^2 \mathfrak{V}_2 \boldsymbol{\lambda})(\boldsymbol{x}) &= \mathcal{T} \boldsymbol{\lambda}(\boldsymbol{x}') \\
&+ \text{p.f.}\, \boldsymbol{n}(\boldsymbol{x}') \cdot \int_\Gamma \mathbf{grad}_{\boldsymbol{x}} \mathrm{div}_{\boldsymbol{x}} \{ G_k(\boldsymbol{x} - \boldsymbol{y}') \big(\lambda_1(\boldsymbol{y}') \boldsymbol{e}_1(\boldsymbol{y}') \\
&\qquad + \lambda_2(\boldsymbol{y}') \boldsymbol{e}_2(\boldsymbol{y}') \big) \} \big|_{\boldsymbol{x}=\boldsymbol{x}'} ds_{\boldsymbol{y}'} \boldsymbol{n}(\boldsymbol{x}') \\
&+ \text{p.f.}\, \boldsymbol{e}_1(\boldsymbol{x}') \cdot \int_\Gamma \mathbf{grad}_{\boldsymbol{x}} \mathrm{div}_{\boldsymbol{x}} \{ G_k(\boldsymbol{x} - \boldsymbol{y}') \big(\lambda_1(\boldsymbol{y}') \boldsymbol{e}_1(\boldsymbol{y}') \\
&\qquad + \lambda_2(\boldsymbol{y}') \boldsymbol{e}_2(\boldsymbol{y}') \big) \} \big|_{\boldsymbol{x}=\boldsymbol{x}'} ds_{\boldsymbol{y}'} \boldsymbol{e}_1(\boldsymbol{x}') \\
&+ \text{p.f.}\, \boldsymbol{e}_2(\boldsymbol{x}') \cdot \int_\Gamma \mathbf{grad}_{\boldsymbol{x}} \mathrm{div}_{\boldsymbol{x}} \{ G_k(\boldsymbol{x} - \boldsymbol{y}') \big(\lambda_1(\boldsymbol{y}') \boldsymbol{e}_1(\boldsymbol{y}') \\
&\qquad + \lambda_2(\boldsymbol{y}') \boldsymbol{e}_2(\boldsymbol{y}') \big) \} \big|_{\boldsymbol{x}=\boldsymbol{x}'} ds_{\boldsymbol{y}'} \boldsymbol{e}_2(\boldsymbol{x}'), \ \boldsymbol{x}' \in \Gamma,
\end{aligned}
$$

where $\boldsymbol{e}_1(\boldsymbol{x}')$, $\boldsymbol{e}_2(\boldsymbol{x}')$ and $\boldsymbol{n}(\boldsymbol{x}')$ form a basis of a local orthogonal coordinate system.

The tangential differential operator $\mathcal{T}_{\mathfrak{V}_2} \boldsymbol{\lambda} = -\frac{1}{k^2} \mathcal{T} \boldsymbol{\lambda}$ and $\mathcal{T} \boldsymbol{\lambda}$ is explicitly given by (9.4.26):

$$
\mathcal{T} = \frac{1}{2} \sum_{0 \le j \le 1} \sum_{|\alpha| = 1-j} (-i)^{|\alpha|+1} \frac{1}{\alpha!} \frac{\partial^\alpha}{(\partial \xi')^\alpha} \boldsymbol{a}^0_{0-j} \big(\boldsymbol{\varrho}', 0, -(0,1) \big) D^\alpha_{\varrho'} .
$$

$\boldsymbol{a}^0_0(\boldsymbol{\xi})$ is given in (12.4.6) and $\boldsymbol{a}^0_{-1} = \mathbf{0}$. With $\boldsymbol{\xi} = (\xi_1', \xi_2', 0)$ we find

$$
\Big(\frac{\partial}{\partial \xi_1} \boldsymbol{a}^0_0 \Big) \frac{\partial}{\partial x_1} \boldsymbol{\lambda} = \begin{pmatrix} 2\xi_1 & \xi_2 & \xi_3 \\ \xi_2 & 0 & 0 \\ \xi_3 & 0 & 0 \end{pmatrix} \big((\boldsymbol{\varrho}', 0), (0,1) \big) \frac{\partial}{\partial x_1} \boldsymbol{\lambda} = \begin{pmatrix} 0 & 0 & 1 \\ 0 & 0 & 0 \\ 1 & 0 & 0 \end{pmatrix} \frac{\partial}{\partial x_1} \lambda_1
$$

$$
\Big(\frac{\partial}{\partial \xi_2} \boldsymbol{a}^0_0 \Big) \frac{\partial}{\partial x_2} \boldsymbol{\lambda} = \begin{pmatrix} 0 & \xi_1 & 0 \\ \xi_1 & 2\xi_2 & \xi_3 \\ 0 & \xi_3 & 0 \end{pmatrix} \big((\boldsymbol{\varrho}', 0); (0,1) \big) \frac{\partial}{\partial x_2} \boldsymbol{\lambda} = \begin{pmatrix} 0 & 0 & 0 \\ 0 & 0 & 1 \\ 1 & 1 & 0 \end{pmatrix} \frac{\partial}{\partial x_2} \lambda_2
$$

and

$$
\boldsymbol{\mathcal{T}}_{\mathfrak{V}_2} \boldsymbol{\lambda} = -\frac{1}{2} \frac{1}{k^2} \Big(\frac{\partial \lambda_1}{\partial x_1} + \big(\frac{\partial \lambda_2}{\partial x_2} \Big) \boldsymbol{n}(\boldsymbol{x}') .
$$

Then

$$
\begin{aligned}
\lim_{\boldsymbol{x} \to \boldsymbol{x}' \in \Gamma} (\mathfrak{V}_2 \boldsymbol{\lambda})(\boldsymbol{x}) &= -\frac{1}{2} \frac{1}{k^2} \mathrm{div}_\Gamma \boldsymbol{\lambda}(\boldsymbol{x}') \boldsymbol{n}(\boldsymbol{x}') \\
&- \frac{1}{k^2} \text{ p.f.} \int_\Gamma \mathbf{grad}_{\boldsymbol{x}} \mathrm{div}_{\boldsymbol{x}} \{ G_k(\boldsymbol{x} - \boldsymbol{y}') \boldsymbol{\lambda}(\boldsymbol{y}') \} \big|_{\boldsymbol{x}=\boldsymbol{x}'} ds_{\boldsymbol{y}'} .
\end{aligned} \tag{12.4.7}
$$

This result is in agreement with the relation (5.5) of Section 6.5 in the book by E. Meister [293].

For the double layer potential in (12.3.1), i.e.,

$$(\mathfrak{W}\varphi)(\boldsymbol{x}) := -\int\limits_{\Gamma} \varphi(\boldsymbol{y}) \times \mathbf{grad}_{\boldsymbol{y}} G_k(\boldsymbol{x}-\boldsymbol{y}) ds_{\boldsymbol{y}}$$

$$= -\mathbf{curl}_{\boldsymbol{x}} \int\limits_{\Gamma} G_k(\boldsymbol{x}-\boldsymbol{y})\varphi(\boldsymbol{y}) ds_{\boldsymbol{y}} \quad \text{for } \boldsymbol{x} \in \Omega, \tag{12.4.8}$$

we consider it as a double layer Newton potential in the form

$$(\mathfrak{W}\varphi)(\boldsymbol{x}) = -\mathbf{curl}_{\boldsymbol{x}} \int\limits_{\mathbb{R}^3} G_k(\boldsymbol{x}-\boldsymbol{y})\big(\varphi(\boldsymbol{y}) \otimes \delta_\Gamma\big) d\boldsymbol{y}\,,\ \boldsymbol{x} \in \mathbb{R}^3 \setminus \Gamma.$$

Then the symbol of $\mathfrak{W}$,

$$\sigma_{\mathfrak{W}}(\boldsymbol{\xi}) = -i\boldsymbol{\xi} \times \frac{1}{|\boldsymbol{\xi}|^2 - k^2}\boldsymbol{I},$$

has the asymptotic expansion

$$\sigma_{\mathfrak{W}}(\boldsymbol{\xi}) = -i\frac{\boldsymbol{\xi}}{|\boldsymbol{\xi}|^2}\Big\{\sum_{\ell=0}^{\infty}\Big(\frac{k^2}{|\boldsymbol{\xi}|^2}\Big)^2\Big\} \times \boldsymbol{I} \tag{12.4.9}$$

$$= \begin{pmatrix} 0 & i\xi_3 & -i\xi_2 \\ -i\xi_3 & 0 & i\xi_1 \\ i\xi_2 & -i\xi_1 & 0 \end{pmatrix} \frac{1}{|\boldsymbol{\xi}|^2}\Big\{\sum_{\ell=0}^{\infty}\Big(\frac{k^2}{|\boldsymbol{\xi}|^2}\Big)^\ell\Big\} \quad \text{for } |\boldsymbol{\xi}| > |k|,$$

$$= \begin{pmatrix} 0 & i\xi_3 & -i\xi_2 \\ -i\xi_3 & 0 & i\xi_1 \\ i\xi_2 & -i\xi_1 & 0 \end{pmatrix} \frac{1}{|\boldsymbol{\xi}|^2} + \mathcal{O}(|\boldsymbol{\xi}|^{-3}k^2) = \boldsymbol{a}^0_{-1}(\boldsymbol{\xi}) + \mathcal{O}(|\boldsymbol{\xi}|^{-3}k^2).$$

The principal part defines a pseudodifferential operator in $\mathcal{L}^{-1}_{c\ell}(\mathbb{R}^3)$ and the remainder is in $\boldsymbol{\mathcal{L}}^{-3}_{c\ell}(\mathbb{R}^3)$. The principal symbol $\boldsymbol{a}^0_{-1}(\boldsymbol{\xi})$ satisfies Hörmander's transmission condition (9.3.40) in Theorem 9.3.11 and extension is guaranteed. $\boldsymbol{a}^0_{-1}(\boldsymbol{\xi})$ also satisfies the parity condition (9.4.2) in Lemma 9.4.1 with parity $\sigma = 1$. Consequently, the Tricomi condition (9.3.18) is satisfied in view of Lemma 9.3.5 and $\mathfrak{W}$ on Γ defines a pseudodifferential boundary integral operator in $\boldsymbol{\mathcal{L}}^0_{c\ell}(\Gamma)$. According to Theorem 9.4.6, the tangential differential operator $\mathcal{T}_{\mathfrak{W}}$ is given by Equation (9.4.26).

$$\mathcal{T}_{\mathfrak{W}}\varphi = \frac{1}{2}(-i)\boldsymbol{a}^0_{-1}\big(\varrho', 0, ; (0,1)\big)\varphi(\boldsymbol{x}') \tag{12.4.10}$$

$$= \frac{1}{2} \begin{matrix} \\ \boldsymbol{e}_1 \\ \boldsymbol{e}_2 \\ \boldsymbol{n} \end{matrix} \begin{matrix} \begin{matrix} \boldsymbol{e}_1 & \boldsymbol{e}_2 & \boldsymbol{n} \end{matrix} \\ \begin{pmatrix} 0 & i & 0 \\ -i & 0 & 0 \\ 0 & 0 & 0 \end{pmatrix} \end{matrix} (-i) \begin{pmatrix} \varphi_1(\boldsymbol{x}') \\ \varphi_2(\boldsymbol{x}') \\ 0 \end{pmatrix}$$

$$= \frac{1}{2}\big(\varphi_1(\boldsymbol{x}')\boldsymbol{e}_2 - \varphi_2(\boldsymbol{x}')\boldsymbol{e}_1\big) = -\frac{1}{2}\boldsymbol{n} \times (\boldsymbol{e}_1\varphi_1 + \boldsymbol{e}_2\varphi_2)$$

since $j = 0$ and $\alpha = 0$, Hence,

$$\lim_{\Omega \ni \boldsymbol{x} \to \boldsymbol{x}' \in \Gamma} (\mathfrak{W}\varphi)(\boldsymbol{x}) = -\frac{1}{2}\boldsymbol{n}(\boldsymbol{x}') \times \varphi(\boldsymbol{x}') - \boldsymbol{W}_k\varphi(\boldsymbol{x}')$$
$$\text{where } \boldsymbol{W}_k\varphi(\boldsymbol{x}') = \text{p.v.} \int_\Gamma \varphi(\boldsymbol{y}) \times \mathbf{grad}_{\boldsymbol{y}} G_k(\boldsymbol{x}' - \boldsymbol{y}) ds_{\boldsymbol{y}}. \tag{12.4.11}$$

Without loss of generality we consider the representation formula (12.3.2) for the solution of the homogeneous equation with $\boldsymbol{f} = \mathbf{0}$:

$$\begin{aligned}\mathbf{E}(\boldsymbol{x}) = &-\int_\Gamma G_k(\boldsymbol{x} - \boldsymbol{y})\boldsymbol{\lambda}(\boldsymbol{y}) ds_{\boldsymbol{y}} \\ &- \frac{1}{k^2}\mathbf{grad}_{\boldsymbol{x}} \int_\Gamma G_k(\boldsymbol{x} - \boldsymbol{y})\text{div}_\Gamma \boldsymbol{\lambda}(\boldsymbol{y}) ds_{\boldsymbol{y}} \\ &- \mathbf{curl}_{\boldsymbol{x}} \int_\Gamma G_k(\boldsymbol{x} - \boldsymbol{y})\varphi(\boldsymbol{y}) ds_{\boldsymbol{y}} \\ = &\ \mathfrak{V}_1(\boldsymbol{\lambda}) + \mathfrak{V}_2(\boldsymbol{\lambda}) + \mathfrak{W}(\varphi)\end{aligned} \tag{12.4.12}$$

where $\boldsymbol{\lambda}(\boldsymbol{y}) = (\boldsymbol{n} \times \mathbf{curl}\,\mathbf{E})(\boldsymbol{y})$ and $\varphi(\boldsymbol{y}) = (\boldsymbol{n} \times \mathbf{E})(\boldsymbol{y})$ are the Cauchy data on Γ for **PE** in Ω, with potentials $\mathfrak{V}_1$ (12.4.3) $\mathfrak{V}_2$ (12.4.4) and $\mathfrak{W}$ (12.4.8).

In what follows, the limit is taken as $\Omega \ni \boldsymbol{x} \to \boldsymbol{x}' \in \Gamma$ of (12.4.12) and we will arrive at a first boundary integral equation

$$\begin{aligned}\pi_t\mathbf{E}(\boldsymbol{x}') &= \pi_t \lim_{\boldsymbol{x} \to \boldsymbol{x}' \in \Gamma} \{\mathfrak{V}_1\boldsymbol{\lambda}(\boldsymbol{x}) + \mathfrak{V}_2\boldsymbol{\lambda}(\boldsymbol{x}) + \mathfrak{W}\varphi(\boldsymbol{x})\} \\ &= -\pi_t\mathbf{V}_k\boldsymbol{\lambda}(\boldsymbol{x}') + \pi_t\Big(-\frac{1}{2}\frac{1}{k^2}\text{div}_\Gamma\boldsymbol{\lambda} \cdot \boldsymbol{n}(\boldsymbol{x}')\Big) \\ -\frac{1}{k^2}&\mathbf{grad}_\Gamma V_k \text{div}_\Gamma \boldsymbol{\lambda}(\boldsymbol{x}') + \pi_t\big(-\frac{1}{2}\boldsymbol{n}(\boldsymbol{x}') \times \varphi(\boldsymbol{x}') - (\mathbf{W}_k\varphi)(\boldsymbol{x}')\big) \\ &= -\big\{\pi_t\mathbf{V}_k\boldsymbol{\lambda}(\boldsymbol{x}') + \frac{1}{k^2}\mathbf{grad}_\Gamma V_k\text{div}_\Gamma\boldsymbol{\lambda}(\boldsymbol{x}')\big\} \\ &\quad -\frac{1}{2}(\boldsymbol{n} \times \varphi)(\boldsymbol{x}') - \pi_t\boldsymbol{W}_k\varphi(\boldsymbol{x}') \quad \text{for } \boldsymbol{x}' \in \Gamma\end{aligned} \tag{12.4.13}$$
$$\text{where } V_k g(\boldsymbol{x}') := \int_\Gamma G_k(\boldsymbol{x}' - \boldsymbol{y})g(\boldsymbol{y}) ds_{\boldsymbol{y}},\ \boldsymbol{x}' \in \Gamma.$$

By applying $\boldsymbol{n} \times \mathbf{curl}\ \mathbf{E}$ to (12.4.11) and taking the limit, we obtain a second boundary integral equation,

$$\gamma_t\mathbf{curl}_\Gamma\mathbf{E} = \gamma_t\big\{\lim \mathbf{curl}\ \mathfrak{V}_1\boldsymbol{\lambda} + \lim \mathbf{curl}\ \mathfrak{V}_2\boldsymbol{\lambda} + \lim \mathbf{curl}\,\mathfrak{W}\varphi\big\}.$$

Again, the limit is taken as $\Omega \ni \boldsymbol{x} \to \boldsymbol{x}' \in \Gamma$. The first term is

$$\gamma_t \lim \mathbf{curl}\,\mathfrak{V}_1\boldsymbol{\lambda} = \gamma_t \lim \int_\Gamma (\mathbf{grad}_{\boldsymbol{y}} G_k(\boldsymbol{x}-\boldsymbol{y})) \times \boldsymbol{\lambda}(\boldsymbol{y}) ds_{\boldsymbol{y}} = \gamma_t \lim \mathfrak{W}\boldsymbol{\lambda}$$
$$= -\frac{1}{2}\gamma_t(\boldsymbol{n}\times\boldsymbol{\lambda})(\boldsymbol{x}') - \gamma_t \boldsymbol{W}_k\boldsymbol{\lambda}(\boldsymbol{x}'),$$
$$= \frac{1}{2}\pi_t\boldsymbol{\lambda}(\boldsymbol{x}') - \gamma_t \boldsymbol{W}_k\boldsymbol{\lambda}(\boldsymbol{x}').$$

The second term is

$$\pi_t\mathfrak{V}_2 = \gamma_t \lim \mathbf{curl}\,\mathfrak{V}_2\boldsymbol{\lambda}$$
$$= \frac{1}{k^2}\gamma_t \lim \mathbf{curl}_{\boldsymbol{x}} \int_\Gamma \mathbf{grad}_{\boldsymbol{y}} G_k(\boldsymbol{x}-\boldsymbol{y}) \operatorname{div}_\Gamma \boldsymbol{\lambda}(\boldsymbol{y}) ds_{\boldsymbol{y}}$$
$$= -\frac{1}{k^2}\gamma_t \lim \mathbf{curl}_{\boldsymbol{x}}\mathbf{grad}_{\boldsymbol{x}} \int_\Gamma G_k(\boldsymbol{x}-\boldsymbol{y}) \operatorname{div}_\Gamma \boldsymbol{\lambda}(\boldsymbol{y}) ds_{\boldsymbol{y}} = \mathbf{0}.$$

For the third term we have the relation

$$\gamma_t \lim \mathbf{curl}_{\boldsymbol{x}}\mathfrak{W}\boldsymbol{\varphi}(\boldsymbol{x}) = -\gamma_t \lim \mathbf{curl}_{\boldsymbol{x}}\mathbf{curl}_{\boldsymbol{x}} \int_\Gamma G_k(\boldsymbol{x}-\boldsymbol{y})\boldsymbol{\varphi}(\boldsymbol{y}) ds_{\boldsymbol{y}}$$
$$= \gamma_t \lim \mathbf{curl}_{\boldsymbol{x}} \text{ p.v.} \int_\Gamma \mathbf{grad}_{\boldsymbol{y}} G_k(\boldsymbol{x}'-\boldsymbol{y}) \times \boldsymbol{\varphi}(\boldsymbol{y}) ds_{\boldsymbol{y}}$$
$$= \gamma_t \text{ p.f.} \int_\Gamma \{\mathbf{curl}_{\boldsymbol{x}}(\mathbf{grad}_{\boldsymbol{y}} G_k(\boldsymbol{x}-\boldsymbol{y}) \times \boldsymbol{\varphi}(\boldsymbol{y})\}|_{\boldsymbol{x}=\boldsymbol{x}'} ds_{\boldsymbol{y}}$$
$$= -\gamma_t \mathbf{curl}\,\boldsymbol{W}_k\boldsymbol{\varphi}(\boldsymbol{x}').$$

Collecting terms, we obtain the **second boundary integral equation**:

$$\gamma_t\, \mathbf{curl}\ \mathbf{E}(\boldsymbol{x}') = \frac{1}{2}\pi_t\boldsymbol{\lambda}(\boldsymbol{x}') - \gamma_t \boldsymbol{W}_k\boldsymbol{\lambda}(\boldsymbol{x}') - \gamma_t \mathbf{curl}\,\boldsymbol{W}_k\boldsymbol{\varphi}(\boldsymbol{x}'). \tag{12.4.14}$$

In both boundary integral equations, we need to substitute $\boldsymbol{\lambda}$ and $\boldsymbol{\varphi}$ by the Cauchy data

$$\boldsymbol{\lambda} = \gamma_t \mathbf{curl}\ \mathbf{E} \quad \text{and} \quad \boldsymbol{\varphi} = \boldsymbol{n} \times \pi_t \mathbf{E},$$

respectively.

In the same manner as for the Laplace equation we are able to define a Calderon projector for the differential equation (12.1.4),

$$\mathbf{curl}\ \mathbf{curl}\ \mathbf{E} - k^2\mathbf{E} = \mathbf{0}$$

in terms of the Cauchy data $\pi_t\mathbf{E}$ and $\gamma_t\mathbf{curl}\ \mathbf{E}$ in the form

$$\begin{pmatrix} \pi_t\mathbf{E} \\ \gamma_t\mathbf{curl}\ \mathbf{E} \end{pmatrix} = \begin{pmatrix} \frac{1}{2}\boldsymbol{I} - \pi_t\widetilde{\boldsymbol{W}}_k, & -\{\pi_t\mathbf{V}_k + \frac{1}{k^2}\mathbf{grad}_{\boldsymbol{x}} V_k \operatorname{div}_\Gamma\} \\ -\gamma_t\, \mathbf{curl}\ \widetilde{\boldsymbol{W}}_k, & \frac{1}{2}\boldsymbol{I} - \gamma_t\boldsymbol{W}_k \end{pmatrix} \begin{pmatrix} \pi_t\mathbf{E} \\ \gamma_t\mathbf{curl}\ \mathbf{E} \end{pmatrix} \tag{12.4.15}$$

where

$$\boldsymbol{W}_k\boldsymbol{u} := \text{p.v.} \int\limits_{\Gamma} \boldsymbol{u}(\boldsymbol{y}) \times \textbf{grad}_{\boldsymbol{y}} G_k(\boldsymbol{x}' - \boldsymbol{y}) ds_{\boldsymbol{y}}, \qquad (12.4.16)$$

$$\widetilde{\boldsymbol{W}}_k\boldsymbol{u} = \boldsymbol{W}_k(\boldsymbol{n} \times \boldsymbol{u}) \quad \text{for } \boldsymbol{x}' \in \Gamma. \qquad (12.4.17)$$

Equation (12.4.15) defines an alternative Calderon projector which is similar to the Calderon projector which is given by (6.10.6), see also (6.4.1)–(6.4.4). With the availability of the two boundary integral equations (12.4.13) and (12.4.14) we now compute their symbols in the next section.

12.5 Symbols of boundary integral operators

For

$$\begin{aligned}
\mathbf{V}_k\boldsymbol{\lambda} = -\gamma_0\mathfrak{V}_1\boldsymbol{\lambda} &= \int\limits_{\Gamma} G_k(\boldsymbol{x}' - \boldsymbol{y})\boldsymbol{\lambda}(\boldsymbol{y}) ds_{\boldsymbol{y}} \\
&= \frac{1}{4\pi} \int\limits_{\Gamma} \sum_{p=0}^{\infty} \frac{\big(ik(|\boldsymbol{x}' - \boldsymbol{y}|)\big)^p}{p!|\boldsymbol{x}' - \boldsymbol{y}|} \boldsymbol{\lambda}(\boldsymbol{y}) ds_{\boldsymbol{y}} \quad \text{with } \boldsymbol{x}' \in \Gamma.
\end{aligned}$$

Fourier transform of the Schwartz kernel's pseudohomogeneous expansion defines the symbol expansion

$$\sigma_{\mathbf{V}_k}(\boldsymbol{\xi}') = \frac{1}{4\pi} \sum_{p=0}^{\infty} \alpha_p |\boldsymbol{\xi}'|^{-p-1} \qquad (12.5.1)$$

where $\alpha_p = 2^{1-p}\Gamma\big(\frac{1-p}{2}\big) \big/ \Gamma\big(\frac{p+1}{2}\big)p!$. Its principal symbol is

$$\boldsymbol{a}_0^0(\boldsymbol{\xi}') = \frac{1}{2}|\boldsymbol{\xi}'|^{-1}.$$

For the second simple layer potential operator we find the pseudohomogeneous kernel expansion

$$\begin{aligned}
\lim \mathfrak{V}_2\boldsymbol{\lambda} &= -\frac{1}{2}\frac{1}{k^2}\boldsymbol{n}(\boldsymbol{x}')\text{div}_{\Gamma}\boldsymbol{\lambda}(\boldsymbol{x}') - \frac{1}{k^2}\textbf{grad}_{\Gamma,\boldsymbol{x}}\mathbf{V}_k\text{div}_{\Gamma}\boldsymbol{\lambda} \qquad (12.5.2) \\
&= -\frac{1}{k^2}\frac{1}{2}\boldsymbol{n}(\boldsymbol{x}')\text{div}_{\Gamma}\boldsymbol{\lambda}(\boldsymbol{x}') \\
&\quad + \frac{1}{k^2}\frac{1}{4\pi}\int\limits_{\Gamma} \Big(\textbf{grad}_{\Gamma,\boldsymbol{y}} \sum_{p=0}^{\infty} \frac{(ik|\boldsymbol{x}' - \boldsymbol{y}|)^p}{p!|\boldsymbol{x}' - \boldsymbol{y}|}\Big)\boldsymbol{\lambda}(\boldsymbol{y}) ds_{\boldsymbol{y}}
\end{aligned}$$

whose two–dimensional Fourier transform yields

$$\lim \widehat{\mathfrak{V}_2 \boldsymbol{\lambda}}(\boldsymbol{\xi}') = \frac{1}{k^2}\frac{1}{2}\boldsymbol{n}(\boldsymbol{x}')(i\xi_1\widehat{\lambda}_1 + i\xi_2\widehat{\lambda}_2)$$
$$+\frac{1}{k^2}\frac{1}{4\pi}\begin{pmatrix} i\xi_1 \\ i\xi_2 \end{pmatrix}\sum_{p=0}^{\infty}\alpha_p|\boldsymbol{\xi}'|^{-p-1}(\widehat{\lambda}_1, \widehat{\lambda}_2).$$

Hence, the symbol reads

$$\begin{aligned}\sigma_{\gamma_0\mathfrak{V}_2}(\boldsymbol{x}',\boldsymbol{\xi}') &= \frac{1}{k^2}\frac{1}{2}\boldsymbol{n}(\boldsymbol{x}')(i\xi_1, i\xi_2) \\ &\quad -\frac{1}{k^2}\frac{1}{4\pi}\Big(\sum_{p=0}^{\infty}\alpha_p|\boldsymbol{\xi}'|^{-p-1}\Big)\begin{pmatrix} \xi_1^2 & 0 \\ o & \xi_2^2 \end{pmatrix}.\end{aligned} \tag{12.5.3}$$

Finally, the double layer potential

$$\lim_{\boldsymbol{x}\to\boldsymbol{x}'\in\Gamma}\mathfrak{W}\varphi(\boldsymbol{x}) = -\lim_{\boldsymbol{x}\to\boldsymbol{x}'\in\Gamma}\mathbf{curl}_{\boldsymbol{x}}\int_{\Gamma} G_k(\boldsymbol{x}-\boldsymbol{y})\varphi(\boldsymbol{y})ds_{\boldsymbol{y}}$$
$$= -\frac{1}{2}\boldsymbol{n}(\boldsymbol{x}')\times\varphi(\boldsymbol{x}') + \text{p.v.}\int_{\Gamma}(\mathbf{grad}_{\boldsymbol{y}}G_k(\boldsymbol{x}'-\boldsymbol{y}))\times\varphi(\boldsymbol{y})ds_{\boldsymbol{y}}$$

which has the pseudohomogeneous expansion

$$\lim_{x\to x'\in\Gamma}\mathfrak{W}\varphi(\boldsymbol{x}) = \frac{1}{2}\boldsymbol{n}(\boldsymbol{x}')\times\varphi(\boldsymbol{x}')$$
$$+\frac{1}{4\pi}\text{p.v.}\int_{\Gamma}\sum_{p=0}^{\infty}\Big(\frac{\partial}{\partial r}\frac{(ikr)^p}{p!r}\Big)\frac{1}{r}\Big|_{r=|\boldsymbol{x}'-\boldsymbol{y}|}(\boldsymbol{y}-\boldsymbol{x}')\times\varphi(\boldsymbol{y})ds_{\boldsymbol{y}}$$
$$= -\frac{1}{2}\boldsymbol{n}(\boldsymbol{x}')\times\varphi(\boldsymbol{x}') + \frac{1}{4\pi}\sum_{p=0}^{\infty}\text{p.v.}\int_{\Gamma}\Big(\frac{\partial}{\partial r}\frac{(ikr)}{p!r}\Big)\frac{1}{r}\Big|_{r=|\boldsymbol{x}'-\boldsymbol{y}|}\times$$
$$\{(y_1-x_1')\varphi_2 - (y_2-x_2')\varphi_1\}\boldsymbol{n}(\boldsymbol{x}')ds_{\boldsymbol{y}}$$

since $\varphi(\boldsymbol{y})$ is a tangential field. The symbol of $\gamma_0\mathfrak{W}$:

$$\begin{aligned}\sigma_{\gamma_0\mathfrak{W}}(\boldsymbol{x}',\boldsymbol{\xi}')\widehat{\varphi}(\boldsymbol{\xi}') &= -\frac{1}{2}\boldsymbol{n}(\boldsymbol{x}')\times\widehat{\varphi}(\boldsymbol{\xi}') \\ &\quad +\sum_{p=0}^{\infty}\frac{1}{4\pi}\alpha_p|\boldsymbol{\xi}'|^{-p-1}\begin{pmatrix} -i\xi_2 & 0 \\ 0 & i\xi_1 \end{pmatrix}\widehat{\varphi}(\boldsymbol{\xi}')\boldsymbol{n}(\boldsymbol{x}').\end{aligned} \tag{12.5.4}$$

With the symbols of $\gamma_0\mathfrak{V}_1, \gamma_0\mathfrak{V}_2$ and $\gamma_0\mathfrak{W}$ available we return to the capacity operators defined by the Calderon projector in (6.10.6) and in (12.4.15).

12.6 Symbols of the Capacity Operators

For the interior domain Ω, the operator $\mathbf{T_E}$ in (6.10.2) on the boundary Γ is given by

$$\begin{aligned}
\mathbf{T_E}(\boldsymbol{n}\times\mathbf{H})(\boldsymbol{x}') &= -(i\omega\mu)\boldsymbol{n}(\boldsymbol{x}')\times\big\{\lim_{\boldsymbol{x}\to\boldsymbol{x}'\in\Gamma}\mathfrak{V}_1(\boldsymbol{n}\times\mathbf{H})(\boldsymbol{x})\\
&\quad+\lim_{\boldsymbol{x}\to\boldsymbol{x}'\in\Gamma}\mathfrak{V}_2(\boldsymbol{n}\times\mathbf{H})(\boldsymbol{x})\big\}\\
&= -(i\omega\mu)\big\{\boldsymbol{n}(\boldsymbol{x}')\times\mathbf{V}_k(\boldsymbol{n}\times\mathbf{H})\\
&\quad+\boldsymbol{n}(\boldsymbol{x}')\times\frac{1}{k^2}\,\mathbf{grad}_{\,\Gamma,\boldsymbol{x}}\mathbf{V}_k\mathrm{div}_\Gamma(\boldsymbol{n}\times\mathbf{H})\big\}.
\end{aligned}$$

$$\begin{aligned}
\sigma_{\mathbf{T_E}}(\boldsymbol{x}',\boldsymbol{\xi}') &= i\omega\mu\{\boldsymbol{n}(\boldsymbol{x}')\times\sigma_{\gamma\mathfrak{V}_1}(\boldsymbol{x}',\boldsymbol{\xi}')+\boldsymbol{n}(\boldsymbol{x}')\times\sigma_{\gamma\mathfrak{V}_2}(\boldsymbol{x}',\boldsymbol{\xi}')\}\\
&= i\omega\mu\Big\{\boldsymbol{n}(\boldsymbol{x}')\times\big(-\sigma_{\mathbf{V}_k}(\boldsymbol{x}')\big)\\
&\quad+\boldsymbol{n}(\boldsymbol{x}')\times\Big(-\frac{1}{k^2}\sigma_{\mathbf{V}_k}(\boldsymbol{\xi}')\Big)\begin{pmatrix}\xi_1^2 & 0\\ 0 & \xi_2^2\end{pmatrix}\Big\}\boldsymbol{I}\\
&= i\omega\mu\begin{pmatrix}0 & -1\\ 1 & 0\end{pmatrix}\Big\{\sigma_{V_k}(\boldsymbol{\xi}')+\frac{1}{k^2}\sigma_{\mathbf{V}_k}(\boldsymbol{\xi}')\begin{pmatrix}\xi_1^2 & 0\\ 0 & \xi_2^2\end{pmatrix}\Big\} \qquad (12.6.1)\\
&= -i\omega\mu\begin{pmatrix}0 & -1\\ 1 & 0\end{pmatrix}\frac{1}{4\pi}\sum_{p=0}^{\infty}\alpha_p|\boldsymbol{\xi}'|^{p-1}\\
&\quad-i\omega\mu\begin{pmatrix}0 & -1\\ 1 & 0\end{pmatrix}\frac{1}{k^2}\frac{1}{4\pi}\sum_{p=0}^{\infty}\alpha_p|\boldsymbol{\xi}'|^{p-1}\begin{pmatrix}\xi_1^2 & 0\\ 0 & \xi_2^2\end{pmatrix}
\end{aligned}$$

For $\mathbf{T_H}(\boldsymbol{n}\times\mathbf{E})$ (see (6.9.5)) the symbol can be computed in the same manner

$$\begin{aligned}
\sigma_{\mathbf{T_H}}(\boldsymbol{x}',\boldsymbol{\xi}') &= -(\sigma-i\omega\varepsilon)\begin{pmatrix}0 & -1\\ 1 & 0\end{pmatrix}\frac{1}{4\pi}\sum_{p=0}^{\infty}\alpha_p|\boldsymbol{\xi}'|^{p-1} \qquad (12.6.2)\\
&\quad-(\sigma-i\omega\varepsilon)\begin{pmatrix}0 & -1\\ 1 & 0\end{pmatrix}\frac{1}{k^2}\frac{1}{4\pi}\sum_{p=0}^{\infty}\alpha_p|\boldsymbol{\xi}'|^{p-1}\begin{pmatrix}\xi_1^2 & 0\\ 0 & \xi_2^2\end{pmatrix}.
\end{aligned}$$

For the double layer Newton potential's boundary values one obtains for $\Omega\ni\boldsymbol{x}\to\boldsymbol{x}'\in\Gamma$:

$$\begin{aligned}
\boldsymbol{n}(\boldsymbol{x}')\times\lim\mathfrak{W}(\boldsymbol{n}\times\mathbf{E})(\boldsymbol{x}) &= -\frac{1}{2}\boldsymbol{n}(\boldsymbol{x}')\times\boldsymbol{n}(\boldsymbol{x}')\times\big(\boldsymbol{n}(\boldsymbol{x}')\times\mathbf{E}\big)\\
&\quad+\boldsymbol{n}(\boldsymbol{x}')\times\int_\Gamma(\mathbf{grad}_{\boldsymbol{y}}G_k(\boldsymbol{x}'-\boldsymbol{y}))\times(\boldsymbol{n}\times\mathbf{E})ds_{\boldsymbol{y}}\\
&= \frac{1}{2}\big(\boldsymbol{n}(\boldsymbol{x}')\times\mathbf{E}(\boldsymbol{x}')\big)-\boldsymbol{R}(\boldsymbol{n}\times\mathbf{E}).
\end{aligned}$$

From (12.5.4) we find that the principal symbol of $\boldsymbol{R}$ vanishes:

$$\sigma^0_{\boldsymbol{R}}(\boldsymbol{x}',\boldsymbol{\xi}')\widehat{\varphi} = \boldsymbol{n}(\boldsymbol{x}') \times \sum_{p=0}^{\infty} \frac{1}{4\pi} \alpha_{\Gamma} |\boldsymbol{\xi}'|^{p-1}(-i\xi_2\widehat{\varphi}_1 + i\xi_1\widehat{\varphi}_2)\boldsymbol{n}(\boldsymbol{x}') = \mathbf{0}. \quad (12.6.3)$$

On the other hand, note that the operator $\boldsymbol{R}$ in (6.9.3) is for $\boldsymbol{x}' \in \Gamma$ given by

$$\boldsymbol{R}(\boldsymbol{n} \times \mathbf{E})(\boldsymbol{x}') = \text{p.v.}\, \boldsymbol{n}(\boldsymbol{x}') \times \mathbf{curl}_{\boldsymbol{x}} \mathbf{V}_k(\boldsymbol{n} \times \mathbf{E}).$$

It has the pseudohomogeneous kernel expansion

$$-\boldsymbol{n}(\boldsymbol{x}') \times \mathbf{curl}_{\boldsymbol{x}} \mathbf{V}_k = -\boldsymbol{n}(\boldsymbol{x}') \times \mathbf{curl}_{\boldsymbol{x}} \frac{1}{4\pi} \sum_{p=0}^{\infty} \frac{(ik|\boldsymbol{x}' - \boldsymbol{y}|)^p}{p!|\boldsymbol{x}' - \boldsymbol{y}|}, \ \boldsymbol{x}' \in \Gamma,$$

with the symbol expansion

$$\sigma_{\boldsymbol{R}}(\boldsymbol{x}',\boldsymbol{\xi}') = -\boldsymbol{n}(\boldsymbol{x}') \times \begin{pmatrix} 0 & i\xi_1 \\ -i\xi_2 & 0 \end{pmatrix} \sum_{p=0}^{\infty} \frac{1}{4\pi} \alpha_p |\boldsymbol{\xi}'|^{p-1}.$$

The symbol is of order 0 and the inverse Fourier transform on the two–dimensional manifold Γ seems to define ψdo_0 on Γ. However, the operator can be written as

$$\begin{aligned} \boldsymbol{R}(\boldsymbol{n} \times \mathbf{E})(\boldsymbol{x}') &= -\boldsymbol{n}(\boldsymbol{x}') \times \int\limits_{\Gamma \setminus \{\boldsymbol{x}'\}} \frac{\partial G_k(\boldsymbol{x}' - \boldsymbol{y})}{\partial n_{\boldsymbol{x}}} (\boldsymbol{n} \times \mathbf{E})(\boldsymbol{y}) ds_{\boldsymbol{y}} \quad \text{for } \boldsymbol{x}' \in \Gamma, \\ &= -\boldsymbol{n}(\boldsymbol{x}') \times \boldsymbol{K}'_k(\boldsymbol{n} \times \mathbf{E})(\boldsymbol{x}') \end{aligned} \quad (12.6.4)$$

with the double layer potential operator $\boldsymbol{K}'_k$ of the Helmholtz equation (see Section 2.1). For smooth Γ, the Schwartz kernel has a pseudohomogenous expansion of order -1, and on the two–dimensional manifold Γ it defines a ψdo_{-1} mapping $\mathbf{H}^s_t(\Gamma)$ into $\mathbf{H}^{s+1}_t(\Gamma)$ which is **compactly imbedded** in $\mathbf{H}^s_t(\Gamma)$.

In summery, the symbols of the capacity operators in the Calderon projectors in Ω, (6.9.6) are given in Equations (12.6.1), (12.6.2), (12.6.3). In the exterior domain $\Omega^c = \mathbb{R}^3 \setminus \overline{\Omega}$, the boundary integral operators have a similar form. The Calderon projector in (12.4.15) will provide two boundary integral equations for the missing parts of the Cauchy data, i.e. $\pi_t\mathbf{E}$ or $\gamma_t\mathbf{curl}\,\mathbf{E}$. If the corresponding integral equation is solved, the solution of $\boldsymbol{P}\mathbf{E} = \mathbf{0}$ in Ω, as the boundary value problem is then obtained by the representation formula (12.3.2).

12.7 Boundary Integral Operators for the Fundamental Boundary Value Problems

The symbols of the boundary integral operators in (12.4.12) are given by

$$\begin{aligned}\sigma_{\pi_t \mathbf{E}}(\boldsymbol{x}', \boldsymbol{\xi}') &= \sigma_{-\pi_t \mathbf{V}_k}(\boldsymbol{x}', \boldsymbol{\xi}') + \sigma_{\pi_t \gamma \mathfrak{V}_2}(\boldsymbol{x}', \boldsymbol{\xi}') + \sigma_{\pi_t \gamma \mathfrak{W}}(\boldsymbol{x}', \boldsymbol{\xi}') \\ &= -\frac{1}{4\pi}\Big(\sum_{p=0}^{\infty} \alpha_p |\boldsymbol{\xi}'|^{-p-1}\Big)\begin{pmatrix}1 & 0\\ 0 & 1\end{pmatrix} \\ &\quad - \frac{1}{k^2}\frac{1}{4\pi}\Big(\sum_{p=0}^{\infty} |\boldsymbol{\xi}'|^{-p-1}\Big)\begin{pmatrix}\xi_1^2 & 0\\ 0 & \xi_2^2\end{pmatrix} - \frac{1}{2}\begin{pmatrix}0 & -1\\ 1 & 0\end{pmatrix}.\end{aligned}$$

For equation (12.4.14), then one finds

$$\begin{aligned}\sigma_{\gamma_t \mathbf{curl}\, \mathbf{E}}(\boldsymbol{x}', \boldsymbol{\xi}') &= \sigma_{\gamma_t \gamma W}(x', \xi') + \sigma_{\gamma_t \mathbf{curl}\, \gamma W}(x', \xi') \\ &= \frac{1}{2}\begin{pmatrix}1 & 0\\ 0 & 1\end{pmatrix} - \frac{1}{4\pi}\sum_{p=0}^{\infty} \alpha_p |\boldsymbol{\xi}'|^{1-p}\begin{pmatrix}1 & 0\\ 0 & 1\end{pmatrix}.\end{aligned}$$

For the first boundary integral equation (12.4.14), namely

$$\pi_t \mathbf{E} = \frac{1}{2}\pi_t \mathbf{E} - \pi_t \widetilde{\boldsymbol{W}}_k \circ \pi_t \mathbf{E} - \pi_t \big[\mathbf{V}_k + \frac{1}{k^2}\mathbf{grad}_{\boldsymbol{x}} \mathbf{V}_k \mathrm{div}_\Gamma\big](\gamma_t \mathbf{curl}\, \mathbf{E}),$$

it is a boundary integral equation of the second kind for $\pi_t \mathbf{E}$ if $\gamma_t \mathbf{curl}\, \mathbf{E}$ is given on Γ. In terms of symbols, the second kind integral equation for $\pi_t \mathbf{E}$ reads

$$\begin{aligned}\frac{1}{2}\widehat{\pi_t \mathbf{E}} &= -\pi_t\big\{\sigma_{\mathbf{V}_k}(\boldsymbol{x}', \boldsymbol{\xi}') + \frac{1}{k^2}\sigma_{\mathbf{grad}_{\boldsymbol{x}} \mathbf{V}_k}(\boldsymbol{x}', \boldsymbol{\xi}')\mathrm{div}_\Gamma(\widehat{\boldsymbol{n} \times \mathbf{curl}\, \mathbf{E}})\big\} \\ &\quad + \mathcal{O}(|\boldsymbol{\xi}'|^{-2})(\widehat{\pi_t \mathbf{E}}, \gamma_t \widehat{\mathbf{curl}\, \mathbf{E}})\end{aligned}$$

for given $\gamma_t \mathbf{curl}\, \mathbf{E}$ on Γ. This corresponds to a classical Fredholm integral equation of the **second** kind. On the other hand, in the second boundary integral equation (12.4.14),

$$-\gamma_t \mathbf{curl}\, \widetilde{\boldsymbol{W}}_k(\pi_t \mathbf{E}) = (\frac{1}{2}I + \gamma_t \boldsymbol{W}_k)(\gamma_t \mathbf{curl}\, \mathbf{E})$$

the unknown $\pi_t \mathbf{E}$ satisfies a boundary integral equation with the corresponding symbols:

$$\begin{aligned}&-\frac{1}{4\pi}\boldsymbol{a}_0^0|\boldsymbol{\xi}'|\begin{pmatrix}1 & 0\\ 0 & 1\end{pmatrix}\widehat{\pi_t \mathbf{E}} \\ &\quad = \big\{\frac{1}{2}\boldsymbol{I} + \gamma_t \sigma_{\boldsymbol{W}_k}(\boldsymbol{x}', \boldsymbol{\xi}')\big\}\gamma_t \widehat{\mathbf{curl}\, \mathbf{E}} + \mathcal{O}|\boldsymbol{\xi}'|^{-1}(\widehat{\pi_t \mathbf{E}}, \gamma_t \widehat{\mathbf{curl}\, \mathbf{E}}).\end{aligned}$$

This is a hypersingular integral equation for $\pi_t \mathbf{E}$ with a coercive principal part and is a Fredholm mapping of index zero.

For the boundary value problem with $\pi_t \mathbf{E}$ on Γ given, the second equation in (12.4.15) will be a **classical Fredholm integral equation** for the unknown ($\gamma_t \mathbf{curl}\, \mathbf{E}$), whereas the first equation (12.4.13) is a **Fredholm equation of the first kind** with Fredholm index zero.

Either one of the boundary integral equations in (12.4.14) can be used for finding the missing Cauchy datum.

A similar analysis can be obtained for the system (12.4.15). For the **EFIE** equations in (6.12.10)

$$\pi_t \mathbf{V}_k \mathbf{J} + \frac{1}{k^2} \mathbf{grad}_\Gamma \{V_k \mathrm{div}_\Gamma \mathbf{J}\} = \boldsymbol{g} \quad \text{for } \boldsymbol{x} \in \Gamma,$$

the Hodge decomposition of $\mathbf{J}$ has the form

$$\mathbf{J} = \mathbf{grad}_\Gamma \varphi + \mathbf{curl}_\Gamma \psi$$

where $\varphi \in \{\varphi \in H^1_*(\Gamma)$ such that $\Delta_\Gamma \varphi \in H^{-1/2}(\Gamma)$ with $\langle \varphi, 1\rangle = 0\}$ and $\psi \in H^{1/2}(\Gamma)$. In terms of φ and ψ we have the equation as

$$\pi_t \mathbf{V}_k (\mathbf{grad}_\Gamma \varphi + \mathbf{curl}_\Gamma \psi) + \frac{1}{k^2} (\mathbf{grad}_\Gamma V_k (\Delta_\Gamma \varphi) = \boldsymbol{g}. \tag{12.7.1}$$

In local coordinates in directions of $\boldsymbol{e}_1$ and $\boldsymbol{e}_2$ in the tangential plane, equation (12.7.1) can be written as

$$\pi_t \mathbf{V}_k (\mathbf{grad}_\Gamma \varphi + \mathbf{curl}_\Gamma \psi) - \frac{1}{k^2} \int\limits_\Gamma (\mathbf{grad}_y G_k(\boldsymbol{x}, \boldsymbol{y})) \Delta_\Gamma \varphi ds_{\boldsymbol{y}} = \boldsymbol{g}. \tag{12.7.2}$$

An existence proof for (12.7.2) has been established in Theorem 6.12.6 which will be revisited in Section 12.10 below.

12.8 Coerciveness and Strong Ellipticity

From the symbol $\sigma_{\boldsymbol{P}}(\boldsymbol{x}', \boldsymbol{\xi}')$ in (12.2.1) of the differential operator $\boldsymbol{P}$ with the determinant (12.2.2) we see that the principal part of the operator has vanishing determinant. Hence, $\boldsymbol{P}$ is **not elliptic** in the sense of Petrovski. For $k = 0$, the determinant of the complete symbol vanishes, thus $\boldsymbol{P}$ is not elliptic either in the sense of Agmon, Douglis, Nirenberg. However, for $k \neq 0$ the inverse symbol matrix of $\boldsymbol{P}$ exists and determines a fundamental solution $\boldsymbol{\Gamma}(\boldsymbol{x}, \boldsymbol{y})$, see (12.2.7).

Moreover, if solutions of the Maxwell system are sought in the subspace of $\mathbf{H}^1(\Omega)$ satisfying $\mathrm{div}\, \mathbf{E} = 0$ i.e., as a solenoidal field, then in that subspace Gårding's inequality is valid and now the system is **strongly elliptic**.

According to M.S.Agranovich (see [5]) the Maxwell system has **"hidden ellipticity"**.

For the Maxwell equations (12.1.1) in Ω and $\operatorname{div}\mathbf{E} = 0$ in Ω, it motivates us to consider the Vektor–Helmholtz equation

$$\mathbf{curl\ curl\ E} - \mathbf{grad}\operatorname{div}\mathbf{E} - k^2\mathbf{E} = \boldsymbol{f} \quad \text{in } \Omega.$$

Costabel considers in [83] (see also [63, p. 60], [96]) the weak variational formulation,

$$\begin{aligned} \boldsymbol{a}_0(\mathbf{E}, \mathbf{E}^t) &= \int\limits_\Omega \{\mathbf{curl\ E} \cdot \mathbf{curl\ E}^t + \operatorname{div}\mathbf{E}\operatorname{div}\mathbf{E}^t - k^2\mathbf{E}\cdot\mathbf{E}^t\} ds_{\boldsymbol{x}} \\ &= \int\limits_\Gamma (\operatorname{div}\mathbf{E})\boldsymbol{n}\cdot\mathbf{E}^t ds - \int\limits_\Gamma \mathbf{E}^t\cdot(\boldsymbol{n}\times\mathbf{curl\ E})ds \end{aligned} \tag{12.8.1}$$

for all test functions $\mathbf{E}^t \in \mathbf{H}^1_n$ or $\mathbf{E}^t \in \mathbf{H}^1_t(\Omega)$. Then $\boldsymbol{a}_0(\mathbf{E}, \mathbf{E})$ satisfies Gårding's inequality

$$\begin{aligned} \boldsymbol{a}_0(\mathbf{E}.\mathbf{E}) &= \int\limits_\Omega \{|\mathbf{curl\ E}|^2 + (\operatorname{div}\mathbf{E})^2 - k^2|\mathbf{E}|^2\}dx \\ &\geq \|\mathbf{E}\|^2_{\mathbf{H}^1_n(\Omega)} - (k^2+1)\|\mathbf{E}\|^2_{L_2(\Omega)} \end{aligned}$$

for all $\mathbf{E}$ in the subspace

$$\mathbf{H}^1_n(\Omega) = \{\boldsymbol{u} \in \mathbf{H}^1(\Omega) \quad \text{with } \gamma_t(\boldsymbol{u}) = \mathbf{0}\},$$

and also satisfies Gårding's inequality

$$\boldsymbol{a}_0(\mathbf{E}, \mathbf{E}) \geq \|\mathbf{E}\|^2_{\mathbf{H}^1_t(\Omega)} - (k^2+1)\|\mathbf{E}\|^2_{L_2(\Omega)}$$

on the second subspace

$$\mathbf{H}^1_t(\Omega) := \{\boldsymbol{u} \in \mathbf{H}^1 \quad \text{with } \gamma_n(\boldsymbol{u}) = 0\} \quad \text{of } \mathbf{H}^1(\Gamma).$$

Hence, the boundary value problems

$$\boldsymbol{P}\mathbf{E} = \mathbf{0} \quad \text{for } \mathbf{E} \in \mathbf{H}^1_t(\Omega)$$

with the natural boundary condition

$$\operatorname{div}\mathbf{E}|_\Gamma = \boldsymbol{g} \in \mathbf{H}^{-1/2}(\Gamma) \quad \text{on } \Gamma \tag{12.8.2}$$

and

$$\boldsymbol{P}\mathbf{E} = \mathbf{0} \quad \text{in } \Omega \quad \text{on the subspace } \mathbf{H}^1_n(\Omega)$$

with the second natural boundary condition

$$\boldsymbol{n} \times \mathbf{curl}\, \mathbf{E} = \boldsymbol{g} \in \mathbf{H}^{-1/2}(\Gamma) \qquad (12.8.3)$$

are uniquely solvable if k^2 is not an eigenvalue of the corresponding homogeneous problems in the subspaces. Therefore Costabel modifies the bilinear forms, where the new natural boundary conditions are not the original standard ones, but the modified bilinear forms now satisfy Gårding's inequalities on the whole space $\mathbf{H}^1(\Gamma)$.

Solenoidal solution fields are also considered by R.C. MacCamy and E.P. Stephan in [276], [277], [278] [166].

For the scattering problem by a perfect conductor from the Stratton–Chu representation formula in Ω^c,

$$\mathbf{E}(\boldsymbol{x}) = i\omega\mu\mathfrak{V}_k(\mathbf{J})(\boldsymbol{x}) - \mathbf{grad}_{\boldsymbol{x}}\mathfrak{V}_k(M)(\boldsymbol{x}) \quad \text{for } \boldsymbol{x} \in \Omega^c \qquad (12.8.4)$$

with $\mathfrak{V}_k\boldsymbol{\lambda}(\boldsymbol{x}) = \int\limits_\Gamma G_k(\boldsymbol{x}-\boldsymbol{y})\boldsymbol{\lambda}(\boldsymbol{y})ds_{\boldsymbol{y}}$, **EFIE** in the form of the boundary condition $\boldsymbol{n} \times \mathbf{E} = -\boldsymbol{n} \times \mathbf{E}^{\text{inc}}$ on Γ, leads to the Equation

$$i\omega\mu\pi_t\mathbf{V}_k\mathbf{J} - \mathbf{grad}_\Gamma V_k M = -\pi_t\mathbf{E}^{\text{inc}} \quad \text{on } \Gamma, \qquad (12.8.5)$$

where $\mathbf{J} = \boldsymbol{n} \times \mathbf{H}$ and $M = \mathbf{E}\cdot\boldsymbol{n}$ are considered as unknowns. Since $\varrho = 0$ in Ω^c and μ and ε are constant, $\mathbf{E}$ is solenoidal, i.e.

$$\operatorname{div}\mathbf{E} = 0 \quad \text{in } \Omega^c$$

which can be used as an additional equation in Ω^c on Γ. Applying div to equation (12.8.4) yields

$$i\omega\mu \operatorname{div}\mathfrak{V}_k(\mathbf{J})(\boldsymbol{x}) + k^2\mathfrak{V}_k(M)(\boldsymbol{x}) = 0 \quad \text{for } \boldsymbol{x} \in \Omega^c$$

in view of $(\Delta_{\boldsymbol{x}} + k^2)G_k(\boldsymbol{x}-\boldsymbol{y}) = 0$ for $\boldsymbol{x} \in \Omega^c$. Taking the limit $\Omega^c \ni \boldsymbol{x} \to \boldsymbol{x}' \in \Gamma$ then yields

$$i\omega\mu V_k(\operatorname{div}_\Gamma\mathbf{J}) + k^2V_k(M) = 0 \quad \text{on } \Gamma. \qquad (12.8.6)$$

This is an additional boundary integral equation for $\mathbf{J}$ and M. In principle, Equations (12.8.5) and (12.8.6) on Γ provide us a system of boundary integral equations for $\mathbf{J}$ and M. However, we may consider a modified system of equations as follows:

Taking the first equation (12.8.5) and applying the surface divergence $\operatorname{div}_\Gamma$ to it yields

$$i\omega\mu\operatorname{div}_\Gamma\pi_t\mathbf{V}_k\mathbf{J} - \Delta_\Gamma V_k M = -\operatorname{div}_\Gamma\pi_t\mathbf{E}^{\text{inc}} \qquad (12.8.7)$$

with Δ_Γ the Laplace–Beltrami operator. Subtracting (12.8.6) from (12.8.7) gives

$$i\omega\mu\{\operatorname{div}_\Gamma\mathbf{V}_k\mathbf{J} - V_k\operatorname{div}_\Gamma\mathbf{J}\} - (\Delta_\Gamma + k^2)V_k M = -\operatorname{div}_\Gamma(\pi_t\mathbf{E}^{\text{inc}}). \qquad (12.8.8)$$

This modified equation (12.8.8) together with the original equation (12.8.5), i.e.

$$i\omega\mu\pi_t\mathbf{V}_k(\mathbf{J}) - \mathbf{grad}_\Gamma V_k(M) = -\pi_t\mathbf{E}^{\text{inc}} \qquad (12.8.9)$$

now forms a **strongly elliptic** system of boundary integral equations on Γ with the triangular principal symbol (see MacCamy and Stephan [277])

$$\sigma(\boldsymbol{\xi}') = i\omega\mu \begin{pmatrix} |\boldsymbol{\xi}'|^2\frac{i}{|\boldsymbol{\xi}'|}, & 0 & 0 \\ i\xi_1|\boldsymbol{\xi}'|^{-1}, & |\boldsymbol{\xi}'|^{-1} & 0 \\ i\xi_2|\boldsymbol{\xi}'|^{-1}, & 0 & |\boldsymbol{\xi}'|^{-1} \end{pmatrix} \qquad (12.8.10)$$

since the commutator

$$\{\mathrm{div}_\Gamma\mathbf{V}_k(\mathbf{J}) - V_k\mathrm{div}_\Gamma\mathbf{J}\}$$

has the principal symbol $\mathbf{0}$, and therefore is of order -1 on Γ mapping $\mathbf{H}^s(\Gamma)$ continuously into $\mathbf{H}^{s+1}(\Gamma)$ and as a mapping $\mathbf{H}^s(\Gamma)$ into $\mathbf{H}^s(\Gamma)$ defines a compact operator for $s > 0$.

As a consequence, the original **EFIE** (12.8.5) together with the modified equation (12.8.5) for the unknowns $\mathbf{J} \in \left(\mathbf{H}^{-1/2}(\Gamma)\right)^2$ and $M \in H^{1/2}(\Gamma)$ is Fredholm with index zero. Hence, uniqueness implies existence of a solution. Uniqueness is guaranteed if k^2 is not an eigenvalue of the interior Dirichlet problem of the Maxwell system. The triangular form of the symbols in (12.8.10) is more amenable than solving the original system from the numerical point of view.

12.9 Gårding's inequality for the sesquilinear form $\mathcal{A}$ in (6.12.23)

Motivated by the work of Buffa, Costabel and Schwab [51] who proved the solvability of Equation (6.12.12) by using a variational saddle point problem for four unknown functions, based on the result by Hildebrandt and Wienholtz [180], we now use our concept of strong ellipticity, Theorem 7.2.7 (see also i.e. Stephan and Wendland [405]), which also refers to Hildebrandt and Wienholtz [180] and tacitly takes care of the invariance of linear systems of differential equations with respect to their isomorphic linear transformations, in order to establish Gårding's inequality for the sesquilinear form $\mathcal{A}$ in (6.12.23). Now, we are using local orthonormal coordinates near Γ with the base $(\boldsymbol{e}_1, \boldsymbol{e}_2, \boldsymbol{n})$ and $\mathbf{grad}_\Gamma\varphi = (\partial_1\varphi)\boldsymbol{e}_1 + (\partial_2\varphi\boldsymbol{e}_2)$, $\mathbf{curl}_\Gamma\psi = (\partial_2\psi)\boldsymbol{e}_1 - (\partial_1\psi)\boldsymbol{e}_2$.

Theorem 12.9.1. *The principal part of the sesquilinear form $\mathcal{A}$ in* (6.12.23), *i.e.*

$$\mathcal{A}_0(\varphi,\psi;\overline{\varphi^t},\overline{\psi^t}) = -\frac{1}{k^2}\langle V_0\Delta_\Gamma\varphi\Delta_\Gamma\overline{\varphi}^t\rangle_\Gamma + \big\langle \pi_t V_0(\mathbf{grad}_\Gamma\varphi + \mathbf{curl}_\Gamma\psi), (\mathbf{grad}_\Gamma\overline{\varphi^t} + \mathbf{curl}_\Gamma\overline{\psi^t})\big\rangle_\Gamma$$

with $\varphi,\varphi^t \in \mathcal{H}(\Gamma) = \{\alpha \in H^1_*(\Gamma) : \Delta_\Gamma\alpha \in H^{-1/2}(\Gamma)$ *with* $\langle\alpha,1\rangle_\Gamma = 0\}$ *and* $(\varphi,\psi;\varphi^t,\psi^t)$ *with* $\varphi,\varphi^t \in \mathcal{H}(\Gamma)$ *and* $\psi,\psi^t \in H^{1/2}(\Gamma)$ *is strongly elliptic satisfying* (7.2.62) *with*

$$\boldsymbol{\Theta} = \begin{pmatrix} -1 & 0 & 0\\ 0 & 1 & 0\\ 0 & 0 & 1\end{pmatrix}, \boldsymbol{\xi} \in \mathbb{R}^2 \quad \textit{and } \boldsymbol{\zeta} \in \mathbb{C}^3:$$

$$\boldsymbol{\zeta}^T \begin{pmatrix} -1 & 0 & 0\\ 0 & 1 & 0\\ 0 & 0 & 0\end{pmatrix} \begin{pmatrix} -\frac{1}{k^2}|\boldsymbol{\xi}| & 0 & 0\\ 0 & \frac{1}{|\boldsymbol{\xi}|}(i\xi_1\boldsymbol{e}_1 + i\xi_2\boldsymbol{e}_2) & 0\\ 0 & 0 & \frac{1}{|\boldsymbol{\xi}|}(i\xi_2\boldsymbol{e}_1 - i\xi_1\boldsymbol{e}_2)\end{pmatrix} :$$

$$\begin{pmatrix} |\boldsymbol{\xi}|^2 & 0 & 0\\ 0 & -(i\xi_1\boldsymbol{e}_1 + i\xi_2\boldsymbol{e}_2) & 0\\ 0 & 0 & -(i\xi_2\boldsymbol{e}_1 - i\xi_1\boldsymbol{e}_2)\end{pmatrix}\overline{\zeta} \tag{12.9.1}$$

$$= \zeta_1|\boldsymbol{\xi}|^3\frac{1}{k^2}\overline{\zeta_1} + \zeta_2|\boldsymbol{\xi}|\overline{\zeta_2} + \zeta_3|\boldsymbol{\xi}|\overline{\zeta_3} \geq c_0|\boldsymbol{\zeta}|^2 \quad \textit{for } |\boldsymbol{\xi}| = 1$$

and with $c_0 = \min\left\{\frac{1}{k^2},1\right\} > 0$.

The system of differential operators on Γ:

$$\boldsymbol{\mathcal{L}}(\varphi,\psi) = -\frac{1}{k^2}V_0\Delta_\Gamma\varphi + \pi_t V_0(\mathbf{grad}_\Gamma\varphi + \mathbf{curl}_\Gamma\psi)$$

is strongly elliptic with $\boldsymbol{\Theta} = \begin{pmatrix} -1 & 0 & 0\\ 0 & 1 & 0\\ 0 & 0 & 1\end{pmatrix}$ and $\sigma = 0$ according to Theorem 7.2.7. [1]

Proof: For the proof we use the sesquilinear form $\boldsymbol{\mathcal{A}}$ defined by (6.12.23), which includes explicitly the principal part of $\boldsymbol{\mathcal{A}}$. The difference

$$\boldsymbol{\mathcal{R}} := \boldsymbol{\mathcal{A}} - \mathbf{B}_0 + k^{-2}\boldsymbol{C}_0$$

is generated by operators of lower order and continuous from $\mathcal{H}\times\mathbf{H}_{||}^{-1/2}(\Gamma)\times\mathbf{H}_{||}^{-1/2}(\Gamma)^2$ into $\{\varphi \in H^{3/2}(\Gamma) : \Delta_\Gamma\varphi \in H^1(\Gamma), \langle\varphi,1\rangle = 0\}\times\mathbf{H}^1_{||}(\Gamma)\times\mathbf{H}^1_{||}(\Gamma)$. This image is compactly imbedded into the space of arguments due to the

[1] In Theorem 7.2.7, the operator $\mathcal{L}$ is denoted by A and α_0 by γ.

Rellich lemma [362], Theorem 4.1.6. Hence, $\boldsymbol{\mathcal{R}}$ is represented by a compact linear operator. Therefore it suffices to consider the principal part of $\boldsymbol{\mathcal{A}}$ given in (6.12.23). With the tangential local base $(\boldsymbol{e}_1, \boldsymbol{e}_2, \boldsymbol{n})$, corresponding local coordinates and the matrix $\boldsymbol{\Theta}$, the principal symbol is given in (12.9.1). Hence, the principal part is strongly elliptic as defined by Stephan and Wendland [405], a simplified version of the conditions by Hildebrandt and Wienholtz [180]. Moreover, the modified sesquilinear form $\boldsymbol{\mathcal{A}}_{0+}$ of the principal part of $\boldsymbol{\mathcal{A}}$,

$$\boldsymbol{\mathcal{A}}_{0+} := \left\langle (-\Delta_\Gamma\varphi, \mathbf{grad}\,_\Gamma\varphi, \mathbf{curl}\,_\Gamma\psi)\boldsymbol{\Theta}\begin{pmatrix} k^{-2}V_0\Delta_\Gamma\overline{\varphi}^t \\ \pi_t\mathbf{V}_0\,\mathbf{grad}\,_\Gamma\overline{\varphi}^t \\ \pi_t\mathbf{V}_0\,\mathbf{curl}\,_\Gamma\overline{\psi}^t \end{pmatrix}\right\rangle_\Gamma$$

satisfies the coerciveness inequality

$$\begin{aligned} &\left\langle (-\Delta_\Gamma\varphi, \mathbf{grad}\,_\Gamma\varphi, \mathbf{curl}\,_\Gamma\psi)\boldsymbol{\Theta}\begin{pmatrix} k^{-2}V_0\Delta_\Gamma\overline{\varphi} \\ \pi_t\mathbf{V}_0\,\mathbf{grad}\,_\Gamma\overline{\varphi} \\ \pi_t\mathbf{V}_0\,\mathbf{curl}\,_\Gamma\overline{\psi} \end{pmatrix}\right\rangle_\Gamma \\ &\geq c_1^V\{k^{-2}\|\Delta_\Gamma\varphi\|^2_{H^{-1/2}(\Gamma)} + \|\mathbf{grad}\,_\Gamma\varphi\|^2_{H_{\|}^{-1/2}(\Gamma)} + \|\mathbf{curl}\,_\Gamma\psi\|^2_{H_{\|}^{-1/2}(\Gamma)}\} \end{aligned} \tag{12.9.2}$$

with the coerciveness constant $c_1^V > 0$ of the scalar single layer potential operator V_0 in (5.6.32) which is independent of k.

$$\begin{aligned} \mathrm{Re}\Big\{&\big\langle(\Delta_\Gamma\varphi, (\mathbf{grad}\,_\Gamma\varphi + \mathbf{curl}\,_\Gamma\psi)), (k^{-2}V_0\Delta_\Gamma\overline{\varphi}, \pi_t\mathbf{V}_0(\mathbf{grad}\,_\Gamma\overline{\varphi} + \mathbf{curl}\,_\Gamma\overline{\psi})\big\rangle \\ &+\big\langle(\mathbf{grad}\,_\Gamma\varphi + \mathbf{curl}\,_\Gamma\psi)\,,\ \boldsymbol{C}_1(\mathbf{grad}\,_\Gamma\overline{\varphi} + \mathbf{curl}\,_\Gamma\overline{\psi})\big\rangle\Big\} \\ \geq c_0\|&\Delta_\Gamma\varphi\|^2_{H^{-1/2}(\Gamma)} + \|\mathbf{grad}\,_\Gamma\varphi\|^2_{H_{\|}^{-1/2}(\Gamma)} + \|\mathbf{curl}\,_\Gamma\psi\|\|^2_{H_{\|}^{-1/2}(\Gamma)} \end{aligned}$$

with the compact mapping $\boldsymbol{C}_1(\bullet,\bullet) = \boldsymbol{C}(\bullet,\bullet) - \pi_t\mathbf{V}_0\mathbf{grad}\,_\Gamma\overline{\varphi}$. This completes the proof of Theorem 12.9.1. ∎

Remark 12.9.1: As shown in Lemma 6.12.3, in the space $\varphi, \varphi^t \in \mathcal{H}(\Gamma) \wedge \mathbf{grad}\,_\Gamma\varphi \in L_2(\Gamma)$, and $\mathbf{grad}\,_\Gamma\varphi^t \in H_{\|}^{-1/2}(\mathrm{div}_\Gamma, \Gamma)$, the sesquilinear form $\langle\pi_t\mathbf{V}_k\mathbf{grad}\,_\Gamma\overline{\varphi}^t\rangle$ is compact and therefore can be neglected in (12.9.2). However, (12.9.2) can still be used for computational purposes.

The saddle point formulation by Buffa, Costabel and Schwab [51]. For the EFIE saddle point formulation (6.12.25) in

$$\boldsymbol{\mathcal{X}} = H^1(\Gamma)/\mathbb{C} \times H^{1/2}(\Gamma)/\mathbb{C} \times H_*^{-1/2} \times H^1(\Gamma/\mathbb{C})$$

with

$$\boldsymbol{\Theta} = \begin{pmatrix} 0 & 0 & 0 & -1 \\ 0 & 1 & 0 & 0 \\ 0 & 0 & -1 & 0 \\ -1 & 0 & 0 & 0 \end{pmatrix}$$

the principal part of the system (6.12.26)

$$\begin{aligned}
&\mathfrak{B}_0\big((\varphi,\psi,m,\lambda)\,,\,\boldsymbol{\Theta}(\overline{\varphi},\overline{\psi},\overline{m},\overline{\lambda})\big)\\
&= \frac{i}{\varepsilon\omega}\langle m,\pi_t,V_0\overline{m}\rangle + i\omega\mu\langle \mathbf{curl}\,_\Gamma\psi,\pi_t\mathbf{V}_0\overline{\mathbf{curl}\,_\Gamma\psi}\rangle\\
&\quad-\langle \mathbf{grad}\,_\Gamma\lambda,\overline{\mathbf{grad}\,_\Gamma\lambda}\rangle - \langle \mathbf{grad}\,_\Gamma\varphi,\overline{\mathbf{grad}\,_\Gamma\varphi}\rangle
\end{aligned}$$

has the symbol

$$\begin{aligned}
\boldsymbol{\sigma}(\boldsymbol{\xi}',\boldsymbol{\zeta}) &= i\{\frac{1}{\varepsilon\omega}\zeta_1\frac{1}{|\boldsymbol{\xi}'|}\overline{\zeta}_1 - \omega\mu\zeta_3(i\xi_2'-i\xi_1')\frac{1}{|\boldsymbol{\xi}'|}(i\xi_2'-i\xi_1')\overline{\zeta}_3\}\\
&\quad-\{\zeta_4(i\xi_1'+i\xi_2')\cdot(i\xi_1'+i\xi_2')\overline{\zeta}_4+\zeta_2(i\xi_1'+i\xi_2')\cdot(i\xi_1'+i\xi_2')\overline{\zeta}_2\}\\
&= i\{\frac{1}{\epsilon\omega}\,\frac{1}{|\boldsymbol{\xi}'|}|\zeta_1|^2+\omega\mu|\boldsymbol{\xi}'|\;|\zeta_3|^2\}+\{|\boldsymbol{\xi}'|^2|\zeta_4|^2+|\boldsymbol{\xi}'|^2|\zeta_2|^2\}
\end{aligned}$$

Hence real and imaginary parts of $\mathfrak{B}_0$ are both strongly elliptic and satisfy the coerciveness inequalities (6.12.28).

12.10 Existence Theorem 6.12.6 revisited

Motivated by the work of Buffa, Costabel and Schwab [51] we now use the pseudodifferential equations corresponding to (6.12.18) for defining a system of boundary integral equations which is strongly elliptic and admits a unique solution if k^2 is not an eigenvalue of the interior Dirichlet problem.

Applying Fourier transform we obtain to the system of boundary integral equations (6.12.18), the equations

$$\begin{aligned}
|\boldsymbol{\xi}'|^{-1}i\xi_1\widehat{\varphi}+|\boldsymbol{\xi}'|^{-1}i\xi_2\widehat{\psi}-\frac{1}{k^2}|\boldsymbol{\xi}'|i\xi_1\widehat{\varphi}&=\widehat{g}_1,\\
|\boldsymbol{\xi}'|^{-1}i\xi_2\widehat{\varphi}-|\boldsymbol{\xi}'|^{-1}i\xi_1\widehat{\psi}-\frac{1}{k^2}|\boldsymbol{\xi}'|i\xi_2\widehat{\varphi}&=\widehat{g}_2.
\end{aligned}\tag{12.10.1}$$

This system also can be written in terms of the symbol matrix as

$$\begin{pmatrix}|\boldsymbol{\xi}'|^{-1}i\xi_1-\frac{1}{k^2}|\boldsymbol{\xi}'|i\xi_1, & |\boldsymbol{\xi}'|^{-1}i\xi_2\\ |\boldsymbol{\xi}'|^{-1}i\xi_2-\frac{1}{k^2}|\boldsymbol{\xi}'|i\xi_2, & -|\boldsymbol{\xi}'|^{-1}i\xi_1\end{pmatrix}\begin{pmatrix}\widehat{\varphi}\\ \widehat{\psi}\end{pmatrix}=\begin{pmatrix}\widehat{g}_1\\ \widehat{g}_2\end{pmatrix}.$$

The determinant of the symbol matrix is given by

$$-\Big(1-\frac{1}{k^2}|\boldsymbol{\xi}'|^2\Big)>0 \quad\text{for}\quad |\boldsymbol{\xi}'|^2>k^2.$$

Hence, the system defines a strongly elliptic system of boundary integral operators on Γ. The principal part is given by

$$\begin{pmatrix} -\frac{1}{k^2}|\boldsymbol{\xi}'|i\xi_1, & |\boldsymbol{\xi}'|^{-1}i\xi_2 \\ -\frac{1}{k^2}|\boldsymbol{\xi}'|i\xi_2, & -|\boldsymbol{\xi}'|^{-1}i\xi_1 \end{pmatrix} \begin{pmatrix} \widehat{\varphi} \\ \widehat{\psi} \end{pmatrix} + \mathcal{O}(|\boldsymbol{\xi}'|^0)\widehat{\varphi} + \mathcal{O}(|\boldsymbol{\xi}'|^{-1})\widehat{\psi} = \begin{pmatrix} \widehat{g}_1 \\ \widehat{g}_2 \end{pmatrix}.$$

and multiplying the first row with $k^2 i\xi_1$ and the second row with $k^2 i\xi_2$ yields after summation

$$|\boldsymbol{\xi}'|^3\widehat{\varphi} = k^2(i\xi_1\widehat{g}_1 + i\xi_2\widehat{g}_2) + k^2\mathcal{O}(|\boldsymbol{\xi}'|)\widehat{\varphi} + k^2\mathcal{O}(|\boldsymbol{\xi}'|^0)\widehat{\psi}.$$

Similarly, by multiplying the first row with $-i\xi_2$, the second row with $i\xi_1$ and summation one finds

$$|\boldsymbol{\xi}'|\widehat{\psi} = (-i\xi_2\widehat{g}_1 + i\xi_1\widehat{g}_2 + \mathcal{O}(|\boldsymbol{\xi}'|^1)\widehat{\varphi} + \mathcal{O}(|\boldsymbol{\xi}'|^0)\widehat{\psi}$$

Hence,

$$\begin{aligned} \widehat{\varphi} &= k^2|\boldsymbol{\xi}'|^{-3}(i\xi_1\widehat{g}_1 + i\xi_2\widehat{g}_2) + k^2\mathcal{O}(|\boldsymbol{\xi}'|^{-2})\widehat{\varphi} + k^2\mathcal{O}(i|\boldsymbol{\xi}|^{-3})\widehat{\psi}, \\ \widehat{\psi} &= |\boldsymbol{\xi}'|^{-1}(-i\xi_2\widehat{g}_1 + i\xi_1\widehat{g}_2) + \mathcal{O}(|\boldsymbol{\xi}'|^0)\widehat{\varphi} + \mathcal{O}(|\boldsymbol{\xi}'|^{-1}\widehat{\psi} \end{aligned}$$

which indicates how to solve the boundary integral equations (6.12.18) and (6.12.21) or (12.10.1) for φ and ψ.

By using a local parametrization in the vicinity of $\boldsymbol{x}'$, we have

$$\mathbf{J} = \mathbf{grad}_{\Gamma}\varphi + \mathbf{curl}_{\Gamma}\psi$$

and **EFIE** takes the form

$$\begin{aligned} &\pi_t\mathbf{V}\Big\{\mathbf{grad}_{\Gamma}\varphi + \mathbf{curl}_{\Gamma}\psi\Big\} \\ &+\frac{1}{k^2}\Big\{\mathbf{grad}_{\Gamma}V\mathrm{div}_{\Gamma}(\mathbf{grad}_{\Gamma}\varphi + \mathbf{curl}_{\Gamma}\psi)\Big\} = \boldsymbol{g} \end{aligned} \tag{12.10.2}$$

Applying div_Γ to (12.10.2) and then testing with φ^t gives

$$\langle\varphi^t, \mathrm{div}_\Gamma\mathbf{V}\{\mathbf{grad}_{\Gamma}\varphi+\mathbf{curl}_{\Gamma}\psi\}\rangle + \frac{1}{k^2}\langle\varphi^t, \mathrm{div}_\Gamma\mathbf{grad}_{\Gamma}V\Delta_\Gamma\varphi\rangle = \langle\varphi^t, \mathrm{div}_\Gamma\boldsymbol{g}\rangle.$$

Integration by parts yields

$$-\big\langle\mathbf{grad}_{\Gamma}\varphi^t, V\{\mathbf{grad}_{\Gamma}\varphi + \mathbf{curl}_{\Gamma}\psi\}\big\rangle + \frac{1}{k^2}\langle\Delta_\Gamma\varphi^t, V\Delta_\Gamma\varphi\rangle = \langle\varphi^t, \mathrm{div}_\Gamma\boldsymbol{g}\rangle. \tag{12.10.3}$$

Combining Appendix $\big(\mathbf{V}(7)\big)$ with $\big(\boldsymbol{IV}(5)\big)$ and $\big(\mathbf{V}(4)\big)$ gives

$$\langle\mathbf{curl}_{\Gamma}\psi^t, \boldsymbol{v}\rangle = \langle\psi^t, \mathrm{curl}_\Gamma\boldsymbol{v}\rangle.$$

Hence, applying curl_Γ to (12.10.2) and then testing with ψ^t gives with $\mathrm{curl}_\Gamma\,\mathbf{grad}_{\Gamma} = 0$ the equation

$$\langle\mathbf{curl}_{\Gamma}\psi^t, \mathbf{V}\mathbf{curl}_{\Gamma}\psi\rangle + \langle\mathbf{curl}_{\Gamma}\psi^t, \mathbf{V}\mathbf{grad}_{\Gamma}\varphi\rangle = \langle\psi^t, \underset{\Gamma}{\mathrm{curl}}\,\boldsymbol{g}\rangle. \tag{12.10.4}$$

After multiplication of (12.10.3) with k^2 and replacing φ^t by $\overline{\varphi}^t$ we obtain the sesquilinear form

$$\langle \Delta_\Gamma, \overline{\varphi}^t, V\Delta_\Gamma\varphi\rangle - k^2\langle \mathbf{grad}\,_\Gamma\overline{\varphi}^t, V\mathbf{grad}\,_\Gamma\varphi\rangle - k^2\langle \mathbf{grad}\,_\Gamma\overline{\varphi}^t, V\mathbf{curl}\,_\Gamma\psi\rangle$$
$$= k^2\langle\overline{\varphi}^t, \mathrm{div}_\Gamma \boldsymbol{g}\rangle. \qquad (12.10.5)$$

It is easy to see that the sesquilinear form on the left is continuous for φ and $\varphi^t \in H^{3/2}(\Gamma)$ and $\psi \in H^{1/2}(\Gamma)$. Moreover, due to (5.6.32) the principal part is coercive:

$$\mathrm{Re}\,\langle \Delta_\Gamma\overline{\varphi}, V_\Gamma\Delta_\Gamma\varphi\rangle \geq c_v \parallel \Delta_\Gamma\varphi \parallel^2_{H^{-1/2}(\Gamma)}$$

and for

$$\varphi \in H^{3/2}(\Gamma)\,,\ \mathbf{grad}\,_\Gamma\varphi \in \mathbf{H}^{1/2}(\Gamma) \Subset (H^{-1/2}(\Gamma))^3$$

which is, due to Rellich's theorem, compactly imbedded into $\mathbf{H}^{-1/2}(\Gamma)$. Finally, the commutator $V\circ\mathbf{curl}\,_\Gamma - \mathbf{curl}\,_\Gamma\circ V$ is a pseudodifferential operator of degree -1 on Γ, hence, $\psi \in H^{1/2}(\Gamma)$ is mapped into $H^{3/2}(\Gamma)$ which, due to Rellich, is compactly imbedded into $H^{1/2}(\Gamma)$. Moreover; $\mathbf{curl}\,_\Gamma\circ\mathbf{grad}\,_\Gamma = \mathbf{0}$.

For the second equation (12.10.4), in the sesquilinear form, i.e.

$$\langle \mathbf{curl}\,_\Gamma\overline{\psi}^t, \mathbf{V}\,\mathbf{curl}\,_\Gamma\psi\rangle + \langle \mathbf{curl}\,_\Gamma\overline{\psi}^t, \mathbf{V}\mathbf{grad}\,_\Gamma\varphi\rangle = \langle\overline{\psi}^t, \mathrm{curl}_\Gamma\boldsymbol{g}\rangle \qquad (12.10.6)$$

we also use the coerciveness of $\mathbf{V}$,

$$\mathrm{Re}\,\langle \mathbf{curl}\,_\Gamma\overline{\psi}, \mathbf{V}\,\mathbf{curl}\,_\Gamma\psi\rangle \geq c_{\mathrm{v}} \parallel \mathbf{curl}\,_\Gamma\psi \parallel^2_{\mathbf{H}^{-1/2}(\Gamma)}\,.$$

Again, the commutator $\mathbf{curl}\,_\Gamma\circ\mathbf{V} - \mathbf{V}\circ\mathbf{curl}\,_\Gamma$ is a pseudodifferential operator of degree -1 on Γ whereas $\mathrm{curl}_\Gamma\circ\mathbf{grad}\,_\Gamma = \mathbf{0}$.

In conclusion we see that the system of equations (12.10.3) and (12.10.4) satisfies Gårding's inequality

$$\mathrm{Re}\Big\{\langle \Delta_\Gamma\varphi, V\Delta_\Gamma\varphi\rangle - k^2\langle \mathbf{grad}\,_\Gamma\overline{\varphi}, \mathbf{V}\,\mathbf{grad}\,_\Gamma\varphi\rangle - k^2\langle \mathbf{grad}\,_\Gamma\overline{\varphi}, \mathbf{V}\mathbf{curl}\,_\Gamma\psi\rangle$$
$$+ \langle \mathbf{curl}\,_\Gamma\overline{\psi}, \mathbf{V}\,\mathbf{curl}\,_\Gamma\psi\rangle + \langle \mathbf{curl}\,_\Gamma\overline{\psi}, \mathbf{V}\mathbf{grad}\,_\Gamma\varphi\rangle\Big\}$$
$$= \mathrm{Re}\,\big\{\langle \Delta_\Gamma\overline{\varphi}, V\Delta_\Gamma\varphi\rangle + \langle \mathbf{curl}\,_\Gamma\overline{\psi}, V\mathbf{curl}\,_\Gamma\psi\rangle + \mathfrak{C}_{k^2}\big((\overline{\varphi},\varphi)\,;\,(\overline{\psi},\psi)\big)$$
$$\geq c(k)\big\{\|\varphi\|^2_{H^{3/2}(\Gamma)/\mathbb{C}} + \|\psi\|^2_{H^{3/2}(\Gamma)/\mathbb{C}}\big\} + \mathfrak{C}_{\boldsymbol{k^2}}\big((\overline{\varphi},\varphi);(\overline{\psi},\psi)\big)$$

where $\mathfrak{C}_{\boldsymbol{k^2}}\big((\overline{\varphi},\varphi)\,;\,(\overline{\psi},\psi)\big)$ denotes a sesquilinear form generated by a compact linear operator from $H^{3/2}(\Gamma)/\mathbb{C}\times H^{1/2}(\Gamma)/\mathbb{C}$ into $\big(H^{1/2}(\Gamma)\big)^2$. Hence, for given $\boldsymbol{g}\in\mathbf{H}^{1/2}(\Gamma)$, the variational equations (12.10.5), (12.10.6) with arbitrary test functions $\varphi^t\in H^{3/2}(\Gamma)\,,\ \psi^t\in H^{1/2}(\Gamma)$ are Fredholm with index zero. Since for $\varphi = c_1\,,\ \psi = c_2$ the homogeneous equation (12.10.2) ($\boldsymbol{g}=\mathbf{0}$) is satisfied, $\varphi_0 = c_1\in\mathbb{C}$ and $\psi_0 = c_2\in\mathbb{C}$ define the null space of (12.10.2) whereas the right hand side $\mathrm{div}_\Gamma\boldsymbol{g}$ in (12.10.3) and $\mathrm{curl}_\Gamma\,\boldsymbol{g}$ in (12.10.4) automatically satisfy the orthogonality conditions $\langle c_1, \mathrm{div}_\Gamma\boldsymbol{g}\rangle = -\langle\mathbf{grad}\,_\Gamma c_1, \boldsymbol{g}\rangle = 0$ and $\langle c_2, \mathrm{curl}_\Gamma\,\boldsymbol{g}\rangle = \langle\mathrm{curl}_\Gamma\,c_2, \boldsymbol{g}\rangle = 0$.

To conclude this section, as long as k^2 is not an eigenvalue of the interior Dirichlet problem in Ω, the sesquilinear equation (12.10.6) admits a unique solution $(\varphi, \psi) \in H^{3/2}(\Gamma)/\mathbb{C} \times H^{1/2}(\Gamma)/\mathbb{C}$ and the scattered solution in Ω^c is given by (12.10.5) as

$$\mathbf{E}(\boldsymbol{x}) = \mathbf{E}^i(\boldsymbol{x}) + i\omega\mu^+ \int_\Gamma G_k(\boldsymbol{x} - \boldsymbol{y})\{\mathbf{grad}_\Gamma\varphi + \mathbf{curl}_\Gamma\psi\}ds_{\boldsymbol{y}}$$
$$- \frac{i}{\omega\varepsilon}\int_\Gamma \mathbf{grad}_{\boldsymbol{y}} G_k(\boldsymbol{x} - \boldsymbol{y})\Delta_\Gamma\varphi ds_{\boldsymbol{y}}\,, \; \boldsymbol{x} \in \Omega^c.$$

12.11 Concluding Remarks

The class of pseudodifferential operators is essentially the smallest algebra of operators which contains all differential operators, all fundamental solutions of elliptic differential operators and all integral operators with pseudohomogeneous kernel expansions. The linear pseudodifferential operators can be characterized by generalized Fourier multipliers, known as *symbols*. The development of the theory of pseudodifferential operators has made it possible to provide a unified treatment of differential and integral operators. These materials are subjects of Chapter 7 and Chapter 8. However, for a quick glance of the materials, we refer the reader to our paper [206] with emphasis on boundary integral equations recast as pseudodifferential equations.

Chapter 12 is a newly added chapter in this second edition. This chapter together with Chapters 10 and 11 are dealing with boundary integral operators arising in potential flow, acoustics, elasticity, Stokes flow, and electromagnetics, which serve as concrete examples illustrating the basic ideas how one may apply the theory of pseudodifferential operators and their calculus to obtain basic properties for the corresponding boundary integral operators.

A. Differential Operators in Local Coordinates with Minimal Differentiability

As we mentioned earlier, the special coordinate transformation (3.3.7) contains $n(\sigma')$ which is defined by derivations of Γ involving one order of differentiability more than that of Γ. Therefore, here we use a slightly different coordinate–transformation which requires no more differentiability than that of the surface Γ as was introduced via the representation in Section 3.3, namely

$$x = x_{(r)} + T_{(r)}(\tau', a_{(r)}(\tau') + \tau_n) = \Psi_{(r)}(\tau) \tag{A.1}$$

and its inverse

$$\tau := (\tau', \tau_n) = \Phi(x)$$

where

$$\begin{aligned} \tau' &= (T_{(r)}^{-1}(x - x_{(r)}))', \\ \tau_n &= \Phi_n(x) := (T_{(x)}^{-1}(x - x_{(r)}))_n - a_{(r)}\left(\left(T_{(r)}^{-1}(x - x_{(r)})\right)'\right) \\ &\quad \text{for } x \in B_{(r)} . \end{aligned} \tag{A.2}$$

In what follows, without loss of generality we assume that $T_{(r)}$ is an *orthonormal matrix* $((T_{jk}))$ associated with the point $x_{(r)}$ by the column vectors

$$T_{\bullet n} = \boldsymbol{n}(x_{(r)})$$

and $T_{\bullet\lambda}, \lambda = 1, \ldots, n-1$, where the latter form an orthonormal basis of the tangent plane to Γ at $x_{(r)}$. Then the inverse of $T_{(r)}$ is given by its transpose with entries

$$(T_{(r)}^{-1})_{jk} = T_{kj} \,. \tag{A.3}$$

For the special choice of the local coordinates in (A.1), the parameters σ from the previous sections and the parameters τ are related by a diffeomorphism given by the equations

$$\begin{aligned} \tau_\kappa &= \sum_{\ell=1}^{n} T_{\ell\kappa}(y_\ell(\sigma') - x_{(r)\ell} + \sigma_n n_\ell(\sigma')) \text{ for } \kappa = 1, \ldots, n-1 \,, \\ \tau_n &= -a_{(r)}(\tau') + \sum_{\ell=1}^{n} T_{\ell n}\left(y_\ell(\sigma') - x_{(r)\ell} + \sigma_n n_\ell(\sigma')\right) . \end{aligned}$$

G. C. Hsiao und W. L. Wendland, *Boundary Integral Equations*,
Applied Mathematical Sciences 164,
https://doi.org/10.1007/978-3-030-71127-6

For the local coordinates (A.1), the Jacobian of the corresponding transformation is given by

$$\frac{\partial x_k}{\partial \tau_\mu} = T_{k\mu} + T_{kn}\frac{\partial a_{(r)}}{\partial \tau_\mu} \quad \text{for } \mu = 1,\dots,n-1\,, \quad \frac{\partial x_k}{\partial \tau_n} = T_{kn} \tag{A.4}$$

and, hence, the coefficients of the Riemannian metric are given by

$$\begin{aligned} g_{\mu\lambda} &= \delta_{\mu\lambda} + \frac{\partial a_{(r)}}{\partial \tau_\mu}\frac{\partial a_{(r)}}{\partial \tau_\lambda} \\ & \qquad\qquad\qquad\qquad \text{for } \lambda,\mu = 1,\dots,n-1\,, \\ g_{\mu n} &= \frac{\partial a_{(r)}}{\partial \tau_\mu}\,, \qquad g_{nn} = 1\,. \end{aligned} \tag{A.5}$$

Since the inverse transformation is explicitly given by (A.2), we find with (A.3)

$$\begin{aligned} \frac{\partial \tau_\lambda}{\partial x_q} &= T_{q\lambda}\,, \\ \frac{\partial \tau_n}{\partial x_q} &= T_{qn} - \sum_{\varrho=1}^{n-1} T_{q\varrho}\frac{\partial a_{(r)}}{\partial \tau_\varrho} \end{aligned} \tag{A.6}$$

for $\lambda = 1,\dots,n-1$ and $q = 1,\dots,n$. Therefore,

$$g^{\ell q} = \sum_{j=1}^{n} \frac{\partial \tau_\ell}{\partial x_j}\frac{\partial \tau_q}{\partial x_j} \tag{A.7}$$

and we obtain with (A.6)

$$\begin{aligned} g^{\lambda\nu} &= \delta^{\lambda\nu}\,,\; g^{\lambda n} = -\frac{\partial a_{(r)}}{\partial \tau_\lambda} \text{ for } \lambda,\nu = 1,\dots,n-1\,, \\ g^{nn} &= 1 + b^2 \text{ where } b^2 = \sum_{\varrho=1}^{n-1}\left(\frac{\partial a_{(r)}}{\partial \tau_\varrho}\right)^2 ;\text{ and } g = 1\,. \end{aligned} \tag{A.8}$$

For the Riemannian metric of Γ, we find

$$\gamma_{\mu\lambda} = \delta_{\mu\lambda} + \frac{\partial a_{(r)}}{\partial \tau_\mu}\frac{\partial a_{(r)}}{\partial \tau_\lambda} \text{ for } \lambda,\mu = 1,\dots,n-1; \tag{A.9}$$

and with elementary algebra

$$\gamma = 1 + b^2 \tag{A.10}$$

and

$$\gamma^{\lambda\nu} = \delta^{\lambda\nu} - \frac{1}{1+b^2}\frac{\partial a_{(r)}}{\partial \tau_\lambda}\frac{\partial a_{(r)}}{\partial \tau_\nu}\,. \tag{A.11}$$

The transformation of differential operators can be obtained from

$$\frac{\partial u}{\partial x_q} = \sum_{\ell=1}^{n} \frac{\partial \tau_\ell}{\partial x_q} \frac{\partial u}{\partial \tau_\ell}$$

by using (A.6) or Lemma 3.4.1; in either case we find

$$\begin{aligned} \frac{\partial u}{\partial x_q} &= \sum_{\mu=1}^{n-1} T_{q\mu} \frac{\partial u}{\partial \tau_\mu} + \left(T_{qn} - \sum_{\lambda=1}^{n-1} T_{q\lambda} \frac{\partial a_{(r)}}{\partial \tau_\lambda} \right) \frac{\partial u}{\partial \tau_n} \\ &= \sum_{\mu=1}^{n-1} T_{q\mu} \frac{\partial u}{\partial \tau_\mu} + \frac{\partial \Phi_n}{\partial x_q} \frac{\partial u}{\partial \tau_n} . \end{aligned} \tag{A.12}$$

Since Γ is given implicitly by $\Phi_n(x) = 0$, the normal vector $\boldsymbol{n}(x)$ can be extended to a vector–field $\underset{\sim}{\boldsymbol{n}}(x)$ in $B_{(r)} \subset \mathbb{R}^n$ by

$$\underset{\sim}{\boldsymbol{n}}(x) := \nabla \Phi_n(x) / |\nabla \Phi_n(x)| . \tag{A.13}$$

In terms of $\underset{\sim}{\boldsymbol{n}}(x)$, then (A.12) can be written as

$$\frac{\partial u}{\partial x_q} = \sum_{\mu=1}^{n-1} T_{q\mu} \frac{\partial u}{\partial \tau_\mu} + \underset{\sim}{n}_q(x) |\nabla \Phi_n(x)| \frac{\partial u}{\partial \tau_n} . \tag{A.14}$$

Correspondingly, we define

$$\frac{\partial u}{\partial \underset{\sim}{n}} := \sum_{q=1}^{n} \underset{\sim}{n}_q(x) \frac{\partial u}{\partial x_q} ;$$

and we find with (A.14) the relation

$$\frac{\partial u}{\partial \underset{\sim}{n}} := \sum_{\ell=1}^{n} \sum_{\mu=1}^{n-1} \underset{\sim}{n}_\ell(x) T_{\ell\mu} \frac{\partial u}{\partial \tau_\mu} + \left| \nabla \Phi_n(x) \right| \frac{\partial u}{\partial \tau_n} .$$

Hence, $|\nabla \Phi_n(x)| \frac{\partial u}{\partial \tau_n}$ can be inserted into (A.14), and we finally obtain

$$\begin{aligned} \frac{\partial u}{\partial x_q} &= \sum_{\lambda=1}^{n-1} \Big(T_{q\lambda} - \underset{\sim}{n}_q(x) \sum_{\ell=1}^{n} \underset{\sim}{n}_\ell(x) T_{\ell\lambda} \Big) \frac{\partial u}{\partial \tau_\lambda} + \underset{\sim}{n}_q(x) \frac{\partial u}{\partial \underset{\sim}{n}} \\ &=: \Big(\underset{\sim}{\mathcal{D}}_q + \underset{\sim}{n}_q \frac{\partial}{\partial \underset{\sim}{n}} \Big) u \end{aligned} \tag{A.15}$$

where

$$\underset{\sim}{\mathcal{D}}_q := \sum_{\lambda=1}^{n-1} \Big(T_{q\lambda} - \sum_{\ell=1}^{n} \underset{\sim}{n}_\ell(x) T_{\ell\lambda} \Big) \frac{\partial}{\partial \tau_\lambda} \tag{A.16}$$

is again a tangential differential operator as in (3.4.33). We proceed in the same manner as for the operator $(\mathcal{D} + \boldsymbol{n}\frac{\partial}{\partial n})^\alpha$ in (3.4.41) by defining

$$D^\alpha = \Big(\underset{\sim}{\mathcal{D}} + \underset{\sim}{n}(x)\frac{\partial}{\partial \underset{\sim}{n}}\Big)^\alpha u = \sum_{k=0}^{|\alpha|} \underset{\sim\alpha,k}{\mathcal{C}} \frac{\partial^k u}{\partial \underset{\sim}{n}^k} \tag{A.17}$$

where

$$\underset{\sim\alpha,k}{\mathcal{C}} u = \sum_{|\gamma|\leq|\alpha|-k} \underset{\sim\alpha,\gamma,k}{c}(x)\Big(\frac{\partial}{\partial\tau'}\Big)^\gamma u\,. \tag{A.18}$$

It turns out that the coefficients $\underset{\sim\alpha,\gamma,k}{c}$ and operators $\underset{\sim\alpha,k}{\mathcal{C}}$ can be defined by the same recursion formulae as for $c_{\alpha,\gamma,k}$ and $\mathcal{C}_{\alpha,k}$ in (3.4.43)–(3.4.46), (3.4.48). However, instead of (3.4.47), we now have

$$\begin{aligned}\underset{\sim\beta,\varrho,k}{c}(x) &= \sum_{\lambda=1}^{n-1}\Big(T_{\ell\lambda} - \underset{\sim\ell}{n}(x)\sum_{q=1}^{n}\underset{\sim q}{n}(x)T_{q\lambda}\Big)\Big(\frac{\partial \underset{\sim\alpha,\varrho,k}{c}}{\partial\tau_\lambda} + \underset{\sim\alpha,(\varrho-(\delta_{j\lambda})),k}{c}\Big)\\ &\quad + \underset{\sim\ell}{n}\Big(\frac{\partial \underset{\sim\alpha,\varrho,k}{c}}{\partial \underset{\sim}{n}} + \underset{\sim\alpha,\varrho,k-1}{c}\Big)\,.\end{aligned} \tag{A.19}$$

The covering of Γ by $B_{(r)} \subset \mathbb{R}^n$ and the local representation (A.1) in $B_{(r)}$ is associated with the partition of unity $\alpha_{(r)}(x)$ introduced in Section 3.3 which satisfies

$$\sum_{r=1}^{p}\alpha_{(r)}(x) = 1 \text{ in some neighbourhood of } \Gamma$$

and $\operatorname{supp}\alpha_{(r)} \Subset B_{(r)}$, $r = 1,\ldots,p$. Then all of our local transformations are valid in $B_{(r)}$ having corresponding tangential and normal differential operators which are different for different r. This now will be indicated by the additional index r.

The general differential operator P in (3.6.1) admits the form

$$\begin{aligned}Pu(x) &= \sum_{r=1}^{p} P(\alpha_{(r)}u)(x) = \sum_{r=1}^{p} P_{(r)}u(x)\\ &= \sum_{r=1}^{p}\sum_{|\beta|\leq 2m} a_{(r)\beta}(x)D^\beta u(x)\end{aligned}$$

with coefficients $a_{(r)\beta}(x)$ defined by the Leibniz product rule in $P(\alpha_{(r)}u)$ and the original coefficients $a_\beta(x)$ of P. In particular, we have $\operatorname{supp}(a_{(r)\beta}) \Subset B_{(r)}$. Then P again can be partitioned along Γ:

$$\begin{aligned}Pu(x) &= \sum_{r=1}^{p}\sum_{k=0}^{2m}\underset{\sim(r)k}{\mathcal{P}}\frac{\partial^k u}{(\partial \underset{\sim(r)}{n})^k}\\ &= \sum_{r=1}^{p}\sum_{|\alpha|\leq 2m}\sum_{k=0}^{|\alpha|}\underset{\sim(r)\alpha,k}{\mathcal{C}}\frac{\partial^k u}{(\partial \underset{\sim(r)}{n})^k}\,.\end{aligned}$$

For $x \in \Gamma$ we have $\tau_n = 0$ and

$$\frac{\partial^k u}{(\partial \underset{\sim}{n}_{(r)})^k} = \frac{\partial^k u}{\partial n^k}\Big|_\Gamma .$$

Hence,

$$Pu(x) = \sum_{k=0}^{2m} \underset{\sim}{\mathcal{P}}_k \frac{\partial^k u}{\partial n^k} \quad \text{for } x \in \Gamma , \tag{A.20}$$

where now

$$\begin{aligned}
\underset{\sim}{\mathcal{P}}_k &= \sum_{r=1}^{p} \sum_{|\beta|\le 2m} a_{(r)\beta}(x) \underset{\sim}{\mathcal{C}}_{(r)\beta,k}(x) \\
&= \sum_{r=1}^{p} \sum_{|\gamma|\le 2m-k} \sum_{|\alpha|\le 2m} a_{(r)\alpha}(x) \underset{\sim}{c}_{(r)\alpha,\gamma,k}(x) \left(\frac{\partial}{\partial \tau'_{(r)}}\right)^{\gamma} \quad \text{for } x \in \Gamma
\end{aligned}$$

are tangential operators. In particular, we again have

$$\underset{\sim}{\mathcal{P}}_{2m}(x) = \sum_{|\alpha|=2m} a_\alpha(x) \boldsymbol{n}^\alpha(x) \text{ for } x \in \Gamma .$$

The operator

$$\underset{\sim}{\mathcal{P}}_0 = \sum_{r=1}^{p} \sum_{|\gamma|\le 2m} \sum_{|\alpha|\le 2m} a_{(r)\alpha}(y) \underset{\sim}{c}_{(r)\alpha,\gamma,0}(y) \left(\frac{\partial}{\partial \tau'_{(r)}}\right)^{\gamma}$$

for $y = x_{(r)} + T_{(r)}(\tau'_{(r)}, a_{(r)}(\tau'_{(r)})) \in \Gamma$ corresponds to the Beltrami operator associated with P on Γ; however, with respect to local parameters $\tau'_{(r)}$.

If $\mathrm{supp}(\alpha_{(r)}) \cap \mathrm{supp}(\alpha_{(r')}) \neq \emptyset$, then we have two different sets of parameters $\tau'_{(r)}$ and $\tau'_{(r')}$ belonging to the same point y on Γ, which are related by the transformation

$$(\tau'_{(r')}, a_{(r')}(\tau_{(r')})) = T^{-1}_{(r')}(x_{(r)} - x_{(r')} + T_{(r)}(\tau'_{(r)}, a_{(r)}(\tau'_{(r)}))) .$$

Hence, the differential operators in $\underset{\sim}{\mathcal{P}}_k$ can eventually be expressed solely in either terms depending on $\tau'_{(r)}$ or $\tau'_{(r')}$, respectively, and we find:

$$Pu(x) = \sum_{k=0}^{2m} \underset{\approx}{\mathcal{P}}_k \frac{\partial^k u}{\partial \underset{\sim}{n}^k} ,$$

where this expansion is to be understood *locally.*

Note, that with this approach one only needs $\Gamma \in C^{2m}$, but not more.

In particular, one finds jump relations corresponding to those in Lemma 3.7.1, namely

$$\frac{\partial^k}{\partial \underset{\sim}{n}^k}(u\chi_\Omega) = \left(\frac{\partial^k u}{\partial \underset{\sim}{n}^k}\right)\chi_\Omega - \sum_{\ell=0}^{k=1}\left(\frac{\partial}{\partial \underset{\sim}{n}}\right)^{k-\ell-1}\left\{\frac{\partial^\ell u}{\partial n^\ell}\Big|_\Gamma \otimes \delta_\Gamma\right\} \tag{A.21}$$

for $k \geq 1$. Again, we find the second Green formula, now in the form,

$$\int_\Omega uP^\top \varphi dy - \int_\Omega \varphi Pudy \tag{A.22}$$

$$= -\sum_{\ell=0}^{2m-1}\sum_{p=0}^{2m-\ell-1}\int_\Gamma \Big\{\Big(\frac{\partial'}{\partial n}\Big)^p \underset{\approx p+\ell+1}{\mathcal{P}^\top} \varphi\Big\}(\gamma_\ell u)ds$$

which is slightly different from formula (3.7.8). In exactly the same manner as in Section 3.7 one may derive representation formulae as (3.7.10)–(3.7.14) by choosing for φ the fundamental solution $E(x, \bullet)$.

B. Vector Identities, Integration Formulae and Surface Differential Operators

In this appendix we summary some of the frequently used vector Identities, Gauss's and Green's formulae, surface differential operators as well as Stokes' Formula and its consequences.

I. Vector and vector field identities

(1) $\mathbf{A} \times (\mathbf{B} \times \mathbf{C}) = (\mathbf{A} \cdot \mathbf{C})\mathbf{B} - (\mathbf{A} \cdot \mathbf{B})\mathbf{C}$
(2) $\nabla \times (\mathbf{B} \times \mathbf{C}) = (\nabla \cdot \mathbf{C})\mathbf{B} - (\nabla \cdot \mathbf{B})\mathbf{C} + (\mathbf{C} \cdot \nabla)\mathbf{B} - (\mathbf{B} \cdot \nabla)\mathbf{C}$
(3) $\mathbf{A} \cdot (\mathbf{B} \times \mathbf{C}) = \mathbf{C} \cdot (\mathbf{A} \times \mathbf{B}) = \mathbf{B} \cdot (\mathbf{C} \times \mathbf{A})$
(4) $\nabla \cdot (\mathbf{B} \times \mathbf{C}) = \mathbf{C} \cdot (\nabla \times \mathbf{B}) - \mathbf{B} \cdot (\nabla \times \mathbf{C})$
(5) $\nabla(\mathbf{A} \cdot \mathbf{B}) = \mathbf{B} \times (\nabla \times \mathbf{A}) + \mathbf{A} \times (\nabla \times \mathbf{B}) + (\mathbf{A} \cdot \nabla)\mathbf{B} + (\mathbf{B} \cdot \nabla)\mathbf{A}$
(6) $\nabla \cdot (\nabla \times \mathbf{A}) = 0$
(7) $\nabla \times (\nabla \times \mathbf{A}) = -\Delta\mathbf{A} + \nabla(\nabla \cdot \mathbf{A})$

II. Integration formulae. In the following, $\Omega \subset \mathbb{R}^3$ is a bounded domain with $\Gamma := \partial\Omega$, the boundary of Ω.

(1) $\int\limits_\Omega \nabla \cdot \boldsymbol{u} dx = \int\limits_\Gamma \boldsymbol{u} \cdot \boldsymbol{n}\, ds.$

(2) $\int\limits_\Omega \nabla \times \boldsymbol{u}\, dx = \int\limits_\Gamma \boldsymbol{n} \times \boldsymbol{u}\, ds.$

(3) $\int\limits_\Omega \nabla\varphi dx = \int\limits_\Gamma \varphi\boldsymbol{n}\, ds.$

(4) $-\int\limits_\Omega \Delta\boldsymbol{u} \cdot \boldsymbol{v}\, dx = \int\limits_\Omega \nabla\boldsymbol{u} : \nabla\boldsymbol{v}\, dx - \int\limits_\Gamma \frac{\partial\boldsymbol{u}}{\partial n} \cdot \boldsymbol{n}\, ds$

(5) $-\int\limits_\Omega \Delta\boldsymbol{u} \cdot \boldsymbol{v}\, dx = \int\limits_\Omega \{\mathbf{curl}\, \boldsymbol{u} \cdot \mathbf{curl}\, \boldsymbol{v} + (\mathrm{div}\, \boldsymbol{u})(\mathrm{div}\, \boldsymbol{v})\}\, dx$
$- \int\limits_\Gamma \{\mathbf{curl}\, \boldsymbol{u} \cdot (\boldsymbol{n} \times \boldsymbol{v}) + (\mathrm{div}\, \boldsymbol{u})(\boldsymbol{n} \cdot \boldsymbol{v})\} ds$

(6) $\int\limits_\Omega \{\mathbf{curl}\, \boldsymbol{u} \cdot \boldsymbol{v} - \boldsymbol{u} \cdot \mathbf{curl}\, \boldsymbol{v}\} dx = \int\limits_\Gamma (\boldsymbol{n} \times \boldsymbol{u}) \cdot \boldsymbol{v}\, ds$

(7) $\int\limits_\Omega \boldsymbol{v} \cdot (\mathbf{curl}\,\mathbf{curl}\, \boldsymbol{u})\, dx = \int\limits_\Omega (\mathbf{curl}\, \boldsymbol{v}) \cdot (\mathbf{curl}\, \boldsymbol{u})\, dx + \int\limits_\Gamma \boldsymbol{n} \cdot (\mathbf{curl}\, \boldsymbol{u} \times \boldsymbol{v})\, ds.$

(8) $\int\limits_\Omega \{\boldsymbol{v} \cdot (\mathbf{curl}\,\mathbf{curl}\, \boldsymbol{u}) - \boldsymbol{u} \cdot (\mathbf{curl}\,\mathbf{curl}\, \boldsymbol{v})\}\, dx$
$= \int\limits_\Gamma \{\boldsymbol{u} \times \mathbf{curl}\, \boldsymbol{v} - \boldsymbol{v} \times \mathbf{curl}\, \boldsymbol{u}\} \cdot \boldsymbol{n}\, ds.$

(9) Helmholtz indentity: Given $\boldsymbol{A}$ with compact support in $L^2(\mathbb{R}^3)^3$ then there exits a scalar field φ and a vector field $\boldsymbol{v}$ such that $\boldsymbol{A}(x) = \mathbf{grad}\,\varphi + \mathbf{curl}\, \boldsymbol{v}$

G. C. Hsiao und W. L. Wendland, *Boundary Integral Equations*,
Applied Mathematical Sciences 164,
https://doi.org/10.1007/978-3-030-71127-6

$$A(x) = \frac{1}{4\pi}\mathbf{curl}\int_{\mathbb{R}^3}\frac{1}{|x-y|}\mathbf{curl}\ A(y)\,dy - \frac{1}{4\pi}\mathbf{grad}\int_{\mathbb{R}^3}\frac{1}{|x-y|}\operatorname{div} A(y)\,dy.$$

(See Dautray and Lions [96, p.313].)

III. Surface differential operators. The differential operators on the surface Γ can be defined either by restriction or by extension. For scalar v and vector $\boldsymbol{v}$ functions in $\mathbb{R}^3$ defined on the tubular neighborhood of Γ, the surface operators are defined as follows:

(1) $\mathbf{grad}_\Gamma v := \mathbf{grad}\ v - \boldsymbol{n}\frac{\partial v}{\partial n} = -\boldsymbol{n}\times(\boldsymbol{n}\times\nabla v) = \pi_t(\mathbf{grad}\ v)$
$\pi_t \boldsymbol{u} := -\boldsymbol{n}\times(\boldsymbol{n}\times\boldsymbol{u})$
(2) $\operatorname{div}_\Gamma \boldsymbol{v} := \operatorname{div}\boldsymbol{v} - \boldsymbol{n}\cdot\frac{\partial \boldsymbol{v}}{\partial n} = -(\boldsymbol{n}\times(\boldsymbol{n}\times\nabla))\cdot\boldsymbol{v}$
(3) $\mathbf{curl}_\Gamma \boldsymbol{v} := \nabla\times\boldsymbol{v} - \boldsymbol{n}\times\frac{\partial \boldsymbol{v}}{\partial n} = -(\boldsymbol{n}\times(\boldsymbol{n}\times\nabla))\times\boldsymbol{v}$

Now for scalar function φ and vector function $\boldsymbol{u}$ in $\mathbb{R}^3$ defined on the surface Γ, we associate with φ and $\boldsymbol{u}$ the extensions (or lifting) $\tilde{\varphi}$ and $\tilde{\boldsymbol{u}}$ defined on the tubular neighborhood of Γ in Ω such that $\tilde{\varphi}|_\Gamma = \varphi$ and $\rceil_\Gamma =$. Then we define

(4) $\mathbf{grad}_\Gamma \varphi := \mathbf{grad}\ \tilde{\varphi}|_\Gamma$ (Surface gradient)
(5) $\mathbf{curl}_\Gamma \varphi := -\gamma_t\,(\mathbf{grad}_\Gamma \varphi)$ (Surface vector curl) ; $\gamma_t \boldsymbol{u} := \boldsymbol{n}\times\boldsymbol{u}$
(6) $\operatorname{div}_\Gamma \boldsymbol{u} := \operatorname{div}\tilde{\boldsymbol{u}}|_\Gamma$ (Surface divergence)
(7) $\operatorname{curl}_\Gamma \boldsymbol{u} := \boldsymbol{n}\cdot\mathbf{curl}\ \tilde{\boldsymbol{u}}|_\Gamma$ (Surface scalar curl)

(See [331, (2.5.145), (2.5.149), (2.5.189), (2.5.190)] for (4) –(7)).
We remark that surface divergence $\operatorname{div}_\Gamma$ and surface scalar $curl_\Gamma$ can also be defined by duality.

IV. Stokes Theorem and its consequences. In the following, S denotes a surface in $\mathbb{R}^3$ with boundary curve $C = \partial S$, and $\Gamma = \partial\Omega$ denotes the closed surface in $\mathbb{R}^3$

(1) $\int_S \ \cdot ds = \oint_C \ \cdot d$ (Stokes Theorem) .

As consequences, we have

(2) $-\int_S \mathbf{curl}_\Gamma \varphi\, ds = \oint_C \varphi d\boldsymbol{r}$ (= $\mathbf{0}$ for a closed surface)
(3) $\int_S \operatorname{div}_\Gamma \boldsymbol{u} ds = \oint_C (\boldsymbol{n}\times\boldsymbol{u})\cdot d\boldsymbol{r}$ (= 0 for a closed surface, if **u** is tangential)
(4) $\int_S \operatorname{curl}_\Gamma \ ds = \oint_C \boldsymbol{u}\cdot d\boldsymbol{r}$ (= 0 for a closed surface, see e.g. [33, p.503])
(5) $\int_\Gamma \mathbf{grad}_\Gamma\varphi\cdot\boldsymbol{v}\,ds + \int_\Gamma \varphi\operatorname{div}_\Gamma \boldsymbol{v}\,ds = 0.$
(6) $\int_\Gamma \mathbf{curl}_\Gamma \varphi\cdot\boldsymbol{v}\,ds - \int_\Gamma \varphi\operatorname{curl}_\Gamma \boldsymbol{v}\,ds = 0.$
(7) $-\int_\Gamma (\Delta_\Gamma w)u\,ds = \int_\Gamma \mathbf{grad}_\Gamma w\cdot\mathbf{grad}_\Gamma u\,ds = \int_\Gamma \mathbf{curl}_\Gamma w\cdot\mathbf{curl}_\Gamma u\,ds$

(8) $-\int_\Gamma \Delta_\Gamma (\boldsymbol{w} \cdot \boldsymbol{v})\, ds = \int_\Gamma \operatorname{div}_\Gamma \boldsymbol{w} \operatorname{div}_\Gamma \boldsymbol{v}\, ds + \int_\Gamma \mathbf{curl}_\Gamma \boldsymbol{w} \cdot \mathbf{curl}_\Gamma \boldsymbol{v}\, ds$

(See, e.g. [331, Theorem 2.5.19] for (5) - (8)).

(9) $\int_\Gamma (\operatorname{curl}_\Gamma \boldsymbol{u})\varphi\, ds = \int_\Gamma (\boldsymbol{n} \times \boldsymbol{u}) \cdot \mathbf{grad}_\Gamma \varphi ds$, which implies
$-\operatorname{div}_\Gamma (\boldsymbol{n} \times \boldsymbol{u}) = \operatorname{curl}_\Gamma \boldsymbol{u}$

V. Surface differential identities

(1) $\Delta_\Gamma u = \operatorname{div}_\Gamma \mathbf{grad}_\Gamma u = -\operatorname{curl}_\Gamma \mathbf{curl}_\Gamma u$
(2) $\Delta_\Gamma \boldsymbol{v} = \mathbf{grad}_\Gamma \operatorname{div}_\Gamma \boldsymbol{v} - \mathbf{curl}_\Gamma \operatorname{curl}_\Gamma \boldsymbol{v}$
(3) $\operatorname{div}_\Gamma \mathbf{curl}_\Gamma u = 0$
(4) $\operatorname{curl}_\Gamma \mathbf{grad}_\Gamma u = 0$
(5) $\operatorname{div}_\Gamma (\boldsymbol{n} \times \boldsymbol{v}) = -\operatorname{curl}_\Gamma \boldsymbol{v}$
(6) $\operatorname{div}_\Gamma (\pi_t \boldsymbol{v}) = \operatorname{curl}_\Gamma (\boldsymbol{n} \times \boldsymbol{v})$ and hence
$\operatorname{div}_\Gamma \boldsymbol{v} = \operatorname{curl}_\Gamma (\boldsymbol{n} \times \boldsymbol{v})$, if $\boldsymbol{v} \in \boldsymbol{L}_t^2(\Gamma)$.
(7) $\mathbf{curl}_\Gamma \varphi = \mathbf{grad}_\Gamma \varphi \times \boldsymbol{n}$.

References

1. Adams, R.A.: Sobolev Spaces. Academic Press, New York 1975.
2. Adrian, S.: *Rapidly converging boundary integral equation solvers in computational Electromagnetics.* Doctorial Thesis, Bretagne–Pays de la Loire École doctorale Math STIC and Technische Universität München 2018.
3. Agmon, S.: Lectures on Elliptic Boundary Value Problems. D.van Nostrand, Princeton, NJ. 1965.
4. Agranovich, M.S.: *Spectral properties of elliptic pseudo–differential operators on a closed curve.* Functional Analysis Appl. **13** (1979) 279–281.
5. Agranovich, M.S.: Sobolev spaces, their Generalization, and Elliptic Problems in Smooth and in Lipschitz Dmains. Springer 2015
6. Agranovich, M.S.: *Elliptic operators on closed manifolds.* In: Encyclopaedia of Mathematical Sciences. Vol. 63, Partial Differential Equations VI. (Yu.V. Egorov, M.A. Shubin eds.) Springer–Verlag, Berlin 1994, pp. 1–130.
7. Ahner, J.F. and Hsiao, G.C.: *On the two–domensional exterior boundary value problems of elasticity.* SIAM J. Appl. Math. **31** (1976) 677–685.
8. Ammari, H.: Propagation d'ondes électromagnétiques. Habilitation à Diriger des Recherches Mathematiques, Univ. de Paris VI, 1999.
9. Ammari, H.and Nedelec, J.C.: *Coupling of finite and boundary element methods for time–harmonic Maxwell equations.* Part I in: Mathematical Aspects of Boundary Element Methods (Bonnet et al. eds.) CRC Press UK 1999. pp.11–22. PartII: A symmetric Formulation. Operator Theory: Advances and Applications **110** (1999), 23–32.
10. Amrouche, C., Bernardi, C., Dauge, M. and Girault, V: *Vector potentials in three-dimensional non-smooth domains,* Math. Meth. Appl. Sci. **21**: 823-864 (1998).
11. Amrouche,C., , Ciarlet, P.G. and Mardare, C.: *On a lemma of Jacques–Louis Lions and its relation to othe fundamental results.* J. Math. Pures Appl. **104** (2015) 207–226.
12. Ammari, H., Buffa, A. and Nedelec, J.–C. : *A justification of eddy current problems,* SIAM J. Appl. Math.**60** (2000) 1805–1823.
13. Angell, T.S., Kleinman, R.E and Kral, J.: *Layer potentials on boundaries with corners and edges.* Čas. Pešt. Mat, **113** (1988) 387–401.
14. Angell, T.S. and Kirsch, A.: Optimization Methods in Electromagnetic Radiation, Springer, New York 2004.
15. Angell, T.S., Hsiao, G.C. and Wen, L.: *On the two–dimensional inverse scattering problems in electromagnetics,* Appl. Anal. **82** (2007) 483–497.
16. Atkinson, K.E.: The Numerical Solutions of Integral Equations of the Second Kind. Cambridge University Press 1997.
17. Aubin, J.–P.: Approximation of Elliptic Boundary Value Problems. John Wiley & Sons, New York 1972.

G. C. Hsiao und W. L. Wendland, *Boundary Integral Equations*,
Applied Mathematical Sciences 164,
https://doi.org/10.1007/978-3-030-71127-6

18. Avantaggiati, A.: *Sulle matrici fondamentali principali per una classe di sistemi ellitici e ipoellitici.* Ann. Mat. Pura Appl. **65** (1964) 191–238.
19. Bach, M.: *On the treatment of propagating mode-1 cracks by variational inequalities.* In: The Mathematics of Finite Elements and Applications X (J. Whiteman ed.) Elsevier, Oxford 2000, pp. 377–386.
20. Bach,M., Nazarov, N. and Wendland, W.: *Stable propagation of a mode-1 planar crack in an isotropic space. Comparison of the Irwin and the Griffith approaches.* In: Current Problems of Analysis of Analysis and Mathematical Physics. Aracne, Rome 2000, pp. 167–189.
21. Balanis, C.A.: Advanced Engineering Electromagnetics, Wiley, New York, 1989.
22. Balanis, C.A.: Antenna Theory: Analysis and Design, 3rd edition, Wiley, 2005.
23. Balaš, J., Sladek J. and Sladek, V.: Stress Analysis by Boundary Element Methods. Elsevier, Amsterdam 1989.
24. Barkeshli, K.: Advanced Electromagnetics and Scattering Theory, Springer, 2015.
25. Bear, J.: Hydraulic of Groundwater. Mac Graw Hill, London 1979.
26. Bellman, R.: Introduction to Matrix Analysis, McGraw–Hill, New York 1960.
27. Bendali, A.: *Numerical analysis of the exterior boundary value problem for the time–harmonic Maxwell equations by a boundary finite element method. Part I: The continuous problem.* Math. Comp. **43** (1984) 29–46.
28. Bendali, A.: *Numerical analysis of the exterior boundary value problem for the time–harmonic Maxwell equations by a boundary finite element method. Part II: The discrete problem.* Math. Comp. **43** (1984) 47–68.
29. Benitez, F.G. (ed.): Fundamental Solutions in Boundary Elements. SAND (Camas) Sevilla 1997.
30. Berger, M. and Gostiaux, B.: Géométrie differentielle: Variétés, Courbes et Surfaces. Presses Universitaires de France, Paris 1987.
31. Biró, O. and Preis, K.: *On the use of the magnetic potential in the finite element analysis of the three–dimensionaleddy currents.* IEEE Trans. Magn. **25** (1988)
32. Bishop, R.L. and Goldberg, S.I.: Tensor Analysis on Manifolds. Dover Publ. Inc., New York 1968.
33. Bladel, J.: Electromagnetic Fields, Second Edition. Wiley Intersience 2007.
34. Boffi, D., Brezzi, F. and Fortin, M.: Mixed Finite Element Methods and Applications. Vol. 44 of Springer Series in Computaional Mathematics, Springer, Heidelberg, 2013.
35. Bogovskii, M.E-: *Solution of the first boundary value problem for an equation of continuity of an incompressible medium.* Sov. Math. Dokl. **20** (1979) 1094–1098.
36. Bonnemay, P.: Equations intégrales pur l'élasticité plane. Thèse de 3ème cycle, Univ. de Paris VI, 1979.
37. Bonnet, M.: Boundary Integral Equation Methods for Solids and Fluids. John Wiley & Sons, Chichester 1995.
38. Bossavit, A: Computational Electromagnetism. Academic Press, London 1998.
39. Boutet de Monvel, L.: *Comportement d'un opérateur pseudo–differentiel sur une variété à bord.* J. Analyse Fonct. **17** (1966) 241–304.
40. Boutet de Monvel, L.: *Opérateurs pseudo–differentiels et problèmes aux limites elliptiques.* Ann. Inst. Fourier **19** (1969) 169–268.
41. Boutet de Monvel, L.: *Boundary problems for pseudo–differential operators.* Acta Math. **126** (1971) 11–51.
42. Brakhage, H. and Werner, P.: *Über des Dirichletsche Aussenraumproblem für die Helmholtzsche Schwingungsgleichung.* Arch. Math. **16** (1965) 325–329.
43. Brebbia, C.A.: Boundary Element Methods. Springer-Verlag, New York 1983.

44. Brebbia, C.A., Telles, J.C.F. and Wrobel, L.C.: Boundary Element Techniques. Springer–Verlag, Berlin 1984.
45. Brezzi, F. and Fortin, M.: Mixed and Hybrid Finite Element Methods. Springer–Verlag New York 1991.
46. Bruhn, G. and Wendland, W.L.: *Über die näherungsweise Lösung von linearen Funktionalgleichungen.* Internat. Schriftenreihe Numerische Mathematik (L. Collatz et al eds.) Birkhäuser–Verlag, Basel **7** (1967) 136–164.
47. Buffa, A.: Some Numerical and Theoretical Problems in Computational Electromagnetism. Doctoral Thesis, Univ. of Pavia 2000.
48. Buffa, A.: Trace theorems on non-smooth boundaries for functional spaces related to Maxwell equations: an overview. In: Computational Electromagnetics, T.L. Bath et al. (Eds), pp.23-34, Springer 2003.
49. Buffa, A. and Ciarlet, P. Jr.: *On traces for functional spaces related to Maxwell's equations I. An integration by parts formula in Lipschitz polyhedra.* Math. Methods Appl. Sci. **24** (2001) 1–30.
50. Buffa, A. and Ciarlet, P. Jr.: *On traces for functional spaces related to Maxwell's equations II. Hodge decompositions on the boundary of Lipschitz polyhedra and applications.* Math. Methods Appl. Sci. **24** (2001) 31–48.
51. Buffa, A., Costabel, M. and Schwab, C.: *Boundary element methods for Maxwell equations on non–smooth domains.* Numer. Math. **92** (2002) 679–710.
52. Buffa, A., Costabel, M. and Sheen, D.: *On traces for H(curl, Ω) in Lipschitz domains,* J. Math. Anal. Appl. **276**, 845-867 (2002).
53. Buffa, A., Hiptmair, R., von Petersdorff, T. and Schwab,C.: *Boundary element methods for Maxwell transmission problems in Lipschitz domains.* Numer. Math. **95** (2003) 459–496.
54. Buffa, A. and Hiptmair, R.: *Galerkin boundary element methods for electromagnetic scattering.* In: Computational Methods in Wave Propagation. Springer–Verlag, 2003.
55. Burago, Yu. and Maz'ya, V.G.: *Potential theory and function theory for irregular regions.* Zap. Naučn. Sem. LOMI **3** (1967) 1–152; Seminars in Mathematics, V.A. Steklov Inst. Leningrad; Consultants Bureau, New York 1969.
56. Burago, Yu., Maz'ya, V.G. and Sapozhnikova, V.D.: *On the double layer potential for nonregular domains.* Dokl. Akad. Nauk SSSR **147** (1962) 523–525; Sov. Math. Dokl. **3** (1962) 1640–1642.
57. Bureau, F.: *Divergent integrals and partial differential equations.* Commun. Pure Appl. Math. **8** (1955) 143–202.
58. Cakoni, F., Hsiao, G.C. and Wendland, W.L.: *On the boundary integral equation method for a mixed boundary value problem of the biharmonic equation.* Complex Variables **50** (2005) 681–696.
59. Calderón, A.P.: *Boundary value problems for elliptic equations.* Outlines of the Joint Soviet–American Symposium on Partial Differential Equations. August, 1963, Novosibirsk.
60. Calderon, A.P. and Zygmund A.: *On singular integrals.* Amer. J. Math. **78** (1956) 289–309.
61. Calderon, A.P. and Zygmund A.: *Singular integral operators and differential equations.* Amer. J. Math. **79** (1957) 901–921.
62. Carleman, T.: Über das Neumann–Poincarésche Problem für ein Gebiet mit Ecken. Doctoral Dissertation, Uppsala 1916.
63. Cessenat, M.: Mathematical Methods in Electromagnetism. World Scientific, Singapore 1996.
64. Chazarain, J. and Piriou, A.: Introduction to the Theory of Linear Partial Differential Equations. North–Holland, Amsterdam 1982.

65. Chen, G. and Zhou, J.: Boundary Element Methods. Academic Press, London 1992.
66. Chenais, D: *Un résultat de compasité d'un ensemble de partie de* $\mathbb{R}^n$. Comp. Rend. Acad. Sci. Paris **277** (1973) 905–907.
67. Christiansen, S.H.: Discrete Fredholm properties and convergence estimates for the EFIE, Tech. Report CMAP Ecole Polytechnique Palaiseau 2001.
68. Ciarlet, P.G.: Mathematical Elasticity, Vol.I: Three–Dimensional Elasticity. North–Holland, Amsterdam 1988.
69. Ciarlet, P.G. and Ciarlet, P.Jr.: *Another approach to linearized elasticity and a new proof of Korn's inequality.* Math. Models and Appl. Sci. **15** (2005) 259–271.
70. Ciavaldini, J.F., Pogu, M. and Tournemine, G.: *Existence and regularity of stream functions for subsonic flows past profiles with a sharp trailing edge.* Arch. Rat. Mech. Anal. **93** (1986) 1–14.
71. Claeys, X. and and Hiptmair, R.: *First kind boundary integral equations for the Hodge–Helmholtz equation.* SIAM J. Appl. Math. (2019).
72. Claeys, X. and and Hiptmair, R.: *First–kind Galerkin boundary element methods for the Hodge–Laplacian in three dimensions.* Math. Meth. Appl. Sci., submitted 2019.
73. Clements, D.L.: Boundary Value Problems Governed by Second Order Systems. Pitman Publ. London 1981.
74. Coclici, C.A. and Wendland, W.L.: *On the treatment of the Kutta–Joukowski condition in transonic flow computations.* Z. Angew. Math. Mech. **79** (1999) 507–534.
75. Colton, D. and Kress, R.: Integral Equation Methods in Scattering Theorie. John Wiley & Sons, New York 1983.
76. Colton, D. and Kress, R: *Karp's theorem in electromagnetic scattering theory.* Proc. AMS **104** (3) (1988) 764–769.
77. Colton, D. and Kress, R: Inverse Acoustic and Electromagnetic Scattering Theory. Springer–Verlag, Berlin 1992, 3rd. Edn. Springer–Verlag, New York 2013.
78. Computational Electromagnetism. (de Castro, A.B. and Valli, A. eds.) Springer, Lecture Notes in Math. **2148**, Springer New York 2015.
79. Costabel, M.: Starke Elliptizität von Randintegralgleichungen erster Art. Habilitation Thesis, Technical University Darmstadt 1984.
80. Costabel, M.: *Boundary integral operators on Lipschitz domains: elementary results.* SIAM J. Math. Anal. **19** (1988) 613–626.
81. Costabel, M.: *Coupling of FEM and BEM for the numerical solution of the problem of electromagnetic scattering from an inhomogeneous body:* In: Computational Mechanics 1988 (S.N. Atluri, G. Yagawa eds.) Vol.1 5v.1-5v.4, Springer–Verlag Berlin.
82. Costabel, M.: *A remark on the regularity of solutions of Maxwell equations in Lipschitz domains.* Math. Methods Appl. Sci. **12** (4) (1990) 365–368.
83. Costabel, M.: *A coercive bilinear form for Maxwell's equations-* J. Math. Anal. Appl. **157** (1991), 527–541.
84. Costabel, M.: *Some historical remarks on the positivity of boundary integral operators.* In: Boundary Element Analysis (M. Schanz, O. Steinbach eds.) Springer Berlin 2007, pp. 1–27.
85. Costabel, M., Dauge, M. and Duduchava, R.: *Asymptotics without logarithmic terms for crack problems.* Comm. PDE **28** (2003) 1673–1687.
86. Costabel, M. and McIntosh, A.: *On Bogovskii and regularized Poincaré integral operators for the Rham complexes on Lipschitz domains.* Math. Z. **265** (2010) 297–320.

87. Costabel, M. and Saranen, J.: *Boundary element analysis of a direct method for the biharmonic equation.* Operator Theory: Advances and Applications **41** (1989) 77-95.
88. Costabel, M. and Stephan, E.: *A direct boundary integral method for transmission problems.* JMAA. **106** (1986) 367–413.
89. Costabel, M. and Stephan, E.: *Strongly elliptic integral equations for electromagnetic transmission problems.* Proc. Royal Soc. Edinburgh **109**A (1988) 271–296.
90. Costabel, M. and Wendland, W.L.: *Strong ellipticity of boundary integral operators.* J. Reine Angew. Mathematik **372** (1986) 34–63.
91. Costabel, M., Stephan, E. and Wendland, W.L.: *On boundary integral equations of the first kind for the bi–Laplacian in a polygonal domain.* Ann. Scuola Norm. Sup. Pisa, Ser. IV, **10** (1983) 197–242.
92. Crouch, S.L. and Starfield, A.M.: Boundary Element Methods in Solid Mechanics. George Allen & Unwin, London 1983.
93. Cruse, T.A.: Boundary Element Analysis in Computational Mechanics. Kluwer Academic Publ., Dordrecht 1988.
94. Dassios, G. and Kleinman, R.: Low Frequency Scattering, Oxford Science Publications, 2000
95. Dautray, R. and Lions, J.L.: Mathematical Analysis and Numerical Methods for Science and Technology. Vol. 1: Physical Origins and Classical Methods. Springer–Verlag, Berlin 1990.
96. Dautray, R. and Lions, J.L.: Mathematical Analysis and Numerical Methods for Science and Technology. Vol. 3: Springer–Verlag, Berlin 1990.
97. Dautray, R. and Lions, J.L.: Mathematical Analysis and Numerical Methods for Science and Technology. Vol. 4: Integral Equations and Numerical Methods. Springer–Verlag, Berlin 1990.
98. De la Bourdonnaye, A.: *Decomposition de* $TH_{\mathrm{div}}^{-1/2}(\Gamma)$ *et nature de l'opérateur de Steklov–Poincaré du problème extérieur de l'électromagnétisme.* C.R. Acad. Sci. Paris, Série I.t.316 (1993) 369–372.
99. De La Bourdonnaye, A.: *Some formulations coupling finite element and integral equation and electromagnetism.* Numer. Math. **69**(1995) 257–268.
100. Dieudonné, J.: Eléments d'Analyse. Vol. 8, Gauthier-Villars, Paris 1978.
101. Dindos, M. and Mitrea, M.: *The stationary Navier–Stokes system in nonsmooth manifolds: The Poisson problem in Lipschitz and* C^1 *domains.* Ration. Mach. Anal. **174** (2004) 1–47.
102. Ding, Z.: *A proof of the trace theorem of Sobolev spaces on Lipschitz domains.* Proc. AMS **124** (1996) 591–600.
103. Douglis, A. and Nirenberg, L.: *Interior estimates for elliptic systems of partial differential equations.* Comm. Pure Appl. Math. **8** (1955) 503–538.
104. Duduchava, R.: *The Green formula and layer potentials.* Integral Eqns. Operator Theory **41** (2001) 127–178.
105. Duduchava, R. and Speck, F.–O.: *Pseudodifferential operators on compact manifolds with Lipschitz boundary.* Math. Nachr. **160** (1993) 149–191.
106. Duduchava, R., Sändig, A.M. and Wendland, W.L.: *Interface cracks in anisotropic composites.* Math. Methods Appl. Sci. **22** (1999) 1413–1446.
107. Duduchava, R. and Wendland, W.L.: *The Wiener–Hopf method for systems of pseudodifferential equations with an application to crack problems.* Integral Eqns. Operator Theory **23** (1995) 294–335.
108. Duffin, R.J. and Noll, W.: *On exterior boundary value problems in elasticity.* Archive Rat. Mech. Anal. **2** (1958/1959) 191–196.

109. Egorov, Yu.V. and Shubin, M.A.: *Linear partial differential equations. Elements of modern theory.* In: Encyclopaedia of Mathematical Sciences, Partial Differential Equations II, Vol.31. (Yu. V. Egorov, M.A. Shubin eds.) Springer–Verlag Berlin 1994, pp. 1–120.
110. Ejdel'man, S.D.: Parabolic Systems. North–Holland 1969.
111. Ehrenpreis, L.: *Solution of some problems of division. Part I. Division by a polynomial of derivation.* Amer. J. Math. **76** (1954) 883–903.
112. Eskin, G.I.: Boundary Value Problems for Elliptic Pseudodifferential Equations. Amer. Math. Soc., Providence, Rhode Island 1981.
113. Essential Numerical Methods in Electromagnetics (Ciarlet, P.G. et al. eds.) Elevier, Amsterdam 2011.
114. Fabes, E.B., Kenig, C.E. and Verchota, G.C.: *The Dirichlet problem for the Stokes system in Lipschitz domains.* Duke Mathematical J. **57** (1988) 769–793.
115. Fabes, E.B., Kenig, C.E. and Verchota, G.C.: *Boundary value problems for the system of elastostatics and Lipschitz domains.* Duke Mathematical J. **57** (1988) 795–818.
116. Farwig, R. Simader, C., Sohr, H. and Varnhorn, W.: *General properties of the Helmholtz decomposition in spaces of L^q–type.* Contemporary Mathematics **666** Amer. Math. Soc. (2016) 163–177.
117. Feng Kang: *Finite element method and natural boundary reduction.* In: Proc. International Congress of Mathematicians. Warszawa 1983, pp.1439–1453.
118. Fichera, G.: *Linear elliptic equations of higher order in two independent variables and singular integral equations, with applications to anisotropic inhomogeneous elasticity.* 1961 Partial Differential Equations and Continuum Mechanics, pp.50–80. Univ. of Wisconsin Press, Madison, Wis.
119. Fichera G.: *Existence theorems in elasticity.* In: Handbuch der Physik, Vol.VIa-2 (S. Flügge, C. Truesdell eds.) Springer–Verlag, Berlin 1972, pp. 347–389.
120. Fichera, G.: *Sul problema misto per le equazioni lineari alle derivate parziali del secondo ordine di tipo elletico.* Rev. Roumaine Pure Appl. **9** (1964) 3–9.
121. Fichera, G.: Linear Elliptic Differential Systems and Eigenvalue Problems. Lecture Notes in Mathematics **8**, Springer-Verlag, Berlin-New York 1965.
122. Filippi, P.: *Integral equations in acoustics.* In: Theoretical Acoustics and Numerical Techniques. CISM Courses and Lectures **277**. Springer–Verlag, Wien–New York 1983, pp. 1–49.
123. Fischer, T., Hsiao, G.C. and Wendland, W.L.: *Singular perturbations for the exterior three–dimensional slow viscous flow problem.* J. Math. Anal. Appl. **110** (1985) 583–603.
124. Fischer, T., Hsiao, G.C. and Wendland, W.L.: *On two–dimensional slow viscous flows past obstacles in a half–plane.* Proc. Royal Soc. Edinburgh **A 104** (1986) 205–215.
125. Folland, G.B.: Lectures on Partial Differential Differential Equations. Tata Institute Fundamental Research, Bombay, Springer-Verlag, Berlin 1983.
126. Folland, G.B.: Introduction to Partial Differential Equations. Princeton Univ. Press, New Jersey 1995.
127. Freeden, W. and Michel, V.: Multiscale Potential Theory. Birkhäuser, Boston 2004.
128. Friedlander, F.G.: Introduction to the Theory of Distributions. Cambridge Univ. Press, Cambridge 1982.
129. Friedman, A.: Partial Differential Equations. Holt, Rinehardt and Winston, Inc., New York 1969.

130. Fulling, S.A. and Kennedy, G.: *The resolvent parametrix of a general elliptic linear differential operator: A closed form for the intrinsic symbol.* Transac. Amer. Math. Soc. **310** (1988) 583–617.
131. Gagliardo, E.: *Caratterizzazioni delle trace sulla frontiera relative ad alcune di funzioni in n variabili.* Rend. Seminario Mat. Universitá di Padova (1957) **27** 284–305).
132. Gagliardo,E.: *Proprietà di alcune classi di funzioni in piu variabili.* Ricerche Mat. **7** (1958) 102–137.
133. Gaier, D.: Konstruktive Methoden der konformen Abbildung. Springer–Verlag, Berlin 1964.
134. Gaier, D.: *Integralgleichungen erster Art und konforme Abbildungen.* Math. Zeitschr. **147** (1976) 113–139.
135. Gakhov, F.D.: Boundary Value Problems. Pergamon Press, New York 1966.
136. Galdi, C.P: An Introduction to the Mathematical Theory of the Navier–Stokes Equations. Vol. I. Springer–Verlag, New York 1994.
137. Gårding, L.: *The Dirichlet problem.* Math. Intelligencer **2** (1979) 43–53.
138. Gatica, G.N.: A Simple Introduction to the Mixed Finite Element Method Theory and Applications. Springer New York 2014.
139. Gatica, G. and Hsiao, G.C.: Boundary–Field Equation Method for a Class of Nonlinear Problems. Pitman Res. Notes in Math. **331**. Addison–Wesley Longman, Edinburgh, Gate Harlow 1995.
140. Gaul, L., Kögel, M. and Wagner, M.: Boundary Element Methods for Engineers and Scientists. Springer, Berlin 2003.
141. Gauss, C.F.: *Allgemeine Lehrsätze in Beziehung auf die im verkehrten Verhältnisse des Quadrats der Entfernung wirkenden Anziehungs– und Abstoßungs–Kräfte.* (1839), Werke **5**, 2nd Ed. Göttingen 1877, 194–242.
142. Gauss, C.F.: Atlas des Erdmagnetismus. (1840), Werke **12**, Göttingen 1929, 326–408.
143. Gelfand, I.M. and Shilov, G.E.: Generalized Functions, Vol.1. Academic Press, New York 1964.
144. Giraud, G.: *Sur certains problèmes concernant des systèmes d'équations du type elliptique.* Comp. Rendus Acad. Sci. Paris **192** (1931) 471–473.
145. Giraud, G.: *Equations à intégrales principales, étude suivie d'une application.* Ann. Ec. Norm. Super. (3) **51** fasc.3 (1934) and fasc.4 (1936) 251–372.
146. Girauld, V. and Raviart, P.–A.: Finite Element Approximation of the Navier–Stokes Equations. Lecture Notes in Mathematics Nr. 749, Springer–Verlag Berlin 1981.
147. Girault, V. and Raviart, P. A.: Finite Element Methods for Navier–Stokes Equations. Theory and Algorithms, Springer, Berlin 1986.
148. Giroire, M.J.: Etudes de Quelques Problèmes aux Limites Extérieurs et Résolution par Equations Intégrales. Thesis Dr. d'Etat. Paris VI, 1987.
149. Giroire, J. and Nedelec, J.C.: *Numerical solution of an exterior Neumann problem using a double layer potential.* Math. Comp. **32** (1978) 973–990.
150. Glauert, H.: The Elements of Aerofoil and Airscrew Theory. Cambridge University Press, Cambridge 1926.
151. Gohberg, I. and Feld'man, I.A.: Convolution Equations and Projection Mathods for their Solution. Amer. Math. Soc., Providence, Rhode Island 1974.
152. Gohberg, I. and Krupnik, N.: One–dimensional Linear Integral Equations. Birkhäuser–Verlag, Basel 1992.
153. Goldstein, S.: Lectures on Fluid Mechanics. AMS, Providence, Rhode Island 1976.
154. Greco, D.: *Le matrici fondamentali di alcuni sistemi di equazioni lineari a derivate parziali di tipo ellitico.* Ricerche di Mat. Napoli **8** (1959) 197–221.

155. Green, G.: An Essay on the Applicability of Mathematical Analysis of the Theories of Elasticity and Magnetism. Nottingham, U.K. 1828.
156. Grisvard, P.: Elliptic Problems in Nonsmooth Domains. Pitman, London 1985.
157. Große, N. and Schneider, C.: *Sobolev spaces on Riemannian manifolds with bounded geometry: General coordinates and traces* . Math. Nachr. 1–28 (2013)
158. Grubb, G.: *Properties of normal boundary problems for elliptic even–order system.* Ann. Sc. Sup. Pisa (4) **1** (1974) 1–61.
159. Grubb, G.: Functional Calculus of Pseudo-Differential Boundary Problems. Birkhäuser, Boston 1986.
160. Grubb, G.: *Pseudodifferential boundary problems and applications.* Jber. d. Dt. Math.–Verein. **99** (1997) 110–121.
161. Grusin, V.V.: *Fundamental solutions of hypoelliptic equations.* Usp. Mat. Nauk **16** n. 4 (1961) 147–153 (Russian).
162. Günter, N.M.: Potential Theory and its Application to Basic Problems of Mathematical Physics. Gestekhizdat, Moscow 1953.
163. Guiggiani, M.: *Hypersingular formulation for boundary stress evaluation.* Engrg. Analysis with Boundary Elements **13** (1994) 169–179.
164. Guo,H.: *Surface Crouzeix–Raviart element for the Laplace–Beltrami equation.* Numer. Math. **144** (2020) 527–551.
165. Gurtin, M.E.: *The linear theory of elasticity.* In: Handbuch der Physik Vol.VIa/2 (S. Flügge ed.), Springer–Verlag, Berlin 1972, pp. 1–295.
166. Gwinner, J. and Stephan, E.P.: Advanced Boundary Element Methods. Springer Nature 2018.
167. Hackbusch, W.: Integralgleichungen. B.G. Teubner, Stuttgart 1989.
168. Hadamard, J.: *Recherches sur les solutions fondamentales e l'intégration des équations linéaires aux dérivées partielles (deuxièm mémoire).* Ann. Éc. Norm. Sup. Sér. 3, **22** (1905) 101–141 (Œuvre **3**, 1195–1235).
169. Han, Houde: *Boundary integro differential equations of elliptic boundary value problems and their numerical solution.* Sci. Sin. Ser. A**31** (1988) 1153–1165.
170. Han, Houde: *The boundary–integro–differential equations of three–dimensional Neumann problem in linear elasticity.* Numer. Math. **68** (1994) 269–281.
171. Harriharan, S.I. and MacCamy, R.C.: *Integral equation procedures for eddy current problems.* J. Compt. Physics **45** (1982) 80–99.
172. Harrington, R.F.: Time-harmonic Electromagnetic Fields, McGraw-Hill, 1961.
173. Hartmann, F.: Introduction to Boundary Elements. Springer-Verlag, Berlin 1989.
174. Hartmann, F. and Zotemantel, R.: *The direct boundary element method in plate bending.* Intern. J. Numer. Methods Engrg. **23** (1986) 2049–2069.
175. Hellwig, G.: *Remarks on a group of Hilbert theorems on systems of elliptic partial differential equations.* Math. Zeitschr. **55** (1952) 276–283.
176. Hellwig, G.: *The boundary value problem of a linear elliptic system.* Math. Zeitschr. **56** (1952) 388–408.
177. Hellwig, G.: Partial Differential Equations. Blaisdell Publ. Comp., New York 1964.
178. Hess, J.L. and Smith, A.M.O.: *Calculation of potential flow about arbitrary bodies.* In: Progress in Aeronautical Sciences **8** (D. Kuchemann ed.) Pergamon Press, Oxford 1966, pp. 1–138.
179. Hilbert, D.: Grundzüge einer allgemeinen Theorie der linearen Integralgleichungen. Teubner, Leipzig 1912.
180. Hildebrandt, S.H. and Wienholtz, E.: *Constructive proofs of representation theorems in seperable Hilbert spaces.* Comm. Pure Appl. Maths. XVII (1964) 369–373.

181. Hiptmair, R.: Symmetric coupling for eddy current problems, SIAM J. Numer. Anal. **40** (1) (2002) 41–65.
182. Hiptmair, R.: *Coupling of finite elements and boundary elements in electromagnetic scattering.* SIAM J. Numer. Anal. **41** (2003) 919–944.
183. Hiptmair, R. and Schwab, C.: *Natural BEM for the electric field integral equation on polyhedra.* Res. Report 2001–04, Sem. on Appl. Math. ETH Zürich 2001.
184. Hörmander, L.: Pseudo–Differential Operators. Comm. Pure Appl. Math. XVIII (1965) 501–517.
185. Hörmander, L.: *Pseudo–differential operators and non–elliptic boundary problems.* Ann. of Math. **83** (1966) 129–209.
186. Hörmander, L.: *Fourier integral operators I.* Acta Math. **127** (1971) 79–183.
187. Hörmander, L: Linear Partial Differential Operators, Springer–Verlag, Berlin 1976.
188. Hörmander, L: The Analysis of Linear Partial Differential Operators, I–IV, Springer–Verlag Berlin 1985.
189. Hofmann, St., Mitrea, M. and Taylor, M.: Singular Integrals and Elliptic Boundary Value Problems on Regular Semmes–Kenig–Toro Domains. Internat. Math. Research Notes. Oxford Univ. Press, Oxford 2009.
190. Hofmann, S., Mitrea, M. and Taylor, M.: *Symbol calculus for operators of layer potential type on Lipschitz surface with VMO normals and related pseudodifferential operator calculus.* Anal. PDE **8** (1) (2015) 115–181.
191. Hsiao, G.C.: *A Neumann series representation for solutions to the exterior boundary value problem of elasticity.* In: Function Theoretic Methods for Partial Differential Equations, Lecture Notes in Mathematics **561** Springer–Verlag Berlin 1976, pp. 252–260.
192. Hsiao, G.C.: *On the stability of integral equations of the first kind with logarithmic kernels.* Archive Rat. Mech. Anal. **94** (1986) 179–192.
193. Hsiao, G.C.: *Mathematical foundations for the boundary–field equations methods in acoustics and electromagnetic scattering.* In: Analytical and Computational Methods in Scattering and Applied Mathematics. Notes in Math. **47** (2000) Chapman & Hall 149–163.
194. Hsiao, G.C. and Kleinman, R.E.: Mathematical foundation for the error estimation in numerical solutions of integral equations in electromagnetics, IEEE Transactions on Antennas and Propagation , vol. 45, No.3, 316–328, (1977)
195. Hsiao, G.C., Kleinman, R.E. and D.Q. Wang, D.Q.: *Applications of boundary integral equation methods in 3D,* J. Computational and Applied Mathematics, vol. 104, pp. 89-110, 1999.
196. Hsiao, G.C., Kopp, P. and Wendland, W.L.: *A Galerkin–collocation method for some integral equations of the first kind.* Computing **25** (1980) 89–130.
197. Hsiao, G.C., Kopp, P. and Wendland, W.L.: *Some applications of a Galerkin–collocation method for boundary integral equations of the first kind.* Math. Methods Appl. Sci. **6** (1984) 280–325.
198. Hsiao, G.C. and MacCamy, R.C.: *Solution of boundary value problems by integral equations of the first kind.* SIAM Rev.**15** (1973) 687–705.
199. Hsiao, G.C. and MacCamy, R.C.: *Singular perturbations for two–dimensional viscous flow problems.* In: Lecture Notes in Math. **942** (W. Eckhaus, E.J. de Jager eds.) Springer–Verlag, Berlin 1982, pp.229–244.
200. Hsiao, G.C., Monk, P. and Nigam, N.: *Error analysis of a finite element–integral equation scheme for approximate the time–harmonic Maxwell system,* SIAM J. Numer. Anal. **40** (2002) 198–219.
201. Hsiao, G.C. and Wendland, W.L.: *A finite element method for an integral equation of the first kind.* J. Math. Anal. Appl. **58** (1977) 449–481.

202. Hsiao, G.C. and Wendland, W.L.: *Boundary element methods for exterior problems in elasticity and fluid mechanics.* In: The Mathematics of Finite Elements and Applications IV (J. Whiteman ed.) Academic Press, London 1988, pp.323–341.
203. Hsiao, G.C. and Wendland, W.L.: *Boundary integral equations in low frequency acoustics.* J. Chinese Inst. Engineers **23** (2000) 369–375.
204. Hsiao, G.C. and Wendland, W.L.: *Boundary element methods: foundation and error analysis.* In: Encyclopaedia of Computational Mechanics, Vol.1 (E. Stein et al eds.) John Wiley & Sons Publ., Chichester 2004, pp. 339–373.
205. Hsiao, G.C. and Wendland, W.L.: *On a boundary integral method for some exterior problems in elasticity.* Proc. Tbilissi Univ. (Trudy Tbiliskogo Ordena Trud. Krasn. Znam. Gosud. Univ.) UDK 539.3, Mat. Mech. Astron. **257**, 18 (1985) 31–60.
206. Hsiao, G. C. and Wendland, W. L.: *Boundary integral equations recast as pseudodifferential equations.* Rendiconti, Accademia Nazionale delle Scienze detta dei XL, Memorie de Matematica e Applicazioni 124° (2006), Vol. XXX, fasc. 1, pagg. 29-48
207. Hsiao, G.C. and Wendland, W.L.: *The boundary integral equation method for almost incompressible elastic materials, revisited.* Rendiconti Acad. Naz. Science XL. Memorie di Matematicae Applicazione **124** (2006) XXX(1) 41–64.
208. Hsiao, G.C. and Wendland, W.L.: *A characterization of the Calderon projector for the biharmonic equation.* Proc. ISAAC Conference 2005. WSPC (2009) 23–38.
209. Hsiao, G.C., Stephan, E.P. and Wendland, W.L.: *On the integral equation method for the plane mixed boundary value problem with the Laplacian.* Math. Methods Appl. Sci. **1** (1979) 265–321.
210. Hsiao, G.C., Stephan, E.P. and Wendland, W.L.: *On the Dirichlet problem in elasticity for a domain exterior to an arc.* J. Comp. Appl. Math. **34** (1991) 1–19.
211. Ilyinsky, A.S and Smirnow, Y.G.: Electromagnetic Wave Diffraction by Conducting Screens. VSP, Utrecht, The Netherlands, 1998. (Russian Original IPRZR 1996)
212. Jameson, A.: *Aerodynamics.* In: Encyclopaedia of Computational Mechanics, Vol 3 (E. Stein et al eds.) John Wiley & Sons Publ., Chichester 2004, pp. 325–406.
213. Jaswon, M.A, and Symm, G.T.: Integral Eduation Methods in Potential Theory and Elastostatics. Academic Press, New York 1977.
214. Jeggle, H.–G.: Konvergenz und Fehlerabschätzung für die diskretisierte Integralgleichung des Neumannschen Problems. Doctoral Thesis, TH Darmstadt 1966.
215. Jeggle, H–G.: Nichtlineare Funktionalanalysis. B.G. Teubner, Stuttgart 1979.
216. Jerison, D.S. and Kenig, C.: *The inhomogeneous Dirichlet problem in Lipschitz domains.* J. Fund. Anal. **130** (1995)-219.
217. Jiang, R. and Kauranen, A.: *Korn's inequality and John domains.* Calculus of Variations and Partial Differential Equations **56** (2017) 109.
218. Jin, J.M.: The Finite Element Method in Electromagnetics, John Wiley and Sons. IEEE–Press 1993.
219. John, F.: Plane Waves and Spherical Means Applied to Partial Differential Equations. Interscience Publ., New York 1955.
220. Jones, D.S.: Methods in Electromagnetic Wave Propagation, Clarendon Press 1979 (Second Edition, Oxford, 1994).

221. Jost, G.: *Integral equations with modified fundamental solution in time-harmonic electromagnetic scattering* , IMA J. Applicat. Mathem., vol. 40 pp. 129-143, 1988.
222. Kantorowitsch, L.V and Akilow, G.P.: Functional Analysis in Normed Spaces. Pergamon Press, New York 1964.
223. Kato, T.: Perturbation Theory for Linear Operators. Springer–Verlag, Berlin 1966.
224. Kellog, O.D.: Foundations of Potential Theory. Springer–Verlag, Berlin 1929.
225. Kieser, R.: Über einseitige Sprungrelationen und hypersinguläre Operatoren in der Methode der Randelemente. Doctoral Thesis, Universität Stuttgart 1990.
226. Kirchhoff, G.: *Zur Theorie freier Flüssigkeitsstrahlen.* Crelles J. f. d. Reine u. Angewandte Mathematik **70** (1869) 295–307.
227. Kirsch, A. and Hettlich, F.: The Mathematical Theory of Time–Harmonic Maxwell's Equations. Springer International Publishing, Switzerland 2015.
228. Kleinman, R. E. and Kress, R.: *On the condition number of integral equations in acoustics using modified fundamental solutions.* IMA. J. Appl. Math. **31** (1983), 79–90.
229. Klingenberg, W.: A Course in Differential Geometry. Springer–Verlag, Heidelberg 1978.
230. Knöpke, B.: *The hypersingular integral equation for bending moments* m_{xx} , m_{yy} *and* m_{xy} *of the Kirchhoff plate.* Comp. Mechanics **14** (1996) 1–12.
231. König, H.: *An explicit formula for fundamental solutions of linear partial differential equations with constant coefficients.* Proc. Amer. Math. Soc. **120** (1994) 1315–1318.
232. Kohr, M.: *The Dirichlet problems for the Stokes resolvent equations in bounded and exterior domains in* $\mathbb{R}^n$. Math. Nachr. **280** (2007) 534–559.
233. Kohr, M. and Pop, I.: Viscous Incompressible Flow. WIT Press, Southampton 2004.
234. Kohr, M. and Wendland, W.L.: *Variational boundary integral equations for the Stokes system.* Applicable Anal. **85** (2006) 1343–1372.
235. Komech, A.I.: *Linear partial differential equations.* In: Linear Partial Differential Equations II. Encyclopaedia of Mathematical Sciences Vol.31 (Yu.V. Egorov, M.A. Shubin eds.) Springer–Verlag, Berlin 1994, pp. 121–255.
236. Kondratiev, V.A. and Oleinik, O.A.: *On Korn's inequalities.* Comp. Rendus Acad. Sci., Paris **308** (1989) 483–487.
237. Kong, J.A.: Theory of Electromagnetic Waves, John Wiley & Sons, New York, 1975.
238. Koshlyakov, N.S., Smirnov, M.M. and Gliner, E.N.: Differential Equations of Mathematical Physics. North Holland, Amsterdam 1964.
239. Král, J.: *The Fredholm method in potential theory.* Trans. Amer. Math. Soc. **125** (1966) 511–547.
240. Král, J.: *Boundary regularity and normal derivatives of logarithmic potentials.* Proc. Roy. Soc. Edinburgh **A 106** (1987), 241–258.
241. Král, J.: Integral Operators in Potential Theory. Lecture Notes in Mathematics, **823**, Springer–Verlag, Berlin 1980.
242. Král, J. and Wendland, W.L.: *Some examples concerning applicability of the Fredholm-Radon method in potential theory.* Apl. Mat. **31** (1986), 293–308.
243. Král, J. and Wendland, W.L.: *On the applicability of the Fredholm–Radon method in potential theory and the panel method.* In: Panel Methods in Fluid Mechanics with Emphasis in Aerodynamics (eds. J. Ballmann et al.) Notes in Fluid Mechanics, Vieweg, Braunschweig **21** (1988) 120–136.
244. Kress, R.: Grundzüge einer Theorie der verallgemeinerten harmonischen Vektorfelder. Meth. Verf. Math. Physik **2** (1069) 49–83.

245. Kress, R.: *Die Behandlung zweier Randwertprobleme für die vektorielle Poissongleichung nach einer Integralgleichngsmethode.* Archive Rat. Mech. Analysis **39** (1070) 206–226.
246. Kress, R.: *Potentialtheoretische Randwertprobleme bei Tensorfeldern beliebigeer Dimension und beliebigen Ranges.* Methoden Verf. Math. Physik **2** (1969) and Archive Rat. Mech. Analysis **47** (1972) 59–80.
247. Kress, R.: Linear Integral Equations. Springer–Verlag, Berlin 1998.
248. Kriezis, E.E., Tsiboukis, T.D. and Panas, S.M.: Eddy currents: theory and applications, Proceedings of IEEE , **80**, (10) (1992) 1559-1589.
249. Kufner, A., John, O. and Fučik, S.: Function Spaces. Academia, Praha, Noordhoff, Groningen 1977.
250. Kufner, A. and Kadlec, J.: Fourier Series. Iliffe Books, London 1971.
251. Kupradze, V.D.: Randwertaufgaben der Schwingungstheorie und Integralgleichungen. Dt. Verlag d. Wissenschaften, Berlin 1956.
252. Kupradze, V.D.: Potential Methods in the Theory of Elasticity. Israel Program Scientific Transl., Jerusalem 1965.
253. Kupradze, V.D., Gegelia, T.G. Basheleishvili, M.O. and Burchuladze, T.V.: Three-Dimensional Problems of the Mathematical Theory of Elasticity and Thermoelasticity. North Holland, Amsterdam 1979.
254. Kurz and Auchmann: Differential forms and boundary integral equations for Maxwell–type problems. In: U. Langer et al. (Eds.): Fast Boundary Element Methods, LNACM 63, pp.1-62. Springer-Verlag Berlin 2012
255. Kythe, P.K: Fundamental Soluitons for Differential Operators and Applications. Birkhäuser, Boston 1996.
256. Ladyženskaya, O.A.: The Mathematical Theory of Viscous Incompressible Flow. Gordon and Breach Science Publishers, New York 1969.
257. Ladyženskaya, O.A.: The Boundary Value Problems of Mathematical Physics. Springer–Verlag, Berlin 1985.
258. Lamb, H.: Hydronamics. University Press, Cambridge 1932.
259. Lamp, U., Schleicher, K. and Wendland, W.L.: *The fast Fourier transform and the numerical solution of one–dimensional boundary integral equations.* Numer. Math. **47** (1985) 15–38.
260. Langenberg, K.J.: *A thorough look at the nonuniqueness of the electromagnetic scattering integral equation solutions as compared to the scalar acoustic ones,* Radio Sci. **38** (2) 8021, doi: 10.1029/2001RS002558, 2003.
261. Lehman, R.S.: *Developments at an analytic corner of solutions of elliptic partial differential equations.* J. Mathematics and Mechanics **8** (1959) 727–760.
262. Leis, R.: Vorlesungen über partielle Differentialgleichungen zweiter Ordnung. BI–Verlag, Mannheim 1967.
263. Leis, R.: Initial Boundary Value Problems in Mathematical Physics. B.G. Teubner, Stuttgart and John Wiley & Sons, Chichester 1986.
264. Leonard, P.J. and Rodger, D.: *Finite element scheme for transient 3D eddy current.* IEEE Trans. Magn. **24** 1988.
265. Levi, E.E.: *Sulle equazioni lineari totalmente ellittiche alle derivate parziali.* Rend. Circ. Mat. Palermo **24** (1907) 275–317.
266. Levi, E.E.: *I problemi dei valori al contorno per le equazioni lineari totalmente ellitiche alle derivate parziali.* Mem. Soc. It. dei Sc. XL **16** (1909) 1–112.
267. Liggett, J.A. and Liu, P.L.F.: The Boundary Integral Equation Method for Porous Media Flow. George Allen & Unwin, London 1983.
268. Lions, J.L.: Espaces intermédaires entre espaces hilbertiens et applications. Bull. Math. Soc. Sc. Math. Phys. R.P. Roumaine **2** (1958) 419–432.
269. Lions, J.L.: Equations Differentielles. Operationuelles et Problemès aux Limites. Springer–Verlag, Berlin 1961.

270. Lions, J.L. and Magenes, E.: Non–Homogeneous Boundary Value Problems and Applications, Vol. I. Springer–Verlag, Berlin 1972.
271. Ljubič, Ju. I.: *On the existence "in the large" of fundamental solutions of second order elliptic equations.* Math. Sbornik **57** (1962) 45–58 (Russian).
272. Lopatinskii, Ya. B.: *Fundamental solutions of a system of differential equations of elliptic type.* Ukrain. Mat. Ž. **3** (1951) 290–316.
273. Lopatinskii, Ya. B.: *A fundamental system of solutions of an elliptic system of differential equations.* Ukrain. Mat. Z. **3** (1951) 3–38 (Russian).
274. MacCamy, R.C.: *Low frequency oscillations.* Quart. Appl. Math. **23** (1965) 247–255.
275. MacCamy, R.C.: *Low frequency expansions for two–dimensional interface scattering problems.* SIAM J. Appl. Mathematics **57** (1997) 1687–1701.
276. MacCamy, R.C. and Stephan, E.P.: *A simple layer potential method for three–dimensional eddy current problems.* In: Lecture Notes in Mathematics (A. Dold, B. Eckmann eds.) **964** (1982) Springer–Verlag Berlin, 477–484.
277. MacCamy, R.C. and Stephan, E.P.: *A boundary element method for an exterior problem for three–dimensional Maxwell's equations.* Appl. Analysis **16** (2) (1983) 141–163.
278. MacCamy, R.C. and Stephan, E.P.: *Solution procedures for the three–dimensional eddy–current problems,* J. Math. Anal. Appl. **101** (1984), 348–379.
279. MacCamy, R.C. and Stephan, E.P.: *A skin effect approximation for eddy current problems.* Arch. Rational Mech. Anal. **90** (1985) 87–98.
280. Malgrange, B.: *Existence et approximation des sulutions des équations aux dérivées partielles et des équations de convolution.* Ann. Inst. Fourier (Grenoble) **6** (1955/56) 271–355.
281. Manolis, G.D. and Beskos, D.E.: Boundary Element Methods in Elastodynamics. Unwin Hyman Ltd., London 1988.
282. Martensen, E.: *Berechnung der Druckverteilung an Gitterprofilen in ebener Potentialströmung mit einer Fredholmschen Integralgleichung.* Archive Rat. Mech. Anal. **3** (1959) 235–270.
283. Martensen, E.: Potentialtheorie. B.G.Teubner, Stuttgart 1968.
284. Maue, A.W.: *Über die Formulierung eines allgemeinen Diffraktionsproblems mit Hilfe einer Integralgleichung.* Zeitschr. f. Physik **126** (1949) 601–618.
285. Mautz, J. R. and Harrington, R. F.: *H-field, E–field, and combined–field solutions for conducting bodies of revolutions,* Arch. Elektron Übertragungstech, Electron, Commun, **32** (4) (1978) 159-164, .
286. Maxwell, J.C.: *A dynamical theory of electromagnetic field.* Phil. Transactions Royal Soc. of London **155** (1865) 459–512.
287. Maxwell Equations: Analysis and Numerics. (Langer, U., Pauly, D. and Repin, S. eds.) Series on Comp. and Appl. Mathematics. De Gruyter, Berlin 2019.
288. Maz'ya, V.G.: Sobolev Spaces. Springer–Verlag, Berlin 1985
289. Maz'ya, V.G.: Boundary Integral Equations. In Encyclopaedia of Mathematical Sciences. Vol 27, Analysis IV. (Maz'ya, V.G., Nikolskiĭ eds.). Springer-Verlag, Berlin 1991, pp. 127–228.
290. McLean, W.: Strongly Elliptic Systems and Boundary Integral Equations. Cambridge Univ. Press, Cambridge UK 2000.
291. McLean, W. and Wendland, W.L.: *Trigonometric approximation of solutions of periodic pseudodifferential operators.* Operator Theory: Advances and Appl. **41** (1989) 359–383.
292. Meddahi, S. and Selgas, V: *A mixed FEM and BEM coupling for a three–dimensional eddy current problem.* M2AN **37** (2) (2003) 291–318.
293. Meister, E.: Partielle Differentialgleichungen. Akademie–Verlag, Berlin 1996.

294. Meyers, N. and Serrin, J.: $H = W$. Proc. Nat. Acad. Sci. USA **51** (1964) 1055–1056.
295. Mikhailov, S.E.: *Analysis of united boundary–domain integro–differential and integral equations for a mixed BVP with variable coefficient.* Math. Methods in Applied Sciences **29** (2006) 715–739.
296. Mikhailov, S.E.: *About traces, extensions and co–normal derivative operators on Lipschitz domains.* In: Integral Methods in Science and Egineering: Techniques And Applications. (C. Constanda, S. Potapenko eds.) Birkhäuser, Boston 2007, pp. 151–162.
297. Mikhailov, S.E.: *Traces, extensions and co–normal derivatives for elliptic systems on Lipschitz domains.* J. Math. Anal. Appl. **378** (2011) 324–342.
298. Mikhlin, S.G.: *Le problème fondamental biharmonique à deux dimensions.* Comp. Rendus Acad. Sci., Paris **197** (1933) 608.
299. Mikhlin, S.G.: *Plane problem of the theory of elasticity.* Trudy Seismol. Inst. Akad. Nauk SSSR **65** (1935), pp. 372–396 (Russian).
300. Mikhlin, S.G.: *Problems of equivalence in the theory of singular integral equations.* Matem. Sbornik **3** (45) (1938) 121–140.
301. Mikhlin, S.G.: Integral Equations and their Applications. Pergamon Press, Oxford 1957.
302. Mikhlin, S.G.: Multidimensional Singular Integrals and Integral Equations. Pergamon Press, Oxford 1965.
303. Mikhlin, S.G.: Mathematical Physics, an Advanced Course. North Holland, Amsterdam 1970.
304. Mikhlin, S.G., Morozov, N.F. and Paukshto, M.V.: The Integral Equations of the Theory of Elasticity. B.G.Teubner, Stuttgart 1995.
305. Mikhlin, S.G. and Prössdorf, S.: Singular Integral Operators. Springer–Verlag, Berlin 1986.
306. Millman, R.S. and Parker, G.D.: Elements of Differential Geometry. Prentice Hall Inc., Englewood Cliffs N.J. 1977.
307. Miranda, C: Partial Differential Equations of Elliptic Type. Springer–Verlag, Berlin 1970.
308. Mitrea, D., Mitrea, M. and Pipher, J.: *Vector potential theory on non–smooth domains in $\mathbb{R}^3$ and applications to electromagnetic scattering.* J. Fourier Anal. Appl. **3** (2) (1967) 131–192.
309. Mitrea, D., Mitrea, M. and Taylor, M.: Layer Potentials, the Hodge Laplacian, and Global Boundary Problems in Nonsmooth Riemannian Manifolds. Memoirs of the Amer. Math. Soc. **150** Nr. 173, 2001.
310. Mitrea, M. and Wright, M.: *Boundary Value Problems for the Stokes System in Arbitrary Lipschitz Domains.* Astérisque **344**, 2012.
311. Mitrea, D., Mitrea,I., Mitrea, M. and Taylor, M.: The Hodge–Laplacian, Vol. 64 of De Gruyter, Studies in Mathematics, De Gruyter, Berlin 2016.
312. Miyazaki, Y.: *New proofs of the trace theorem of Sobolev spaces.* Proc. Japan Acad. Ser. A. **84** (2008) 112–116.
313. Moioda, A. and Perugia, I.: *A space–time Trefftz discontinuous Galerkin method of the acoustic wave equation in the first–order formulation.* Numer. Math. **138** (2018) 389–435.
314. Monk, P.: Finite Element Methods for Maxwell's Equations. Clarendon Press, Oxford 2003.
315. Morrey, C.B.: *Second order elliptic systems of differential equations.* Proc. Nat. Acad. USA **39** (1953) 201–206.
316. Morrey, C.B.: *Second order elliptic systems of differential equations.* Ann. Math. Studies **33** (1954) 101–159.

317. Müller, C.: Foundations of the Mathematical Theory of Electromagnetic Waves. Springer–Verlag, Berlin 1969.
318. Murat, F.: *Compacité par compensation II.* In:Proc. Int. Meeting on Recent Methods in Nonlinear Analysis (E. Giorgi, E. Magenes, U. Mosco eds.), Pitagora Editrice, Bologna, 1979, pp. 245–256.
319. Muskhelishvili, N.I.: Some Basic Problems of the Mathematical Theory of Elasticity. Noordhoff, Groningen 1953.
320. Muskhelishvili, N.I.: Singular Integral Equations. Noordhoff, Groningen 1958.
321. Natroshvili, D.: Estimation of Green's Tensors of the Theory of Elasticity and some of their Applications. Izdat. Tbiliskogo Universiteta, Tbilisi 1978 (Russian).
322. Natroshvili, D.: Investigations of Boundary Value and Initial–Boundary Value Problems of the Mathematical Theory of Elasticity and Thermoelasticity for Homogeneous Anisotropic Media Using the Potential Methods. Dr. of Science Thesis, Tbilisi 1984 (Russian).
323. Natroshvili, D., Džagmaidze, A.Ja. and Svanadze, M.Ž.: Some Problems in the Linear Theory of Elastic Mixtures. Izd. Tbiliskogo Universiteta, Tbilisi 1986 (Russian).
324. Natroshvili, D. and Wendland, W.L.: *Boundary variational inequalities in the theory of interface cracks.* In: Operator Theory: Advances and Appl. **147** (2004) 387–402.
325. Nazarov, S.A.: *The polynomial property of self–adjoint elliptic boundary value problems and algebraic description of their attributes.* Uspekhi Mat. Nauk **54**:5 (1999) 77–142.
326. Nečas, J.: Les Méthodes Directes en Théorie des Equations Elliptiques. Masson, Paris 1967. English Translation: Direct Methods in the Theory of Elliptic Equations. Springer–Verlag Berlin–Heidelberg 2011.
327. Nedelec, J.C.: *Curved finite element methods for solutions of singular integral equations on surfaces in* $\mathbb{R}^3$. Comput. Methods Appl. Mech. Engrg. **9** (1976) 191–216.
328. Nedelec, J.C.: Approximation des Equationes Intégrales en Mécanique et en Physique. Lecture Notes, Centre de Mathématiques Appliquées, Ecole Polytechnique, Palaiseau, France, 1977.
329. Nedelec, J.C.: *Approximation par potential de double cuche du problème de Neumann extérieur.* Comp. Rendus Acad. Sci. Paris, Sér. I. Math. **286** (1978) 616-619.
330. Nedelec, J.C.: *Integral equations with non integrable kernels.* Integral Eqs. Operator Theory **5** (1982) 562–572.
331. Nedelec, J.C.: Acoustic and Electromagnetic Equations, Integral Representations for Harmonic Problems. Springer–Verlag, New York 2001.
332. Nedelec, J.C. and Planchard, J.: *Une méthode variationelle d'élements pour la résolution numérique d'un problème extérieur dans* $\mathbb{R}^3$. RAIRO, Anal. Numer. **7** (1973) 105–129.
333. Neittaanmäki, P. and Roach, F. G.: *Weighted Sobolev spaces and exterior problems for the Helmholtz equation.* Proc. Roy. Soc. London Ser. A **410** (1987), No. 1839, 373–383.
334. Netuka, I.: *Smooth surfaces with infinite cyclic variation.* Casopis pro pěstováni matematiky **96** (1971) 86–101.
335. Neumann, C.: *Zur Theorie des logarithmischen und des Newtonschen Potentials.* Ber. Verh. Königl. Sächsische Ges. d. Wiss. zu Leipzig **22** (1870) 45–56, 264–321.
336. Neumann, C.: Untersuchungen über das logarithmische und Newtonsche Potential. Teubner–Verlag, Leipzig 1877.

337. Neumann, C.: Über die Methode des arithmetischen Mittels. Hirzl, Leipzig 1887 (erste Abhandlung), 1888 (zweite Abhandlung).
338. Nitsche, J.: On Korn's second inequality. RAIRO, Anal. Numer. **15** (1981) 237–248.
339. Of, G.: BETI–Gebietszerlegungsmethoden mit schnellen Randelementverfahren und Anwendungen. Doctoral Thesis, Universität Stuttgart 2006.
340. Of, G., Steinbach, O. and Wendland, W.L.: *Applications of a fast multipole Galerkin boundary element method in linear elastostatics.* Computing and Visualization in Sciences **8** (2005) 201–209.
341. Ortner, V.N.: *Regularisierte Faltung von Distributionen,* Teil 1: *Zur Berechnung von Fundamentallösungen.* ZAMP **31** (1980) 133–154.
342. Ortner, V.N.: *Regularisierte Faltung von Distributionen,* Teil 2: *Eine Tabelle von Fundamentallösungen.* ZAMP **31** (1980) 155–173.
343. Parasjuk, L.S.: *Fundamental solutions of elliptic systems of differential equations with discontinuous coefficients.* Dopovidi Akad. Nauk Ukrain. RSR (1963) 986–989 (Ukrainian).
344. Park, B.-J.: *On the boundedness of pseudo–differential operators on Triebel–Lizorkin and Besov spaces.* J. Math. Anal. Appl. **461** (2018) 544–576.
345. Peetre, J,: *Another approach to elliptic boundary value problems* Comm. Pure Appl. Math. **14**, 711–731, 1961.
346. Petersen, B.E.: Introduction to the Fourier Transform and Pseudodifferential Operators. Pitman, London 1983.
347. Picard, R. and Des McGhee: Partial Differential Equations. Walter de Gruyter, Berlin 2011.
348. Picard, R., Weck, N. and Witsch, k.-J.: *Time–harmonic Maxwell equations in the exterior of perfectly conducting irregular obstacles.* Analysis (Munich) **21**(3), (2001) 231–263.
349. Plemelj, J.: Potentialtheoretische Untersuchungen. B.G. Teubner–Verlag, Leipzig 1911.
350. Poincaré, H.: *La méthode de Neumann et le problème de Dirichlet.* Acta Math. **20** (1896) 59–142.
351. Pomp, A.: The Boundary-Domain Integral Method for Elliptic Systems with an Application to Shells. Lecture Notes in Mathematics, **1683**. Springer-Verlag, Berlin 1998.
352. Power, H. and Wrobel, L.C.: Boundary Integral Methods in Fluid Mechanics. Comp. Mech. Publ., Southampton 1995.
353. Pozrikidis, C.: Boundary Integral and Singularity Methods for Linearized Viscous Flow. Cambridge Univ. Press, Cambridge 1992.
354. Prössdorf, S.: *Linear integral equations.* In: Encyclopaedia of Mathematical Sciences. Vol 27, Analysis IV, Maz'ya, V.G., Nikolskii (eds.). Springer–Verlag, Berlin 1991, pp.1–125.
355. Prössdorf, S. and Rathsfeld, A.: *On strongly elliptic singular integral operators with piecewise continuous coefficients.* Integral Equations and Operator Theory **8** (1985) 825–841.
356. Prössdorf, S. and Schmidt, G.: *A finite element collocation method for singular integral equations.* Math. Nachr. **100** (1981) 33–60.
357. Prössdorf, S. and Schmidt, G.: *A finite element collocation method for systems of singular integral equations.* In: Complex Analysis and Applications '81, Sofia 1984, pp. 428–439.
358. Prössdorf, S. and Silbermann, B.: Projektionsverfahren und die näherungsweise Lösung singulärer Gleichungen. Teubner–Verlag, Leipzig 1977.
359. Prössdorf, S. and Silbermann, B.: Numerical Analysis for Integral and Related Operators. Birkhäuser–Verlag, Basel 1991.

360. Radon, J.: *Über die Randwertaufgabe beim logarithmischen Potential.* Sitzungsber. Akad. Wiss. Wien, Math.-nat. Kl. Abt. IIa **128** (1919) 1123–1167.
361. Reidinger, B. and Steinbach, O.: *A symmetric boundary element method for the Stokes problem in multiple connected domains.* Math. Methods Appl. Sci. **26** (2003) 77–93.
362. Rellich, F.: *Ein Satz über mittlere Konvergenz.* Göttinger Nachr. Math. Phys. Klasse (1930) 30–35.
363. Rellich, F.: *Darstellungen der Eigenwerte von $\Delta u + \lambda u = 0$ durch ein Randintegral.* Math. Z. **46** (1940) 635–636.
364. Rellich, F.: *Über das asymptotische Verhalten der Lösungen von $\Delta u + \lambda u = 0$ in unendlichen Gebieten.* Jber. Deutsch. Math. Verein **53** (1943) 57–65.
365. Rodriguez, R.: *Numerical approximation of Maxwell equations in low–frequency regime.* In: Computational Electromagnetism (A. Bermudez de Castro, A. Valli eds.) Lecture Notes in Mathematics **2148** Springer (2014) pp. 59–110.
366. Rodriguez, A. A. and Valli, A: Eddy Current Approximation of Maxwell Equations, Theory, Algorithms and Applications, Springer–Verlag Italia 2010.
367. Rudin, W.: Functional Analysis. Mc Graw–Hill, New York 1973.
368. Rumsey, V.H.: *Reaction concept in electromagnetic theory.* Physical Review **94** (1954) 1438–1491.
369. Saranen, J. and Vainikko, G.: Integral and Pseudodifferential Equations with Numerical Approximation. Springer–Verlag, Berlin 2002.
370. Saranen, J. and Wendland, W.L.: *The Fourier series representation of pseudo–differential operators on closed curves.* Complex Variables **8** (1987) 55–64.
371. Sauter, St. and Schwab, Ch.: Randelementmethoden. B.G. Teubner, Stuttgart 2004.
372. Sayas, F.–J., Brown, Th.S. and Hassell, M.E.: Variational Techniques for Elliptic Partial Differential Equations. CRC Press, Boca Raton, London, New York 2019.
373. Schanz, M. and Steinbach, O. (Eds.): Boundary Element Analysis. Springer–Verlag, Berlin 2006.
374. Schatz, A., Thomée, V. and Wendland, W.L.: Mathematical Theory of Finite and Boundary Elements. Birkhäuser Verlag, Basel 1990.
375. Schechter, M.: *Integral inequalities for partial differential operators and functions satisfying general boundary conditions.* Comm. Pure Appl. Math. **12** (1959) 37–88.
376. Schechter, M.: Principles of Functional Analysis. Academic Press, New York 1971.
377. Schechter, M.: Modern Methods in Partial Differential Equations. McGraw–Hill, New York 1977.
378. Schrohe, E. and Schulze, B.–W.: *Boundary value problems in Boutet de Monvel's calculus for manifolds with conical singularities I and II.* In: Advances in Partial Differential Equations, Pseudodifferential Calculus and Mathematical Physics. Vol.I, pp.97–209 (1994) and Vol.II, pp.70–205 (1995) Akademie–Verlag Berlin.
379. Schulze, B.–W.: Boundary Value Problems and Singular Pseudo–Differential Operators. John Wiley & Sons, Chichester 1998.
380. Schwab, C. and Wendland, W.L.: *Kernel properties and representations of boundary integral operators.* Math. Nachr. **156** (1992) 187–218.
381. Schwab, C. and Wendland,W.L.: *On the extraction technique in boundary integral equations.* Math. Comp. **68** (1999) 91–121.
382. Schwartz, L.: Théorie des Distributions. Hermann, Paris 1950.

383. Seeley, R.T.: *Extension of C^∞ functions defined in a half–space.* Proc. Amer. Math. Soc.**15** (1964) 625–626.
384. Seeley, R.T.: *Singular integrals and boundary value problems.* Amer. J. Math. **88** (1966) 781–809.
385. Seeley, R.T.: *Topics in pseudo–differential operators.* In: *Pseudodifferential Operators* (L. Nirenberg ed.), Centro Internazionale Mathematico Estivo (C.I.M.E.), Edizioni Cremonese , Roma 1969, pp. 169–305.
386. Sellier, A.: *Hadamard's finite part concept in dimension $n \geq 2$, distributional definition, regularization forms and distributional derivatives.* Proc. R. Soc. London A **445** (1994) 69–98.
387. Sellier, A.: *Hadamard's finite part concept in dimension $n \geq 2$; definition and change of variables, associated Fubini's theorem, derivation.* Math. Proc. Camb. Phil. Soc. **122** (1997) 131–148.
388. P.P. Silvester and R.L. Ferrari: Finite Elements for Electrical Engineers, Third Edition, Cambridge Univeristy Press, 1996.
389. Simader, C.G. and Sohr, M.: *Necessary and sufficiant conditions for the existence of Helmholtz decompositions in general domains.* Annali dell' Unicersitá di Ferrara, **60** (2014): Scienzematematiche, 245–262.
390. Sloan, I. and Wendland, W.L.: *Qualocation methods for elliptic boundary integral equations.* Numer. Math. **79** (1998) 451–483.
391. Sloan, I. and Wendland, W.L.: *Spline qualocation methods for variable–coefficient elliptic equations on curves.* Numer. Math. **83** (1999) 497–533.
392. Smirnov, V.I.: A Course of Higher Mathematics. Vol IV, Pergamon and Addison-Wesley, New York 1964.
393. Sobolev, S.L.: Equations of Mathematical Physics. Gostokhizdat. Moscow 1954. (Russian)
394. Sokolnikoff, I.S.: Tensor Analysis, Theory and Applications. Mac Graw Hill, New York 1951.
395. Sommerfeld, A.: Partial Differential Equations in Physics. Academic Press, New York 1969.
396. Somigliana, C.: *Sui sistemi simmetrici di equazioni a derivate parziali.* Ann. Mat. Pura Appl. **22** (1894) 143–156.
397. Staelin, D.H., Morgenthaler, A.W. and Kong, J.A.: Electromagnetic Waves. Prentice Hall, New Jersey 1994.
398. Steinbach, O.: *A robust boundary element method for nearly incompressible linear elasticity.* Numer. Math. **95** (2003) 553–562.
399. Steinbach, O.: Numerische Verfahren für elliptische Randwertprobleme. B.G. Teubner, Stuttgart 2003.
400. Steinbach, O.: *A note on the ellipticity of the single layer potential in two–dimensional linear elastostatics.* J. Math. Anal. Appl. **294** (2004) 1–6.
401. Steinbach, O. and Wendland, W.L.: *On C. Neumann's method for second order elliptic systems in domains with non–smooth boundaries.* J. Math. Anal. Appl. **262** (2001) 733–748.
402. Stephan, E.P.: *Boundary integral equations for magnetic screens in $\mathbb{R}^3$.* Proc. Royal Soc. Edinburgh **102A** (1986) 189–210.
403. Stephan, E.P.: Boundary Integral Equations for Mixed Boundary Value Problems, Screen and Transmission Problems in $\mathbb{R}^3$. Habilitationsschrift, TH Darmstadt 1984.
404. Stephan, E.P.: *A boundary integral equation procedure for the mixed boundary value problem of the vector Helmholtz equation.* Appl. Anal. **35** (1990) 59–72.

405. Stephan, E.P. and Wendland, W.L.: *Remarks to Galerkin and least squares methods with finite elements for general elliptic problems.* Manuscripta Geodaetica **1** (1976) 93–123.
406. Stephan, E.P. and Wendland, W.L.: *An augmented Galerkin procedure for the boundary integral method applied to two–dimensional screen– and crack problems.* Applicable Analysis **18** (1984) 183–220.
407. Sterz, O. and Schwab, C.: *A scalar boundary integrodifferential equation for eddy current problem using impedance boundary condition.* Comput. Visual. Sci. **3** (2001) 129–136,
408. Stratton, J.A.: Electromagnetic Theory. McGraw–Hill, New York 1941.
409. Stratton, J.A. and Chu, I.J.: *Diffraction theory of electromagnetic waves.* Physical Review **56** (1939) 99–107.
410. Stummel, F.: Rand– und Eigenwertaufgaben in Sobolewschen Räumen. Lecture Notes in Mathematics **102**, Springer–Verlag, Berlin 1969.
411. Szabo, I.: Höhere Technische Mechanik. Springer–Verlag, Berlin 1956.
412. Taira, K: Diffusion Processes and Partial Differential Equations. Academic Press, Boston 1988.
413. Tartar, L.: *On the characterization of traces of a Sobolov space used for Maxwell's equations.* In: Proc. Conf. in honour of M. Artota, Bordeaux 1997.
414. Tartar, L.: An Introduction to Sobolev Spaces and Interpolation Spaces. Springer–Verlag Berlin 2007.
415. Taylor, M.E.: Pseudodifferential Operators. Princeton Univ. Press, Princeton 1981.
416. Taylor, M.E.: Partial Differential EquationsI. Basic Theory. Springer, New York 1996.
417. Temam, R.: Navier–Stokes Equations. North–Holland Publ., Amsterdam 1979.
418. Treves, F.: Linear Partial Differential Equations with Constant Coefficients. Gordon and Breach, New York 1966.
419. Treves, F.: Topological Vector Spaces, Distributions and Kernels. Academic Press, New York 1967.
420. Treves, F.: Pseudodifferential and Fourier Integral Operators. Vol. 1: Pseudodifferential Operators. Vol. 2: Fourier Integral Operators. Plenum Press, New York 1980.
421. Triebel, H.: Höhere Analysis. Dt. Verlag d. Wissenschaften, Berlin 1972.
422. Tychonoff, A.N. and Samarski, A.A.: Partial Differential Equations of Mathematical Physics. Vol I and II, Holden–Day, San Francisco 1964 and 1967.
423. Ursell, F.: *On the exterior problems of acoustics.* Proc. Cambridge Philos. Soc., I **74** (1973) 117–125; II **84** (1978) 545–548.
424. Varnhorn, W.: The Stokes Equations. Akademie Verlag, Berlin 1994.
425. Vekua, I.N.: New Methods for Solving Elliptic Equations. North Holland, Amsterdam 1968 (Russian 1948, Moscow).
426. Vishik, M.J.: *On strongly elliptic systems of differential equations.* Mat. Sb. **25** (1951) 615–676 (Russian).
427. Vishik, M.J. and Eskin, G.T.: *Elliptic equations in convolution in a bounded domain and their applications.* Russian Math. Surveys **22** (1967) 13–75 (Uspekhi Mat. Nauk **22** (1967) 15–76).
428. Wagner, P.: Parameterintegration zur Berechnung von Fundamentallösungen. Polska Akademia Nauk, Inst. Matematyczny, Dissertationes Mathematicae, CCXXX, Warszawa 1984.
429. Wagner, P.: *On the explicit calculation of fundamental solutions.* J. Math. Anal. Appl. **297** (2004) 404–418.

430. Weber, Ch. and Werner, P.: *A local compactness theorem for Maxwell's equations.* Math. Meth. in the Appl. Sci. **2** (1980) 12–25.
431. Weggler, L.: *Stabilized boundary element methods for low–frequency electromagnetic scattering.* Math. Meth. Appl. Sci. **35** (2012) 574–597.
432. Wendland, W.L.: *Die Behandlung von Randwertaufgaben im R_3 mit Hilfe von Einfach– und Doppelschichtpotentialen.* Numer. Math. **11** (1968) 380–404.
433. Wendland, W.L.: *Martin Costabel's version of the trace theorem revisited.* Math. Methods Appl. Sci. **37** (2014) 1924–1955.
434. Wendland, W.L. and Zhu, J.: *The boundary element method for three–dimensional Stokes flows exterior to an open surface.* Math. Comput. Modelling **15** (1991) 19–41.
435. Werner, P.: *Zur mathematischen Theorie akustischer Wellenfelder.* Archive Rat. Mech. Anal. **6** (1960) 321-260.
436. Werner, P.: *On the behaviour of stationary electromagnetic wave fields for small frequencies.* J. Math. Anal. Appl. **15** (1966) 447–496.
437. Wielandt, H.: *Das Iterationsverfahren bei nicht selbstadjungierten linearen Eigenwertaufgaben.* Math. Z. **50** (1944) 93–143.
438. Wilcox, C.H.: Scattering Theory for the d'Alembert Equation in Exterior Domains. Lecture Notes in Mathematics **442**, Springer-Verlag, Berlin 1975.
439. Wloka, Y.: Partielle Differentialgleichungen. B.G. Teubner, Stuttgart 1982.
440. Wloka, J.T., Rowley, B. and Lawruk, B.: Boundary Value Problems for Elliptic Systems. Cambridge Univ. Press, Cambridge 1995.
441. Yaghjian, A. D. : *Augmented electric–and magnetic integral equations,* Radio Sci., **16** (1981) 987-1001 .
442. Yu, De–hao: Natural Boundary Integral Method and its Applications. Kluwer Acad. Publ., Dordrecht 2002.
443. Zeidler, E.: Nonlinear Functional Analysis and its Applications I: Fixed Point Theorems. Springer–Verlag Berlin 1985.
444. Zeidler, E.: Nonlinear Functional Analysis and its Applications. Vol. II/A: Linear Monotone Operators. Springer–Verlag, Berlin 1990.

Index

G. C. Hsiao und W. L. Wendland, *Boundary Integral Equations*,
Applied Mathematical Sciences 164,
https://doi.org/10.1007/978-3-030-71127-6

The manufacturer's authorised representative in the EU is Springer Nature Customer Service Centre GmbH, Europaplatz 3, 69115 Heidelberg, Germany. If you have any concerns regarding our products, please contact ProductSafety@springernature.com

Printed and bound by CPI Group (UK) Ltd, Croydon, CR0 4YY
22/09/2026
02220979-0019